# Adenoviral Vectors for Gene Therapy

# Adenoviral Vectors for Gene Therapy

## Second Edition

*Edited by*

## *David T. Curiel*

ELSEVIER

AMSTERDAM • BOSTON • HEIDELBERG • LONDON
NEW YORK • OXFORD • PARIS • SAN DIEGO
SAN FRANCISCO • SINGAPORE• SYDNEY • TOKYO
Academic Press is an imprint of Elsevier

Academic Press is an imprint of Elsevier
125 London Wall, London EC2Y 5AS, UK
525 B Street, Suite 1800, San Diego, CA 92101-4495, USA
50 Hampshire Street, 5th Floor, Cambridge, MA 02139, USA
The Boulevard, Langford Lane, Kidlington, Oxford OX5 1GB, UK



First Edition 2002



**British Library Cataloguing-in-Publication Data**
A catalogue record for this book is available from the British Library

**Library of Congress Cataloging-in-Publication Data**
A catalog record for this book is available from the Library of Congress

ISBN: 978-0-12-800276-6

For information on all Academic Press publications
visit our website at https://www.elsevier.com/



*Acquisition Editor:* Linda Versteeg-Buschman
*Editorial Project Manager:* Halima Williams
*Production Project Manager:* Karen East and Kirsty Halterman
*Designer:* Alan Studholme

Typeset by TNQ Books and Journals
www.tnq.co.in

# Contents

# List of Contributors

**Yadvinder S. Ahi** HIV Drug Resistance Program, National Cancer Institute, Frederick National Laboratory for Cancer Research, Frederick, MD, USA

**Steven M. Albelda** Thoracic Oncology Research Group, Pulmonary, Allergy, and Critical Care Division, Department of Medicine, Perelman School of Medicine, University of Pennsylvania, Philadelphia, PA, USA

**Yasser A. Aldhamen** Department of Microbiology and Molecular Genetics, Michigan State University, East Lansing, MI, USA

**Ramon Alemany** IDIBELL-Institut Català d'Oncologia, L'Hospitalet de Llobregat, Barcelona, Spain

**Marta M. Alonso** Department of Medical Oncology, Clínica Universidad de Navarra, University of Navarra, Pamplona, Spain

**P.M. Alves** iBET, Instituto de Biologia Experimental e Tecnológica, Oeiras, Portugal; Instituto de Tecnologia Química e Biológica, Universidade Nova de Lisboa, Oeiras, Portugal

**Andrea Amalfitano** Department of Microbiology and Molecular Genetics, Michigan State University, East Lansing, MI, USA; College of Osteopathic Medicine, Michigan State University, East Lansing, MI, USA

**Rachael Anatol** Office of Cellular, Tissue, and Gene Therapies, Center for Biologics Evaluation and Research, Food and Drug Administration, Silver Spring, MD, USA

**C.A. Anderson** Merck Research Laboratories, Merck & Co., Inc., West Point, PA, USA

**Svetlana Atasheva** Lowance Center for Human Immunology, Departments of Pediatrics and Medicine, Emory University, Atlanta, GA, USA

**Michael A. Barry** Division of Infectious Diseases, Department of Internal Medicine, Mayo Clinic, Rochester, MN, USA; Department of Immunology, Mayo Clinic, Rochester, MN, USA; Department of Molecular Medicine, Mayo Clinic, Rochester, MN, USA

**Raj K. Batra** UCLA School of Medicine, Division of Pulmonary and Critical Care Medicine, GLA-VAHCS, Los Angeles, CA, USA; Jonsson Comprehensive Cancer Center, UCLA, Los Angeles, CA, USA

**A.J. Bett** Merck Research Laboratories, Merck & Co., Inc., West Point, PA, USA

**A. Bout** Crucell NV, Leiden, The Netherlands

**K. Brouwer** Crucell NV, Leiden, The Netherlands

**Nicola Brunetti-Pierri** Telethon Institute of Genetics and Medicine, Pozzuoli, Italy; Department of Translational Medicine, Federico II University, Naples, Italy

**Andrew P. Byrnes** Division of Cellular and Gene Therapies, FDA Center for Biologics Evaluation and Research, Silver Spring, MD, USA

**Shyambabu Chaurasiya** Department of Oncology, Faculty of Medicine and Dentistry, University of Alberta, Edmonton, AB, Canada

**L. Chen** Merck Research Laboratories, Merck & Co., Inc., West Point, PA, USA

**A.S. Coroadinha** iBET, Instituto de Biologia Experimental e Tecnológica, Oeiras, Portugal; Instituto de Tecnologia Química e Biológica, Universidade Nova de Lisboa, Oeiras, Portugal

**Igor P. Dmitriev** Department of Radiation Oncology, School of Medicine, Washington University, St. Louis, MO, USA

**Hildegund C.J. Ertl** Wistar Institute, Philadelphia, PA, USA

**P. Fernandes** iBET, Instituto de Biologia Experimental e Tecnológica, Oeiras, Portugal; Instituto de Tecnologia Química e Biológica, Universidade Nova de Lisboa, Oeiras, Portugal; Autolus, London, UK

**Juan Fueyo** Department of Neuro-Oncology, The University of Texas MD Anderson Cancer Center, Houston, Texas, USA; Department of Neurosurgery, The University of Texas MD Anderson Cancer Center, Houston, Texas, USA

**S.M. Galloway** Merck Research Laboratories, Merck & Co., Inc., West Point, PA, USA

**Thomas A. Gardner** Department of Urology, Indiana University Medical Center, Indianapolis, IN, USA; Department of Microbiology and Immunology, Indiana University Medical Center, Indianapolis, IN, USA

**Candelaria Gomez-Manzano** Department of Neuro-Oncology, The University of Texas MD Anderson Cancer Center, Houston, Texas, USA; Department of Genetics, The University of Texas MD Anderson Cancer Center, Houston, Texas, USA

**Urs F. Greber** Institute of Molecular Life Sciences, University of Zurich, Zurich, Switzerland

**Diana Guimet** Department of Molecular Genetics and Microbiology, School of Medicine, Stony Brook University, Stony Brook, NY, USA

**Michael Havert** Office of Cellular, Tissue, and Gene Therapies, Center for Biologics Evaluation and Research, Food and Drug Administration, Silver Spring, MD, USA

**Patrick Hearing** Department of Molecular Genetics and Microbiology, School of Medicine, Stony Brook University, Stony Brook, NY, USA

**Masahisa Hemmi** Laboratory of Biochemistry and Molecular Biology, Graduate School of Pharmaceutical Sciences, Osaka University, Osaka, Japan

**R.B. Hill** Merck Research Laboratories, Merck & Co., Inc., West Point, PA, USA

**Mary M. Hitt** Department of Oncology, Faculty of Medicine and Dentistry, University of Alberta, Edmonton, AB, Canada

**Ying Huang** Office of Cellular, Tissue, and Gene Therapies, Center for Biologics Evaluation and Research, Food and Drug Administration, Silver Spring, MD, USA

**Ilan Irony** Office of Cellular, Tissue, and Gene Therapies, Center for Biologics Evaluation and Research, Food and Drug Administration, Silver Spring, MD, USA

**Hong Jiang** Department of Neuro-Oncology, The University of Texas MD Anderson Cancer Center, Houston, Texas, USA

**Sergey A. Kaliberov** Department of Radiation Oncology, School of Medicine, Washington University, St. Louis, MO, USA

**Chinghai H. Kao** Department of Urology, Indiana University Medical Center, Indianapolis, IN, USA; Department of Microbiology and Immunology, Indiana University Medical Center, Indianapolis, IN, USA

**Dayananda Kasala** Department of Bioengineering, College of Engineering, Hanyang University, Seongdong-gu, Seoul, Republic of Korea

**D. Kaslow** Merck Research Laboratories, Merck & Co., Inc., West Point, PA, USA

**Benjamin B. Kasten** Department of Radiology, The University of Alabama at Birmingham, Birmingham, AL, USA

**Johanna K. Kaufmann** German Cancer Research Center (DKFZ), Heidelberg, Germany

**Jay K. Kolls** Richard King Mellon Foundation Institute for Pediatric Research, Children's Hospital of Pittsburgh of UPMC, Pittsburgh, PA, USA; Department of Pediatrics, School of Medicine, University of Pittsburgh, Pittsburgh, PA, USA

**Johanna P. Laakkonen** Department of Biotechnology and Molecular Medicine, A.I. Virtanen Institute for Molecular Sciences, University of Eastern Finland, Kuopio, Finland

**R. Lardenoije** Crucell NV, Leiden, The Netherlands

**J. Lebron** Merck Research Laboratories, Merck & Co., Inc., West Point, PA, USA

**B.J. Ledwith** Merck Research Laboratories, Merck & Co., Inc., West Point, PA, USA

**J. Lewis** Merck Research Laboratories, Merck & Co., Inc., West Point, PA, USA

**Erik Lubberts** Department of Immunology, Erasmus MC, University Medical Center, Rotterdam, The Netherlands; Department of Rheumatology, Erasmus MC, University Medical Center, Rotterdam, The Netherlands

**Stefania Luisoni** Institute of Molecular Life Sciences, University of Zurich, Zurich, Switzerland

**S.V. Machotka** Merck Research Laboratories, Merck & Co., Inc., West Point, PA, USA

**S. Manam** Merck Research Laboratories, Merck & Co., Inc., West Point, PA, USA

**D. Martinez** Merck Research Laboratories, Merck & Co., Inc., West Point, PA, USA

**Suresh K. Mittal** Department of Comparative Pathobiology, College of Veterinary Medicine and Purdue University Center for Cancer Research, Purdue University, West Lafayette, IN, USA

**Hiroyuki Mizuguchi** Laboratory of Biochemistry and Molecular Biology, Graduate School of Pharmaceutical Sciences, Osaka University, Osaka, Japan

**Edmund Moon** Thoracic Oncology Research Group, Pulmonary, Allergy, and Critical Care Division, Department of Medicine, Perelman School of Medicine, University of Pennsylvania, Philadelphia, PA, USA

**Stephen J. Murphy** Molecular Medicine Program, Mayo Clinic, Rochester, MN, USA

**Dirk M. Nettelbeck** German Cancer Research Center (DKFZ), Heidelberg, Germany

**Philip Ng** Department of Molecular and Human Genetics, Baylor College of Medicine, Houston, TX, USA

**W.W. Nichols** Merck Research Laboratories, Merck & Co., Inc., West Point, PA, USA

**Raymond John Pickles** Cystic Fibrosis/Pulmonary Research and Treatment Center, University of North Carolina at Chapel Hill, Chapel Hill, NC, USA

**Sudhanshu P. Raikwar** Department of Veterinary Medicine and Surgery, College of Veterinary Medicine, University of Missouri and Harry S. Truman Veterans' Memorial Hospital, Columbia, MO, USA

**Paul N. Reynolds** Department of Thoracic Medicine and Lung Research Laboratory, Royal Adelaide Hospital, Adelaide

**Jillian R. Richter** Department of Radiology, The University of Alabama at Birmingham, Birmingham, AL, USA

**Yisel Rivera-Molina** Department of Neuro-Oncology, The University of Texas MD Anderson Cancer Center, Houston, Texas, USA

**Qian Ruan** PaxVax Inc., San Diego, CA, USA

**C. Russo** Merck Research Laboratories, Merck & Co., Inc., West Point, PA, USA

**Carl Scandella** Carl Scandella Consulting, Bellevue, WA, USA

**Paul Shabram** PaxVax Inc., San Diego, CA, USA

**Anurag Sharma** Department of Pediatrics, Weill Cornell Medical College, New York, NY, USA

**Sherven Sharma** UCLA/Wadsworth Pulmonary Immunology Laboratory, Division of Pulmonary and Critical Care Medicine, GLA-VAHCS, Los Angeles, CA, USA

**Dmitry M. Shayakhmetov** Lowance Center for Human Immunology, Departments of Pediatrics and Medicine, Emory University, Atlanta, GA, USA

**A.C. Silva** iBET, Instituto de Biologia Experimental e Tecnológica, Oeiras, Portugal; Instituto de Tecnologia Química e Biológica, Universidade Nova de Lisboa, Oeiras, Portugal

**Phoebe L. Stewart** Department of Pharmacology and Cleveland Center for Membrane and Structural Biology, Case Western Reserve University, Cleveland, OH, USA

**Hideyo Ugai** Cancer Biology Division, Department of Radiation Oncology, School of Medicine, Washington University, St. Louis, MO, USA

**D. Valerio** Crucell NV, Leiden, The Netherlands

**M. van der Kaaden** Crucell NV, Leiden, The Netherlands

**Gary Vellekamp** Vellekamp Consulting LLC, Montclair, NJ, USA

**Sai V. Vemula** Laboratory of Molecular Virology, Center for Biologics Evaluation and Research, Food and Drug Administration, Silver Spring, MD, USA

**Richard G. Vile** Molecular Medicine Program, Mayo Clinic, Rochester, MN, USA

**R. Vogels** Crucell NV, Leiden, The Netherlands

**Stefan Worgall** Department of Pediatrics, Weill Cornell Medical College, New York, NY, USA; Department of Genetic Medicine, Weill Cornell Medical College, New York, NY, USA

**Lily Wu** Jonsson Comprehensive Cancer Center, UCLA, Los Angeles, CA, USA; Department of Urology, UCLA School of Medicine, Los Angeles, CA, USA; Department of Pediatrics, UCLA School of Medicine, Los Angeles, CA, USA

**Enric Xipell** Department of Medical Oncology, Clínica Universidad de Navarra, University of Navarra, Pamplona, Spain

**Seppo Ylä-Herttuala** Department of Biotechnology and Molecular Medicine, A.I. Virtanen Institute for Molecular Sciences, University of Eastern Finland, Kuopio, Finland; Department of Medicine, University of Eastern Finland, Kuopio, Finland; Gene Therapy Unit, Kuopio University Hospital, Kuopio, Finland

**Chae-Ok Yun** Department of Bioengineering, College of Engineering, Hanyang University, Seongdong-gu, Seoul, Republic of Korea

**Kurt R. Zinn** Department of Radiology, The University of Alabama at Birmingham, Birmingham, AL, USA

**D. Zuidgeest** Crucell NV, Leiden, The Netherlands

# Adenovirus Structure

**1**

*Phoebe L. Stewart*
Department of Pharmacology and Cleveland Center for Membrane and Structural Biology, Case Western Reserve University, Cleveland, OH, USA

## 1. Historical Perspective on Adenovirus Structure

The structure of the adenovirus virion is quite complex and our understanding of it has been evolving from before 1965. Early negative stain electron micrographs of adenovirus revealed an icosahedral capsid with 252 capsomers and long fibers protruding from the vertices.[1] Later these capsomers were identified as 240 hexons and 12 pentons, with the pentons at the fivefold vertices of the capsid. The pentons each have five neighboring capsomers and the hexons each have six neighboring capsomers. As the adenoviral molecular components were identified and their stoichoimetries characterized, it became apparent that the hexons and pentons were different proteins. The hexons are trimeric proteins and the pentons are formed by two proteins, a pentameric penton base and a trimeric fiber.[2] Subsequently, X-ray crystallography provided atomic resolution structures of hexon,[3] penton base,[4] fiber,[5,6] and adenovirus protease,[7] which is involved in virion maturation. In addition to the three major protein components of the capsid (hexon, penton base, and fiber), there are four minor capsid proteins (proteins IIIa, VI, VIII, IX).[8,9] The minor proteins are also referred to as cement proteins as they serve to stabilize the capsid. They also play important roles in the assembly, disassembly, and cell entry of the virus. Atomic resolution structures have not yet been determined for the minor proteins isolated from the adenovirus capsid. However, cryo-electron microscopy (cryoEM) has provided moderate structural information on the density of the minor proteins in the context of the virion.[10–13] In 2010, atomic resolution structures of adenovirus were determined by cryoEM and X-ray crystallography.[14,15] Despite these two atomic, or near atomic, resolution (3.5–3.6 Å) structures, controversies remained regarding the structure and assignment of the minor capsid proteins. In 2014, a refined crystal structure of adenovirus at 3.8 Å resolution revised the minor capsid protein structures and locations.[16]

The adenoviral genome is relatively large, with ~30–40 kb.[8] It is notable in that large deletions and insertions can be tolerated, a feature that contributes to the enduring popularity of adenovirus as a gene delivery vector.[17] Within the core of the virion there are five proteins associated with the double-stranded DNA genome (proteins V, VII, mu, IVa2, and terminal binding protein).[9] The structure of the genome and how it is packaged with its associated proteins in the core of the virion is not well understood. Early negative stain EM and ion etching studies suggested that the core is organized as 12 large spherical nucleoprotein assemblies, termed adenosomes.[18,19] However, cryoEM and crystallographic structures of adenovirus show that the core does not follow the strict icosahedral symmetry of the capsid.[14–16]

**Adenoviral Vectors for Gene Therapy.**

Adenovirus was one of the first samples imaged during the development of the cryoEM technique[20] and was among the first set of viruses to have its structure determined by the cryoEM single particle reconstruction method.[21] Since then cryoEM structures have been determined for multiple types of adenovirus and adenovirus in complex with various host factors.[10–12,14,22–29] Docking of crystal structures of capsid proteins into the cryoEM density and difference imaging have been useful approaches for dissecting the complex nature of the capsid. An early example of difference imaging was applied in two dimensions to scanning transmission electron microscopy (STEM) images of the group-of-nine hexons and this work helped to elucidate the position of protein IX within the icosahedral facet.[30] Difference imaging in three dimensions led to an early tentative assignment for the positions of the minor capsid proteins within the capsid based on copy number and approximate mass.[13] As higher resolution cryoEM structures were determined, some of these initial assignments were revised.[10–12] Visualization of α-helices was achieved with a 6Å resolution cryoEM structure.[12] This structure facilitated more accurate docking of hexon and penton base crystal structures and produced a clearer difference map and more detailed density for the minor capsid proteins. Secondary structure prediction for the minor capsid proteins was used to tentatively assign density regions to minor capsid proteins. Determination of an atomic resolution (3.6Å) structure by cryoEM was facilitated by the use of a high-end FEI Titan Krios electron microscope.[14] Micrographs for this dataset were collected on film and scanned for digital image processing. The final dataset included 31,815 individual particle images. The resolution was estimated by reference-based Fourier shell correlation coefficient and supported by observation of both α-helical and β-strand density. Density was also observed for some of the side chains, particularly bulky amino acids. The assignments for the minor capsid protein locations were assumed to be the same as interpreted from the 6Å resolution cryoEM structure.[12] Atomic models were produced for minor capsid proteins IIIa, VIII, and IX from the atomic resolution cryoEM density map using bulky amino acids as landmarks.[14]

Attempts to crystallize intact adenovirus began in 1999 and proceeded for more than 10 years before the first atomic resolution crystal structure was published.[15,31] Several factors hampered early crystallization efforts, including the long protruding fiber, the instability of virions at certain pH values, the tendency of adenovirus particles to aggregate, and relatively low yields from standard virus preparations. Use of a vector based on human adenovirus type 5 (HAdV5), but with the short fiber from type 35 (Ad5.F35, also called Ad35F), helped to solve some of the production and crystallization difficulties. This vector was also used for several moderate resolution cryoEM structural studies.[11,12] Collection of diffraction data for atomic resolution structure determination spanned several years. Even though crystals were flash-cooled in liquid nitrogen, they were still highly radiation sensitive and only 2–5% of the crystals diffracted to high resolution. Diffraction data from nearly 900 crystals were collected but only a small subset of these data was used to generate the dataset. The best crystals diffracted well to 4.5Å resolution and weakly to 3.5Å at synchrotron sources. The initial phase information was derived from a pseudo-atomic capsid of adenovirus generated from fitting the crystallographic structures of hexon and penton base into a cryoEM structure of Ad5.F35 at 9Å resolution.[11] In 2010, partial atomic models were built for some of the minor capsid proteins.[15]

After collection of more diffraction data and additional refinement a refined crystal structure was published with more complete models for minor capsid proteins IIIa, VI, VIII, and IX and surprisingly for a portion of the core protein V.[16] To compensate for the relatively modest resolution (3.8 Å) of the structure, a method was devised to evaluate the reliability of assigned amino acid sequences to the experimental electron density. This gives credence to the latest assignments for the locations of the minor capsid proteins within the capsid. It is important to recognize that adenovirus is one of the largest biomolecular assemblies with an atomic resolution structure determined by X-ray crystallography (>98,000 nonhydrogen atoms used in refinement of the asymmetric unit). With an assembly of this size and complexity and with less than ideal resolution data, assigning the locations of the minor capsid proteins is quite a challenging task.

There are over 60 HAdV types categorized in seven species (human adenovirus A–G). Species D adenoviruses are the most numerous, many of which were identified during the AIDS epidemic.[32] AIDS patients and other immunocompromised patients are particularly susceptible to adenovirus. Adenovirus causes acute respiratory illness, epidemic keratoconjunctivitis, acute hemorrhagic cystitis, hepatitis, myocarditis, and gastroenteritis in humans. Adenoviruses have also been characterized from the five major classes of vertebrate species, mammals, birds, reptiles, amphibians, and fish.[33] Structural studies of human and animal adenoviruses have contributed to our understanding of the molecular complexity within the Adenoviridae family.

## 2. Hexon Structure and Capsid Packing

The icosahedral capsid of adenovirus is composed of 240 trimeric hexons and 12 pentameric penton bases at the vertices with associated fibers. For HAdV2, hexons account for the majority (>83%) of the protein mass in the capsid.[34] The first hexon crystal structure was that of HAdV2.[3] At that time in 1986, the hexon subunit was the longest polypeptide whose structure was determined by X-ray crystallography with 967 residues per hexon monomer. Higher resolution (2.2 and 2.5 Å) crystal structures of HAdV2 and HAdV5 hexons are now available (PDB-ID: 1P2Z; PDB-ID: 1P30).[34] The hexon crystal structure revealed that although it is a trimeric protein, the base of the molecule is shaped as a hexagon, which is optimal for close packing within the capsid. The hexagonal base of the hexon trimer is formed by two viral jellyroll domains in each hexon monomer, with each jellyroll situated at a point of the hexagon. The topology of the jellyrolls is similar to that of icosahedral RNA viruses, although the architectural roles of the jellyrolls in forming the icosahedral capsids of these viruses are different.[3] Intriguingly, the hexon fold is the same as that of the major capsid protein P3 of the bacteriophage PRD1.[35]

The top of the hexon trimer is trimeric in shape with three protruding towers. Each tower is formed by intertwined loops from all three hexon monomers. The intertwining within the hexon trimer is so extensive that an accessory protein, called the 100k protein, is required to help fold the hexon trimer.[36,37] Hexon has a large subunit interface and each subunit of hexon clasps its neighboring subunit, resulting in a highly stable trimeric structure.[38]

Comparison of hexon sequences from multiple adenovirus types led to the finding of multiple hypervariable regions within the hexon.[39] Originally it was thought that only some of these regions were mapped to the top of the hexon. However, determination of the HAdV2 crystal structure at 2.5 Å resolution led to an atomic model with 25% of the sequence reassigned compared to the earlier HAdV5 crystal structure.[38] Later both the HAdV2 and the HAdV5 crystal structures were refined with newer protocols.[34] The hexons from these two adenovirus types are highly homologous (86% identity) and their refined structures are very similar. The revised HAdV2 and HAdV5 hexon crystal structures place all of the hypervariable loops near the exposed top of the hexon trimer.

Hexon sequences from different viral types also revealed a high level of sequence conservation within a particular human species (~88%), reduced conservation between types of different human species (79–81%), and less conservation between types of different animal species (66–68%).[39] The majority of the differences found in hexon sequences are within the hypervariable loops and these loops are often the targets of neutralizing antibodies.[40] Following vaccination and natural infection, neutralizing antibodies are produced to both hexon and fiber, although the response to hexon appears to be dominant.[41] The flexibility and sequence tolerance of the hexon hypervariable loops have made them useful as insertion sites for modification of the adenovirus capsid.[42]

On each hexon trimer between the three protruding towers that project from the outer viral surface is a central depression. CryoEM structures of adenovirus in complex with vitamin K-dependent blood coagulation factor X indicate that the hexon depression is the binding site for the GLA (γ-carboxyglutamic acid rich) domain of the factor.[23,26,29] Specifically, a single threonine residue (T425) of HAdV5 is critical for the interaction with factor X, as mutation of this residue in the context of the virion abrogates binding to factor X.[23] Injection of mice intravenously with this virus mutant indicated that it does not infect hepatocytes efficiently, whereas wild-type and other virus mutants with single or double hexon mutations are efficient in this regard. Factor X plays a role in mediating Ad-hepatocyte transduction in vivo after intravenous administration. The adenovirus/factor X complex utilizes an alternative cellular uptake pathway and the adenovirus-bound factor X interacts with heparan sulfate proteoglycan on macrophages.[23,43]

The hexons are arranged with 12 trimers in each of the 20 facets of the icosahedron. There are four unique positions for the hexon trimer within the asymmetric unit of the capsid (Figure 1). The asymmetric unit is the smallest repeating unit of the capsid and corresponds to one-third of an icosahedral facet. Although different conventions have been used for numbering the hexons, the most common convention labels the hexons next to the penton base as position 1, the hexons next to the icosahedral twofold axes as position 2, the hexons next to the icosahedral threefold axes as position 3, and the fourth remaining site as position 4 (Figure 1(A)). The hexons next to the penton base, which are also referred to as the peripentonal hexons, have been observed to dissociate separately from the other hexons.[44] The remaining hexons dissociate in groups-of-nine hexons. These nine hexons (three each in positions 2, 3, and 4) form the central part of each icosahedral facet. The group-of-nine hexons are held together by the minor capsid protein, protein IX.[30]

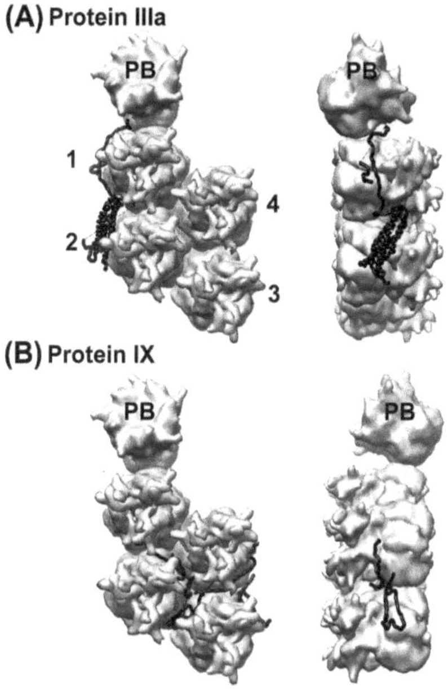


**Figure 1** Structure and location of the outer capsid proteins as assigned in the refined adenovirus crystal structure.[16] (A) The enlarged asymmetric unit, with four independent hexon trimers (1–4) and a complete penton base (PB), is shown as a 5 Å surface representation (light gray) together with the ordered portion of protein IIIa (black). Protein IIIa is chain O in PDB-ID: 4CWU. (B) The enlarged asymmetric unit together with the ordered portions of four copies of protein IX (black). Only the N-terminal portions of protein IX are ordered. The four copies of protein IX in the asymmetric unit are chains P, Q, R, and S in PDB-ID: 4CWU. Top and side views are shown in both panels. Dashed lines represent disordered regions. This figure was made with UCSF Chimera.[126]

## 3. Penton Base Structure and Integrin-Binding RGD Loop

The penton base is a pentameric protein that is shaped as a pentagon and packs nicely at each vertex of the capsid within a ring of five peripentonal hexons. The penton base of human and animal adenovirus types is typically highly conserved with ~70% homology between the sequences of any two types.[4] Negative stain electron micrographs of the adenovirus penton, composed of penton base and fiber, showed a pentameric structure with the fiber shaft protruding from the center.[45] CryoEM structures of dodecahedra composed of 12 HAdV3 penton bases or complete pentons showed subtle changes in the penton base structure with fiber binding.[46]

The crystal structure of the penton base was first determined for an N-terminally truncated form of the HAdV2 protein that formed regular dodecahedral particles with 12 complete pentamers.[4] Two structures were determined at the same time, one of penton base alone and one with an N-terminal fragment of the fiber protein revealing how the fiber interacts with the penton base (PDB-ID: 1X9P; PDB-ID: 1X9T).[4] The crystal structure of the HAdV2 penton revealed that the top of the penton base has grooves between the subunits that serve as binding sites for a conserved motif near the N-terminal end of fiber.[4] There is a symmetry mismatch between the trimeric fiber and the pentameric penton base, meaning that only three of the five grooves are occupied in each penton base of the assembled virion.[2] The pentameric form of the penton base buries a significant portion of the total surface area of each monomer. Mainly hydrophobic surfaces are buried in formation of the pentamer. The oligomeric penton base is composed of tilted monomers that form an assembly with an overall right-handed twist.

The pentagonal shape at the basal end of the molecule is formed by one jellyroll in each monomer. Intriguingly, the jellyroll within penton base is topologically related to the jellyroll in hexon. In addition to the jellyroll motif each monomer has an upper insertion domain, which protrudes from the outer capsid surface. The upper insertion domain is formed by two long insertion loops between strands of the jellyroll. One insertion loop contains the hypervariable Arg–Gly–Asp (RGD) region. This region is the most variable in sequence and length among adenovirus types. The RGD loop for HAdV2 is ~80 aa and is glycine- and alanine-rich. Most of the loop is flexible as no density is observed for residues 298–375 in the X-ray structure.[4] The second insertion, called the variable loop, forms a flexible β ribbon projecting from the top of penton base. In HAdV2 this loop is formed by residues 142–169, but in other adenovirus types it can be up to 10 residues longer.

The sequences of a penton base region including the RGD loop, variable loop, and surrounding residues from 51 human adenovirus types were used for phylogenetic analysis and structural prediction.[47] As expected, the phylogenetic analysis demonstrated clustering of the adenovirus types according to their species. In addition, clustering of the species B types supported the concept of dividing species B types into subspecies B1 and B2. Structural models for the various penton base proteins were built based on the crystallographic structure of the HAdV2 penton base. The divergence of the jellyroll motif compared to the HAdV2 penton base structure was predicted to be only 9.8–15.5%, whereas the divergence of the upper insertion domain was in the range of 37.3–38.8%.

Most, but not all, types of adenovirus have an RGD motif in one of the two surface loops of the penton base.[48] This motif is required for interactions with cellular integrins. Clustering of integrins on the host cell surface is promoted by interaction with penton base RGD loops and this leads to activation of signaling pathways that result in rapid internalization of the virus into clathrin-coated pits and endosomes.[49] The enteric adenovirus types HAdV40 and HAdV41 of species F lack the RGD motif on their penton base and do not utilize integrins for cell entry.[50,51]

Moderate resolution cryoEM structures have been determined for HAdV2 and HAdV12 in complex with soluble forms of $\alpha v\beta 5$ integrin.[22,27] Modeling with integrin crystal structures indicates that only a maximum of four integrins can bind per penton base. This is consistent with the surface plasmon resonance measurement of 4.2 integrin molecules per HAdV2 penton base at close to saturation.[22] The spacing of the RGD protrusions on the penton base (~60 Å) appears to be too close to allow five integrin heterodimers to bind to one penton base. Modeling shows that there is room to bind four integrin heterodimers to one penton base, but significant flexibility within the penton base RGD loops is required to accommodate this binding configuration. It was hypothesized that the strain arising from this symmetry-mismatched interaction might lead to a conformational change in the penton base and promote partial release of penton base pentamers from the capsid.[27]

The flexibility of the penton base RGD loops was first demonstrated by a cryoEM structure of HAdV2 in complex with a Fab fragment from a monoclonal antibody that binds a peptide region of penton base including RGD.[52] The HAdV2 penton base crystal structure is missing quite a large peptide region of 78 residues in the RGD loop due to disorder.[4] Alignment of penton base sequences from human adenovirus types indicates that HAdV12 has one of the shorter RGD loops, with just 15 residues corresponding to the missing 78 residues in HAdV2.[52] However, even the shorter HAdV12 RGD loop is flexible as indicated by the cryoEM structures of HAdV12 in complex with $\alpha v\beta 5$ integrin.[22,27]

Submission of two penton base sequences, those of HAdV5 and HAdV19c, to the ProteinDisOrder System (PrDOS) prediction webserver indicated that these RGD loops are predicted to be intrinsically disordered.[24] The significance of having an intrinsically disordered RGD loop might be related to increasing the binding rate constant of penton base to integrins on the cell surface. It has been demonstrated that the binding of intrinsically disordered proteins to structured targets with strong electrostatic interactions enhances the binding rate constants by several orders of magnitude.[53]

The penton base RGD loops have been implicated in binding human alpha defensins, which are peptides of the innate immune system.[24] Human alpha defensin 5 (HD5) can inhibit cell entry of adenoviral types from species A, B1, B2, C, and E, whereas species D and F types are resistant.[28] CryoEM structures of adenovirus/defensin complexes led to a model in which the RGD loops of sensitive adenoviral types wrap around HD5 monomers or dimers at the interface between penton base and fiber and stabilize the penton base/fiber complex.[24,28] This stabilization effect is thought to prevent release of the adenoviral membrane lytic factor, protein VI, and therefore adenovirus cannot escape from the endosome and is degraded by the host cell in the lysosomal pathway.

## 4. Fiber Structure and Receptor Interactions

The fiber is composed of three distinct regions: a short penton base interaction region near the N terminus, a shaft domain with a variable number of repeats, and a distal knob domain, which interacts with various receptors. The first atomic resolution structural information for the fiber was for the knob domain of HAdV5 (PDB-ID: 1KNB).[6] The crystal structure revealed an eight-stranded antiparallel $\beta$-sandwich structure in each monomer. The trimeric knob has a large buried surface area, indicating that the trimer is probably the most prevalent form of the fiber in solution. Crystal structures have now been determined for fiber knobs of numerous human adenovirus types, including HAdV3, HAdV7, HAdV11, HAdV12, HAdV14, HAdV16, HAdV21, HAdV35, and HAdV37.[54–61] In addition, crystal structures have been determined for canine and porcine fiber knobs.[62,63] These structures all reveal the same overall fold for the knob domain.

Sequence alignment of the shaft domain of multiple adenovirus types showed a common 15-residue repeat pattern.[64] The fold of this repeat pattern was revealed in a crystal structure of the knob domain plus four repeat units of the shaft from the HAdV2 protein (PDB-ID: 1QIU).[5] The fiber shaft fold represents a new structural motif for fibrous proteins, named the triple β-spiral. This fold is characterized by an extended β-strand running parallel to the fiber axis, a turn with a conserved glycine or proline, a second β-strand, and a following solvent-exposed loop of variable length. This structural motif is also found in the shaft domain of the reovirus sigma-1 protein.[65]

The structure of a short peptide region near the N terminus of fiber, termed as the universal fiber motif, was revealed in the crystal structure of the HAdV2 penton base with a 21-residue fiber peptide (PDB-ID: 1X9T).[4] The universal fiber motif is a mostly hydrophobic peptide region (FNPVYPY) that binds at the top of penton base at the subunit interface. All of the interactions observed between the fiber peptide and the penton base involve the conserved motif of the fiber and highly conserved residues of the penton base with the exception of one residue (Lys-387 of HAdV2 penton base). This indicates that it is likely that there is a universal mode of association between the N-terminal fiber motifs and the penton bases of various adenovirus types. The interactions between the fiber N-terminal region and the penton base were confirmed in a model of the HAdV5 fiber built by homology modeling and fitting of models within a 3.6 Å resolution cryoEM structure of the intact HAdV5 virion.[66]

The fiber knob is responsible for interaction with a variety of host cell attachment receptors, including coxsackie-adenovirus receptor (CAR), CD46 (membrane cofactor protein), sialic acid-containing oligosaccharides, GD1a glycan, and desmoglein-2 (DSG-2).[67–69] Numerous crystal structures have been determined with fiber knobs of various adenoviral types in complex with CAR,[54,70] CD46,[56,71,72] and sialic acid-containing molecules.[55,63,67] Whereas CAR and CD46 bind on the side of the trimeric fiber knob, sialic acid for the most part binds at the top of the fiber knob near the threefold symmetry axis. One exception to this, however, is the structure of the canine adenovirus type 2 (CAdV2) knob in complex with sialic acid.[63] This structure shows a distinct binding site for sialic acid, still on the top of the knob but more toward the periphery. The observation that CAR and CD46 bind on the side of the adenoviral fiber knobs, while sialic acid binds on the top of the knobs from human adenoviruses, suggests that there may be situations in which one fiber binds two different attachment receptors. This possibility is supported by a crystal structure of the HAdV37 fiber knob in complex with both a CAR domain and sialyl-D-lactose.[63]

## 5. Atomic Resolution Cryo-Electron Microscopy and X-ray Crystallographic Adenovirus Structures

Of course to truly appreciate the structure of adenovirus it is necessary to obtain an atomic resolution structure of the intact virion. In 2010, atomic resolution structures were published as determined by both cryoEM[14] and X-ray crystallography.[15] Four years later in 2014, a refined crystal structure was published.[16] Interpretation of the cryoEM structure was aided by the known structures of the major capsid proteins. In addition, the resolution was sufficient to observe density for bulky side chain, and de novo atomic models were created for several of the minor capsid proteins (PDB-ID: 3IYN). Solving of the crystal structure at 3.5 Å resolution was aided by a pseudo-atomic capsid produced by fitting the coordinates of isolated capsid proteins into a cryoEM density map.[11] The first crystal structure provides atomic descriptions of hexon and penton base together with partial models for some of the minor capsid proteins (PDB-ID: 1VSZ).

Both structures represent tremendous achievements given the large size of adenovirus, 150 MDa, and the complexity of the capsid with over 100,000 nonhydrogen atoms per asymmetric unit. The problem is that with this size and level of complexity, and with only partial side-chain densities apparent, the assignment of the minor capsid proteins is ambiguous and the two structures differ in their interpretations. The cryoEM structure is of HAdV5 and the crystal structures are of the HAdV5-based vector, Ad5.F35. In terms of molecular composition they should only vary in their fibers, with Ad5.F35 containing the shorter HAdV35 fiber. Given that the structure of penton base and a fiber fragment indicated a universal mode of association between fibers and penton bases of various adenovirus types,[4] it would seem to be a safe assumption that, except for the fibers and possible crystal packing effects, the structure of icosahedral capsid would be the same between HAdV5 and Ad5.F35.

A refined crystal structure at 3.8 Å resolution was published in 2014 with more complete atomic models for the minor capsid proteins (PCD-ID: 4CWU).[16] Ideally an atomic resolution crystal structure would be at high enough resolution to observe density for all, or most, of the side chains so that the assignment of density regions to specific amino acid sequences would be unambiguous. Unfortunately, this was not the case. Therefore a strategy was designed to evaluate and score assigned sequences to features in the experimental density map. This involved grouping the 20 amino acids into six groups based on side-chain size. Scores were assigned based on how well the sequence matched the density. Comparisons were made after shifting the amino acid sequence by one residue at a time. In addition, the N to C direction of each polypeptide chain was reversed and the scores recalculated to confirm that the best match for the density was chosen. This careful analysis of the X-ray density lends support to the assignments of the minor capsid proteins made by Reddy and Nemerow.[16]

## 6. Hexons in the Atomic Resolution Adenovirus Structures

Comparison of the hexon coordinates within the cryoEM and crystallographic atomic resolution adenovirus structures is complicated by the fact that the authors chose a different set of four unique hexons to include in the asymmetric unit, or basic repeating unit of the capsid.[14–16] The nomenclature of the four hexons is the same in all structures, with hexon 1 next to the penton base, hexon 2 next to the icosahedral twofold axes, hexon 3 next to the icosahedral threefold axes, and hexon 4 at the remaining position in the asymmetric unit. However, the four representative hexons of the cryoEM structure were chosen to surround the four-helix bundle at a facet edge, while the four hexons of the X-ray structure are all on the same side of the four-helix bundle.

Within the crystal structure of Ad5.F35 all 12 of the independent hexon subunits have virtually identical folds with a ~1 Å root mean square deviation on superimposition.[15] The main differences between the hexon subunits within the Ad5.F35 crystal structure are found at the N- and C-termini. Both the cryoEM and the crystal structures of adenovirus report coordinates for a few extra residues at the N- and C-termini of hexon, compared to the crystal structure of the isolated HAdV5 hexon.[34] However, the details of the hexon N- and C-terminal tail structures differ somewhat. Both the cryoEM and the crystal structures provide coordinates for some of the residues in the hexon hypervariable loops, which were disordered in the isolated hexon structure.[34] When packed in the adenovirus capsid the hypervariable loops mediate interhexon interactions and interactions with other capsid proteins. Selection and superimposition of a matching set of four hexons from the full icosahedral capsids of both the cryoEM and the crystal structures reveal some differences in interpretation for the hexon hypervariable loop structures.

## 7. Conformational Differences of the Penton Base in the Atomic Resolution Adenovirus Structures

The crystal structure of the isolated HAdV2 penton base was determined with an N-terminal truncation missing the first 48 residues because the full-length protein was easily degraded.[4] The coordinates for the isolated HAdV2 penton base (PDB-ID: 1X9T) begin with residue 52. In the HAdV5 atomic resolution cryoEM structure additional residues (aa 37–51) are traced in the N-terminal tail of the penton base.[14] The HAdV2 and HAdV5 penton base proteins are highly homologous (98% identity) and the overall fold is nearly identical. In the cryoEM structure the N-terminal residues of the HAdV5 penton base are observed to interact with a minor capsid protein below the penton base and then turn inward to connect to the genomic core. However, the N-terminal extensions of the penton base are not identified in the crystal structure of Ad5.F35.[15,16]

One of the more obvious differences between the cryoEM and the crystallographic adenovirus structures is the overall conformation of the penton base.[14–16] In the cryoEM structure the conformation matches the crystal structure of the isolated penton base in complex with an N-terminal fiber peptide,[4] whereas in the X-ray structure of adenovirus the penton base has a more expanded conformation and a larger central pore. In the isolated penton base structure the central pore of the pentamer has a maximum diameter of 28 Å, which is too narrow to accommodate the fiber shaft. In the X-ray structure of the intact Ad5.F35 virion, the penton base pore has an expanded pore diameter of 50 Å and density assigned to the fiber shaft is observed within the pore.[15] In the HAdV5 cryoEM structure, density for a short portion of the fiber shaft is observed on top of the penton base, consistent with the structure of the isolated penton base.[14] It is possible that crystal packing forces helped to induce the altered conformation of the penton base in the Ad5.F35 crystal structure.

The observation of two conformations for the penton base is intriguing. Conformational flexibility of the penton base may play a role in early events in viral cell entry and may be necessary for the programmed disassembly of the virion.[73] It is known that the minor capsid protein VI is membrane lytic and that it is released from the capsid in the endosome during viral cell entry.[74] In the mature adenovirus virion, protein VI is packaged on the inner capsid surface.[16] A conformational change in the capsid, such as dissociation of the penton base, may lead to release of protein VI at the appropriate time during cell entry. A cryoEM study of the adenovirus–integrin interaction led to the hypothesis that strain arising from the symmetry mismatch between four integrin heterodimers and the fivefold penton base might lead to a conformational change in the penton base and promote its release from the capsid.[27]

## 8. Alternate Assignments for the Four-Helix Coiled Coil

Both the cryoEM and the X-ray structures of adenovirus show four-helix coiled coils at the facet edges (Figure 1(A)).[14–16] Density at this location in the capsid was assigned to a portion of protein IIIa in an early cryoEM analysis of the molecular architecture of adenovirus.[13] This assignment to protein IIIa was based on mass and copy number per capsid. At higher resolution, this density resolved into a four-helix coiled coil, which led to an alternate assignment of this density as the C-terminal domain of protein IX.[12] This new assignment was based on the fact that the C-terminal domain of protein IX is strongly predicted to form a coiled coil and this region was the only observed coiled coil within the icosahedral capsid. This assignment implied that the N-terminal domains of IX form trimers, cementing together hexons within a facet,[30] and the C-terminal domains form four-helix bundles at the facet edges.

The assignment of protein IX to the coiled-coil density at the facet edge seemed to be supported by two moderate resolution cryoEM tagging studies.[75,76] In a tagging study by Marsh et al., an engineered adenovirus with enhanced green fluorescent protein (EGFP) fused to the C terminus of protein IX was examined by cryoEM.[76] The cryoEM structure at 22 Å resolution showed extra density assigned to EGFP at the facet edges hovering above the coiled-coil regions, although these regions were not resolved into separate helices. In a second tagging study by Fabry et al., a 12 residue peptide (called SY12) was engineered at the C terminus of protein IX.[75] A cryoEM structure of the engineered adenovirus at 11 Å resolution showed extra density at both ends of the cylinder of density, representing the coiled-coil region at the facet edge. In addition, anti-SY12 Fab fragments were added and a cryoEM structure of the complex was determined at 22 Å resolution. This structure showed apparent Fab density at both ends of the cylinder of density at the facet edge, indicating that the bundle includes antiparallel helices.

In the atomic resolution cryoEM structure of the intact virion,[14] the four-helix coiled coils were interpreted as the C-terminal domains of protein IX, as assigned earlier by Saban et al.[12] and as indicated by the cryoEM tagging studies.[75,76] The higher resolution cryoEM structure enabled chain tracing within the coiled-coil region with density apparent for several large side chains, including arginines and lysines, which aligned with the atomic model for the C-terminal domain of protein IX.

In the cryoEM-derived atomic model of the four-helix bundle, the helices were linked by a ladder of hydrophobic residues (leucines and valines). Chain tracing indicated that three of the helices were parallel and the fourth was antiparallel. The antiparallel helix was traced as coming from a protein IX N-terminal domain within an adjacent facet. Support for this assignment was provided by the observation that when the cryoEM density was contoured with a low-density threshold, connections were observed between most of the protein IX N-terminal domains and the helices within the coiled coil. The coiled coil is held in place by an interaction with a projecting loop (aa 251–256) on the side of the hexon in position 4 within the capsid.

In the first X-ray structure of the intact virion it was noted that two of the helices in the four-helix bundle appeared to be connected at one end.[15] This density connection between two helices suggested that the four helices might be from a domain of a single protein. This observation, in combination with the lack of clear side-chain density for the helical residues in the coiled coil, led the authors to consider the possibility that this density might be a domain of protein IIIa as originally proposed.[13] In the refined X-ray structure of the virion the assignment of the four-helix bundle to protein IIIa is confirmed with as much certainty as possible given the resolution of the density map (3.8 Å).[16]

## 9. Protein IIIa Structure

As discussed above, there have been differing assignments for the location of protein IIIa within the capsid. Protein IIIa is the largest cement protein in the capsid (63 kDa) and it is present in 60 copies per virion.[2] It is known to play a role in viral assembly and maturation as temperature-sensitive mutants of protein IIIa are defective for assembly.[77,78] Secondary structure prediction indicates that protein IIIa is highly α-helical with at least 16 predicted helices. Analysis of a cryoEM structure of Ad5.F35 at 6 Å resolution in which α-helices within the capsid were resolved resulted in the assignment of protein IIIa to a cluster of helices below the penton base on the inside of the capsid.[12] A cryoEM labeling study of protein IIIa seemed to support this assignment as it indicated that the N terminus of protein IIIa is located beneath the vertex complex between the penton base and the peripentonal hexons.[79]

The cluster of helices below the penton base was also observed in the atomic resolution cryoEM structure of adenovirus.[14] The backbone fold of a large portion of protein IIIa (aa 7–300) was traced into this density below the vertex. It was reported that side-chain densities were visualized for ~85% of the residues. However, the densities were not distinctive enough to identify individual amino acids and therefore the large side chains were used more as "landmarks" for guiding the building of an atomic model (PDB-ID: 3IYN).

In the refined X-ray structure of adenovirus protein IIIa is assigned to the four-helix bundle on the exterior of the capsid (PDB-ID: 4CWU) (Figure 1(A)).[16] Three segments of protein IIIa are resolved with good certainty. The N-terminal region of protein IIIa (aa 48–102) is observed to extend toward the penton base at the vertex. Another short stretch within the N-terminal region (aa 9–25) is also traced, but the assignment of this region is less certain. Two segments in the middle of protein IIIa (103–209 and 252–355) are traced within the four-helix bundle at the facet edge. In the atomic model the helical bundle is formed by two long helix–turn–helix motifs with one disordered connection (aa 210–253). In addition, the large C-terminal region of protein IIIa (aa 356–585) is disordered.

Mass spectrometry indicates that the C-terminal 15 residues of HAdV5 protein IIIa (aa 571–585) are cleaved by adenovirus protease,[80] as had been predicted for protein IIIa of HAdV2.[77,81] Reddy and Nemerow surmise that the C-terminal region of protein IIIa remains on the capsid exterior near the icosahedral twofold axis. One remaining puzzle about protein IIIa is how the C-terminal tails are cleaved by the adenovirus protease, which is packaged in the core of the virion.

## 10. Protein IX Structure

Protein IX is known to help stabilize the virion, as virions lacking protein IX have poor thermostability.[82,83] Recently protein IX has gained prominence as a convenient site of ligand addition for both vector retargeting and fluorescence labeling.[84] The location of the N-terminal domain of protein IX was established by STEM of capsid dissociation fragments called groups-of-nine hexons, or GONs.[30] Four trimeric regions were observed stabilizing the hexon array. Initially these regions were thought to represent the location of the full-length protein IX. However, later it was shown that only the conserved N-terminal domain of protein IX (aa 1–39) is necessary for stabilization of the Ad capsid.[85,86] Volume analysis in a cryoEM study of the Ad5.F35 vector at 9 Å resolution indicated that the locations identified by STEM for protein IX were likely to correspond to only the N-terminal viral interaction domains.[11]

The atomic resolution cryoEM and X-ray crystal structures of adenovirus show density for the N-terminal region of protein IX.[14–16] In the first X ray structure only coordinates for the Cα backbone atoms were deposited (PDB-ID: 1VSZ).[15] In the cryoEM structure density was visualized for ~85% of the side chains in the N-terminal domain and coordinates were deposited for the majority of the residues in this domain (PDB-ID: 3IYN).[14] Similarly, in the refined X-ray structure coordinates for the N-terminal domain of protein IX were deposited (PDB-ID: 4CWU).[16] However, the N to C direction of the polypeptide backbone is reversed in these two atomic models.

In the refined X-ray structure the best match/confidence scores are obtained for protein IX compared to the scores for the other cement proteins, lending confidence to the X-ray-derived atomic model for protein IX. The protein IX N-terminal regions form triskelion shapes between hexon trimers in a group-of-nine hexons in the middle of each icosahedral facet. In each facet one triskelion sits at the icosahedral threefold axis in the middle of the facet, and three additional triskelions sit at local threefold axes. In the asymmetric unit with just four hexon trimers, one triskelion at a local threefold axis is observed along with one-third of the triskelion at the icosahedral threefold axis (Figure 1(B)). The polypeptide orientation of the refined X-ray atomic model places the N-termini of protein IX at the distal ends of the triskelion and the middle of the protein IX sequence (~aa 77) at the center of the triskelion.

The protein IX C-terminal domain has a heptad-repeat motif typical of a helix bundle.[87] High-resolution (4–5 Å) cryoEM structures of two bovine adenovirus intermediates showed three-helix coiled coils above the trimeric regions formed by the N-terminal domains of protein IX.[88] No coiled coils are observed in these locations in the human adenovirus structures. In fact, no density at all is observed for the C-terminal domains of protein IX in the refined X-ray structure.[16] The fact that the linker region between the conserved N-terminal region and the predicted C-terminal coiled coil is significantly shorter in bovine adenovirus type 3 (BAdV3) (~24 aa) than in HAdV5 (~42 aa) may explain why a protein IX coiled coil is only observed for BAdV3 and not for human adenoviruses.

A moderate resolution cryoEM structure of the canine adenovirus CAdV2 showed cylinders of density above the protein IX triskelions in the same place as the coiled coils in the BAdV3 structures.[89] As for BAdV3, the linker between the N- and the C-terminal domains of protein IX is significantly smaller in CAdV2 (~15 aa) than in HAdV5 (~42 aa). To help support the assignment of the cylinders to the C-terminal domain of protein IX, Schoehn et al. determined a cryoEM structure of CAdV2 with GFP fused to the C terminus of protein IX.[89] As expected, extra density assigned to GFP was observed above the cylinders. It seems reasonable to conclude that the relatively long linker in HAdV5 protein IX may prevent formation of a rigid coiled-coil bundle extending directly above the N-terminal triskelion region of protein IX.

Given the homology among the N-terminal domains of protein IX among human, bovine, and canine adenovirus, it also seems reasonable to assume that all of these domains have the same fold in the context of intact virions. Assuming that the refined X-ray atomic model is correct,[16] this means that the middle of the protein IX sequence is appropriately placed to have a coiled-coil form above the protein IX triskelion if the linker is short enough. This is apparently the case for both BAdV3 and CAdV2 but not for any of the human adenovirus types that have been studied by cryoEM or X-ray crystallography, including HAdV2, HAdV5, and HAdV12.

## 11. Core Protein V Structure

One unexpected finding in the refined X-ray atomic model of adenovirus is the positioning of a portion of core protein V on the inner capsid surface (Figure 2(A)).[16] An atomic model was built for 72 residues of protein V (aa 208–219 and 236–295) out of a total of 368 residues. This region of protein V interacts with protein VI below the peripentonal hexons. This positioning is consistent with cross-linking experiments that indicated that proteins V and VI interact within the virion.[90,91] The ordered region of protein V is also observed to interact with the copy of protein VIII that is closest to the vertex. The complex of proteins V, VI, and VIII is observed to stabilize the peripentonal hexons and link them to the adjacent group-of-nine hexons.[16]

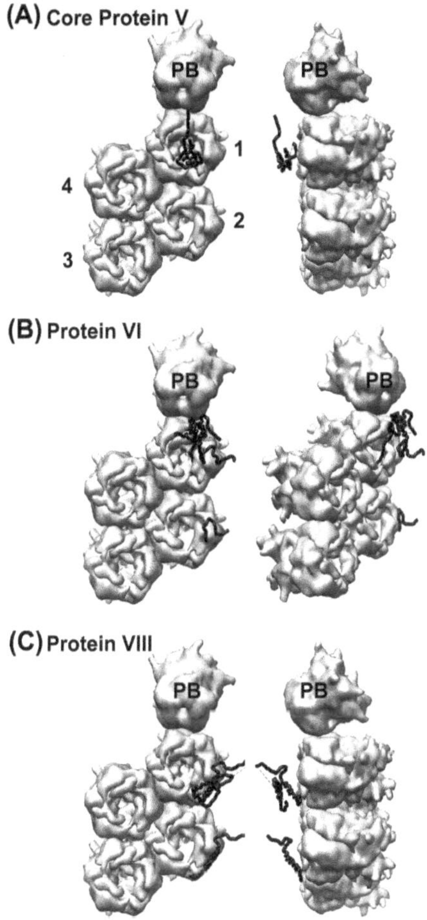


**Figure 2** Structure and location of the inner capsid proteins as assigned in the refined adenovirus crystal structure.[16] (A) The enlarged asymmetric unit, with four independent hexon trimers (1–4) and a complete penton base (PB), is shown as in Figure 1 but viewed from the inside of the capsid together with the ordered portion of core protein V (black). Protein V is chain T in PDB-ID: 4CWU. (B) The enlarged asymmetric unit together with the ordered portions of two copies of protein VI (black). The two copies of protein VI in the asymmetric unit are chains U and V in PDB-ID: 4CWU. (C) The enlarged asymmetric unit together with the ordered portions of two copies of protein VIII (black). The two copies of protein VIII in the asymmetric unit are chains X and Y in PDB-ID: 4CWU. Top and side views are shown in panels A and C. Top and a 45° tilted views are shown in panel B. Dashed lines represent disordered regions. This figure was made with UCSF Chimera.[126]

## 12. Protein VI Structure

Protein VI has multiple functions in the adenovirus lifecycle including regulation of hexon import into the nucleus during adenovirus assembly,[92] disruption of the endosomal membrane during cell entry,[74] and provision of a peptide cofactor for adenovirus protease.[93,94] During the production of progeny virions in host cells, the viral structural proteins are produced in the cytoplasm while the viral genome is replicated and new viral particles are assembled in the nucleus. Wodrich et al. showed that protein VI shuttles between the nucleus and the cytoplasm and links hexon to the nuclear import machinery via an importin alpha/beta-dependent mechanism.[92] Protein VI contains nuclear import and export signals in a short C-terminal segment, which is proteolytically removed by the adenoviral protease during virus maturation. Wiethoff et al. showed that the N-terminal domain of protein VI has a predicted amphipathic α-helix that is required for membrane lytic activity.[74] Release of protein VI from the virion is thought to occur in the endosome during cell entry. In 1993, two groups showed that an 11-residue peptide cleaved from the C-terminus of the precursor form of protein VI serves as a cofactor for the protease.[93,94]

A direct association between protein VI and hexon has been demonstrated[95,96] and protein VI has also been shown to bind DNA.[97] Therefore a location for protein VI on the inner capsid surface of the virion in the vicinity of the viral genome seems most likely. Also consistent with an internal capsid location is the fact that both the N- and the C-terminal peptide regions of protein VI are cleaved by adenovirus protease. There are ~369 copies of protein VI per virion,[80] which corresponds to ~1.5 copies of protein VI per hexon trimer. Saban et al. first noted density bound within the hexon cavities on the inner capsid surface and tentatively assigned it to protein VI.[11] No coordinates for protein VI were deposited with the atomic resolution cryoEM structure or the first X-ray structure of adenovirus.[14,15]

The refined X-ray structure of adenovirus provided the first atomic model for protein VI (Figure 2(B)).[16] Three regions of protein VI were traced (aa 6–31, 34–79, and 87–157). One copy of protein VI is found within the hexon cavity of each peripentonal hexon. The fold of protein VI appears to be distinct and is predominantly α-helical. The predicted amphipathic α-helix of protein VI[74] does not form an α-helix in the refined X-ray structure. However, it may adopt a helical conformation on interaction with the endosomal membrane.[98] One of the three traced regions (aa 6–31) corresponds to the 33-residue N-terminal propeptide that is cleaved by adenovirus protease. The refined X-ray structure shows that after cleavage the ends of the newly formed fragments are separated by ~24 Å. The majority of the residues in the 33-residue N-terminal propeptide are found with the peripentonal hexon cavity. The propeptide interactions with hexon are consistent with hydrogen–deuterium exchange mass spectrometry results that indicate that the N-terminal propeptide associates with peripentonal hexons.[99] The new structural results are also in agreement with the measured high affinity of the precursor form of protein VI to hexon.[100]

## 13. Protein VIII Structure

The assignment of protein VIII to two hammer-like regions per asymmetric unit on the inner capsid surface was first made by Fabry et al.[10] These two regions were also observed in the 6 Å resolution cryoEM structure[12] and in the atomic resolution cryoEM and X-ray structures of adenovirus.[14–16] Adenovirus protease cleaves protein

VIII in two places resulting in three fragments. The refined X-ray structure provides coordinates for fragment 1 in both copies of protein VIII within the asymmetric unit (Figure 2(C)).[16] These coordinates mostly agree with the atomic resolution cryoEM coordinates for fragment 1.[14] The refined X-ray structure also includes coordinates for fragment 3 in one copy of protein VIII, although these coordinates differ significantly from the cryoEM coordinates for fragment 3. No density was observed for fragment 2 (aa 112–157) in the refined X-ray structure and it is possible that this fragment is released from the virion after proteolytic processing.

One copy of protein VIII within the asymmetric unit is below the peripentonal hexons. At this position, protein VIII interacts with proteins V and VI and helps to stabilize the interaction between the peripentonal hexons. The second copy of protein VIII is near the icosahedral twofold axis and interacts with protein VI that is bound to the inner side of hexon in position 2 of the asymmetric unit. Both copies of protein VIII are at the edge of a group-of-nine hexons and help to stabilize adjacent facets of hexons.

## 14. Adenovirus Protease

The adenovirus protease catalyzes the maturational processing of six structural proteins in adenovirus and this step is essential for the production of infectious virus particles.[101,102] These six structural proteins are the precursor forms of proteins IIIa, VI, VII, VIII, mu, and terminal protein (TP).[101,103,104] Three of these are capsid proteins (IIIa, VI, and VIII) and the other three (VII, mu, and TP) are proteins associated with the viral DNA in the core of the virion. Adenovirus protease is also responsible for cleaving the presumed scaffolding protein L1-52K.[105] There are ~50 copies of protease packaged within the core of the virion.[106] Since it plays a critical role in the viral life cycle, adenovirus protease has been proposed as a target for the design and development of antiviral agents to protect against adenovirus infections.[107]

Structures of active[7,108] and inactive[109] forms of adenovirus protease have been determined. The structures confirm the idea proposed earlier that adenovirus protease represents a distinct class of the cysteine proteases.[110] Adenovirus protease was categorized as a cysteine protease on the basis of biochemical and mutagenesis studies.[94,111] Common active site cysteine protease inhibitors are active against adenovirus protease.[106] Originally the sequence of adenovirus protease was unrelated to any other protease sequence in the databases until a weak similarity was found with ubiquitin-like proteinase 1 (Ulp1), which is required for cell-cycle progression in yeast.[112] Recently, two other viral proteases have been added to the adenovirus protease family. These are from vaccinia virus[113] and African swine fever virus.[114] A few other proteins have been found to be homologous to adenovirus protease, including two paralogous gene products in *Chlamydia*,[115] a virulence factor in *Yersinia pestis*, YopJ,[116] and a protease involved in the regulation of chromosome condensation in *Saccharomyces cerevisiae*.[117]

When adenovirus protease is compared to papain, the archetypical cysteine protease, the order of the catalytic Cys and His residues in the primary sequence is different, with His54 followed by Cys122 in adenovirus protease and Cys25 followed by His159 in papain.[118–120] Remarkably, however, when the structure of the active form of adenovirus protease is superimposed on that of papain, the active site Cys–His–Glu triplet and the oxyanion hole are found in similar arrangements in both proteins.[7] This alignment of catalytic elements despite quite different protein folds strongly suggests that adenovirus protease employs the same catalytic mechanism as papain.[121]

Purified recombinant adenovirus protease is inactive and two cofactors are needed for maximal activity. One cofactor is an 11-residue peptide cleaved from the C terminus of the precursor form of protein VI (pVIc),[93,94] and another is the viral DNA.[93] The sequence of pVIc is GVQSLKRRRCF. Adenovirus protease is likely to cleave its own peptide cofactor from the precursor protein, as there is a consensus cleavage site immediately preceding pVIc in the precursor of protein VI (IVGL/GVQS). The structures of the active form of the enzyme show that the pVIc peptide forms a sixth strand of a β-sheet and a disulfide linkage with Cys-104 of adenovirus protease. This disulfide bond forms both in vitro[7,108] and in vivo[122] in the virus particle. The specificity constant ($k_{cat}/K_m$) of adenovirus protease is increased by both cofactors.[93,123–125] The viral DNA increases the specificity constant by 110-fold. The pVIc peptide increases it 1130-fold. Both cofactors together result in maximum activity, with an increase in the specificity constant of 15,800-fold.

The fold of adenovirus protease is of the $\alpha+\beta$ type, with five β-strands, six α-helices, and two 3–10 helices. These elements form two domains, an α-helical domain and a β-strand domain, with the active site at the domain interface. Comparing the structure of the inactive form with that of the active form in complex with the pVIc peptide reveals why protease is inactive without the pVIc peptide cofactor.[109] The major structural difference between the two forms is in the β-sheet domain.[109] In the structure of the active form with the pVIc cofactor, the general base His-54 Nδ1 is close enough (3.9 Å) to Cys-122 Sγ to be nucleophilic. In contrast, in the structure of the inactive form, His-54 Nδ1 is 7.0 Å away from Cys-122 Sγ and too far to act as a nucleophile. There are multiple conformational changes that must occur between the inactive and the active forms of the proteins, including reduction in flexibility of a critical loop region. The activation of adenovirus protease by pVIc is proposed to occur along a 62-amino acid pathway of contiguous conformational changes.[109]

## 15. Concluding Remarks

Our knowledge of the molecular composition and structure of adenovirus has greatly increased since the early electron micrographs of the virus taken from before 1965.[1] X-ray crystal structures of isolated hexon,[3,34,38] penton base,[4] and fragments of the fiber[5,6] have helped us to appreciate these basic molecular building blocks of the virus. Crystal structures of the active and inactive form of adenovirus protease have led to an understanding of how the protease is activated by a peptide cofactor.[7,108,109]

Both cryoEM and X-ray crystallography have contributed to our understanding of the architecture of the intact adenovirus virion. However, the large size and high level of complexity of adenovirus have led to many twists and turns along the path to understanding. The first cryoEM difference map of adenovirus, which was generated by docking multiple copies of the hexon crystal structure into the cryoEM density and subtracting the hexon density, revealed minor capsid proteins stabilizing the capsid on both the exterior surface and the inner capsid surface facing the viral core.[13] More details about these minor capsid proteins were revealed in moderate resolution cryoEM structures.[10–12] Molecular mass, copy number, and predicted α-helical content were used to make tentative assignments of density to the minor capsid proteins in efforts to describe the architecture of the virus. CryoEM and molecular tagging studies were also used in attempts to pinpoint the locations of the minor capsid proteins within the virion.[75,76,79] The size and complexity of adenovirus have made assignment of the minor proteins a daunting challenge.

In 2010, two "atomic resolution" structures were published, one by cryoEM at 3.6Å resolution,[14] and another by X-ray crystallography at 3.5Å resolution.[15] Although both represent tremendous achievements, unfortunately the resolution of both structures was less than ideal for tracing the chains of the minor capsid proteins (IIIa, VI, VIII, IX). The authors had to rely on the presence of density for bulky side chains. This task was especially difficult in the absence of atomic resolution structures of the minor proteins in their isolated forms separate from the virion. Therefore, perhaps not surprisingly, these two structures differed in their interpretations of the locations of the minor capsid proteins.

Four years later, in 2014, with a refined X-ray crystal structure of adenovirus, an attempt was made to set the record straight on the locations and chain tracings of the minor capsid proteins.[16] However, again less than ideal resolution of the density map (3.8Å) made tracing of the minor capsid proteins challenging. A strategy to evaluate the correspondence of the sequence alignment with the somewhat crude side-chain density information available from the density map was developed. This analysis provides a measure of confidence to the new assignments. The minor capsid assignments of the refined X-ray structure are presented in this chapter. These assignments include protein IIIa at the facet edges and protein IX N-terminal regions between the hexons on the exterior of the capsid. On the inner capsid surface facing the viral core, a portion of core protein V is found to interact with protein VI and protein VIII below the peripentonal hexons and a second copy of protein VI is found to interact with a second copy of protein VIII below hexon in position 2 within the asymmetric unit.

Density for the C-terminal domain of protein IX was not resolved in the refined X-ray structure of a human adenovirus vector.[16] In cryoEM structures of canine[89] and bovine[88] adenovirus, density was observed for the C-terminal domain directly above the N-terminal protein IX triskelions. In the case of the bovine structures, the C-terminal domains were resolved as coiled coils[88] as predicted for this domain.[87] One remaining puzzle is where the C-terminal domains of protein IX are in human adenovirus. The longer linker between N- and C-terminal domains in human adenoviruses might lead to completely random positions for the C-terminal domains. Alternatively, perhaps an occasional C-terminal helix of protein IX sits down on the helical cluster of protein IIIa at the facet edge. This possibility might explain the cryoEM tagging results that seem to support the assignment of the coiled coil at the facet edge to protein IX.

One observation on adenovirus structure that is not in question is the incredible complexity of this virus. Another amazing feature of adenovirus is the multifunctionality of many of its "structural" proteins. Many of the proteins that form the icosahedral capsid of adenovirus play additional key roles during the viral life cycle. As we learn more about the structure of the virion, this information will undoubtedly help to guide the development of new adenoviral vectors.

## References


1. Rowe WP, Hartley JW. A general review of the adenoviruses. *Ann NY Acad Sci* 1962; **101**:466–74.
2. van Oostrum J, Burnett RM. Molecular composition of the adenovirus type 2 virion. *J Virol* 1985;**56**:439–48.
3. Roberts MM, White JL, Grutter MG, Burnett RM. Three-dimensional structure of the adenovirus major coat protein hexon. *Science* 1986;**232**:1148–51.
4. Zubieta C, Schoehn G, Chroboczek J, Cusack S. The structure of the human adenovirus 2 penton. *Mol Cell* 2005;**17**:121–35.
5. van Raaij MJ, Mitraki A, Lavigne G, Cusack S. A triple beta-spiral in the adenovirus fibre shaft reveals a new structural motif for a fibrous protein. *Nature* 1999;**401**:935–8.
6. Xia D, Henry LJ, Gerard RD, Deisenhofer J. Crystal structure of the receptor-binding domain of adenovirus type 5 fiber protein at 1.7 Å resolution. *Structure* 1994;**2**:1259–70.
7. Ding J, McGrath WJ, Sweet RM, Mangel WF. Crystal structure of the human adenovirus proteinase with its 11 amino acid cofactor. *EMBO J* 1996;**15**:1778–83.
8. Berk AJ. Adenoviridae: the viruses and their replication. In: Knipe DM, Howley PM, editors. *Fields Virology*. 5th ed. Philadelphia, PA: Lippincott Williams & Wilkins; 2007. p. 2355–94.
9. Russell WC. Adenoviruses: update on structure and function. *J Gen Virol* 2009;**90**:1–20.
10. Fabry CM, Rosa-Calatrava M, Conway JF, Zubieta C, Cusack S, Ruigrok RW, et al. A quasi-atomic model of human adenovirus type 5 capsid. *EMBO J* 2005;**24**:1645–54.
11. Saban SD, Nepomuceno RR, Gritton LD, Nemerow GR, Stewart PL. CryoEM structure at 9 Å resolution of an adenovirus vector targeted to hematopoietic cells. *J Mol Biol* 2005;**349**:526–37.
12. Saban SD, Silvestry M, Nemerow GR, Stewart PL. Visualization of α-helices in a 6-angstrom resolution cryoelectron microscopy structure of adenovirus allows refinement of capsid protein assignments. *J Virol* 2006;**80**:12049–59.
13. Stewart PL, Fuller SD, Burnett RM. Difference imaging of adenovirus: bridging the resolution gap between X-ray crystallography and electron microscopy. *EMBO J* 1993;**12**:2589–99.
14. Liu H, Jin L, Koh SB, Atanasov I, Schein S, Wu L, et al. Atomic structure of human adenovirus by cryo-EM reveals interactions among protein networks. *Science* 2010;**329**:1038–43.
15. Reddy VS, Natchiar SK, Stewart PL, Nemerow GR. Crystal structure of human adenovirus at 3.5 Å resolution. *Science* 2010;**329**:1071–5.
16. Reddy VS, Nemerow GR. Structures and organization of adenovirus cement proteins provide insights into the role of capsid maturation in virus entry and infection. *Proc Natl Acad Sci USA* 2014;**111**:11715–20.

17. Vorburger SA, Hunt KK. Adenoviral gene therapy. *Oncologist* 2002;**7**:46–59.
18. Brown DT, Westphal M, Burlingham BT, Winterhoff U, Doerfler W. Structure and composition of the adenovirus type 2 core. *J Virol* 1975;**16**:366–87.
19. Newcomb WW, Boring JW, Brown JC. Ion etching of human adenovirus 2: structure of the core. *J Virol* 1984;**51**:52–6.
20. Adrian M, Dubochet J, Lepault J, McDowall AW. Cryo-electron microscopy of viruses. *Nature* 1984;**308**:32–6.
21. Stewart PL, Burnett RM, Cyrklaff M, Fuller SD. Image reconstruction reveals the complex molecular organization of adenovirus. *Cell* 1991;**67**:145–54.
22. Chiu CY, Mathias P, Nemerow GR, Stewart PL. Structure of adenovirus complexed with its internalization receptor, $\alpha_V\beta5$ integrin. *J Virol* 1999;**73**:6759–68.
23. Doronin K, Flatt JW, Di Paolo NC, Khare R, Kalyuzhniy O, Acchione M, et al. Coagulation factor X activates innate immunity to human species C adenovirus. *Science* 2012;**338**:795–8.
24. Flatt JW, Kim R, Smith JG, Nemerow GR, Stewart PL. An intrinsically disordered region of the adenovirus capsid is implicated in neutralization by human alpha defensin 5. *PLoS One* 2013;**8**:e61571.
25. Irons EE, Flatt JW, Doronin K, Fox TL, Acchione M, Stewart PL, et al. Coagulation factor binding orientation and dimerization may influence infectivity of adenovirus-coagulation factor complexes. *J Virol* 2013;**87**:9610–9.
26. Kalyuzhniy O, Di Paolo NC, Silvestry M, Hofherr SE, Barry MA, Stewart PL, et al. Adenovirus serotype 5 hexon is critical for virus infection of hepatocytes in vivo. *Proc Natl Acad Sci USA* 2008;**105**:5483–8.
27. Lindert S, Silvestry M, Mullen TM, Nemerow GR, Stewart PL. Cryo-electron microscopy structure of an adenovirus-integrin complex indicates conformational changes in both penton base and integrin. *J Virol* 2009;**83**:11491–501.
28. Smith JG, Silvestry M, Lindert S, Lu W, Nemerow GR, Stewart PL. Insight into the mechanisms of adenovirus capsid disassembly from studies of defensin neutralization. *PLoS Pathog* 2010;**6**:e1000959.
29. Waddington SN, McVey JH, Bhella D, Parker AL, Barker K, Atoda H, et al. Adenovirus serotype 5 hexon mediates liver gene transfer. *Cell* 2008;**132**:397–409.
30. Furcinitti PS, van Oostrum J, Burnett RM. Adenovirus polypeptide IX revealed as capsid cement by difference images from electron microscopy and crystallography. *EMBO J* 1989;**8**:3563–70.
31. Reddy VS, Natchiar SK, Gritton L, Mullen TM, Stewart PL, Nemerow GR. Crystallization and preliminary X-ray diffraction analysis of human adenovirus. *Virology* 2010;**402**:209–14.
32. Robinson CM, Singh G, Lee JY, Dehghan S, Rajaiya J, Liu EB, et al. Molecular evolution of human adenoviruses. *Sci Rep* 2013;**3**:1812.
33. Benko M, Harrach B. Molecular evolution of adenoviruses. *Curr Top Microbiol Immunol* 2003;**272**:3–35.
34. Rux JJ, Kuser PR, Burnett RM. Structural and phylogenetic analysis of adenovirus hexons by use of high-resolution X-ray crystallographic, molecular modeling, and sequence-based methods. *J Virol* 2003;**77**:9553–66.
35. Benson SD, Bamford JK, Bamford DH, Burnett RM. Viral evolution revealed by bacteriophage PRD1 and human adenovirus coat protein structures. *Cell* 1999;**98**:825–33.
36. Cepko CL, Sharp PA. Assembly of adenovirus major capsid protein is mediated by a nonvirion protein. *Cell* 1982;**31**:407–15.

37. Hong SS, Szolajska E, Schoehn G, Franqueville L, Myhre S, Lindholm L, et al. The 100K-chaperone protein from adenovirus serotype 2 (Subgroup C) assists in trimerization and nuclear localization of hexons from subgroups C and B adenoviruses. *J Mol Biol* 2005;**352**:125–38.
38. Rux JJ, Burnett RM. Type-specific epitope locations revealed by X-ray crystallographic study of adenovirus type 5 hexon. *Mol Ther* 2000;**1**:18–30.
39. Crawford-Miksza L, Schnurr DP. Analysis of 15 adenovirus hexon proteins reveals the location and structure of seven hypervariable regions containing serotype-specific residues. *J Virol* 1996;**70**:1836–44.
40. Bradley RR, Maxfield LF, Lynch DM, Iampietro MJ, Borducchi EN, Barouch DH. Adenovirus serotype 5-specific neutralizing antibodies target multiple hexon hypervariable regions. *J Virol* 2012;**86**:1267–72.
41. Bradley RR, Lynch DM, Iampietro MJ, Borducchi EN, Barouch DH. Adenovirus serotype 5 neutralizing antibodies target both hexon and fiber following vaccination and natural infection. *J Virol* 2012;**86**:625–9.
42. Wu H, Han T, Belousova N, Krasnykh V, Kashentseva E, Dmitriev I, et al. Identification of sites in adenovirus hexon for foreign peptide incorporation. *J Virol* 2005;**79**:3382–90.
43. Bradshaw AC, Parker AL, Duffy MR, Coughlan L, van Rooijen N, Kahari VM, et al. Requirements for receptor engagement during infection by adenovirus complexed with blood coagulation factor X. *PLoS Pathog* 2010;**6**:e1001142.
44. van Oostrum J, Smith PR, Mohraz M, Burnett RM. The structure of the adenovirus capsid. III. Hexon packing determined from electron micrographs of capsid fragments. *J Mol Biol* 1987;**198**:73–89.
45. Ruigrok RW, Barge A, Albiges-Rizo C, Dayan S. Structure of adenovirus fibre. II. Morphology of single fibres. *J Mol Biol* 1990;**215**:589–96.
46. Schoehn G, Fender P, Chroboczek J, Hewat EA. Adenovirus 3 penton dodecahedron exhibits structural changes of the base on fibre binding. *EMBO J* 1996;**15**:6841–6.
47. Madisch I, Hofmayer S, Moritz C, Grintzalis A, Hainmueller J, Pring-Akerblom P, et al. Phylogenetic analysis and structural predictions of human adenovirus penton proteins as a basis for tissue-specific adenovirus vector design. *J Virol* 2007;**81**:8270–81.
48. Nemerow GR, Pache L, Reddy V, Stewart PL. Insights into adenovirus host cell interactions from structural studies. *Virology* 2009;**384**:380–8.
49. Li E, Stupack D, Bokoch GM, Nemerow GR. Adenovirus endocytosis requires actin cytoskeleton reorganization mediated by Rho family GTPases. *J Virol* 1998;**72**:8806–12.
50. Albinsson B, Kidd AH. Adenovirus type 41 lacks an RGD α(v)-integrin binding motif on the penton base and undergoes delayed uptake in A549 cells. *Virus Res* 1999;**64**:125–36.
51. Davison AJ, Telford EA, Watson MS, McBride K, Mautner V. The DNA sequence of adenovirus type 40. *J Mol Biol* 1993;**234**:1308–16.
52. Stewart PL, Chiu CY, Huang S, Muir T, Zhao Y, Chait B, et al. Cryo-EM visualization of an exposed RGD epitope on adenovirus that escapes antibody neutralization. *EMBO J* 1997;**16**:1189–98.
53. Zhou HX, Pang X, Lu C. Rate constants and mechanisms of intrinsically disordered proteins binding to structured targets. *Phys Chem Chem Phys* 2012;**14**:10466–76.
54. Bewley MC, Springer K, Zhang YB, Freimuth P, Flanagan JM. Structural analysis of the mechanism of adenovirus binding to its human cellular receptor, CAR. *Science* 1999;**286**:1579–83.

55. Burmeister WP, Guilligay D, Cusack S, Wadell G, Arnberg N. Crystal structure of species D adenovirus fiber knobs and their sialic acid binding sites. *J Virol* 2004;**78**:7727–36.
56. Cupelli K, Muller S, Persson BD, Jost M, Arnberg N, Stehle T. Structure of adenovirus type 21 knob in complex with CD46 reveals key differences in receptor contacts among species B adenoviruses. *J Virol* 2010;**84**:3189–200.
57. Pache L, Venkataraman S, Nemerow GR, Reddy VS. Conservation of fiber structure and CD46 usage by subgroup B2 adenoviruses. *Virology* 2008;**375**:573–9.
58. Pache L, Venkataraman S, Reddy VS, Nemerow GR. Structural variations in species B adenovirus fibers impact CD46 association. *J Virol* 2008;**82**:7923–31.
59. Persson BD, Muller S, Reiter DM, Schmitt BB, Marttila M, Sumowski CV, et al. An arginine switch in the species B adenovirus knob determines high-affinity engagement of cellular receptor CD46. *J Virol* 2009;**83**:673–86.
60. Wang H, Liaw YC, Stone D, Kalyuzhniy O, Amiraslanov I, Tuve S, et al. Identification of CD46 binding sites within the adenovirus serotype 35 fiber knob. *J Virol* 2007;**81**:12785–92.
61. Wang H, Yumul R, Cao H, Ran L, Fan X, Richter M, et al. Structural and functional studies on the interaction of adenovirus fiber knobs and desmoglein 2. *J Virol* 2013;**87**:11346–62.
62. Guardado-Calvo P, Munoz EM, Llamas-Saiz AL, Fox GC, Kahn R, Curiel DT, et al. Crystallographic structure of porcine adenovirus type 4 fiber head and galectin domains. *J Virol* 2010;**84**:10558–68.
63. Seiradake E, Henaff D, Wodrich H, Billet O, Perreau M, Hippert C, et al. The cell adhesion molecule "CAR" and sialic acid on human erythrocytes influence adenovirus in vivo biodistribution. *PLoS Pathog* 2009;**5**:e1000277.
64. Chroboczek J, Ruigrok RW, Cusack S. Adenovirus fiber. *Curr Top Microbiol Immunol* 1995;**199**(Pt 1):163–200.
65. Chappell JD, Prota AE, Dermody TS, Stehle T. Crystal structure of reovirus attachment protein sigma1 reveals evolutionary relationship to adenovirus fiber. *EMBO J* 2002;**21**:1–11.
66. Liu H, Wu L, Zhou ZH. Model of the trimeric fiber and its interactions with the pentameric penton base of human adenovirus by cryo-electron microscopy. *J Mol Biol* 2011;**406**:764–74.
67. Nilsson EC, Storm RJ, Bauer J, Johansson SM, Lookene A, Angstrom J, et al. The GD1a glycan is a cellular receptor for adenoviruses causing epidemic keratoconjunctivitis. *Nat Med* 2011;**17**:105–9.
68. Wang H, Li ZY, Liu Y, Persson J, Beyer I, Moller T, et al. Desmoglein 2 is a receptor for adenovirus serotypes 3, 7, 11 and 14. *Nat Med* 2011;**17**:96–104.
69. Zhang Y, Bergelson JM. Adenovirus receptors. *J Virol* 2005;**79**:12125–31.
70. Seiradake E, Lortat-Jacob H, Billet O, Kremer EJ, Cusack S. Structural and mutational analysis of human Ad37 and canine adenovirus 2 fiber heads in complex with the D1 domain of coxsackie and adenovirus receptor. *J Biol Chem* 2006;**281**:33704–16.
71. Persson BD, Reiter DM, Marttila M, Mei YF, Casasnovas JM, Arnberg N, et al. Adenovirus type 11 binding alters the conformation of its receptor CD46. *Nat Struct Mol Biol* 2007;**14**:164–6.
72. Persson BD, Schmitz NB, Santiago C, Zocher G, Larvie M, Scheu U, et al. Structure of the extracellular portion of CD46 provides insights into its interactions with complement proteins and pathogens. *PLoS Pathog* 2010;**6**:e1001122.
73. Greber UF, Willetts M, Webster P, Helenius A. Stepwise dismantling of adenovirus 2 during entry into cells. *Cell* 1993;**75**:477–86.

74. Wiethoff CM, Wodrich H, Gerace L, Nemerow GR. Adenovirus protein VI mediates membrane disruption following capsid disassembly. *J Virol* 2005;**79**:1992–2000.
75. Fabry CM, Rosa-Calatrava M, Moriscot C, Ruigrok RW, Boulanger P, Schoehn G. The C-terminal domains of adenovirus serotype 5 protein IX assemble into an antiparallel structure on the facets of the capsid. *J Virol* 2009;**83**:1135–9.
76. Marsh MP, Campos SK, Baker ML, Chen CY, Chiu W, Barry MA. Cryoelectron microscopy of protein IX-modified adenoviruses suggests a new position for the C terminus of protein IX. *J Virol* 2006;**80**:11881–6.
77. Boudin ML, D'Halluin JC, Cousin C, Boulanger P. Human adenovirus type 2 protein IIIa. II. Maturation and encapsidation. *Virology* 1980;**101**:144–56.
78. Chroboczek J, Viard F, D'Halluin JC. Human adenovirus 2 temperature-sensitive mutant 112 contains three mutations in the protein IIIa gene. *Gene* 1986;**49**:157–60.
79. San Martin C, Glasgow JN, Borovjagin A, Beatty MS, Kashentseva EA, Curiel DT, et al. Localization of the N-terminus of minor coat protein IIIa in the adenovirus capsid. *J Mol Biol* 2008;**383**:923–34.
80. Lehmberg E, Traina JA, Chakel JA, Chang RJ, Parkman M, McCaman MT, et al. Reversed-phase high-performance liquid chromatographic assay for the adenovirus type 5 proteome. *J Chromatogr B Biomed Sci Appl* 1999;**732**:411–23.
81. Anderson CW. The proteinase polypeptide of adenovirus serotype 2 virions. *Virology* 1990;**177**:259–72.
82. Caravokyri C, Leppard KN. Constitutive episomal expression of polypeptide IX (pIX) in a 293-based cell line complements the deficiency of pIX mutant adenovirus type 5. *J Virol* 1995;**69**:6627–33.
83. Colby WW, Shenk T. Adenovirus type 5 virions can be assembled in vivo in the absence of detectable polypeptide IX. *J Virol* 1981;**39**:977–80.
84. Parks RJ. Adenovirus protein IX: a new look at an old protein. *Mol Ther* 2005;**11**:19–25.
85. Rosa-Calatrava M, Grave L, Puvion-Dutilleul F, Chatton B, Kedinger C. Functional analysis of adenovirus protein IX identifies domains involved in capsid stability, transcriptional activity, and nuclear reorganization. *J Virol* 2001;**75**:7131–41.
86. Vellinga J, van den Wollenberg DJ, van der Heijdt S, Rabelink MJ, Hoeben RC. The coiled-coil domain of the adenovirus type 5 protein IX is dispensable for capsid incorporation and thermostability. *J Virol* 2005;**79**:3206–10.
87. Lupas A, Van Dyke M, Stock J. Predicting coiled coils from protein sequences. *Science* 1991;**252**:1162–4.
88. Cheng L, Huang X, Li X, Xiong W, Sun W, Yang C, et al. Cryo-EM structures of two bovine adenovirus type 3 intermediates. *Virology* 2014;**450–451**:174–81.
89. Schoehn G, El Bakkouri M, Fabry CM, Billet O, Estrozi LF, Le L, et al. Three-dimensional structure of canine adenovirus serotype 2 capsid. *J Virol* 2008;**82**:3192–203.
90. Chatterjee PK, Vayda ME, Flint SJ. Interactions among the three adenovirus core proteins. *J Virol* 1985;**55**:379–86.
91. Everitt E, Lutter L, Philipson L. Structural proteins of adenoviruses. XII. Location and neighbor relationship among proteins of adenovirion type 2 as revealed by enzymatic iodination, immunoprecipitation and chemical cross-linking. *Virology* 1975;**67**:197–208.
92. Wodrich H, Guan T, Cingolani G, Von Seggern D, Nemerow G, Gerace L. Switch from capsid protein import to adenovirus assembly by cleavage of nuclear transport signals. *EMBO J* 2003;**22**:6245–55.
93. Mangel WF, McGrath WJ, Toledo DL, Anderson CW. Viral DNA and a viral peptide can act as cofactors of adenovirus virion proteinase activity. *Nature* 1993;**361**:274–5.
94. Webster A, Hay RT, Kemp G. The adenovirus protease is activated by a virus-coded disulphide-linked peptide. *Cell* 1993;**72**:97–104.

95. Matthews DA, Russell WC. Adenovirus protein-protein interactions: hexon and protein VI. *J Gen Virol* 1994;**75**(Pt 12):3365–74.
96. Matthews DA, Russell WC. Adenovirus protein-protein interactions: molecular parameters governing the binding of protein VI to hexon and the activation of the adenovirus 23K protease. *J Gen Virol* 1995;**76**(Pt 8):1959–69.
97. Russell WC, Precious B. Nucleic acid-binding properties of adenovirus structural polypeptides. *J Gen Virol* 1982;**63**(Pt 1):69–79.
98. Drin G, Antonny B. Amphipathic helices and membrane curvature. *FEBS Lett* 2010;**584**:1840–7.
99. Snijder J, Benevento M, Moyer CL, Reddy V, Nemerow GR, Heck AJ. The cleaved N-terminus of pVI binds peripentonal hexons in mature adenovirus. *J Mol Biol* 2014; **426**:1971–9.
100. Graziano V, McGrath WJ, Suomalainen M, Greber UF, Freimuth P, Blainey PC, et al. Regulation of a viral proteinase by a peptide and DNA in one-dimensional space: I. binding to DNA AND to hexon of the precursor to protein VI, pVI, of human adenovirus. *J Biol Chem* 2013;**288**:2059–67.
101. Weber JM. Adenovirus endopeptidase and its role in virus infection. *Curr Top Microbiol Immunol* 1995;**199**(Pt 1):227–35.
102. Weber JM, Tihanyi K. Adenovirus endopeptidases. *Methods Enzymol* 1994;**244**:595–604.
103. McGrath WJ, Abola AP, Toledo DL, Brown MT, Mangel WF. Characterization of human adenovirus proteinase activity in disrupted virus particles. *Virology* 1996;**217**:131–8.
104. Webster A, Russell S, Talbot P, Russell WC, Kemp GD. Characterization of the adenovirus proteinase: substrate specificity. *J Gen Virol* 1989;**70**(Pt 12):3225–34.
105. Perez Berna AJ, Mangel WF, McGrath WJ, Graziano V, Flint J, San Martin C. Processing of the ll 52/55k protein by the adenovirus protease: a new substrate and new insights into virion maturation. *J Virol* 2014;**88**:1513–24.
106. Brown MT, McGrath WJ, Toledo DL, Mangel WF. Different modes of inhibition of human adenovirus proteinase, probably a cysteine proteinase, by bovine pancreatic trypsin inhibitor. *FEBS Lett* 1996;**388**:233–7.
107. Pang YP, Xu K, Kollmeyer TM, Perola E, McGrath WJ, Green DT, et al. Discovery of a new inhibitor lead of adenovirus proteinase: steps toward selective, irreversible inhibitors of cysteine proteinases. *FEBS Lett* 2001;**502**:93–7.
108. McGrath WJ, Ding J, Didwania A, Sweet RM, Mangel WF. Crystallographic structure at 1.6-Å resolution of the human adenovirus proteinase in a covalent complex with its 11-amino-acid peptide cofactor: insights on a new fold. *Biochim Biophys Acta* 2003;**1648**:1–11.
109. Baniecki ML, McGrath WJ, Mangel WF. Regulation of a viral proteinase by a peptide and DNA in one-dimensional space: III. Atomic resolution structure of the nascent form of the adenovirus proteinase. *J Biol Chem* 2013;**288**:2081–91.
110. Rawlings ND, Barrett AJ. Families of cysteine peptidases. *Methods Enzymol* 1994;**244**:461–86.
111. Tihanyi K, Bourbonniere M, Houde A, Rancourt C, Weber JM. Isolation and properties of adenovirus type 2 proteinase. *J Biol Chem* 1993;**268**:1780–5.
112. Li SJ, Hochstrasser M. A new protease required for cell-cycle progression in yeast. *Nature* 1999;**398**:246–51.
113. Ansarah-Sobrinho C, Moss B. Role of the I7 protein in proteolytic processing of vaccinia virus membrane and core components. *J Virol* 2004;**78**:6335–43.
114. Andres G, Alejo A, Simon-Mateo C, Salas ML. African swine fever virus protease, a new viral member of the SUMO-1-specific protease family. *J Biol Chem* 2001;**276**:780–7.

115. Stephens RS, Kalman S, Lammel C, Fan J, Marathe R, Aravind L, et al. Genome sequence of an obligate intracellular pathogen of humans: Chlamydia trachomatis. *Science* 1998;**282**:754–9.
116. Orth K, Xu Z, Mudgett MB, Bao ZQ, Palmer LE, Bliska JB, et al. Disruption of signaling by Yersinia effector YopJ, a ubiquitin-like protein protease. *Science* 2000;**290**:1594–7.
117. Strunnikov AV, Aravind L, Koonin EV. Saccharomyces cerevisiae SMT4 encodes an evolutionarily conserved protease with a role in chromosome condensation regulation. *Genetics* 2001;**158**:95–107.
118. Grierson AW, Nicholson R, Talbot P, Webster A, Kemp G. The protease of adenovirus serotype 2 requires cysteine residues for both activation and catalysis. *J Gen Virol* 1994;**75**(Pt 10):2761–4.
119. Jones SJ, Iqbal M, Grierson AW, Kemp G. Activation of the protease from human adenovirus type 2 is accompanied by a conformational change that is dependent on cysteine-104. *J Gen Virol* 1996;**77**(Pt 8):1821–4.
120. Rancourt C, Tihanyi K, Bourbonniere M, Weber JM. Identification of active-site residues of the adenovirus endopeptidase. *Proc Natl Acad Sci USA* 1994;**91**:844–7.
121. Polgar L. Mercaptide-imidazolium ion-pair: the reactive nucleophile in papain catalysis. *FEBS Lett* 1974;**47**:15–8.
122. McGrath WJ, Aherne KS, Mangel WF. In the virion, the 11-amino-acid peptide cofactor pVIc is covalently linked to the adenovirus proteinase. *Virology* 2002;**296**:234–40.
123. Baniecki ML, McGrath WJ, McWhirter SM, Li C, Toledo DL, Pellicena P, et al. Interaction of the human adenovirus proteinase with its 11-amino acid cofactor pVIc. *Biochemistry* 2001;**40**:12349–56.
124. Mangel WF, Toledo DL, Brown MT, Martin JH, McGrath WJ. Characterization of three components of human adenovirus proteinase activity in vitro. *J Biol Chem* 1996;**271**:536–43.
125. McGrath WJ, Baniecki ML, Li C, McWhirter SM, Brown MT, Toledo DL, et al. Human adenovirus proteinase: DNA binding and stimulation of proteinase activity by DNA. *Biochemistry* 2001;**40**:13237–45.
126. Goddard TD, Huang CC, Ferrin TE. Software extensions to UCSF chimera for interactive visualization of large molecular assemblies. *Structure* 2005;**13**:473–82.

# Biology of Adenovirus Cell Entry: Receptors, Pathways, Mechanisms

**2**

*Stefania Luisoni, Urs F. Greber*
Institute of Molecular Life Sciences, University of Zurich, Zurich, Switzerland

## 1. Introduction

The family of *Adenoviridae* is composed of a large number of pathogens that cause respiratory, ocular, blood-borne, and intestinal infections and can heavily affect immune-compromised individuals.[1,2] Adenoviruses (AdVs) are widely used vectors for virotherapies (http://www.wiley.com/legacy/wileychi/genmed/clinical/). These nonenveloped nanoparticles can be grown to high titers and purified to near homogeneity under good manufacturing practice.[3,4] Adenoviruses are popular gene delivery agents because they transduce both dividing and quiescent cells efficiently, have high physical and genetic stability, and occur in a large number of variants yielding high application flexibility. The infection biology of certain AdV types has been elucidated in sufficient detail to grant clinical applications. In particular, deep mechanistic insight has been gained with human AdV (HAdV) types of the species B and C, and members of the HAdV-C are widely used in oncolytic therapy. HAdVs E4 and B7 have been used successfully as a vaccine in enteric applications for military personnel.[5,6] The therapeutic potential of AdV-based vectors for clinical applications is broad and includes novel vaccination strategies and oncolysis.[7,8] A major challenge for the field has been to better understand features that make AdVs so highly immunogenic and distinct features that render AdV vectors the most popular cancer cell killers. Such insights will provide a basis to improve vectors: for example, to increase their circulation time and better target them to cells of interest in clinical applications.[9]

## 2. Entry Pathways: Impact of Capsid Proteins

Virus entry determines the efficacy of virotherapies, disease onset, and progression. Adenovirus entry is best characterized in epithelial cells, although entry into immune cells is of emerging importance (for reviews, see Refs 10, 11). Adenovirus entry occurs in sequential steps, including virus attachment to target cells, signaling, endocytosis, endosomal escape, transport through the cytosol, separation of the viral genome from the capsid, and delivery of the genome into the nucleus (reviewed in Refs 12–16). A schematic summary of generic HAdV entry is depicted in Figure 1. Even minor defects or enhancements in the entry process, for example by soluble factors such as serum factors, have an impact on viral gene expression or innate immunity responses against the incoming virus.

Adenoviruses are icosahedral particles of pseudo $T=25$ triangulation, in which each facet has 12 trimers of the major coat protein hexon. The hexons contain exposed

**Adenoviral Vectors for Gene Therapy.**

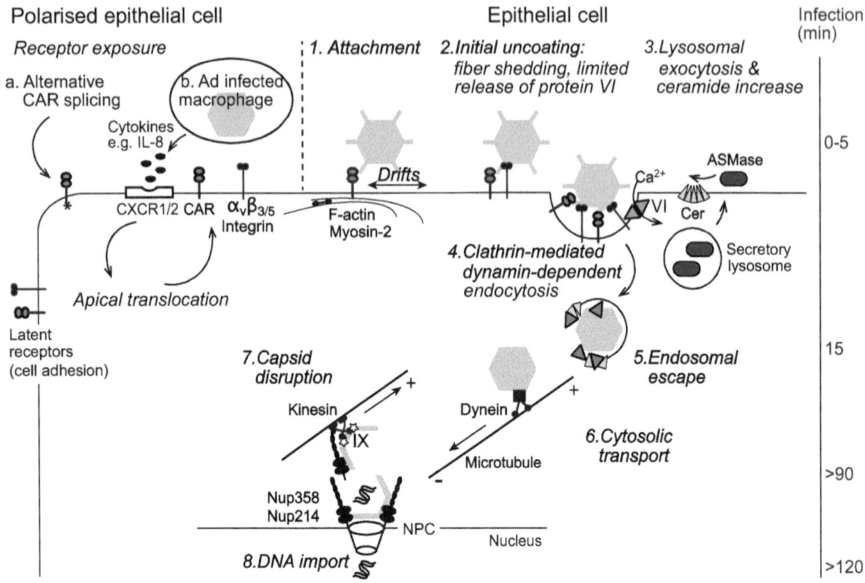


**Figure 1 Schematic depiction of HAdV-C2/5 entry into epithelial cells.** For infection of polarized epithelial cells, which lack HAdV-C2/5 receptors on the luminal apical side, alternatively spliced forms of Coxsackie virus B AdV receptor (CAR) can be transported apically in low amounts.[255] In addition, innate immune responses of macrophages and other immune cells stimulate the expression of cytokines and chemokines, such as IL-8, and thereby lead to apical localization of CAR and integrin receptors.[55,56] Interleukin-8 binds to the CXCR1/2 receptor and triggers translocation of CAR and integrin receptors to the apical surface of epithelial cells. (1) This allows virus attachment to CAR exposed on the cell surface, followed by drifting motions and retrograde flow mediated by F-actin and myosin-2 whereas particles are still outside the cell.[34,256] Eventually, virus binds to integrins, which are immobile and have a lower affinity to the virus than CAR.[63,257] (2) Mechanical forces generated by receptor dynamics initiate the stepwise uncoating process.[172] Fiber shedding and the release of limited amounts of the membrane lytic protein VI from the virus occur on the cell surface.[34,35,258] This is likely aided by interaction between integrins and penton base, inducing a clockwise untwisting of the pentamer and thereby loosening the viral capsid.[144] (3) Transient protein VI-mediated permeabilization of the plasma membrane and $Ca^{2+}$ influx elicit lysosomal exocytosis, a cellular response to membrane injury.[40] Secreted lysosomal acid sphingomyelinase (ASMase) generates ceramides. (4) Virus endocytosis requires integrin signaling and is clathrin-mediated and dynamin-dependent. (5) As virus is endocytosed, additional protein VI is released and interacts with ceramide lipids to enhance endosomal breakage. Thereby, the virus is released to the cytosol and avoids passage through acidic compartments.[40,173] (6) Partially uncoated HAdV-C2/5 binds to cytosolic dynein via hexon and travels on microtubules toward the microtubule organizing center.[21,225,226,259] (7) Virions dock on the cytosolic side of the nuclear pore complex (NPC) by engaging with the nucleoporin Nup214.[232,243,244,246] Simultaneously, protein XI and Nup358 bind to kinesin-1, which exerts forces on the capsid and the NPC and thereby sets free the viral DNA genome and displaces a fraction of the nucleoporins.[246] (8) Viral DNA in association with the condensing protein VII is imported into the nucleus.[229,232,260,261]

hypervariable regions (HVRs), which are subject to immune recognition by both neutralizing antibodies and virus-specific T-cell responses.[17] They also interact with host proteins, including blood factors and cytoplasmic motors.[18–21] A trimeric fiber is anchored at each vertex to a pentameric penton base, which binds to integrins and controls virus uptake and cell signaling (for reviews, see Refs 13, 22, 23). The capsid is held together with a range of cementing proteins, as indicated by an X-ray crystal structure and a cryo-electromagnetic (EM) structure.[24,25] Neither the function nor the location of these cementing proteins is entirely clear, however. In particular, the cryo-EM and the X-ray structures of AdV are not in complete agreement with respect to the location of the cementing proteins. For example, the cryo-EM structure localizes protein IIIa beneath the vertices, in close association with protein VIII,[24] whereas the X-ray structure assigns protein IIIa to a four-helix bundle on the outer surface of the coat, aligned along the icosahedral edges.[25] The X-ray structure places the inner cementing proteins VI, VIII, and the deoxyribonucleic acid (DNA) core protein V near the vertex, and protein IX outside the virion. The vertex anchors the fiber protein, which protrudes away from the capsid and attaches virus to cell surface proteins or sugars (Table 1).[26,27]

The capsid harbors a single copy of a linear double-stranded viral DNA. The capsid is a metastable structure that progressively uncoats in response to cellular cues. This feature ensures diverse functions of the capsid, such as endosomal escape, DNA shielding during cytoplasmic transport, and finally, DNA release for nuclear import (for a review, see Ref. 16). Meta-stability of the capsid is a key feature and is largely controlled by the viral cysteine protease, L3-p23 (AdV protease, AVP), which cleaves the precursors of protein IIIa, VI, VII, VIII, X, terminal protein, and the scaffold protein L1-52/55K (reviewed in Ref. 28). L1-52/55K is important for the assembly of the particle because it provides a bridge between the inner wall of the coat and the viral DNA. Its cleavage by AVP spring-loads the particle to a metastable state capable of responding to cellular cues for uncoating.[29,30] Uncoating is the process by which the viral genome is exposed and then separated from the protein or lipid coat. For example, the temperature-sensitive mutant HAdV-C2_TS1 (TS1) has a packaging-defective AVP and fails to open the coat and release the DNA. Interestingly, the surface of TS1 is indistinguishable from that of wild-type HAdV-C2, although TS1 has more mass in the vertex region than HAdV-C2.[31,32] Because the vertex is the weakest part of the icosahedral coat,[30,33] it is possible that limited proteolysis of vertex proteins renders the particle metastable. It is likely that TS1 cannot interpret the mechanical cues it receives from the differential movements of the viral receptors on the cell surface (described in Fig. 1), unlike HAdV-C2, which is responsive to these cues and exposes the membrane lytic protein VI.[34–36] This is compatible with the notion that TS1 binds to cells and is endocytosed, but fails to uncoat and does not deliver its genome to the nucleus.[34,37–40]

## 3. Attachment Factors and Signaling Receptors

To identify a target cell, AdVs rely on their affinity for attachment factors and signaling receptors. Some receptors are readily available on the site of entry whereas others such as Coxsackie virus B AdV receptor (CAR), desmoglein-2 (DSG2),

**Table 1** **Host Factors Binding to Adenoviruses, and Their Endocytic Uptake Pathways**

| Binding Factor | Expression | Exposure | Serotype | Affinity | Avidity | Virus Binding Domain | Pathway |
|---|---|---|---|---|---|---|---|
| CAR | Broad | Tight junctions; low apical abundance of CAR[ex8] | 2,4,5,9,12, 15,19,31,41,52 and others, except species B,D and G[101,142,262] | **2,5**,9,**12** **41L**[64,262] | | Fiber knob | Clathrin–dynamin; lipid rafts mediated for HAdV-C5 or fiber knob of canine Ad2 in neurons[263] |
| CD46 | Broad | Apical and basolateral[264] | 3,7,11,14,16, 21,35,49,50[265] | **3**,7,**11**,14, **16,21,35** | **3,7**[77] | Fiber knob | Macropinocytosis[81,82] |
| DSG2 | Broad | Desmosome | 3,7,11,14 | 3[91] | **3**[90] | Fiber knob | ? |
| SA | Broad | Cell surface | 8,19,37,52 | 37,52[99,101] | **37**[103] | Fiber knob, interactions of +/– charges | Caveolin for HAdV-D37 in corneal cells[204] |
| HSPG | Broad | Cell surface, ECM | 2,5[113] 3[266] | | **3** | KKTK motive in fiber shaft<br>Penton base, fiber knob[111,266] | ? |
| Integrins | Broad | Basolateral surface | Most types, except 40,41, CAdV-2 | **2,9**[136,143] | **2,5**[257,267] | RGD domain in penton | Clathrin–dynamin; macropinocytosis[38] |
| SR | Macrophages, other cell types[114,268] | Cell surface | 5[18,118] | | | Hexon hypervariable regions (HVRs) 1, 2, 5, 7 | Macropinocytosis? |
| FX | Blood | Soluble | 2,3,5,6,7,11, 13,16,18,35, 37,46,49,50; except 20,29,25,17, 26,28,44,48[a] | **5**[153] | **5,35**[153] | T423, E424, T425 of HVR7 and K10 in FX GLA domain[19] | Residues R93, K96, R125, R165, K169, K236, and R240 in SP domain of FX bind to HSPG.[150] Endocytic pathway? |

| | | | | | | | |
|---|---|---|---|---|---|---|---|
| Lf | Body fluids | Soluble | 1,2,5,6 | | | IIIa and hexon: Far-Western blots of soluble biotinylated-bovine Lf[269] | CAR-independent[160,165] |
| DPPC | Lung surfactant | Soluble or intracellular secretory organelles[270] | 2,5 | | 2 | Hexon: Flotation assay and sodium dodecyl sulfate–polyacrylamide gel electrophoresis[269] | CAR-independent[167] |
| α-Defensins HD5, HNP1 | Extracellular fluids, nasal, lung, female urogenital tract epithelia[146] | Soluble | 2,3,5,7,11, 12,14,16,35 4,19,23,25 37,41,51 | 5[271,272] | | Between penton base and fiber[271] | CAR-mediated uptake, inhibition of uncoating Enhanced infection[271] |
| Secreted Ig | Blood, extracellular fluids | Soluble | Most serotypes, including MAdV-1 | | | Variable | Virus neutralization by C3b coating; Ig-mediated phagocytosis[154,273,274] |

Cell-bound and soluble factors binding to adenoviruses are depicted in light gray and dark gray, respectively. Affinities were derived from binding strengths of soluble virus factors (trimeric fiber knob, penton base, hexon) to host receptors. Avidity represents synergistic but not additive interactions of receptors or cells with highly repetitive ligands such as viruses, penton dodecahedrons (PtDd), oligomeric viral proteins. High affinity or avidity are defined here as $K_D < 1$ μM, indicated in bold. Information about expression of CAR, DSG2, and CD46 can be found at:
http://www.proteinatlas.org/ENSG00000117335-CD46/tissue.
http://atlasgeneticsoncology.org/Genes/GC_DSG2.html.
http://www.proteinatlas.org/ENSG00000154639-CXADR/tissue.
41-L, long fiber of HAdV-F41; 52-S, short fiber of HAdV-G52; Coxsackie virus B AdV receptor (CAR).
[a]$K_D$ is indicated for HAdV-C5, HAdV-B35, and HAdV-D25. The other serotypes are grouped together if they have similar binding properties.

or integrins, are scarce or constrained to the basolateral surface of polarized epithelia. Besides attachment, CAR, integrins, CD46, or DSG2 themselves have a crucial role in virus endocytosis, uncoating, and signaling. These aspects determine productive entry, gene expression, and infection. For most of the other factors proposed to mediate AdV attachment to cells, a specific role in infection has not been identified. Hence, their subsequent discussion will be descriptive.

## 3.1 Accessing the CAR

The CAR is an immunoglobulin family protein implicated in cell adhesion, migration, and growth. It is highly expressed in the myocardium, intestines, pancreas, lung, liver, and kidney and is present in low amounts in adult brain or cancer cells.[27,41,42] CAR connects neighboring cells through homophilic contacts at the lateral and basal surfaces of polarized cells.[43,44] It is not readily exposed to viruses that approach the apical side of polarized epithelia. Nonetheless, it is a high-affinity attachment factor for most HAdVs with the exception of the B, D, and G types.[45–50] It also binds to canine AdV-2 (CAdV-2) and avian AdV CELO.[51]

Three mechanisms have been implicated in rendering CAR accessible to HAdVs and enabling infection of the respiratory, gastrointestinal, and ocular tracts. The first is disruption of tight junctions by a preexisting condition such as physical or chemical stress in asthma, cystic fibrosis, and chronic obstructive pulmonary disease.[52–54] Second, HAdV-C2/5 subverts the innate immune response from macrophages to increase apical CAR availability.[55,56] HAdV-C2/5 is taken up by alveolar macrophages and triggers the release of proinflammatory chemokines and cytokines. Among these, interleukin 8 (IL-8 or cysteine-X [any amino acid]–cysteine type chemokine receptor [CXCL-8]) induces the translocation of CAR and integrin co-receptors to the apical surface of polarized epithelia.[55] CAR relocalization to the apical membrane depends on CXCR-1/2 and Src tyrosine kinases, and mimics a cell migration phenotype. The third mechanism occurs when IL-8 enhances the expression of an alternatively spliced CAR isoform, $CAR^{ex8}$, which traffics in low amounts to the apical surface owing to its C-terminal tail.[56,57] $CAR^{ex8}$ expression is associated with activation of protein kinase B (or AKT) and S6-kinase and inhibition of glycogen synthase kinase-3β. Virus binding to CAR enhances signaling through p42/44 extracellular kinase-1,2 and increases expression of the entry receptors $\beta_{1/3}$ integrins.[58,59] Signaling cascades are reinforced during viral gene expression, when the viral immediate early protein E1A stabilizes the messenger RNAs encoding for proinflammatory cytokines, including IL-8.[60] That CAR has a role in inflammatory processes is supported by the observation that in an inflamed mucosal tissue, this receptor is liberated from its homophilic contacts on polarized cells and engages in activation of immune cells such as neutrophils or gamma-delta T cells.[56,61]

In addition to a well-characterized function in entry, CAR has a role in virus egress from infected epithelia. At late stages of infection when viruses are assembled in the nucleus, structural viral proteins are produced in large amounts, and some of them, including fibers and penton base, exit the infected cell by unconventional secretion.[43,62] The released fibers bind to the lateral surface of CAR, akin to homophilic CAR-CAR binding at basolateral cell adhesion sites of polarized cells.[63] This binding dissolves cell adhesions, most likely involving avidity effects owing to the trimeric nature of the fiber knob.[64] This shows that CAR is a versatile entry and egress receptor for a large number of mastadenoviruses, including human, simian, canine, and chicken AdVs.

## 3.2 Using CD46 in High- and Low-Affinity Modes

Many types of the species B and D, HAdVs attach to the membrane cofactor protein CD46 (see Refs 13, 26). CD46 is an inhibitor of complement activation and is present on all nucleated cells.[65] It is localized apically in polarized epithelial lung tissues,[66] although it can also be expressed basolaterally in other tissue types.[67,68] Virus binding to CD46 occurs top-down on tilted CD46, which suggests that straight fiber shafts projecting from the virus are suitable for this interaction. The trimeric fiber knob attaches to a short consensus repeat in the extracellular variable loop of CD46.[69–76] The knobs of HAdV species B have evolved affinities to CD46 ranging from low nanomolar for species B2, such as HAdV-B11 or B35, to several micromolar for species B1, such as HAdV-B3 or B7. The low apparent affinities are compensated for by avidity mechanisms, which allow the B1 HAdV to bind to CD46-expressing cells with nanomolar affinity.[77] We term this "low-affinity, high-avidity binding." This type of binding is possible for receptors that are highly expressed.

CD46 protects from homologous complement-mediated cell killing, maintains the homeostasis of epithelial barriers, and promotes formation of cellular cytoskeleton.[78,79] Human adenovirus attachment to CD46 triggers receptor clustering and macropinocytosis.[80–82] Ligands of CD46, including fiber knobs of HAdV-B35 and HAdV-B11p, rapidly downregulate surface expression of CD46.[83–85] Together with degradation of CD46, this enhances complement-mediated lysis, alters antigen presentation on major histocompatibility complex (MHC) I and II, and decreases IL-12 secretion in activated human peripheral blood mononuclear cells.[86–88] In summary, CD46 is a versatile receptor for both species B1, which preferentially infect respiratory epithelia, and B2 HAdVs, which predominantly infect the kidney and urinary tracts. CD46 captures HAdV-B and -D with a range of affinities and may function as a docking site for viruses on the apical domain of polarized cells. It will be interesting to determine whether highly pathogenic HAdV-B such as B14 also use CD46. Together, AdVs exploit two different strategies for cell attachment. The first involves high-affinity receptors and ensures binding even when only scarce amounts of receptors are available. The second strategy relies on high avidity toward abundant but low-affinity attachment sites and may provide opportunities for expanding viral tropism and pathogenicity.

## 3.3 Desmoglein-2

The species B1 HAdVs not only bind to CD46 but also exploit another receptor, DSG2.[77,89–91] The interaction between DSG2 and HAdVs is likely of low affinity and is enhanced by avidity. This is suggested by the observation that only intact virus or penton dodecahedron (PtDd) complexes composed of penton base and fibers, but not trimeric fiber knob, measurably attach to DSG2. DSG2 participates in the formation of desmosomes and is largely restricted to the basolateral side of polarized epithelia.[92]

The role of DSG2 in a primary infection in vivo remains unclear, in part because mouse DSG does not bind to HAdV-B3.[89] It is speculated that HAdVs might exploit DSG2 for virus spreading, because cells infected with species B or D viruses release high amounts of PtDd before cell lysis. These assemblies consist of fibers and penton bases,[93,94] interact with DSG2, and initiate a signaling cascade that mimics epithelial to mesenchymal transition.[90] Epithelial to mesenchymal transition widens cellular junctions and exposes basolateral factors, thereby facilitating lateral spreading of viral progeny.[93]

In summary, whereas CD46 is a unique receptor for HAdV-B16, HAdV-B21, HAdV-B35, and HAdV-B50,[89,95] HAdV-B3, and HAdV-B7 bind to DSG2 and engage with CD46 in an avidity-enhanced mechanism.[77] HAdV-B14 and B11 use both receptors with comparable affinities.[95] It is conceivable that DSG2 and CD46 are independent attachment sites for HAdV-B types: that is, DSG2 at desmosomal junctions and CD46 on apical and basolateral surfaces.[66,96,97] It is possible that apical CD46 leads to activation or degradation of DSG2, and thereby opens cell–cell contact sites for viruses to access the basolateral pool of receptors, including DSG2 and integrins. Surprisingly, despite the nearly undetectable apical binding, HAdV-B35 efficiently transduced polarized Caco-2 from the apical side.[98] It will be interesting to determine the underlying mechanisms of this process.

## 3.4 Sialic Acid

Sialic acid (SA) and its variants are low-affinity attachment sites (dissociation constant ($K_D$) between 19 and 37 μM) for epidemic keratoconjunctivitis causing HAdV-D8, HAdV-D19, and HAdV-D37, and the gastroenteric HAdV-G52.[99–101] Sialic acid is highly abundant on the apical side of polarized epithelial tissues, and might therefore serve as an initial anchor for primary infections despite the low binding affinity. The family of sialic acid encompasses over 50 related sugars with variable chemical modifications, linkages, and tissue expression.

HAdVs exploit the ubiquitous Neu5Ac, which is similar to the glycans of GD1a gangliosides.[102] HAdV-D8, HAdV-D19, and HAdV-D37 harbor three conserved positively charged binding sites (pKa=9) on the fiber knob that interact with the negatively charged Neu5Ac (pKa=2.6).[103,104] HAdV-G52 contains a long and a short fiber. Although the long fiber binds to CAR, attachment to some cells also depends on the short fiber, which interacts with sialic acid.[101] Sialic acid is also highly abundant on erythrocytes and may be involved in AdV-mediated hemagglutination and in determining the biodistribution of HAdVs and their vectors.[105]

## 3.5 Heparan Sulfate

Heparan sulfate–containing proteoglycans (HSPG) are linear chains of polymerized disaccharides (approximately 80 residues) attached to either the extracellular matrix

(ECM) or cell surface components, where they serve to reinforce cell adhesion.[106] Interaction with fibronectin regulates the formation of focal adhesions and stress fibers.[107]

HSPG are involved in a multitude of cellular processes, including blood coagulation[108] and liver lipid metabolism.[109] They act as attachment sites for HAdVs that are free or coated with other soluble factors (see also Section 3.8). They are reported to have low affinity for HAdV species C serotypes, and possibly HAdV-B3 and HAdV-B35, which may facilitate downstream interaction of the viruses with integrins.[110–112] Whereas binding of HAdV-C5 occurs through a positively charged lysine–lysine–tyrosine–lysine (KKTK) motif in the fiber shaft,[113] HAdV-B3 apparently uses fiber knob.[111] Mechanisms of HAdV-B35 binding remain unclear and physiological consequences are difficult to elucidate.

## 3.6 Scavenger Receptors and Other Attachment Factors

Scavenger receptors (SR) encompass eight classes of proteins with low sequence homology. They have the propensity to bind negatively charged molecules, are highly expressed on macrophages, and expose an electrostatic positive patch implicated in ligand binding.[114,115] SR have been implicated in the clearance of HAdV vectors by Kupffer cells in the liver.[116–118] Kupffer cells are considered detrimental to systemic adenoviral therapies because they clear a large fraction of the injected HAdV-C5–based vectors.[119–121] They channel the vector to a degradation pathway, which leads to necrosis and inflammation.[122,123] Interestingly, HAdV-C6 or chemically shielded HAdV vectors evade clearance by Kupffer cells.[117,124]

The scavenger receptor A-II (SR-A) mediates macrophage uptake of HAdV-C5–based vectors, as shown, for example, by gain-of-function experiments.[18,125–127] This occurs through external negatively charged HVRs 1, 2, 5, and 7 of hexon, which interact with SR-A. In addition, respiratory tract infections with HAdV are likely to involve alveolar macrophages, which tune the cytokine profiles such that epithelial cells expose HAdV receptors CAR and integrins (see Section 3.1).[55] Further investigations are required to elucidate the physiological roles of SR and other attachment factors in natural and therapeutic infections of model organisms and humans. It remains to be seen whether such factors include the previously proposed candidate receptors CD80/CD86 and MHC I.[128–131]

## 3.7 Internalization Co-receptors: Integrins

Integrins are internalization and signaling co-receptors for most HAdVs, except, for example, HAdV-F40 and HAdV-F41, which lack an arginine-glycine-aspartate (RGD) motive in penton base.[13,26] Integrins anchor cells to the ECM, transduce signals to and from the cell, and regulate cell survival, migration, and differentiation.[132] They are located on the basolateral side of polarized epithelia but can be relocated apically in the context of infection. Integrins are in a dynamic equilibrium between active ligand-bound and inactive ligand-free states. Distinct integrin conformations are regulated by binding of extracellular divalent cations (outside-in) and cytosolic proteins (inside-out).[133]

Interaction of integrins with the endocytic machinery critically determines integrin turnover.[134]

Integrins occur as 24 α/β heterodimers of 18 α and 8 β subunits. Interaction between HAdV, $\alpha_v\beta_3$, and $\alpha_v\beta_5$ are best characterized, but HAdVs have also been shown to engage with $\alpha_v\beta_1$, $\alpha_M\beta_2$, and $\alpha_3\beta_1$.[135–140] Human AdV binding to integrins requires an RGD sequence in penton base and mimics attachment of integrins to RGD-containing ECM proteins such as laminin, fibronectin, and vitronectin.[141] HAdV-C2 internalization, but not binding to CAR-positive cells, was inhibited by soluble penton base or synthetic RGD peptides.[136] Nevertheless, in the absence of CAR, integrins can serve as attachment receptors.[139,142] Remarkably, HAdV-D9 pentameric penton base and $\alpha_V\beta_3$ interact with a $K_D$ in the nanomolar range, and one penton base of HAdV-A12 engages with a maximum of four $\alpha_v\beta_5$s.[143,144]

## 3.8 Extracellular Factors Influencing Viral Tropism

Body fluids contain factors that bind to AdVs, confer alternative viral tropism, and influence the outcome of both natural infection and systemic gene therapy. For example, dipalmitoylphosphatidylcholine (DPPC) in lung surfactant or lactoferrin (Lf) in secretory fluids apparently enhance infection of polarized cells regardless of CAR availability. A range of blood coagulation factors, such as factor X (FX), FVII, FIX, and protein C binds to HAdV-C5 and its derived vectors.[145] For example, FX shields circulating viruses from immunoglobulin (Ig) M and complement-mediated neutralization and influences virus liver tropism. Extracellular released defensins inhibited uncoating and delayed infection with some serotypes and enhanced infection with others, such as HAdV-D and F (Table 1).[146] This suggests that defensins modulate attachment of HAdV to the cell surface.

### 3.8.1 Coagulation Factors

Vector bioavailability in systemic HAdV therapies is restricted by rapid neutralization through Kupffer cells or intrinsic viral liver tropism. A soluble ligand suspected to orchestrate liver targeting in mice and nonhuman primates is blood coagulation FX.[147–149] Factor X is a vitamin K–dependent coagulation factor that circulates in the bloodstream, attaches to intravenously injected HAdV-C5–based vectors, and bridges them to HSPG via the FX γ-carboxyl-glutamic acid–rich (GLA) domain.[150] Cryo-EM experiments identified the solvent-exposed hexon HVR7, particularly amino acids T423-E424-T425 and E451, as essential for binding to FX and hepatocyte transduction.[19,151,152]

Not all HAdV species bind to FX and binding affinities depend on the amino acid sequence of HVRs. For example, HAdV-C5 and HAdV-C2 are strong binders (for HAdV-C5, the $K_D$ is about 2 nM), whereas HAdV-B3 and HAdV-B35 are weak binders.[153] Although several reports have pinpointed that FX is detrimental for AdV-based therapies, it has emerged that FX is also beneficial because it shields HAdV from rapid neutralization by IgM and the complement system.[154] For further reading on coagulation factors and liver targeting, see, for example, Lopez-Gordo et al.[155]

### 3.8.2 Immunoglobulins and Complement

Ablation of FX binding to HAdV-C5 by the vitamin K inhibitor warfarin renders capsids accessible to circulating IgM and the complement components C1q and C4.[154] These factors not only trigger an antiviral cytokine response but also directly coat viruses to prevent capsid binding to target cells, including hepatocytes.[154,156,157] Accordingly, high levels of HAdV-specific IgM efficiently blocked liver transduction in mice.[127] Intriguingly, the endosomal escape-defective mutant TS1 binds to antibodies and elicits early complement cascade activation in vitro but not in vivo.[158] These data suggest complex dynamics of complement activation in vivo. Complement activation is influenced by cell damage signals or other events associated with infection. Immunoglobulin coating of viruses also confers tropism to specialized immune cells that internalize viruses by phagocytosis[11] (see also Section 4.3).

### 3.8.3 Lactoferrin

Lf is a globular glycoprotein present in mucosa and body fluids such as tears, nasal fluids, and saliva.[159] Using tear fluids, it was found that Lf enhances infection of cultured human cells with HAdV-C5 but not the eye disease associated HAdV-D37.[160] It was suggested that Lf acts as a bridging factor between HAdV and the cell. An Lf-mediated increase in HAdV transduction was apparently independent of CAR. At high concentrations, however, Lf inhibited HAdV infection. This suggested that Lf competed with the HAdV–Lf complex for binding to an as-yet unidentified receptor.[161,162] A candidate receptor for the HAdV–Lf complexes is the lactoferrin receptor (LfR), which is expressed on human epithelial cells and T cells.[163,164] Another receptor candidate is the lectin DC-SIGN, which recognizes the mannose-type N-linked glycans on Lf and triggers transduction of monocytes and dendritic cells by HAdV5-Lf.[165,166] It will be interesting to determine how Lf binds the viruses and enhances or inhibits infection of CAR-negative immune cells or polarized cells.

### 3.8.4 Dipalmitoylphosphatidylcholine

Major macromolecules in alveolar surfactant are lipids, proteins, and polysaccharides. One of the lipid species, disaturated DPPC, was found to enhance infection of cultured lung epithelial A549 cells with HAdV-C2, as judged by transgene expression.[167,168] Purified viruses or hexons are cofractionated with liposomes. The authors interpreted the data to mean that DPPC would enhance HAdV-C2 infection of cells in the absence of virus receptors, and perhaps gate virus entry from the apical side of polarized epithelial cells. Mechanistic data will be required to further support this idea.

## 4. Endocytosis

HAdVs enter cells by endocytosis, a vital process that occurs in all cells. Endocytosis is the uptake process for fluids, solutes, membrane-associated proteins, and membranes as well as receptor ligands. It determines cell fate and well-being, including the engulfment of nutrients, sensing of the environment, and tuning of signaling.[169–171] Endocytosis occurs through the clathrin-mediated pathway, macropinocytosis, phagocytosis, caveolae-dependent uptake, and the CLIC/GEEC, IL-2, Arf6-dependent and flotillin-dependent pathways. These pathways require subsets of effector molecules and are subject to distinct regulations. AdVs evolved to exploit multiple pathways of internalization. Endocytic uptake of HAdV-C2/5 is highly effective in receptor-positive epithelial cells, where > 80% of bound viruses are endocytosed with the half-time of 5–10 min.[38,81,82,172,173] This is a great advantage for the virus because rapid endocytosis clears viral traces from the cell surface and limits the risk of premature detection of the infection by the immune system.

## 4.1 Clathrin-Mediated Endocytosis

Clathrin-mediated endocytosis (CME) is initiated at specialized membrane domains rich in phosphatidyl-inositol-4,5 bisphosphate ($PI4,5P_2$). There, the endocytic cargo is recognized by cytosolic adapters that recruit adapter proteins, clathrin, and other factors for vesicle formation and scission.[174,175] Human AdV-C2 and HAdV-C5 gain access into epithelial cells by clathrin- and dynamin-mediated endocytosis. Initial EM studies revealed a clathrin coat on endocytic pits and vesicles enclosing incoming HAdV.[176–179] Further studies with HAdV-C2/5 employing dominant-negative mutants revealed the requirement of clathrin, dynamin, and Eps15 for endocytosis and infection.[38,180,181] The endocytic process of HAdV is enhanced by β-integrins.[135,136,182] Many β-integrins harbor an asparagine-proline-any amino acid-tyrosine (NPXY) or NPXY-like motive in their cytoplasmic domain, which binds to cargo selectors such as NUMB and Dab2. NUMB directly interacts with the adapter AP-2 and clathrin. The site of interaction surrounds the focal contacts, where NUMB is involved in integrin trafficking during cell migration.[183] Dab2 is another adapter molecule that initiates CME by bridging β-integrins with clathrin and $PI4,5P_2$. Interestingly, Dab2 supports CME in AP-2–deficient cells.[184,185] This suggests a certain redundancy, perhaps ensuring efficient endocytosis under diverse conditions.

Membrane shaping might require actin, depending on the cell type. Actin is not required for endocytosis in nonpolarized cells, yet in polarized cells it has been implicated in the invagination of membrane domains on the apical but not the basolateral side. Actin is thought to provide a force to counteract membrane tension and complete membrane bending and pit constriction.[186] F-Actin and the clathrin lattice are connected by huntingtin interacting protein 1 related (HIP1R).[186] During HAdV entry, actin dynamics are tuned by signaling through integrins involving PI3K, p130CAS, and the actin remodeling GTPases Rac1 and Cdc42.[135,187] The clathrin-coated pits are finally subjected to closure and scission by dynamin 2 (Dyn2) in epithelial and immune cells. The expression of Dyn2_K44A, a dominant negative form of Dyn2 defective in GTP binding, blocks HAdV-C2/5 at the cell surface. Supported with RNA interference of Dyn2, these data imply that the major pathway of HAdV-C2/5 entry into epithelial cells is dynamin dependent.[38,180]

## 4.2 Macropinocytosis

Macropinocytosis is the engulfment and uptake process of large amounts of fluids and membranes. It occurs by actin-dependent membrane protrusion and retraction, and results in large intracellular vacuoles (>0.2 μM), referred to as macropinosomes, which can mature to late endosomes.[188] Constitutive macropinocytosis contributes to antigen presentation in macrophages and dendritic cells, and is widely exploited by pathogens to modulate the immune response.[11,189]

HAdV-B3 and HAdV-B35 subvert macropinocytosis for infection of hematopoietic and nonpolarized epithelial cells.[81,82] They cluster CD46 and $\alpha_V$ integrins and activate cytosolic Rac1, a small GTPase coordinating actin remodeling and plasma membrane blebbing. Downstream of Rac1, p21-activated kinase (Pak1) activates the C-terminal binding protein 1 of E1A (CtBP1).[190] CtBP1 and Pak1 colocalize with macropinosomes that contain the fluid phase-marker dextran, viruses, CD46, and $\alpha_V$ integrins.[81] Remarkably, activation of CtBP1 might influence infection beyond membrane trafficking. CtBP1 is a transcriptional regulator and possibly inhibits the AdV transactivator E1A during viral gene expression.[191,192] Incoming HAdV-B that use CD46 have been shown to activate TLR9.[193] This could result in the upregulation of CtBP1-repressed genes and thereby promote an antiviral state in the infected cell.[194–196] This scenario suggests that CtBP1-dependent macropinocytosis is a defense reaction against HAdV. In support of this hypothesis, HAdV-C2/5 induce macropinocytosis, although they do not engage with CD46. For these viruses, macropinocytosis is not the infectious pathway and is slightly delayed compared with their main uptake pathway, CME.[180] All of these observations highlight the broad significance of macropinocytosis for infection, ranging from a virus uptake pathway to stimulating innate immunity.

## 4.3 Phagocytosis

Phagocytosis mostly accounts for pathogen neutralization and antigen presentation: for example, when specialized immune cells internalize opsonized viruses.[11,197] Immunoglobulins contain constant (Fc) and variable domains, coat HAdV, and trigger phagocytosis after attaching to Fc receptors (FcR) on antigen-presenting cells, macrophages, and dendritic cells.[198] For example, IgG-clustered HAdV-C5 particles are internalized by dendritic cells via the Fc$\gamma$R2 and Fc$\gamma$R3 receptors.[199]

Human AdVs exploit the Fc-mediated pathway for infection. An example is the uptake mechanism of HAdV-C2-CARexFc clusters. CARexFc is a fusion protein carrying the extracellular domain of CAR and the constant region of a human IgG.[200,201] The clusters, which are several micrometers in diameter, transduced CAR-negative hematopoietic cells by phagocytosis through the Fc receptor Fc$\gamma$R1 (CD64). Intriguingly, other studies found that HAdVs in complex with IgG, IgM, or IgA were delivered to the cytosol and elicited a virus-neutralizing innate immune response triggered by the cytosolic FcR TRIM21.[198,202,203] This highlights that a cytosolic antibody-mediated barrier defends the host against incoming viruses.

### 4.4 Caveolar Endocytosis

It has been reported that the keratoconjunctivitis-causing HAdV-D37 enters corneal cells by caveolar endocytosis.[204] Corneal cells have high levels of constitutive endocytosis and form multiple flask-shaped pits lacking a clathrin coat, classified as caveolin-containing pits. Interestingly, corneal cells of caveolin knockout mice are less susceptible to HAdV-D37 entry and HAdV-D37-dependent activation of Src kinase and CXCL-1 production. It remains to be shown whether caveolar uptake controls HAdV-D37 infection or whether caveolae function as a signaling hub in controlling HAdV-D37 infection.

## 5. Endosomal Escape

HAdVs penetrate cellular membranes from endosomes. Upon initial steps of uncoating and endocytosis, the inner capsid protein VI is externalized and disrupts the limiting membrane. This strategy avoids detrimental lysis of the plasma membrane, limits the damage to intracellular endosomes, and ensures clearance of viral traces from the cell surface.

### 5.1 Protein VI for Membrane Lysis

Endosomal escape is efficient, with a success rate of 80–95% for HAdV-C5[81,173,205] and occurs rapidly. Human AdV-C2 and HAdV-B3 are found in the cytosol within 15 min after cell attachment.[173] Endosomal penetration of HAdV is coupled to the stepwise nature of uncoating, as indicated by studies of HAdV-C2/5 and the mutant TS1. The TS1 mutant is not infectious. It remains trapped in endosomes, and is degraded in lysosomes.[206] Underlying reasons are an uncoating defect of TS1, which precludes exposure of the membrane lytic factor VI. This defect occurs because TS1 fails to respond to mechanical cues from receptors on the plasma membrane. Mechanical forces occur when viruses bind to CAR and undergo acto-myosin–dependent drifts that are counteracted by immobile integrins. These events trigger fiber release followed by externalization of VI from a location within the coat.[24,144,207,208] TS1 is further defective at releasing the tight interactions between the DNA core and the inner side of the coat.[30–32,34,275]

Electron microscopy analysis revealed that wild-type AdVs are largely intact when they reach the cytosol,[209] which suggests that endosomes are lysed rather than perforated with discrete pores. In support of this notion, a decrease in total endosomes after infection was measured.[210] It was also observed that HAdVs enhance cytosolic or nuclear co-delivery of macromolecules, such as 70-kDa fluorescein isothiocyanate dextrans or bovine serum albumin coupled to a nuclear localization peptide.[180,211,212] In addition, the ribotoxin α-sarcin was found to intoxicate cells in a vATPase–dependent manner when added to cells together with HAdV-C5.[36,213] Any co-delivery assay is an indirect measurement of virus penetration because it measures leakage of any endocytic organelle formed during virus and toxin co-uptake. In contrast to co-delivery assays, virus penetration can be addressed directly at the single-cell, single-virus level. Direct measurement of virus penetration combined with infection studies has shown that HAdV-C2/5 and HAdV-B3 penetrate endosomal membranes independently of the vacuolar endosomal $H^+$-ATPase inhibitor bafilomycin A1 or the protonophore niclosamide.[173,210,214,215] Remarkably, however, virus penetration was sensitive to lysosomotropic agents, which suggests that the ionic milieu beyond protons is important for HAdV penetration.[172,173]

Membrane lysis requires the activity of the viral membrane lytic factor protein VI.[216] In vitro studies revealed a role of VI in lysis of model membranes such as liposomes, notably independent of low pH.[36] Liposomes exposed to VI collapsed and formed tubular structures. Fragmentation is enhanced in the presence of a lipid that induces positive curvature, lyso-PC.[205] Membrane disruption depends on the N-terminal 54 amino acids of VI.[36] Particularly, amino acids 36–53 of VI likely fold into an alpha helical structure with amphipathic properties that allow association of VI in parallel to the lipid bilayer, as suggested by tryptophan depth profiling experiments.[217]

That the amphipathic N-terminal domain is important for virus penetration was shown by virus mutagenesis experiments. The mutation of alanine to glutamine at position 40 (VI-L40Q) reduced not only the membrane lytic activity in vitro but also cell infection.[216] Another mutation in the alpha helical domain, a glycine at position 48 (G48C), resulted in aberrant disulfide bonds and the formation of VI dimers, and decreased membrane lysis.[218] Intriguingly, the N-terminal amino acids 34–50 of protein VI do not form a secondary structure within the intact capsid.[18] It is conceivable that in endosomes, protein VI encounters cellular cues, such as lipids or specific ions, that favor folding of its lytic domain into a membrane active amphipathic alpha helix. The lytic mechanism might resemble that of antimicrobial amphipathic alpha helical peptides,[219] some of which insert their hydrophobic domain into the hydrocarbon backbone of membranes and traverse the lipid bilayer.[220]

## 5.2 A New Concept: Hijacking Membrane Repair for Endosome Lysis

Endosomal lysis is not thought to be a physiological event. However, although membrane penetration strongly depends on viral proteins, host factors contribute in multiple ways to this process. For example, cues from the receptors and mechanical forces are indispensable for VI exposure.[34] In addition, new data indicate that incoming viruses release few copies of VI at the cell surface, and thereby induce transient membrane permeabilization.[40] This triggers the influx of calcium ions and lysosomal exocytosis, a plasma membrane repair response associated with the increase of ceramide lipids. Ceramide lipids enhance virus endocytosis and interact with protein VI in endosomes, thereby enhancing endosomal lysis. This provides evidence for a positive feedback loop between limited virus uncoating, protein VI-mediated membrane lysis, and localized lipid production. It is likely that besides ceramide, other endosomal lipids and host factors influence virus uncoating and membrane stability, such as ion fluxes, cytosolic factors, or lipid and protein modulating enzymes.[205,221]

# 6. Targeting the Nucleus

HAdVs engage with cytoskeleton-dependent motor proteins to transit the cytoplasm from the site of endosomal escape to the nucleus.

## 6.1 Transport through the Cytoplasm

Long-range cytoplasmic transport is largely mediated by microtubules (MT) and associated motors. Microtubules are polarized filaments composed of α–β tubulin heterodimers. In nonpolarized cells they nucleate at the microtubule organization center (MTOC) near the nuclear periphery.[222] In polarized epithelial cells, microtubules nucleate at the apical side and elongate toward the basal site, where the positive end is situated.[223] Motor proteins carry cargos across cells along microtubules. For this, motors engage in either minus end–directed motions toward the MTOC (dynein), or plus end–directed motions toward the periphery (kinesin).[224] HAdVs have been shown to traffic in bidirectional modes on MT.[225–228]

Adenoviruses exploit the MT motor complex dynein–dynactin to reach the nucleus of epithelial cells. They reach the nucleus as partially dismantled DNA-positive capsids 30–60 min after entry.[225,227–232] HAdV binding to the dynein motor likely occurs through hexon, as supported by stochastic simulation modeling of virus motions on microtubules.[233] In addition, low pH-primed hexons directly interact with the light intermediate chain 1 and the 74-kDa intermediate chain of dynein, presumably through the HVR1 loop.[21,234] HVR1 is susceptible to cleavage by the bacterial protease dispase, and this cleavage abrogated hexon binding to dynein in vitro. However, HAdV-B/C does not pass through acidic compartments and is insensitive to inhibitors of endosomal acidification.[38,173,205] Cues other than low pH are therefore likely to tune the conformation of hexon and promote binding to dynein–dynactin.

Dynein-dependent motions are subject to regulation. The dynein cofactor dynactin regulates motor processivity and enhances HAdV motions.[225,226,235] In addition, dynein-dependent motions are enhanced by virus signaling from the cell surface, for example, through the cAMP-dependent protein kinase A (PKA), the Rho family GTPases RhoA and Rac1, and MAP kinase p38.[59,236,237] It is possible that PKA and p38 signaling stabilizes the MT network.[238] Alternatively, signaling may enhance dynein–dynactin binding to hexon and boost the fast microtubule-dependent motions.[225,234,239]

## 6.2 Deoxyribonucleic Acid Uncoating and Nuclear Import

How viral capsids transit from MT to the NPC is poorly understood. Yet, imaging by fluorescence and electron microscopy has shown that the viral DNA is in the context of a hexon-containing coat when it reaches the NPC. Accordingly, small compounds or antihexon antibodies blocked the translocation of cytosolic virus to the NPC.[240–242]

A key factor for the association of the incoming virus with the NPC is the nuclear export factor chromosome region of maintenance 1 (CRM1). Inhibition of CRM1 by leptomycin B, ratjandone A, or RNA interference strikingly arrested incoming viruses at the MTOC.[240] It is possible that CRM1-mediated nuclear export is required for HAdV-C2/5 to dock to the NPC, that it facilitates a special type of transport from the MTOC to the NPC, or that it discharges viruses from MT proximal to the NPC. Docking of the partly uncoated HAdV capsids to the NPC depends on the nucleoporin Nup214.[243,244] Depletion of Nup214 blocked HAdV-C5 nuclear targeting, whereas expression of the N-terminal 137 amino acids of Nup214 restored virus binding.

The maximal transport diameter of the NPC is about 40 nm, which excludes incoming AdV from the nucleus.[245] To overcome this size restriction, the DNA is uncoated before being imported into the nucleus. For HAdV-C2/5, this occurs through the recruitment of conventional kinesin-1.[246] When AdVs are docked to the NPC, both protein XI and Nup358 engage with kinesin-1 via the kinesin light and heavy chain. Kinesin motions provide pulling forces that disrupt capsids and release viral DNA. Consequently, the nucleoporins Nup358, Nup68, and Nup214 are displaced to the cytoplasm with parts of the uncoated capsids. Partly disassembled NPCs show increased permeation of fluorescent dextrans, and are probably permissive to viral DNA import. Further cellular factors are implicated in the import of viral DNA, including histone H1, heat-shock cognate protein 70, and importins $\alpha$, $\beta1$, $\beta2$, and 7.[243,247–249] Tracking of single viral DNA using copper(I)-catalyzed azide-alkyne cyclo-addition (click) chemistry and super-resolution microscopy revealed that, surprisingly, only 6–48% of viral genomes were successfully imported into the nucleus, and 25% of viral DNA was spread throughout the cytosol with large cell-to-cell variability.[232] Future work employing single-genome tracking will further elaborate on mechanisms and implications of nuclear import and misdelivery of HAdV genomes (see also Ref. 250).

## 7. Conclusions and Perspectives

Adenoviruses have been known for more than 50 years. Yet, there are many open questions regarding how these viruses interact with their host during entry. We estimate that close to 100 host factors are involved in AdV entry. In addition, unresolved questions include the nature of virus, for example, the location of the minor cementing proteins, and the mode of genome encapsidation. The latter may affect the mechanisms of viral DNA unpacking during entry.

Important questions regarding virus–host interactions include how the virus overcomes the hurdles toward long-lasting gene expression from adenovirus vectors. Answers to these questions are the basis for taking best advantage of the highly immunogenic nature of certain human AdVs and killing cancer cells. The aims involve strategies to de-target the virus from immune cells and the liver, to shield it from blood factors, target it to relevant cells, and finally arm it with genes of interest. Whereas re-targeting efforts are advanced, shielding is still in its infancy. It awaits the development of tunable coats that can respond to local cues of target cells: for example, by exposing the virus to the uncoating cues that determine efficient entry and genome delivery. Promising and versatile shielding and re-targeting strategies have exploited designed ankyrin repeats that can be selected for virtually any target at high affinity and selectivity.[251,252] Alternatively, single-chain variable fragments, immunoglobulins from camelids, and derivatives from the latter have also been developed.[253] Regarding immunity, key problems for the near future are to elucidate immunogenic features of AdVs and determine the mechanisms of how they occur in vitro and in vivo.[254] A major challenge is to translate the findings from in vitro studies and animal studies to humans. It will be interesting to see whether clinical trials with oncolytic AdVs can close this gap or create new puzzles that require new technologies to be resolved.

## Acknowledgments

We thank Maarit Suomalainen and Justin Flatt for comments on the manuscript. The work was supported by the Swiss National Science Foundation (grant 31003A_141222/1 to UFG) and the Kanton Zurich.

## References

1. Matthes-Martin S, Boztug H, Lion T. Diagnosis and treatment of adenovirus infection in immunocompromised patients. *Expert Rev Anti Infect Ther* 2013;**11**(10):1017–28.
2. Ghebremedhin B. Human adenovirus: viral pathogen with increasing importance. *Eur J Microbiol Immunol* 2014;**4**(1):26–33.
3. Verma IM, Weitzman MD. Gene therapy: twenty-first century medicine. *Annu Rev Biochem* 2005;**74**:711–38.
4. Raty JK, Lesch HP, Wirth T, Yla-Herttuala S. Improving safety of gene therapy. *Curr Drug Saf* 2008;**3**(1):46–53.
5. Top Jr FH, Dudding BA, Russell PK, Buescher EL. Control of respiratory disease in recruits with types 4 and 7 adenovirus vaccines. *Am J Epidemiol* 1971;**94**(2):142–6.
6. Hoke Jr CH, Hawksworth A, Snyder Jr CE. Initial assessment of impact of adenovirus type 4 and type 7 vaccine on febrile respiratory illness and virus transmission in military basic trainees, March 2012. *MSMR* 2012;**19**(3):2–4.
7. Alemany R, Balague C, Curiel DT. Replicative adenoviruses for cancer therapy. *Nat Biotechnol* 2000;**18**(7):723–7.
8. Capone S, D'Alise AM, Ammendola V, Colloca S, Cortese R, Nicosia A, et al. Development of chimpanzee adenoviruses as vaccine vectors: challenges and successes emerging from clinical trials. *Expert Rev Vaccines* 2013;**12**(4):379–93.
9. Appaiahgari MB, Vrati S. Adenoviruses as gene/vaccine delivery vectors: promises and pitfalls. *Expert Opin Biol Ther* 2015;**15**(3):337–51.
10. Fejer G, Freudenberg M, Greber UF, Gyory I. Adenovirus triggered innate signalling pathways. *Eur J Microbiol Immunol* December 2011;**1**(4):279–88.
11. Mercer J, Greber UF. Virus interactions with endocytic pathways in macrophages and dendritic cells. *Trends Microbiol* 2013;**21**(8):380–8.
12. Meier O, Greber UF. Adenovirus endocytosis. *J Gene Med* 2003;**5**(6):451–62.
13. Wolfrum N, Greber UF. Adenovirus signalling in entry. *Cell Microbiol* 2013;**15**(1):53–62.
14. Stewart PL, Nemerow GR. Cell integrins: commonly used receptors for diverse viral pathogens. *Trends Microbiol* 2007;**15**(11):500–7.
15. Greber UF, Arnberg N, Wadell G, Benko M, Kremer EJ. Adenoviruses - from pathogens to therapeutics: a report on the 10th International Adenovirus Meeting. *Cell Microbiol* 2013;**15**(1):16–23.

16. Suomalainen M, Greber UF. Uncoating of non-enveloped viruses. *Curr Opin Virol* 2013;**3**:27–33.
17. Deal C, Pekosz A, Ketner G. Prospects for oral replicating adenovirus-vectored vaccines. *Vaccine* 2013;**31**(32):3236–43.
18. Khare R, Reddy VS, Nemerow GR, Barry MA. Identification of adenovirus serotype 5 hexon regions that interact with scavenger receptors. *J Virol* 2012;**86**(4):2293–301.
19. Doronin K, Flatt JW, Di Paolo NC, Khare R, Kalyuzhniy O, Acchione M, et al. Coagulation factor X activates innate immunity to human species C adenovirus. *Science* 2012;**338**(6108):795–8.
20. Kelkar S, De BP, Gao G, Wilson JM, Crystal RG, Leopold PL. A common mechanism for cytoplasmic dynein-dependent microtubule binding shared among adeno-associated virus and adenovirus serotypes. *J Virol* 2006;**80**(15):7781–5.
21. Bremner KH, Scherer J, Yi J, Vershinin M, Gross SP, Vallee RB. Adenovirus transport via direct interaction of cytoplasmic dynein with the viral capsid hexon subunit. *Cell Host Microbe* 2009;**6**(6):523–35.
22. Nemerow GR, Stewart PL. Antibody neutralization epitopes and integrin binding sites on nonenveloped viruses. *Virology* 2001;**288**(2):189–91 [Review].
23. Greber UF. Signalling in viral entry. *Cell Mol Life Sci* 2002;**59**:608–26.
24. Liu H, Jin L, Koh SB, Atanasov I, Schein S, Wu L, et al. Atomic structure of human adenovirus by cryo-EM reveals interactions among protein networks. *Science* 2010; **329**(5995):1038–43.
25. Reddy VS, Nemerow GR. Structures and organization of adenovirus cement proteins provide insights into the role of capsid maturation in virus entry and infection. *Proc Natl Acad Sci USA* 2014;**111**(32):11715–20.
26. Arnberg N. Adenovirus receptors: implications for targeting of viral vectors. *Trends Pharmacol Sci* 2012;**33**(8):442–8.
27. Excoffon KJ, Bowers JR, Sharma P. 1. Alternative splicing of viral receptors: a review of the diverse morphologies and physiologies of adenoviral receptors. *Recent Res Dev Virol* 2014;**9**:1–24.
28. Mangel WF, San Martin C. Structure, function and dynamics in adenovirus maturation. *Viruses* 2014;**6**(11):4536–70.
29. Greber UF. Virus assembly and disassembly: the adenovirus cysteine protease as a trigger factor. *Rev Med Virol* 1998;**8**:213–22.
30. Ortega-Esteban A, Perez-Berna AJ, Menendez-Conejero R, Flint SJ, San Martin C, de Pablo PJ. Monitoring dynamics of human adenovirus disassembly induced by mechanical fatigue. *Sci Rep* 2013;**3**:1434.
31. Silvestry M, Lindert S, Smith JG, Maier O, Wiethoff CM, Nemerow GR, et al. Cryo-electron microscopy structure of adenovirus type 2 temperature-sensitive mutant 1 reveals insight into the cell entry defect. *J Virol* 2009;**83**(15):7375–83.
32. Perez-Berna AJ, Marabini R, Scheres SH, Menendez-Conejero R, Dmitriev IP, Curiel DT, et al. Structure and uncoating of immature adenovirus. *J Mol Biol* 2009;**392**(2):547–57.
33. Snijder J, Uetrecht C, Rose RJ, Sanchez-Eugenia R, Marti GA, Agirre J, et al. Probing the biophysical interplay between a viral genome and its capsid. *Nat Chem* 2013; **5**(6):502–9.
34. Burckhardt CJ, Suomalainen M, Schoenenberger P, Boucke K, Hemmi S, Greber UF. Drifting motions of the adenovirus receptor CAR and immobile integrins initiate virus uncoating and membrane lytic protein exposure. *Cell Host Microbe* 2011;**10**(2):105–17.
35. Wodrich H, Henaff D, Jammart B, Segura-Morales C, Seelmeir S, Coux O, et al. A capsid-encoded PPxY-motif facilitates adenovirus entry. *PLoS Pathog* 2010;**6**(3):e1000808.

36. Wiethoff CM, Wodrich H, Gerace L, Nemerow GR. Adenovirus protein VI mediates membrane disruption following capsid disassembly. *J Virol* 2005;**79**(4):1992–2000.
37. Hannan C, Raptis LH, Dery CD, Weber J. Biological and structural studies with an adenovirus type 2 temperature-sensitive mutant defective for uncoating. *Intervirology* 1983;**19**:213–23.
38. Gastaldelli M, Imelli N, Boucke K, Amstutz B, Meier O, Greber UF. Infectious adenovirus type 2 transport through early but not late endosomes. *Traffic* 2008;**9**(12):2265–78.
39. Nakano MY, Greber UF. Quantitative microscopy of fluorescent adenovirus entry. *J Struct Biol* 2000;**129**(1):57–68.
40. Luisoni S, Suomalainen M, Boucke K, Tanner LB, Wenk MR, Guan XL, et al. Co-option of membrane wounding enables virus penetration into cells. *Cell Host Microbe* 2015;**18**(1):75–85.
41. Philipson L, Pettersson RF. The coxsackie-adenovirus receptor–a new receptor in the immunoglobulin family involved in cell adhesion. *Curr Top Microbiol Immunol* 2004;**273**:87–111.
42. Zhang Y, Bergelson JM. Adenovirus receptors. *J Virol* 2005;**79**(19):12125–31.
43. Walters RW, Freimuth P, Moninger TO, Ganske I, Zabner J, Welsh MJ. Adenovirus fiber disrupts CAR-mediated intercellular adhesion allowing virus escape. *Cell* 2002;**110**(6):789–99.
44. Coyne CB, Bergelson JM. CAR: a virus receptor within the tight junction. *Adv Drug Deliv Rev* 2005;**57**(6):869–82.
45. Tan PK, Michou AI, Bergelson JM, Cotten M. Defining CAR as a cellular receptor for the avian adenovirus CELO using a genetic analysis of the two viral fibre proteins. *J Gen Virol* 2001;**82**(Pt 6):1465–72.
46. Bergelson JM, Cunningham JA, Droguett G, Kurt-Jones EA, Krithivas A, Hong JS, et al. Isolation of a common receptor for Coxsackie B viruses and adenoviruses 2 and 5. *Science* 1997;**275**:1320–3.
47. Roelvink PW, Lizonova A, Lee JGM, Li Y, Bergelson JM, Finberg RW, et al. The coxsackievirus-adenovirus receptor protein can function as a cellular attachment protein for adenovirus serotypes from subgroups A, C, D, E, and F. *J Virol* 1998;**72**(10):7909–15.
48. Kirby I, Davison E, Beavil AJ, Soh CP, Wickham TJ, Roelvink PW, et al. Identification of contact residues and definition of the CAR-binding site of adenovirus type 5 fiber protein. *J Virol* 2000;**74**(6):2804–13.
49. Freimuth P, Springer K, Berard C, Hainfeld J, Bewley M, Flanagan J. Coxsackievirus and adenovirus receptor amino-terminal immunoglobulin V-related domain binds adenovirus type 2 and fiber knob from adenovirus type 12. *J Virol* 1999;**73**(2):1392–8.
50. Tomko RP, Xu R, Philipson L. HCAR and MCAR: the human and mouse cellular receptors for subgroup C adenoviruses and group B coxsackieviruses. *Proc Natl Acad Sci USA* 1997;**94**:3352–6.
51. Soudais C, Boutin S, Hong SS, Chillon M, Danos O, Bergelson JM, et al. Canine adenovirus type 2 attachment and internalization: coxsackievirus-adenovirus receptor, alternative receptors, and an RGD-independent pathway. *J Virol* 2000;**74**(22):10639–49.
52. Frickmann H, Jungblut S, Hirche TO, Groß U, Kuhns M, Zautner AE. Spectrum of viral infections in patients with cystic fibrosis. *Eur J Microbiol Immunol* 2012;**2**(3):161–75.
53. Gern JE, Pappas T, Visness CM, Jaffee KF, Lemanske RF, Togias A, et al. Comparison of the etiology of viral respiratory illnesses in inner-city and suburban infants. *J Infect Dis* 2012;**206**(9):1342–9.
54. Hayashi S, Hogg JC. Adenovirus infections and lung disease. *Curr Opin Pharmacol* 2007;**7**(3):237–43.

55. Lutschg V, Boucke K, Hemmi S, Greber UF. Chemotactic antiviral cytokines promote infectious apical entry of human adenovirus into polarized epithelial cells. *Nat Commun* 2011;**2**:391.
56. Kotha PL, Sharma P, Kolawole AO, Yan R, Alghamri MS, Brockman TL, et al. Adenovirus entry from the apical surface of polarized epithelia is facilitated by the host innate immune response. *PLoS Pathog* 2015;**11**(3):e1004696.
57. Excoffon KJ, Gansemer ND, Mobily ME, Karp PH, Parekh KR, Zabner J. Isoform-specific regulation and localization of the coxsackie and adenovirus receptor in human airway epithelia. *PLoS One* 2010;**5**(3):e9909.
58. Farmer C, Morton PE, Snippe M, Santis G, Parsons M. Coxsackie adenovirus receptor (CAR) regulates integrin function through activation of p44/42 MAPK. *Exp Cell Res* 2009;**315**(15):2637–47.
59. Suomalainen M, Nakano MY, Boucke K, Keller S, Greber UF. Adenovirus-activated PKA and p38/MAPK pathways boost microtubule-mediated nuclear targeting of virus. *Embo J* 2001;**20**(6):1310–9.
60. Ferrari R, Gou D, Jawdekar G, Johnson SA, Nava M, Su T, et al. Adenovirus small E1A employs the lysine acetylases p300/CBP and tumor suppressor Rb to repress select host genes and promote productive virus infection. *Cell Host Microbe* 2014;**16**(5):663–76.
61. Verdino P, Witherden DA, Havran WL, Wilson IA. The molecular interaction of CAR and JAML recruits the central cell signal transducer PI3K. *Science* 2010;**329**(5996):1210–4.
62. Trotman LC, Achermann DP, Keller S, Straub M, Greber UF. Non-classical export of an adenovirus structural protein. *Traffic* 2003;**4**:390–402.
63. Bewley MC, Springer K, Zhang YB, Freimuth P, Flanagan JM. Structural analysis of the mechanism of adenovirus binding to its human cellular receptor, CAR. *Science* 1999;**286**(5444):1579–83.
64. Lortat-Jacob H, Chouin E, Cusack S, van Raaij MJ. Kinetic analysis of adenovirus fiber binding to its receptor reveals an avidity mechanism for trimeric receptor-ligand interactions. *J Biol Chem* 2001;**276**(12):9009–15.
65. Cole JL, Housley Jr GA, Dykman TR, MacDermott RP, Atkinson JP. Identification of an additional class of C3-binding membrane proteins of human peripheral blood leukocytes and cell lines. *Proc Natl Acad Sci USA* 1985;**82**(3):859–63.
66. Sinn PL, Williams G, Vongpunsawad S, Cattaneo R, McCray Jr PB. Measles virus preferentially transduces the basolateral surface of well-differentiated human airway epithelia. *J Virol* 2002;**76**(5):2403–9.
67. Maisner A, Zimmer G, Liszewski MK, Lublin DM, Atkinson JP, Herrler G. Membrane cofactor protein (CD46) is a basolateral protein that is not endocytosed. Importance of the tetrapeptide FTSL at the carboxyl terminus. *J Biol Chem* 1997;**272**(33):20793–9.
68. McLaughlin BJ, Fan W, Zheng JJ, Cai H, Del Priore LV, Bora NS, et al. Novel role for a complement regulatory protein (CD46) in retinal pigment epithelial adhesion. *Invest Ophthalmol Vis Sci* 2003;**44**(8):3669–74.
69. Sirena D, Lilienfeld B, Eisenhut M, Kaelin S, Boucke K, Beerli RR, et al. The human membrane cofactor CD46 is a receptor for species B adenovirus serotype 3. *J Virol* 2004;**78**:4454–62.
70. Sakurai F, Murakami S, Kawabata K, Okada N, Yamamoto A, Seya T, et al. The short consensus repeats 1 and 2, not the cytoplasmic domain, of human CD46 are crucial for infection of subgroup B adenovirus serotype 35. *J Control Release* 2006;**113**(3):271–8.
71. Cupelli K, Muller S, Persson BD, Jost M, Arnberg N, Stehle T. Structure of adenovirus type 21 knob in complex with CD46 reveals key differences in receptor contacts among species B adenoviruses. *J Virol* 2010;**84**(7):3189–200.

72. Pache L, Venkataraman S, Reddy VS, Nemerow GR. Structural variations in species B adenovirus fibers impact CD46 association. *J Virol* 2008;**82**(16):7923–31.
73. Persson BD, Muller S, Reiter DM, Schmitt BB, Marttila M, Sumowski CV, et al. An arginine switch in the species B adenovirus knob determines high-affinity engagement of cellular receptor CD46. *J Virol* 2009;**83**(2):673–86.
74. Fleischli C, Verhaagh S, Havenga M, Sirena D, Schaffner W, Cattaneo R, et al. The distal short consensus repeats 1 and 2 of the membrane cofactor protein CD46 and their distance from the cell membrane determine productive entry of species B adenovirus serotype 35. *J Virol* 2005;**79**(15):10013–22.
75. Fleischli C, Sirena D, Lesage G, Havenga MJ, Cattaneo R, Greber UF, et al. Species B adenovirus serotypes 3, 7, 11 and 35 share similar binding sites on the membrane cofactor protein CD46 receptor. *J Gen Virol* 2007;**88**(Pt 11):2925–34.
76. Sirena D, Ruzsics Z, Schaffner W, Greber UF, Hemmi S. The nucleotide sequence and a first generation gene transfer vector of species B human adenovirus serotype 3. *Virology* 2005;**343**(2):283–98.
77. Trinh HV, Lesage G, Chennamparampil V, Vollenweider B, Burckhardt CJ, Schauer S, et al. Avidity binding of human adenovirus serotypes 3 and 7 to the membrane cofactor CD46 triggers infection. *J Virol* 2012;**86**:1623–37.
78. Cardone J, Al-Shouli S, Kemper C. A novel role for CD46 in wound repair. *Front Immunol* 2011;**2**:28.
79. Liszewski MK, Post TW, Atkinson JP. Membrane cofactor protein (MCP or CD46): newest member of the regulators of complement activation gene cluster. *Annu Rev Immunol* 1991;**9**:431–55.
80. Crimeen-Irwin B, Ellis S, Christiansen D, Ludford-Menting MJ, Milland J, Lanteri M, et al. Ligand binding determines whether CD46 is internalized by clathrin-coated pits or macropinocytosis. *J Biol Chem* 2003;**278**(47):46927–37.
81. Amstutz B, Gastaldelli M, Kälin S, Imelli N, Boucke K, Wandeler E, et al. Subversion of CtBP1 controlled macropinocytosis by human adenovirus serotype 3. *EMBO J* 2008;**27**(7):956–66.
82. Kalin S, Amstutz B, Gastaldelli M, Wolfrum N, Boucke K, Havenga M, et al. Macropinocytotic uptake and infection of human epithelial cells with species B2 adenovirus type 35. *J Virol* 2010;**84**(10):5336–50.
83. Gustafsson DJ, Andersson EK, Hu YL, Marttila M, Lindman K, Strand M, et al. Adenovirus 11p downregulates CD46 early in infection. *Virology* 2010;**405**(2):474–82.
84. Sakurai F, Akitomo K, Kawabata K, Hayakawa T, Mizuguchi H. Downregulation of human CD46 by adenovirus serotype 35 vectors. *Gene Ther* 2007;**14**(11):912–9.
85. Russell S. CD46: a complement regulator and pathogen receptor that mediates links between innate and acquired immune function. *Tissue Antigens* 2004;**64**(2):111–8.
86. Iacobelli-Martinez M, Nepomuceno RR, Connolly J, Nemerow GR. CD46-utilizing adenoviruses inhibit C/EBPbeta-dependent expression of proinflammatory cytokines. *J Virol* 2005;**79**(17):11259–68.
87. Gerlier D, Trescol-Biémont MC, Varior-Krishnan G, Naniche D, Fugier-Vivier I, Rabourdin-Combe C. Efficient major histocompatibility complex class II-restricted presentation of measles virus relies on hemagglutinin-mediated targeting to its cellular receptor human CD46 expressed by murine B cells. *J Exp Med* 1994;**179**(1):353–8.
88. Schneider-Schaulies J, Schnorr JJ, Schlender J, Dunster LM, Schneider-Schaulies S, ter Meulen V. Receptor (CD46) modulation and complement-mediated lysis of uninfected cells after contact with measles virus-infected cells. *J Virol* 1996;**70**(1):255–63.

89. Wang H, Li ZY, Liu Y, Persson J, Beyer I, Moller T, et al. Desmoglein 2 is a receptor for adenovirus serotypes 3, 7, 11 and 14. *Nat Med* 2011;**17**(1):96–104.
90. Wang H, Li Z, Yumul R, Lara S, Hemminki A, Fender P, et al. Multimerization of adenovirus serotype 3 fiber knob domains is required for efficient binding of virus to desmoglein 2 and subsequent opening of epithelial junctions. *J Virol* 2011;**85**(13):6390–402.
91. Wang H, Yumul R, Cao H, Ran L, Fan X, Richter M, et al. Structural and functional studies on the interaction of adenovirus fiber knobs and desmoglein 2. *J Virol* 2013;**87**(21):11346–62.
92. Delva E, Tucker DK, Kowalczyk AP. The desmosome. *Cold Spring Harb Perspect Biol* 2009;**1**(2):a002543.
93. Lu ZZ, Wang H, Zhang Y, Cao H, Li Z, Fender P, et al. Penton-dodecahedral particles trigger opening of intercellular junctions and facilitate viral spread during adenovirus serotype 3 infection of epithelial cells. *PLoS Pathog* 2013;**9**(10):e1003718.
94. Fender P, Hall K, Schoehn G, Blair GE. Impact of human adenovirus type 3 dodecahedron on host cells and its potential role in viral infection. *J Virol* 2012;**86**(9):5380–5.
95. Tuve S, Wang H, Ware C, Liu Y, Gaggar A, Bernt K, et al. A new group B adenovirus receptor is expressed at high levels on human stem and tumor cells. *J Virol* 2006;**80**(24):12109–20.
96. Blau DM, Compans RW. Entry and release of measles virus are polarized in epithelial cells. *Virology* 1995;**210**(1):91–9.
97. Strauss R, Sova P, Liu Y, Li ZY, Tuve S, Pritchard D, et al. Epithelial phenotype confers resistance of ovarian cancer cells to oncolytic adenoviruses. *Cancer Res* 2009;**69**(12):5115–25.
98. Lecollinet S, Gavard F, Havenga MJ, Spiller OB, Lemckert A, Goudsmit J, et al. Improved gene delivery to intestinal mucosa by adenoviral vectors bearing subgroup B and d fibers. *J Virol* 2006;**80**(6):2747–59.
99. Nilsson EC, Storm RJ, Bauer J, Johansson SM, Lookene A, Angstrom J, et al. The GD1a glycan is a cellular receptor for adenoviruses causing epidemic keratoconjunctivitis. *Nat Med* 2011;**17**(1):105–9.
100. Seiradake E, Henaff D, Wodrich H, Billet O, Perreau M, Hippert C, et al. The cell adhesion molecule "CAR" and sialic acid on human erythrocytes influence adenovirus in vivo biodistribution. *PLoS Pathog* 2009;**5**(1):e1000277.
101. Lenman A, Liaci AM, Liu Y, Ardahl C, Rajan A, Nilsson E, et al. Human adenovirus 52 uses sialic acid-containing glycoproteins and the coxsackie and adenovirus receptor for binding to target cells. *PLoS Pathog* 2015;**11**(2):e1004657.
102. Arnberg N, Edlund K, Kidd AH, Wadell G. Adenovirus type 37 uses sialic acid as a cellular receptor. *J Virol* 2000;**74**(1):42–8.
103. Arnberg N, Kidd AH, Edlund K, Nilsson J, Pring-Akerblom P, Wadell G. Adenovirus type 37 binds to cell surface sialic acid through a charge-dependent interaction. *Virology* 2002;**302**(1):33–43.
104. Burmeister WP, Guilligay D, Cusack S, Wadell G, Arnberg N. Crystal structure of species D adenovirus fiber knobs and their sialic acid binding sites. *J Virol* 2004;**78**(14):7727–36.
105. Pring-Akerblom P, Heim A, Trijssenaar FE. Molecular characterization of hemagglutination domains on the fibers of subgenus D adenoviruses. *J Virol* 1998;**72**(3):2297–304.
106. Bernfield M, Götte M, Park PW, Reizes O, Fitzgerald ML, Lincecum J, et al. Functions of cell surface heparan sulfate proteoglycans. *Annu Rev Biochem* 1999;**68**:729–77.
107. Morgan MR, Humphries MJ, Bass MD. Synergistic control of cell adhesion by integrins and syndecans. *Nat Rev Mol Cell Biol* 2007;**8**(12):957–69.

108. Bourin MC, Lindahl U. Glycosaminoglycans and the regulation of blood coagulation. *Biochem J* 1993;**289**(Pt 2):313–30.
109. Mahley RW, Ji ZS. Remnant lipoprotein metabolism: key pathways involving cell-surface heparan sulfate proteoglycans and apolipoprotein E. *J Lipid Res* 1999;**40**(1):1–16.
110. Dechecchi MC, Melotti P, Bonizzato A, Santacatterina M, Chilosi M, Cabrini G. Heparan sulfate glycosaminoglycans are receptors sufficient to mediate the initial binding of adenovirus types 2 and 5. *J Virol* 2001;**75**(18):8772–80.
111. Tuve S, Wang H, Jacobs JD, Yumul RC, Smith DF, Lieber A. Role of cellular heparan sulfate proteoglycans in infection of human adenovirus serotype 3 and 35. *PLoS Pathog* 2008;**4**(10):e1000189.
112. Gout E, Schoehn G, Fenel D, Lortat-Jacob H, Fender P. The adenovirus type 3 dodecahedron's RGD loop comprises an HSPG binding site that influences integrin binding. *J Biomed Biotechnol* 2010;**2010**:541939.
113. Smith TA, Idamakanti N, Rollence ML, Marshall-Neff J, Kim J, Mulgrew K, et al. Adenovirus serotype 5 fiber shaft influences in vivo gene transfer in mice. *Hum Gene Ther* 2003;**14**(8):777–87.
114. Mukhopadhyay S, Plüddemann A, Gordon S. Macrophage pattern recognition receptors in immunity, homeostasis and self tolerance. *Adv Exp Med Biol* 2009;**653**:1–14.
115. Canton J, Neculai D, Grinstein S. Scavenger receptors in homeostasis and immunity. *Nat Rev Immunol* 2013;**13**(9):621–34.
116. Xu Z, Tian J, Smith JS, Byrnes AP. Clearance of adenovirus by Kupffer cells is mediated by scavenger receptors, natural antibodies, and complement. *J Virol* 2008;**82**(23):11705–13.
117. Prill JM, Espenlaub S, Samen U, Engler T, Schmidt E, Vetrini F, et al. Modifications of adenovirus hexon allow for either hepatocyte detargeting or targeting with potential evasion from Kupffer cells. *Mol Ther J Am Soc Gene Ther* 2011;**19**(1):83–92.
118. Piccolo P, Vetrini F, Mithbaokar P, Grove NC, Bertin T, Palmer D, et al. SR-A and SREC-I are Kupffer and endothelial cell receptors for helper-dependent adenoviral vectors. *Mol Ther* 2013;**21**(4):767–74.
119. Alemany R, Suzuki K, Curiel DT. Blood clearance rates of adenovirus type 5 in mice. *J Gen Virol* 2000;**81**(Pt 11):2605–9.
120. Ganesan LP, Mohanty S, Kim J, Clark KR, Robinson JM, Anderson CL. Rapid and efficient clearance of blood-borne virus by liver sinusoidal endothelium. *PLoS Pathog* 2011;**7**(9):e1002281.
121. Di Paolo NC, van Rooijen N, Shayakhmetov DM. Redundant and synergistic mechanisms control the sequestration of blood-born adenovirus in the liver. *Mol Ther* 2009;**17**(4):675–84.
122. Manickan E, Smith JS, Tian J, Eggerman TL, Lozier JN, Muller J, et al. Rapid Kupffer cell death after intravenous injection of adenovirus vectors. *Mol Ther* 2006;**13**(1):108–17.
123. Liu Q, Muruve DA. Molecular basis of the inflammatory response to adenovirus vectors. *Gene Ther* 2003;**10**(11):935–40.
124. Khare R, May SM, Vetrini F, Weaver EA, Palmer D, Rosewell A, et al. Generation of a Kupffer cell-evading adenovirus for systemic and liver-directed gene transfer. *Mol Ther* 2011;**19**(7):1254–62.
125. Haisma HJ, Kamps JA, Kamps GK, Plantinga JA, Rots MG, Bellu AR. Polyinosinic acid enhances delivery of adenovirus vectors in vivo by preventing sequestration in liver macrophages. *J Gen Virol* 2008;**89**(Pt 5):1097–105.
126. Haisma HJ, Boesjes M, Beerens AM, van der Strate BW, Curiel DT, Plüddemann A, et al. Scavenger receptor A: a new route for adenovirus 5. *Mol Pharm* 2009;**6**(2):366–74.
127. Khare R, Hillestad ML, Xu Z, Byrnes AP, Barry MA. Circulating antibodies and macrophages as modulators of adenovirus pharmacology. *J Virol* 2013;**87**(7):3678–86.

128. Short JJ, Vasu C, Holterman MJ, Curiel DT, Pereboev A. Members of adenovirus species B utilize CD80 and CD86 as cellular attachment receptors. *Virus Res* 2006;**122**(1–2): 144–53.
129. Hong SS, Karayan L, Tournier J, Curiel DT, Boulanger PA. Adenovirus type 5 fiber knob binds to MHC class I alpha2 domain at the surface of human epithelial and B lymphoblastoid cells. *EMBO J* 1997;**16**(9):2294–306.
130. McDonald D, Stockwin L, Matzow T, Blair Zajdel ME, Blair GE. Coxsackie and adenovirus receptor (CAR)-dependent and major histocompatibility complex (MHC) class I-independent uptake of recombinant adenoviruses into human tumour cells. *Gene Ther* 1999;**6**(9):1512–9.
131. Davison E, Kirby I, Elliott T, Santis G. The human HLA-A*0201 allele, expressed in hamster cells, is not a high-affinity receptor for adenovirus type 5 fiber. *J Virol* 1999;**73**(5):4513–7.
132. Byron A, Morgan MR, Humphries MJ. Adhesion signalling complexes. *Curr Biol* 2010;**20**(24):R1063–7.
133. Moser M, Legate KR, Zent R, Fässler R. The tail of integrins, talin, and kindlins. *Science* 2009;**324**(5929):895–9.
134. Bridgewater RE, Norman JC, Caswell PT. Integrin trafficking at a glance. *J Cell Sci* 2012;**125**(Pt 16):3695–701.
135. Li E, Brown SL, Stupack DG, Puente XS, Cheresh DA, Nemerow GR. Integrin alpha(v) beta1 is an adenovirus coreceptor. *J Virol* 2001;**75**(11):5405–9.
136. Wickham TJ, Mathias P, Cheresh DA, Nemerow GR. Integrins alpha v beta 3 and alpha v beta 5 promote adenovirus internalization but not virus attachment. *Cell* 1993;**73**(2):309–19.
137. Mathias P, Wickham T, Moore M, Nemerow G. Multiple adenovirus serotypes use alpha v integrins for infection. *J Virol* 1994;**68**(10):6811–4.
138. Mathias P, Galleno M, Nemerow GR. Interactions of soluble recombinant integrin alphav beta5 with human adenoviruses. *J Virol* 1998;**72**(11):8669–75.
139. Huang S, Kamata T, Takada Y, Ruggeri ZM, Nemerow GR. Adenovirus interaction with distinct integrins mediates separate events in cell entry and gene delivery to hematopoietic cells. *J Virol* 1996;**70**(7):4502–8.
140. Salone B, Martina Y, Piersanti S, Cundari E, Cherubini G, Franqueville L, et al. Integrin alpha3beta1 is an alternative cellular receptor for adenovirus serotype 5. *J Virol* 2003; **77**(24):13448–54.
141. Humphries JD, Byron A, Humphries MJ. Integrin ligands at a glance. *J Cell Sci* 2006; **119**(Pt 19):3901–3.
142. Roelvink PW, Kovesdi I, Wickham TJ. Comparative analysis of adenovirus fiber-cell interaction: adenovirus type 2 (Ad2) and Ad9 utilize the same cellular fiber receptor but use different binding strategies for attachment. *J Virol* 1996;**70**(11):7614–21.
143. Veesler D, Cupelli K, Burger M, Gräber P, Stehle T, Johnson JE. Single-particle EM reveals plasticity of interactions between the adenovirus penton base and integrin $\alpha V\beta 3$. *Proc Natl Acad Sci USA* 2014;**111**(24):8815–9.
144. Lindert S, Silvestry M, Mullen TM, Nemerow GR, Stewart PL. Cryo-electron microscopy structure of an adenovirus-integrin complex indicates conformational changes in both penton base and integrin. *J Virol* 2009;**83**(22):11491–501.
145. Parker AL, Waddington SN, Nicol CG, Shayakhmetov DM, Buckley SM, Denby L, et al. Multiple vitamin K-dependent coagulation zymogens promote adenovirus-mediated gene delivery to hepatocytes. *Blood* 2006;**108**(8):2554–61.
146. Wilson SS, Wiens ME, Smith JG. Antiviral mechanisms of human defensins. *J Mol Biol* 2013;**425**(24):4965–80.

147. Kalyuzhniy O, Di Paolo NC, Silvestry M, Hofherr SE, Barry MA, Stewart PL, et al. Adenovirus serotype 5 hexon is critical for virus infection of hepatocytes in vivo. *Proc Natl Acad Sci USA* 2008;**105**(14):5483–8.
148. Vigant F, Descamps D, Jullienne B, Esselin S, Connault E, Opolon P, et al. Substitution of hexon hypervariable region 5 of adenovirus serotype 5 abrogates blood factor binding and limits gene transfer to liver. *Mol Ther* 2008;**16**(8):1474–80.
149. Alba R, Bradshaw AC, Mestre-Francés N, Verdier JM, Henaff D, Baker AH. Coagulation factor X mediates adenovirus type 5 liver gene transfer in non-human primates (*Microcebus murinus*). *Gene Ther* 2012;**19**(1):109–13.
150. Duffy MR, Bradshaw AC, Parker AL, McVey JH, Baker AH. A cluster of basic amino acids in the factor X serine protease mediates surface attachment of adenovirus/FX complexes. *J Virol* 2011;**85**(20):10914–9.
151. Alba R, Bradshaw AC, Parker AL, Bhella D, Waddington SN, Nicklin SA, et al. Identification of coagulation factor (F)X binding sites on the adenovirus serotype 5 hexon: effect of mutagenesis on FX interactions and gene transfer. *Blood* 2009;**114**(5):965–71.
152. Irons EE, Flatt JW, Doronin K, Fox TL, Acchione M, Stewart PL, et al. Coagulation factor binding orientation and dimerization may influence infectivity of adenovirus-coagulation factor complexes. *J Virol* 2013;**87**(17):9610–9.
153. Waddington SN, McVey JH, Bhella D, Parker AL, Barker K, Atoda H, et al. Adenovirus serotype 5 hexon mediates liver gene transfer. *Cell* 2008;**132**(3):397–409.
154. Xu Z, Qiu Q, Tian J, Smith JS, Conenello GM, Morita T, et al. Coagulation factor X shields adenovirus type 5 from attack by natural antibodies and complement. *Nat Med* 2013;**19**(4):452–7.
155. Lopez-Gordo E, Denby L, Nicklin SA, Baker AH. The importance of coagulation factors binding to adenovirus: historical perspectives and implications for gene delivery. *Expert Opin Drug Deliv* 2014;**11**(11):1795–813.
156. Cichon G, Boeckh-Herwig S, Schmidt HH, Wehnes E, Muller T, Pring-Akerblom P, et al. Complement activation by recombinant adenoviruses. *Gene Ther* 2001;**8**(23):1794–800.
157. Shayakhmetov DM, Li ZY, Ni S, Lieber A. Interference with the IL-1-signaling pathway improves the toxicity profile of systemically applied adenovirus vectors. *J Immunol* 2005;**174**(11):7310–9.
158. Tian J, Xu Z, Smith JS, Hofherr SE, Barry MA, Byrnes AP. Adenovirus activates complement by distinctly different mechanisms in vitro and in vivo: indirect complement activation by virions in vivo. *J Virol* 2009;**83**(11):5648–58.
159. Weinberg ED. Human lactoferrin: a novel therapeutic with broad spectrum potential. *J Pharm Pharmacol* 2001;**53**(10):1303–10.
160. Johansson C, Jonsson M, Marttila M, Persson D, Fan XL, Skog J, et al. Adenoviruses use lactoferrin as a bridge for CAR-independent binding to and infection of epithelial cells. *J Virol* 2007;**81**(2):954–63.
161. Arnold D, Di Biase AM, Marchetti M, Pietrantoni A, Valenti P, Seganti L, et al. Antiadenovirus activity of milk proteins: lactoferrin prevents viral infection. *Antiviral Res* 2002;**53**(2):153–8.
162. Pickles RJ, McCarty D, Matsui H, Hart PJ, Randell SH, Boucher RC. Limited entry of adenovirus vectors into well-differentiated airway epithelium is responsible for inefficient gene transfer. *J Virol* 1998;**72**(7):6014–23.
163. Bi BY, Leveugle B, Liu JL, Collard A, Coppe P, Roche AC, et al. Immunolocalization of the lactotransferrin receptor on the human T lymphoblastic cell line Jurkat. *Eur J Cell Biol* 1994;**65**(1):164–71.

164. Ghio AJ, Carter JD, Samet JM, Reed W, Quay J, Dailey LA, et al. Metal-dependent expression of ferritin and lactoferrin by respiratory epithelial cells. *Am J Physiol* 1998;**274**(5 Pt 1): L728–36.
165. Adams WC, Bond E, Havenga MJ, Holterman L, Goudsmit J, Karlsson Hedestam GB, et al. Adenovirus serotype 5 infects human dendritic cells via a coxsackievirus-adenovirus receptor-independent receptor pathway mediated by lactoferrin and DC-SIGN. *J Gen Virol* 2009;**90**(Pt 7):1600–10.
166. Günther PS, Mikeler E, Hamprecht K, Schneider-Schaulies J, Jahn G, Dennehy KM. CD209/DC-SIGN mediates efficient infection of monocyte-derived dendritic cells by clinical adenovirus 2C isolates in the presence of bovine lactoferrin. *J Gen Virol* 2011;**92**(Pt 8): 1754–9.
167. Balakireva L, Schoehn G, Thouvenin E, Chroboczek J. Binding of adenovirus capsid to dipalmitoyl phosphatidylcholine provides a novel pathway for virus entry. *J Virol* 2003;**77**(8):4858–66.
168. Jobe AH, Ikegami M, Yei S, Whitsett JA, Trapnell B. Surfactant effects on aerosolized and instilled adenoviral-mediated gene transfer. *Hum Gene Ther* 1996;**7**(6):697–704.
169. Doherty GJ, McMahon HT. Mechanisms of endocytosis. *Annu Rev Biochem* 2009; **78**:857–902.
170. Mercer J, Schelhaas M, Helenius A. Virus entry by endocytosis. *Annu Rev Biochem* 2010;**79**:803–33.
171. Sigismund S, Confalonieri S, Ciliberto A, Polo S, Scita G, Di Fiore PP. Endocytosis and signaling: cell logistics shape the eukaryotic cell plan. *Physiol Rev* 2012;**92**(1):273–366.
172. Greber UF, Willetts M, Webster P, Helenius A. Stepwise dismantling of adenovirus 2 during entry into cells. *Cell* 1993;**75**(3):477–86.
173. Suomalainen M, Luisoni S, Boucke K, Bianchi S, Engel DA, Greber UF. A direct and versatile assay measuring membrane penetration of adenovirus in single cells. *J Virol* 2013;**87**(22):12367–79.
174. Doherty GJ, McMahon HT. Mediation, modulation, and consequences of membrane-cytoskeleton interactions. *Annu Rev Biophys* 2008;**37**:65–95.
175. Schmid SL, Frolov VA. Dynamin: functional design of a membrane fission catalyst. *Annu Rev Cell Dev Biol* 2011;**27**:79–105.
176. Svensson U. Role of vesicles during adenovirus 2 internalization into HeLa cells. *J Virol* 1985;**55**(2):442–9.
177. Patterson S, Russell WC. Ultrastructural and immunofluorescence studies of early events in adenovirus-HeLa cell interactions. *J Gen Virol* 1983;**64**(Pt 5):1091–9.
178. Chardonnet Y, Dales S. Early events in the interaction of adenoviruses with HeLa cells. I. Penetration of type 5 and intracellular release of the DNA genome. *Virology* 1970;**40**(3):462–77.
179. Pastan I, Seth P, FitzGerald D, Willingham M. In: Notkins AL, Oldstone MBA, editors. *Adenovirus entry into cells: some new observations on an old problem*. New York: Springer Verlag; 1986. p. 141–6.
180. Meier O, Boucke K, Hammer SV, Keller S, Stidwill RP, Hemmi S, et al. Adenovirus triggers macropinocytosis and endosomal leakage together with its clathrin-mediated uptake. *J Cell Biol* 2002;**158**(6):1119–31.
181. Wang K, Huang S, Kapoor-Munshi A, Nemerow G. Adenovirus internalization and infection require dynamin. *J Virol* 1998;**72**(4):3455–8.
182. Bai M, Harfe B, Freimuth P. Mutations that alter an Arg-Gly-Asp (RGD) sequence in the adenovirus type 2 penton base protein abolish its cell-rounding activity and delay virus reproduction in flat cells. *J Virol* 1993;**67**(9):5198–205.

183. Nishimura T, Kaibuchi K. Numb controls integrin endocytosis for directional cell migration with aPKC and PAR-3. *Dev Cell* 2007;**13**(1):15–28.
184. Maurer ME, Cooper JA. The adaptor protein Dab2 sorts LDL receptors into coated pits independently of AP-2 and ARH. *J Cell Sci* 2006;**119**(Pt 20):4235–46.
185. Motley A, Bright NA, Seaman MN, Robinson MS. Clathrin-mediated endocytosis in AP-2-depleted cells. *J Cell Biol* 2003;**162**(5):909–18.
186. Boulant S, Kural C, Zeeh JC, Ubelmann F, Kirchhausen T. Actin dynamics counteract membrane tension during clathrin-mediated endocytosis. *Nat Cell Biol* 2011;**13**(9):1124–31.
187. Li E, Stupack D, Bokoch GM, Nemerow GR. Adenovirus endocytosis requires actin cytoskeleton reorganization mediated by Rho family GTPases. *J Virol* 1998;**72**(11):8806–12.
188. Swanson JA. Shaping cups into phagosomes and macropinosomes. *Nat Rev Mol Cell Biol* 2008;**9**(8):639–49.
189. Kerr MC, Teasdale RD. Defining macropinocytosis. *Traffic* 2009;**10**(4):364–71.
190. Liberali P, Kakkonen E, Turacchio G, Valente C, Spaar A, Perinetti G, et al. The closure of Pak1-dependent macropinosomes requires the phosphorylation of CtBP1/BARS. *EMBO J* 2008;**27**(7):970–81.
191. Chinnadurai G. CtBP family proteins: more than transcriptional corepressors. *Bioessays* 2003;**25**(1):9–12.
192. Berk AJ. Recent lessons in gene expression, cell cycle control, and cell biology from adenovirus. *Oncogene* 2005;**24**(52):7673–85.
193. Iacobelli-Martinez M, Nemerow GR. Preferential activation of Toll-like receptor nine by CD46-utilizing adenoviruses. *J Virol* 2007;**81**(3):1305–12.
194. Schuierer M, Hilger-Eversheim K, Dobner T, Bosserhoff AK, Moser M, Turner J, et al. Induction of AP-2alpha expression by adenoviral infection involves inactivation of the AP-2rep transcriptional corepressor CtBP1. *J Biol Chem* 2001;**276**(30):27944–9.
195. Zhao H, Granberg F, Elfineh L, Pettersson U, Svensson C. Strategic attack on host cell gene expression during adenovirus infection. *J Virol* 2003;**77**(20):11006–15.
196. Granberg F, Svensson C, Pettersson U, Zhao H. Adenovirus-induced alterations in host cell gene expression prior to the onset of viral gene expression. *Virology* 2006;**353**(1):1–5.
197. Aderem A, Underhill DM. Mechanisms of phagocytosis in macrophages. *Annu Rev Immunol* 1999;**17**(6755):593–623.
198. Bidgood SR, Tam JC, McEwan WA, Mallery DL, James LC. Translocalized IgA mediates neutralization and stimulates innate immunity inside infected cells. *Proc Natl Acad Sci USA* 2014;**111**(37):13463–8.
199. Mercier S, Rouard H, Delfau-Larue MH, Eloit M. Specific antibodies modulate the interactions of adenovirus type 5 with dendritic cells. *Virology* 2004;**322**(2):308–17.
200. Ebbinghaus C, Al-Jaibaji A, Operschall E, Schoffel A, Peter I, Greber UF, et al. Functional and selective targeting of adenovirus to high-affinity Fcgamma receptor I-positive cells by using a bispecific hybrid adapter. *J Virol* 2001;**75**(1):480–9.
201. Meier O, Gastaldelli M, Boucke K, Hemmi S, Greber UF. Early steps of clathrin-mediated endocytosis involved in phagosomal escape of Fcgamma receptor-targeted adenovirus. *J Virol* 2005;**79**(4):2604–13.
202. Hauler F, Mallery DL, McEwan WA, Bidgood SR, James LC. AAA ATPase p97/VCP is essential for TRIM21-mediated virus neutralization. *Proc Natl Acad Sci USA* 2012;**109**(48):19733–8.
203. Mallery DL, McEwan WA, Bidgood SR, Towers GJ, Johnson CM, James LC. Antibodies mediate intracellular immunity through tripartite motif-containing 21 (TRIM21). *Proc Natl Acad Sci USA* 2010;**107**(46):19985–90.
204. Yousuf MA, Zhou X, Mukherjee S, Chintakuntlawar AV, Lee JY, Ramke M, et al. Caveolin-1 associated adenovirus entry into human corneal cells. *PLoS One* 2013;**8**(10):e77462.

205. Maier O, Marvin SA, Wodrich H, Campbell EM, Wiethoff CM. Spatiotemporal dynamics of adenovirus membrane rupture and endosomal escape. *J Virol* 2012;**86**(19):10821–8.
206. Greber UF, Webster P, Weber J, Helenius A. The role of the adenovirus protease on virus entry into cells. *EMBO J* 1996;**15**(8):1766–77.
207. Reddy VS, Natchiar SK, Stewart PL, Nemerow GR. Crystal structure of human adenovirus at 3.5 A resolution. *Science* 2010;**329**(5995):1071–5.
208. San Martin C. Latest insights on adenovirus structure and assembly. *Viruses* 2012; **4**(5):847–77.
209. Morgan C, Rosenkranz HS, Mednis B. Structure and development of viruses as observed in the electron microscope. V. Entry and uncoating of adenovirus. *J Virol* 1969;**4**(5): 777–96.
210. Brabec M, Schober D, Wagner E, Bayer N, Murphy RF, Blaas D, et al. Opening of size-selective pores in endosomes during human rhinovirus serotype 2 in vivo uncoating monitored by single-organelle flow analysis. *J Virol* 2005;**79**(2):1008–16.
211. Prchla E, Plank C, Wagner E, Blaas D, Fuchs R. Virus-mediated release of endosomal content in vitro: different behavior of adenovirus and rhinovirus serotype 2. *J Cell Biol* 1995;**131**(1):111–23.
212. Farr GA, Zhang LG, Tattersall P. Parvoviral virions deploy a capsid-tethered lipolytic enzyme to breach the endosomal membrane during cell entry. *Proc Natl Acad Sci USA* 2005;**102**(47):17148–53.
213. Otero MJ, Carrasco L. Proteins are cointernalized with virion particles during early infection. *Virology* 1987;**160**(1):75–80.
214. Jurgeit A, McDowell R, Moese S, Meldrum E, Schwendener R, Greber UF. Niclosamide is a proton carrier and targets acidic endosomes with broad antiviral effects. *PLoS Pathog* 2012;**8**(10):e1002976. http://dx.doi.org/10.1371/journal.ppat.
215. Perez L, Carrasco L. Involvement of the vacuolar H(+)-ATPase in animal virus entry. *J Gen Virol* 1994;**75**:2595–606.
216. Moyer CL, Wiethoff CM, Maier O, Smith JG, Nemerow GR. Functional genetic and biophysical analyses of membrane disruption by human adenovirus. *J Virol* 2011; **85**(6):2631–41.
217. Maier O, Galan DL, Wodrich H, Wiethoff CM. An N-terminal domain of adenovirus protein VI fragments membranes by inducing positive membrane curvature. *Virology* 2010;**402**(1):11–9.
218. Moyer CL, Nemerow GR. Disulfide-bond formation by a single cysteine mutation in adenovirus protein VI impairs capsid release and membrane lysis. *Virology* 2012;**428**(1):41–7.
219. Last NB, Schlamadinger DE, Miranker AD. A common landscape for membrane-active peptides. *Protein Sci* 2013;**22**(7):870–82.
220. Shai Y. Mechanism of the binding, insertion and destabilization of phospholipid bilayer membranes by alpha-helical antimicrobial and cell non-selective membrane-lytic peptides. *Biochim Biophys Acta* 1999;**1462**(1–2):55–70.
221. Greber UF. How cells tune viral mechanics–insights from biophysical measurements of influenza virus. *Biophys J* 2014;**106**(11):2317–21.
222. Raynaud-Messina B, Merdes A. Gamma-tubulin complexes and microtubule organization. *Curr Opin Cell Biol* 2007;**19**(1):24–30.
223. Müsch A. Microtubule organization and function in epithelial cells. *Traffic* 2004;**5**(1):1–9.
224. Hancock WO. Bidirectional cargo transport: moving beyond tug of war. *Nat Rev Mol Cell Biol* 2014;**15**(9):615–28.
225. Suomalainen M, Nakano MY, Boucke K, Keller S, Stidwill RP, Greber UF. Microtubule-dependent minus and plus end-directed motilities are competing processes for nuclear targeting of adenovirus. *J Cell Biol* 1999;**144**:657–72.

226. Engelke MF, Burckhardt CJ, Morf MK, Greber UF. The dynactin complex enhances the speed of microtubule-dependent motions of adenovirus both towards and away from the nucleus. *Viruses* 2011;**3**(3):233–53.
227. Warren JC, Cassimeris L. The contributions of microtubule stability and dynamic instability to adenovirus nuclear localization efficiency. *Cell Motil Cytoskelet* 2007;**64**(9):675–89.
228. Kelkar SA, Pfister KK, Crystal RG, Leopold PL. Cytoplasmic dynein mediates adenovirus binding to microtubules. *J Virol* 2004;**78**(18):10122–32.
229. Greber UF, Suomalainen M, Stidwill RP, Boucke K, Ebersold M, Helenius A. The role of the nuclear pore complex in adenovirus DNA entry. *EMBO J* 1997;**16**:5998–6007.
230. Leopold PL, Kreitzer G, Miyazawa N, Rempel S, Pfister KK, Rodriguez-Boulan E, et al. Dynein- and microtubule-mediated translocation of adenovirus serotype 5 occurs after endosomal lysis. *Hum Gene Ther* 2000;**11**(1):151–65.
231. Greber UF, Way M. A superhighway to virus infection. *Cell* 2006;**124**(4):741–54.
232. Wang IH, Suomalainen M, Andriasyan V, Kilcher S, Mercer J, Neef A, et al. Tracking viral genomes in host cells at single-molecule resolution. *Cell Host Microbe* 2013;**14**(4):468–80.
233. Gazzola M, Burckhardt CJ, Bayati B, Engelke M, Greber UF, Koumoutsakos P. A stochastic model for microtubule motors describes the in vivo cytoplasmic transport of human adenovirus. *PLoS Comp Biol* 2009;**5**(12):e1000623.
234. Scherer J, Vallee RB. Conformational changes in the adenovirus hexon subunit responsible for regulating cytoplasmic dynein recruitment. *J Virol* 2015;**89**(2):1013–23.
235. Schroer TA, Sheetz MP. Two activators of microtubule-based vesicle transport. *J Cell Biol* 1991;**115**(5):1309–18.
236. Tibbles LA, Spurrell JCL, Bowen GP, Liu Q, Lam M, Zaiss AK, et al. Activation of p38 and ERK signaling during adenovirus vector cell entry lead to expression of the C-X-C chemokine IP-10. *J Virol* 2002;**76**(4):1559–68.
237. Warren JC, Rutkowski A, Cassimeris L. Infection with replication-deficient adenovirus induces changes in the dynamic instability of host cell microtubules. *Mol Biol Cell* 2006;**17**(8):3557–68.
238. Giannakakou P, Nakano M, Nicolaou KC, O'Brate A, Yu J, Blagosklonny MV, et al. Enhanced microtubule-dependent trafficking and p53 nuclear accumulation by suppression of microtubule dynamics. *Proc Natl Acad Sci USA* 2002;**99**(16):10855–60.
239. Helmuth JA, Burckhardt CJ, Koumoutsakos P, Greber UF, Sbalzarini IF. A novel supervised trajectory segmentation algorithm identifies distinct types of human adenovirus motion in host cells. *J Struct Biol* 2007;**159**(3):347–58.
240. Strunze S, Trotman LC, Boucke K, Greber UF. Nuclear targeting of adenovirus type 2 requires CRM1-mediated nuclear export. *Mol Biol Cell* 2005;**16**(6):2999–3009.
241. Smith JG, Cassany A, Gerace L, Ralston R, Nemerow GR. A neutralizing antibody blocks adenovirus infection by arresting microtubule-dependent cytoplasmic transport. *J Virol* 2008;**82**(13):6492–500.
242. Duffy MR, Parker AL, Kalkman ER, White K, Kovalskyy D, Kelly SM, et al. Identification of novel small molecule inhibitors of adenovirus gene transfer using a high throughput screening approach. *J Control Release* 2013;**170**(1):132–40.
243. Trotman LC, Mosberger N, Fornerod M, Stidwill RP, Greber UF. Import of adenovirus DNA involves the nuclear pore complex receptor CAN/Nup214 and histone H1. *Nat Cell Biol* 2001;**3**:1092–100.
244. Cassany A, Ragues J, Guan T, Begu D, Wodrich H, Kann M, et al. Nuclear import of adenovirus DNA involves direct interaction of hexon with an N-terminal domain of the nucleoporin Nup214. *J Virol* 2014;**89**(3).

245. Pante N, Kann M. Nuclear pore complex is able to transport macromolecules with diameters of about 39 nm. *Mol Biol Cell* 2002;**13**(2):425–34.
246. Strunze S, Engelke MF, Wang I-H, Puntener D, Boucke K, Schleich S, et al. Kinesin-1-mediated capsid disassembly and disruption of the nuclear pore complex promote virus infection. *Cell Host Microbe* 2011;**10**:210–23.
247. Saphire ACS, Guan TL, Schirmer EC, Nemerow GR, Gerace L. Nuclear import of adenovirus DNA in vitro involves the nuclear protein import pathway and hsc70. *J Biol Chem* 2000;**275**(6):4298–304.
248. Hindley CE, Lawrence FJ, Matthews DA. A role for transportin in the nuclear import of adenovirus core proteins and DNA. *Traffic* 2007;**8**(10):1313–22.
249. Wodrich H, Cassany A, D'Angelo MA, Guan T, Nemerow G, Gerace L. Adenovirus core protein pVII is translocated into the nucleus by multiple import receptor pathways. *J Virol* 2006;**80**(19):9608–18.
250. Flatt JW, Greber UF. Misdelivery at the nuclear pore complex—stopping a virus dead in its Tracks. *Cells* 2015;**4**(3):277–96.
251. Dreier B, Honegger A, Hess C, Nagy-Davidescu G, Mittl PR, Grutter MG, et al. Development of a generic adenovirus delivery system based on structure-guided design of bispecific trimeric DARPin adapters. *Proc Natl Acad Sci USA* 2013;**110**(10):E869–77.
252. Plückthun A. Designed ankyrin repeat proteins (DARPins): binding proteins for research, diagnostics, and therapy. *Annu Rev Pharmacol Toxicol* 2015;**55**:489–511.
253. Kaliberov SA, Kaliberova LN, Buggio M, Tremblay JM, Shoemaker CB, Curiel DT. Adenoviral targeting using genetically incorporated camelid single variable domains. *Lab Invest* 2014;**94**(8):893–905.
254. Hendrickx R, Stichling N, Koelen J, Kuryk L, Lipiec A, Greber UF. Innate immunity to adenovirus. *Hum Gene Ther* 2014;**25**:265–84.
255. Kolawole AO, Sharma P, Yan R, Lewis KJ, Hostetler HA, Excoffon KJ. The PDZ1 and PDZ3 domains of MAGI-1 regulate the eight exon isoform of the coxsackievirus and adenovirus receptor. *J Virol* 2012;**86**(17):9244–54.
256. Burckhardt CJ, Greber UF. Virus movements on the plasma membrane support infection and transmission between cells. *PLoS Pathog* 2009;**5**(11):e1000621.
257. Chiu CY, Mathias P, Nemerow GR, Stewart PL. Structure of adenovirus complexed with its internalization receptor, alpha(v)beta 5 integrin. *J Virol* 1999;**73**(8):6759–68.
258. Nakano MY, Boucke K, Suomalainen M, Stidwill RP, Greber UF. The first step of adenovirus type 2 disassembly occurs at the cell surface, independently of endocytosis and escape to the cytosol. *J Virol* 2000;**74**(15):7085–95.
259. Leopold PL, Ferris B, Grinberg I, Worgall S, Hackett NR, Crystal RG. Fluorescent virions - dynamic tracking of the pathway of adenoviral gene transfer vectors in living cells. *Hum Gene Ther* 1998;**9**(3):367–78.
260. Walkiewicz MP, Morral N, Engel DA. Accurate single-day titration of adenovirus vectors based on equivalence of protein VII nuclear dots and infectious particles. *J Virol Methods* 2009;**159**(2):251–8.
261. Puntener D, Engelke MF, Ruzsics Z, Strunze S, Wilhelm C, Greber UF. Stepwise loss of fluorescent core protein V from human adenovirus during entry into cells. *J Virol* 2011;**85**(1):481–96.
262. Kirby I, Lord R, Davison E, Wickham TJ, Roelvink PW, Kovesdi I, et al. Adenovirus type 9 fiber knob binds to the coxsackie B virus-adenovirus receptor (CAR) with lower affinity than fiber knobs of other CAR-binding adenovirus serotypes. *J Virol* 2001;**75**(15):7210–4.
263. Salinas S, Zussy C, Loustalot F, Henaff D, Menendez G, Morton PE, et al. Disruption of the coxsackievirus and adenovirus receptor-homodimeric interaction triggers lipid

microdomain- and dynamin-dependent endocytosis and lysosomal targeting. *J Biol Chem* 2014;**289**(2):680–95.

264. Maisner A, Liszewski MK, Atkinson JP, Schwartz-Albiez R, Herrler G. Two different cytoplasmic tails direct isoforms of the membrane cofactor protein (CD46) to the basolateral surface of Madin-Darby canine kidney cells. *J Biol Chem* 1996;**271**(31):18853–8.
265. Andersson EK, Mei YF, Wadell G. Adenovirus interactions with CD46 on transgenic mouse erythrocytes. *Virology* 2010;**402**(1):20–5.
266. Vives RR, Lortat-Jacob H, Chroboczek J, Fender P. Heparan sulfate proteoglycan mediates the selective attachment and internalization of serotype 3 human adenovirus dodecahedron. *Virology* 2004;**321**(2):332–40.
267. Lyle C, McCormick F. Integrin alphavbeta5 is a primary receptor for adenovirus in CAR-negative cells. *Virol J* 2010;**7**:148.
268. Kelley JL, Ozment TR, Li C, Schweitzer JB, Williams DL. Scavenger receptor-A (CD204): a two-edged sword in health and disease. *Crit Rev Immunol* 2014;**34**(3):241–61.
269. Pietrantoni A, Di Biase AM, Tinari A, Marchetti M, Valenti P, Seganti L, et al. Bovine lactoferrin inhibits adenovirus infection by interacting with viral structural polypeptides. *Antimicrob Agents Chemother* 2003;**47**(8):2688–91.
270. Schmitz G, Müller G. Structure and function of lamellar bodies, lipid-protein complexes involved in storage and secretion of cellular lipids. *J Lipid Res* 1991;**32**(10):1539–70.
271. Smith JG, Silvestry M, Lindert S, Lu W, Nemerow GR, Stewart PL. Insight into the mechanisms of adenovirus capsid disassembly from studies of defensin neutralization. *PLoS Pathog* 2010;**6**(6):e1000959.
272. Flatt JW, Kim R, Smith JG, Nemerow GR, Stewart PL. An intrinsically disordered region of the adenovirus capsid is implicated in neutralization by human alpha defensin 5. *PLoS One* 2013;**8**(4):e61571.
273. Tam JC, Bidgood SR, McEwan WA, James LC. Intracellular sensing of complement C3 activates cell autonomous immunity. *Science* 2014;**345**(6201):1256070.
274. Vaysburd M, Watkinson RE, Cooper H, Reed M, O'Connell K, Smith J, et al. Intracellular antibody receptor TRIM21 prevents fatal viral infection. *Proc Natl Acad Sci USA* 2013;**110**(30):12397–401.
275. Ortega-Esteban A, Bodensiek K, San Martin C, Suomalainen M, Greber UF, de Pablo PJ, et al. Fluorescence tracking of genome release during mechanical unpacking of single viruses. *ACS Nano* 2015. http://dx.doi.org/10.1021/acsnano.5b03020.

# Adenovirus Replication

3

*Diana Guimet, Patrick Hearing*
Department of Molecular Genetics and Microbiology, School of Medicine,
Stony Brook University, Stony Brook, NY, USA

## 1. Introduction

The biology of adenoviruses (Ad) has been studied since before 1965, unraveling the profound range of virus–host interactions including the complexity of viral regulation of gene expression and Ad inhibition of host antiviral activities. Human Ad was first isolated from adenoid tissue in the 1950s as viral agents associated with respiratory infections.[1] Ad can establish acute and persistent infections. Most Ad infections are associated with mild disease occurring mainly in children, but Ad is increasingly being recognized as a significant viral pathogen in immunocompromised individuals.[2] Patients undergoing immunosuppressive therapy, including organ transplant patients and hematopoietic stem cell transplant recipients, and AIDS patients are included in this category. Each Ad serotype can infect a great variety of tissues and cells; however, a distinct disease pattern is observed for Ads that belong to different subgroups. For example, subgroup A and F Ads (see below) cause gastrointestinal infections, and subgroup B and C Ads cause upper respiratory tract infections, which may be accompanied by acute respiratory disease. Interestingly, some Ads of subgroup D cause a distinct disease, epidemic keratoconjunctivitis. Over 100 Ads have been identified and characterized in a wide range of vertebrate species. The virion is a nonenveloped, icosahedral capsid with a diameter of ~80–90 nm, containing a linear double-stranded DNA genome of ~36 kbp for human Ads. The size of Ad genomes varies from ~30 to 40 kbp.

In the early 1960s, researchers demonstrated that some human Ads cause tumors in rodents, which led to a surge in studies of the molecular biology, genetics, and physiology of Ads that continues to this day. Additionally, recombinant Ad vectors are being utilized in ~25% of current human gene therapy trials. Of particular excitement is the use of oncolytic Ad as a new class of anticancer agents with great therapeutic potential.[3] Extensive insight into the biology of Ad has opened the door for engineered Ads that target cancer cells.

## 2. Classification

Ads belong to the *Adenoviridiae* family. Currently there are five genera: *Mastadeno virus* and *Aviadenovirus* originate from mammals or birds, respectively; *Atadeno virus* and *Siadenovirus* have a broader range of hosts including birds and frog; and

**Adenoviral Vectors for Gene Therapy.**

*Ichtadenovirus* are identified in fish. Human Ads comprise more than 60 different serotypes classified into seven subgroups, designated A–G, that are organized based on their capacity to agglutinate red blood cells of human, rat, and monkey as well as on their oncogenicity in rodents. Recently, the field of DNA sequencing has taken the classification to a new level, and new sequence availability has allowed for more detailed phylogenetic analyses.[4]

## 3. Adenovirus Genome Organization

The Ad genome (Figure 1) is flanked by inverted terminal repeats (ITR) of ~100 bp that contain the origins of viral DNA replication. These sequences are followed by the viral packaging sequences at the left end of the genome, which direct viral DNA encapsidation. Ad terminal protein (pTP) is covalently linked to the 5′ ends of the genome and plays an important role in the initiation of viral DNA replication. The capsid is composed of three major proteins (II, III, and IV) and five minor proteins (IIIa, IVa2, VI, VIII, and IX). Proteins V, VII, and X/mu are associated with the DNA and form the core within the virion. These proteins are believed to condense the Ad DNA and mediate interactions between the core and the capsid. Ad protease is required for the maturation of the assembled particle to form fully infectious virus.[5]

The human Ad genome encodes ~40 proteins, which are classified as either early or late, based on their expression before or after DNA replication, respectively. The genome (Figure 1) contains the immediate-early region E1A, four early transcription units (E1B, E2, E3, and E4), and a set of "delayed early" units encoding proteins IX and IVa2. The Ad major late promoter (MLP) directs the synthesis of a single late pre-mRNA that is alternatively polyadenylated and alternatively spliced to generate five families of late mRNAs, regions L1–L5. Two additional small late transcripts are produced, virus-associated (VA) RNAI and VA RNAII. The transcription units of the Ad genome are transcribed from both strands of the chromosome.

Ads are excellent example of viruses that efficiently use limited genetic space and information to maximize their protein production for optimal virus propagation. Expression of the viral genes is temporally regulated at many different levels to produce a stepwise, logical progression of gene expression in order to take full advantage

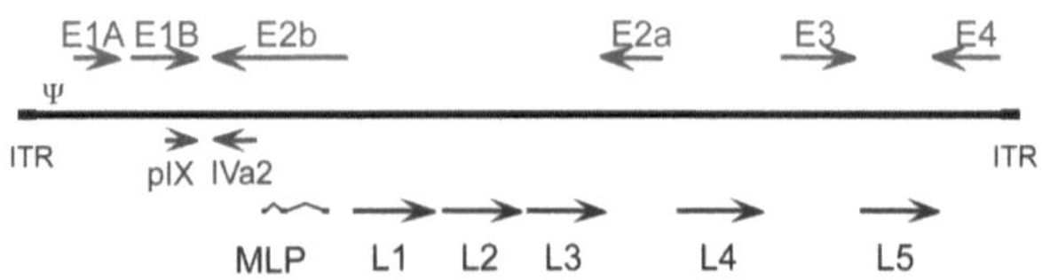


**Figure 1** Schematic of the human adenovirus genome. The Ad genome is depicted by a bold line with the inverted terminal repeats (ITR) and packaging sequences (Ψ) shown. Arrows above the genome indicate early genes, and arrows below the genome indicate intermediate and late genes.

of the cellular machinery to direct virus production. As an example, the late viral genes are not expressed to a full extent until after viral replication takes place and, even then, late gene expression is tightly controlled at both the transcriptional and the posttranscriptional level.

## 4. Virus Infection

The Ad fiber protein binds to the coxsackie-adenovirus receptor, which is the primary receptor for both Ad5 and coxsackie B virus.[6] Attachment to the receptor occurs in concert with the binding of the RGD peptide on the penton base to cellular integrins ($\alpha_v\beta_3$ and $\alpha_v\beta_5$). Ad may enter cells using heparin sulfate proteoglycans as an alternative receptor, likely through interaction with blood factors such as factor IX, factor X, or complement component C4-binding protein.[7] CD46 also is used as a receptor for subgroup B human Ad.[6] Ad is internalized by receptor-mediated endocytosis via clathrin-coated pits.[8] Endosome acidification alters virus topology and the capsid components partly disassemble. The partially degraded virion is transported through the cytoplasm to the nucleus along the microtubule network and the capsid is further disassembled en route.[8] On reaching the nuclear pore complex, the protein VII-coated Ad DNA enters the nucleus.[8] Current published findings suggest that it is only protein VII-wrapped DNA that enters the nucleus, escorted by histone H1.[5] Protein VII protects the viral DNA from activating the DNA damage response.[9] At this point, the DNA is still highly condensed and must undergo extensive remodeling to decondense before transcription of early genes can begin.[5] At the beginning of viral gene expression, histones are found bound to the viral DNA, particularly histone H3.3, along with protein VII.[5] Viral chromatin remodeling continues throughout the life cycle. Viral DNA replication and assembly of progeny virions occur entirely within the nucleus of infected cells. Ad DNA associates with newly synthesized pVII and the DNA–protein complex subsequently is packaged into the Ad capsid.[5] The Ad life cycle takes 24–36 h, with a single virus-infected cell producing $\sim 10^4$ daughter virions.

## 5. Early Gene Expression

When the viral genome enters the nucleus, early gene expression is directed toward achieving three main objectives. First, the host cell is stimulated to enter the S phase of the cell cycle and provide the virus with the necessary intracellular niche for optimal virus replication. Second, Ads devote a considerable part of their coding capacity to immune evasion functions that facilitate virus propagation and spread as well as gene products that counteract host antiviral responses. Third, viral proteins must be produced and used in concert with cellular proteins to carry out viral DNA replication. Ad encodes ~25 early gene products. The early genes are expressed in a temporal and coordinated manner.

## 5.1 Early Region 1A

The first early region expressed after Ad infection is the immediate-early transcription unit E1A since it requires only cellular transcription factors for its expression. The E1A gene products in turn activate transcription from the other early promoters. The E1A gene comprises two exons, and several E1A polypeptides are produced following alternative splicing of a primary RNA transcript (Figure 2). The most abundant E1A proteins are referred to as the E1A 243 amino acid (243 aa) and 289 amino acid (289 aa) gene products. The E1A 243 aa and 289 aa proteins act as major regulators of early viral transcription as well as important modulators of host cell gene expression and proliferation.[10] The E1A 243 aa and 289 aa proteins share two conserved regions within exon 1, referred to as CR1 and CR2, as well as another conserved region (CR4) at the C terminus in exon 2 (Figure 2). The two proteins differ only in a 46-residue internal exon segment present in the 289 aa protein, referred to as conserved region 3 (CR3). This region is important for the transcriptional transactivation properties of the E1A 289 aa protein.[10]

The E1A proteins exert their effects by interactions with numerous cellular proteins, many of which are involved in transcriptional regulation (Figure 2). The E1A proteins interact with a number of important cellular proteins including: (1) the retinoblastoma tumor suppressor family members pRb, p107, and p130 via CR1 and CR2; (2) transcriptional coactivators p300/CBP, PCAF, GCN5, TRRAP, and p400 via amino terminal sequences and CR1; (3) additional transcription factors such as TATA-binding protein, members of the ATF family, and the RNA polymerase II mediator complex via CR3; and (4) the transcriptional repressor CtBP via CR4.[10] The expression of E1A alone is sufficient to induce immortalization of primary rodent cells. E1A fully transforms such cells in conjunction with other oncogenes such as the Ad E1B proteins or activated Ras.[11] The expression of E1A also is sufficient to induce S phase progression in quiescent cells.[12] E1A activates gene expression via the

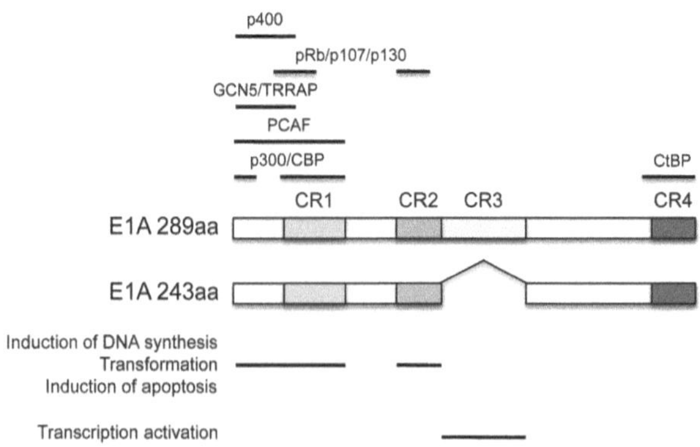


**Figure 2** Functional map of E1A proteins. The coding sequences of the E1A 289 aa and 243 aa proteins are shown with conserved regions depicted (CR1, CR2, CR3, and CR4). Binding sites for cellular proteins are indicated by bars above the E1A proteins. E1A functional activities are listed below the proteins.

E2F family of transcription factors. E2F transcription factors play a major role in the expression of cellular genes important for cell cycle progression. E2Fs both positively and negatively regulate gene expression.[13] In general, E2F-1, -2, and -3a/b activate gene expression, whereas E2F-4 and -5 repress gene expression. E2F-6, -7, and -8 may function as dominant-negative effectors. Activating E2Fs recruit histone acetyl transferases (HATs), and other transcriptional activators, to activate transcription. In contrast, repressing E2Fs recruit histone deacetylases (HDACs) and other transcriptional repressors to repress transcription. These latter complexes are formed via the interaction of repressing E2Fs with members of the retinoblastoma gene family.[13] Rb family binding to E2Fs is controlled through phosphorylation by cyclin-dependent kinases (Cdks). The phosphorylation of Rb family proteins by Cdks in the G1 phase of the cell cycle results in their dissociation from E2Fs and derepression of E2F-responsive genes.[13] The activation of E2F complexes results in the promotion of the S phase of the cell cycle via the expression of cellular genes that promote cell cycle progression. E1A acts to subvert the tight control of E2Fs by binding directly to Rb family proteins via an LXCXE motif in E1A common to other DNA tumor virus transforming proteins.[13] E1A sequesters Rb family members and frees E2Fs to activate viral and cellular gene expression. Both E1A 243 aa and 289 aa products direct the release of Rb family members from E2Fs, and both E1A proteins promote cellular transformation, in part, via this mechanism. p300/CBP, PCAF, and GCN5 are all HATs, while TRRAP and p400 serve as scaffolding proteins to bridge the interactions of HATs with other transcriptional regulators.[14] The binding of E1A to these proteins promotes cellular DNA synthesis and E1A mutants that cannot interact with these effectors are defective for transformation.[10] E1A binds to p400 through the N-terminal domain.[14] p400 is related to the yeast chromatin-modifying proteins SWI2/SNF2 and forms a complex with TRRAP. An E1A mutant that is defective for p400 binding also is defective for transformation. p400 is part of a larger HAT complex, termed TIP60, that contains TRRAP, GCN5, PCAF, TIP48, and TIP49.[14]

All the aforementioned results demonstrate the complex regulatory circuit that governs the regulation of cellular proliferation and how disruption of this carefully coordinated system by the N-terminal domains of E1A leads to profound effects on the cell. The C-terminal exon of the E1A proteins contains a nuclear localization signal and a binding site (CR4) for the transcriptional corepressor CtBP.[15,16] E1A exon 2 exhibits transcriptional regulatory activities that antagonize exon 1 functions. E1A exon 2 negatively regulates E1A functions in transformation and tumorigenesis.[15,16] CtBP functions as a transcriptional corepressor when tethered to a promoter region and binds to a number of HDACs including HDAC-1, -4, and -5. The C-terminal region of E1A also binds cellular proteins Dyrk and Foxk that may contribute to, or regulate, E1A activities.[15,16] E1A also plays a role in the induction of apoptosis in infected cells. Sustained, unregulated E2F activity triggers cellular checkpoint signaling and causes an increase in the level of the tumor suppressor. Activated p53 induces gene expression by binding specific promoter sequences, which activates genes that are involved in a number of cellular processes of p53.[17] p53 can induce cell cycle arrest, thus inhibiting progression of cell division; p53 also can induce cell death by the induction of apoptosis.[17] The activation of p53, and induction of cellular apoptosis, would be deleterious to Ad replication. Therefore, Ad has evolved several proteins encoded by the E1B and E4 transcription units that repress p53 activity and inhibit apoptosis.

## 5.2 Early Region 1B

Early region 1B (E1B) encodes the E1B-19K and E1B-55K proteins. The major roles of these proteins in Ad infection are to inhibit apoptosis and further modify the intracellular environment in order to make the cell more hospitable to viral protein production and viral DNA replication. Viruses with mutations in either E1B protein are significantly reduced in virus yield due to cell death by apoptosis prior to the completion of the replication cycle. The E1B-55K protein is essential for a variety of important functions in the viral life cycle including the inhibition of the induction of p53-dependent apoptosis.[18] The E1B-55K protein binds to the N-terminal transactivation domain of p53 and inhibits p53-induced transcription. E1B-55K also disrupts the interaction of p53 with the HAT PCAF and interferes with p53 acetylation.[18] Interestingly, the E1B-55K protein promotes cell transformation independently of repression of p53 transcriptional activity. This may relate to the ability of E1B-55K to inhibit other apoptotic activities in the cell including the proapototic cellular activity Daxx.[18] The E1B-55K protein is modified by the small, ubiquitin-like protein SUMO-1 and sumoylation of E1B-55K is required for its role in transformation.[19] Finally, E1B-55K acts in a complex with another Ad early protein, E4-ORF6, to promote the proteasome-dependent degradation of p53, among other cellular protein targets.[20]

The E1B-19K protein also is involved in the inhibition of apoptosis. E1B-19K acts to block apoptotic pathways that do not rely on p53, such as the tumor necrosis factor α (TNFα) and Fas ligand cell death pathways.[21] E1B-19K is a functional homologue of a cellular suppressor of apoptosis, Bcl-2. E1B-19K acts in the same manner as Bcl-2 and predominantly inhibits apoptosis by binding proapoptotic activities Bax and Bak.[21] E1B-19K also plays a role in the inhibition of TNFα-induced apoptosis by blocking the oligomerization of death-inducing complexes involving Fas-associated death domain (FADD).[21] FADD is a protein that is activated by binding Fas via death domains, thus its name (Fas-associated death domain). The exact function of E1B 19K in FADD regulation is not well understood.

## 5.3 Early Region 2

The early region 2 (E2) transcription unit encodes proteins that are required for viral DNA replication: DNA-binding protein (DBP), precursor terminal protein (pTP), and Ad DNA polymerase (Ad-Pol). DBP binds cooperatively to single-stranded DNA to stimulate both the initiation and the elongation of viral DNA replication.[22] DBP contains two major domains: the N-terminal globular core domain and the C-terminal part, which harbors most of the biological functions of DBP including nucleic acid binding and DNA replication.[22] The outermost C-terminal part of DBP is sufficient for all DNA replication functions and forms a protruding C-terminal arm and contains a hook. This hook is involved in the formation of a DBP chain where a C-terminal arm hooks into another DBP to form a multiprotein complex. This multimerization is the driving force for ATP-independent DNA unwinding by DBP during replication elongation.[22] During the initiation of viral DNA replication, DBP also stimulates the formation of a covalent linkage between the pTP and the trinucleotide primer CAT by lowering the $K_m$ value of the reaction, possibly via a direct interaction with the pTP–Pol complex.[22] Indirectly, DBP stimulates replication initiation by increasing the binding of NFI (see below) to the origin of replication. Other roles for DBP during DNA replication include enhancing the processivity of Ad-Pol, which is achieved by cooperative binding to the displaced strand during replication, thereby protecting it from nuclease digestion and facilitating strand displacement.[22]

pTP binds both single-stranded and double-stranded DNA and has several functions in Ad DNA replication.[23] pTP forms a heterodimer with Ad-Pol, which constitutes the preinitiation complex for Ad DNA replication.[23] The C-terminal region of pTP contains the Ad-Pol-binding region, while the N-terminal portion is involved in DNA binding. pTP binds to the entrance of the primer-binding groove of Ad-Pol, with its priming part located at the polymerase active site, and allows for protein-primed DNA synthesis to begin.[23] Besides helping to stabilize the pTP–Pol complex, other functions of pTP include attachment of viral DNA to the nuclear matrix, which is important for efficient early transcription and possibly for DNA replication.[24] The presence of pTP also renders the viral DNA inaccessible to 5′ exonucleases and prevents binding of end-recognizing proteins that could inhibit DNA replication.[23]

Ad-Pol belongs to a group of Pol-α DNA polymerases that employs a protein primer for DNA replication.[23] Ad-Pol has DNA-binding activity, DNA polymerase activity, and participates in the initiation of Ad DNA replication. The polymerase and exonuclease active sites of Ad-Pol are spatially distinct: the exonuclease activity of Ad-Pol is located at the N-terminal part of the protein and the polymerase activity is located at the C terminus. The molecular architecture of Ad-Pol is similar to that of RB69 DNA polymerases, a model enzyme for the family B polymerases.[23] When bound to DNA, Ad-Pol covers a region of 14–15 nt. Compared to free Ad-Pol, the pTP–Pol complex has reduced polymerase and exonuclease activity.[23] This is most likely due to the fact that pTP binds at the entrance of the primer-binding groove of Ad-Pol, which results in a competition between the pTP and the DNA located at the primer-binding groove.[23] After dissociation, Ad-Pol becomes a more active and processive enzyme.

## 5.4 Early Region 3

The early region 3 (E3) region of the Ad2 and Ad5 genome contains seven expressed open reading frames, most of which encode proteins with immunomodulatory functions (Figure 3). The E3 region is dispensable for viral replication in culture since E3 genes primarily are involved in the evasion of host immune defenses. E3 encodes integral membrane proteins that subvert host defense mechanisms by modulating immune response mechanisms including inhibition of antigen presentation suppression of natural killer (NK) cell activation, downregulation of apoptosis receptors, and interference with TNFα receptor-induced activities.[25,26] It is important that the infected cell remains viable during the extended period of infection, so antiapoptotic mechanisms are critical for survival. TNFα and Fas are members of the tumor necrosis

| E3 protein | Function(s) |
|---|---|
| 12.5K | unknown |
| 6.7K | inhibits TRAIL apoptosis |
| gp19K | blocks MHC class I presentation, inhibits CTL killing |
| ADP | cell killing, promotes virus release |
| RIDα/RIDβ | Receptor Internalization and Degradation complex, blocks TNFα, FasL, and TRAIL apoptosis |
| 14.7K | Inhibits TNFα, FasL, and TRAIL apoptosis, blocks arachadonic acid release |

**Figure 3** Schematic of Ad5 E3 proteins. The different proteins encoded by the E3 region are indicated by bars. The functions ascribed to different E3 proteins are listed below the diagram.

factor receptor (TNFR) superfamily that contain death domains that participate in protein–protein interactions, leading to the activation of proapoptotic caspases. At high concentrations, TNFα can inhibit the replication of certain viruses, including Ad, by inducing lysis and/or apoptosis of infected cells.[25,26] Binding of a death ligand in the TNF family, such as TNFα, Fas ligand (FasL), and TRAIL to its cognate death receptor, TNFR1, FAS, and TRAIL receptors 1 and 2 (TNFR1, TNFR2), leads to complex protein–protein interactions that result in a cascade of caspase-mediated proteolytic cleavages and the activation of transcription factors such as NF-κB and AP1. This triggers events that ultimately lead to the destruction of the cell via apoptosis. The receptors for TRAIL, Fas ligand, and TNFα play a role in killing infected cells, but Ad E3 proteins have the ability to inhibit such killing to prolong infection.[25,26] Another viral strategy encoded by E3 aims to inhibit viral antigen presentation. Cytotoxic T cells (CTLs) recognize antigenic peptides presented by major histocompatibility complex (MHC) class I antigens on the surface of infected cells. Assembly of MHC class I antigens occurs in the endoplasmic reticulum (ER) and this process is assisted by chaperones. MHC class I molecules carry peptides that are produced by the proteasome and translocate across the ER membrane by the transporter associated with antigen presentation.[25,26] After recognition, CTLs release perforins and granzymes that promote killing of the infected cell. Alternatively, CTL can induce apoptosis by the Fas pathway. By avoiding antigen presentation, the continued replication of Ad in infected host cells is possible.

E3-19K is a membrane glycoprotein localized in the ER and contains three segments: a luminal portion, a long cytoplasmic tail, and a transmembrane segment. E3-19K functions to counter the recognition of infected cells by both innate and adaptive cellular immune responses. E3-19K forms a complex with MHC class I molecules and inhibits transport of newly synthesized MHC molecules to the cell surface, preventing peptide presentation and suppressing recognition by T cells.[25,26] The function of E3-19K is based on two activities: MHC-I-binding activity by use of the luminal portion of the protein, and the ability to localize to the ER where it retains the MHC-I molecules. The latter activity is mediated by two structural elements: the ER retention signal contained in the transmembrane segment and the ER retrieval signal in the cytoplasmic tail. The retrieval signal mediates retrograde transport of E3-19K, and associated MHC-I, from the *cis*-Golgi to the ER where it is retained. Recently, an additional functional element has been characterized in the transmembrane domain (TMD) of E3-19K.[27] The TMD, together with the ER retrieval function, plays a role in efficient ER localization and transport inhibition of MHC-I. A potential caveat to human leukocyte antigen downregulation by E3-19K is the potential to render Ad-infected cells vulnerable to NK cell recognition. NK cells are a heterogeneous population of cells expressing a wide range of activating and inhibitory receptors, including NKG2. MHC class I chain-related proteins A and B (MICA and MIBC) are two of the seven human cellular NKG2D ligands (NKG2DLs). Induction of NKG2DLs occurs in response to stress, such as virus infection, and has been shown to be downregulated by E1A, which renders infected cells susceptible to NK cell attack.[28] However, while infection enhances the synthesis of NKG2DLs, MICA and MICB expression on the cell surface is suppressed by E3-19K.[29] This is a newly discovered function of E3-19K, which sequesters both MICA and MICB within the ER to allow Ad to successfully evade NK activation.[27]

The knowledge of the mechanism by which E3-19K protein targets MHC class I molecules for retention in the ER is derived mainly from studies of Ad2. Interestingly, there are low levels of amino acid conservation between E3-19K proteins of different Ad serotypes, suggesting that proteins from different serotypes have distinct MHC-I regulatory properties. This may lead to differing abilities of Ad serotypes to cause persistent infections.[30]

The E3-10.4–14.5K complex, named receptor internalization and degradation (RID), modulates a selective set of plasma membrane receptors involved in apoptosis and growth control.[25,26] E3-10.4K, or the RIDα subunit, is expressed as two isoforms: in one, the signal peptide is cleaved, whereas in the other form, it remains attached and serves as a second membrane anchor. The two isoforms form a disulfide-linked dimer and associate with E3-14.5K, or the RIDβ subunit. RIDβ is a type I transmembrane protein that is O-glycosylated and phosphorylated. Both viral proteins contain sequence elements in their cytoplasmic tails that include conserved transport motifs. RIDβ has three putative tyrosine motifs in the cytoplasmic tail and RIDα has a dileucine sorting motif, both of which trigger rapid internalization from the cell surface and aid in sorting to endosomal and lysosomal compartments, as well as mediate trafficking to the *trans*-Golgi network.[25,26] The YXXϕ motif is important for endocytosis of the RID complex, while the dileucine motif plays a role in targeting the complex to a recycling pathway, diverting it away from a degradation pathway. Therefore, RID targets specific cell surface receptors for degradation, but at some point along the endocytic pathway, RID drops off its target molecule and recycles back to the cell surface. The RID complex is localized predominantly to the plasma membrane, but neither RID subunit alone can reach the cell surface; RIDα alone is primarily localized in the Golgi, and RIDβ alone is localized in the ER and Golgi.[25,26]

E3–RID removes epidermal growth factor receptor (EGFR) from the cell surface by diverting internalized receptors to a degradation compartment and protects the infected cells from ligand-induced apoptosis.[25,26] E3–RID also plays a role in protecting human lymphocytes from apoptosis induced by ligation of Fas, a mechanism important for regulating lymphocyte populations.[25,26] On ligand engagement with and subsequent trimerization of Fas, Fas associates with death domain protein FADD via the death domain present in both proteins. In turn, the death effector domains present in FADD and procaspase 8 interact, resulting in the cleavage and production of caspase 8 eventually leading to cellular apoptosis. E3–RID downregulates surface Fas by a mechanism similar to that of EGFR, via endocytosis of the receptors into endosomes followed by transport to and degradation within lysosomes. This blocks events immediately downstream of Fas ligation, like Fas-FADD association and caspase-8 cleavage.[26,31] It is possible that expression of RID facilitates long-term infection by preventing Fas-mediated deletion of persistently infected lymphocytes. The apparently overlapping functions of E3–RID can be separated at the molecular level, leading to distinct mechanisms of action. For example, the RIDα subunit has an extracellular domain that contains sequences that are important for downregulation of EGFR, but not Fas,[32] suggesting that RID has specific mechanism for the downregulation of receptors.

TNFR1 is a proinflammatory receptor that activates both NF-κB and AP1 transcription factors in parallel. Both transcription factors are involved in proapoptotic and proinflammatory functions, both of which hinder the survival of Ad in infected cells. The E3–RID complex is sufficient to inhibit TNF signal transduction through TNFR, and the inhibitory effects extend to both the AP-1 pathway and the NF-κB pathway.[33] E3–RID inhibition of signal transduction through TNFR is accomplished by elimination of TNFR1 from the cell surface by clathrin-mediated endocytosis. E3–RID downregulates TNFR1 via an AP2 clathrin-mediated pathway and associates with TNFR1 in the same complex.[34] Once again, the mechanisms of action for downregulation of TNFR1 and Fas by E3–RID are different: it is likely that E3–RID associates with TNFR1 on the plasma membrane and Fas in the endosome (similar to where E3–RID associates with EGFR). Additional studies have shown that the E3–RID complex also interferes with the activation of the NF-κB pathway by interfering with a critical phosphorylation step and preventing NF-κB from entering the nucleus and becoming active.[35]

TNFα-induced apoptosis requires activation of one or more cytokines or chemokines. Chemokines play important roles where injury and infection are present for the early phase of inflammation. Interleukin-1 beta (IL-1β) is an example of such chemokine that is produced in response to inflammation, or induced by lipopolysaccharide (LPS), a component of the cell wall of Gram-negative bacteria. Both IL-1β and LPS share significant portions of the signaling pathways downstream of the respective receptors, IL-1R and Toll-like receptor (TLR4), including the adaptor molecules that are required to stimulate signaling and both signaling pathways are strongly inhibited by E3–RID. However, E3-RID expression has differential effects on IL-1R and TLRF4 signaling.[36] The inhibition of TLR4 signaling by E3–RID does not involve receptor downregulation. This suggests a new, yet unidentified mechanism by which E3–RID disrupts chemokine expression and signal transduction without having a direct effect on the receptor. In addition, TNFα-induced apoptosis requires the activation of cytosolic phospholipase A2 (cPLA2), an enzyme responsible for the production of inflammatory mediators. In the presence of submicromolar levels of $Ca^{2+}$, cPLA2 translocates to membranes where it cleaves arachidonic acid (AA), a potent mediator of inflammation, from membrane phospholipids. TNF-induced release of AA is inhibited by E3–RID (and E3-14.7K, discussed below), by preventing TNFα-induced translocation of cPLA2 to membranes of infected cells.[26,31]

E3–RID is one of three proteins encoded by Ad that independently inhibits TRAIL-induced apoptosis of infected cells.[37] The other two proteins are E1B-19K and E3-14.7K. T lymphocytes and NK cells use TRAIL to induce apoptosis in virus-infected and tumor cells. TRAIL induces apoptosis through two receptors: TRAIL-R1 (or death receptor 4) and TRAIL-R2 (death receptor 5).[25,26] E3–RID induces the internalization of TRAIL-R1 from the cell surface where it is internalized in endosomes/lysosomes and is eventually degraded in lysosomes.[37] One study showed that E3–RID is necessary and sufficient for downregulation of TRAIL-R1,[37] while another study showed that E3–RID may need E3-6.7K (discussed below) for this function.[38] Later, it was shown that both E3–RID and E3-6.7K are necessary for internalization and degradation of TRAIL-R2, whereas only E3–RID is required for TRAIL-R1 downregulation.[25,26]

E3-6.7K is a small hydrophobic integral membrane glycoprotein encoded by subgroup C Ad.[39] A small portion of the E3-6.7K protein is localized on the plasma membrane and forms a complex with the E3–RIDβ protein, and this complex is sufficient to induce downmodulation of TRAIL-R1 (and possibly TRAIL-R2) from the cell surface and inhibit TRAIL-mediated apoptosis.[38,39] E3-6.7K was shown to maintain ER $Ca^{2+}$ homeostasis and inhibit the induction of apoptosis, reduce the levels of AA induced by TNFα, and protect cells against apoptosis induced through Fas and TNFα, in addition to TRAIL receptors. E3-6.7K translocates across the membrane of the ER in a posttranslational, ribosome-independent, ATP-dependent manner. In addition, it has the ability to adopt more than one membrane topology.[40] It contains a single hydrophobic stretch that initiates membrane insertion and acts as the TMD, containing a signal anchor sequence. E3-6.7K has defied much of the current understanding of membrane protein conformation.

The nonmembrane E3-14.7K protein has a large proportion of amino acids with charged residues giving the protein hydrophilic properties. It is localized to both the cytosol and the nucleus. E3-14.7K inhibits TNFα-induced apoptosis by an independent mechanism.[26,41] TNFα-mediated apoptosis is initiated by ligand-induced recruitment of TNFα receptor-associated death domain, FADD, and caspase-8 to the death domain of TNFR1, thereby establishing the death-inducing signaling complex, DISC. E3-14.7K inhibits apoptosis by targeting key factors in this process. For example, caspase 8 cleaves key structural components of cells, ultimately leading to apoptosis and cPLA2 activation. E3-14.7K can bind and inhibit the function of caspase 8, and ultimately inhibit cPLA2 levels and the release of AA.[26,41] A number of cellular E3-14.7K-interacting proteins, named FIPs, are known to bind E3-14.7K directly. The FIPs in turn interact with other cellular proteins that are involved in membrane trafficking and morphogenesis, NF-κB signal transduction pathway, cell cycle control, and trafficking from and to the nucleus and cytoplasm.[26,41] New molecular mechanisms implemented by E4-14.7K to escape immunosurveillance have been discovered from 2005 to 2015. For example, E3-14.7K targets TNFR1 endocytosis and directly prevents TNFα-induced DISC formation.[42] Additionally, E3-14.7K is a potent inhibitor of NF-κB transcription activity following TLR or TNFR signaling. E3-14.7K can inhibit transcriptional activity at a point distal to the initial signaling pathway: specifically, it directly binds to the p50 subunit of NF-κB and prevents NF-κB DNA binding.[43] E3-14.7K has also been shown to independently inhibit TRAIL-induced apoptosis in certain cell types,[37] and inhibit STAT1 function by preventing its phosphorylation and nuclear translocation.[44]

To elucidate the molecular mechanism for E3-14.7K-mediated cell death protection, the biophysical properties of the protein have been characterized. The C-terminal two-thirds of the protein is highly structured and binds its putative cellular receptors. This C-terminal domain retains the capacity to interact with FIP-1 and the death effector domain of caspase 8, and binds zinc. Additionally, another FIP found is optineurin (OPTN).[45] OPTN has several functions including regulation of receptor endocytosis, vesicle trafficking, antiviral signaling, and regulation of the NF-κB pathway. Binding of OPTN to E3-14.7K potentially recruits both proteins to the TNFR1 complex, but OPTN is dispensable for E3-14.7K-mediated protection against TNFα-induced cytotoxicity.[46]

Recently, a new E3 protein, E3-49K, was discovered with subgroup D Ad.[47] Overall, E3 represents one of the most divergent regions of Ad, and some E3 genes seem to be unique to a particular subgroups. Ad19a and Ad65 contain the E3-49K gene that is absent in Ad of other subgroups, but is present in all subgroup D Ad examined. This protein is a glycosylated type I transmembrane protein of approximately 80–100 kDa molecular weight, the largest E3 protein discovered. The protein is localized in the Golgi-*trans*-Golgi network in early endosomes, and in lysosomes during the late phase of infection. E3-49K is cleaved and the large ectodomain is secreted.[47] This finding is significant, because to date, E3-49K is the only known E3 protein that is shed or secreted and offers an excellent example of immunomodulatory activities that are expressed by a single Ad subgroup. Interestingly, the protein targets noninfected leukocytes by binding to the protein phosphatase CD45, which suppresses leukocyte activation. E3-49K suppressed the functions for both NK cells and T cells.[47]

Finally, the E3-11.6K protein, now known as adenovirus death protein (ADP), plays a role in efficient cell lysis and subsequent release of virus from the infected cell. ADP is a nuclear membrane and Golgi glycoprotein that is produced in small amounts from scarce E3 mRNAs at early times postinfection, but is greatly amplified at late stages of infection.[48] ADP expression at late stages of infection is regulated by the L4-22K protein, a master regulator of late gene expression (discussed below).[49] During the late stages of infection, ADP is synthesized from mRNAs that contain the Ad tripartite leader (TPL). Over the course of infection, the abundance of ADP increases in the Golgi and ER, but ultimately ADP accumulates in the nuclear membrane late in infection. Extensive structure–function studies have identified specific domains important for protein processing, exit from the Golgi, and subcellular localization. Specifically, the luminal domain of ADP is important for protein stability and efficiency of cell lysis, while the cytoplasmic–nucleoplasmic domain is important for protein localization. In addition, ADP undergoes a complex process of N- and

O-linked glycosylation and proteolytic cleavage.[50] Just as it is critical for viruses to block apoptosis and evade the immune system, viruses must eventually be released from the cell in order to further amplify virus production. Mutation or deletion of ADP leads to a small plaque phenotype due to impaired virus release,[51] showing that ADP plays an important role in killing infected cells and for virion release. ADP may cause virus release by disrupting nuclear membrane integrity. Interestingly, the requirement for ADP for virus spread can be readily compensated by disrupting the functions of the E1B-19K protein, suggesting that the two viral proteins act as antagonists to influence viral spread.[48] Both caspase-dependent and caspase-independent mechanisms of cell killing have been observed when ADP is overexpressed,[52] but the mechanism is not well understood. However, overexpression of ADP has been reported to induce a pattern of cell death that exhibited some characteristics of apoptosis,[53] while other studies show that ADP has a role in the induction of necrosis-like cell death.[54]

## 5.5 Early Region 4

Whether they are expressed early or late during infection, a common theme among the Ad transcription units is that they encode multiple proteins of related functions. However, Ad early region 4 (E4) is the only transcription unit that produces proteins of relatively disparate functions. E4 encodes at least seven proteins according to analysis of open reading frames (ORF) and spliced mRNAs (Figure 4). The gene products exhibit a wide range of activities. Proteins expressed from the E4 region have been shown to be important for transcriptional regulation, viral DNA replication, viral mRNA transport and splicing, shutoff of host cell protein synthesis, oncogenic transformation, and the regulation of apoptosis. Subgroup D Ads are unique in their ability to induce estrogen-dependent mammary tumors in animals.[55] The primary oncogenic determinant of these viruses is the E4-ORF1 protein,[56] rather than the E1A and E1B proteins described above. Based on sequence similarity,

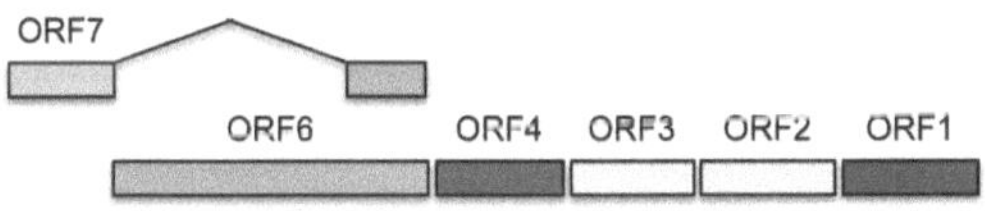


| E4 protein | Function(s) |
|---|---|
| ORF1 | binds PDZ proteins, stimulates proliferation/transformation |
| ORF2 | unknown |
| ORF3 | inhibits PML-NB activities, blocks the DNA damage response, inhibits p53 |
| ORF4 | regulates PP2A, induces cell killing |
| ORF6 | forms E3 ubiquitin ligase complex and induces degradation of cellular substrates, blocks the DNA damage response, inhibits p53, host cell shutoff |
| ORF6/7 | regulates E2Fs |

**Figure 4** Schematic of E4 proteins. The different proteins encoded by the E4 region are indicated by bars. The functions ascribed to different E4 proteins are listed below the diagram.

the E4-ORF1 protein appears to have evolved from a cellular dUTP pyrophosphatase gene, although E4-ORF1 does not possess this enzymatic activity.[56] Rather, E4-ORF1 appears to have utilized the structural aspects of dUTP pyrophosphatases in order to form homotrimers. The tumorigenic property of E4-ORF1 depends on a C-terminal sequence motif referred to as a PDZ domain. Ad9 E4-ORF1 contains a class I PDZ-binding domain motif following the consensus sequence (S/T)–X–(V/I/L)–COOH (X is any amino acid) at the C terminus of the protein.[56] PDZ domains are involved in protein–protein interactions and, in the case of Ad9 E4-ORF1, mediate the binding of E4-ORF1 to cellular proteins including Dlg1, MUPP1, PATJ, MAGI-1, and ZO-2.[56] PDZ proteins function as scaffolds to target signaling complexes to specific sites at the plasma membrane, and the PDZ proteins that bind to Ad9 E4-ORF1 have tumor suppressor activities. The binding of Ad9 E4-ORF1 to PDZ-containing proteins mediates the oncogenic function of this viral gene product. MUPP1, PATJ, MAGI-1, and ZO-2 localize to tight junctions at sites of cell–cell contact with epithelial cells. In epithelial cells, Ad9 E4-ORF1 prevents localization of PATJ and ZO-2 to tight junctions and, as such, disrupts tight junctions, resulting in a loss of apicobasal polarity.[56] Tight junction disruption and the loss of apicobasal polarity are common features of epithelial cancers. An additional function of Ad9 E4-ORF1 is activation of PI-3 kinase and this activity also is involved in the oncogenic properties of this protein. Activation of PI-3 kinase is associated with many human cancers and the interaction of Ad9 E4-ORF1 with Dlg1 may mediate this process.[56] Finally, the E4-ORF1 protein recently was shown to induce Myc activation and promote cellular anabolic glucose metabolism to promote viral replication.[57]

The E4-ORF3 protein is highly conserved among different Ad and is multifunctional. The E4-ORF3 and E4-ORF6 proteins both bind to the E1B-55K product, although with different outcomes. E4-ORF6 enhances the inhibition of p53 by E1B-55K, whereas E4-ORF3 transiently relieves the repression of p53 by E1B-55K.[58] E4-ORF3 has been shown to localize with discrete nuclear structures alternatively known as PML nuclear bodies (PML-NB), PML oncogenic domains (PODs), or ND10.[59] PML-NB exist as multiprotein complexes that exhibit a discreet, punctate appearance in the nucleus of a cell. E4-ORF3 is necessary and sufficient to cause redistribution of these protein complexes into long, track-like structures. PML-NB have been implicated in a number of cellular processes including transcriptional regulation, the regulation of apoptosis, DNA damage repair, protein modification, and an antiviral response.[60] PML-NB have also been shown to react to stresses such as heat shock and heavy metals as well as interferon, suggesting a role in cellular defense mechanisms. The E4-ORF3 protein of subgroup C Ad (e.g., Ad2 and Ad5) inhibits a cellular DNA damage response. E4-ORF3 directs the reorganization of the Mre11–Rad50–Nbs1 complex (MRN complex) into PML-containing tracks.[58] The MRN complex serves as a sensor of DNA damage and is recruited to the ends of damaged DNA. This serves to trigger effector cascades that lead to cell cycle arrest and the repair of the DNA damage. In the context of Ad infection, this process, if unabated, results in the end-to-end ligation of viral genomes effectively inhibiting viral DNA replication. E4-ORF3 interferes with this process by sequestering MRN proteins in the nucleus and blocking their function.[58] The E4-ORF3 protein also blocks p53 signaling by inducing heterochromatin formation at p53-induced promoters, thereby blocking p53 DNA binding and transactivation.[61]

The E4-ORF4 protein is a multifunctional regulator. First, E4-ORF4 binds to the B55 subunit of the serine/threonine phosphatase PP2A.[62,63] By binding this subunit, the trimeric form of PP2A is activated, which results in the dephosphorylation of target proteins such as mitogen-activated protein (MAP) kinases that are important in signal transduction pathways. Increased PP2A activity leads to decreased phosphorylation and inactivation of certain transcription factors, such as E4F, through direct interaction or through the inactivation of MAP kinases. E4-ORF4 expression also results in decreased E1A phosphorylation at MAP kinase consensus sites that are important for E4 transactivation.[62,63] By decreasing the activity of E1A and E4F, E4-ORF4 regulates the expression of the E4 region itself and thus may suppress the oncogenic potential of the E1A proteins. Second, E4-ORF4 is able to induce p53-independent apoptosis in transformed cells.[62,63] Oncogenic transformation of cells sensitizes them to E4-ORF4-induced cell killing. Depending on the cell type, E4-ORF4-induced apoptosis utilizes the classical pathway involving caspases or a nonclassical pathway that is caspase independent.[62,63] The binding to and regulation of PP2A by E4-ORF4 are essential for the induction of cell death. E4-ORF4-dependent apoptosis also requires modulation of Src-family kinases.[62,63] Thus, by an alternative mechanism, E4-ORF4 represses the oncogenic potential of the Ad E1 proteins. E4-ORF4 may be useful in the future as a therapeutic agent to target human cancers for apoptotic cell death.

The E4-ORF6 protein binds to and inhibits p53, providing Ad yet another defense for p53 effects within the cell.[10,55,58] E4-ORF6 augments the transformed phenotype of Ad E1-transformed cells through the downregulation of p53 expression.[10,55,58] E4-ORF6 forms a direct protein complex with the E1B-55K protein. The E4-ORF6/E1B-55K complex recruits a CUL5-containing E3 ubiquitin ligase complex to target p53, and other cellular proteins including proteins involved in DNA damage repair, for polyubiquitination and proteasome-dependent degradation.[10,55,58] By this mechanism, the E4-ORF6/E1B-55K complex counteracts the induction of p53 stability provided by E1A. Other targets of this virus-induced E3 ligase activity include Mre11, Rad50, and DNA ligase IV.[10,55,58]

The E4-ORF6/7 protein is produced from a spliced mRNA that encodes the amino terminus of E4-ORF6 linked to the unique E4-ORF7 sequence. E4-ORF6/7 forms stable homodimers that contribute to viral DNA synthesis by enhancing the production of E2 products. E4-ORF6/7 binds free E2F and induces cooperative and stable binding of E2F/DP heterodimers to inverted E2F-binding sites in the Ad E2 early promoter.[64] E4-ORF6/7 induces expression from the cellular E2F-1 promoter and is able to functionally compensate for E1A in Ad infection by displacing Rb family members from E2Fs.[65] Further, the E4-6/7 protein alters the subcellular localization of E2F family members and directs E2F-4 from the cytoplasm to the nucleus.[66] Thus, E4-6/7 displays functional redundancy with the E1A proteins in terms of activating E2F family members.

## 6. Viral DNA Replication

Ad DNA replication is a very efficient process. Three viral proteins (pTP, Ad-Pol, and DBP) and three cellular transcription factors (NFI, OCT-1, and NFII) are required for efficient Ad replication. The core origin of replication, a conserved region between nucleotides 9 and 18 at the ends of the viral genome, binds the pTP–Pol complex. Adjacent to the core origin is the auxiliary origin bound by NFI and OCT-1. NFI and OCT-1 enhance initiation by several hundred fold.[22,67] Between the core region and the auxiliary region is an A/T-rich region, also important for replication.

NFI is part of a family of related proteins that plays an important role in transcription regulation of a large variety of cellular and viral genes.[67] The conserved DNA-binding domain and dimerization region of NFI are required for Ad DNA replication. NFI binds as a dimer to the consensus sequence TGG(C/A)($N_5$)GCCAA located at the origin of replication nucleotides 25 to 38. DBP enhances the binding of NFI to the auxiliary origin by causing a structural change in the DNA that allows for increased flexibility of the NFI-binding site. NFI stabilizes binding of the pTP–Pol complex to the origin by directly interacting with the pTP–Pol complex.[67] NFI influences the kinetics of replication by increasing the number of active initiation complexes. Specifically, NFI facilitates formation of the preinitiation complex by inducing a bend of 60° in DNA on binding to the Ad origin, which in addition to the origin-binding site requires the A/T-rich region preceding this site.[67] OCT-1 belongs to the family of octamer-binding transcription factors containing a POU DNA-binding domain.[67] OCT-1 binds at nucleotides 39–47 adjacent to the NFI-binding sequence and interacts with pTP via its POU domain. This interaction recruits pTP to the origin and tethers it there. Both NFI and OCT-1 act in concert to enhance initiation, bending the DNA and facilitating optimal assembly of the preinitiation complex, thereby strongly stimulating replication.[67]

After the preinitiation complex is formed, the stress induced by bending assists in the unwinding of the origin. Ad replication begins, and employs a protein primer for replication initiation. During initiation, pTP presents its Ser-580 residue to the Ad-Pol active site, and influences its catalytic activity. After binding with an incoming dCTP nucleotide, Ad-Pol covalently couples the first dCTP residue with the OH group of Ser-580.[22,67] After coupling of the first dCTP, a pTP–CAT intermediate is formed using an integral GTA triplet at positions 4–6 as a template. DBP stimulates the formation of the pTP–CAT intermediate by lowering the $K_m$ for the reaction by influencing the Ad-Pol active site, either via direct interaction with Ad-Pol or by changing the template conformation.[22,67] Subsequently, the pTP–CAT intermediate jumps back three bases and becomes paired with template residues 1–3 at the beginning of the template strand. A common feature among protein-primed replication systems is replication initiation at an internal start site, rather than at the genome termini. A similar sliding back mechanism was first identified in bacteriophages. This mechanism allows for error corrections during initiation that cannot yet be repaired by the proofreading ability of Ad-Pol because when bound to pTP, the exonuclease activity of Ad-Pol is inhibited.[22,67] During or shortly after the three-nucleotide jump, Ad-Pol dissociates from pTP, increasing the rate of polymerization and the proofreading activity of Ad-Pol.

During elongation, the pTP–CAT primer is efficiently and processively elongated by Ad-Pol by strand displacement. During elongation, DBP enhances the rate of processivity of Ad-Pol and modifies the sensitivity to nucleotide analogs indirectly, by modifying the DNA structure.[22,67] DBP helps to unwind short stretches of double-stranded DNA in an ATP-independent manner. A third cellular factor, NFII, is necessary for elongation and ensures synthesis of genome-length DNA. NFII is a type I DNA topoisomerase.[22,67] During elongation, a new duplex genome is formed and the nontemplate strand is displaced. The ITRs of the displaced nontemplate strand may anneal together to restore the functional origins and can serve as substrates for new rounds of replication.

## 7. Virus-Associated RNA Genes

Ad encodes two ~160 nt VA RNAs (VA RNAI and VA RNAII) transcribed by cellular RNA polymerase III. These are single-stranded RNA molecules that fold into a well-conserved secondary structure that can be divided into a terminal stem, an apical stem, and a central domain. The VA RNAs counter cellular antiviral defense mechanisms in order to allow efficient synthesis of viral proteins.[68] VA RNAI blocks the activity of the RNA-dependent protein kinase (PKR) an interferon-inducible protein kinase activated in infected cells as part of the antiviral response.[68] Ad infection leads to the production of double-stranded RNA, which activates PKR. Activated PKR phosphorylates the α subunit of eukaryotic initiation factor 2 (eIF-2), which eventually leads to the shutoff of protein synthesis. Ad VA RNAI binds to PKR and blocks PKR activation to maintain protein synthesis. In addition, VA RNAI stabilizes ribosome-associated viral mRNAs, which could lead to enhanced levels of protein synthesis. The apical stem–loop is required for the binding to PKR and the central domain is involved in the inhibition of PKR activation.

## 8. Late Gene Expression

Ad replication initiates a transcription cascade that drives late gene expression from the MLP. The MLP shows basal transcriptional activity at early times of infection, with an efficiency comparable to other early viral promoters, but it is highly active late after infection. During the replication cycle, there is a switch from early-to-late gene expression in which the MLP is fully activated.[69] This activation requires viral DNA replication, and at least one viral protein, IVa2. Transcription of IVa2 is first activated as a result of viral DNA synthesis-dependent titration of a cellular transcriptional repressor that binds to the IVa2 promoter.[70] The synthesis of IVa2 in turn leads to maximally efficient transcription from the MLP. The late-specific increase of MLP activity requires additional *cis*-acting sequence elements located downstream of the MLP start site, termed downstream elements, DE1 and DE2.[70] DE1 is bound by a heterodimer of IVa2 (DEF-A) while DE2 is bound by DEF-B, which is suggested to be a heterodimer of IVa2 and the L4-22K protein.[71] The L4-22K protein transactivates the MLP consistent with this idea.[72,73]

The MLP drives transcription of the major late transcription unit, MLTU, that results in the production of five different groups of mature mRNAs (Figure 1; L1 to L5). Ad late proteins act as capsid structural proteins, promote assembly, direct genome packaging, and serve regulatory functions. The MLP produces all late mRNAs by alternative splicing and alternative polyadenylation of a primary transcript. Prior to DNA replication, the MLP is active at low levels with transcription proceeding only as far as the L3 region and mRNA production restricted to the L1-52/55K and i-leader proteins.[69] Following DNA replication, the MLP is fully activated and transcription proceeds to the L4 and L5 regions. After polyadenylation, Ad late primary transcripts are spliced so that each mature mRNA contains the untranslated TPL sequence composed of leaders 1, 2, and 3. There is a fourth exon, called the i-leader, that is sometimes inserted between leaders 2 and 3 in early L1 52/55K transcripts, but the i-leader is not present in other late mRNAs. L1 mRNAs that contain the i-leader coding region are less stable that those that lack it.[74] The TPL enhances translation of mRNAs during the late phase of Ad infection and can also increase the efficiency of mRNA export from the nucleus.[75] The L4-100K protein is the first late viral protein to be synthesized and it promotes translation of TPL mRNAs by ribosome shunting and inhibits cellular protein synthesis.[76] Ribosome shunting involves the loading of the 40S ribosome subunits to the 5′ end of the capped mRNA, followed by its direct translocation to the downstream initiation codon, directed by shunting elements in the TPL.

The cellular initiation factor eIF4F is a protein complex composed of phosphorylated cap-binding protein eIF4E, eIF4E kinase MnK1, eIF4A, poly(A)-binding protein, and eIF4G. L4-100K binds to the carboxyl terminus of eIF4G and competitively displaces MnK1 from cap initiation complexes, preventing eIF4E phosphorylation.[76] Late in infection, the modified L4-100K cap initiation complex associates with higher specificity to mRNAs that contain the TPL. The L4-100K–TPL complex enhances association with initiation factor eIF4G and poly(A)-binding protein, and together, the complexes are recruited to Ad late mRNAs, which utilize the complex to promote TPL-directed translation by ribosome shunting.[76,77] The mechanism behind this function of L4-100K is partially understood. L4-100K is tyrosine-phosphorylated, which was shown to be essential for efficient ribosome shunting and late protein synthesis, but not involved in binding eIF4G.[76] Additionally, L4-100K is arginine-methylated, a modification that plays an essential role in modulating protein–protein/RNA interactions by L4-100K.[78]

Following full activation of the MLP, transcription continues to the L4 and L5 regions. The L4 poly(A) site usage is prominent and it is only after an increase in expression of L4 gene products that the L2, L3, and L5 poly(A) sites are used, generating the complete late viral gene expression profile.[79] The L4 region encodes four proteins: L4-100K, protein pVIII, a structural protein of the viral capsid, L4-33K, a viral splicing factor, and L4-22K. Besides its role in DNA encapsidation, described below, L4-22K is associated with other functions pertaining to the regulation of Ad late gene expression.[49,72,73,80] The timing of L4 expression fits well with the idea that its products function to regulate late viral gene expression.[72,79] However, the essential role of L4 family proteins at this stage in the viral life cycle creates a paradox since their expression is achieved only as a consequence of the activation of late-phase expression. Morris et al. identified a novel, intermediate-phase Ad5 promoter, the L4 promoter (L4P), that directs the expression of the L4-22K and L4-33K proteins independent from the MLP,[81] providing an apparent solution to this paradox. L4P is active in its natural context and the amount of L4-22K protein expressed via L4P is sufficient to induce the early-to-late transition in MLTU activity. L4P is strongly activated by replication of the viral genome and is also partly activated by E1A, E4-ORF3, and IVa2. This promoter also is significantly activated by the cellular stress response regulator, p53.[82] In addition, the two products of L4P activation, L4-22K and L4-33K, inhibit p53 activation of the promoter, suggesting the presence of a negative feedback mechanism controlling L4P.

Following alternative polyadenylation, the TPL is spliced to one of many alternative 3′ splice sites, generating the cytoplasmic mRNAs that encode the Ad late proteins.[69] The individual 3′ acceptor splice and poly(A) sites within the main MLTU body have different efficiencies of usage that change as infection proceeds. The L4-33K protein serves as a viral splicing factor in the L1 unit.[69] The L1 unit has a common 5′ donor splice site that can be joined to one of two alternative 3′ acceptor splice sites forming the L1-52/55K or IIIa mRNAs. During early times of infection, the L1-52/55K acceptor site is preferentially used, while the IIIa acceptor splice site is not active until late times of infection. The L4-33K protein is required for the early-to-late shift in the L1 alternative splicing pattern and plays a positive role in regulating L1 splice site selection.[69] The role of L4-33K as an alternative RNA splicing factor has been confirmed in the context of viral infection, with primary targets identified as the proteins IIIa and pVI, with a lesser effect on fiber.[83]

Although the role of L4-33K in alternative splicing is well understood, the role of L4-22K in the regulation of Ad late gene expression is a very young field of study, and a deeper understanding regarding the mechanisms of regulation is still needed. Morris and Leppard established that L4-22K has an important role in regulating the pattern of MLTU gene expression and this role is independent of the effect of L4-33K on late mRNA splicing.[72] Analysis of the phenotypes of L4-22K mutant viruses confirmed that L4-22K is important for the transition of early-to-late viral gene expression[49] and that it acts at the level of late gene mRNA production.[80] This function is specific to a unique set of mRNA transcripts, with L4-33K and pVIII pre-mRNAs identified as the primary targets for L4-22K regulation. Specifically, L4-22K is required for efficient splicing of the L4-33K mRNA transcript and subsequent expression of the protein. The current working model of late gene expression places L4-22K, activated early in infection by L4P,[81] as the master regulator of late gene expression: L4-22K regulates accumulation of different L4 transcripts, including its own, thereby regulating the levels of L4 gene products.[49,72,80] As L4-22K protein levels increase, L4-33K pre-mRNA splicing will be stimulated, which would effectively reduce the accumulation of L4-22K mRNAs since L4-33K itself regulates Ad late pre-mRNA splicing.[69] Additionally, L4-22K is believed to suppress early gene expression during the late stage of infection, specifically E1A, although it is not clearly understood if this is also a posttranscriptional mechanism.[49] This finely tunes the temporal switch of gene expression patterns during the late phase of infection.

# 9. Viral DNA Packaging

The assembly of mature Ad virus particles is a multistep process that takes place in the nucleus. There are four different Ad particle forms identified by CsCl equilibrium centrifugation: empty capsids devoid of viral DNA, light assembly intermediates that contain the left end of the genome, heavy assembly intermediates that contain the full-length viral genome and precursor forms of certain capsid proteins, and mature virus particles that contain the full-length genome with proteolytically processed capsid proteins. Efficient packaging is required for the formation of mature virus particles and a complex of DNA and proteins at the packaging domain plays a key role in this process. The current model of Ad DNA packaging includes a multiprotein complex for packaging similar to the molecular motors of double-stranded DNA bacteriophages, which insert their genomes into preformed capsids driven by an ATP-hydrolyzing molecular machinery. The IVa2 protein was shown to bind ATP,[84] leading to the speculation that IVa2 is the ATPase providing the power stroke of the Ad packaging machinery.

Ad DNA packaging is dependent on a *cis*-acting region located at the left end of the viral genome (nucleotides 230–380) termed the packaging domain.[85] To achieve optimal packaging activity, the packaging domain must be near an end of the genome. The role of the packaging domain is to target the genome to an immature procapsid via proteins that bind to its sequences. The packaging sequences contain seven AT-rich repeats, termed A1 thru A7, that have the consensus sequence 5′-TTTG–$N_8$–CGNG-3′.[85] Repeats A1, A2, A5, and A6 are the most important repeats functionally.[85] The identification of the core nucleotide sequences involved in packaging allowed for the production of simplified packaging domains consisting of multiple copies of one A repeat or several A repeats, with A1 and A2 commonly used as a synthetic sequence that functions as wild type.[86] Several viral proteins have been described for their critical function in DNA packaging, including IVa2 and L4-22K. IVa2 and L4-22K mutant viruses produce empty particles containing no viral DNA.[49,87,88] In vitro, IVa2 binds to the CG motif of the A1 repeat, and L4-22K binds to the TTTG motif and forms a complex with IVa2 on the A2 repeat. Additional IVa2 and L4-22K proteins bind to adjacent packaging repeats.[71,88] In vivo studies showed that the L4-22K and IVa2 proteins are dependent on each other for binding to the packaging domain.[49] The binding of these proteins to the packaging domain may result in a complex initiating the formation of a portal entry site in an immature virus particle that leads to DNA encapsidation.

Detailed characterization of the assembly of IVa2 and L4-22K proteins onto packaging sequences shows that L4-22K binding promotes cooperative assembly of IVa2 onto packaging sequences.[89,90] The critical roles of both of these viral proteins in DNA encapsidation, and other aspects of infection, have prompted several structure–function studies. The Ad IVa2 protein has highly conserved sequences that resemble Walker boxes A and B associated with ATPases, and these elements are required for Ad packaging to occur.[87] The putative helix–turn–helix motif at the extreme C terminus of IVa2 is responsible for DNA-binding activity specifically related to packaging, but not its activation of the MLP.[91] The L4-22K protein has a unique C-terminal region that is highly conserved among human Ad serotypes.[80] Within this region is a conserved pair of cysteine residues that are required for efficient packaging, but not for regulation of late gene expression; therefore these residues uncouple the functions of L4-22K in Ad DNA packaging from the regulation of viral gene expression.[80]

A working model for Ad DNA packaging suggests that IVa2 and L4-22K bind directly to packaging sequences and together they recruit the L1-52/55K and L1-IIIa proteins to promote encapsidation. L1-52/55K is a nuclear phosphoprotein that is present in empty capsids and assembly intermediates, but it is not found in mature virions, suggesting a scaffolding role for this protein. In the absence of L1-52/55K, only empty capsids are formed,[92] further strengthening the notion that this protein is critical for virus packaging. IVa2 and L1-52/55K proteins bind to the packaging sequence in vivo, but independent of each other.[86,93] L1-52/55K also interacts with IVa2 via the N-terminal 173 amino acids of L1-52/55K.[93] Additionally, the L1-IIIa protein likely interacts with the L1-52/55K protein and associates with viral packaging sequences,[94] indicating that the complex of proteins on the packaging sequence is more intricate than initially believed.

Outside of the conventional model of packaging that includes the four viral proteins described above, other cellular and viral factors have been implicated and extensively studied. Several cellular DNA-binding proteins that bind packaging repeats were identified, but were not found to be relevant to the packaging process. These cellular proteins include COUP-TF, OCT-1, and P complex containing CCAAT displacement protein.[95] The packaging domain overlaps with the transcriptional enhancer region of E1A; thus, these cellular proteins as well as Ad packaging proteins may be important for E1A transcription. Additionally, L4-33K is suggested to play a role in virus assembly[96–98] and genome packaging.[83] A virus deficient in L4-33K produces only empty capsids, but the protein does not bind to packaging sequences or influence the interaction of other Ad proteins with the packaging sequences;[83] therefore, a better understanding of this protein's role in this aspect of the viral life cycle is needed.

## 10. Conclusion

Adenoviruses have a life cycle that represents a complex interplay between virus and host with a series of events that are temporarily regulated in order to optimize production of progeny virus. Ads devote a considerable part of their genome coding capacity to immune evasion functions and to encode proteins that counteract innate cellular responses to infection. At the same time, Ads are excellent at utilizing the cellular machinery to carry out viral functions. Since Ads significantly manipulate normal host cell functions to promote infection, studies into Ad basic biology have contributed to the fields of cell biology, DNA replication, gene expression, translation, immunology, and cancer. There are still numerous unanswered questions about Ad biology and many unsolved problems as the growing cases of serious Ad infections in immunologically compromised individuals rise. There is still no virus-specific therapy available. Additionally, with the widespread use of Ad gene therapy vectors, it is becoming increasingly important to gain a better understanding of the molecular mechanisms of infection.

# References


1. Rowe W, Huebner RJ, Gilmore LK, Parrott RH, Ward TG. Isolation of a cytopathogenic agent from human adenoids undergoing spontaneous degeneration in tissue culture. *Proc Soc Exp Biol Med* 1953;**84**:570–3.
2. Echavarría M. Adenoviruses in immunocompromised hosts. *Clin Microbiol Rev* 2008;**21**:704–15.
3. Yamamoto M, Curiel DT. Current issues and future directions of oncolytic adenoviruses. *Mol Ther* 2010;**18**(2):243–50.
4. Davison AJ, Benko M, Harrach B. Genetic content and evolution of adenoviruses. *J Gen Virol* 2003;**84**(Pt. 11):2895–908.
5. Giberson AN, Davidson AR, Parks RJ. Chromatin structure of adenovirus DNA throughout infection. *Nucl Acids Res* 2012;**40**(6):2369–76.
6. Russell WC. Adenoviruses: update on structure and function. *J Gen Virol* 2009;**90** (Pt. 1):1–20.
7. Waddington SN, McVey JH, Bhella D, Parker AL, Barker K, Atoda H, et al. Adenovirus serotype 5 hexon mediates liver gene transfer. *Cell* 2008;**132**(3):397–409.
8. Wolfrum N, Greber UF. Adenovirus signalling in entry. *Cell Microbiol* 2013;**15**(1):53–62.
9. Karen KA, Hearing P. Adenovirus core protein VII protects the viral genome from a DNA damage response at early times after infection. *J Virol* 2011;**85**(9):4135–42.
10. Berk AJ. Recent lessons in gene expression, cell cycle control, and cell biology from adenovirus. *Oncogene* 2005;**24**(52):7673–85.
11. Ruley HE. Adenovirus early region 1A enables viral and cellular transforming genes to transform primary cells in culture. *Nature* 1983;**304**(5927):602–6.
12. Frisch SM, Mymryk JS. Adenovirus-5 E1A: paradox and paradigm. *Nat Rev Mol Cell Biol* 2002;**3**(6):441–52.
13. DeCaprio JA. How the Rb tumor suppressor structure and function was revealed by the study of adenovirus and SV40. *Virology* 2009;**384**(2):274–84.
14. Turnell AS, Mymryk JS. Roles for the coactivators CBP and p300 and the APC/C E3 ubiquitin ligase in E1A-dependent cell transformation. *Br J Cancer* 2006;**95**(5):555–60.
15. Chinnadurai G. Opposing oncogenic activities of small DNA tumor virus transforming proteins. *Trends Microbiol* 2011;**19**(4):174–83.
16. Yousef AF, Fonseca GJ, Cohen MJ, Mymryk JS. The C-terminal region of E1A: a molecular tool for cellular cartography. *Biochem Cell Biol* 2012;**90**(2):153–63.
17. Levine AJ. The common mechanisms of transformation by the small DNA tumor viruses: the inactivation of tumor suppressor gene products: p53. *Virology* 2009;**384**(2):285–93.
18. Blackford AN, Grand RJ. Adenovirus E1B 55-kilodalton protein: multiple roles in viral infection and cell transformation. *J Virol* 2009;**83**(9):4000–12.
19. Wimmer P, Schreiner S, Dobner T. Human pathogens and the host cell SUMOylation system. *J Virol* 2012;**86**(2):642–54.
20. Schreiner S, Wimmer P, Dobner T. Adenovirus degradation of cellular proteins. *Future Microbiol* 2012;**7**(2):211–25.
21. White E. Mechanisms of apoptosis regulation by viral oncogenes in infection and tumorigenesis. *Cell Death Differ* 2006;**13**(8):1371–7.
22. de Jong RN, van der Vliet PC, Brenkman AB. Adenovirus DNA replication: protein priming, jumping back and the role of the DNA binding protein DBP. *Curr Top Microbiol Immunol* 2003;**272**:187–211.
23. de Jong RN, Meijer LA, van der Vliet PC. DNA binding properties of the adenovirus DNA replication priming protein pTP. *Nucleic Acids Res* 2003;**31**(12):3274–86.

24. Schaack J, Ho WY, Freimuth P, Shenk T. Adenovirus terminal protein mediates both nuclear matrix association and efficient transcription of adenovirus DNA. *Genes Dev* 1990;**4**:1197–208.
25. Fessler SP, Delgado-Lopez F, Horwitz MS. Mechanisms of E3 modulation of immune and inflammatory responses. *Curr Top Microbiol Immunol* 2004;**273**:113–35.
26. Lichtenstein DL, Toth K, Doronin K, Tollefson AE, Wold WS. Functions and mechanisms of action of the adenovirus E3 proteins. *Int Rev Immunol* 2004;**23**(1–2):75–111.
27. Sester M, Ruszics Z, Mackley E, Burgert HG. The transmembrane domain of the adenovirus E3/19K protein acts as an endoplasmic reticulum retention signal and contributes to intracellular sequestration of major histocompatibility complex class I molecules. *J Virol* 2013;**87**(11):6104–17.
28. Routes JM, Ryan S, Morris K, Takaki R, Cerwenka A, Lanier LL. Adenovirus serotype 5 E1A sensitizes tumor cells to NKG2D-dependent NK cell lysis and tumor rejection. *J Exp Med* 2005;**202**(11):1477–82.
29. McSharry BP, Burgert HG, Owen DP, Stanton RJ, Prod'homme V, Sester M, et al. Adenovirus E3/19K promotes evasion of NK cell recognition by intracellular sequestration of the NKG2D ligands major histocompatibility complex class I chain-related proteins A and B. *J Virol* 2008;**82**(9):4585–94.
30. Garnett CT, Talekar G, Mahr JA, Huang W, Zhang Y, Ornelles DA, et al. Latent species C adenoviruses in human tonsil tissues. *J Virol* 2009;**83**(6):2417–28.
31. Windheim M, Hilgendorf A, Burgert HG. Immune evasion by adenovirus E3 proteins: exploitation of intracellular trafficking pathways. *Curr Top Microbiol Immunol* 2004;**273**:29–85.
32. Zanardi TA, Yei S, Lichtenstein DL, Tollefson AE, Wold WS. Distinct domains in the adenovirus E3 RIDalpha protein are required for degradation of Fas and the epidermal growth factor receptor. *J Virol* 2003;**77**(21):11685–96.
33. Fessler SP, Chin YR, Horwitz MA. Inhibition of tumor necrosis factor (TNF) signal transduction by the adenovirus group C RID complex involves downregulation of surface levels of TNF receptor 1. *J Virol* 2004;**78**(23):13113–21.
34. Chin YR, Horwitz MS. Adenovirus RID complex enhances degradation of internalized tumour necrosis factor receptor 1 without affecting its rate of endocytosis. *J Gen Virol* 2006;**87**(11):3161–7.
35. Friedman JM, Horwitz MS. Inhibition of tumor necrosis factor alpha-induced NF-kappa B activation by the adenovirus E3-10.4/14.5K complex. *J Virol* 2002;**76**(11):5515–21.
36. Delgado-Lopez F, Horwitz MS. Adenovirus RIDalphabeta complex inhibits lipopolysaccharide signaling without altering TLR4 cell surface expression. *J Virol* 2006;**80**(13): 6378–86.
37. Tollefson AE, Toth K, Doronin K, Kuppuswamy M, Doronina OA, Lichtenstein DL, et al. Inhibition of TRAIL-induced apoptosis and forced internalization of TRAIL receptor 1 by adenovirus proteins. *J Virol* 2001;**75**(19):8875–87.
38. Benedict CA, Norris PS, Prigozy TI, Bodmer JL, Mahr JA, Garnett CT, et al. Three adenovirus E3 proteins cooperate to evade apoptosis by tumor necrosis factor-related apoptosis-inducing ligand receptor-1 and -2. *J Biol Chem* 2001;**276**(5):3270–8.
39. Lichtenstein DL, Doronin K, Toth K, Kuppuswamy M, Wold WS, Tollefson AE. Adenovirus E3-6.7K protein is required in conjunction with the E3-RID protein complex for the internalization and degradation of TRAIL receptor 2. *J Virol* 2004;**78**(22):12297–307.
40. Moise AR, Grant JR, Lippé R, Gabathuler R, Jefferies WA. The adenovirus E3-6.7K protein adopts diverse membrane topologies following posttranslational translocation. *J Virol* 2004;**78**(1):454–63.

41. Burgert HG, Blusch JH. Immunomodulatory functions encoded by the E3 transcription unit of adenoviruses. *Virus Genes* 2000;**21**(1–2):13–25.
42. Schneider-Brachert W, Tchikov V, Merkel O, Jakob M, Hallas C, Kruse ML, et al. Inhibition of TNF receptor 1 internalization by adenovirus 14.7K as a novel immune escape mechanism. *J Clin Invest* 2006;**116**(11):2901–13.
43. Carmody RJ, Maguschack K, Chen YH. A novel mechanism of nuclear factor-kappaB regulation by adenoviral protein 14.7K. *Immunol* 2006;**117**(2):188–95.
44. Spurrell E, Gangeswaran R, Want P, Cao F, Gao D, Feng B, et al. STAT1 interaction with E3-14.7K in monocytes affects the efficacy of oncolytic adenovirus. *J Virol* 2014;**88**(4):2291–300.
45. Horwitz MS. Function of adenovirus E3 proteins and their interactions with immunoregulatory cell proteins. *J Gene Med* 2004;**6**(Suppl. 1):S172–83.
46. Klingseisen L, Ehrenschewender M, Heigl U, Wajant H, Hehigans T, Schütze S, et al. E3-14.7K is recruited to TNF-receptor 1 and blocks TNF cytolysis independent from interaction with optineurin. *PLoS One* 2012;**7**(6):e38348.
47. Windheim M, Southcombe JH, Kremmer E, Chaplin L, Urlaub D, Falk CS, et al. A unique secreted adenovirus E3 protein binds to the leukocyte common antigen CD45 and modulates leukocyte functions. *Proc Natl Acad Sci USA* 2013;**110**(50):E4884–93.
48. Subramanian T, Vijayalingam S, Chinnadurai G. Genetic identification of adenovirus type 5 genes that influence viral spread. *J Virol* 2006;**80**(4):2000–12.
49. Wu K, Orozco D, Hearing P. The adenovirus L4-22K protein is multifunctional and is an integral component of crucial aspects of infection. *J Virol* 2012;**86**(19):10474–83.
50. Tollefson AE, Scaria A, Ying B, Wold WS. Mutations within the ADP (E3-11.6K) protein alter processing and localization of ADP and the kinetics of cell lysis of adenovirus-infected cells. *J Virol* 2003;**77**(14):7764–78.
51. Doronin K, Toth K, Kuppuswamy M, Krajcsi P, Tollefson AE, Wold WS. Overexpression of the ADP (E3-11.6K) protein increases cell lysis and spread of adenovirus. *Virology* 2003;**305**(2):378–87.
52. Zou A, Atencio I, Huang WM, Horn M, Ramachandra M. Overexpression of adenovirus E3-11.6K protein induces cell killing by both caspase-dependent and caspase-independent mechanisms. *Virology* 2004;**326**(2):240–9.
53. Yun CO, Kim E, Koo T, Kim H, Lee YS, Kim JH. ADP-overexpressing adenovirus elicits enhanced cytopathic effect by induction of apoptosis. *Cancer Gene Ther* 2005;**12**(1):61–71.
54. Abou El Hassan MA, van der Meulen-Muileman I, Abbas S, Kruyt FA. Conditionally replicating adenoviruses kill tumor cells via a basic apoptotic machinery-independent mechanism that resembles necrosis-like programmed cell death. *J Virol* 2004;**78**(22):12243–51.
55. Endter C, Dobner T. Cell transformation by human adenoviruses. *Curr Top Microbiol Immunol* 2004;**273**:163–214.
56. Javier RT, Rice AP. Emerging theme: cellular PDZ proteins as common targets of pathogenic viruses. *J Virol* 2011;**85**(22):11544–56.
57. Thai M, Graham NA, Braas D, Nehil M, Komisopoulou E, Kurdistani SK, et al. Adenovirus E4ORF1-induced MYC activation promotes host cell anabolic glucose metabolism and virus replication. *Cell Metab* 2014;**19**(4):694–701.
58. Weitzman MD, Ornelles DA. Inactivating intracellular antiviral responses during adenovirus infection. *Oncogene* 2005;**24**(52):7686–96.
59. Doucas V, Ishov AM, Romo A, Juguilon H, Weitzman MD, Evans RM, et al. Adenovirus replication is coupled with the dynamic properties of the PML nuclear structure. *Genes Dev* 1996;**10**(2):196–207.

60. Geoffroy MC, Chelbi-Alix MK. Role of promyelocytic leukemia protein in host antiviral defense. *J Interferon Cytokine Res* 2011;**31**(1):145–58.
61. Soria C, Estermann FE, Espantman KC, O'Shea CC. Heterochromatin silencing of p53 target genes by a small viral protein. *Nature* 2010;**466**(7310):1076–81.
62. Branton PE, Roopchand DE. The role of adenovirus E4orf4 protein in viral replication and cell killing. *Oncogene* 2001;**20**(54):7855–65.
63. Kleinberger T. Induction of transformed cell-specific apoptosis by the adenovirus E4orf4 protein. *Prog Mol Subcell Biol* 2004;**36**:245–67.
64. Huang MM, Hearing P. The adenovirus early region 4 open reading frame 6/7 protein regulates the DNA binding activity of the cellular transcription factor, E2F, through a direct complex. *Genes Dev* 1989;**3**:1699–710.
65. O'Connor RJ, Hearing P. The E4-6/7 protein functionally compensates for the loss of E1A expression in adenovirus infection. *J Virol* 2000;**74**(13):5819–24.
66. Schaley JE, Polonskaia M, Hearing P. The adenovirus E4-6/7 protein directs nuclear localization of E2F-4 via an arginine-rich motif. *J Virol* 2005;**79**(4):2301–8.
67. Van der Vliet PC. Adenovius DNA replication. *Curr Top Microbiol Immunol* 1995;**199**:1–30.
68. Ma Y, Mathews MB. Structure, function, and evolution of adenovirus-associated RNA: a phylogenetic approach. *J Virol* 1996;**70**(8):5083–99.
69. Akusjarvi G. Temporal regulation of adenovirus major late alternative RNA splicing. *Front Biosci* 2008;**13**:5006–15.
70. Young CSY. The structure and function of the adenovirus lajor late promoter. *Curr Top Microbiol Immunol* 2003;**272**:213–50.
71. Ostapchuk P, Anderson ME, Chandrasekhar S, Hearing P. The L4 22-kilodalton protein plays a role in packaging of the adenovirus genome. *J Virol* 2006;**80**(14):6973–81.
72. Morris SJ, Leppard K,N. Adenovirus serotype 5 L4-22K and L4-33K proteins have distinct functions in regulating late gene expression. *J Virol* 2009;**83**(7):3049–58.
73. Backström E, Kaufmann KB, Lan X, Göran A. Adenovirus L4-22K stimulates major late transcription by a mechanism requiring the intragenic late-specific transcription factor-binding site. *Virus Res* 2010;**151**:220–8.
74. Soloway PD, Shenk T. The adenovirus type 5 i-leader open reading frame functions in cis to reduce the half-life of L1 mRNAs. *J Virol* 1990;**64**:551–8.
75. Huang W, Flint SJ. The tripartite leader sequence of subgroup c adenovirus major late mRNAs can increase the efficiency of mRNA export. *J Virol* 1998;**72**(1):225–35.
76. Xi Q, Cuesta R, Schneider RJ. Regulation of translation by ribosome shunting through phosphotyrosine-dependent coupling of adenovirus protein 100k viral mRNAs. *J Virol* 2005;**79**(9):5676–83.
77. Xi Q, Cuesta R, Schneider RJ. Tethering of eIF4G to adenoviral mRNAs by viral 100k protein drives ribosome shunting. *Genes Dev* 2004;**18**(16):1997–2009.
78. Koyuncu OO, Dobner T. Arginine methylation of human adenovirus type 5 L4 100-kilodalton protein is required for efficient virus production. *J Virol* 2009;**83**(10):4778–90.
79. Farley DC, Brown J,L, Leppard K,N. Activation of the early-late switch in adenovirus type 5 major late transcription unit expression by L4 gene products. *J Virol* 2004;**78**(4):1782–91.
80. Guimet D, Hearing P. The adenovirus L4-22K protein has distinct functions in the post-transcriptional regulation of gene expression and encapsidation of the viral genome. *J Virol* 2013;**87**(13):7688–99.
81. Morris SJ, Scott GE, Leppard KN. Adenovirus late-phase infection is controlled by a novel L4 promoter. *J Virol* 2010;**84**(14):7096–104.
82. Wright J, Leppard KN. The human adenovirus 5 L4 promoter is activated by cellular stress response protein p53. *J Virol* 2013;**87**(21):11617–25.

83. Wu K, Guimet D, Hearing P. The adenovirus L4-33K protein regulates both late gene expression patterns and viral DNA packaging. *J Virol* 2013;**87**:6739–47.
84. Ostapchuk P, Hearing P. Adenovirus IVa2 protein binds ATP. *J Virol* 2008;**82**(20):10290–4.
85. Ostapchuk P, Hearing P. Control of adenovirus packaging. *J Cell Biochem* 2005;**96**:25–35.
86. Ostapchuk P, Jihong Y, Auffarth E, Hearing P. Functional interaction of the adenovirus IVa2 protein with adenovirus type 5 packaging sequences. *J Virol* 2005;**79**(5):2831–8.
87. Ostapchuk P, Almond M, Hearing P. Characterization of empty adenovirus particles assembled in the absence of a functional adenovirus IVa2 protein. *J Virol* 2011;**85**(11):5524–31.
88. Zhang W, Low JA, Christensen JB, Imperiale MJ. Role for the adenovirus IVa2 protein in packaging of viral DNA. *J Virol* 2001;**75**(21):10446–54.
89. Ewing SG, Byrd SA, Christensen JB, Tyler RE, Imperiale M. Ternary complex formation on the adenovirus packaging sequence by the IVa2 and L4 22-kilodalton proteins. *J Virol* 2007;**81**(22):12450–7.
90. Yang T-C, Karl MN. Cooperative heteroassembly of the adenoviral L4-22K and IVa2 proteins onto the viral packaging sequence DNA. *Biochem* 2012;**51**:1357–68.
91. Christensen JB, Ewing SG, Imperiale MJ. Identification and characterization of a DNA binding domain on the adenovirus IVa2 protein. *Virology* 2012;**433**(1):124–30.
92. Gustin KE, Imperiale M. Encapsidation of viral DNA requires the adenovirus L1 52/5-kilodalton protien. *J Virol* 1998;**72**:7860–70.
93. Perez-Romero P, Gustin KE, Imperiale MJ. Dependence of the encapsidation function of the adenovirus L1 52/55-kilodalton protein on its ability to bind the packaging sequence. *J Virol* 2006;**80**(4):1965–71.
94. Ma H-C, Hearing P. Adenovirus structural protein IIIa is involved in the serotype specificity of viral DNA packaging. *J Virol* 2011;**85**(15):7849–55.
95. Erturk E, Ostapchuk P, Wells SI, Yang J, Gregg K, Nepveu A, et al. Binding of CCAAT displacement protein CDP to adenovirus packaging sequences. *J Virol* 2003;**77**(11):6255–64.
96. Fessler SP, Young CSH. The role of the L4-33K gene in adenovirus infection. *Virology* 1999;**263**:507–16.
97. Finnen RL, Biddle JF, Flint SJ. Truncation of the human adenovirus type 5 L4-33K protein: evidence for an essential role of the carboxy-terminus in the viral infectious cycle. *Virology* 2001;**289**:388–99.
98. Kulshreshtha V, Babiuk LA, Tikoo SK. Role of bovine adenovirus-3 33K protein in viral replication. *Virology* 2004;**323**:59–69.

# Adenoviral Vector Construction I: Mammalian Systems

**4**

*Shyambabu Chaurasiya, Mary M. Hitt*
Department of Oncology, Faculty of Medicine and Dentistry, University of Alberta, Edmonton, AB, Canada

## 1. Introduction

Rowe and colleagues first discovered adenovirus (Ad) in 1953 while trying to culture human adenoid tissue in the laboratory.[1] Following the discovery of human Ad, nonhuman Ads have been isolated from a number of species including dog, mouse, chimpanzee, and pigs as well as other mammalian and avian species.[2,3] After their discovery, Ads were extensively studied as a model system to understand basic eukaryotic cellular processes such as DNA replication, transcription, RNA splicing, and translation.[4] The study of Ad led Sharp and colleagues to discover the existence of introns and the process of mRNA splicing.[5] During late 1960s it was found that adenoviruses can recombine during growth in culture. This finding ultimately set the stage for the use of Ad as a vector for gene delivery to cells both in vitro and in vivo.[6–8]

Ads have many features that make them a suitable vector for gene therapy including: (1) the viral genome is relatively easy to manipulate by recombinant DNA technology; (2) scaling up and purification of the recombinant virus for use in the clinic are relatively easy; (3) the virus infects both quiescent and dividing cells with high efficiency; (4) recombinant viruses are fairly stable as the viral genome does not undergo rearrangement at a high rate; (5) in permissive cells the virus replicates to high levels producing up to 10,000 plaque-forming units (pfu) per infected cell; and (6) high levels of transgene expression are achieved. Moreover, the viral genome is maintained as an episome in the infected cell and rarely integrates into the cellular genome. This increases the safety of adenoviral vectors as the risk of insertional mutagenesis is quite low. However, because of the episomal nature of the vector genome, transgene expression is transient in dividing cells (reviewed in Sadeghi and Hitt[9]). These features have made Ad a vector of choice for gene therapy, which is evident from the fact that adenoviral vectors have been used in almost a quarter of all the gene therapy clinical trials performed to date.[10]

### 1.1 Adenovirus Biology

More than 100 serotypes of Ad are known, 51 among which are isolated from humans. Based on sequence homology and their ability to agglutinate red blood cells, the 51 serotypes of human Ads have been classified into six groups: A to F.[3] The serotypes most widely studied and most commonly used as vectors for gene therapy are Ad2

**Adenoviral Vectors for Gene Therapy.**

and Ad5, both of which belong to group C.[2,11] This chapter will focus mainly on the biology of these two serotypes of Ad. The adenovirion is a nonenveloped icosahedral particle about 70–90 nm in size containing a linear double-stranded DNA genome of approximately 36 kilobase pairs (kbp). The facets of the icosahedral capsid of the virion are composed mainly of trimers of hexon protein, and some other minor proteins. The vertices of the capsid are composed of penton bases anchoring the fiber proteins that are responsible for the primary attachment of the virion to the cell surface.

The first event in virus infection is the binding of fiber protein to the coxsackievirus adenovirus receptor (CAR) on the cell surface. This is followed by a secondary interaction between virion penton and $\alpha_v\beta_3$ and $\alpha_v\beta_5$ integrins, leading to internalization of the virion by clathrin-dependent endocytosis.[12,13] The levels of primary (CAR) and secondary (integrins) receptors present on the cell surface determine the efficiency with which the cell will be infected with adenovirus.[14] After internalization, the acidic environment of the endosome leads to escape of the virion to the cytoplasm. Here the virion is trafficked by dynein along microtubules toward the nucleus.[15] During translocation toward the nucleus, the virion undergoes sequential disassembly and the viral genome is ultimately imported to the nucleus through the nuclear pore complex. Viral DNA replication begins 6–8 h postinfection and it takes 24–36 h for the virus to complete its life cycle.[16]

The viral genome is flanked by inverted terminal repeats (ITRs) of 90–140 bp that are required in *cis* for the replication of the viral genome.[17,18] The ITRs are covalently bound by terminal protein.[4] In addition to the ITRs, the packaging signal ($\psi$) is also required in *cis* for proper folding and packaging of the viral genome into the capsid.[19] The viral genome is divided into noncontiguous, overlapping early and late transcription regions: E1A, E1B, E2, E3, and E4 are early genes whereas L1 to L5 are late genes.[2,20] The products of early genes as well as the replication of viral DNA are prerequisites for the expression of late genes.[21]

E1A, the first transcription unit to be expressed, produces two major proteins following differential mRNA processing. These proteins are required for the transcriptional activation of other early genes (E1B, E2, E3, and E4) and also to induce an S-phase-like state in the infected cells.[22] The E1A proteins bind to retinoblastoma protein (pRb), allowing the release of E2F and ultimately forcing the infected cells to enter into S-phase.[23] Because of the crucial role of E1A in viral replication, E1A is often deleted in order to make the virus replication deficient. The two major products of the E1B transcription unit are involved in blocking host mRNA transport, promoting viral mRNA transport, and blocking E1A-induced apoptosis to prevent premature death of the infected cells.[24,25] The E1B product (E1B-55 kDa) directly binds to the p53 protein to block E1A-induced apoptosis. E1A and E1B are considered oncogenes as they have the ability, when used in combination, to transform human and rodent cells in vitro.[2,26,27]

The two transcription units in the E2 region encode proteins required for the replication of viral DNA.[28] E2a encodes the 72-kDa DNA-binding protein whereas E2b encodes the viral DNA polymerase and terminal protein precursor (pTP). The E3 region encodes at least seven proteins, most of which are involved in subversion of the host immune system to allow a more robust infection. For example, E3-gp19K blocks the presentation of viral antigens by major histocompatibility complex (MHC) class I, thus preventing lysis of the infected cells by cytotoxic T lymphocytes.[28] The E3 region is nonessential for virus replication in vitro. At least six proteins are encoded by the E4 region. The products of the E4 region have diverse functions including facilitation of viral DNA replication, enhancement of late gene expression, and downregulation of host protein synthesis.[28] This region can also play a role in promoting the transforming ability of E1A.[29]

All the late region genes (L1–L5) are expressed from a common promoter called major late promoter. The primary major late transcript undergoes alternative splicing to produce individual transcripts. The products of late genes are mainly structural in function.[28]

### 1.2 Adenovirus Vectors

Different regions of the viral genome can be replaced with transgene(s) to generate mammalian gene transfer vectors. As described above, E1A-encoded proteins are crucial for the expression of both early and late viral genes and hence for replication of the virus. Deletion of the E1A region not only makes the virus replication deficient but also increases the cloning capacity of the vector. The packageable viral genome is limited in length to 105% of the wild-type genome size; thus one can insert only up to 1.8 kb in the vector without deletion of any viral sequences.[30] However, deletion of the E1 region allows insertion of transgenes up to 5.1 kb in size. Because E3-encoded proteins are nonessential for virus replication in vitro, the E3 region is often removed from Ad vectors. Deletion of E3 together with E1 can further increase the cloning capacity, accommodating insertion of foreign genes up to 8.2 kb in size.[31] Ad vectors deleted in E1, both with and without E3 deletion, are referred to as first-generation vectors.[32] First-generation vectors are the most commonly used Ad vectors for the purpose of gene therapy. In this chapter we will focus on the construction of first-generation Ad vectors.

## 2. Cell Lines for Propagating Adenovirus Vectors

Human Ads can undergo productive replication only in cells of primate, pig, and cotton rat origin.[33–35] Adenovirus infection of nonpermissive cells (e.g., cells from mouse, hamster, or rat other than cotton rat) results in abortive replication or occasional transformation of the cells due to rare integration of viral E1 sequences into the cellular genome.[36] Propagation of human Ad vectors is generally carried out in human cells that complement the E1 deletion in the vector. The first E1-complementing human cell line was developed by Graham and colleagues in their studies on E1-induced transformation.[37] In their landmark study, they used their novel technique of calcium phosphate coprecipitation to introduce sheared DNA from Ad5 into human embryonic kidney (HEK) cells.[38] The HEK cells transformed with sheared DNA from Ad5 were called HEK-293 cells. This cell line has been widely distributed since its isolation before 1980. According to the ISI Web of Science, the original paper describing the isolation of this cell line[37] has been cited nearly 3500 times. HEK-293 cells contain the "left end" of the Ad5 viral genome (1–4344 bp), including early region E1, integrated into chromosome 19.[39] These cells have been extensively used for the construction and propagation of E1-deleted nonreplicating Ad vectors. Additionally, the HEK-293 cell line has been widely used for diverse transfection-related studies because of the high efficiency of transfection and high level of transgene expression. The high expression levels are thought to result from promiscuous activation of the transfected promoter by E1A; and blockage of apoptosis, induced to varying degrees by different transfection procedures, by E1B.[40] Although HEK-293 cells were long considered to be kidney epithelial cells, evidence suggests that they may have been derived from a neural cell in the complex embryonic kidney cell culture.[41]

Several investigators have attempted to stably express E1 proteins in established human cancer cell lines such as A549, for the purpose of propagating E1-deleted Ad vectors. However, limited success has been achieved with this strategy, partly because growth of established cells is not dependent on E1 expression and also because it is difficult to isolate E1-expressing cells due to E1A-mediated toxicity. Although some encouraging data have been published in generating such cells, the use of these cells for construction and/or propagation of Ad vectors has been very limited.[42,43] An advantage of HEK-293 cells is that growth of the cells is dependent on the expression of E1 and hence constant levels of E1 expression are maintained over time.

One difficulty of propagating Ad vectors is the potential for Ad sequences carried by the propagating cell line to recombine with residual E1 (or immediately downstream) sequences in the vector, regenerating a wild-type E1 region. This type of homologous recombination would give rise to replication-competent adenovirus (RCA)-contaminated vector stocks, which would be especially problematic if the wild-type virus had a growth advantage over the recombinant vector. The PER.C6 cell line was specifically established to avoid RCA contamination during the propagation of Ad vectors to produce clinical grade stocks.[42] PER.C6 cells were derived from human embryonic retina cells by transforming with a minimal E1 region of Ad5. These are discussed in more detail in a later chapter of this book. Likewise, a system based on E1-transformed amniocyte-derived primary cells has been developed for rescue and propagation of Ad vectors with no overlapping E1 sequences.[44] In this chapter, we will focus on the construction and propagation of Ad vectors in HEK-293 cells.

## 2.1 Propagation of Adenovirus Vectors Encoding Toxic Transgenes for Cancer Gene Therapy

Cancer is a disease caused by the accumulation of many genetic mutations that allow the cells to undergo uncontrolled division. Unlike gene augmentation therapy where the goal is to restore a defective gene, the goal of many cancer gene therapies is to kill the cancer cells. One approach to killing cancer cells through gene therapy is by delivering proapoptotic or toxic genes to cancer cells. High levels of transgene expression are usually desirable in the target cells; however, construction and propagation of Ad vectors encoding such genes are challenging as the transgene expression can induce toxicity in the packaging cells, reducing vector yield.[45] In some cases the toxicity in packaging cells is so severe that the cells die after transfection with the vector DNA resulting in total failure to obtain the viral vector. In other cases the toxicity places a strong selective pressure on the resulting viral vector to reduce or completely eliminate transgene activity. This selective pressure may give rise to revertants or to mutations within the transgene expression cassette leading to reduction or complete ablation of transgene expression. The replicative advantage of these revertant/mutant viruses over the desired vector would reduce the feasibility of large-scale vector production.[45]

Different approaches have been proposed to address this challenge. The most common approaches involve differential regulation of transgene expression at the transcriptional level in the packaging and target cells (Figure 1). The use of tissue/tumor-specific promoters, such as the human telomerase (hTERT)[46] promoter, prostate specific antigen (PSA)[47] promoter, and carcinoembryonic antigen (CEA)[48] promoter to control the transgene, is a common strategy for achieving high levels of transgene expression in target cells with minimal expression in the packaging cells (Figure 1(A)). In our lab we have previously shown that the upstream sequence of the mammoglobin gene, a gene that is expressed at high levels in breast cancer and at very low levels in nonmammary cells, could be used to target transgene expression to breast cancer cells.[49] The use of tissue/tumor-specific promoters not only makes the construction and propagation of the vectors easier but also increases the cancer specificity and hence the overall safety of the vector. Moreover, in vivo in immune-competent animals, nonselective viral promoters such as the cytomegalovirus (CMV) immediate early promoter and the SV40 promoter are prone to silencing by TNF-$\alpha$ and interferon-$\gamma$; hence transgene expression is not long-lasting.[50–52] Studies comparing the duration of transgene expression driven by viral (CMV or SV40) promoters to that driven by cellular promoters have found that transgene expression lasts longer when driven by cellular promoters not only in the case of first generation Ad vectors but also in the case of helper-dependent Ad vectors, which are devoid of all viral coding sequences.[53] One drawback of this type of targeting is that tissue/tumor-specific eukaryotic promoters are usually inferior to viral promoters in terms of expression intensity.[54] Incorporation of additional elements could increase the expression intensity of the tissue/tumor-specific promoters. For instance, in our laboratory we have shown that addition of two enhancer elements upstream of the minimal mammoglobin promoter greatly increases the expression intensity of the promoter without compromising the tissue specificity.[55] Several other strategies have been used to improve the expression intensity of the tissue/tumor-specific promoter, the discussion of which is beyond the scope of this chapter. Readers are encouraged to see an excellent review on this topic by Papadakis et al.[53]

An alternative strategy to silence toxic transgenes during vector propagation is to insert a DNA sequence containing a strong transcription-terminating sequence between the promoter and the transgene (Figure 1(B)). The inserted sequence is flanked by *loxP* recognition sequences of the site-specific Cre recombinase. The presence of the inserted sequence should completely block transgene expression in the packaging cells. For therapeutic use, coinfection of target cells with another Ad vector encoding Cre causes excision of the *loxP*-flanked sequence, inducing expression of the transgene.[56] However, the requirement of an additional vector expressing the recombinase makes this system less suitable, especially for in vivo studies.

A third strategy for the construction and propagation of vectors encoding toxic genes is through the use of exogenously regulated expression systems, which have the distinct advantage of allowing pharmacological control of transgene expression

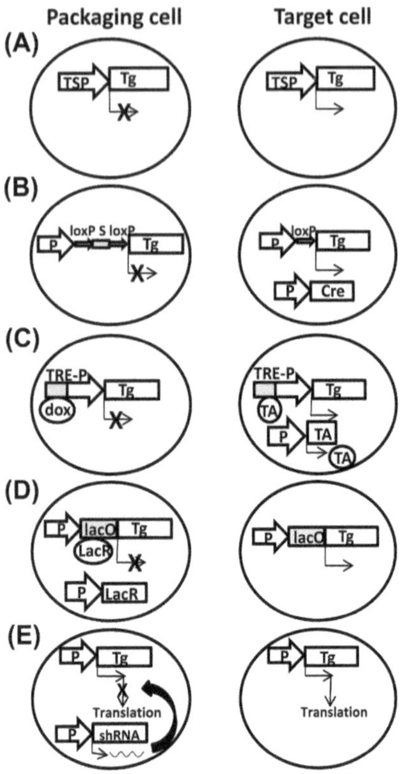


**Figure 1** Strategies for rescue of Ad vectors encoding toxic transgene(s). (A) Control of transgene expression by cell-specific regulation. Tissue/tumor-specific promoters can be used to achieve low levels of transgene expression in packaging cells and high levels in the target cell type. (B) Control of transgene expression by Cre–*loxP*-mediated regulation. An exogenous sequence, containing strong transcription termination sequence(s) and flanked by *loxP* sites in direct orientation, is inserted in between the promoter and the transgene open reading frame to inhibit transcription of the transgene in packaging cells. Coinfection of target cells with this vector and another Ad vector encoding Cre recombinase causes the excision of the exogenous sequence, allowing expression of the transgene. (C) Control of transgene expression using the Tet-OFF system to silence the transgene in packaging cells. In this system, the promoter is fused to a tetracycline response element (TRE) and a transactivator is encoded either by the same vector or by a different vector. Tetracycline (or an analog such as doxycycline) prevents the transactivator from binding to the TRE, and as a result, the transgene remains silent. Transgene expression is activated in target cells in the absence of doxycycline. (D) Control of transgene expression using *lac* repressor regulation. The *lac* operator sequence, to which the *lac* repressor protein binds, is inserted in close proximity to the promoter driving transgene expression. Packaging cells are modified to express the *lac* repressor protein which suppresses transgene expression. Absence of the *lac* repressor protein in target cells allows expression of the transgene (see Section 2.1). (E) Control of transgene expression by RNA silencing. Packaging cells are modified to express an shRNA that targets the transgene transcript, preventing its expression. Absence of the shRNA in the target cells allows high-level expression of the transgene. (P, promoter; TSP, tissue/tumor-specific promoter; Tg, transgene; S, exogenous stuffer sequence with termination sequences; dox, doxycycline; TRE-P, tetracycline response element fused to promoter; lacO, *lac* operator sequence; LacR, lac repressor protein; TA, transactivator.)

both in vitro and in vivo[53] (Figure 1(C)). The tetracycline (tet) on/off system, based on the highly sensitive prokaryotic tetracycline resistance operon, is probably the most commonly used regulatable expression system. In this system, the transgene is placed under the control of a tet-response element (tet operator), and a transactivator (TA) is encoded either by the same vector or by a different vector. The TA is a fusion of a tet repressor with the activation domain of a transcription factor, such as VP16 from herpes virus. This system requires constant administration of tetracycline analogs such as doxycycline to prevent TA binding to the response element, hence the switch is "off" (tet-OFF).[53] Based on this system, Gu et al. constructed an adenoviral vector that carries the apoptotic *Bax* gene transcriptionally controlled by the tet-OFF transactivator protein, which is encoded by the same vector under the control of the hTERT promoter.[57] Expression of *Bax* can be inhibited by the addition of doxycycline, which acts by inhibiting the transactivator protein. Although propagation requires continuous administration of doxycycline to silence the transgene, no drug is needed to induce *Bax* expression in clinical applications, which is an advantage. An alternative approach is to use the (tet-ON) system,[53] in which a mutant "reverse" tet repressor binds to the response element and activates transgene transcription only in the presence of doxycycline (or other tetracycline analogs).[58] Sipo et al. used the tet-ON system to construct and propagate an Ad vector encoding the apoptotic gene FasL in which the FasL gene was driven by the tet operator fused to the CMV promoter and the reverse TA was encoded by a different Ad vector.[59]

Another commonly used regulatable expression system makes use of the prokaryotic *lac* operon repressor protein that binds to the *lac* operator sequence and suppresses gene expression (reviewed in Rubinchik et al.[45]) (Figure 1(D)). In this system, operator binding sites are placed in close proximity to the promoter driving the transgene. Binding of the *lac* repressor to the operator sequences prevents binding of RNA polymerase II to the promoter and hence represses transcription of the transgene. Packaging cell lines can be engineered to stably express the repressor protein, which would ensure that transgene expression is suppressed during virus production. However, absence of the repressor protein in target cells allows high levels of transgene expression. Zhao et al. used this system to obtain high titers of an Ad vector encoding the cytolytic HIV-1 *env* protein.[60] Matthews et al. used a related system to construct and propagate an adenoviral vector that encodes the rabies virus glycoprotein following unsuccessful attempts using standard HEK-293 cells to rescue the virus.[61]

Posttranscriptional gene silencing using HEK-293 cells stably transfected with shRNA against the transgene has also been used to grow adenoviral vectors to high titer (Figure 1(E)). Wang et al. used this strategy to produce a vector encoding hIcon, an antiangiogenic protein. Interestingly, although hIcon is not directly toxic to cells, the authors suggest that transgene silencing reduced nutrient consumption during vector propagation, thus allowing higher virus yields.[62] Alternatively, packaging cells stably expressing antiapoptotic genes have been shown to increase yields of vectors encoding apoptotic genes. Bruden et al. found that stable expression of the antiapoptotic gene *CrmA*, encoding a poxvirus serpin, in E1-complementing packaging cells such as HEK-293 or AE25, dramatically increased the yield of adenoviral vectors encoding apoptotic genes such as Fas ligand, Fas-associated protein with death domain, caspase-8, or Fas/APO1.[63]

# 3. Construction of First-Generation Adenoviral Vectors

## 3.1 Early Methods

In 1973, Graham and colleagues showed that purified DNA from Ad5 and also from simian virus 40, when coprecipitated with calcium phosphate, can be taken up by human cells, resulting in the production of infectious virus particles.[38] This observation that purified viral DNA could be infectious laid the foundation for the studies manipulating the Ad genome for the construction of recombinant Ad vectors. Early methods of modified Ad construction mainly used two approaches: (1) in vitro ligation of viral DNA following cleavage with restriction enzymes[64,65] and (2) homologous recombination between viral DNAs in cotransfected cells.[66] For the first of these approaches, Stow devised a technique that employed in vitro ligation between purified virion DNA and plasmid DNA containing the left end of the Ad genome.[67] The E1 shuttle plasmid and purified viral DNA (from the Ad5 mutant dl309 that has a unique *Xba*I site in the E1 region[68]) were both digested with *Xba*I and then ligated together in vitro. The ligation product was then used to transfect HEK-293 cells, which resulted in the production of recombinant Ad virions.[67] This study elegantly showed that infectious virus could be reconstructed using a cloned subgenomic Ad sequence to shuttle precise E1 modifications into recombinant virus. However, due to the location of the *Xba*I site, most of the E1 region is retained in the recombinant, and few other unique restrictions sites are available in the Ad genome, so this strategy is not ideal for construction of gene therapy vectors.

At about the same time, Kapoor and Chinnadurai developed a system to rescue mutations into the Ad E1 region by in vivo homologous recombination between the "left" end Ad sequences cloned into a plasmid and purified Ad virion DNA.[69] This "left end" shuttle plasmid could be easily manipulated in vitro to incorporate the desired mutations in E1. The overlapping sequence in the shuttle plasmid and the cotransfected viral DNA allowed homologous recombination to take place in HEK-293 cells, resulting in the generation of recombinant virions with alterations in E1. This system obviates the need for unique restriction enzyme sites since it does not involve ligation of two DNA molecules. However, the viral DNA must be cleaved in the left end before cotransfection in order to reduce contamination with nonrecombinant parental virus. The most commonly used sites for cleaving the viral DNA are the unique *Xba*I in Ad5 dl309 mutant and the unique *Cla*I site in the wild-type Ad5. Despite cleaving the viral DNA with these enzymes, contamination with the parental virus remains an issue. Both the *Xba*I and the *Cla*I sites are located at the very left end of the genome; hence there is a very small size difference between undigested viral DNA and viral DNA digested with *Xba*I or *Cla*I. Because of the small size difference it is difficult, using agarose gel electrophoresis, to confirm whether the digestion is complete. The undigested parental DNA generates virus more efficiently than the DNAs, which require recombination. In addition, the small fragment produced by *Xba*I or *Cla*I digestion can potentially be carried over during the transfection and the fragments may religate in the cell to generate wild-type or parental virus. This could potentially be another source of nonrecombinant virus contamination.[70,71]

Later, Mizuguchi and Kay proposed an alternate strategy to replace E1 sequences in the Ad genome with transgene expression cassettes.[72] Plasmids were constructed containing the entire viral genome with and without the E3 region and with three unique restriction enzyme sites (I-*Ceu*I, *Swa*I, and PI-*Sce*I) in place of El. Transgene expression cassettes, flanked by an I-*Ceu*I site at one end and a PI-*Sce*I site at the other, were ligated to the genomic plasmid following digestion with these two restriction enzymes. Ligation in the presence of *Swa*I reduces the recovery of nonrecombinant parental plasmids. The modified genomic plasmid is then linearized to release viral sequences, and used to transfect HEK-293 cells to produce recombinant virus. This strategy addresses both the problems associated with lack of unique restriction enzyme sites in the viral genome and the problem associated with high levels of wild-type or parental virus contamination.[72] However, construction of each vector involves manipulation, cloning, and scale-up of separate plasmids >30 kb in size, which can be difficult in some cases.

These methods for the construction of recombinant Ads rely on the use of the viral genome either alone or in combination with a shuttle plasmid. However, the large viral DNA genome is not only time-consuming and laborious to isolate and purify, but also difficult to manipulate genetically. Some of these approaches can also lead to high levels of contamination with the nonrecombinant parental virus. It is not uncommon for the parental virus to outgrow the recombinant virus, making the rescue of the recombinant vectors further problematic. Given the potential of Ad vectors to be used in gene therapy and other purposes, an efficient method for the construction of Ad vectors with minimal wild-type or parental contamination was needed.

## 3.2 The Two-Plasmid Rescue System

McGrory et al. (1984) developed a two-plasmid rescue system to overcome the limitations faced by the earlier approaches of Ad vector construction.[73] The two-plasmid rescue system is based on the ability of two plasmids to undergo recombination in mammalian cells. The two-plasmid rescue system has gone through many modifications, including the switch from homologous recombination strategy to a site-specific recombination strategy, to make the method more efficient and to reduce wild-type or parental virus contamination. In addition to the two-plasmid rescue system, other methods have been developed for efficient construction of Ad vectors using bacterial systems or in vitro ligation, which are discussed in other chapters in this book. The remainder of this chapter will focus on the two-plasmid rescue system and modifications in this method for high-efficiency Ad vector construction.

### 3.2.1 Development of the Two-Plasmid Rescue System

A study by Berkner and Sharp in 1983 demonstrated that recombinant Ads could be produced in cotransfected HEK-293 cells by homologous recombination between cloned fragments of viral DNA.[74] Rescue of infectious Ads was dependent on cleavage of at least one of the plasmids at the junction of the ITR and the plasmid DNA, releasing the ITR. Unlike earlier methods that used purified virion DNA for the construction of Ad recombinants, this method used only noninfectious plasmids, thus avoiding the need to isolate virion DNA. Furthermore, since full-length viral DNA was not used, nonrecombinant parental virus could not be generated, which was a major problem with the earlier methods.

The linear genome of Ad is thought to replicate in a semiconservative manner, with replication starting at either end of the genome. However, Ruben et al. (1983) showed that up to 10% of Ad DNA molecules in an infected cell are joined head-to-tail due, at least in part, to the formation of covalently closed circles.[75] In purified virus stocks, viral DNA does not exist in a circular form, but circular viral DNA can be detected intracellularly before the onset of viral DNA replication.[75] This finding suggested that the full-length viral genome could be cloned and maintained as a bacterial plasmid. Subsequently, in 1984 Graham cloned the entire Ad5 genome as a plasmid (pFG140) with plasmid sequences containing the β-lactamase gene and a bacterial origin of replication inserted at the *Xba*I site at nt 1339.[76] This Ad genomic plasmid could be amplified in *Escherichia coli* in the presence of ampicillin. Moreover, this plasmid was shown to be nearly as efficient as purified virion DNA in generating infectious virus following transfection into HEK-293 cells.

Another important finding, made in 1987 by Ghosh-Choudhury et al., was the discovery that protein IX (pIX) is essential for the generation of infectious virus.[77] They constructed an Ad5 genomic plasmid that was similar to pFG140 except that it had a deletion of the gene encoding protein IX. Unlike pFG140, this plasmid was noninfectious. To confirm the essential nature of pIX, they cotransfected HEK-293 cells with the pIX-deleted genomic plasmid and a plasmid encoding the left end of the Ad genome, including the pIX gene. All viruses recovered from the cotransfection carried the pIX gene as a result of homologous recombination between the two plasmids. Later studies determined that pIX plays important roles in packaging of full-length viral genome and also in the stability of the viral icosahedrons.[77,78]

These findings laid the foundation for the development of the first two-plasmid rescue system by McGrory et al. for the construction of Ad vectors in which a transgene expression cassette replaces the E1 region.[73] In this study, the authors inserted a sequence into pFG140 to increase its size to 40 kb, which is beyond the packaging capacity of Ad. The resulting plasmid, pJM17, was noninfectious, but could serve as a template for replication in HEK-293 cells. For the second component of the system, they constructed a shuttle plasmid containing the left end of Ad5 with foreign DNA of up to 5.4 kb in place of E1. Infectious recombinant Ad vectors bearing the foreign DNA sequence were generated following cotransfection of HEK-293 cells with pJM17 and the shuttle plasmid. Since both plasmids are noninfectious, in principle, only recombinant E1-substituted vectors should be generated, resulting from in vivo homologous recombination between the overlapping Ad sequences in the genomic and the shuttle plasmids. This system was highly successful as it was able to overcome some of the limitations associated with the earlier methods of Ad vector construction, such as ease of transgene insertion and significant contamination with parental virus. However, the pJM17 genomic plasmid was able to generate a low level of infectious virus in HEK-293 cells even in the absence of a cotransfected shuttle plasmid. Infectivity of pJM17 was discovered to be due to the spontaneous deletion of sequences from the plasmid backbone resulting in reduction in size of the genomic plasmid to within the packaging constraints of Ad.[73] This posed the risk of parental virus contamination in recombinant virus preparations; hence modification of this method was needed.

In 1994, Bett et al. modified the two-plasmid rescue method by constructing an improved genomic Ad plasmid, pBHG10.[79] Like pJM17, pBHG10 contains essentially the entire Ad5 genome with some important modifications. First, it has a deletion of 3180 bp in the E1 region, removing E1A and the packaging signal ($\psi$) required for packaging the viral genome into the capsid. Removal of the $\psi$ sequence renders the plasmid noninfectious. The second modification was the deletion of ~2.7 kb from the nonessential E3 region and addition of a *Pac*I restriction enzyme site in its place. Shuttle plasmids bearing the left end of viral DNA, including the ITR and packaging signal but with a deletion in E1 from 339 to 3533 bp, were also constructed. A linker containing multiple cloning sites (MCPs) was introduced in the shuttle plasmids in place of E1 to allow easy insertion of a transgene. This modified system introduced two improvements in the two-plasmid system developed by McGrory et al.[73] First, the combined E1 and E3 deletions in this system increased the cloning capacity of the resulting recombinant vectors to allow insertion of up to ~8 kb of foreign DNA. Second, one can insert foreign DNA into either the E1 or the E3 region using this system. Insertion of a transgene into the E3 region is facilitated by the unique *Pac*I site in the large pBHG10 plasmid. Insertion can be expedited by using the kanamycin-resistant pABS.4 plasmid (Microbix Biosystem Inc., Mississauga, ON, Canada). This plasmid contains a *Swa*I-flanked kanamycin resistance gene within a MCS flanked by *Pac*I sites. For E3 insertions, the transgene is first cloned into the MCS of pABS.4. The resulting plasmid is then digested with *Pac*I and the fragment bearing the transgene and the kanamycin resistance gene is then inserted into the *Pac*I site in pBHG10. The resulting large plasmid is then used to transform *E. coli* and positive clones bearing the E3 insertion are selected based on their resistance to both ampicillin and kanamycin. Finally, prior to cotransfection for vector construction, the kanamycin resistance gene is removed from the genomic plasmid by digestion with *Swa*I. The increased cloning capacity and versatility of the method as well as the absence of parental virus contamination made this version of the two-plasmid rescue system very popular for the construction of nonreplicating Ad vectors. Like other Ad vector rescue systems developed by that time, the efficiency of vector rescue was fairly low, typically requiring cotransfection of 12 to 30 60-mm dishes to ensure rescue of around 10 independent isolates of the recombinant vector.

### 3.2.2 *Fine-Tuning of the Two-Plasmid Rescue System*

A possible explanation for the low efficiency of vector rescue by the two-plasmid system developed by McGrory et al. is that homologous recombination frequencies are simply not high enough. The observation that the infectious plasmid pFG140 has a plaque-forming efficiency ~100-fold higher than that of a typical cotransfection for vector rescue supports this hypothesis. Ng et al.[80] proposed that recombination mediated by the site-specific Cre recombination system would be more efficient than homologous recombination for the rescue of recombinant Ad vectors. Therefore, they inserted *loxP* sites into pBHG10 upstream of the pIX gene and into the shuttle plasmid after the transgene expression cassette. Cotransfection of Cre-expressing HEK-293 cells with these two *loxP*-containing plasmids allowed vector rescue with an efficiency ~30-fold higher than that mediated by homologous recombination. The efficiency of virus rescue was increased even further by replacement of the single ITR with two ITRs fused head-to-head (referred to here as an ITR junction) in the shuttle plasmid. Several reasons were proposed to explain the observed enhancement in virus rescue by replacing a single ITR with an ITR junction in the shuttle plasmid. First, in contrast to plasmids with a single ITR, plasmids containing an ITR junction should increase in copy number following cotransfection of this shuttle plasmid and the Ad genomic plasmid. The ITR junction can serve as an origin of viral DNA replication,[76] with the Ad genomic plasmid providing all the *trans*-acting viral factors essential for viral DNA replication. Recognition of ITR junction-containing plasmids as templates for the viral replication machinery results in production and amplification of linear shuttle plasmid DNA flanked by the ITRs. Since both the genomic plasmid and the shuttle plasmid contain ITR junctions, both should increase in copy number in cotransfected HEK293 cells. This increase in the pool of substrates for recombination should enhance the rescue of recombinant virus. Second, the replicating linear shuttle DNA might serve as a better substrate for recombination with linear Ad genomic DNA than the nonreplicating circular shuttle DNA. Third, recombination between a linear shuttle DNA and the Ad genomic DNA should generate a packageable, infectious genome in a single step. In contrast, generation of a packageable genome from a circular shuttle plasmid and a linear genomic DNA is likely a two-step process. In the first step the circular plasmid integrates into the linear genomic DNA at the *loxP* site. However, this is nonpackageable because the packaging signal (from the shuttle plasmid) in the recombinant DNA is far from the terminus of the molecule. Also the size of this DNA molecule exceeds the virion packaging limit. Generation of infectious DNA would require additional step(s) to eliminate extraneous sequences from the recombinant molecule. Together, the replacement of homologous recombination with Cre-mediated recombination and the replacement of a single ITR with an ITR junction in the shuttle plasmid increased the efficiency of virus rescue by ~100-fold compared to the efficiency of the earlier two-plasmid systems.[73,79]

This two-plasmid system developed by Ng et al.[80] improved the efficiency of virus rescue but virus could be rescued only in HEK-293, or other E1-complementing cells, that also express Cre. Therefore, to improve the utility of the system, a Cre expression cassette was inserted into the genomic plasmid within the plasmid backbone so that the Cre cassette is not incorporated in the final recombinant vector. This insertion eliminates the requirement for Cre-expressing cells for virus rescue. The efficiency of virus rescue obtained using this Cre-expressing genomic plasmid in parental HEK-293 cells was found to be comparable to that obtained using the previous generation of Ad genomic plasmid in Cre-expressing HEK-293 cells.[81]

The Cre/*loxP*-mediated recombination system for the generation of the Ad vectors (Figure 2) elegantly addressed most of the limitations of previously developed rescue systems. Although suitable for rescue of most Ad vectors, this system precludes the use of *loxP* sites anywhere else in the genome, for example, for the purpose of regulating transgene expression[82] or inhibiting vector packaging.[83] To circumvent this problem, Ng et al. designed an alternate site-directed recombination system based on the yeast flippase (FLP) recombinase.[84] The yeast FLP-recombinase expression cassette replaced

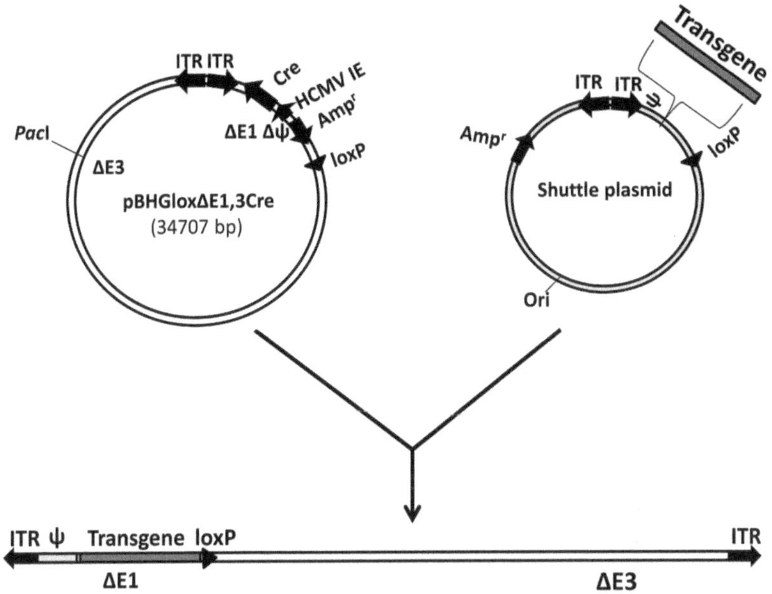


**Figure 2** Two-plasmid rescue system for the construction of Ad vectors. The genomic plasmid used in this system contains most of the Ad5 genome, flanked at the "left" end by a *loxP* site and at the "right" end by two ITR sequences fused head-to-head. The plasmid backbone of the genomic plasmid contains a Cre recombinase expression cassette. The shuttle plasmid used in this system contains a transgene expression cassette flanked at the "left" end by two fused ITR sequences and a packaging signal, and at the "right" end by a *loxP* site. Cotransfection of HEK-293 cells with the two plasmids generates recombinant vectors following *loxP*-specific recombination mediated by the Cre recombinase. (ITR, inverted terminal repeat; HCMV-IE, human cytomegalovirus immediate early promoter; Amp$^r$, gene conferring ampicillin resistance; *Pac*I, restriction site for insertions replacing the E3 region; ψ, packaging signal.)

the Cre expression cassette in the genomic plasmid, and *frt* sites replaced the *loxP* sites in the shuttle and genomic plasmids. No significant difference was observed in the efficiency of virus rescue between the Cre-mediated and the FLP-mediated recombination systems.

## 4. Steps Involved in Adenovirus Vector Construction

Here we will describe the steps involved in Ad vector construction using the Cre/*loxP*-based two-plasmid rescue system developed by Ng et al.[81] Typically, a foreign expression cassette is inserted in the shuttle plasmid for rescue in the E1 region of the vector. For E3 insertions, the expression cassette is cloned directly, or via the pABS.4 transfer plasmid, into the larger (genomic) plasmid. Foreign expression cassette(s) can be inserted into the E1 and/or E3 regions in either parallel or antiparallel orientation relative to the E1 or E3 transcription units. Generally, a higher level of gene expression

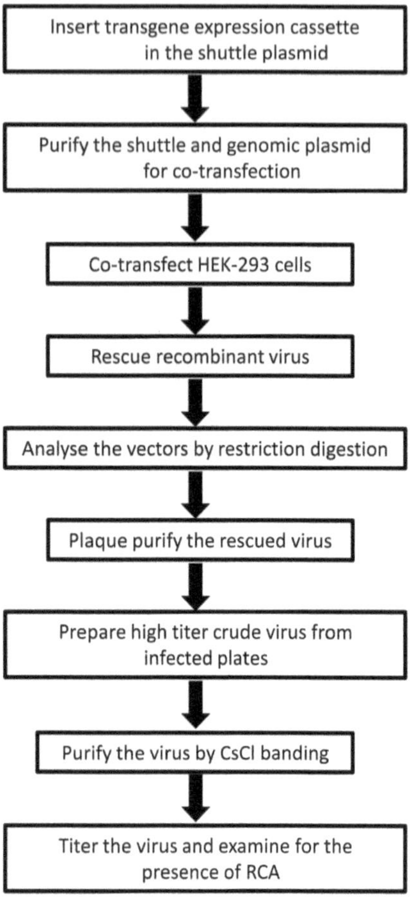


**Figure 3** Flowchart for steps involved in the construction, propagation, purification, and characterization of first-generation Ad vectors.

is achieved when the transgenes are in parallel orientation to the viral transcription units they replace; however, the overall expression levels also depend on the type of promoter and sequence of the insert itself.[31] The following sections describe methods to rescue, purify, and titer the recombinant Ad vectors (Figure 3).

## 4.1 Preparation of Adenovirus Genomic and Shuttle Plasmid DNA for Cotransfection

Based on the size of transgene and desired application of the vector to be generated, the genomic plasmid can be selected from a variety of plasmids available from Microbix Biosystem Inc. (Mississauga, ON, Canada). The genomic plasmids pBHGloxΔE1Cre and pBHGfrtΔE1FLP (formerly designated as pBHGloxE3Cre and pBHGfrtE3FLP) retain an intact E3 region, and thus have a reduced cloning capacity compared to the E3-deleted plasmids. Foreign expression cassettes of up to 5 kb can be rescued into vectors using these plasmids owing to the size constraints of Ad. Genomic plasmids (pBHGloxΔE1,3Cre and pBHGfrtΔE1,3FLP) have deletions of 2653 bp in the E3 region, allowing the rescue of up to 8 kb foreign sequence in the vector. Although these plasmids offer the highest cloning capacity, the vectors generated from them have slightly reduced growth (around twofold) compared to vectors that retain more of the E3 region.[31] The unique *Pac*I site, as in the original pBHG10-based system, can be used to insert a transgene cassette in the E3-deleted region of pBHGloxΔE1,3Cre or pBHGfrtΔE1,3FLP if the transgene is desired in the E3 region of the recombinant vector. The desired genomic plasmid is amplified in *E. coli* and purified for cotransfection. High-speed plasmid purification kits (e.g., from Qiagen) may be adequate for small quantities of plasmid DNA; however, for large quantities of genomic plasmid DNA, CsCl gradient purification procedures may be preferable.

Shuttle plasmids for the construction of Ad vectors based on Cre- or FLP-mediated recombination are available from Microbix Biosystem Inc. The E1 insertion plasmids pDC311 and pDC312 allow rescue of transgene cassettes into the vector via Cre-mediated recombination; pDC511 and pDC512 allow vector rescue via FLP-mediated recombination. The transgene cassette, including promoter, transgene, and polyadenylation (poly(A)) signal, is inserted into the pUC-based shuttle plasmid at the MCS. To further simplify cloning, shuttle plasmids are available that contain the immediate early promoter of murine cytomegalovirus (MCMV) and the poly(A) sequence from SV40. The MCMV promoter drives high-level expression in most cell types of both human and murine origin, in contrast to the HCMV promoter that is less active in murine tissues.[85] pDC315 and pDC316 are based on the Cre recombination system, and pDC515 and pDC516 are based on the FLP recombination system. The choice of shuttle plasmid depends on which MCS orientation is most convenient for transgene insertion, which site-specific recombination system is desired for vector rescue, and whether high levels of transgene expression are required for the intended application. Prior to use in cotransfections, the shuttle plasmid containing the transgene cassette should be amplified in *E. coli* and purified using standard plasmid purification kits (e.g., from Qiagen).

## 4.2 Cotransfection of HEK-293 Cells with Genomic and Shuttle Plasmid

Cotransfection of HEK-293 cells with the two plasmids gives rise to recombinant Ad that are observed as plaques on the cell monolayer. Individual plaques are then isolated for expansion of the vector; therefore it is desirable to have plaques that are well separated from each other. The number of plaques formed depends on many factors, including the transfection efficiency, state of cells, quality of plasmid DNA, and the amount of plasmid DNA used. In the Cre- or FLP-mediated recombination system, an average of ~40 plaques are formed per 60-mm dish of HEK-293 cells transfected with 2 μg of shuttle plasmid and 2 μg of genomic plasmid. The infectious Ad genomic plasmid pFG140 can be used as a control for transfection efficiency and plaque formation. Under optimal conditions, pFG140 should yield up to ~100 plaques per 0.5 μg DNA. It is a good practice to transfect with a range of plasmid concentrations to obtain a high number of plaques without risk of cross-contamination between the plaques.

To prepare for cotransfection, low passage (<p40) HEK-293 cells are grown in 60-mm dishes using complete MEMF11 supplemented with 10% FBS. Cultures at ~70–80% confluency are best for cotransfection. In general, a nearly confluent 150-mm dish of HEK-293 cells can be split into eight 60-mm dishes, which would be ready for transfection the next day. For the construction of one recombinant vector, generally 16 cultures in 60-mm dishes are prepared which are enough to perform the cotransfections in quadruplicate using three different concentrations of experimental plasmids and one concentration of pFG140. The next day, 1 h prior to cotransfection, the medium in each of the 60-mm dishes is replaced with freshly prepared medium. To prepare the DNA, 0.08 mg salmon sperm DNA, used as carrier DNA, is added to 8 ml of HEPES-buffered saline and then vortexed for 1 min to shear the DNA. The sheared salmon sperm is divided among four polystyrene tubes labeled A, B, C, and D. Each of these tubes is sufficient for cotransfection of four dishes of HEK-293 cells. To tubes A, B, and C, add 2, 8, and 20 μg, respectively, of both shuttle and genomic plasmids. To the control tube D add 2 μg of the infectious plasmid pFG140. The tubes are mixed well, and then 0.1 ml 2.5 M $CaCl_2$ is added to each tube dropwise with gentle mixing. Finally, 0.5 ml of the resulting suspension from each tube is added dropwise to the medium in separate 60-mm dishes (four dishes per tube). One day later, the medium is replaced with an agarose overlay (0.5% in MEMF11). Plaques (round turbid areas in the transparent monolayer) are generally visible within a week post cotransfection, and can be isolated between day 10 and day 14. Well-isolated plaques are collected by repeated stabs through the agarose at the site of the plaque using a sterile cotton-plugged Pasteur pipette or a 1 ml pipet tip. The agarose pieces from each plaque isolate are transferred to a vial containing 0.5 ml phosphate buffered saline (PBS) supplemented with $Mg^{2+}$ and $Ca^{2+}$ (PBS++) and 10% glycerol and then stored at –80 °C. Isolation of 5 to 20 plaques for further analysis is recommended.

## 4.3 Analysis of the Rescued Recombinant Adenovirus Vectors

The vector isolates obtained by picking plaques are amplified in HEK-293 cells to verify recombinants by analysis of viral DNA and to generate seed stocks for subsequent vector production. Briefly, HEK-293 cells are grown to near confluence in 60-mm dishes. (Note: It takes a longer time for complete cytopathic effect (CPE) to appear if the cells are too confluent or older.) The virus plaque suspensions are freeze–thawed three times, and then half of each suspension is used to infect separate 60-mm dishes of cells. The reader is discouraged from amplifying pFG140-based virus at the same time as nonreplicating vectors due to the risk of cross-contamination and potential growth advantage of pFG140-based virus. Cultures are incubated until most of the cells are rounded up and detached from the dish (complete CPE). Semiadherent cells are collected by gentle pipetting and combined with nonadherent cells. (Note: If infections are harvested too soon, it will be difficult to observe vector DNA bands above the background of cellular DNA when the DNA is analyzed by gel electrophoresis.) Approximately 3.5 ml of cell suspension from each dish is transferred to a vial containing 0.5 ml sterile glycerol and stored at −80 °C for use as a vector seed stock. The remaining 1.5 ml of cell suspension is centrifuged briefly in a microfuge tube to pellet the cells. The supernatant is aspirated, leaving behind 0.1 ml in each tube to aid in resuspending the pellet. A solution of Pronase–sodium dodecyl sulfate (SDS) is added to the infected cells, and viral and cellular proteins are degraded by overnight digestion. Viral DNA is purified from the lysate by ethanol precipitation and resuspended in 50 μl buffer.

For the analysis of viral DNA, 5 μl of the DNA from the infected cells is digested with *Hind*III and then the resulting fragments are separated by agarose gel electrophoresis and stained with ethidium bromide. Viral DNA bands should be clearly visible under UV light, above a background smear of cellular DNA. *Hind*III digested wild-type Ad DNA can be run alongside that of the recombinant vector for the purpose of comparison. It should be noted that *Hind*III digestion of human DNA produces a band at 1.8 kb. To further verify the candidate recombinant Ad vector, the extracted DNA can be digested with other restriction enzymes and analyzed by agarose gel electrophoresis. When using the AdMax kit (Microbix Biosystem Inc.), virtually all the plaques obtained should be correct. However, it is good laboratory practice to carry out at least one round of plaque purification to ensure that all the recombinants in a high-titer vector stock have the same genome, having descended from a single infectious virus particle.

## 4.4 Plaque Purification of Recombinant Adenovirus Vector

Plaque assays are commonly used both to purify and to determine the titer of adenovirus vectors. In both cases, confluent HEK-293 cells in 60-mm dishes are infected with virus stock serially diluted in PBS++, at a range of $10^{-2}$ to $10^{-6}$ for partially purified virus or $10^{-5}$ to $10^{-10}$ for highly purified and concentrated virus. An agarose overlay is applied to the cell monolayer after infecting the monolayer with the virus. The agarose overlay immobilizes viruses and prevents cross-contamination among plaques. Plaques should be visible by day 4 postinfection. At day 10 postinfection, well-isolated plaques are collected and correct recombinants verified by DNA analysis as described in Section 4.3.

## 4.5 Preparation of High-Titer Virus Stock (Crude Lysate)

High-titer Ad virus stocks can be prepared by concentrating infected HEK-293 cells as the virus is not released from the cells until very late in infection when the cell lyses. For preparing high-titer stocks, cells can be infected either in monolayer (HEK-293) or suspension culture (HEK-293N3S). Suspension culture is more amenable for large-scale vector production, due to the ease of infected cell collection. However, complete CPE, used frequently to determine the appropriate time to harvest the infected cells, is easier to visualize in monolayer cultures. In this section, we will describe the protocol for achieving high-titer virus stock from both monolayer and suspension cultures.

### 4.5.1 Preparation of High-Titer Crude Virus Stocks from Monolayer Culture

It is desirable to infect the cells at a multiplicity of infection (MOI) of 1–10 pfu/cell. However, as a close approximation, we generally dilute an infected cell lysate (such as that generated in Section 4.3) 1:8 in PBS++ and use 1 ml to infect each of eight 150-mm dishes of near confluent HEK-293 cells. Cultures should be examined every day for the appearance of CPE. When most of the cells are rounded up (but not all detached), scrape the adherent cells and combine with the cells in suspension. Pellet the infected cells, and combine pellets from all eight plates, resuspending in 8 ml of PBS++ with 10% glycerol. Aliquot and store at −80 °C. Perform three freeze–thaw cycles to release the virus prior to use for infection or further amplification. To purify virus by CsCl banding (see Section 4.6 below), this crude lysate should be used for one further round of amplification in forty 150-mm dishes, and the final infected cell pellet resuspended in 15 ml 0.1 M Tris, pH 8, and stored at −80 °C.

### *4.5.2 Preparation of High-Titer Crude Virus Stocks from Suspension Culture*

Suspension cultures of HEK-293N3S,[86] a derivative of HEK-293 cells, can be used for large-scale preparation of Ad vectors. HEK-293N3S cells are most conveniently maintained as a semiadherent monolayer culture until expansion is desired. Three 150-mm dishes of nearly confluent HEK-293N3S cells are sufficient to establish a 500-ml suspension culture in Joklik's modified MEM supplemented with 10% horse serum. This suspension culture can be expanded by diluting 1:2 or 1:3 when the cell density reaches $\sim 5 \times 10^5$ cells/ml. A 4 l culture is generally sufficient to prepare enough vector for CsCl gradient purification. To infect HEK-293N3S cells, the culture is centrifuged and resuspended in 0.1 vol of fresh medium, and then inoculated with virus (MOI of 1–20 pfu/cell). After gentle stirring for 1 h, the culture is brought to its original volume with fresh medium and incubation continued. Unlike infections in monolayer cultures, CPE in infected suspension cultures cannot be simply visualized under a microscope. In order to determine the optimal time for harvesting the infected cells, a small sample is taken from the suspension culture daily and examined for the presence of inclusion bodies by orcein staining. Late in infection, inclusion bodies appear as densely stained nuclear structures that result from the accumulation of a large amount of viral products. Uninfected cells should be used as a negative control for staining. When the inclusion bodies are visible in 80–90% cells (usually at day 3, depending on the MOI used), cells are harvested by centrifugation and resuspended in 20 ml PBS++ supplemented with 10% glycerol and stored at −80 °C. For purification by CsCl banding, the final infected cell pellet should be resuspended in 15 ml 0.1 M Tris, pH 8, and stored at −80 °C.

## *4.6 Purification of High-Titer Adenovirus Vector by CsCl Banding*

CsCl gradient purification is commonly used to purify and to concentrate adenovirus. Although crude virus stocks can be used for some in vitro experiments, the virus must be purified for other experiments, particularly in vivo work. The CsCl banding described here can be used for the purification of crude lysate from a 4 l suspension culture (HEK-293N3S) or thirty to forty 150-mm dishes of monolayer cultures (HEK-293), which have similar virus yields. Infection and collection of crude lysates from suspension and monolayer cultures are described in Section 4.5. The infected cell lysate is subjected to three freeze–thaw cycles, and then sodium deoxycholate is added to a final concentration of 3.75%. After 30 min at room temperature, the solution should be highly viscous, and all virus particles should be released from the cells. The lysate is digested with DNase I to reduce viscosity, and then clarified by centrifugation. The supernatant is carefully layered over a three-stage CsCl step gradient prepared by layering CsCl solutions at densities of 1.25d and 1.35d (each at about half the volume of the lysate) over a 1.5d cushion of CsCl solution. Spin at 20,000*g* for 1 h. Collect the virus band at the interface between 1.35d and 1.25d, pool all tubes from the same virus preparation, and recentrifuge overnight. The virus, visible as a turbid band, should be collected in the smallest possible volume and then dialyzed against 10 mM Tris, pH 8, or desalted by column chromatography (e.g., PD-10) and glycerol added to a final concentration of 10%. Depending on the application, other storage buffers may also be appropriate.[87] Store the purified virus in small aliquots at −80 °C.

## 4.7 Characterization of Adenovirus Vectors

After preparation of the viral vector, the DNA structure should be confirmed, the titer of virus particles and infectious units should be determined, transgene expression should be ascertained, and the stock must be tested for the presence of RCA.

The identity of the recombinant vector can be verified by restriction enzyme analysis as described in Section 4.3, using 0.025 ml purified virus as starting material instead of infected cells. The vector preparation can be titrated using the classical plaque assay or using commercially available kits. For titration, plaque assays (described in Section 4.4) should be carried out with a broad range of virus dilutions ($10^{-4}$ to $10^{-10}$). Plaques are usually counted 10 days postinfection. Alternatively, virus titers can be determined using the Adeno-X™ rapid titer kit (Clontech, Cat. No. 632250), which detects the viral hexon protein within infected cells. This assay has the advantage of being significantly faster (~48 h) than the plaque assay (~10 days). Determination of viral particle concentration and the test for RCA are described below.

### 4.7.1 Determination of Particles to Plaque-Forming Units Ratio

In addition to determining the concentration of infectious vector (pfu/ml), it is necessary to determine the concentration of virus particles, including noninfectious particles, especially if the vector is to be used in humans. In fact, the FDA recommends that patient doses be calculated on the basis of virus particles rather than the infectious particles.[88] This recommendation is based on two important facts. First, the determination of virus particle is based on physical measurement, and hence is more precise than the determination of infectious particles. Second, a primary toxicity of Ad vectors is from the innate immune response directed against the viral coat, which is dependent on particle number and largely independent of transgene expression. There are many methods to determine the concentration of virus particles such as anion exchange high-performance liquid chromatography (HPLC), measurement of virion DNA using a DNA-binding dye (e.g., PicoGreen), reverse-phase HPLC analysis of viral protein components, and spectrophotometric analysis after solubilizing the vector.[88] To determine the concentration of virus particles by spectrophotometric analysis, purified virus is diluted in Tris–EDTA buffer supplemented with 0.1% SDS and heated to 56 °C for 10 min, and the $OD_{260}$ is determined using a UV spectrophotometer. Based on the extinction coefficient of wild-type Ad as determined by Maizel et al.,[89] the concentration of viral particles is calculated as follows:

$$\text{Particles/ml} = (OD_{260})\ (\text{dilution factor})\ (1.1 \times 10^{12}).$$

The particle:pfu ratio is between 20:1 and 80:1 for most Ad vector preparations.

### 4.7.2 Replication-Competent Adenovirus Assay

HEK-293 cells,[39] for many years the only cell line that would support growth of E1-deleted Ad vectors, are transformed with the left end of Ad5 (viral nucleotide sequence 1–4344) that includes the E1 region.[39] During the propagation of Ad vectors in HEK-293 cells, viral sequences in the HEK-293 cells may recombine with viral sequences in the vector, producing E1-positive RCA. Although the frequency with which the recombination occurs is not known, RCA is likely to replicate faster than many E1-deleted vectors in HEK-293 cells. Therefore, prolonged propagation of the vector may increase the proportion of RCA in the vector preparation, and should be avoided. To minimize RCA contamination, one should scale up vector production from a single plaque to large-scale culture in as few steps as possible. RCA contamination is considered a safety issue especially if the vector will be used clinically. The FDA recommends that there be no more than one RCA in $10^9$ infectious adenovirus virions in a clinical stock.[88] Several different approaches have been developed for the detection of RCA, including Southern blot hybridization, quantitative polymerase chain reaction, and biological assay.[88]

The biological assay used for RCA detection in our lab is based on the induction of CPE in the non-E1-complementing A549 cell line following infection with the test vector. Infection with Ad, even in the absence of RCA, will frequently result in death of the initially infected A549 monolayer due to toxicity of viral proteins in the inoculum. Therefore, the RCA test is carried out in two stages. In the first stage, 150-mm dishes of A549 cells are infected with $10^6$, $10^7$, or $10^8$ pfu vector (one dish per virus amount). Greater amounts of vector can be tested if detection of RCA at higher sensitivity is required. One week after infection, or sooner if most of the culture shows CPE, the infected cultures (monolayer plus medium) are harvested. The harvested cultures are taken through three freeze–thaw cycles, and then 1 ml of each lysate is used to infect a fresh dish of A549 cells. This second round of infection is observed for 3 weeks, replacing medium every 5 days. Any signs of CPE in this second stage would indicate the presence of RCA. If no CPE is observed then the original inoculum of vector used to generate the lysate would have been free from RCA. If CPE is observed in any plate, then viral DNA can be recovered from that plate and analyzed by agarose gel electrophoresis after digestion with appropriate restriction enzymes for further confirmation (as described in Section 4.3).

## 5. High-Efficiency Construction of Adenovirus Vectors for Generating Adenovirus-Based cDNA Expression Libraries

The human genome project revealed the presence of an estimated 25,000 genes and a much larger number of proteins encoded by these genes.[90] The functions of the majority of these gene products are unknown. It is important to identify the activities of the genes, especially those whose products play a role in human disease, as these genes could be potential targets for therapy. However, identification of the full spectrum of a gene's function is very difficult partly due to complex interactions of the gene product with other proteins and factors that vary depending on the specific cell type or developmental stage under study. Thus, there is a need for a highly efficient mammalian expression system that would facilitate cloning and direct determination of gene function on a genomic scale in cell-based assays.

The ability of Ad to efficiently transduce a wide variety of cell types including primary cells makes it an ideal vector for cDNA delivery in functional assays. Other desirable features for an expression library vector are that the method of construction should yield only recombinant vectors (with minimal wild-type or parental vector contamination) and the method should produce a large number of clones.[91] Although several methods of Ad vector construction, including the two-plasmid rescue systems discussed in earlier sections, generate recombinant vectors that are free from contamination with parental virus, the efficiency of vector rescue with these methods is not sufficient for generating a cDNA expression library of high complexity. Fewer than 100 plaques per microgram of vector DNA are obtained at best, using the conventional methods of Ad vector construction.[91]

Mammalian cells are speculated to contain ~$10^5$ mRNA species; thus, at least $10^6$ independent clones must be produced for adequate representation of all transcripts in the cDNA library. Construction of such a complex population of recombinant Ad vectors by conventional methods would not be feasible. The low efficiency of transgene rescue following transfection with a plasmid-derived vector genome is thought to be partly due to low infectivity of cloned viral DNA. It has been found that plasmid-derived viral DNA is ~1000-fold less efficient than virion DNA in producing infectious virus. This huge difference is due to the absence of terminal protein (TP) in the cloned viral genome. In the virion, TP is bound covalently to both ends of the linear Ad genome and plays an important role in enhancing infectivity and template efficiency for viral DNA replication.[92,93] Miyake et al. developed an efficient method for construction of Ad vectors that employs viral DNA termini complexed to TP as a substrate in the generation of recombinants.[94] In this study, they inserted the transgene at a unique site in the full-length viral genome carried in a cosmid backbone. The cosmid vector was used to cotransfect HEK-293 cells together with TP-bound virion DNA cleaved at several sites with a restriction enzyme to reduce recovery of nonrecombinant virions. The use of DNA–TP complexes greatly increased the efficiency of vector generation: several hundred plaques were formed per microgram of viral DNA–TP complex. However, only a fraction of the resulting clones were the desired recombinants, suggesting that fragmentation of the Ad-TP donor genome is not sufficient to prevent its contamination of the recombinant pool.

In order to more easily identify and recover recombinant clones, reporter genes can be either incorporated in the parental viral genome, such that it is lost following transgene rescue and can be used for negative selection, or incorporated in the shuttle plasmid, such that it is rescued along with the transgene and can be used for positive selection. Schaack et al. employed the *E. coli LacZ* gene for positive selection of Ad clones expressing their gene of interest.[70] The use of reporter genes for the selection of positive clones is useful but such screening would be time- and reagent-consuming if making a library of several thousand clones. This selection system is further hampered by the fact that often recombinant vectors have a growth disadvantage relative to the parental virus, which makes the isolation of recombinant clones from a library difficult unless the clones are positively selected for growth.

Elahi et al. (2002) developed a positive selection system that should be compatible with generation of a large number of recombinant Ad clones.[91] In this study they made use of the essential late Ad protease (PS) gene. Ad deleted of the PS gene can undergo only one round of DNA replication in HEK-293 cells. For this strategy, the authors made a shuttle plasmid bearing the left end of Ad and containing, in place of E1, a bicistronic expression cassette incorporating the PS gene and the transgene. HEK-293 cells were transfected with this shuttle plasmid following infection with a PS-deleted full-length Ad. Because the PS-deleted viral genome cannot go beyond one round of replication, only recombinant vectors that have acquired the PS gene can result in productive infection. With this interesting and scalable system, virtually all of the recovered viruses are recombinant vectors with a diversity predicted to be as high as one million clones.

Hatanaka et al. (2003) developed a Cre–*loxP*-based recombination system for the generation of an Ad cDNA expression library.[95] Unlike all the strategies discussed above, the recombination event in this method takes place in vitro. First, a cDNA library was constructed in an Ad shuttle plasmid background, with a *loxP* site just downstream of the cDNA cassette. A pool of linearized shuttle plasmids was added in vitro to a complex of TP and viral DNA deleted of the left end and the mixture was treated with Cre recombinase. The resulting recombinant DNA was then used to transfect HEK-293 cells to obtain a library of infectious vectors. Using this system the authors were able to isolate cDNA for CD2 (present at a frequency of less than 1 in 3000 T cell transcripts) from human T cells.

All the methods discussed above provided evidence that construction of Ad cDNA expression libraries is feasible. However, one common drawback in all the approaches for creating Ad-based cDNA expression libraries is that they are technically demanding and time-consuming. In 2006, Hillgenberg et al. modified the previously developed Cre–*loxP*-based Ad construction methods to generate ~$10^6$ independent clones of recombinant Ad[96] in a short time. In this system the shuttle plasmid carries the viral 5′ ITR, complete viral packaging signal, the cDNA expression cassette, and a single *loxP* site. A mixture of shuttle plasmids is used to transfect HEK-293 cells expressing Cre recombinase that have been infected with donor Ad attenuated by partial deletion of the packaging signal, which is flanked by *loxP* sites. Site-specific recombination causes excision of the packaging signal from the donor virus rendering it completely nonpackageable. A second recombination between the *loxP* site in shuttle plasmid and the *loxP* site in the donor virus rescues the transgene and packaging signal, resulting in an infectious recombinant vector. Individual clones are then identified and purified by plaque assay. The residual donor viruses are counterselected during the amplification of recombinant vectors because of their impaired growth. This rapid, efficient, and elegant construction system should prove very useful for the production of cDNA expression libraries of sufficient complexity for identification of gene function in cell-based assays.

## 6. Conclusion

As described in this chapter, the two-plasmid rescue system using mammalian cells, particularly HEK-293, is one of the earliest and most commonly used methods for the construction of Ad vectors. The two-plasmid rescue system depends on recombination between a shuttle and a genomic plasmid. Replacement of homologous recombination with site-specific recombination as a means to rescue the transgene into the vector has greatly increased the efficiency of recombinant virus production. With advancements made in the construction process, as discussed in the last section of this chapter, it is now feasible to construct Ad-based human cDNA libraries in a short time.

## Acknowledgments

We wish to thank P. Ng and F.L. Graham for their contributions. This work was supported by grants from the Canadian Institutes of Health Research (CIHR) and Canadian Breast Cancer Foundation, Prairies/NWT Chapter.

## References

1. Rowe WP, Huebner RJ, Gilmore LK, Parrott RH, Ward TG. Isolation of a cytopathogenic agent from human adenoids undergoing spontaneous degeneration in tissue culture. *Proc Soc Exp Biol Med* 1953;**84**:570–3.
2. Shenk T. Adenoviridae: the viruses and their replication. In: Fields BN, Knipe DM, Howley PM, editors. *Virology*. 3rd ed. Philadelphia: Lippincott-Raven Publishers; 1996. p. 2111–48.
3. Wadell G, Hammarskjold ML, Winberg G, Varsanyi TM, Sundell G. Genetic variability of adenoviruses. *Ann NY Acad Sci* 1980;**354**:16–42.
4. McConnell MJ, Imperiale MJ. Biology of adenovirus and its use as a vector for gene therapy. *Hum Gene Ther* 2004;**15**:1022–33.
5. Berget SM, Moore C, Sharp PA. Spliced segments at the 5' terminus of adenovirus 2 late mRNA. *Proc Natl Acad Sci USA* 1977;**74**:3171–5.
6. Lewis Jr AM, Baum SG, Prigge KO, Rowe WP. Occurrence of adenovirus-SV40 hybrids among monkey kidney cell adapted strains of adenovirus. *Proc Soc Exp Biol Med* 1966;**122**:214–8.

7. Pierce WE, Rosenbaum MJ, Edwards EA, Peckinpaugh RO, Jackson GG. Live and inactivated adenovirus vaccines for the prevention of acute respiratory illness in naval recruits. *Am J Epidemiol* 1968;**87**:237–46.
8. Lewis Jr AM, Rowe WP. Isolation of two plaque variants from the adenovirus type 2-simian virus 40 hybrid population which differ in their efficiency in yielding simian virus 40. *J Virol* 1970;**5**:413–20.
9. Sadeghi H, Hitt MM. Transcriptionally targeted adenovirus vectors. *Curr Gene Ther* 2005;**5**:411–27.
10. *J Genet Med* 2014. Accessed at:www.wiley.co.uk/genmed/clinical.
11. Douglas JT. Adenoviral vectors for gene therapy. *Mol Biotechnol* 2007;**36**:71–80.
12. Defer C, Belin MT, Caillet-Boudin ML, Boulanger P. Human adenovirus-host cell interactions: comparative study with members of subgroups B and C. *J Virol* 1990;**64**:3661–73.
13. Wang K, Guan T, Cheresh DA, Nemerow GR. Regulation of adenovirus membrane penetration by the cytoplasmic tail of integrin beta5. *J Virol* 2000;**74**:2731–9.
14. Ambriovic-Ristov A, Gabrilovac J, Cimbora-Zovko T, Osmak M. Increased adenoviral transduction efficacy in human laryngeal carcinoma cells resistant to cisplatin is associated with increased expression of integrin alphavbeta3 and coxsackie adenovirus receptor. *Int J Cancer J Int Cancer* 2004;**110**:660–7.
15. Kelkar SA, Pfister KK, Crystal RG, Leopold PL. Cytoplasmic dynein mediates adenovirus binding to microtubules. *J Virol* 2004;**78**:10122–32.
16. Liu H, Naismith JH, Hay RT. Adenovirus DNA replication. *Curr Top Microbiol Immunol* 2003;**272**:131–64.
17. Wolfson J, Dressler D. Adenovirus-2 DNA contains an inverted terminal repetition. *Proc Natl Acad Sci USA* 1972;**69**:3054–7.
18. Garon CF, Berry KW, Rose JA. A unique form of terminal redundancy in adenovirus DNA molecules. *Proc Natl Acad Sci USA* 1972;**69**:2391–5.
19. Ostapchuk P, Hearing P. Minimal cis-acting elements required for adenovirus genome packaging. *J Virol* 2003;**77**:5127–35.
20. Horwitz MS. Adenoviridae: the viruses and their replication. In: Fields BN, Knipe DM, et al., editors. *Virology*. 2nd ed. New York: Raven Press; 1990. p. 1679–721.
21. Thomas GP, Mathews MB. DNA replication and the early to late transition in adenovirus infection. *Cell* 1980;**22**:523–33.
22. Moran E, Mathews MB. Multiple functional domains in the adenovirus E1A gene. *Cell* 1987;**48**:177–8.
23. Akusjarvi G, Pettersson Ulf, Roberts Richard J. Structure and function of the adenovirus-2 genome. *Adenovirus DNA Viral Genome Expr* 1986:53–96.
24. Pilder S, Moore M, Logan J, Shenk T. The adenovirus E1B-55K transforming polypeptide modulates transport or cytoplasmic stabilization of viral and host cell mRNAs. *Mol Cell Biol* 1986;**6**:470–6.
25. Moran E. Interaction of adenoviral proteins with pRB and p53. *FASEB J Off Publ Fed Am Soc Exp Biol* 1993;**7**:880–5.
26. Graham FL. Transformation by and oncogenicity of human adenoviruses. In: Ginsberg HS, editor. *The adenoviruses*. New York (NY): Plenum Press; 1984. p. 339–98.
27. Green M, Wold WS, Brackmann KH, Cartas MA. Identification of families of overlapping polypeptides coded by early "transforming" gene region 1 of human adenovirus type 2. *Virology* 1979;**97**:275–86.
28. Jared D, Evans PH. Adenovirus replication. In: Curiel DT, editor. *Adenoviral vectors for gene therapy*. (USA): Academic Press; 2002. p. 39–70.
29. Moore M, Horikoshi N, Shenk T. Oncogenic potential of the adenovirus E4orf6 protein. *Proc Natl Acad Sci USA* 1996;**93**:11295–301.

30. Bett AJ, Prevec L, Graham FL. Packaging capacity and stability of human adenovirus type 5 vectors. *J Virol* 1993;**67**:5911–21.
31. Ng P, Graham FL. Adenoviral vector construction I: mammalian systems. In: Curiel DT, Douglas JT, editors. *Adenoviral vectors for gene therapy*. 1st ed. San Diego (California): Academic Press; 2002. p. 71–104.
32. Danthinne X, Imperiale MJ. Production of first generation adenovirus vectors: a review. *Gene Ther* 2000;**7**:1707–14.
33. Lubeck MD, Davis AR, Chengalvala M, Natuk RJ, Morin JE, Molnar-Kimber K, et al. Immunogenicity and efficacy testing in chimpanzees of an oral hepatitis B vaccine based on live recombinant adenovirus. *Proc Natl Acad Sci USA* 1989;**86**:6763–7.
34. Torres JM, Alonso C, Ortega A, Mittal S, Graham F, Enjuanes L. Tropism of human adenovirus type 5-based vectors in swine and their ability to protect against transmissible gastroenteritis coronavirus. *J Virol* 1996;**70**:3770–80.
35. Pacini DL, Dubovi EJ, Clyde Jr WA. A new animal model for human respiratory tract disease due to adenovirus. *J Infect Dis* 1984;**150**:92–7.
36. Trentin JJ, Yabe Y, Taylor G. The quest for human cancer viruses. *Science* 1962;**137**:835–41.
37. Graham FL, Smiley J, Russell WC, Nairn R. Characteristics of a human cell line transformed by DNA from human adenovirus type 5. *J Gen Virol* 1977;**36**:59–74.
38. Graham FL, van der Eb AJ. A new technique for the assay of infectivity of human adenovirus 5 DNA. *Virology* 1973;**52**:456–67.
39. Louis N, Evelegh C, Graham FL. Cloning and sequencing of the cellular-viral junctions from the human adenovirus type 5 transformed 293 cell line. *Virology* 1997;**233**:423–9.
40. Berk AJ. Recent lessons in gene expression, cell cycle control, and cell biology from adenovirus. *Oncogene* 2005;**24**:7673–85.
41. Shaw G, Morse S, Ararat M, Graham FL. Preferential transformation of human neuronal cells by human adenoviruses and the origin of HEK 293 cells. *FASEB J Off Publ Fed Am Soc Exp Biol* 2002;**16**:869–71.
42. Fallaux FJ, Bout A, van der Velde I, van den Wollenberg DJ, Hehir KM, Keegan J, et al. New helper cells and matched early region 1-deleted adenovirus vectors prevent generation of replication-competent adenoviruses. *Hum Gene Ther* 1998;**9**:1909–17.
43. Imler JL, Chartier C, Dreyer D, Dieterle A, Sainte-Marie M, Faure T, et al. Novel complementation cell lines derived from human lung carcinoma A549 cells support the growth of E1-deleted adenovirus vectors. *Gene Ther* 1996;**3**:75–84.
44. Schiedner G, Hertel S, Kochanek S. Efficient transformation of primary human amniocytes by E1 functions of Ad5: generation of new cell lines for adenoviral vector production. *Hum Gene Ther* 2000;**11**:2105–16.
45. Rubinchik S, Norris JS, Dong JY. Construction, purification and characterization of adenovirus vectors expressing apoptosis-inducing transgenes. *Methods Enzym* 2002;**346**:529–47.
46. Shay JW, Bacchetti S. A survey of telomerase activity in human cancer. *Eur J Cancer* 1997;**33**:787–91.
47. Gotoh A, Ko SC, Shirakawa T, Cheon J, Kao C, Miyamoto T, et al. Development of prostate-specific antigen promoter-based gene therapy for androgen-independent human prostate cancer. *J Urol* 1998;**160**:220–9.
48. Schrewe H, Thompson J, Bona M, Hefta LJ, Maruya A, Hassauer M, et al. Cloning of the complete gene for carcinoembryonic antigen: analysis of its promoter indicates a region conveying cell type-specific expression. *Mol Cell Biol* 1990;**10**:2738–48.
49. Shi CX, Long MA, Liu L, Graham FL, Gauldie J, Hitt MM. The human SCGB2A2 (mammaglobin-1) promoter/enhancer in a helper-dependent adenovirus vector directs high levels of transgene expression in mammary carcinoma cells but not in normal nonmammary cells. *Mol Ther J Am Soc Gene Ther* 2004;**10**:758–67.

50. Acsadi G, O'Hagan D, Lochmüller H, Prescott S, Larochelle N, Nalbantoglu J, et al. Interferons impair early transgene expression by adenovirus-mediated gene transfer in muscle cells. *J Mol Med Berl* 1998;**76**:442–50.
51. Harms JS, Splitter GA. Interferon-gamma inhibits transgene expression driven by SV40 or CMV promoters but augments expression driven by the mammalian MHC I promoter. *Hum Gene Ther* 1995;**6**:1291–7.
52. Qin L, Ding Y, Pahud DR, Chang E, Imperiale MJ, Bromberg JS. Promoter attenuation in gene therapy: interferon-gamma and tumor necrosis factor-alpha inhibit transgene expression. *Hum Gene Ther* 1997;**8**:2019–29.
53. Papadakis ED, Nicklin SA, Baker AH, White SJ. Promoters and control elements: designing expression cassettes for gene therapy. *Curr Gene Ther* 2004;**4**:89–113.
54. Nettelbeck DM, Jerome V, Muller R. A strategy for enhancing the transcriptional activity of weak cell type-specific promoters. *Gene Ther* 1998;**5**:1656–64.
55. Shi CX, Graham FL, Hitt MM. A convenient plasmid system for construction of helper-dependent adenoviral vectors and its application for analysis of the breast-cancer-specific mammaglobin promoter. *J Gene Med* 2006;**8**:442–51.
56. Okuyama T, Fujino M, Li XK, Funeshima N, Kosuga M, Saito I, et al. Efficient Fas-ligand gene expression in rodent liver after intravenous injection of a recombinant adenovirus by the use of a Cre-mediated switching system. *Gene Ther* 1998;**5**:1047–53.
57. Gu J, Zhang L, Huang X, Lin T, Yin M, Xu K, et al. A novel single tetracycline-regulative adenoviral vector for tumor-specific Bax gene expression and cell killing in vitro and in vivo. *Oncogene* 2002;**21**:4757–64.
58. Rendahl KG, Quiroz D, Ladner M, Coyne M, Seltzer J, Manning WC, et al. Tightly regulated long-term erythropoietin expression in vivo using tet-inducible recombinant adeno-associated viral vectors. *Hum Gene Ther* 2002;**13**:335–42.
59. Sipo I, Hurtado Picó A, Wang X, Eberle J, Petersen I, Weger S, et al. An improved Tet-On regulatable FasL-adenovirus vector system for lung cancer therapy. *J Mol Med Berl* 2006;**84**:215–25.
60. Zhao C, Crews CJ, Derdeyn CA, Blackwell JL. Lac-regulated system for generating adenovirus 5 vaccine vectors expressing cytolytic human immunodeficiency virus 1 genes. *J Virol Methods* 2009;**160**:101–10.
61. Matthews DA, Cummings D, Evelegh C, Graham FL, Prevec L. Development and use of a 293 cell line expressing lac repressor for the rescue of recombinant adenoviruses expressing high levels of rabies virus glycoprotein. *J Gen Virol* 1999;**80**(Pt 2):345–53.
62. Wang L, Qi X, Shen R, Sun Y, Tuveson DA. An shRNA silencing a non-toxic transgene reduces nutrient consumption and increases production of adenoviral vectors in a novel packaging cell. *J Cell Physiol* 2009;**219**:365–71.
63. Bruder JT, Appiah A, Kirkman 3rd WM, Chen P, Tian J, Reddy D, et al. Improved production of adenovirus vectors expressing apoptotic transgenes. *Hum Gene Ther* 2000;**11**:139–49.
64. Carlock LR, Jones NC. Transformation-defective mutant of adenovirus type 5 containing a single altered E1a mRNA species. *J Virol* 1981;**40**:657–64.
65. Solnick D. An adenovirus mutant defective in splicing RNA from early region 1A. *Nature* 1981;**291**:508–10.
66. Chinnadurai G, Chinnadurai S, Brusca J. Physical mapping of a large-plaque mutation of adenovirus type 2. *J Virol* 1979;**32**:623–8.
67. Stow ND. Cloning of a DNA fragment from the left-hand terminus of the adenovirus type 2 genome and its use in site-directed mutagenesis. *J Virol* 1981;**37**:171–80.
68. Jones N, Shenk T. Isolation of adenovirus type 5 host range deletion mutants defective for transformation of rat embryo cells. *Cell* 1979;**17**:683–9.

69. Kapoor QS, Chinnadurai G. Method for introducing site-specific mutations into adenovirus 2 genome: construction of a small deletion mutant in VA-RNAI gene. *Proc Natl Acad Sci USA* 1981;**78**:2184–8.
70. Schaack J, Langer S, Guo X. Efficient selection of recombinant adenoviruses by vectors that express beta-galactosidase. *J Virol* 1995;**69**:3920–3.
71. Munz PL, Young CS. End-joining of DNA fragments in adenovirus transfection of human cells. *Virology* 1991;**183**:160–9.
72. Mizuguchi H, Kay MA. Efficient construction of a recombinant adenovirus vector by an improved in vitro ligation method. *Hum Gene Ther* 1998;**9**:2577–83.
73. McGrory WJ, Bautista DS, Graham FL. A simple technique for the rescue of early region I mutations into infectious human adenovirus type 5. *Virology* 1988;**163**:614–7.
74. Berkner KL, Sharp PA. Generation of adenovirus by transfection of plasmids. *Nucleic Acids Res* 1983;**11**:6003–20.
75. Ruben M, Bacchetti S, Graham F. Covalently closed circles of adenovirus 5 DNA. *Nature* 1983;**301**:172–4.
76. Graham FL. Covalently closed circles of human adenovirus DNA are infectious. *EMBO J* 1984;**3**:2917–22.
77. Ghosh-Choudhury G, Haj-Ahmad Y, Graham FL. Protein IX, a minor component of the human adenovirus capsid, is essential for the packaging of full length genomes. *EMBO J* 1987;**6**:1733–9.
78. Rosa-Calatrava M, Grave L, Puvion-Dutilleul F, Chatton B, Kedinger C. Functional analysis of adenovirus protein IX identifies domains involved in capsid stability, transcriptional activity, and nuclear reorganization. *J Virol* 2001;**75**:7131–41.
79. Bett AJ, Haddara W, Prevec L, Graham FL. An efficient and flexible system for construction of adenovirus vectors with insertions or deletions in early regions 1 and 3. *Proc Natl Acad Sci USA* 1994;**91**:8802–6.
80. Ng P, Parks RJ, Cummings DT, Evelegh CM, Sankar U, Graham FL. A high-efficiency Cre/loxP-based system for construction of adenoviral vectors. *Hum Gene Ther* 1999;**10**:2667–72.
81. Ng P, Parks RJ, Cummings DT, Evelegh CM, Graham FL. An enhanced system for construction of adenoviral vectors by the two-plasmid rescue method. *Hum Gene Ther* 2000;**11**:693–9.
82. Anton M, Graham FL. Site-specific recombination mediated by an adenovirus vector expressing the Cre recombinase protein: a molecular switch for control of gene expression. *J Virol* 1995;**69**:4600–6.
83. Parks RJ, Chen L, Anton M, Sankar U, Rudnicki MA, Graham FL. A helper-dependent adenovirus vector system: removal of helper virus by Cre-mediated excision of the viral packaging signal. *Proc Natl Acad Sci USA* 1996;**93**:13565–70.
84. Ng P, Cummings DT, Evelegh CM, Graham FL. Yeast recombinase FLP functions effectively in human cells for construction of adenovirus vectors. *BioTechniques* 2000;**29**:524–6. 528.
85. Addison CL, Hitt M, Kunsken D, Graham FL. Comparison of the human versus murine cytomegalovirus immediate early gene promoters for transgene expression by adenoviral vectors. *J Gen Virol* 1997;**78**(Pt 7):1653–61.
86. Graham FL. Growth of 293 cells in suspension culture. *J Gen Virol* 1987;**68**(Pt 3):937–40.
87. Croyle MA, Cheng X, Wilson JM. Development of formulations that enhance physical stability of viral vectors for gene therapy. *Gene Ther* 2001;**8**:1281–90.
88. Hutchins B, Sajjadi N, Seaver S, Shepherd A, Bauer SR, Simek S, et al. Working toward an adenoviral vector testing standard. *Mol Ther J Am Soc Gene Ther* 2000;**2**:532–4.
89. Maizel Jr JV, White DO, Scharff MD. The polypeptides of adenovirus. I. Evidence for multiple protein components in the virion and a comparison of types 2, 7A, and 12. *Virology* 1968;**36**:115–25.

90. National Human Genome Research Institute, NIH. 2012. Accessed at: http://www.genome.gov/12011238.
91. Elahi SM, Oualikene W, Naghdi L, O'Connor-McCourt M, Massie B. Adenovirus-based libraries: efficient generation of recombinant adenoviruses by positive selection with the adenovirus protease. *Gene Ther* 2002;**9**:1238–46.
92. Stillman BW. The replication of adenovirus DNA with purified proteins. *Cell* 1983;**35**:7–9.
93. Pronk R, van der Vliet PC. The adenovirus terminal protein influences binding of replication proteins and changes the origin structure. *Nucleic Acids Res* 1993;**21**:2293–300.
94. Miyake S, Makimura M, Kanegae Y, Harada S, Sato Y, Takamori K, et al. Efficient generation of recombinant adenoviruses using adenovirus DNA-terminal protein complex and a cosmid bearing the full-length virus genome. *Proc Natl Acad Sci USA* 1996;**93**:1320–4.
95. Hatanaka K, Ohnami S, Yoshida K, Miura Y, Aoyagi K, Sasaki H, et al. A simple and efficient method for constructing an adenoviral cDNA expression library. *Mol Ther J Am Soc Gene Ther* 2003;**8**:158–66.
96. Hillgenberg M, Hofmann C, Stadler H, Loser P. High-efficiency system for the construction of adenovirus vectors and its application to the generation of representative adenovirus-based cDNA expression libraries. *J Virol* 2006;**80**:5435–50.

# Adenoviral Vector Construction II: Bacterial Systems

**5**

*Masahisa Hemmi, Hiroyuki Mizuguchi*
Laboratory of Biochemistry and Molecular Biology, Graduate School of Pharmaceutical Sciences, Osaka University, Osaka, Japan

## 1. Introduction

Adenoviruses (Ad) are nonenveloped icosahedral particles with a diameter of 70–100 nm.[1] Each virion consists of a DNA core surrounded by a protein shell composed of 252 subunits called capsomeres (240 subunits, hexons, and 12 subunits, pentons). So far, more than 65 serotypes of human Ad have been identified and classified into seven distinct subgroups (A–F), many of which are associated with respiratory, gastrointestinal, or ocular disease.[1–3] Two of them, Ad types 2 and 5, which belong to subgroup C, have been the most extensively studied both genetically and biochemically, and the findings of these studies have contributed to our knowledge of viral and cellular gene expression and regulation, DNA replication, cell cycle control, and other biological processes. Because of the extensive knowledge of the genetic and biological characteristics of Ad types 2 and 5, these types are generally used to prepare recombinant Ad vectors.

Human Ad contains a linear, approximately 36 kb, double-stranded DNA genome that encodes over 70 gene products.[1] The viral genome contains five early transcription units (E1A, E1B, E2, E3, E4), two early delayed (intermediate) transcription units (protein IX (pIX) and IVa2), and five late units (L1–L5), which mostly encode structural proteins for the capsid and the internal core. Inverted terminal repeats (ITR) at the ends of the viral chromosome function as replication origins. The E1A gene is the first transcription unit to be activated shortly after infection, and is essential to the activation of other promoters and replication of the viral genome. In the first generation Ad vectors, the E1 (E1A and E1B) gene is replaced by the foreign gene and the virus is propagated in E1-transcomplementing cell lines, such as 293,[4] 911,[5] or PER.C6 cells.[6] The E3 region encodes products associated with host defense mechanisms, which are not required for viral replication in vitro, and thus the E3 region not only is often deleted to enlarge the packageable size limit for foreign genes but also is replaced with foreign genes. Since up to 3.2 and 3.1 kb of the E1 and E3 region, respectively, can be deleted,[7] and approximately 105% of the wild-type genome can be packaged into the virus without affecting the viral growth rate and titer,[8] the E1/E3-deleted Ad vectors allow the packaging of approximately 8.1–8.2 kb of foreign DNA.[7]

Recombinant Ad vectors have been extensively used to deliver foreign genes to a variety of cell types and tissues both in vitro and in vivo.[9–12] They can be easily grown to high titer and can efficiently transfer genes into both dividing and nondividing cells. The viral genome persists as an episome in the nucleus of the transduced cells. Since

**Adenoviral Vectors for Gene Therapy.**

these vectors do not replicate extrachromosomally and rarely integrate into the host chromosome, Ad vector-mediated gene expression is variable and dependent on factors such as cellular turnover and immune responses directed against transduced cells. Ad vectors have not only become promising vectors for gene therapy, but also important tools for gene transfer into mammalian cells. The construction of Ad vectors, however, is a time-consuming and labor-intensive procedure, and improvements have been made to several of the systems involved in order to facilitate the process. In this paper, we review the advances made so far in the methods for generating recombinant Ad vectors.

# 2. Construction of First Generation Adenovirus Vectors

## 2.1 Early Methods for Adenovirus Vector Construction

Initially the methods for constructing recombinant Ad vectors required direct manipulation of whole Ad DNA purified from virions. There were two principal methods. One of these methods, the in vivo homologous recombination of Ad DNAs,[13] included cotransfection of Ad DNA restriction fragments with overlapping homologous sequences that were cleaved from Ad types 2 and 5, resulting in recombination in 293 cells and the production of infectious recombinant DNAs. The other method was in vitro ligation of Ad DNAs cleaved by restriction enzymes.[14,15] However, both methods had the major limitation that it was very difficult to manipulate large Ad DNAs. In the early 1980s, this limitation was partially overcome by modifications of the left end of the Ad genome. In brief, Stow et al.[16] used in vitro ligation between a cloned subgenomic Ad fragment and the viral DNA, both of which had unique *XbaI* sites. The shuttle plasmid containing the modified left end of the viral DNA and the purified viral DNA were cleaved with *XbaI* and ligated in vitro. Then, by transfecting the ligation product into 293 cells, the recombinant Ad vectors bearing the modified E1A region were generated. On the other hand, Kapoor and Chinnadurai[17] employed homologous recombination in 293 cells. They also constructed a shuttle plasmid bearing a modified E1 region and cotransfected the shuttle plasmid with the viral DNA, which was cleaved with *XbaI* in advance, into 293 cells.

Although these methods were useful, the requirement for manipulating whole viral DNAs to generate recombinant Ad vectors was a major technical issue. Purification of viral DNAs was time-consuming and tedious. Moreover, the use of these techniques caused the contamination of parental nonrecombinant Ad viruses, such that the desired recombinant Ad vectors also had to be purified. This was because the parental Ad viruses had a growth advantage over the recombinant vectors and inhibited the propagation of the vectors in 293 cells. Thus, easier and more efficient methods were urgently needed at the time.

From the late 1980s to early 1990s, other standard methods to make E1-deleted Ad vectors were developed; these included a homologous recombination method in 293 cells[7,18] and an in vitro ligation method.[19–21] In 1994, Bett et al. developed a homologous recombination method in the E1-complementing cell lines (293 cells) to generate recombinant Ad vectors[7] (Figure 1(A)). Their method used two plasmids containing

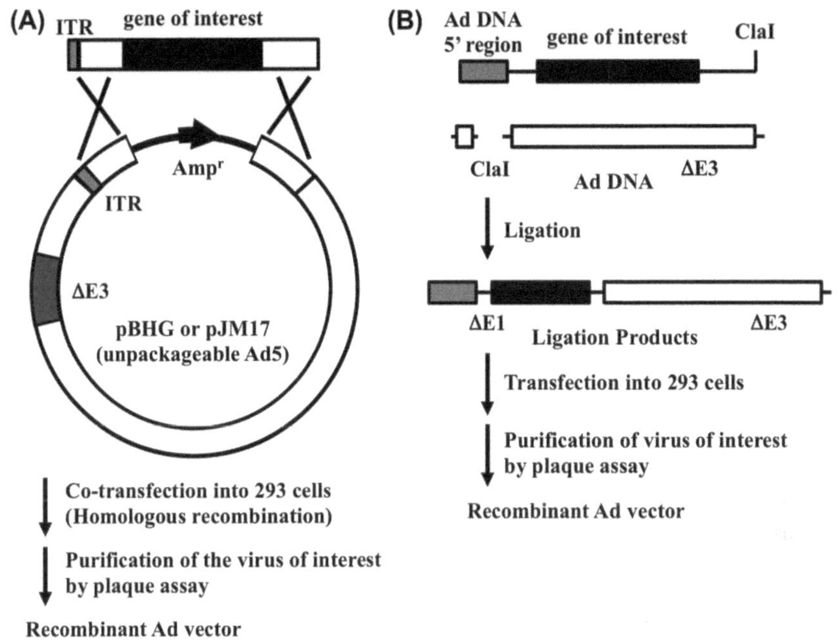


**Figure 1** Early methods of constructing first generation adenovirus vectors. (A) Homologous recombination method in 293 cells. (B) In vitro ligation method.

overlapping fragments that recombine. The first plasmid contained most of the viral genome in circular form, but lacked the DNA packaging signals and the E1 region. The second plasmid contained the left ITR, packaging signal, and sequence overlapping the first plasmid. After the gene of interest had been introduced into the second plasmid, the two plasmids were cotransfected into 293 cells, and the virus produced by recombination in 293 cells was isolated through plaque purification. This method was widely used at the time, and it greatly contributed to the widespread use of the Ad vectors. The major limitations of these approaches were the low frequency of the recombination event and the tedious and time-consuming plaque purification procedure required to select the recombinant virus of interest; the purification procedure was required because most of the virus produced was of the wild type (in most cases 20–70%) due to recombination with the Ad sequence integrated into the chromosome of the 293 cells. In addition, the potential instability of the large Ad genome due to the presence of a head-to-head ITR junction was another restriction.[22] Several other systems that overcome these limitations have been developed and are described in Chapter 4.

On the other hand, the in vitro ligation method also used whole viral DNA genomes and a plasmid containing the left end of the Ad genome with the left ITR, the packaging signal, and the E1A enhancer sequence (map unit: 0–1.3) (Figure 1(B)). This system employed ClaI, which is a unique site located in the E1 region of the Ad type 5 genome (map unit: 2.6). After the gene of interest was inserted into the region downstream of the viral sequence of the plasmid, the ClaI-digested fragment containing the left viral sequence and gene of interest was ligated with the Ad genome digested by ClaI, replacing a portion of the viral E1A gene. The ligated DNAs were then directly transfected into 293 cells to generate the recombinant virus. This method, however, was not very efficient and required purification of the recombinant virus by plaque assay, because wild-type and transgene null viruses resulting from incomplete restriction digestion and self-religation were also generated. Moreover, the E1 region was not completely removed, limiting the space for insertion of a foreign gene. As a result, this method is rarely used today. More efficient and simplified in vitro ligation systems will be reviewed below.

## 2.2 *Homologous Recombination Method in* Escherichia coli

A new method called the AdEasy system[23] has been developed for generating Ad vectors based on the homologous recombination of two plasmids using bacteria[24,25] (Figure 2(A)). Although most bacteria do not recombine transformed DNA readily and are not widely used for plasmid manipulation by homologous recombination, this method takes advantage of the highly efficient homologous recombination machinery of specialized *E. coli* (BJ5183, recBCsbcBC). The vector plasmid contains the full-length (or the E3-deleted) Ad genome flanked by the PacI site, which is an 8-bp recognition restriction enzyme (rare cutter), an ampicillin resistance gene, and a plasmid origin of replication. The shuttle plasmid contains the left region of the viral genome including the left ITR, packaging signal, and overlapping sequence downstream of the E1 region. The vector plasmid cut by ClaI (a unique restriction enzyme), which is located in the E1 deletion region, and the shuttle plasmid, in which the gene of interest is cloned, are cotransformed with recBCsbcBC *E. coli*. The recombination event occurs through overlapping of the fragments of each plasmid. The plasmid is isolated after culturing of the independent *E. coli* clones for a short time (usually fewer than 8 h). Since recBCsbcBC *E. coli* is not suitable for large-scale preparation of the plasmid, the recovered plasmid is retransformed and cultured with more conventionally used strains of *E. coli* (e.g., DH5α). The plasmid is then isolated, and positive clones are selected by restriction analysis. Transfection of linearized plasmids digested by PacI, which cuts at the end of the left and right ITRs, into 293 cells generates the recombinant Ad vectors. Unlike the homologous recombination method in 293 cells, the generation of the wild-type virus is extremely low due to the transfection of homogeneous DNAs in which the E1 gene has already been replaced by the foreign gene. Thus, the time-consuming plaque purification procedure is not absolutely required to produce the virus. In 2001, the AdEasy system[24] was simply improved.[26] Before the transformation of a shuttle plasmid and a vector plasmid, a viral backbone plasmid is preselected by electroporation of the vector plasmid into BJ5183 followed by the picking up of a single colony. Thus, this two-step transformation protocol allows a higher success rate during the selection for recombinants by eliminating defective and nonreplicative viral backbone plasmids in advance.

A modified system using homologous recombination in bacteria has also been reported.[27] In this system, the incP-plasmid, which is capable of replicating in the polA

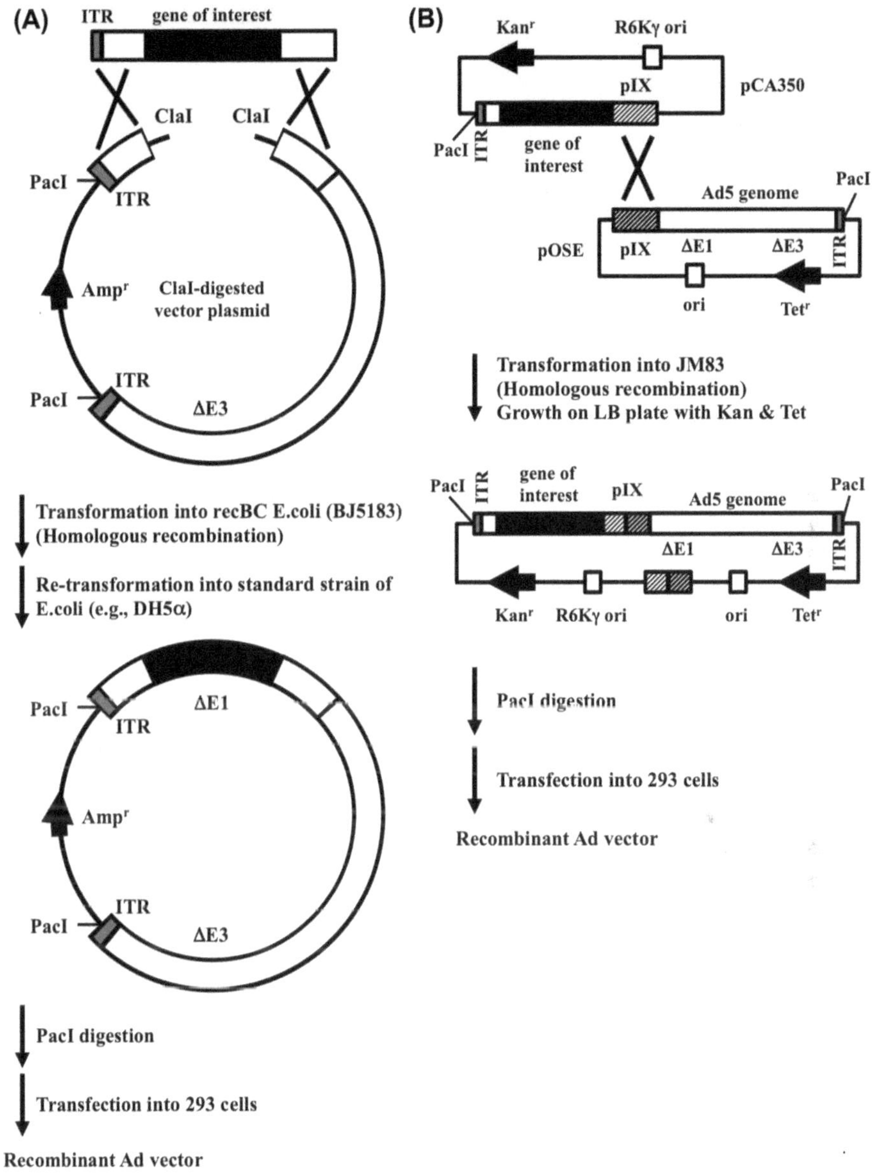


**Figure 2** Methods of constructing first generation adenovirus vectors by homologous recombination in *E. coli*. (A) Homologous recombination method. (B) Improved homologous recombination method.

mutant of *E. coli* and accommodating a large insert and the SacB gene of *Bacillus subtilis*, which is a lethal marker for *E. coli* in the presence of sucrose, is used to improve the selection of positive recombinant Ad plasmids. Another modified approach was taken by using a common *E. coli* strain.[28] Since BJ5183 do not produce a sufficient amount of recombinant plasmid to allow for restriction analysis, the recombinant plasmid must be retransformed into another strain to observe the expected construction. To avoid this complicated process, the *E. coli* strain Top10F′ was used. Top10F′ permits the homologous recombination to yield recombinant plasmids and the production of large-scale plasmid DNAs.

In 2004, Mullan et al. improved the *E. coli* homologous recombination system.[29] Conventional methods manipulating recombinant Ad vector construction in *E. coli* generally entail a screening step to identify the correct plasmid. Mullan and colleagues described that the recombination process yielding an Ad plasmid requires only one step, and always leads to the formation of only the desired recombinant Ad clone. Thus, no screening is needed to identify the objective clone encoding the desired recombinant Ad DNA. They introduced a gene of interest into a conditionally replicating shuttle plasmid, pCA350. pCA350 contains the left end of an Ad genome encompassing the pIX-coding region, an R6Kγ conditional bacterial origin of replication, and a kanamycin resistance cassette. The R6Kg plasmid cannot replicate in the absence of specific replication initiator proteins known as π proteins, which are encoded by the *pir* gene. A recombinant pCA350 was transformed into a π protein-expressing *E. coli* (e.g., λPir +TAM1) bearing a pOSE plasmid series containing the right end of an Ad genome commencing from the pXI-coding region, a bacterial origin of replication, and a tetracycline resistance cassette, leading to homologous recombination in pXI-coding regions contained in both plasmids on the LB with kanamycin and tetracycline. After the recombination, only a cointegrated recombinant vector along with a specific bacterial origin of replication, kanamycin resistance cassette, and tetracycline resistance cassette could replicate and form colonies. Thus, the resulting recombinant was truly clonal, obviating the need for screening. In addition, the individual plasmids used in the homologous recombination process are incapable of generating contaminating Ad by themselves. Mullan et al. reported that this approach successfully generated more than 200 recombinant Ad vectors, and other groups have also employed this construction system[30,31] (Figure 2(B)).

## 2.3 Improved in vitro Ligation Method

About a decade ago, Mizuguchi and Kay developed a two plasmid in vitro ligation method that does not require a recombination step to produce Ad vectors[32,33] (Figure 3(A)). Since Ad contains a large genome, unique and useful restriction sites are limited. This makes in vitro manipulation of Ad DNA difficult and explains why the method of simple in vitro ligation based on plasmid construction has never been developed. Mizuguchi and Kay overcame this limitation by using the rare-cutting enzymes, I-CeuI and PI-SceI. I-CeuI[34] and PI-SceI[35] are intron-encoded endonucleases[36] that do not cut the Ad genome, and their sequence specificities are at least 9–10 and 11 bp, respectively. The vector plasmid contains a complete E1/E3/E4-deleted Ad type 5 genome with three unique restriction sites, I-CeuI, SwaI, and PI-SceI, in an E1 deletion site and an ampicillin resistance gene. SwaI is also a rare-cutting restriction enzyme with a sequence specificity of 8 bp. The shuttle plasmid contains a multicloning site between the I-CeuI and the PI-SceI sites and a kanamycin resistance gene.

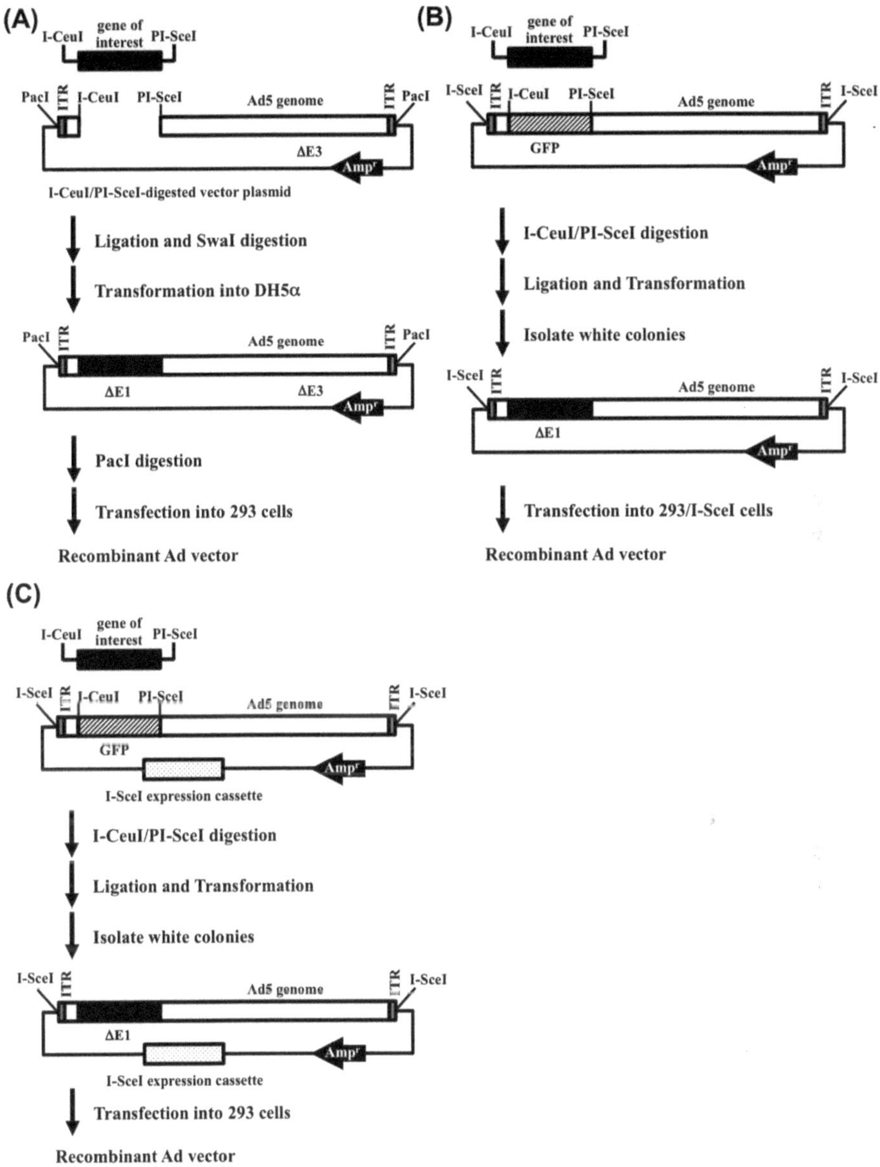


**Figure 3** Methods of constructing first generation adenovirus vectors by in vitro ligation. (A) Improved in vitro ligation method. (B and C) Modified "improved in vitro ligation method," using I-SceI expressing 293 cells (293/I-SceI cells) in (B) and a vector plasmid with I-SceI expression cassette in (C).

A variety of genes can be inserted in any site of the multicloning site. In this section, we introduce their efficient and improved in vitro ligation techniques. The system basically involves two cloning steps in *E. coli* followed by transduction of the linearized recombinant Ad DNA into 293 cells (Figure 3(A)).

First, in 1998, Mizuguchi and Kay generated four basic viral backbones, pAdHM1-4 in bacterial plasmids[32]; all four backbones are the E1 deleted and the viral DNA is flanked by PacI sites at both ends (pAdHM3, 4). In these backbones, pAdHM4 has the E3 deletion. They also constructed pHM3, which contains the pUC18-derived multicloning site between the I-CeuI and the PI-SceI sites. After a gene of interest is inserted into the multicloning site of a shuttle plasmid, the gene can be easily introduced into the E1 deletion site of a vector plasmid at the I-CeuI and PI-SceI sites. Basically, the recombinant shuttle plasmid and vector plasmid are then digested with I-CeuI and PI-SceI, and the mixture is directly ligated without gel purification of either the transgene expression cassette sequence or the vector viral sequence. SwaI digestion of the ligation products is performed in order to prevent production of a plasmid containing a parental Ad genome (null vector). By transformation into standard strains of *E. coli*, such as chemical-competent DH5α (electro-competent DH5α can also be used), and growth in ampicillin, only the ligated Ad plasmid DNAs with inserts are selected (more than 90% of the transformants have the correct insert). A recombinant Ad vector can be generated by the transfection of the PacI-digested, linearized plasmid into 293 cells, resulting in a homogeneous population of the recombinant virus. The time-consuming plaque purification procedure is not absolutely required to produce the virus as in homologous recombination methods. Thus, recombinant Ad vectors are produced by simple molecular biology techniques, without the need for homologous recombination, in mammalian cells, bacteria, or yeast (Figure 3(A)).

Next, they simplified this system by deleting the time-consuming process of gel purification of the cloned DNA fragments.[33] A shuttle plasmid that has a kanamycin resistance cassette instead of an ampicillin resistance cassette was constructed. Since the vector plasmid contained an ampicillin resistance cassette, the isolation of the recombinant Ad fragment from a gel was obviated. Moreover, in the same study they also reported a series of shuttle plasmids with different promoters, such as a CMV or RSV promoter, and a series of vector plasmids for producing the E1-, E1/E3-, and E1/E3/E4-deleted vectors. pAdHM10 has a larger E3 deletion than pAdHM4, while, in addition to the larger deletion in the E3 region, pAdHM12 has the E4 deletion. Up to 4.9, 7.6, 8.1, and 10.9 kb of exogenous DNA can be inserted into pAdHM3, 4, 10, and 12, respectively. These modifications allow any laboratory to construct Ad vectors easily for the delivery of a variety of genes (Figure 3(A)).

Around the same time, a different in vitro ligation method employing the same logic was developed, called the pAdvantage system.[37] The viral backbone (pAdvantage) was based on Ad serotype 2, which contained an I-CeuI site in the E1 deletion region and the ITRs flanked by SnaBI sites. In this method, however, only one unique restriction enzyme site is introduced into the vector plasmid containing most of the viral sequence, making it less efficient than the method described above.

The improved in vitro ligation method has several advantages over the homologous recombination method in bacteria: only simple molecular biology cloning techniques are required, no special *E. coli* strain is needed, and no additional transformation is required (the *E. coli* system formerly required a two-step transformation using two different *E. coli* strains).

In 2003, Gao et al. devised an advanced method for the generation of Ad vectors,[38] which was based on the method of Mizuguchi and Kay[32,32] (Figure 3(B) and 3(C)). The method included two types of modification. The first step consisted of isolating bacterial transformants containing the correct recombinant vectors. They incorporated a convenient green–white selection step into the cloning process, reducing the numbers of transformant colonies to be screened by plasmid DNA purification and restriction analysis. An *Aequorea victoria* green fluorescent protein (GFP) expression cassette was inserted into the E1 region of a viral backbone. The objective clones for an Ad vector are produced by removing the GFP expression cassette and inserting a gene of interest, while the background clones still contain the cassette. We can distinguish recombinants (no fluorescence; white colony) from background transformants (GFP fluorescence; green colony) on an ampicillin plate. Using this method, Gao and colleagues reported that a correct recombinant clone is usually guaranteed by picking up less than three white colonies. Thus, this green–white selection simplifies the cloning of Ad vectors and improves the cloning efficiency for the construction of Ad vectors. The second step consisted of transfecting and rescuing clones of recombinant Ad vectors in packaging cells. They extracted the Ad vector genome from a circular viral plasmid in 293 cells after transfection. Rare cutter I-SceI sites were inserted into the viral backbone outside the ITRs of an Ad genome. I-SceI is an intron-encoding endonuclease present in the mitochondria of yeast *Saccharomyces cerevisiae*. In addition, I-SceI expression is compatible with eukaryotic cell growth.[39,40] The circular viral plasmid containing I-SceI sites was transfected into 293 cells expressing I-SceI endonuclease (293/I-SceI cells), resulting in the efficient release of the viral linear genome in 293 cells. Alternatively, a transcription unit expressing I-SceI endonuclease was incorporated into the viral backbone, which also contains the viral genome flanked by I-SceI sites. The transfection of this self-cleaving plasmid into normal 293 cells led to the expression of I-SceI endonuclease, followed by the efficient release of the linear DNAs. Using this I-SceI-mediated cleavage, they demonstrated that circular viral DNAs can be efficiently transfected into 293 cells and that the cleavage in the cells improves the rescue efficiency of recombinant Ad vectors (Figure 3(B) and 3(C)).

## 2.4 Homologous Recombination Method in Yeast

Another method was devised to overcome the difficulty of manipulating Ad DNAs. Yeast has highly efficient homologous recombination capacity. In 1994, Ketner et al. used *S. cerevisiae* to construct a yeast artificial chromosome (YAC) that contains a complete sequence of the linear viral genome.[41] This system could allow multiple mutation inserts to be introduced into the viral genome. An infectious Ad vector was generated after introducing the modified Ad genome, digested from the YAC vector, into 293 cells. However, a major restriction of this method was the low yield of DNA.

In 2003, Hokanson et al. removed this restriction by taking a hybrid approach that combined the strengths of the yeast and *E. coli* systems.[42] This approach exploited the facts that viral DNAs are easily modified in yeast and efficiently amplified in bacteria. The recombinant Ad genome was generated through recombination between a

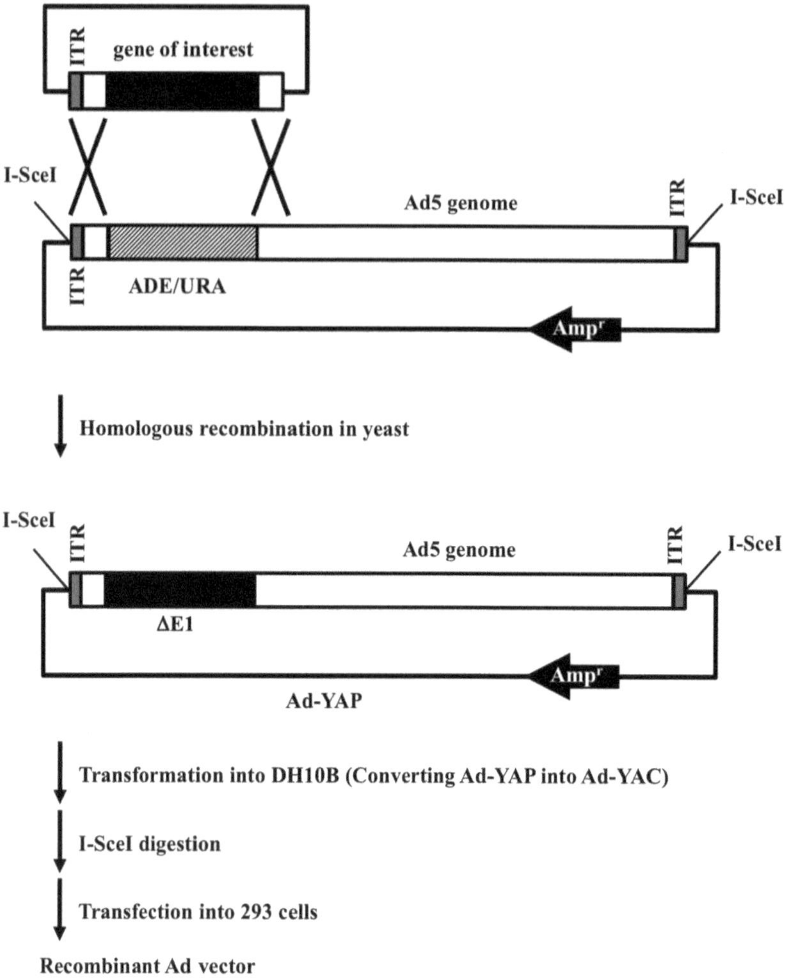


**Figure 4** A method of constructing first generation adenovirus vectors by homologous recombination in yeast. ADE/URA, yeast selectable markers.

shuttle plasmid and a yeast artificial plasmid (YAP) encoding the full-length Ad DNA with yeast and bacterial origins of replication, resulting in an Ad-YAP. The modified YAP vector was introduced into bacteria, thereby converting the Ad-YAP to a plasmid artificial chromosome, which was then amplified. After introducing the Ad genome cleaved from the Ad-YAP into 293 cells, a recombinant Ad vector with a gene of interest was generated (Figure 4).

An advantageous feature of this yeast system was that it allowed efficient manipulation of large DNA fragments.[43] Nonetheless, the system was also quite complicated. It required the use of an additional host (yeast), including yeast culture and manipulation.

### 2.5 Transposon-Mediated Recombination Method

In 2000, a transposon-mediated recombination system was developed, called the admid system.[44] Recombinant Ad vectors were generated by Tn7-mediated, site-specific transposition in *E. coli* strain DH10B. This Tn7-mediated recombination method had been used to construct recombinant baculoviruses.[45] The authors prepared two plasmids: (i) a low copy number *E. coli* plasmid (admid) containing a full-length Ad genome with a β-galactosidase-coding sequence and the Tn7 attachment site (lacZattTn7) replacing the E1 region and (ii) a transfer plasmid with a mini-Tn7 containing a gene of interest flanked by Tn7R and Tn7L. These two plasmids were stably maintained in DH10B. After the transposition of the mini-Tn7 into the lacZattTn7 site, transposed recombinant Ad vectors were readily identified by their β-galactosidase phenotype. The authors reported that this admid system accommodates DNA inserts up to 5.8 kb in basic E1/E3-deleted Ad vectors and the transposition efficiency is about 25%. Transfection of the PacI-digested admid recombinant into 293 cells resulted in the production of recombinant Ad vectors.

### 2.6 Other Methods

Other methods to construct Ad vectors have been reported: a cre–lox-mediated recombination method,[46–50] a method combining homologous recombination in mammalian cells and bacteria,[51] a method based on cosmid construction,[52–55] and others.[56] Several variations of the cre–lox-mediated recombination method have been developed in 293 cells,[47–50] which overcomes some of the inefficiencies of the homologous recombination method in mammalian cells.[48,49] In addition, cre–lox-mediated recombination has been used to construct the whole Ad vector genome in a test tube, eliminating the need for cellular recombination.[46] The usefulness of these systems depends on the experimental carrier used by the individual investigator, although additional steps are required, such as lambda packaging for cosmid construction or purification of the virus of interest by plaque assay.

## 3. Construction of the E1/E3-Substituted Adenovirus Vectors

Most first generation Ad vectors in current use are of the E1-substitution type. Sometimes, however, it may be preferable to insert foreign genes into the E3 deletion region as well as into the E1 deletion region. For example, when heterologous gene expression cassettes inserted into the E1-deletion region are coexpressed, promoter interference sometimes occurs, i.e., transcription from one promoter suppresses transcription from another.[57–59] Ad vectors containing foreign genes that can be introduced into both the E1 and the E3 deletion regions eliminate such problems, because each of the genes can be efficiently expressed.

Some of the systems described in Section 2 allow for the insertion of foreign genes into both the E1 and the E3 deletion regions. In one method, developed by Graham et al., a unique PacI site in the E3 deletion region of the plasmid containing the Ad genome was used to clone the gene of interest into the E3 deletion region, and another gene of interest was then inserted into the E1 deletion region by homologous recombination in 293 cells[7,60] (Figure 5(A)). However, insertion of foreign genes into the E3 deletion region is sometimes difficult because the vector plasmid contains long palindromic sequences of the ITR, which induces plasmid instability in *E. coli* (by contrast, the vector plasmids in Sections 2.2 and 2.3 do not contain palindromic sequences). In another method, the insertion relied on homologous recombination in bacteria[25] (Figure 5(B)). This method used a shuttle plasmid containing the Ad sequence around the E3 region. After the gene of interest was inserted into the E3 deletion region of the shuttle plasmid, the linearized shuttle plasmid and SpeI (or SrfI)-digested vector plasmid were cotransformed into recBCsbcBC *E. coli*, by the same procedure as that used for the cloning of foreign genes into the E1 deletion region described in Section 2.2 (the SpeI and SrfI sites located around the E3 region are unique to the Ad genome).

Our group developed a modified system to clone the gene of interest into both the E1 and the E3 deletion regions based on simple plasmid construction using in vitro ligation[61] (Figure 5(C)). To do this, a unique restriction site, the Csp45I, ClaI, or I-SceI sites, was introduced into the E3 deletion region of the Ad vector plasmid containing the unique I-CeuI, SwaI, and PI-SceI sites in the E1 deletion region. Shuttle plasmids containing a multicloning site flanked by Csp45I, ClaI, or I-SceI sites were also constructed to assist introduction of the gene of interest into the E3 deletion region. Csp45I and ClaI produce compatible cohesive ends. Thus, if the gene of interest does not have both the Csp45I and the ClaI sites, the recombinant plasmid is produced from recleavable ligation products by Csp45I or ClaI. When the gene of interest contains both the Csp45I and the ClaI sites, the rare cutter I-SceI site is ideal for use as an alternative cloning site.[39,40] In this method, Ad vectors containing heterologous genes in the E1 and E3 deletion region are generated by a procedure similar to that described in Section 2.3.[62]

A major advantage of a binary transgene expression system in which heterologous genes can be inserted into both the E1 and the E3 regions is that gene products that interact with each other can be expressed in a single vector. A typical example is a tetracycline (tet)-controllable expression system, which allows for regulatable transgene expression.[63–65] In these experiments, two kinds of the E1-substituted Ad vectors were cotransfected: one expressing a tet-responsive transcriptional activator and the other, driven by a tet-responsive promoter, expressing a gene of interest. In their system, by inserting the gene of interest with a tetracycline-regulatable promoter and tetracycline-responsive transcriptional activator gene into the E1 and E3 deletion regions, respectively, Ad vectors containing a tetracycline-controllable expression system could be generated.[61] This approach had the advantage of eliminating extra labor to generate vectors and did not require cotransduction.

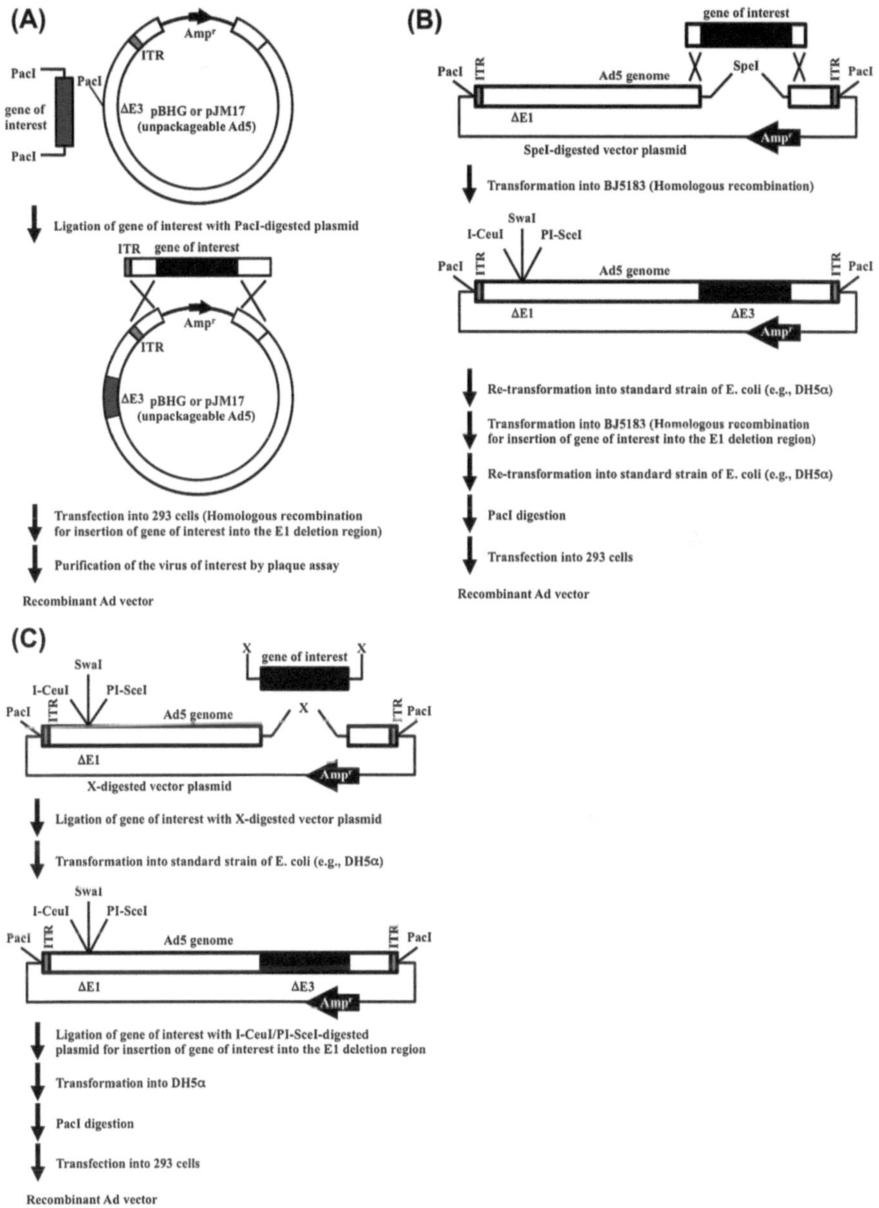


**Figure 5** Methods of constructing adenovirus vectors containing genes of interest in both the E1 and the E3 deletion region. (A) Homologous recombination method in 293 cells. (B) Homologous recombination method in *E. coli*. (C) Improved in vitro ligation method.

# 4. Construction of Capsid-Mutant Adenovirus Vectors

One of the hurdles confronting Ad-mediated gene transfer is that gene transfer with Ad vectors is inefficient in cells lacking the primary receptor, the coxsackievirus and adenovirus receptor (CAR)[66–70]; such cells include many advanced tumor cells, skeletal muscle cells, smooth muscle cells, peripheral blood cells, hematopoietic stem cells, dendritic cells, and so on. A high dose of vector is required to achieve efficient gene transfer to these cell types. This in turn increases unwanted side effects, such as vector-associated immunogenic toxicities. Another hurdle confronting Ad vector-mediated gene transfer is the nonspecific distribution of the vectors in tissue after in vivo gene transfer because of the relatively broad expression of CAR, αv integrin (the secondary receptor), and heparan sulfate (the third receptor). This property imposes an increased risk of toxicity due to vector dissemination to nontargeted cells, such as antigen-presenting cells (e.g., macrophages and dendritic cells). This occurs even when Ad vectors are locally administered to the tissue of interest. Vector targeting to a specific tissue or cell type would enhance gene therapy efficacy and permit the delivery of lower doses, which should result in reduced toxicity.

## 4.1 Construction of Fiber-Mutant Adenovirus Vectors

Genetic modification of the Ad capsid, such as its fiber, pIX, or hexon, is an attractive strategy for altering the Ad tropism. Among these options, modification of the fiber proteins has been the most widely studied. Fiber proteins consist of three distinct domains: the tail, shaft, and knob. Each domain has distinct functions in host cell infection. Ad infection of susceptible cells requires two distinct steps. In the first step, the initial high-affinity binding of the virus to the CAR on the cell surface occurs via the trimeric subunits of the C-terminal knob domain of the fiber protein.[66,70] In the second step, interaction between the RGD motif of the penton bases with the secondary host cell receptors, αvβ3 and αvβ5 integrins, expressed on most cell types facilitates internalization via receptor-mediated endocytosis.[71–73] Therefore, the interaction of the fiber knob with CAR on the cell is the key mediator by which the Ad vector enters the cells. Thus, modification of fiber protein is an attractive strategy for overcoming the limitations imposed by the CAR dependence of Ad infection, and in several studies performed around the year 2000, two approaches were adopted to accomplish this. One was the addition of foreign peptides to the C-terminal end of the fiber knob,[74–77] and the other was the insertion of foreign peptides into the HI loop of the fiber knob.[78–81] Both approaches allowed Ad tropism to be expanded by binding of the foreign ligand to the cellular receptor.

Curiel et al. first reported characterization of an Ad vector containing a heterologous peptide epitope in the HI loop of the fiber knob.[78,79] They constructed a fiber-modified vector by the homologous recombination method in bacteria (Figure 6(A)). To do this, they constructed a shuttle plasmid containing the fiber-coding region and a vector plasmid. The shuttle plasmid contained a mutated-fiber gene in which a unique EcoRV site was incorporated in place of the HI loop-coding region. The vector plasmid was constructed so that it contained a unique SwaI site in the fiber-coding region. Oligonucleotides corresponding to the peptide of interest were first inserted into the

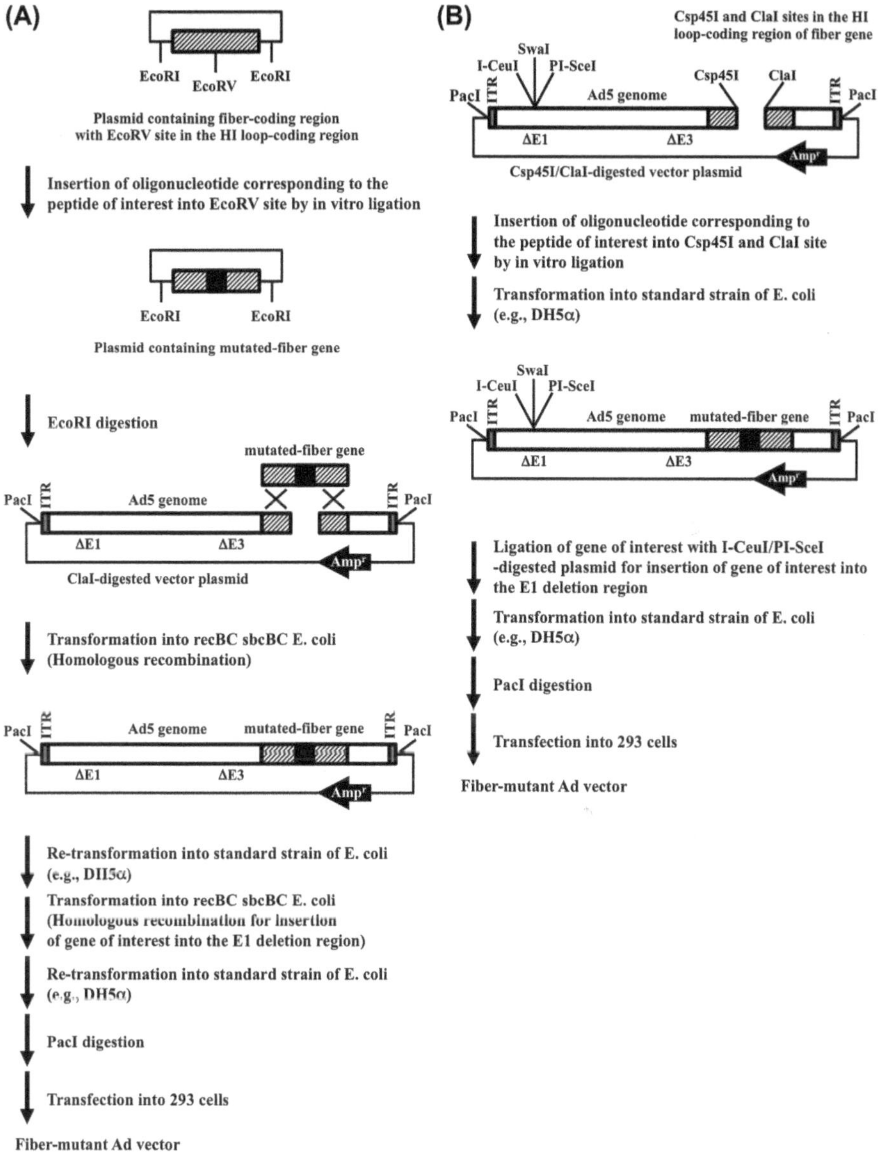


**Figure 6** Methods of constructing fiber-mutant adenovirus vectors. (A) Homologous recombination method in *E. coli*. (B) Improved in vitro ligation method.

EcoRV site of the shuttle plasmid. The EcoRI-digested linearized shuttle plasmid containing the mutated-fiber gene and SwaI-digested linearized vector plasmid were then cotransformed with recBCsbcBC *E. coli* for homologous recombination. The following steps were performed by a method similar to that described in Section 2.2 (see also Figure 2(A)). This method required at least five transformations, including transformation of the plasmid into different strains of bacteria to produce the fiber-mutant Ad vector that expresses the foreign gene.

We developed a method of constructing fiber-modified Ad vectors by using simple in vitro ligation[81] (Figure 6(B)). The vector plasmid contains a complete E1/E3-deleted Ad genome and extra 12-bp foreign DNAs, which are the recognition sequences produced by Csp45I and ClaI, in the HI loop-coding region of the fiber knob. Oligonucleotides corresponding to the peptide of interest and containing a Csp45I and ClaI recognition site are ligated into the Csp45I- and ClaI-digested vector plasmid. The foreign transgene expression cassette is inserted into the E1 deletion site of the vector plasmid by the improved in vitro ligation method described in Section 2.3 (see also Figure 3(B)). The fiber-mutant Ad vector is produced by transfection of the PacI-digested recombinant vector plasmid into 293 cells. As a result, only a two-step, simple in vitro ligation and transformation using a standard strain of *E. coli* are required to construct fiber-mutant Ad vectors containing the gene of interest. In this method, two to three additional amino acids flanking the peptide of interest are introduced into the mutated fiber by the additional nucleotides contained within the Csp45I and ClaI recognition sites. However, these additional amino acids do not exert any effect on the function of the peptide of interest.[81] Moreover, on the basis of this procedure, we constructed a vector plasmid that allowed the peptide of interest to be expressed in Ad vectors in either the HI loop or the C-terminus of the fiber knob, or both.[82] This plasmid contains a complete E1/E3-deleted Ad genome with I-CeuI, SwaI, and PI-SceI sites in the E1 deletion region, a Csp45I site between T546 and P547 of the fiber protein, which is the HI loop-coding region of the fiber protein, and a ClaI site between E581 (the last amino acid) and the stop codon of the fiber protein of the Ad genome. By using our system, the foreign DNA-coding peptide of interest could easily be cloned into two regions of the fiber knob by a simple in vitro ligation method: the HI loop-coding region with Csp45I/ClaI or Csp45I only, and the C-terminus with ClaI.

We and other groups reported that Ad vectors containing the RGD peptide motif (CDCRGDCFC), which binds with high affinities to the αvβ3 and αvβ5 integrins on the cell surface[83–85] on the fiber knob, mediate not only CAR-dependent gene delivery but CAR-independent, RGD-integrin (αvβ3 and αvβ5)-dependent gene delivery as well.[78,80,81] The virus containing the RGD peptide on the fiber knob was able to infect human glioma cells lacking CAR expression about 100–1000 times more efficiently than the virus containing a wild-type fiber.[81] Since αvβ3 and αvβ5 integrins are expressed on most types of cells, except some blood cells, Ad vectors containing RGD peptides on the fiber knob mediate efficient gene transfer into CAR-deficient cells.

Although Ad vectors can transduce most cells because of the insufficient expression of CAR, they mediate less efficient gene transfer in some of the important target tissues (cells) for gene therapy, including differentiated airway epithelium, skeletal muscle, smooth muscle, peripheral blood cells, and hematopoietic stem cells.[68,69,86] Many reports have shown that fiber-mutant Ad vectors containing appropriate foreign peptides on the fiber knob may transduce these cells efficiently and may be a powerful tool for gene transfer into mammalian cells.[87–89]

## 4.2 Construction of Other Capsid Protein-Mutant Adenovirus Vectors

The C-terminal region of pIX and the hypervariable region (HVR) 5 of hexon loop L1 are other candidate locations for capsid modification. pIX is a minor structural protein contained in the Ad virion.[90] The attractive characteristic of ligand insertion into the pIX region is that the C-terminus of pIX tolerates the insertion of large ligands.[91–93] On the other hand, hexons are the most abundant capsid proteins and compose each geometrical face of the capsid. As hexons are mostly targeted by neutralizing antibodies,[94] hexon modification was reported to escape from neutralizing antibodies as well as to modify the tropism.[95] The HVR of the hexon is a candidate location for incorporating foreign peptides without affecting the normal function of Ad type 5 as a gene transfer vector (i.e., viral growth, virus formation, virion stability, CAR-mediated infectivity).[96]

One attractive point of pIX or hexon modification is that 240 or 720 molecules of foreign peptides per virion are displayed at pIX or the hexon, respectively, while only 36 molecules are displayed at the fiber (note that the fiber and hexon are composed of trimeric subunits). Therefore, pIX- or hexon-modified Ad vectors containing heterologous peptides might be more effective than fiber-modified Ad vectors. While the pIX- or hexon-modified Ad vectors described above were generated by conventional methods, such as the homologous recombination system in *E. coli*[91–93] or a cosmid system,[95] we devised a simple method for constructing pIX- or hexon-modified Ad vectors by using in vitro ligation-based plasmid construction.[97] By expanding on our previous idea[82] (Figure 6(B)), we constructed new vector plasmids containing a unique XbaI site in the coding regions of the C-terminal of pIX or HVR5 of the hexon, so that heterologous peptide sequences could be inserted into pIX or the hexon.

As the vector system shown in this section enables easy construction of capsid-modified Ad vectors displaying a peptide of interest, it has great potential for gene therapy and gene transfer experiments.

## 5. Construction of Small-Interfering RNA-Expressing Adenovirus Vectors

RNA interference (RNAi), which mediates the sequence-specific suppression of gene expression in a wide variety of eukaryotes by double-stranded RNA homologies to the target gene,[98] is a powerful tool for the knockdown of gene expression. Transduction of synthetic small-interfering RNA (siRNA; 19 to 29 nucleotides of RNA) or the promoter-based expression of siRNA in the cells results in sequence-dependent degradation of target mRNA and subsequent reduction of target gene expression. Most promoter-based RNAi systems express short hairpin RNA (shRNA), which is then trimmed by Dicer, generating functional siRNA. Polymerase III-based promoters, such as the small nuclear RNA U6 promoter or the human RNase P RNA H1 promoter, are widely used for the expression of shRNA (siRNA),[98] although polymerase II-based promoters are also used.[99,100] The promoter-based method has an advantage in that both viral and nonviral vectors can be used for delivery of the siRNA expression unit, whereas only nonviral vectors are used for the delivery of synthetic siRNA.

To construct Ad vectors expressing siRNA by the conventional recombination methods described above (the improved in vitro ligation method or the homologous recombination method in *E. coli*), shRNA-coding oligonucleotides are introduced downstream of the polymerase III (or polymerase II)-based promoter cloned in a shuttle plasmid.

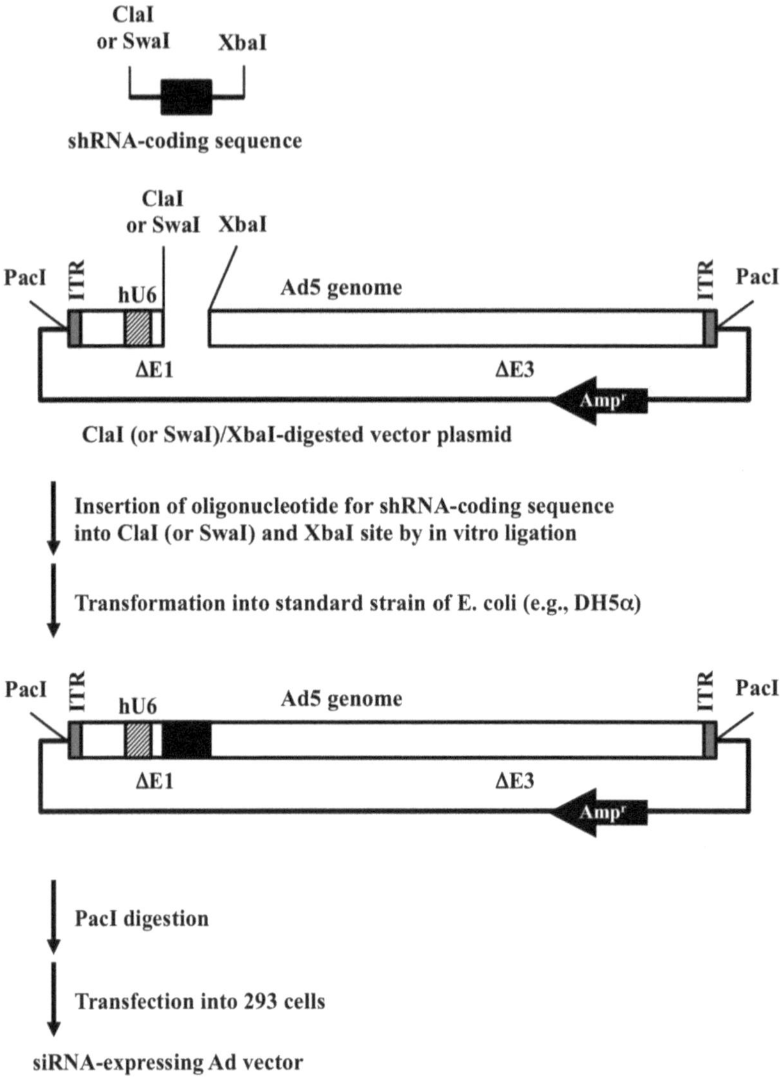


**Figure 7** A method of constructing small-interfering RNA-expressing adenovirus vectors.

An shRNA (siRNA) expression cassette, which is cloned in the shuttle plasmid, is then introduced into the E1 deletion region of the Ad vector plasmid, which clones a full Ad genome, by a simple in vitro ligation or homologous recombination in *E. coli*. The resulting plasmid is then linearized and transfected into 293 cells, generating an Ad vector expressing siRNA. Therefore, two-step *E. coli* transformation and plasmid manipulation are required for the improved in vitro ligation method, whereas three-step *E. coli* transformation and plasmid manipulation are required in the homologous recombination method in *E. coli*. Ad vector-mediated delivery of an siRNA expression unit provides a valuable tool for both gene function studies and therapeutic applications.[101,102]

We developed a simple method for generating Ad vectors expressing siRNA, in which shRNA-coding oligonucleotides could be directly introduced into an Ad vector plasmid containing the human U6 (hU6) promoter sequence[103] (Figure 7). In brief, the hU6 promoter sequence containing unique restriction enzyme sites (ClaI/XbaI or SwaI/XbaI) at the transcription start region was introduced into a shuttle plasmid. The shuttle plasmids were then transferred into the E1 deletion region of one of the pAdHM plasmid series by the in vitro ligation method, resulting in new vector plasmids containing ClaI/XbaI or SwaI/XbaI sites at the region downstream of the hU6 promoter sequence (pAdHM-hU6). To generate a recombinant vector plasmid for Ad vectors expressing siRNA, oligonucleotides for shRNA against a target gene are synthesized, annealed, and ligated with ClaI/XbaI or SwaI/XbaI-digested pAdHM-hU6. The oligonucleotides must be designed so that the recombinant vector plasmid containing the shRNA-coding sequence is redigested with XbaI, but not with ClaI or SwaI. By designing oligonucleotides like the one described above, the generation of self-ligated plasmids can be avoided by digestion of the ligation products with ClaI or SwaI. Using this method, only one-step *E. coli* transformation is required to generate an Ad vector plasmid containing an siRNA expression cassette. This system should be useful for RNAi-based experiments,[104,105] and might facilitate the development of an siRNA-expressing Ad vector library for functional screening.

## 6. Conclusion

Many systems have been developed to generate Ad vectors, each with its own advantages and disadvantages, depending on the experimental carrier used by individual investigators. A major advantage of the homologous recombination method in bacteria is that any mutation can be introduced into the whole Ad genome, at a unique restriction site around the region to be mutated. The improved in vitro ligation method requires only routine molecular reagents and techniques and allows any laboratory to construct Ad vectors for gene transfer studies. Progress in the technology for the generation of Ad vectors should make this vector more attractive for gene therapy, gene transfer experiments, and studies of gene function in basic research.

## References

1. Shenk T. Adenoviridae. In: Fields BN, Knipe DM, Howley PM, editors. *Fields virology*, Vol. 2. Philadelphia, PA: Lippincott-Raven Publishing; 1996. p. 2111–48.
2. Russell WC. Adenoviruses: update on structure and function. *J Gen Virol* 2009;**90**:1–20.
3. Yu P, Ma C, Nawaz M, Han L, Zhang J, Du Q, et al. Outbreak of acute respiratory disease caused by human adenovirus type 7 in a military training camp in Shaanxi, China. *Microbiol Immunol* 2013;**57**:553–60.
4. Graham FL, Smiley J, Russell WC, Nairn R. Characteristics of a human cell line transformed by DNA from human adenovirus type 5. *J Gen Virol* 1977;**36**:59–74.

5. Fallaux FJ, Kranenburg O, Cramer SJ, Houweling A, Van Ormondt H, Hoeben RC, et al. Characterization of 911: a new helper cell line for the titration and propagation of early region 1-deleted adenoviral vectors. *Hum Gene Ther* 1996;**7**:215–22.
6. Fallaux FJ, Bout A, van der Velde I, van den Wollenberg DJ, Hehir KM, Keegan J, et al. New helper cells and matched early region 1-deleted adenovirus vectors prevent generation of replication-competent adenoviruses. *Hum Gene Ther* 1998;**9**:1909–17.
7. Bett AJ, Haddara W, Prevec L, Graham FL. An efficient and flexible system for construction of adenovirus vectors with insertions or deletions in early regions 1 and 3. *Proc Natl Acad Sci USA* 1994;**91**:8802–6.
8. Bett AJ, Prevec L, Graham FL. Packaging capacity and stability of human adenovirus type 5 vectors. *J Virol* 1993;**67**:5911–21.
9. Kozarsky KF, Wilson JM. Gene therapy: adenovirus vectors. *Curr Opin Genet Dev* 1993;**3**:499–503.
10. Kovesdi I, Brough DE, Bruder JT, Wickham TJ. Adenoviral vectors for gene transfer. *Curr Opin Biotechnol* 1997;**8**:583–9.
11. Benihoud K, Yeh P, Perricaudet M. Adenovirus vectors for gene delivery. *Curr Opin Biotechnol* 1999;**10**:440–7.
12. Yeh P, Perricaudet M. Advances in adenoviral vectors: from genetic engineering to their biology. *FASEB J* 1997;**11**:615–23.
13. Chinnadurai G, Chinnadurai S, Brusca J. Physical mapping of a large-plaque mutation of adenovirus type 2. *J Virol* 1979;**32**:623–8.
14. Carlock LR, Jones NC. Transformation-defective mutant of adenovirus type 5 containing a single altered E1a mRNA species. *J Virol* 1981;**40**:657–64.
15. Solnick D. An adenovirus mutant defective in splicing RNA from early region 1A. *Nature* 1981;**291**:508–10.
16. Stow ND. Cloning of a DNA fragment from the left-hand terminus of the adenovirus type 2 genome and its use in site-directed mutagenesis. *J Virol* 1981;**37**:171–80.
17. Kapoor QS, Chinnadurai G. Method for introducing site-specific mutations into adenovirus 2 genome: construction of a small deletion mutant in VA-RNAI gene. *Proc Natl Acad Sci USA* 1981;**78**:2184–8.
18. Miyake S, Makimura M, Kanegae Y, Harada S, Sato Y, Takamori K, et al. Efficient generation of recombinant adenoviruses using adenovirus DNA-terminal protein complex and a cosmid bearing the full-length virus genome. *Proc Natl Acad Sci USA* 1996;**93**:1320–4.
19. Berkner KL, Sharp PA. Generation of adenovirus by transfection of plasmids. *Nucleic Acids Res* 1983;**11**:6003–20.
20. Gilardi P, Courtney M, Pavirani A, Perricaudet M. Expression of human alpha 1-antitrypsin using a recombinant adenovirus vector. *FEBS Lett* 1990;**267**:60–2.
21. Rosenfeld MA, Siegfried W, Yoshimura K, Yoneyama K, Fukayama M, Stier LE, et al. Adenovirus-mediated transfer of a recombinant alpha 1-antitrypsin gene to the lung epithelium in vivo. *Science* 1991;**252**:431–4.
22. Ghosh-Choudhury G, Haj-Ahmad Y, Brinkley P, Rudy J, Graham FL. Human adenovirus cloning vectors based on infectious bacterial plasmids. *Gene* 1986;**50**:161–71.
23. Luo J, Deng Z-L, Luo X, Tang N, Song W-X, Chen J, et al. A protocol for rapid generation of recombinant adenoviruses using the AdEasy system. *Nat Protoc* 2007;**2**:1236–47.
24. He TC, Zhou S, da Costa LT, Yu J, Kinzler KW, Vogelstein B. A simplified system for generating recombinant adenoviruses. *Proc Natl Acad Sci USA* 1998;**95**:2509–14.
25. Chartier C, Degryse E, Gantzer M, Dieterle A, Pavirani A, Mehtali M. Efficient generation of recombinant adenovirus vectors by homologous recombination in *Escherichia coli*. *J Virol* 1996;**70**:4805–10.

26. Zeng M, Smith SK, Siegel F, Shi Z, Van Kampen KR, Elmets CA, et al. AdEasy system made easier by selecting the viral backbone plasmid preceding homologous recombination. *Biotechniques* 2001;**31**:260–2.
27. Crouzet J, Naudin L, Orsini C, Vigne E, Ferrero L, Le Roux A, et al. Recombinational construction in *Escherichia coli* of infectious adenoviral genomes. *Proc Natl Acad Sci USA* 1997;**94**:1414–9.
28. Renaut L, Bernard C, D'Halluin JC. A rapid and easy method for production and selection of recombinant adenovirus genomes. *J Virol Methods* 2002;**100**:121–31.
29. Mullan B, Dugué C, Moutard V, Raoux D, Tremp G, Denèfle P, et al. Robust functional gene validation by adenoviral vectors: one-step *Escherichia coli*-derived recombinant adenoviral genome construction. *Gene Ther* 2004;**11**:1599–605.
30. Bouquet C, Lamandé N, Brand M, Gasc J-M, Jullienne B, Faure G, et al. Suppression of angiogenesis, tumor growth, and metastasis by adenovirus-mediated gene transfer of human angiotensinogen. *Mol Ther* 2006;**14**:175–82.
31. Frau E, Magnon C, Opolon P, Connault E, Opolon D, Beermann F, et al. A gene transfer comparative study of HSA-conjugated antiangiogenic factors in a transgenic mouse model of metastatic ocular cancer. *Cancer Gene Ther* 2007;**14**:251–61.
32. Mizuguchi H, Kay MA. Efficient construction of a recombinant adenovirus vector by an improved in vitro ligation method. *Hum Gene Ther* 1998;**9**:2577–83.
33. Mizuguchi H, Kay MA. A simple method for constructing E1- and E1/E4-deleted recombinant adenoviral vectors. *Hum Gene Ther* 1999;**10**:2013–7.
34. Marshall P, Lemieux C. Cleavage pattern of the homing endonuclease encoded by the fifth intron in the chloroplast large subunit rRNA-encoding gene of *Chlamydomonas eugametos*. *Gene* 1991;**104**:241–5.
35. Gimble FS, Thorner J. Homing of a DNA endonuclease gene by meiotic gene conversion in *Saccharomyces cerevisiae*. *Nature* 1992;**357**:301–6.
36. Perlman PS, Butow RA. Mobile introns and intron-encoded proteins. *Science* 1989;**246**:1106–9.
37. Souza DW, Armentano D. Novel cloning method for recombinant adenovirus construction in *Escherichia coli*. *Biotechniques* 1999;**26**:502–8.
38. Gao G, Zhou X, Alvira MR, Tran P, Marsh J, Lynd K, et al. High throughput creation of recombinant adenovirus vectors by direct cloning, green-white selection and I-Sce I-mediated rescue of circular adenovirus plasmids in 293 cells. *Gene Ther* 2003;**10**:1926–30.
39. Colleaux L, D'Auriol L, Galibert F, Dujon B. Recognition and cleavage site of the intron-encoded omega transposase. *Proc Natl Acad Sci USA* 1988;**85**:6022–6.
40. Monteilhet C, Perrin A, Thierry A, Colleaux L, Dujon B. Purification and characterization of the in vitro activity of I-Sce I, a novel and highly specific endonuclease encoded by a group I intron. *Nucleic Acids Res* 1990;**18**:1407–13.
41. Ketner G, Spencer F, Tugendreich S, Connelly C, Hieter P. Efficient manipulation of the human adenovirus genome as an infectious yeast artificial chromosome clone. *Proc Natl Acad Sci USA* 1994;**91**:6186–90.
42. Hokanson CA, Dora E, Donahue BA, Rivkin M, Finer M, Mendez MJ. Hybrid yeast-bacteria cloning system used to capture and modify adenoviral and nonviral genomes. *Hum Gene Ther* 2003;**14**:329–39.
43. Gagnebin J, Brunori M, Otter M, Juillerat-Jeanneret L, Monnier P, Iggo R. A photosensitising adenovirus for photodynamic therapy. *Gene Ther* 1999;**6**:1742–50.
44. Richards CA, Brown CE, Cogswell JP, Weiner MP. The admid system: generation of recombinant adenoviruses by Tn7-mediated transposition in *E. coli*. *Biotechniques* 2000;**29**:146–54.

45. Luckow VA, Lee SC, Barry GF, Olins PO. Efficient generation of infectious recombinant baculoviruses by site-specific transposon-mediated insertion of foreign genes into a baculovirus genome propagated in *Escherichia coli*. *J Virol* 1993;**67**:4566–79.
46. Aoki K, Barker C, Danthinne X, Imperiale MJ, Nabel GJ. Efficient generation of recombinant adenoviral vectors by Cre-lox recombination in vitro. *Mol Med* 1999;**5**:224–31.
47. Hardy S, Kitamura M, Harris-Stansil T, Dai Y, Phipps ML. Construction of adenovirus vectors through Cre-lox recombination. *J Virol* 1997;**71**:1842–9.
48. Ng P, Parks RJ, Cummings DT, Evelegh CM, Sankar U, Graham FL. A high-efficiency Cre/loxP-based system for construction of adenoviral vectors. *Hum Gene Ther* 1999;**10**:2667–72.
49. Ng P, Parks RJ, Cummings DT, Evelegh CM, Graham FL. An enhanced system for construction of adenoviral vectors by the two-plasmid rescue method. *Hum Gene Ther* 2000;**11**:693–9.
50. Tashiro F, Niwa H, Miyazaki J. Constructing adenoviral vectors by using the circular form of the adenoviral genome cloned in a cosmid and the Cre-loxP recombination system. *Hum Gene Ther* 1999;**10**:1845–52.
51. Anderson RD, Haskell RE, Xia H, Roessler BJ, Davidson BL. A simple method for the rapid generation of recombinant adenovirus vectors. *Gene Ther* 2000;**7**:1034–8.
52. Danthinne X. New vectors for the construction of double recombinant adenoviruses. *J Virol Methods* 1999;**81**:11–20.
53. Danthinne X, Werth E. New tools for the generation of E1- and/or E3-substituted adenoviral vectors. *Gene Ther* 2000;**7**:80–7.
54. Fu S, Deisseroth AB. Use of the cosmid adenoviral vector cloning system for the in vitro construction of recombinant adenoviral vectors. *Hum Gene Ther* 1997;**8**:1321–30.
55. Kojima H, Ohishi N, Yagi K. Generation of recombinant adenovirus vector with infectious adenoviral genome released from cosmid-based vector by simple procedure allowing complex manipulation. *Biochem Biophys Res Commun* 1998;**246**:868–72.
56. Okada T, Ramsey WJ, Munir J, Wildner O, Blaese RM. Efficient directional cloning of recombinant adenovirus vectors using DNA-protein complex. *Nucleic Acids Res* 1998;**26**:1947–50.
57. Cullen BR, Lomedico PT, Ju G. Transcriptional interference in avian retroviruses–implications for the promoter insertion model of leukaemogenesis. *Nature*; **307**:241–45.
58. Emerman M, Temin HM. Genes with promoters in retrovirus vectors can be independently suppressed by an epigenetic mechanism. *Cell* 1984;**39**:449–67.
59. Emerman M, Temin HM. Quantitative analysis of gene suppression in integrated retrovirus vectors. *Mol Cell Biol* 1986;**6**:792–800.
60. Bramson J, Hitt M, Gallichan WS, Rosenthal KL, Gauldie J, Graham FL. Construction of a double recombinant adenovirus vector expressing a heterodimeric cytokine: in vitro and in vivo production of biologically active interleukin-12. *Hum Gene Ther* 1996;**7**:333–42.
61. Mizuguchi H, Kay MA, Hayakawa T. In vitro ligation-based cloning of foreign DNAs into the E3 and E1 deletion regions for generation of recombinant adenovirus vectors. *Biotechniques* 2001;**30**:1112–4, 1116.
62. Anderson BD, Nakamura T, Russell SJ, Peng K-W. High CD46 receptor density determines preferential killing of tumor cells by oncolytic measles virus. *Cancer Res* 2004;**64**:4919–26.
63. Gossen M, Bujard H. Tight control of gene expression in mammalian cells by tetracycline-responsive promoters. *Proc Natl Acad Sci USA* 1992;**89**:5547–51.
64. Harding TC, Geddes BJ, Murphy D, Knight D, Uney JB. Switching transgene expression in the brain using an adenoviral tetracycline-regulatable system. *Nat Biotechnol* 1998;**16**:553–5.

65. Corti O, Sabaté O, Horellou P, Colin P, Dumas S, Buchet D, et al. A single adenovirus vector mediates doxycycline-controlled expression of tyrosine hydroxylase in brain grafts of human neural progenitors. *Nat Biotechnol* 1999;**17**:349–54.
66. Bergelson JM, Cunningham JA, Droguett G, Kurt-Jones EA, Krithivas A, Hong JS, et al. Isolation of a common receptor for Coxsackie B viruses and adenoviruses 2 and 5. *Science* 1997;**275**:1320–3.
67. Miller CR, Buchsbaum DJ, Reynolds PN, Douglas JT, Gillespie GY, Mayo MS, et al. Differential susceptibility of primary and established human glioma cells to adenovirus infection: targeting via the epidermal growth factor receptor achieves fiber receptor-independent gene transfer. *Cancer Res* 1998;**58**:5738–48.
68. Pickles RJ, McCarty D, Matsui H, Hart PJ, Randell SH, Boucher RC. Limited entry of adenovirus vectors into well-differentiated airway epithelium is responsible for inefficient gene transfer. *J Virol* 1998;**72**:6014–23.
69. Zabner J, Freimuth P, Puga A, Fabrega A, Welsh MJ. Lack of high affinity fiber receptor activity explains the resistance of ciliated airway epithelia to adenovirus infection. *J Clin Invest* 1997;**100**:1144–9.
70. Tomko RP, Xu R, Philipson L. HCAR and MCAR: the human and mouse cellular receptors for subgroup C adenoviruses and group B coxsackieviruses. *Proc Natl Acad Sci USA* 1997;**94**:3352–6.
71. Bai M, Harfe B, Freimuth P. Mutations that alter an Arg-Gly-Asp (RGD) sequence in the adenovirus type 2 penton base protein abolish its cell-rounding activity and delay virus reproduction in flat cells. *J Virol* 1993;**67**:5198–205.
72. Wickham TJ, Mathias P, Cheresh DA, Nemerow GR. Integrins alpha v beta 3 and alpha v beta 5 promote adenovirus internalization but not virus attachment. *Cell* 1993;**73**:309–19.
73. Wickham TJ, Filardo EJ, Cheresh DA, Nemerow GR. Integrin alpha v beta 5 selectively promotes adenovirus mediated cell membrane permeabilization. *J Cell Biol* 1994;**127**:257–64.
74. Bouri K, Feero WG, Myerburg MM, Wickham TJ, Kovesdi I, Hoffman EP, et al. Polylysine modification of adenoviral fiber protein enhances muscle cell transduction. *Hum Gene Ther* 1999;**10**:1633–40.
75. Gonzalez R, Vereecque R, Wickham TJ, Vanrumbeke M, Kovesdi I, Bauters F, et al. Increased gene transfer in acute myeloid leukemic cells by an adenovirus vector containing a modified fiber protein. *Gene Ther* 1999;**6**:314–20.
76. Wickham TJ, Tzeng E, Shears LL, Roelvink PW, Li Y, Lee GM, et al. Increased in vitro and in vivo gene transfer by adenovirus vectors containing chimeric fiber proteins. *J Virol* 1997;**71**:8221–9.
77. Yoshida Y, Sadata A, Zhang W, Saito K, Shinoura N, Hamada H. Generation of fiber-mutant recombinant adenoviruses for gene therapy of malignant glioma. *Hum Gene Ther* 1998;**9**:2503–15.
78. Dmitriev I, Krasnykh V, Miller CR, Wang M, Kashentseva E, Mikheeva G, et al. An adenovirus vector with genetically modified fibers demonstrates expanded tropism via utilization of a coxsackievirus and adenovirus receptor-independent cell entry mechanism. *J Virol* 1998;**72**:9706–13.
79. Krasnykh V, Dmitriev I, Mikheeva G, Miller CR, Belousova N, Curiel DT. Characterization of an adenovirus vector containing a heterologous peptide epitope in the HI loop of the fiber knob. *J Virol* 1998;**72**:1844–52.
80. Reynolds P, Dmitriev I, Curiel D. Insertion of an RGD motif into the HI loop of adenovirus fiber protein alters the distribution of transgene expression of the systemically administered vector. *Gene Ther* 1999;**6**:1336–9.

81. Mizuguchi H, Koizumi N, Hosono T, Utoguchi N, Watanabe Y, Kay MA, et al. A simplified system for constructing recombinant adenoviral vectors containing heterologous peptides in the HI loop of their fiber knob. *Gene Ther* 2001;**8**:730–5.
82. Koizumi N, Mizuguchi H, Utoguchi N, Watanabe Y, Hayakawa T. Generation of fiber-modified adenovirus vectors containing heterologous peptides in both the HI loop and C terminus of the fiber knob. *J Gene Med* 2003;**5**:267–76.
83. Arap W, Pasqualini R, Ruoslahti E. Cancer treatment by targeted drug delivery to tumor vasculature in a mouse model. *Science* 1998;**279**:377–80.
84. Pasqualini R, Koivunen E, Ruoslahti E. Alpha v integrins as receptors for tumor targeting by circulating ligands. *Nat Biotechnol* 1997;**15**:542–6.
85. Koivunen E, Wang B, Ruoslahti E. Phage libraries displaying cyclic peptides with different ring sizes: ligand specificities of the RGD-directed integrins. *Biotechnology (N Y)* 1995;**13**:265–70.
86. Wickham TJ. Targeting adenovirus. *Gene Ther* 2000;**7**:110–4.
87. Kawabata K, Sakurai F, Yamaguchi T, Hayakawa T, Mizuguchi H. Efficient gene transfer into mouse embryonic stem cells with adenovirus vectors. *Mol Ther* 2005;**12**:547–54.
88. Ulasov IV, Tyler MA, Han Y, Glasgow JN, Lesniak MS. Novel recombinant adenoviral vector that targets the interleukin-13 receptor alpha2 chain permits effective gene transfer to malignant glioma. *Hum Gene Ther* 2007;**18**:118–29.
89. Belousova N, Krendelchtchikova V, Curiel DT, Krasnykh V. Modulation of adenovirus vector tropism via incorporation of polypeptide ligands into the fiber protein. *J Virol* 2002;**76**:8621–31.
90. Ghosh-Choudhury G, Haj-Ahmad Y, Graham FL. Protein IX, a minor component of the human adenovirus capsid, is essential for the packaging of full length genomes. *EMBO J* 1987;**6**:1733–9.
91. Dmitriev IP, Kashentseva EA, Curiel DT. Engineering of adenovirus vectors containing heterologous peptide sequences in the C Terminus of capsid protein IX. *J Virol* 2002;**76**:6893–9.
92. Le LP, Everts M, Dmitriev IP, Davydova JG, Yamamoto M, Curiel DT. Fluorescently labeled adenovirus with pIX-EGFP for vector detection. *Mol Imaging* 2004;**3**:105–16.
93. Meulenbroek RA, Sargent KL, Lunde J, Jasmin BJ, Parks RJ. Use of adenovirus protein IX (pIX) to display large polypeptides on the virion–generation of fluorescent virus through the incorporation of pIX-GFP. *Mol Ther* 2004;**9**:617–24.
94. Sumida SM, Truitt DM, Lemckert AAC, Vogels R, Custers JHHV, Addo MM, et al. Neutralizing antibodies to adenovirus serotype 5 vaccine vectors are directed primarily against the adenovirus hexon protein. *J Immunol* 2005;**174**:7179–85.
95. Roberts DM, Nanda A, Havenga MJE, Abbink P, Lynch DM, Ewald BA, et al. Hexon-chimaeric adenovirus serotype 5 vectors circumvent pre-existing anti-vector immunity. *Nature* 2006;**441**:239–43.
96. Wu H, Han T, Belousova N, Krasnykh V, Kashentseva E, Dmitriev I, et al. Identification of sites in adenovirus hexon for foreign peptide incorporation. *J Virol* 2005;**79**:3382–90.
97. Kurachi S, Koizumi N, Sakurai F, Kawabata K, Sakurai H, Nakagawa S, et al. Characterization of capsid-modified adenovirus vectors containing heterologous peptides in the fiber knob, protein IX, or hexon. *Gene Ther* 2007;**14**:266–74.
98. Scherer LJ, Rossi JJ. Approaches for the sequence-specific knockdown of mRNA. *Nat Biotechnol* 2003;**21**:1457–65.
99. Xia H, Mao Q, Paulson HL, Davidson BL. siRNA-mediated gene silencing in vitro and in vivo. *Nat Biotechnol* 2002;**20**:1006–10.

100. Shinagawa T, Ishii S. Generation of Ski-knockdown mice by expressing a long double-strand RNA from an RNA polymerase II promoter. *Genes Dev* 2003;**17**:1340–5.
101. Hosono T, Mizuguchi H, Katayama K, Xu Z-L, Sakurai F, Ishii-Watabe A, et al. Adenovirus vector-mediated doxycycline-inducible RNA interference. *Hum Gene Ther* 2004;**15**:813–9.
102. Uprichard SL, Boyd B, Althage A, Chisari FV. Clearance of hepatitis B virus from the liver of transgenic mice by short hairpin RNAs. *Proc Natl Acad Sci USA* 2005;**102**:773–8.
103. Mizuguchi H, Funakoshi N, Hosono T, Sakurai F, Kawabata K, Yamaguchi T, et al. Rapid construction of small interfering RNA-expressing adenoviral vectors on the basis of direct cloning of short hairpin RNA-coding DNAs. *Hum Gene Ther* 2007;**18**:74–80.
104. Yamashiro T, Kuge H, Zhang J, Honke K. Calcineurin mediates the angiotensin II-induced aldosterone synthesis in the adrenal glands by up-regulation of transcription of the CYP11B2 gene. *J Biochem* 2010;**148**:115–23.
105. Motegi Y, Katayama K, Sakurai F, Kato T, Yamaguchi T, Matsui H, et al. An effective gene-knockdown using multiple shRNA-expressing adenovirus vectors. *J Control Release* 2011;**153**:149–53.

# Upstream Bioprocess for Adenovirus Vectors

6

*P. Fernandes*[1,2,3], *A.C. Silva*[1,2], *A.S. Coroadinha*[1,2], *P.M. Alves*[1,2]
[1]iBET, Instituto de Biologia Experimental e Tecnológica, Oeiras, Portugal; [2]Instituto de Tecnologia Química e Biológica, Universidade Nova de Lisboa, Oeiras, Portugal; [3]Autolus, London, UK

## 1. Adenovirus Biology

### 1.1 Structure and Genome

Adenoviruses are nonenveloped, icosahedral viruses of 60–90 nm with a linear double-stranded DNA genome of ~36 kbp. The genome can be divided into two sets of transcriptional units: early genes that are expressed before the onset of viral DNA replication and late genes that are preferentially expressed after viral DNA replication (Figure 1). This classification also defines the early and late phase of the infectious cycle. E1A is the first gene to be expressed following infection and is involved in the transcription of the other viral early genes. The mRNAs of the major late transcription unit are grouped into five families (L1–L5) that are dependent on the activation of the major late promoter during viral DNA replication.

### 1.2 Infection and Replication Cycle

The sequential uptake process relies on an initial contact with cellular receptors responsible for attachment[1] and internalization of the virus particle.[2] Internalization occurs by receptor-mediated endocytosis.[3] Once in the cytosol, the virion is transported via microtubuli toward the nucleus. Meanwhile, the particle is dismantled by an ordered elimination of structural proteins so that when it reaches the nuclear membrane, only the core particle is left. Adenovirus uncoating culminates with the release of the viral DNA into the nucleus via nuclear pore complexes.[4] The early genes are responsible for expressing mainly nonstructural, regulatory proteins.[5] These proteins alter the expression of host proteins that are necessary for DNA synthesis, activate other early genes (such as the virus-encoded DNA polymerase), and avoid premature death of the infected cell by the host-immune defenses. During viral genome replication, late-phase transcription is activated.[5] This infection phase is mainly focused on producing capsid proteins and packaging the replicated viral genomes. Structural proteins are assembled into virions, viral DNA is packaged, and viruses are released from the cell as a result of virus-induced cell lysis (Figure 2).

**Adenoviral Vectors for Gene Therapy.**

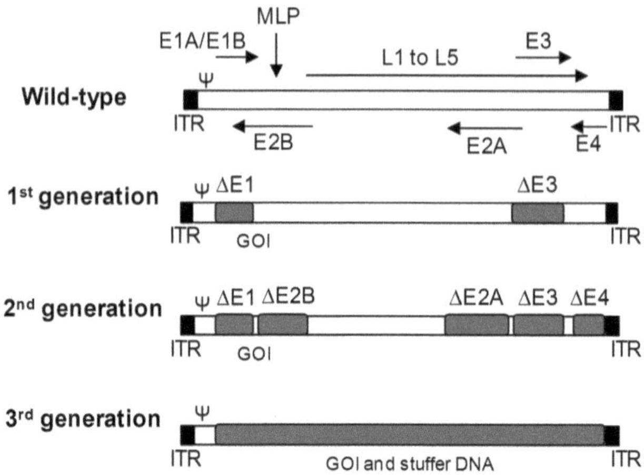


**Figure 1** Schematic representation of wild-type adenovirus genome and the different generations of nonreplicating AdV. E1 to E4, early-region transcript units; L1 to L5, late region transcript unit; ITR, inverted terminal repeats; MLP, major late promoter; Ψ, packaging signal, GOI, gene of interest.

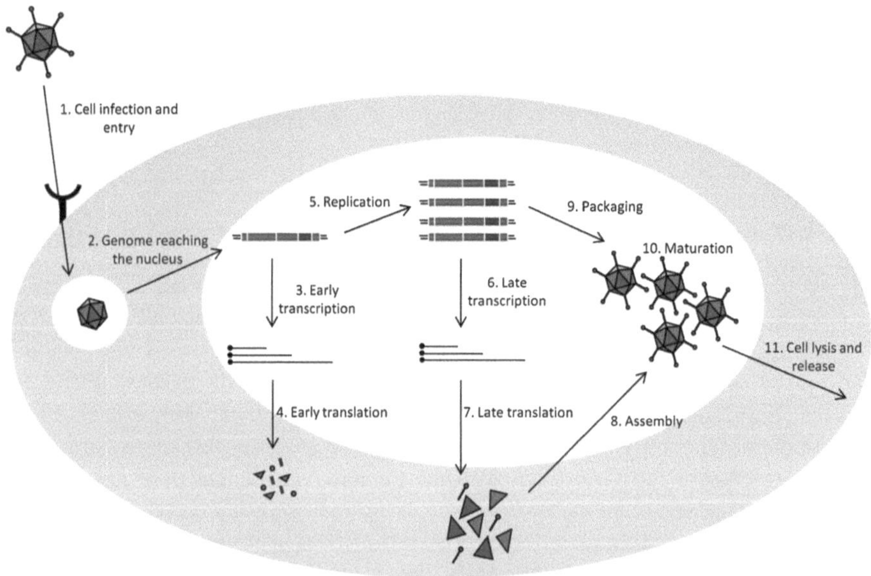


**Figure 2** Adenovirus replication cycle. Briefly, after cell binding (1), the mature viral particle is internalized and transported toward the nucleus (2). During this transport, the particle is dismantled by an ordered elimination of some structural proteins and culminates with the delivery of the viral genome into the nucleus via nuclear pore complexes. Once in the nucleus, early genes are expressed (3, 4) and viral DNA replication starts (5). Late phase of infection is then activated and structural proteins expressed (6, 7). These proteins are likely assembled into empty virions (8), followed by the packaging of viral genome (9). Finally, viral particles are subjected to a maturation process becoming infectious (10) and cell lysis is accomplished (11).

# 2. Manufacturing of Adenovirus Vectors for Gene Therapy

Viral vectors are currently the most efficient tools for in vivo gene transfer. Due to their in vivo efficiency, adenovirus vectors (AdV) are used more often than any other vector in clinical trials (http://www.wiley.com/legacy/wileychi/genmed/clinical/), in which the human adenovirus vector (HAdV) serotypes 5 and 2 are the most characterized ones among all other serotypes of the same family.[6,7] Adenovirus-wide cell tropism in quiescent and nonquiescent cells, its inability to integrate the host genome, and its high production titer[8] make AdV very good candidates for human gene therapy.

## 2.1. Adenovirus Vectors and Producer Cells

The development of AdV is based on modifications of the wild-type genome mainly by removing viral gene(s) and adding expression cassettes with the gene(s) of interest. While these viral genes are essential for virus replication, to propagate AdV it is necessary to establish producer cell lines, that is, cells expressing the viral elements deleted from the vector genome. The AdV that are nonreplicating in a clinical setting and strictly used for gene transfer can be categorized as first-, second-, or third-generation vectors, depending on the extension of viral genes removed (Figure 1). The conditionally replicating vectors are another category of AdV and are designed to specifically target, propagate, and deliver gene(s) in cancer cells.

### 2.1.1 First-Generation Vectors

The majority of AdV used for gene therapy or many other therapeutic purposes are nonreplicating, in which the E1 region is deleted from the genome often in combination with E3, providing space for the insertion of expression cassettes. Such adenovirus vectors are denominated as first-generation or E1-deleted (ΔE1) vectors. The E1 region can be divided in E1A and E1B subunits and codes for products involved in the activation of other early and late gene expressions and in making host cells more amenable to initiate virus propagation (i.e., inhibition of antiviral response and apoptosis).[9–15] Therefore, to produce ΔE1 vectors a producer cell line containing adenovirus E1 sequences is necessary to complement these functions. On the other hand, the E3 region, which is involved in antagonizing host defense mechanisms, is not essential for viral amplification in vitro.[16] Together, the deletion of E1 and E3 permits a transgene insert capacity of up to 8.2 kb. HEK 293 represents the traditional cell line used to trans-complement the lack of E1 and produce AdV. Because of the significant homology of first-generation adenovirus vectors with the HEK 293 DNA, the main disadvantage in using HEK 293 is its potential to generate replication-competent adenoviruses (RCA), raising safety concerns in a therapeutic product.[17–21] To avoid this, several cell lines were established under the rational of reducing the homology of viral DNA sequences incorporated in cell lines with viral vector genomes (reviewed in Kovesdi and Hedley).[22] Following this effort, PER.C6 represents the first cell line in which the generation of RCA is abolished and typical adenovirus yields are ensured.[23]

The HEK 293 cell line was generated with sheared HAdV-5 DNA; however, the current strategy for generating producer cell lines for E1 complementation is through the incorporation of the contiguous sequences of E1A and E1B into the cell line genomes. Strategies to minimize any recombination between viral genome and cell DNA were further developed and involve the integration of E1A and E1B into the producer cell genome at separate locations.[24,25] E1-complementing cell lines are the backbones for the remaining producer cells used for other adenovirus vectors.

### 2.1.2 Conditionally Replicating Vectors

Conditionally replicating adenovirus vectors, also denominated oncolytic adenoviruses, have been employed for cancer treatment to specifically target and replicate in cancer cells.[26–28] The lytic nature of adenovirus replication directly kills tumor-infected cells, releasing the associated antigens. On the other hand, progeny viruses can be spread throughout a tumor, further infecting and destroying other cancer cells.

The construction of oncolytic adenoviruses is based on the deletion or modification of viral gene functions that are critical to viral replication in normal cells, but dispensable in tumor cells. This includes insertion of mutations in the E1A region,[29,30] or deletion of E1B[31,32] from the wild-type genome, to target cancer cells with defects in the retinoblastoma (Rb) and p53 pathways, respectively, as these pathways are defective in most human tumors. Another strategy for targeting oncolytic adenoviruses replication in cancer cells involves the control of E1 transcription by using tumor- or tissue-specific promoters such as prostate-specific enhancer/promoter for prostate cancer,[33] or E2F-I for cancer cells with a defective Rb pathway.[34] In addition to these modifications, these vectors can further incorporate genes encoding immune stimulatory factors to boost the antitumor immunity (reviewed in Choi and Yun),[28] giving rise to the so-called armed vectors and to one of the most promising gene delivery systems for cancer therapy. Production of oncolytic adenoviruses is similar to that of first-generation vectors. Cell lines already established for first-generation vectors are used for the manufacture of these vectors and include HeLa,[35] A549,[36] HEK 293, and PER.C6.[37]

### 2.1.3 Second-Generation Vectors

Although the removal of the E1 region renders the virus replication defective, the delivery of high doses of first-generation vectors and/or the presence of E1-like factors in many cells can lead to the expression of other viral proteins in vivo.[22,38–40] This can induce a strong immune response, reducing the efficacy of these vectors. To circumvent that, further deletions in E2 and/or E4 regions were explored, leading to the establishment of second-generation vectors. The E2 region (E2A and E2B subunits) codes for three proteins essential for viral replication: DNA-binding protein transcribed from E2A subunit, terminal protein, and viral DNA polymerase from E2B subunit.[41] E4 products modulate transcription, the cell cycle, cell signaling, and DNA repair and are essential for productive virus infection,[42] but only one of the ORF3 or ORF6 is required for successful virus production in cell culture.[43,44]

Given the toxicity associated with E2 and E4 viral products, most of the cell lines for second-generation vectors rely on the use of an inducible system (reviewed in Kovesdi and Hedley).[22] Similar to first-generation vectors, the majority of E1/E2 and E1/E4 complementary producer cells lines use HEK 293 as the parental cell line.

Despite the sophisticated systems available for second-generation vector manufacturing, including vector construct and incorporation of inducible systems in the producer cell lines, the use of these vectors remains without advances. In general, reduced production yields are obtained for vectors with multiple deletions. In fact, when compared to first-generation vectors, the yields of these vectors can be reduced, and no major improvement in toxicity is observed in vivo.[22,45–50] Moreover, the transgene expression of these vectors is also described as unstable, which probably made third-generation vectors a more suitable choice for increased safety and long-term transgene expression. (reviewed in Segura et al.).[51]

### 2.1.4 Third-Generation Vectors

Third-generation vectors, also known as gutted or gutless, high-capacity, or helper-dependent vectors (HDVs), are the most advanced AdV. They are devoid of all viral coding genes, only harboring on their genome the essential *cis*-acting elements: inverted terminal repeats and packaging signal. This allows the insertion of therapeutic gene or genes up to ~37 kb. Additional stuffer DNA is included to render vector genome size similar to that of the wild type and maintain viral particle stability.[52] Apart from their increased safety, the use of these vectors results in long-term transgene expression (reviewed in Segura et al.).[51] To produce these vectors, while E1 functions are provided by the producer cell line, the remaining functions are provided by a helper vector (HV). In fact, the need for an HV increases the complexity of the production system, and it is the main disadvantage of these type of vectors. Adding HV to the production system implicates the production of both HDV and HV, raising the need to remove HV contamination from the final product. To date, the most elegant way to prevent HV propagation has been the approach described by Parks et al. in which the HV packaging signal is flanked by loxP sites, and under the expression of Cre recombinase the HV genome is cleaved (hampering its encapsidation) and production of HV particles is minimized.[53] Thus, besides E1, producer cell lines for HDV must further express Cre recombinase. Following the same approach, the FLPe recombinase system is also used to minimize HV contamination.[54,55] Similar to E1 trans-complementing cell lines, the most common producer cell lines using recombinase system are derived from HEK 293 cells, although PER.C6 cells are also used for that purpose (reviewed in Kovesdi and Hedley).[22] Despite these efforts in the HDV production system, two major bottlenecks are still found when considering the use of HDV in patients: (i) HV contamination is still not fully eliminated and (ii) HDV production yield still faces the need for multiple amplification steps and inconsistency in infectious particles yields (discussed in Section 2.3). Despite these bottlenecks, because of the advantages of increased cassette incorporation, improved expression, and lower immunity it is worth promoting their further development.[51]

### 2.1.5 Novel Adenovirus Vectors

While most research on vector development is based on the utilization of human adenovirus serotypes 5 and 2, over 80% of the adult population have been naturally exposed to these viruses.[56] Therefore, preexisting humoral and cellular immunity may preclude efficient gene transfer when these AdV are used.[57–62] Apart from limitations in therapeutic efficacy,[63] immune responses against the vector may result in a number of undesirable side effects, including liver toxicity[64] and systemic inflammatory response syndrome due to repeated vector administration.[65] Eliminating liver tropism and the epitopes involved in viral proteins recognition by neutralizing antibodies has been proposed to reduce the immune response. Alternatively, vectors based on low seroprevalence AdVs can be also be used to circumvent the preexisting immunity. In addition, the diversity of AdV protein isoforms and their variety of ligand–receptor interactions found in the different serotypes can also be the basis to target different cell types.[62] Adenovirus vectors derived from alternative human and nonhuman serotypes, to which the human population has a lower or no prevalence of neutralizing antibodies, are currently being investigated.

Human subgroup B adenoviruses, and in particular serotypes 11 and 35, are being used to develop AdV (reviewed in Kovesdi and Hedley).[22] Typically, E1 (both E1A and E1B)-deleted vectors from these serotypes cannot replicate in regular E1 producer cell lines already established for serotype 5 vectors, although E1A-deleted vectors can be propagated in PER.C6 cells.[66] These vectors are therefore produced by modifying the typical first-generation cell lines already available[67–70] or replacing vectors genes by HAdV5 sequences to permit viral propagation on unmodified cell lines, such as PER.C6.[71]

Several nonhuman adenoviruses derived from bovine, simian, porcine, ovine, murine, and canine sources have been used as backbone to develop E1-deleted vectors for gene therapy or vaccine purposes (reviewed in Lopez-Gordo et al.).[62] Table 1 summarizes the main nonhuman adenoviruses used as vectors and the corresponding cell lines. For the majority of nonhuman vectors, specialized E1 producer cells have been developed to propagate these vectors, such as bovine kidney or fetal retinal cells expressing bovine or human adenovirus E1 sequences,[72] or canine kidney cells expressing the canine adenovirus type 2 (CAV-2) E1.[73,74] Chimpanzee-derived vectors of subgroup E can be produced on HEK 293 cells already developed.[75,76] However, similar to HAdV11 and HAdV35, chimpanzee-derived vectors of subgroup B cannot be propagated on these cells, and a chimeric strategy is used to allow propagation in HEK 293 cells.[77,78] CAV-2-derived vectors are probably the best described and advanced nonhuman vectors.[79] The ability to preferentially transduce neurons combined with a remarkable capacity of axonal transport makes CAV-2 vectors candidates for the treatment of neurodegenerative diseases.[80] Furthermore, third-generation CAV-2 vectors were already developed and tested in animal models, confirming their potential for gene transfer-based therapies.[81,82] In accordance, new producer cell lines[74] and scalable bioprocess[83,84] were recently developed to facilitate the manufacture of CAV-2 vectors at larger scales and the regulatory approval for clinical grade production.

**Table 1 Adenoviruses Used as Vectors and the Corresponding Cell Lines Used for Vector Production[22,73–78,83,126–144]**

| Host | | Serotype | Species | Genus | Main Producer Cell Lines |
|---|---|---|---|---|---|
| Human | | HAdV-5, HAdV-2 | Human adenovirus C | *Mastadenovirus* | HEK 293, PER. C6 (reviewed in (22)) |
| | | HAdV-11, HAdV-35 | Human adenovirus B | | |
| Nonhuman | Canine | CAdV-2 (or CAV-2) | Canine adenovirus A | | DK (73), MDCK (74, 83) |
| | Bovine | BAdV-3 | Bovine adenovirus B | | FBRT (126, 127), MDBK (128) |
| | Porcine | PAdV-3 | Porcine adenovirus A | | FPRT (129-131) |
| | Simian | SAdV-21* | Human adenovirus B | | HEK 293 (75-78, 132) |
| | | SAdV-22 to SAdV25 | Human adenovirus E | | |
| | | SAdV-7 | Human adenovirus G | | |
| | Murine | MAdV-1 | Murine adenovirus A | | L929 (133), A549 (134), 3T6 (135, 136) |
| | Fowl | FAdV-1 | Fowl adenovirus A | *Aviadenovirus* | CH-SAH (137), LMH (138-141) |
| | | FAdV-10 | Fowl adenovirus C | | |
| | | FAdV-9 | Fowl adenovirus D | | |
| | | FAdV-8 | Fowl adenovirus E | | |
| | Ovine | OAdV-7 | Ovine adenovirus D | *Atadenovirus* | CSL503 (142-144) |

HEK293, human embryonic kidney cells; PER.C6, human retinal cells; DK, dog kidney; MDCK, Madin–Darby canine kidney; FBRT, fetal bovine retinal cells; MDBK, Madin–Darby bovine kidney; FPRT, fetal porcine retinal cells; L929, murine fibrosarcoma cells; A549, carcinomic human alveolar basal epithelial cells; 3T6, murine embryonal fibroblast cells; CH-SAH, chicken hepatoma cells; LMH, chicken hepatocarcinoma cells; CSL503, sheep fetal lung cells; * chimera.

## 2.2 Upstream Process for Adenovirus Vectors

An adenovirus production process starts by growing the cell line of choice to the desired cell density for infection followed by the inoculation of an adenovirus stock to initiate the infection and virus production cycle (Figure 3).[85] The need for significant amounts of clinical grade AdV, which in some cases may reach

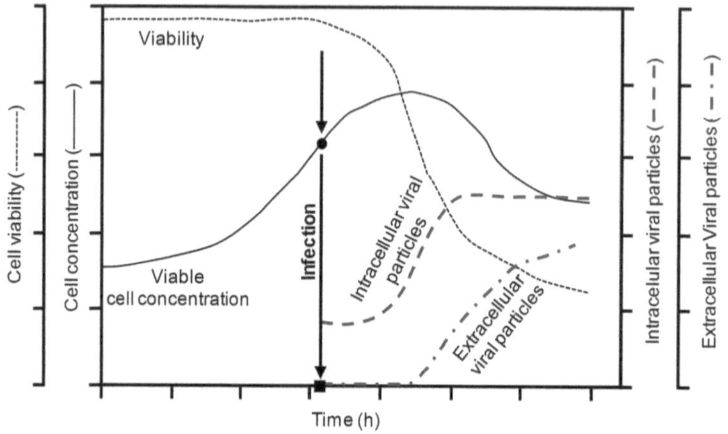


**Figure 3** Typical profile of an adenovirus production process, including cell growth, viability, and infection kinetics. After inoculation, cells grow exponentially and are usually infected at the end of exponential growth phase (represented by an arrow). Viral particles are assembled intracellularly. Due to the lytic nature of adenovirus propagation, cell lysis is induced after the production of viral particles. As a result, cell concentration and viability decrease and viral particles are released to the supernatant. (●) Cell concentration at infection (CCI); (■) time of infection (TOI). (Adapted from Altaras et al.[85])

$10^{13}$ total particles/patient ($10^{11}$ infectious particles/patient), requires efficient and robust processes for production and purification at a large scale compliant with good manufacturing practices (GMP).[8] From a scalable point of view, the production of AdV encompasses several bioengineering aspects that must be carefully undertaken to maximize bioprocess yields and reduce production costs. In this section, process considerations for the adenovirus production will be discussed, mostly focusing on first-generation vectors, as the main bioprocess advances have been achieved with these vectors.

### *2.2.1 Cell Culture and Adenovirus Production Process*

Adenovirus production processes can be performed in static or stirred cultures. From a scalable point of view, stirred cultures are preferred. The use of microcarriers represents the most common approach for transferring adherent cells to stirred culture systems. The initial steps of microcarriers preparation and cell seed from adherent cultures represent the main drawback when using this system,[86] since cell manipulation is cumbersome when working at larger scales. Most of the production processes for adenoviruses reported in the literature use cells growing in suspension (reviewed in Silva et al.).[87] Although adaptation of cells to grow in suspension can be time-consuming at the expense of lowering cell-specific productivity, suspension cell lines are preferred as they greatly facilitate large-scale productions. Furthermore, the adaptation of cells to suspension involves transference to serum-free culture medium, which represents an additional advantage for biopharmaceutical production.

In general, the medium is first chosen and optimized for its ability to support cell growth. During the infection and vector production phase, the specific consumption rate of several medium components is increased, and must be taken into account when defining the final medium formulation. Also, cells cultured with media containing animal serum should be avoided. The undefined composition and high batch-to-batch variability of serum, together with its potential source of contaminations raises safety concerns and hinders the standardization of cell culture processes for the production of biopharmaceuticals.[88] Almost all media manufacturers commercialize serum-free formulations designed and optimized for specific cell lines and/or final product.

The majority of scalable adenovirus manufacturing processes are performed in stirred-tank bioreactors both for microcarriers and for suspension cultures.[87] Technological advances in disposable equipments, such as single-use Wave and stirred-tank bioreactors, have been shown by several companies (e.g., GE Healthcare, Sartorius, PBS). The use of such systems poses several advantages for GMP production by avoiding the need for cleaning and sterilization validation and alleviating facility requirements. Therefore, the increasing use of disposable equipment for adenovirus manufacturing is anticipated.

### 2.2.2 Adenovirus Seed Stocks

AdV viral particles must be rescued from the initial vector genome construct. This is usually performed by linearization of vector genome from the plasmid, and transfection of producer cells. Viruses are then harvested once cytopathic effect (CPE) is evident or, as alternative, 2–3 days after transfection. This procedure is typically performed under static culture conditions. When working with producer cells easily transfectable and first-generation vectors, the rescue of viral particles after this transfection step is relatively high, and one or two more production rounds are performed to amplify the amount of viral vectors and establish a purified viral seed stock. In such cases, scalable production processes are usually established once a purified viral seed stock is obtained. When developing viral vectors that typically present low productivities, such as HDV, process development is defined from the transfection step to the final viral seed stock production to maximize working-scale and volumetric productivity in all steps and minimize the number of amplifications required to produce sufficient viral material[85] (discussed in Section 2.3.).

For GMP production, the purified viral seed stock must be certified and tested to confirm the absence of adventitious agents (see Section 3.1). On the other hand, the use of well-established purified viral seed stocks is advised even for research purposes as it ensures reproducibility between different production batches. Therefore, determination of viral particles, physical to infectious particles ratio, and selection of best storage conditions are carefully undertaken to ensure that the properties of particles from the viral seed stock are maintained.[89–93]

### 2.2.3 The Infection Process and Harvest Strategy

The production process starts once producer cells are growing and then infected afterward and the newly produced adenoviruses are harvested at the end of the process (Figures 2 and 3). Defining the best process-related parameters, such as multiplicity of infection (MOI), cell concentration at infection (CCI), time of infection (TOI), and time of harvest (TOH), is the determinant to ensure an optimal production process. Because adenovirus infections are relatively fast and lytic, processes are established as single-round infections with MOI>1 to ensure that all cells are infected. Although MOI>1 needs to be established, infecting the cells with more virus than those optimally required may have a negative effect on cell viability, compromising virus productivity. TOI or CCI is usually selected when cells are at the exponential growth phase. For adherent, static cultures, cells are infected when 60–80% confluence is achieved, while for stirred cultures the typical cell density range used for adenovirus production is $0.5–1 \times 10^6$ cells/mL in batch mode with medium exchange at infection. To produce adenoviruses above this cell density and maintain cell-specific productivity, fed-batch or perfusion modes must be added to the bioprocess (discussed in Section 2.2.5.).

After infection, the cell growth is arrested within 24 hpi. Viral DNA replication and virus assembly occur between 10–24 hpi and 20–48 hpi, respectively. As virus production progresses, cell viability decreases after 24 hpi. Typically, at 48 hpi, when the volumetric virus productivity reaches its plateau, the cell viability is around 40–80%. At this point, a percentage of the virus (between 10 and 50%) has already been released from lysed cells, but the rest of the virus remains intracellular. The cultivation process can proceed further with no significant increase in virus production, but with a significant increase in virus found in the culture medium. Harvest timing and method can be tailored to the entire bioreaction bulk, or only to the intra- or extracellular fractions. The remaining cells are lysed either by freeze–thaw cycles[94,95] or using a detergent.[96,97]

### 2.2.4 Productivity of Adenovirus Vector Manufacturing

One of the most attractive features for AdV, namely ΔE1-deleted vectors, is the high production yields obtained when compared to other viral vectors. Typically, titers of ΔE1-deleted vectors range from $10^3$ to $10^4$ IP per cell, whereas genome-containing particles are one log higher (reviewed in Silva et al.).[87] Considering the usual CCI used for adenovirus production, this corresponds to a volumetric productivity of $10^9$–$10^{10}$ IP/mL or $10^{10}$–$10^{11}$ PP/mL. It is also worth noting that such yields are dependent on AdV construct and culture conditions. Some production processes for HDV are described as holding relatively low specific ($10^2$ IP/cell) and volumetric ($10^8$ IP/mL) titers[98,99] or inconsistent PP:IP ratios[100]. In addition, cell-specific productivity under batch mode is limited to CCI below $1 \times 10^6$ cells/mL; producing adenoviruses above this cell concentration results in lower specific yields (see Section 2.2.5). Finally, a cell's physiological state can also impact the infectivity of the newly produced adenoviruses. Indeed, when producing adenoviruses with MOI higher than the optimal, the resulting PP:IP ratio tends to increase.

### 2.2.5 Adenovirus Production at High Cell Density

In a batch operation mode, the cell density at infection is a very important parameter as it impacts virus production. In fact, the narrow range for an optimal infection cell density between 0.5 and $1.0 \times 10^6$ cells/mL is well documented (reviewed elsewhere).[85] A drop in specific productivity at higher cell densities occurs even though medium formulations allow cell growth up to $5 \times 10^6$ cells/mL. This is often referred to as the "cell density effect." While this phenomenon is not totally understood, the decrease in productivity is thought to be due to limitations in a nutrient critical for virus propagation, accumulation of inhibitory by-products, or a combination of both.[101] Following this hypothesis, the typical approach for maintaining cell-specific productivity at increased cell densities consists of exchanging medium at infection time. This operation mode, widely used for viral vector production, enables maximum cell-specific productivity up to $2 \times 10^6$ cells/mL. The main limitation of such approach is the incorporation of a medium exchange/cell separation step. While at scales of few liters, this is relatively easy to perform by centrifugation, when dealing with hundreds or thousands of liters such manipulation is highly cumbersome. At these scales, fed-batch or perfusion are more feasible approaches. Several fed-batch approaches were devoted to adenoviral production based on the control or addition of glutamine, glucose, and amino acids.[102–104] Despite the simplicity of implementation, most of the fed-batch strategies toward maintaining cell-specific productivity have failed.

Perfusion processes are capable of achieving high cell densities, through a continuous renewal of fresh medium, providing new nutrients and diluting/removing inhibiting by-products. Therefore, perfusion must be such that nutrient supply and by-product removal rates are sufficient to ensure robust cell performance at increasing cell concentration. The perfusion rates described for adenovirus production can range from 0.5 to 3 culture volumes per day (V/day), with 2 V/day the rate mostly used.[35,105–107] When compared to low cell density batch productions, specific productivity of HEK 293 cells infected at $3–6 \times 10^6$ cells/mL was maintained, which constituted a 5-fold increase in the final production yield.[105,107] The highest cell density tested to successfully produce adenoviruses under perfusion mode is described in Yuk et al.[35] using Hela cells at $10^7$ cells/mL.

Other efforts were made to better understand the cell density effect. The specific metabolic demands of producer cells during growth and virus production have been analyzed through a metabolic flux analysis approach, showing that a favorable metabolic state for adenovirus production should have an increase in glycolytic and TCA fluxes,[107,108] and in ATP production rates on infection. This state is also extended to perfusion modes.[107] Furthermore, it was demonstrated that a decrease in the proportion of cells in the S phase was related to a decrease in specific productivity at high cell densities.[109] Following this observation, the synchronization of HEK 293 cells chemically or by lowering temperature led to an enrichment of cells in the S phase and up to a 7.3-fold increase in AdV cell-specific titer.[110] Despite these insights, perfusion modes still remain the best strategy for successfully producing AdV at high cell densities.

### 2.2.6 Physical Parameters and Process Monitoring

Physical parameters such as temperature, pH, and dissolved oxygen (DO) are usually well established when manufacturing AdV at larger scales. The impact of such parameters on AdV production has been previously reviewed[85,87]; therefore only optimal values used to produce AdV are discussed.

Production of adenoviruses is typically set to a cell growth temperature of 37 °C, although previous works showed that by decreasing the temperature to 35 °C an improvement in virus production can be obtained. The common pH values for adenoviruses have been established as 7.2–7.3. On the other hand, production of canine adenoviruses in a stirred-tank bioreactor at pH 7.4 showed production yields similar to those of human adenoviruses. Despite the lack of reports addressing the effect of DO on adenovirus production, most of the controlled processes set DO to values higher than 30%.[86,87]

Small-scale productions are typically monitored with offline sampling. While such offline methods can be used to monitor some parameters, such as cell concentration and pH, online methods are very useful for stirred-tank bioreactor production at a large scale, especially to select TOI and TOH. The cell/infection status can be monitored through a variety of noninvasive online measurements. In particular, cell density can be estimated by the oxygen uptake rate and the capacitance levels,[107,111,112] as they can also be used to monitor infection kinetics and confirm the maximum productivity point and harvest timing.

## 2.3 Production Process for Helper-Dependent Vectors

Despite the potential of HDV, extensive use of such vectors for gene transfer experiments has been restrained by two main bottlenecks: production yields and HV contamination. Similar to first-generation vectors, HDV particles must be rescued from linearized plasmid after transfecting producer cells. To provide all the viral elements required for HDV replication, cells must also be infected with HV (see Section 2.1.3). Typically, low yields from this initial step are obtained and/or multiple rounds of HDV production are needed until the desired vectors titer is achieved. Reducing the number of amplifications is a primary condition for establishing a robust production process. This also minimizes the possibility of recombination between HV/HDV and producer cell line, avoiding the occurrence of RCA, packaging-competent or recombinant HV, which have a propagation advantage when compared to HDV.[113,114] Therefore, the most significant advances for scalable production of HDV are designed early in the transfection step to maximize volumetric productivity and set amplification rounds to its minimum.[99]

The importance of a recombinase system, such as Cre/loxP, with high levels of Cre to avoid HV propagation is unquestionable.[115] In fact, considering that the remaining HV contaminant was due to limiting Cre levels that permitted HV to escape packaging signal excision,[115] major advances were made to increase the levels of recombinase during HDV production.[100] Moreover, the definition of optimal MOI ratio showed a critical impact either in maximizing HDV propagation or in reducing HV contamination. The use of high HV MOI, besides being unnecessary and even negative for HDV production,[98] implicates the development of alternative designs to attain higher levels of Cre than those supported by the cell line to reduce HV contamination.[100] Further, this also leads to the accumulation of higher levels of excised helper DNA molecules (after Cre recombinase excision), increasing the chance of recombination between viral vectors. In fact, some authors showed that these helper DNA molecules, being prone to rearrangements, contributed to the generation of recombination between viral vectors, in which HV with rearranged genomes had a growth advantage.[114] In addition, some studies show a relatively high and/or great inconsistency in maintaining PP:IP ratios among different preparations of the same HDV.[100,116] Special attention must be paid to this, as the PP:IP ratio of clinical grade adenoviruses is limited by Food and Drug Administration (see Section 3.4.).

## 2.4 Downstream Process for Adenovirus Vectors

Since downstream processing is not in the scope of this chapter, only a brief overview is presented. Traditional purifications of AdV are performed by cesium chloride (CsCl) density gradient ultracentrifugation. While useful for preparations at laboratory scales, scalable purification of AdV relies on membrane and chromatographic processes.[85,117,118] The main aim of downstream processing is to eliminate contaminants, either process related (e.g., bovine serum albumin, Benzonase, extractables, and leachables) or product related (e.g., host cell proteins, DNA, proteoglycans, and glycosaminoglycans); other product-related impurities include free proteins, aggregates, and empty capsids.[119] The ultimate goal is to obtain a product with high purity, potency, and quality, which can meet the stringent guidelines of the regulatory authorities, such as the FDA and the EMA. After harvesting and prior purification two main steps are employed to production bulk: cell lysis, to release intracellular adenoviruses and increase the yield, and genomic DNA breakdown, to facilitate DNA removal. Purification can then be divided into three major steps: clarification, concentration/purification, and polishing. A suitable clarification step should remove cell debris and large aggregates and can be performed by centrifugation, widely used at laboratory scales, continuous flow centrifugation, and/or microfiltration at industrially relevant scales.[85,117,120,121] The concentration/purification step aims to reduce the stream volume and facilitate the upfront equipment and materials. In this step, low-molecular-weight proteins, fragmented DNA, and other impurities are further removed and ultrafiltration or chromatography columns are usually employed.[85,117,122] The step of polishing is usually performed using chromatographic processes and applied to remove remaining impurities closely related to the product of interest.[85,122] The final step for the production of GMP-grade adenovirus product is a filtration using a 0.2 μm sterile membrane. In the final product, host cell DNA and protein levels should be below the specifications set by the European Pharmacopoeia (Ph. Eur 5.2.3) and the World Health Organization (WHO Expert Committee on Biological Standardization, WHO Technical Report Series 878, 47th report, 1998).[118]

# 3. Concerns in the Manufacturing of Adenovirus Vectors for Clinical Product Release

For vaccines intended for human use, and before the release of adenovirus vectored products, tests including those for potency, general safety, sterility, purity, identity, and constituent materials are performed. In addition, screenings for bacterial endotoxins and pyrogens are added when the product is intended for use by injection (US. Pharmacopeial Convention and USP-NF bulletin, http://www.usp.org). Every biopharmaceutical has its own characteristics that are considered when developing and qualifying these tests. Therefore this section is focused on the tests used in the manufacturing process of AdV extraneous agents, RCAs, vector genetic stability, and quantity/potency.

## 3.1 Extraneous Agents

Viral inactivation and clearance steps are generally performed in the production of inactivated vaccines. These steps help in removing potential extraneous agents from the final product. However, these procedures cannot be applied in AdV for gene therapy, as their bioactivity is critical for gene transfer. Thus safety strategies need to be implemented to lower the risk of introduction and carryover of contaminants, such as extraneous agents, adventitious viruses, and transmissible spongiforme encephalopathy-causing agents. According to good manufacturing practices, a track record of all the materials used in the construction and manufacture of AdV must be kept. Master cell banks and master viral seeds must be extensively tested to confirm the absence of adventitious viruses according to the ICH Q5A and Q5D guidelines. The use of animal-derived components should be avoided as much as possible, as it represents a potential source of contamination.[88] In situations where the use of animal components is unavoidable, excellent traceability and testing are performed to discard any risk of entry and transfer of extraneous agents to the manufacturing process. Screening for adventitious agents has relied on the use of in vitro infectivity assays, in vivo studies, and specific polymerase chain reaction (PCR) tests. More recently, massive parallel sequencing has also been applied to this end.[123]

## 3.2 Replication-Competent Adenovirus

Contamination of clinical batches with RCA is an important safety concern. Possible consequences include increased local inflammatory responses and tissue damage due to uncontrolled systemic replication in immune-compromised individuals. Although the issue of RCA formation by homologous recombination was solved by the introduction of new cell lines such as PER.C6 (see Section 2.1), regulatory agencies still require testing to confirm their absence in clinical batches.[118] The maximum contamination level is set to one RCA per $3 \times 10^{10}$ VP and is based on the current FDA guidelines for HAdV5 vectors (FDA Gene Therapy Letter, 2000). The detection of replicating virus is based on the screening for a CPE after inoculating noncomplementing cell lines. CPE-positive results are then confirmed by a more specific assay such as PCR or immunofluorescence.[124]

## 3.3 Vector Genetic Stability

To confirm product stability throughout the manufacturing process, the genetic stability of the viral vector is tested. To demonstrate this, viral vectors are propagated for a number of passages beyond the level used in production, generally five passages (more information at US. Pharmacopeial Convention and USP-NF bulletin, http://www.usp.org). Mutation frequency in replication-deficient adenovirus is considered rare; however, such analysis is important for evaluating that transgene region and corresponding expression are maintained, and critical when considering the use of oncolytic adenoviruses, which, unlike gene transfer-strict vectors, are intended to further propagate in a clinical setting.[37] Extended propagation permits the detection of any recombinant or mutant vector that has a replication advantage over the target vector. For this purpose, PCR detection combined with sequence analysis of PCR fragments can be used for screening any variations in transgene sequence that might occur.[118]

### 3.4 Quantity and Potency

Quantification of viral particles is important for monitoring yields during process development and also in the final product to control the amount of viral protein injected within an acceptable safety window. Typically, absorbance measurements are used to quantify physical particles based on the correlations between adenovirus preparations and absorbance at 260 nm described by Maizel et al.[145]. While these measurements require purified and concentrated preparations, alternative methods have been developed to allow adenovirus quantification during process development and, simultaneously, improve the limit of detection.[85]

Adenovirus vectors exert their clinical effect by transducing target cells. Therefore, quantification of infectious particles is required during the overall production process, for product release, and stability assessments of viral preparations. In fact, a parameter considered to represent the quality of preparations is the ratio between the amount of physical and infectious particles. For gene therapy programs using adenoviruses, the PP:IP ratio must be $<30:1$ (FDA Gene Therapy Letter, 2000).

Standard plaque assays or tissue culture infectious dose ($TCID_{50}$) assays using complementing cell lines can be employed to quantify infectious titers. To improve the accuracy and precision of these assays, alternative methods can be applied that are based on quantitative PCR techniques or on the detection of the transgene expression.

## 4. Conclusion and Future Directions in Adenovirus

Bioprocess the development of a high-yield AdV production process requires an integrated approach, in which the interplay of producer cell line specifications, cell culture characteristics, bioprocess parameters, and viral construction should be considered to fully maximize AdV production and clinical efficiency.

While bioprocess advances have been traditionally focused on first-generation HAdV5, special attention is now being paid to the next generation vectors, including alternative human and nonhuman serotypes or chimeras. Intrinsically linked to new vectors is the design of complementing cell lines suitable for production. In some cases, this implicates the development of specialized producer cell lines. Alternatively, when developing new vectors (namely those from human serotypes), it is worth considering a chimeric approach, that is, incorporate further vector modifications that would allow manufacturing on the already established producer cell lines. This would accelerate vector development and overcome licensing/regulatory considerations when using new cell lines. Regardless of the strategy used, the ideal production system must be designed to fulfill both vector attributes (appropriate tropism, low PP:IP ratio, limited preexisting immunity in the target population) and vector productivity. Also included in the next generation vectors are HDVs. Despite some effort, the large-scale manufacturing is considered underdeveloped, possibly because HDV productivities and PP:IP ratios are still difficult to manage from a bioprocess and clinical point of view. Nevertheless, considerable progress in reducing HV contamination[98,100] and reduce HDV amplification steps[99] has been recently achieved. Moreover, understanding bottlenecks from an adenovirus life cycle perspective would be an important approach for identifying and hopefully overcoming the current limitations found in the manufacturing of these vectors.

The currently disposable and single-use bioreactors available are key process modifications that will contribute significantly in lowering the cost of goods and streamline adenovirus production processes. On the other hand, final manufacturing processes and costs are dependent on the success of adenovirus production at high cell density, which still remains underdeveloped. A better understanding of viral amplification and cell physiology will contribute to the elaboration of an improved viral cell system for high viral-specific productivities at high cell density. For instance, increasing knowledge of the metabolic requirements for cell growth and viral production by metabolic flux analysis might allow improvements of viral productivities and surpass the "cell density effect" by the manipulation of energy metabolism, as already shown for other systems.[125]

## Acknowledgments

This work was supported by Fundação para a Ciência e a Tecnologia (FCT)–Portugal, through the projects PTDC/EBB-BIO/118615/2010 and PTDC/EBB-BIO/119501/2010, and European Commission through FP7 project BrainCAV (222992). Paulo Fernandes and Ana Carina Silva acknowledge FCT for the Ph.D. Grants SFRH/BD/70810/2010 and SFRH/BD/45786/2008, respectively.

## References

1. Soudais C, Boutin S, Hong SS, et al. Canine adenovirus type 2 attachment and internalization: coxsackivirus-adenovirus receptor, alternative receptors, and an RGD-independent pathway. *J Virol* 2000;**74**:10639–49.
2. Wickham TJ, Mathias P, Cheresh DA, Nemerow GR. Integrins alpha v beta 3 and alpha v beta 5 promote adenovirus internalization but not virus attachment. *Cell* 1993;**73**:309–19.
3. Nemerow GR, Stewart PL. Role of alpha(v) integrins in adenovirus cell entry and gene delivery. *MMBR* 1999;**63**:725–34.

4. Greber UF, Suomalainen M, Stidwill RP, Boucke K, Ebersold MW, Helenius A. The role of the nuclear pore complex in adenovirus DNA entry. *EMBO J* 1997;**16**: 5998–6007.
5. Russell WC. Update on adenovirus and its vectors. *J Gen Virol* 2000;**81**:2573–604.
6. McConnell MJ, Imperiale MJ. Biology of adenovirus and its use as a vector for gene therapy. *Hum Gene Ther* 2004;**15**:1022–33.
7. Tatsis N, Ertl HC. Adenoviruses as vaccine vectors. *Mol Ther* 2004;**10**:616–29.
8. Dormond E, Perrier M, Kamen A. From the first to the third generation adenoviral vector: what parameters are governing the production yield? *Biotechnol Adv* 2009;**27**:133–44.
9. Berk AJ. Functions of adenovirus E1A. *Cancer Surv* 1986;**5**:367–87.
10. Rao L, Debbas M, Sabbatini P, Hockenbery D, Korsmeyer S, White E. The adenovirus E1A proteins induce apoptosis, which is inhibited by the E1B 19-kDa and Bcl-2 proteins. *Proc Natl Acad Sci USA* 1992;**89**:7742–6.
11. Gallimore PH, Turnell AS. Adenovirus E1A: remodelling the host cell, a life or death experience. *Oncogene* 2001;**20**:7824–35.
12. Moran E. DNA tumor virus transforming proteins and the cell cycle. *Curr Opin Genet Dev* 1993;**3**:63–70.
13. Berk AJ. Adenovirus promoters and E1A transactivation. *Annu Rev Genet* 1986;**20**:45–79.
14. Debbas M, White E. Wild-type p53 mediates apoptosis by E1A, which is inhibited by E1B. *Genes Dev* 1993;**7**:546–54.
15. Schmitz ML, Indorf A, Limbourg FP, Stadtler H, Traenckner EB, Baeuerle PA. The dual effect of adenovirus type 5 E1A 13S protein on NF-kappaB activation is antagonized by E1B 19K. *Mol Cell Biol* 1996;**16**:4052–63.
16. Doerfler W, Böhm P. *The molecular repertoire of adenoviruses*. Springer; 1995.
17. Lochmuller H, Jani A, Huard J, et al. Emergence of early region 1-containing replication-competent adenovirus in stocks of replication-defective adenovirus recombinants (delta E1+delta E3) during multiple passages in 293 cells. *Hum Gene Ther* 1994;**5**:1485–91.
18. Hehir KM, Armentano D, Cardoza LM, et al. Molecular characterization of replication-competent variants of adenovirus vectors and genome modifications to prevent their occurrence. *J Virol* 1996;**70**:8459–67.
19. Smith JG, Eck SL. Molecular characterization of an adenoviral vector resulting from both homologous and nonhomologous recombination. *Cancer Gene Ther* 1999;**6**:475–81.
20. Zhu J, Grace M, Casale J, et al. Characterization of replication-competent adenovirus isolates from large-scale production of a recombinant adenoviral vector. *Hum Gene Ther* 1999;**10**:113–21.
21. Murakami P, Pungor E, Files J, et al. A single short stretch of homology between adenoviral vector and packaging cell line can give rise to cytopathic effect-inducing, helper-dependent E1-positive particles. *Hum Gene Ther* 2002;**13**:909–20.
22. Kovesdi I, Hedley SJ. Adenoviral producer cells. *Viruses* 2010;**2**:1681–703.
23. Fallaux FJ, Bout A, van der Velde I, et al. New helper cells and matched early region 1-deleted adenovirus vectors prevent generation of replication-competent adenoviruses. *Hum Gene Ther* 1998;**9**:1909–17.
24. Farson D, Tao L, Ko D, et al. Development of novel E1-complementary cells for adenoviral production free of replication-competent adenovirus. *Mol Ther* 2006;**14**:305–11.
25. Howe JA, Pelka P, Antelman D, et al. Matching complementing functions of transformed cells with stable expression of selected viral genes for production of E1-deleted adenovirus vectors. *Virology* 2006;**345**:220–30.

26. Liu TC, Galanis E, Kirn D. Clinical trial results with oncolytic virotherapy: a century of promise, a decade of progress. *Nat Clin Pract Oncol* 2007;**4**:101–17.
27. Choi JW, Lee JS, Kim SW, Yun CO. Evolution of oncolytic adenovirus for cancer treatment. *Adv Drug Deliv Rev* 2012;**64**:720–9.
28. Choi IK, Yun CO. Recent developments in oncolytic adenovirus-based immunotherapeutic agents for use against metastatic cancers. *Cancer Gene Ther* 2013;**20**:70–6.
29. Heise C, Hermiston T, Johnson L, et al. An adenovirus E1A mutant that demonstrates potent and selective systemic anti-tumoral efficacy. *Nat Med* 2000;**6**:1134–9.
30. Lamfers ML, Grill J, Dirven CM, et al. Potential of the conditionally replicative adenovirus Ad5-Delta24RGD in the treatment of malignant gliomas and its enhanced effect with radiotherapy. *Cancer Res* 2002;**62**:5736–42.
31. Bischoff JR, Kirn DH, Williams A, et al. An adenovirus mutant that replicates selectively in p53-deficient human tumor cells. *Science* 1996;**274**:373–6.
32. Freytag SO, Rogulski KR, Paielli DL, Gilbert JD, Kim JH. A novel three-pronged approach to kill cancer cells selectively: concomitant viral, double suicide gene, and radiotherapy. *Hum Gene Ther* 1998;**9**:1323–33.
33. Yu DC, Chen Y, Dilley J, et al. Antitumor synergy of CV787, a prostate cancer-specific adenovirus, and paclitaxel and docetaxel. *Cancer Res* 2001;**61**:517–25.
34. Bristol JA, Zhu M, Ji H, et al. In vitro and in vivo activities of an oncolytic adenoviral vector designed to express GM-CSF. *Mol Ther* 2003;**7**:755–64.
35. Yuk IH, Olsen MM, Geyer S, Forestell SP. Perfusion cultures of human tumor cells: a scalable production platform for oncolytic adenoviral vectors. *Biotechnol Bioeng* 2004;**86**:637–42.
36. Longley Jr R, Radzniak L, Santoro M, et al. Development of a serum-free suspension process for the production of a conditionally replicating adenovirus using A549 cells. *Cytotechnology* 2005;**49**:161–71.
37. Working PK, Lin A, Borellini F. Meeting product development challenges in manufacturing clinical grade oncolytic adenoviruses. *Oncogene* 2005;**24**:7792–801.
38. Nevins JR. Mechanism of activation of early viral transcription by the adenovirus E1A gene product. *Cell* 1981;**26**:213–20.
39. Gaynor RB, Berk AJ. *Cis*-acting induction of adenovirus transcription. *Cell* 1983;**33**:683–93.
40. Imperiale MJ, Kao HT, Feldman LT, Nevins JR, Strickland S. Common control of the heat shock gene and early adenovirus genes: evidence for a cellular E1A-like activity. *Mol Cell Biol* 1984;**4**:867–74.
41. Swaminathan S, Thimmapaya B. Regulation of adenovirus E2 transcription unit. In: Doerfler W, Böhm P, editors. *The molecular repertoire of adenoviruses III*. Berlin Heidelberg: Springer; 1995. p. 177–94.
42. Weitzman MD. Functions of the adenovirus E4 proteins and their impact on viral vectors. *Front Biosci* 2005;**10**:1106–17.
43. Bridge E, Ketner G. Redundant control of adenovirus late gene expression by early region 4. *J Virol* 1989;**63**:631–8.
44. Huang MM, Hearing P. Adenovirus early region 4 encodes two gene products with redundant effects in lytic infection. *J Virol* 1989;**63**:2605–15.
45. Gao GP, Yang Y, Wilson JM. Biology of adenovirus vectors with E1 and E4 deletions for liver-directed gene therapy. *J Virol* 1996;**70**:8934–43.
46. Lusky M, Christ M, Rittner K, et al. In vitro and in vivo biology of recombinant adenovirus vectors with E1, E1/E2A, or E1/E4 deleted. *J Virol* 1998;**72**:2022–32.
47. O'Neal WK, Zhou H, Morral N, et al. Toxicological comparison of E2a-deleted and first-generation adenoviral vectors expressing alpha1-antitrypsin after systemic delivery. *Hum Gene Ther* 1998;**9**:1587–98.

48. Christ M, Louis B, Stoeckel F, et al. Modulation of the inflammatory properties and hepatotoxicity of recombinant adenovirus vectors by the viral E4 gene products. *Hum Gene Ther* 2000;**11**:415–27.
49. Brough DE, Hsu C, Kulesa VA, et al. Activation of transgene expression by early region 4 is responsible for a high level of persistent transgene expression from adenovirus vectors in vivo. *J Virol* 1997;**71**:9206–13.
50. Wang Q, Greenburg G, Bunch D, Farson D, Finer MH. Persistent transgene expression in mouse liver following in vivo gene transfer with a delta E1/delta E4 adenovirus vector. *Gene Ther* 1997;**4**:393–400.
51. Segura MM, Alba R, Bosch A, Chillon M. Advances in helper-dependent adenoviral vector research. *Curr Gene Ther* 2008;**8**:222–35.
52. Parks RJ, Graham FL. A helper-dependent system for adenovirus vector production helps define a lower limit for efficient DNA packaging. *J Virol* 1997;**71**:3293–8.
53. Parks RJ, Chen L, Anton M, Sankar U, Rudnicki MA, Graham FL. A helper-dependent adenovirus vector system: removal of helper virus by Cre-mediated excision of the viral packaging signal. *Proc Natl Acad Sci USA* 1996;**93**:13565–70.
54. Ng P, Beauchamp C, Evelegh C, Parks R, Graham FL. Development of a FLP/frt system for generating helper-dependent adenoviral vectors. *Mol Ther* 2001;**3**:809–15.
55. Umana P, Gerdes CA, Stone D, et al. Efficient FLPe recombinase enables scalable production of helper-dependent adenoviral vectors with negligible helper-virus contamination. *Nat Biotechnol* 2001;**19**:582–5.
56. Garnett CT, Erdman D, Xu W, Gooding LR. Prevalence and quantitation of species C adenovirus DNA in human mucosal lymphocytes. *J Virol* 2002;**76**:10608–16.
57. Bangari DS, Mittal SK. Current strategies and future directions for eluding adenoviral vector immunity. *Curr Gene Ther* 2006;**6**:215–26.
58. Paillard F. Advantages of non-human adenoviruses versus human adenoviruses. *Hum Gene Ther* 1997;**8**:2007–9.
59. Thomas CE, Birkett D, Anozie I, Castro MG, Lowenstein PR. Acute direct adenoviral vector cytotoxicity and chronic, but not acute, inflammatory responses correlate with decreased vector-mediated transgene expression in the brain. *Mol Ther* 2001;**3**:36–46.
60. Yang Y, Su Q, Wilson JM. Role of viral antigens in destructive cellular immune responses to adenovirus vector-transduced cells in mouse lungs. *J Virol* 1996;**70**:7209–12.
61. Sumida SM, Truitt DM, Lemckert AA, et al. Neutralizing antibodies to adenovirus serotype 5 vaccine vectors are directed primarily against the adenovirus hexon protein. *J Immunol* 2005;**174**:7179–85.
62. Lopez-Gordo E, Podgorski II, Downes N, Alemany R. Circumventing antivector immunity: potential use of nonhuman adenoviral vectors. *Hum Gene Ther* 2014;**25**:285–300.
63. Kuriyama S, Tominaga K, Kikukawa M, et al. Inhibitory effects of human sera on adenovirus-mediated gene transfer into rat liver. *Anticancer Res* 1998;**18**:2345–51.
64. Vlachaki MT, Hernandez-Garcia A, Ittmann M, et al. Impact of preimmunization on adenoviral vector expression and toxicity in a subcutaneous mouse cancer model. *Mol Ther* 2002;**6**:342–8.
65. Raper SE, Chirmule N, Lee FS, et al. Fatal systemic inflammatory response syndrome in a ornithine transcarbamylase deficient patient following adenoviral gene transfer. *Mol Genet Metab* 2003;**80**:148–58.
66. Seshidhar Reddy P, Ganesh S, Limbach MP, et al. Development of adenovirus serotype 35 as a gene transfer vector. *Virology* 2003;**311**:384–93.
67. Vogels R, Zuijdgeest D, van Rijnsoever R, et al. Replication-deficient human adenovirus type 35 vectors for gene transfer and vaccination: efficient human cell infection and bypass of preexisting adenovirus immunity. *J Virol* 2003;**77**:8263–71.

68. Gao W, Robbins PD, Gambotto A. Human adenovirus type 35: nucleotide sequence and vector development. *Gene Ther* 2003;**10**:1941–9.
69. Sirena D, Ruzsics Z, Schaffner W, Greber UF, Hemmi S. The nucleotide sequence and a first generation gene transfer vector of species B human adenovirus serotype 3. *Virology* 2005;**343**:283–98.
70. Stone D, Ni S, Li ZY, et al. Development and assessment of human adenovirus type 11 as a gene transfer vector. *J Virol* 2005;**79**:5090–104.
71. Havenga M, Vogels R, Zuijdgeest D, et al. Novel replication-incompetent adenoviral B-group vectors: high vector stability and yield in PER.C6 cells. *J Gen Virol* 2006;**87**:2135–43.
72. van Olphen AL, Tikoo SK, Mittal SK. Characterization of bovine adenovirus type 3 E1 proteins and isolation of E1-expressing cell lines. *Virology* 2002;**295**:108–18.
73. Kremer EJ, Boutin S, Chillon M, Danos O. Canine adenovirus vectors: an alternative for adenovirus-mediated gene transfer. *J Virol* 2000;**74**:505–12.
74. Fernandes P, Santiago VM, Rodrigues AF, Tomas H, Kremer EJ, Alves PM, et al. Impact of E1 and Cre on adenovirus vector amplification: developing MDCK CAV-2-E1 and E1-Cre transcomplementing cell lines. *PLoS One* 2013;**8**:e60342.
75. Farina SF, Gao GP, Xiang ZQ, Rux JJ, Burnett RM, Alvira MR, et al. Replication-defective vector based on a chimpanzee adenovirus. *J Virol* 2001;**75**:11603–13.
76. Roy S, Gao G, Lu Y, Zhou X, Lock M, Calcedo R, et al. Characterization of a family of chimpanzee adenoviruses and development of molecular clones for gene transfer vectors. *Hum Gene Ther* 2004;**15**:519–30.
77. Roy S, Zhi Y, Kobinger GP, Figueredo J, Calcedo R, Miller JR, et al. Generation of an adenoviral vaccine vector based on simian adenovirus 21. *J Gen Virol* 2006;**87**:2477–85.
78. Roy S, Clawson DS, Lavrukhin O, Sandhu A, Miller J, Wilson JM. Rescue of chimeric adenoviral vectors to expand the serotype repertoire. *J Virol Methods* 2007;**141**:14–21.
79. Bru T, Salinas S, Kremer EJ. An update on canine adenovirus type 2 and its vectors. *Viruses* 2010;**2**:2134–53.
80. Soudais C, Laplace-Builhe C, Kissa K, Kremer EJ. Preferential transduction of neurons by canine adenovirus vectors and their efficient retrograde transport in vivo. *Faseb J* 2001;**15**:2283–5.
81. Soudais C, Skander N, Kremer EJ. Long-term in vivo transduction of neurons throughout the rat CNS using novel helper-dependent CAV-2 vectors. *FASEB J* 2004;**18**:391–3.
82. Lau AA, Rozaklis T, Ibanes S, et al. Helper-dependent canine adenovirus vector-mediated transgene expression in a neurodegenerative lysosomal storage disorder. *Gene* 2012;**491**:53–7.
83. Fernandes P, Peixoto C, Santiago VM, Kremer EJ, Coroadinha AS, Alves PM. Bioprocess development for canine adenovirus type 2 vectors. *Gene Ther* 2013;**20**:353–60.
84. Segura MM, Puig M, Monfar M, Chillon M. Chromatography purification of canine adenoviral vectors. *Hum Gene Ther methods* 2012;**23**:182–97.
85. Altaras NE, Aunins JG, Evans RK, Kamen A, Konz JO, Wolf JJ. Production and formulation of adenovirus vectors. *Adv Biochem Eng Biotechnol* 2005;**99**:193–260.
86. Silva A, Fernandes P, Sousa MQ, Alves P. Scalable production of adenovirus vectors. In: Chill6n M, Bosch A, editors. *Adenovirus*. Humana Press; 2014. p. 175–96.
87. Silva AC, Peixoto C, Lucas T, et al. Adenovirus vector production and purification. *Curr Gene Ther* 2010;**10**:437–55.
88. Falkner E, Appl H, Eder C, Losert UM, Schoffl H, Pfaller W. Serum free cell culture: the free access online database. *Toxicol Vitro* 2006;**20**:395–400.
89. Croyle MA, Roessler BJ, Davidson BL, Hilfinger JM, Amidon GL. Factors that influence stability of recombinant adenoviral preparations for human gene therapy. *Pharm Dev Technol* 1998;**3**:373–83.

90. Croyle MA, Cheng X, Wilson JM. Development of formulations that enhance physical stability of viral vectors for gene therapy. *Gene Ther* 2001;**8**:1281–90.
91. Obenauer-Kutner LJ, Ihnat PM, Yang TY, et al. The use of field emission scanning electron microscopy to assess recombinant adenovirus stability. *Hum Gene Ther* 2002;**13**:1687–96.
92. Evans RK, Nawrocki DK, Isopi LA, et al. Development of stable liquid formulations for adenovirus-based vaccines. *J Pharm Sci* 2004;**93**:2458–75.
93. Cruz PE, Silva AC, Roldao A, Carmo M, Carrondo MJ, Alves PM. Screening of novel excipients for improving the stability of retroviral and adenoviral vectors. *Biotechnol Prog* 2006;**22**:568–76.
94. Huyghe BG, Liu X, Sutjipto S, et al. Purification of a type 5 recombinant adenovirus encoding human p53 by column chromatography. *Hum Gene Ther* 1995;**6**:1403–16.
95. Blanche F, Cameron B, Barbot A, et al. An improved anion-exchange HPLC method for the detection and purification of adenoviral particles. *Gene Ther* 2000;**7**:1055–62.
96. Zhang S, Thwin C, Wu Z, Cho, T. US Patent 6,194,191 B1, 2001.
97. Goerke AR, To BC, Lee AL, Sagar SL, Konz JO. Development of a novel adenovirus purification process utilizing selective precipitation of cellular DNA. *Biotechnol Bioeng* 2005;**91**:12–21.
98. Dormond E, Perrier M, Kamen A. Identification of critical infection parameters to control helper-dependent adenoviral vector production. *J Biotechnol* 2009;**142**:142–50.
99. Dormond E, Meneses-Acosta A, Jacob D, et al. An efficient and scalable process for helper-dependent adenoviral vector production using polyethylenimine-adenofection. *Biotechnol Bioeng* 2009;**102**:800–10.
100. Gonzalez-Aparicio M, Mauleon I, Alzuguren P, et al. Self-inactivating helper virus for the production of high-capacity adenoviral vectors. *Gene Ther* 2011;**18**:1025–33.
101. Nadeau I, Kamen A. Production of adenovirus vector for gene therapy. *Biotechnol Adv* 2003;**20**:475–89.
102. Nadeau I, Garnier A, Cote J, Massie B, Chavarie C, Kamen A. Improvement of recombinant protein production with the human adenovirus/293S expression system using fed-batch strategies. *Biotechnol Bioeng* 1996;**51**:613–23.
103. Nadeau I, Gilbert PA, Jacob D, Perrier M, Kamen A. Low-protein medium affects the 293SF central metabolism during growth and infection with adenovirus. *Biotechnol Bioeng* 2002;**77**:91–104.
104. Ferreira TB, Ferreira AL, Carrondo MJ, Alves PM. Effect of re-feed strategies and non-ammoniagenic medium on adenovirus production at high cell densities. *J Biotechnol* 2005;**119**:272–80.
105. Henry O, Dormond E, Perrier M, Kamen A. Insights into adenoviral vector production kinetics in acoustic filter-based perfusion cultures. *Biotechnol Bioeng* 2004;**86**:765–74.
106. Cortin V, Thibault J, Jacob D, Garnier A. High-titer adenovirus vector production in 293S cell perfusion culture. *Biotechnol Prog* 2004;**20**:858–63.
107. Henry O, Perrier M, Kamen A. Metabolic flux analysis of HEK-293 cells in perfusion cultures for the production of adenoviral vectors. *Metab Eng* 2005;**7**:467–76.
108. Nadeau I, Jacob D, Perrier M, Kamen A. 293SF metabolic flux analysis during cell growth and infection with an adenoviral vector. *Biotechnol Prog* 2000;**16**:872–84.
109. Zhang CH, Ferreira TB, Cruz PE, Alves PM, Haury M, Carrondo MJT. The importance of 293 cell cycle phase on adenovirus vector production. *Enzyme Microb Technol* 2006;**39**:1328–32.
110. Ferreira TB, Perdigao R, Silva AC, et al. 293 cell cycle synchronisation adenovirus vector production. *Biotechnol Prog* 2009;**25**:235–43.

111. Garnier A, Cote J, Nadeau I, Kamen A, Massie B. Scale-up of the adenovirus expression system for the production of recombinant protein in human 293S cells. *Cytotechnology* 1994;**15**:145–55.
112. Monica TJ, Montgomery T, Ayala JL, et al. Monitoring adenovirus infections with on-line and off-line methods. *Biotechnol Prog* 2000;**16**:866–71.
113. Sandig V, Youil R, Bett AJ, et al. Optimization of the helper-dependent adenovirus system for production and potency in vivo. *Proc Natl Acad Sci USA* 2000;**97**:1002–7.
114. Ahn M, Gamble A, Witting SR, et al. Vector and helper genome rearrangements occur during production of helper-dependent adenoviral vectors. *Hum Gene Ther Methods* 2013;**24**:1–10.
115. Ng P, Evelegh C, Cummings D, Graham FL. Cre levels limit packaging signal excision efficiency in the Cre/loxP helper-dependent adenoviral vector system. *J Virol* 2002;**76**:4181–9.
116. Kreppel F, Biermann V, Kochanek S, Schiedner G. A DNA-based method to assay total and infectious particle contents and helper virus contamination in high-capacity adenoviral vector preparations. *Hum Gene Ther* 2002;**13**:1151–6.
117. Vicente T, Mota JP, Peixoto C, Alves PM, Carrondo MJ. Rational design and optimization of downstream processes of virus particles for biopharmaceutical applications: current advances. *Biotechnol Adv* 2011;**29**:869–78.
118. Vellinga J, Smith JP, Lipiec A, et al. Challenges in manufacturing adenoviral vectors for global vaccine product deployment. *Hum Gene Ther* 2014;**25**:318–27.
119. Lee DS, Kim BM, Seol DW. Improved purification of recombinant adenoviral vector by metal affinity membrane chromatography. *Biochem Biophys Res Commun* 2009;**378**:640–4.
120. Puig M, Piedra J, Miravet S, Segura MM. Canine adenovirus downstream processing protocol. *Methods Mol Biol* 2014;**1089**:197–210.
121. Jungbauer A. Continuous downstream processing of biopharmaceuticals. *Trends Biotechnol* 2013;**31**:479–92.
122. Segura MM, Kamen AA, Garnier A. Overview of current scalable methods for purification of viral vectors. *Methods Mol Biol* 2011;**737**:89–116.
123. Onions D, Kolman J. Massively parallel sequencing, a new method for detecting adventitious agents. *Biologicals* 2010;**38**:377–80.
124. Marzio G, Kerkvliet E, Bogaards JA, et al. A replication-competent adenovirus assay for E1-deleted Ad35 vectors produced in PER.C6 cells. *Vaccine* 2007;**25**:2228–37.
125. Carinhas N, Bernal V, Monteiro F, Carrondo MJ, Oliveira R, Alves PM. Improving baculovirus production at high cell density through manipulation of energy metabolism. *Metab Eng* 2010;**12**:39–52.
126. Bangari DS, Sharma A, Mittal SK. Bovine adenovirus type 3 internalization is independent of primary receptors of human adenovirus type 5 and porcine adenovirus type 3. *Biochem Biophys Res Commun* 2005;**331**:1478–84.
127. Bangari DS, Shukla S, Mittal SK. Comparative transduction efficiencies of human and nonhuman adenoviral vectors in human, murine, bovine, and porcine cells in culture. *Biochem Biophys Res Commun* 2005;**327**:960–6.
128. Mittal SK, Prevec L, Graham FL, Babiuk LA. Development of a bovine adenovirus type 3-based expression vector. *J Gen Virol* 1995;**76**:93–102.
129. Reddy PS, Idamakanti N, Hyun BH, Tikoo SK, Babiuk LA. Development of porcine adenovirus-3 as an expression vector. *J Gen Virol* 1999;**80**(Pt 3):563–70.
130. Reddy PS, Idamakanti N, Babiuk LA, Mehtali M, Tikoo SK. Porcine adenovirus-3 as a helper-dependent expression vector. *J Gen Virol* 1999;**80**(Pt 11):2909–16.

131. Bangari DS, Mittal SK. Porcine adenoviral vectors evade preexisting humoral immunity to adenoviruses and efficiently infect both human and murine cells in culture. *Virus Res* 2004;**105**:127–36.
132. Pichla-Gollon SL, Drinker M, Zhou X, Xue F, Rux JJ, Gao GP, et al. Structure-based identification of a major neutralizing site in an adenovirus hexon. *J Virol* 2007;**81**:1680–9.
133. Charles PC, Guida JD, Brosnan CF, Horwitz MS. Mouse adenovirus type-1 replication is restricted to vascular endothelium in the CNS of susceptible strains of mice. *Virology* 1998;**245**:216–28.
134. Lenaerts L, Verbeken E, De Clercq E, Naesens L. Mouse adenovirus type 1 infection in SCID mice: an experimental model for antiviral therapy of systemic adenovirus infections. *Antimicrob Agents Chemother* 2005;**49**:4689–99.
135. Beard CW, Spindler KR. Analysis of early region 3 mutants of mouse adenovirus type 1. *J Virol* 1996;**70**:5867–74.
136. Nguyen T, Nery J, Joseph S, Rocha C, Carney G, Spindler K, et al. Mouse adenovirus (MAV-1) expression in primary human endothelial cells and generation of a full-length infectious plasmid. *Gene Ther* 1999;**6**:1291–7.
137. Corredor JC, Nagy E. A region at the left end of the fowl adenovirus 9 genome that is non-essential in vitro has consequences in vivo. *J Gen Virol* 2010;**91**:51–8.
138. Michou AI, Lehrmann H, Saltik M, Cotten M. Mutational analysis of the avian adenovirus CELO, which provides a basis for gene delivery vectors. *J Virol* 1999;**73**:1399–410.
139. Francois A, Eterradossi N, Delmas B, Payet V, Langlois P. Construction of avian adenovirus CELO recombinants in cosmids. *J Virol* 2001;**75**:5288–301.
140. Tan PK, Michou AI, Bergelson JM, Cotten M. Defining CAR as a cellular receptor for the avian adenovirus CELO using a genetic analysis of the two viral fibre proteins. *J Gen Virol* 2001;**82**:1465–72.
141. Greenall SA, Tyack SG, Johnson MA, Sapats SI. Antibody fragments, expressed by a fowl adenovirus vector, are able to neutralize infectious bursal disease virus. *Avian Pathol: J WVPA* 2010;**39**:339–48.
142. Vrati S, Macavoy ES, Xu ZZ, Smole C, Boyle DB, Both GW. Construction and transfection of ovine adenovirus genomic clones to rescue modified viruses. *Virology* 1996;**220**:200–3.
143. Xu ZZ, Hyatt A, Boyle DB, Both GW. Construction of ovine adenovirus recombinants by gene insertion or deletion of related terminal region sequences. *Virology* 1997;**230**:62–71.
144. Loser P, Hillgenberg M, Arnold W, Both GW, Hofmann C. Ovine adenovirus vectors mediate efficient gene transfer to skeletal muscle. *Gene Ther* 2000;**7**:1491–8.
145. Maizel Jr JV, White DO, Scharff MD. The polypeptides of adenovirus. I. Evidence of multiple protein components in the virion and a comparison of types 2, 7A, and 12. *Virology* 1968; **36**(1):115–25.

# Propagation of Adenoviral Vectors: Use of PER.C6™ Cells

7

*W.W. Nichols[1], R. Lardenoije[2], B.J. Ledwith[1], K. Brouwer[2], S. Manam[1], R. Vogels[2], D. Kaslow[1], D. Zuidgeest[2], A.J. Bett[1], L. Chen[1], M. van der Kaaden[2], S.M. Galloway[1], R.B. Hill[1], S.V. Machotka[1], C.A. Anderson[1], J. Lewis[1], D. Martinez[1], J. Lebron[1], C. Russo[1], D. Valerio[2], A. Bout[2]*
[1]Merck Research Laboratories, Merck & Co., Inc., West Point, PA, USA; [2]Crucell NV, Leiden, The Netherlands

## 1. Introduction

### 1.1 Scope of the Paper

Gene therapy aims at the introduction of gene(s) into somatic cells of humans for therapeutic purposes. The success of gene therapy is therefore dependent on the efficiency by which a therapeutic gene can be transferred to the patient's target tissues. In many cases, viruses are exploited for gene transfer purposes, and in particular gene transfer vectors derived from adenoviruses (adenoviral vectors) are often used to achieve this (for review see Ref. 1).

The reason for this is that adenoviral vectors:

- efficiently transfer genes to many different cell types,
- can be propagated on well-defined production systems to high yields, and
- are very stable, which makes purification and long-term storage possible, thereby making pharmaceutical production feasible.

This contribution will focus on the production systems for clinical lots of adenoviral vectors. Particular attention will be paid to the generation and use of complementation cell lines that carry the E1 genes. Particular emphasis will be on the PER.C6™ cell line, which was developed to prevent generation of replication competent adenovirus (RCA) during propagation of E1-deleted adenoviral vectors. In addition, safety issues with respect to the use of the cell line for making clinical-grade material will be addressed.

### 1.2 Adenoviruses

Human adenovirus was isolated for the first time in 1953 from cultured adenoidal tissue.[2,3] Since then, 51 different serotypes have been isolated from various tissues and excretions of humans, of which serotypes 42–51 were obtained from immunocompromised individuals.[4–6] A serotype is defined on the basis of its immunological distinctiveness as judged by quantitative neutralization with animal antisera (horse and rabbit).

**Adenoviral Vectors for Gene Therapy.**

If neutralization shows a certain degree of cross-reaction between two viruses, distinctiveness of serotype is assumed if (1) the hemagglutinins are unrelated, as shown by lack of cross-reaction on hemagglutination–inhibition, or (2) substantial biophysical/biochemical differences in DNA exist.[7]

Human adenoviruses are subdivided into six different groups (A–F), which are based mainly on differences in hemagglutination, restriction enzyme analysis, and DNA homology.[8] The adenoviruses were found to be associated with different disease patterns (see e.g., Refs 9,10). In addition to the human adenoviruses, some 40 different serotypes have been isolated from various animal species.[11]

All adenoviruses possess a DNA molecule that is surrounded by a capsid consisting essentially of hexon, penton base, and fiber proteins. The virion has an icosahedral symmetry and, depending on the serotype, a diameter of 60–90 nm.

The well-characterized adenovirus serotypes 2 and 5 have a linear double-stranded DNA genome of approximately 36,000 base pairs (bp) (Figure 1). Other adenoviruses have genome sizes ranging from 30 to 38 kbp. The genome contains, at both its ends, identical inverted terminal repeats (ITRs) of approximately 90–140 bp with the exact length depending on the serotype. The viral origins of replication are within the ITRs exactly at the genome ends. Sequences required for encapsidation (Ψ) of the viral genome are located in a region of approximately 400 bp downstream of the left ITR.

The structure of the adenoviral genome is described on the basis of the adenovirus genes expressed following infection of human cells, which are called early (E) and late (L), according to whether transcription of these regions takes place prior to or after onset of DNA replication.

Infection of a target cell starts by interaction of the fiber with a receptor on the surface of the cell. Many, but not all,[12] adenoviruses use the coxsackie–adenovirus receptor for this,[13,14] which is present on the cell surface. Integrins act as secondary receptors by binding to the viral penton-base protein. Subsequently, the virus is

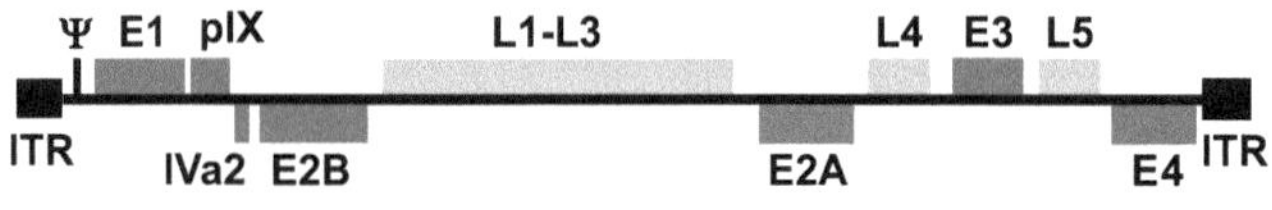


**Figure 1** Map of the adenovirus genome. The 36 kbp (base pairs) (for adenovirus type 5) double-stranded DNA molecule is usually divided into 100 map units (mu). The early (E) and late (L) regions are indicated on the map. The ITR (inverted terminal repeats) sequences are identical, inverted, terminal repeats of approximately 100 bp, depending on the serotype, which are required for replication. Ψ is a stretch of sequences involved in packaging of the viral DNA into particles. E1 comprises the E1A and E1B region, both encoding two proteins, which are described in detail in Section 1.3. E2A encodes the DNA-binding protein, E2B, the precursor terminal protein and DNA polymerase. E3 encodes a number of proteins that are predominantly involved in modulating the host's immune response against adenoviral-infected cells. E4 proteins (6 in total) are involved in modulation of gene expression and viral replication, mainly through interactions with the host cell. IVa2 (transcriptional activator of major late promoter) and pIX (essential for assembly of the virion) are intermediate proteins. L1–L5 encode the late proteins, which are mainly capsid proteins, including penton (L2), hexon (L3), hexon-assembly (L4), and fiber (L5) protein.

internalized by receptor-mediated endocytosis. The adenoviruses escape from the endocytic vesicles (or receptosomes) by virtue of a change in the configuration of the virion surface due to the low pH in these vesicles. As a consequence, the virus particles are released in the cytoplasm of the cell, where they are further degraded,[15] with the DNA ending up in the nucleus, where a complex with histone proteins is formed, which may attach to the nuclear matrix for replication.[16]

The adenovirus DNA is usually not integrated into the host cell chromosomal DNA but remains episomal (extra-chromosomal) unless transformation or tumorigenesis has occurred.

## 1.3 Adenovirus Replication

As indicated before, a productive adenovirus infection is divided into two distinct phases: the early (E) and the late (L) phase. In the early phase, the so-called early genes (E1, E2, E3, and E4) of adenovirus are expressed to prepare the host cell for virus replication. During the late phase, actual viral DNA replication and production of viral structural proteins takes place, leading to the formation of new viral particles.

Adenovirus replication requires both host-cell proteins and viral proteins (for reviews see Refs 8,16). The cellular proteins needed for replication are nuclear factors I, II, and III,[16] which are involved in initiation of viral DNA replication and elongation, as well as in increasing the efficiency of replication.

Adenovirus DNA replication starts with expression of the "immediate early" E1 genes. The E1 region comprises two different transcription units, E1A and E1B. The main functions of the E1A gene products are (1) to induce quiescent cells to enter the cell cycle and resume cellular DNA synthesis, and (2) to transcriptionally activate the E1B gene and the other early regions (E2, E3, and E4). The E1A region encodes two major RNA products, 12S and 13S, which are generated by one transcription unit and which differ in size due to alternative splicing. The RNAs encode acidic proteins of 243 and 289 amino acids, respectively (for adenovirus 5). These are phosphorylated proteins present in the nucleus of the cells. In addition, during lytic infection mRNA's of 9S, 10S, and 11S are produced, but these proteins were found to be not essential for adenoviral replication.[17,18] The function of these proteins has not yet been resolved.

The E1B region codes for one 22S mRNA, which is translated into two proteins, with molecular weights (for adenovirus 5) of 21 kDa and 55 kDa. E1B proteins assist E1A in redirecting the cellular functions to allow viral replication. The E1B 55-kDa protein forms a complex with the E4 open reading frame 6 (ORF6) 34-kDa protein, which is localized in the nucleus.[19,20] Its main function is to inhibit the synthesis of host proteins and to facilitate the expression of viral genes. In addition, it also blocks the p53 tumor-suppressor protein, thereby inhibiting apoptosis.[21] The E1B 21-kDa protein is important for quenching the cytotoxic effects to the target cells induced by E1A proteins. It has anti-apoptotic functions similar to the human Bcl-2 protein, which is important for preventing premature death of the host cell before the virus life cycle has been completed.[22] Mutant viruses incapable of expressing the E1B 21-kDa gene product exhibit a shortened infection cycle that is accompanied by excessive degradation of host cell chromosomal DNA (*deg*-phenotype) and in an enhanced cytopathic effect (*cyt*-phenotype).[23]

The E2 region encodes three different proteins that function in viral DNA replication: an Ad-specific DNA polymerase, the precursor terminal protein (pTP), and the DNA-binding protein.[16] The DNA-binding protein, which is encoded by the E2A gene, binds to single-stranded DNA and is involved in unwinding duplex DNA. It might also be involved in the regulation of transcription. The precursor of the terminal protein (pTP) and the DNA polymerase, which are present as a heterodimer, are encoded by the E2B region. The pTP is attached to the adenoviral DNA and is cleaved by the viral protease late in infection. It has a function in protection of the DNA from nucleolytic breakdown and in attaching the adenoviral DNA to the nuclear matrix, which may localize the viral genome to areas of the nucleus in which high concentrations of replication and transcription factors are present. The polymerase is involved in the synthesis of new DNA strands.

None of the E3 products is required for virus replication. They do, however, play an important role in virus multiplication in vivo, since they protect virus-infected cells from being eradicated by the host's immune response (for review see Ref. 9).

Several differentially spliced mRNAs are synthesized from the E4 region during infection, and six different polypeptides have been identified in infected cells.[24] These proteins are involved in modulation of gene expression and viral replication, mainly through interactions with the host cell.

The E4 ORF3 as well as E4 ORF6 encoded proteins are involved in posttranscriptional processes that increase viral late protein synthesis. They do so by facilitating the cytoplasmic accumulation of the mRNAs encoding these proteins and by expansion of the pool of late RNAs in the nucleus, most likely by influencing splicing. In addition, the E4 ORF6 encoded protein forms a complex with the E1B 55-kDa protein that selectively increases the rate of export of viral late mRNAs from the nucleus. The complex is located in so-called viral inclusion bodies, the region where viral DNA replication, viral late gene transcription, and RNA processing occur.[25] The E4 ORF6 protein, either alone or in a complex with the E1B 55-kDa protein, binds the cellular protein p53 thereby blocking its potential to activate the transcription of tumor-suppressing genes.[26,27]

E4 ORF1 sequences are related to dUTPase enzymes. It has been hypothesized that this gene has a role in stimulating quiescent cells.[24]

The E4 ORF4 protein binds to protein phosphatase 2A, which results in hypophosphorylation of some proteins, including the adenovirus E1 proteins. This perhaps limits cytotoxic effects of E1A and may lead to a more productive infection. It is also in line with the observation that E4 ORF4 mutants are more effective than wild-type viruses in killing nonpermissive rodent cells.[28] E4 ORF4 also induces apoptosis in transformed cells like 293 cells.[29]

The E4 ORF6/7 modulates the activity of the cellular transcription factor E2F, which may subsequently activate cellular genes, which are important for the S-phase.[30]

The functions of E4 ORF1, ORF2, and ORF3/4 during lytic infection are less clear and are dispensable for growth of the virus in laboratory cell lines.

After onset of DNA replication, expression of the late genes L2–L5, which are all under control of one promoter, is switched on. These genes encode the structural components of the virus particles, including L2 the penton, L3 the hexon, L4 the hexon assembly, and L5 the fiber protein. These proteins form the new virus particles into which the adenoviral DNA becomes entrapped. Depending on the serotype, 10,000–100,000 progeny adenovirus particles can be generated in a single cell.

The adenoviral replication causes lysis of the cells.

# 2. Cells Expressing E1 of Adenovirus

## 2.1 Transformation of Cells by E1 of Adenovirus

In the previous section of this chapter, the function of adenoviral gene products in the replication of adenovirus has been described. There is extensive influence of adenoviral proteins on a large number of cellular functions. In the absence of lytic viral replication, adenoviral genes may have a profound effect on cellular functions: the most striking being transformation by the adenoviral E1A and E1B proteins. Clearly, these proteins interfere with the regulatory mechanism of cellular proliferation.

Human adenoviruses have a narrow host range for productive infections, and can be propagated in cells of human, chimpanzee,[31] pig,[32] and cotton rats.[33] In rodent cells, e.g., from rat (with the exception of the cotton rat), hamster, or mouse, they bring about an abortive infection, which occasionally leads to transformation.[34] In the transformed cells, the adenoviral DNA is integrated into the genome and at least the genes of the viral E1 region are expressed (reviewed in Ref. 35).

The viruses that were used for such studies were mainly adenovirus serotypes 2, 5, and 12. The various Ad serotypes differ in their ability to induce tumors upon inoculation into newborn hamsters; for example, Ad type 5 (Ad5) is non-oncogenic,[36] whereas Ad12 is highly oncogenic.[34] However, all Ad serotypes or their DNA can transform rodent cells.[37,38] Ad5E1 transformed cells can only form tumors in immunodeficient mice and rats, whereas Ad12E1 transformed cells are oncogenic both in immunodeficient and in immunocompetent animals,[35] which correlates with the ability of Ad12E1 to repress expression of major histocompatibility complex class I genes.[39]

In culture, both rodent cells, e.g., from rat, mouse, or hamster, and human cells can be transformed by Ad DNA, although human cells, including fibroblasts and epithelial cells, are relatively refractory to transformation. Adenovirus-DNA-transformed human cell lines have been made from cultures of human embryonic kidney,[40,41] human embryonic retina,[42–45] human embryonic lung,[43] and, recently, from human amniocytes.[46]

As described before, the E1 region consists of two transcriptional units, E1A and E1B. For complete morphological transformation, both regions are needed, but the E1A region by itself can immortalize rodent cells[47] and occasionally human cells,[42] albeit with very low efficiency. Expression of E1A usually results in induction of programmed cell death (apoptosis), which can be prevented by coexpression of E1B.[48] The E1A associates with a number of cellular proteins, including the tumor-suppressor gene product pRb, as well as p107, p130, cyclins A and E, cyclin-dependent kinase 2 (cdk2), and p300 (reviewed in Refs 49–51). Most of these proteins are involved in cell-cycle control, and, with the exception of p300, regulate the activity of the transcription factor E2F.[50] The E1A proteins do not exert their activity in initiation of transcription by direct, sequence-specific binding to DNA, but rather do so by binding to cellular transcription factors.

The E1B 55-kDa[19] and 21-kDa[52] proteins cooperate independently with E1A in transformation and are required to inhibit the apoptotic response initiated by E1A. The 55-kDa E1B protein inhibits apoptosis by blocking the function of the p53 tumor-suppressor protein, which mediates E1A-induced apoptosis.[21] The 21-kDa E1B protein inhibits apoptosis in a way similar to the cellular Bcl-2 protein.[22]

## 2.2 E1 Expressing Cell Lines for Adenoviral Vector Production

Most adenoviral vectors currently used in gene therapy have a deletion in the E1 region, where novel genetic information can be introduced.[53] The E1 deletion renders the recombinant virus replication defective, which is a prerequisite for most of the clinical applications. In order to be able to produce E1-deleted recombinant adenoviral vectors, complementing cell lines have to be used that express the E1 proteins of adenovirus. One of the main challenges here is to express sufficient levels of the E1 protein to achieve this. However, adenovirus E1 proteins, and in particular E1A proteins, are very toxic to cells. E1A has a profound effect on the transcription of many cellular genes, which leads to alteration of the morphology and growth of the cells and may lead to apoptosis.

A few examples have been reported in literature, where cells have been immortalized (but not transformed) with E1A only. This has been described both for rodent[47] and for human cells.[42] It is not known whether cells that express E1A only are able to complement adenoviral vectors that are deleted for both E1A and E1B.

Attempts have also been made to express E1 proteins in established cell lines such as A549. Growth of established cells is not dependent on E1 expression, and toxicity of E1 proteins made it difficult to isolate clones that show stable expression of the E1 proteins, although a few papers report encouraging results.[45,54,55] To the best of our knowledge, there is limited use of such cells and therefore this chapter will deal mainly with the group of E1 expressing cells that use the transforming capacity of the adenoviral E1 genes.

Typical examples are the cell lines derived from human embryonic kidney,[40,41] human embryonic retina (HER),[42–45] and human amniocytes.[46] The advantage of using E1 for immortalization is that such cells are dependent on E1 expression for growth, and therefore the levels of E1 expression are remarkably constant over time.

The vast majority of cell lines that were made by immortalization and transformation of primary cells were made to study immortalization and transformation and were not made for propagation of E1-deleted adenoviral vectors. The only documented cell line based on the E1 immortalization principle, which was made specifically for use in gene therapy, is the PER.C6™ cell line[45] and the amniocyte-derived cell line.[46]

These cell lines have been tailor-made for the manufacture of clinical lots of adenoviral vectors, with special attention to avoiding generation of RCA (see below). In addition, proper documentation and adequate safety testing are pivotal to ensure manufacture of safe batches of adenoviral vectors. As PER.C6™ is the only cell currently used for making clinical lots of adenoviral vectors, a description of the generation of PER.C6™ is given below. Also, the performance of the cell line in production of recombinant adenovirus as well as results of safety and genetic testing is provided.

# 3. PER.C6™ Prevents RCA during Vector Production

## *3.1 RCA*

The majority of preparations of E1-deleted adenoviral vectors were produced on 293 cells. This cell line was generated in Leiden in the group of Prof Van der Eb, by transfection of E1 sequences of adenovirus type 5 into primary HEK cells.[40] The aim of this experiment was to study the transforming potential of adenoviral E1 sequences, and the DNA used for it was sheared adenoviral DNA.[40] Precise mapping of the adenoviral sequences present in this cell line indicated that the cell line had integrated bases 1–4137 of the adenoviral DNA.[56]

Adenoviral vectors carry a deletion in the E1 region that runs from approximately 400 to 3500 nucleotides of the adenoviral genome. This means that there is a substantial sequence overlap between the E1 sequences present in the cell line and the adenoviral vector DNA (see Figure 2). This sequence overlap may result in homologous recombination between the sequences. Due to a double cross-over, the E1 region present in the cellular chromosome may end up into the E1-deleted adenoviral vector[57] (Figure 2). The resulting virus is E1 positive, and therefore capable of replicating independently in cells that do not contain E1 sequences in the chromosome. Several reports have described the occurrence of RCA in adenoviral vector batches produced on 293 cells.[45,54,57–59]

RCA in clinical preps is unwanted, both from manufacturing as well as from safety point of view.

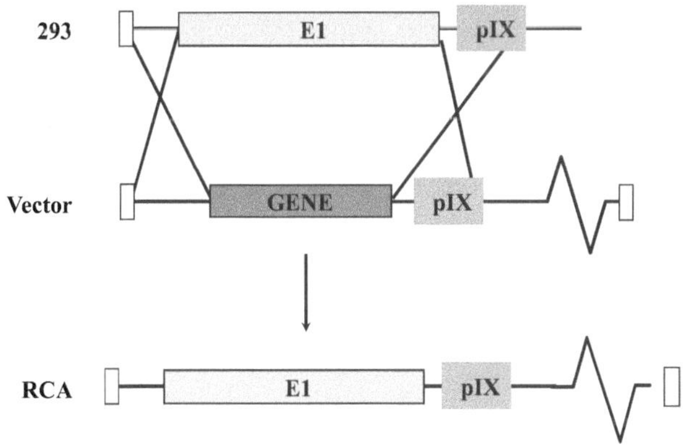


**Figure 2** Mechanism of generation of RCA in 293 cells. Adenoviral vectors contain sequences that overlap with sequences present in the genome of 293 cells, indicated by the crossing lines. Due to the sequence homology, cross-over events can occur, which leads to exchange of DNA. E1 sequences replace the transgene in the adenoviral vectors, resulting in E1-containing adenoviruses that are replication competent.

Its appearance in batches is a chance process and is therefore unpredictable and difficult to control. This is a significant problem for good manufacturing practice (GMP), in particular if large-scale batches have to be prepared.

It is also unwanted from safety point of view, as upon coinfection of a cell, RCA causes the E1-deleted adenoviral vector to replicate in an uncontrollable way. It causes shedding of the vector.[60] In addition, RCA has been shown to cause inflammatory responses.[58,61] Therefore, RCA generation during production of E1-deleted adenoviral vectors has to be circumvented.

## 3.2 PER.C6™: Absence of Sequence Overlap Eliminates RCA Generation

The strategy to prevent RCA occurrence was to eliminate sequence overlap between the E1 sequences present in the cellular genome and the adenoviral vector.[45] A potential hurdle to do this is the way the E1B and pIX gene are regulated. Both E1B and pIX use the same poly(A) sequences.[62] Furthermore, the pIX gene is not expressed upon transfection in cultured cells[63] but can only be expressed if present in an adenoviral genome. Therefore, an RCA-free packaging system should consist of two components: (1) an adenoviral vector that is deleted for E1A and E1B but contains the pIX expression cassette; and (2) a cell line that expresses E1A and E1B and is devoid of pIX sequences.

### 3.2.1 E1 Construct Used for Making PER.C6™

To create the novel cell line, the aim was to use only a minimal number of sequences derived from human adenovirus type 5, i.e., the E1 protein coding sequences only, to prevent sequence overlap with E1-deleted recombinant adenovirus vectors (rAV). The E1 promoter and poly(A) sequences were therefore obtained from non-adenovirus sources. The E1 promoter was replaced by the human phosphoglycerate kinase (PGK) promoter (see below), which is a known house-keeping promoter,[64] and the poly(A) sequences were isolated from Hepatitis B virus.[65,66] The construct pIG.E1A.E1B contains, in addition to the E1A and E1B coding regions, sequences upstream of the E1A gene, including E1A enhancer elements, the cap sequence, and untranslated E1A sequences were retained in the construct. These elements were included since earlier studies indicated that this results in efficient expression of the E1A gene.[67]

A map of the construct, designated as pIG.E1A.E1B, is presented in Figure 3.

Despite removal of the splice site at position 3509 of the adenoviral genome,[62] which is highly conserved, and truncation of the E1B transcript, high expression levels

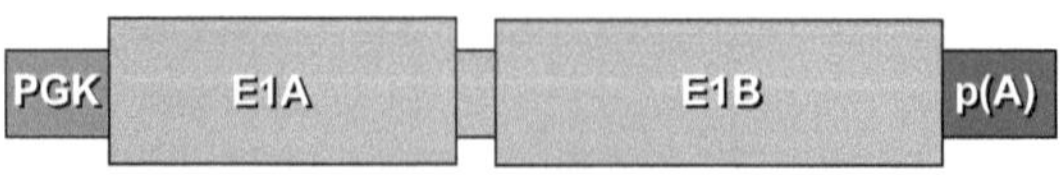


**Figure 3** The E1 construct used to generate PER.C6™. The pIG.E1A.E1B construct contains adenovirus type 5 sequences 459–3510. E1A expression is driven by the human phosphoglycerate kinase (PGK) promoter. E1B transcription is terminated by hepatitis-B derived poly(A) sequences.

of both E1B 21 kDa and E1B 55 kDa were obtained.[45] In fact, the expression of the E1B proteins was even higher than in 293 and 911 cells, whereas equal expression levels of E1A were observed.[45]

To prevent sequence overlap with E1 present in PER.C6™ cells, adenoviral vectors were constructed that carry a deletion of the complete E1 region. These vectors were shown to propagate efficiently in PER.C6™ (see below) and were found to express the pIX gene.[45]

### 3.2.2 Generation of PER.C6™

The primary cells selected for making a new E1 complementing cell line were human embryonic retinoblasts. The choice for retinoblasts[42] was based on the observation that Ad12 could transform hamster retinal cells in vitro[68] and induce retinoblastomas following intraocular injection into newborn baboons.[69] It has been described that these cells can be immortalized relatively easily by E1 of human Ad5[42,43,70] and Ad12.[71] In addition, the 911 cells, which are derived from human embryonic retinoblasts, are very efficient in production of rAV and easy to use,[70] thus providing a second argument for the use of primary human embryonic retinoblasts as the source of primary cells to make a novel cell line.

Primary human embryonic retinoblasts (HER cells) have a limited life span. Such cells can be cultured only a few passages after which the cells senesce. Transfection of HER cells with E1 constructs results in immortalization and transformation of the cells, reflected by focus formation in the cultures. This is easily recognized by both macroscopic and microscopic examination of the cultures. Such foci can be isolated and cultured further. Therefore, the pIG.E1A.E1B construct was transfected into primary HER cells, and PER.C6™ cells were isolated as described in detail before.[45]

After propagation of the cells to passage number 29, a research Master Cell Bank (MCB) was laid down, which was extensively characterized and tested for safety (including sterility testings) (see below).

Immortalization of primary cells with E1 sequences of adenovirus guarantees (1) a stable expression of E1 proteins, as the cells need E1 expression for growth, and (2) that no external selection marker is needed to distinguish E1 expressing from nonexpressing cells. Therefore, the strategy was to transfect primary cells with E1 sequences of adenovirus. Human adenovirus serotype 5 was taken as the donor for E1 sequences, as this serotype is most commonly used in the rAV arena.

## 3.3 Frequency of RCA Occurrence

In order to test whether PER.C6™ cells are able to propagate adenoviral vectors without concomitant generation of RCA, E1-deleted adenoviral vectors were propagated on 293 cells and on PER.C6™ cells. The adenoviral vectors used did not have any sequence overlap with E1 sequences in PER.C6™. The batches of vector were analyzed for the presence of RCA, using cell-culture-based assays, as described before.[45,57] The results (summarized in Table 1) clearly indicate that adenoviral vectors, when propagated on 293 cells, get contaminated with RCA. On the other

**Table 1 Frequency of RCA Occurrence in 293 Cells and in PER.C6™ Cells**

| Helper cell | No. of productions | No. of cells per production | No. of RCA positive batches | |
|---|---|---|---|---|
| | | | 2.5E9 IU | 2.5E10 IU |
| 293 | 22 | 1E8–3E9 | 13/22 | ND |
| PER.C6 | 8 | 1E8–3E9 | 0/8 | 0/2 |
| PER.C6 | 3 | 1E10–3E10 | ND | 0/3 |

Batches of E1-deleted adenoviral vectors, propagated on either 293 and PER.C6™ cells, were tested for the presence of RCA at a level of sensitivity of either 1 RCA in 2.5E9 infectious units (IU) or 1 RCA in 2.5E10 infectious IU of E1-deleted adenoviral vector.
The number of batches that were produced on either cell line, as well as the number of cells used for the production, is indicated as well.

hand, the data provided in Table 1 clearly demonstrate that PER.C6™ cells support RCA-free propagation of E1-deleted adenoviral vectors, even if large-scale batches (produced on 1–3E10 PER.C6 cells) were tested for RCA in a very sensitive assay (1 RCA/2.5 E10 infectious units).

In a separate experiment, an E1-and E3-deleted Ad5 vector was derived and propagated in PER.C6™ cells. A master virus seed (MVS), prepared from passage 12, was used to generate eight virus-production lots (passage 13). The unprocessed virus harvest (vector-infected suspension culture) of the MVS and the virus-production lots were tested for RCA. In brief, test articles were frozen and thawed then assayed by inoculation onto the human-lung-carcinoma (A549, ATCC CCL 185) cell line for approximately 1–2 h at 37 °C, after which the inoculum was removed and the culture was re-fed with medium. Cultures were passaged three times to amplify any putative RCA present, with incubation times ranging from 4 to 7 days for the early passages and from 2 to 5 days for the final passage. The cultures were examined for cytopathic effects at each passage. The virus-production scale was approximately 20 l, and a 60-ml volume (diluted to 600 ml to avoid toxicity and interference with detection of RCA) was tested for RCA for each lot. The testing volume was selected on the basis of a worst-case calculation to insure the testing of at least three dose equivalents of virus. Earlier virus-production studies suggested that the freeze-thaw extract would contain at least $5 \times 10^9$ particles/ml (or $10^{11}$ particles/20 ml). Thus, at least $3 \times 10^{11}$ Ad5 particles (three dose equivalents) would be tested. Assuming a random (Poisson) distribution of RCA, if there were an average of one RCA per $1 \times 10^{11}$ particles (20) ml), one would predict a probability of not detecting it by testing only $1 \times 10^{11}$ particles to be $e^{-1}$ or 0.3679 (36.79% chance). By testing $3 \times 10^{11}$ particles (60 ml), the (binomial) probability of not detecting 1 RCA/$1 \times 10^{11}$ particles is reduced to $e^{-3}$ or 0.04979 (4.98% chance). Mathematically, this is equivalent to three independent tests of 20 ml each (60 ml total).

No RCA was detected in the MVS or in any of eight virus-production lots assayed. Using the ratio of particle/TCID50 determined for purified virus (15.6 particles/infectious units), the virus-production lots were estimated to have an average of $1.9 \times 10^{10}$ particles/ml. It was estimated that the mean probability of not detecting at least one RCA in a dose of $10^{11}$ particles of virus-production lots was 0.000887%. Besides having directly tested the infected cell suspension of the MVS for RCA, the repeated inability to detect RCA in the various clinical batches bodes well for the RCA-free nature of the MVS. For the clinical production runs, 1 mL of MVS is used to inoculate each of 100 roller bottles. This means a total of 800 ml of MVS have been used for these "clinical lots." Following the same calculation scheme as above, if there were 1 RCA per 20 ml of the MVS, there would be $e^{-1 \times 5}$ or 0.00674 probability (0.674% chance) of not transmitting an RCA when preparing a single clinical batch. Moreover, cumulatively across the eight clinical production runs, there would be only $(e^{-1 \times 5})^8$ or a $4.25 \times 10^{-18}$ probability ($4.25 \times 10^{-16}$% chance) of not transmitting RCA in the preparation of eight lots. In conclusion, the 60-ml freeze-thaw sample used for RCA testing provided adequate assurance for the detection of RCA in virus-production lots, at a level of 1 RCA per $10^{11}$ dose. However, for testing of future Ad5 vector lots, we plan to use a clarified lysate. In this case, the probability estimated for detection of RCA will be based on more direct measurement of virus concentration.

In summary, eliminating overlap between E1 sequences in the cell and the E1-deleted adenoviral vector eliminates RCA.

# 4. Production of Adenoviral Vectors

## 4.1 Vector Stability

When constructing E1-deleted adenovirus vectors, a number of choices must be made on the structure of the vector backbone and on the composition of the transgene. One must determine if the size of the E1 deletion will be adequate to accommodate the size of the transgene, or if additional deletions, such as in the E3 region, will be needed. One must also decide on the placement of the transgene within the genome (E1 vs E3) and the orientation of the transgene (E1 parallel vs E1 antiparallel). Finally, one must decide on the composition of the transgene in terms of the transcriptional regulation elements that are utilized (promoter and polyadenylation signals).

All of these parameters make constructing adenoviral vectors, which express the transgene to the desired level, are genetically stable and propagate well enough to allow high-level production, a somewhat empirical process. The net genome size of the vector, the deletions used, transgene orientation, the composition of the transgene, and the transgene product itself can all effect the growth and productivity of the vector. The degree to which vector and transgene structure can affect genomic stability and productivity is illustrated by our experience with Ad5 Vector 1 (Figure 4). Vector 1 contains an E1 deletion into which the transgene was introduced in the E1 antiparallel orientation. The transgene is composed of our gene of interest flanked by the immediate early gene promoter and intron A from the human cytomegalovirus, and the bovine growth hormone polyadenylation signal sequence. In addition to the deletion of the E1 region, the vector has an E3 deletion.[72]

When the genetic stability of Vector 1 was assessed after serial passage in PER.C6™, it was found to be unstable. Restriction analysis of purified viral DNA recovered from passages 12 to19 indicated that the virus population contained genetic variants (Figure 5). Over this passage series, the proportion and number of variants appeared to increase. An analyses of the novel restriction fragments and close to

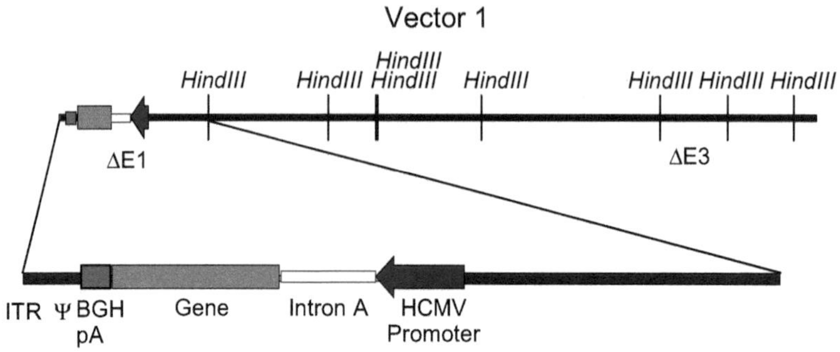


**Figure 4** Genetic structure of Ad5 Vectors 1.

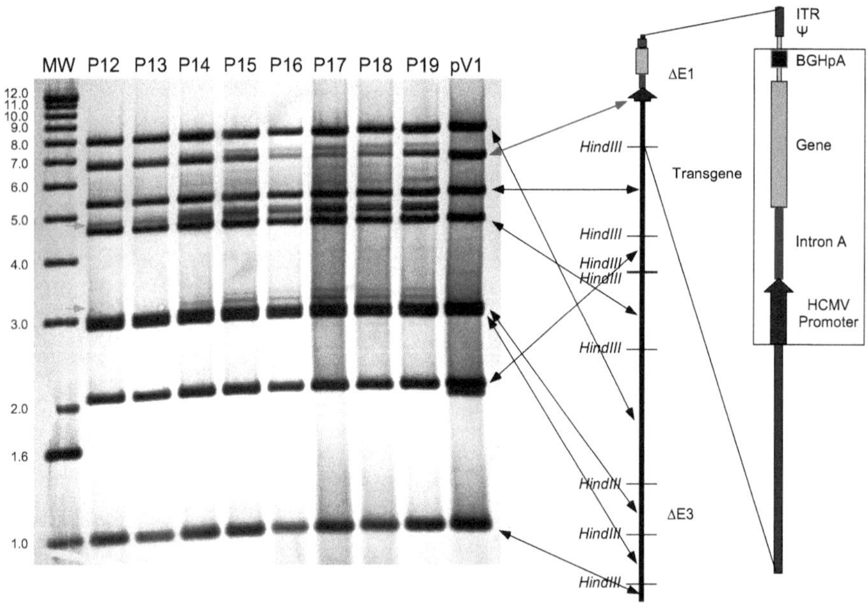


**Figure 5** Genetic structure of serially passaged Vector 1. Viral DNA was purified from passages 12 to 19 of Vector 1 digested with *Hin*dIII and end-label with $P^{32}$-dATP. The end-labeled restriction fragments were then size-fractionated by gel electrophoresis and detected by autoradiography. pV1, the plasmid used to derive Vector 1 is shown for comparison. The position in the Vector 1 genome to which the restriction fragments correspond is indicated on the right. The reduction and upward shift in the 6.6kb transgene containing restriction fragment (red (light gray in print versions) arrow) is due to amplification of the 107bp sequence in the packaging region. Novel bands seen at approximately 4.8 and 3.2kbp (blue (dark gray in print versions) arrows) are due to deletions in the transgene in association with amplification in the packaging region.

1000 individually recovered, circularized viral genomes indicated that two genetic mechanisms could account for all of the observed **restriction fragment length polymorphisms**: (1) deletions of the transgene expression cassette, particularly in the region of the hCMV promotor and intron A, and, in two instances, deletion of only adenovirus sequence; and (2) amplification (two to four repeats) of a 107-bp sequence in the region containing the viral packaging elements. No rearrangements or insertions in the E3 region were detected.

The genetic analysis of Vector 1 has led to the development of highly stable vectors that can be easily propagated in PER.C6™ cells, suggesting that the genetic instability can be overcome by vector design and is not necessarily related to the use of PER. C6™ cells.

## 4.2 The Production Process

To make E1-deleted adenoviral vectors for human gene therapy, a scalable process suitable for commercial manufacturing under GMP conditions was developed.

One of the key factors in the development of cell-culture-based production processes is the culture system. In particular, if scaling of the process is needed, culture of the cells in a bioreactor is highly desired. For robust and scalable systems suspension growth of the required cell line is extremely advantageous. PER.C6™ cells can be cultured both as adherent cells as well as in suspension culture. For suspension growth-specific, well-defined, serum-free (SF) media have been developed (e.g., ExCell 525; JRH Biosciences). These media do not contain any protein that is derived from human or animal tissue or specimen. This results not only in many fewer contaminants to be removed during downstream processing but also a favorable safety profile with respect to pathogens which might be introduced by animal-/human-derived components.

The serum-free culture medium (SF-medium) supports the growth of PER.C6™ cells to densities of 1.5–2.5 × 1E6 cells/ml in routine T-flask and roller bottle cultures. In perfused bioreactor systems, cell densities up to 1E7 cell/ml are easily obtained.

An overview of the process of production of E1-deleted recombinant adenoviruses is presented in Figure 6 and is summarized below.

After thawing of a vial of PER.C6™, expansion in a T-flask containing SF-medium is done, followed by transfer of the suspension culture to roller bottles. Then these roller bottles are cultured until sufficient cells are generated to inoculate a bioreactor.

In the standard batch-wise production process (e.g., in 2- or 20-l bioreactor), half of the bioreactor working volume is inoculated at $0.5 \times 10^6$ cell/ml. Then PER.C6™ is grown in 2 days to $2 \times 10^6$ cell/ml and diluted once to $1 \times 10^6$ cell/ml by adding the same volume of fresh medium. Then the seed virus is added and temperature is lowered from 37 to 35 °C, followed by harvest after 3 days by pelleting. The latter is necessary if the purification process consists of ultracentrifugation with CsCl density gradients. After these 3 days the virus particles become suspended utilizing cell lysis. The batch process is very robust but not economical since only low cell densities can be obtained due to the rapid consumption of nutrients from the medium. When high-cell densities are required, a perfusion system can be used. Nutrients are

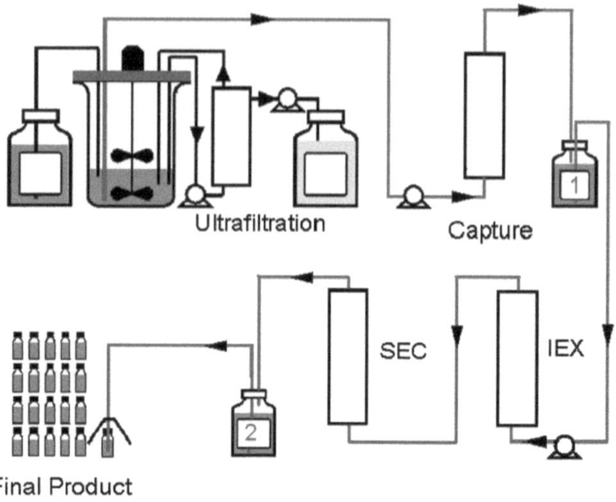


**Figure 6** Overview of the process of production of E1-deleted recombinant adenovirus. The process is described is Section 4.2 of this book chapter.

replaced and metabolites removed by perfusion of fresh medium. A suitable perfusion system can be obtained with hollow fiber modules. These modules are operated externally on the bioreactor and can therefore easily be replaced when malfunction occurs. Hollow-fiber technology also has the opportunity for virus retention, easy scale-up, and its potential application as a first step in the virus isolation. To take full advantage of high-density cultures the virus replication should last longer than 3 days to enable the utilization of all cells present because a repeated infection can occur with newly released particles from lysed cells. A typical example of a 20-l bioreactor run is presented in Figure 7. Because a large part of the total produced virus will be in suspension, the volume of such a culture is too large to enable purification by ultracentrifugation. Hollow fiber ultrafiltration and chromatography are methods of choice for virus isolation and purification. With these systems directly connected to the bioreactor, thereby ensuring a closed system, all viruses can be isolated from the culture medium. After capture of the virus, the bulk product can be further purified utilizing ion exchange chromatography and/or size exclusion chromatography systems. The obtained product is of high purity and infectivity. Final formulation can be done by ultrafiltration bringing the product at the final concentration in the required buffer.

## 4.3 Yields of Adenoviral Vectors

The yields of virus obtained after propagation in PER.C6™ cells in 20-l suspension cultures ranges from $0.6\times10^{11}$ vp/ml to $1.1\times10^{11}$ vp/ml culture medium with an average yield of $0.8\times10^{11}$ vp/ml ($n=5$). The cell density during infection was approximately $3\times10^6$ cells/ml. The calculated virus yield per cell is therefore $0.2\times10^5$–$0.4\times10^5$ vp/cell. As the cultures are inoculated at a multiplicity of infection of 40 vp/cell, an

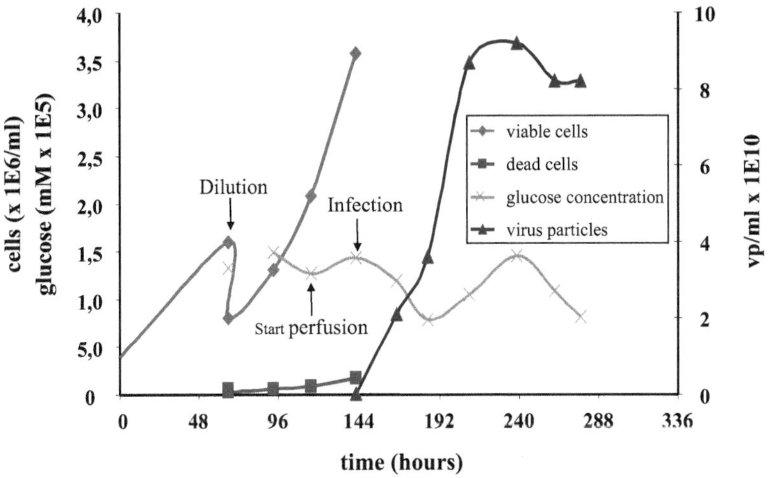


**Figure 7** Example of production of E1-deleted adenoviral vectors in PER.C6™ in a 20 l bioreactor. PER.C6 cells are seeded at a density of 0.5E6 cells/ml, in ExCell 525 culture medium. Perfusion is started 48 h later, at a rate of 1 bioreactor volume/24 h. The glucose concentration remains constant during perfusion. Under these conditions, cell densities of $1 \times 10^7$ cells/ml are obtained.

amplification factor of 500 was achieved. The loss during isolation and purification can be held to 70–80%. This figure was consistently obtained in multiple runs, for three different adenoviral vectors.

Similar yields of E1-deleted adenoviral vectors obtained on PER.C6™ have been obtained by others.[73]

## 4.4 Scale of Adenoviral Vector Production

The estimated scale of the required bioreactor and cell-line stability is calculated as follows. The cell density used for virus production in perfusion mode is $3–6 \times 10^6$ cell/ml. Therefore, assuming at least 20,000 virus particles per cell yield, the overall expected yield in the crude bioreactor harvest is $2 \times 10^4$ vp/cell $\times 5 \times 10^6$ cell/ml $= 1 \times 10^{11}$ vp/ml. Further, after optimization, maximum expected loss of virus particles after downstream processing by column chromatography is 75%. Therefore, from a 20-l perfusion bioreactor $1 \times 10^{11}$ vp/ml $\times 0.25$ (recovery) $\times 10^4$ ml $= 5 \times 10^{14}$ vp can be obtained. This gives $5 \times 10^{14}$ vp/$1 \times 10^{10}$ vp/dose $= 50{,}000$ doses (assuming $1 \times 10^{10}$ vp/dose). When during product development, 40% of the batch is retained for QC and archiving purposes: 3000 patients can receive $50{,}000 \times 0.6/3000 = 10$ doses each. Therefore, using the currently developed technology, this 20-l bioreactor is sufficient for the generation of material for the first clinical studies. However, to be able to do process development on a larger scale, a larger vessel is required for full commercial production. Full production scale is expected to be about 5 times larger, and therefore a 100-l bioreactor is expected to be the maximum volume required for application with single doses up to $10^{10}$ vp. To propagate the cells from a working cell bank ampoule, containing 5E6 cells, to a 5E6 cell/ml culture in a 100 l bioreactor would take 17 cell doublings. So a reliable production process would require a cell line, which is at least stable over 20 cell doublings. PER.C6™ was shown to be stable with respect to E1 expression for at least 98 cell doublings.

# 5. Safety Considerations of PER.C6™

## *5.1 QC Testing of PER.C6™ Cells for Use in the Manufacture of Biologicals and Vaccines*

The safety of vaccines and biologicals manufactured in continuous cell lines of animal or human origin is of *paramount importance* and *must be ensured* by the manufacturer through a program of quality control (QC) testing applied to the product before release for human administration. This QC testing is intended to (1) ensure the *identity* of the product (2) ensure the *safety and sterility* of the product by demonstrating the absence of adventitious microbial agents, and (3) ensure the *safety and sterility* of the product by demonstrating the absence of adventitious viral agents.

The program for QC testing applied to a biological product, formalized as a *Release Protocol*, is developed as a responsibility of a Department of BioAnalytical Development. The *Release Protocol* is developed through an evaluation and integration of (1) relevant compendial literature and precedents, (2) the origin of the cell line used for production and its development as a MCB, (3) the sourcing and quality control testing of raw materials of animal origin used in manufacture, and (4) the method of cGMP manufacture of the bulk and intermediate and final product considering among other considerations the quality of environment in which bioprocessing is conducted, the method of manufacture in particular the isolation of the culture system from operators and the consistency of preparation.

The *Release Protocol* prescribes the QC testing to be applied to not only final products but importantly, MCBs, master virus seeds and other bioprocess inputs, raw materials of animal origin, as well as intermediate bulk products developed during downstream processing purification and formulation. The *Release Protocol* specifies testing methods and volumes to be tested relying upon bacterial broth and agar cultures, embryonated eggs, small animals, and in vitro cell culture in a variety of primary and continuous cell lines of mammalian or human origin. These methods are well known to be sensitive to the detection of a variety of bacterial and viral agents and when applied in concert provide a comprehensive and sensitive analytical approach upon which to ensure product safety. More recently, with the development of exquisitely sensitive polymerase chain reaction (PCR) methods for the detection of agents, which are refractory to animal or cell culture, these classical propagation methods are commonly supplemented with agent-specific testing, using PCR and polymerase-enhanced reverse transcriptase assays. The general methods of testing to ensure product safety are presented in illustrated form in Figure 8.

| *Method* | *Criteria for Evaluation* |
|---|---|
|  ***Sterility*** *Inoculation of Broth and Agar Culture and Cell Cultures with Observation of 14-21 days* | ***Turbidity, Colony Formation Cytoplasmic Fluoresence*** |
| 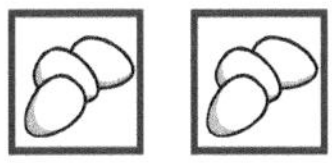 ***In Vivo Testing in Eggs*** *Injection of Eggs by Amniotic, Allantoic or Yolk Sac Routes with Observation for 7-14 days* | ***Viability Gross morphology Hemagluttination*** |
|  ***In Vivo Testing in Animals*** *Injection of Adult or Suckling mice, Guinea pigs or Rabbits by IM, IP, or SC Routes with Observation for 7-60 days* | ***Viability Fitness Evidence of Disease*** |
|  ***In Vitro Testing in Cell Culture*** *Inoculation of Primary or Continuous Cell Lines of Human, Primate or Animal Origin with Observation for 14-28 days* | ***Evidence of Cytopathology Hemadsorption Hemagluttination*** |
| 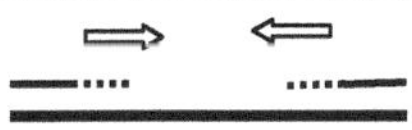 ***Testing for Specific Virus Agents*** *Use of Sequence Specific Primers for PCR Amplification or PERT, or TEM* | ***Evidence of Gene-Specific Product*** ***Evidence of Enzymatic Activity of RT*** |

**Figure 8** Testing methods for the demonstration of product safety.

### 5.1.1 QC Testing for the Release of PER.C6™ Master Cell Bank

The development of a PER.C6™ research Master Cell Bank (rMCB), A068-016, to support manufacture of biologics has been previously described. The *Release Protocol* to ensure the (1) *Identity*, (2) *Sterility*, and (3) *Viral Safety* of the rMCB is presented

Table 2 **Release Protocol for Crucell rMCB A068-016**

| Test | Method |
|---|---|
| ***Identity*** | *Isoenzyme analysis* |
| ***Sterility*** | *broth and agar for cultivation of bacteria, fungi, mycoplasma*<br>*In vitro indicator cells for detection using Hoechst stain* |
| ***Viral safety***<br>*In-vivo eggs* | *Eggs (allantoic and yolk sac)* |
| ***Viral safety***<br>*In-vitro cell culture* | *MRC-5, HeLa, Vero, bovine cells* |
| ***Viral safety***<br>*Agent-specific*<br>*Testing using PCR* | *HBV, HCV, EBV, HHV6, HIV-1, HIV-2, HTLV-1 HTLV-2 AAV, B19, SV40* |
| ***Viral safety***<br>*Agent-specific*<br>*Testing for retroviruses* | *PERT, $S^+L^-$, XC testing* |

in Table 2. The QC testing was conducted by contract at Inveresk Research (Tranent, Scotland) and at MicroSafe (Leiden, The Netherlands).

### *5.1.2 QC Testing for the Release of a PER.C6™ Working Cell Bank*

The *Release Protocol* of research Working Cell Bank (rWCB), A068-043W, according to the panel of testing is presented in Table 3. The QC testing was conducted by contract at Inveresk Research and at MicroSafe. This testing included tests for (1) *Identity* (2) *Sterility*, and (3) *Viral Safety* in cells of human and simian origin.

### *5.1.3 Development of a Master Cell Bank at the Merck Research Laboratories*

Cryopreserved vials of the rWCB were obtained from Crucell by the Merck Research Laboratories and expanded under conditions of cGMP manufacture to create a MCB for future manufacturing use. This MCB has been released for use in the propagation of recombinant adenovirus according to a *Release Protocol* presented in Table 4. The preponderance of this QC testing was conducted by Q-One BioTech in Glasgow, Scotland.

This Release Protocol for the rWCB provides persuasive demonstration of the (1) *Identity* (2) *Sterility, and* (3) *Viral Safety* of the PER.C6™ MCB. This *Release Protocol* specifies animal testing in small animals to supplement the egg safety testing applied to the rWCB, expands the variety of primary and continuous cell lines used for viral safety using in vitro cell culture, and greatly broadens the variety of agent-specific testing using PCR-based testing and biochemical testing for retroviruses. The human cell line 293 was included in the panel of tissue-culture cell lines in an attempt to detect the presence of any defective adventitious virus that requires the presence of E1 in the host cell. The direct assay for reverse transcriptase, as well as the detection of reverse transcriptase (RT) in co-cultivation supernatant fluids, was done with the highly sensitive PCR-based reverse transcriptase assay. The supplemental PCR tests

Table 3 **Release Protocol for Crucell rWCB A068-043W**

| Test | Method |
|---|---|
| ***Identity*** | *Isozyme* |
| ***Sterility*** | *broth and agar for cultivation of bacteria, fungi, mycoplasma* |
| | *In-vitro cell culture testing for mycoplasma* |
| ***Viral safety*** | *Vero, MRC-5, PER.C6* |
| *In-vitro cell culture* | |
| ***Viral safety*** | *Adeno-associated virus* |
| *Agent-specific testing* | |
| *Using PCR* | |

Table 4 **Protocol for Release for the a PER.C6 Master Cell Bank**

| Test | Method |
|---|---|
| ***Identity*** | *Isozyme analysis* |
| | *DNA fingerprinting* |
| | *PCR-based test for E1* |
| ***Sterility*** | *broth & agar for cultivation of bacteria, fungi, mycoplasma* |
| | *In-vitro cell culture testing for mycoplasma* |
| ***Viral safety*** | *Eggs (allantoic and yolk sac)* |
| *In-vivo eggs* | |
| ***Viral safety*** | *Guinea pig* |
| *In-vivo animals* | *Adult and suckling mouse* |
| ***Viral safety*** | *VERO, MRC-5, 293, RK* Vero* |
| *In-vitro cell culture* | *Bovine turbinate, porcine kidney* |
| ***Viral safety*** | *transmission electron microscopy PERT for RT* |
| ***Agent specific testing*** | *Raji, RD, H9 Cell-cocultivation for retroviruses* |
| | *PCR for HBV, HCV* |
| | *CMV, EBV, HHV6, HHV7, HHV8* |
| | *HIV-1, HIV-2, HTLV-1 &HTLV-2, SiFV* |
| | *SFV, AAV, B19, bovine polyoma, SV40* |

were included with due consideration for the human origin of the cell line and the use of bovine serum for the derivation of the cell line. The tumorigenic potential of the cell line was tested beyond the anticipated manufacturing cell-passage level.

Satisfactory results were obtained from all QC testing. The results of the testings are presented in Table 5.

## 5.2 *Tumorigenicity*

### 5.2.1 *Tumorigenicity Studies of PER.C6™ Cells*

Three tumorigenicity studies were carried out on the PER.C6™ cell line. The results of these studies are summarized in Table 6. In the first study, nude (nu/nu) mice were

## Table 5 Summary of Testing of PER.C6 Research Master Cell Bank (Passage Number 29)

| Test | Specification | Result |
|---|---|---|
| Sterility (EP) | Negative | Negative |
| Mycoplasma (broth, agar and DNA staining) | Negative | Negative |
| In-vitro virology for adventitious viruses (28 days, with cytopathic effect and hemadsorption) on Vero, MRC-5, HeLa and PER.C6 cells (PTC) | Negative | Negative |
| Specific viruses: | Negative | Negative |
| Human immunodeficiency virus types | Negative | Negative |
| 1 and 2 | Negative | Negative |
| Human T-lymphotropic virus types 1 and 2 | Negative | Negative |
| Human hepatitis B + C | Negative | Negative |
| Human cytomegalovirus | Negative | Negative |
| Human parvovirus B 19 | Negative | Negative |
| Human herpes virus 6 | Negative | Negative |
| Simian virus 40 | Negative | Negative |
| Adeno-associated virus | | |
| Epstein–Barr virus | | |
| Bovine viruses (BVD, IBR and PI3) | Negative | Negative |
| In-vivo virology in suckling mice (i.c.and i.p.), and embryonated eggs, allantoic and yolk sac injections (PTC) | Negative | Negative |
| Isoenzyme test for human origin | Confirmed | Confirmed |
| In-vivo virology (adult mice, guinea pigs and suckling mice) and transmission electron microscopy (TEM) | Absence of adventitious microbial contamination | Free from infectious adventitious microbial contamination |
| Reverse transcriptase assay | Negative | Negative |
| $S^{+}L^{-}$ focus forming assay and XC Plaque assay | Negative | Negative |
| Tumorigenicity in nude mice | Report result | Tumorigenic |
| Restriction analysis | No evidence of mutation or rearrangements | No evidence of mutation or rearrangements |
| Sequencing | Report sequence | Sequence reported |
| Copy number determination | Report result | 5–6 copies Per haploid genome |
| DNA profiling rMCB (passage 29) and late passage cells (passage 98) | Late passage banding pattern resembles rMCB | Late passage banding pattern resembles rMCB |
| Karyotyping/Chromosomal analysis | Report chromosome numbers | Modal number 86<br>Range 68-106 |

**Table 5 Continued**

| Test | Specification | Result |
|---|---|---|
| Fluorescent product enhanced reverse transcriptase assay | Negative | Negative |
| $S^+L^-$ focus forming assay and XC plaque assay | Negative | Negative |
| Multicolor In situ hybridization (MFISH) | Report integration site | Chromosome 14 |
| Copy number determination (fiber FISH analysis) | Report results | 13.6 ± 6.1<br>5 copies/Haploid genome |
| Prions:<br>Determination of prions<br>Sequence analysis | No evidence for infectious $PrP^{Sc}$ | Confirmed |

**Table 6 Tumorigenicity of PER.C6™ cells**

**A. Day 28 tumorigenicity of PER.C6™ and KB cells in nude mice**

| Cell type | No. of cells | Male | Female |
|---|---|---|---|
| KB | $1 \times 10^7$ | 10/10 | 10/10 |
| PER.C6™ | $1 \times 10^7$ | 9/10 | 7/10 |

**B. 84-day tumorigenicity study of PER.C6™ and HeLa cells**

| Cell type | No. of cells | Day 21 | Day 42 | Day 84 |
|---|---|---|---|---|
| HeLa | $1 \times 10^6$ | NA | 10/10 | NA |
| PER.C6™ | $1 \times 10^7$ | 5/10 | 5/10 | 1/10 |
| Medium control | --- | 1/10[a] | 0/10 | 0/10 |

**C. Titration tumorigenicity study of PER.C6™ cells in nude mice**

| Cell type | No. of cells | Day 21 | Day 42 | Day 84 |
|---|---|---|---|---|
| PER.C6™ | $1 \times 10^3$ | 0/10 | 0/10 | 0/10 |
| PER.C6™ | $1 \times 10^5$ | 0/10 | 0/10 | 0/10 |
| PER.C6™ | $1 \times 10^7$ | 5/10 | 9/10 | 7/10[b] |
| Medium | – | 0/10 | 0/10 | 0/10 |

Details of the experiment are presented in Section 5.2.
[a]Benign lung adenoma.
[b]seven animals sacrificed, with tumors on day 56 and leaving 0/3 at day 84.

injected subcutaneously with $10^7$ PER.C6™ cells. Positive control animals were injected subcutaneously with $10^7$ KB cells. KB is a known tumor-producing cell line derived from an epidermoid carcinoma (American Type Culture Collection; CCL-121). The study was of a 28-day duration, at which point all animals were necropsied and examined grossly and histologically. All of the positive control animals had growing nodules, and 8 of 10 male mice and 7 of 10 female mice receiving PER.C6™ cells had growing nodules, thus producing a positive test (Table 6(A)).

At the time of the first study, 21 or 28 days was the duration that was usually used. Subsequently the Center for Biologics Evaluation and Research (CBER) of the Food and Drug Administration had suggested the observation period be extended to 84 days. This was to give more time for slow-growing tumors to appear and for non-tumorigenic nodules to regress or disappear. Therefore, the tumorigenicity study on the PER.C6™ cells was repeated.

The second study was performed in nude (nu/nu) mice over an 84-day period. Thirty nude mice were injected subcutaneously with $10^7$ PER.C6™ cells in 0.2 ml of SF medium. As a positive control, 10 mice were injected subcutaneously with $10^6$ HeLa cells in 0.2 ml of SF medium. As a negative control, 30 mice were injected with 0.2 ml of medium. The mice were palpated at the injection site every 3–7 days and any nodules found were measured in two dimensions. The PER.C6™ cell test arm and the negative control arm had 10 mice necropsied 21, 42, and 84 days post injections. The positive control arm was necropsied at 42 days post injection. Gross and histological examinations were performed on all injection sites and nodules if they appeared. During the initial days after injection, palpable nodules were present at the subcutaneous injection sites in all animals inoculated with PER.C6™ cells. Between post-injection days 5 and 14, the detectable masses disappeared from the injection sites. However, in several of these mice, the masses subsequently reappeared by around day 21 and continued to enlarge until the animals were necropsied. Of the mice injected with PER.C6™ cells, 5 of 10 sacrificed on day 21, 5 of 10 sacrificed on day 42, and 1 of 10 sacrificed on day 84 (actually sacrificed on day 49 due to tumor size) had gross or microscopic evidence of a tumor (Table 6(B)). Histologically, these recurrent nodules were composed of sheets of large pleomorphic cells with numerous, sometimes abnormal, mitotic figures. These masses compressed the surrounding tissues but were not invasive. No tumors were observed outside the injection sites. The interpretation of the test is that PER.C6™ cells are positive for tumorigenicity.

In view of the positive tumorigenicity results obtained following injection of $10^7$ PER.C6™ cells, a titration study was performed in which nude mice were injected with PER.C6™ cells at doses of $10^7$, $10^5$, or $10^3$ cells per animal. Mice were necropsied 21, 42, or 84 days post injection. No animals receiving $10^3$ PER.C6™ cells had palpable masses at the injection site from the first palpation day until necropsy. None of these animals had gross or microscopic evidence of nodules or tumor cell collections at any necropsy time point. Two of the thirty mice receiving $10^5$ PER.C6™ cells had palpable nodules on post injection Day 3. These masses disappeared by day 7 and did not recur. Gross and histological examination of the injection sites was negative at all necropsy time points. In the mice that received $10^7$ PER.C6™ cells, 29 of 30 animals had palpable nodules on day 3—some of which disappeared or became smaller but most of these recurred and grew progressively until necropsy. At necropsy, 5 of 10 mice on day 21 had tumors, 9 of 10 mice sacrificed on day 42 had tumors, and 7 of 10 in the group scheduled for day 84 had tumors but were sacrificed on day 56 because of tumor size (Table 6(C)). The histological and gross features of the PER.C6™ cell tumors were similar to those described for the previous study (above). No metastatic nodules were found. Thus, the tumorigenicity studies of PER.C6™ cells were positive at $10^7$ cells per animal and negative at $10^5$ and $10^3$ cells per animal. This would indicate that not all of the PER.C6™ cells are tumorigenic and/or a critical mass of tumorigenic cells is necessary for tumor formation.

### 5.2.2 *Tumorigenicity Studies of Residual DNA from PER.C6™ Cells*

In view of the positive tumorigenicity studies with $10^7$ PER.C6™ cells, the oncogenic potential of residual DNA from these cells was tested in both nude mice and newborn hamsters. For these studies, DNA was isolated from passage 61 PER.C6™ cells using standard procedures. The DNA preparation was shown to be of high molecular weight (average size ~100 Kbp) and devoid of significant protein or RNA impurities.

In the nude mice study, 20 female nude (nu/nu) mice were injected subcutaneously with 225 μg of PER.C6™ DNA (in a volume of 0.25 ml). For negative controls, two groups of 20 female mice each were injected subcutaneously with 0.25 ml of vehicle. Approximately 5 months after injection, the mice were necropsied and examined histologically for tumor growth. None of the mice in this study exhibited gross or microscopic evidence of tumors at the injection site. One treated mouse had a lymphoma at a distant site. However, nude mice—particularly females—are known to have a high incidence of spontaneous lymphoma,[74–77] and the occurrence of a single lymphoma in twenty treated mice is consistent with the spontaneous incidence.

Although the lymphoma was almost certainly a spontaneous event, a PCR study was performed on the lymphoma DNA to determine if there was any evidence for the presence of the adenovirus E1 region—the transforming agent of PER.C6™ cells. The study was negative, with a sensitivity of approximately one copy of E1 per 750 tumor cells. Previously, E1 expression has been shown to be necessary to maintain the transformed state of 293 cells, which, like PER.C6™ cells, were transformed by E1.[78] The results of the PCR analysis support the conclusion that the lymphoma was a spontaneous event, not induced by PER.C6™ DNA.

A second tumorigenicity study using DNA from PER.C6™ cells was carried out in newborn hamsters. Between 18 and 36 h after birth, female and male hamsters (28 in total) were injected subcutaneously with approximately 100 μg of PER.C6™ DNA (in a volume of 110 μl). Two groups of controls hamsters (50, mixed sex, per group) were injected with 100 μl of vehicle. Several pups in each group were lost due to maternal cannibalism, reducing the group sizes to 20 (11 female, 9 male) in the PER.C6™ DNA group, 40 (19 female, 21 male) in control group-1, and 45 (27 female, 18 male) in control group-2. After weaning, the hamsters were palpated on a weekly basis. The hamsters were necropsied approximately 5 months after injection and examined grossly and histologically for tumor growth. One female hamster in control group-2 died approximately 21 weeks after injection of a malignant ovarian teratoma. No evidence of tumors was found in the 20 hamsters that were injected with PER.C6™ DNA.

### 5.2.3 Concerns About Using a Tumorigenic Cell Substrate

The basis for concern about using a tumorigenic cell substrate to produce a vaccine includes three theoretical possibilities. First, DNA from the cells carrying a putative-activated oncogene or cancer-causing mutation could be integrated into the recipients' genome and produce a tumor. Second, a transforming protein in the cells could be transmitted and result in a tumor. Third, an adventitious tumor virus may be present and could be transmitted to the recipient and produce a tumor.

Concerning residual DNA from a tumorigenic cell substrate, there have now been several reports demonstrating that DNA extracted from tumorigenic cell lines or tumors growing in vivo—and even purified activated oncogenes—do not produce tumors when injected into animals at levels up to 1000 μg of DNA.[79–86] The negative results obtained with PER.C6™ DNA in nude mice and newborn hamsters are consistent with these findings. In the case of the PER.C6™ studies, the amount of DNA injected (~100 or 225 μg) represents a >$10^6$-fold excess compared to the amount of residual DNA present in a dose of vaccine produced on this cell substrate. Others have calculated that 100 pg of residual DNA from tumorigenic cells would be equal to less than a billionth of a tumor-producing dose.[79–86]

The second concern of transforming proteins or growth factors has been considered by a WHO study group to be significant only if those are continually produced by cells or have continued administration.[79,80] The study group did not consider the presence of contaminating known growth factors, in the concentrations that they would be found, to constitute a serious risk in biological products prepared from continuous cell lines.

The third category of concern, viruses or other adventitial agents, does present a potential risk. This risk is greatest when primary cells are used because of the frequent need for newly acquired cells that require repeats of the extensive testing for adventitial agents. Human diploid cell lines and continuous tumorigenic cell lines are thoroughly and routinely tested for a wide variety of known and unknown adventitial agents in a series of in vitro and in vivo assays, thus providing adequate assurance that adventitial agents will not be transmitted.

## 5.3 Prion-Related Issues

It is now generally accepted that an abnormal form of the cell surface glycoprotein PrP, or prion protein, is the main infectious agent in transmissible spongiform encephalopathies like scrapie, bovine spongiform encephalopathy (BSE), and Creutzfeldt–Jakob disease (CJD) (Ref. 87 and reviewed in Ref. 88). The abnormal form of PrP, called $PrP^{sc}$ or PrP-res, is characterized by a remarkable resistance to denaturing agents and to degradation by Proteinase K (Prot K). Diagnostic tests take advantage of this unusual stability that allows a distinction between $PrP^{c}$ and $PrP^{sc}$ using antibodies that recognize both forms of PrP (e.g., Ref. 89).

Human prion diseases occur in sporadic, acquired, or inherited forms with different clinical and pathological phenotypes (reviewed in Ref. 90). In 1996 a new variant of CJD (vCJD) was reported in relatively young patients with clinical features different from the known CJD forms.[91] It was also found by strain typing that the prion protein of these patients was indistinguishable from the one that causes BSE thus raising the question whether vCJD could be acquired by consumption of meat from cattle suffering from BSE.[92,93] The possibility of transmission of $PrP^{sc}$ from bovine to human raises safety issues for cultured cell lines used for the production of human drugs.

Therefore, PER.C6™ cells were carefully examined for the PrP phenotype (see below) as well as genotype. It has been found that specific mutations in the PrP gene are associated with hereditary forms of human prion disease (reviewed in Refs 88,90). Furthermore, a common methionine/valine polymorphism at codon 129 of the PrP gene appears to be associated with phenotypic variability and susceptibility to sporadic and iatrogenic CJD. The vast majority of patients suffering from sCJD, and also from vCJD, were found to be homozygous for 129M whereas patients heterozygous at codon 129 were strikingly underrepresented.[94–96] To examine whether the PER. C6™ PrP gene contains any of the known mutations associated with susceptibility to prion diseases, the PER.C6™ PrP gene was sequenced. For these sequencing studies genomic DNA from PER.C6™ cells was isolated and used to amplify the PrP gene sequences by PCR. The resulting PCR product was cloned into a vector, and the PrP gene in each of 13 PrP-containing clones was sequenced by BaseClear (Leiden, The Netherlands). Five of these clones contained sequences coding for the 129M $PrP^c$ protein, while the other eight contained the 129V $PrP^c$ sequence, demonstrating the heterozygosity at this position. To confirm this observation, the resulting PCR product was also sequenced. As expected, a double peak (g/a) was observed in the 129 codon at a position defining it as a valine (if the nucleotide is a guanine) or as methionine (if the nucleotide is an adenine). The PER.C6™ PrP gene sequence was then compared to the wild-type sequence published in Genbank (accession number M12899) and was found to be identical to the wild-type gene; thus, ruling out the possibility that these cells possessed a hereditary mutation that would be predisposing for prion diseases. The sequence also revealed that PER.C6™ cells are heterozygous for methionine/valine at codon 129.

$PrP^c$ is constitutively expressed in adult brain[89,97,98] and at lower levels in other tissues like liver and spleen.[99] PrP expression has also been found in a variety of rodent and human cell lines. Our studies on PER.C6™ and 293 cells have shown that these cells also express the cellular form of PrP. A validated Western blot analysis of Prot K-treated protein extracts of PER.C6™ cells, and their parental HER cells have failed to detect any Prot K-resistant forms of PrP at passages 33 and 36 of PER.C6™ cells and passage 6 of their parental HER cells.

In addition to the sequencing of the prion gene and testing for the presence of abnormal prion protein in the PER.C6™ cells at an early and late passage level of the culture, serum and trypsin batches that were used were traced to see if any were derived in the United Kingdom.

Finally, it has been possible to adopt the PER.C6™ cells to SF suspension so that bovine serums can be completely avoided in the future if desired.

The above-mentioned characteristics of PER.C6™ make it a safe manufacturing cell line in this respect.

## 5.4 Genetic Characterization of PER.C6™ Cells

### 5.4.1 Sequence Analysis of E1

The integrity of the E1A and E1B coding regions present in PER.C6™ was tested by sequence analysis. This was done by bidirectional sequencing of PCR fragments generated from these regions, and the sequence of these fragments was compared to the original pIG.E1A.E1B sequence, the construct that was initially used in transfection.

No mutations, deletions, or insertions were detected between the sequence of the PCR fragments and pIG.E1A.E1B, indicating that no genetic alterations were introduced in the E1A and E1B regions during transfection and subsequent culture of the cells.

### 5.4.2 Site of Integration of E1

The chromosomal integration site of the plasmid pIG.E1A.E1B in PER.C6™ was determined by using the multicolor in situ hybridization (MFISH) technique in combination with the principle of combined binary ratio labeling (COBRA).[100] This technique combines 24-color COBRA–MFISH using chromosome-specific painting probes for all human chromosomes with plasmid probe (pIG.E1A.E1B) visualization (25th color).

The pIG.E1A.E1B integration site was determined using PER.C6™ cells that are derived from the research MCB (passage number 29). Cells were analyzed at passage numbers 31, 41, 55, and 99. Two-hundred-and-fifty metaphases and interphases were studied.

pIG.E1A.E1B integration was only detected on chromosome 14 (Figure 9) and in both sister chromatids of the chromosome in all PER.C6™ passage numbers screened. 75–80% of the 47 metaphases and 203 interphases consisted of integration of pIG.E1A.E1B in one chromosome 14, whereas 20–25% consisted of integration in two chromosomes 14.[101]

### 5.4.3 Copy Number of the E1 Construct

The number of copies of pIG.E1A.E1B present in the PER.C6™ chromosome was studied by southern blot analysis, dot bot analysis, and fiber FISH analysis.[101]

Southern hybridization revealed the presence of several integrated copies of pIG.E1A.E1B in the genome of PER.C6™.[45]

In addition, dot blot analysis showed a pIG.E1A.E1B plasmid copy number of $19 \pm 3$ (research MCB) and $24 \pm 16$ (extended cell bank, passage number 99) per genome.

From the results, it was concluded that PER.C6™ consists of five to six copies of pIG.E1A.E1B per haploid genome.

Fiber FISH enables physical length measurements of in situ-hybridized DNA probes on linearized DNA fibers with a resolution equal to the theoretical length of a

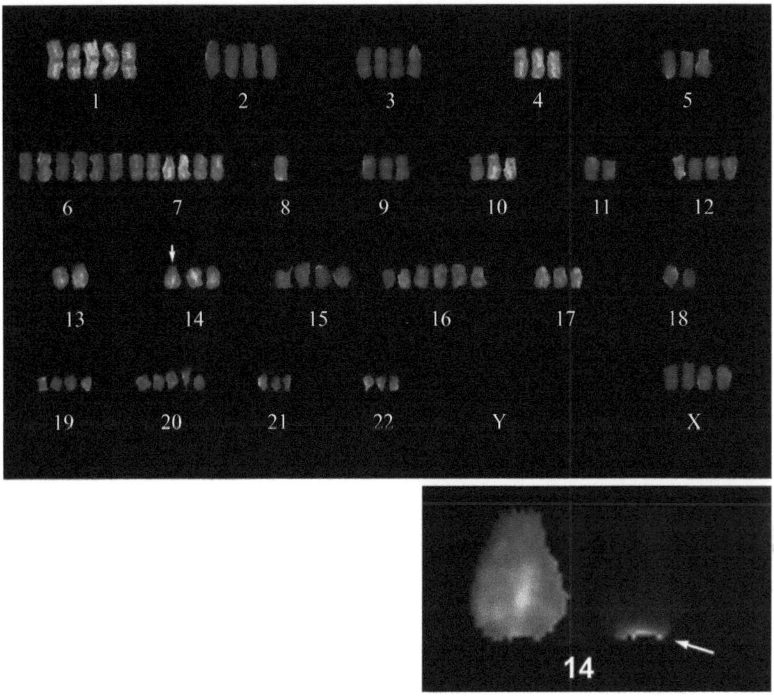


**Figure 9** Chromosomal localization of pIG.E1A.E1B in PER.C6™ cells. Twenty five-color COBRA FISH with 24 human chromosome-specific painting probes combined with integrated plasmid probe DNA on PER. C6 metaphase chromosomes. One out of three chromosomes 14 contains the integrated E1 construct; this chromosome is shown as an enlargement.

linearized DNA molecule according the model of Watson and Crick (1 kb is 0.34 μm). Therefore, fiber FISH was conducted to measure the length of the integrated construct in the PER.C6™ cell line at passage number (pn) 31, 41, and 99. Twenty fibers were measured. It was determined that pIG.E1A.E1B was integrated in tandem copies in chromosome 14 of PER.C6™. The copy number of these in tandem integrations was determined to be: pn31: $13.6 \pm 6.1$; pn41: $18 \pm 4.5$, and pn99: $20.1 \pm 7.9$.

One integration site was detected per cell, which resulted in approximately five copies per haploid genome.

In conclusion, one pIG.E1A.E1B integration site is detected per PER.C6™ cell, which results in approximately five copies per haploid genome.

### *5.4.4 Chromosome Analysis*

PER.C6™ cells from cellular passage 44 and 66 were harvested for chromosome analysis to determine the modal chromosome number and the karyotype in a sample of metaphase plates. Cells were harvested, and slides were prepared and stained using a standard giemsa banding technique. At each passage level, the number of chromosomes in 50 metaphase plates was counted. Also, full karyotypes were prepared from each passage level.

At passage level 44, the chromosome number ranged from 43 to 160. The mean number of chromosomes was 72 and the modal number was 61. All metaphase plates examined had structural chromosomal changes and rearrangements. A marker chromosome 19 with additional material in the long arm ($19q^+$) was the most common alteration and was found in 14 of the 20 metaphase plates that were karyotyped.

At passage 66, the chromosome number ranged from 42 to 112. The mean number of chromosomes was 63 and the modal number was 64. All metaphase plates karyotyped again were found to have structural changes. The $19q^+$ was again the most common change, observed in 15 of 20 karyotypes. There was also a marker chromosome 11 with extra material in the short arm ($11p^+$) in 14 of the 20 karyotypes and a marker chromosome 9 with additional material in the short arm (9p+) in 8 of the 20 karyotypes.

Several of the markers differed at the two passage levels, but the most common marker, 19q+, was the same. The continuing changes seen as passage level increases are typical of heteroploid continuous cell lines.

### 5.4.5 DNA Fingerprinting

PER.C6™ cells were also analyzed on two occasions by DNA fingerprinting. DNA profile analysis of PER.C6™ indicated no changes in the banding pattern obtained between the research MCB (pn 29) and an extended cell bank that was laid down at passage number 99. On a second occasion, a consistent DNA fingerprint was obtained between pn 45 and pn 67. There was no evidence of cross-contamination with other cell lines.

## 6. Conclusions

At the present time, the PER.C6™ cell line is the best substrate for the production of adenovirus vectors for gene therapy or vaccines. This conclusion is based on the ability to obtain good yields and safety considerations.

The major safety considerations are the possibility of:

1. the production of RCA;
2. a tumorigenic risk from the transformed cell line;
3. the presence of abnormal prions; and
4. contamination by adventitial agents.

As described in this chapter, the lack of any overlap between the genome of the adenoviral vectors that carry the E1 deletion and the adenoviral E1 sequences carried in the PER.C6™ cells makes homologous recombination impossible, thereby preventing the formation of RCA.

It is well known that many transformed cell lines can produce tumors when injected into immunodeficient animals. As described, PER.C6™ cells produce tumors in nude mice when $10^7$ cells are injected. They do not produce tumors, however, when $10^5$ or $10^3$ cells are injected. Since it is not anticipated that there will be any PER.C6™ cells in a final product, this leaves the question of possible tumorigenicity of residual PER.

C6™ cellular DNA. Studies in nude mice and newborn hamsters in which DNA from PerC6 cells was injected were negative for tumor production.

The possibility of the presence of abnormal prions that could produce a neurodegenerative disease was also considered. This could occur if the PER.C6™ cells had a mutation in a prion gene or if the cells were contaminated with abnormal prions such as in bovine spongioform encephalopathy. As far as possible, all serum and trypsin batches used from the time of origin of the culture were traced and no contact of serum from British sources was identified. The PER.C6™ cell line was also adapted to SF suspension cultures.

The prion protein gene of PER.C6™ cells was sequenced, and no mutations were found, and the cell line was shown to be heterozygous for the 129 M/V polymorphism. The cell line was also analyzed for the presence of abnormal prions at an early and late passage and an early passage of the HER parental line and none were found. In total, these studies indicate that the risk of a prion disease from the use of PER.C6™ cells is vanishingly small.

Finally, extensive studies for known and unknown adventitial agents have been documented and are negative.

While there can be no absolute elimination of risk, this body of studies indicates a minimal if any risk from the use of this cell substrate for the production of adenovirus vectors. As new studies are developed, they will also be applied to ensure that no hazards are present. It has often been pointed out that a continuous cell line such as PER. C6™ permits extensive analysis for adventitial agents and other safety concerns and thus is less hazardous than short-lived primary cell cultures for which testing must be repeated for each newly established culture.

## References


1. Russell W. Update on adenovirus and its vectors. *J Gen Virol* 2000;**81**:2573–604.
2. Rowe W, Huebner R, Gilmore L, Parrott R, Ward T. Isolation of a cytopathogenic agent from human adenoids undergoing spontaneous degeneration in tissue culture. *Proc Soc Exp Biol Med* 1953;**84**:570–3.
3. Hilleman M, Werner J. Recovery of new agents from patients with acute respiratory illness. *Proc Soc Exp Biol Med* 1954;**85**:183–8.
4. Hierholzer J, Wigand R, Anderson L, Adrian T, Gold J. Adenoviruses from patients with AIDS: a plethora of serotypes and a description of five new serotypes of subgenus D (types 43-47). *J Infect Dis* 1988;**158**:804–13.
5. Schnurr D, Dondero ME. Two new candidate adenovirus serotypes. *Intervirology* 1993;**36**:79–83.
6. Jong Jd, Wermenbol A, Verweij-Uijterwaal M, Slaterus K, Dillen PW-V, Doornum GV, et al. Adenoviruses from human immunodeficiency virus-infected individuals, including two strains that represent new candidate serotypes Ad50 and Ad51 of species B1 and D, respectively. *J Clin Microbiol* 1999;**37**:3940–5.
7. Francki R, Fauquet C, Knudson D, Brown F. Classification and nomenclature of viruses. Fifth report of the international committee on taxonomy of viruses. *Arch Virol* 1991;(Suppl. **2**):140–4.

8. Horwitz MS. Adenoviridae and their replication. In: Fields BN, Knipe DM, editors. *Virology*. New York: Raven Press, Ltd; 1990. p. 1679–740.
9. Wold WSM, Tollefson AE, Hermiston TW. Strategies of immune modulation by adenoviruses. In: McFadden G, editor. *Viroceptors, virokines and related immune modulators encoded by DNA viruses*. Heidelberg: Springer-Verlag; 1995. p. 147–85.
10. Wadell G. Molecular epidemiology of adenoviruses. *Curr Top Microbiol Immunol* 1984;**110**:191–220.
11. Ishibashi M, Yasue H. In: Ginsberg H, editor. *The adenoviruses*. London, New York: Plenum Press; 1984. p. 497–561.
12. Arnberg N, Kidd A, Edlund K, Olfat F, Wadell G. Initial interactions of subgenus D adenoviruses with A549 cellular receptors: sialic acid versus αv integrins. *J Virol* 2000;**74**:7691–3.
13. Roelvink P, Lizonova A, Lee J, Li Y, Bergelson J, Finberg R, et al. The coxsackievirus-adenovirus receptor protein can function as a cellular attachment protein for adenovirus serotypes from subgroups A, C, D, E and F. *J Virol* 1998;**72**:7909–15.
14. Bergelson J, Cunningham J, Droguett G, Kurt-Jones E, Krithivas A, Hong J, et al. Isolation of a common receptor for coxsackie B viruses and adenoviruses 2 and 5. *Science* 1997;**275**:1320–3.
15. Greber UF, Willetts M, Webster P, Helenius A. Stepwise dismantling of adenovirus 2 during entry into cells. *Cell* 1993;**75**:477–86.
16. Vlietvander P. Adenovirus DNA replication. In: Doerfler W, Bohm P, editors. *The molecular repertoire of adenoviruses*, vol. II. Berlin: Springer Verlag; 1995.
17. Ulfendahl PJ, Linder S, Kreivi JP, Nordqvist K, Sevensson C, Hultberg H, et al. A novel adenovirus-2 E1A mRNA encoding a protein with transcription activation properties. *EMBO J* 1987;**6**:2037–44.
18. Stephens C, Harlow E. Differential splicing yields novel adenovirus 5 E1A mRNAs that encode 30 kd and 35 kd proteins. *EMBO J* 1987;**6**:2027–35.
19. Sarnow P, Hearing P, Anderson CW, Halbert DN, Shenk T, Levine AJ. Adenovirus early region 1B 58,000-dalton tumor antigen is physically associated with an early region 4 25,000-dalton protein in productively infected cells. *J Virol* 1984;**49**:692–700.
20. Rubenwolf S, Schutt M, Nevels H, Dobner T. Structural analysis of the adenovirus type 5 E1B 55-kilodalton-E4ORF6 protein complex. *J Virol* 1997;**71**:1115–23.
21. Debbas M, White E. Wild-type p53 mediates apoptosis by E1A which is inhibited by E1B. *Genes Dev* 1993;**7**:546.
22. Tarodi B, Subramanian T, Chinnadurai G. Functional similarity between adenovirus E1B 19K gene and Bcl2 oncogene – mutant complementation and suppression of cell death induced by DNA damaging agents. *Int J Oncol* 1993;**3**:467–72.
23. Telling GC, Perera S, Szatkowski OM, Williams J. Absence of an essential regulatory influence of the adenovirus E1B 19-kilodalton protein on viral growth and early gene expression in human diploid WI38, HeLa, and A549 cells. *J Virol* 1994;**68**:541–7.
24. Leppard KN. E4 gene function in adenovirus, adenovirus vector and adeno-associated virus infections. *J Gen Virol* 1997;**78**:2131–8.
25. Pombo A, Ferreira J, Bridge E, Carmo Fonseca M. Adenovirus replication and transcription sites are spatially separated in the nucleus of infected cells. *EMBO J* 1994;**13**: 5075–85. ISSN:0261–4189.
26. Dobner T, Horikoshi N, Rubenwolf S, Shenk T. Blockage by adenovirus E4orf6 of transcriptional activation by the p53 tumor suppressor. *Science* 1996;**272**:1470–3.
27. Moore M, Horikoshi N, Shenk T. Oncogenic potential of the adenovirus ORF6 protein. *Proc Natl Acad Sci USA* 1996;**93**:11295–303.

28. Muller U, Kleinberger T, Shenk T. Adenovirus E4orf4 protein reduces phosphorylation of c-Fos and E1A proteins while simultaneously reducing the level of AP-1. *J Virol* 1992;**66**:5867–78.
29. Shtrichman R, Kleinberger T. Adenovirus type 5 E4 open reading frame 4 protein induces apoptosis in transformed cells. *J Virol* 1998;**72**:2975–82.
30. Johnson D, Ohtani K, Nevins J. Autoregulatory control of E2F1 expression in response to positive and negative regulators of cell cycle progression. *Genes Dev* 1994;**8**:1514–25.
31. Lubeck MD, Davis AR, Chengalvala M, Natuk RJ, Morin JE, Molnar KK, et al. Immunogenicity and efficacy testing in chimpanzees of an oral hepatitis B vaccine based on live recombinant adenovirus. *Proc Natl Acad Sci USA* 1989;**86**:6763–7.
32. Torres JM, Alonso C, Ortega A, Mittal S, Graham F, Enjuanes L. Tropism of human adenovirus type 5-based vectors in swine and their ability to protect against transmissible gastroenteritis coronavirus. *J Virol* 1996;**70**:3770–80.
33. Pacini DL, Dubovi EJ, Clyde WA. A new animal model for human respiratory tract disease due to adenovirus. *J Infect Dis* 1984;**150**:92–7.
34. Trentin J, Yale Y, Taylor G. The quest for human cancer viruses. *Science* 1962;**137**:835–41.
35. Bernards R, van der Eb AJ. Adenovirus: transformation and oncogenicity. *Biochim Biophys Acta* 1984;**783**:187–204.
36. Flint SJ. Transformation by adenoviruses. In: Tooze J, editor. *Molecular biology of tumor viruses, II. DNA tumor viruses*. Cold Spring Harbor, NY: Cold Spring Harbor Laboratory; 1980. p. 547–76.
37. Freeman A, Black P, VanderPool E, Henry P, Austin J, Huebner R. *Proc Natl Acad Sci USA* 1967;**58**:1205–12.
38. Graham FL, van der Eb AJ, Heijneker HL. Size and location of the transforming region in human adenovirus type 5 DNA. *Nature* 1974;**251**:687–91.
39. Schrier P, Bernards R, Vaessen R, Houweling A, van der Eb AJ. Expression of class I major histocompatibility antigens switched off by highly oncogenic adenovirus 12 in transformed rat cells. *Nature* 1983;**305**:771–5.
40. Graham FL, Smiley J, Russell WC, Nairn R. Characteristics of a human cell line transformed by DNA from adenovirus type 5. *J Gen Virol* 1977;**36**:59–72.
41. Whittaker JL, Byrd PJ, Grand RJA, Gallimore PH. Isolation and characterization of four adenovirus type 12-transformed human embryo kidney lines. *Mol Cell Biol* 1984;**4**:110–6.
42. Gallimore PH, Grand RJA, Byrd PJ. Transformation of human embryo retinoblasts with simian virus 40, adenovirus and ras oncogenes. *Anticancer Res* 1986;**6**:467–99.
43. Vaessen RTMJ, Houweling A, Israel A, Kourilsky P, van der Eb AJ. Adenovirus E1A-mediated regulation of class I MHC expression. *EMBO J* 1986;**5**:335–41.
44. Fallaux FJ, Hoeben RC, Cramer SJ, van den Wollenberg DJM, Ormondt Hv, van der Eb AJ. Human cell line 911 is a good alternative for the production and titration of adenovirus-5-based vectors lacking early region 1. *Gene Ther* 1994;**1**:S12.
45. Fallaux FJ, Bout A, Velde Iv d, van den Wollenberg DJM, Hehir K, Keegan J, et al. New helper cells and matched early region 1-deleted adenovirus vectors prevent generation of replication-competent adenoviruses. *Hum Gene Ther* 1998;**9**:1909–17.
46. Schiedner G, Hertel S, Kochanek S. Efficient transformation of primary human amniocytes by E1 functions of Ad5: generation of new cell lines for adenoviral vector production. *Hum Gene Ther* 2000;**11**:2105–16.
47. Houweling A, Elsen Pv d, van der Eb AJ. Partial transformation of primary rat cells by the left-most 4.5% of adenovirus 5 DNA. *Virology* 1980;**105**:537–50.
48. Teodoro J, Branton P. Regulation of apoptosis by viral gene products. *J Virol* 1997;**71**:1739–46.

49. Dyson N, Harlow E. Adenovirus E1A targets key regulators of cell proliferation. *Cancer Surv* 1992;**12**:161–95.
50. Peeper DS, Zantema A. Adenovirus-E1A proteins transform cells by sequestering regulatory proteins. *Mol Biol Rep* 1993;**17**:197–207.
51. Peeper DS, van der Eb AJ, Zantema A. The G1/S cell-cycle checkpoint in eukaryotic cells. *Biochim Biophys Acta* 1994;**1198**:215–30.
52. White E, Cipriani R, Sabbatini P, Denton A. Adenovirus E1B 19-kilodalton protein overcomes the cytotoxicity of E1A proteins. *J Virol* 1991;**65**:2968.
53. Bout A. Towards safe and effective adenoviral vectors for human gene therapy. *Eur Biopharm Rev* December 1999:94–9.
54. Imler JL, Chartier C, Dreyer D, Dieterlé A, Sainte-Marie M, Faure T, et al. Novel complementation cell lines derived from human lung carcinoma A549 cells support the growth of E1-deleted adenovirus vectors. *Gene Ther* 1996;**3**:75–84.
55. Brough DE, Lizonova A, Kovesdi I. Stable cell lines for complementation of adenovirus early regions E1, E2A and E4. In: *Abstract book CSH conference on gene Therapy*. 1996. p. 42.
56. Louis N, Evelegh C, Graham FL. Cloning and sequencing of the cellular-viral junctions from the human adenovirus type 5 transformed 293 cell line. *Virology* 1997;**233**:423–9.
57. Hehir KM, Armentano D, Cardoza LM, Choquette TL, Berthelette PB, White GA, et al. Molecular characterization of replication-competent variants of adenovirus vectors and genome modifications to prevent their occurrence. *J Virol* 1996;**70**:8459–67.
58. Lochmüller H, Jani A, Huard J, Prescott S, Simoneau M, Massie B, et al. Emergence of early region 1-containing replication-competent adenovirus in stocks of replication-defective adenovirus recombinants (ÐE1+ÐE3) during multiple passages in 293 cells. *Hum Gene Ther* 1994;**5**:1485–92.
59. Zhu J, Grace M, Casale J, Chang A, Musco M, Bordens R, et al. Characterization of replication-competent adenovirus isolates from large-scale production of a recombinant adenoviral vector. *Hum Gene Ther* 1999;**10**:113–21.
60. Imler JL, Bout A, Dreyer D, Dieterle A, Schultz H, Valerio D, et al. Trans-complementation of E1-deleted adenovirus: a new vector to reduce the possibility of codissemination of wild-type and recombinant adenoviruses. *Hum Gene Ther* 1995;**6**:711–21.
61. Hermens WTJMC, Verhaagen J. Adenoviral-vector-mediated gene expression in the nervous system of immunocompetent Wistar and T cell-deficient nude rats: preferential survival of transduced astroglial cells in nude rats. *Hum Gene Ther* 1997;**8**:1049–63.
62. Ormondt Hv, Galibert F. Nucleotide sequences of adenovirus DNAs. In: *Current topics in microbiology and immunology*, vol. 110. Heidelberg: Springer-Verlag; 1984. p. 73–142.
63. Vales LD, Darnell JJ. Promoter occlusion prevents transcription of adenovirus polypeptide IX mRNA until after DNA replication. *Genes Dev* 1989;**3**:49–59.
64. Singer-Sam J, Keith DH, Tani K, Simmer RL, Shively L, Lindsay S, et al. Sequence of the promoter region of the gene for X-linked 3-phosphoglycerate kinase. *Gene* 1984;**32**:409–17.
65. Valerio D, Duyvesteyn MGC, Dekker BMM, Weeda G, Bervens TM, Voorn Lv d, et al. Adenosine deaminase: characterization and expression of a gene with a remarkable promoter. *EMBO J* 1985;**4**:437–43.
66. Simonsen C, Levinson A. Analysis of processing and polyadenylation signals of the hepatitis B virus surface antigen gene by using simian virus 40-Hepatitis B virus chimeric plasmids. *Mol Cell Biol* 1983;**3**:2250–8.
67. Roberts BE, Miller JS, Kimelman D, Cepko CL, Lemischka IR, Mulligan RC. Individual adenovirus type 5 early region 1A products elicit distinct alterations of cellular morphology and gene expression. *J Virol* 1985;**56**:404–13.

68. Albert D, Rabson A, Dalton A. In vitro neoplastic transformation of uveal and retinal tissues by oncogenic DNA viruses. *Invest Ophthalmol* 1968;**7**:357–65.
69. Mukai N, Kalter S, Cummins L, Mathews V, Nishida T, Nakajima T. Retinal tumor induction in the baboon by human adenovirus 12. *Science* 1980;**210**:1023–5.
70. Fallaux FJ, Kranenburg O, Cramer SJ, Houweling A, Ormondt Hv, Hoeben RC, et al. Characterization of 911: a new helper cell line for the titration and propagation of early-region-1-deleted adenoviral vectors. *Hum Gene Ther* 1996;**7**:215–22.
71. Byrd P, Brown K, Gallimore P. Malignant transformation of human embryo retinoblasts by cloned adenovirus 12 DNA. *Nature* 1982;**298**:69–71.
72. Bett AJ, Haddara W, Prevec L, graham FL. An efficient and flexible system for construction of adenovirus vectors with insertions or deletions in early regions 1 and 3. *Proc Natl Acad Sci USA* 1994;**91**:8802–6.
73. Blanche F, Cameron B, Barbot A, Ferrero L, Guillemin T, Guyot S, et al. An improved anion-exchange HPLC method for the detection and purification of adenoviral particles. *Gene Ther* 2000;**7**:1055–62.
74. Sadoff D, Giddens Jr W, DiGiacomo R, Vogel A. Neoplasms in NIH Type II athymic (nude) mice. *Lab Anim Sci* 1988;**38**:407–12.
75. Furmanski P, Rich M. Neoplasms of the hematopoietic system. In: Foster H, Small J, Fox J, editors. *The mouse in biomedical research.* London: Academic Press; 1982. p. 351–71.
76. Gershwin M, Yoshiyuki P, Castles J, Ikeda R, Ruebner B. Anti-μ induces lymphoma in germfree congenitally athymic (nude) but not heterozygous (nu/+) mice. *J Immunol* 1983;**131**:2069–73.
77. Frith C, Ward J, Frederickson T, Harleman J. Neoplastic lesions of the hematopoietic system. In: Mohr U, Dungsworth D, Capen C, Carlton W, Sundberg J, Ward J, editors. *Pathobiology of the aging mouse.* Washington DC: ILSI Press; 1996. p. 219–35.
78. Quinlan MP. Expression of antisense E1A in 293 cells results in altered cell morphologies and cessation of proliferation. *Oncogene* 1993;**8**:257–65.
79. WHO Report. *Acceptability of cell substrates for production of biologicals* Report of WHO study group. Geneva: World Health Organization; 1987 (WHO Technical Report Series no. 747)(see tables 1 and 2).
80. WHO Report. *WHO requirements for the use of animal cells as in vitro substrates for the production of biologicals**. *(Requirements for biological substances no. 50).* 1998. *Reproduced from WHO Technical Report Series no. 878.
81. Grachev V. Advances in biotechnical processes. In: Mizraki A.*Viral vaccines*, vol. 14. Wiley Liss; 1990.
82. Mufson R, Gesner T. Lack of tumorigenicity of cellular DNA and oncogene DNA in newborn Chinese Hamsters. In: Hopps H, Petricciani J, editors. *Abnormal cells, new products and risk.* 1985. p. 168–9.
83. Levinson A, Svedersky L, Palladino Jr M. Tumorigenic potential of DNA derived form mammalian cell lines. In: Hopps H, Petricciani J, editors. *Abnormal cells, new producst and risks.* 1985. p. 161–5.
84. Palladino M, Levinson A, Svedersky L, Objeski J. Safety issues related to the use of recombinant DNA-derived cell culture products. I cellular components. *Dev Biol Stand* 1987;**66**:13–22.
85. Wierenga D, Cogan J, Petricciani J. Administration of tumor cell chromatin to immunosuppressed and non-immunosuppressed non-human primates. *Biologicals* 1995;**23**:221–4.
86. Lower J. Risk of tumor induction in vivo by residual cellular DNA: quantitative considerations. *J Med Virol* 1990;**31**:50–3.

87. Prusiner S. Novel proteinaceous infectious particle causes scrapie. *Science* 1982;**216**: 136–44.
88. Prusiner S. Prions. *Proc Natl Acad Sci USA* 1998;**95**:13363–83.
89. Oesch B, Westaway D, Walchli M, McKinley M, Kent S, Aebersold R, et al. A cellular gene encodes scrapie PrP 27-30 protein. *Cell* 1985;**40**:735–46.
90. Ironside J. Prion diseases in man. *J Pathol* 1998;**186**:227–34.
91. Will R, Ironside J, Zeidler M, Cousens S, Estibeiro K, Alperovitch A, et al. A new variant of Creutzfeldt-Jakob disease in the U.K. *Lancet* 1996;**347**:921–5.
92. Bruce M, Will R, Ironside J, McConnell I, Drummond D, Suttie A, et al. Transmissions to mice indicate that 'new variant' CJD is caused by the BSE agent. *Nature* 1997;**389**:498–501.
93. Collinge J, Sidle K, Meads J, Ironside J, Hill A. Molecular analysis of prion strain variation and the aetiology of 'new variant' CJD. *Nature* 1996;**383**:685–90.
94. Palmer M, Dryden A, Hughes J, Collinge J. Homozygous prion protein genotype predisposes to sporadic Creutzfeld-Jakob disease. *Nature* 1991;**352**:340–2.
95. Collinge J, Palmer M, Dryden A. Genetic predisposition to iatrogenic Creutzfeld-Jakob disease. *Lancet* 1991;**337**:1441–2.
96. Parchi P, Castellani R, Capellari S, Ghetti B, Young K, Chen S, et al. Molecular basis of phenotypic variability in sporadic Creutzfeldt-Jakob disease. *Ann Neurol* 1996;**39**:767–78.
97. Chesebro B, Race R, Wehrly K, Nishio J, Bloom M, Lechner D, et al. Identification of scrapie prion-protein mRNA in scrapie-infected and uninfected brain. *Nature* 1985;**315**:331–5.
98. Meyer R, McKinley M, Bowman K, Braunfeld M, Barry R, Prusiner S. Separation and properties of cellular and scrapie prion protein. *Proc Natl Acad Sci USA* 1986;**83**:2310–4.
99. Caughey B, Race R, Chesebro B. Detection of prion protein mRNA in normal and scrapie-infected tissues and cell lines. *J Gen Virol* 1988;**69**:711–6.
100. Tanke HJ, Wiegant J, van Gijlswijk RP, Bezrookove V, Pattenier H, Heetebrij RJ, et al. New strategy for multi-colour fluorescence in situ hybridisation: COBRA: COmbined Binary RAtio labelling. *Eur J Hum Genet* 1999;**7**:2–11.
101. Wiegant J, Brouwer K, Bezrookove V, Raap A, Bout A, Tanke H. Molecular cytogenetic characterization of PER.C6 cells, in preparation.

# Purification of Adenovirus

8

*Paul Shabram[1], Gary Vellekamp[2], Qian Ruan[1], Carl Scandella[3]*
[1]PaxVax Inc., San Diego, CA, USA; [2]Vellekamp Consulting LLC, Montclair, NJ, USA; [3]Carl Scandella Consulting, Bellevue, WA, USA

## 1. Introduction

Since the late 1950s, adenoviruses have been purified using classical methods of density gradient ultracentrifugation. These methods were efficient and could supply the quantities of highly purified viral particles necessary for research. The need for larger quantities has arisen with the advent of the use of adenoviral vectors for gene therapy trials. In this chapter, we discuss the techniques for extracting adenoviral particles from a complex milieu. Selecting the best technique for purification requires an understanding of the physical nature of the particles as well as the nature of contaminants. Knowledge of these properties is essential for developing a purification process that is sufficient to supply the commercial market for a therapeutic adenovirus.

### *1.1 Physical Characteristics of the Adenovirus Particle in Solution*

There are numerous adenoviruses that possess specific tropism for many species of animals including human, bovine, ovine, equine, canine, porcine, murine, and simian adenovirus subgenera. Although many of these adenoviruses are capable of delivering a transgene to human tissues, the development of clinical adenoviral agents most often employs human adenoviral vectors derived from human adenovirus serotypes 2 or 5. Consequently, for the purposes of this discussion, the information given here refers to human adenovirus types 2 and 5.

#### *1.1.1 Particle Size*

Size and shape are key factors involved in purifying any macromolecule. These factors are equally applicable to adenoviruses, which are much larger than most biomolecules commonly purified. Adenoviruses are icosahedral in shape with fiber-like extensions from each of the 12 vertices. The adenovirus comprises DNA, protein, and carbohydrate. The viral DNA is packaged in a highly organized protein coat termed the "capsid." Negative staining electron microscopy of the adenovirus capsid was used to estimate a diameter of 73 nm along a fivefold symmetry axis with a vertex-to-vertex diameter of about 83 nm.[1] Freeze fracture studies demonstrate a slightly larger capsid diameter. Fibers extending from each of 12 vertices increase the vertex-to-vertex diameter by about 40 nm for human adenovirus serotype 5. Oliver et al.[2] employed photon correlation spectroscopy to characterize the adenovirus type 5 particles in

**Adenoviral Vectors for Gene Therapy.**

solution, reporting a molecular weight of $1.67\times10^{8}$ Da and a corresponding particle diameter of 98 nm.

### 1.1.2 Diffusion of Adenovirus Particles

The adenovirus particle diffuses very slowly in solution. The diffusion coefficient for Ad-5 is $4.46\times10\,m^{12}/s^{2}$ in serum-containing media at 37 °C.[2,3] Figure 1 compares the diffusion constants of some well-known macromolecules to adenovirus. The large size and corresponding slow diffusion of adenovirus particles in solution require consideration in mixing because, given the density of the particle and its slow rate of diffusion, Brownian motion cannot be relied on to disperse the particle. If left alone the particles in solution would take weeks to reach equilibrium. However, gravity will intervene and cause the particles to sediment to the bottom of the container. Consequently, adenovirus solutions require greater agitation than protein solutions to disperse the particles evenly.

The slow diffusion rate also complicates analytical methods. For example, the interaction between the virus and a cell takes much longer than protein–cell interactions because of the slow diffusion rate. Typical biological methods for quantifying particles depend on Brownian motion for bringing about virus–cell interaction. Without accounting for this slow diffusion, the titer of the material tested may be underestimated. This will be discussed in more detail below (see Section 3.2.2 of this chapter).

### 1.1.3 Capsid Surface

The surface of the adenovirus capsid is of particular interest when selecting a separation technique as there are many binding methods available. The adenovirus capsid consists of 252 capsomeres. Two hundred and forty of these capsomeres are hexons

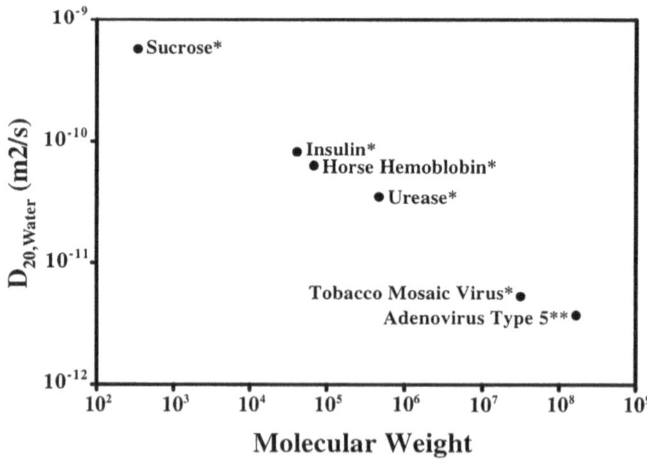


**Figure 1** The diffusion coefficients of macromolecules are related to their sizes. Adenoviral particles diffuse very slowly in solution. This slow diffusion rate affects mixing, separation, and analytical methods.

and 12 are pentons. Hexons are trimers of protein II and are the main structural component of the capsid. Pentons, constructed from penton base and fiber proteins, are prominent features as these structures protrude and add to the hydrodynamic radius of the virus particle. In addition, protein IIIa and protein IX may also contribute to surface characteristics. Protein IIIa is essential and two of these proteins are found at the junction joining adjacent facets together like stitching. Protein IX is not essential for capsid assembly but enhances the stability of the virus at higher temperatures. Four protein IX trimers stabilize a group of nine hexons (ninemers) that have assembled into a facet.[4]

The hexon, which composes about 50% of the total virion protein, dominates the charge characteristics of the particle. Hexon capsomeres possess an isoelectric point (pI) near pH 6.[1] At physiological pH, the capsomere would be expected to bear a negative charge. The complete adenovirus structure would also be expected to show a net anionic surface charge under physiological conditions. It is generally advisable to avoid exposing the proteins to solutions at or near the pI because many proteins change conformation, degrade, or precipitate when titrated through it.

Ninemers, hexons, and complete virions precipitate from solution when titrating the solution from pH 7 to pH 5.[5,6] If the particle is titrated through the pI rapidly and further lowered, losses occur but active particles can be recovered from solutions such as acetic acid. The virus is also stable up to around pH 8; the exact threshold is dependent on the composition of the solution. Exposing particles to a pH greater than 8 generally leads to a loss of activity and particle disruption.

### *1.1.4 Hydration of the Adenovirus Particle*

The degree of hydration of the particle is an important consideration for both purification and stability of the adenovirus. Sedimentation by ultracentrifugation using Schlieren optics[2] suggested that the adenovirus particle contains a "hard core" (water excluded) of a diameter of 76 nm, similar to that obtained using negatively stained electron microscopy. The difference between the "hard core" diameter of 76 nm and the light scattering diameter of 98 nm is significant because it suggests that the particle contains water. The amount of water represented by the difference suggests that the virus particle contains 2.3 g of water for every gram of virus.[2] These observations are consistent with measurements of other viruses.[7,8]

While proteins are stabilized by the incorporation of a few water molecules ("waters of hydration"), the amount of water suggested by these studies is more than 21 million water molecules per virion. This amount far exceeds water typically bound to proteins. This degree of hydration corresponds to a theoretical density calculation of about 1.4 g/ml, which is very close to the observed density of 1.34 g/ml. The additional water should not be surprising because the virus exists in an aqueous environment. With these data in mind, however, the particle would be expected to show unique properties.

The adenovirus particle is generally considered rigid. However, the degree of hydration suggests that a certain amount of flexibility should be considered. Because the particle is not encapsulated in a membrane small ions may have ready access to the core (see Section 4). Changes in ionic strength may induce conformational changes in the capsid proteins; some of these may be beneficial for separation and some may be catastrophic. The particle may also be sensitive to rapid changes in salt concentrations. One might also predict that hydrophobic solvents should be avoided.

## 1.2 Features of the Milieu

Effective virus purification capitalizes on the differences between the physical properties of the adenovirus relative to the components of the mixture from which it is being isolated. The exact composition of the milieu varies with the cell culture process and, to a lesser extent, every batch. In general, the large-scale purification of adenovirus requires the isolation of the virus from infected cell lysate taken from a bioreactor. This mixture consists of a formulated medium sometimes containing bovine serum, and less frequently antifoaming agents, or anticlumping agents (pluronics). Significant amounts of additives, however, present difficult challenges for any recovery procedure. Efficient large-scale production requires high cell densities which in turn require high gas exchange rates. This can cause severe foaming and necessitate the addition of agents to control it. Other additives such as anticlumping agents and lipids adapt the media for large-scale cell culture. Cell lysis, necessary to release the adenovirus from the host cell, results in the additional release of DNA, protein, lipids, carbohydrates, and other cellular components. Culture conditions, media components, cell-derived contaminants, and additives may have a significant impact on downstream processing.

### 1.2.1 Culture Conditions

Adenoviruses are produced by infection of cell lines in culture with a viral seed stock. The particular cell line used requires a highly developed cell culture method to achieve maximum yield. Flat stock culture, although useful for small-scale work, is generally not sufficient for larger scale applications. Some of the cell lines used in flat stock culture have resisted attempts to adapt them to the suspension and serum-free conditions preferred for large-scale processes. A compromise is struck by the culture of attachment-dependent cells using microcarriers in a bioreactor. These microcarrier-based processes introduce yet another component that must be separated from the adenovirus. Similarly, if serum is utilized, it will be necessary to consider the effective removal of its components.

### 1.2.2 Construct-Induced Contaminants and Considerations

The viral construct may also contribute to the milieu as the viral DNA backbone may lead to the contribution of many more contaminants. For downstream purification, higher titers favor better recovery and cleaner preparations because recovery and purification are enrichment processes. Even with maximum productivity, however, adenoviral particles represent a small fraction of molecular entities produced by the end of the culture process. Therefore, factors affecting the end titer can also affect the process.

The majority of adenoviral vectors for gene therapy are serotype 5 and have been rendered deficient for replication in most cells. With the exception of replication-competent adenoviruses, most vectors have been crippled to eliminate their replication in normal human cells. In general, when compared to wild-type virus, deletions or mutations in the early genes tend to attenuate viral replication in all cell lines. Attenuation for replication is typically achieved by large deletions in the immediate early region E1. These vectors require specialized packaging cell lines for efficient production. Cell lines such as HEK 293[9] or PER.C6[10] have been transformed with adenoviral DNA and provide sufficient E1 function *in trans* to enable replication.

In addition to E1 deletions, many vectors possess deletions for much of the E3 region. The deleted E3 genes are considered nonessential for viral replication and these deletions allow for larger transgene packaging capacity. Other deletions have been made to reduce the frequency of recombination during culture. Some vectors may have additional early gene deletions (e.g., in E4) as well as a deletion of protein IX encoding sequences.[11] Elimination of certain essential genes from the virus requires that the cell line be able to complement these protein functions *in trans* to package the virus.[12]

A significant unwanted by-product of adenoviral replication is DNA. Wild-type human adenoviruses are able to replicate in a variety of both quiescent and proliferating human cells due to the function of adenoviral immediate early genes. E1a proteins can be observed within an hour after infection, as cellular transcription factors are sufficient to transcribe the Ela genes. El expression initiates the adenoviral life cycle by altering the cell cycle machinery to induce cellular DNA replication even in quiescent cells. Viral and cellular proteins activate subsequent viral transcription. New copies of viral DNA are synthesized and viral production proceeds in a replication cascade. By the end of viral DNA replication, a large amount of DNA is present in the infected cell.

The purpose of a gene therapy vector is to convey a therapeutic effect by the delivery and expression of therapeutic genes. Many of these transgenes have a significant effect on the cells and adenoviral life cycle. Some genes, such as the retinoblastoma protein and the cyclin-dependent kinase inhibitor p21, directly affect the levels of activated E2F. E2F is a cellular transcription factor that is necessary for the transactivation of the adenoviral E2 promoter and thus the expression of viral DNA replication proteins. Other transgene products, such as pro-apoptotic proteins, can overcome the adenoviral block to apoptosis leading to early cell death and can interfere with adenoviral particle assembly. Secreted pleiotropic transgene products, such as growth factors, can trigger undesired effects in the packaging cell and severely attenuate production. Before adenoviral DNA can be coated with viral core proteins, the DNA is available for transcription. In this way, some adenoviral genes, especially those encoding capsid components, are not expressed until DNA replication occurs.[13] During this phase, the expression of transgene product is enhanced by the replication cycle itself by expanding the copy number of the transgene with each copy becoming available for transcription. Strong exogenous promoters may also sequester transcription factors and the cellular protein synthesis machinery can become clogged with transgene expression leading to attenuated adenoviral protein production. If the sequence of events leading to particle maturation is disturbed, even by an imbalance in protein production, large quantities of viral proteins, incomplete particle assemblies, transgene products, abnormal cellular structures, and in some cases extreme amounts of extracellular proteins can be added to the milieu. These complications can significantly impede purification and add to the analytical requirements.

### 1.3 Summary of Characteristics

The attributes of the adenoviral particle, the culture process, components of the media, and properties of the vector itself have significant impacts on the design of a purification process. Table 1 outlines the salient aspects of the particle and the lysate.

## 2. Recovery and Purification of Adenoviral Particles

Purification must take place in the context of a complicated lysate. Table 2 summarizes the key features of the adenovirus particle and suggests recovery techniques that may be employed. Together, the features of the particle and the milieu point to a sequence of process steps that yield purified adenovirus (Table 1). Because the particles are produced in cells, the first recovery step is harvesting of the infected cells. The next step requires lysis of the cell to release the virus. The cell lysate contains cell debris so a clarification step is necessary to protect downstream steps. A substantial amount of DNA is present and must be eliminated early in the process. The clarified lysate is too crude for high-resolution purification so an initial purification is needed. Once the preparation has been simplified by the initial purification, a fine separation step removes the remaining contamination. The purification may utilize salts or buffers that are undesirable for use in the clinic. These components may need to be exchanged for a final formulation. The following sections provide techniques to accomplish this sequence.

### 2.1 Harvest Methods

Cells grown and infected in large-scale flat stock culture eventually detach from the surface. Alternatively, trypsin may be used to detach cells before the onset of a cytopathic effect (CPE). Cells free in the medium may be collected by centrifugation or filtration. Cells grown in suspension are also harvested in the same manner. Cells grown on microcarriers may be harvested by allowing the cells to settle so that the spent medium can be decanted. Infected cells may be removed from the microcarriers by trypsinization or processed while still on the carrier. Infected cells remain suspended and therefore can be decanted with the spent medium.

In order to maximize the harvest yield, one should consider the point at which harvest occurs. The life cycle of the adenovirus was thought to terminate at the time a CPE is observed. Analysis using anion exchange high-performance liquid chromatography (AEHPLC; Section 3.6 of this chapter), however, demonstrated that the peak

Table 1 **Properties of Adenoviral Particles**

| Property | Data | Consideration | Issues |
|---|---|---|---|
| Diffusion | $4.46 \times 10^{-12}\ m^2/s$ | Filtration, centrifugation, chromatography, freezing, and thawing | • Slow diffusion rate may provide a means for separation on filters and chromatography resins.<br>• Centrifugation not counteracted by diffusion. Mixing will be problematic especially during freeze–thaw operations.<br>• Slow diffusion may decrease concentration-dependent aggregation.<br>• Assays are hindered |
| Water content | High compared to proteins | Fragility, salt concentration, small ions, solvents, freezing, and thawing | • Particle may be swelled with water.<br>• High salt and solvents may result in degradation.<br>• Sensitivity to shearing forces. |
| Viral genome alterations | Viral, cellular, and transgene expression | Early gene alterations and deletions overexpression of transgene products | • Cell lysate may contain many incorrectly assembled particles, adenoviral protein structures, and unusual cellular structures that are similar in size to adenoviral particles.<br>• Transgene product maybe a contaminant that could confound potency assays.<br>• High titers result in less difficult purification |
| Lysate | Complex | Contaminants | • Particle assay for crude materials.<br>• DNA, lipids, BSA, antifoaming agents, anticlumping agents, and other contaminants may bind to the particle and copurify or foul filters and resins. |

**Table 2** **Physical Properties of Adenovirus Particles that Can Be Exploited for Purification**

| Property | Data | Method | Issues |
|---|---|---|---|
| Density | 1.34 g/ml | Density gradient ultracentrifugation | • Classical method–small scale. |
| Size | ~100 nm | Filtration<br>Size exclusion chromatography | • Some cell debris similar size, lipids, and cell culture additives may interference.<br>• Particle larger than resin pore sizes.<br>• Could be used for buffer exchange. |
| Surface | Protein: ionic, hydrophobic, specific surface chemistry | Ion exchange, hydrophobic interaction, affinity chromatography ligand binding filtration<br>Reversed-phase chromatography, solvent extraction | • Scalable methods employed for protein purification.<br>• Well-established literature to predict behavior of cell culture components. Particle size larger than pore sizes.<br>• Solvents may be problematic. |

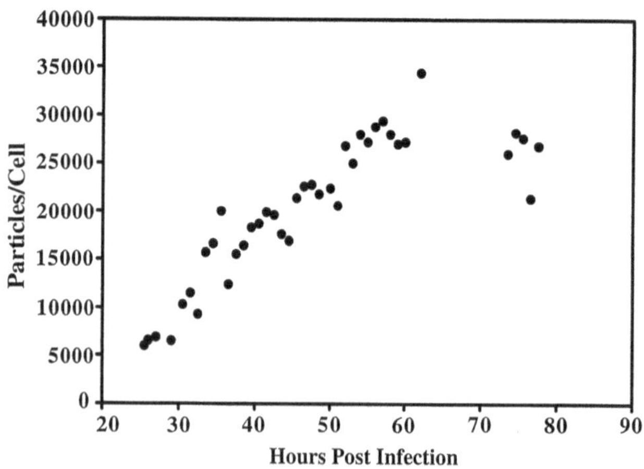


**Figure 2** Adenoviral particle production can be followed using anion exchange high-performance liquid chromatography as discussed in Section 3.6. The per cell productivity was monitored during an SL bioreactor run with a replication-deficient type 5 adenovirus grown in HEK 293 cells. The completion of particle production occurred before obvious signs of cytopathic effects are manifested.

of particle production occurs before cells begin to detach. Figure 2 shows the particle count taken at various time points during a bioreactor process. In most bioreactor runs the particle concentration drops slightly from the peak, but in some cases the particle concentration falls to as low as 10% of the peak value in just a few hours.

This phenomenon underscores the need to monitor the process carefully to obtain maximum yield.

## 2.2 Lysis Methods

Following recovery of the infected cells, the adenovirus must be released from the cell. This is accomplished by lysing the cell. There are many methods available for both large- and small-scale cell lysis. The most useful of these methods are discussed below.

### 2.2.1 Freeze Thaw

Cells burdened with a full viral load are fragile and easily disrupted. Freezing and thawing the infected cells achieve the release of virus. This method is attractive at small scale because it does not require specialized equipment. It is less attractive at large scale because the freezing and thawing of a large sample are difficult to control. Consequently other methods are preferred for large scale.

The freeze–thaw lysis method requires that one observe how the solution freezes and how it thaws. During the freezing process, solutes, such as salts, proteins, and free viral particles, depress the freezing point of the solution. Small ice crystals of pure water begin to form. The solutes excluded from the ice tend to concentrate in spaces between the ice crystals. These areas of concentrated solutes experience freezing point depression. If the freezing process is too slow, the virus may be found in highly concentrated bands. Once the thawing process begins low molecular weight solutes are free to diffuse away rapidly, but adenovirus particles remain roughly in the same place, owing to their large size and slow diffusion constant. Given that nearly all proteins precipitate from solution at a critical concentration, one would expect that the virus particle would also be similarly limited. At temperatures above 0 °C, frequent collisions among particles lead to an aggregation cascade. While freeze–thaw releases more than 90% of the virus in three cycles under favorable conditions, improper control may lead to greater than 50% loss.

Consideration for damage and loss of the particles represents the greatest concern. Damage may not be obvious at first when a structure as large as a virus is involved; rather, damage suffered during early recovery steps may manifest itself as reduced stability of the purified virus. Particles are packed into the cell in a tight array; once released the viral particle can aggregate. Collisions leading to aggregation in the cell may be limited by mediating proteins, which impede movement and hold the particle in the soluble array.

The control of pH and salt concentration is also critical for freeze–thaw. Some buffers such as phosphate do not maintain buffering capacity when the solution freezes. This may be because sodium phosphates precipitate at low temperatures.

### 2.2.2 Homogenization

A useful method for cell disruption used for recovery of recombinant proteins involves passage of bacterial or yeast cells through a small orifice under pressures up to 20,000 psi. On passage through the orifice, the cells expand and rupture as they experience a sudden drop in pressure. The French press has been used for this purpose at laboratory scale for many years.[14] Large-scale processing using this principle may be

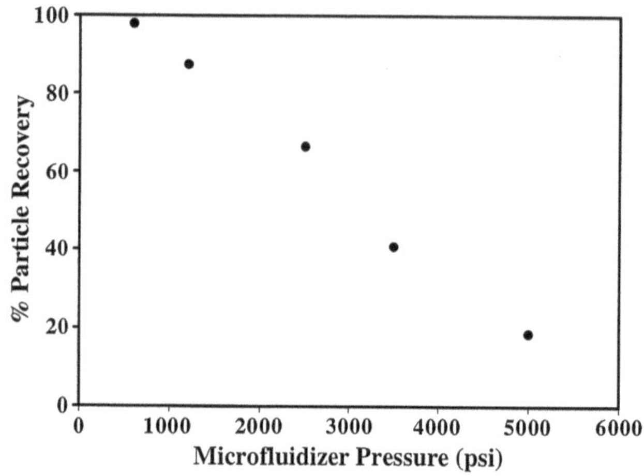


**Figure 3** Methods to lyse infected cells vary from repeated cycles of freezing and thawing to microfluidization. The cell can be lysed by a sudden pressure drop generated by several different means. This experiment used anion exchange HPLC to measure the recovery of particles over a range of pressures in a microfluidizer. Samples were also submitted for titer and showed the same trend. Particle recovery declines at differential pressures above 600 psi. Similar results were obtained using a nitrogen bomb apparatus (data not shown).

accomplished using a Gaulin homogenizer (APV Gaulin, Wilmington, MA). Lysis of infected mammalian cells can also be achieved using a similar device. However at the pressures under which this method is normally used to rupture bacterial cells viral loss is observed. To protect the virus the pressures must be minimized. Figure 3 illustrates this effect by comparing the recovery of adenovirus particles after lysis at different fluidization pressures in a microfluidizer.

### *2.2.3 Sonication*

Sonication is widely used for breaking small quantities of cells for research[14] because it is rapid and convenient for small samples. This method results in excellent release of virus from infected cells. Disadvantages of this method include the generation of heat, production of free radicals, and attendant chemical damage, as well as the lack of equipment suitable for large-scale applications.

### *2.2.4 Simultaneous Harvest and Lysis*

Using a continuous flow centrifuge bacterial and yeast cells can be readily harvested at large scale by the use of continuous flow centrifuges. The use of these devices to harvest intact mammalian cells is less certain. These centrifuges are configured to concentrate and harvest cells by different means. To aid in the separation these centrifuges often are fitted with a series of conical discs inside the rotor. All systems expose the cells and liquid to a centrifugal field, allowing the cells to be concentrated in the interior of a hollow rotor. Supernatant, which has a lower buoyant density, flows out of the rotor. The "pellet," which remains liquid, collects in the rotor, exits through restrictions on the outer edge of the rotor, or is ejected in a discharge cycle as the rotor opens. This third method is to briefly open the rotor on the outer edge while it is spinning. If the rotor is opened for just a fraction of a second, the pellet is discharged and little supernatant is lost. Generally, these centrifuges develop between 10,000g and 20,000 g. Fragile cells, such as mammalian cells, are particularly at risk of lysis in both the discharge- and the nozzle-type disc stack machines because of the shear forces and rapid pressure changes. For the same reason these systems are ideal for large-scale concentration and lysis of infected cells.[15] Shear from the rotating discs provide some cellular disruption and the rapid pressure drop from discharge finishes the job. The resulting lysate may be further clarified.

### 2.2.5 Lysis during Filtration

Another method used for lysis is cross-flow filtration. Cross-flow filtration, also known as tangential flow filtration, separates particles in solution by passing the solution along the surface of a membrane. Liquid passes through the membrane because of the pressure differential across the membrane. Particles and solutes are retained if they are larger than the "cutoff size" of the membrane. Cross-flow systems are configured to allow for recirculation of material along the membrane surface. In this way, larger components are retained (retentate) and smaller components are collected in the passed liquid (permeate). Membranes may be configured in flat plate, spiral wound, or hollow fiber systems. Cutoff sizes vary from particles visible to the eye down to molecules as small as 300 molecular weight. A typical clarification operation can be achieved by using either a 0.45 μm or a 0.2 μm nominal cutoff size. Infected cells in this system are exposed to both shear and rapid pressure drops. The advantage of these systems is that the viral particles, which are slightly less than 0.1 μm in diameter, are not allowed to traverse the filter until they have been released from the cell. Cellular debris is retained and, therefore, separated from the virus.

### 2.2.6 Lysis with Detergent

Detergent lysis is a simple and robust alternative lysis method. Nonionic detergents (such as Tween, Triton, and Brij) are commonly used to solubilize membranes and, as a result, release adenovirus from cells with varying kinetics and efficiency. Whereas detergent will inactivate enveloped virus, adenovirus, a nonenveloped virus, is stable in the detergents. It is recommended that the concentration of detergent and other lysis buffer components, as well as the time, be optimized for lysis. Tween and Triton have been widely used with success in the range of 0.05–1%. Detergent lysis is simple to carry out at both the lab and the manufacturing scale. It provides an additional benefit of providing some level of viral clearance for enveloped virus even at low concentrations. However, additional process steps (such as density gradient or ion exchange chromatography) are recommended for removing the detergent for final formulation if there are concerns of residual detergent in the finished product. Dialysis is ineffective with detergents that readily form micelles because the micelles may be too large to pass through the membranes. Micelle formation may be strongly affected by salts in the media.

For adherent cell culture (flat stock or microcarriers), the spent medium can be easily removed and the adenoviruses can be treated with harvest buffer containing the detergent for a period of time sufficient to lyse the cells. The adenoviruses will be released from the cells into the buffers and the cellular debris can be removed by coarse filtration. For suspension cell culture, the detergent lysis process can be carried out either by direct spiking of the detergent into the culture or by adding a harvest buffer containing the detergent after medium removal (centrifuge or microfiltration). The mixing time, temperature, and agitation speed need to be optimized to achieve the highest recovery. To greatly reduce the storage volume at large scale, a microfiltration step (concentrating the infected cells) prior to adding detergent or an ultrafiltration step after the harvest may be desirable.

## 2.3 Clarification

The lysis procedure releases the virus from the packed array inside the cell into the medium. All procedures produce cell debris, which must be removed before purification. A common procedure for lysis includes concentration of the intact cells followed by lysis. This method works well at the laboratory scale and may be necessary for production with viral vectors that exhibit poor per cell productivity. With optimized culture methods overconcentration becomes a concern. Highly concentrated lysates exhibit significant losses during lysis and clarification. The loss seems to be associated with aggregation of the virus particle.

### 2.3.1 Centrifugation

Centrifugation at low relative centrifugal force (RCF, 1000 × g min) is sufficient for cell debris removal. Centrifugation in a swinging bucket centrifuge is a common method and can be efficient for volumes below 5 l. The disadvantages include performing this operation using an aseptic technique and the possible generation of an aerosol of virus particles. Centrifugation at higher RCF can lead to the loss of virus. Figure 4 shows relative yield loss over time of particles after centrifugation for 5 min at increasing RCF.

### 2.3.2 Filtration

Filtration is another clarification method. Either cross-flow filtration or dead end single pass filtration can be used to remove debris.[16] A study employing a variety of membranes with varying compositions is necessary for optimizing yield. Some membrane materials bind proteins and may also bind adenovirus. Membranes composed of polyethersulfone possess low protein binding characteristics. Adenovirus will pass through these membranes with excellent yields if the concentration of the adenoviral particles is kept below $5 \times 10^{11}$ particles/ml. Most filters are rated by their performance in passing dyes or particles of standard sizes. However, manufacturers generate membrane pores in different ways. Some membranes possess pore sizes similar to the cutoff size; whereas most possess pores much larger than the cutoff. These types of membranes rely on the torturous path the solute must follow to get through the membrane. Proteins and other contaminants can interfere with filtration by forming a barrier that effectively

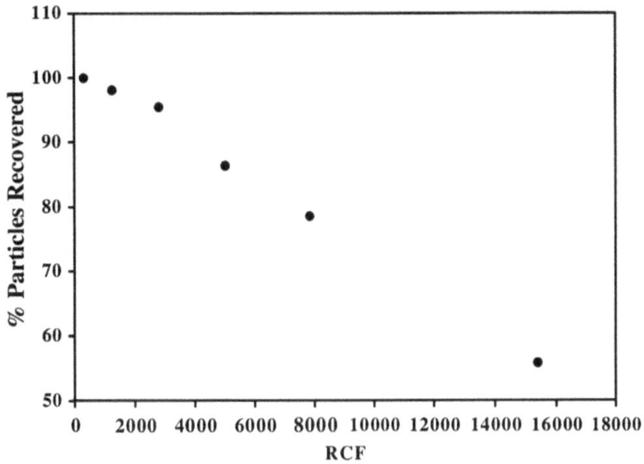


**Figure 4** The density of adenoviral particles is significantly higher than most lysate components. Samples of infected cell lysate were spun in an Eppendorf Microfuge Model 5415c at 4 °C for 5 min at different speeds. Recovery of particles was measured using anion exchange HPLC. Yields were plotted against relative centrifugal force (RCF). Particle loss occurred when the sample was subject to RCF greater than 3000*g* min.

reduces the pore size. Adenovirus can be filtered through some 0.22 μm membranes with better than 90% yield, but the membrane must be selected by experimentation. Fortunately, adenoviral preparations may be sterilized by 0.22 μm filtration with filters available from many manufacturers such as Millipore, Gelman, or Sartorius.

### 2.3.3 Expanded Bed Chromatography

Expanded bed chromatography removes cell debris using an upward flowing chromatography column partially filled with large or dense beads.[17–19] "Expanded bed" differs from "fluidized bed" in that the suspended bed is stabilized by a gradient of bead densities so that mixing is minimized. Resin of this type is available from GE Healthcare Life Sciences. In this mode of operation, the column is fed from the bot tom at a flow rate sufficient to suspend the beads throughout the column but not cause the resin to pack at the top. Supernatant exits through the top frit. After absorption of the entire volume to be clarified, the direction of flow reverses and the column is packed, washed, and eluted. The main advantage of this technique is that it accomplishes debris removal and column chromatography in one unit operation. Drawbacks include the need for a special column and a special resin, limiting flexibility. The flow rate for loading the column is restricted to a rather narrow range by the requirements to keep the bed suspended but not packed at the top of the column.

### 2.3.4 Digestion of DNA

Lysis releases a large amount of DNA both in large and in small fragments. Some of the DNA associates with DNA binding proteins and are found in large structures. DNA (and RNA) digestion is necessary because this contaminant promotes aggregation and complicates downstream processing. The adenovirus particle possesses sufficient surface area such that significant amounts of DNA bind to the capsid despite the anionic surface charge of the particle. Much of this DNA is viral and if not separated can be taken up by the target cell causing abnormal replication. In some cases, E1 can be cotransfected with an E1-deficient particle and can give rise to the generation of replication-competent adenovirus. Consequently, it is important to eliminate as much of the exogenous DNA as is practical.

Fortunately, nucleases such as Benzonase® (Merck KGaA, Darmstadt, Germany) are available in highly pure forms are able to digest the majority of the complicating nucleic acids. The enzymes function best at 37 °C near pH 7 in the presence of magnesium ion. The salt concentration is also critical as high salt inhibits the enzymes. Aggregation of the adenovirus, which depends on collisions, occurs more rapidly at warmer temperatures, whereas the enzymes function poorly below 15 °C. The lysate must be cleared of debris because the quantity of host cell and free viral DNA competes for enzyme and results in incomplete digestion. A compromise is to perform a modest clarification (see above), adjust the salt concentration by dilution, buffer the solution to maintain physiological pH, add magnesium ion, and then perform the digestion at room temperature. The final concentration of nuclease can be increased to accelerate the process.

## 2.4 Purification

Once the clarified lysate is free of exogenous DNA substantial purification can proceed by a variety of techniques. At small scale both chromatographic or density gradient centrifugation methods are effective. Large scale, however, favors chromatography.

### 2.4.1 Ultracentrifugation

Meselson et al.[20] presented a method for determining the molecular weight and partial specific volume of macromolecules by density gradient centrifugation. This technique has been particularly useful for macromolecules such as DNA and viruses. Salts, such as $Cs_2SO_4$ and CsCl, form density gradients when subjected to a strong centrifugal field. Macromolecules separated from contaminates on the basis of their respective buoyant densities collect in bands at their own density if the sample is centrifuged to equilibrium.

Adenoviruses can be purified using this technique because the buoyant density of the particles is approximately 1.34 g/ml. A typical purification scheme is a three step process where the infected cell is lysed and the DNA is digested. The sample is then applied to a step gradient of CsCl in a tube where the density of the bottom layer of CsCl is around 1.4 g/ml and the top layer is around 1.25 g/ml (both layers are buffered with Tris to approximately pH 8). After spinning at approximately 150,000 × g for 1–2 h, the virus separates from cellular debris and collects in a band between the CsCl layers. The band is collected by puncturing the tube and drawing the material out with a syringe. This collected band is then mixed with CsCl at 1.35 g/ml, placed in a centrifuge tube, and subjected to 200,000 × g overnight. The intact virus separates from DNA, proteins, and defective particles and is collected as before. CsCl is then removed by dialysis.

This method is easy to perform and yields high-purity virus preparations. Unfortunately, the process time required is long. CsCl must be removed from the final product, and specialized equipment is required. The main disadvantage is that this process cannot be performed at the large scales demanded for pivotal clinical trial or market use.

### 2.4.2 Purification by Chromatography

Column chromatography is by far the most versatile and powerful method for purification of viruses. The methods described above serve to prepare the viral preparation for chromatography by freeing it from cells, cell debris, and interfering substances. The clarified lysate must be in a buffer suitable for application to a chromatography column. Modes of chromatography applicable to viruses include ion exchange, reversed phase, hydrophobic interaction, size exclusion (gel filtration), immobilized metal chelate, affinity, etc. For each mode, one could choose from many commercially available resins and mobile phases. The selection of an optimal sequence of chromatography steps has been made easier by commercial instruments that are able to perform a systematic search of columns and gradient conditions. Books and review articles describe these modes of chromatography and offer strategies for selecting the best combination.[21–27] Specific purification methods for adenoviruses are also found in the literature.[28]

Fundamental differences distinguish analytical and preparative chromatography.[29] Analytical runs are performed by injecting a small amount of sample onto a column with high resolving power. Such columns typically have very small particle size beads (3–10 μm), high theoretical plate numbers, and rapid run times under high pressure. The result of the run is judged by the appearance of the chromatogram; that is, the peaks should be symmetrical, narrow, and well resolved. Preparative runs, in contrast, are carried out by applying a large sample load (usually near the maximum for the column) using columns of 1–500 l bed volume packed with larger beads (20–90 μm or larger). The cost of the packing becomes more significant for such large columns. The result of the preparative run is measured by the ability to recover pure fractions from the column with high yield. This means that an analytical technique is needed to judge the purity and yield of the fractions. AEHPLC and reversed-phase HPLC (RPHPLC)[30–32] serve as the two most powerful analytical techniques (see Section 3 of this chapter). Without such information, the outcome of a preparative separation is not known because the chromatograms for many preparative separations are complex and difficult to interpret.

Method scouting is conveniently done on small columns in high-pressure systems (HPLC, FPLC) in order to speed development and conserve material. Media with smaller bead sizes may be used provided larger bead sizes are also available. Preliminary screening should identify two modes of chromatography able to resolve the virus of interest from the contaminants. The first column step usually employs a resin with high binding capacity and/or high selectivity for the product; otherwise, a very large column may be needed. Anion exchange chromatography is often selected as the first step. Adenoviruses do not adsorb to cation exchange resins at physiological pH, but this type of resin may be used as a first step. While the adenovirus passes through cation exchange columns, some protein contaminants will be removed from the viral solution. For adenovirus, this method is not needed. Resins for each step are then selected based on resolution, recovery, speed, and cost, and possibly other factors such as freedom from extractable materials and availability of documentation required for the production of clinical material for human trials under current Good Manufacturing Practices (cGMP, Part 21CFR). The mobile phase is selected and the gradient optimized. Sample volume and concentration influence resolution in column chromatography so both of these parameters must be optimized as well. All of this work can be done on relatively small columns. The product produced at a small scale should meet all of the purity requirements desired for the final product.

Scale-up of chromatography steps is performed by maintaining the media bed height and linear flow rate of the mobile phase while increasing cross-sectional area (hence column volume) of the column. Fine-tuning of the process is usually done at the production scale; only minor adjustments should be needed.

Column packing instructions depend on the particle size and nature of the resin.[25] The resin manufacturer's instructions should be followed and then checked by measuring theoretical plates, $n$, for the column with acetone (UV detection) or sodium acetate or sodium chloride (conductivity detection). The theoretical plate number can be measured from the chromatogram by the formula[29]

$$n = 5.55\left(\frac{t_r}{w_{1/2}}\right),$$

where $t$ is the retention time and $w_{1/2}$ is the width of the peak at half of its maximum height. Sometimes the width is measured at the baseline. In this case, the constant in the equation changes from 5.54 to 16. The plate number increases with column length. Often it is useful to correct the plate number of column length, yielding a parameter known as "height equivalent theoretical plate" (HETP), given by

$$HETP = \frac{L\,(cm)}{n},$$

where $L$ is the length of the column in centimeters. These parameters offer a simple way to monitor column performance over time. Care should be taken to avoid introducing air into the column because air pockets degrade performance and may necessitate repacking of the column.

Columns should be cleaned after use and stored in a suitable bacteriostatic environment[25] following the manufacturer's directions. Most process resins can be cleaned and sanitized with sodium hydroxide solutions in the range of 0.1–1 N (exception: silica-based materials dissolve at alkaline pH).

#### 2.4.2.1 Ion Exchange Chromatography of Adenovirus

Ion exchange chromatography offers a powerful method for adenoviral fractionation because of its high capacity and resolution. Ion exchange chromatography exploits the charge that proteins carry on their surface. The net charge of these groups varies with pH and the amino acids exposed in the protein surface.[27,33] Adenoviral capsids are highly anionic in nature, making anion exchange ideal for purifying them. Anion exchange resins carry positively charged groups such as diethylaminoethyl (DEAE) or quaternary amino ethyl, which bind anionic proteins in a manner that depends on pH. Elution may also be accomplished by changing pH to eliminate the ionic interaction with the protein. For proteins, it is helpful to know the isoelectric point of interest and how the protein charge varies with pH. These properties can be measured by isoelectric focusing and electrophoretic titration.[25] Good binding and elution characteristics are often obtained about 1–1.5 pH units above the isoelectric point for anion exchange or an equal increment below the isoelectric point for cation exchange.

The predominate capsid protein is the hexon, which possesses an isoelectric point near pH 6. As noted above the particle is only stable in a narrow pH range near pH 7. Ion exchange resins also bind protons (cation exchange) or hydroxyl ions (anion exchange). Increasing salt concentration may lead to large changes in pH because salt competes with protons or hydroxyl ions for binding sites on the resin. One should be alert to the possibility of pH changes, possibly as much as one pH unit during chromatography. Maintaining the pH with the correct buffer at sufficient concentration is important for stabilizing the pH during elution.

Several ion exchange resins should be tested for binding capacity and resolution at a constant flow rate because resins with the same functional group may differ considerably in these properties owing to differences in their backbone, density of substitution, or other factors. After selection of the resin and mobile phase the other critical parameters can be optimized: sample load and volume, flow rates for absorption and elution, and elution gradient.

Hexon is a noncovalent trimer that is anionic at pH 7. The capsid is composed of 240 of these capsomeres and gives the particle a large number of negative charges on the surface at neutral pH. Proteins bind to ion exchange resins in low salt (5–50 mM NaCl) and elute with high salt (0.1–1 M). Concerted binding of capsomeres in the capsid to the resin allows the particle to adsorb at higher salt concentrations than those used to elute endotoxins and most proteins. This allows easy separation of viral particles from proteins. Most chromatographic resins are optimized for different classes of ligands by making the resin particle with various pore sizes. Proteins and other ligands have access to a substantial amount of resin surface area inside the pores. Adenoviral particles do not have access and are limited to the outer surface of the resin. Large fragments of DNA, however, are also highly anionic but with a higher charge density. Consequently, DNA elutes at higher salt concentrations than adenovirus. These properties result in an order of binding and elution for the constituents of a clarified lysate. At buffered salt concentrations as high as 350 mM NaCl, the particle binds to the column whereas nearly no free proteins bind under these conditions. Large DNA–protein complexes such as incorrectly assembled particles and some cellular structures bind under these conditions. DNA binds tightly. Using a linear salt gradient the order of elution will generally be proteins first, followed by complex contaminants, viral particles, other cellular derived structures, and lastly DNA that has escaped digestion. The elution of these components results in excellent separation between peaks (Figure 5). Purification yields are as great as 99% but can be lower if the peak must be trimmed to improve purity.

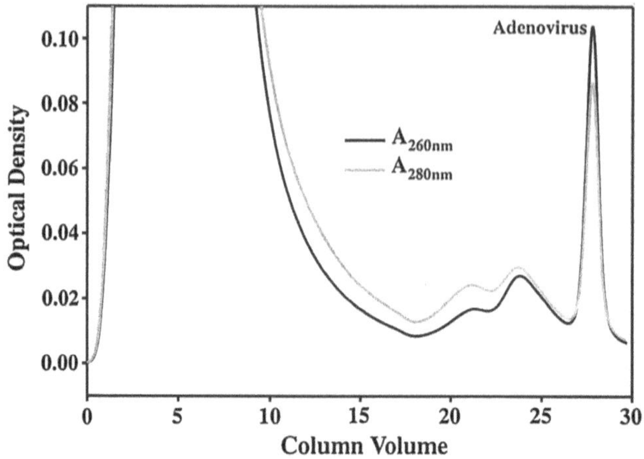


**Figure 5** Anion exchange chromatography is the most robust method for recovering and purifying adenoviral particles from crude stocks. Monitoring the optical density at 280 and 260 nm allows the chromatographer to easily recognize the fractions containing adenovirus by taking a ratio of $A_{260}$ to $A_{280}$. The ratio for pure virus is around 1.25. The chromatogram above is plotted as the optical density vs. the number of column volumes of materials that have been pumped through the column. In this example, DEAE Fractogel 650M (EM Sciences) was used. The column was buffered in 50 mM HEPES, pH 7.5, at room temperature throughout the process. Infected cell lysate was loaded onto the column with an adjusted salt concentration of approximately 350 mM NaCl. The load produced significant absorbance as the majority of contaminants passed through the column. More contaminants were eluted during a post loading wash with equilibration buffer. Elution was with a linear salt gradient from 350 to 600 mM NaCl. The adenovirus peak is well resolved at the end of the chromatogram. The approximate salt concentration of the collected peak was about 450 mM NaCl. Column cleaning was achieved with 1 M NaCl and 0.5 N NaOH (data not shown). The column height for this chromatography was about 5 cm. The chromatography looks the same, however, at 10 cm bed height. Scale-up produces the same chromatogram if the column diameter is increased using the same bed height and the flow rate is adjusted accordingly.

**2.4.2.1.1 Disposable Ion-Exchange Membrane Absorber** Disposable membrane absorbers (such as Mustang Q® from Pall Scientific and Sartobind Q® from Sartorius) have become popular for the ion exchange chromatography process, especially during clinical product production. The disposable membrane absorber technology exploits same principle as that of traditional resin chromatography. However, the membrane absorber technology provides the advantage of: (1) faster flow rates, operate at flow rates between 10 and 40 membrane volumes per minute, making them much faster than conventional columns which typically operate at 0.5–3 column volumes per minute; (2) higher binding capacity; (3) no packing or cleaning validation required for clinical usage; (4) no resin lifetime issues or storage issues.

Manufacturers specifically design these membrane absorbers with the same bed height for the scale-down model and large-scale products. Therefore, a linear scale-up process can be simply achieved by maintaining the linear flow rate. The ease of linear scale-up ensures a shortened process development time by significantly reducing the reoptimization required between scale-up steps.

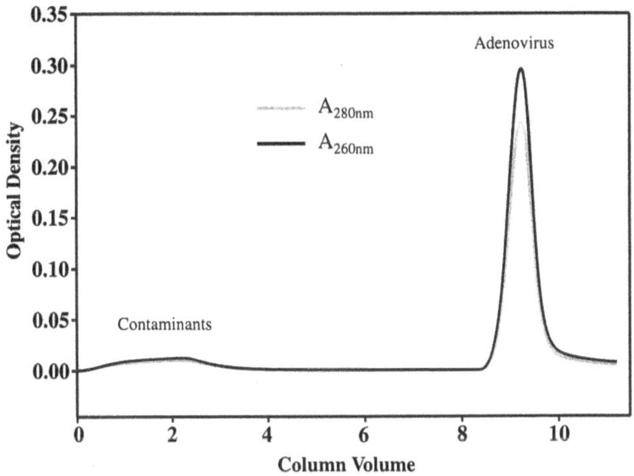


**Figure 6** High-resolution techniques, such as zinc metal affinity chromatography, are needed to complete the purification of adenoviral particles. TosoHaas AF chelate 650 M immobilized metal affinity resin was charged with divalent zinc. The column was equilibrated at room temperature with 450 mM NaCl in 50 mM HEPES at pH 7.5. DEAE adenovirus fractions (Figure 5) were loaded onto the column followed by a wash with equilibration buffer. Remaining contaminants eluted from the column during the load and wash. The adenovirus was eluted from the column with a 500 mM glycine step gradient. After elution, the column was stripped with EDTA followed by 1 M NaCl and then 0.5 N NaOH (data not shown).

### 2.4.2.2 Immobilized Metal Affinity Chromatography of Adenovirus

In 1975 Porath showed that metal ions could be linked to a column in a 1:1 complex with a chelating ligand, iminodiacetic acid, bound to the column.[34,35] These columns had unique properties for fractionation and provide orthogonal methods for purification. This technique is referred to as "immobilized metal affinity chromatography" or IMAC. Beaded agarose is the most common support and iminodiacetic acid remains the most popular chelating ligand. Such columns can be charged with a variety of divalent metal ions; $Zn^{2+}$ and $Cu^{2+}$ are preferred for protein chromatography.[36] Adenovirus particles bind readily to $Zn^{2+}$ charged resin whereas $Cu^{2+}$ is not as efficient. Excess metal ion is removed by washing before applying protein. Bound metal ion forms a coordination complex leaving some coordination sites free to interact with proteins. Protein binding typically occurs through histidine residues,[23,36] which occupy the free coordination sites. However, coordination with epsilon amino groups is also probable. Elution can be achieved either by changing pH or by adding competitors such as imidazole or glycine, for the binding sites. Ethylenediaminetetraacetic acid (EDTA) can be used to elute the column, as EDTA strips the metal from the column and the protein.

IMAC works well as a polishing step for purification because it removes residual host cell contaminants. Because IMAC can be operated under high salt conditions, fractions from an initial recovery column, such as anion exchange eluate, can be loaded directly on the equilibrated and charged column. The buffer selected must not strongly chelate metals, of course. The yields typically fall in the range of 60–80% and the purity is greater than that with CsCl ultracentrifugation methods. Figure 6 shows a chromatogram of adenovirus type 5 purified with zinc charged IMAC and eluted with a step gradient of glycine.

Modifications to the fiber can also affect the chromatography. Interestingly, "fiber-less" adenovirus, vectors that have been altered to express truncated fiber protein, bind normally to anion exchange resin, but do not bind at all to IMAC charged with zinc. Presumably the residues serving as zinc binding sites have been removed from the surface of the fiber by mutation. Alternatively, the adenovirus fiber can be engineered to provide histidine repeats that will bind to $Ni^{2+}$ very effectively.[37]

#### 2.4.2.3 Reversed-Phase Chromatography

Reversed-phase HPLC can be used for analysis of adenovirus particles or as a polishing step after initial purification by anion exchange chromatography. The recoveries for the preparative method run in the range of 20%, which is poor compared to other methods. The lower recovery may be related to the presence of organic solvents in the mobile phase as high molecular weight proteins tend to denature or precipitate. Recoveries may be improved through careful selection of column, solvent, ion pairing agent, and pH.

#### 2.4.2.4 Hydrophobic Interaction Chromatography of Adenovirus

The discovery of hydrophobic interaction chromatography (HIC) resulted from an attempt to make affinity columns.[38] This fortunate accident uncovered a unique mode of protein chromatography. On the surface HIC resembles reversed-phase chromatography[27,39,40] in that the protein binds to the column through hydrophobic interactions in an aqueous solvent. Both resin types consist of a stationary phase with a hydrophobic surface. Thereafter the two techniques diverge.[23] HIC resins are typically constructed from polysaccharide or polymeric material. Reversed-phase resins are typically bonded silicas. HIC resins have a lower density of substitution and they tend to be less hydrophobic than reversed-phase media.

Typically, the conformational changes are driven by high salts (such as ammonium sulfate). The salt presents an ionic environment that is favorable to hydrophilic surfaces. Hydrophobic surfaces are driven together so that exposure to the environment is reduced. In this way, the proteins are partially "salted out" and adsorb to the resin. The use of ammonium sulfate is relatively gentle because most proteins are stabilized in the presence of high concentrations of ammonium sulfate. High salt, however, may destabilize the particle (see Section 4 below), possibly because of its high water content or because the capsid proteins are twisted into destabilized conformations. The loading material may be adjusted to high salt prior to application to the column. Alternatively, small amounts may be applied to the column repeatedly, washing with equilibration buffer, or the material could also be diluted in equilibration buffer inline with the load. The advantage of the two later methods is that protein precipitation occurs slowly from the time of salt addition. Limiting the time of exposure to high salt may improve the chromatography and mitigate yield loss. Elution is achieved by reducing the salt concentration with a reverse gradient.

With the exception of the direction of the gradient, HIC columns are optimized and operated along the same lines as ion exchange columns. Residence time on the column should be minimized because of the possibility of denaturation.[23] Yields of virus from this type of chromatography typically range between 20% and 60%.

#### 2.4.2.5 Size Exclusion Chromatography of Adenovirus

Size exclusion, also called gel filtration or gel permeation, is the only mode of chromatography that is not intended to involve binding of proteins to the resin.[41,42] The pore structure of the resin provides a molecular sieve, where smaller molecules can access the entire volume of the pores and large molecules are excluded from the pores. If a mixture of proteins differing in size is applied to a size exclusion column, the largest proteins will emerge first and smallest last. Molecules above a certain size do not penetrate the pores at all. Normally, these are the first to elute from the column; the volume at which they elute is termed the "excluded volume." Adenovirus particles may not elute in the excluded volume of the column. Instead, the particles may elute as much as a column volume beyond the excluded volume. Late elution from size exclusion chromatography often results from charge interactions between the sample and the column. This effect may be reversed by raising the salt concentration. Similarly, molecules below a certain size all elute at the "included volume."

All other species elute between the excluded volume and the included volume. This property limits the resolving power of size exclusion chromatography because the number of peaks that can fit into the volume allowed is small. A further limitation of size exclusion chromatography is that resolution deteriorates if sample volume exceeds 4–5% of the bed volume of the column. Another limitation is that resolution deteriorates with increasing sample viscosity. The maximum protein concentration allowing good resolution is usually in the range of 5–10 mg/ml protein. Taken together, these two factors mean that a size exclusion column has 1–5% as much protein capacity as an ion exchange column of the same size! Hence, size exclusion has limited utility for purifying adenoviral particles or proteins and is usually reserved as a last step.

Sample load volume, maximum protein concentration, and flow rate should be determined by experiment. A careful column packing technique is critical for good results. Plate number should be measured for a new column and at regular intervals during use. Despite these limitations, size exclusion chromatography has an important place in the arsenal. It is gentle and rapid so yields are nearly quantitative. Additionally, it provides an opportunity to exchange the buffer to the desired formulation because size exclusion chromatography is compatible with a wide range of aqueous buffers.

## *2.5 Buffer Exchange*

Every process confronts the problems of removing low MW species and/or concentrating the desired fractions. Dialysis[43–45] or gel filtration (Section 2.4 above) may be used to remove small molecules or exchange buffers when the sample volume is in the range of 1 l or less. Several types of ultrafiltration devices are available for concentration of proteins on a laboratory scale, including pressurized stirred cells and filters driven by centrifugal force or other means.

Concentration of protein solutions at a process scale is usually done by ultrafiltration using a tangential flow filter.[16,27,46] Buffer exchange and removal of low MW species are usually done by diafiltration; both concentration and diafiltration may be done on the same device and combined as a unit operation. Diafiltration is more efficient than dialysis in that less buffer and less time is needed to achieve a given level of solute removal. The concentration of solute remaining in the retentate ($C_R$) after $V$ volumes of continuous diafiltration that began with the solute concentration ($C_0$) is given by[47]

$$C_R = C_0 e^{-V(1-\sigma)},$$

where $\sigma$ represents the rejection of the solute by the membrane. For solutes freely permeable to the membrane, $\sigma=0$. Under these conditions, diafiltration with 3 vol of buffer reduces the concentration of solute by 95%. In contrast, 20 vol of buffer would be required for the same result by standard dialysis. In either case, buffer exchange of viral particles can be achieved. Care must be taken to avoid foaming or excessive shear. Special attention should be devoted to pH control. The process should be monitored to avoid overconcentration and possible loss of product through precipitation.

## 3. Analytical Methods for Process Development and Process Tracking

Analytical methods are as important for purification as the process steps themselves. Analytical methods are essential for following the process and assessing the purity of adenovirus particles throughout the process. The methods must be rapid, reliable, and informative about the quantity and quality of adenovirus particles. They should be sensitive enough to detect subtle changes. As with the process techniques, characteristics of the virus are useful for selecting analytical techniques. In this section we have highlighted several assays not only because of their usefulness for a process, but because an understanding of what these assays mean is critical in producing vector with the quality required for use in humans.

### 3.1 Plaque-Forming Titer Assays

Plaque-forming assays have been in use as biological assays since early in the twentieth century. A common method for many viruses, this type of assay has been employed with adenovirus since they were discovered. The plaque assay is performed using many variations but generally consists of diluting the virus preparation to a point that a thin layer placed over sensitive cells will result in a countable number of infection events. This is usually accomplished in a petri dish or in six-well plates. Once cells have been exposed, the viral solution is removed and a layer of warm agar applied on top of the cells. After 1–2 weeks of incubation, the cells are stained with a dye, such as neutral red, and plaques of lysed cells become visible. It is assumed that because of the extreme dilution, each plaque is the result of a single viral particle infecting a cell. The plaque arises following the replication of that viral particle and subsequent infection of adjacent cells by virus progeny. The number of plaques in a well divided by the inoculum volume and corrected for dilution yields a titer. This method is simple but relatively insensitive. If the cells are robust, the inoculum layer may be very thin and the exposure reasonably long. The slow diffusion rate of the particle and the formation of a meniscus in the well limit the method to sampling about 10% of the virions in a sample.

## 3.2 Adenovirus 96 Well Titer Plate Assay

A convenient method for estimating adenoviral titers uses cells (of an appropriate cell line) in the wells of a 96-well plate. The cells are plated such that they reach 50% confluence after 1 day of growth. The sample is then diluted so that the final particle concentration falls between 5 and 10 particles/ml. Several different dilutions are prepared. It is best to perform an initial dilution in the original sample tube using the whole sample because the freeze–thaw process concentrates the virus into bands that will not disperse without substantial mixing. The largest error in dilution usually occurs with this first dilution. The initial dilution should be limited so that the concentration after the initial dilution can be verified using a particle assay. Thereafter, dilution of the sample is not problematic. To infect the cells the entire medium is first removed from the seeded wells. Each different dilution of virus is pipetted into at least 10 wells per dilution. When using a diffusion-adjusted calculation (see Section 3.2.2. below) the wells should be filled to the top. It is common in practice, however, to inoculate 1 to 200 µm of medium with 10 to 50 µm of virus solution. For diffusion-corrected calculation the infection time is limited to an hour or less (15 min is convenient) and then the virus solution is replaced with medium. For methods using Spearman–Karber titer calculations, such as the Lynn Titerpint analysis,[48] analysts typically leave the virus solution on the plate for the duration of the assay. Incubation of cells takes place at 37 °C, 7–10% $CO_2$, and 90–100% humidity for varying times depending on the method of detection. One method is to fix the cells after 3 to 5 days with methanol and acetone followed by staining with a FITC-conjugated anti-adenovirus antibody. This method of detection requires microscopic examination of each well under ultraviolet light. A well is counted if one or more fluorescent cells are positive. If a cytopathic effect is used for detection, 1 to 2 weeks of incubation will be required. The dilutions that produce fewer than 100% positive wells are used in the titer calculations.

### 3.2.1 Spearman–Karber Analysis

Spearman–Karber analysis based on Finney[49] essentially converts data such that graphing the data as log dilution verses positive wells approaches a straight line. Spearman–Karber performs an interpolation to a midpoint. Thus, Spearman–Karber gives a log dilution where 50% of the wells would have been positive. Titer is expressed as a negative log of the dilution. The Lynn program transforms the number by taking the reciprocal of the dilution (10 raised to the power of the Spearman–Karber number) and divides by the inoculum amount ostensibly to obtain the inoculum concentration. However, this value is expressed as $ED_{50}$ (or $TCID_{50}$) per milliliter and is often substituted for a virus concentration. This assumes that everything that was put into the well is measured by the assay and the fact that positive wells follow a Poisson distribution is ignored. A more appropriate analysis accepts the Poisson distribution in that even if the average number of virions per well is one, not every well would get a virion; but the Poisson distribution is applicable for this kind of assay. The Poisson distribution is given by

$$p = \frac{S^r e^{-S}}{r!},$$

where $p$=probability or fraction of positive wells, $S$=the event density or average virion per well, and $r$=the number of virions in a particular well. For the number of wells that get zero particles, $r$=0, $s^r$=1, and $r!$=1 so that the fraction of positive wells is given by $1-e^{-s}$. The Spearman–Karber analysis gives us a convenient way to take all the data into account, calculate a standard error, and then apply the Poisson distribution to obtain a concentration. Since the value obtained gives the dilution where the fraction of positive wells is 0.5, $p$=0.5. Solving for $S$ yields $S$=0.69 virion per well. This is always the case when 50% of the wells are positive. If, for example the inoculum were 50 μm, then the concentration of the inoculum would be 13.8 particles per milliliter. Using the dilution obtained by the Spearman–Karber number the original concentration can be calculated. As in the plaque assay, the key assumption is that no virion in a well escapes detection. In the time frame of these assays, this is clearly not possible. A more precise analysis must take into account Brownian motion (diffusion).

### 3.2.2 Diffusion-Normalized Calculation

In the older animal virus literature the methodology used for measuring infectious titer was simple: a thin layer of a viral preparation was placed over the target cells for as long as practical. It was intuitively understood that the infection process was diffusion limited. This methodology would help minimize underrepresenting the titer. Done properly, these assays may underestimate by 10- to 100-fold, but the values obtained in a given experiment were useful in a relative sense. However, many of the cells used for replication-deficient adenoviruses cannot be maintained with very low media levels. Unfortunately, the adenoviral particle, or any particle of similar size, diffuses very slowly in solution. Adding more media mitigated the sensitivity of the cells, but resulted in substantially greater underestimations of titer. This is because the probability of infecting a given cell is dependent on the concentration of the virus and the time of exposure.

The discrepancy between the adenoviral titers and the more precise particle assays (see below) has furthered the concept of particle to infectious unit ratio, or PIU. In the case of adenoviral vectors, this concept is based on the supposition that, in a population of intact and otherwise complete particles, most are not infective. However, a search of the literature does not support this supposition. To the contrary, work by Nyberg–Hoffman et al.[3] used a model derived from Fick's laws of diffusion to demonstrate that most, if not all, adenovirus particles are indeed infective. That work, and others,[50] showed the importance of diffusion mechanics for virus binding and demonstrated the dependence of titer determination on often ignored experimental conditions.

The defective particle misconception neglects the important factor that most of the added particles in a well never contact the target cell during the critical period of the assay. Changes in assay conditions such as particle concentration or exposure time can have a dramatic effect on the results. A significant outcome of the misconception has been the requirement by regulatory agencies to demand that there be at least one infective particle per 100 total particles (FDA Guidelines). Substantial resources have been spent in attempts to purify away the "defective particles." Unfortunately, in many cases these particles are merely experimental artifacts and thus cannot be removed by purification. Other ramifications of this misconception have been discussed in the literature.[50,51]

Diffusion can be accounted for by the use of diffusion-normalized analysis.[3] This analysis takes into account the diffusion of the particle under the conditions of the assay by deriving normalization equations from Fick's laws of diffusion. For a titer plate assay the equation is given by

$$V = -\frac{ln\left(1 - \frac{p_w}{n}\right)}{A_w C_w I \sqrt{t}} \times Dilution\ Factor,$$

where $p_w$ is the number of positive wells per dilution, $n$ is the total number of wells per dilution, $A_w$ is the area of the bottom of the well in square centimeters, $C_w$ is the confluence of the well at the time of infection, $I$ is a constant incorporating the diffusion coefficient and is equal to $2.38 \times 10^{-4}$ cm/particles $\cdot$ $s^{1/2}$, and $t$ is the exposure time in seconds. From the equation, one can see that $p_w$ must be less than $n$ and greater than zero. Optimally, the number of positive wells should be between 20% and 80% of the total wells in a dilution. This method yields titers that are up to 50% of the particle concentration.

## 3.3 Flow Cytometry

Fluorescence activated cell sorting (FACS) offers another sensitive method for assessing infective titer.[52] Sufficient quantities of permissive cells for analysis can be grown and infected in six-well plates. Infected cells are harvested, fixed, and stained with a FITC-conjugated anti-adenovirus antibody. Infected cells are brightly stained whereas uninfected cells are not. FACS analysis determines the fraction of cells infected at the time of harvest. The fraction of infected cells at the time of infection is diluted at the time of harvest because uninfected cells continue to divide. By performing cell counts at infection and at harvest times, one can then calculate the proportion of cells infected at infection time by using the fact that the number of infected cells does not increase during the incubation period. Multiplying the proportion of infected cells by the number of cells at infection time yields the number of cells originally infected. Because the total number of cells at infection time is also known, the proportion of infected cells at infection time is also known. This value can be used to calculate a titer.

Cells are infected with a virus concentration high enough to obtain 5–10% of cells to stain positive. Exposure time ranges from 30 s to 60 min. Incubation is up to 50 h to avoid secondary infections. The titer is given by the diffusion-adjusted equation,

$$V = -\frac{ln\,(1-F)}{I\varphi\sqrt{t}},$$

where the average cell area (two-dimensional footprint), symbolized by $\varphi$, is a variable that must be determined for each cell line. Subclones of HEK 293 cells, for example, can display morphology differences from the parent stock. Cell area can be determined using image analysis to analyze micrographs of cells. HEK 293 cells obtained from ATCC and at low passage number possess a cell area of approximately $6.3 \times 10^{-6}\,cm^2$. Other subclones of 293 cells can have larger or smaller areas. $I$ is $2.38 \times 10^{-4}$ cm/particles·$s^{1/2}$ of as above. $F$ is the final adjusted fraction of positive cells detected. For Ad-5 this method yields titers that are 50–80% of the particle concentration. This is probably the best value that can be obtained considering efficiency factors, such as the fiber length for the particular serotype, the degree of cooperativity for multiple copies of early immediate genes, the activation level of E2F transcription factor, and that no step in the infection process is 100% efficient.

## 3.4 Particle Concentration Determination by Ultraviolet Absorbance

The most common method for measuring particle concentration is to disrupt the particles using sodium dodecyl sulfate (SDS) followed by absorbance measurement at 260 nm. Maizel[53] determined that the absorptivity of adenovirus was $1.1 \times 10^{12}$ particles/ml/absorbance unit at 260 nm. This method is convenient and rapid but not without limitations. The sample must be pure and free of particulates and aggregates in order to obtain an accurate reading. The buffer formulation of the sample can affect the reading as salt concentration partially determines the concentration at which SDS may form micelles. This method cannot distinguish disrupted particles from intact particles. Contaminating DNA may increase the absorbance and lead to an overestimation of particle concentration. Absorbance readings at several wavelengths provide a check on the validity of the assay. Absorbance in the longer ultraviolet and visible regions indicates light scattering. The ratio of 260 nm (DNA) to 280 nm (protein) should fall between 1.2 and 1.3. A ratio outside this range indicates contamination.

## 3.5 Analytical Reverse-Phase High-Performance Liquid Chromatography

RPHPLC first achieved prominence as an analytical technique because of its wide applicability and ability to resolve a large number of components in a single chromatographic run.[54] This excellent technique also works well as a preparative method for some proteins, mainly those of lower molecular weights (<30,000).[25,55] RPHPLC is the dominant method for the purification of peptides and protein of MW < 10,000.[56] For example, human insulin is produced at a level of tons per year using RPHPLC. Application to larger proteins is limited by the denaturing tendency of organic solvents. For this reason RPHPLC is considered a denaturing technique. The binding of proteins to reversed-phase columns results from a hydrophobic interaction between the exposed regions of the protein and the hydrophobic surface of the stationary phase (the resin). Denaturation by the mobile phase exposes hydrophobic regions buried within the protein. Elution is achieved by applying a gradient of increasing concentration of organic solvent, usually acetonitrile. Proteins tend to elute as broad, asymmetrical peaks unless an ion pairing agent, such as trifluoroacetic acid (TFA) is included. TFA is thought to bind to the positive charges on the protein, masking negative charges on the resin matrix and providing additional hydrophobic surface to interact with the column.[54]

RPHPLC is a powerful analytical tool to use for adenoviral samples because it resolves the proteins contained in the virus particle. This method, described in Lehmberg et al.,[31] has provided detailed insights into the nature of adenovirus preparations. The method consists of injecting the sample onto a C4 reversed-phase column that has been equilibrated in 20% acetonitrile at a constant TFA concentration of 0.1%. The column temperature, a critical parameter for RPHPLC, is kept at 40 °C. These conditions dissociate the particle into proteins that in turn bind to the column. The column is then eluted with an acetonitrile gradient beginning at 20% and ending at 60%. The TFA concentration is maintained at 0.1%. The absorbance is monitored at 214 nm and can be monitored at 260 and 280 nm. The resultant chromatogram gives a characteristic fingerprint of the adenovirus proteins. Mass spectrometry and N-terminal sequencing have identified 14 major peaks. These proteins can be recovered quantitatively, enabling this assay to be used as a quantitative method for determining particle concentration. The relative peak areas of the proteins can be compared to a known standard to assess particle quality.

## 3.6 Analytical Anion Exchange High-Performance Liquid Chromatography

The anionic nature of the adenoviral particle lends itself to analysis by anion exchange HPLC.[30,32] This method is nondestructive and yields a wealth of information. Before this method, monitoring the production and purification of adenovirus particles was limited to infectious titer assays. For the purposes of process development and monitoring, infectious titer assays were too slow, resulting in low throughput. They were not sensitive enough to distinguish small differences among samples. Protein analysis methods, such as sodium dodecyl sulfate–polyacrylamide gel electrophoresis, could not quantitate virus in crude samples because most of the viral protein is not incorporated into whole virus.[1] Absorbance techniques to assess viral particles based on the UV absorption of DNA were not applicable to crude samples. This is because free viral and host cell DNA are present in the sample. These limitations are overcome by AEHPLC.

A l-ml Resource Q anion exchange column (GE Healthcare Life Sciences) is convenient for HPLC analysis of samples: Lysates are prepared by treatment with nuclease. Semipure or pure virus can be injected directly because they do not require nuclease treatment. As long as the column is not reequilibrated during injection, the assay is independent of the injection volume. After sample loading, the column is washed with equilibration buffer (HEPES or Tris, at approximately pH 7.5) followed by linear salt gradient elution. The chromatography can be monitored on a standard ultraviolet detector. Significantly more information can be gleaned by the use of a photodiode array detector scanning from 210 to 300 nm. The retention time of the adenovirus peak varies with the serotype of the virus.

The AEHPLC chromatogram reveals information about purity, particle integrity, and particle quantity. Figure 7 shows the chromatograms of purified adenovirus type 5 and infected cell lysate. Other serotypes give slightly different peak elution times. Provided that the HPLC system remains below the pressure limits of the resin, the chromatography can be performed in fewer than 6 min. Monitoring the process by anion exchange HPLC enables the production staff to rapidly obtain a picture of the progress of the purification.

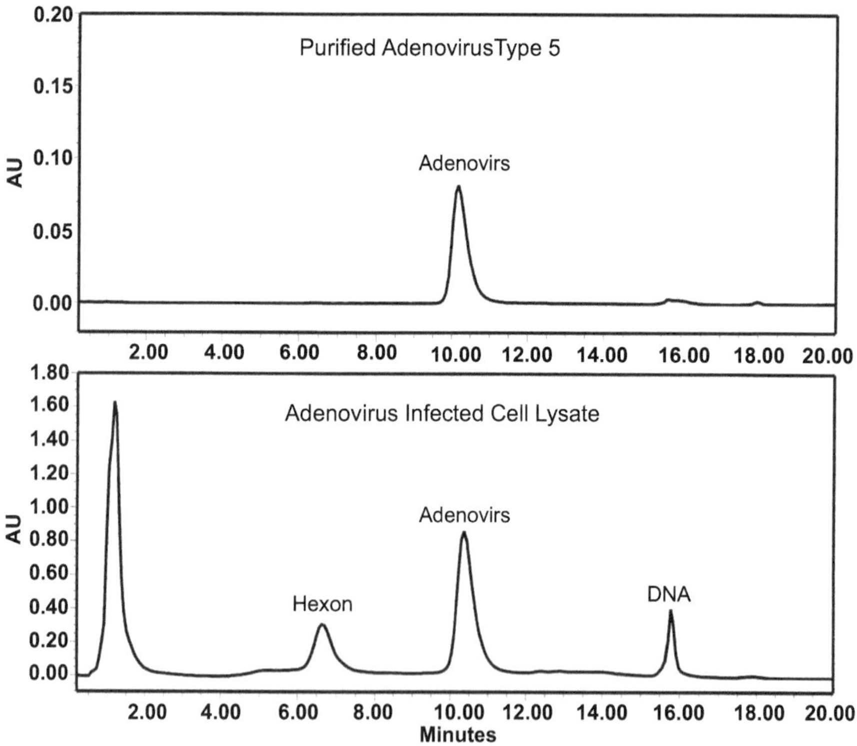


**Figure 7** The most useful method for analyzing both crude and pure samples of adenovirus is anion exchange HPLC. Pure adenovirus elutes in a nearly symmetrical peak. The adenoviral peak from lysate elutes with baseline separation from contaminants such as hexon and undigested DNA. The method is detailed in Shabram et al.[30]

## 4. Formulation and Stability

Process development must determine product stability over a wide range of conditions. The time course of the purification must be scrutinized for excessive run times and delays because rapid processing favors high yields. A well-designed process includes holding points selected such that the viral preparation can be safely stored in the event of a planned or unplanned delay. Stability studies are critical for identifying those steps at which the virus is at risk. Additionally the lessons gleaned from these studies point to a stabilizing formulation needed for clinical trials and beyond.

Aggregation of adenovirus particles is well known to those in the field. The adenovirus demonstrates low stability at 4 °C. This phenomenon has yet to be explained, although there are examples of cold-sensitive enzymes in the literature.[57,58] The instability seems to be related to the tendency of adenovirus particles to aggregate. Aggregation requires that particles bind to each other after a collision. As the collisions proceed the aggregating particle grows. A systematic analysis of aggregation was published in1917 by Smoluchowski.[59] Using Smoluchowski's coagulation model the aggregation frequency of adenoviral particle can be roughly estimated at one per ever 50collisions! This may seem surprisingly frequent until one realizes that collisions between particles are relatively rare due to the slow diffusivity of the particle. It also explains the observation that aggregation is dependent on particle concentration.

Aggregation may also be the end result of damage that occurs early in the production process but does not manifest itself until the sample is concentrated. Aggregation can be mitigated by changing the conditions of the formulation such that aggregation events are not favored. Collision frequency can be reduced by increasing the viscosity of the solution. Figure 8 shows the effect of concentration on the stability of virus in phosphate-buffered saline with 2% sucrose at 4 °C. Aggregation was measured as disappearance on AEHPLC. Aggregation accelerated at concentrations greater than $5\times10^{11}$. This is likely due to the ability of glycerol to cause preferential hydration of protein surfaces,[60,61] leading to a tighter association of the capsid subunit structure.

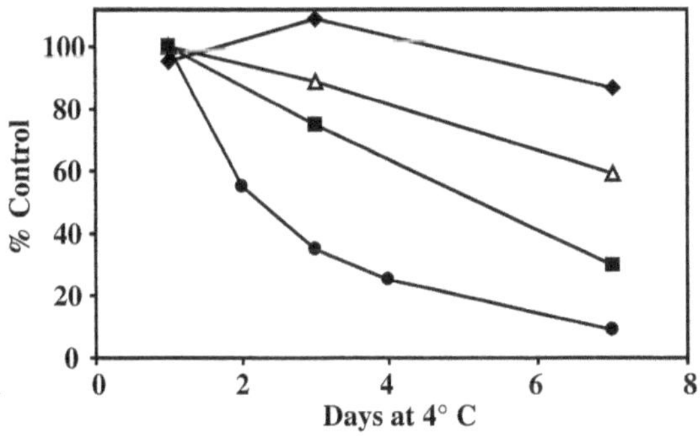


**Figure 8** Adenovirus stability is affected by concentration. Recombinant adenovirus serotype 5 was diluted to $4\times10^{11}$ (♦); $5\times10^{11}$ (Δ); $6\times10^{11}$ (■); or $8\times10^{11}$ (•); at 4 °C. On the indicated days aliquots were assayed for virus concentration with the anion exchange HPLC assay. The buffer was 20 mM sodium phosphate, pH 8, 100 mM NaCl, 2 mM $MgCl_2$, and 2% sucrose.

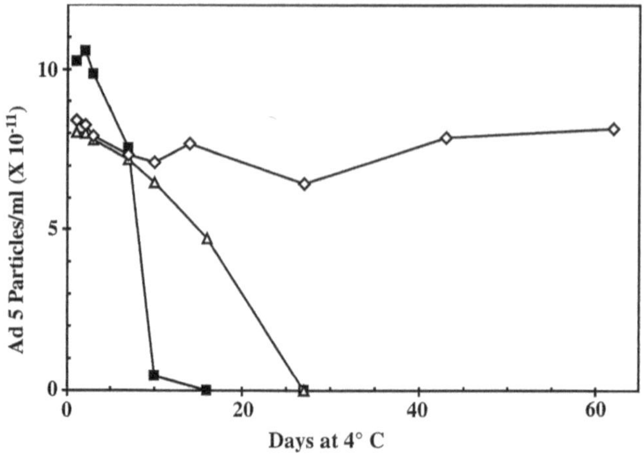


**Figure 9** Glycerol stabilizes the adenovirus in solution. Recombinant adenovirus at 10 particles/ml was stored at 4 °C in aliquots: undiluted (■); diluted 20% with water (Δ); or diluted with 50% glycerol to a final concentration of 10% (v/v) glycerol (◊). On the indicated days aliquots were assayed for virus concentration with the anion exchange HPLC assay. The buffer was 20 mM sodium phosphate, pH 8, 100 mM NaCl, 2 mM $MgCl_2$, and 2% sucrose.

Additionally, an increase in viscosity reduces the number of collisions between particles. The addition of glycerol increased the virus stability (Figure 9).

Salt concentration plays a significant role in the stability of the particle in that stability is reduced in salt solutions above 300 mM. Stability at these salt levels was increased in the presence of glycerol. The putative damage may be due to an anion-specific effect and probably follows the Hoffmeister series. Studies with potassium chloride, sodium chloride, and cesium chloride showed that whereas chloride concentrations above 1.5 M were not destabilizing, concentrations between approximately 0.4 and 1.0 M could be harmful regardless of the cation. In contrast high concentrations of sodium sulfate, sodium phosphate, or potassium phosphate were not destabilizing.

A major consideration is the exposure to pH outside the physiological range. As noted earlier, phosphate buffer at pH 7.2 lacks buffering capacity during freeze thaw. Hence, the rate of freezing and thawing is critical. In general, fast freezing and fast thawing improve the stability of the material. Buffers such as HEPES or Tris maintain buffering capacity during freeze thaw and therefore are preferable for stabilizing pH. One might suspect that the lower stability at 4 °C may be related to pH stability and therefore the buffering capacity of the formulation. Whereas phosphate may be problematic for freeze thaw it is interesting that substituting Tris for phosphate does not affect the stability at 4 °C.

Freezing and thawing were noted earlier as a potential risk for the viral preparation. Cryoprotection agents are often used to mitigate the risk by disturbing ice crystal formation and providing for an amorphous frozen solid. Typically carbohydrates are used to accomplish this. Sucrose and mannitol are often found in formulations where the freezing process is critical. Typically, mannitol provides slightly better protection with proteins than sucrose and is preferred for its superior cake formation in a lyophilized product. With adenovirus, however, the opposite is true. Sucrose provides moderate protection but mannitol has a clear negative effect.

## 5. Conclusions

The use of adenovirus vectors for gene therapy has placed increased demands on the technology for production, purification, and characterization of virus particles. Some of the classic technology has been reexamined and improved. A new class of methods based on column chromatography has added a powerful set of tools to this array. In large part, the chromatographic methods are based on modes of chromatography and resins originally developed for protein purification. With proper consideration for the size and other characteristics of adenovirus particles column chromatography may be applied with considerable success. Column chromatography is now a preferred method for adenovirus purification because of its versatility and ability to purify large amounts of virus to a high state of purity while retaining biological activity.

## Acknowledgments

The authors would like to acknowledge the indispensable help of Rich Murphy, Beth Hutchins, Daniel Giroux, and Muralidhara Ramachandra for input, advice, and editing of this manuscript. The work presented here is a compilation of effort by many individuals over the last seven years. In particular, we would like to thank and acknowledge: Bernie Huyghe, Daniel Giroux, Mark Hom, Xioa-Dong Lui, Suganto Sutjipto, Emo Pungor, Estuardo Aguilar-Cordova, Douglas Cornell, Mary Nunnaly, Kai Tam, Ann Goudreau, Nancy Connelly, Cassandra Nyberg-Hoffman, Jing Zhou, Fred Porter, Andreas Frei, Barry Sugarman, Margarita Nodelman, Josefina Beltran, Anastasia Sofianos, Karli Watson, Thomas Schluep, Susan Miller, Shu-Fen Wen, Tattanahalli Nagabhushan, and Michael Shepard.

## References


1. Philipson L. Structure and assembly of adenoviruses. *Curr Top Microb Immunol* 1984;**109**:1–52.
2. Oliver CJ, Shortridge KF, Belyavin G. Diffusion coefficient and molecular weight of type 5 adenovirus by photon-correlation spectroscopy. *Biochim Biophys Acta* 1976;**437**: 589–98.
3. Nyberg-Hoffman C, Shabram P, Li W, Giroux D, Aguilar-Cordova E. Sensitivity and reproducibility in adenoviral infectious titer determination. *Nat Med* 1997;**3**:808–11.
4. Stewart PL, Fuller SD, Burnett RM. Difference imaging of adenovirus: bridging the resolution gap between X-ray crystallography and electron microscopy. *EMBO J* 1993;**12**:2589–99.
5. Pettersson U, Philipson L, Hoglund S. Structural proteins of adenoviruses. I. Purification and characterization of the adenovirus type 2 hexon antigen. *Virology* 1967;**33**:575–90.
6. Boulanger PA, Flamencourt P, Biserte G. Isolation and comparative chemical study of structural proteins of the adenoviruses 2 and 5: hexon and fiber antigens. *Eur J Biochem* 1969;**10**:116–31.
7. Pusey PN, Koppel DE, Schaefer DW, Camerini-Otero RD, Koenig SH. Intensity fluctuation spectroscopy of laser light scattered by solutions of spherical viruses: R17, Q beta, BSV, PM2, and T7. I. Light-scattering technique. *Biochemistry* 1974;**13**:952–60.

8. Sakaki Y, Maeda T, Oshima T. Bacteriophage phiNS II: a lipid- containing phage of acidophilic thermophilic bacteria. IV. Sedimentation coefficient, diffusion coefficient, partial specific volume, and particle weight of the phage. *J Biochem (Tokyo)* 1979;**85**:1205–11.
9. Graham FL, Smiley J, Russell WC, Nairn R. Characteristics of a human cell line transformed by DNA from human adenovirus type 5. *J Gen Virol* 1977;**36**:59–74.
10. Fallaux FJ, Bout A, van der Velde I, van den Wollenberg DJ, Hehir KM, Keegan J, et al. New helper cells and matched early region 1-deleted adenovirus vectors prevent generation of replication-competent adenoviruses. *Hum Gene Ther* 1998;**9**:1909–17.
11. Hehir KM, Armentano D, Cardoza LM, Choquette TL, Berthelette PB, White GA, et al. Molecular characterization of replication-competent variants of adenovirus vectors and genome modifications to prevent their occurrence. *J Virol* 1996;**70**:8459–67.
12. Branton PE, editor. *Early gene expression*. Austin, TX: R.G. Landes Company; 1999.
13. Vales LD, Darnell JE. Promoter occlusion prevents transcription of adenovirus polypeptide IX mRNA until after DNA replication. *Genes Dev* 1989;**3**:49–59.
14. Penefsky HS, Tzagoloff A. Extraction of water soluble enzymes and proteins from membranes. *Method Enzymol* 1971;**22**:204–19.
15. Monica T. *Viral vectors and vaccines*. Lake Tahoe, NV: Williamsburg Bioprocessing Foundation; 2000.
16. McGregor WC. Membrane separation in biotechnology. In: McGregor WC, editor. *Bioprocessing technology*. New York: Marcel Dekker; 1986.
17. Chase HA. The use of affinity adsorbents in expanded bed adsorption. *J Mol Recognit* 1998;**11**:217–21.
18. Hjorth R. Expanded-bed adsorption in industrial bioprocessing: recent developments. *Trends Biotechnol* 1997:230–5.
19. Thommes J. Fluidized bed adsorption as a primary recovery step in protein purification. *Adv Biochem Eng Biotechnol* 1987;**58**:185–230.
20. Meselson M, Stahl FW, Vinograd J. Equilibrium sedimentation of macromolecues in density gradients. *Proc Nat Acad Sci* 1957;**43**:581–8.
21. Burgess RR. Protein purification: micro to macro. In: *UCLA symposia on molecular and cellular biology*. New York: Alan R. Liss; 1987.
22. Deutscher MP, editors. *Guide to protein purification. Methods in enzymology*, vol. 180. New York: Academic Press; 1990.
23. Gagnon P. *Purification tools for monoclonal antibodies*. Tuscon: Validated Biosystems; 1996.
24. Jacoby WB, editors. *Enzyme purification. Methods in enzymology*, vol. 104. New York: Academic Press; 1984. p. 528.
25. Janson JC. Protein purification. In: Janson JC, editor. *Protein purification: principles, high resolution methods, and applications*. New York: VCH Publishers, Inc.; 1989.
26. Kenny A, Powell S. Practical protein chromatography. In: *Methods in Molecular Biology*, vol. 11. 1992. p. 1–327.
27. Wheelwright SM. *Protein purification: design and scale up of downstream processing*. Munich: Hanser Publishers; 1991.
28. Huyghe BG, Liu X, Sutjipto S, Sugarman BJ, Horn MT, Shepard HM, et al. Purification of a type 5 recombinant adenovirus encoding human p53 by column chromatography. *Hum Gene Ther* 1995;**6**:1403–16.
29. Snyder LR, Kirkland JJ. *Introduction to modem liquid chromatography*. New York: John Wiley & Sons; 1979.

30. Shabram PW, Giroux DD, Goudreau AM, Gregory RJ, Horn MT, Huyghe BG, et al. Analytical anion-exchange HPLC of recombinant type-5 adenoviral particles. *Hum Gene Ther* 1997;**8**:453–65.
31. Lehmberg E, Traina JA, Chakel JA, Chang RJ, Parkman M, McCaman MT, et al. Reversed-phase high-performance liquid chromatographic assay for the adenovirus type 5 proteome. *J Chromatogr B Biomed Sci Appl* 1999;**1732**:411–23.
32. Blanche F, Cameron B, Barbot A, Ferrero L, Guillemin T, Guyot S, et al. An improved anion-exchange HPLC method for the detection and purification of adenoviral particles. *Gene Ther* 2000;**7**:1055–62.
33. Karlsson RM. Ion exchange chromatography. In: Janson JC, Ryden L, editors. *Protein purification: principles, high resolution methods, and applications*. New York: VCH Publishers, Inc.; 1989.
34. Porath J, Olin B. Immobilized metal ion affinity adsorption and immobilized metal ion affinity chromatography of biomaterials. Serum protein affinities for gel-immobilized iron and nickel ions. *Biochemistry* 1983;**22**:1621–30.
35. Porath J, Carlsson J, Olsson I, Belfrage G. Metal chelate affinity chromatography, a new approach to protein fractionation. *Nature* 1975;**258**:598–9.
36. Kagedal L. *Immobilized metal ion affinity chromatography*. New York: VCH Publishers; 1989.
37. Douglas JT, Miller CR, Kim M, Dmitriev I, Mikheeva G, Krasnykh V, et al. A system for the propagation of adenoviral vectors with genetically modified receptor specificities. *Nat Biotechnol* 1999;**7**:470–5.
38. Shaltiel S. Hydrophobic chromatography. In: Jacoby WB, editor. *Enzyme purification and related techniques*. New York: Academic Press; 1984.
39. Eriksson KO. Hydophobic interaction chromatography. In: Janson JC, Ryden L, editors. *Protein purification: principles, high resolution methods, and applications*. New York: VCH Publishers, Inc.; 1989.
40. Kennedy RM. Hydrophobic chromatography. In: Deutscher MP, editor. *Guide to protein purification*. New York: Academic Press; 1990.
41. Fischer L. *Gel filtration chromatography*. Amsterdam: Elsevier; 1980.
42. Hagel L. Gel filtration. In: Janson JC, Ryden L, editors. *Protein purification: principles, high resolution methods, and applications*. New York: VCH Publishers, Inc.; 1989.
43. Coligan JE, Dunn BM, Ploegh HL, Speicher DW, Wingfield PT. In: Chanda VB, editor. *Current protocols*. New York: John Wiley & Sons; 1996.
44. Craig LC. Techniques for the study of peptides and proteins by dialysis and diffusion. In: Colowick SP, Kaplan NO, editors. *Methods in enzymology*. New York: Academic Press; 1967. p. 870–905.
45. McPhie P. Dialysis. In: Colowick SP, Kaplan NO, editors. *Methods in enzymology*. New York: Academic Press; 1971. p. 23–32.
46. Wiseman A. *Handbook of enzyme biotechnology*. New York: Halsted Press/John Wiley and Sons; 1985.
47. Cheryan M. *Ultrfiltration handbook*. Lancaster: Technomic Publishing Co; 1986.
48. Lynn DE. A BASIC computer program for analyzing endpoint assays. *Biotechniques* 1992;**12**:880–1.
49. Finney DJ. *Probit analysis*. Cambridge, UK: Cambridge University Press; 1962.
50. Andreadis S, Lavery T, Davis HE, LeDoux JM, Yarmush ML, Morgan JR. Toward a more accurate quantitation of the activity of recombinant retroviruses: alternatives to titer and multiplicity of infection. [corrected and republished article originally printed in *J Virol* February 2000;74(3): 1258–66] *J Virol* 2000;**74**:3431–9.

51. Shabram P, Aguilar-Cordova E. Multiplicity of infection/multiplicity of confusion. *Mol Ther* 2000;**2**:420–1.
52. Musco ML, Cui S, Small D, Nodelman M, Sugarman B, Grace M. Comparison of flow cytometry and laser scanning cytometry for the intracellular evaluation of adenoviral infectivity and p53 protein expression in gene therapy. *Cytometry* 1998;**33**:290–6.
53. Maizel JV, White D0, Scharff MD. The polypeptides of adenovirus. I. Evidence for multiple protein components in the virion and a comparison of types 2, 7A, and 12. *Virology* 1968;**36**:115–25.
54. Snyder LR, Kirkland JJ, Glajch JL. *Practical HPLC method development*. New York: John Wiley & Sons; 1997.
55. Hearn MTW. Reversed phase high performance liquid chromatography. In: Jakoby WB, editor. *Enzyme purification and related techniques, part C*. New York: Academic Press; 1984. p. 190–212.
56. Rivier J. Reversed-phase high-performance liquid chromatography: preparative purification of synthetic peptides. *J Chromatogr* 1984;**288**:303–28.
57. Lee ML, Muench KH. Prolyl transfer ribonucleic acid synthetase of *E. coli*. *J Biol Chern* 1969;**244**:223–30.
58. Frieden C. Protein-protein interaction and enzymatic activity. *Ann Rev Biochem* 1971;**40**: 653–96.
59. Smoluchowski MV. Versuch einer mathematischentheorie der koagulationskinetic kolloider losungen. *Z Phy Chern* 1917;**92**:129–68.
60. Gekko K, Timasheff SN. Mechanism of protein stabilization by glycerol: preferential hydration in glycerol-water mixtures. *Biochemistry* 1981;**20**:4667–76.
61. Gekko K, Timasheff SN. Thermodynamic and kinetic examination of protein stabilization by glycerol. *Biochemistry* 1981;**20**:4677–86.

# Targeted Adenoviral Vectors I: Transductional Targeting

9

*Igor P. Dmitriev, Sergey A. Kaliberov*
Department of Radiation Oncology, School of Medicine, Washington University, St. Louis, MO, USA

## 1. Introduction

Human adenovirus[1] (Ad) has been used extensively to derive replication-incompetent gene delivery vectors to correct genetic disorders and develop candidate vaccines for a variety of infectious diseases and cancer immunotherapy, and as conditionally replicative Ad (CRAd) agents for cancer virotherapy. Adenovirus vectors have been used in 22% of all gene therapy clinical trials, followed by retroviral vectors (19%) and naked/plasmid deoxyribonucleic acid (DNA) (18%).[2,3] A major factor limiting the effectiveness of current-generation Ad vectors is their inability to accomplish specific gene delivery to cells of interest. Indeed, a National Institutes of Health report identified the "need for vector targeting" as a central objective for the field of gene therapy.[4] Extensive studies of interactions between Ad capsid proteins and host cells in vitro revealed that efficient Ad infection requires the presence of sufficient levels of receptors responsible for virus attachment to the cellular membrane and internalization. Adenovirus attachment to the cell is mediated by fiber binding with its C-terminal knob domain to a primary cellular receptor. Subsequent interaction of $\alpha_v\beta_{3/5}$-integrins, the secondary cellular receptor, with an Arg-Gly-Asp (RGD) sequence within a protein loop extended from the penton base is required to trigger endocytosis resulting in virus internalization. Most Ad of species B have been shown to use human membrane cofactor CD46 as the predominant attachment receptor[5] whereas the coxsackievirus group B and Ad receptor (CAR)[6] has been identified as the primary high-affinity receptor for many representatives of species A, C, D, E, and F.[7–9] Therefore, levels of CD46 and CAR expression determine the infection efficacy of Ad serotype 35 of species B and Ad serotype 2 (Ad2) or 5 (Ad5) (both of species C), respectively, which are mostly used for vector construction purposes. Thus, an unfavorable expression pattern of primary Ad receptor in a clinical context would result in an insufficient level of infection of target cells while leading to ectopic virus sequestration by non-target tissues. The delineation of key steps of the Ad cellular entry pathway in vitro, in which cell attachment is distinct from subsequent virus internalization, suggested that Ad recognition of cognate primary receptor represents a rate-limiting step, which could be intervened in an effort to redirect virus-cell binding via an alternative cellular receptor to confer susceptibility to Ad vector infection. Transductional targeting strategies seek to redirect Ad binding to appropriate nonnative receptors to increase the efficiency of gene transfer to the cell type selected to achieve therapeutic intervention.

**Adenoviral Vectors for Gene Therapy.**

# 2. Adapter-Mediated Ad Vector Targeting Approach

Efforts to redirect Ad vectors via receptors overexpressed on the cells that are refractory to Ad infection mainly focus on incorporating targeting ligands by means of chemical conjugation or genetic modification of viral capsid proteins and using bispecific adapter molecules to mediate virus recognition of target cells. The use of bispecific protein adapters was originally proposed to bridge viral particle and cell surface molecule to overcome inefficient virus infectivity owing to the scarcity of Ad attachment receptor[7–9] or its localization on inaccessible parts of the cell.[10–17] This goal was originally addressed by the development of bispecific antibody (bsAb) conjugates, which are able to bind both the viral capsid protein and the cell surface receptor, allowing indirect linkage between viral particles and cellular receptor (Figure 1).

## *2.1 Use of Ab Conjugates for Ad Targeting*

To construct bsAb adapters, Wickham et al. used monoclonal antibody (mAb) against an FLAG peptide, which was genetically incorporated in place of the deleted RGD sequence in penton base protein, chemically conjugated to mAb with specificities for $\alpha_V$-integrin receptors or human CD3 to redirect the AdFLAG vector to endothelial and smooth muscle cells or T cells, respectively.[18,19] Although successfully demonstrating the feasibility of in vitro virus retargeting via non–Ad receptors displayed on human venule endothelial cells, intestinal smooth muscle cells, and resting T cells that are normally refractory, this approach was later abandoned, apparently because of reduced virus viability resulting from RGD sequence deletion.[20] An alternative

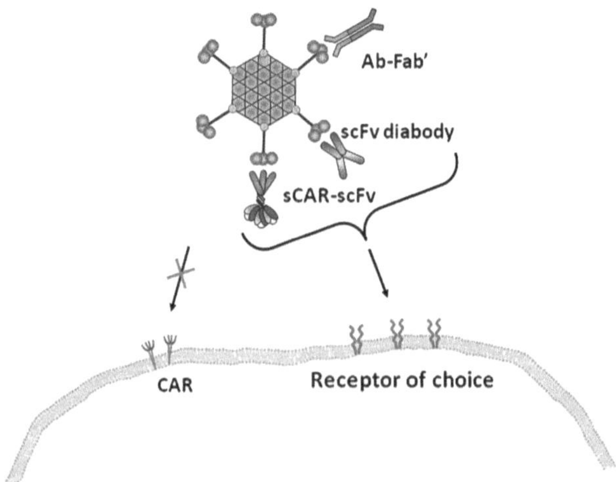


**Figure 1 Strategies of Ad vector targeting using bispecific adapters.** Adenovirus retargeting from various cell types can be achieved using bispecific adapter proteins. Bispecific adapters consist of Ad fiber knob-binding moiety fused to alternative receptor-binding ligand including Fab antibodies, scFv Ab, and biological ligands. Targeting adapters allow efficient CAR-independent transduction of cells of interest.

approach to provide adapter binding to Ad capsid was proposed by Douglas et al.,[21] based on the use of neutralizing mAb 1D6.14, which blocks binding of the Ad5 fiber knob to CAR. The feasibility of Ad retargeting via a nonviral receptor was demonstrated by conjugating the Fab' fragments of mAb 1D6.14 to folate to allow virus linkage to the folate receptor, which is overexpressed on the surface of a variety of malignant cells. This Fab–folate conjugate was complexed with Ad5 vector carrying the luciferase reporter gene and was shown to redirect virus infection of target cells via the folate receptor at a high efficiency. When complexed with Ad5 carrying the gene for herpes simplex virus thymidine kinase, the Fab–folate conjugate mediated the specific killing of cells that overexpress the folate receptor. This work demonstrated the feasibility of employing an adapter approach both to ablate endogenous viral tropism and to introduce novel tropism in vitro.

Use of the Ab-based Ad5 vector targeting approach was further explored to circumvent the lack of CAR expression and improve gene transfer specifically to tumor cells by generating a bispecific Ab conjugate consisting of anti-knob Fab' fragments conjugated to mAb 425,[22] which was derived against the epidermal growth factor receptor (EGFR), a tumor-associated marker negligibly expressed in normal mitotically quiescent tissues.[23] Targeting of Ad5 vector via EGFR using this Ab conjugate led to enhanced gene transfer relative to untargeted Ad in 7 of 12 human glioma cell lines and 6 of 8 primary glioma samples derived from tumors of various histologies.[24] Furthermore, EGFR retargeting showed marked transduction enhancement in both squamous cell carcinoma of the head and neck cell lines and primary tumor tissue compared with normal tissue from the same patient.[25] These studies illustrated that Ad targeting via EGFR overcomes cell deficiency in CAR expression to achieve an increase in gene transfer efficiency in tumor cell types, and therefore suggests that a bispecific adapter approach could augment Ad vector potency for cancer gene therapy applications.

Based on these essential findings, an adapter approach was employed to promote Ad-mediated gene transfer in dendritic cells (DC) to assess its targeting utility in the context of important therapeutic applications proposed for genetically modified DCs. To this end, Tillman et al.[26] tested Fab' fragment (1D6.14) chemically conjugated to mAb G28-5 agonistically binding DC's receptor CD40.[27] The CD40 receptor is attractive for DC targeting because it has an important role in inducing DC maturation and priming cytotoxic T cells.[28] Ad5 vector retargeting via the CD40 pathway using this bispecific construct dramatically enhanced gene transfer to monocyte-derived DCs (MoDCs) established from peripheral blood of normal human volunteer donors and induced both their phenotypic and functional maturation as demonstrated by increased T cell stimulation in an allogeneic mixed leukocyte reaction and by enhanced interleukin (IL)-12p70 release.[26] To explore the potential of an adapter-mediated targeting approach to enhance the efficacy of DC-based vaccinations in vivo, Tillman et al.[29] employed a similar Fab' conjugated with mAb FGK45[30] against mouse CD40 (mCD40) along with Ad vector encoding a tumor antigen. To this end, AdE7 vector expressing the human papillomavirus type-16 (HPV-16) E7 oncogene, which represents an attractive target for antigen-specific immunity of cervical cancer, was coupled with Fab-anti-murine CD40 and then was used to load bone marrow–derived DCs

(BMDCs) ex vivo. It was shown that subcutaneous injection of BMDCs infected with CD40-targeted AdE7 provided superior protection against HPV-16–induced tumor challenge and improved prophylaxis against outgrowth of established tumors relative to BMDCs infected by untargeted Ad. This study illustrated that Ad-modified DCs may be used in repeated vaccination to establish antigen-specific and $CD8^+$ T cell–dependent protection. These findings suggested that Ad-based DC loading with tumor antigens can elicit productive antitumor immunity and that the enhancement of gene transfer and DC maturation mediated by CD40-targeted Ad complex may facilitate this process.

To further demonstrate the clinical utility of adapter-mediated DC targeting, de Gruijl et al.[31] evaluated CD40-targeted Ad vectors performance in the context of three-dimensional human tissue under physiological and clinically highly relevant conditions. To this end, a human skin explant model was used to test transduction efficiency of cutaneous DC after intradermal injection of Ad5 vector preincubated with antiknob Fab'-G28-5 conjugate.[26] Significantly enhanced transduction efficiency and selectivity and an increased activation state of migrating DC were achieved while extending antigen-specific cytotoxic T lymphocyte (CTL)-stimulatory ability for up to 1 week after the start of migration, in contrast to DC transduced by untargeted Ad. Because DC targeting in vivo might obviate the need for the in vitro culture of autologous DC for adoptive transfer, CD40-targeted Ad vectors constitute a promising new vaccine modality for tumor immunotherapy.

To determine whether an adapter-mediated Ad-targeting approach could maintain fidelity upon systemic vascular administration, Reynolds et al.[32] used a bispecific Ab conjugate to target Ad infection specifically to angiotensin-converting enzyme (ACE), which is preferentially expressed on pulmonary capillary. Administration of ACE-targeted vector complexes via tail vein injection into rats resulted in at least a 20-fold increase in both Ad genome localization and luciferase transgene expression in the lungs whereas luciferase activity in the liver was reduced by over 80% compared with the untargeted vector. This study showed that an adapter-mediated Ad targeting can indeed alter the biodistribution profile of an Ad vector given systemically, thus providing encouraging implications for the further development of targetable, injectable Ad vectors that may enable gene therapy for pulmonary vascular disease. The use of ACE-targeting adapter combined with endothelial-specific transgene expression driven by flt-1 promoter resulted in a synergistic 300,000-fold improvement in the selectivity of luciferase expression for lung versus the usual site of vector sequestration, the liver.[33]

The use of adapter-mediated Ad retargeting toward tumor cells was demonstrated using mAb against the epithelial cell adhesion molecule (EpCAM)[34] conjugated with antifiber knob Fab' fragments.[35] The EpCAM-targeted Ad vectors complexed with this bispecific Ab conjugate showed an improved transduction of primary tumor cells and cell lines established from gastric and esophageal adenocarcinoma compared with normal gastric epithelium.[36] Using a similar approach, chemical conjugation of the antiknob Fab' was achieved with basic fibroblast growth factor (FGF2)[37] in an effort to develop a new treatment approach for Kaposi sarcoma.[38] Of note, use of FGF2-targeted Ad complexes achieved direct therapeutic goals in a murine orthotopic model of human ovarian carcinoma relevant to a current human clinical cancer gene therapy scheme.[39,40]

## 3. Recombinant Ad Targeting Adapters

Further refinement of adapter approach was accomplished by engineering recombinant proteins consisting of a neutralizing single-chain fragment variable (scFv) Ab S11 against Ad fiber knob fused with human EGF[41] or scFv 425 against EGFR[42] to improve Ad5 infection efficiency in cancer cells. Recombinant adapter molecules such as these have advantages for Ad retargeting, because use of the chemical conjugation of Ab molecules increases the difficulties of producing Ad retargeting complexes, which makes this approach relatively complex and expensive to develop. To further improve vector targeting specificity, the use of native tropism-ablated Ad, which was previously constructed to contain both CAR- and $\alpha_V$-integrin-binding mutated residues,[43,44] was tested using bispecific scFv adapters targeted toward human EGFR or EpCAM.[45] An elegant study by van Beusechem et al. demonstrated that these native tropism-ablated Ad vectors complexed with bispecific scFv efficiently and selectively targeted both alternative receptors on the surface of human cancer cell lines and primary human tumor specimens. Moreover, EGFR-targeted doubly ablated vectors were selective for human brain tumors versus the surrounding normal brain tissue, resulting in a 5- to 38-fold improved tumor-to-normal brain targeting index compared with nonablated control vectors.[45] Application of EpCAM-targeted double-ablated Ad vector for gastric cancer gene therapy showed a favorable ration of tumor over normal tissue transduction.[46] Of note, the transduction efficiency mediated by EpCAM-targeted native tropism-ablated Ad complexes reached levels similar to or exceeding those achieved with native Ad control for EpCAM-expressing primary human gastric tumors, whereas transduction of gastric epithelium and liver tissue was reduced at least 10-fold.

To achieve targeted genetic modification of hepatic stellate cells (HSCs), Reetz et al.[47] designed a peptide of the nerve growth factor (NGFp) with specific affinity for the p75 neurotrophin receptor (p75NTR) present on HSCs. Coupling of this NGFp to Ad particles was done via chemical conjugation using bifunctional polyethylene glycol (PEG) or by coating with a fusion protein composed of scFv S11 and p75NTR. Coupling of NGFp to Ad via S11 or PEGylation resulted in markedly reduced liver tropism and enhanced gene transfer to HSCs, whereas Ad GFP-S11-NGFp transduced activated HSCs better than Ad GFP-PEG-NGFp. This study contributed to the development of gene transfer system targeted to activated HSCs based on systemically applied Ad vector modified with NGFp.

These successful examples of employing bispecific adapters to achieve receptor-specific Ad gene transfer rationalized further development of the recombinant adapter molecule design. In this regard, Dmitriev et al. proposed using the soluble extracellular CAR domain (sCAR) fused to human EGF as a targeting ligand to engineer a novel class of adapters capable of blocking CAR-dependent Ad tropism while promoting infection of CAR-deficient cell types overexpressing EGFR including human mammary gland, ovarian, epidermoid, squamous, and pancreatic carcinoma cells.[48,49] A similar approach was applied to engineer sCAR ectodomain fused to the Fc region of the human immunoglobulin G1 protein to target Ad vector via high-affinity $Fc_{\gamma}$ receptor I while achieving up to a 250-fold increase in transgene expression in CAR-negative human monocytic cell lines expressing the target receptor (CD64).[50] Using noninvasive optical imaging to monitor firefly luciferase (luc) luciferin-dependent bioluminescent activity, Liang et al. showed that systemic vascular administration of Ad5-luc vector coated with the newly generated sCAR-EGF protein resulted in significantly reduced ectopic luc expression in the liver and markedly facilitated luc expression in tumor xenografts displaying elevated EGFR levels compared with sCAR-6His-coated Ad5-luc control.[51] This demonstration of both liver untargeting and tumor retargeting of Ad vector mediated by bispecific recombinant adapter suggested that sCAR-EGF–coated virions could maintain fidelity after systemic delivery, thus providing encouraging implications for the development of targetable, injectable Ad vector systems that may enable gene therapy for cancer. To assess the use of an adapter approach for Ad targeting to colon, lung, and breast epithelial tumors that express carcinoembryonic antigen (CEA), Li et al. used noninvasive optical imaging of bioluminescent luc activity provided by Ad complexed with a bispecific sCAR-MFE protein containing an scFv MFE-23 against CEA.[52] The use of sCAR-MFE adapter resulted in Ad vector retargeting to CEA-positive epithelial tumor cells in cell culture, subcutaneous tumor xenografts, and hepatic tumor grafts while showing greater than 90% reduction of Ad-directed luc expression in the liver after systemic vector administration.

Use of recombinant adapter molecules eliminates chemical conjugation and provides a high degree of flexibility for ligand substitution, and consequently expands the targeting capabilities of Ad vectors. These considerations warranted further development of the adapter-mediated Ad targeting approach to improve its potency in the context of systemic applications. One development endeavor was to design bispecific recombinant molecules that have higher binding affinity to viral capsid to maintain fidelity of virus–adapter complexes subsequent to systemic delivery. In this regard, both structural analysis of fiber knob bound to CAR D1 domain[53] and identification of a conserved CAR-binding site on the fiber protein[43] suggested an avidity mechanism when three CAR molecules could simultaneously bind per one fiber knob trimer, which was supported by kinetic analysis of Ad2 knob binding to the CAR D1 domain.[54] Based on these considerations, it was hypothesized that trimeric sCAR-ligand molecules could achieve high-affinity linkage to fiber knob and promote ligand-mediated binding to target receptors.

To test this hypothesis, Kashentseva et al. engineered the sCARfC6.5 adapter protein consisting of sCAR, phage T4 fibritin-derived polypeptide, and C6.5 scFv against c-erbB-2 oncoprotein to confer Ad targeting capability on cancer cells expressing the c-erbB-2/HER-2/neu oncogene.[55] It was demonstrated that incorporation of fibritin polypeptide provided trimerization of sCAR fusion proteins that resulted in increased affinity to Ad fiber knob and augmented the ability to block CAR-dependent Ad infection, compared with monomeric sCAR protein. As illustrated in cancer cell lines that overexpress c-erbB-2, targeted Ad, complexed with sCARfC6.5 adapter protein, provided 1.5- to 17-fold enhancement of gene transfer compared with Ad alone and up to

130-fold increase compared with untargeted Ad complexed with sCARfibritin control protein. In a parallel study, Kim et al. employed an isoleucine GCN4 trimerization domain to improve sCAR binding to fiber knob[56] while engineering recombinant adapters containing a cyclic RGD peptide (cRGD) or the receptor-binding domain of apolipoprotein E to achieve efficient gene transfer in human diploid fibroblasts in vitro. Whereas the trimerized sCAR devoid of targeting ligand provided efficient blocking of ectopic liver gene transfer in normal C57BL/6 mice, addition of either ligand failed to retarget the liver in vivo. To apply gene therapy treatment for hepatic colorectal cancer (CRC) metastatic tumors, which often express both cyclooxygenase-2 (COX-2) and CEA, Li et al. coupled the use of COX-2 promoter for transcriptional control with transductional targeting mediated by a trimerized sCARfMFE adapter containing anti-CEA scFv.[57] This study demonstrated that the use of both transcriptional control and sCARfMFE adapter allowed retargeting of Ad-mediated expression of the herpes simplex virus type 1 thymidine kinase (HSV1-tk) therapeutic gene from normal liver tissue to hepatic CRC tumors after systemic virus injection, which increased the therapeutic efficacy of ganciclovir treatment for hepatic CRC tumors while reducing its hepatic toxicity. These results indicate that trimerized sCAR-ligand proteins can markedly improve Ad targeting potency in vivo owing to its high-affinity binding to fiber knob, which efficiently blocks CAR-dependent viral tropism while conferring a novel cell-binding specificity mediated by trimeric ligand moiety via an alternative tumor-associated receptor.

The sCAR-derived adapters have also been exploited to confer Ad targeting abilities toward dendritic cells (DCs) to orchestrate immune responses in an effort to develop vaccines and potent anticancer immunotherapy. The current procedure of ex vivo loading of autologous DCs with tumor-associated antigen (TAA) and their activation for clinical application is laborious and expensive, and remains poorly standardized. The use of viral vectors represents an attractive alternative approach to loading resident DCs in vivo by targeted TAA delivery and simultaneous activation. The feasibility of sCAR-mediated Ad targeting to DCs was demonstrated by Pereboev et al., by generating sCAR fusion with scFv against human CD40, which was derived using the G28-5 hybridoma cell line[27] and demonstrating highly efficient transduction of immature MoDCs.[58] Using this sCAR-G28 adapter, Asiedu et al.[59] showed that improved transduction of mature rhesus monkey MoDCs with Ad expressing transforming growth factor (TGF)-β1 could significantly suppress alloimmune responses and inhibit proliferation of CD4 and CD8 responder T cells. These results and work by Clement et al.[60] illustrated that adapter-mediated Ad targeting can promote TGF-β1 gene expression in nonhuman primate mature MoDCs to function as alloantigen-specific cellular immunosuppressants, an approach that has the potential to facilitate induction of allograft tolerance in vivo. The study by Brando et al.[61] showed that the bispecific scFv S11-G28 adapter can serve as well to significantly enhance Ad transduction efficiency of human MoDCs while increasing the ability of MoDC to activate CTL in an antigen-specific manner.

Further development of the CD40-targeting approach was achieved using the adapter molecule, CFm40L, which was designed by fusing ectodomains of CAR and mCD40 ligand (mCD40L) via a trimerization motif.[62] Incorporation of the trimerization motif served to increase fiber knob binding avidity[55] while maintaining the native trimeric CD40L conformation necessary for efficient mCD40 binding and function,[28] which is compatible with its human counterpart owing to the high degree of homology between mouse and human tumor necrosis factor–like CD40L domains.[63] Pereboev et al. showed that gene transfer to mouse BMDC using CFm40L-targeted Ad was over four orders of magnitude more efficient than that for the untargeted Ad5 control, resulting in transduction of 70% of the BMDC compared with undetectable transduction using Ad5 control. Most important, CD40-targeted Ad induced in vivo phenotypical DC maturation, upregulated IL-12 expression, and elicited superior Th and CTL responses against the β-galactosidase model antigen in Balb/c mice. Results of this study demonstrated that Ad-mediated gene transfer to DC can be significantly enhanced using nonnative transduction pathways such as the CD40 pathway, which may have important applications in genetic vaccination for cancer and infectious diseases. To study the effects of adapter-mediated Ad targeting via the CD40 pathway in vivo, Huang et al.[64] compared biodistribution and immune responses after intravenous (i.v.), intradermal (i.d.), and intranasal (i.n.) administration of CFm40L-coated Ad5 and untargeted Ad5 control in Balb/c mice. The CD40-targeted Ad5 injected i.v. revealed increased transgene expression in the lung and thymus, which normally do not sequester significant amounts of virus after systemic administration. After i.d. injection, CD40-targeted Ad showed about 300-fold lower gene transfer signals detected mainly in local draining lymph nodes and skin compared with control Ad5. Of note, undesirable ectopic sequestration of untargeted Ad5, which was detected in brain tissue that showed the second highest gene expression level after the lung, was largely ablated using CD40-targeted Ad complexes. Moreover, CD40 targeting elicited more sustained antigen-specific cellular immune responses (up to 17-fold) against nucleocapsid protein of SARS-CoV, which was used as a model antigen, at later time points (30 days after boosting) after i.d. and i.n. application, but also significantly hampered humoral responses irrespective of the administration route. This study demonstrated that CFm40L adapter-mediated Ad targeting can profoundly alter the patterns of virus biodistribution and immune responses against the transgene after local and systemic administration.

Preclinical evidence of therapeutic use of an adapter-mediated DC targeting approach for cancer immunotherapy was obtained by Hangalapura et al.[65] using Ad encoding the full-length melanoma antigen recognized by T cell-1 (MART-1) coupled with the CFm40L adapter.[62] It was demonstrated that this CD40-targeted Ad-MART-1vector enhanced transduction of conventional and plasmacytoid DC subsets, but not B cells, in suspensions of human melanoma-draining sentinel lymph nodes ex vivo resulting in reduction of regulatory T cell (Tregs) frequencies while facilitating expansion of functional MART-1–specific $CD8^+$ T cells. Further study by Hangalapura et al.[66] demonstrated enhanced transduction and maturation of cultured BMDCs with CFm40L-coupled Ad-GFP-$TRP2_{aa180-188}$ vector encoding the immunodominant H-2Kb–binding epitope of tyrosinase-related protein 2 (TRP2) fused to eGFP compared with untargeted control. The BMDCs transduced with DC-targeted vector ex vivo induced stronger $TRP2_{aa180-188}$–specific $CD8^+$ T-cell responses in peripheral blood while resulting in improved prophylactic vaccination efficacy in the aggressive and poorly immunogenic murine B16F10 melanoma model.[67] To assess the effect of CD40 targeting on the induction of immunity against weakly immunogenic TAAs, CFm40L-coupled Adgp100 vector encoding a full-length human gp100[68] was employed for i.d. vaccination. These studies revealed that CD40-targeted Adgp100 significantly enhanced the induction of a $gp100_{25-33}$-specific $CD8^+$ T cell response and antitumor efficacy in both prophylactic and therapeutic vaccination settings, which translated into an improved survival of tumor-bearing animals receiving a CFm40L-Adgp100 vaccine. These results thus clearly showed enhanced antitumor efficacy afforded by the CFm40L-mediated in vivo targeting of Ad5-based vaccines encoding weakly immunogenic TAAs to DCs. Taken together, these studies support the use of CFm40L-coupled Ad vectors for in vivo DC targeting to accomplish high-efficacy CTL priming while breaking immune tolerance against TAAs to achieve therapeutic anticancer efficacy in preclinical and clinical studies.

To test whether the CD40-targeting strategy can improve the outcomes of prostate cancer immunotherapy, Williams et al.[69] developed a murine model of prostate cancer by generating derivatives of the mouse RM-1 prostate cancer cell line expressing human prostate–specific membrane antigen (PSMA).[70] To maximize antigen presentation in target cells, both major histocompatibility complex class I and transporter associated with antigen processing protein expression was induced in RM-1 cells by transduction with Ad5-IFN-gamma vector expressing interferon-gamma.[71] Administering DCs infected ex vivo using CD40-targeted Ad5-huPSMA coupled with CFm40L adapter, as well as direct intraperitoneal injection of the vector–adapter complexes, resulted in high levels of tumor-specific CTL responses against RM-1-PSMA cells pretreated with Ad5-IFN-gamma, thus significantly improving the therapeutic antitumor efficacy. These data suggested that DC-targeted Ad delivery of PSMA mediated by CFm40L adapter may be effective clinically for prostate cancer immunotherapy.

The adapter approach was explored to improve Ad vector utility for T lymphocyte–based therapies.[72] To surmount T lymphocyte resistance to Ad infection, Beatty et al.[73] proposed designing sCAR ectodomain fusion with murine interleukin 2 (sCAR-mIL-2) that targets Ad to the murine IL-2 receptor (IL-2R). Interleukin-2R is T lymphocyte specific and highly expressed in therapeutic T lymphocyte populations such as $CD4^+Foxp3^+$ regulatory T lymphocytes and activated $CD4^+$ and $CD8^+$ T lymphocytes.[74] This study showed the use of Ad5 vector coupled with an sCAR-mIL-2 adapter to infect a murine T-cell line, CTLL-2, and activated primary murine T lymphocytes allowed a nine- and fourfold improvement in reporter gene expression levels compared with Ad5 vector alone, respectively. These findings have broad application for the study of T cell biology and genetic modification of T cells for therapeutic use.

The technologies of designed ankyrin repeat proteins (DARPins) and ribosome display were employed to develop a DARPin that binds the Ad5 fiber knob domain with low nanomolar affinity. In particular, Dreier et al.[75] reported a novel design of bispecific adapter protein that chelated the knob in a bivalent or trivalent fashion while providing binding specificity for HER-2, an established cell-surface biomarker of human cancers. This study showed that the efficacy of gene transfer by the adapter–Ad complex increased accordingly with the functional affinity of these molecules, enabling efficient virus transduction at low stoichiometric adapter-to-fiber ratios. In principle, DARPins can be generated against any target, which makes this versatile adapter approach useful for developing a broad range of disease-specific Ad vector applications. The most recent refinement of DARPin technology allowed the development of a series of adapters that bind the Ad5 fiber knob with such high affinity that they remain fully bound for more than 10 days while blocking Ad native receptor tropism and mediating interaction with a surface receptor of choice.[76] By solving the crystal structure of the complex of the trimeric knob with three bound DARPins at 1.95-A resolution, Dreier et al. used computer modeling to devise a trimeric protein of extraordinary kinetic stability. Specifically, the capsid protein SHP from the lambdoid phage 21 served to bind the knob like a trimeric clamp fused with DARPins of varying specificities, thus allowing Ad5-mediated gene transfer in a HER-2-, EGFR-, or EpCAM-dependent manner with transduction efficiencies comparable to or even exceeding those of Ad5 alone. With these adapters, efficiently produced in *Escherichia coli*, Ad can be conferred new receptor specificities using receptor-binding ligands available for many cell types of choice, which suggests the means to engineer practical and effective Ad targeting approaches.

## 3.1 Combination of Genetic Capsid Modification and Adapter-Mediated Ad Targeting

To achieve a strong association between viral particles and adapter proteins, several groups proposed combining genetic capsid modification with a targeting adapter approach. To this end, the Ad5 fiber capsid protein was genetically fused to the C-terminal biotin acceptor peptide (BAP).[77] Adenovirus 5 particles bearing this BAP were metabolically biotinylated during vector production by the endogenous biotin ligase in 293 cells to produce covalently biotinylated virions. The resulting biotinylated vector could be retargeted to new receptors by conjugation to biotinylated antibodies using tetrameric avidin ($K_d = 10^{-15}$ M). Campos et al.[78] used a panel of metabolically biotinylated Ad vectors to directly compare targeted transduction mediated through the fiber, protein IX, and hexon capsid proteins using a variety of biotinylated ligands including mAb, transferrin, EGF, and cholera toxin B. This study clearly demonstrated that effective cell targeting could be achieved only when biotinylated fiber protein served for receptor-binding ligand conjugation. In contrast, protein IX and hexon-mediated ligand conjugation with the same ligands failed to provide vector targeting, likely because of aberrant trafficking at the cell surface or inside targeted cells. These data suggested that Ad targeting will likely be the most efficient through fiber modification rather than pIX or hexon protein. Using Ad5 vector containing metabolically biotinylated fiber proteins, Chen et al.[79] showed retargeting to primary cultured human corneal epithelial cells, which was mediated by conjugation with biotinylated EGF, providing up to ninefold increased transduction of EGFR-expressing corneal epithelial progenitor cells while reducing transduction of differentiated corneal epithelial cells. A biotin–avidin linkage was also used to conjugate Ad vectors to ligands that bind with high affinity to ChemR23, $\alpha_v\beta_3$-integrins, and DC-SIGN receptors[80] to improve the efficacy of human MoDCs transduction, maturation, and ability to stimulate cytokine production by autologous memory $CD8^+$ T cells against the vector-encoded immunodominant human cytomegalovirus pp65 protein compared with untargeted virus. This study expanded the range of receptors that could be employed for DC targeting to facilitate the development of Ad-based vaccines.

An alternative targeting strategy was proposed to combine genetic incorporation of the immunoglobulin (Ig) binding domain of *Staphylococcus aureus* protein A into the Ad fiber protein with targeting ligands fused to the Ig Fc domain to form vector-ligand targeting complexes.[81–83] Korokhov et al.[82] showed that targeting ligands containing the Fc domain and either an anti-CD40 scFv or CD40L form stable complexes with Ad vector incorporating the so-called Cd of *S. aureus* protein A, which resulted in significant augmentation of gene delivery to MoDCs target cells. Using a similar approach of genetic fiber modification to insert a synthetic 33–amino acid IgG-binding domain (Z33) derived from protein A, Volpers et al.[81] demonstrated up to a 77-fold increased gene transfer efficacy in differentiated primary human muscle cells, which was achieved by preincubation of the AdFZ33 vector with mAb directed against neuronal cell adhesion molecule or $\alpha_7$-integrin. This versatile Ad targeting strategy was employed by Kawashima et al.[84] to demonstrate highly efficient gene transfer in biliary cancer cells using AdFZ33 vector combined with mAb against EpCAM or EGFR compared with the control antibody or without antibody. This study showed that AdFZ33 vector, which was constructed to express uracil phosphoribosyl transferase, complexed with anti-EpCAM or anti-EGFR mAb, remarkably enhanced the sensitivity of biliary cancer cells to 5-fluorouracil but not cells lacking EpCAM or EGFR expression including normal hepatocytes and thus resulting in significantly suppressed growth of biliary cancer xenografts in nude mice. Employment of this versatile IgG-binding Ad vector approach holds promise to solve the problem of structural and biosynthetic compatibility between viral capsid proteins and targeting ligands by allowing direct use of the available repertoire of mAb against cell surface antigens for Ad targeting to a variety of cellular receptors.

Use of the bispecific adapter approach has established several key concepts with respect to the goal of Ad vector retargeting. (1) It was clearly shown that Ad5-based vectors can provide effective gene transfer via CAR-independent cell entry pathways. Thus, virus interaction with its primary attachment receptor does not appear to be essential to attain the effective cell entry. (2) Achievement of CAR-independent infection via alternative cellular receptors allows augmented levels of gene transfer. Indeed, redirecting Ad5 infection via nonviral receptors allows improving the susceptibility of target cells in vitro and in vivo. (3) The targeting use of adapter molecules depends on interaction with viral capsomers. In this regard, the targeting ability of bispecific molecules appears to be the most efficient through Ad fiber interaction rather than pIX or hexon protein.

## 4. Adenovirus Targeting Using Genetic Modification of Capsid Proteins

As discussed, molecular adapters have allowed modification of Ad tropism and key proof-of-principle demonstrations of targeted gene transfer in both in vitro and in vivo delivery contexts. However, the genetic capsid modification approach is the preferred

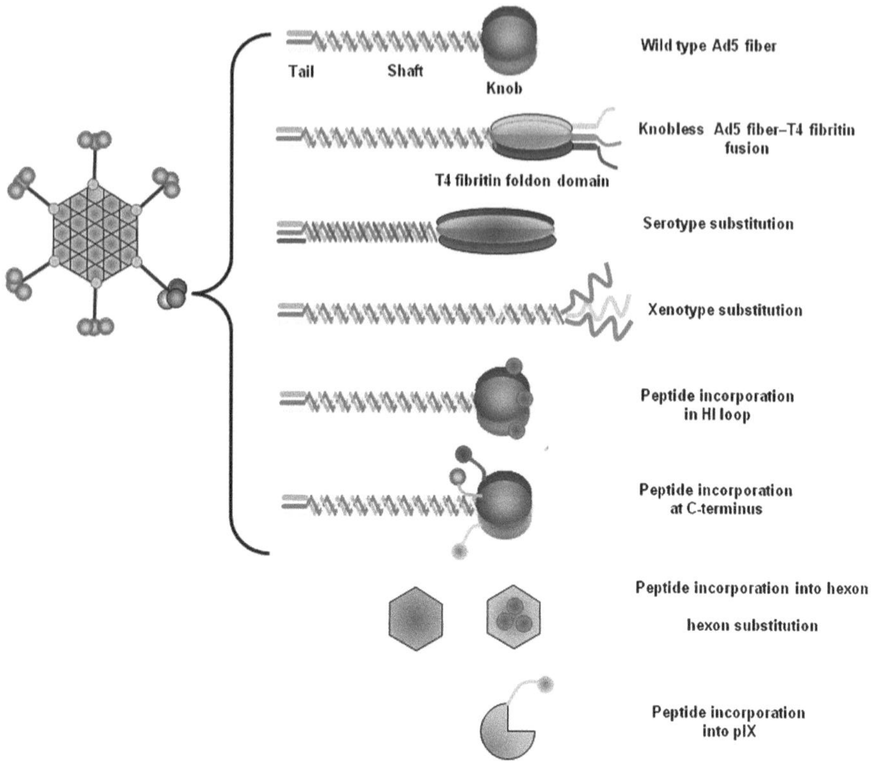


**Figure 2 Schematic outline of Ad capsid modifications using genetic incorporation of heterologous ligands.** Adenovirus targeting strategies employing manipulation of the Ad genome included peptide incorporation to the HI loop and C-terminus of the knob domain, serotype chimerism, fiber/knob replacement, and combinatorial approaches. Chimeric Ads composed of fiber/knob domains from alternative serotypes, fiber-xenotyped Ad vectors displayed fibers from nonhuman Ad species, and knobless Ad fiber–fibritin fusion. Genetic incorporation of peptides into the HVR of hexon and C-terminus of the pIX.

configuration for clinical applications of targeted Ad vectors. Methods to alter Ad vector tropism have capitalized on the knowledge that viral capsid proteins including fiber, hexon, penton base, and pIX are the key determinants of specificity of Ad infection. On the basis of these considerations, several approaches have been developed to alter Ad5 tropism using genetic capsid modifications (Figure 2).

## 5. Employment of Chimeric and Mosaic Fibers

In the first instance, the method of Ad retargeting is based on use of Ad capsid pseudotyping using fiber substitution with fiber from different serotypes. These chimeric fibers are primarily derived from viruses that employ different receptors for cell binding including CD46,[85] CD80/CD86,[86] and desmoglein-2.[87,88] In early studies, the use of a method to construct an Ad5/3 vector containing chimeric fibers composed of the tail and shaft domains of Ad5 and the knob domain of Ad serotype 3 (Ad3) was established.[89] More recent studies demonstrated high transduction efficacy of Ad5/3-based vectors in a variety of Ad5-refractory tumor cell types with low CAR expression including renal cell carcinoma,[90] ovarian cancer,[91] melanoma,[92] and prostate cancer cells.[93] Several fiber chimeric Ad vectors have been developed to achieve tumor- or tissue-specific gene delivery by employing fibers derived from Ad35, Ad40, or Ad19p.[94–98] Over the past decade, additional approaches to Ad retargeting were developed using fibers from nonhuman Ad species. A number of fiber-xenotyped Ad5 vectors were developed based on chimeric fibers with knob domains from aviadenovirus or atadenovirus,[99] canine Ad serotype 2 (CAV-2),[100] canine Ad serotype 1 (CAV-1),[101] and porcine Ad serotype 3 and 4 (PAd3 and PAdV-4).[102,103] In addition to fibers from nonhuman Ad vectors, a fiber-mosaic Ad5 vector encoding two different fibers including the wild-type Ad5 and the receptor-binding molecule of Dearing (T3D) reovirus serotype 3 was constructed. Use of fiber-like σ1 attachment protein provided enhanced infectivity in tissues with low CAR expression and tropism expansion via infection of cells expressing sialic acid and junction adhesion molecule 1.[104]

Ad targeting must embody the concept that tumors are complex tissues that are composed of many interdependent cellular components, including malignant cells, cancer stem (or stem-like) cells, and tumor-associated stromal elements. In this regard, a fiber-mosaic strategy of capsid modification in which single viral particles can incorporate two distinct fiber species was evaluated using mosaic Ad5 vectors with fibers derived from wild-type Ad5 fiber and FF6H protein consisting of the amino-terminal segment of Ad5 fiber sequence genetically fused with the carboxy-terminal portion of the phage T4 fibritin protein, followed by the linker and the 6-His ligand.[105] In another study, Murakami et al. generated a fiber-mosaic Ad vector displaying both Ad5 fiber and a chimeric fiber protein composed of the Ad5 tail domain and the Ad3 shaft and knob domains. The capacity of the dual-fiber Ad vector to transduce distinct cell types in a mixed cell population was demonstrated in vitro. This fiber profile allows the expanded tropism required for an inclusion targeting strategy, which is based on the use of fiber-mosaic viral particles that can infect cells efficiently with a distinctive receptor's repertoire.[106] More recent studies have demonstrated that employment of fibers derived from both Ad5 and Ad3 increased oncolytic potency of CRAd. An experimental therapy study using a human pancreatic tumor xenograft model demonstrated that employment of complex mosaicism increased efficacy of the combination of oncolytic virotherapy with chemotherapy.[107]

## 6. Employment of Targeting Peptides in Fiber Modification

Several strategies have been developed to alter tropism of Ad5-based vectors to achieve a cell-specific gene delivery by employing fiber modifications using genetic incorporation of targeting motifs. Generally, retargeting strategies have focused on

Ad fiber modifications, because it is the major determinant of Ad tropism. It was shown that insertion of an integrin-binding RGD motif or polylysine peptides into the C-terminus of the fiber knob significantly reduced the transduction efficiency of CAR-positive cells by Ad vectors.[108] In early studies, the use of an exposed HI-loop structure connecting β-sheets H and I in the Ad knob domain as an alternate location for the cysteine-constrained RGD-4C (CDCRGDCFC) peptide insertion was demonstrated by Dmitriev et al.[109] Based on evidence that RGD and polylysine (pK7) motifs bind to different cell surface proteins, cellular integrins, and heparan sulfate–containing receptors, respectively, double-modified Ad fiber knob with RGD and pK7 motifs have been shown to enhance Ad5 infection via CAR-independent pathways and improved gene transfer efficiency.[110] Promising data have been demonstrated in studies in which the phage display technique was used to determine specific binding peptides. The display of polypeptide repertoires on the surface of filamentous phages as well as peptide incorporated Ad libraries was shown to be a valuable method for isolating unique peptides that can be employed for Ad targeting. In vivo selection of phage display libraries and Ad libraries displaying random peptides on the fiber knob techniques were successfully employed to acquire a number tumor-homing peptides with a targeting specificity related to angiogenic blood vessels (CDCRGDCFC and CNGRCVSGCAGRC),[111] tumor lymphatic vessels (CGNKRTRGC),[112] pancreatic cancer (SYENFSA),[113] and renal cell carcinoma (HITSLLS).[98] Phage display technology was used to isolate an HVGGSSV peptide that binds specifically to tax-interacting protein-1 receptor in irradiated tumors.[114] Although these modifications allowed for tropism alterations, in many instances the application of this approach has been limited by incompatibility between the capsid and ligand. In addition, targeting motifs fused to the fiber protein demonstrated decreased binding functionality or an impaired proper protein tertiary structure.[115]

The means of transductional retargeting of the Ad was accomplished by fiber modification approaches allowing incorporation of large and complex targeting moieties and retaining trimerization of the fiber. These modifications are based on the concept of chimeric knobless fiber by replacing the native Ad fiber knob protein with an alternative protein capable of providing trimerization functions and allowing the incorporation of targeting peptides.[116] Initial studies involved generating knobless Ad vectors with trimerization motifs derived from Moloney murine leukemia virus,[117] bacteriophage T4 fibritin,[118] or trimerization motifs derived from reoviral σ1 protein[119] were introduced in place of knob domain followed by the C-terminal Myc-epitope or 6His-tag. All of these fiber-modified vectors were shown to mediate receptor-specific transduction in vitro through interaction with surface-expressed Abs.

Whereas a wide range of targeting moieties have been employed for recombinant Ad vectors, the restricted repertoire of available targeting peptides that are functionally compatible with insertion in the fiber protein has led to the consideration of various Ab species for Ad retargeting purposes. Furthermore, the biosynthesis of candidate Ab species designed for Ad incorporation must be compatible with Ad capsid protein synthesis and assembly. To this point, available Ab species have not proved to be biologically compatible with cytosolic Ad capsid synthesis and assembly, resulting in loss of binding affinities. This loss of binding specificity, in the instance of incorporated scFv, is likely because Ad capsid proteins are normally synthesized in the cytosol with assembly in the nucleus, whereas scFv molecules are typically routed through the rough endoplasmic reticulum.[120] In this context, the redox state of the cytosol likely results in improper scFv folding that perturbs the structural configuration required for antigen recognition, leading to loss of binding specificity. Despite the demonstrated utility of stabilized scFv with molecular scaffold motifs designed to resist the deleterious effect of the cytosol redox state for Ad retargeting,[116,121] the limited available repertoire of target specificities of this class of scFv practically restricts this approach.

Recent studies validated the use of fiber-based targeting moieties using synthetically constructed monobodies representing single-domain Ab mimics based on the tenth human fibronectin type III domain (10Fn3) scaffold to achieve selectivity of gene transfer using tropism-modified Ad.[122] In contrast to these synthetically constructed monobodies, Kaliberov et al. considered the use of alternate antibody species that might embody a stability profile compatible with the cytosolic biosynthesis of Ad capsid proteins.[123] The discovery of unconventional immunoglobulins derived from the serum of animals in the camelid family (camels and alpacas) that consist of only the two heavy-chains (hcAbs) as the basis of antigen recognition and binding has made possible their use for Ad-mediated gene therapy.[124] Unlike conventional immunoglobulins, hcAbs contain a single variable domain (VHH) linked to two constant domains.[125] Cloned and isolated single-domain antibodies have shown effective targeting in model systems and a remarkable stability profile compared with conventional immunoglobulins and scFvs.[126,127] It was shown that expression of anti-CEA VHH genetically incorporated into a deknobbed Ad5 fiber-fibritin protein did not disrupt the trimerization capability of the Ad fiber and retain antigen recognition functionality. The ability of an anti-CEA VHH fused to fiber-fibritin chimera to provide specific and efficient targeting of the CEA-expressing cancer cells for Ad-mediated gene transfer was also demonstrated.[123]

## 7. Employment of Alternative Capsid Sites for Ligand Incorporation

Despite the demonstrated use of fiber modification for Ad retargeting, this approach has been limited by incompatibility between the fiber protein and ligand that leads to impaired antigen recognition functionality. Another approach has focused on the development of retargeting Ad using other capsid proteins besides fiber. In early studies, small peptides were incorporated into Ad capsid proteins, such as peptide epitope from the hemagglutinin protein of influenza virus within a penton base.[128]

Evidence shows that the hexon is the most abundant Ad capsid protein, which makes the hexon an attractive site for the presentation of targeting moieties. The tendency of i.v. administered Ad5 to localize in the liver represents a major factor limiting current strategies to accomplish targeting of Ad vector. The major pathway of liver transduction involves interactions of Ad capsid proteins with circulating blood cells and with plasma proteins including several components of complement pathway and blood coagulation zymogens.[129,130] Although not universally accepted, liver uptake of Ad is mediated by high-affinity interaction between the major protein in the Ad5 capsid, hexon, and γ-carboxylated glutamic acid domain of coagulation factor X (FX). The Ad5–FX complex attaches to hepatocytes through binding of the serine protease domain of FX to cell surface heparan sulfate proteoglycans.[131–133] It was shown that different Ad serotypes interact with FX with distinct affinities. For instance, human Ad serotypes 26, 35, and 48 bind to FX with relatively low affinity compared with Ad5.[131,134–136] More recently, it was shown that ablation of FX binding to Ad5 with modified hexon protein resulted in decreased liver tropism.[130,137,138]

Comparisons of the hexon sequence among different Ad serotypes revealed several unique serotype-specific sequences: hypervariable regions (HVR1–9) at loops 1 and 2, which are exposed on the exterior surface of the hexon molecule. Incorporation of an αv-specific DCRGDCF ligand in the HVR5 of hexon resulted in enhanced transduction of cells with low levels of CAR expression.[139] In another study, the 6-His epitope was incorporated in HVR2 and HVR5.[140] It was shown that HVRs 2, 3, 5, 6, and 7 are amenable to insertion of a 6-His motif. In addition, anti–6-His Ab recognized Ad vectors with 6-His inserted into HVRs 2 and 5.[110] A subsequent study demonstrated that HVR5 of hexon was capable to accommodate a peptide up to 36 amino acids (aa) in length[141] as well as the 71-aa BAP protein[142] with minimal adverse effects on virion stability. It was later shown that substitution of HVR7 of the Ad5 hexon with HVR7 from Ad3 resulted in decreased liver tropism and dramatically altered biodistribution of gene expression. Systemic administration of AdH5/H3CMVLuc, AdH5RoboLuc, and AdH5/H3RoboLuc in C57BL6J mice produced Luc expression in the liver that was 59- and 431- and over 240,000-fold, respectively, lower than wild-type AdH5CMVLuc. The results of these studies suggest that the combination of liver detargeting using a genetic modification of hexon with an endothelium-specific transcriptional control element produces an additive effect in the improvement of Ad5 biodistribution.[143]

The minor capsid protein IX (pIX), which is present in 240 copies in the Ad capsid, was exploited as an anchor for heterologous C-terminal extensions of up to 113 aa in length, which included 75 Å α-helical spacers between pIX protein and peptide ligands. The MYC-tagged-pIX molecules were readily accessible to anti-MYC Ab.[144] In early studies, use of pIX for genetic incorporation of targeting ligands was established by Dmitriev et al.[145] In this study, Ad vectors containing modified pIX carrying a C-terminal Flag epitope along with a heparan sulfate binding motif consisting of either eight consecutive lysines or a polylysine sequence were constructed. The pIX variants were efficiently incorporated into the capsid of Ad particles. Using an anti-Flag Ab, it was shown that modified pIXs are incorporated into virions and display Flag-containing C-terminal sequences on the capsid surface. The incorporation of a polylysine motif into the pIX ectodomain resulted in significant augmentation of Ad fiber knob-independent infection of CAR-deficient cell types.[145] Using this strategy, Ad retargeting was achieved by incorporating large targeting moieties, including eGFP,[146] HSV1-tk,[147] and metallothionein.[148,149]

The use of pIX protein as a platform for presenting scFv or sdAb molecules for Ad retargeting was evaluated. The 13R4 scFv directed against β-galactosidase, which was selected for its capacity to fold correctly in a reducing environment such as the cytoplasm, was fused with pIX using a 75-Å-spacer sequence.[121] In another study, a single-chain T cell receptor directed against cancer/testis antigen melanoma-associated antigen (MAGE)-A1 in complex with the human leukocyte antigen (HLA) class I molecule of haplotype HLA-A1 was fused with the C terminus of the pIX. Generated particles specifically transduced melanoma cells expressing the HLA-A1/MAGE-A1 target complex with at least 10-fold higher efficiency than control viruses.[150] However, because of the nature of the Ad capsid proteins synthesis and virion assembly, even the endoplasmic reticulum–targeted pIX-scFv proteins were incorporated into the Ad capsid at a low level that was not sufficient to retarget virus infection. In contrast, it was shown that expression of anti-EGFRvIII sdAb on the Ad capsid through fusion to pIX can be used to redirect Ad infection.[120]

## 8. Conclusion

It is widely acknowledged that improving the therapeutic potential of Ad vectors requires elimination of the natural viral tropism and introduction of a novel mechanism of selective cell recognition that would allow directed virus localization to the target tissue. The strategies described above including the use of bispecific adapter molecules and the genetic incorporation of targeting ligands into capsid proteins were extensively developed to redirect Ad5 infection via nonnative pathways. Targeted Ad vectors hold the promise to expand the types of diseases that can be treated by gene therapy and to make the therapeutic applications of Ad vectors more effective. The increased specificity achieved by targeting virus infection to cells of interest will ultimately allow lower and safer doses of Ad vectors to be provided when regional or systemic delivery is contemplated in the future.

The nature of the virus–host interactions that dictate the fate of systemically administered Ad vectors has come under considerable scrutiny in recent years. Recent studies focused on the biology of interactions between Ad capsid components and host blood factors and their influence on systemic virus biodistribution revealed the ability of the vitamin K–dependent coagulation factors VII, IX, X, and protein C to bind trimeric hexon in the viral capsid and facilitate CAR-independent infection of hepatic cells after intravascular Ad5 vector administration. These efforts serve to highlight the complexity of virus–host interplay in the artificial blood-borne environment and have identified modifications of the fiber and hexon proteins that significantly decrease infection and virus-induced toxicity in the liver. Thus, it is recognized that the infection pathway of systemically administered Ad5 is mediated via multiple mechanisms involving blood factors rather than direct virus interaction with cellular receptors. On this basis, it becomes increasingly apparent that engineering of capsid proteins to overcome ectopic sequestration in the liver coupled with virus retargeting via a nonnative infection pathway represents a rational strategy to direct Ad vector localization to the tissue of interest subsequent to systemic vascular administration. In this regard, genetic engineering of the Ad fiber protein appears the most straightforward way to generate targeted Ad vectors with novel tropism.

Despite major advancements illustrating the potential of genetic Ad targeting in vitro, efforts to employ high-affinity ligands including growth factors and scFvs have mostly been unsuccessful, frustrating targeting of Ad vectors to many attractive cellular markers. On the basis of these deliberations, the use of alternate Ab species that might embody a stability profile compatible with the cytosolic biosynthesis of Ad capsid proteins was considered. Camelid hcAbs possess characteristics ideal for an Ad retargeting strategy: (1) cytosolic stability allowing functional incorporation into the Ad capsid and (2) compatibility with phage biopanning selection to allow target cell specificity. Based on these useful attributes, a number of targeted Ad vectors using genetic incorporation of sdAb into fiber-fibritin or pIX proteins have been developed. This finding provides an important technical approach allowing practical linkage of capsid modification of Ad vector and ligand-based strategies for targeting gene delivery.

Whereas single-component vector systems have been favored for employment in the context of human clinical trials, rigorous analysis of the pharmacodynamics and systemic stability of vector–adapter complexes could provide the rationale for clinical translation. In this respect, previous in vivo studies using various Ad5 fiber knob-binding adapters have provided compelling evidence of reduced ectopic liver transduction and receptor-specific vector delivery to target organs or tumors. The utility of adapter molecules constructed using an anti-Ad5 knob scFv or the sCAR ectodomain is obviously limited to Ad5 and other CAR-binding Ad serotypes. This provides a rationale for the development of a new class of protein adapters capable of Ad vector targeting by virtue of binding to alternative capsid epitopes. The use of such a serotype-independent targeting modality could provide the technical means for testing the ability of vectors derived from representatives of various Ad species to localize to the tissue of interest while overcoming ectopic organ sequestration.

Thus, novel Ad tropism modification maneuvers that embody the concepts of detargeting and retargeting by combining elements of genetic capsid modification and adapter-based approaches have encouraging implications for further development of advanced delivery vehicles.

## References

1. Shenk T. Adenoviridae: the viruses and their replication. In: Fields BN, Knipe DM, Howley PM, editors. *Fields virology*. 3rd ed. Philadelphia: Lippincott – Raven Publishers; 1996. p. 2111–48.
2. Gene therapy clinical trials worldwide. *J Gene Med* 2014. Available at: http://www.wiley.com//legacy/wileychi/genmed/clinical/.
3. Ginn SL, Alexander IE, Edelstein ML, Abedi MR, Wixon J. Gene therapy clinical trials worldwide to 2012 – an update. *J Gene Med* February 2013;**15**(2):65–77.
4. Fox JL. Orkin-Motulsky panel calls for gene therapy basic research. *Gene Ther* January 1996;**3**(1):pre-1.
5. Gaggar A, Shayakhmetov DM, Lieber A. CD46 is a cellular receptor for group B adenoviruses. *Nat Med* November 2003;**9**(11):1408–12.

6. Bergelson JM, Cunningham JA, Droguett G, Kurt-Jones EA, Krithivas A, Hong JS, et al. Isolation of a common receptor for Coxsackie B viruses and adenoviruses 2 and 5. *Science* February 28, 1997;**275**(5304):1320–3.
7. Hidaka C, Milano E, Leopold PL, Bergelson JM, Hackett NR, Finberg RW, et al. CAR-dependent and CAR-independent pathways of adenovirus vector-mediated gene transfer and expression in human fibroblasts. *J Clin Invest* February 1999;**103**(4):579–87.
8. Leon RP, Hedlund T, Meech SJ, Li S, Schaack J, Hunger SP, et al. Adenoviral-mediated gene transfer in lymphocytes. *Proc Natl Acad Sci USA* October 27, 1998;**95**(22):13159–64.
9. Nalbantoglu J, Pari G, Karpati G, Holland PC. Expression of the primary coxsackie and adenovirus receptor is downregulated during skeletal muscle maturation and limits the efficacy of adenovirus-mediated gene delivery to muscle cells. *Hum Gene Ther* April 10, 1999;**10**(6):1009–19.
10. Cohen CJ, Shieh JT, Pickles RJ, Okegawa T, Hsieh JT, Bergelson JM. The coxsackievirus and adenovirus receptor is a transmembrane component of the tight junction. *Proc Natl Acad Sci USA* December 18, 2001;**98**(26):15191–6.
11. Coyne CB, Bergelson JM. CAR: a virus receptor within the tight junction. *Adv Drug Deliv Rev* April 25, 2005;**57**(6):869–82.
12. Pickles RJ, Fahrner JA, Petrella JM, Boucher RC, Bergelson JM. Retargeting the coxsackievirus and adenovirus receptor to the apical surface of polarized epithelial cells reveals the glycocalyx as a barrier to adenovirus-mediated gene transfer. *J Virol* July 2000;**74**(13):6050–7.
13. Raschperger E, Thyberg J, Pettersson S, Philipson L, Fuxe J, Pettersson RF. The coxsackie- and adenovirus receptor (CAR) is an in vivo marker for epithelial tight junctions, with a potential role in regulating permeability and tissue homeostasis. *Exp Cell Res* May 15, 2006;**312**(9):1566–80.
14. Sharma P, Kolawole AO, Wiltshire SM, Frondorf K, Excoffon KJ. Accessibility of the coxsackievirus and adenovirus receptor and its importance in adenovirus gene transduction efficiency. *J Gen Virol* January 2012;**93**(Pt 1):155–8.
15. Stonebraker JR, Wagner D, Lefensty RW, Burns K, Gendler SJ, Bergelson JM, et al. Glycocalyx restricts adenoviral vector access to apical receptors expressed on respiratory epithelium in vitro and in vivo: role for tethered mucins as barriers to lumenal infection. *J Virol* December 2004;**78**(24):13755–68.
16. Walters RW, Grunst T, Bergelson JM, Finberg RW, Welsh MJ, Zabner J. Basolateral localization of fiber receptors limits adenovirus infection from the apical surface of airway epithelia. *J Biol Chem* April 9, 1999;**274**(15):10219–26.
17. Walters RW, van't Hof W, Yi SM, Schroth MK, Zabner J, Crystal RG, et al. Apical localization of the coxsackie-adenovirus receptor by glycosyl-phosphatidylinositol modification is sufficient for adenovirus-mediated gene transfer through the apical surface of human airway epithelia. *J Virol* August 2001;**75**(16):7703–11.
18. Wickham TJ, Lee GM, Titus JA, Sconocchia G, Bakacs T, Kovesdi I, et al. Targeted adenovirus-mediated gene delivery to T cells via CD3. *J Virol* October 1997;**71**(10):7663–9.
19. Wickham TJ, Segal DM, Roelvink PW, Carrion ME, Lizonova A, Lee GM, et al. Targeted adenovirus gene transfer to endothelial and smooth muscle cells by using bispecific antibodies. *J Virol* October 1996;**70**(10):6831–8.
20. Bai M, Harfe B, Freimuth P. Mutations that alter an Arg-Gly-Asp (RGD) sequence in the adenovirus type 2 penton base protein abolish its cell-rounding activity and delay virus reproduction in flat cells. *J Virol* September 1993;**67**(9):5198–205.
21. Douglas JT, Rogers BE, Rosenfeld ME, Michael SI, Feng M, Curiel DT. Targeted gene delivery by tropism-modified adenoviral vectors. *Nat Biotechnol* November 1996;**14**(11):1574–8.

22. Snelling L, Miyamoto CT, Bender H, Brady LW, Steplewski Z, Class R, et al. Epidermal growth factor receptor 425 monoclonal antibodies radiolabeled with iodine-125 in the adjuvant treatment of high-grade astrocytomas. *Hybridoma* April 1995;**14**(2):111–4.
23. Wong AJ, Bigner SH, Bigner DD, Kinzler KW, Hamilton SR, Vogelstein B. Increased expression of the epidermal growth factor receptor gene in malignant gliomas is invariably associated with gene amplification. *Proc Natl Acad Sci USA* October 1987;**84**(19):6899–903.
24. Miller CR, Buchsbaum DJ, Reynolds PN, Douglas JT, Gillespie GY, Mayo MS, et al. Differential susceptibility of primary and established human glioma cells to adenovirus infection: targeting via the epidermal growth factor receptor achieves fiber receptor-independent gene transfer. *Cancer Res* December 15, 1998;**58**(24):5738–48.
25. Blackwell JL, Miller CR, Douglas JT, Li H, Reynolds PN, Carroll WR, et al. Retargeting to EGFR enhances adenovirus infection efficiency of squamous cell carcinoma. *Arch Otolaryngol Head Neck Surg* August 1999;**125**(8):856–63.
26. Tillman BW, de Gruijl TD, Luykx-de Bakker SA, Scheper RJ, Pinedo HM, Curiel TJ, et al. Maturation of dendritic cells accompanies high-efficiency gene transfer by a CD40-targeted adenoviral vector. *J Immunol* June 1, 1999;**162**(11):6378–83.
27. Francisco JA, Gilliland LK, Stebbins MR, Norris NA, Ledbetter JA, Siegall CB. Activity of a single-chain immunotoxin that selectively kills lymphoma and other B-lineage cells expressing the CD40 antigen. *Cancer Res* July 15, 1995;**55**(14):3099–104.
28. van Kooten C, Banchereau J. CD40–CD40 ligand. *J Leukoc Biol* January 2000;**67**(1):2–17.
29. Tillman BW, Hayes TL, DeGruijl TD, Douglas JT, Curiel DT. Adenoviral vectors targeted to CD40 enhance the efficacy of dendritic cell-based vaccination against human papillomavirus 16-induced tumor cells in a murine model. *Cancer Res* October 1, 2000;**60**(19):5456–63.
30. Rolink A, Melchers F, Andersson J. The SCID but not the *RAG-2* gene product is required for Sμ–Sε heavy chain class switching. *Immunity* October 1996;**5**(4):319–30.
31. de Gruijl TD, Luykx-de Bakker SA, Tillman BW, van den Eertwegh AJ, Buter J, Lougheed SM, et al. Prolonged maturation and enhanced transduction of dendritic cells migrated from human skin explants after in situ delivery of CD40-targeted adenoviral vectors. *J Immunol* November 1, 2002;**169**(9):5322–31.
32. Reynolds PN, Zinn KR, Gavrilyuk VD, Balyasnikova IV, Rogers BE, Buchsbaum DJ, et al. A targetable, injectable adenoviral vector for selective gene delivery to pulmonary endothelium in vivo. *Mol Ther* December 2000;**2**(6):562–78.
33. Reynolds PN, Nicklin SA, Kaliberova L, Boatman BG, Grizzle WE, Balyasnikova IV, et al. Combined transductional and transcriptional targeting improves the specificity of transgene expression in vivo. *Nat Biotechnol* September 2001;**19**(9):838–42.
34. Edwards DP, Grzyb KT, Dressler LG, Mansel RE, Zava DT, Sledge Jr GW, et al. Monoclonal antibody identification and characterization of a Mr 43,000 membrane glycoprotein associated with human breast cancer. *Cancer Res* March 1986;**46**(3):1306–17.
35. Haisma HJ, Pinedo HM, Rijswijk A, der Meulen-Muileman I, Sosnowski BA, Ying W, et al. Tumor-specific gene transfer via an adenoviral vector targeted to the pan-carcinoma antigen EpCAM. *Gene Ther* August 1999;**6**(8):1469–74.
36. Heideman DA, Snijders PJ, Craanen ME, Bloemena E, Meijer CJ, Meuwissen SG, et al. Selective gene delivery toward gastric and esophageal adenocarcinoma cells via EpCAM-targeted adenoviral vectors. *Cancer Gene Ther* May 2001;**8**(5):342–51.
37. Rogers BE, Douglas JT, Ahlem C, Buchsbaum DJ, Frincke J, Curiel DT. Use of a novel cross-linking method to modify adenovirus tropism. *Gene Ther* December 1997;**4**(12):1387–92.

38. Goldman CK, Rogers BE, Douglas JT, Sosnowski BA, Ying W, Siegal GP, et al. Targeted gene delivery to Kaposi's sarcoma cells via the fibroblast growth factor receptor. *Cancer Res* April 15, 1997;**57**(8):1447–51.
39. Printz MA, Gonzalez AM, Cunningham M, Gu DL, Ong M, Pierce GF, et al. Fibroblast growth factor 2-retargeted adenoviral vectors exhibit a modified biolocalization pattern and display reduced toxicity relative to native adenoviral vectors. *Hum Gene Ther* January 1, 2000;**11**(1):191–204.
40. Rancourt C, Rogers BE, Sosnowski BA, Wang M, Piche A, Pierce GF, et al. Basic fibroblast growth factor enhancement of adenovirus-mediated delivery of the herpes simplex virus thymidine kinase gene results in augmented therapeutic benefit in a murine model of ovarian cancer. *Clin Cancer Res* October 1998;**4**(10):2455–61.
41. Watkins SJ, Mesyanzhinov VV, Kurochkina LP, Hawkins RE. The 'adenobody' approach to viral targeting: specific and enhanced adenoviral gene delivery. *Gene Ther* October 1997;**4**(10):1004–12.
42. Haisma HJ, Grill J, Curiel DT, Hoogeland S, van Beusechem VW, Pinedo HM, et al. Targeting of adenoviral vectors through a bispecific single-chain antibody. *Cancer Gene Ther* June 2000;**7**(6):901–4.
43. Roelvink PW, Mi Lee G, Einfeld DA, Kovesdi I, Wickham TJ. Identification of a conserved receptor-binding site on the fiber proteins of CAR-recognizing adenoviridae. *Science* November 19, 1999;**286**(5444):1568–71.
44. Einfeld DA, Schroeder R, Roelvink PW, Lizonova A, King CR, Kovesdi I, et al. Reducing the native tropism of adenovirus vectors requires removal of both CAR and integrin interactions. *J Virol* December 2001;**75**(23):11284–91.
45. van Beusechem VW, Grill J, Mastenbroek DC, Wickham TJ, Roelvink PW, Haisma HJ, et al. Efficient and selective gene transfer into primary human brain tumors by using single-chain antibody-targeted adenoviral vectors with native tropism abolished. *J Virol* March 2002;**76**(6):2753–62.
46. Heideman DA, van Beusechem VW, Offerhaus GJ, Wickham TJ, Roelvink PW, Craanen ME, et al. Selective gene transfer into primary human gastric tumors using epithelial cell adhesion molecule-targeted adenoviral vectors with ablated native tropism. *Hum Gene Ther* September 20, 2002;**13**(14):1677–85.
47. Reetz J, Genz B, Meier C, Kowtharapu BS, Timm F, Vollmar B, et al. Development of adenoviral delivery systems to target hepatic stellate cells in vivo. *PLoS One* 2013;**8**(6):e67091.
48. Dmitriev I, Kashentseva E, Rogers BE, Krasnykh V, Curiel DT. Ectodomain of coxsackievirus and adenovirus receptor genetically fused to epidermal growth factor mediates adenovirus targeting to epidermal growth factor receptor-positive cells. *J Virol* August 2000;**74**(15):6875–84.
49. Wesseling JG, Bosma PJ, Krasnykh V, Kashentseva EA, Blackwell JL, Reynolds PN, et al. Improved gene transfer efficiency to primary and established human pancreatic carcinoma target cells via epidermal growth factor receptor and integrin-targeted adenoviral vectors. *Gene Ther* July 2001;**8**(13):969–76.
50. Ebbinghaus C, Al-Jaibaji A, Operschall E, Schoffel A, Peter I, Greber UF, et al. Functional and selective targeting of adenovirus to high-affinity Fcγ receptor I-positive cells by using a bispecific hybrid adapter. *J Virol* January 2001;**75**(1):480–9.
51. Liang Q, Dmitriev I, Kashentseva E, Curiel DT, Herschman HR. Noninvasive of adenovirus tumor retargeting in living subjects by a soluble adenovirus receptor-epidermal growth factor (sCAR-EGF) fusion protein. *Mol Imaging Biol* Nov-Dec 2004;**6**(6):385–94.

52. Li HJ, Everts M, Pereboeva L, Komarova S, Idan A, Curiel DT, et al. Adenovirus tumor targeting and hepatic untargeting by a coxsackie/adenovirus receptor ectodomain anti-carcinoembryonic antigen bispecific adapter. *Cancer Res* June 1, 2007;**67**(11):5354–61.
53. Bewley MC, Springer K, Zhang YB, Freimuth P, Flanagan JM. Structural analysis of the mechanism of adenovirus binding to its human cellular receptor, CAR. *Science* November 19, 1999;**286**(5444):1579–83.
54. Lortat-Jacob H, Chouin E, Cusack S, van Raaij MJ. Kinetic analysis of adenovirus fiber binding to its receptor reveals an avidity mechanism for trimeric receptor-ligand interactions. *J Biol Chem* March 23, 2001;**276**(12):9009–15.
55. Kashentseva EA, Seki T, Curiel DT, Dmitriev IP. Adenovirus targeting to c-erbB-2 oncoprotein by single-chain antibody fused to trimeric form of adenovirus receptor ectodomain. *Cancer Res* January 15, 2002;**62**(2):609–16.
56. Kim J, Smith T, Idamakanti N, Mulgrew K, Kaloss M, Kylefjord H, et al. Targeting adenoviral vectors by using the extracellular domain of the coxsackie-adenovirus receptor: improved potency via trimerization. *J Virol* February 2002;**76**(4):1892–903.
57. Li HJ, Everts M, Yamamoto M, Curiel DT, Herschman HR. Combined transductional untargeting/retargeting and transcriptional restriction enhances adenovirus gene targeting and therapy for hepatic colorectal cancer tumors. *Cancer Res* January 15, 2009;**69**(2):554–64.
58. Pereboev AV, Asiedu CK, Kawakami Y, Dong SS, Blackwell JL, Kashentseva EA, et al. Coxsackievirus-adenovirus receptor genetically fused to anti-human CD40 scFv enhances adenoviral transduction of dendritic cells. *Gene Ther* September 2002;**9**(17):1189–93.
59. Asiedu C, Dong SS, Pereboev A, Wang W, Navarro J, Curiel DT, et al. Rhesus monocyte-derived dendritic cells modified to over-express TGF-β1 exhibit potent veto activity. *Transplantation* September 15, 2002;**74**(5):629–37.
60. Clement A, Pereboev A, Curiel DT, Dong SS, Hutchings A, Thomas JM. Converting nonhuman primate dendritic cells into potent antigen-specific cellular immunosuppressants by genetic modification. *Immunol Res* 2002;**26**(1–3):297–302.
61. Brandao JG, Scheper RJ, Lougheed SM, Curiel DT, Tillman BW, Gerritsen WR, et al. CD40-targeted adenoviral gene transfer to dendritic cells through the use of a novel bispecific single-chain Fv antibody enhances cytotoxic T cell activation. *Vaccine* June 2, 2003;**21**(19–20):2268–72.
62. Pereboev AV, Nagle JM, Shakhmatov MA, Triozzi PL, Matthews QL, Kawakami Y, et al. Enhanced gene transfer to mouse dendritic cells using adenoviral vectors coated with a novel adapter molecule. *Mol Ther* May 2004;**9**(5):712–20.
63. Karpusas M, Hsu YM, Wang JH, Thompson J, Lederman S, Chess L, et al. 2 A crystal structure of an extracellular fragment of human CD40 ligand. *Structure* October 15, 1995;**3**(10):1031–9.
64. Huang D, Pereboev AV, Korokhov N, He R, Larocque L, Gravel C, et al. Significant alterations of biodistribution and immune responses in Balb/c mice administered with adenovirus targeted to CD40(+) cells. *Gene Ther* February 2008;**15**(4):298–308.
65. Hangalapura BN, Oosterhoff D, Aggarwal S, Wijnands PG, van de Ven R, Santegoets SJ, et al. Selective transduction of dendritic cells in human lymph nodes and superior induction of high-avidity melanoma-reactive cytotoxic T cells by a CD40-targeted adenovirus. *J Immunother* September 2010;**33**(7):706–15.
66. Hangalapura BN, Oosterhoff D, de Groot J, Boon L, Tuting T, van den Eertwegh AJ, et al. Potent antitumor immunity generated by a CD40-targeted adenoviral vaccine. *Cancer Res* September 1, 2011;**71**(17):5827–37.
67. Fidler IJ. Biological behavior of malignant melanoma cells correlated to their survival in vivo. *Cancer Res* January 1975;**35**(1):218–24.

68. Tuettenberg A, Jonuleit H, Tuting T, Bruck J, Knop J, Enk AH. Priming of T cells with Ad-transduced DC followed by expansion with peptide-pulsed DC significantly enhances the induction of tumor-specific CD8+ T cells: implications for an efficient vaccination strategy. *Gene Ther* February 2003;**10**(3):243–50.
69. Williams BJ, Bhatia S, Adams LK, Boling S, Carroll JL, Li XL, et al. Dendritic cell based PSMA immunotherapy for prostate cancer using a CD40-targeted adenovirus vector. *PLoS One* 2012;**7**(10):e46981.
70. Ghosh A, Heston WD. Tumor target prostate specific membrane antigen (PSMA) and its regulation in prostate cancer. *J Cell Biochem* February 15, 2004;**91**(3):528–39.
71. Martini M, Testi MG, Pasetto M, Picchio MC, Innamorati G, Mazzocco M, et al. IFN-γ-mediated upmodulation of MHC class I expression activates tumor-specific immune response in a mouse model of prostate cancer. *Vaccine* April 30, 2010;**28**(20):3548–57.
72. Morgan RA, Dudley ME, Rosenberg SA. Adoptive cell therapy: genetic modification to redirect effector cell specificity. *Cancer J* 2010 July–August;**16**(4):336–41.
73. Beatty MS, Curiel DT. Augmented adenovirus transduction of murine T lymphocytes utilizing a bi-specific protein targeting murine interleukin 2 receptor. *Cancer Gene Ther* August 2013;**20**(8):445–52.
74. Malek TR. The biology of interleukin-2. *Annu Rev Immunol* 2008;**26**:453–79.
75. Dreier B, Mikheeva G, Belousova N, Parizek P, Boczek E, Jelesarov I, et al. Her2-specific multivalent adapters confer designed tropism to adenovirus for gene targeting. *J Mol Biol* January 14, 2011;**405**(2):410–26.
76. Dreier B, Honegger A, Hess C, Nagy-Davidescu G, Mittl PR, Grutter MG, et al. Development of a generic adenovirus delivery system based on structure-guided design of bispecific trimeric DARPin adapters. *Proc Natl Acad Sci USA* March 5, 2013;**110**(10):E869–77.
77. Parrott MB, Adams KE, Mercier GT, Mok H, Campos SK, Barry MA. Metabolically biotinylated adenovirus for cell targeting, ligand screening, and vector purification. *Mol Ther* October 2003;**8**(4):688–700.
78. Campos SK, Barry MA. Comparison of adenovirus fiber, protein IX, and hexon capsomeres as scaffolds for vector purification and cell targeting. *Virology* June 5, 2006;**349**(2):453–62.
79. Chen Z, Mok H, Pflugfelder SC, Li DQ, Barry MA. Improved transduction of human corneal epithelial progenitor cells with cell-targeting adenoviral vectors. *Exp Eye Res* October 2006;**83**(4):798–806.
80. Maguire CA, Sapinoro R, Girgis N, Rodriguez-Colon SM, Ramirez SH, Williams J, et al. Recombinant adenovirus type 5 vectors that target DC-SIGN, ChemR23 and $\alpha(v)\beta_3$ integrin efficiently transduce human dendritic cells and enhance presentation of vectored antigens. *Vaccine* January 30, 2006;**24**(5):671–82.
81. Volpers C, Thirion C, Biermann V, Hussmann S, Kewes H, Dunant P, et al. Antibody-mediated targeting of an adenovirus vector modified to contain a synthetic immunoglobulin g-binding domain in the capsid. *J Virol* February 2003;**77**(3):2093–104.
82. Korokhov N, Mikheeva G, Krendelshchikov A, Belousova N, Simonenko V, Krendelshchikova V, et al. Targeting of adenovirus via genetic modification of the viral capsid combined with a protein bridge. *J Virol* December 2003;**77**(24):12931–40.
83. Noureddini SC, Krendelshchikov A, Simonenko V, Hedley SJ, Douglas JT, Curiel DT, et al. Generation and selection of targeted adenoviruses embodying optimized vector properties. *Virus Res* [Research Support, N.I.H., Extramural, Research Support, U.S. Gov't, Non-P.H.S.] March 2006;116(1–2):185–95.
84. Kawashima R, Abei M, Fukuda K, Nakamura K, Murata T, Wakayama M, et al. EpCAM- and EGFR-targeted selective gene therapy for biliary cancers using Z33-fiber-modified adenovirus. *Int J Cancer* September 1, 2011;**129**(5):1244–53.

85. Sirena D, Lilienfeld B, Eisenhut M, Kalin S, Boucke K, Beerli RR, et al. The human membrane cofactor CD46 is a receptor for species B adenovirus serotype 3. *J Virol* May 2004;**78**(9):4454–62.
86. Short JJ, Vasu C, Holterman MJ, Curiel DT, Pereboev A. Members of adenovirus species B utilize CD80 and CD86 as cellular attachment receptors. *Virus Res* December 2006;**122**(1–2):144–53.
87. Tuve S, Wang H, Ware C, Liu Y, Gaggar A, Bernt K, et al. A new group B adenovirus receptor is expressed at high levels on human stem and tumor cells. *J Virol* December 2006;**80**(24):12109–20.
88. Wang H, Li ZY, Liu Y, Persson J, Beyer I, Moller T, et al. Desmoglein 2 is a receptor for adenovirus serotypes 3, 7, 11 and 14. *Nat Med* January 2011;**17**(1):96–104.
89. Krasnykh VN, Mikheeva GV, Douglas JT, Curiel DT. Generation of recombinant adenovirus vectors with modified fibers for altering viral tropism. *J Virol* October 1996;**70**(10):6839–46.
90. Haviv YS, Blackwell JL, Kanerva A, Nagi P, Krasnykh V, Dmitriev I, et al. Adenoviral gene therapy for renal cancer requires retargeting to alternative cellular receptors. *Cancer Res* August 1, 2002;**62**(15):4273–81.
91. Kanerva A, Mikheeva GV, Krasnykh V, Coolidge CJ, Lam JT, Mahasreshti PJ, et al. Targeting adenovirus to the serotype 3 receptor increases gene transfer efficiency to ovarian cancer cells. *Clin Cancer Res* January 2002;**8**(1):275–80.
92. Rivera AA, Davydova J, Schierer S, Wang M, Krasnykh V, Yamamoto M, et al. Combining high selectivity of replication with fiber chimerism for effective adenoviral oncolysis of CAR-negative melanoma cells. *Gene Ther* December 2004;**11**(23):1694–702.
93. Murakami M, Ugai H, Belousova N, Pereboev A, Dent P, Fisher PB, et al. Chimeric adenoviral vectors incorporating a fiber of human adenovirus 3 efficiently mediate gene transfer into prostate cancer cells. *Prostate* March 1, 2010;**70**(4):362–76.
94. Shinozaki K, Suominen E, Carrick F, Sauter B, Kahari VM, Lieber A, et al. Efficient infection of tumor endothelial cells by a capsid-modified adenovirus. *Gene Ther* January 2006;**13**(1):52–9.
95. Preuss MA, Glasgow JN, Everts M, Stoff-Khalili MA, Wu H, Curiel DT. Enhanced gene delivery to human primary endothelial cells using tropism-modified adenovirus vectors. *Open Gene Ther J* January 1, 2008;**1**:7–11.
96. Granio O, Ashbourne Excoffon KJ, Henning P, Melin P, Norez C, Gonzalez G, et al. Adenovirus 5-fiber 35 chimeric vector mediates efficient apical correction of the cystic fibrosis transmembrane conductance regulator defect in cystic fibrosis primary airway epithelia. *Hum Gene Ther* March 2010;**21**(3):251–69.
97. Rodriguez E, Romero C, Rio A, Miralles M, Raventos A, Planells L, et al. Short-fiber protein of ad40 confers enteric tropism and protection against acidic gastrointestinal conditions. *Hum Gene Ther Methods* August 2013;**24**(4):195–204.
98. Diaconu I, Denby L, Pesonen S, Cerullo V, Bauerschmitz GJ, Guse K, et al. Serotype chimeric and fiber-mutated adenovirus Ad5/19p-HIT for targeting renal cancer and untargeting the liver. *Hum Gene Ther* June 2009;**20**(6):611–20.
99. Renaut L, Colin M, Leite JP, Benko M, D'Halluin JC. Abolition of hCAR-dependent cell tropism using fiber knobs of Atadenovirus serotypes. *Virology* April 10, 2004;**321**(2):189–204.
100. Glasgow JN, Kremer EJ, Hemminki A, Siegal GP, Douglas JT, Curiel DT. An adenovirus vector with a chimeric fiber derived from canine adenovirus type 2 displays novel tropism. *Virology* June 20, 2004;**324**(1):103–16.

101. Stoff-Khalili MA, Rivera AA, Glasgow JN, Le LP, Stoff A, Everts M, et al. A human adenoviral vector with a chimeric fiber from canine adenovirus type 1 results in novel expanded tropism for cancer gene therapy. *Gene Ther* December 2005;**12**(23):1696–706.
102. Bangari DS, Mittal SK. Porcine adenovirus serotype 3 internalization is independent of CAR and $\alpha_v\beta_3$ or $\alpha_v\beta_5$ integrin. *Virology* February 5, 2005;**332**(1):157–66.
103. Kim JW, Glasgow JN, Nakayama M, Ak F, Ugai H, Curiel DT. An adenovirus vector incorporating carbohydrate binding domains utilizes glycans for gene transfer. *PLoS One* 2013;**8**(2):e55533.
104. Tsuruta Y, Pereboeva L, Glasgow JN, Luongo CL, Komarova S, Kawakami Y, et al. Reovirus σ1 fiber incorporated into adenovirus serotype 5 enhances infectivity via a CAR-independent pathway. *Biochem Biophys Res Commun* September 16, 2005;**335**(1):205–14.
105. Pereboeva L, Komarova S, Mahasreshti PJ, Curiel DT. Fiber-mosaic adenovirus as a novel approach to design genetically modified adenoviral vectors. *Virus Res* September 15, 2004;**105**(1):35–46.
106. Murakami M, Ugai H, Wang M, Belousova N, Dent P, Fisher PB, et al. An adenoviral vector expressing human adenovirus 5 and 3 fiber proteins for targeting heterogeneous cell populations. *Virology* November 25, 2010;**407**(2):196–205.
107. Kaliberov SA, Kaliberova LN, Buchsbaum DJ, Curiel DT. Experimental virotherapy of chemoresistant pancreatic carcinoma using infectivity-enhanced fiber-mosaic oncolytic adenovirus. *Cancer Gene Ther* July 2014;**21**(7):264–74.
108. Wickham TJ, Tzeng E, Shears 2nd LL, Roelvink PW, Li Y, Lee GM, et al. Increased in vitro and in vivo gene transfer by adenovirus vectors containing chimeric fiber proteins. *J Virol* November 1997;**71**(11):8221–9.
109. Dmitriev I, Krasnykh V, Miller CR, Wang M, Kashentseva E, Mikheeva G, et al. An adenovirus vector with genetically modified fibers demonstrates expanded tropism via utilization of a coxsackievirus and adenovirus receptor-independent cell entry mechanism. *J Virol* December 1998;**72**(12):9706–13.
110. Wu H, Seki T, Dmitriev I, Uil T, Kashentseva E, Han T, et al. Double modification of adenovirus fiber with RGD and polylysine motifs improves coxsackievirus-adenovirus receptor-independent gene transfer efficiency. *Hum Gene Ther* September 1, 2002;**13**(13):1647–53.
111. Arap W, Pasqualini R, Ruoslahti E. Cancer treatment by targeted drug delivery to tumor vasculature in a mouse model. *Science* January 16, 1998;**279**(5349):377–80.
112. Laakkonen P, Porkka K, Hoffman JA, Ruoslahti E. A tumor-homing peptide with a targeting specificity related to lymphatic vessels. *Nat Med* July 2002;**8**(7):751–5.
113. Yamamoto Y, Hiraoka N, Goto N, Rin Y, Miura K, Narumi K, et al. A targeting ligand enhances infectivity and cytotoxicity of an oncolytic adenovirus in human pancreatic cancer tissues. *J Control Release* October 28, 2014;**192**:284–93.
114. Hariri G, Yan H, Wang H, Han Z, Hallahan DE. Radiation-guided drug delivery to mouse models of lung cancer. *Clin Cancer Res* October 15, 2010;**16**(20):4968–77.
115. Magnusson MK, Hong SS, Henning P, Boulanger P, Lindholm L. Genetic retargeting of adenovirus vectors: functionality of targeting ligands and their influence on virus viability. *J Gene Med* 2002 Jul-Aug;**4**(4):356–70.
116. Hedley SJ, Auf der Maur A, Hohn S, Escher D, Barberis A, Glasgow JN, et al. An adenovirus vector with a chimeric fiber incorporating stabilized single chain antibody achieves targeted gene delivery. *Gene Ther* January 2006;**13**(1):88–94.
117. van Beusechem VW, van Rijswijk AL, van Es HH, Haisma HJ, Pinedo HM, Gerritsen WR. Recombinant adenovirus vectors with knobless fibers for targeted gene transfer. *Gene Ther* November 2000;**7**(22):1940–6.

118. Krasnykh V, Belousova N, Korokhov N, Mikheeva G, Curiel DT. Genetic targeting of an adenovirus vector via replacement of the fiber protein with the phage T4 fibritin. *J Virol* May 2001;**75**(9):4176–83.
119. Schagen FH, Wensveen FM, Carette JE, Dermody TS, Gerritsen WR, van Beusechem VW. Genetic targeting of adenovirus vectors using a reovirus σ1-based attachment protein. *Mol Ther* May 2006;**13**(5):997–1005.
120. Poulin KL, Lanthier RM, Smith AC, Christou C, Risco Quiroz M, Powell KL, et al. Retargeting of adenovirus vectors through genetic fusion of a single-chain or single-domain antibody to capsid protein IX. *J virology* [Research Support, Non-U.S. Gov't] October 2010;**84**(19):10074–86.
121. Vellinga J, de Vrij J, Myhre S, Uil T, Martineau P, Lindholm L, et al. Efficient incorporation of a functional hyper-stable single-chain antibody fragment protein-IX fusion in the adenovirus capsid. *Gene Ther* April 2007;**14**(8):664–70.
122. Matsui H, Sakurai F, Katayama K, Abe Y, Machitani M, Kurachi S, et al. A targeted adenovirus vector displaying a human fibronectin type III domain-based monobody in a fiber protein. *Biomaterials* May 2013;**34**(16):4191–201.
123. Kaliberov SA, Kaliberova LN, Buggio M, Tremblay JM, Shoemaker CB, Curiel DT. Adenoviral targeting using genetically incorporated camelid single variable domains. *Lab Invest* August 2014;**94**(8):893–905.
124. Revets H, De Baetselier P, Muyldermans S. Nanobodies as novel agents for cancer therapy. *Expert Opin Biol Ther* January 2005;**5**(1):111–24.
125. Hamers-Casterman C, Atarhouch T, Muyldermans S, Robinson G, Hamers C, Songa EB, et al. Naturally occurring antibodies devoid of light chains. *Nature* [Research Support, Non-U.S. Gov't] June 3, 1993;**363**(6428):446–8.
126. Cortez-Retamozo V, Backmann N, Senter PD, Wernery U, De Baetselier P, Muyldermans S, et al. Efficient cancer therapy with a nanobody-based conjugate. *Cancer Res* April 15, 2004;**64**(8):2853–7.
127. Cortez-Retamozo V, Lauwereys M, Hassanzadeh Gh G, Gobert M, Conrath K, Muyldermans S, et al. Efficient tumor targeting by single-domain antibody fragments of camels. *Int J Cancer* March 20, 2002;**98**(3):456–62.
128. Einfeld DA, Brough DE, Roelvink PW, Kovesdi I, Wickham TJ. Construction of a pseudoreceptor that mediates transduction by adenoviruses expressing a ligand in fiber or penton base. *J Virol* November 1999;**73**(11):9130–6.
129. Shayakhmetov DM, Gaggar A, Ni S, Li ZY, Lieber A. Adenovirus binding to blood factors results in liver cell infection and hepatotoxicity. *J Virol* June 2005;**79**(12):7478–91.
130. Alba R, Bradshaw AC, Parker AL, Bhella D, Waddington SN, Nicklin SA, et al. Identification of coagulation factor (F)X binding sites on the adenovirus serotype 5 hexon: effect of mutagenesis on FX interactions and gene transfer. *Blood* July 30, 2009;**114**(5):965–71.
131. Parker AL, Waddington SN, Nicol CG, Shayakhmetov DM, Buckley SM, Denby L, et al. Multiple vitamin K-dependent coagulation zymogens promote adenovirus-mediated gene delivery to hepatocytes. *Blood* October 15, 2006;**108**(8):2554–61.
132. Shayakhmetov DM, Gaggar A, Ni S, Li ZY, Lieber A. Adenovirus binding to blood factors results in liver cell infection and hepatotoxicity. *J Virol* [Research Support, N.I.H., Extramural, Research Support, U.S. Gov't, P.H.S.] June 2005;**79**(12):7478–91.
133. Zinn KR, Szalai AJ, Stargel A, Krasnykh V, Chaudhuri TR. Bioluminescence imaging reveals a significant role for complement in liver transduction following intravenous delivery of adenovirus. *Gene Ther* [Research Support, U.S. Gov't, Non-P.H.S., Research Support, U.S. Gov't, P.H.S.] October 2004;**11**(19):1482–6.
134. Waddington SN, McVey JH, Bhella D, Parker AL, Barker K, Atoda H, et al. Adenovirus serotype 5 hexon mediates liver gene transfer. *Cell* February 8, 2008;**132**(3):397–409.

135. Kalyuzhniy O, Di Paolo NC, Silvestry M, Hofherr SE, Barry MA, Stewart PL, et al. Adenovirus serotype 5 hexon is critical for virus infection of hepatocytes in vivo. *Proc Natl Acad Sci USA* April 8, 2008;**105**(14):5483–8.
136. Parker AL, McVey JH, Doctor JH, Lopez-Franco O, Waddington SN, Havenga MJ, et al. Influence of coagulation factor zymogens on the infectivity of adenoviruses pseudotyped with fibers from subgroup D. *J Virol* [Research Support, Non-U.S. Gov't] April 2007;81(7):3627–31.
137. Alba R, Bradshaw AC, Coughlan L, Denby L, McDonald RA, Waddington SN, et al. Biodistribution and retargeting of FX-binding ablated adenovirus serotype 5 vectors. *Blood* October 14, 2010;**116**(15):2656–64.
138. Short JJ, Rivera AA, Wu H, Walter MR, Yamamoto M, Mathis JM, et al. Substitution of adenovirus serotype 3 hexon onto a serotype 5 oncolytic adenovirus reduces factor X binding, decreases liver tropism, and improves antitumor efficacy. *Mol Cancer Ther* [Research Support, N.I.H., Extramural, Research Support, Non-U.S. Gov't] September 2010;**9**(9):2536–44.
139. Vigne E, Mahfouz I, Dedieu JF, Brie A, Perricaudet M, Yeh P. RGD inclusion in the hexon monomer provides adenovirus type 5-based vectors with a fiber knob-independent pathway for infection. *J Virol* June 1999;**73**(6):5156–61.
140. Wu H, Han T, Belousova N, Krasnykh V, Kashentseva E, Dmitriev I, et al. Identification of sites in adenovirus hexon for foreign peptide incorporation. *J Virol* March 2005;**79**(6):3382–90.
141. McConnell MJ, Danthinne X, Imperiale MJ. Characterization of a permissive epitope insertion site in adenovirus hexon. *J Virol* June 2006;**80**(11):5361–70.
142. Campos SK, Barry MA. Rapid construction of capsid-modified adenoviral vectors through bacteriophage lambda Red recombination. *Hum Gene Ther* November 2004;**15**(11):1125–30.
143. Kaliberov SA, Kaliberova LN, Hong Lu Z, Preuss MA, Barnes JA, Stockard CR, et al. Retargeting of gene expression using endothelium specific hexon modified adenoviral vector. *Virology* December 2013;**447**(1–2):312–25.
144. Vellinga J, Rabelink MJ, Cramer SJ, van den Wollenberg DJ, Van der Meulen H, Leppard KN, et al. Spacers increase the accessibility of peptide ligands linked to the carboxyl terminus of adenovirus minor capsid protein IX. *J Virol* April 2004;**78**(7):3470–9.
145. Dmitriev IP, Kashentseva EA, Curiel DT. Engineering of adenovirus vectors containing heterologous peptide sequences in the C terminus of capsid protein IX. *J Virol* July 2002;**76**(14):6893–9.
146. Le LP, Everts M, Dmitriev IP, Davydova JG, Yamamoto M, Curiel DT. Fluorescently labeled adenovirus with pIX-EGFP for vector detection. *Mol Imaging* April 2004;**3**(2):105–16.
147. Li J, Le L, Sibley DA, Mathis JM, Curiel DT. Genetic incorporation of HSV-1 thymidine kinase into the adenovirus protein IX for functional display on the virion. *Virology* August 1, 2005;**338**(2):247–58.
148. Mathis JM, Bhatia S, Khandelwal A, Kovesdi I, Lokitz SJ, Odaka Y, et al. Genetic incorporation of human metallothionein into the adenovirus protein IX for non-invasive SPECT imaging. *PLoS One* 2011;**6**(2):e16792.
149. Liu L, Rogers BE, Aladyshkina N, Cheng B, Lokitz SJ, Curiel DT, et al. Construction and radiolabeling of adenovirus variants that incorporate human metallothionein into protein IX for analysis of biodistribution. *Mol Imaging* 2014;**13**.
150. de Vrij J, Uil TG, van den Hengel SK, Cramer SJ, Koppers-Lalic D, Verweij MC, et al. Adenovirus targeting to HLA-A1/MAGE-A1-positive tumor cells by fusing a single-chain T-cell receptor with minor capsid protein IX. *Gene Ther* July 2008;**15**(13):978–89.

# Targeted Adenoviral Vectors III: Transcriptional Targeting

**10**

*Sudhanshu P. Raikwar*[1], *Chinghai H. Kao*[2,3], *Thomas A. Gardner*[2,3]
[1]Department of Veterinary Medicine and Surgery, College of Veterinary Medicine, University of Missouri and Harry S. Truman Veterans' Memorial Hospital, Columbia, MO, USA; [2]Department of Urology, Indiana University Medical Center, Indianapolis, IN, USA; [3]Department of Microbiology and Immunology, Indiana University Medical Center, Indianapolis, IN, USA

## 1. Introduction—Rationale of Transcriptional Targeting

Gene therapy is an innovative approach aimed to introducing genetic material into an organism for therapeutic intent. Still in its infancy, this novel concept has witnessed a fundamental preclinical success with numerous ongoing clinical trials to confirm these findings. Critical to the success of gene therapy trials are issues relating to specific delivery of physiologically active biomolecules at therapeutically significant concentrations. Initially this was achieved by using direct intralesional injections of vectors to localize the delivery to the target tissue and universal promoters to maximized expression at that site. Over the past several years we and several other investigators have investigated the potential of tumor-specific promoters to transcriptionally regulate gene expression in the laboratory and in clinical trials. The safety demonstrated by these trials using tumor/tissue-specific promoters has led to the recent approval of a trial administering a conditionally replicative adenovirus systemically for the treatment of metastatic prostate cancer.

In order for gene therapy to be widely applicable, there is an urgent need to develop new generation of viral vectors capable of achieving these goals of targeted delivery and controlled gene expression at the target site. The aim of this chapter is to discuss various potential strategies that have been utilized to achieve tissue-/tumor-specific expression using adenoviral vectors. A better understanding of tissue-specific gene expression necessitates basic review of the eukaryotic transcription process at the molecular level. Consequently, we begin by examining the molecular architecture of DNA and its relationship with the transcriptional mechanism.

## 2. Regulation of Transcription in Eukaryotes

To fully understand the complexity underlining transcriptional targeting, a brief review of the mammalian transcriptional process follows:

**Adenoviral Vectors for Gene Therapy.**

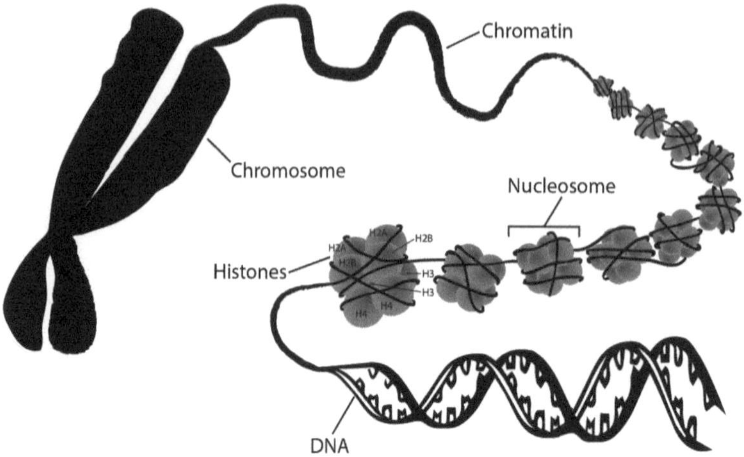


**Figure 1 DNA organization:** The double helical DNA is tightly complexed with a core of eight histones to form nucleosomes. Nucleosomes bound with H1 histone form chromatosomes which in turn form highly condensed fibers to form compact chromatids that are finally packaged into chromosomes.

## *2.1 Molecular Organization of DNA (Figure. 1)*

During interphase the genetic material in association with proteins is dispersed throughout the nucleus in the form of chromatin. At the onset of mitosis, chromatin condensation takes place, and during prophase it undergoes further compression into recognizable chromosomes. The associated proteins are basic, positively charged (lysine and arginine containing) histones and less positively charged nonhistones including high-mobility group proteins. Histones play a key role in chromatin structural organization and are subject to various posttranslational modifications like acetylation, phosphorylation, and ubiquitination. Histones constitute nearly half of all the chromatin protein by weight and can be divided into six types: H1, H2A, H2B, H3, H4, and H5. DNA is incorporated into a 100-Å nucleosomal fiber comprising of two molecules each of H2A, H2B, H3, and H4, which form the core histone octamer along with one linker histone H1 or H5. The nucleosome core particle consists of 146 base pairs (bp) of DNA while the core histone octamer interacts with about 200 bp of DNA. While the histones function by interacting with DNA to form nucleosome, the nonhistone proteins are responsible for performing diverse functions including tissue-specific transcription. The 100-Å nucleosomal fiber is arranged into a higher order structure termed as a 300-Å supercoiled filament or solenoid. Evidence indicates that certain nonhistone proteins including topoisomerase II bind to chromatin every 60–100 kilobases and tether the supercoiled, 300-Å filament into structural loops. Further interaction with other nonhistone proteins leads to gathering of loops into rosettes, which in association with additional nonhistones undergo condensation, forming a scaffold. This is known as radical loop-scaffold model of compaction. Special, irregularly spaced repetitive base sequences associate with nonhistone proteins to define chromatin loops. These stretches of DNA are known as scaffold-associated regions. In order to be competent for transcription, the 300-Å chromatin filament must undergo decondenzation.

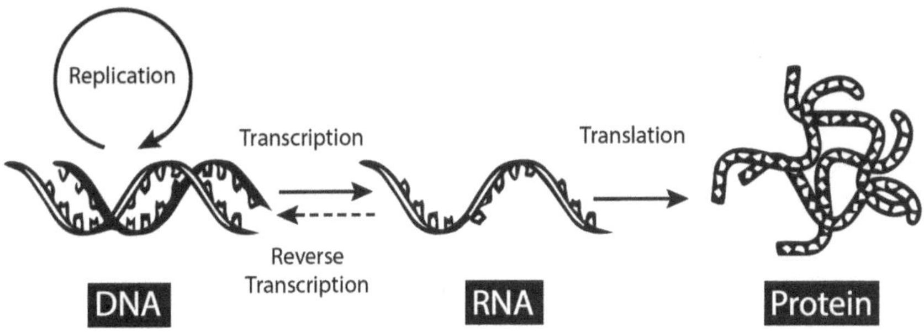


**Figure 2 Central dogma:** According to the central dogma the flow of genetic information is deterministic and unidirectional and refers to transcription of DNA to RNA and translation of RNA to protein. The genetic information cannot be transferred back from protein to either protein, RNA or DNA. Exceptions to central dogma include (a) prions where genetic information is transferred between prion proteins and (b) retroviruses which transcribe the RNA into DNA by the enzyme reverse transcriptase.

## 2.2 *The Central Dogma (Figure. 2)*

According to the central dogma, the genetic information flows from (1) DNA to DNA during genomic replication and (2) DNA to protein during gene expression. Gene expression can be simply defined as a phenomenon by which the genetic information stored in DNA is transferred to a protein. It involves two distinct processes. The process by which cells convert genetic information from DNA to RNA is called **transcription** and the decoding of RNA information to generate a specific sequence of amino acids is called **translation**. In addition, the flow of genetic information from RNA to DNA has been demonstrated in the case of retroviruses. Thus, the flow of genetic information from DNA to RNA is sometimes reversible; however, this flow is unidirectional from RNA to protein and irreversible since, normally, the genetic information within the mRNA intermediate is not altered. However, a few exceptions in the form of RNA editing seem to challenge the present concept. RNA editing has been shown to alter the information content of the gene transcripts by changing the structures of individual bases and by inserting or deleting uridine monophosphate residues.

Gene expression in eukaryotes is a spatial and temporally regulated process. Gene expression is regulated at multiple levels including transcription, posttranscriptional processing, nucleocytoplasmic transport, mRNA stability, translation, posttranslational modification, and intracellular trafficking of the protein.

## 2.3 *Transcription (Figure. 3)*

In eukaryotes, transcription occurs in the nucleus with the help of RNA polymerase to generate a single-stranded RNA molecule that is complementary in base sequence to the DNA template strand. There are three different types of RNA polymerases for the transcription of different types of genes. **RNA polymerase I** functions to transcribe rRNA genes to generate a large rRNA primary transcript which undergoes processing within the nucleolus to generate a 28S rRNA, a 5.8S rRNA, and an 18S rRNA. **RNA polymerase II** transcribes all of the protein coding genes into primary transcripts

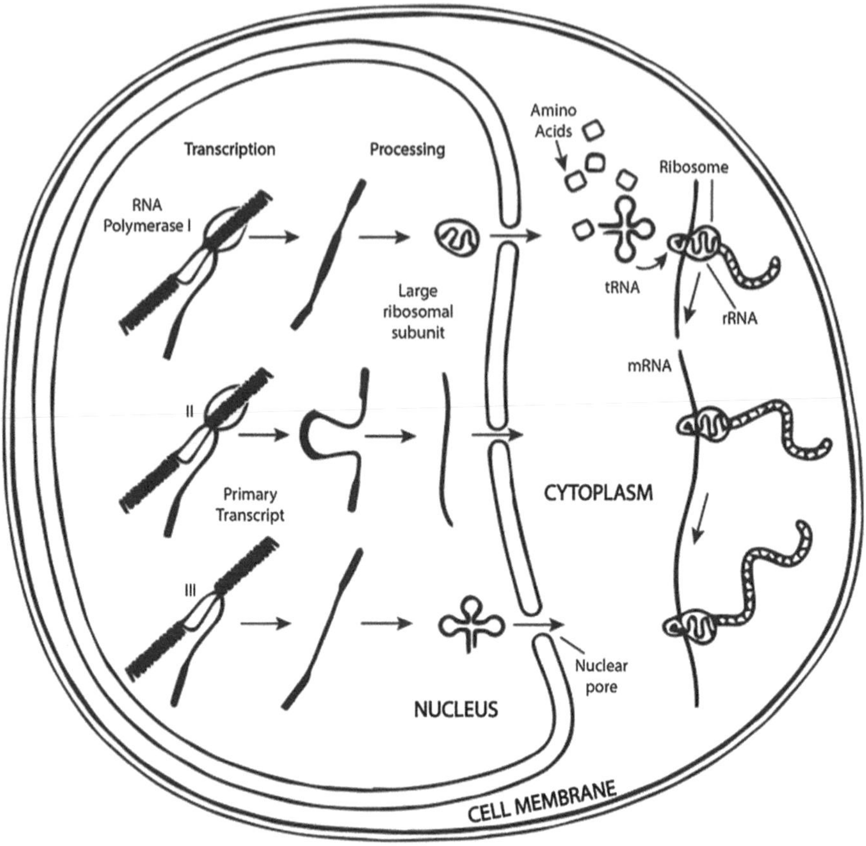


**Figure 3 Eukaryotic transcriptional machinery:** Eukaryotic transcription is governed by RNA polymerases I, II and III. RNA polymerase I transcribes polycistronic preribosomal RNA which is the precursor for the 5.8S, 18S and 28S rRNAs. RNA polymerase II transcribes all protein coding genes and synthesizes mRNA. Extensive RNA processing in the nucleus including addition of 5' cap on mRNA, intron removal by RNA splicing and polyadenylation occurs prior to RNA export to the cytoplasm. RNA polymerase II and III transcribe microRNAs as well as ncRNAs. RNA polymerase III transcribes the genes for tRNA, 5SRNA and U6 snRNA.

called pre-mRNAs that upon posttranscriptional processing generate mRNAs. While **RNA polymerase III** is known to transcribe tRNAs, 5S rRNA, and small nuclear RNAs (snRNAs).

There are two main types of *cis*-acting elements in all polII-transcribed genes: promoters and enhancers. The promoter, which is in close proximity to the protein coding region, consists of nucleotide sequences spanning approximately −40 and +50 relative to a transcription initiation site. A typical core promoter consists of four distinct elements: **(1)** A unique sequence called Goldberg-Hogness or the TATA box has a consensus sequence TATAAAA and is located about −25 to −30 nucleotides upstream of the transcription initiation site. The TATA box alone is sufficient for independently directing a low-level polII mediated transcription. **(2)** An initiator element that is functionally analogous to TATA box that directly overlaps the transcription start site and has a loose consensus sequence PyPyA+1NT/APyPy. **(3)** The downstream core promoter element is located approximately at position +30 downstream of the initiation site and acts in conjunction with the initiator element to direct transcription initiation. **(4)** The TFIIB recognition element, which has the consensus sequence $^{G}/_{C}{}^{G}/_{C}{}^{G}/_{A}$CGCC and is located from −32 to −38 upstream of the TATA box. Another *cis*-acting element called CAAT box has a consensus sequence GGCCAATCT and is located near position −70 to −80 relative to transcription initiation site. Mutagenesis studies suggest the critical role of CAAT box in modulating promoter's ability to facilitate transcription.

Besides, polII promoters often contain two conserved sequences, the SP1 or GC box (GGGCGG) at about position −110 and the octamer box (ATTTGCAT); however, their positions are variable and they may occur either singly or in multiple copies. These consensus sequences are known to influence the efficiency of the promoter in initiating transcription. In addition, the regulatory regions termed enhancers are located farther upstream, downstream, or within the gene. The activity of the enhancers is independent of the location, orientation, and the gene type. Although, they may not be involved directly in template binding, they are capable of modulating highly efficient transcription initiation.

The promoter regions are normally sequestered within the nucleosome and thus are rarely able to bind to basal transcription factors and RNA polymerase thereby leading to transcriptional repression or silencing. In order for transcription initiation to occur, the sequestered promoter must be exposed so that it can readily bind basal factors, and this is achieved by chromatin remodeling. The DNA in highly compacted chromatin is relatively resistant to nuclease DNaseI digestion. Thus, sensitivity of the DNA to DNaseI reveals the degree of chromatin condensation and is directly proportional to the transcriptional activity of a particular gene. Chromatin remodeling by acetylation and deacetylation of the histone proteins represents a major regulatory mechanism during gene activation and repression, respectively. The acetylation of histones by histone acetylase causes neutralization of the lysine basic charge, which in turn causes relaxation of contacts between the histones and the DNA. Thus, acetylated histones are preferentially found in active or potentially active genes where the chromatin is less tightly packed. Further, treatment of cultured cells with compounds like sodium butyrate, which enhances histone acetylation, leads to activation of previously silenced cellular genes.

The normal chromatin in the nucleosomal conformation can be converted into highly condensed heterochromatin, which is transcriptionally inactive, by the addition of methyl groups to a series of cytosine residues in the CpG di-nucleotides found in tissue-specific genes. Thus, methylation and demethylation may play a crucial role in tissue-specific gene regulation. Locus control regions (LCRs) are specialized regulatory sequences located several kilo base pairs upstream of the gene and capable of modulating transcription of gene clusters by influencing the chromatin structure. An assembled LCR-transcription factor complex is called an enhanceosome, and if any of the components of this complex are missing, transcriptional activation of the gene cluster cannot occur. Insulators or boundary elements are regulatory sequences located in the vicinity of junctions between condensed and decondensed chromatin, which represent transcriptionally active and inactive loci, respectively. Insulators do not enhance transcription and are responsible for position-independent effects, but can prevent transcription when placed between an enhancer and a promoter.

## 2.4 Mechanism of Transcription

Eukaryotic transcription by RNA polII involves five stages (1) formation of the pre-initiation complex, (2) initiation, (3) promoter clearance, (4) elongation, and (5) termination. RNA polII cannot interact directly with the promoter to initiate transcription but requires recruitment to the promoter by interacting with transcription factors. Transcription initiation is precisely controlled by the binding of a variety of *trans*-acting proteins termed transcription factors to the promoter and the enhancer. Transcription factors that assist the binding of RNA polymerase II to the promoter and initiate low levels of transcription are called basal factors, while other transcription factors are termed activators and repressors by binding to the enhancers. The transcription factors that bind to the TATA box are known as TATA-binding protein (TBP) and are essential to the initiation of transcription from all polII genes. A number of other basal factors that associate with TBP are called TBP-associated factors ($TAF_{II}$s) and they help in the assembly and binding of the complex to the promoter, which in turn leads to transcription initiation.

The first event in the formation of pre-initiation complex involves recognition of TATA box by a multi subunit transcription factor IID (TFIID) complex. A complex consisting of TBP and $TAF_{II}$s called TFIID specifically binds to the TATA box to induce conformational changes that favor the binding of other transcription factors like TFIIA and TFIIB both of which can interact directly with TFIID. TFIIB serves two critical roles in transcription initiation: (1) It acts as a bridge and recruits TFIIF/RNA polII to the promoter and (2) It aids in the selection of the transcriptional start site. TFIIB interacts asymmetrically with TFIID-DNA and contacts the phosphodiester backbone of DNA both upstream and downstream of TATA box. The position of the amino terminus of TFIIB in the DNA-TFIID-TFIIB complex is located near the transcription start site, which might explain the role of TFIIB in stabilizing the melting of the promoter prior to RNA synthesis.

Following the assembly of the DNA–TFIID–TFIIA–TFIIB complex, RNA polII is recruited to the promoter by TFIIF. TFIIF has two subunits: (1) The larger subunit RAP74 has an ATP-dependent DNA helicase activity which may catalyze the local unwinding of the DNA to initiate transcription and (2) The smaller subunit RAP38 by which it binds tightly to the RNA polII. This is followed by binding of TFIIE to the DNA downstream from the transcriptional start point. Two other factors, TFIIH and TFIIJ are recruited to the initiation complex but their locations in the complex are unknown. The interaction of the pre-initiation complex with the core promoter alone is not sufficient to initiate transcription. A sequence of events beginning with the phosphorylation of the carboxy-terminal domain of RNA polII by TFIIF followed by ATP hydrolysis set the stage for DNA melting, initiation of synthesis, and promoter clearance. Most of the TFII factors dissociate before RNA polII leaves the promoter. The carboxy-terminal domain coordinates the processing of RNA with transcription.

The general process of transcription initiation is similar to that catalyzed by bacterial RNA polymerase. Binding of the RNA polII generates a closed complex, which is converted at a later stage to an open complex in which the DNA strands have been separated. TFIIE and TFIIH are involved in an extension of the unwound region of the DNA to allow polymerase to begin transcription elongation. Several elongation factors including TFIIF, SII, SIII, ELL, and P-TEFb function to suppress or prevent premature pausing of RNA polII as it traverses the DNA template. Early in the elongation process when the growing RNA chains are about 30 nucleotides long, the 5′ ends of the pre-mRNAs are modified by the addition of 7-methyl guanosine caps. The 7-methyl guanosine cap contains an unusual 5′–5′ triphosphate linkage, and two methyl groups are added posttranscriptionally. The 7-methyl guanosine caps are recognized by protein factors involved in the initiation of translation and also help by protecting the growing RNA chains from degradation by nucleases.

The 3′ ends of the RNA transcripts are produced by endonucleolytic cleavage of the primary transcripts rather than by the termination of transcription. The transcription termination occurs at multiple sites located 1000 to 2000 nucleotides downstream from the site that will eventually become the 3′ end of the mature transcript. The endonucleolytic cleavage occurs 11 to 30 nucleotides downstream from the conserved consensus sequence AAUAAA, which is located near the end of the transcription unit. Following endonucleolytic cleavage, the enzyme ploy-A polymerase adds a stretch of about 200 nucleotides long poly-A tail to the 3′ ends of the transcript in a process termed polyadenylation.

## 2.5 Structural Motifs (Figure. 4)

The transcription factors are modular in nature and contain characteristic structural motifs. The DNA-binding domain, as the name implies, binds to the DNA sequences present in the promoters and enhancers while the *trans* activation domain is responsible for the activation of transcription via protein–protein interactions. The DNA-binding domains have characteristic three-dimensional motifs, which result from associations between amino acids present within the polypeptide chains. Thus far, at least five types of DNA-binding motifs have been extensively characterized. These include, (1) helix-turn-helix (HTH) motif, (2) leucine zipper motif, (3) helix-loop-helix (HLH) motif, (4) zinc finger motif, and (5) steroid-hormone-binding motif.

1. **Helix-turn-helix motif (HTH)** was first discovered as the DNA-binding domain of phage repressor proteins. It is characterized by a geometric conformation that consists of two α-helical regions separated by a turn of several amino acids, which enable it to bind to DNA. Unlike other DNA-binding motifs, HTH cannot function alone but as part of larger DNA-binding domain; it fits well into the major groove of the DNA. The HTH motif has been identified in a stretch of 180-bp sequence called homeobox, which specifies a 60-amino-acid-homeodomain sequence in a large number of eukaryotic transcription factors involved in the developmentally regulated genes.
2. **Leucine zipper motif** consists of a stretch of amino acids with a leucine residue in every seventh position. The leucine-rich regions form a α-helix with leucine residue protruding at every other turn, and when two such molecules dimerize, the leucine residues zip together. The dimmer contains two α-helical regions adjacent to the zipper, which bind to phosphate

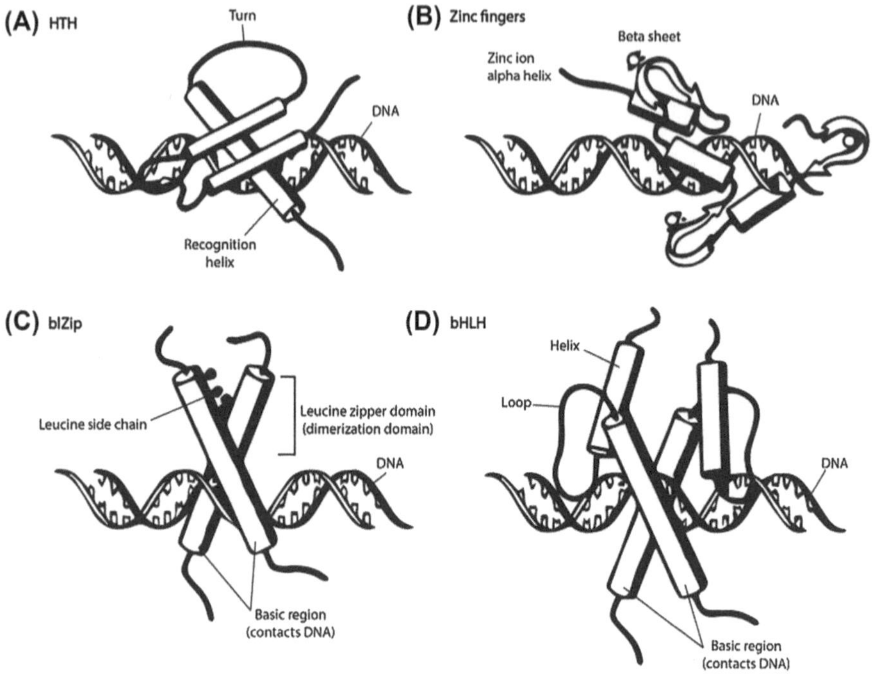


**Figure 4 Structural motifs:** The DNA-binding domains of activator proteins interact with DNA through specific amino acids which have high-affinity to specific nucleotide sequence. (A) Helix-turn-helix motif is found in phage repressors and contains one helix in the wide groove of DNA and the other helix at an angle across DNA. (B) Zinc finger motif comprises a DNA binding domain and has the consensus sequence Cys-X2-4-Cys-X3-Phe-X3-Leu-X2-His-X3-His which is stabilized by metal ions including zinc. (C) Leucine zippers are the dimerization domains of bZIP class of transcription factors which have high binding affinity for ACGT motifs. (D) Transcription factors containing Helix-loop-helix motif are dimeric with each helix containing basic amino acid residues that facilitate DNA binding.

residues and specific bases in DNA giving it a scissors like appearance. The transcription factor AP1 has two major components: Jun and Fos polypeptides encoded by c-jun and c-fos genes, respectively. Both Jun and Fos contain leucine zippers in their dimerization domains. A Jun leucine zipper can interact with another Jun leucine zipper to form a homodimer or with a Fos leucine zipper to form a heterodimer; however, Fos leucine zipper is unable to interact with another Fos leucine zipper to form a homodimer. Neither Jun nor Fos alone can bind to DNA and thus in their monomeric forms, they are unable to act as transcription factors. However, Jun–Jun homodimers or Jun–Fos heterodimers are both transcription factors and bind to DNA with same target specificity but with different affinities. The ability to form homo or heterodimers greatly increases the repertoire of potential transcription factors a cell can assemble from a limited number of gene products.

3. **Helix-loop-helix (HLH) motif** consists of a stretch of 40–50 amino acids containing two amphipathic α-helices separated by 12–28-amino-acid-long non-helical loop. The proteins bearing HLH form both homodimers and heterodimers by means of interactions between the hydrophobic residues on the corresponding faces of the two helices. The HLH proteins that contain a stretch of highly basic amino acids adjacent to the HLH motif are termed as

bHLH proteins. These bHLH proteins are of two types: Class A consists of proteins that are ubiquitously expressed (e.g., Mammalian E12/E47), while, Class B consists of proteins that are expressed in a tissue-specific manner (e.g., mammalian MyoD, myogenin, Myf-5).

4. **Zinc finger motif** was first recognized in the Xenopus RNA polIII transcription factor TFIIIA. There are several types of zinc finger proteins, however, the classic zinc finger consists of about 23 amino acids with a loop of 12–14 amino acids between the Cys and His residues and a 7–8-amino-acid linker between the loops. The consensus sequence of a typical zinc finger is Cys-$X_{2-4}$-Cys-$X_3$-Phe-$X_3$-Leu-$X_2$-His-$X_3$-His. The interspersed cysteine and histidine residues covalently bind a single zinc ion to form a tetrahedral structure thereby folding the amino acids into loops. The crystal structure analysis of DNA bound by zinc fingers suggests that the C-terminal part of each finger forms α-helices that bind DNA while the N-terminal part forms a β-sheet. Three α-helices fit into one turn of the major groove and each α-helix makes two sequence-specific contacts with DNA. A zinc finger transcription factor may contain anywhere from 2 to 13 zinc fingers. Thus zinc finger binds to DNA and also controls the specificity of dimerization. Therefore, a zinc finger motif offers a novel strategy to design artificial sequence-specific DNA-binding protein, aimed at regulating specific gene expression.

Recent studies indicate that it is possible to engineer gene switches based on zinc finger protein for precise and specific regulation of gene expression. Beerli et al.[1] have utilized zinc finger domains to design a polydactyl protein specifically recognizing 9 or 18 bp sequences in the 5′ untranslated region of the erbB-2/HER-2 promoter. They have evaluated the efficacy of gene regulation by converting the polydactyl finger into a transcriptional repressor by fusion with Kruppel-associated box (KRAB), ERF repressor domain (ERD) or mSIN3 interaction domain (SID) repressor domains. Transcriptional activators were generated by fusion with HSV VP16 activation domain or with a tetrameric repeat of VP16's minimal activation domain, termed VP64. Their results indicate that both gene repression and activation can be achieved by targeting designed proteins to a single site within the transcribed region of a gene. Kang and Kim[2] have examined the ability of designer zinc finger transcription factors to regulate transcription in vitro using an ecdysone inducible system. They constructed a 268/NRE chimeric peptide by linking the three-finger peptide from Zif268, which recognizes the site 5′-GCGTGGGCG-3′, and the three-finger NRE peptide (a variant of the Zif268 peptide) that binds specifically to part of a nuclear hormone response element, 5′-AAGGGTTCA-3′. By incorporating a 19-bp binding site for the 268/NRE near the transcriptional start site in the luciferase reporter vectors, >99% repression of activated transcription was observed in vivo. Earlier studies have shown that 268/NRE peptide binds to the 19-bp recognition sequence about 6000-fold more tightly than the Zif268 peptide (Kim and Pabo[3]). Imanishi et al.[4] have utilized zinc fingers to create six-zinc finger proteins Sp1ZF6(Gly)n by connecting two DNA-binding domains of transcription factor Sp1 with flexible polyglycine peptide linkers. These peptides were capable of inducing specific DNA-bending by binding to two GC boxes and may provide an optimized approach to control gene expression by changing the DNA-bending direction.

Corbi et al.[5] have engineered a novel gene "Jazz" that encodes for a three-zinc finger peptide capable of binding the 9-bp DNA sequence, 5′-GCTGCTGCG-3′, present in the promoter region of human and murine utrophin gene. Chimeric transcription factors Gal4-Jazz and Sp1-Jazz were able to drive the expression of luciferase from the human utrophin promoter. Moore[6] have addressed the issue of zinc finger DNA-binding specificity by altering the way in which zinc finger arrays are constructed. Their results suggest that by linking three two-finger domains rather than two three-finger units, far greater target specificity and binding with picomolar affinity can be achieved through increased discrimination against mutated or closely related sequences. Taken together, the overall results suggest the potential utility of zinc finger based designer transcription factors in achieving regulation of gene specific expression in diverse applications including gene therapy, functional genomics and transgenic organisms.

### 2.6 Regulation of Adenoviral DNA Transcription Process

The adenovirus is a double-stranded DNA virus that has evolved to infect a host cell, transport its DNA into the nucleus of the host, replicate its DNA, use the host transcriptional apparatus to produce necessary structural proteins for replication, assemble itself, and destroy the host to release the newly formed infectious particles to perpetuate the process further. This process has been described in detail in Chapter 3.

## 3. Approaches of Transcriptional Regulation

### 3.1 Prior Rationale Universal Promoters

Several universal promoters have been utilized to attempt to maximize gene expression. The LTR, CMV, and RSV promoter were isolated from Maloney retrovirus, cytomegalovirus, and Rous sarcoma virus, respectively. These promoter elements were used because of the universal transcriptional activation over a broad host range. This universal transcription allowed for excellent but non-discriminatory gene transcription and subsequent transgene expression. Because of the high levels of gene expression within several DNA constructs (i.e., viruses, cosmids, plasmids, etc.) these promoters are still used daily throughout the scientific community to test hypothesis, which require uniform and high-level gene transcription. These were the promoters utilized in the first wave of gene therapy clinical trials, which focused on maximal gene expression in local injection techniques used to control the region of gene expression achieved. The LTR promoter was used to control hsv-TK expression in a retroviral vector by placing retroviral producer cells into residual brain tumors to confer TK expression to the brain tumor, which could lead to conversion of a prodrug and subsequent tumor cell death. The CMV promoter was used in a replication deficient adenovirus to deliver p53 gene expression after intralesional delivery to both lung and head and neck tumors of patients and is still under clinical investigation. The RSV promoter was employed to express hsv-TK after intralesional delivery in patients with several different tumor types.

### 3.2 Current Rationale of Tissue-Specific Promoters (Figure. 5)

A major challenge-facing gene therapy is to generate vectors capable of achieving tissue-or tumor-specific expression. Initial gene therapy strategies utilized universal promoters that demonstrated gene transfer, but were associated with toxicity associated with non-specific gene transduction (see Section 3.2). Tissue-specific promoters offer

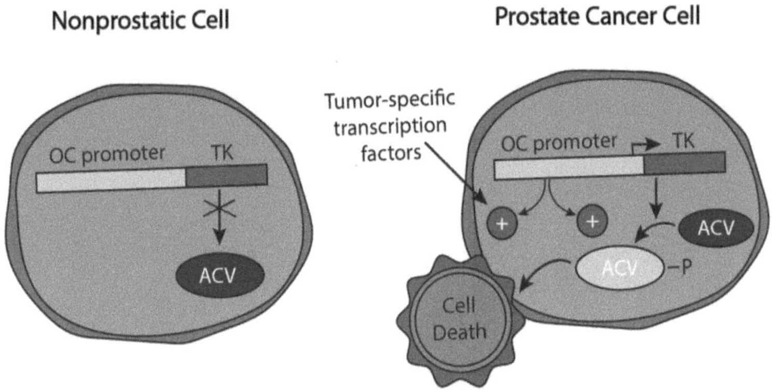


**Figure 5 Tissue specific promoter:** Replication defective adenoviral vector Ad-OC-TK engineered to express osteocalcin promoter-driven TK gene achieves precise and highly specific targeted killing of only prostate cancer cells in the presence acyclovir (ACV) while sparing nonprostatic cells. Ad-OC-TK trial has shown promising results in metastatic prostate cancer patients. OC, osteocalcin.

a novel approach to develop transcriptionally targeted viral vectors with enhanced potential for human gene therapy applications as described below. Several important characteristics are required to develop a tissue-/tumor-specific strategy for a particular disease. Fortunately, the recent explosion of our understanding of molecular events that are present in a variety of disease processes has simplified the identification of suitable promoters. Additionally the completion of the genome project and the utilization of microarray technology have enhanced the development of tissue- or tumor-specific promoters by allowing for the identification of novel but specific molecules associated with a particular disease (e.g., cancer). The advancements in molecular cloning techniques (e.g., PCR) have allowed the investigator to extract regulatory sequences from genomic DNA and evaluate each component through site-directed mutagenesis analysis in plasmid expression vectors. Additionally, the development of luciferase and green fluorescent protein as well as other quantifiable transgenes has enabled the investigator to test the tissue- or tumor-specific nature of a particular promoter.

To illustrate the concept and utility of a tissue-/tumor-specific promoter, five such promoters have been selected from Table 1. The basic rationale for selection, in-vitro and in-vivo laboratory investigation, and the clinical testing associated with each will be briefly reviewed below.

### 3.2.1 Carcinoembryonic Antigen Promoter

#### 3.2.1.1 Rationale

Carcinoembryonic antigen (CEA) is a 180-kDA cell surface glycoprotein overexpressed in 90% of gastrointestinal malignancies, including colon, gastric, rectal, and pancreatic tumors, 70% of lung cancers, and about 50% of breast cancers.[7] Thompson[8] initially reported on the molecular cloning of the CEA gene from a human genomic library. Subsequently, Schrewe[9] also isolated and characterized a cosmid clone containing

**Table 1 Gene Therapy Applications of Tissue-specific Promoters for Transcriptional Targeting**

| Promoter | Tissue-Specificity | Transgene | Vector | References |
|---|---|---|---|---|
| AFP | HCC | HSV-TK | Adenoviral | Bilbao et al., Ishikawa et al.[121] |
| | | CD, IL-2 | Adenoviral | Kanai et al.,[27] Bui et al.[29] |
| Albumin | Liver | Factor VIII | Adenoviral | Balague et al., Brann et al. |
| α-actin | Muscle | GHRH | Nonviral | Draglia-Akli et al. |
| lactalbumin | Breast cancer | CD | Adenoviral | Anderson et al. |
| lactoglobulin | Breast cancer | HSV-TK | Adenoviral | Anderson et al. |
| globin | Erythroid cells | β-globin | Retroviral | Li et al.[123] |
| c-erbB2 | Breast and pancreatic cancer | HSV-TK | Adenoviral | Vassaux et al. |
| CEA | Breast, pancreatic, lung and colorectal carcinoma | HSV-TK, cre | Adenoviral | Tanaka et al.,[19] Kijima et al. |
| | | H-ras mutant | Adenoviral | Takeuchi et al.[15] |
| Egr-1 | Radiation induced | TNF-α, LacZ | Adenoviral | Staba et al., Manome et al. |
| E-selectin | Tumor endothelium | TNF-α | Retroviral | Jaggar et al. |
| GFAP | Glial cells | FasL | Adenoviral | Morelli et al. |
| | | TH | Retroviral | Cortez et al. |
| Grp78 (BIP) | Anoxic/acidic tumor tissue | HSV-TK | Adenoviral | Gazi et al. |
| | | HSV-TK | Retroviral | Chen et al. |
| hAAT | Hepatocytes | FactorIX | Nonviral | Miao et al. |
| HGH | Pituitary | HSV-TK, | Adenoviral | Lee et al.[48] |
| HGPH-α | Pituitary | HSV-TK, | Adenoviral | Lee et al.[48] |
| HIF-1α/HRE | Hypoxia inducible | Erythropoietin | Nonviral | Rinsch et al. |
| HSP | Heat induced | p53, TNF-α | Nonviral | Luna et al. |
| L-Plastin | Epithelial tumors | LacZ | Adenoviral | Chung et al. |
| MBP | Oligodendrocytes | Caspase 8 | Adenoviral | Shinoura et al. |
| | | GFP | AAV | Chen et al. |
| MCK | Undifferentiated muscle | LacZ | Adenoviral | Hauser et al. |
| MMTV-LTR | Breast cancer | Antisense c-myc | Retroviral | Steiner et al. |

| | | | | |
|---|---|---|---|---|
| MUC1 (DF3) | Breast cancer | E1A | Adenoviral | Kurihara T. et al. |
| | | HSV-TK | Retroviral | Ring et al. |
| Nestin | Glioma, Glioblastoma | Cre, LacZ | Adenoviral | Kurihara H. et al. |
| NSE | Neurons | FasL, | Adenoviral | Morelli et al. |
| | | BDNF | AAV | Klein et al. |
| Osteocalcin | Osteosarcoma, prostate | HSV-TK | Adenoviral | Koeneman et al. |
| PEPCK | Hepatocytes | Insulin | Adenoviral | Lu et al. |
| PSA | Prostate | Nitroreductase | Adenoviral | Latham et al.[44] |
| | | HSV-TK, PNP | Adenoviral | Gotoh et al.,[47] Martiniello-Wilks et al.[46] |
| Preproenkephalin | CNS | LacZ | HSV | Kaplitt et al. |
| Probasin | Prostate | E1A | Adenoviral | Yu et al.[128] |
| Prolactin | Pituitary lactotrophic cells | HSV-TK | Adenoviral | Southgate et al. |
| SLPI | Ovarian, cervical carcinoma | HSV-TK | Nonviral | Robertson et al. |
| SM22α | Smooth muscle cells | LacZ | Adenoviral | Kim et al. |
| Surfactant protein C | Respiratory epithelium | HSV-TK | Adenoviral | Harrod et al. |
| Tyrosinase | Melanocytes | Luc, PNP | Nonviral | Park et al. |
| Tyrosine Hydroxylase | Sympathetic nervous system | LacZ | HSV | Wang et al. |

the entire coding region of the CEA gene including its promoter. The CEA promoter region encompasses 400 bp upstream of the translational start site and is known to confer tissue-specific CEA expression. Hauck and Stanners[10] have demonstrated that the CEA promoter region located between −403 and −124 bp upstream of the translational initiation site is capable of directing high levels of gene expression in CEA-expressing human colon cancer CRC cells. Chen et al.[11] have identified the CEA promoter region to lie between −123 and −28 bp upstream from transcriptional start site and have shown the presence of SP1 and upstream stimulatory factor binding sites. According to Richards[12] the CEA promoter is located between −90 and +69 bp upstream from the transcriptional start site and the essential sequences of the CEA promoter reside between −90 and −17 bp upstream of the transcriptional start site of the CEA gene. Cao et al.[13] have compared the CEA core promoter regions between −135 and +69 bp isolated from human colorectal carcinoma and normal adjacent mucosa and found that both the sequences were identical and without any mutations. Taking advantage of this fact various studies have suggested potential utility of the CEA promoter for restricted expression of heterologous genes (Osaki,[14] Hauck and Stanners,[10] Richards[12]).

#### 3.2.1.2 In Vitro and In Vivo Experiments with CEA Promoter

Takeuichi[15] demonstrated that an adenoviral vector-encoding CEA promoter-driven N116Y-dominant negative H-Ras mutant was capable of suppressing liver metastasis by human pancreatic cancer cell line PCI-43 in a nude mice model. Lan et al.[16,17] have demonstrated successful adenoviral mediated transduction of *Eschershia coli* cytosine deaminase (CD) in vitro as well as in an immunodeficient in vivo model of MKN45 gastric carcinoma. As compared to an adenoviral vector, in which CD expression is driven by CAG promoter, the expression of CD under the control of CEA promoter was confined to tumor xenografts. However, the reduction in tumor burden by AdCEA-CD/5FC, although significant, was not as superior as that induced by AdCA-CD/5FC. In fact the CEA promoter was shown to be 200 times less active as compared to the CAG promoter.

Similar results have been described in mice-bearing xenografts that were transfected with CEA-CD constructs and subsequently treated with 5-FC (DiMaio[18] and Richards[12]). Tanaka and collogues[19] have used CEA promoter sequence located between −424 and −2 bp upstream of translational start site to generate an adenoviral vector expressing HSV-TK and examined its efficacy to kill CEA-producing cancer cells in vitro and in vivo. By employing intratumoral Ad-CEA-TK injection and GCV administration, the growth of the tumors was inhibited by 20% as compared to untreated tumors. Brand et al.[20] have used CEA promoter (−296 to +102 with respect to transcriptional start site) to drive the expression of HSV-TK in an adenoviral vector. Their results indicate that CEA promoter was active in several human and rat tumor-derived cell lines but not in rat primary hepatocytes and in mouse liver, while the CMV promoter was highly active in all cell types. Although the CEA promoter driven TK expression was less, it was sufficient to kill 100% of cancer cells indicating a significant bystander effect. Treatment of subcutaneous tumors in SCID mice with Ad-CEA-TK was able to significantly reduce tumor growth, and the tail vein injection of a high dose of this virus caused no side effects in the liver.

Kijima et al.[20] have utilized a novel Cre-Lox system-based strategy to achieve enhanced antitumor effect against CEA-producing human lung and colon cancer cell lines. Their strategy involved generation of two recombinant adenoviral vectors: one expressing Cre recombinase gene under the control of the CEA promoter, while the second adenoviral vector is designed to express HSV-TK gene from the CAG promoter only after Cre excises the neomycin-resistance gene (inserted between the CAG promoter and HSV-TK) in a loxP site-specific manner (Cre recombinase derived from bacteriophage P1 mediates site-specific excisional deletion of a DNA sequence that is flanked by a pair of loxP sites composed of 34 nucleotides.). This novel approach requires simultaneous co-infection of a cell by the two adenoviral vectors. Using this approach, CEA-producing human cancer cell line was rendered 8.4-fold more sensitive to GCV as compared to infection by Ad-CEA-TK alone. Intratumoral injection of Ad-CEA-Cre along with Ad-lox-TK followed by GCV treatment almost completely eradicated CEA-producing tumors in an athymic subcutaneous tumor model, whereas intratumoral injection of Ad-CEA-TK with GCV treatment showed reduced tumor growth.

### 3.2.2 α-Fetoprotein Promoter

#### 3.2.2.1 Rationale

The human α-fetoprotein (AFP) gene is developmentally regulated and is expressed at high levels in the fetal liver, but its transcription declines rapidly after birth and is barely detectable in adult life (Belanger,[21] Nahon[22]). However, overexpression of AFP gene is a characteristic feature of human hepatocellular carcinoma. The AFP gene is about 20 kb long and contains 15 exons and 14 introns (Sakai[23]). The cap site is located at position 44 nucleotides upstream of the translation initiation site, and the TATA box is located 27 nucleotides upstream from the cap site and is flanked by sequences with dyad symmetry. Other sequences in the 5′ untranslated region include a CCAAC pentamer, a 14-bp enhancer-like sequence, a 9-bp sequence homologous to the glucocorticoid-responsive element, a 90-bp direct repeat and several alternating purine/pyrimidine sequences.

The AFP promoter is 200-bp region upstream of the transcriptional start site. It is regulated by hepatocyte nuclear factor 1 (HNF1), nuclear factor 1 (NF1), and CCAAT/enhancer-binding protein (C/EBP). The human AFP enhancer is located between −4.9 and −3.0 kb upstream of the transcriptional start site and consists of at least two functional domains designated A and B, which have binding sites for at least four transcription factors, including HNF1, HNF3, HNF4, and C/EBP. The domain B is located at −3.7 to −3.3 kb upstream of the transcriptional start site and is solely responsible for typical enhancer effects, but maximum enhancer activity is observed together with domain A located at −5.1 to −3.7 kb. A hepatoma-specific nuclear factor termed AFP1 is known to bind to an AT-rich sequence, TGATTAATAATTACA, in the B domain of the human AFP enhancer. The AFP enhancer plays a critical role in enhancing AFP gene expression in the fetal liver as well as in hepatocellular carcinoma. The AFP silencer which is a negative *cis*-acting element with a consensus sequence, 5′-CTTCATAACCTAATACTT-3′, has been identified (Nakabayashi[24]). Two transcriptional silencer elements have been identified: proximal silencer, which contains a single copy of the consensus sequence at −0.31 kb, and the distal silencer at −1.75 kb, which carries four copies of the consensus sequence. Of the two silencers, the distal silencer exhibits a higher suppressive activity than the proximal silencer. The silencer activity is manifested only when the silencer is located downstream of the enhancer and upstream of the promoter. An inverse correlation exists between the silencer activity and the AFP expression levels in hepatocellular carcinoma cell lines, thereby suggesting the role of the silencer in downregulating the level of AFP expression.

#### 3.2.2.2 In Vitro and In Vivo Experiments with the AFP Promoter

Because of its tissue-specific nature, the AFP promoter has been used in adenoviral vectors for transcriptional targeting of suicide genes in AFP-producing hepatocellular carcinoma cells in vitro as well as in vivo. Kaneko[25] developed adenoviral vectors containing either 4.9 kb AFP promoter (Av1AFPTK1) or RSV promoter (Av1TK1) to express HSV-TK gene. In vitro and in vivo cell-specific killing was observed in AFP-producing HuH7 hepatocellular carcinoma cells transduced with Av1AFPTK1 and treated with GCV. In contrast to HuH7 tumors, AFP-nonproducing hepatocellular carcinoma SK-Hep-1 did not show complete regression when treated with Av1AFPTK1. Av1TK1 was able to cause complete regression in SK-Hep-1 tumors. Using a similar approach, Kanai[26] developed adenoviral vectors by incorporating AFP enhancer domains A and B (−4.0 to −3.3 kb) and 0.17 kb AFP promoter to drive the expression of HSV-TK. These vectors conferred cell-specific killing in AFP-producing HuH-7 and HepG2 cell lines but not in AFP-nonproducing HLE and HLF cell lines. Kanai et al.[27] have also reported on the development of adenoviral vectors in which the expression of *E. coli* CD is driven by AFP promoter. These vectors were capable of causing regression of HCC xenografts following treatment with 5FC. Arbuthnot et al.[28] have analyzed in vitro and in vivo cell-specific expression of the nuclear β-galactosidase using adenoviral vectors containing transcriptional elements derived from either rat AFP or the human insulin-like growth factor II genes. Their results indicate hepatoma cell-specific expression using AFP promoter, however, primary hepatoma cells were poorly infected by these adenoviral vectors. Bui et al.[29] have compared adenoviral vector-mediated expression of IL-2 under the transcriptional control of murine AFP promoter and CMV promoters for the treatment of established human hepatocellular xenografts in CB-17/SCID mice. Intratumoral injection of these adenoviral vectors resulted in growth retardation and regression in a majority of animals but with wider therapeutic index and less systemic toxicity using the AFP vector. Using the AFP promoter and cre-lox based approach Sato et al. were able to achieve strictly tissue-specific expression of LacZ in AFP-producing cells in vitro as well as in vivo in nude mice-bearing AFP-producing tumor xenografts.

### *3.2.3 Prostate-Specific Antigen Promoter*

#### 3.2.3.1 Rationale

The gene for prostate-specific antigen (PSA), a member of the glandular kallikrein family, was independently characterized by Riegman[30,31] and Lundwall[32] from a human genomic library. The gene contains five exons and is located on the long arm of chromosome 19, in the region q13.3-qter (Riegman[31]). The gene is 7130 bp long and includes 633 bp of 5′ and 639 bp of 3′ flanking sequence. The promoter region contains a variant TATA box (TTTATA) at position −28 to −23, a GC box at −53 to −48, and a CACCC box at −129 to −125. An imperfect palindromic sequence AGAACAGCAAGTGCT closely related to reverse complement of the consensus sequence for steroid hormone receptor binding TGTACANNNTGTC/TCT is found at position −170 to −156. In addition GGGAGGG and CAGCCTC repeats are located in the region −123 to −72. Expression of PSA is primarily detected in human prostate (Wang,[33] Wang,[34] Gallee[35]). Further, PSA expression has been shown to be androgen-responsive (Riegman[36]). This is achieved by several transcription factors that are involved in regulating prostate-specific antigen gene.

Two functionally active androgen receptor-binding sites or androgen-response elements have been identified at positions −170 (ARE-I) and −394 (ARE-II) (Riegman,[36] Cleutjens [Wang, 1997] Cleutjens[37–39] et al., 1996, 1997). Cleutjens et al. (1997) have identified a complex, androgen-regulated 440-bp enhancer (−4366 to −3874) which contains a high-affinity AR-binding site, ARE-III **5′-GGAGGAACATATTGTATCGAT-3′** at position −4200. In subsequent studies a 6-kb PSA promoter fragment has been shown to confer prostate-specific and androgen-regulated expression of β-galactosidase in transgenic mice (Cleutjens.[40]) Pang et al.[41] have identified an 822-bp PSA gene regulatory sequence (PSAR) which when combined with the PSA promoter (PCPSA-P) exhibited an enhanced luciferase activity in LNCaP cells. Upon stimulation with 10–100 nM dihydrotestosterone, a more than 1000-fold increase in expression was observed as compared to androgen negative controls. Their studies further suggest that this 822-bp sequence alone could serve as a promoter thereby indicating that the complete PSA promoter contains two functional domains: a proximal promoter and a distal promoter, which can also function as an enhancer.

Yeung et al.[42] have identified two cis-acting elements within the 5.8-kb PSA promoter that are essential for the androgen-independent activity of PSA promoter in prostate cancer cells. Their studies provide evidence that androgen-independent activation of PSA promoter in androgen-independent prostate cancer cell line C4-2 involves two distinct regions: a 440-bp AREc and a 150-bp pN/H, which are responsible for upregulation of the PSA promoter activity by employing two different pathways. AREc confers high-basal PSA promoter activity in C4-2 cells, while pN/H is a strong AR-independent positive-regulatory element of the PSA promoter in both LNCaP and C4-2 cells. Further a 17-bp RI fragment within the pN/H region was identified as the key *cis*-element, which interacts with a 45-kDa prostate cancer cell-specific transcription factor to mediate androgen and AR-independent transcriptional activation of the PSA promoter. By juxtaposing AREc and pN/H a chimeric PSA promoter has been created that exhibits twofold to threefold higher activity than wild-type PSA promoter in both LNCaP and C4-2 cells. Oettgen et al.[43] have identified a novel prostate epithelial-specific Ets transcription factor, PDEF, that is involved in PSA gene regulation and acts as a co-regulator of AR. PDEF acts as an androgen-independent transcriptional activator of the PSA promoter. It also directly interacts with the DNA-binding domain of AR and enhances androgen-mediated activation of the PSA promoter. Thus, strong tissue specificity of the PSA promoter makes it an ideal candidate for prostate cancer gene therapy. Latham et al.[44] have compared tissue-specific expression of luciferase reporter vectors by employing PSA, human glandular kallikrein (hKLK2), and CMV promoters in PSA-positive LNCaP and PSA-negative CoLo320, DG75, A2780, and Jurkat cells. Their studies revealed that minimal 628-bp PSA and hKLK2 promoters showed only low-level androgen-independent expression in both PSA positive and -negative cell lines. Tandem duplication of the PSA promoter slightly increased expression in LNCaP cells. Addition of CMV enhancer upstream of the PSA or hKLK2 promoter led to substantially enhanced and nonspecific luciferase expression in all the cell lines. By placing a 1455-bp PSA enhancer sequence upstream of either the PSA or hKLK2 promoter a 20-fold increase in tissue-specific luciferase expression was observed. Tandem duplication of the PSA enhancer increased the expression 50-fold higher than either promoter, while retaining tissue specificity. The expression from all the enhancer constructs was 100-fold above the basal levels upon induction with androgen dihydrotestosterone.

#### 3.2.3.2 In Vitro and In Vivo Experiments with the PSA Promoter

These enhancer sequences were incorporated in adenoviral vectors to express EGFP and nitroreductase. The results indicate low-level expression of EGFP by PSA enhancer promoter in LNCaP and no expression in non-PSA-producing EJ cells when compared with CMV promoter-driven EGFP. However, PSA enhancer promoter was able to express comparable levels of nitroreductase in tissue-specific manner in LNCaP cell alone. These transduced LNCaP cells upon treatment with CB1954 exhibited cytotoxicity. A replication competent adenoviral vector CN706 in which the E1A gene is under the transcriptional control of the PSA enhancer/promoter has been shown to exhibit selective toxicity toward PSA-expressing prostate cancer cells (Rodriguez.[45]) Martinello-Wilks et al.[46] have examined the efficacy of adenoviral vectors with a 630-bp PSA promoter-driven HSV-TK and *E. coli* PNP genes for their ability to kill an AI prostate cancer cell line PC-3 tumor xenografts in a nude mouse model. Both HSV-TK and *E. coli* PNP-expressing adenoviral vectors were able to achieve significant tumor regression in vivo following GCV or 6MPDR treatment. Gotoh et al.[47] have developed transcriptionally targeted recombinant adenoviral vectors by incorporating either 5837 bp long or 642 bp short PSA promoter elements to drive the expression of HSV-TK. The long PSA promoter was shown to have superior activity over the short promoter and was more active in C4-2 cells than in LNCaP cells. In vitro expression of TK conferred marked killing of C4-2 cells upon acyclovir treatment. Administration of this virus in an in vivo subcutaneous C4-2 tumor model, followed by acyclovir treatment revealed significant inhibition of tumor burden. Lee et al.[48] have demonstrated tissue-specific growth suppression of PSA positive and negative cell lines by transfecting PSA promoter–enhancer-driven p53 tumor-suppressor gene. Recently, human prostate cancer- and tissue-specific genes P503, P540S, and P510S have been identified using a combination of cDNA library subtraction and high-throughput microarray screening (Xu[49]). It would be interesting to characterize the promoter region of these genes and use them in developing transcriptionally targeted adenoviral vectors.

### *3.2.4 Osteocalcin Promoter (Figure. 6)*

#### 3.2.4.1 Rationale

Osteocalcin (OC) (bone γ-carboxyglutamic acid (Gla) containing protein (BGP)) is a 50-amino acid, 5.8-kDa, major noncollagenous protein found in adult bone and has

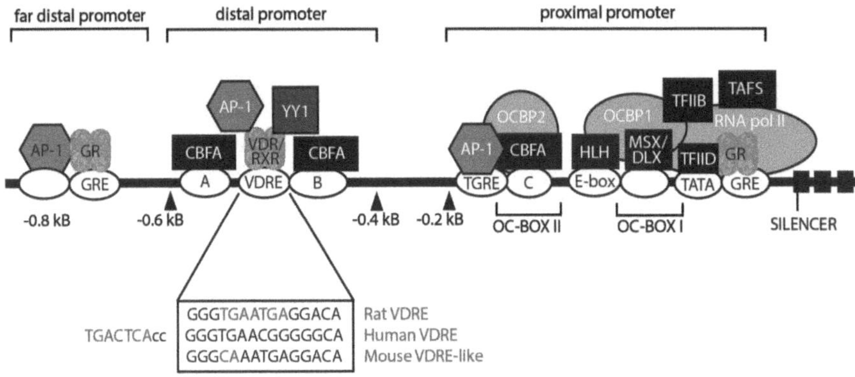


**Figure 6 Schematic representation of osteocalcin (OC) promoter:** The organization of OC promoter is very complex and includes far distal, distal, proximal promoter, OC box I, OC box II as well as vitamin D response element which is induced by vitamin D3. A novel replication competent adenoviral vector Ad-hOC-E1 in which adenoviral E1a and E1b genes are driven by a single bidirectional human OC promoter has enhanced tumor regression in androgen dependent and androgen-independent prostate cancer cells.
*Source*: Ref. 130.

been shown to be transcriptionally regulated by 1,25-dihydroxyvitamin $D_3$ (Pan and Price,[50] Price[51]). The human, rat, and murine OC genes have been cloned and each consists of four exons and three introns (Kerner,[52] Lian,[53] Desbois[54,55]). Montecino et al.[56] have reported that the key promoter elements are located in two DNascI hypersensitive sites. The proximal hypersensitive site (−170 to −70) includes sequence motifs that specifically interact with basal transcription factors such as Msx (Hoffman,[57] Towler[58,59]), HLH protein Id-1 (Tamura and Noda,[60]), AP-1 (Banerjee[61]), a bone-specific nuclear matrix-associated protein, NMP-2 (Bidwell,[62] Merriman[63]), and a member of the AML family of transcription factors (Geoffroy,[64] Banerjee[65]). The distal hypersensitive domain (−600 to −400) contains the vitamin D responsive element (VDRE, −465 to −437), which interacts with the VDR-RXRα complex in a ligand-dependent manner (Markose,[66] Demay,[67] Breen[68]). Montecino et al.[69] have demonstrated that the promoter segment −343 to −108 is critical for inducing both proximal nuclease hypersensitivity and basal transcriptional activity, and the DnaseI hypersensitivity at −600 is not essential for vitamin D-dependent transcriptional upregulation. Two additional NMP-2 sites (site A: −604 to −599, and site B: −440 to −435) have been identified in the sequences flanking the distal DNase I hypersensitive domain that might support specific interactions between the nuclear matrix and the OC gene promoter (Bidwell,[62] Merriman.[63]) Analysis of the 5′ flanking sequence of rat OC gene reveals a modular organization of the promoter consisting of TATAAAA sequence between −31 and −25 and CCAAT sequence between −92 and −88 (Yoon.[70]) Lian et al.[53] have identified a 24-nucleotide regulatory sequence, 5′ATGACCCCCAATTAGTCCTGGCAG-3′, in the proximal promoter region with a CAAT motif as a central element and have designated this sequence as an OC box since only two nucleotide substitutions are found in the rat and human OC genes in this region. Hoffman et al.[57] have reported that the OC box is located at nucleotide positions between −99 and −76 and TATA box containing consensus GRE between −44 and −31. The stimulation of OC gene expression by 1,25-dihydroxyvitamin $D_3$ is associated with sequence-specific binding of nuclear factors to a 26-bp sequence 5′-CTGGGTGAATGAGGACATTACTGACC-3′ located between −462 and −437. This sequence contains a region of hyphenated dyad symmetry and shares homology with consensus steroid-responsive elements. The promoter region has been shown to contain two sites of an E-box motif (a consensus-binding site for HLH proteins) termed as OCE1 (CACATG at −102) and OCE2 (CAGCTG at −149) (Tamura and Noda[60]). Mutagenesis studies have indicated that osteoblastic-specific gene transcription is regulated via the interaction between certain E-box binding transcription factors in osteoblasts and the OCE1 sequence in the promoter region of the OC gene. Banerjee et al.[61] have demonstrated that an AML-1 binding sequence within the proximal promoter (nt −138 to −130) contributes to 75% of the level of OC gene expression. The promoter region is not GC rich and does not contain consensus sequence for the SP1 binding site.[71] Theofan et al.[72] have performed a detailed analysis of the BGP promoter region. Three regulatory elements that share partial homology with the consensus sequence for the glucocorticoid-responsive element (GRE) have been located at nucleotide positions −356, −178 and −68, respectively. In addition, two sequences related to the consensus sequence for the metal ion-responsive element (MRE) have been identified at position −190 and −143. An octanucleotide sequence TGCAGTCA is located directly adjacent 3′ to the second MRE. Two other sequences that share homology with the cAMP-responsive element are found at −437 (TGAGGACA) and −392 (TCACGGCA). The BGP promoter region also contains several pairs of inverted repeat sequences that form regions of dyad symmetry. Three particularly long regions of imperfect dyad symmetry are located between −523 to −504, −234 to −214, and −51 to −28. An octanucleotide palindromic sequence from −134 to −127 partially overlaps both a putative MRE- and cAMP-responsive element. A short sequence GCAG or its complement CTGC is repeated 17 times. A region of alternating purines and pyrimidines at location −90 to −81 from the CAT box has the potential to form a Z-DNA structure which may be important in gene regulation. A 7-bp OC silencer element 5′-TGGCCCT-3′ has been located between +29 and +35 position in the first exon of the human OC gene while two silencer elements 5′-CCTCCT-3′ (nt +106 to +111 and +135 to +140) and 5′-TTTCTTT-3′ (nt +118 to +124) have been located in the first intron of the rat OC gene (Goto,[73] Kearns[74]).

#### 3.2.4.2 In Vitro and In Vivo Experiments with the OC Promoter

Ko et al.[75] have developed an OC promoter-driven TK-expressing recombinant adenoviral vector to achieve tissue-specific killing of osteosarcoma cells in experimental animal model. Administration of this vector followed by acyclovir treatment led to a significant growth inhibition of osteosarcoma in experimental animal model. Cheon et al.[76] have used a chemogene therapy approach by combining OC promoter-driven TK expression and acyclovir plus methotrexate treatment regimen in nude mice bearing either subcutaneous human osteosarcoma (MG-63) or rat osteosarcoma (ROS). Their results indicate that osteosarcoma tumor growth was more efficiently inhibited due to synergistic effects of combined methotrexate and acyclovir treatment. Shirakawa et al.[77] have further demonstrated the potential utility of an adenoviral OC promoter mediated suicide gene therapy for osteosarcoma pulmonary metastasis in nude mice. Hou et al.[78] have demonstrated osteoblast-specific gene expression in adherent bone marrow cells using a 1.7-kb rat OC-CAT gene. Recipient mice were shown to be positive for osteoblast-specific expression following bone marrow transplantation.

Using a replication defective adenovirus Ad-OC-TK we completed a phase I clinical trial that demonstrated the expected safety profile and gene transfer that we expected. Eleven men with recurrent or metastatic prostate cancer were enrolled in a phase I intralesional dose escalating trial combining two Ad-OC-TK injections with 3 weeks of valacyclovir administration. In summary, this was well tolerated at all doses reaching a maximum of 5×10^10 pfu (or 1×10^12 vp) in patients in the high-dose group. Viral distribution studies revealed that after intralesional administration the patients demonstrated a measurable viremia for 2–3 days. Despite the presence of viral particles at these time points no patient demonstrated hepatotoxity with valacyclovir administration. This is in direct contrast to intralesional delivery of Ad-RSV-TK to the prostatic recurrence in which patients experience hepatotoxicity upon prodrug administration. Finally, comparison of biopsy specimens prior to the first(day 0) and second (day 7) injection and at the end of the study (day 30) revealed successful gene transfer at day 7 by immunohistochemical staining for hsv-TK and some evidence of tumor destruction by day 30. This expected and encouraging results have led us to propose a phase I trial to test the transcriptional ability of the OC promoter to regulate adenoviral replication in a similar format.

## 3.3 MN/CA9 Promoter

### 3.3.1 Rationale

The human MN/CA9 gene has been isolated, sequenced and characterized by Opavsky.[79] This gene is a member of the carbonic anhydrase family, which codes for a diverse group of catalysts of the reversible conversion of carbon dioxide to carbonic acid. MN/CA9 expression has been detected in several types of carcinomas including renal, ovarian and cervical as well as in normal gastric mucosa (Der and Stanbridge,[80] Zavada,[81] Liao,[82] Pastorekova[83]). The complete genomic sequence of MN/CA9 gene including the 5′-flanking region encompasses 10.9 kb with a coding sequence comprising of 11 exons. The MN/CA9 protein contains 459 amino acids with a molecular weight ranging from 54 to 58 kDa. MN displays CA activity and binds zinc (Pastorek[84]). The nucleotide sequence close to 5′ end shows 91.4% sequence homology to the U3 region of the long terminal repeats (LTR) of the human HERV-K endogenous retroviruses (Ono,[85]). This LTR like sequence is 222 bp long with an A-rich tail at its 3′ end. Analysis of the MN/CA9 promoter region between −507 and +1 upstream of the transcription initiation site indicates that despite the presence of 60% GC residues the additional features of TATA-less promoters are absent but presence of consensus sequences for AP1, AP2 and p53 transcription factor binding sites has been demonstrated (Locker and Buzard,[86] Imagawa,[87] El-Deiry[88]). Functional characterization of the 3.5 kb MN 5′ upstream region by deletion analysis led to the identification of −173 to +31 fragment as the MN promoter. The promoter region lacks the CpG-rich islands that are typical for TATA-less promoters but contains two non-overlapping consensus initiator sequences required for the promoter activity.

### 3.3.2 In Vitro and In Vivo Experiences with MN

Initial in vitro studies with this promoter driving luciferase expression demonstrated tumor specificity for both renal cell carcinoma and cervical carcinoma. Based on the expression assays we have constructed an oncolytic adenovirus with MN promoter with has demonstrated 40–100-fold increased killing in human renal cell carcinomas compare to control cell lines not expressing this promoter activity. We are currently evaluating this oncolytic vector in animal models of human renal cell carcinoma.

## 3.4 Inducible Transcription

The ability to precisely regulate spatial and temporal expression of a particular gene is likely to have a significant impact in the field of human gene therapy. In order to be effective, such an approach must necessarily fulfill several criteria including: (1) biological safety, (2) ease of administration, (3) low basal expression, (4) high and gene-specific inducibility, (5) reversibility, and (6) must be preferably of human origin to minimize immunogenicity. A wide variety of inducible systems for regulating gene expression have been developed. These include the use of metal response promoter (Searle[89]), heat shock promoter (Fuqua[90]), the glucocorticoid inducible promoter (Hirt[91]), IPTG inducible lac repressor/operator system (Figge,[92] Baim[93]), tetracycline inducible system (Gossen and Bujard,[94]), RU 486 inducible system (Wang[95]), ecdysone inducible system (No[96]), FK506/rapamycin inducible system (Rivera[97]) hypoxia-inducible-factor-1system (Dachs[98]), radiation inducible system (Scott[99]) and the tamoxifen inducible system (Putzer[100]). It is beyond the scope of this chapter to provide an in-depth information on all of the above mentioned inducible systems, consequently we would like to focus on those inducible systems that might have a higher potential for human gene therapy applications.

### 3.4.1 Tetracycline Inducible System

The tet-inducible system originally developed by Bujard and co-workers (Gossen and Bujard,[94] Gossen[101]) is widely used to regulate gene expression. The tet-inducible system is based on the tetracycline resistance operon of *E. coli*. The system utilizes the specificity of the tet repressor (tetR) for the tet operator sequence (tetO), the sensitivity of tetR to tetracycline and the potent transactivator function of herpes simplex virus protein VP16. The system is based upon the concept of negatively regulating the transcription of the bacterial resistance gene by tetR protein binding to tetO DNA sequences. Addition of tetracycline or doxycycline causes derepression by binding to the (tetR) protein thereby allowing transcription to proceed. This has been achieved by employing a tet transactivator (tTA) which is a chimeric tetracycline-repressed transactivator generated by fusing carboxy terminal of tetR protein to the carboxy-terminal 127 amino acids of VP16. The tTA when bound to tetracycline is prevented from binding to seven copies of tetO sequences, which are juxtaposed upstream to a minimal human cytomegalovirus promoter thereby selectively turning off the transcription of the gene in question. Removal of tetracycline results in binding of tTA to the tetO sequences in the tet-inducible promoter following which the VP16 moiety of tTA transactivates the target gene by promoting assembly of a transcriptional initiation complex thereby selectively turning on the gene expression. A recent modification of this system allows for selective induction of gene expression in the presence of tetracycline. In this strategy, a mutated tetR called reverse tTA (rtTA) has been generated by incorporating four amino acid changes into tTA thereby facilitating rtTA to bind to tetO sequence in the presence of tetracycline. Another variation involves fusion of tTA with the KRAB repressor domain of the human zinc finger protein Kox1. Upon binding to tetO sequences, this protein is capable of blocking transcription as far as 3 kb downstream (Deuschle[102]). A further variation has revealed that by placing two minimal promoters in the opposite orientation on either side of the tetO sequences it is possible to simultaneously regulate the expression of two genes from a single plasmid (Baron[103]). Massie et al.[104] have used tet-inducible system to generate a recombinant adenoviral vector encoding a deletion in the R1 subunit of the herpes simplex virus type 2 ribonucleotide reductase. Topical and tetracycline inducible gene expression in transgenic mice carrying a gene under tet-inducible promoter has been achieved by adenovirus mediated gene transfer and expression of tTA (Ghersa[105]). Rubinchik et al.[106] have developed a tet-inducible, double recombinant adenoviral vector expressing a fusion of murine FasL and green fluorescent protein. In this virus the tet-responsive element and the transactivator element are built into the opposite ends of the same vector to avoid enhancer interference. The in vitro expression of FasL-GFP in various cell lines could be conveniently regulated by tetracycline or doxycycline in a dose dependent manner.

### 3.4.2 FK506/Rapamycin-Inducible System

Latest in the armamentarium of inducible gene expression systems are the chemical dimerizers that rely upon drug dependent recruitment of a transactivation domain to a basal promoter to drive the expression of the therapeutic gene. The strategy is based upon generating a genetic fusion comprising of heterologous DNA-binding domain and activation domain with the drug binding domain thereby enabling a bivalent drug to crosslink the two proteins and reconstitute an active transcription factor. This is achieved by using small cell-permeable immunosuppressive molecules FK506, rapamycin and cyclosporine to bind members of the immunophilin family. The FK506 molecule binds tightly to the cellular protein FKBP12 while FK1012 a synthetic dimer of FK506 causes dimerization of several chimeric proteins containing FKBP12 (Liang[107]). Another synthetic compound FKCsA created by fusion between FK506 and cyclosporine A binds with high affinity to FKBP12 and cyclophilin and has been used for inducible transcription of exogenous genes (Belshaw[108]). However, the most promising results have been obtained using the heterodimerizer rapamycin which binds simultaneously to the human proteins FKBP and FRAP (Standaert,[109] Brown[110]). In this system, transcriptional activation is achieved through rapamycin induced reconstitution of a transcription factor complex formed by coupling of (1) a unique DNA-binding domain ZFHD genetically fused to FKBP and (2) activation domain of the p65 subunit of nuclear factor kappa B (NFκB), fused with the rapamycin binding domain of FRAP. This novel approach has been successfully utilized for stable in vivo delivery of secreted alkaline phosphate, murine erythropoietin and human growth hormone using eukaryotic expression vectors, adenoviral, retroviral and adeno-associated viral vectors (Magari,[111] Ye,[112] Rivera[113]). One of the limitations of this approach is the growth inhibitory and immunosuppressive activity of rapamycin which is due to the inhibition of endogenous FRAP activity (Brown[110]). This limitation can be overcome by nonimmunosuppressive analogs (rapalogs) of rapamycin by incorporating mutations in the FRAP domain that accommodate modified drugs (Liberles,[114] Clackson[115]). Considerable progress has also been made in designing novel synthetic dimerizers of ligand for human FKBP12 and mutated FKBP (Amara,[116] Clackson,[115] Rollins[117]). These studies are suggestive of the potential utility of this novel approach for human gene therapy applications.

### *3.4.3 RU 486*

Wang et al.[95] have developed a novel regulated transcriptional activator consisting of a truncated ligand binding domain of the human progesterone receptor, the DNA-binding domain of yeast transcriptional activator GAL4, and a C-terminal fragment of the herpes simplex VP16 transcriptional regulator protein. This novel transcriptional activator binds with high affinity to the synthetic progesterone antagonist RU 486 but binds very poorly to progesterone. In conjunction with the target gene containing four copies of the consensus GAL4 binding site, the gene expression was only activated in the presence of RU 486 (Wang,[95] Wang [Wang, 1997a]). Wang [Wang, 1997b] have also developed an inducible repressor system by substituting the KRAB transcriptional repressor domain for the VP16 transactivation domain. In addition to RU 486, this system can be activated by other synthetic progesterone antagonists at low concentration. The efficacy of this system has been demonstrated using an ex vivo transplantation approach in which the cells containing stably integrated chimeric regulator GLVP and a target gene (tyrosine hydroxylase) were grafted in rats. One of the caveats of this system is the low but distinctive basal activity of the GAL4 responsive promoter in the absence of RU 486. Consequently, this system has been refined by designing a synthetic transcription factor which contains a 35 amino acid truncation of the progesterone receptor rather than the 42 amino acid truncation (Delort and Capecchi,[118]). This system exhibits twofold to threefold lower basal activities as compared to the earlier version.

## 4. Enhanced Control of Transgene Expression

### *4.1 Safety Improvements*

Prior to initiating our clinical trial with Ad-OC-TK, we performed a distribution study measured TK activity in a variety of organs harvested 3 days after intravenous (IV) injection of Ad-CMV-TK ($2\times10^9$ pfu) or Ad-OC-TK ($2\times10^9$ pfu) with three mice per group. TK enzymatic activity was detected only in the Ad-CMV-TK group (liver and spleen only), but not the Ad-OC-TK group. Next a comparative study in which 10 C57/bl mice received one IV injection of $2\times10^9$ pfu of Ad-OC-TK or Ad-CMV-TK and intraperitoneal (IP) GCV. Significant mortality with severe hepatic histopathology was observed in the Ad-CMV-TK/GCV group (90% mortality) while the Ad-OC-TK/GCV administration did not affect survival of the treated animals (100% survival). These data and the above tissue distribution studies support the hypothesis that, in syngeneic hosts, the OC promoter is tissue specific for tumors, since Ad-OC-TK inhibits tumor growth as effectively as do RSV-TK and CMV-TK, but without the generalized toxicity observed with these universal promoters. These findings paralleled the formal GLP toxicology study in mice and our toxicology profile in our clinical trial. Others have demonstrated the lethal effects of both universal promoter hsv-TK viruses in mice (), rats (), and hepatotoxicity in humans () after intraprostatic injections.

### 4.2 Potency Concerns

The initial concern with a tumor-specific promoter is that the magnitude of the transgene expression would be decrease because of the specificity of the promoter. To address this issue we compared the in vivo growth inhibition associated with intralesional administration of Ad-OC-TK with that of Ad-CMV-TK, using a rat osteosarcoma (ROS 17/2.8) subcutaneous model. Ten athymic nude mice were injected with 1 x $10^6$ ROS cells per site in four subcutaneous locations. After establishment of tumor growth at greater than 5 mm diameter, Ad-CMV-TK or Ad-OC-TK was injected intralesionally into 5 animals (or 20 tumors) each. After viral injection, the animals received IP GCV (3 mice, 12 tumors) or phosphate-buffered saline (PBS; 2 mice, 8 tumors) for a 2-week period. The animals received one additional adenoviral injection 7 days after the first. The tumors were measured weekly and the animals were sacrificed after the second week of GCV or PBS administration. Both Ad-OC-TK and Ad-CMV-TK forms of therapy demonstrated a greater growth-inhibitory effect than was observed with PBS administration. The growth inhibition was superior with the Ad-OC-TK adenovirus. Therefore, the OC promoter has high intrinsic activity rivaling that of the strong universal CMV promoter, at least in ROS cells.

## 5. Future Directions

### 5.1 Enhancement of Weak but Specific Promoters

A wide variety of highly tissue-specific promoters have been evaluated for achieving transcriptional targeting, however, their applicability has been hampered due to weak transcriptional activity. Enhancement of weak tissue-specific promoters can be achieved by employing several different strategies. One of the simplest approach involves (1) deletion of those sequences from the promoter that do not contribute to tissue specificity or transcriptional activity and (2) incorporating multiple copies of the enhancer and positive regulatory elements. This approach has been successfully used in the case of PSA promoter (Pang,[41] tyrosinase promoter (Siders[119,120]) and CEA promoter (Richards[12])).

Another approach involves generation of activating point mutations within the promoter region as has been in the case of AFP promoter (Ishikawa[121]) and the MDR 1 promoter (Stein[122]). Yet another strategy involves selective combination of multiple positive regulatory and tissue-specific elements to achieve enhancement of weak promoters. This strategy has shown promising results in augmenting melanoma-specific gene expression when tyrosinase promoter, either alone or in combination with single or dual, tandem melanocyte specific enhancer was used to drive the expression of luciferase and *E. coli* purine nucleoside phosphorylase gene.

Transient expression studies indicated 5–500-fold increase in luciferase activity following incorporation of either single or tandem enhancer elements. In another example, when 5–20 muscle-specific transcriptional elements were randomly assembled and linked to minimal chicken α-actin promoter, sixfold higher activity was observed as compared to the CMV promoter (Li[123]). In case of adenoviral vectors it might be possible to selectively increase specific expression from exogenous promoters by co-expression of modified VAI genes. Using this approach, Eloit[124] were able to achieve 12.5–502-fold increased reporter gene expression. The fact that activity of certain E2F responsive promoters in tumor cells exceeds that achieved in mitotically active normal cells has been exploited for tumor selective transgene expression using an adenoviral vector in a malignant glioma model (Parr[125]). A novel approach involves development of dual specificity promoters that are both cell type specific and cell cycle regulated. In this approach the transgene is under the transcriptional control of an artificial heterodimeric transcription factor whose DNA-binding domain is expressed from a tissue-specific promoter, whereas the transactivating subunit is transcribed from a cell cycle regulated promoter (). The feasibility of this approach has been successfully tested in a transient transfection system (Jerome and Muller,[126] Nettelbeck[127]).

Transcriptional targeting of viral replication for selective killing of tumor cells can be achieved by deletion of adenoviral E1B/55 kDa protein which is essential for viral replication but is dispensable in p53 deficient tumor cells. An alternate approach involves generation of a replication competent adenoviral vector in which E1A or E1A and E1B genes are under the transcriptional control of a tumor-specific promoters like PSA, kallikrein-2 or AFP (Rodriguez,[45] Yu,[128] Hallenbeck[129]).

## 5.2 Improving Specificity with Multiple Promoter Segments

Several investigators have placed combinations of promoter sequences in tandem to derive more specific transgene expression. The authors of the following chapter have both laboratory and clinical experience with this approach and this topic is well covered in their chapter.

## 5.3 Tumor-Specific Oncolysis

Several different approaches have been designed to achieved cancer cell specific adenoviral replication and subsequent tumor lysis. Based on our previous work in the laboratory and the clinic, we design an adenoviral vector that would only replicate in cell, which could activate the OC promoter. We have recently received approval for OBA using the OC promoter which transcriptionally regulates adenoviral replication for the treatment of men with metastatic and recurrent prostate cancer. This approach is thoroughly reviewed in the next chapter.

### 5.4 Combined Targeting Approaches

The two preceding chapters demonstrate elegant methods to achieve transductional targeting. These approaches will allow for concentration of adenovirus at metastatic tumor deposits after a systemic administration. In collaboration with these investigators we have begun to combine both transductional and transcriptional targeting to allow for both tumor-specific concentration and tumor-specific oncolysis. This approach combines many of the individual strides achieved in adenoviral gene therapy in the past decade and holds great promise for the future of adenoviral cancer gene therapy.

## 6. Summary

In summary, we believe that the success of gene therapy and its general applicability to medicine will be partially linked to the development effective transcriptional targeting strategies. The main purpose of this chapter was to illustrate to the reader the benefits of transcriptional targeting and how this approach can be used to generate tumor- or tissue-specific gene expression. The main example of the OC promoter was used because of our laboratories significant investigation in this promoter.

## Acknowledgment

With thanks to my daughter 'Apoorva Raikwar - Latham Fellow, Department of Microbiology, Roy J. and Lucille A. Carver College of Medicine, University of Iowa, Iowa City, USA' for providing the figures for this chapter — Sudhanshu Raikwar.

## References

1. Beerli RR, Segal DJ, Dreier B, Barbas CF. Toward controlling gene expression at will: specific regulation of the erbB-2/HER-2 promoter by using polydactyl zinc finger proteins constructed from modular building blocks. *Proc Natl Acad Sci USA* 1998;**95**:14628–33.
2. Kang JS, Kim JS. Zinc finger proteins as designer transcription factors. *J Biol Chem* 2000;**275**:8742–8.
3. Kim JS, Pabo CO. Getting a handhold on DNA: design of poly-zinc finger proteins with femtomolar dissociation constants. *Proc Natl Acad Sci USA* 1998;**95**:2812–7.
4. Imanishi M, Hori Y, Nagaoka M, Sugiura Y. DNA-bending finger: artificial design of 6-zinc finger peptides with polyglycine linker and induction of DNA bending. *Biochemistry* 2000;**39**:4383–90.
5. Corbi N, Libri V, Fanciulli M, Tinsley JM, Davies KE, Passananti C. The artificial zinc finger coding gene 'Jazz' binds the utrophin promoter and activates transcription. *Gene Ther* 2000;**7**:1076–83.
6. Moore M, Klug A, Choo Y. From the cover: improved DNA binding specificity from polyzinc finger peptides by using strings of two-finger units. *Proc Natl Acad Sci USA* 2001;**98**:1437–41.

7. Gold P, Shuster J, Freedman SO. Carcinoembryonic antigen (CEA) in clinical medicine: historical perspectives, pitfalls and projections. *Cancer* 1978;**42**:1399–405.
8. Thompson JA, Pande H, Paxton RJ, Shively L, Padma A, Simmer RL, et al. Molecular cloning of a gene belonging to the carcinoembryonic antigen gene family and discussion of a domain model. *Proc Natl Acad Sci USA* 1987;**84**:2965–9.
9. Schrewe H, Thompson J, Bona M, Hefta LJ, Maruya A, Hassauer M, et al. Cloning of the complete gene for carcinoembryonic antigen: analysis of its promoter indicates a region conveying cell type-specific expression. *Mol Cell Biol* 1990;**10**:2738–48.
10. Hauck W, Stanners CP. Transcriptional regulation of the carcinoembryonic antigen gene. Identification of regulatory elements and multiple nuclear factors. *J Biol Chem* 1995;**270**:3602–10.
11. Chen CJ, Li LJ, Maruya A, Shively JE. In vitro and in vivo footprint analysis of the promoter of carcinoembryonic antigen in colon carcinoma cells: effects of interferon gamma treatment. *Cancer Res* 1995;**55**:3873–82.
12. Richards CA, Austin EA, Huber BE. Transcriptional regulatory sequences of carcinoembryonic antigen: identification and use with cytosine deaminase for tumor-specific gene therapy. *Hum Gene Ther* 1995;**6**:881–93.
13. Cao G, Kuriyama S, Gao J, Mitoro A, Cui L, Nakatani T, et al. Comparison of carcinoembryonic antigen promoter regions isolated from human colorectal carcinoma and normal adjacent mucosa to induce strong tumor-selective gene expression. *Int J Cancer* 1998;**78**:242–7.
14. Osaki T, Tanio Y, Tachibana I, Hosoe S, Kumagai T, Kawase I, et al. Gene therapy for carcinoembryonic antigen-producing human lung cancer cells by cell type-specific expression of herpes simplex virus thymidine kinase gene. *Cancer Res* 1994;**54**:5258–61.
15. Takeuchi M, Shichinohe T, Senmaru N, Miyamoto M, Fujita H, Takimoto M, et al. The dominant negative H-ras mutant, N116Y, suppresses growth of metastatic human pancreatic cancer cells in the liver of nude mice. *Gene Ther* 2000;**7**:518–26.
16. Lan KH, Kanai F, Shiratori Y, Okabe S, Yoshida Y, Wakimoto H, et al. Tumor-specific gene expression in carcinoembryonic antigen—producing gastric cancer cells using adenovirus vectors. *Gastroenterology* 1996;**111**:1241–51.
17. Lan KH, Kanai F, Shiratori Y, Ohashi M, Tanaka T, Okudaira T, et al. In vivo selective gene expression and therapy mediated by adenoviral vectors for human carcinoembryonic antigen-producing gastric carcinoma. *Cancer Res* 1997;**57**:4279–84.
18. DiMaio JM, Clary BM, Via DF, Coveney E, Pappas TN, Lyerly HK. Directed enzyme pro-drug gene therapy for pancreatic cancer in vivo. *Surgery* 1994;**116**:205–13.
19. Tanaka T, Kanai F, Lan KH, Ohashi M, Shiratori Y, Yoshida Y, et al. Adenovirus-mediated gene therapy of gastric carcinoma using cancer-specific gene expression in vivo. *Biochem Biophys Res Commun* 1997;**231**:775–9.
20. Brand K, Loser P, Arnold W, Bartels T, Strauss M. Tumor cell-specific transgene expression prevents liver toxicity of the adeno-HSVtk/GCV approach. *Gene Ther* 1998;**5**:1363–71.
21. Belanger L, Baril P, Guertin M, Gingras MC, Gourdeau H, Anderson A, et al. Oncodevelopmental and hormonal regulation of alpha 1-fetoprotein gene expression. *Adv Enzyme Regul* 1983;**21**:73–99.
22. Nahon JL, Danan JL, Poiret M, Tratner I, Jose-Estanyol M, Sala-Trepat JM. The rat alpha-fetoprotein and albumin genes. Transcriptional control and comparison of the sequence organization and promoter region. *J Biol Chem* 1987;**262**:12479–87.
23. Sakai M, Morinaga T, Urano Y, Watanabe K, Wegmann TG, Tamaoki T. The human alpha-fetoprotein gene. Sequence organization and the 5′ flanking region. *J Biol Chem* 1985;**260**:5055–60.

24. Nakabayashi H, Hashimoto T, Miyao Y, Tjong KK, Chan J, Tamaoki T. A position-dependent silencer plays a major role in repressing alpha-fetoprotein expression in human hepatoma. *Mol Cell Biol* 1991;**11**:5885–93.
25. Kaneko S, Hallenbeck P, Kotani T, Nakabayashi H, McGarrity G, Tamaoki T, et al. Adenovirus-mediated gene therapy of hepatocellular carcinoma using cancer-specific gene expression. *Cancer Res* 1995;**55**:5283–7.
26. Kanai F, Shiratori Y, Yoshida Y, Wakimoto H, Hamada H, Kanegae Y, et al. Gene therapy for alpha-fetoprotein-producing human hepatoma cells by adenovirus-mediated transfer of the herpes simplex virus thymidine kinase gene. *Hepatology* 1996;**23**:1359–68.
27. Kanai F, Lan KH, Shiratori Y, Tanaka T, Ohashi M, Okudaira T, et al. In vivo gene therapy for alpha-fetoprotein-producing hepatocellular carcinoma by adenovirus-mediated transfer of cytosine deaminase gene. *Cancer Res* 1997;**57**:461–5.
28. Arbuthnot PB, Bralet MP, Le Jossic C, Dedieu JF, Perricaudet M, Brechot C, et al. In vitro and in vivo hepatoma cell-specific expression of a gene transferred with an adenoviral vector. *Hum Gene Ther* 1996;**7**:1503–14.
29. Bui LA, Butterfield LH, Kim JY, Ribas A, Seu P, Lau R, et al. In vivo therapy of hepatocellular carcinoma with a tumor-specific adenoviral vector expressing interleukin-2. *Hum Gene Ther* 1997;**8**:2173–82.
30. Riegman PH, Vlietstra RJ, van der Korput JA, Romijn JC, Trapman J. Characterization of the prostate-specific antigen gene: a novel human kallikrein-like gene. *Biochem Biophys Res Commun* 1989;**159**:95–102.
31. Riegman PH, Vlietstra RJ, Klaassen P, van der Korput JA, Geurts van Kessel A, Romijn JC, et al. The prostate-specific antigen gene and the human glandular kallikrein-1 gene are tandemly located on chromosome 19. *FEBS Lett* 1989;**247**:123–6.
32. Lundwall A. Characterization of the gene for prostate-specific antigen, a human glandular kallikrein. *Biochem Biophys Res Commun* 1989;**161**:1151–9.
33. Wang MC, Valenzuela LA, Murphy GP, Chu TM. Purification of a human prostate specific antigen. *Invest Urol* 1979;**17**:159–63.
34. Wang MC, Papsidero LD, Kuriyama M, Valenzuela LA, Murphy GP, Chu TM. Prostate antigen: a new potential marker for prostatic cancer. *Prostate* 1981;**2**:89–96.
35. Gallee MP, van Vroonhoven CC, van der Korput HA, van der Kwast TH, ten Kate FJ, Romijn JC, et al. Characterization of monoclonal antibodies raised against the prostatic cancer cell line PC-82. *Prostate* 1986;**9**:33–45.
36. Riegman PH, Vlietstra RJ, van der Korput HA, Romijn JC, Trapman J. Identification and androgen-regulated expression of two major human glandular kallikrein-1 (hGK-1) mRNA species. *Mol Cell Endocrinol* 1991;**76**:181–90.
37. Cleutjens CB, Steketee K, van Eekelen CC, van der Korput JA, Brinkmann AO, Trapman J. Both androgen receptor and glucocorticoid receptor are able to induce prostate-specific antigen expression, but differ in their growth-stimulating properties of LNCaP cells. *Endocrinology* 1997;**138**:5293–300.
38. Cleutjens KB, van der Korput HA, van Eekelen CC, van Rooij HC, Faber PW, Trapman J. An androgen response element in a far upstream enhancer region is essential for high, androgen-regulated activity of the prostate-specific antigen promoter. *Mol Endocrinol* 1997;**11**:148–61.
39. Cleutjens KB, van Eekelen CC, van der Korput HA, Brinkmann AO, Trapman J. Two androgen response regions cooperate in steroid hormone regulated activity of the prostate-specific antigen promoter. *J Biol Chem* 1996;**271**:6379–88.
40. Cleutjens KB, van der Korput HA, Ehren-van Eekelen CC, Sikes RA, Fasciana C, Chung LW, et al. A 6-kb promoter fragment mimics in transgenic mice the prostate-specific and androgen-regulated expression of the endogenous prostate-specific antigen gene in humans. *Mol Endocrinol* 1997;**11**:1256–65.

41. Pang S, Dannull J, Kaboo R, Xie Y, Tso CL, Michel K, et al. Identification of a positive regulatory element responsible for tissue-specific expression of prostate-specific antigen. *Cancer Res* 1997;**57**:495–9.
42. Yeung F, Li X, Ellett J, Trapman J, Kao C, Chung LW. Regions of prostate-specific antigen (PSA) promoter confer androgen-independent expression of PSA in prostate cancer cells. *J Biol Chem* 2000;**275**:40846–55.
43. Oettgen P, Finger E, Sun Z, Akbarali Y, Thamrongsak U, Boltax J, et al. PDEF, a novel prostate epithelium-specific ets transcription factor, interacts with the androgen receptor and activates prostate-specific antigen gene expression. *J Biol Chem* 2000;**275**:1216–25.
44. Latham JP, Searle PF, Mautner V, James ND. Prostate-specific antigen promoter/enhancer driven gene therapy for prostate cancer: construction and testing of a tissue-specific adenovirus vector. *Cancer Res* 2000;**60**:334–41.
45. Rodriguez R, Schuur ER, Lim HY, Henderson GA, Simons JW, Henderson DR. Prostate attenuated replication competent adenovirus (ARCA) CN706: a selective cytotoxic for prostate-specific antigen-positive prostate cancer cells. *Cancer Res* 1997;**57**:2559–63.
46. Martiniello-Wilks R, Garcia-Aragon J, Daja MM, Russell P, Both GW, Molloy PL, et al. In vivo gene therapy for prostate cancer: preclinical evaluation of two different enzyme-directed prodrug therapy systems delivered by identical adenovirus vectors. *Hum Gene Ther* 1998;**9**:1617–26.
47. Gotoh A, Ko SC, Shirakawa T, Cheon J, Kao C, Miyamoto T, et al. Development of prostate-specific antigen promoter-based gene therapy for androgen-independent human prostate cancer. *J Urol* 1998;**160**:220–9.
48. Lee SE, Jin RJ, Lee SG, Yoon SJ, Park MS, Heo DS, et al. Development of a new plasmid vector with PSA-promoter and enhancer expressing tissue-specificity in prostate carcinoma cell lines. *Anticancer Res* 2000;**20**:417–22.
49. Xu LL, Srikantan V, Sesterhenn IA, Augustus M, Dean R, Moul JW, et al. Expression profile of an androgen regulated prostate specific homeobox gene NKX3.1 in primary prostate cancer. *J Urol* 2000;**163**:972–9.
50. Pan LC, Price PA. The effect of transcriptional inhibitors on the bone gamma-carboxyglutamic acid protein response to 1,25-dihydroxyvitamin D3 in osteosarcoma cells. *J Biol Chem* 1984;**259**:5844–7.
51. Price PA, Williamson MK. Primary structure of bovine matrix Gla protein, a new vitamin K-dependent bone protein. *J Biol Chem* 1985;**260**:14971–5.
52. Kerner SA, Scott RA, Pike JW. Sequence elements in the human osteocalcin gene confer basal activation and inducible response to hormonal vitamin D3. *Proc Natl Acad Sci USA* 1989;**86**:4455–9.
53. Lian J, Stewart C, Puchacz E, Mackowiak S, Shalhoub V, Collart D, et al. Structure of the rat osteocalcin gene and regulation of vitamin D-dependent expression. *Proc Natl Acad Sci USA* 1989;**86**:1143–7.
54. Desbois C, Hogue DA, Karsenty G. The mouse osteocalcin gene cluster contains three genes with two separate spatial and temporal patterns of expression. *J Biol Chem* 1994;**269**:1183–90.
55. Desbois C, Seldin MF, Karsenty G. Localization of the osteocalcin gene cluster on mouse chromosome 3. *Mamm Genome* 1994;**5**:321–2.
56. Montecino M, Pockwinse S, Lian J, Stein G, Stein J. DNase I hypersensitive sites in promoter elements associated with basal and vitamin D dependent transcription of the bone-specific osteocalcin gene. *Biochemistry* 1994;**33**:348–53.
57. Hoffmann HM, Catron KM, van Wijnen AJ, McCabe LR, Lian JB, Stein GS, et al. Transcriptional control of the tissue-specific, developmentally regulated osteocalcin gene requires a binding motif for the Msx family of homeodomain proteins. *Proc Natl Acad Sci USA* 1994;**91**:12887–91.

58. Towler DA, Rutledge SJ, Rodan GA. Msx-2/Hox 8.1: a transcriptional regulator of the rat osteocalcin promoter. *Mol Endocrinol* 1994;**8**:1484–93.
59. Towler DA, Bennett CD, Rodan GA. Activity of the rat osteocalcin basal promoter in osteoblastic cells is dependent upon homeodomain and CP1 binding motifs. *Mol Endocrinol* 1994;**8**:614–24.
60. Tamura M, Noda M. Identification of a DNA sequence involved in osteoblast-specific gene expression via interaction with helix-loop-helix (HLH)-type transcription factors. *J Cell Biol* 1994;**126**:773–82.
61. Banerjee C, Hiebert SW, Stein JL, Lian JB, Stein GS. An AML-1 consensus sequence binds an osteoblast-specific complex and transcriptionally activates the osteocalcin gene. *Proc Natl Acad Sci USA* 1996;**93**:4968–73.
62. Bidwell JP, Van Wijnen AJ, Fey EG, Dworetzky S, Penman S, Stein JL, et al. Osteocalcin gene promoter-binding factors are tissue-specific nuclear matrix components. *Proc Natl Acad Sci USA* 1993;**90**:3162–6.
63. Merriman HL, van Wijnen AJ, Hiebert S, Bidwell JP, Fey E, Lian J, et al. The tissue-specific nuclear matrix protein, NMP-2, is a member of the AML/CBF/PEBP2/runt domain transcription factor family: interactions with the osteocalcin gene promoter. *Biochemistry* 1995;**34**:13125–32.
64. Geoffroy V, Ducy P, Karsenty G. A PEBP2 alpha/AML-1-related factor increases osteocalcin promoter activity through its binding to an osteoblast-specific cis-acting element. *J Biol Chem* 1995;**270**:30973–9.
65. Banerjee C, Stein JL, Van Wijnen AJ, Frenkel B, Lian JB, Stein GS. Transforming growth factor-beta 1 responsiveness of the rat osteocalcin gene is mediated by an activator protein-1 binding site. *Endocrinology* 1996;**137**:1991–2000.
66. Markose ER, Stein JL, Stein GS, Lian JB. Vitamin D-mediated modifications in protein-DNA interactions at two promoter elements of the osteocalcin gene. *Proc Natl Acad Sci USA* 1990;**87**:1701–5.
67. Demay MB, Gerardi JM, DeLuca HF, Kronenberg HM. DNA sequences in the rat osteocalcin gene that bind the 1,25- dihydroxyvitamin D3 receptor and confer responsiveness to 1,25- dihydroxyvitamin D3. *Proc Natl Acad Sci USA* 1990;**87**:369–73.
68. Breen EC, van Wijnen AJ, Lian JB, Stein GS, Stein JL. In vivo occupancy of the vitamin D responsive element in the osteocalcin gene supports vitamin D dependent transcriptional upregulation in intact cells. *Proc Natl Acad Sci USA* 1994;**91**:12902–6.
69. Montecino M, Frenkel B, Lian J, Stein J, Stein G. Requirement of distal and proximal promoter sequences for chromatin organization of the osteocalcin gene in bone-derived cells. *J Cell Biochem* 1996;**63**:221–8.
70. Yoon KG, Rutledge SJ, Buenaga RF, Rodan GA. Characterization of the rat osteocalcin gene: stimulation of promoter activity by 1,25-dihydroxyvitamin D3. *Biochemistry* 1988;**27**:8521–6.
71. Briggs MR, Kadonaga JT, Bell SP, Tjian R. Purification and biochemical characterization of the promoter-specific transcription factor, Sp1. *Science* 1986;**234**:47–52.
72. Theofan G, Haberstroh LM, Price PA. Molecular structure of the rat bone Gla protein gene and identification of putative regulatory elements. *DNA* 1989;**8**:213–21.
73. Goto K, Heymont JL, Klein-Nulend J, Kronenberg HM, Demay MB. Identification of an osteoblastic silencer element in the first intron of the rat osteocalcin gene. *Biochemistry* 1996;**35**:11005–11.
74. Kearns AE, Goto K, Gianakakos G, Lippmann W, Demay MB. Transcriptional repression of the rat osteocalcin gene: role of two intronic CCTCCT motifs. *Endocrinology* 1999;**140**:4120–6.
75. Ko SC, Cheon J, Kao C, Gotoh A, Shirakawa T, Sikes RA, et al. Osteocalcin promoter-based toxic gene therapy for the treatment of osteosarcoma in experimental models. *Cancer Res* 1996;**56**:4614–9.

76. Cheon J, Ko SC, Gardner TA, Shirakawa T, Gotoh A, Kao C, et al. Chemogene therapy: osteocalcin promoter-based suicide gene therapy in combination with methotrexate in a murine osteosarcoma model. *Cancer Gene Ther* 1997;**4**:359–65.
77. Shirakawa T, Ko SC, Gardner TA, Cheon J, Miyamoto T, Gotoh A, et al. In vivo suppression of osteosarcoma pulmonary metastasis with intravenous osteocalcin promoter-based toxic gene therapy. *Cancer Gene Ther* 1998;**5**:274–80.
78. Hou Z, Nguyen Q, Frenkel B, Nilsson SK, Milne M, van Wijnen AJ, et al. Osteoblast-specific gene expression after transplantation of marrow cells: implications for skeletal gene therapy. *Proc Natl Acad Sci USA* 1999;**96**:7294–9.
79. Opavsky R, Pastorekova S, Zelnik V, Gibadulinova A, Stanbridge EJ, Zavada J, et al. Human MN/CA9 gene, a novel member of the carbonic anhydrase family: structure and exon to protein domain relationships. *Genomics* 1996;**33**:480–7.
80. Der CJ, Stanbridge EJ. A tumor-specific membrane phosphoprotein marker in human cell hybrids. *Cell* 1981;**26**:429–38.
81. Zavada J, Zavadova Z, Pastorekova S, Ciampor F, Pastorek J, Zelnik V. Expression of MaTu-MN protein in human tumor cultures and in clinical specimens. *Int J Cancer* 1993;**54**:268–74.
82. Liao SY, Brewer C, Zavada J, Pastorek J, Pastorekova S, Manetta A, et al. Identification of the MN antigen as a diagnostic biomarker of cervical intraepithelial squamous and glandular neoplasia and cervical carcinomas. *Am J Pathol* 1994;**145**:598–609.
83. Pastorekova S, Parkkila S, Parkkila AK, Opavsky R, Zelnik V, Saarnio J, et al. Carbonic anhydrase IX, MN/CA IX: analysis of stomach complementary DNA sequence and expression in human and rat alimentary tracts. *Gastroenterology* 1997;**112**:398–408.
84. Pastorek J, Pastorekova S, Callebaut I, Mornon JP, Zelnik V, Opavsky R, et al. Cloning and characterization of MN, a human tumor-associated protein with a domain homologous to carbonic anhydrase and a putative helix- loop-helix DNA binding segment. *Oncogene* 1994;**9**:2877–88.
85. Ono M. Molecular cloning and long terminal repeat sequences of human endogenous retrovirus genes related to types A and B retrovirus genes. *J Virol* 1986;**58**:937–44.
86. Locker J, Buzard G. A dictionary of transcription control sequences. *DNA Seq* 1990;**1**:3–11.
87. Imagawa M, Chiu R, Karin M. Transcription factor AP-2 mediates induction by two different signal- transduction pathways: protein kinase C and cAMP. *Cell* 1987;**51**:251–60.
88. el-Deiry WS, Kern SE, Pietenpol JA, Kinzler KW, Vogelstein B. Definition of a consensus binding site for p53. *Nat Genet* 1992;**1**:45–9.
89. Searle PF, Stuart GW, Palmiter RD. Building a metal-responsive promoter with synthetic regulatory elements. *Mol Cell Biol* 1985;**5**:1480–9.
90. Fuqua SA, Blum-Salingaros M, McGuire WL. Induction of the estrogen-regulated "24K" protein by heat shock. *Cancer Res* 1989;**49**:4126–9.
91. Hirt RP, Fasel N, Kraehenbuhl JP. Inducible protein expression using a glucocorticoid-sensitive vector. *Methods Cell Biol* 1994;**43**:247–62.
92. Figge J, Wright C, Collins CJ, Roberts TM, Livingston DM. Stringent regulation of stably integrated chloramphenicol acetyl transferase genes by *E. coli* lac repressor in monkey cells. *Cell* 1988;**52**:713–22.
93. Baim SB, Labow MA, Levine AJ, Shenk T. A chimeric mammalian transactivator based on the lac repressor that is regulated by temperature and isopropyl beta-D-thiogalactopyranoside. *Proc Natl Acad Sci USA* 1991;**88**:5072–6.
94. Gossen M, Bujard H. Tight control of gene expression in mammalian cells by tetracycline- responsive promoters. *Proc Natl Acad Sci USA* 1992;**89**:5547–51.

95. Wang Y, O'Malley BW, Tsai SY. A regulatory system for use in gene transfer. *Proc Natl Acad Sci USA* 1994;**91**:8180–4.
96. No D, Yao TP, Evans RM. Ecdysone-inducible gene expression in mammalian cells and transgenic mice. *Proc Natl Acad Sci USA* 1996;**93**:3346–51.
97. Rivera VM, Clackson T, Natesan S, Pollock R, Amara JF, Keenan T, et al. A humanized system for pharmacologic control of gene expression. *Nat Med* 1996;**2**:1028–32.
98. Dachs GU, Patterson AV, Firth JD, Ratcliffe PJ, Townsend KM, Stratford IJ, et al. Targeting gene expression to hypoxic tumor cells. *Nat Med* 1997;**3**:515–20.
99. Scott SD, Marples B, Hendry JH, Lashford LS, Embleton MJ, Hunter RD, et al. A radiation-controlled molecular switch for use in gene therapy of cancer. *Gene Ther* 2000;**7**:1121–5.
100. Putzer BM, Stiewe T, Crespo F, Esche H. Improved safety through tamoxifen-regulated induction of cytotoxic genes delivered by Ad vectors for cancer gene therapy. *Gene Ther* 2000;**7**:1317–25.
101. Gossen M, Freundlieb S, Bender G, Muller G, Hillen W, Bujard H. Transcriptional activation by tetracyclines in mammalian cells. *Science* 1995;**268**:1766–9.
102. Deuschle U, Meyer WK, Thiesen HJ. Tetracycline-reversible silencing of eukaryotic promoters. *Mol Cell Biol* 1995;**15**:1907–14.
103. Baron U, Freundlieb S, Gossen M, Bujard H. Co-regulation of two gene activities by tetracycline via a bidirectional promoter. *Nucleic Acids Res* 1995;**23**:3605–6.
104. Massie B, Couture F, Lamoureux L, Mosser DD, Guilbault C, Jolicoeur P, et al. Inducible overexpression of a toxic protein by an adenovirus vector with a tetracycline-regulatable expression cassette. *J Virol* 1998;**72**:2289–96.
105. Ghersa P, Gobert RP, Sattonnet-Roche P, Richards CA, Merlo Pich E, Hooft van Huijsduijnen R. Highly controlled gene expression using combinations of a tissue-specific promoter, recombinant adenovirus and a tetracycline-regulatable transcription factor. *Gene Ther* 1998;**5**:1213–20.
106. Rubinchik S, Ding R, Qiu AJ, Zhang F, Dong J. Adenoviral vector which delivers FasL-GFP fusion protein regulated by the tet-inducible expression system. *Gene Ther* 2000;**7**:875–85.
107. Liang X, Hartikka J, Sukhu L, Manthorpe M, Hobart P. Novel, high expressing and antibiotic-controlled plasmid vectors designed for use in gene therapy. *Gene Ther* 1996; **3**:350–6.
108. Belshaw PJ, Spencer DM, Crabtree GR, Schreiber SL. Controlling programmed cell death with a cyclophilin-cyclosporin-based chemical inducer of dimerization. *Chem Biol* 1996;**3**:731–8.
109. Standaert RF, Galat A, Verdine GL, Schreiber SL. Molecular cloning and overexpression of the human FK506-binding protein FKBP. *Nature* 1990;**346**:671–4.
110. Brown EJ, Albers MW, Shin TB, Ichikawa K, Keith CT, Lane WS, et al. A mammalian protein targeted by G1-arresting rapamycin-receptor complex. *Nature* 1994;**369**:756–8.
111. Magari SR, Rivera VM, Iuliucci JD, Gilman M, Cerasoli F. Pharmacologic control of a humanized gene therapy system implanted into nude mice. *J Clin Invest* 1997;**100**:2865–72.
112. Ye X, Rivera VM, Zoltick P, Cerasoli F, Schnell MA, Gao G, et al. Regulated delivery of therapeutic proteins after in vivo somatic cell gene transfer. *Science* 1999;**283**:88–91.
113. Rivera VM, Ye X, Courage NL, Sachar J, Cerasoli F, Wilson JM, et al. Long-term regulated expression of growth hormone in mice after intramuscular gene transfer. *Proc Natl Acad Sci USA* 1999;**96**:8657–62.
114. Liberles SD, Diver ST, Austin DJ, Schreiber SL. Inducible gene expression and protein translocation using nontoxic ligands identified by a mammalian three-hybrid screen. *Proc Natl Acad Sci USA* 1997;**94**:7825–30.

115. Clackson T, Yang W, Rozamus LW, Hatada M, Amara JF, Rollins CT, et al. Redesigning an FKBP-ligand interface to generate chemical dimerizers with novel specificity. *Proc Natl Acad Sci USA* 1998;**95**:10437–42.
116. Amara JF, Clackson T, Rivera VM, Guo T, Keenan T, Natesan S, et al. A versatile synthetic dimerizer for the regulation of protein-protein interactions. *Proc Natl Acad Sci USA* 1997;**94**:10618–23.
117. Rollins CT, Rivera VM, Woolfson DN, Keenan T, Hatada M, Adams SE, et al. A ligand-reversible dimerization system for controlling protein-protein interactions. *Proc Natl Acad Sci USA* 2000;**97**:7096–101.
118. Delort JP, Capecchi MR. TAXI/UAS: a molecular switch to control expression of genes in vivo. *Hum Gene Ther* 1996;**7**:809–20.
119. Siders WM, Halloran PJ, Fenton RG. Transcriptional targeting of recombinant adenoviruses to human and murine melanoma cells. *Cancer Res* 1996;**56**:5638–46.
120. Siders WM, Halloran PJ, Fenton RG. Melanoma-specific cytotoxicity induced by a tyrosinase promoter-enhancer/herpes simplex virus thymidine kinase adenovirus. *Cancer Gene Ther* 1998;**5**:281–91.
121. Ishikawa H, Nakata K, Mawatari F, Ueki T, Tsuruta S, Ido A, et al. Utilization of variant-type of human alpha-fetoprotein promoter in gene therapy targeting for hepatocellular carcinoma. *Gene Ther* 1999;**6**:465–70.
122. Stein U, Walther W, Shoemaker RH. Vincristine induction of mutant and wild-type human multidrug- resistance promoters is cell-type-specific and dose-dependent. *J Cancer Res Clin Oncol* 1996;**122**:275–82.
123. Li X, Eastman EM, Schwartz RJ, Draghia-Akli R. Synthetic muscle promoters: activities exceeding naturally occurring regulatory sequences. *Nat Biotechnol* 1999;**17**:241–5.
124. Eloit M, Adam M, Gallais I, Fournier A. High level of transgene expression in cell cultures and in the mouse by replication-incompetent adenoviruses harboring modified VAI genes. *J Virol* 1997;**71**:5375–81.
125. Parr MJ, Manome Y, Tanaka T, Wen P, Kufe DW, Kaelin WG, et al. Tumor-selective transgene expression in vivo mediated by an E2F-responsive adenoviral vector. *Nat Med* 1997;**3**:1145–9.
126. Jerome V, Muller R. Tissue-specific, cell cycle-regulated chimeric transcription factors for the targeting of gene expression to tumor cells. *Hum Gene Ther* 1998;**9**:2653–9.
127. Nettelbeck DM, Jerome V, Muller R. A dual specificity promoter system combining cell cycle-regulated and tissue-specific transcriptional control. *Gene Ther* 1999;**6**:1276–81.
128. Yu DC, Chen Y, Seng M, Dilley J, Henderson DR. The addition of adenovirus type 5 region E3 enables calydon virus 787 to eliminate distant prostate tumor xenografts. *Cancer Res* 1999;**59**:4200–3.
129. Hallenbeck PL, Chang YN, Hay C, Golightly D, Stewart D, Lin J, et al. A novel tumor-specific replication-restricted adenoviral vector for gene therapy of hepatocellular carcinoma. *Hum Gene Ther* 1999;**10**:1721–33.
130. Chung LWK, et al. New targets for therapy in prostate cancer: Modulation of stromal-epithelial interactions. *Urology* 2003;**62**(Suppl. 5A):44–54.

# Adenoviral Vector Targeting via Mitigation of Liver Sequestration

11

*Michael A. Barry*
Division of Infectious Diseases, Department of Internal Medicine, Mayo Clinic, Rochester, MN, USA; Department of Immunology, Mayo Clinic, Rochester, MN, USA; Department of Molecular Medicine, Mayo Clinic, Rochester, MN, USA

## 1. Introduction

Certain applications can apply adenoviruses (Ads) by injection into a single tissue (i.e., vaccination into the muscle, intratumoral injection). In contrast, many diseases are systemic in character, so Ads need to be injected by a route that can distribute them to many sites in the body. Examples include treating metastatic cancer and cell-autonomous genetic diseases that affect all cells in the body. In other cases, tissues such as endothelial cells or liver hepatocytes that are in direct or near-direct contact with the blood are the therapeutic target. In these cases, intravascular (i.v.) virus injection is the most obvious and robust method to hit these targets.

Before reaching the ultimate target cell, intravenously injected virions encounter a multitude of host factors and barriers that can reduce or ablate viral efficacy (reviewed in Ref. 1). These barriers reduce the amount of vector that can reach cells in need of therapy. These barriers also invite the use of increasing doses of viruses that can lead to significant immunologic and toxic side effects.

**First things first: there is not one Ad: There is a whole "virome" of Ads.** Many discuss the biology of Ads as "Ad does this" and "Ad does that," when in reality they should be saying "Ad serotype 5 (Ad5) does this." There are more than 60 human Ad serotypes and an even wider variety found in other species.[1–3] Across the human Ad "virome" there is approximately 45% genetic diversity at the genome level.[4] Therefore, we will try to avoid the trap of describing of extrapolating the biology of this vast virome of many Ads to the lessons learned for only one: Ad5.

**Adenovirus hexon proteins.** Other chapters delve into the biology and structure of Ads. Here we will focus on a less explored protein of these viruses: the hexon. Many of the interactions that determine the fate of systemically delivered Ad occur on the surface of its icosahedron.[1] The bulk of the icosahedron is made up of 720 monomers of hexon per virion. Hexon trimerizes to create a tower-like structure when viewed from the side (Figure 1(A)).[5,6] Each hexon monomer has seven serotype-specific hypervariable regions (HVR)[7] located on the surface of the hexon trimer and the icosahedron (Figure 1(B)).[8] This location allows the HVRs of hexon to interact with neutralizing antibodies, receptors, proteins, and cells. There are 5040 HVRs per icosahedron providing a complex repeating three-dimensional array as an excellent

**Adenoviral Vectors for Gene Therapy.**

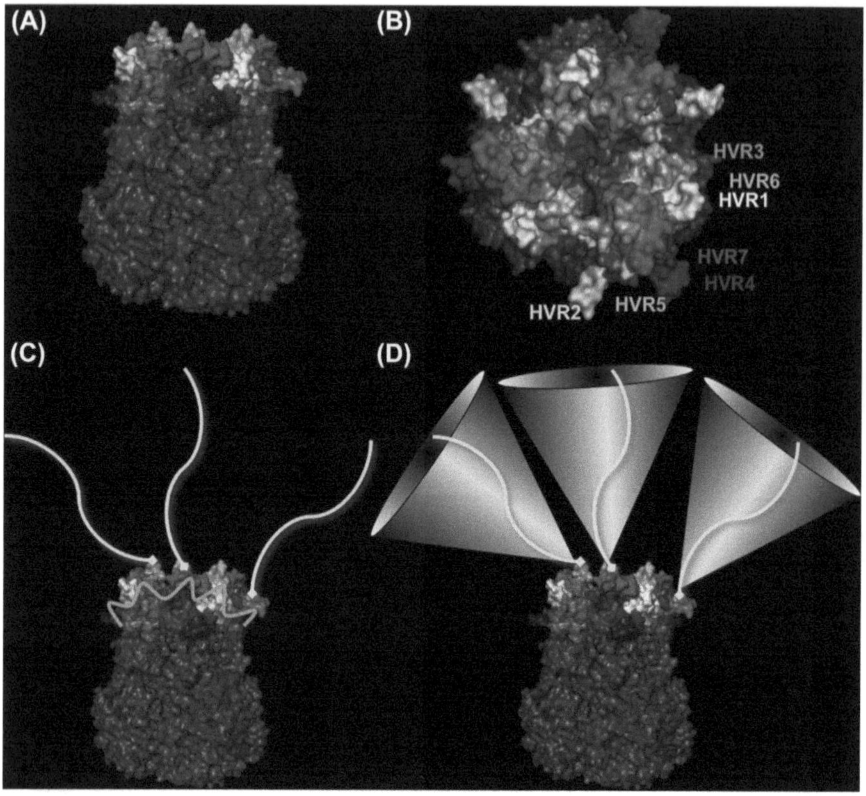


**Figure 1 Adenovirus hexon structure.** (A) Side view of a hexon trimer. (B) View of the exposed surface of a hexon trimer with individual HVRs indicated. (C) Side view of a hexon trimer with three linear PEG molecules shown attached to the hexon surface and on HPMA molecule "stitched" to the surface. (D) Theoretical representation of cones of PEG shielding produced by gyration of these linear polymers.

surface for the deposition of low- or high-affinity host factor proteins, for recognition by innate immune cell pattern recognition, and for phagocytosis. Notably, some of these HVRs are innately large and flexible to the extent that they are "unstructured" by X-ray diffraction or cryo-electron microscopy structural predictions (e.g., Ad5 hexon) (Figure 2).[5,6] For example, in Figure 1 most of HVR1 is actually missing because it is unstructured in the X-ray reconstruction. Therefore, it is unclear how robust efforts to model hexon and proteins[9–11] are if the largest feature on the hexon is unknown. In contrast, the structures of all HVRs can be more accurately assigned and modeled for Ads with smaller and more rigid HVRs such as Ad26 (data not shown).

The specific sequences and size of many of the HVRs vary substantially between Ads across the virome and within Ad subtypes or species (Figure 2). For example, archetype vector human species C Ad5 has a large 29–amino acid HVR1 with a considerable number of charged residues (1 positive and 14 negative for a net of −13). Another species C virus, Ad6, also has a larger, but less negative 32–amino acid HVR1

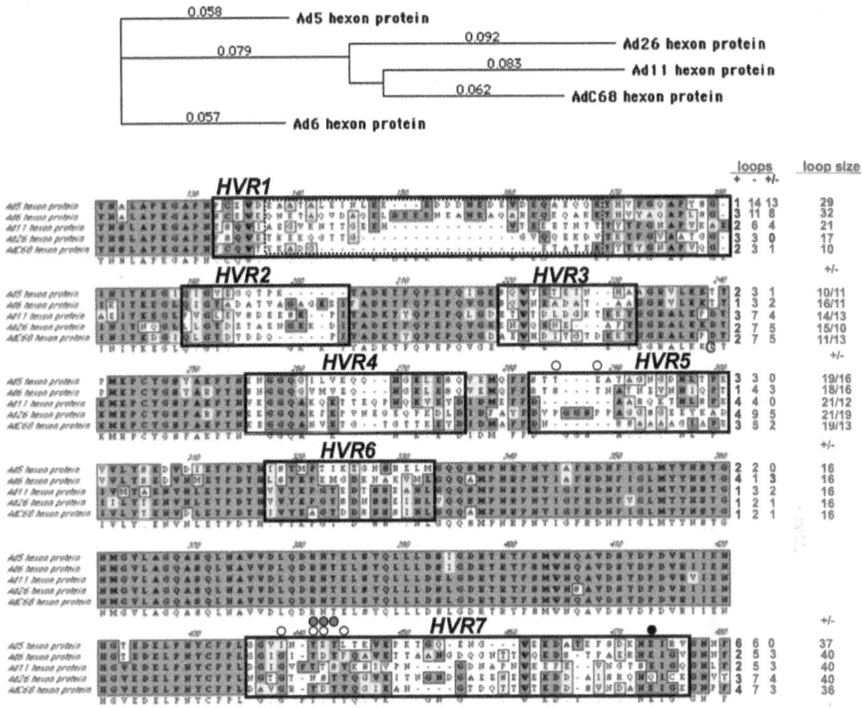


**Figure 2 Sequence alignment of HVRs of hexons of selected Ads.** The hexon regions from the indicated Ads were aligned by ClustalW using MacVector. A phylogenetic dendrogram is shown at top. On the sequence alignment, HVRs are shown boxed. At the right of the alignment, number of positive and negative charged amino acids within the boxed HVRs is shown. A circle above the sequence indicates glutamic acid that is conserved in FX binding Ads.

(net charge of −8). In contrast, human species B Ad11 and human species D Ad26, and chimpanzee AdC68 have smaller HVR1s (21, 17, and 10 amino acids, respectively) with markedly less charge (−4, 0, and −1 net charge, respectively). Similar charge and lesser size variations are observed in the other HVRs (Figure 2). It is therefore not surprising that a display of 5040 highly charged HVRs on the Ad nanoparticle invites low- and high-affinity interactions with proteins and cells.

**Adenovirus interactions with the endothelium.** When a needle is used to inject Ads into the bloodstream, it tears through the vascular wall, damaging the endothelial layer. This activates a natural clotting cascade locally at the site of injection. In the high velocity of the bloodstream, the vast majority of injected virus likely avoids interactions with this damaged area and instead speeds downstream. Within the blood, though, Ads are exposed to and can be sequestered by nearly $6\times10^{13}$ endothelial cells with up to 7 $m^2$ of surface area in humans.[12] This endothelial cell "sink" for Ads scales down to approximately $2\times10^{11}$ endothelial cells with 0.025 $m^2$ of surface area in a 25-g mouse. To put this in context, the average mouse's liver has 10 times *less* hepatocytes than the total number of endothelial cells.[13]

Adenovirus 5 can infect endothelial cells through penton base RGD and αv–integrin interactions.[14–17] Adenovirus 5 can also induce frank breaches in blood vessel walls after high-dose injection.[18] Most Ad serotypes have RGDs in their pentons, and so are likely also to be sequestered by these cells and provoke similar side effects. Adenoviruses can enter endothelial cells by productive infection pathways, but they can also be pinocytosed or phagocytosed by liver sinusoidal endothelial cells (see below).

**First interactions in the blood.** After breaching the endothelium, a bolus or slow drip of virus is injected into the complex environment of the blood. Fifty-four percent of blood is plasma, 45% is red blood cells, and approximately 1% is made up of white blood cells.[19] The good news is that most of plasma is actually water, so 50% of blood (i.e., the water) represents no danger to injected adenoviruses. The bad news is that the remaining soluble and cellular components in the blood can inactivate or profoundly change the virus and its pharmacokinetics (Figure 1).[20–23] Subtract water from the equation and Ads have a 50/50 chance to be inactivated or retargeted by blood cells or by soluble blood factors.

**Blood cells.** Most in vitro tests of Ads with cells confront the viruses with cell concentrations on the order of a million cells per milliliter in an artificial medium. In contrast, the blood confronts these viruses with more than 50 million cells per milliliter. In rank concentration order, these cells include red blood cells, platelets, and white blood cells at 50, 5, and 0.1 million cells per milliliter, respectively.

Adenovirus interactions with red blood cells have been known since soon after their discovery.[24] The fact that human Ad serotypes differentially hemagglutinate rat, monkey, and other red blood cells[24,25] reinforces the concept that the in vivo pharmacology of all Ads cannot be extrapolated from only one Ad, Ad5. For example, human species A Ads[11,17,29] have little agglutination. Species B Ads[3,7,10,13,15,20,32,33] are completely agglutinated by monkey red blood cells. Benchmark species C Ads[1,2,5,6] are partially agglutinated by rat erythrocytes. Species D Ads[6,8,9,12,14,16,18,19,21–28,30,31,34–37,40–45] are completely agglutinated by rat erythrocytes.[26] These differences in hemagglutination are attributed in large part to differences in receptor binding by the different Ad serotypes via interactions with the Coxsackie and Ad receptor (CAR), CD46, and integrins.[2] These are likely mediated by high-affinity interactions with viral fiber or penton base proteins. However, the role of low-affinity hexon interactions or hexon-blood factor–mediated interactions is likely underestimated, because these in vitro assays generally do not include normal plasma proteins in the mix (see below).

Of course, these in vitro binding assays are not only adenovirus species-specific, they are also host species-specific. For example, archetype Ad5 vector does not bind or hemagglutinate mouse red blood cells, it binds strongly to rat and human red blood cells.[14,27] Indeed, when Ad5 is mixed with human blood, 98% is cell associated and not in the plasma fraction.[27] In contrast, most Ad5 is found in the plasma in mouse blood.[27] Adenovirus 5 binds human neutrophils and monocytes in vitro in addition to red blood cells.[27]

**Platelets.** Platelets are small nonnucleated cells in the blood that are present at concentrations of approximately 10% that of red blood cells. As thrombocytes, they have a primary role in blood clotting at the surface of damaged endothelial surfaces.[14] One of the toxicities associated with intravenous Ad5 use is severe thrombocytopenia, which is usually acute and can be triggered even at low doses.[14] Adenovirus 5 binds αIIbβ3 integrins on platelets via its RGD motif in its penton base.[28] This results in Ad and the platelets being targeted to reticuloendothelial cells with the net effect of both the virus and the cells being destroyed.[29] This depletes therapeutic Ad5 for systemic distribution, but it can also provoke dangerous disseminated intravascular coagulation.[30] This is our understanding for Ad5. Because most Ads display an RGD motif in their penton base, it is likely that many serotypes interact with platelets. In vitro tests of Ad–platelet interactions likely underestimate the role of hexons in these interactions, because most in vitro tests are not performed in plasma, but are instead performed in artificial media or buffers that lack factors that bind Ad hexons.

**Plasma.** Plasma makes up 55% of the blood fluid. Plasma is saturated with proteins with average total concentrations of 100 mg/ml. Many plasma proteins such as albumin may actually stabilize the virus as crowding agents. However, there are a number of proteins in plasma that bind to Ad5, most of which interact with hexon. Below, we discuss these known hexon-binding proteins and the need to evade them by hexon engineering approaches.

**Hexon-binding proteins in the blood: neutralizing antibodies.** It is well known that adenovirus neutralizing or binding antibodies can drastically affect the fate of systemically administered Ads. Adenoviruses were discovered and numbered in order of observation in humans, so Ads with lower numbers tend to be more commonly observed in patients. This means that most humans are already immune to many of our favorite Ads, including Ad5.[31] Indeed, one could arguably not have picked a worse virus for use in humans than Ad5 considering that 27–100% of humans are already immune to the virus.[31] Therefore, one goal is to recruit less seroprevalent Ads for use as therapeutics to avoid preexisting immunity.

One might assume that most neutralizing antibodies bind the receptor binding fiber protein. In actuality, most of Ad-neutralizing antibodies actually bind hexon.[32] Indeed, this targeting has likely driven the evolution of HVRs in hexon to avoid these antibodies. As their name suggests, the HVRs are hypervariable and define in large part differences between Ads in the same viral species or between serotypes (Figure 2).

Preexisting antibodies are challenging in patients. However, each Ad therapy is also a "one-off" treatment as far as the immune system is concerned. Injection of naked Ad virions will provoke robust antibody and cellular immune responses against the injected viral capsid proteins and against any Ad proteins that are expressed in infected cells. These include preexisting neutralizing antibodies that can bind and inactivate the virus. These vector-induced antibodies can strongly attenuate a second use of the virus if the same serotype of Ad is used.[33]

One approach to evade neutralizing antibodies is to "serotype switch" the vector. For example, mice administered with Ad2 serotype vectors generate potent neutralizing antibodies against Ad2 that drastically reduce transgene expression if Ad2 is used again.[34] However, if an Ad2 vector is used for the first round of transduction and then an Ad5 vector is used for the second round, there is little reduction in transduction because the Ad2-specific antibodies do not overtly neutralize this different serotype.[34] This approach has been demonstrated as a robust approach for Ad vectors for human immunodeficiency virus vaccines.[35–41] Another approach is to swap HVRs from low-seroprevalence Ads into Ad5.[42,43] Another is to chemically or genetically reengineer the hexons of Ad5 or other serotypes to avoid preexisting neutralizing antibodies in humans.

**Hexon-binding proteins in the blood: natural antibodies.** Neutralizing antibodies are an obvious confounder for systemic delivery of Ads. Less known but more ubiquitous are natural antibodies that exist without any prior Ad exposure.[44] Antibodies are known to opsonize bacterial and parasitic pathogens. However, they can also have a key role in macrophage-dependent clearance of foreign particles including Ads.[45] For Ad5, these appear to be mediated by natural antibodies that are not the product of memory immune responses, but that are instead encoded in the germline. These immunoglobulin M (IgM) antibodies bind to the repetitive structure of the 720 hexons on Ad5.[46–48] This binding either retargets Ad5 to liver Kupffer cells where both the virus and the cells are destroyed or activates complement to further modify the virion surface[46–48] (see below). Notably, this effect is serotype-specific. If the hexon HVRs of Ad5 are replaced with those from another species C virus, Ad6, natural antibodies still bind the virus.[48] However, unlike Ad5, the Ad6 hexon avoids recognition by scavenger receptors and Kupffer cells.[48] Therefore, it is likely that the even more diverse "virome" of Ad serotypes will avoid this effect.

**Hexon-reactive proteins in the blood: complement.** Early work demonstrated that several complement proteins bind to Ad5.[49,50] More recently, these observations have been integrated with the role of natural antibodies and their effects on Ad5 pharmacology.[45–47,51] In the presence of IgM, complement can neutralize Ad5 by reducing the ability of the virus to bind cells.[46] Natural antibodies appear to activate conversion of complement protein C3 to C3b. C3b has a reactive thioester group that covalently reacts with the Ad5 virion surface.[46] Considering that the thioester group reacts nonspecifically with proteins, any Ad serotype that binds IgM natural antibodies will likely run the risk of being neutralized by C3b deposition on its hexon surface.

**Hexon-binding proteins in the blood: vitamin K–dependent blood clotting factors.** The biology of Ad5 interactions with CAR and $\alpha_v\beta_{1,3,5}$ integrins was well established based on in vitro studies.[52,53] Unfortunately, these interactions may have only a secondary role after intravenous injection of Ad5, at least when considering infection in the liver. When CAR-binding motifs in the Ad5 fiber and penton RGD ligands were knocked out in Ad5, this had relatively little effect on liver transduction after i.v. injection in mice.[54]

This effect was resolved in part by a body of work by several groups that determined that Ad5 binds vitamin K–dependent blood coagulation factors after i.v. injection.[55–57] It was demonstrated that clotting factors FVII, FIX, FX, and protein C, but not FXI, bind Ad5.[56–58] Blood factor X (FX) binds with highest affinity (subnanomolar) to Ad5 and this binding occurs on its hexon protein.[56,57,59] In the presence of FX, Ad5 efficiently transduces liver hepatocytes. In its absence, liver transduction is markedly reduced. Thus endogenous host ligands appear to have a more profound effect on viral tropism in vivo than virally encoded CAR and penton ligands that were defined in vitro.

It was initially thought that FX binding acted as a bridge that bound the hexon and retargeted the virus to heparin sulfate proteoglycans on hepatocytes.[55] This clearly occurs in vitro and Shayakhmetov's group[10,11] still proposes that it occurs. Support for this model comes from data showing that whereas FVII and FX both bind Ad5 hexon, only FX is effective at bridging to heparin sulfate proteoglycans.[11] Factor X can bridge because its heparin-binding protease domains are exposed, whereas FVII cannot because it is sequestered by dimerization on the virus surface.

Whereas FX may enhance transduction by this mechanism, FX may influence in vivo Ad tropism by more complex mechanisms. Recent data suggest that FX binding to Ad5 hexon may protect it from natural antibody and complement-mediated destruction at Kupffer cells.[46,47] In this model championed by Andrew Byrnes, FX protects i.v. injected Ad5, allowing it to reach liver hepatocytes in sufficient amounts to mediate detectable gene delivery.[46] In the absence of FX, more Ad5 is intercepted by natural antibodies and complement to target the virus to macrophages for phagocytic destruction. Indeed, it is speculated that FX may actually have a role as an evolutionarily evolved pathogen detection system to active innate immune responses.[10]

Not all Ad serotypes bind vitamin K–dependent blood factors.[56,57,59] Therefore, absorption in the liver and the balance between sequestration by Kupffer cells, liver sinusoidal endothelial cells (LSECs), and hepatocytes for different serotypes will likely be different for the diverse serotypes.

**All dressed up and somewhere to go.** Most Ads are likely rapidly bound by soluble factors and cells soon after injection into the bloodstream. Most interactions with cells likely occur by binding to fiber and penton base proteins. Lower-affinity hexon or FX–hexon bridging interactions may also occur with cells, but this may be less likely in the high shear of the bloodstream under physiologic conditions. Whereas fiber and CAR dominate cellular interactions in the blood, the multivalent patterned array of 720 copies of the hexon appears to be an attractive scaffold for binding by a number of soluble proteins in the blood. Adenovirus 5 and other Ad serotypes are likely decorated with a mosaic of antibodies, FX, and perhaps complement soon after i.v. administration. The degree and balance of factors binding these viruses likely determine their pharmacokinetics and ultimate fate: either reaching the intended therapeutic target or being destroyed by the natural protective process of the body. Adenovirus 5 and other viruses can be sequestered by the lung, kidney, and spleen. However, this chapter focuses on their sequestration by the liver and engineering efforts to avoid this.

**The liver as a pharmacologic "sink" for systemic Ads.** As researchers geared up to hit systemic targets, we fantasized that we could add novel receptor moieties to the fiber or penton base proteins and have the vectors "target" any cell in the body. Indeed, these retargeted vectors worked well in tissue culture dishes but almost always failed in vivo after i.v. injection.

Insight into this problem was gained as a number of groups began studying the pharmacology of these 150-MDa "drugs". In mice, approximately 98% of systemically delivered Ad5 is absorbed by the liver.[60] If one is targeting the liver, innate Ad5 pharmacology highly supports this therapeutic goal. However, if one aims to hit nearly any other cell in the body, the liver represents a dramatic pharmacologic dead end for Ad5 and likely most Ads (Figure 3). We focus below on the liver as a dead end for Ads, but stipulate that Ads are also absorbed by other tissues.

It was observed that injecting increasing amounts of Ad5 did not translate into linear increases in liver transduction.[61] When doses were increased, a certain threshold could

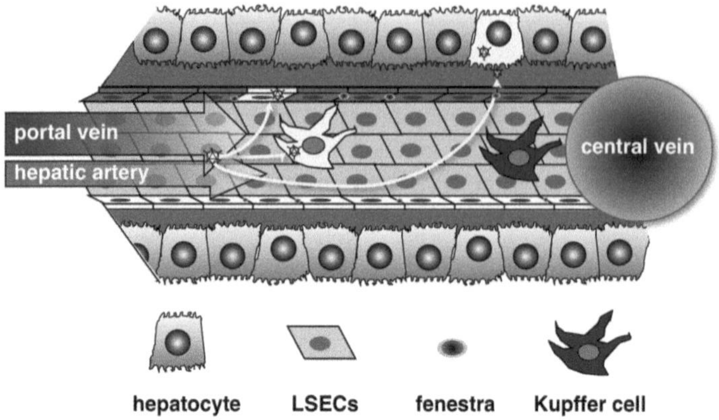


**Figure 3 Diagram of interactions of Ad with cells in the liver.**

be reached in which transduction became linear with Ad5 dose. This so-called "threshold effect" is a saturable phenomenon and varies in different mouse strains.[47,48,62]

**Kupffer cells.** Any cell in the liver could formally be involved in absorbing Ads from the blood. Approximately 65% of the liver is made up of hepatocytes, 25% endothelial cells, and 7% Kupffer cells; stellate and other cells make up the rest (Figure 3).[12] Whereas all of these cells can act as pharmacologic "sinks" for Ads, liver Kupffer cells were one of the most likely candidates for Ad5 absorption.[60] Kupffer cells are the resident macrophage in the liver.[63] These phagocytic cells line the lumen of the sinusoids that carry blood into the parenchyma of the liver, where they serve as sentinels to remove particulates from the blood stream including viral particles (Figure 3). Kupffer cell can phagocytose particles up to 2 µm in diameter,[64] so about 100-nm Ad particles are well within their capacity.

A role for liver Kupffer cells has been shown in several studies. Liver transduction at doses below the threshold can be amplified up to 100-fold by "predosing", in which an irrelevant Ad or another particulate reagent such as $GdCl_2$, clodronate liposomes, or polyinosinic acid is injected hours before therapeutic virus injection.[22,65–67] This predosing destroys or inactivates liver Kupffer cells, allowing transduction to occur in a linear dose-dependent fashion. Adenosine 5 or other serotypes do not appear to transduce Kupffer cells efficiently compared with hepatocytes, although low-level gene delivery can be observed. Instead, Ad5 and Kupffer cells engage in the biological equivalent of mutually assured destruction, because bulk Ad entry and viral uncoating kills Kupffer cells within 10 min of Ad5 uptake.[68] These aggregates of partially uncoated Ad and dying Kupffer cells fragment and detach from the liver vasculature and these fragments lodge in the lung, where they may exacerbate systemic side effects.[68,69]

**Liver sinusoidal endothelial cells.** Studies in wild-type and scavenger receptor knockout mice demonstrate the role of Kupffer cells in Ad5 destruction; however, studies also suggest that there are redundant mechanisms for Ad removal from the systemic circulation.[16] A second layer of Ad sequestration occurs in the larger number of LSECs (Figure 3). Liver sinusoidal endothelial cells can pinocytose particles less than 0.23 μm in diameter, allowing uptake of Ad virions.[63] Because pinocytosis is receptor independent, LSECs can likely sequester many Ad serotypes regardless of their receptor specificity. Liver sinusoidal endothelial cells also express scavenger receptors, allowing a second mechanism of potentially destructive sequestration. In addition, Ad5 and many Ad serotypes also appear able to infect or transduce endothelial cells by normal infection routes that may not be overtly destructive to the virus, but may damage the endothelium.[18] From a mass-action standpoint, there are three to four times as many LSECs as Kupffer cells, so even less efficient uptake by these cells may significantly alter the fate of systemically administered Ads.

**Hepatocytes.** Once Ads evade antibodies, complement, blood cells, Kupffer cells, and LSECs, they can then be absorbed by hepatocytes if the virus can extravasate through fenestrae in the endothelial wall of the liver sinusoids (Figure 3). There are approximately 10 times as many hepatocytes as Kupffer cells, so this cell "sink" in the liver for Ads is substantial. After being protected by FX, the virally encoded receptor binding ligands on fiber and CAR (or other Ad receptors) can now come into play, likely by mechanisms that were originally defined in vitro. If FX protection is not effective or if natural antibodies and complement are not evaded, the receptor binding lessons learned in tissue culture are largely irrelevant.

This balance between Kupffer cell destruction and hepatocyte infection is best reflected in species C Ads including Ad1, 2, 5, and 6. Although many assume that these highly related viruses behave identically, in reality their ability to transduce hepatocytes varies over two orders of magnitude, with prototype Ad2 being the worst and less studied Ad6 being the best.[70] Archetype Ad5 is good at liver transduction but Ad6 is actually three times better. All four of these viruses bind FX, so this is not the source of variation. Instead, Ad6 appears better primarily owing to its innate ability to avoid Kupffer cells.[43] This effect is seated solely in the Ad6 hexon as replacement of Ad5's HVRs with those from Ad6 transfers Kupffer cell evasion and increased liver transduction.[43] Therefore, CAR, integrin, and other receptor binding functions that are encoded by the virus appear as a result of blood factor binding effects, at least for species C Ads. Binding of natural antibodies and complement to Ads increases absorption of these viruses by Kupffer cells and perhaps LSECs. Conversely, FX binding to Ads can protect them from natural antibodies and complement.

More than half of Ad serotypes do not bind FX, so some could experience more Kupffer cell sequestration than FX binding viruses. This suggests other Ad serotypes may play by a different set of rules than species C human Ads. For example, wild-type replication-competent adenoviruses including species B Ad11 and 35 and species D Ad26 and 48 do not bind FX, they cause markedly lower liver damage than Ad5, and they do not predose Kupffer cells[71] (data not shown). Consistent with this, replacement of the HVRs of Ad5 in a replication-defective vector with those from Ad48 produces a virus with reduced hepatocyte transduction and lower uptake in Kupffer cells (KCs).[72] However, this Ad5/48 mosaic vector triggers strong proinflammation in the liver, which provokes rapid liver damage, even at relatively low doses of vector. In contrast, replication-defective Ad48 did not elevate transaminases or induce inflammation.[72] These data suggest that different serotypes interact with liver cells in varied ways. Therefore, extrapolating Ad5 lessons to all Ads is a mistake. These data also suggest that combining features from different serotypes may improve systemic functions (i.e., Ad5 with Ad6 hexon) or increase side effects (i.e., Ad5 with Ad48 hexon).

**Strategies to detarget the liver for systemic therapy.** Three cells in the liver appear to be the primary pharmacologic sinks for Ads after intravenous injection: Kupffer cells, LSECs, and hepatocytes. Below, we discuss select engineering efforts to reduce uptake into these cells.

**Pharmacological engineering.** Some of the earliest and sometimes most effective approaches involve not reengineering Ads, but reengineering the host with drugs or agents.

**Pharmacologic elimination of cells.** One approach is to eliminate problematic cells. This approach is feasible for Kupffer cells, because they do not serve a fundamental functional role in the liver. Eliminating LSECs or hepatocytes is, of course, not feasible, because removing either would have lethal consequences. Kupffer cells can be removed most simply by predosing with an Ad that kills these cells 4h before injecting a therapeutic dose of another Ad.[22,65] Nonadenovirus predosing agents include $GdCl_2$ and clodronate liposomes.[65,73,74] Whereas predosing strategies can improve systemic therapy, destruction of these proinflammatory phagocytic cells can also increase potentially dangerous inflammatory responses. These can also provoke strong side effects after Kupffer cell fragments and their predosing agents lodge in the lung and elsewhere.[68,69,75]

**Pharmacologic inhibition of recognition by cells.** Phagocytosis by Kupffer cells and/or LSECs can be blocked by polyinosinic acid[66,67] or with antibodies that block scavenger receptors.[76] Integrin-mediated Ad uptake into LSECs or hepatocytes can be blocked by RGD peptides or RGD-containing proteins.[77] Coxsackie and Ad receptor or CD46-mediated uptake into hepatocytes can be performed by blocking these with free knob proteins or blocking antibodies.[78] This approach would obviously be problematic if the therapeutic Ad still relies using these interactions to infect target cells.

**Pharmacologic elimination of bridging proteins.** Natural antibodies enhance phagocytosis of Ads by targeting them to Kupffer cells and perhaps other cells. Eliminating these in animals is feasible primarily by use of immunodeficient mice such as SCID or Rag mice that lack all immunoglobulins or that lack IgM.[45–48,51] It may be feasible to reduce levels of natural antibodies in humans by strategies used to treat IgM monoclonal gammopathy of undetermined significance that include steroids and plasmapheresis. Alternately, antibodies such as rituximab that target CD20 on B cells may deplete these B cells.[79] Complement C3b covalently reacts with the Ad5 virion surface.[46] Whereas C3 can be blocked with agents such as cobra venom factor,[46] the cost–benefit of such an intervention would have to be exceptional. The development of more specific complement inhibitors may have better utility.

By far, the most success in this category has been to eliminate FX and vitamin K–dependent clotting factors to reduce hepatocyte infection in animals.[22,58] Although this can be effective in mice, the doses of warfarin used are orders of magnitude higher than are used in humans to reduce clotting after heart attack. Considering this and the fact that warfarin is also the active ingredient in mouse poisons, warfarin will likely not be a first choice to reduce hepatocyte infection by FX-binding Ads. In addition, warfarin was originally thought to act by blocking FX retargeting of Ad to heparin sulfate proteoglycans on hepatocytes. Considering that FX may instead simply protect Ads from natural antibody-mediated destruction in Kupffer cells,[46] warfarin may simply increase Kupffer cell destruction and death and may increase side effects.

**Combining Kupffer cell and hepatocyte pharmacologic detargeting.** Based on these results, we tested whether detargeting Kupffer cells or hepatocytes was more important during systemic oncolytic therapy in a real therapeutic model.[22] When predosing or warfarin was used alone for i.v. injection of oncolytic Ad5, each had the desired individual pharmacologic effect, but neither alone augmented the ability to kill distant tumors. Instead, if the two were combined, this enhanced oncolytic efficacy after a single round of i.v. therapy in mice bearing human tumors.[22] Avoiding just one cell in the liver was not sufficient for systemic therapy. Instead, one needed to avoid both Kupffer cells and hepatocytes. Unfortunately, we did not have the tools to avoid LSECs, but subsequent work by Shayakhmetov's group showed that these, too, represent a significant pharmacologic dead end for Ads.[16]

**Chemical engineering of Ads.** There is a good body of data for the use of chemical modification of Ad5 to reduce binding of neutralizing antibodies, blood factors, and complement to the virion. In most cases, these approaches have been tested with archetype Ad5,[33,60,80–95] but other serotypes such as Ad6 have been tested.[96]

**Semirandom polymer modification of Ads.** In most examples, either polyethylene glycol (PEG) or *N*-(2-hydroxypropyl)methacrylamide (HPMA) has been covalently attached to the surface of Ad5.[33,60,80–95] In most cases, these polymers react with primary amines. For Ads, this usually translates into nearly random alkylation of the virus on exposed lysine residues. Polyethylene glycol is generally a linear polymer with one reactive group on one end and so forms a "hairy" surface on hexon (gray strands projecting from hexon) (Figure 1(C)). *N*-(2-Hydroxypropyl)methacrylamide has multiple reactive groups along its polymer length allowing it to conjugate at several sites on hexon to form polymer "stitches" (wavy strand on hexon) (Figure 1(C)).

When 5 kDa PEG is used, this translates into covalent attachment of approximately 15,000 PEGs to the virion surface.[91] This profoundly reduces in vitro transduction by Ad5, but surprisingly does not reduce liver transduction in mice.[91] Use of larger PEGs can block liver transduction,[97,98] likely by preventing extravasation through liver fenestrae (Figure 3). However, this can be highly variable in different mouse strains and with different sources of PEG.

Any lysine on any exposed capsomer can be attacked by PEG under these conditions. Penton base and its RGD motif are largely unaffected by PEGylation.[91] Lysines on the extended Ad5 fiber are directly PEGylated and these conjugation events ablate CAR binding.[89,91] This is consistent with observations that PEGylation markedly reduces Ad5 in vitro transduction.

Considering that there are 36 fibers, 60 penton bases, and 720 hexons per virion, the vast majority of covalent modification does not occur on fiber or penton, but instead targets the surface of hexon. Coating Ad with HPMA prevents binding of blood FX to Ad hexon and binding to erythrocytes.[99] In contrast, 5 kDa PEGylation does not ablate FX binding to Ad5.[14] Although one would expect PEGylation to block this, this failure is ironically likely related to the ability of PEG to "shield" the virion surface. Covalent conjugation of 5 kDa PEG to hexon results in the attachment of a string-like, 35-nm linear polymer to the icosahedron surface (green strands, Figure 1(C)). To put this in context, each 5-kDa PEG is about the same length as Ad5 fiber itself. Once PEG is attached, it "shields" this site from interactions with proteins as desired. However, this PEG will also shield the hexon from further reactions with other PEG molecules (gray cones of shielding, Figure 1(D)). Therefore, as each PEG is covalently attached, it likely creates a "shadow" in which other PEGs cannot react and protect adjacent amino acids or surfaces. If so, long 35-nm PEG molecules may not be able to reach the hexon surface, but smaller, more globular, 4×9 nm FX may still be able to bind. Therefore, linear PEG may still allow FX binding, whereas more complex HPMA may not, for better or for worse.

In contrast, IgM natural antibodies weigh in at 970 kDa, or more than 15 times the mass of FX. Immunoglobulin Ms are also 31 nm,[100] which makes them nearly as wide as Ad5 fiber is long. One might predict that only one IgM might be able to bind one facet of Ad. If so, even spare PEGylation of the hexon surface might block natural antibodies but not smaller FX. This is consistent with the fact that random PEGylation of Ad5 reduces its uptake by Kupffer cells, but retains liver transduction. This is also consistent with observations that 5 kDa PEG blocks C3a complement activation in vitro and in vivo.[51]

Integrating these data on detargeting the liver, random PEGylation inhibits binding by scavenger receptors, reduces uptake of Ad5 by phagocytic cells, and detargets Ad5 from liver Kupffer cells.[91,97,101] Random PEGylation also blocks uptake of Ad5 by SREC, the endothelial cell scavenger receptor,[101] and reduces Ad5 binding and activation of endothelial cells in vitro and in vivo.[14] Finally, random PEGylation generally preserves hepatocyte transduction unless very large PEG conjugates are used.[91,97,98,101]

*N*-(2-Hydroxypropyl)methacrylamide likely blocks uptake of Ad by all of these liver cells.[92,99] However, HPMA also appears to block CAR binding like PEG.[95] This combined with blocking FX binding produces a virus with little ability to infect any cell, liver or otherwise. This is a disadvantage for applications in which you would like the vector to use native Ad transduction. Conversely, this can be used as an advantage to detarget Ad5 and retarget it with peptide ligands with HPMA[95] or with PEG.[84,88,102,103]

All of these data make chemical engineering of Ads sound promising, but nothing is perfect. It is true that randomly PEGylated Ads are still able to transduce cells for vaccine, gene therapy, and oncolytic applications using native Ad5 receptor binding or FX-mediated transduction.[14,91,97,98,104,105] However, loss of CAR binding after PEGylation blunts the level of gene-based vaccination with Ad5.[105] When PEGylated Ad5 was tested as a systemic oncolytic virus against lymph node carcinoma of prostate (LNCaP) human prostate tumors in immunodeficient mice, the PEGylated virus had equal efficacy as unmodified Ad5.[98] However, LNCaP tumors are relatively easy to cure with Ad5. In contrast, in our more recent studies against harder-to-cure DU145 human prostate tumors in mice, PEGylation markedly reduced Ad5 efficacy after single i.v. injection.[106] PEGylated Ads were also less effective after single i.v. injection against HAK tumors in immunocompetent hamsters. These data suggest that loss of receptor binding and/or FX bridging by random polymer modification may reduce vector efficacy in more stringent therapeutic systems or in humans.

**Retargeting Ad by genetic engineering of hexon and other capsomers.** David Curiel's group led many early explorations of displaying cell targeting ligands on different Ad5 capsomer proteins including fiber, IX, hexon, and IIIa.[107–110] For hexon, to our knowledge, the earliest retargeting effort involved insertion of the archetype ligand RGD into Ad5 HVR5.[111] This was formally not a "re" targeting approach, but really just improved display of the ligand, because Ad5 already has RGD on its penton base. In this work, the authors demonstrated that RGD display on hexon increased Ad5 transduction on certain cells. After this, Curiel's group inserted a His6 tag into HVR2, HVR3, HVR5, HVR6, and HVR7 of Ad5.[109] This work showed that His6 could be tolerated in these HVRs by the virus, but that only HVR2 and 5 were displayed sufficiently to bind anti-His6 antibody. Unfortunately, when these His6-displaying viruses were tested for retargeting to an artificial anti-His6 receptor on cells, none of the HVRs displaying vectors worked.[109] This called into question whether hexon was a viable display scaffold for high-affinity cell targeting ligands. Although RGD worked, it is small and relatively low-affinity, and perhaps a special case.

To explore this question, our group used metabolic biotinylation[112–114] to determine how well different Ad capsomers could display ligands and how different ligands behave on each capsomers. Metabolic biotinylation takes advantage of endogenous mammalian or co-expressed bacterial biotin ligase enzymes to covalently biotinylated biotin acceptor peptides (BAPs).[112–114] Biotin is an appealing tag because avidin can be used to bridge two biotinylated moieties with the strongest affinity known in nature: $10^{-15}$ M or one-quarter the strength of a carbon–carbon covalent bond. To test this for Ad targeting, we inserted BAPs into fiber, IX, and the HVR5 of hexon of replication-defective Ad5.[115–118] As these viruses are produced, the endogenous biotin ligase in 293 cells covalently biotinylates the tagged capsomer. After purification, we showed that the BAPs on fiber, IX, and hexon are all accessible on the surface of the virion and could be bound by avidin.[115–118]

Adenovirus 5–fiber–BAP was efficiently retargeted by avidin bridging with a wide array of biotinylated ligands including oligonucleotides, carbohydrates, proteins, toxins, and a variety of monoclonal antibodies.[115–119] Adenovirus–fiber–BAP mediated 40- to 60-fold increases in transduction in vitro. In contrast, the same ligands on Ad5–IX–BAP increased transduction twofold at best. On Ad5–hexon–BAP these same ligands increased transduction at most sixfold.[117,118]

We expected that displaying 720 or 240 ligands on hexon and IX would be better than displaying only 36 on fiber, so these results were surprising. It is possible that engaging too many receptors at the cell surface via 240 or 720 capsomers may inhibit the ability of the virus to be internalized or may lead it down an unproductive entry path. Alternately, this failure may result from the use of high-affinity rather than low-affinity ligands. With these high-affinity ligands, IX and hexon–BAP vectors appear to become trapped on the targeted receptors in endosomes or on the cells surface and be unable to proceed to the nucleus.[117] We speculate that this failure reflects the biology of the capsomers. Fiber is shed from virions in the endosome, so the virion releases any ligand bound to any targeted receptor. In contrast, IX and hexon remain associated until reaching the nuclear membrane. Therefore, a high-affinity ligand linked to IX or hexon has no way to release its receptor or the virion, and so may become trapped on the receptor in endosomes. These endosomes may recycle to the cell surface along with the virus.

Inserting RGD into hexon may avoid some of these problems because this is a small ligand whose affinity may well be low when displayed in HVR5. Therefore, it can function by avidity, but can also be more readily released from receptors in the endosome during acidification after entry. Therefore, we believe that IX and hexon can be useful retargeting platforms, but that they may work only with low-affinity ligands or ligands that can release receptors. A case in point can be seen in the use of transferrin as a ligand. Transferrin is unique in that it naturally releases its receptor after acidification in endosomes. For Ad5–IX–BAP, we demonstrated that most ligands failed, but that the only ligand that worked well was transferrin.[117] It is likely that transferrin has become one of the most popular ligands for genetic-chemical targeting with IX and hexon-modified vectors because of this novel ligand-receptor biology.[90,120] Therefore, we predict that low-affinity ligands and ligands that can release receptors may be most effective for hexon retargeting efforts. More recent proof of principle for this has been provided by insertion of a transforming growth factor-β receptor targeting peptide in the Ad5 hexon for pancreatic cancer.[121]

**Retargeting Ad by engineering hexon binding partners.** A number of groups have applied bridging strategies using a fiber binding protein fused to cell targeting ligands (reviewed in Ref. 1,122,123). Similar efforts for hexon have lagged because of the lack of binding partners. Our group made use of the subnanomolar affinity of FX for hexon to retarget oncolytic Ad5.[124] We fused the γ-glutamic acid (GLA) domain of FX that binds Ad5 hexon to a series of single-chain (ScFv) monoclonal antibodies that target different cellular receptors. We showed in vitro that GLA-anti-Her-2 or anti-EGFR could bind to hexon and functionally retarget Ad5 to increase infection of breast and ovarian cancer cells.[124] We also showed that GLA-anti-ABCG2 could target this stem cell receptor. Finally, we showed in vivo in mice that GLA-anti-EGFR ScFv could be expressed from oncolytic Ad5 to increase infection of peritoneal and subcutaneous ovarian tumors by Ad5, and that this increased oncolytic efficacy.[124]

While retargeting with GLA worked, we had hoped that binding GLA to hexon would also detarget hepatocytes. Unfortunately it did not. At the time, the field believed that FX was acting as a bridging molecule to retarget Ad5 to heparin sulfate proteoglycans on hepatocytes. It was thought that GLA bound Ad5 and the protease domain of FX bound hepatocytes. We therefore expected that blocking this interaction with GLA-ScFv would block bridging and hepatocyte infection. In the current model in which FX is not bridging, but is instead protecting Ad5 from IgM and complement, it is possible that our GLA-ScFv may also protect the virion while allowing efficient hepatocyte infection or that FX may simply outcompete GLA-ScFv on the surface. Regardless, our bridging strategy failed to detarget Ad5 from the liver.

**Detargeting Ad by genetic engineering.** Some of the off-target infection by Ads is mediated by interactions of virally encoded ligands. One approach to avoiding these interactions is to genetically delete these ligands from the virus to detarget these interactions. Examples include mutating fiber to ablate CAR or CD46 binding[54,125,126] and mutating penton base to block integrin binding.[54,127,128] Early work knocking out CAR and integrin binding had modest effects on Ad5 tropism in the liver. Subsequent work by Shayakhmetov's group showed that one must detarget all three cells in the liver: Kupffer cells, LSECs, and hepatocytes.[16] This work combined pharmacologic detargeting with the use of an RGD-deleted Ad. This RGD deletion was critical to detargeting LSECs. Although this may detarget LSEC, most retargeted Ads do not function without using the integrin entry pathway, so RGD deletion may not be viable for retargeting unless you can provide an alternate entry receptor. Put another way, the ultimate detargeted vector may simply be dead.

**Detargeting hexon by peptide insertion into hexon.** These efforts target relatively high-affinity ligands on fiber and penton. It is generally thought that hexon is not involved in direct cell receptor binding, but this is likely inaccurate because even low-affinity interactions can become important when multiplied by 720 copies of a protein as a binding surface. Consider again that Ads can have a considerable concentration of charged amino acids concentrated on their surface-exposed HVRs (Figure 2). It is therefore likely that 720 low-affinity interactions on hexon may compete with 36 high-affinity interactions on fiber. This suggests that hexon interactions likely have a bigger role than may be currently appreciated and that these will vary among Ad serotypes.

The first demonstration of genetic engineering to detarget Ad5 from hepatocytes was actually a byproduct of the first efforts to retarget Ad by inserting RGD into hexon.[111] When this RGD vector was tested by i.v. injection, it mediated markedly less Ad liver transduction by reducing interactions with FX.[59] This work also showed that substituting other GA repeat peptides or the HVRs from Ad2, 19, and 30 also reduced liver transduction. These were great observations helping the field consider trying to detarget hepatocytes by genetic engineering.

After we heard of liver detargeting by HVR5-RGD, we wondered whether our hexon–BAP virus might also be detargeted. Whereas the RGD virus had an 11–amino acid insertion into HVR5, the BAP insertion is relatively huge, weighing in at 8 kDa.[116,118] Shayakhmetov's group showed that the BAP in HVR5 reduced FX binding 10,000-fold and strongly detargeted liver hepatocytes.[56] To test this in a therapeutic application, we inserted the BAP into HVR5 of oncolytic Ad5.[129] We found that liver infection by this replication-competent Ad5–hexon–BAP was four orders of magnitude lower than Ad5. This detargeting decreased liver damage and allowed 10-fold higher doses of the hexon–BAP oncolytic to be injected i.v. than Ad5. Adenovirus–hexon–BAP had equal efficacy against Hep3B hepatocellular carcinoma tumor–bearing nude mice after i.v. injection, but had markedly lower toxicity.[129] These data provided first proof of principle for the use of genetic detargeting of hexon in a real therapeutic model.

**Swapping and mutating HVRs in hexons to block FX binding.** In parallel to the hexon–BAP work, Andrew Baker's group genetically modified Ad5 hexon replacing its HVR5 or HVR7 with HVRs from non-FX binding Ad26 or by mutating selected amino acids.[9] Replacement or point mutations in Ad5 HVR5 reduced FX binding partially. Hypervariable region 7 replacement or point mutations reduced FX binding and FX-mediated transduction, but the single-point mutant in HVR7 E451Q did not entirely abolish FX binding. Compound mutations were needed.

Andrew Baker's group next replaced the Ad5 hexon HVRs with those from Ad48.[72] Adenovirus 48 does not bind FX, so this modification was expected to detarget the liver. When this virus was injected i.v., it did indeed have low liver transduction. This was accompanied by increased uptake into Kupffer cells, increased inflammation, and increased IL-6 responses[72] like Ad–hexon–BAP. This lends further support to a model in which FX binding to Ad5 hexon reduces its destruction in Kupffer cells. They followed this with a reciprocal hexon swap by introducing Ad5 HVRs into non–FX-binding Ad26.[130] They showed that incorporation of HVR 1, 2, 3, 5, 6, and 7 from Ad5 into Ad26 made this virus an FX binder and sensitized it to natural antibodies. Addition of Ad5 HVRs to Ad26 increased hepatocyte transduction, which was abrogated by FX depletion.

Much of the thinking was that FX was redirecting Ad5 to hepatocytes by retargeting it to heparin sulfate proteoglycans. In Andrew Byrnes' model, FX protects Ad5 from IgM and complement.[46] One can interpret the results with RGD and BAP viruses because these insertions in hexon may simply block FX binding and Ad5 is now unprotected from IgM and complement. Consistent with this, Ad5–hexon–BAP drove 10-fold increases in IL-6 responses compared with Ad5.[129]

Later work by Shayakhmetov's group tested mutating residues 423–425 (TET) in Ad5 hexon. They showed that the mutant T425A entirely blocked FX binding and markedly reduce hepatocyte transduction.[10] They also showed that this point mutant generated markedly reduced systemic and splenic inflammatory responses. They argue that FX should be thought of as an innate immune response sensor and suggested a more complex and divergent mechanism for how Ad5 interacts with Kupffer cells, macrophages, and FX.

One question is, if RGD and BAP block FX binding to hexon, why do they not also block natural antibodies? It could be that FX and natural antibodies bind different specific sites on hexon. We hypothesize that FX and natural antibody IgM actually do not bind in specific sites, but instead engage in less specific charge–charge interactions. This seems possible, because the GLA domain of FX is simply a surface that displays 6 to 12 $Ca^{2+}$ molecules, which it uses to bind negatively charged head groups on membranes, not specific receptors.[131] Likewise, natural antibody IgMs do not bind specific sites, but instead bind repeating structures such as nucleic acids, phospholipids, carbohydrates, and viral capsids.[132,133]

Although there have been efforts to model these interactions,[9–11] it is unclear how accurate any molecular modeling can be when the largest and most charged HVR (HVR1) is unstructured and excluded from any of these calculations. In contrast, non-FX binding Ads such as Ad26 and 48 have hexons with small HVR1s. This suggests to us that much of the action for FX and natural antibody binding may involve HVR1 domains perhaps in concert with HVR5 or 7. Alternately, the charge on HVR1 may act as a landing platform for the Gla domain of FX and then allow it to shift into a higher-affinity lock-and-key interaction. However, it is undeniable that small-point mutations can block FX binding.[9–11] Whether they do this by a direct lock-and-key mechanism as has been suggested remains to be determined. Large insertions in the right place(s) or whole HVR swaps may do the trick.

**Swapping and mutating HVRs in hexons that retain FX binding but detarget Kupffer cells and macrophages.** In most of these engineering cases, FX binding and Kupffer cell destruction are reciprocal. Factor X protects Ad5 from IgM and Kupffer destruction. Remove FX binding by HVR swap or RGD or BAP insertion and the virus becomes destroyed in Kupffer cells. There is one exception to this effect that we observed with species C Ad6. As mentioned above, Ad6 is more efficient than Ad5 for liver transduction and as a systemic oncolytic in mice and hamsters.[43,48,70,71,134] This improved efficacy appears to be related to the ability of Ad6 hexon to evade scavenger receptors and IgM-mediated destruction in Kupffer cells.[43,48] This change in pharmacology is seated in the Ad6 HVRs, because replacement of Ad5's HVRs with Ad6's transfers this phenotype to Ad5.[43]

In contrast to HVR swaps between Ad5 and Ad26 and 48, where FX binding was inversely related to destruction in Kupffer cells, these two effects are detached in Ad5 and Ad6 hexon chimeras. Adenovirus 6 still binds FX, but actually with 10-fold lower affinity.[43] It still binds IgM, yet it evades Kupffer cell destruction. This effect is most notable in BALB/c mice that have highest levels of IgMs and is hard to observe in other mouse strains with less immunoglobulin.[48] Rather than reducing liver transduction, the Ad6 hexon actually increased gene delivery by Ad5. In this case, Ad6 hexon still binds FX and IgM but evades recognition by scavenger receptors and Kupffer cells by an as yet undetermined mechanism. This suggests that FX biology and Kupffer cell targeting are separable phenotypes and that one may be able to detarget these events separately.

**Targeted genetic-chemical modification of Ads.** In the chemical engineering section above, PEG and HPMA polymers were reacted randomly with primary amines on the Ad surface. This can result in knockout of useful functions (such as CAR binding). This can also result in overmodification (the ultimate detargeted vector is actually totally inactive and worthless). An alternate approach to random modification is targeted polymer modification by "genetic and chemical" engineering.[90] This approach makes use of the fact that Ad5 has few exposed cysteines on its surface and one can purchase maleimide-activated PEG molecules that react relatively specifically with these cysteines. Cysteines are genetically introduced into an exposed site on an Ad capsomer and this site can be conjugated with any maleimide-reactive moiety.[90,101,135]

For Ad5, this approach was first applied to insert a cysteine in its fiber[90] and later in the HVR5 of its hexon protein.[136] When HVR5 was conjugated to maleimide–5-kDa PEG, targeted PEGylation did not reduce in vitro virus transduction in contrast to random PEGylation. This HVR-targeted PEG also this blocked FX binding and decreased transduction of hepatocytes. We followed this work by introducing cysteines separately into each of the seven HVRs of Ad5 hexon to mutate them conditionally and evaluate their role in Ad biology.[101] We showed that targeted PEGylation of any HVR did not did not reduce CAR-mediated transduction in vitro, in marked contrast to the negative effects of random PEGylation. Targeted PEGylation of all tested HVRs reduced recognition by scavenger receptor SRA-II that is expressed on Kupffer cells. In contrast, only PEGylation of HVRs 1, 3, and 7 reduced recognition by the endothelial cell scavenger receptor SREC. Modification of HVR2 and 6 did not.[101] When tested in vivo after i.v. injection in mice, targeted PEGylation of HVRs 1, 2, 5, and 7 increased liver transduction up to 20-fold after i.v. injection.[101] PEGylation of HVR3 and 4 had no increases. These data suggest that HVR1, 2, 5, and 7 of Ad5 may be involved in Kupffer cell and perhaps LSEC recognition and subsequent destruction. This suggests again that FX and IgM binding are separable events.

These data demonstrate that this conditional genetic-chemical mutation strategy is a useful tool to investigate the interactions of Ads with host tissues. Although our work suggests that conditional mutation of HVRs 1, 2, 5, and 7 with and without PEG still allows FX to bind but detarget Kupffer cells, we have to remember that large 5-kDa PEG molecules do not saturate every surface. Therefore, the large PEGs may be a blunt tool to probe these interactions. Indeed, when one labels Ad5 with smaller chemical moieties such as the fluorophore IR800,[137] one can drastically reduce liver gene delivery without affecting liver sequestration.

This also suggests that targeted PEGylation that spares need Ad functions may be superior to random PEGylation for clinical translation. The only negatives of this approach will be the need to obtain Food and Drug Administration or other regulatory approvals for not just one article (Ad), but for two (Ad and PEG) and ensuring that a reproducible product can be routinely generated.

## References

1. Barry MA, Hofherr SE, Chen CY, Senac JS, Hillestad ML, Shashkova EV. Systemic delivery of therapeutic viruses. *Curr Opin Mol Ther* 2009;**11**(4):411–20.
2. Bailey A, Mautner V. Phylogenetic relationships among adenovirus serotypes. *Virology* 1994;**205**(2):438–52.
3. Barry MA, Weaver EA, Chen CY. Mining the adenovirus "Virome" for systemic oncolytics. *Curr Pharm Biotechnol* 2011.
4. Weaver EA, Palmer D, Ng P, Barry MA. Characterization of species C human adenovirus serotype 6 (Ad6). *Virology*.
5. Liu H, Jin L, Koh SB, Atanasov I, Schein S, Wu L, et al. Atomic structure of human adenovirus by cryo-EM reveals interactions among protein networks. *Science (New York, NY)* 2010;**329**(5995):1038–43.
6. Reddy VS, Natchiar SK, Stewart PL, Nemerow GR. Crystal structure of human adenovirus at 3.5 A resolution. *Science (New York, NY)* 2010;**329**(5995):1071–5.
7. Mizuta K, Matsuzaki Y, Hongo S, Ohmi A, Okamoto M, Nishimura H, et al. Stability of the seven hexon hypervariable region sequences of adenovirus types 1-6 isolated in Yamagata, Japan between 1988 and 2007. *Virus Res* 2009;**140**(1–2):32–9.
8. Rux JJ, Burnett RM. Type-specific epitope locations revealed by X-ray crystallographic study of adenovirus type 5 hexon. *Mol Ther* 2000;**1**(1):18–30.
9. Alba R, Bradshaw AC, Parker AL, Bhella D, Waddington SN, Nicklin SA, et al. Identification of coagulation factor (F)X binding sites on the adenovirus serotype 5 hexon: effect of mutagenesis on FX interactions and gene transfer. *Blood* 2009.
10. Doronin K, Flatt JW, Di Paolo NC, Khare R, Kalyuzhniy O, Acchione M, et al. Coagulation factor X activates innate immunity to human species C adenovirus. *Science (New York, NY)* 2012;**338**(6108):795–8.
11. Irons EE, Flatt JW, Doronin K, Fox TL, Acchione M, Stewart PL, et al. Coagulation factor binding orientation and dimerization may influence infectivity of adenovirus-coagulation factor complexes. *J Virol* 2013;**87**(17):9610–9.
12. Cines DB, Pollak ES, Buck CA, Loscalzo J, Zimmerman GA, McEver RP, et al. Endothelial cells in physiology and in the pathophysiology of vascular disorders. *Blood* 1998;**91**(10):3527–61.
13. Sohlenius-Sternbeck AK. Determination of the hepatocellularity number for human, dog, rabbit, rat and mouse livers from protein concentration measurements. *Toxicol In Vitro* 2006;**20**(8):1582–6.

14. Hofherr SE, Mok H, Gushiken FC, Lopez JA, Barry MA. Polyethylene glycol modification of adenovirus reduces platelet activation, endothelial cell activation, and thrombocytopenia. *Hum Gene Ther* 2007;**18**(9):837–48.
15. Othman M, Labelle A, Mazzetti I, Elbatarny HS, Lillicrap D. Adenovirus-induced thrombocytopenia: the role of von Willebrand factor and P-selectin in mediating accelerated platelet clearance. *Blood* 2007;**109**(7):2832–9.
16. Di Paolo NC, van Rooijen N, Shayakhmetov DM. Redundant and synergistic mechanisms control the sequestration of blood-born adenovirus in the liver. *Mol Ther* 2009;**17**(4):675–84.
17. Liu Q, Zaiss AK, Colarusso P, Patel K, Haljan G, Wickham TJ, et al. The role of capsid-endothelial interactions in the innate immune response to adenovirus vectors. *Hum Gene Ther* 2003;**14**(7):627–43.
18. Morral N, O'Neal WK, Rice K, Leland MM, Piedra PA, Aguilar-Cordova E, et al. Lethal toxicity, severe endothelial injury, and a threshold effect with high doses of an adenoviral vector in baboons. *Hum Gene Ther* 2002;**13**(1):143–54.
19. *Ganong's review of medical physiology*. New York: McGraw-Hill Medical; 2010.
20. Ikeda K, Wakimoto H, Ichikawa T, Jhung S, Hochberg FH, Louis DN, et al. Complement depletion facilitates the infection of multiple brain tumors by an intravascular, replication-conditional herpes simplex virus mutant. *J Virol* 2000;**74**(10):4765–75.
21. Brown BD, Cantore A, Annoni A, Sergi LS, Lombardo A, Della Valle P, et al. A microRNA-regulated lentiviral vector mediates stable correction of hemophilia B mice. *Blood* 2007;**110**(13):4144–52.
22. Shashkova EV, Doronin K, Senac JS, Barry MA. Macrophage depletion combined with anticoagulant therapy increases therapeutic window of systemic treatment with oncolytic adenovirus. *Cancer Res* 2008;**68**(14):5896–904.
23. Carlisle RC, Di Y, Cerny AM, Sonnen AF, Sim RB, Green NK, et al. Human erythrocytes bind and inactivate type 5 adenovirus by presenting coxsackievirus-adenovirus receptor and complement receptor 1. *Blood* 2009;**113**(9):1909–18.
24. Rosen L. A hemagglutination-inhibition technique for typing adenoviruses. *Am J Hyg* 1960;**71**:120–8.
25. Norrby E. The structural and functional diversity of Adenovirus capsid components. *J Gen Virol* 1969;**5**(2):221 36.
26. Fields B. *Field's virology*. New York: Raven Press Books, Ltd; 1985.
27. Lyons M, Onion D, Green NK, Aslan K, Rajaratnam R, Bazan-Peregrino M, et al. Adenovirus type 5 interactions with human blood cells may compromise systemic delivery. *Mol Ther* 2006;**14**(1):118–28.
28. Eggerman TL, Mondoro TH, Lozier JN, Vostal JG. Adenoviral vectors do not induce, inhibit, or potentiate human platelet aggregation. *Hum Gene Ther* 2002;**13**(1):125–8.
29. Stone D, Liu Y, Shayakhmetov D, Li ZY, Ni S, Lieber A. Adenovirus-platelet interaction in blood causes virus sequestration to the reticulo-endothelial system of liver. *J Virol* 2007.
30. Raper SE, Chirmule N, Lee FS, Wivel NA, Bagg A, Gao GP, et al. Fatal systemic inflammatory response syndrome in a ornithine transcarbamylase deficient patient following adenoviral gene transfer. *Mol Genet Metabol* 2003;**80**(1–2):148–58.
31. Abbink P, Lemckert AA, Ewald BA, Lynch DM, Denholtz M, Smits S, et al. Comparative seroprevalence and immunogenicity of six rare serotype recombinant adenovirus vaccine vectors from subgroups B and D. *J Virol* 2007;**81**(9):4654–63.
32. Sumida SM, Truitt DM, Lemckert AA, Vogels R, Custers JH, Addo MM, et al. Neutralizing antibodies to adenovirus serotype 5 vaccine vectors are directed primarily against the adenovirus hexon protein. *J Immunol* 2005;**174**(11):7179–85.

33. Croyle MA, Chirmule N, Zhang Y, Wilson JM. "Stealth" adenoviruses blunt cell-mediated and humoral immune responses against the virus and allow for significant gene expression upon readministration in the lung. *J Virol* 2001;**75**(10):4792–801.
34. Parks R, Evelegh C, Graham F. Use of helper-dependent adenoviral vectors of alternative serotypes permits repeat vector administration. *Gene Ther* 1999;**6**(9):1565–73.
35. Pinto AR, Fitzgerald JC, Giles-Davis W, Gao GP, Wilson JM, Ertl HC. Induction of CD8(+) T cells to an HIV-1 antigen through a prime boost regimen with heterologous E1-deleted adenoviral vaccine carriers. *J Immunol* 2003;**171**(12):6774–9.
36. Barouch DH, Pau MG, Custers JH, Koudstaal W, Kostense S, Havenga MJ, et al. Immunogenicity of recombinant adenovirus serotype 35 vaccine in the presence of pre-existing anti-Ad5 immunity. *J Immunol* 2004;**172**(10):6290–7.
37. Lemckert AA, Sumida SM, Holterman L, Vogels R, Truitt DM, Lynch DM, et al. Immunogenicity of heterologous prime-boost regimens involving recombinant adenovirus serotype 11 (Ad11) and Ad35 vaccine vectors in the presence of anti-ad5 immunity. *J Virol* 2005;**79**(15):9694–701.
38. McCoy K, Tatsis N, Korioth-Schmitz B, Lasaro MO, Hensley SE, Lin SW, et al. Effect of preexisting immunity to adenovirus human serotype 5 antigens on the immune responses of nonhuman primates to vaccine regimens based on human- or chimpanzee-derived adenovirus vectors. *J Virol* 2007;**81**(12):6594–604.
39. Weaver EA, Nehete PN, Buchl SS, Senac JS, Palmer D, Ng P, et al. Comparison of replication-competent, first generation, and helper-dependent adenoviral vaccines. *PLoS One* 2009;**4**(3):e5059.
40. Weaver EA, Nehete PN, Nehete BP, Buchl SJ, Palmer D, Montefiori DC, et al. Protection against mucosal SHIV challenge by peptide and helper-dependent adenovirus vaccines. *Viruses* 2009;**1**(3):920.
41. Weaver EA, Nehete PN, Nehete BP, Yang G, Buchl SJ, Hanley PW, et al. Comparison of systemic and mucosal immunization with helper-dependent adenoviruses for vaccination against mucosal challenge with SHIV. *PLoS One* 2013;**8**(7):e67574.
42. Roberts DM, Nanda A, Havenga MJ, Abbink P, Lynch DM, Ewald BA, et al. Hexon-chimaeric adenovirus serotype 5 vectors circumvent pre-existing anti-vector immunity. *Nature* 2006;**441**(7090):239–43.
43. Khare R, May SM, Vetrini F, Weaver EA, Palmer D, Rosewell A, et al. Generation of a Kupffer cell-evading adenovirus for systemic and liver-directed gene transfer. *Mol Ther* 2011;**19**(7):1254–62.
44. Ehrenstein MR, Notley CA. The importance of natural IgM: scavenger, protector and regulator. *Nat Rev Immunol* 2010;**10**(11):778–86.
45. Xu Z, Tian J, Smith JS, Byrnes AP. Clearance of adenovirus by Kupffer cells is mediated by scavenger receptors, natural antibodies, and complement. *J Virol* 2008;**82**(23):11705–13.
46. Xu Z, Qiu Q, Tian J, Smith JS, Conenello GM, Morita T, et al. Coagulation factor X shields adenovirus type 5 from attack by natural antibodies and complement. *Nat Med* 2013;**19**(4):452–7.
47. Qiu Q, Xu Z, Tian J, Moitra R, Gunti S, Notkins AL, et al. Impact of natural IgM concentration on gene therapy with adenovirus type 5 vectors. *J Virol* 2015;**89**(6):3412–6.
48. Khare R, Hillestad ML, Xu Z, Byrnes AP, Barry MA. Circulating antibodies and macrophages as modulators of adenovirus pharmacology. *J Virol* 2013;**87**(7):3678–86.
49. Jiang H, Wang Z, Serra D, Frank MM, Amalfitano A. Recombinant adenovirus vectors activate the alternative complement pathway, leading to the binding of human complement protein C3 independent of anti-ad antibodies. *Mol Ther* 2004;**10**(6):1140–2.

50. Shayakhmetov DM, Gaggar A, Ni S, Li ZY, Lieber A. Adenovirus binding to blood factors results in liver cell infection and hepatotoxicity. *J Virol* 2005;**79**(12):7478–91.
51. Tian J, Xu Z, Smith JS, Hofherr SE, Barry MA, Byrnes AP. Adenovirus activates complement by distinctly different mechanisms in vitro and in vivo: indirect complement activation by virions in vivo. *J Virol* 2009;**83**(11):5648–58.
52. Greber UF, Willetts M, Webster P, Helenius A. Stepwise dismantling of adenovirus 2 during entry into cells. *Cell* 1993;**75**(3):477–86.
53. Wickham TJ, Mathias P, Cheresh DA, Nemerow GR. Integrins avb3 or avb5 promote adenovirus internalization but not virus attachment. *Cell* 1993;**73**:309–19.
54. Smith T, Idamakanti N, Kylefjord H, Rollence M, King L, Kaloss M, et al. In vivo hepatic adenoviral gene delivery occurs independently of the coxsackievirus-adenovirus receptor. *Mol Ther* 2002;**5**(6):770–9.
55. Baker AH, McVey JH, Waddington SN, Di Paolo NC, Shayakhmetov DM. The influence of blood on in vivo adenovirus bio-distribution and transduction. *Mol Ther* 2007;**15**(8):1410–6.
56. Kalyuzhniy O, Di Paolo NC, Silvestry M, Hofherr SE, Barry MA, Stewart PL, et al. Adenovirus serotype 5 hexon is critical for virus infection of hepatocytes in vivo. *Proc Natl Acad Sci USA* 2008;**105**(14):5483–8.
57. Waddington SN, McVey JH, Bhella D, Parker AL, Barker K, Atoda H, et al. Adenovirus serotype 5 hexon mediates liver gene transfer. *Cell* 2008;**132**(3):397–409.
58. Parker AL, Waddington SN, Nicol CG, Shayakhmetov DM, Buckley SM, Denby L, et al. Multiple vitamin K-dependent coagulation zymogens promote adenovirus-mediated gene delivery to hepatocytes. *Blood* 2006;**108**(8):2554–61.
59. Vigant F, Descamps D, Jullienne B, Esselin S, Connault E, Opolon P, et al. Substitution of hexon hypervariable region 5 of adenovirus serotype 5 abrogates blood factor binding and limits gene transfer to liver. *Mol Ther* 2008;**16**(8):1474–80.
60. Alemany R, Suzuki K, Curiel DT. Blood clearance rates of adenovirus type 5 in mice. *J Gen Virol* 2000;**81**(Pt 11):2605–9.
61. Tao N, Gao GP, Parr M, Johnston J, Baradet T, Wilson JM, et al. Sequestration of adenoviral vector by Kupffer cells leads to a nonlinear dose response of transduction in liver. *Mol Ther* 2001;**3**(1):28–35.
62. Snoeys J, Lievens J, Wisse E, Jacobs F, Duimel H, Collen D, et al. Species differences in transgene DNA uptake in hepatocytes after adenoviral transfer correlate with the size of endothelial fenestrae. *Gene Ther* 2007;**14**(7):604–12.
63. Bedard MB, Parkkari M, Weaver B, Riendeau J, Dahlquist M. Assessment of driving performance using a simulator protocol: validity and reproducibility. *Am J Occup Ther* 2010;**64**(2):336–40.
64. Elvevold K, Smedsrod B, Martinez I. The liver sinusoidal endothelial cell: a cell type of controversial and confusing identity. *Am J Physiol* 2008;**294**(2):G391–400.
65. Lieber A, He CY, Meuse L, Schowalter D, Kirillova I, Winther B, et al. The role of Kupffer cell activation and viral gene expression in early liver toxicity after infusion of recombinant adenovirus vectors. *J Virol* 1997;**71**(11):8798–807.
66. Haisma HJ, Kamps JA, Kamps GK, Plantinga JA, Rots MG, Bellu AR. Polyinosinic acid enhances delivery of adenovirus vectors in vivo by preventing sequestration in liver macrophages. *J Gen Virol* 2008;**89**(Pt 5):1097–105.
67. Haisma HJ, Boesjes M, Beerens AM, van der Strate BW, Curiel DT, Pluddemann A, et al. Scavenger receptor A: a new route for adenovirus 5. *Mol Pharm* 2009;**6**(2):366–74.
68. Manickan E, Smith JS, Tian J, Eggerman TL, Lozier JN, Muller J, et al. Rapid Kupffer cell death after intravenous injection of adenovirus vectors. *Mol Ther* 2006;**13**(1):108–17.

69. Smith JS, Tian J, Muller J, Byrnes AP. Unexpected pulmonary uptake of adenovirus vectors in animals with chronic liver disease. *Gene Ther* 2004;**11**(5):431–8.
70. Weaver EA, Hillestad ML, Khare R, Palmer D, Ng P, Barry MA. Characterization of species C human adenovirus serotype 6 (Ad6). *Virology* 2011;**412**(1):19–27.
71. Shashkova EV, May SM, Barry MA. Characterization of human adenovirus serotypes 5, 6, 11, and 35 as anticancer agents. *Virology* 2009. Epub Sep. 17.
72. Coughlan L, Bradshaw AC, Parker AL, Robinson H, White K, Custers J, et al. Ad5:Ad48 hexon hypervariable region substitutions lead to toxicity and increased inflammatory responses following intravenous delivery. *Mol Ther* 2012;**20**(12):2268–81.
73. Wang S, Baum BJ, Kagami H, Zheng C, O'Connell BC, Atkinson JC. Effect of clodronate on macrophage depletion and adenoviral-mediated transgene expression in salivary glands. *J Oral Pathol Med* 1999;**28**(4):145–51.
74. Schiedner G, Hertel S, Johnston M, Dries V, van Rooijen N, Kochanek S. Selective depletion or blockade of Kupffer cells leads to enhanced and prolonged hepatic transgene expression using high-capacity adenoviral vectors. *Mol Ther* 2003;**7**(1):35–43.
75. Smith JS, Tian J, Lozier JN, Byrnes AP. Severe pulmonary pathology after intravenous administration of vectors in cirrhotic rats. *Mol Ther* 2004;**9**(6):932–41.
76. Piccolo P, Vetrini F, Mithbaokar P, Grove NC, Bertin T, Palmer D, et al. SR-A and SREC-I are Kupffer and endothelial cell receptors for helper-dependent adenoviral vectors. *Mol Ther* 2013;**21**(4):767–74.
77. Haisma HJ, Bellu AR. Pharmacological interventions for improving adenovirus usage in gene therapy. *Mol Pharm* 2011;**8**(1):50–5.
78. Zinn KR, Douglas JT, Smyth CA, Liu HG, Wu Q, Krasnykh VN, et al. Imaging and tissue biodistribution of 99mTc-labeled adenovirus knob (serotype 5). *Gene Ther* 1998;**5**(6):798–808.
79. Ramchandren S, Lewis RA. An update on monoclonal gammopathy and neuropathy. *Curr Neurol Neurosci Rep* 2012;**12**(1):102–10.
80. Cattaneo R, Miest T, Shashkova EV, Barry MA. Reprogrammed viruses as cancer therapeutics: targeted, armed and shielded. *Nat Rev* 2008;**6**(7):529–40.
81. Kreppel F, Kochanek S. Modification of adenovirus gene transfer vectors with synthetic polymers: a scientific review and technical guide. *Mol Ther* 2008;**16**(1):16–29.
82. Chillon M, Lee JH, Fasbender A, Welsh MJ. Adenovirus complexed with polyethylene glycol and cationic lipid is shielded from neutralizing antibodies in vitro. *Gene Ther* 1998;**5**(7):995–1002.
83. O'Riordan CR, Lachapelle A, Delgado C, Parkes V, Wadsworth SC, Smith AE, et al. PEGylation of adenovirus with retention of infectivity and protection from neutralizing antibody in vitro and in vivo. *Hum Gene Ther* 1999;**10**(8):1349–58.
84. Romanczuk H, Galer CE, Zabner J, Barsomian G, Wadsworth SC, O'Riordan CR. Modification of an adenoviral vector with biologically selected peptides: a novel strategy for gene delivery to cells of choice [see comments]. *Hum Gene Ther* 1999; **10**(16):2615–26.
85. Croyle MA, Yu QC, Wilson JM. Development of a rapid method for the PEGylation of adenoviruses with enhanced transduction and improved stability under harsh storage conditions. *Hum Gene Ther* 2000;**11**(12):1713–22.
86. Croyle MA, Chirmule N, Zhang Y, Wilson JM. PEGylation of E1-deleted adenovirus vectors allows significant gene expression on readministration to liver. *Hum Gene Ther* 2002;**13**(15):1887–900.
87. Cheng X, Ming X, Croyle MA. PEGylated adenoviruses for gene delivery to the intestinal epithelium by the oral route. *Pharm Res* 2003;**20**(9):1444–51.

88. Lanciotti J, Song A, Doukas J, Sosnowski B, Pierce G, Gregory R, et al. Targeting adenoviral vectors using heterofunctional polyethylene glycol FGF2 conjugates. *Mol Ther* 2003;**8**(1):99–107.
89. Ogawara K, Rots MG, Kok RJ, Moorlag HE, Van Loenen AM, Meijer DK, et al. A novel strategy to modify adenovirus tropism and enhance transgene delivery to activated vascular endothelial cells in vitro and in vivo. *Hum Gene Ther* 2004;**15**(5):433–43.
90. Kreppel F, Gackowski J, Schmidt E, Kochanek S. Combined genetic and chemical capsid modifications enable flexible and efficient de- and retargeting of adenovirus vectors. *Mol Ther* 2005;**12**(1):107–17.
91. Mok H, Palmer DJ, Ng P, Barry MA. Evaluation of polyethylene glycol modification of first-generation and helper-dependent adenoviral vectors to reduce innate immune responses. *Mol Ther* 2005;**11**(1):66–79.
92. Fisher KD, Stallwood Y, Green NK, Ulbrich K, Mautner V, Seymour LW. Polymer-coated adenovirus permits efficient retargeting and evades neutralising antibodies. *Gene Ther* 2001;**8**(5):341–8.
93. Sakuma S, Lu ZR, Kopeckova P, Kopecek J. Biorecognizable HPMA copolymer-drug conjugates for colon-specific delivery of 9-aminocamptothecin. *J Control Release* 2001;**75**(3):365–79.
94. Parker AL, Fisher KD, Oupicky D, Read ML, Nicklin SA, Baker AH, et al. Enhanced gene transfer activity of peptide-targeted gene-delivery vectors. *J Drug Target* 2005;**13**(1):39–51.
95. Stevenson M, Hale AB, Hale SJ, Green NK, Black G, Fisher KD, et al. Incorporation of a laminin-derived peptide (SIKVAV) on polymer-modified adenovirus permits tumor-specific targeting via alpha6-integrins. *Cancer Gene Ther* 2007;**14**(4):335–45.
96. Nguyen TV, Barry ME, Barry MA. *Polymer shielded adenovirus serotype 6 (Ad6) for systemic virotherapy against human prostate cancers.*
97. Hofherr SE, Shashkova EV, Weaver EA, Khare R, Barry MA. Modification of adenoviral vectors with polyethylene glycol modulates in vivo tissue tropism and gene expression. *Mol Ther* 2008;**16**(7):1276–82.
98. Doronin K, Shashkova EV, May SM, Hofherr SE, Barry MA. Chemical modification with high molecular weight polyethylene glycol reduces transduction of hepatocytes and increases efficacy of intravenously delivered oncolytic adenovirus. *Hum Gene Ther* 2009;**20**(9):975–88.
99. Subr V, Kostka L, Selby-Milic T, Fisher K, Ulbrich K, Seymour LW, et al. Coating of adenovirus type 5 with polymers containing quaternary amines prevents binding to blood components. *J Control Release* 2009;**135**(2):152–8.
100. Czajkowsky DM, Shao Z. The human IgM pentamer is a mushroom-shaped molecule with a flexural bias. *Proc Natl Acad Sci USA* 2009;**106**(35):14960–5.
101. Khare R, Reddy VS, Nemerow GR, Barry MA. Identification of adenovirus serotype 5 hexon regions that interact with scavenger receptors. *J Virol* 2012;**86**(4):2293–301.
102. Takahashi S, Mok H, Parrott MB, Marini 3rd FC, Andreeff M, Brenner MK, et al. Selection of chronic lymphocytic leukemia binding peptides. *Cancer Res* 2003;**63**(17):5213–7.
103. Menezes KM, Mok HS, Barry MA. Increased transduction of skeletal muscle cells by fibroblast growth factor-modified adenoviral vectors. *Hum Gene Ther* 2006;**17**(3):314–20.
104. Hofherr S, Senac JS, Chen CY, Palmer D, Ng P, Barry MA. Short-term rescue of neonatal lethality in a mouse model of propionic acidemia by gene therapy. *Hum Gene Ther* 2008.
105. Weaver EA, Barry MA. Effects of shielding adenoviral vectors with polyethylene glycol on vector-specific and vaccine-mediated immune responses. *Hum Gene Ther* 2008;**19**(12):1369–82.

106. Nguyen TV, Barry ME, Turner MA, Crosby CM, Trujillo MA, Morris JC, et al. *Comparison of detargeting strategies for systemic therapy with oncolytic adenovirus serotype 5.*
107. Michael S, Hong J, Curiel D, Engler J. Addition of a short peptide ligand to the adenovirus fiber protein. *Gene Ther* 1995;**2**:660–9.
108. Dmitriev IP, Kashentseva EA, Curiel DT. Engineering of adenovirus vectors containing heterologous peptide sequences in the C terminus of capsid protein IX. *J Virol* 2002;**76**(14):6893–9.
109. Wu H, Han T, Belousova N, Krasnykh V, Kashentseva E, Dmitriev I, et al. Identification of sites in adenovirus hexon for foreign peptide incorporation. *J Virol* 2005;**79**(6):3382–90.
110. San Martin C, Glasgow JN, Borovjagin A, Beatty MS, Kashentseva EA, Curiel DT, et al. Localization of the N-terminus of minor coat protein IIIa in the adenovirus capsid. *J Mol Biol* 2008;**383**(4):923–34.
111. Vigne E, Mahfouz I, Dedieu JF, Brie A, Perricaudet M, Yeh P. RGD inclusion in the hexon monomer provides adenovirus type 5-based vectors with a fiber knob-independent pathway for infection. *J Virol* 1999;**73**(6):5156–61.
112. Parrott MB, Barry MA. Metabolic biotinylation of recombinant proteins in mammalian cells and in mice. *Mol Ther* 2000;**1**(1):96–104.
113. Parrott MB, Barry MA. Metabolic biotinylation of secreted and cell surface proteins from mammalian cells. *Biochem Biophys Res Comm* 2001;**281**(4):993–1000.
114. Barry MA, Campos SK, Ghosh D, Adams KE, Mok H, Mercier GT, et al. Biotinylated gene therapy vectors. *Expert Opin Biol Ther* 2003;**3**(6):926–40.
115. Parrott MB, Adams KE, Mercier GT, Mok H, Campos SK, Barry MA. Metabolically biotinylated adenovirus for cell targeting, ligand screening, and vector purification. *Mol Ther* 2003;**8**(4):688–700.
116. Campos SK, Barry MA. Rapid construction of capsid-modified adenoviral vectors through bacteriophage lambda red recombination. *Hum Gene Ther* 2004;**15**(11):1125–30.
117. Campos SK, Parrott MB, Barry MA. Avidin-based targeting and purification of a protein IX-modified, metabolically biotinylated adenoviral vector. *Mol Ther* 2004;**9**(6):943–55.
118. Campos SK, Barry MA. Comparison of adenovirus fiber, protein IX, and hexon capsomeres as scaffolds for vector purification and cell targeting. *Virology* 2006;**349**(2):453–62.
119. Chen Z, Mok H, Pflugfelder SC, Li DQ, Barry MA. Improved transduction of human corneal epithelial progenitor cells with cell-targeting adenoviral vectors. *Exp Eye Res* 2006;**83**(4):798–806.
120. Laakkonen JP, Engler T, Romero IA, Weksler B, Couraud PO, Kreppel F, et al. Transcellular targeting of fiber- and hexon-modified adenovirus vectors across the brain microvascular endothelial cells in vitro. *PLoS One* 2012;**7**(9):e45977.
121. Lucas T, Benihoud K, Vigant F, Schmidt CQ, Bachem MG, Simmet T, et al. Hexon modification to improve the activity of oncolytic adenovirus vectors against neoplastic and stromal cells in pancreatic cancer. *PLoS One* 2015;**10**(2):e0117254.
122. Campos SK, Barry MA. Current advances and future challenges in Adenoviral vector biology and targeting. *Curr Gene Ther* 2007;**7**(3):189–204.
123. Khare R, Chen CY, Weaver EA, Barry MA. Advances and future challenges in adenoviral vector pharmacology and targeting. *Curr Gene Ther* 2011;**11**(4):241–58.
124. Chen CY, May SM, Barry MA. Targeting adenoviruses with factor x-single-chain antibody fusion proteins. *Hum Gene Ther* 2010;**21**(6):739–49.
125. Roelvink PW, Mi Lee G, Einfeld DA, Kovesdi I, Wickham TJ. Identification of a conserved receptor-binding site on the fiber proteins of CAR-recognizing adenoviridae. *Science (New York, NY)* 1999;**286**(5444):1568–71.

126. Wang H, Liaw YC, Stone D, Kalyuzhniy O, Amiraslanov I, Tuve S, et al. Identification of CD46 binding sites within the adenovirus serotype 35 fiber knob. *J Virol* 2007;**81**(23):12785–92.
127. Einfeld DA, Schroeder R, Roelvink PW, Lizonova A, King CR, Kovesdi I, et al. Reducing the native tropism of adenovirus vectors requires removal of both CAR and integrin interactions. *J Virol* 2001;**75**(23):11284–91.
128. Shayakhmetov DM, Eberly AM, Li ZY, Lieber A. Deletion of penton RGD motifs affects the efficiency of both the internalization and the endosome escape of viral particles containing adenovirus serotype 5 or 35 fiber knobs. *J Virol* 2005;**79**(2):1053–61.
129. Shashkova EV, May SM, Doronin K, Barry MA. Expanded anticancer therapeutic window of hexon-modified oncolytic adenovirus. *Mol Ther* 2009;**17**(12):2121–30.
130. Ma J, Duffy MR, Deng L, Dakin RS, Uil T, Custers J, et al. Manipulating adenovirus hexon hypervariable loops dictates immune neutralisation and coagulation factor X-dependent cell interaction in vitro and in vivo. *PLoS Pathog* 2015;**11**(2):e1004673.
131. Sunnerhagen M, Forsen S, Hoffren AM, Drakenberg T, Teleman O, Stenflo J. Structure of the Ca(2+)-free Gla domain sheds light on membrane binding of blood coagulation proteins. *Nat Struct Biol* 1995;**2**(6):504–9.
132. Panda S, Ding JL. Natural antibodies bridge innate and adaptive immunity. *J Immunol* 2015;**194**(1):13–20.
133. Notkins AL. Polyreactivity of antibody molecules. *Trends Immunol* 2004;**25**(4):174–9.
134. Chen CY, Weaver EA, Khare R, May SM, Barry MA. Mining the adenovirus virome for oncolytics against multiple solid tumor types. *Cancer Gene Ther* 2011;**18**(10):744–50.
135. Espenlaub S, Corjon S, Engler T, Fella C, Ogris M, Wagner E, et al. Capsomer-specific fluorescent labeling of adenoviral vector particles allows for detailed analysis of intracellular particle trafficking and the performance of bioresponsive bonds for vector capsid modifications. *Hum Gene Ther* 2010;**21**(9):1155–67.
136. Prill JM, Espenlaub S, Samen U, Engler T, Schmidt E, Vetrini F, et al. Modifications of adenovirus hexon allow for either hepatocyte detargeting or targeting with potential evasion from Kupffer cells. *Mol Ther* 2011;**19**(1):83–92.
137. Hofherr SE, Adams KE, Chen CY, May S, Weaver EA, Barry MA. Real-time dynamic imaging of virus distribution in vivo. *PLoS One* 2011;**6**(2):e17076.

# Molecular Design of Oncolytic Adenoviruses

12

*Ramon Alemany*
IDIBELL-Institut Català d'Oncologia, L'Hospitalet de Llobregat, Barcelona, Spain

## 1. Introduction

Among different oncolytic viruses, adenovirus has several traits that facilitate the design of oncolytic or tumor-selective recombinants. Most serotypes of adenovirus use the fiber (nt 31037–32782 of human Ad5 reference material) to interact with well-defined protein receptors (CAR, CD46, or desmoglein 2) at the cell membrane, allowing for receptor-targeting strategies. Later, in the late phase of the infectious cycle, the newly synthesized fiber will not only pass to the progeny but also will be secreted to the extracellular medium in large amounts, and this can be used to open intercellular junctions on adjacent cells to facilitate virus spread. The penton base RGD (nt 15174–15183) motif is involved in a second step of cell internalization by binding to integrins that drive the virus to the endocytic pathway, and this internalization step can also be modulated in recombinant adenoviruses. The protein surface of the virus, mainly formed by the hexon (nt 18842–21700), contributes to surface charge and to other interactions with host antibodies and other proteins (such as FX clotting factor or scavenger receptors) that can also be modulated to improve systemic tumor targeting. The existence of multiple serotypes of adenovirus may be used to create chimerical viruses that evade preexisting immunity issues. Adenovirus has a natural tropism to infect and replicate in epithelial cells, the origin of most solid tumors. The viral DNA expression occurs in the nuclei of infected cells and it follows a timely orchestrated series of activation steps, initiated by the expression of early 1a (*E1a*) genes. This allows the control viral replication replacing the E1a promoter by a tumor-selective promoter. Viral progeny accumulates in the nuclei of infected cells and its release can be accelerated using different genetic modifications. Finally, the adenoviral genome can be armed with transgenes at different insertion sites of the double-stranded linear 36 kb genome. These transgenes can be controlled by exogenous promoters or by endogenous viral promoters that will express the transgene in the appropriate phase of the viral life cycle. With genetic recombineering techniques in yeast[1] or bacteria[2] that allow the modification of any nucleotide position of the plasmid-cloned genome in a two-step replacement of positive–negative selectable genes, restriction enzymes sites are no longer a limitation in the design of oncolytic adenoviruses.

**Adenoviral Vectors for Gene Therapy.**

# 2. Genetic Modifications to Achieve Tumor-Selective Replication

## 2.1 Deletion of Viral Functions Dispensable in Tumor Cells

*E1A* (nt 560–1545) is the first viral gene expressed from the virus genome. Differential splicing produces a shorter E1A-12S (243 aa) and a large E1A-13S (289 aa) protein with several conserved regions among serotypes (CR1 spans residues 41 to 80, CR2 residues 122 to 139 LTCHEAGF (nt 923–946), and CR3 residues 140 to 188, this later only present in E1A-13S). These conserved regions are involved in interactions with cellular proteins to induce the transcription of cellular and viral genes. CR1 and CR2 mediate binding to pRB family proteins (pRB, p107, and p130) dissociating them from the E2F transcription factor. The released "free" E2F activates cellular genes involved in cell cycle and DNA synthesis and also the E2 viral genes. On the other hand, E1A proteins use the CR1 and the amino-terminal domain to bind to nuclear lysine acetylases (CBP and p300) to form a complex that stimulates the transcriptional activity of E2F. Removing the E2F inhibitor pRB and stimulating the E2F activators p300 are two complementary functions of E1A proteins necessary to pass the G2/M phases of the cell cycle, but each of them separately is enough to induce DNA synthesis. In cycling cells, pRB phosphorylation by cyclin D-CDK4/6 and cyclin E/A-CDK2 changes its conformation and dissociates it from E2F. In tumor cells pRB is inactivated by constitutive phosphorylation or deletion. Therefore, when cycling or tumor cells are infected with adenovirus, the CR2 of E1A becomes dispensable. AdD24 and dl922-947 are two oncolytic viruses based on such CR2 deletions.[3,4] (See Figure 1 for deletions and insertions commonly used to design oncolytic adenoviruses.) Arresting normal cells or transferring pRB into pRB-mutant tumor cells reduces the production levels (burst size) of such mutants up to 3 logs. However, focusing on their potency in tumor cells, a certain degree (1 log) of attenuation in several tumor cell lines has

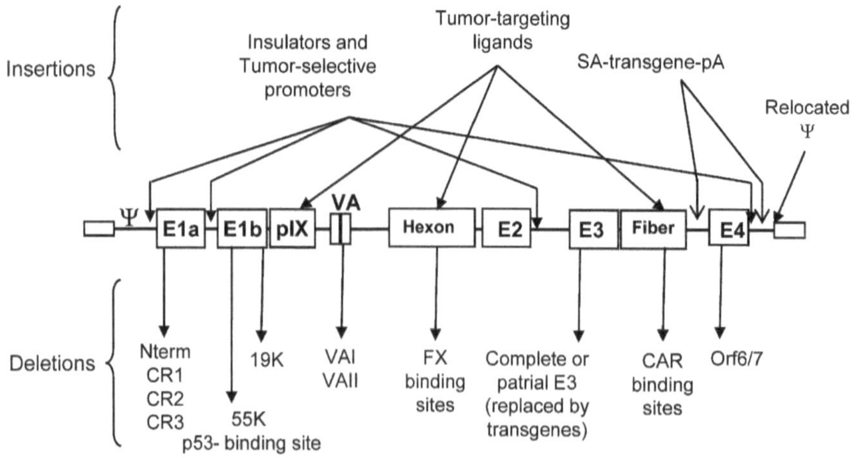


**Figure 1** Insertion and deletions commonly used to design oncolytic adenoviruses. SA, splicing acceptor; pA, polyadenylation signal.

also been observed, and it could be associated with the role of the p300–E1A–pRB complex to repress cell genes that inhibit viral replication.[5] If safety/toxicity is a concern, further deletion of CR1 to impair p300 binding increases the attenuation of these mutants in normal cells[6] especially under proliferating conditions,[7] but their yield in tumor cells (potency) may also be lower.[4] In addition, E4orf6/7 (nt 34072–32909, note reverse orientation) binds free E2F to activate the E2 promoter but it also can help to dissociate pRB–E2F complexes. Therefore, the E4orf6/7 deletion could increase the selectivity of CR2 mutants but at the expense of some potency.

CR3 is needed for efficient transactivation of other early viral promoters, mainly by binding to the TATA-binding protein and displacing p53 that causes transcriptional repression. Deletion of the CR3 of E1A or the splice variant E1A-13S (such in Ad mutant dl520 that has a deletion of nt 1107 to nt 1117 eliminating the splice donor site of E1A-13S) renders the Ad unable to transactivate early Ad genes in normal cells. However, in tumor cells this defect can be bypassed to a certain degree by the accumulation of the transcription factor YB-1 in the cell nucleus, which is able to activate the late E2 promoter.[8]

E1b-19K (or E1b-21K, nt 1714–2244) is a homolog of Bcl2 that blocks apoptosis in infected cells. Wild-type adenovirus does not cause apoptosis because it is blocked by E1b-19K. In normal cells, lack of this protein causes premature cell death by apoptosis on infection and the mutant Ad cannot complete the life cycle. However, in certain tumor cells with overactive antiapoptotic pathways, apoptosis on infection of E1b-19K-defective mutants (E1B19K can be partly deleted using *EcoNI* nt 1710 and *BstEII* nt 1915) is prevented or delayed to the late phase of the viral cycle. Such a delayed apoptosis results in a more efficient release of virus from infected cells and increases viral spread.[9] In fact, the selection of large plaques after random mutagenesis of the viral genome often selects for E1b-19K mutants.[10] Nevertheless, the function of E1b-19K cannot be complemented unless the antiapoptotic pathways of the cells are highly active.[11] The proapoptotic cell death induced by E1b-19K mutants makes them especially suitable for combination with apoptosis-inducing chemotherapies.[12] On the other hand, E1b-19K-defective mutants induce greater inflammatory responses that could induce greater antiviral and antitumor immune responses and may affect the oncolytic outcome.[13]

E1b-55K (nt 2019–3509) forms a complex with E4orf6 and p53 to promote ubiquitin-mediated degradation of p53. This prevents early apoptosis of infected cells and induces the S phase that favors virus replication. Functional loss of p53 in tumor cells would render E1B-55K dispensable. On this basis, an E1b-55K deletion mutant was proposed as a tumor-selective or oncolytic adenovirus for p53-defective tumors.[14] The main caveat of this design is that E1b-55K has other functions besides p53 inactivation, such as the shutoff of RNA nuclear export and the nuclear localization of YB1 to activate the late E2 promoter. Certain tumor cells can inactivate p53, counteract protein shutoff, and accumulate nuclear YB1 simultaneously, but in most cases a significant loss of E1b-55K mutant yields is observed. Obtaining E1b-55K mutations that only affect the p53 binding site would overcome this limitation.[15] E4-orf3 (nt 34699–34349) can block the expression of p53-dependent genes, and therefore its deletion should also be considered to attenuate the virus in p53+ normal cells.

Tumor cells are less sensitive to the protein translation shutoff induced by interferon (IFN). This IFN resistance is the basis for the oncolytic selectivity of different IFN-sensitive RNA viruses. On the other hand, IFN-resistant viruses can be designed to be tumor selective if the viral functions that counteract IFN are eliminated. This can be achieved in adenovirus by deletion of VA-RNAs (VA-I nt 10620–10779; VA-II nt 10876–11038) that block the IFN-inducible protein kinase R (PKR).[16] However, other factors may contribute to the replication potency of these mutants as the function of VA-RNAs is not just blocking PKR. For example, VA-RNAs produce microRNAs to inhibit cellular genes to promote efficient virus growth.[17] The simultaneous inhibitory and activating functions of most viral RNAs and proteins (E1A, E1B, E4) make it very difficult in designing defective Ad mutants to target a specific molecular defect of tumor cells. In this regard, the use of tumor-selective promoters seems more predictable for achieving selective replication while not affecting the replication potency in tumor cells.

## 2.2 Insertion of Tumor-Selective Promoters

As E1A 12S and 13S proteins are the first ones expressed on infection and control the expression of the rest of early viral genes, replacing the E1a promoter by an exogenous tumor-selective or tissue-selective promoter is the main strategy for achieving a transcriptionally controlled oncolytic adenovirus. Structural elements of the E1a promoter that can affect the inserted exogenous promoter are the following: An SP1-binding site (GGGTGG) in the left ITR (nt 1–103) at nt 52–57, ATF (activating transcription factor)-binding sites (TGACGT) in the ITR at nt 64–69 and nt 96–101, and downstream of the ITR at nt 170–175 and nt 456–461, the enhancer I elements (AGGAAGT at nt 199 and nt 296, and CGGATGT at nt 155 and nt 229), two E2F-binding sites (TTTCGCG) at nt 212 and nt 275, the enhancer II at nt 249 to nt 282 (that stimulate transcription of E1a and also other early viral genes), and OCT1-binding site at nt 303–316, the TATA box at nt 468–475, the cap site at nt 499, and the first E1a codon (ATG) at nt 560. The exogenous promoter has been commonly inserted between the cap site and this codon. Among these regulatory elements, those present in the ITR and the enhancers I/II that overlap with the A repeats of the packaging signal cannot be removed without deleterious replication or packaging effects. The SP1 and ATF sites at the ITR are able to originate transcripts from the ITR itself.[18] The readthrough over the inserted exogenous promoter would affect its transcriptional activity, and to avoid it a pA signal can be inserted downstream of the ITR. The interference of enhancers I/II can be avoided by moving the whole packaging signal to the right end of the genome, upstream to the right ITR (nt 35832–35934).

E1b, E2, and E4 promoters have also been replaced with tumor-selective promoters. The E1b promoter is composed of a GC box at nt 1654 and a TATA box at nt 1672, and it is affected by enhancers near the E1A stop codon (nt 1545). As E1B-19K activates E1A, E2, E3, and E4 promoters and E1B-55K activates the E2 late promoter, control of E1b helps to obtain tumor-selective replication. An easy way to control E1a and E1b is to connect them using an internal ribosome entry site.[19] The dual regulation of E1a and E1b seems to add 100-fold selectivity to the single E1a regulation,[20] but for certain combinations it did not add selectivity.[21] E4 proteins, mainly ORF6/7, transactivate E2 and can induce E1A-independent replication. Replacing the E4 promoter (nt 35575–35818) can increase up to 10,000-fold the selective replication of E1A-controlled viruses.[22] E2 has an early promoter highly dependent on E1A transactivation composed of a TATA box (CTTAA, nt 27069–27073), and E2F palindrome site (TTTCTCAAATTTAAGCGCGAAAA; nt 27091–27113; formed by two E2F boxes TTTCGCG and CGCGAAA), and one ATF site (ACGTCAT; nt 27116–27122), and a late promoter composed of a TATA box, two Sp1 sites, and three CCAAT boxes repressed by E1A and activated by E1B-55K jointly with YB-1. Among both E2 promoters, in designing oncolytic adenoviruses the E2 early promoter can be replaced with tumor-selective promoters. When using tumor-selective Tcf-binding sites (AGATCAAAGG) it has been found that the adjacent AP1 (TGACTAAC, nt 27469–27476) and ATF (TCGTCAC; nt 27503–27509) sites belonging to the E3 promoter had to be mutated (preserving the overlapping L4-pVIII ORF) for proper control.[21] In theory, as E2 encodes proteins essential for viral replication (DNA polymerase, terminal protein, and DNA binding protein) the E2 promoter control could not be bypassed by cellular factors, as may occur with E1a and E4 proteins that control gene expression or the cell cycle, but replacing E1a or E4 promoters has yielded more selective replication (up to 10,000-fold with E1a or E4 compared to 100-fold with E2).

For a proper regulation in the viral genome, the exogenous promoters should be insulated from transcriptional sites and enhancers by means of transcription stop signals (poly signal AATAAA) and insulators (such as the HS4, H19, DM1, and synthetic insulators).[23]

# 3. Genetic Modifications to Enhance Oncolytic Potency

## 3.1 Modifications to Increase the Release of Virus from Infected Cells

Research on improving the design of oncolytic Ad has moved progressively from selectivity to potency. Random selection of faster spreading adenoviruses has led to mutations in E1b-19K (nt 1714–2244), iLeader (nt 7978–8427), or E3-19K (nt 28729–29211) proteins that show a common mechanism of action based on a faster release of virus particles from infected cells.[24] This reveals that progeny release from infected cells is a rather inefficient process. Despite that faster release is often at the expense of lower virus production yields (burst size), it clearly increases oncolytic potency in xenograft models, indicating that spread is a more limiting factor for efficacy than virus yields. However, the advantage of faster released mutants is harder to prove in the presence of an immune system. As noted above, the loss of E1-19K function promotes early apoptosis, which induces a faster release of viruses. On the other hand, the mechanisms responsible for the faster release of E3-19K C-terminal mutants (e.g., the "T1" or "445A" mutation, which inserts an additional "A" at the AAAA stretch located at nt 29170–29173 and shifts the E3-19K ORF), which express E3-19K at the plasma membrane instead of the endoplasmic reticulum, could be associated with an ionic permeabilization or viroporin function of this protein.[25] The molecular pathway involved in the fast release associated with the C-terminal truncation of the iLeader protein (e.g., G at nt 8374 mutated to T, which changes aa E to a stop codon) is still unknown.[26] Another way to accelerate release is overexpressing the adenovirus death protein (ADP; nt 29485–29763).[27] ADP is the only *E3* gene expressed at the late phase. The absence of this protein in E3-deleted viruses was found to decrease their oncolytic potency and causes a small plaque phenotype. Conversely, its overexpression by deleting other *E3* genes or by inserting a second copy of the major late promoter (MLP) before the ADP ORF enhances viral release. When E3-deleted viruses are mutagenized and propagated in vitro, compensating E1B-19K deletions arise. Therefore, for E3-deleted oncolytic viruses, the E1b-19K deletion is of special advantage.

## 3.2 Arming Oncolytic Adenoviruses with Transgenes

A different strategy for increasing the efficacy of oncolytic adenovirus has been to insert exogenous transgenes. In general terms, oncolytic adenoviruses have been "armed" with transgenes for several purposes. Often, the aim is to increase viral cytotoxicity using prodrug-activating genes (such as thymidine kinase, carboxypeptidase, nitroreductase, or cytosine deaminase/uracyl phosphoribosyltransferase), fusogenic proteins (such as GALV), toxins (such onconase), or immunostimulatory genes (such as *IL2*, *IL4*, *IL12*, *IL18*, *IL24*, *IFN-a*, *RANTES*, *CD40L*, *HSP-70*, *THF*, and *GM-CSF*). A conceptually different aim is to arm the virus with genes that foster virus replication, spread, or yields. For this later aim, oncolytic adenoviruses have been armed with genes that increase viral replication or progeny release (such as p53 at the late phase[28]) or enzymes that degrade the extracellular matrix (such as relaxin, metalloproteases, and hyaluronidase).[29] Eventually, a combination of several genes would be ideal, but the maximum genome size that can fit within the capsid is 38 kb, that is, 2 kb over the wild-type size. Although the transgene can be inserted as an expression cassette with its own promoter and poly(A) signal, as usually adenoviral vectors are used for gene therapy or vaccination, the design of oncolytic adenoviruses often uses endogenous viral promoters to express the transgene. This saves space and allows a better tuning of the expression with the replication cycle of the virus. In addition, it avoids creating directed or inverted repeats by repetition of promoter or regulatory sequences that result in rearrangements and genome instability. Transgenes can also cause aberrant splicing with viral genes, and preferred consensus splicing donor sites (e.g., AGGTGAGT and AGGTAAGT) in the transgene should be mutated at the internal (underlined) GT nucleotides without changing the aa sequence of the protein.

Transgenes have been inserted at different sites of the Ad genome.[30] A good option for obtaining early expression without adding additional promoters is to link the transgene to E1A using an internal ribosome entry site[31] or a 2A ribosome skipping sequence. Different 2A sequences have been used (e.g., from the porcine teschovirus 1 (ATNFSLLKQAGDVEENPGP)[32] or from the Thosea asigna virus (EGRGSLLTCGDVEENPGP)),[33] usually with a GSG linker between the upstream protein and the 2A signal. The 2A linker strategy has the advantage of requiring less cloning space (66 bp of 2A vs 600 bp of the IRES) but it leaves the entire 2A sequence except the last proline at the C terminus of the viral protein and the proline residue at the N terminus of the exogenous protein. It should be determined that such tails do not affect the function of the 2A-linked proteins, as seems to be the case for E1a, pIX, and fiber. In addition, ribosome skipping by 2A is not 100% effective and a significant percentage of fusion protein remains. Substitution of *E3* genes, except for ADP, leads also to early gene expression. For a late gene expression the transgene can be linked to the MLP using a splicing acceptor signal upstream of the transgene and locating it at different positions downstream of the MLP.[30] In E3-deleted viruses, one possible location is at the E3-deleted region itself.[34] To avoid deleting viral genes, insertions between the fiber poly(A) signal (AATAAA nt 32778–32783; that overlaps with the fiber stop codon) and the E4 poly(A) (TTTATT, nt 32,811–32816) or between the E4 promoter (nt 35575–35818) and the right ITR (nt 35832–35,934) have been used by various groups. Different splicing acceptor signals used derive from the adenovirus genes, such as the IIIa gene,[35,36] Ad41 long fiber gene,[37] Ad40 long fiber gene,[34] or from a consensus splicing signal including a branch point, a polypyrimidine track, and a splice acceptor sequence (named BPS).[30,33] Between the splicing signal and the transgene, a Kozak consensus signal (CCACC) can be inserted to increase the expression level of the transgene. A pA signal should be added after the stop codon of the protein, which can be overlapping (TAAtaaa) if the stop codon is changed to TAA. A left-to-right transcriptional direction is preferred and seems to increase expression levels.[30] The location after the fiber leads to higher transgene expression but the location upstream to the left ITR leads to a tighter replication-dependent expression, which is more adequate for toxic proteins,[38] and it seems also the most appropriate location to express shRNAs by RNA polymerase III.[39] The use of IRES or a ribosome skipping 2A sequence can also be used to fuse the transgene to fiber, hexon, pIX, or other viral proteins.[28,31] Comparing oncolytic adenoviruses with luciferase downstream of the fiber with 2A or IRES or next to the right ITR with a splicing acceptor indicated that IRES mediated the strongest expression but the splicing acceptor preserved the highest replication dependent or tumor-selective expression.[31] Given the unpredictable effects of the remaining 2A sequences, the long size of IRES, and the good results obtained with the splice acceptor signal, this later seems to be the best choice to arm oncolytic adenoviruses.

To obtain more than 2 kb of cloning space, deletion of dispensable *E3* or *E4* genes has been considered. E3 encodes for seven proteins: 12.5K (nt 27858–28181), 6.7K (nt 28547–28736), 19K (nt 28735–29215), ADP (nt 29491–29770), RIDa (10.4K), RIDb (14.5K), and 14.7K, from left to right. These proteins are nonessential for viral replication in vitro and they have been classically replaced with antigen-expression cassettes (with exogenous promoters) in replication-competent adenovirus vaccines. For oncolytic adenoviruses, partial E3 substitutions preserve the E3 promoter and different E3 genes (only certain open-reading frames are substituted).[40] Except for ADP, all other E3 proteins have immunomodulatory functions and are expressed at the early phase of the viral cycle (8–12 h postinfection, before genome replication). The 6.9K/19K have overlapping reading frames: the 6.9K stop codon TGA is at the second nucleotide of the start codon (ATG) of 19K (…ccaag*ATG*Attaggtac…). The 6.9K/19K substitution (nt 28532 Nhe site–29355 Mun I site) offers early gene expression and wild-type level of other *E3* genes at the expense of a possible higher immunoclearance because the anti-TNF protection given by 6.7K and the anti-MHC-I retention activity of 19K are lost.[41] A 19K-only substitution can render also very high levels of expression and it can be designed adding a TAA stop codon for 6.7K, a Kozak sequence, and the start codon of the exogenous protein (ccaagTAAccATG…), followed by the rest of the transgene ORF replacing the E3-19K ORF.[42] The ADP substitution decreases viral release, but as a side effect the infected cell viability and protein synthesis are extended over time, and a slight increase in virus production is observed.[41] In addition, ADP is the only *E3* gene expressed at the late phase, which implies that expression is tumor selective in oncolytic adenoviruses. Finally, the substitution of RID genes and 14.7K (which together form the E3B region) has some unexpected favorable traits such as higher levels of expression, late phase expression, ADP overexpression, and faster cytotoxicity, at the expense of the anti-TNF activity provided by these three proteins.[43] The immunomodulatory activity of E3 requires an immunocompetent model to fairly test the effect of their substitution. In this setting, viruses with an intact *E3* region showed higher replication and antitumor responses.[44] In general terms, the insertion of transgenes in E3 has shown more interferences with adjacent viral genes and promoters, and less reliable expression and genome stability compared to insertion at E1, after fiber, or next to the right ITR.

## 4. Modification of Capsid Proteins to Achieve Tumor Targeting, Enhance Infectivity, and Display Antigens

Adenovirus has a nonenveloped icosahedral capsid with hexon forming the 20 facets and fibers at the 12 vertices. Penton base proteins connect the fiber to the facets and protein IX between hexon cements the capsid. These four proteins are the major determinants of the pharmacokinetics (clearance from blood), organ and tissue biodistribution, and cellular tropism of the virus. Human adenovirus type 5 in nature infects through the respiratory track, and viremia, although it can occur if the infection is not blunted by the immune system, does not account for the natural histopathology of the infection. Therefore, capsid interactions in blood have been studied recently by those interested in the systemic delivery of the virus as a therapeutic agent.[45] To achieve the maximum delivery of virus particles to tumor sites intravenously, several obstacles must be overcome: avoid neutralizing interactions with antibodies, blood factors, and blood cells; avoid extravasation in nontargeted organs (mainly liver and spleen); avoid uptake by macrophages (mainly Kupffer cells); induce extravasation and penetration in tumors; and promote the efficient entry of virus into tumor cells. All these aspects of the systemic tumor-targeting challenge can be approached with capsid modifications.

Most humans (85%) have been preexposed to adenovirus type 5 and have neutralizing antibodies. These target mainly the exposed hypervariable regions (HVR) (or loops) of the hexon and neutralize the virus by blocking postentry steps of the virus cellular trafficking. Replacing the HVR from one serotype to another can avoid this antibody-mediated neutralization.[46] For CAR-binding Ads, HVR1 seems the most important region for the replacement, in agreement with the major role of HVR1 to bind dinein and mediate the transport of the virus to the nuclear pores after leaving the endocytic vesicles. Serotypes that do not enter through CAR (such as Ad48, which uses CD46) are useful for avoiding neutralizing antibodies and other capsid interactions but they follow a different uptake pathway. Replacing the HVR1 of Ad5 with that of Ad48 impairs dinein binding and nuclear transport.[47] Therefore, to avoid neutralizing antibodies while preserving trafficking, the HVR1 of a CAR-binding Ad should be replaced with the HVR1 of another CAR-binding Ad serotype. Specific residues of certain serotypes located at the hexon HVR are also involved in capsid binding to FX; therefore, replacing them with HVR from nonbinding serotypes avoids this interaction, but again, the efficiency of nuclear transport mediated by the modified hexons should be carefully evaluated. Further, the benefit of FX ablation is controversial as FX protects virus from complement binding and degradation and reduces the virus-induced inflammation, and it may contribute to infection of tumor cells. Negative residues of HVR1, 2, 5, and 7 of Ad5 also bind to scavenger receptors in macrophages and endothelial cells, and replacing them with less charged residues from other serotypes (such as Ad6) can inhibit this uptake and clearance—a clear advantage, as long as dinein binding is not affected.

The interaction of fiber with CAR promotes Ad5 binding to human erythrocytes (mouse erythrocytes do not express CAR). This interaction can be abrogated using CAR-ablating fiber mutations such as the replacement of aa Y fiber aa 477 (nt 32465–32467) with aa A (Y477A mutation). However, as the virus will remain at the surface of the erythrocyte (integrins would be required for entry) and it is still infective,[48] if CAR-binding shows an association–dissociation reversible equilibrium, erythrocyte binding may not preclude tumor targeting and, conversely, erythrocytes could ferry viruses to tumors. A similar unpredictable effect can happen with platelets, whose integrins bind the RGD motif of the penton base and, with their tumor tropism and without viral uptake, they could promote tumor targeting.

While most hexon modifications aim to promote a passive tumor targeting by avoiding undesired neutralizing and clearance interactions, some modifications aim at an active tumor targeting, seeking new receptor interactions. The modification of the hexon with high-affinity ligands for tumor cell receptors has failed because, contrary to the fiber, the hexon remains attached to the capsid during the transport to the nuclear pore. If the interaction with the receptor is of high affinity, the hexon remains stuck to the receptor and this precludes the transport to the nuclei (so-called "sticky virus"). A competition for the interaction with dinein can also explain the deficient nuclear transport of the hexon-modified viruses. A different outcome occurs with low-affinity ligands (e.g., peptides such as RGD[49]) or with nonreceptor-based membrane interactions (such as the hexon modification with the protein transduction domain TAT from HIV[50]) that promote virus infection in hard-to-infect cells.[51]

The fiber is the main target for promoting an active tumor targeting by high-affinity ligands. The major challenge in the insertion of ligands to the fiber has been the structural constraints of the trimeric fiber, where the fiber C-terminal knob exposed to the solvent is crucial for fiber trimerization. Insertion ligands at the fiber target the exposed HI loop or the C terminus of the knob, or are based on the replacement of the entire knob. The relatively long, flexible, and exposed HI loop can accept exogenous peptides up to 83 aa long. Among different peptides, the integrin-binding RGD4C has been one of the more useful ligands for promoting infectivity.[52] Although wild-type Ad5 has RGD at the penton base and uses integrins for virus entry after the fiber has bound to CAR, the presence of the RGD at the fiber allows the use of integrins as a primary receptor independent of CAR binding. As many tumor cells lack CAR but express integrins, this RGD insertion has proven widely useful. RGD has also been used to replace the KKT residues of the fiber shaft that contributed to hepatocyte infectivity on systemic administration.[53] This RGD location at the shaft promoted simultaneously some liver detargeting and tumor targeting. Other interesting peptides that have been inserted at the HI loop to broadly increase the CAR-independent transduction of adenoviruses are derived from protein transduction domains, such as the TAT48-60 peptide of HIV.[54] The HI loop has also been used to insert randomized variants of the Z domain of staphylococcal protein A-named affibodies that may be used to target selected receptors.[55] The C terminus of the fiber (…YIAQUE*) hides below the solvent-exposed trifoil of the knob. This C-terminal end can be extended up to around 80 residues but it needs a long flexible linker (usually formed by serines or alanines, GSx5 or PSASASASAPGSGS) to display the terminal peptide beyond the umbrella formed by the trifoil.[56] Again, this has restricted the C-terminal modification strategy to small peptides, and even then, many of them lose the binding to the target (such as the GYIGSR laminin ligand) or result unviable (such as the laminin-binding SIKVAV, the E-selectin binding peptide TRSDITWDQLWDLMKTS, the 10 amino acid gastrin-releasing peptide, or the 13 amino acid melanocyte-stimulating hormone). This C-terminal location is most appropriate for peptides that only bind the receptor in a free-end conformation, for example, for the insertion of the iRGD peptide that promotes both integrin targeting and neuropilin-mediated extravasation.[57] The C-terminal location is also very useful for inserting ligands that become activated on proteolysis, such as a metalloprotease-activatable TAT domain.[58] However, for ligands that do not need to be terminal (with one free end) such as RGD or TAT, the fiber HI loop has proven superior to the C terminus for targeting.[54] To expose a large peptide, such as a receptor-binding protein or a fragment of an antibody, the peptide must replace the knob. In this design, the trimerization function of the knob has been provided by the fiber–fibritin domain of bacteriophage T4 fibritin or by a 36 aa trimerization motif of the human lung surfactant protein D.[59] Although some in vitro successful reports have been published, stability issues seem to preclude the large-scale production of these recombinants. Even in the most efficient configuration, containing the whole shaft (ending at VGN residues) followed by the 12th coiled–coil segment of fibritin and a (GGGS)x3 linker (Table 1), retargeted fibers (to CD40L) were poorly incorporated in the capsid.[60] In addition, fiber synthesis at the cytoplasm made the use of ligands with disulfide bonds difficult.[59] Fiber-mediated targeting with very specific high-affinity ligands (such and single chain antibodies) has other disadvantages: there is no true tumor-selective receptor, and given the tumor heterogeneity at the receptor level, a very narrow receptor specificity may lead to tumor-resistant or escape variants without the receptor. Adenovirus-infected cells secrete large amounts of fiber to block receptors on adjacent cells to self-limit its propagation in the host.[61]

An alternative to the insertion of exogenous ligands in the Ad5 fiber has been the use of recombinant chimeric fibers where knob,[62] or the knob and shaft,[63] is from a serotype other than Ad5. For knob-only replacement a good fusion point is the flexible motif formed by the TLWT residues between the shaft and the knob often preserved among serotypes.[62] The most common serotypes have been Ad3, which binds to desmoglein 2, and Ad35, which binds to CD46. These two receptors have been found overexpressed in tumors. This overexpression leads to selective tumor targeting in vivo.[64] Fibers with long shafts, such as the Ad5 fiber, bind to blood factors that promote virus

**Table 1 Selected Relevant Sequences for the Designs of Oncolytic Adenoviruses: Amino Acid (When the Sequence is Translated) and Nucleotide Sequences are Shown**

| Sequence id | Sequence: AA(nt) or nt | Ref |
|---|---|---|
| 2A teschovirus | GSGATNFSLLKQAGDVEENPGP (gggtccg-gagccacgaacttctctctgttaaagcaagcaggagacgtg-gaagaaaaccccgggccc) | 32 |
| IIIa SA (3VDE) | agtactaagcggtgatgtttctgatcag | 35,36 |
| BPS | tgcttatcttcctttctctcttcag | 28,31 |
| RGD4C | CDCRGDCFC (tgtgactgccgcggagactgtttctgc) | 52 |
| iRGD cter +pA | CRGDKGPDC-stop (tgccgcggtgacaagggacccgactgctaataaa) | 57 |
| TAT48-60 | GRKKRRQRRRPPQ (ggcaggaagaagcggagacagcgacgaagacctcctcaa) | 50,51 |
| Tat-like peptide | YGSKKRRQRRRGG (tatggcagcaagaagcggagacagcgacgaagaggcggc) | 50,54 |
| MMP-cleavable linker | AKGLYK (gccaagggcctgtacaag) | 58 |
| TAT-blocking sequence cter +pA | GGEEEGEEEEE-stop (ggcggcgaggaagagggc-gaggaggaagaggagtaataaa) | 58 |
| (GGGS)x3 linker | GGGGSGGGGSGGGGS (ggcggaggaggtagcggt-ggaggtggatctgggggaggtggctcc) | 56 |
| T4 fibritin 12th coiled–coil | LTNS IKANETNIASVTQEVN-TAKGNISSLQGDVQALQEA (ttaaccaattcaataaaagctaacgaaactaacattgcatcagt-tacacaagaagtgaatacagctaaaggcaatatatcttctttta-caaggtgatgttcaagctctccaagaagcc) | 60 |
| T4 fibritin foldon | GYIPEAPRDGQAYVRKDGEWVFLSTFLSPA (ggttatattcctgaagctccaagagatgggcaagcttacgttcg-taaagatggcgaatgggtattcctttctacctttttatcaccagca) | 60 |
| Alpha-helical linker for pIX display | EETRARLSKELQAAQARLGAD-MEDVCGRLVQYRGEVQA (gaggagacccgggcccgcctctccaaggagctgcaggccg-cccaggcccgcctgggcgccgacatggaggacgtgtgcggc-cgcctggtgcagtaccgcggggaggtgcaggcc) | 60 |

uptake by Kupffer cells and infection of hepatocytes.[65] As Ad3 and Ad35 have short fibers, a chimera that includes those shafts untargets these cells. An oncolytic virus can be designed to express two complete fibers that assemble at different vertices, forming a mosaic capsid.[66] The fiber from the porcine Ad4 is trimeric in the N-terminal half but becomes momomeric for the rest. This unique structure may be ideal for designing a chimeric fiber that allows the insertion of large targeting ligands.[67]

The adenovirus capsid can be modified to display epitopes of antigens. This modification, commonly used in vaccine designs, can be applied to immunostimulatory oncolytic adenoviruses. The modified capsid, even if neutralized by preexisting antibodies and taken up by macrophages, will induce immune responses to the displayed epitopes. The variable and exposed loops of fiber and hexon have been the main targets for antigenic epitope display. A comparison of locations indicated a greater antigenicity when located at the fiber.[68,69] Another Ad protein of particular interest on epitope insertion is E3-19K as it can deliver the epitopes directly to the endoplasmic reticulum to be loaded to the MHC in a TAP-independent way (A. Rodriguez-Garcia and R. Alemany, unpublished). As tumor cells are often TAP deficient to escape immune presentation, this strategy is particularly designed to promote responses against tumor epitopes.

Finally the C terminus of pIX (nt 4028) (…PPNAV*) has been also used to fuse exogenous polypeptides of up to 113 amino acids or even the complete GFP protein.[70,71] A long alpha-helical spacer followed by the flexible (GGGS)x3 is used to help expose the peptide ligands to the virus surface. Propagation and retargeting efficiency of pIX modified ads have been so far modest, suggesting structural limitations on pIX genetic modification or a possible impairment of the nonstructural functions of pIX, such as transcriptional activity and PML nuclear body reorganization. Despite this, viruses have been rescued with large proteins fused to pIX, such as GFP, RFP, luciferase, thymidine kinase, or a hyperstable scFv.[72,73] pIX has also been used to display antigens on the capsid,[74] offering a larger capability compared to hexon and fiber. For retargeting purposes, as pIX does not detach from the capsid on cell entry, it is expected that the same problem of "sticky" viral particles as found with the hexon occurs, and only low-affinity ligands become compatible with virus nuclear transport. For pIX modification capsid stability (at 45C) should be tested, as a minor C-terminal modification of pIX may affect it.[32]

## Acknowledgments

The author is grateful to the students and postdoctoral members of his group for their contribution to the designs described in this chapter. The author's work is supported by grants MICINN BIO2011-30299-C02-01; AECC Barcelona Section 2013 Grant; and EU Project number 29002. FP7-PEOPLE-2011-ITN Advance.

## References

1. Gimenez-Alejandre M, Gros A, Alemany R. Construction of capsid-modified adenoviruses by recombination in yeast and purification by iodixanol-gradient. *Methods Mol Biol* 2012;**797**:21–34.
2. Stanton RJ, McSharry BP, Armstrong M, Tomasec P, Wilkinson GW. Re-engineering adenovirus vector systems to enable high-throughput analyses of gene function. *Biotechniques* 2008;**45**:659–62, 664–8.
3. Fueyo J, Gomez-Manzano C, Alemany R, et al. A mutant oncolytic adenovirus targeting the Rb pathway produces anti-glioma effect in vivo. *Oncogene* 2000;**19**:2–12.
4. Heise C, Hermiston T, Johnson L, et al. An adenovirus E1A mutant that demonstrates potent and selective systemic anti-tumoral efficacy. *Nat Med* 2000;**6**:1134–9.
5. Ferrari R, Gou D, Jawdekar G, et al. Adenovirus small E1A employs the lysine acetylases p300/CBP and tumor suppressor Rb to repress select host genes and promote productive virus infection. *Cell Host Microbe* 2015;**16**:663–76.

6. Balague C, Noya F, Alemany R, Chow LT, Curiel DT. Human papillomavirus E6E7-mediated adenovirus cell killing: selectivity of mutant adenovirus replication in organotypic cultures of human keratinocytes. *J Virol* 2001;**75**:7602–11.
7. Howe JA, Demers GW, Johnson DE, et al. Evaluation of E1-mutant adenoviruses as conditionally replicating agents for cancer therapy. *Mol Ther* 2000;**2**:485–95.
8. Holm PS, Lage H, Bergmann S, et al. Multidrug-resistant cancer cells facilitate E1-independent adenoviral replication: impact for cancer gene therapy. *Cancer Res* 2004;**64**:322–8.
9. Sauthoff H, Heitner S, Rom WN, Hay JG. Deletion of the adenoviral E1b-19kD gene enhances tumor cell killing of a replicating adenoviral vector. *Hum Gene Ther* 2000;**11**:379–88.
10. Subramanian T, Vijayalingam S, Chinnadurai G. Genetic identification of adenovirus type 5 genes that influence viral spread. *J Virol* 2006;**80**:2000–12.
11. Cuconati A, Degenhardt K, Sundararajan R, Anschel A, White E. Bak and Bax function to limit adenovirus replication through apoptosis induction. *J Virol* 2002;**76**:4547–58.
12. Oberg D, Yanover E, Adam V, et al. Improved potency and selectivity of an oncolytic E1ACR2 and E1B19K deleted adenoviral mutant in prostate and pancreatic cancers. *Clin Cancer Res* 2010;**16**:541–53.
13. Radke JR, Grigera F, Ucker DS, Cook JL. Adenovirus E1B 19-kilodalton protein modulates innate immunity through apoptotic mimicry. *J Virol* 2014;**88**:2658–69.
14. Bischoff JR, Kirn DH, Williams A, et al. An adenovirus mutant that replicates selectively in p53-deficient human tumor cells. *Science* 1996;**274**:373–6.
15. Shen Y, Kitzes G, Nye JA, Fattaey A, Hermiston T. Analyses of single-amino-acid substitution mutants of adenovirus type 5 E1B-55K protein. *J Virol* 2001;**75**:4297–307.
16. Cascallo M, Capella G, Mazo A, Alemany R. Ras-dependent oncolysis with an adenovirus VAI mutant. *Cancer Res* 2003;**63**:5544–50.
17. Kondo S, Yoshida K, Suzuki M, Saito I, Kanegae Y. Adenovirus-encoding virus-associated RNAs suppress HDGF gene expression to support efficient viral replication. *PLoS One* 2014;**9**:e108627.
18. Hatfield L, Hearing P. Redundant elements in the adenovirus type 5 inverted terminal repeat promote bidirectional transcription in vitro and are important for virus growth in vivo. *Virology* 1991;**184**:265–76.
19. Zhang J, Ramesh N, Chen Y, et al. Identification of human uroplakin II promoter and its use in the construction of CG8840, a urothelium specific adenovirus variant that eliminates established bladder tumors in combination with docetaxel. *Cancer Res* 2002;**62**:3743–50.
20. Yu DC, Sakamoto GT, Henderson DR. Identification of the transcriptional regulatory sequences of human kallikrein 2 and their use in the construction of calydon virus 764, an attenuated replication competent adenovirus for prostate cancer therapy. *Cancer Res* 1999;**59**:1498–504.
21. Brunori M, Malerba M, Kashiwazaki H, Iggo R. Replicating adenoviruses that target tumors with constitutive activation of the wnt signaling pathway. *J Virol* 2001;**75**:2857–65.
22. Doronin K, Kuppuswamy M, Toth K, et al. Tissue-specific, tumor-selective, replication-competent adenovirus vector for cancer gene therapy. *J Virol* 2001;**75**:3314–24.
23. Majem M, Cascallo M, Bayo-Puxan N, Mesia R, Germa JR, Alemany R. Control of E1A under an E2F-1 promoter insulated with the myotonic dystrophy locus insulator reduces the toxicity of oncolytic adenovirus Ad-Delta24RGD. *Cancer Gene Ther* 2006;**13**:696–705.
24. Yan W, Kitzes G, Dormishian F, et al. Developing novel oncolytic adenoviruses through bioselection. *J Virol* 2003;**77**:2640–50.
25. Gros A, Martinez-Quintanilla J, Puig C, et al. Bioselection of a gain of function mutation that enhances adenovirus 5 release and improves its antitumoral potency. *Cancer Res* 2008;**68**:8928–37.

26. Puig-Saus C, Gros A, Alemany R, Cascallo M. Adenovirus i-leader truncation bioselected against cancer-associated fibroblasts to overcome tumor stromal barriers. *Mol Ther* 2011;**20**:54–62.
27. Toth K, Djeha H, Ying B, et al. An oncolytic adenovirus vector combining enhanced cell-to-cell spreading, mediated by the ADP cytolytic protein, with selective replication in cancer cells with deregulated wnt signaling. *Cancer Res* 2004;**64**:3638–44.
28. Sauthoff H, Pipiya T, Heitner S, et al. Late expression of p53 from a replicating adenovirus improves tumor cell killing and is more tumor cell specific than expression of the adenoviral death protein. *Hum Gene Ther* 2002;**13**:1859–71.
29. Choi JW, Lee JS, Kim SW, Yun CO. Evolution of oncolytic adenovirus for cancer treatment. *Adv Drug Deliv Rev* 2012;**64**:720–9.
30. Jin F, Kretschmer PJ, Hermiston TW. Identification of novel insertion sites in the Ad5 genome that utilize the Ad splicing machinery for therapeutic gene expression. *Mol Ther* 2005;**12**:1052–63.
31. Rivera AA, Wang M, Suzuki K, et al. Mode of transgene expression after fusion to early or late viral genes of a conditionally replicating adenovirus via an optimized internal ribosome entry site in vitro and in vivo. *Virology* 2004;**320**:121–34.
32. Funston GM, Kallioinen SE, de Felipe P, Ryan MD, Iggo RD. Expression of heterologous genes in oncolytic adenoviruses using picornaviral 2A sequences that trigger ribosome skipping. *J Gen Virol* 2008;**89**:389–96.
33. Quirin C, Rohmer S, Fernandez-Ulibarri I, et al. Selectivity and efficiency of late transgene expression by transcriptionally targeted oncolytic adenoviruses are dependent on the transgene insertion strategy. *Hum Gene Ther* 2011;**22**:389–404.
34. Carette JE, Graat HC, Schagen FH, Abou El, Hassan MA, Gerritsen WR, et al. Replication-dependent transgene expression from a conditionally replicating adenovirus via alternative splicing to a heterologous splice-acceptor site. *J Gene Med* 2005;**7**:1053–62.
35. Guedan S, Gros A, Cascallo M, Vile R, Mercade E, Alemany R. Syncytia formation affects the yield and cytotoxicity of an adenovirus expressing a fusogenic glycoprotein at a late stage of replication. *Gene Ther* 2008;**15**:1240–5.
36. Muhlemann O, Kreivi JP, Akusjarvi G. Enhanced splicing of nonconsensus 3′ splice sites late during adenovirus infection. *J Virol* 1995;**69**:7324–7.
37. Fuerer C, Iggo R. 5-Fluorocytosine increases the toxicity of Wnt-targeting replicating adenoviruses that express cytosine deaminase as a late gene. *Gene Ther* 2004;**11**:142–51.
38. Fernandez-Ulibarri I, Hammer K, Arndt MA, et al. Genetic delivery of an immunoRNase by an oncolytic adenovirus enhances anticancer activity. *Int J Cancer* 2015;1**36**:2228–40.
39. Pei Z, Shi G, Kondo S, et al. Adenovirus vectors lacking virus-associated RNA expression enhance shRNA activity to suppress hepatitis C virus replication. *Sci Rep* 2013;**3**:3575.
40. Bauzon M, Castro D, Karr M, Hawkins LK, Hermiston TW. Multigene expression from a replicating adenovirus using native viral promoters. *Mol Ther* 2003;**7**:526–34.
41. Hawkins LK, Johnson L, Bauzon M, et al. Gene delivery from the E3 region of replicating human adenovirus: evaluation of the 6.7 K/gp19 K region. *Gene Ther* 2001;**8**:1123–31.
42. Zhu M, Bristol JA, Xie Y, et al. Linked tumor-selective virus replication and transgene expression from E3-containing oncolytic adenoviruses. *J Virol* 2005;**79**:5455–65.
43. Hawkins LK, Hermiston T. Gene delivery from the E3 region of replicating human adenovirus: evaluation of the E3B region. *Gene Ther* 2001;**8**:1142–8.
44. Wang Y, Hallden G, Hill R, et al. *E3* gene manipulations affect oncolytic adenovirus activity in immunocompetent tumor models. *Nat Biotechnol* 2003;**21**:1328–35.
45. Coughlan L, Alba R, Parker AL, et al. Tropism-modification strategies for targeted gene delivery using adenoviral vectors. *Viruses* 2011;**2**:2290–355.

46. Bradley RR, Maxfield LF, Lynch DM, Iampietro MJ, Borducchi EN, Barouch DH. Adenovirus serotype 5-specific neutralizing antibodies target multiple hexon hypervariable regions. *J Virol* 2012;**86**:1267–72.
47. Scherer J, Vallee RB. Conformational changes in the adenovirus hexon subunit responsible for regulating cytoplasmic dynein recruitment. *J Virol* 2015;**89**:1013–23.
48. Cichon G, Boeckh-Herwig S, Kuemin D, et al. Titer determination of Ad5 in blood: a cautionary note. *Gene Ther* 2003;**10**:1012–7.
49. Vigne E, Mahfouz I, Dedieu JF, Brie A, Perricaudet M, Yeh P. RGD inclusion in the hexon monomer provides adenovirus type 5-based vectors with a fiber knob-independent pathway for infection. *J Virol* 1999;**73**:5156–61. [the above report in]
50. Yu D, Jin C, Leja J, et al. Adenovirus with hexon Tat-protein transduction domain modification exhibits increased therapeutic effect in experimental neuroblastoma and neuroendocrine tumors. *J Virol* 2011;**85**:13114–23.
51. Jin C, Yu D, Cancer M, Nilsson B, Leja J, Essand M. Tat-PTD-modified oncolytic adenovirus driven by the SCG3 promoter and ASH1 enhancer for neuroblastoma therapy. *Hum Gene Ther* 2013;**24**:766–75.
52. Krasnykh V, Dmitriev I, Mikheeva G, Miller CR, Belousova N, Curiel DT. Characterization of an adenovirus vector containing a heterologous peptide epitope in the HI loop of the fiber knob. *J Virol* 1998;**72**:1844–52.
53. Bayo-Puxan N, Gimenez-Alejandre M, Lavilla-Alonso S, et al. Replacement of adenovirus type 5 fiber shaft heparan sulfate proteoglycan-binding domain with RGD for improved tumor infectivity and targeting. *Hum Gene Ther* 2009;**20**:1214–21.
54. Kurachi S, Tashiro K, Sakurai F, et al. Fiber-modified adenovirus vectors containing the TAT peptide derived from HIV-1 in the fiber knob have efficient gene transfer activity. *Gene Ther* 2007;**14**:1160–5.
55. Myhre S, Henning P, Friedman M, Stahl S, Lindholm L, Magnusson MK. Re-targeted adenovirus vectors with dual specificity; binding specificities conferred by two different Affibody molecules in the fiber. *Gene Ther* 2009;**16**:252–61.
56. Wickham TJ, Tzeng E, Shears LL, et al. Increased in vitro and in vivo gene transfer by adenovirus vectors containing chimeric fiber proteins. *J Virol* 1997;**71**:8221–9.
57. Puig-Saus C, Rojas LA, Laborda E, et al. iRGD tumor-penetrating peptide-modified oncolytic adenovirus shows enhanced tumor transduction, intratumoral dissemination and antitumor efficacy. *Gene Ther* 2014;**21**:767–74.
58. Jose A, Rovira-Rigau M, Luna J, et al. A genetic fiber modification to achieve matrix-metalloprotease-activated infectivity of oncolytic adenovirus. *J Control Release* 2014;**192**:148–56.
59. Magnusson MK, Hong SS, Henning P, Boulanger P, Lindholm L. Genetic retargeting of adenovirus vectors: functionality of targeting ligands and their influence on virus viability. *J Gene Med* 2002;**4**:356–70.
60. Noureddini SC, Krendelshchikov A, Simonenko V, et al. Generation and selection of targeted adenoviruses embodying optimized vector properties. *Virus Res* 2006;**116**:185–95.
61. Rebetz J, Na M, Su C, et al. Fiber mediated receptor masking in non-infected bystander cells restricts adenovirus cell killing effect but promotes adenovirus host co-existence. *PLoS One* 2009;**4**:e8484.
62. Rogozhin VN, Logunov DY, Shchebliakov DV, et al. An efficient method for the delivery of the interleukin-2 gene to human hematopoietic cells using the fiber-modified recombinant adenovirus. *Acta Naturae* 2011;**3**:100–6.
63. Shayakhmetov DM, Papayannopoulou T, Stamatoyannopoulos G, Lieber A. Efficient gene transfer into human CD34(+) cells by a retargeted adenovirus vector. *J Virol* 2000;**74**:2567–83.

64. Ni S, Gaggar A, Di Paolo N, et al. Evaluation of adenovirus vectors containing serotype 35 fibers for tumor targeting. *Cancer Gene Ther* 2006;**13**:1072–81.
65. Shayakhmetov DM, Gaggar A, Ni S, Li ZY, Lieber A. Adenovirus binding to blood factors results in liver cell infection and hepatotoxicity. *J Virol* 2005;**79**:7478–91.
66. Takayama K, Reynolds PN, Short JJ, et al. A mosaic adenovirus possessing serotype Ad5 and serotype Ad3 knobs exhibits expanded tropism. *Virology* 2003;**309**:282–93.
67. Kim JW, Glasgow JN, Nakayama M, Ak F, Ugai H, Curiel DT. An adenovirus vector incorporating carbohydrate binding domains utilizes glycans for gene transfer. *PLoS One* 2013;**8**:e55533.
68. Krause A, Joh JH, Hackett NR, et al. Epitopes expressed in different adenovirus capsid proteins induce different levels of epitope-specific immunity. *J Virol* 2006;**80**:5523–30.
69. Sharma A, Krause A, Xu Y, Sung B, Wu W, Worgall S. Adenovirus-based vaccine with epitopes incorporated in novel fiber sites to induce protective immunity against *Pseudomonas aeruginosa*. *PLoS One* 2013;**8**:e56996.
70. Meulenbroek RA, Sargent KL, Lunde J, Jasmin BJ, Parks RJ. Use of adenovirus protein IX (pIX) to display large polypeptides on the virion—generation of fluorescent virus through the incorporation of pIX-GFP. *Mol Ther* 2004;**9**:617–24.
71. Vellinga J, Rabelink MJ, Cramer SJ, et al. Spacers increase the accessibility of peptide ligands linked to the carboxyl terminus of adenovirus minor capsid protein IX. *J Virol* 2004;**78**:3470–9.
72. Glasgow JN, Everts M, Curiel DT. Transductional targeting of adenovirus vectors for gene therapy. *Cancer Gene Ther* 2006;**13**:830–44.
73. Vellinga J, de Vrij J, Myhre S, et al. Efficient incorporation of a functional hyper-stable single-chain antibody fragment protein-IX fusion in the adenovirus capsid. *Gene Ther* 2007;**14**:664–70.
74. Bayer W, Tenbusch M, Lietz R, et al. Vaccination with an adenoviral vector that encodes and displays a retroviral antigen induces improved neutralizing antibody and CD4+ T-cell responses and confers enhanced protection. *J Virol* 2010;**84**:1967–76.

# Conditionally Replicative Adenoviruses—Clinical Trials

13

*Enric Xipell*[4], *Yisel Rivera-Molina*[1], *Candelaria Gomez-Manzano*[1,2], *Hong Jiang*[1], *Marta M. Alonso*[4], *Ramon Alemany*[5,*], *Juan Fueyo*[1,3,*]
[1]Department of Neuro-Oncology, The University of Texas MD Anderson Cancer Center, Houston, Texas, USA; [2]Department of Genetics, The University of Texas MD Anderson Cancer Center, Houston, Texas, USA; [3]Department of Neurosurgery, The University of Texas MD Anderson Cancer Center, Houston, Texas, USA; [4]Department of Medical Oncology, Clínica Universidad de Navarra, University of Navarra, Pamplona, Spain; [5]IDIBELL-Institut Català d'Oncologia, L'Hospitalet de Llobregat, Barcelona, Spain

## 1. Origins of Cancer Virotherapy

The effects of cancer virotherapy were first described more than 100 years ago. In 1912, DePace reported that administration of an attenuated prophylactic rabies virus to a woman bitten by a dog-induced necrosis in cervical cancer.[1] The next advance in virotherapy emerged in later decades. In 1950, tissue culture techniques that made possible the growth of cells and viruses in the laboratory allowed scientists to make rapid progress in characterizing these biological agents, one of which was adenovirus.[2,3]

In 1953, the laboratory of Ward and colleagues demonstrated that an infectious agent present in the adenoid culture obtained from a child could be cultured in vitro, and although the tissue culture could not be maintained for long, they were able to transmit the infective agent to other different cultures including HeLa cells.[2] To determine the origin of this new agent, this group filtered the medium and tried to cultivate the infective agent in bacteriological medium, without success. They concluded that the new cytopathic agent in the culture that could not be trapped by a bacterial filter was a virus. About the same time that adenoviruses were discovered, virotherapy was arriving in the clinical setting.

In 1949, the hepatitis B virus was administered to cancer patients to combat Hodgkin lymphoma, a disease then considered untreatable and lethal.[4] Results of the second clinical assay were published in 1952, when the Egypt 101 virus was used against advanced and unresponsive neoplastic disease.[5,6] Tumor reduction was evident in 10% of patients, a small but significant percentage. However, in both clinical scenarios, the investigators also reported numerous side effects that included fever, malaise, mild encephalitis, and, in the most severe cases, death. Because of the unwanted secondary effects of the previously used viruses, investigators tested safer viruses in the clinic. Huebner, who had worked with Ward's group in identifying adenoviruses, proposed

* Both authors contributed equally.

**Adenoviral Vectors for Gene Therapy.**

using these newly identified agents as tools against cancer. This led to clinical testing in 1956 of replication-competent adenoviruses as anticancer tools in patients with cervical cancer.[7] In addition to the fact that adenovirus, unlike the hepatitis B and Egypt 101 viruses, caused only mild disease, the rationale for the clinical trial was based on the documented ability of adenoviruses to replicate in and kill HeLa cells, a cell line derived from a patient with cervical cancer and maintained in vitro since 1951. In these studies, 30 women with cervical carcinoma were treated with mixtures of several serotypes of wild-type adenoviruses. After adenovirus administration, the patients experienced clinical symptoms similar to adenoviral infection, such as fever and malaise. Importantly, the investigators observed necrosis in the tumor site in more than 50% of patients. However, and of further clinical relevance, tumor regression was never complete and the infection could not prevent regrowth of the tumor and disease progression that eventually resulted in death.[7] Despite the potential benefit of adenovirus inoculation in treating cancer, investigators did not retest the idea of replication-competent adenoviruses in the clinical setting until more than 40 years later.

Over the next 50 years, after these first clinical trials, researchers concentrated on in-depth characterization of the adenovirus infection and replication cycle.[8,9] Such studies were required to achieve the next objective in virotherapy: genetic modification of the adenovirus genome (Figure 1). On the basis of new knowledge of the mechanisms of adenoviral replication, pioneers of vectorology proposed deleting the proteins that are indispensable for viral replication, to make the virus replication deficient, and replacing them with exogenous genes. Thus, the safe adenoviral vector would infect and transduce the ectopic gene without killing the cell by lysis. The first strategy proposed included removing the viral E1 genes, which encode the first proteins expressed after infection and orchestrate the transcription of the rest of the adenoviral genes. The empty space in the adenovirus genome was used to insert a mini-cassette to express a foreign therapeutic or reporter protein in the target cell in a virus-independent way. This strategy required the development of a cell line that was transformed with the *E1* gene, allowing replication of the adenoviral vectors and their production at high titers in any laboratory.[10] With this change, viral particles that lack the ability to express viral genes are obtained.[11]

The Ad-p53 adenoviral vector was one of the first constructs built by taking advantage of the E1-deleted strategy. Ad-p53 is capable of transferring the p53 complementary deoxyribonucleic acid (cDNA), which encodes for a potent apoptosis inducer that selectively kills tumor cells expressing a mutated p53.[12] In vitro data produced by Gomez-Manzano et al.[13] provided the rationale for the development of a clinical trial in patients with brain tumors at the University of Texas MD Anderson Cancer Center. The dose-escalation phase I trial, led by Lang et al.,[12] involved the intratumoral injection of Ad-p53 into 15 patients with recurrent glioma, followed by total resection of the treated tumor 3 days later. This design allowed for an examination of the function of the ectopic p53 in gliomas. Results from the trial showed a lack of toxicity, and thus the maximum tolerated dose was not achieved. However, the Ad-p53 treatment did not induce complete regression in any of the patients and the improvement in survival was modest. Interestingly, histopathologic examination of the tumors revealed transference of the *p53* gene through detecting the expression of p53 protein that could

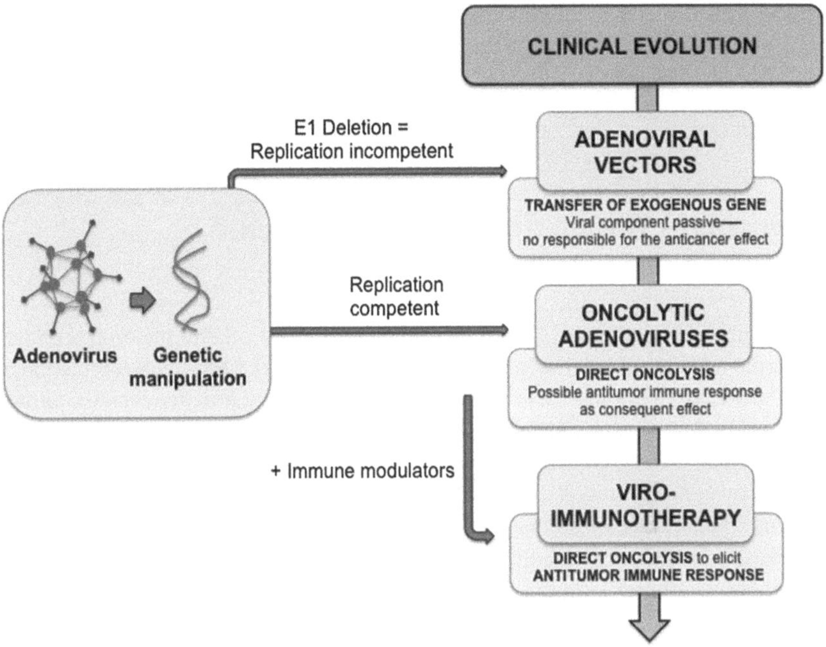


**Figure 1** Evolution in the clinical use of adenoviruses for the treatment of cancer. Adenoviruses have been extensively tested as vectors to transfer exogenous genes to cancer cells. Sophisticated modifications of the adenovirus genome to improve infectivity and selectivity have made possible the production of oncolytic adenoviruses intended to destroy tumors directly by adenovirus-mediated cell lysis. In some instances, the use of adenovirus might elicit not only an antiviral immune response, but also a desirable antitumoral response. The combination of oncolytic adenovirus and immune modulators is predicted to be tested in the near future.

transduce well-characterized transcriptional targets such as p21.[12] The histopathology of the tumors also showed that transduction of the *p53* gene was observed in an area of a few millimeters surrounding the injection site, which emphatically demonstrated that Ad-p53, and for that matter probably any other adenoviral vector, was incapable of transducing enough cells to induce a significant anticancer effect when administered intratumorally.[12]

The Ad-p53 trial and other similar trials using adenoviral vectors showed the need to elucidate ways to improve gene delivery. One way to improve distribution of the adenoviral effect is to enable viral replication. Although there are many other approaches to enhance delivery, the fact that replication-competent viruses may spread through a tumor and therefore be more effective in delivering exogenous genes caused a revival of interest in replication-competent viruses as therapeutic tools. However, wild adenoviruses had already been tested in clinical trials and were shown to be toxic, and therefore the new oncolytic viruses needed to be more potent than wild-type viruses and at the same time safer than wild-type adenoviruses. Basic research into how adenoviral proteins interact with cell proteins and into the mechanisms for adenoviral anchorage on the cell membrane as well as internalization paved the way for the design of a new generation of oncolytic adenoviruses (Figure 1). The first oncolytic adenovirus, dl1520, was designed to exploit the differing status of the p53 protein in normal cells and cancer cells.

## 2. Oncolytic Adenoviruses: The dl1520 and H1011 Concept

Adenoviruses are episomal DNA viruses that require host cells to undergo S phase to activate the adenoviral promoters and the enzymes required for DNA replication and protein synthesis. For these reasons, the early genes of adenovirus encoded proteins that create an appropriate cellular environment for viral replication. However, the initiation of viral DNA replication and the synthesis of adenoviral proteins would be identified by cellular sensors that trigger apoptosis or induce cell cycle arrest and an exit from S phase. One of the main cellular sensors that detect adenovirus replication and induce cell cycle arrest and apoptosis is the protein p53. The p53 protein, like other master regulatory proteins including Rb, is a cellular target for the adenovirus early proteins, specifically the E1B55K protein that binds and inactivates p53.[14]

On the basis of these data, the laboratory of McCormick developed a project based on a mutant adenovirus lacking the expression of E1B55K protein, and which could not inactivate p53. This mutant adenovirus could, at least in theory, replicate in cells with an inactivated mutant p53 but not in normal cells with wild-type p53 protein. Because the p53 protein is mutated in most cancer cells, the E1B55K-deleted adenovirus could specifically target cancer cells. This new oncolytic adenovirus was first called dl1520, and later, ONYX-015.[15]

Preclinical data for dl1520 were attractive,[15] and it was not surprising that the first clinical trial was initiated soon after publication of the study.[15] The dl1520 was tested in several types of cancer[16] but the best clinical results were obtained in phase I and subsequent phase II studies that combined intratumoral injection of dl150 with administration of chemotherapy for patients with head and neck cancer. The phase I study proved the safety of the approach,[17] showing that the most frequent secondary effects were fever and injection site pain without more serious adverse effects.[17] In five patients, injection of the virus reduced the tumor size, but metastatic tumors showed no clinical response.[17] The purposes of the phase II study were to confirm the safety of dl1520 as a treatment against recurrent squamous cell carcinoma of the head and neck in combination with cisplatin and 5-fluorouracil and to evaluate potential clinical responses. The study enrolled 40 patients who received the highest virus dose tested in the phase I trial ($2.3 \times 10^{11}$ plaque-forming units) administered (intratumoral single injection).[18] The study showed again that the most frequent unwanted symptoms were fever and injection site pain. Viral genome was detected in the plasma of 41% of patients 24h after the injection and viremia persisted in 9% of the patients for 9days. Clinical responses were remarkable: 14% of treated patients achieved complete regression, approximately 50% had stable disease, and the rest of the patients

(36%) had disease progression.[18] Unfortunately, despite the results of the phase II study, phase III was canceled; less than 10% of the patients had been enrolled because of a corporate takeover.[19]

Although testing of E1B55K-deleted viruses in the United States came to an unfortunate end, scientists in the People's Republic of China patented an adenovirus, called H101, that incorporated the E1B55K deletion and thus was similar to the virus designed by McCormick. Chinese clinicians initiated a phase III clinical trial to study the effectiveness and toxicity of H101 in combination with a cisplatin plus 5-fluorouracil regimen or a doxorubicin plus 5-fluorouracil regimen. The clinical assay had four different arms involving administration of the two chemotherapies with or without intratumoral administration of H101. This trial enrolled 160 patients, and the results were excellent. The overall response rate was 78.8% in the group treated with cisplatin and 5-fluorouracil and the adenovirus H101. This was the first clinical trial demonstrating the antitumoral efficacy of an adenovirus, albeit in combination with chemotherapy. The success of this phase III trial led to the approval in 2005 of H101, specifically for treatment of nasopharyngeal carcinoma in combination with cisplatin-based chemotherapy. However, this trial was relatively small for a phase III study, compared with phase III of dl1520, which had planned a sample size of more than 450 patients, allowing for strong statistical power to detect important differences in survival between treatment groups.[19] In addition, the standard of care for locally advanced nasopharyngeal carcinoma is the combination of radiotherapy with chemotherapy, which normally results in response rates of 80%. It is unclear why the authors used chemotherapy without radiotherapy as a control group.[19] It was also not completely clear how clinical responses were evaluated, and follow-up of the patients seemed to be difficult or incomplete.[19]

Further studies of the E1B55K-deleted virus concluded that the mechanism of tumor selectivity was not as simple as the targeting of p53.[20] In addition, and relevant for cancer virotherapy, the dl1520 was highly attenuated and replicated poorly in several types of cancers.[8,15,21–23] For these reasons, investigators sought to develop new tumor-selective adenoviruses with higher anticancer potency than E1B55K-deleted viruses. One of these viruses targeted the Rb pathway and was called Delta-24.[24]

## 3. Delta-24

Several cellular proteins have been discovered by their interaction with adenoviral proteins. In this regard, Harlow and collaborators defined the regions of the E1A protein that interact with Rb protein, a master regulator of the cell cycle.[24,25] If inactivation of p53 is key to preventing apoptosis, inactivation of the Rb protein during adenovirus infection is critical to inducing the entry of the cell into S phase. Impairment of the interaction of E1A with Rb would render effective viral replication, which is impossible in normal cells arrested in G0.[24] On the basis of this concept, we constructed a replication-competent adenovirus encompassing a deletion of 24-base pair (bp) nucleotides in the *E1A* gene encoding for the Rb-interacting segment of the E1A protein. The resulting mutant oncolytic virus was called Delta-24.[24]

The Delta-24 deletion in E1A should render the adenovirus deficient in replication in normal cells that are out of the cell cycle in a postmitotic status, but it would not affect the replication of the virus in cancer cells in which the Rb pathway is normally inactivated.[24] Importantly, Heise et al.[26] demonstrated that the E1A mutation did not attenuate the virus and that this mutant adenovirus replicated up to 100 times more efficiently than ONYX-015 in the majority of cancer cells.[24,26]

In the original report of the anticancer effect of Delta-24,[24] we observed that some glioma cell lines were more resistant to the oncolytic effect than others. The difference in the effect resulted from the different levels of expression of adenovirus receptors on the surface of cancer cells. In fact, many cancer cells do not express high levels of the natural receptor for adenovirus, termed coxsackievirus and adenovirus receptor (CAR).[27–29] Therefore, although E1A mutants such as Delta-24 displayed superior anticancer potency compared with their predecessor E1B55K mutant adenovirus, further improvements were required if the goal was to infect most cancer cells in tumors. This could be done using a variety of strategies. In the case of Delta-24, through collaboration between our group and the Curiel laboratory, the fiber of the adenovirus was modified by inserting the sequence of a peptide that allowed for binding to integrins in the HI loop of the adenoviral fiber protein.[28,30] By maintaining an intact fiber knob, the modified Delta-24 could interact with either CAR or integrins for its anchorage on the target cells. Different from its wild-type counterpart, the fiber-modified Delta-24 displayed CAR-independent infectivity.[28] This tropism-enhanced Delta-24 was named Delta-24-RGD.[28,30]

## 4. Delta-24-RGD

Enhancement of the tropism of the Delta-24 adenovirus was performed by adding a specific motif in the HI loop of its fiber protein that consisted of the sequence of a peptide ACDCRGDCFCG (RGD-4C) containing Arg-Gly-Asp[30–32] that has been shown to home to tumor vessels.[31,32] The choice of the RGD peptide was justified based on studies from virology and cellular biology. Adenoviruses encompass the RGD motif as a key part of the penton base proteins that interact with cellular integrins to make possible internalization of the adenovirus into the tumor cell.[33,34] The purpose of this modification was to increase the anchorage of the virus on cancer cells with low expression of CAR, through interaction of the fiber knob directly with $\alpha v\beta 5$ and $\alpha v\beta 3$ integrins, resulting in participation of the integrin-binding domain in both anchorage and internalization in a CAR-independent manner.[28,30,35]

Delta-24-RGD had a more potency and widespread anticancer effect than did Delta-24. Thus, using a system of glioma cell lines, Fueyo et al. demonstrated that Delta-24-RGD could significantly extend the survival of glioma-bearing mice compared with Delta-24: At the same dose, Delta-24-RGD was always significantly more potent.[28] Importantly, Delta-24-RGD could infect, replicate in, and kill glioma stem cells,[36] a population of cells resistant to conventional therapies including chemotherapy and radiotherapy.[31,37]

## 5. Clinical Experience with Delta-24-RGD

The safety of Delta-24-RGD was first tested in a phase I clinical trial of women with ovarian cancer.[38] The purposes of the study were to determine the safety of intraperitoneal delivery of the oncolytic virus and to characterize the biologic effects encountered in these patients with advanced cancer cells. The 21 patients were treated daily for 3 days through an intraperitoneal catheter with doses up to $1 \times 10^{12}$ viral particles (vp). Toxicity effects were limited to fever, fatigue, and abdominal pain, and the investigators did not observe serious adenovirus-related toxicity. Although the adenovirus was safe, there were no partial or complete responses. Examination of the ascites using polymerase chain reaction found Delta-24-RGD in most patients. Importantly, an increase in copy number was observed in 30% of patients 3 days after the inoculation, strongly suggesting active adenovirus replication.

Delta-24-RGD was also tested in patients with recurrent malignant glioma; the results of the trial have been submitted for publication (Lang et al.). The study enrolled 37 patients with recurrent high-grade gliomas, and the objectives included determination of the maximum tolerated dose of Delta-24-RGD administered into the tumor and a description of the biological effects found in tumors that were resected after adenovirus injection. The patients were divided into two groups. The first group received a single intratumoral injection of Delta-24-RGD and was observed clinically with no other intervention. Patients in the second group also were inoculated with Delta-24-RGD, but their tumors were resected 15 days after the injection. The most interesting finding from this study was that a small percentage of patients showed complete responses after the treatment and a greater percentage showed a partial response. Data are currently being examined; further information cannot be provided at this moment. However, that some patients had a complete regression suggests that gliomas could be efficiently treated with oncolytic adenoviruses such as Delta-24-RGD. Data from the clinical trial also suggested that an antitumor immune response might complement the direct antiglioma effect of the viral oncolysis.

Interestingly, Delta-24-RGD is also currently being tested in patients with recurrent malignant gliomas in the Clinica Universidad de Navarra, Spain. In this phase I trial, Delta-24-RGD is administered in combination with temozolomide. The rationale for this combination is based on preclinical studies performed by Alonso and collaborators showing that the E1A protein inactivates the promoter of MGMT, the enzyme that mediates resistance to temozolomide.[39] Preliminary results from this trial show a potential antiglioma effect of Delta-24-RGD when administered directly into the glioma mass.

In addition to these trials, another multicenter trial was started in 2014 to treat glioma patients at the MD Anderson Cancer Center (Houston, Texas), Moffit Cancer Center (Tampa, Florida), and three other institutions in the United States. Patients will be treated with a single injection of Delta-24-RGD or with a combination of Delta-24-RGD and interferon-gamma. Patients have already been enrolled at MD Anderson and Moffit. Unfortunately, it is too early to comment on any result from this trial.

## 6. Delta-24-RGD and Antitumor Immune Response

The paradigm of cancer virotherapy is shifting. If we started our research decades ago thinking that virus-mediated oncolysis could eradicate a solid tumor, we are now learning that another important component in the anticancer effect induced by oncolytic adenoviruses, and probably any other virus, is the triggering of an immune response against the infected and uninfected cancer cells of an inoculated tumor (Figure 1).[39] This is not a straightforward theory and requires further investigation to be confirmed. Whereas adenoviruses are highly immunogenic and the immune response directed against viruses during an infection is deeply understood, the host immune response against cancer cells is not well understood. The shift of the immune response against viruses toward antigens expressed by cancer cells is not completely proven yet, at least in the clinical trial setting. However, we have plenty of sophisticated evidence from in vitro and animal studies suggesting that this may be the case. The infection of a cancer cell by an adenovirus is followed by the trigger of a massive and productive autophagy.[40] The role of autophagy in the infected cells is multifaceted and depends on the experimental conditions. Thus, in vitro, autophagy would never prevent adenovirus replication and it would participate in spreading the virus by playing a role during cell lysis.[40,41] In fact, autophagy is so relevant for an efficient adenovirus-mediated cell lysis that cells deficient in autophagy display a deficient lysis.[40] Thus, autophagy seems to have a provirus role as has been described during infection by other pathogens.[42] However, the role of autophagy in vivo may be different. In vivo, autophagy acts as part of the immune response machinery by degrading, processing, and presenting epitopes to the immune cells.[43] In this regard, autophagy may complement the role of the immunoproteasome in the triggering of an antigen-mediated immune response in cells infected with adenovirus.[43,44] This is true in the case of cells infected with the Epstein–Barr virus, because the Epstein–Barr virus nuclear antigen is predominantly generated in the autolysosomes.[45] Further evidence was provided by studies of Delta-24-RGD in immunocompetent animal models. Jiang and collaborators reported that infection of murine gliomas results in the triggering of a systemic immune response that recognizes not only cells infected with adenoviruses but also cancer cells that have never been infected, which suggests that the adenovirus infection is followed by the presentation of cellular antigens to the immune system.[44] But what are the precise mechanisms that allow adenoviruses to elicit an antitumor immune response against tumor cells? The most supported hypothesis suggests that there is cooperation between the pathogen-associated molecular patterns (PAMPs) and the damage-associated molecular patterns (DAMPs). The hypothesis predicts that although PAMPs and DAMPs are generated by different sets of signals and produced to trigger two different immune responses, under certain circumstances, the two molecular patterns would not compete to elicit the specific immune response, but would cooperate. Strong evidence suggesting that this is the case comes from reports showing that DAMPs and PAMPs can share the same Toll-like receptors.[46] We anticipate that over the next 5 years, virotherapy and immunotherapy specialists will determine the precise mechanism that allows certain patients to transform an antiviral response into the most potent anticancer therapy.

# 7. ICOVIR Platform

To increase the selectivity of replication associated with a deregulated Rb pathway, we developed a series of adenoviruses that combine tumor selectivity based on the deletion of the CR2 domain of E1A (such as in Delta-24 and Delta-24-RGD) with the transcriptional regulation of E1A using an exogenous promoter regulated by this pathway. We selected the promoter from transcription factor E2F1 because it was previously validated by other groups.[47–49] This E2F1 promoter was of particular interest in the context of Delta-24 viruses because it reinforced the selectivity of Delta-24 based on the same Rb pathway commonly altered in tumor cells. Other promoters used to control oncolytic adenovirus replication usually restrict the replication to specific tumor types, such as tyrosinase to melanoma, alpha-fetoprotein to hepatocellular carcinoma, or prostate-specific antigen to prostate cancer. To avoid interference with the enhancers located in the adenovirus ITR and *E1A* promoter, we insulated the E2F1 promoter using DM1 insulators.[50] This virus (named ICOVIR2) was later modified by inserting a Kozak sequence before *E1A* to boost its translation (virus ICOVIR5)[51] and with additional E2F-binding sites (virus ICOVIR7).[52] The DM1 insulators and the full E2F1 promoter enlarged the virus genome to almost the packaging limit (38 kb). Therefore, to insert exogenous transgenes in these oncolytic adenoviruses with the E2F-based promoter, regulation of *E1A* had to be redesigned and only minimal E2F-binding sites were used to control *E1A* expression (virus named ICOVIR15).[53] These binding sites were expanded only 150 bp and allowed us to arm the Rb-dependent oncolytic virus with transgenes such as hyaluronidase to favor the spread of the virus through the tumor stroma (virus named ICOVIR17).[54] We modified such a hyaluronidase-armed, Rb-dependent oncolytic virus with the insertion of an RGD at the fiber shaft instead of on the previous location at the HI loop of the knob (virus known as ICOVIR17K or VCN01).[55] This different localization of the integrin-binding RGD motif resulted in a lower infection of hepatocytes and a greater infection of tumor cells upon systemic administration of the virus.[56] Preclinical work in mice and in hamsters, which allow more permissive human adenovirus replication than do mice, has shown that ICOVIR viruses are selective, cause minor toxicity problems (transaminitis) at up to $5 \times 10^{10}$ vp per animal, and have antitumor activity when given intravenously once.

# 8. Clinical Experience with ICOVIR Viruses

Nonarmed ICOVIR5 (containing the Delta-24 mutation, the RGD at the fiber knob, the insulated E2F1 promoter, and the Kozak sequence before *E1A*) and ICOVIR7 (with four additional E2F sites in the E2F1 promoter) were considered for clinical tests. ICOVIR5 is currently being administered to metastatic melanoma patients by intravenous single administration (Trial NCT01864759). New treatment options based on immunotherapy with checkpoint inhibitor antibodies (anti-CTLA4, anti-PD1, and anti-PD-L1) have become available for this type of patient, which has had an effect on the recruitment of patients, by biasing patient selection in favor of choroidal melanoma (with common hepatic metastases) over cutaneous melanoma. Mild fever, vomiting, asthenia, and thrombocytopenia have been observed at $1 \times 10^{12}$ vp or higher doses. Although no clear signs of efficacy have been observed so far (dose of $1 \times 10^{13}$ vp), biopsy specimens of metastases have been analyzed for five patients, three of which have been positive for the presence of virus genomes, which indicates that systemic adenovirus can reach cutaneous and liver metastases. ICOVIR7 has been administered to 21 patients with a variety of metastatic solid tumors up to a dose of $1 \times 10^{12}$ vp via intratumoral, intravenous, or intracavital single administration.[57] Mild-to-moderate fever, transaminitis, anemia, hyponatremia, pain, and chills were commonly observed, and four patients had partial response or stable disease.

To increase the delivery of the virus to tumors, ICOVIR5 has also been administered within cellular vehicles, in particular in autologous mesenchymal stem cells (MSC). These cells have a tropism toward tumor stroma when injected systemically.[58] Because MSC are expanded in vitro for infection with ICOVIR5, the Rb pathway is inactive and the virus can replicate despite being normal cells. After ex vivo infection and irradiation (which affect cell viability but not virus propagation), infected MSC were infused intravenously into four children with metastatic neuroblastoma.[59] One of these children had a complete response and is still in complete remission, 7 years after treatment (M. Ramírez, personal communication). This patient had a widespread and high tumor burden, and if lysis of tumor cells had been produced by the virus only, viremia and virus infection symptoms would have been observed, which was not the case. Therefore, it is likely that an antitumor immune response was the ultimate reason for the successful outcome. It is tempting to speculate that MSC could have attenuated the strong immune response elicited by the virus to allow a better immune response to tumor antigens.[60] Harnessing the immunodominance of oncolytic virus antigens will likely be crucial to their successful use as immunotherapy tools.

In addition to issues of systemic delivery and antiviral immunodominance, the poor diffusion of the virus within the infected tumor needs to be addressed for successful virotherapy. This poor diffusion represents a particular challenge in tumors with a dense stroma, such as pancreatic adenocarcinoma. We specifically developed an armed version of our Rb-pathway–dependent oncolytic adenoviruses to digest hyaluronic acid, the major stroma component of the growing edge of tumors. Whereas collagen is more prevalent in inner desmoid tumor, the less dense granulation tissue of the growing edge of the tumor is richer in fibronectin and hyaluronic acid. ICOVIR17K (or VCN01) is currently being tested in patients with different types of metastatic tumors by single intravenous administration (Trial NCT02045602) and in patients with nonresectable pancreatic cancer by single intratumoral administration (Trial NCT02045589). In the intravenous route trial, patients with antibody titers against human adenovirus type 5 higher than 1/320 were excluded to improve the possibility of reaching tumors. The extracellular matrix of glioblastoma is also rich in hyaluronic acid, and therefore ICOVIR17 or ICOVIR17K could be suitable for glioblastoma treatment. Taking into account that the evidence for the tumor-targeting properties of MSC injected in proximity to the glioblastoma tumor mass is better established than the evidence for tumor targeting of systemic MSC, the use of MSC to deliver oncolytic viruses to glioblastoma is particularly interesting. This strategy has been validated with ICOVIR17 in preclinical models of glioblastoma.[61]

## 9. Conclusions and Future Directions

Oncolytic adenoviruses have arrived in the clinical setting, and in the People's Republic of China, one E1B mutant adenovirus, termed H101, has been approved for use as conventional therapy for patients with head and neck tumors. This clinical approval was based on the results of a phase III trial performed in China, and although there has been some criticism about the design and number of patients and the exclusion of radiotherapy, the trial confirmed results from a phase II clinical trial using a similar virus and similar patients that was conducted in the United States. Second-generation oncolytic viruses, including Delta-24-RGD and ICOVIR, have also reached the clinical setting for a variety of cancers. On the basis of preliminary, and unpublished, results from these phase I trials, we predict that phase II and phase III trials will soon clarify the role of the host immune system in the antitumor effect elicited by the viruses. According to data now being collected by several groups in the United States and Europe, the role of the immune system activation that should follow a phase of tumor infection and oncolysis would be decisive in inducing partial and complete responses in patients treated with replication-competent adenoviruses. It therefore seems safe to predict that a set of clinical trials in the near future will test the combination of oncolytic adenoviruses with immune modulatory molecules including immune checkpoint inhibitors and stimulators.

## Conflict of Interest

The funders had no role in study design, data collection and analysis, decision to publish, or preparation of the manuscript. CG-M and JF are shareholders of DNAtrix, Inc. RA is shareholder of VCB Biosciences.

## Acknowledgments

This work has been partially supported by the National Institutes of Health (P50CA127001; R01NS069964), the Marnie Rose Foundation, the Will Power Foundation, and the Broach Foundation.

## References

1. Pack GT. Note on the experimental use of rabies vaccine for melanomatosis. *AMA Arch Derm Syphilol* 1950;**62**(5):694–5.
2. Rowe WP, Huebner RJ, Gilmore LK, et al. Isolation of a cytopathogenic agent from human adenoids undergoing spontaneous degeneration in tissue culture. *Proc Soc Exp Biol Med* 1953;**84**(3):570–3.
3. Kelly E, Russell SJ. History of oncolytic viruses: genesis to genetic engineering. *Mol Ther* 2007;**15**(4):651–9.

4. Hoster HA, Zanes Jr RP, Von Haam E. Studies in Hodgkin's syndrome; the association of viral hepatitis and Hodgkin's disease; a preliminary report. *Cancer Res* 1949;**9**(8):473–80.
5. Southam CM, Moore AE. Clinical studies of viruses as antineoplastic agents with particular reference to Egypt 101 virus. *Cancer* 1952;**5**(5):1025–34.
6. Moore AE. Effects of viruses on tumors. *Annu Rev Microbiol* 1954;**8**:393–410.
7. Huebner RJ, Rowe WP, Schatten WE, et al. Studies on the use of viruses in the treatment of carcinoma of the cervix. *Cancer* 1956;**9**(6):1211–8.
8. Babiss LE, Ginsberg HS, Darnell Jr JE. Adenovirus E1B proteins are required for accumulation of late viral mRNA and for effects on cellular mRNA translation and transport. *Mol Cell Biol* 1985;**5**(10):2552–8.
9. Lillie JW, Green MR. Transcription activation by the adenovirus E1a protein. *Nature* 1989;**338**(6210):39–44.
10. Graham FL, Smiley J, Russell WC, et al. Characteristics of a human cell line transformed by DNA from human adenovirus type 5. *J Gen Virol* 1977;**36**(1):59–74.
11. He TC, Zhou S, da Costa LT, et al. A simplified system for generating recombinant adenoviruses. *Proc Natl Acad Sci USA* 1998;**95**(5):2509–14.
12. Lang FF, Bruner JM, Fuller GN, et al. Phase I trial of adenovirus-mediated p53 gene therapy for recurrent glioma: biological and clinical results. *J Clin Oncol* 2003;**21**(13):2508–18.
13. Gomez-Manzano C, Fueyo J, Kyritsis AP, et al. Adenovirus-mediated transfer of the p53 gene produces rapid and generalized death of human glioma cells via apoptosis. *Cancer Res* 1996;**56**(4):694–9.
14. Blackford AN, Grand RJ. Adenovirus E1B 55-kilodalton protein: multiple roles in viral infection and cell transformation. *J Virol* 2009;**83**(9):4000–12.
15. Bischoff JR, Kirn DH, Williams A, et al. An adenovirus mutant that replicates selectively in p53-deficient human tumor cells. *Science* 1996;**274**(5286):373–6.
16. Kirn D. Clinical research results with dl1520 (Onyx-015), a replication-selective adenovirus for the treatment of cancer: what have we learned? *Gene Ther* 2001;**8**(2):89–98.
17. Ganly I, Kirn D, Eckhardt G, et al. A phase I study of Onyx-015, an E1B attenuated adenovirus, administered intratumorally to patients with recurrent head and neck cancer. *Clin Cancer Res* 2000;**6**(3):798–806.
18. Nemunaitis J, Khuri F, Ganly I, et al. Phase II trial of intratumoral administration of ONYX-015, a replication-selective adenovirus, in patients with refractory head and neck cancer. *J Clin Oncol* 2001;**19**(2):289–98.
19. Kirn DH. The end of the beginning: oncolytic virotherapy achieves clinical proof-of-concept. *Mol Ther* 2006;**13**(2):237–8.
20. O'Shea CC, Johnson L, Bagus B, et al. Late viral RNA export, rather than p53 inactivation, determines ONYX-015 tumor selectivity. *Cancer Cell* 2004;**6**(6):611–23.
21. Leppard KN, Shenk T. The adenovirus E1B 55 kd protein influences mRNA transport via an intranuclear effect on RNA metabolism. *EMBO J* 1989;**8**(8):2329–36.
22. Pilder S, Moore M, Logan J, et al. The adenovirus E1B-55K transforming polypeptide modulates transport or cytoplasmic stabilization of viral and host cell mRNAs. *Mol Cell Biol* 1986;**6**(2):470–6.
23. Heise C, Sampson-Johannes A, Williams A, et al. ONYX-015, an E1B gene-attenuated adenovirus, causes tumor-specific cytolysis and antitumoral efficacy that can be augmented by standard chemotherapeutic agents. [see comment] *Nat Med* 1997;**3**(6):639–45.
24. Fueyo J, Gomez-Manzano C, Alemany R, et al. A mutant oncolytic adenovirus targeting the Rb pathway produces anti-glioma effect in vivo. *Oncogene* 2000;**19**(1):2–12.
25. Whyte P, Buchkovich KJ, Horowitz JM, et al. Association between an oncogene and an anti-oncogene: the adenovirus E1A proteins bind to the retinoblastoma gene product. *Nature* 1988;**334**(6178):124–9.

26. Heise C, Hermiston T, Johnson L, et al. An adenovirus E1A mutant that demonstrates potent and selective systemic anti-tumoral efficacy. *Nat Med* 2000;**6**(10):1134–9.
27. Miller CR, Buchsbaum DJ, Reynolds PN, et al. Differential susceptibility of primary and established human glioma cells to adenovirus infection: targeting via the epidermal growth factor receptor achieves fiber receptor-independent gene transfer. *Cancer Res* 1998;**58**(24):5738–48.
28. Fueyo J, Alemany R, Gomez-Manzano C, et al. Preclinical characterization of the antiglioma activity of a tropism-enhanced adenovirus targeted to the retinoblastoma pathway. *J Natl Cancer Inst* 2003;**95**(9):652–60.
29. Bergelson JM, Cunningham JA, Droguett G, et al. Isolation of a common receptor for Coxsackie B viruses and adenoviruses 2 and 5. *Science* 1997;**275**(5304):1320–3.
30. Suzuki K, Fueyo J, Krasnykh V, et al. A conditionally replicative adenovirus with enhanced infectivity shows improved oncolytic potency. *Clin Cancer Res* 2001;**7**(1):120–6.
31. Piccirillo SG, Reynolds BA, Zanetti N, et al. Bone morphogenetic proteins inhibit the tumorigenic potential of human brain tumour-initiating cells. *Nature* 2006;**444**(7120): 761–5.
32. Pasqualini R, Koivunen E, Ruoslahti E. A peptide isolated from phage display libraries is a structural and functional mimic of an RGD-binding site on integrins. *J Cell Biol* 1995;**130**(5):1189–96.
33. Wickham TJ, Mathias P, Cheresh DA, et al. Integrins alpha v beta 3 and alpha v beta 5 promote adenovirus internalization but not virus attachment. *Cell* 1993;**73**(2):309–19.
34. Seth P, editor. *Adenoviruses: basic biology to gene therapy*. Austin, Texas: R.G. Landes; 1999. p. 31–5.
35. Jiang H, McCormick F, Lang FF, et al. Oncolytic adenoviruses as antiglioma agents. *Expert Rev Anticancer Ther* 2006;**6**(5):697–708.
36. Jiang H, Gomez-Manzano C, Aoki H, et al. Examination of the therapeutic potential of Delta-24-RGD in brain tumor stem cells: role of autophagic cell death. *J Natl Cancer Inst* 2007;**99**(18):1410–4.
37. Bao S, Wu Q, McLendon RE, et al. Glioma stem cells promote radioresistance by preferential activation of the DNA damage response. *Nature* 2006;**444**(7120):756–60.
38. Kimball KJ, Preuss MA, Barnes MN, et al. A phase I study of a tropism-modified conditionally replicative adenovirus for recurrent malignant gynecologic diseases. *Clin Cancer Res* 2010;**16**(21):5277–87.
39. Alonso MM, Gomez-Manzano C, Bekele BN, et al. Adenovirus-based strategies overcome temozolomide resistance by silencing the O6-methylguanine-DNA methyltransferase promoter. *Cancer Res* 2007;**67**(24):11499–504.
40. Jiang H, White EJ, Rios-Vicil CI, et al. Human adenovirus type 5 induces cell lysis through autophagy and autophagy-triggered caspase activity. *J Virol* 2011;**86**(10):4720–9.
41. Klein SR, Piya S, Lu Z, et al. C-Jun N-terminal kinases are required for oncolytic adenovirus-mediated autophagy. *Oncogene* 2015 Jan 26. http://dx.doi.org/10.1038/onc.2014.452. [Epub ahead of print].
42. Wileman T. Aggresomes and autophagy generate sites for virus replication. *Science* 2006;**312**(5775):875–8.
43. Levine B, Deretic V. Unveiling the roles of autophagy in innate and adaptive immunity. *Nat Rev Immunol* 2007;**7**(10):767–77.
44. Jiang H, Clise-Dwyer K, Ruisaard KE, et al. Delta-24-RGD oncolytic adenovirus elicits anti-glioma immunity in an immunocompetent mouse model. *PLoS One* 2014;**9**(5): e97407.
45. Paludan C, Schmid D, Landthaler M, et al. Endogenous MHC class II processing of a viral nuclear antigen after autophagy. *Science* 2005;**307**(5709):593–6.

46. Escamilla-Tilch M, Filio-Rodriguez G, Garcia-Rocha R, et al. The interplay between pathogen-associated and danger-associated molecular patterns: an inflammatory code in cancer? *Immunol Cell Biol* 2013;**91**(10):601–10.
47. Johnson L, Shen A, Boyle L, et al. Selectively replicating adenoviruses targeting deregulated E2F activity are potent, systemic antitumor agents. *Cancer Cell* 2002;**1**(4):325–37.
48. Jakubczak JL, Ryan P, Gorziglia M, et al. An oncolytic adenovirus selective for retinoblastoma tumor suppressor protein pathway-defective tumors: dependence on E1A, the E2F-1 promoter, and viral replication for selectivity and efficacy. *Cancer Res* 2003;**63**(7):1490–9.
49. Tsukuda K, Wiewrodt R, Molnar-Kimber K, et al. An E2F-responsive replication-selective adenovirus targeted to the defective cell cycle in cancer cells: potent antitumoral efficacy but no toxicity to normal cell. *Cancer Res* 2002;**62**(12):3438–47.
50. Majem M, Cascallo M, Bayo-Puxan N, et al. Control of E1A under an E2F-1 promoter insulated with the myotonic dystrophy locus insulator reduces the toxicity of oncolytic adenovirus Ad-Δ24RGD. *Cancer Gene Ther* 2006;**13**(7):696–705.
51. Cascallo M, Alonso MM, Rojas JJ, et al. Systemic toxicity-efficacy profile of ICOVIR-5, a potent and selective oncolytic adenovirus based on the pRB pathway. *Mol Ther* 2007;**15**(9):1607–15.
52. Rojas JJ, Cascallo M, Guedan S, et al. A modified E2F-1 promoter improves the efficacy to toxicity ratio of oncolytic adenoviruses. *Gene Ther* 2009;**16**(12):1441–51.
53. Rojas JJ, Guedan S, Searle PF, et al. Minimal RB-responsive E1A promoter modification to attain potency, selectivity, and transgene-arming capacity in oncolytic adenoviruses. *Mol Ther* 2010;**18**(11):1960–71.
54. Guedan S, Rojas JJ, Gros A, et al. Hyaluronidase expression by an oncolytic adenovirus enhances its intratumoral spread and suppresses tumor growth. *Mol Ther* 2010;**18**(7):1275–83.
55. Rodriguez-Garcia A, Gimenez-Alejandre M, Rojas JJ, et al. Safety and efficacy of VCN-01, an oncolytic adenovirus combining fiber HSG-binding domain replacement with RGD and hyaluronidase expression. *Clin Cancer Res* 2015;**21**(6):1406–18.
56. Bayo-Puxan N, Gimenez-Alejandre M, Lavilla-Alonso S, et al. Replacement of adenovirus type 5 fiber shaft heparan sulfate proteoglycan-binding domain with RGD for improved tumor infectivity and targeting. *Hum Gene Ther* 2009;**20**(10):1214–21.
57. Nokisalmi P, Pesonen S, Escutenaire S, et al. Oncolytic adenovirus ICOVIR-7 in patients with advanced and refractory solid tumors. *Clin Cancer Res* 2010;**16**(11):3035–43.
58. Yong RL, Shinojima N, Fueyo J, et al. Human bone marrow-derived mesenchymal stem cells for intravascular delivery of oncolytic adenovirus Δ24-RGD to human gliomas. *Cancer Res* 2009;**69**(23):8932–40.
59. Garcia-Castro J, Alemany R, Cascallo M, et al. Treatment of metastatic neuroblastoma with systemic oncolytic virotherapy delivered by autologous mesenchymal stem cells: an exploratory study. *Cancer Gene Ther* 2010;**17**(7):476–83.
60. Alemany R. Chapter four–Design of improved oncolytic adenoviruses. *Adv Cancer Res* 2012;**115**:93–114.
61. Martinez-Quintanilla J, He D, Wakimoto H, et al. Encapsulated stem cells loaded with hyaluronidase-expressing oncolytic virus for brain tumor therapy. *Mol Ther* 2015;**23**(1):108–18.

# Innate Immune Response to Adenovirus Vector Administration In Vivo

**14**

*Svetlana Atasheva, Dmitry M. Shayakhmetov*
Lowance Center for Human Immunology, Departments of Pediatrics and Medicine, Emory University, Atlanta, GA, USA

The application of adenoviruses (Ad) as vectors for transgene delivery, vaccines, or oncolytic therapy is a promising approach for the treatment of a variety of clinically significant diseases. Adenovirus vectors form highly stable proteinaceous particles that can efficiently deliver foreign genetic material to both dividing and quiescent cells. Other advantages of Ad vectors as gene therapy tools include the capacity to deliver large therapeutic genes, the ease of genetic manipulations, the ability to produce large quantities of clinical-grade virus, and episomal replication of the vector genome. However, the successful application of Ad vectors as a treatment in clinical settings has been hampered by the unexpected complexity of host–vector interactions. Interactions between the Ad vector and the host immune system in humans may result in severe clinical outcomes. One of the harshest setbacks in Ad vector research was caused by the death of a patient involved in a clinical trial in 1999.[1] The patient died owing to severe Ad vector–triggered toxicity, which resulted in multiple organ failure.[2] Clearly, detailed and thorough understanding of the host response to Ad vector administration is a crucial prerequisite for the successful and safe use of Ad vectors in humans.

The availability of a variety of mouse strains that are deficient in genes involved in immune response allows for a thorough examination of the molecular mechanisms of host response in vivo. Although in vitro studies are important and required for certain experiments aiming to reveal cell intrinsic mechanisms of virus interaction with the host, they do not account for the complexity of the host. Therefore, they may lack critical factors that have an important role in guiding virus–host interactions in vivo. Study of Ad infection in vivo shows that Ad vector interactions with a host involve an intricate system of cellular receptors and soluble mediators that interact with different Ad vector capsid proteins.

Intravenous Ad vector delivery for gene transfer purposes, especially at high doses, stimulates strong innate and adaptive immune responses that can prove fatal for the host.[1–4] Experiments in rodents showed that upon systemic Ad administration, vector is rapidly sequestered from the bloodstream by the liver.[5–8] The removal of Ad from circulation temporally coincides with the induction of transcription and release into the blood of proinflammatory cytokines and chemokines, including interleukin (IL)-6, tumor necrosis factor-α (TNF-α), CCL5, CXCL10, IL-8, CCL3, CCL4, and CXCL2.[9–13] Macrophages throughout the body, including tissue residential

**Adenoviral Vectors for Gene Therapy.**

macrophages (e.g., Kupffer cells in the liver), and dendritic cells are the principal source of these cytokines and chemokines.[14] In human gene therapy trials, serum levels of IL-6 and IL-1 were elevated after intravenous Ad injection at high doses ($2\times10^{12}$–$6\times10^{13}$ virus particles).[15–17] Histological evaluation of tissues, including the lung, liver, and spleen, revealed areas of leukocyte and neutrophil infiltration and local necrosis,[10,18] which indicated that most, if not all, tissues in the body respond to intravenous Ad injection in a proinflammatory manner.

Using microarray technology, it was demonstrated that Ad entry into cells alters expression of up to 15% of all messenger ribonucleic acid transcripts in mouse liver tissue or human epithelial cells.[19,20] In mouse and nonhuman primate preclinical animal models, Wilson et al. showed activation of innate responses by transcriptionally defective Ad particles,[11,21] indicating that the induction of innate immune and inflammatory responses occurs shortly after intravenous Ad vector delivery but before the initiation of viral gene expression. In later studies, a severe acute inflammatory response was also observed in a nonhuman primate model after intravenous injection of a helper-dependent Ad vector that lacked all viral genes.[22] These data provide clear evidence that the dose-dependent activation of innate immune and inflammatory responses to Ad occurs primarily as a result of Ad particle interaction with host cells and does not require viral gene expression. The importance of interactions of host cells with the Ad vector capsid before establishing a replication cycle suggests that the outcome of the Ad vector administration depends on early events after injection. Therefore, in this chapter we aim to summarize in vivo data of the molecular mechanisms of the events that occur immediately after intravenous Ad administration.

## 1. Adenovirus Interactions with Blood Cells and Components of Plasma

Systemic delivery of adenoviral vectors for the treatment of different inborn or acquired diseases requires use of an intravascular route of administration. Blood, however, possesses multiple means of protecting the host from viral pathogens. Different soluble factors found in the plasma, such as neutralizing antibodies, natural antibodies, complement, coagulation factors, or defensins, can recognize viral invasion and either directly neutralize the Ad or activate an immune response against the administered virus.[23] Moreover, blood cells themselves represent a defense machinery that can bind and sequester Ad vector in the bloodstream. The interactions of Ad vector with blood components result in the activation of the host immune system, but they also affect vector biodistribution.[23] More than 90% of the administered Ad vector is sequestered by the liver,[5] and blood factors have a significant role in Ad vector hepatotropism.[24] Understanding the mechanisms of the interactions between Ad and blood components immediately after administration is important for understanding the mechanisms of liver sequestration. It is conceivable that this awareness can lead to the development of more efficient targeting of the Ad to extrahepatic tissues and cell types.

**Coagulation factors.** The first record of the influence of blood coagulation factors on in vivo Ad biodistribution came from experiments with perfused mouse livers under asanguineous conditions.[25] We found that efficient hepatocytes transduction occurs only in the presence of the plasma and showed that human Ad 5 (HAd5) directly interacts with coagulation factor IX and other plasma proteins (C4BP, fibrinogen, and PZP).[25] Factor IX is not the only coagulation factor able to support hepatocyte transduction by HAd5. Later, it was found that other vitamin K-dependent blood coagulation factors (FVII, FIX, FX, and protein C) can directly bind to HAd5 and mediate hepatocyte transduction.[26] FX, however, had the most prominent effect on the efficiency of the hepatocytes transduction both in vivo and in vitro compared with other coagulation factors.[26] FVII, FIX, FX, and protein C are all vitamin K-dependent serine proteases with a common domain structure, but their zymogen activity is not required to mediate hepatocyte transduction.[26] Interestingly, these factors also increase HAd5 vector transduction efficiency in tissue culture (reviewed in Ref. 27).

Detailed studies of the interaction of the HAd5 with FX revealed that FX binds to the hexon protein trimer on the HAd5 capsid. The cryo-EM studies showed that FX is bound at the top of each hexon trimer, and the entire virion surface is covered with molecules of FX.[28–31] Furthermore, it was shown that hypervariable regions (HVR) 5 and 7 are involved in hexon–FX interaction.[32] Mutagenesis of HVR5 reduced the binding capacities of the hexon to FX, whereas the mutation in HVR7 T425A completely ablated FX–Ad hexon interactions, which suggests that HVR7 has a crucial role in binding FX.[29]

In addition to the enhanced hepatocytes transduction, the interaction of the hexon with FX has a significant role in triggering the host immune response. The FX binding to HAd5 activates the nuclear factor-κB–dependent inflammatory pathway through the TLR4-TRIF/MyD88-TRAF6 signaling axis, resulting in a release of proinflammatory cytokines and chemokines.[29] In addition, the Ad species that were able to bind FX activated a significantly broader spectrum of chemokines and cytokines than those whose hexon proteins cannot bind FX. Thus, FX that is bound to the surface of the virion can be recognized as a pathogen-associated molecular pattern and trigger a host inflammatory response.[29]

On the other hand, a high concentration of FX in the plasma and its ability to bind to each hexon trimer on the viral surface may be used by the virus to shield the capsid from antibody recognition. Xu et al. showed that FX binding prevents Ad vector from recognition by natural immunoglobulin M (IgM) antibodies, helping to escape inactivation by the complement system.[33]

**Antibodies and complement.** Because of natural infections, most of the human population has preexisting neutralizing antibodies (NAbs) against HAd5, the most commonly used vector for gene therapy. The Abs target all of the three capsid proteins: hexon, penton, and fiber.[34,35] Suggested mechanisms of neutralization include Ab-mediated aggregation of viral capsids, blocking interaction of Ad vector capsid proteins with cellular receptors, and stabilizing the virion structure preventing proper uncoating after internalization.[36] Depending on the target, the amount of NAbs required for neutralization varies from 1.6 to 240 Ab per virion.[36,37] Interestingly, some NAbs can affect postinternalization steps of virus infection. The monoclonal Ab 9C12 prevents

HAd5 infection by affecting capsid association to microtubules or microtubule motors, thus blocking virus trafficking to the nucleus.[38]

As an alternative to direct neutralization by antibodies, Ad vectors in the bloodstream can be targeted by the complement system. Activation of both branches of the complement system, the classic and alternative pathways, can be detected by measuring C3a concentration in the plasma. Injection of Ad vector results in an increased level of C3a within 10 min after administration of the virus, which demonstrates that complement activation is a rapid process.[39] However, complement activation reaches a peak at 30 min after Ad administration and then declines to a normal level in 90 min.[40] The mechanism of complement activation in response to Ad vector administration is not completely understood. In vitro studies demonstrated that C3a activation is absolutely dependent on the presence of natural IgM antibodies.[23] Contrary to in vitro data, in vivo experiments showed that there is no requirement for natural antibodies in complement activation.[40] Also, the ts1 mutant, which cannot escape endosome after internalization, was unable to activate complement, which suggests that C3 in plasma cannot directly recognize Ad vector capsid and be activated on the virus surface under in vivo conditions.[40] Endosome rupture and the reticuloendothelial system were critical for complement activation in vivo.[41] However, in later studies, Xu et al. showed the importance of natural antibodies in complement activation in the absence of coagulation factor X, which shields the Ad capsid in the blood.[33] Clearly, antibodies and complement have a significant role in Ad vector neutralization in the bloodstream; however, to date, the exact molecular mechanisms that mediate in vivo virus neutralization remain unknown.

**Defensins**. Other blood host factors that can interfere with efficient Ad vector delivery are defensins. Defensins are small peptides (18–45 amino acids) expressed by neutrophils, Paneth cells, and epithelial cells. These cationic peptides target a broad range of pathogens, including bacteria, viruses, fungi, and protozoa.[42] The amphiphilic nature of defensins allows them to target membranous pathogens and directly disrupt the membrane. Although Ad vectors do not have a membrane envelope, they are sensitive to defensins.[43] Smith et al. showed that alpha-defensins HNP1 and HD5 have potent anti-Ad activity.[44] The mechanism of HD5 antiadenoviral activity is based on the ability of the defensin to bind to the Ad vector capsid proteins, penton base, and fiber.[45] HD5 binds to the disordered region on the penton base and the negatively charged EDES sequence in the N-terminal fiber region, thus reinforcing the penton and fiber vertex complex. Defensin binding to the Ad capsid stiffens and stabilizes the virion structure, resulting in an inability of capsid to release the penton and protein VI.[46] The latter protein mediates endosomal escape of the Ad vector, and in its absence the Ad vector stays trapped in the endosome.

**Blood cells**. Besides soluble factors, blood cells themselves have the means to interact and inactivate Ad. Erythrocytes represent the largest fraction of the cells in the blood. Interestingly, human but not murine erythrocytes express CAR, the Coxsackie and Ad receptor[47,48] that can interact with the HAd5 fiber knob domain. Lyons et al.[49] analyzed HAd5 interaction with human and mouse blood cells ex vivo and found that after a short incubation with blood cells, over 90% of the applied virus was stably associated with human (but not murine) erythrocytes. Moreover, human erythrocytes express a complement receptor (CRI) that can mediate virus binding in the presence of human or murine plasma.[50] Both of these receptors on the surface of human erythrocytes are efficient in binding and neutralizing Ad virions, thus preventing further virus dissemination and infection of other cells.[47] These data were confirmed by in vivo observations from clinical trials involving cancer patients, which showed that most of the administrated oncolytic Ad vector is associated with erythrocytes or granulocytes.[51]

Cotter et al.[52] described interactions with neutrophils, the primary effector cells of the innate immune system. Using fluorescent and electron microscopy techniques, the authors demonstrated that Ad particles can bind primary human neutrophils and subsequently internalize. Binding to neutrophils depended on the presence of complement and antibodies and was reduced by blocking complement receptor or Fc receptors.[52]

Another blood cell type that has been shown to interact with the Ad vector is platelets. Unlike neutrophils that are present in the circulation in high numbers only during times of acute host inflammatory response, platelets are constantly present in the blood in high numbers ($1.5 \times 10^5$–$4.5 \times 10^5$ platelets/ml) and are an essential part of the blood coagulation system. Several groups showed that Ad vector administration in mice leads to clearing of platelets, which results in thrombocytopenia.[53,54] Othman et al.[53] reported that platelets can internalize Ad, unlike inert latex particles that were able to induce internalization only after platelet activation with adenosine diphosphate (ADP).[55] Moreover, Ad5 induces platelet–leukocyte aggregate formation in mice in a von Willebrand factor–dependent manner. However, the inert latex particles could not induce aggregates formation even in ADP-activated platelets.[53,55]

Moreover, platelets interaction with Ad vector in blood facilitated virus sequestration by the liver reticuloendothelial system. It was proposed that virus–platelet aggregates are trapped in the liver sinusoids and are further taken up and degraded by the Kupffer cells.[56] Depletion of platelets before intravascular Ad administration reduced Ad vector sequestration to different organs, thus demonstrating the importance of this interaction in vivo.[56] The mechanism of platelet recognition of Ad is unclear; however, 78% of human platelets are positive for the Ad5 receptor, CAR.[23] On the other hand, Ad11 associates with platelets to a much higher degree than Ad5, and this interaction occurs in a fiber-independent manner.[57] Together with the observation that platelets cannot be activated by inert latex particles[55] or PEGylated Ad5 vector,[58] these data suggest a novel and still uncharacterized mechanism that likely mediates Ad interactions with circulating blood cells.

## 2. Adenovirus Interactions in the Liver

Undoubtedly, Ad vectors represent a highly attractive platform for gene therapy; however, they also possess several undesirable characteristics. Intravenously delivered Ad vectors display strong hepatotropism: more than 90% of administered Ad vector is sequestered by the liver.[5,8] Interactions between the Ad vector and liver cells lead to clinically significant hepatotoxicity and represent a major limitation for gene delivery to extrahepatic cells and tissues, such as disseminated metastatic tumors. Dissecting

Ad interactions with host factors will further our understanding of the molecular mechanisms that dictate Ad biodistribution and identify targets for manipulation of Ad vectors for future clinical use.

Liver sequestration of intravenously administrated Ad is a remarkably rapid process. The half-life of injected HAd5 in the bloodstream is 2 min.[5] Within 30 min after intravascular injection, the liver traps 99% of the injected Ad, eliminating it from the bloodstream.[5] Kupffer cells, the residential macrophages in the liver, are responsible for most (about 90%) of viral uptake from the blood. Other liver cells, such as hepatocytes and liver sinusoidal endothelial cells (LSECs), uptake the vector to a much lesser degree: 6%, and 2% respectively.[24] The rest of the administered Ad vector is likely trapped in the Disse space. Interestingly, despite taking up massive amounts of the virus, Kupffer cells do not support Ad replication or expression, but degrade the trapped virus.[7] In the liver, 90% of viral DNA is rapidly degraded in the first 24 h. This process requires an intact viral capsid and is independent of the immune status of the mice.[8] Another striking detail of Ad uptake by the liver is that it does not depend on the function of the Ad receptor, CAR.[59,60] Moreover, despite use of different cellular receptors, the amount of Ad5 and Ad11 DNA in the liver at 30 min postadministration was comparable.[57] Unlike Kupffer cells, hepatocytes support vector transduction and express substantial amounts of the transgene, although the level of expression declines over time.[7,61] Different fates of the vector DNA in different cell types suggest that recognition of the vector by these cells is mediated by diverse mechanisms. Indeed, studies show that the Ad vector can interact with an array of different factors in the bloodstream, and these complex interactions dictate the Ad biodistribution in vivo. These factors include cellular receptors, intact viral capsid proteins, and soluble blood factors that may bridge an Ad particle to cellular surface molecules on hepatic cells.[24,26,62]

**Kupffer cells**. Upon intravascular delivery, most of the virus in the liver is sequestered by Kupffer cells.[7,62] Although Kupffer cells can accumulate large amounts of blood-borne Ad, virus entry into Kupffer cells does not lead to their transduction, which suggests that Kupffer cells are a poor host for Ad propagation. However Kupffer cells have a limited capacity for virus uptake, and after reaching the limit, they cannot longer accumulate more virus.[63] As a result, the excess of the virus can transduce other liver cells such as hepatocytes and LSECs.[24] Tao et al. showed that the threshold can be reached with a single dose of the virus by injecting more than $1 \times 10^{11}$ particles per mouse.[7] In addition, the saturation limit can be achieved by two separate administrations of the Ad vector. If a second dose of Ad vector is given after just 5 min, Kupffer cells are unable to uptake the second dose of the injected virus.[7] The inability of Kupffer cells to uptake the second Ad dose leads to enhanced hepatocyte transduction efficiency and higher levels of transgene expression.[9] The same effect of higher hepatocytes transduction can be achieved if Kupffer cells are depleted or functionally blocked.[64]

The mechanisms of virus sequestration by Kupffer cells are not completely understood. Kupffer cells function as blood filters and sense foreign particles by recognizing certain molecular patterns. Kupffer cells can uptake inert latex particles, which suggests that they can discriminate foreign material based on size.[65] However, inactivated virus particles were sequestered by the liver at lower rates than a native virus, which suggests that capsid protein interactions are important for Kupffer cell sequestration.[8] The Ad receptor, CAR, which is required for transduction of cell in vitro, was also dispensable for Kupffer cell uptake.[59] Moreover, the RGD motif within the penton base on the virus capsid that engages $\alpha_3\beta_1$, $\alpha_5\beta_1$, $\alpha_v\beta_1$, $\alpha_v\beta_3$, and $\alpha_v\beta_5$ integrin molecules on the cellular surface[66] was also unnecessary for Kupffer cell uptake.[24]

Since finding that the Ad5 virion can interact with coagulation factor IX and C4BP (one factors of the complement system) in the bloodstream,[25] the search for factors that are required for liver uptake has shifted to blood components. Xu et al. showed that scavenger receptors, natural antibodies, and complement have a significant role in virus uptake by Kupffer cells.[62]

Scavenger receptors have a variety of functions, one of which is recognition and removal of foreign materials from the circulation. They can recognize negatively charged clusters that are usually associated with oxidized lipids, the products of metabolism.[67] It is conceivable that the negatively charged hypervariable region 1 (HVR1) of the hexon protein on the surface of the virion can be recognized by scavenger receptors. However, mice lacking one of the major scavenger receptors, SR-A, demonstrated no difference in Kupffer cells uptake of Ad compared with mice in which Kupffer cells expressed SR-A.[68] Recently it was shown that scavenger receptors ES-1 in Kupffer cell can have a role in Ad uptake.[69] Peptides that target and inactivate these receptors increased hepatocytes transduction efficiency, which suggests a blockage of Kupffer cell uptake of the Ad vector.[70] The importance of natural antibodies in Kupffer cells uptake also gained more evidence. Khare et al. showed that antibodies targeting HVR1, 2, 5, and 7 of the hexon protein on the surface of Ad capsid have a significant role in Kupffer cell uptake.[71,72] Complement importance was also supported by finding that CRIg is the receptor on Kupffer cells that mediates early response to Ad vector administration.[50] Taken together, these data show that the key components for Kupffer cell sequestration of Ad include native hexon protein: in particular, HVR1, 2, 5, and 7; complement; natural antibodies; and scavenger receptors. However, the molecular details of how all of these seemingly distinct components of the virus–host interactions come into play remain unclear.

After uptake, Kupffer cells release an array of proinflammatory cytokines and chemokines. Within 1 h after Ad administration, IL-1α, IL-16, granulocyte macrophage–colony-stimulating factor (GM-CSF), CCL1, CCL4, CCL11, CXCL1, and CXCL2 are elevated in the plasma.[10] However, Kupffer cells themselves respond to the Ad vector uptake drastically; they undergo severe cytosolic disorganization within 15 min of Ad administration and die via necrotic type cell death.[73,74] Di Paolo et al. found that the mechanism of Kupffer cell death does not depend on known principal mediators of canonical necrotic or apoptotic cell death programs. They found that IRF3 is a nonredundant factor required for Ad-mediated Kupffer cell death. Interestingly, neither known IRF3 upstream activators nor IRF3 phosphorylation at S396 that signals its activation as a transcription factor was required for Kupffer cell death.[74] Moreover, IRF3 is known for activation of type I interferon (IFN) signaling; it has transcriptional activities and can activate expression of IFN-dependent genes.[75] However, these processes did not have a role in IRF3-mediated Kupffer cell death after Ad vector administration.

The most striking detail of the Kupffer cell death is timing: 15 min after Ad vector administration, Kupffer cells were positive for propidium iodine in the nuclei, an indicator of cellular membrane disruption.[74] Interestingly, Kupffer cells in IRF3$^{-/-}$ mice did not undergo necrosis despite comparable Ad uptake. Also, the Ad ts1 mutant that can bind the cell surface and internalize but is unable to escape the endosome did not induce Kupffer cell death.[74] The authors also showed that this suicidal mechanism of Kupffer cell death is a protective measure of the host defense against viral invasion. In IRF3$^{-/-}$ mice, Kupffer cells that did not undergo cell death had higher viral load.[74] In summary, Kupffer cells in the liver represent the first line of host defense against blood-borne pathogens by taking up the virus through various mechanisms including scavenger receptors, natural antibody, and the complement system. If the amount of virus exceeds the threshold, Kupffer cells trigger defensive suicide necrotic cell death, killing the virus within them and preventing virus dissemination.

**Hepatocytes**. Sequestration of the Ad by Kupffer cells after its intravascular delivery is not the only mechanism responsible for the uptake of blood-borne Ad in the liver (Figure 1).[24] The virus that escapes Kupffer cell sequestration and is covered by coagulation FX can transduce hepatocytes, the main type of cells in the liver parenchyma.[24,26] Factor X can bind to heparan sulfate proteoglycan molecules that are scattered on the surface of hepatocytes and can mediate attachment of Ad–FX complexes.[76] Unlike Kupffer cells, hepatocytes allow for viral replication and Ad can transcribe its genome and start the replication cycle. The hepatocyte transduction is an efficient process; it is possible to reach expression of virally encoded transgenes in nearly 90% of hepatocytes depending on the virus dose.[30,61] However, the level of transgene expression steeply declines with time, and by 2 weeks after Ad administration no expression can be detected in immune-competent hosts.[77] One reason for the decline in transgene expression is elimination of infected cells by cells of the adaptive immune system, namely cytotoxic lymphocytes (CTLs). In a comprehensive study by Wohlleber et al., it was shown that CTLs target and eliminate infected hepatocytes.[78] Interestingly, the classical priming of CTLs by dendritic cells was not required for elimination of the infected hepatocytes. Instead, LSECs were able to uptake viral antigens and cross-present them to the CTLs, thus priming them to kill infected hepatocytes.[78] No less intricate was the mechanism of killing, which was mediated by TNF-α; however, only infected cells were sensitized to TNF-α. By an unknown mechanism, only virus-infected cells were expressing caspase 3 and its presence enhanced TNF-α signaling, resulting in the elimination of infected cells.[78]

**Liver sinusoidal endothelial cells**. Similar to hepatocytes, LSECs are permissive for Ad replication.[24] The mechanism of Ad vector uptake in LSECs depends on integrins expressed on the cell surface.[24] The RGD motif in Ad penton functions as a moiety that binds cellular integrins during Ad cell entrance.[79,80] Integrins mediate capsid internalization in many cell lines; however, they can also mediate both capsid attachment to the cell surface and internalization of Ad inside the cell.[46] Structural studies demonstrated that engagement of the RGD motif with integrin molecules leads to structural relaxing of the Ad virion, thus allowing for partial capsid disassembly.[46,81] The exact mechanism of LSECs transduction is not completely understood, but the importance of the integrins was shown by Di Paolo et al.[24]

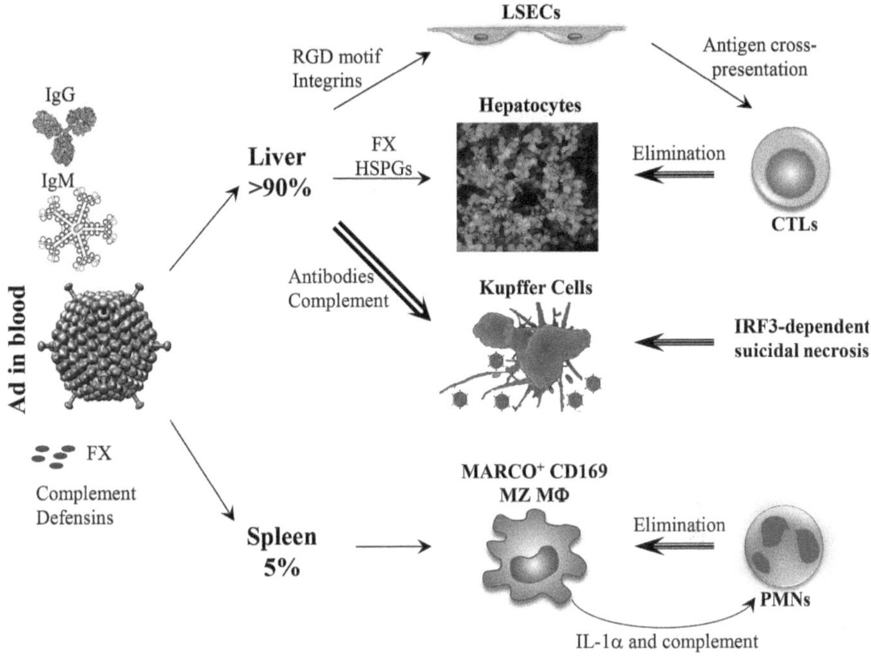


**Figure 1 Model of mechanisms of immune-mediated inactivation of adenovirus after intravascular delivery.** After intravascular Ad administration, soluble blood factors interact with Ad and facilitate clearing of virus particles from the blood. Most of the injected virus is sequestered by liver residential macrophages, Kupffer cells, which undergo cell-autonomous IRF3-dependent necrosis. Hepatocytes absorb blood-borne Ad particles in an FX-dependent manner. Liver sinusoidal endothelial cells can prime cytotoxic T lymphocytes by cross-presenting the virus antigens, leading to elimination of infected hepatocytes. In the spleen, MARCO+ and CD169+ marginal zone macrophages activate IL-1α-IL-1RI signaling and complement that recruit neutrophils, which eliminate virus-containing cells. In both tissues, the functional consequence of host response to Ad is elimination of virus-containing cells and restriction of systemic virus spread. FX, coagulation factor X; LSECs, liver sinusoidal endothelial cells; HSPGs, heparan sulfate proteoglycans; CTLs, cytotoxic T lymphocytes; MZ, marginal zone; MF, macrophages; PMNs, polymorphonuclear neutrophils.

The authors demonstrated that LSECs in mice deficient in β3 integrins were unable to internalize Ad capsid. Accordingly, a virus with deleted RGD motif, Ad5ΔRGD, was not recovered from LSECs.[24]

Di Paolo et al. proposed a model for sequestration of blood-borne Ad in the liver.[24] This model suggests that a defined set of specific molecular mechanisms become engaged in a redundant, synergistic, and orderly manner to ensure the clearance of blood-borne Ad from the circulation. When small amounts of Ad particles appear in the blood, virus trapping by Kupffer cells works as a dominant mechanism, mediating Ad sequestration in the liver. When the Ad dose exceeds the capacity of Kupffer cells, hepatocytes absorb blood-borne Ad particles in a blood factor-dependent manner, serving as a second dominant mechanism mediating sequestration of blood-borne Ad.

However, when the Ad dose is high and both Kupffer cells and blood factor pathways are inactivated, LSECs and the anatomical architecture of liver sinusoids become the third line of defense that sequesters Ad particles in an RGD motif-dependent manner.

Liver sequestration independence from in vitro-identified Ad fiber-specific receptors, the rapid Kupffer cells response to the Ad administration, and the engagement of host factors that recognize Ad prior established replication cycle suggest that mechanisms of virus clearance from the blood are likely decoy pathways that ensure the removal of pathogens from circulation to prevent their dissemination into other tissues.

## 3. Adenovirus Interactions in the Spleen

The spleen is the largest secondary lymphoid organ in the body that is responsible for orchestrating immune responses to blood-borne antigens. Another function of the spleen is removal of old and damaged red blood cells. The spleen consists of two compartments that are functionally and anatomically distinct: the red pulp and the white pulp. The red pulp functions as a blood filter and removes damaged erythrocytes and foreign material. The white pulp is composed of three compartments: the periarteriolar lymphoid sheath, the follicles, and the marginal zone (MZ).[82] Blood that enters the marginal sinus and MZ percolates through the MZ and flows either directly into the venous sinuses or goes into the red pulp. About 5% of the intravenously administered Ad is sequestered by the spleen.[83] In the spleen, the virus is captured by $CD169^+$ and $MARCO^+$ macrophages in the MZ.[68] Within 10 min after Ad administration, $MARCO^+$ macrophages activate transcription of proinflammatory cytokines and chemokines IL-1α, IL-1β, and CXCL1.[68] The alacrity of the response suggests that $MARCO^+$ macrophages are activated by sensing early steps of viral invasion, most likely virus attachment or internalization. Indeed, $MARCO^+$ macrophages did not trigger cytokine transcription in response to Ad5ΔRGD virus administration.[68] Ad5ΔRGD has a deletion of an RGD motif in the penton protein and is incapable of β3 integrin binding.[66] Furthermore, the Ad ts1 mutant that can adhere to the cell surface and can be internalized but is incapable of escaping endosomes was not able to activate a full-scale immune response.[68] These data suggest that $MARCO^+$ macrophages trap the virus from the bloodstream and activate an immune response immediately after virus attachment to cell surface integrins. However, if the signal of virus attachment is coupled with endosomal membrane damage, the immune response can be drastically amplified.[68] Interestingly, $MARCO^+$ macrophages containing the virus undergo drastic cytosol rearrangement and display severe distortions of mitochondria that result in cell elimination within 24 h.[84] Surprisingly, IRF3, the key player in triggering Kupffer cell death in the liver,[74] was not required for $MARCO^+$ macrophage death. Moreover, unlike Kupffer cells, $MARCO^+$ macrophages did not accumulate propidium iodine, which suggests that the cellular membranes stayed intact and necrotic cell death programs were not activated in $MARCO^+$ macrophages upon their interaction with Ad.[84]

Marginal zone macrophages that trap Ad particles from the blood initiate release of IL-1α, which in turn signals through IL-1R and activates production of both CXCL1 and CXCL2 in the spleen.[68] CXC family chemokines, CXCL1 and CXCL2, are among the most potent chemoattractants that stimulate neutrophil migration and activation.[85] Indeed, after Ad administration, bone marrow responded with the release of great numbers of polymorphonuclear leukocytes (PMNs) into the bloodstream. Interestingly, about a third of all released PMNs were retained in the MZ of spleen in close proximity to MACRO+ macrophages.[68] However, the solely chemokine signaling was not sufficient for PMNs retention in the spleen. The authors found that complement component C3 cooperated with IL-1a–dependent signaling in retaining PMNs in close proximity to MACRO+ macrophages.[68] Neutrophils, together with circulating monocytes and tissue macrophages, are professional phagocytic cells that form the effector arm of innate immunity.[86] Although both PMNs and macrophages can efficiently engulf and destroy extracellular pathogens, their effector mechanisms are distinct and nonredundant. The PMN migration and retention in the splenic MZ and their localization in proximity to Ad-containing MARCO+ macrophages result in an elimination of the virus-containing cells from the MZ.[86]

Another type of signaling that is involved in PMN retention in the MZ is TLR4-dependent. It was found that mice that are deficient in TLR4 even with uncompromised IL-1a signaling cannot retain PMNs in the spleen MZ.[29] As a result, the PMNs are unable to clear the virus-containing cells, which leads to higher viral load.[29] The molecular mechanism of TLR4-dependent PMNs retention in the MZ is not clear and requires further investigation.

# 4. Adenovirus Interactions in the Lungs

After intravascular injection of Ad in mice, less than 1% of the injected virus can be recovered from the lungs owing to highly efficient virus sequestration by liver and spleen macrophages.[83] However, if a preexisting liver condition exists, such as liver cirrhosis, which results in redistribution of phagocytic macrophages from the liver to other tissues,[87] lungs become one of the major organs accumulating Ad.[88] In experiments with animals with cirrhotic livers, Smith et al. showed that biodistribution of the injected Ad dramatically shifted toward the lungs instead of the liver, whereas virus accumulation in other organs was not significantly changed.[88] In the lungs of the cirrhotic animals, the virus accumulated in the pulmonary intravascular macrophages (PIMs). In rodents and humans, PIMs are not constitutively present in the lungs. However, they can be induced by different conditions, for example, during liver cirrhosis (reviewed in Ref. 89). Interestingly, redistribution of the virus to the lungs did not result in higher transgene expression in the pulmonary tissue, which suggests that viral DNA in the PIMs was degraded before transcription of the genome.[88] In addition, after Ad administration, cirrhotic rats had higher mortality, most likely because of severe pulmonary hemorrhagic edema, and an increase in IL-6 and TNF-$\alpha$ compared with healthy rats.[90]

Whereas PIMs are located on the vascular side in the lungs and are not constitutively present in many species including humans and rodents, alveolar macrophages are residential lung macrophages that survey the respiratory epithelial surface.[91]

Alveolar macrophages rapidly uptake intratracheally administrated Ad.[92,93] Uptake by the macrophages seems to depend on surfactant proteins present in the alveoli. Alveolar macrophages in mice deficient in major surfactant protein, SP-A, did not efficiently uptake the virus from the lumens.[94] In response to Ad uptake, alveolar macrophages but not airway epithelial or vascular endothelial cells were able to activate production of proinflammatory cytokines (TNF-α, IL-6, MIP-1α, and CXCL2) as early as 30 min after administration.[92] Similar to macrophages in the liver, alveolar macrophages are not permissive for Ad vector replication and do not express a transgene encoded by the virus. Instead, they accumulate the virus and degrade viral DNA.[92] Degradation of viral DNA is a biphasic process. During the early phase, the alveolar macrophages degrade approximately 70% of the Ad DNA. This degradation is independent of the T-cell immune status of mice and occurs within 24 h after virus administration.[93] The remaining Ad DNA degrades slowly during several days or weeks after virus administration. This second phase of degradation depends on the functions of $CD8^+$ T cells that target Ad-infected cells.[95] Unlike alveolar macrophages, other lung cell types such as airway epithelial, alveolar lining, and adventitial cells can be effectively transduced by Ad and express a transgene from 2 to 14 days after Ad delivery.[96] Interestingly, similar to the liver, the absence of alveolar macrophages allows for enhanced transduction of other lung cell types with an increase in viral DNA levels and amplified transgene expression.[93] In addition, immune deficiencies in $CD8^+$ T cells result in prolonged expression of a transgene encoded by the Ad vector.[95]

In summary, intravascular Ad delivery causes a complex response of the host innate immune system. Numerous studies of virus biodistribution after intravascular injection have shown that over 90% of administered virus particles accumulate in the liver and spleen. The mechanisms of Ad vector sequestration depend on the interplay of host factors that can promote vector uptake and transduction or neutralize the virus. Different organs with their residential macrophages respond to Ad vector administration in a similar fashion that is aimed at eliminating invading pathogens. Although the molecular mechanisms of elimination are slightly variable, they all culminate in death and removal of virus-containing cells. After trapping Ad particles from the bloodstream, liver residential macrophages, Kupffer cells, undergo IRF3-dependent necrosis. In the spleen, $MARCO^+$ MZ macrophages activate IL-1α-IL-1RI–dependent proinflammatory cytokine and chemokine production. Cooperation of chemokines CXCL1 and CXCL2 with complement factor C3 results in recruiting and retaining PMNs in the splenic MZ in close proximity to virus-containing $MARCO^+$ macrophages. Activated PMNs eliminate both the virus and virus-containing cells. In the lungs, PIMs or alveolar macrophages sequester Ad and release proinflammatory cytokines attracting and activating $CD8^+$ T cells that eliminate virus-infected cells. Collectively, cooperation of tissue macrophages, PMNs, and CTLs results in effective elimination of the virus from the blood and Ad-transduced cells from the tissues. If Ad-mediated therapy is to be successful, effective strategies need to be devised that prevent Ad interaction with a network of circulating humoral factors and both innate and adaptive arms of the immune system.

## References


1. Raper SE, Yudkoff M, Chirmule N, Gao GP, Nunes F, Haskal ZJ, et al. A pilot study of in vivo liver-directed gene transfer with an adenoviral vector in partial ornithine transcarbamylase deficiency. *Hum Gene Ther* 2002;**13**:163–75.
2. Raper SE, Chirmule N, Lee FS, Wivel NA, Bagg A, Gao GP, et al. Fatal systemic inflammatory response syndrome in a ornithine transcarbamylase deficient patient following adenoviral gene transfer. *Mol Genet Metab* 2003;**80**:148–58.
3. Morral N, O'Neal WK, Rice K, Leland MM, Piedra PA, Aguilar-Cordova E, et al. Lethal toxicity, severe endothelial injury, and a threshold effect with high doses of an adenoviral vector in baboons. *Hum Gene Ther* 2002;**13**:143–54.
4. Lozier JN, Csako G, Mondoro TH, Krizek DM, Metzger ME, Costello R, et al. Toxicity of a first-generation adenoviral vector in rhesus macaques. *Hum Gene Ther* 2002;**13**:113–24.
5. Alemany R, Suzuki K, Curiel DT. Blood clearance rates of adenovirus type 5 in mice. *J Gen Virol* 2000;**81**:2605–9.
6. Kirn D. Clinical research results with dl1520 (Onyx-015), a replication-selective adenovirus for the treatment of cancer: what have we learned? *Gene Ther* 2001;**8**:89–98.
7. Tao N, Gao GP, Parr M, Johnston J, Baradet T, Wilson JM, et al. Sequestration of adenoviral vector by Kupffer cells leads to a nonlinear dose response of transduction in liver. *Mol Ther* 2001;**3**:28–35.
8. Worgall S, Wolff G, Falck-Pedersen E, Crystal RG. Innate immune mechanisms dominate elimination of adenoviral vectors following in vivo administration. *Hum Gene Ther* 1997;**8**:37–44.
9. Lieber A, He CY, Meuse L, Schowalter D, Kirillova I, Winther B, et al. The role of Kupffer cell activation and viral gene expression in early liver toxicity after infusion of recombinant adenovirus vectors. *J Virol* 1997;**71**:8798–807.
10. Muruve DA, Barnes MJ, Stillman IE, Libermann TA. Adenoviral gene therapy leads to rapid induction of multiple chemokines and acute neutrophil-dependent hepatic injury in vivo. *Hum Gene Ther* 1999;**10**:965–76.
11. Zhang Y, Chirmule N, Gao GP, Qian R, Croyle M, Joshi B, et al. Acute cytokine response to systemic adenoviral vectors in mice is mediated by dendritic cells and macrophages. *Mol Ther* 2001;**3**:697–707.
12. Borgland SL, Bowen GP, Wong NC, Libermann TA, Muruve DA. Adenovirus vector-induced expression of the C-X-C chemokine IP-10 is mediated through capsid-dependent activation of NF-κB. *J Virol* 2000;**74**:3941–7.
13. Bowen GP, Borgland SL, Lam M, Libermann TA, Wong NC, Muruve DA. Adenovirus vector-induced inflammation: capsid-dependent induction of the C-C chemokine RANTES requires NF-κB. *Hum Gene Ther* 2002;**13**:367–79.
14. Liu Q, Muruve DA. Molecular basis of the inflammatory response to adenovirus vectors. *Gene Ther* 2003;**10**:935–40.
15. Reid T, Galanis E, Abbruzzese J, Sze D, Andrews J, Romel L, et al. Intra-arterial administration of a replication-selective adenovirus (dl1520) in patients with colorectal carcinoma metastatic to the liver: a phase I trial. *Gene Ther* 2001;**8**:1618–26.
16. Crystal RG, Harvey BG, Wisnivesky JP, O'Donoghue KA, Chu KW, Maroni J, et al. Analysis of risk factors for local delivery of low- and intermediate-dose adenovirus gene transfer vectors to individuals with a spectrum of comorbid conditions. *Hum Gene Ther* 2002;**13**:65–100.

17. Ben-Gary H, McKinney RL, Rosengart T, Lesser ML, Crystal RG. Systemic interleukin-6 responses following administration of adenovirus gene transfer vectors to humans by different routes. *Mol Ther* 2002;**6**:287–97.
18. McCoy RD, Davidson BL, Roessler BJ, Huffnagle GB, Janich SL, Laing TJ, et al. Pulmonary inflammation induced by incomplete or inactivated adenoviral particles. *Hum Gene Ther* 1995;**6**:1553–60.
19. Hartman ZC, Black EP, Amalfitano A. Adenoviral infection induces a multi-faceted innate cellular immune response that is mediated by the toll-like receptor pathway in A549 cells. *Virology* 2007;**358**:357–72.
20. Hartman ZC, Kiang A, Everett RS, Serra D, Yang XY, Clay TM, et al. Adenovirus infection triggers a rapid, MyD88-regulated transcriptome response critical to acute-phase and adaptive immune responses in vivo. *J Virol* 2007;**81**:1796–812.
21. Schnell MA, Zhang Y, Tazelaar J, Gao GP, Yu QC, Qian R, et al. Activation of innate immunity in nonhuman primates following intraportal administration of adenoviral vectors. *Mol Ther* 2001;**3**:708–22.
22. Brunetti-Pierri N, Palmer DJ, Beaudet AL, Carey KD, Finegold M, Ng P. Acute toxicity after high-dose systemic injection of helper-dependent adenoviral vectors into nonhuman primates. *Hum Gene Ther* 2004;**15**:35–46.
23. Baker AH, McVey JH, Waddington SN, Di Paolo NC, Shayakhmetov DM. The influence of blood on in vivo adenovirus bio-distribution and transduction. *Mol Ther* 2007;**15**:1410–6.
24. Di Paolo NC, van Rooijen N, Shayakhmetov DM. Redundant and synergistic mechanisms control the sequestration of blood-born adenovirus in the liver. *Mol Ther* 2009;**17**:675–84.
25. Shayakhmetov DM, Gaggar A, Ni S, Li ZY, Lieber A. Adenovirus binding to blood factors results in liver cell infection and hepatotoxicity. *J Virol* 2005;**79**:7478–91.
26. Parker AL, Waddington SN, Nicol CG, Shayakhmetov DM, Buckley SM, Denby L, et al. Multiple vitamin K-dependent coagulation zymogens promote adenovirus-mediated gene delivery to hepatocytes. *Blood* 2006;**108**:2554–61.
27. Lopez-Gordo E, Denby L, Nicklin SA, Baker AH. The importance of coagulation factors binding to adenovirus: historical perspectives and implications for gene delivery. *Expert Opin Drug Deliv* 2014;**11**:1795–813.
28. Irons EE, Flatt JW, Doronin K, Fox TL, Acchione M, Stewart PL, et al. Coagulation factor binding orientation and dimerization may influence infectivity of adenovirus-coagulation factor complexes. *J Virol* 2013;**87**:9610–9.
29. Doronin K, Flatt JW, Di Paolo NC, Khare R, Kalyuzhniy O, Acchione M, et al. Coagulation factor X activates innate immunity to human species C adenovirus. *Science* 2012;**338**:795–8.
30. Kalyuzhniy O, Di Paolo NC, Silvestry M, Hofherr SE, Barry MA, Stewart PL, et al. Adenovirus serotype 5 hexon is critical for virus infection of hepatocytes in vivo. *Proc Natl Acad Sci USA* 2008;**105**:5483–8.
31. Waddington SN, McVey JH, Bhella D, Parker AL, Barker K, Atoda H, et al. Adenovirus serotype 5 hexon mediates liver gene transfer. *Cell* 2008;**132**:397–409.
32. Alba R, Bradshaw AC, Parker AL, Bhella D, Waddington SN, Nicklin SA, et al. Identification of coagulation factor (F)X binding sites on the adenovirus serotype 5 hexon: effect of mutagenesis on FX interactions and gene transfer. *Blood* 2009;**114**:965–71.
33. Xu Z, Qiu Q, Tian J, Smith JS, Conenello GM, Morita T, et al. Coagulation factor X shields adenovirus type 5 from attack by natural antibodies and complement. *Nat Med* 2013;**19**:452–7.
34. Bradley RR, Lynch DM, Iampietro MJ, Borducchi EN, Barouch DH. Adenovirus serotype 5 neutralizing antibodies target both hexon and fiber following vaccination and natural infection. *J Virol* 2012;**86**:625–9.

35. Sumida SM, Truitt DM, Lemckert AA, Vogels R, Custers JH, Addo MM, et al. Neutralizing antibodies to adenovirus serotype 5 vaccine vectors are directed primarily against the adenovirus hexon protein. *J Immunol* 2005;**174**:7179–85.
36. Wohlfart C. Neutralization of adenoviruses: kinetics, stoichiometry, and mechanisms. *J Virol* 1988;**62**:2321–8.
37. Varghese R, Mikyas Y, Stewart PL, Ralston R. Postentry neutralization of adenovirus type 5 by an antihexon antibody. *J Virol* 2004;**78**:12320–32.
38. Smith JG, Cassany A, Gerace L, Ralston R, Nemerow GR. Neutralizing antibody blocks adenovirus infection by arresting microtubule-dependent cytoplasmic transport. *J Virol* 2008;**82**:6492–500.
39. Baker AH, Nicklin SA, Shayakhmetov DM. FX and host defense evasion tactics by adenovirus. *Mol Ther* 2013;**21**:1109–11.
40. Tian J, Xu Z, Smith JS, Hofherr SE, Barry MA, Byrnes AP. Adenovirus activates complement by distinctly different mechanisms in vitro and in vivo: indirect complement activation by virions in vivo. *J Virol* 2009;**83**:5648–58.
41. Shayakhmetov DM, Di Paolo NC, Mossman KL. Recognition of virus infection and innate host responses to viral gene therapy vectors. *Mol Ther* 2010;**18**:1422–9.
42. Lehrer RI, Lu W. α-Defensins in human innate immunity. *Immunol Rev* 2012;**245**:84–112.
43. Bastian A, Schafer H. Human α-defensin 1 (HNP-1) inhibits adenoviral infection in vitro. *Regul Pept* 2001;**101**:157–61.
44. Smith JG, Nemerow GR. Mechanism of adenovirus neutralization by human α-defensins. *Cell Host Microbe* 2008;**3**:11–9.
45. Smith JG, Silvestry M, Lindert S, Lu W, Nemerow GR, Stewart PL. Insight into the mechanisms of adenovirus capsid disassembly from studies of defensin neutralization. *PLoS Pathog* 2010;**6**:e1000959.
46. Snijder J, Reddy VS, May ER, Roos WH, Nemerow GR, Wuite GJ. Integrin and defensin modulate the mechanical properties of adenovirus. *J Virol* 2013;**87**:2756–66.
47. Carlisle RC, Di Y, Cerny AM, Sonnen AF, Sim RB, Green NK, et al. Human erythrocytes bind and inactivate type 5 adenovirus by presenting Coxsackie virus-adenovirus receptor and complement receptor 1. *Blood* 2009;**113**:1909–18.
48. Seiradake E, Henaff D, Wodrich H, Billet O, Perreau M, Hippert C, et al. The cell adhesion molecule "CAR" and sialic acid on human erythrocytes influence adenovirus in vivo biodistribution. *PLoS Pathog* 2009;**5**:e1000277.
49. Lyons M, Onion D, Green NK, Aslan K, Rajaratnam R, Bazan-Peregrino M, et al. Adenovirus type 5 interactions with human blood cells may compromise systemic delivery. *Mol Ther* 2006;**14**:118–28.
50. He JQ, Katschke Jr KJ, Gribling P, Suto E, Lee WP, Diehl L, et al. CRIg mediates early Kupffer cell responses to adenovirus. *J Leukoc Biol* 2013;**93**:301–6.
51. Escutenaire S, Cerullo V, Diaconu I, Ahtiainen L, Hannuksela P, Oksanen M, et al. In vivo and in vitro distribution of type 5 and fiber-modified oncolytic adenoviruses in human blood compartments. *Ann Med* 2011;**43**:151–63.
52. Cotter MJ, Zaiss AK, Muruve DA. Neutrophils interact with adenovirus vectors via Fc receptors and complement receptor 1. *J Virol* 2005;**79**:14622–31.
53. Othman M, Labelle A, Mazzetti I, Elbatarny HS, Lillicrap D. Adenovirus-induced thrombocytopenia: the role of von Willebrand factor and P-selectin in mediating accelerated platelet clearance. *Blood* 2007;**109**:2832–9.
54. Wolins N, Lozier J, Eggerman TL, Jones E, Aguilar-Cordova E, Vostal JG. Intravenous administration of replication-incompetent adenovirus to rhesus monkeys induces thrombocytopenia by increasing in vivo platelet clearance. *Br J Haematol* 2003;**123**:903–5.

55. Gupalo E, Kuk C, Qadura M, Buriachkovskaia L, Othman M. Platelet-adenovirus vs. inert particles interaction: effect on aggregation and the role of platelet membrane receptors. *Platelets* 2013;**24**:383–91.
56. Stone D, Liu Y, Shayakhmetov D, Li ZY, Ni S, Lieber A. Adenovirus-platelet interaction in blood causes virus sequestration to the reticuloendothelial system of the liver. *J Virol* 2007;**81**:4866–71.
57. Stone D, Ni S, Li ZY, Gaggar A, DiPaolo N, Feng Q, et al. Development and assessment of human adenovirus type 11 as a gene transfer vector. *J Virol* 2005;**79**:5090–104.
58. Hofherr SE, Mok H, Gushiken FC, Lopez JA, Barry MA. Polyethylene glycol modification of adenovirus reduces platelet activation, endothelial cell activation, and thrombocytopenia. *Hum Gene Ther* 2007;**18**:837–48.
59. Alemany R, Curiel DT. CAR-binding ablation does not change biodistribution and toxicity of adenoviral vectors. *Gene Ther* 2001;**8**:1347–53.
60. Einfeld DA, Schroeder R, Roelvink PW, Lizonova A, King CR, Kovesdi I, et al. Reducing the native tropism of adenovirus vectors requires removal of both CAR and integrin interactions. *J Virol* 2001;**75**:11284–91.
61. Shayakhmetov DM, Li ZY, Ni S, Lieber A. Analysis of adenovirus sequestration in the liver, transduction of hepatic cells, and innate toxicity after injection of fiber-modified vectors. *J Virol* 2004;**78**:5368–81.
62. Xu Z, Tian J, Smith JS, Byrnes AP. Clearance of adenovirus by Kupffer cells is mediated by scavenger receptors, natural antibodies, and complement. *J Virol* 2008;**82**:11705–13.
63. Shashkova EV, Doronin K, Senac JS, Barry MA. Macrophage depletion combined with anticoagulant therapy increases therapeutic window of systemic treatment with oncolytic adenovirus. *Cancer Res* 2008;**68**:5896–904.
64. Schiedner G, Hertel S, Johnston M, Dries V, van Rooijen N, Kochanek S. Selective depletion or blockade of Kupffer cells leads to enhanced and prolonged hepatic transgene expression using high-capacity adenoviral vectors. *Mol Ther* 2003;**7**:35–43.
65. Wheeler MD, Yamashina S, Froh M, Rusyn I, Thurman RG. Adenoviral gene delivery can inactivate Kupffer cells: role of oxidants in NF-κB activation and cytokine production. *J Leukoc Biol* 2001;**69**:622–30.
66. Wickham TJ, Mathias P, Cheresh DA, Nemerow GR. Integrins $\alpha_v\beta_3$ and $\alpha_v\beta_5$ promote adenovirus internalization but not virus attachment. *Cell* 1993;**73**:309–19.
67. Canton J, Neculai D, Grinstein S. Scavenger receptors in homeostasis and immunity. *Nat Rev Immunol* 2013;**13**:621–34.
68. Di Paolo NC, Miao EA, Iwakura Y, Murali-Krishna K, Aderem A, Flavell RA, et al. Virus binding to a plasma membrane receptor triggers interleukin-1α-mediated proinflammatory macrophage response in vivo. *Immunity* 2009;**31**:110–21.
69. Piccolo P, Vetrini F, Mithbaokar P, Grove NC, Bertin T, Palmer D, et al. SR-A and SREC-I are Kupffer and endothelial cell receptors for helper-dependent adenoviral vectors. *Mol Ther* 2013;**21**:767–74.
70. Piccolo P, Annunziata P, Mithbaokar P, Brunetti-Pierri N. SR-A and SREC-I binding peptides increase HDAd-mediated liver transduction. *Gene Ther* 2014;**21**:950–7.
71. Khare R, Hillestad ML, Xu Z, Byrnes AP, Barry MA. Circulating antibodies and macrophages as modulators of adenovirus pharmacology. *J Virol* 2013;**87**:3678–86.
72. Khare R, May SM, Vetrini F, Weaver EA, Palmer D, Rosewell A, et al. Generation of a Kupffer cell-evading adenovirus for systemic and liver-directed gene transfer. *Mol Ther* 2011;**19**:1254–62.
73. Manickan E, Smith JS, Tian J, Eggerman TL, Lozier JN, Muller J, et al. Rapid Kupffer cell death after intravenous injection of adenovirus vectors. *Mol Ther* 2006;**13**:108–17.

74. Di Paolo NC, Doronin K, Baldwin LK, Papayannopoulou T, Shayakhmetov DM. The transcription factor IRF3 triggers "defensive suicide" necrosis in response to viral and bacterial pathogens. *Cell Rep* 2013;**3**:1840–6.
75. Takeuchi O. IRF3: a molecular switch in pathogen responses. *Nat Immunol* 2012;**13**:634–5.
76. Kritz AB, Nicol CG, Dishart KL, Nelson R, Holbeck S, Von Seggern DJ, et al. Adenovirus 5 fibers mutated at the putative HSPG-binding site show restricted retargeting with targeting peptides in the HI loop. *Mol Ther* 2007;**15**:741–9.
77. Jover R, Bort R, Gomez-Lechon MJ, Castell JV. Cytochrome P450 regulation by hepatocyte nuclear factor 4 in human hepatocytes: a study using adenovirus-mediated antisense targeting. *Hepatology* 2001;**33**:668–75.
78. Wohlleber D, Kashkar H, Gartner K, Frings MK, Odenthal M, Hegenbarth S, et al. TNF-induced target cell killing by CTL activated through cross-presentation. *Cell Rep* 2012;**2**:478–87.
79. Nemerow GR, Stewart PL. Role of α(v) integrins in adenovirus cell entry and gene delivery. *Microbiol Mol Biol Rev* 1999;**63**:725–34.
80. Wickham TJ, Carrion ME, Kovesdi I. Targeting of adenovirus penton base to new receptors through replacement of its RGD motif with other receptor-specific peptide motifs. *Gene Ther* 1995;**2**:750–6.
81. Veesler D, Cupelli K, Burger M, Graber P, Stehle T, Johnson JE. Single-particle EM reveals plasticity of interactions between the adenovirus penton base and integrin $\alpha_v\beta_3$. *Proc Natl Acad Sci USA* 2014;**111**:8815–9.
82. Cesta MF. Normal structure, function, and histology of the spleen. *Toxicol Pathol* 2006;**34**:455–65.
83. Johnson M, Huyn S, Burton J, Sato M, Wu L. Differential biodistribution of adenoviral vector in vivo as monitored by bioluminescence imaging and quantitative polymerase chain reaction. *Hum Gene Ther* 2006;**17**:1262–9.
84. Di Paolo NC, Baldwin LK, Irons EE, Papayannopoulou T, Tomlinson S, Shayakhmetov DM. IL-1α and complement cooperate in triggering local neutrophilic inflammation in response to adenovirus and eliminating virus-containing cells. *PLoS Pathog* 2014;**10**:e1004035.
85. Kuida K, Lippke JA, Ku G, Harding MW, Livingston DJ, Su MS, et al. Altered cytokine export and apoptosis in mice deficient in interleukin-1 beta converting enzyme. *Science* 1995;**267**:2000–3.
86. Silva MT. When two is better than one: macrophages and neutrophils work in concert in innate immunity as complementary and cooperative partners of a myeloid phagocyte system. *J Leukoc Biol* 2010;**87**:93–106.
87. Holmberg JT, Bergqvist L, Hultberg B, Hagerstrand I, Ihse I, Ryden S. Radiolabelled colloid uptake distribution and pulmonary contents and localization of lysosomal enzymes in cholestatic rats. *Scand J Gastroenterol* 1986;**21**:291–9.
88. Smith JS, Tian J, Muller J, Byrnes AP. Unexpected pulmonary uptake of adenovirus vectors in animals with chronic liver disease. *Gene Ther* 2004;**11**:431–8.
89. Schneberger D, Aharonson-Raz K, Singh B. Pulmonary intravascular macrophages and lung health: what are we missing? *Am J Physiol Lung Cell Mol Physiol* 2012;**302**:L498–503.
90. Smith JS, Tian J, Lozier JN, Byrnes AP. Severe pulmonary pathology after intravenous administration of vectors in cirrhotic rats. *Mol Ther* 2004;**9**:932–41.
91. Hussell T, Bell TJ. Alveolar macrophages: plasticity in a tissue-specific context. *Nat Rev Immunol* 2014;**14**:81–93.
92. Zsengeller Z, Otake K, Hossain SA, Berclaz PY, Trapnell BC. Internalization of adenovirus by alveolar macrophages initiates early proinflammatory signaling during acute respiratory tract infection. *J Virol* 2000;**74**:9655–67.

93. Worgall S, Leopold PL, Wolff G, Ferris B, Van Roijen N, Crystal RG. Role of alveolar macrophages in rapid elimination of adenovirus vectors administered to the epithelial surface of the respiratory tract. *Hum Gene Ther* 1997;**8**:1675–84.
94. Harrod KS, Trapnell BC, Otake K, Korfhagen TR, Whitsett JA. SP-A enhances viral clearance and inhibits inflammation after pulmonary adenoviral infection. *Am J Physiol* 1999;**277**:L580–8.
95. Yang Y, Li Q, Ertl HC, Wilson JM. Cellular and humoral immune responses to viral antigens create barriers to lung-directed gene therapy with recombinant adenoviruses. *J Virol* 1995;**69**:2004–15.
96. Rodman DM, San H, Simari R, Stephan D, Tanner F, Yang Z, et al. In vivo gene delivery to the pulmonary circulation in rats: transgene distribution and vascular inflammatory response. *Am J Respir Cell Mol Biol* 1997;**16**:640–9.

# Antibodies against Adenoviruses

**15**

*Andrew P. Byrnes*
Division of Cellular and Gene Therapies, FDA Center for Biologics Evaluation and Research, Silver Spring, MD, USA

## 1. Introduction

Although adenoviruses are efficient vectors for in vivo gene therapy, antibodies can be a major barrier. Even in animals that have never been exposed to adenoviruses, innate natural immunoglobulin M (IgM) antibodies can significantly inhibit the efficiency of gene therapy with adenovirus vectors.[1] The adaptive immune system is also strongly activated by adenoviruses, leading to capsid-specific antibodies that are often highly inhibitory.[2–5] Numerous studies have examined the in vitro and in vivo impact of antibodies against the three major capsid proteins that are exposed on the surface of the virion: hexon, fiber, and penton. Whereas antibodies to each of these proteins can impair the ability of an adenovirus to infect cells in vitro, antibodies to hexon have the most impact in vivo for both gene therapy and vaccine applications.[6–11]

Serology has historically been an important tool for distinguishing adenovirus serotypes and for diagnosis of adenovirus infections, although deoxyribonucleic acid (DNA)-based methods are now more common. Each adenovirus type is serologically unique, and the major type-specific antigenic determinants are on hexon and fiber. Type-specific antifiber antibodies can be identified using hemagglutination inhibition assays, because these antibodies prevent interaction of fiber with receptors on erythrocytes.[12] Antibodies against hexon are especially useful in classifying adenoviruses, and antihexon antibodies may be either type specific (e.g., human adenovirus 5), species specific (e.g., *human adenovirus C*, consisting of types 1, 2, 5, and 6), or even genus specific (*Mastadenovirus*).[12–15] Type-specific hexon antibodies are directed against the hexon hypervariable regions, which are presented on the outside of the capsid and have significant diversity among serotypes.[16–18] In contrast, species-specific and genus-specific antibodies recognize more conserved parts of hexon, many of which are inaccessible epitopes that are exposed only after hexon is dissociated from the capsid. Type-specific antibodies against hexon can neutralize virus infectivity. In contrast, species-specific and genus-specific antibodies cross-react against hexons from related adenoviruses in binding assays such as enzyme-linked immunosorbent assays (ELISAs) but do not neutralize virus.[19–21]

Engineering adenovirus vectors that can avoid antibodies are an important goal for increasing the efficiency of gene therapies, and there has been considerable progress using genetic and chemical methods that alter the surface of virions to prevent recognition by antibodies. This chapter surveys the most promising of these methods. First, however, the chapter reviews how antibodies inhibit adenoviruses, in particular the mechanisms through which antibodies work in vivo. This knowledge provides the

**Adenoviral Vectors for Gene Therapy.**

foundation for the rational design of more effective adenovirus vectors. Although this chapter focuses on gene therapy, studies of vaccine vectors and wild-type adenovirus infection will also be reviewed because they, too, shed considerable light on how antibodies affect adenovirus gene therapy.

# 2. Mechanisms of Antibody Action

## 2.1 Neutralization

Antibodies can interfere with adenoviral infection through multiple mechanisms, but neutralization is by far the easiest of these mechanisms to measure in vitro. Neutralization assays involve mixing antibody and virus together, then assaying the ability of the virus to infect or transduce cells. Neutralizing antibody titers are widely used as a surrogate measure of anti-adenoviral immunity in vivo. For example, in recipients of adenovirus type 4 and 7 vaccine, neutralizing antibody titers correlate well with protection against infection.[22] Although neutralization is simple to measure, it is a surprisingly complex phenomenon that can involve blocking of virion binding to receptors, aggregation of virions, or novel interactions of antibody-coated virions with proteins in the cytoplasm.

Adenovirus capsids contain three proteins that are accessible to antibodies: hexon, penton, and fiber. Although each of these proteins can be the target of neutralizing antibodies, the efficiency of neutralization varies considerably. Antibodies against penton can interfere with viral entry into cells,[23] likely because penton contributes to viral internalization.[24] However, antipenton antibodies neutralize relatively poorly overall, and virions coated with antipenton antibodies can still enter cells through alternative mechanisms.[23] Antibodies to fiber can neutralize adenovirus by cross-linking virions into aggregates, as well as by inhibiting binding or entry into cells.[23,25,26] In addition, antibodies against fiber and penton neutralize synergistically.[27]

Antihexon antibodies are typically much more potent for neutralization than antibodies against other capsid proteins.[10,11,26,28] Neutralization by antihexon antibodies can occur by a single-hit mechanism that requires as little as 1.4 antibodies per virion.[26,29] Antibodies against hexon clearly work through different mechanisms than antibodies against penton and fiber, because antihexon antibodies fail to aggregate virions and do not block attachment of virions to cells.[26,30] Unlike antipenton or antifiber antibodies, antihexon antibodies can neutralize even if they are added after virions have already attached to cells.[26] Taken together, this evidence strongly suggests that antihexon antibodies act primarily by neutralizing virions inside cells.

How can antibodies against hexon neutralize virions intracellularly? Early studies suggested that antihexon antibodies impair penetration of virus from endosomes into the cytoplasm[26] and that they block the transport of virus along microtubules to the nucleus.[31] It seems likely that antibodies against hexon can interfere with adenovirus infection by multiple mechanisms, especially when high concentrations of antibodies are used. However, groundbreaking studies have identified the predominant mechanism for neutralization by antihexon antibodies: an intracellular Fc receptor—TRIM21—recognizes antibody-coated virions in the cytoplasm and targets the virions for destruction.[32,33] Studies of TRIM21-dependent neutralization have so far focused on adenovirus as a model system, but TRIM21 seems likely to have a similar role in antibody-dependent intracellular neutralization of other nonenveloped viruses, and even certain bacteria.[34]

TRIM21 is a tripartite motif (TRIM)[35] protein that can bind the Fc region of Immunoglobulin G (IgG) with high affinity.[36] TRIM21 is also known as Ro52, and was first identified because it is a common target of autoantibodies in Sjögren syndrome and other systemic autoimmune diseases.[37,38] The interaction between TRIM21 and IgG is highly conserved in mammals, and TRIM21 from one species can bind IgG from a variety of other species.[39] TRIM21 binds to the Fc region of IgA with lower affinity than TRIM21 binds to Fc region of IgG, but IgA directed against hexon can still mediate TRIM21-dependent neutralization.[40]

After antibody-coated adenovirus enters cells, virions rapidly colocalize with TRIM21 in the cytoplasm.[41] How does TRIM21 inhibit infection by antibody-coated adenovirus? Like other TRIM proteins, TRIM21 has multiple domains: the PRYSPRY domain of TRIM21 binds to Fc,[36] and the RING domain is an E3 ubiquitin ligase.[41] The RING domain mediates autoubiquitination of TRIM21 and is essential for TRIM21 to mediate neutralization.[41] Ubiquitination is a signal that targets proteins to the proteasome for degradation, and indeed antibody-coated adenovirus is degraded rapidly inside cells in a manner that requires both the proteasome and TRIM21.[41] Because the proteasome is unable to degrade large proteins without help, TRIM21-dependent neutralization also requires the unfoldase *valosin-containing protein* (VCP).[42,43] Thus, efficient intracellular neutralization requires TRIM21 binding to antibody, TRIM21 autoubiquitination, VCP unfoldase activity, and the protein-degradation activity of the proteasome.

Although TRIM21 is undoubtedly a key mediator of neutralization by antihexon antibodies, there must also be other mechanisms. Antihexon antibodies retain some neutralizing activity even when the Fc region is mutated to abrogate TRIM21 binding.[29] Similarly, antihexon antibodies can neutralize adenovirus in cells from *Trim21* knockout mice, although neutralization in knockout cells is much less effective than in wild-type cells.[40]

As mentioned, neutralization by antihexon antibodies requires only slightly more than a single antibody per virion, but this is under optimal conditions.[26,29] Interestingly, neutralization in cells that express low amounts of TRIM21 requires more antibody per virion than in cells that express high amounts of TRIM21, and upregulation of TRIM21 expression by type I interferon (IFN) substantially increases neutralization by antibody.[29,41] In addition, TRIM21 appears to become saturated at high antibody concentrations or high virus concentrations.[29] The saturable nature of the TRIM21 pathway may help to explain the phenomenon of the persistent fraction, which is the longstanding observation that it is impossible to neutralize 100% of virions even at high concentrations of antibody. In sum, the amount of TRIM21 expression is a critical determinant of neutralization, and thus the effectiveness of neutralization depends not only on antibody, but also on the state of the cell being infected.

During adenoviruses infection, cytoplasmic DNA sensors detect the adenovirus genome and induce type I IFN.[44] However, adenoviruses produce proteins and transcripts that actively interfere with IFN-induced antiviral mechanisms, and thus adenovirus replication has only moderate sensitivity to IFN.[45–48] It is known that IFN synergizes with antibodies to suppress replication of adenovirus and other viruses: IFN and antibody together inhibit replication more than either IFN or antibody alone.[49] The mechanism for this synergy (at least for adenovirus) is now known to be IFN upregulation of TRIM21.[41] Therefore, although IFN is best known for inducing cellular proteins that directly inhibit viral replication, IFN has an additional indirect antiviral effect by inducing TRIM21, which sensitizes cells to antiviral antibodies.

## 2.2 Beyond Neutralization: Other Mechanisms of Antibody Action

Although neutralization assays may be the easiest way to measure the antiviral activity of antibodies, neutralization should not be mistaken for being the only way that antibodies inhibit adenovirus, or assumed to be the most important activity for in vivo antibody function. Antibody-coated virions engage multiple additional antiviral mechanisms, including redirecting virions to plasma membrane Fcγ receptors (FcγR), activation of complement, and augmented proinflammatory responses. These multiple mechanisms are likely to contribute synergistically to the antiviral effects of antibodies in vivo.

Fcγ receptors on the plasma membrane of cells bind to IgG-coated virions and can change both the cellular tropism and the intracellular fate of virions. Immunoglobulin G antibodies retarget adenovirus to FcγR-expressing cells of the innate immune system such as neutrophils, macrophages, and dendritic cells.[50–53] Fcγ receptor–bound virus can also undergo altered trafficking within cells; in macrophages, antibody-bound virus is targeted to phagolysosomes in an FcγR-dependent manner.[53]

Immunoglobulin M and most isotypes of IgG activate the classical complement pathway. Immunoglobulin G antibodies against adenovirus activate complement strongly, even when the IgG is nonneutralizing.[54] Natural IgM activates the classical complement pathway on Ad5 virions that are unable to bind coagulation factor X (FX), which is a protein that normally binds to hexon and protects Ad5 from complement.[55,56] Activation of complement leads to opsonization, in which fragments of complement protein C3 covalently attach to virions.[56–58] This opsonization by C3 can greatly enhance neutralization in vitro.[56,58] In addition, opsonization can retarget virions to plasma membrane complement receptors, leading to uptake or sequestration of virions by new cell types, including neutrophils, rhabdomyosarcoma cells, and erythrocytes.[51,57,59] Such retargeting can have functional consequences in vivo; for example, binding of opsonized virions to complement receptor CR1 on erythrocytes reduces the amount of adenovirus vector that can transduce the liver,[57] and complement contributes to clearance of adenovirus vectors from the circulation by Kupffer cells.[60,61]

Antibodies and antibody-induced complement also activate innate immunity, which may contribute to the toxic effects of adenovirus vectors. One way that antibodies enhance innate immunity is by retargeting virions to cells of the innate immune system such as macrophages.[53] Interestingly, another way in which antibodies activate innate immunity is by engaging TRIM21 inside cells. It has been shown that TRIM21 not only mediates intracellular neutralization, but also stimulates proinflammatory signaling via nuclear factor-κB, activator protein-1, and interferon regulatory factor (IRF) pathways.[40,43,62] This proinflammatory activity of TRIM21 requires both the RING domain and ubiquitination, similar to requirements for TRIM21-mediated neutralization.[62] Downstream of ubiquitination, however, the neutralization and proinflammatory pathways diverge; although TRIM21-dependent neutralization requires the proteasome, TRIM21-dependent upregulation of IRF7 does not require the proteasome.[62] In addition to this antibody/TRIM21-activated proinflammatory pathway, antibody-mediated opsonization of adenovirus by complement can also activate a newly discovered intracellular complement-recognition system that stimulates proinflammatory signaling and enhances neutralization.[58] This system is activated by C3 fragment-coated virions in a TRIM21-independent manner, but the exact molecular mechanism for intracellular C3 recognition is currently unknown.

In sum, antibodies do much more than just neutralize adenovirus; they also retarget adenovirus to Fcγ receptors and activate the classical pathway of complement. In addition, IgG/C3-opsonized virions that reach the cytoplasm trigger several antiviral mechanisms: TRIM21 binds to IgG and causes neutralization and proinflammatory activation, and C3 triggers TRIM21-independent proinflammatory activation.

## 2.3 Do In Vitro Assays Predict In Vivo Effects of Antibodies?

Because of the multiple direct and indirect actions of antibodies on adenoviruses, it can be extraordinarily difficult to isolate the contribution of any single mechanism in vivo. Presumably, many of the mechanisms described above act in parallel. Nevertheless, it is essential to understand the role of each mechanism in vivo; such knowledge is a prerequisite for determining which in vitro assays accurately predict the in vivo behavior of adenovirus vectors. To add a further complication, the relative importance of each mechanism seems likely to be influenced by the in vivo model system being used (viral infection, gene therapy, or vaccine) and perhaps also by the species being studied.

The neutralizing activity of an anti-adenoviral serum typically correlates with its ability to suppress adenovirus infection or transduction in animal models.[63–66] However, revealing exceptions to this general rule illustrate the importance of nonneutralizing actions of antibodies in vivo. Mutation of a hexon surface loop allows AdC68 vectors to escape neutralization by antibodies targeted to this loop, although these antibodies can still bind to virions.[67,68] In vivo, these nonneutralizing antibodies are fully capable of inhibiting muscle transduction and vaccine activity, thus demonstrating a lack of correlation between in vitro neutralization assays and in vivo antiviral activity.[68] The mechanism for this in vivo antiviral activity remains unknown, but it does not depend on FcγR.[68]

As another example in which neutralization assays fail to capture the complexity of antibody behavior in vivo, natural mouse IgM is poor at neutralizing adenovirus,[56] but transfer of natural IgM to antibody-deficient mice decreases liver transduction by adenovirus vectors[69,70] because of enhanced vector clearance by Kupffer cells.[60]

Even when one has identified an antiviral mechanism that is important in vivo, it can be challenging to develop in vitro assays that accurately model the in vivo mechanism. Designing suitable assays often requires a deep understanding of mechanism.

For example, intravenous (iv) injection of Ad5 vectors into mice activates complement.[55] Complement has diverse effects on adenovirus in vivo, altering virion uptake by Kupffer cells, hepatocyte transduction, and both innate and adaptive immune responses.[56,60,61,71,72] Citrate-anticoagulated plasma is widely used for complement assays, and indeed Ad5 activates complement in citrated plasma.[55] However, results from in vitro assays in citrate plasma failed to correlate with in vivo observations of complement activation in mice.[55] The reason for this lack of correlation became apparent when it was found that citrate reduces the plasma calcium concentration to a point that is too low to support calcium-dependent binding of FX to Ad5.[56] Because FX shields Ad5 from complement activation, in vitro studies of complement in citrate plasma (where FX is unable to bind to Ad5) can produce nonphysiological results.

# 3. Consequences of Anti-Adenoviral Antibodies In Vivo

Antibodies are critical for controlling wild-type adenovirus infection and can also severely inhibit adenovirus-based gene therapies and vaccines.[63,64,73] It is well established that preexisting antibodies often suppress the effectiveness of adenovirus vectors, but surprisingly there are also some situations in which such antibodies may be less of a barrier, or may even be beneficial in controlling vector-mediated toxicity.

Although preexisting antibodies are certainly a major hurdle for gene therapy and vaccine applications, they are not the only obstacle presented by the adaptive immune system. T cell responses against adenovirus proteins also impair the effectiveness of adenoviral vectors.[64,74] Unfortunately, T cell epitopes are often highly conserved across human adenovirus serotypes,[9,75,76] which makes anti-adenoviral T cells even harder to avoid than antibodies. Humans who are seronegative for a particular adenovirus serotype often have T cells that recognize that serotype.[77,78] It is important to be aware when reviewing the literature that studies of preexisting anti-adenoviral immunity in animal models often do not discriminate between the effects of antibodies and T cells; preimmunizing animals with vector induces both antibodies and T cells against adenovirus. Experiments that transfer anti-adenoviral antibodies to animals are the best way to examine the effects of preexisting antibodies in isolation.

## *3.1 Impact of Antibodies on Adenoviral Infection*

The pathogenesis of human adenoviruses is difficult to study in animal models because of species-specific restriction factors, but fortunately, mouse adenovirus is an excellent model system that has provided insight into how antibodies help to control adenovirus infection. Mice cannot control mouse adenovirus type 1 (MAV-1) infection without antibodies, and transfer of antibodies against MAV-1 can inhibit infection.[63,79] Early T cell–independent IgM responses appear to be particularly important for viral control.[63] TRIM21 has a central role in the control of MAV-1 infection by antibodies. In vitro, anti-MAV-1 antibodies neutralize MAV-1 in a TRIM21-dependent manner, and they also induce a TRIM21-dependent proinflammatory response.[43] In vivo, *Trim21* knockout mice are poor at controlling infection with MAV-1, and transferring anti-MAV-1 antibodies fails to protect *Trim21* knockout mice against fatal MAV-1 infection.[80] Interestingly, MAV-1 was found to upregulate TRIM21 expression in wild-type mice (presumably mediated through IFN), thereby demonstrating that adenovirus infection sensitizes mice to antiviral antibodies.[80] TRIM21 can mediate both neutralization and proinflammatory activity, and it is not yet known which of these activities is most important for control of MAV-1 in mice. This is a difficult question to address in an in vivo model system, but it is reasonable to speculate that both mechanisms are likely to have a role.

Much less is known about the role of antibodies in controlling adenoviral infection in humans, although it seems safe to assume that antibodies have a major role. Genetic immunoglobulin deficiencies may be associated with an increased incidence and severity of adenovirus infection.[81,82] However, T cells are also essential for control of human adenoviral infection. Reactivation of adenoviral infections is a common and life-threatening consequence of myelosuppressive therapies, especially in pediatric patients.[83,84] Transfer of adenoviral-specific T cells to such patients may resolve ongoing adenoviral infections.[85]

## 3.2 Impact of Antibodies on Gene Therapy Vectors and Oncolytic Adenoviruses

Adenoviruses Most studies of antiviral antibodies focus on high-affinity antibodies that are induced after exposure to a virus, but natural antibodies can also have important effects on gene therapy. Natural antibodies have flexible antigen-binding regions and possess an innate ability to bind to diverse viruses, including adenovirus.[60,86,87] These virus-binding natural antibodies are mostly of the IgM isotype and are important during the earliest phases of an infection, before development of specific, high-affinity antibody responses.[86,88] Transfer of serum from wild-type mice to antibody-deficient mice causes increased clearance of intravenously injected adenovirus vector by Kupffer cells,[60] which reduces hepatocyte transduction by the vector.[69,70] Variation in IgM concentration among mouse strains can affect gene therapy: BALB/c mice have genetically high IgM concentration and low liver transduction compared with C57BL/6 mice.[1] Interestingly, vectors with Ad6 hexon are less influenced by natural IgM than vectors with Ad5 hexon, and therefore Ad6-based vectors are less susceptible to clearance by Kupffer cells.[69,89]

On its own, natural IgM has little neutralizing activity against adenovirus, but it can activate complement.[56] Normally, binding of FX to Ad5 hexon prevents IgM from activating complement, but when FX is blocked, IgM-activated complement can neutralize the vector.[56] The effects of natural IgM antibodies in vivo are particularly dramatic when Ad5 vectors cannot bind FX, in which case natural IgM and complement are extraordinarily inhibitory for liver transduction.[1,56] Adenovirus 5 hexon appears to confer a unique susceptibility to natural IgM and/or complement, because other adenovirus serotypes are less susceptible to this type of neutralization than Ad5.[90]

Both adenovirus infection and exposure to adenovirus vectors stimulate the adaptive immune system to produce specific antibodies against adenovirus. In humans, neutralizing anti-adenovirus antibodies have been identified not only in blood, but also in a variety of other compartments including synovial fluid, peritoneal fluid, and bronchoalveolar lavage.[91–93] Transfer of neutralizing human serum to mice can substantially inhibit liver transduction by intravenously injected adenovirus vectors.[65,73,94] Experiments with chimeric vectors have identified antihexon antibodies as being the most important for inhibiting vector transduction of the liver, although antifiber antibodies are also inhibitory.[10] Transfer of an antihexon monoclonal IgG antibody is sufficient to inhibit muscle transduction by vector in mice.[68] Together, these animal studies show that antihexon IgG antibodies are major barriers to gene therapy and are likely to inhibit delivery of vector to a variety of tissues and organs.

Although there have been numerous clinical trials with adenovirus gene therapy vectors and many of these trials have collected information about antibodies, it is often difficult to measure transduction and efficacy, and so there are currently few data on how preexisting antibodies affect adenovirus gene therapy vectors in humans. As discussed below, such data are easier to obtain and evaluate for vaccine clinical trials.

In animals, preexisting antibodies can affect not only transduction, but also vector-induced toxicity. For example, high doses of iv adenovirus vector can cause shock in several animal models.[95–97] Vector-induced shock in mice can be prevented by transfer of mouse anti-adenovirus serum, but for unclear reasons shock is not prevented by human serum containing anti-adenoviral antibodies.[96] As another example, preimmunization with vector can reduce some types of toxicity in both mice and monkeys, but other types of toxicity and inflammation actually worsen.[98–100] It is particularly notable that preimmunized mice show a large increase in early mortality after vector administration.[99,100] Because these latter studies used preimmunization instead of antibody transfer, it is unclear whether the increased toxicity was caused by antibodies, T cells, or other unidentified factors.

Oncolytic adenovirus vectors are designed to replicate selectively in tumors, but they can produce toxicity in normal tissues as well. Transfer of human adenovirus-immune serum to mice inhibits tumor transduction far less than the serum inhibits liver transduction.[94] Anti-adenovirus antibodies can somewhat diminish anti-tumor efficacy; on the other hand, antibodies greatly inhibit toxicity, in part by blocking liver infection by the vector.[101,102] Altogether, these animal studies suggest that preexisting antibodies may be less of a problem for oncolytic adenoviruses than for nonreplicating vectors. Even though antibodies can attenuate the antitumor activity of replication-competent adenovirus vectors, antibodies also significantly improve tolerability by preventing vector spillover to organs and decreasing vector-induced toxicity.

### 3.3 Impact of Antibodies on Adenovirus-Based Vaccines

The effects of preexisting anti-Ad5 antibodies on Ad5-based vaccines have been studied extensively in both animals and in clinical trials. The picture that emerges is nuanced and difficult to generalize. Although preexisting antibodies can undoubtedly inhibit or even block responses to adenovirus-based vaccines, there are also cases in which preexisting antibodies seem to have no effect on vaccine-induced responses.

An Ad5 vaccine expressing human immunodeficiency virus (HIV) Env was able to induce robust anti-Env T cell responses in mice after transfer of antifiber antibodies but not after transfer of antihexon antibodies.[11] Transfer of antihexon IgG to mice reduces vaccine effectiveness in a dose-dependent manner, and high doses of IgG completely suppress vaccine activity.[11,64] Transfer of neutralizing antibody also biases the T cell immune response toward a central memory phenotype.[66] These studies clearly demonstrate that high titers of antihexon antibodies are sufficient to suppress the activity of adenovirus-based vaccines in animal models.

In humans, the apparent impact of preexisting antibodies on adenovirus-vectors vaccines differs among clinical trials, for unclear reasons. Humans have variable levels of preexisting antibodies and T cells against adenovirus, and the levels do not correlate with each other,[103] which suggests that it is too facile to use neutralizing antibodies as a surrogate measure of preexisting immunity against adenovirus. In addition, it has been shown in animal studies that preexisting anti-adenoviral immunity leads to differences in the homing and localization of vaccine-induced T cells in organs.[104] These differences cannot be detected by assaying T cells in the blood, which raises the question of whether preexisting immunity might cause changes in vaccine-induced responses that are impossible to measure in clinical trials.

In human studies of Ad5-vectored malaria and tuberculosis vaccines, preexisting antibodies had no apparent influence on T cell responses to the vaccine-encoded antigens.[105,106] For Ad5-vectored HIV and Ebola vaccines, however, preexisting antibodies correlated with lower responses against vector-encoded antigens.[107,108] Surprisingly, in a clinical trial of an Ad5-vectored HIV vaccine, vaccine recipients who had preexisting neutralizing anti-Ad5 titers were more likely to become infected with HIV compared with unvaccinated controls.[109,110] Reasons for this disappointing finding are still unclear but may be connected to differences in vector-induced activation of the immune system in the presence or absence of neutralizing antibodies.[111–113] Other studies of Ad5-vectored HIV vaccines have not found an increase in HIV infection in subjects who had preexisting neutralizing antibodies.[114]

## 4. Evading Antibodies

The adenovirus gene therapy and vaccine research communities have invested a great deal of effort in designing novel strategies to avoid preexisting antibodies.[2,4] Most of these strategies rely on modifying the virus capsid or using vectors derived from low-prevalence adenoviruses. However, there have also been attempts to inhibit antibodies with immunosuppression or to minimize antibody impact by altering the route of vector administration.

### *4.1 Rare Human Serotypes and Nonhuman Adenoviruses*

The most popular adenovirus vectors are based on human Ad5, but Ad5 infections are common in childhood, leading to a generally high seroprevalence in adults.[115,116] There is a window in early childhood when the prevalence of Ad5 antibodies is low—after the disappearance of maternal antibodies but before infection with Ad5—and it has therefore been suggested that Ad5-based vectors would be most effective when administered to children.[115,116] However, a more general strategy to avoid preexisting antibodies is to use adenovirus vectors that are derived from low-seroprevalence adenoviruses. Vectors that are derived from rare human adenoviruses can be effective in animal studies, even in the face of high titers of antibodies against Ad5.[73,117] It is important to note that serotype prevalence can vary significantly among human populations, such that a serotype that is considered rare in one part of the world may be more common in other parts of the world.[118–120]

There has also been wide interest in developing vectors based on nonhuman adenoviruses.[121] Examples include porcine,[122,123] canine,[124] bovine,[125] ovine,[126] and nonhuman primate[127–129] adenoviruses. Humans tend to have low amounts of neutralizing antibodies against animal adenoviruses.[122,128–132] However, there are exceptions: neutralizing antibodies against certain chimpanzee adenoviruses are more common in people who live in sub-Saharan Africa, compared with the United States or Thailand.[130] It has been shown that many nonhuman adenovirus vectors retain full activity even in animals that have been preimmunized with human adenoviruses.[125,126,128] An additional potential advantage of bovine and porcine adenoviruses is that they have relatively limited overlap of T cell epitopes with human adenoviruses, at least in mouse studies,[133] and thus it is possible that using such vectors in humans would be less likely to provoke attack by preexisting anti-adenoviral T cells. However, T cells reactive against chimpanzee adenoviruses have been detected in humans.[134]

When selecting a vector, one must consider not only the prevalence of preexisting antibodies, but also other properties of the virus that affect vector activity and toxicity. It is well established that vaccines based on different adenoviruses vary markedly in their immunogenicity.[129,135,136] Similarly, when considering vectors for gene therapy, one must take into account that innate proinflammatory activity and tropism vary significantly among adenoviruses.[137–139]

Studies of seroprevalence in humans typically focus on neutralizing antibodies and often do not measure virus-binding antibodies using a method such as ELISA. The percentage of humans who have antibodies that bind to a given adenovirus serotype is usually higher than the percentage with neutralizing antibodies against the same serotype,[140] and the levels of binding and neutralizing antibodies in individual samples are often poorly correlated.[141] As noted above, even nonneutralizing antibodies can activate complement or inhibit vector transduction in vivo,[56,68] although there are also examples where transfer of nonneutralizing antibodies had no suppressive effect on vaccine-induced immunity.[66] Further work is needed to determine the in vivo impact of nonneutralizing anti-adenoviral antibodies.

Adenovirus vectors may need to be administered more than once, for example, to boost an immune response to a vaccine or to provide multiple gene therapy treatments. Antibody and T cell responses induced by the first injection of vector can greatly decrease the effectiveness of subsequent injections with the same vector.[142] For vaccines, one can avoid this problem by using a heterologous prime-boost, in which one injection is an adenovirus vector and the other is either a plasmid or a nonadenoviral vector.[143] Another option is to use two different human adenovirus serotypes or to switch from a human to a nonhuman adenovirus. These adenovirus serotype switch strategies have shown promise in animal studies for both vaccines[142,144] and gene therapies.[145–151] In some cases, however, exposure to one adenovirus can induce low amounts of cross-reactive antibodies that neutralize another serotype, especially when the adenoviruses are closely related.[7,152,153] The occurrence of cross-reactive antibody responses can be minimized by using adenoviruses that are distantly related and immunologically distinct.[152] Serotype switching is clearly a feasible strategy for vaccines, although the requirement to manufacture and develop two or more products can increase cost and complexity. For gene therapy, serotype switching may also be possible if only a few injections are required.

## 4.2 Hexon Mutation or Pseudotyping

As previously noted, antibodies against hexon are more of a barrier than antibodies against penton and fiber, as demonstrated in vivo for both gene therapy and vaccine applications.[10,11] This observation suggests that one can minimize the detrimental effects of anti-adenoviral antibodies simply by changing the hexon, without needing to change the entire vector to a different serotype. Changes to hexon may be made in several ways: substituting the entire hexon with hexon from another adenovirus; swapping just the surface-exposed hypervariable domains of hexon (which determine the serotype); or targeted mutation of hexon hypervariable domains. Changing the hexon has potential advantages over switching the entire vector to a different serotype, because one can preserve parts of the vector that control vector tropism and immunogenicity. However, it has also been noted that seemingly conservative changes to hexon—intended only to create a vector that avoids antibodies—may alter other vector properties in an unpredictable manner. For example, replacing the Ad5 hexon hypervariable regions with those of Ad48 hexon yields a vector that has different properties from both Ad5 and Ad48, including unexpected proinflammatory effects and toxicity.[139,154]

Engineering changes in hexon can be technically challenging, leading to decreased vector viability or even complete failure to rescue virus.[9,90,155] Full hexon swaps between closely related adenovirus serotypes are more successful than swaps between more distantly related serotypes. It was found that Ad5 hexon could be easily replaced with hexon from other species C viruses (types 1, 2, and 6).[9] However, when more distant hexon swaps were attempted, only one of 15 was successful. Replacing the Ad5 hexon with Ad6 hexon is sufficient to avoid neutralization by Ad5 antibodies in vitro and in vivo.[9] However, Ad2 and Ad5 hexons can induce cross-neutralizing antibodies that interfere with vector delivery,[7] which demonstrates that conservative swaps between closely related hexons may not always be sufficient to avoid antibodies.

Other examples of hexon swaps include Ad5 gene therapy vectors with hexon replaced by Ad3 or Ad12 hexon.[8,155] In both of these cases, iv injection of the hexon-chimeric vector into Ad5-preimmunized mice resulted in successful liver transduction. In another example, grafting the Ad48 hexon hypervariable regions into an Ad5 vaccine vector allowed effective vaccination in the face of preexisting Ad5 immunity.[156] In addition, when this chimeric Ad5/48 vaccine was used in a prime-boost strategy with Ad5, the T cell response to the vector-encoded transgene was higher than when the prime and boost were both performed with the same vector.[156]

## 4.3 Shielding with Polymers and Other Molecules

Coating the surface of virions with polymers can shield vectors against undesirable interactions with antibodies and other proteins, although these vector modifications must be performed in a way that does not negatively affect vector tropism or activity. Polyethylene glycol (PEG) is the most popular polymer for this purpose, in part because of the successful track record of PEGylation for improving the pharmacokinetic properties of a number of protein therapeutics.[157] One common but nonspecific method is to couple PEG to accessible lysine residues on the capsid, resulting in labeling of all three major capsid proteins.[158] Polyethylene glycol can also be coupled in a targeted manner to free cysteine residues that have been genetically engineered into specific locations on the capsid.[159,160] Other polymers such as *N*-[2-hydroxypropyl]methacrylamide (HPMA) can be used to modify adenoviruses in a similar manner.[161]

Polyethylene glycol and HPMA shield adenovirus from antibodies both in vitro and in vivo.[158,162–165] PEGylation also reduces antibody-dependent complement activation by adenovirus both in vitro and in vivo.[55] PEGylated gene therapy vectors can successfully transduce organs even in animals that have developed antibodies as a result of previous administration of unPEGylated vector,[162,166] and PEGylating vaccine vectors improves boosting of responses in Ad-immune animals.[167] Polyethylene glycol can itself induce antibody responses that neutralize PEGylated vector, rendering the PEGylated vector less effective for multiple administrations, but this problem can be mitigated by using different methods to couple PEG chemically to the capsid.[162,166]

Beyond polymers, a more limited number of studies have shown that adenovirus vectors can be shielded from antibodies by liposomes[168,169] or by encapsulation in microspheres.[170]

## 4.4 Route of Administration

Anti-adenoviral antibodies may be less problematic for certain modes of vector administration. Antibodies can be completely avoided if one is transducing cells ex vivo. For vaccines, adenovirus vector-transduced *peripheral blood mononuclear cells* can be used to avoid inhibition by antibodies,[171] although this approach is much more costly and cumbersome than direct administration of vector. Dendritic cells can also be transduced ex vivo and then administered as tumor vaccines.[172]

Although adenovirus vectors are extremely vulnerable to antibodies when vectors are administered by iv or intramuscular injection, a few routes of administration show less susceptibility to antibodies. When boosting responses to adenovirus-vectored vaccines in Ad-immune mice, oral or intranasal boosting can be substantially more effective than intramuscular boosting.[173–175] For adenovirus gene therapy vectors, readministration to the lung is somewhat more successful than readministration to other organs,[176,177] especially when using helper-dependent vectors that induce relatively weak antibody responses.[178] The reasons for these route-dependent differences are unclear, but may be related to low amounts of antibodies in certain compartments such as the lungs.[179]

### 4.5 Immunosuppression

Immunosuppressive treatments may help to diminish the levels of anti-adenoviral antibodies. Because preexisting antibodies are difficult to remove in vivo, however, research on this topic has largely focused on attenuating vector-induced antibody responses, with the goal of improving repeat administration of vector. Treatments that have been found to suppress antibody responses and allow repeated vector administration in animals include B cell depletion,[180] immunosuppressive chemotherapies,[181,182] CD4$^+$ T cell depletion,[183,184] antibodies against CD40 ligand,[185] and tolerization.[186] Immunosuppression carries risks and complicates therapies, so immunosuppression may be less generally applicable than the vector modification or shielding strategies described above.

## 5. Future Directions

The past few years have seen great progress in understanding the varied mechanisms by which antibodies affect adenovirus vectors, especially in vitro. More work needs to be done in vivo to understand which of these antiviral mechanisms have the most impact on gene therapy with adenovirus vectors, and how these mechanisms synergize. Many ingenious new strategies are being developed to avoid antibodies or to shield vectors from antibodies. There also continues to be productive cross-fertilization of ideas and techniques between the adenovirus gene therapy and vaccine fields. Although in many cases vaccine studies (which try to induce immune responses) and gene therapy studies (which generally try to avoid immune responses) have opposing goals, both types of application are generally most effective when they can avoid preexisting antivector immune responses.

One area in most urgent need of research is an understanding of which types of anti-adenoviral mediators—antibodies or T cells—are most important for inhibiting gene therapy applications and how to measure these mediators in a way that captures the most pertinent information. As alluded to several times, measuring neutralizing antibodies is easy but perhaps not always the most relevant parameter. Understanding the underlying mechanisms is critically important for deciding what to measure. An even more difficult issue is how to measure outcomes (vector transduction and activity) in gene therapy clinical trials in a way that allows one to understand how the ultimate success of the therapy is affected by preexisting immune responses against the vector. Animal models have an indispensable role in this process because it is often extremely challenging to measure vector transduction, transgene expression, or efficacy in gene therapy clinical trials. In this regard, results from vaccine clinical trials can be informative for the gene therapy field because vaccine studies usually have surrogate outcome measures—antibody and T cell responses to vector-encoded transgenes—that can aid in understanding the impact of preexisting immunity on vector activity.

Finally, despite the many decades in which adenoviruses have been studied, major discoveries are still being made about the basic biology of these viruses. Adenoviruses are amazingly diverse, and it is clear that many human and animal serotypes have unique properties that either aid or hinder their usefulness for particular gene therapy or vaccine applications. Recent findings of unexpected adenoviral properties (interaction with coagulation factors)[56,187–189] or novel antiviral mechanisms (TRIM21-mediated neutralization)[33] highlight the fact that there is still much to be gained from basic research on adenovirus biology.

## Acknowledgments

I thank Suzanne Epstein, Mike Havert, Andy Harmon, Zhili Xu, and Ritu Moitra for helpful comments and discussion.

## References

1. Qiu Q, Xu Z, Tian J, Moitra R, Gunti S, Notkins AL, et al. Impact of natural IgM concentration on gene therapy with adenovirus type 5 vectors. *J Virol* 2015;**89**(6):3412–6.
2. Thacker EE, Timares L, Matthews QL. Strategies to overcome host immunity to adenovirus vectors in vaccine development. *Expert Rev Vaccines* 2009;**8**(6):761–77.
3. Zaiss AK, Machado HB, Herschman HR. The influence of innate and pre-existing immunity on adenovirus therapy. *J Cell Biochem* 2009;**108**(4):778–90.
4. Ahi YS, Bangari DS, Mittal SK. Adenoviral vector immunity: its implications and circumvention strategies. *Curr Gene Ther* 2011;**11**(4):307–20.
5. Fausther-Bovendo H, Kobinger GP. Pre-existing immunity against Ad vectors. *Hum Vaccin Immunother* 2014;**10**(10):2875–84.
6. Gall J, Kass-Eisler A, Leinwand L, Falck-Pedersen E. Adenovirus type 5 and 7 capsid chimera: fiber replacement alters receptor tropism without affecting primary immune neutralization epitopes. *J Virol* 1996;**70**(4):2116–23.
7. Gall JG, Crystal RG, Falck-Pedersen E. Construction and characterization of hexon-chimeric adenoviruses: specification of adenovirus serotype. *J Virol* 1998;**72**(12):10260–4.
8. Roy S, Shirley PS, McClelland A, Kaleko M. Circumvention of immunity to the adenovirus major coat protein hexon. *J Virol* 1998;**72**(8):6875–9.
9. Youil R, Toner TJ, Su Q, Chen M, Tang A, Bett AJ, et al. Hexon gene switch strategy for the generation of chimeric recombinant adenovirus. *Hum Gene Ther* 2002;**13**(2):311–20.
10. Roy S, Clawson DS, Calcedo R, Lebherz C, Sanmiguel J, Wu D, et al. Use of chimeric adenoviral vectors to assess capsid neutralization determinants. *Virology* 2005;**333**(2):207–14.
11. Sumida SM, Truitt DM, Lemckert AA, Vogels R, Custers JH, Addo MM, et al. Neutralizing antibodies to adenovirus serotype 5 vaccine vectors are directed primarily against the adenovirus hexon protein. *J Immunol* 2005;**174**(11):7179–85.
12. Kjellen L, Pereira HG. Role of adenovirus antigens in the induction of virus neutralizing antibody. *J Gen Virol* 1968;**2**(1):177–85.
13. Norrby E, Wadell G. Immunological relationships between hexons of certain human adenoviruses. *J Virol* 1969;**4**(5):663–70.
14. Russell WC, Patel G, Precious B, Sharp I, Gardner PS. Monoclonal antibodies against adenovirus type 5: preparation and preliminary characterization. *J Gen Virol* 1981;**56**(Pt 2):393–408.
15. Adam E, Lengyel A, Takacs M, Erdei J, Fachet J, Nasz I. Grouping of monoclonal antibodies to adenovirus hexons by their cross-reactivity. *Arch Virol* 1986;**87**(1–2):61–71.

16. Toogood CI, Crompton J, Hay RT. Antipeptide antisera define neutralizing epitopes on the adenovirus hexon. *J Gen Virol* 1992;**73**(Pt 6):1429–35.
17. Crawford-Miksza L, Schnurr DP. Analysis of 15 adenovirus hexon proteins reveals the location and structure of seven hypervariable regions containing serotype-specific residues. *J Virol* 1996;**70**(3):1836–44.
18. Liu M, Tian X, Li X, Zhou Z, Li C, Zhou R. Generation of neutralizing monoclonal antibodies against a conformational epitope of human adenovirus type 7 (HAdv-7) incorporated in capsid encoded in a HAdv-3-based vector. *PLoS One* 2014;**9**(7):e103058.
19. Haase AT, Pereira HG. The purification of adenovirus neutralizing antibody: adenovirus type 5 hexon immunoadsorbent. *J Immunol* 1972;**108**(3):633–6.
20. Willcox N, Mautner V. Antigenic determinants of adenovirus capsids. I. Measurement of antibody cross-reactivity. *J Immunol* 1976;**116**(1):19–24.
21. Willcox N, Mautner V. Antigenic determinants of adenovirus capsids. II. Homogeneity of hexons, and accessibility of their determinants, in the virion. *J Immunol* 1976;**116**(1):25–9.
22. Kuschner RA, Russell KL, Abuja M, Bauer KM, Faix DJ, Hait H, et al. A phase 3, randomized, double-blind, placebo-controlled study of the safety and efficacy of the live, oral adenovirus type 4 and type 7 vaccine, in U.S. military recruits. *Vaccine* 2013;**31**(28):2963–71.
23. Wohlfart CE, Svensson UK, Everitt E. Interaction between HeLa cells and adenovirus type 2 virions neutralized by different antisera. *J Virol* 1985;**56**(3):896–903.
24. Wickham TJ, Mathias P, Cheresh DA, Nemerow GR. Integrins αvβ3 and αvβ5 promote adenovirus internalization but not virus attachment. *Cell* 1993;**73**(2):309–19.
25. Valentine RC, Pereira HG. Antigens and structure of the adenovirus. *J Mol Biol* 1965; **13**(1):13–20.
26. Wohlfart C. Neutralization of adenoviruses: kinetics, stoichiometry, and mechanisms. *J Virol* 1988;**62**(7):2321–8.
27. Gahery-Segard H, Farace F, Godfrin D, Gaston J, Lengagne R, Tursz T, et al. Immune response to recombinant capsid proteins of adenovirus in humans: antifiber and anti-penton base antibodies have a synergistic effect on neutralizing activity. *J Virol* 1998;**72**(3): 2388–97.
28. Wadell G. Sensitization and neutralization of adenovirus by specific sera against capsid subunits. *J Immunol* 1972;**108**(3):622–32.
29. McEwan WA, Hauler F, Williams CR, Bidgood SR, Mallery DL, Crowther RA, et al. Regulation of virus neutralization and the persistent fraction by TRIM21. *J Virol* 2012; **86**(16):8482–91.
30. Varghese R, Mikyas Y, Stewart PL, Ralston R. Postentry neutralization of adenovirus type 5 by an antihexon antibody. *J Virol* 2004;**78**(22):12320–32.
31. Smith JG, Cassany A, Gerace L, Ralston R, Nemerow GR. Neutralizing antibody blocks adenovirus infection by arresting microtubule-dependent cytoplasmic transport. *J Virol* 2008;**82**(13):6492–500.
32. Watkinson RE, McEwan WA, James LC. Intracellular antibody immunity. *J Clin Immunol* 2014;**34**(Suppl. 1):S30–4.
33. McEwan WA, James LC. TRIM21-Dependent intracellular antibody neutralization of virus infection. *Prog Mol Biol Transl Sci* 2015;**129**:167–87.
34. Rakebrandt N, Lentes S, Neumann H, James LC, Neumann-Staubitz P. Antibody- and TRIM21-dependent intracellular restriction of Salmonella enterica. *Pathog Dis* 2014;**72**(2):131–7.
35. Rajsbaum R, Garcia-Sastre A, Versteeg GA. TRIMmunity: the roles of the TRIM E3-ubiquitin ligase family in innate antiviral immunity. *J Mol Biol* 2014;**426**(6): 1265–84.

36. James LC, Keeble AH, Khan Z, Rhodes DA, Trowsdale J. Structural basis for PRYSPRY-mediated tripartite motif (TRIM) protein function. *Proc Natl Acad Sci USA* 2007;**104**(15):6200–5.
37. Chan EK, Hamel JC, Buyon JP, Tan EM. Molecular definition and sequence motifs of the 52-kD component of human SS-A/Ro autoantigen. *J Clin Invest* 1991;**87**(1):68–76.
38. Hudson M, Pope J, Mahler M, Tatibouet S, Steele R, Baron M, et al. Clinical significance of antibodies to Ro52/TRIM21 in systemic sclerosis. *Arthritis Res Ther* 2012;**14**(2):R50.
39. Keeble AH, Khan Z, Forster A, James LC. TRIM21 is an IgG receptor that is structurally, thermodynamically, and kinetically conserved. *Proc Natl Acad Sci USA* 2008;**105**(16):6045–50.
40. Bidgood SR, Tam JC, McEwan WA, Mallery DL, James LC. Translocalized IgA mediates neutralization and stimulates innate immunity inside infected cells. *Proc Natl Acad Sci USA* 2014;**111**(37):13463–8.
41. Mallery DL, McEwan WA, Bidgood SR, Towers GJ, Johnson CM, James LC. Antibodies mediate intracellular immunity through tripartite motif-containing 21 (TRIM21). *Proc Natl Acad Sci USA* 2010;**107**(46):19985–90.
42. Hauler F, Mallery DL, McEwan WA, Bidgood SR, James LC. AAA ATPase p97/VCP is essential for TRIM21-mediated virus neutralization. *Proc Natl Acad Sci USA* 2012;**109**(48):19733–8.
43. Watkinson RE, Tam JC, Vaysburd MJ, James LC. Simultaneous neutralization and innate immune detection of a replicating virus by TRIM21. *J Virol* 2013;**87**(13):7309–13.
44. Lam E, Stein S, Falck-Pedersen E. Adenovirus detection by the cGAS/STING/TBK1 DNA sensing cascade. *J Virol* 2014;**88**(2):974–81.
45. Ackrill AM, Foster GR, Laxton CD, Flavell DM, Stark GR, Kerr IM. Inhibition of the cellular response to interferons by products of the adenovirus type 5 E1A oncogene. *Nucleic Acids Res* 1991;**19**(16):4387–93.
46. Ghadge GD, Malhotra P, Furtado MR, Dhar R, Thimmapaya B. In vitro analysis of virus-associated RNA I (VAI RNA): inhibition of the double-stranded RNA-activated protein kinase PKR by VAI RNA mutants correlates with the in vivo phenotype and the structural integrity of the central domain. *J Virol* 1994;**68**(7):4137–51.
47. Lei M, Liu Y, Samuel CE. Adenovirus VAI RNA antagonizes the RNA-editing activity of the ADAR adenosine deaminase. *Virology* 1998;**245**(2):188–96.
48. Kajon AE, Spindler KR. Mouse adenovirus type 1 replication in vitro is resistant to interferon. *Virology* 2000;**274**(1):213–9.
49. Langford MP, Villarreal AL, Stanton GJ. Antibody and interferon act synergistically to inhibit enterovirus, adenovirus, and herpes simplex virus infection. *Infect Immun* 1983;**41**(1):214–8.
50. Mercier S, Rouard H, Delfau-Larue MH, Eloit M. Specific antibodies modulate the interactions of adenovirus type 5 with dendritic cells. *Virology* 2004;**322**(2):308–17.
51. Cotter MJ, Zaiss AK, Muruve DA. Neutrophils interact with adenovirus vectors via Fc receptors and complement receptor 1. *J Virol* 2005;**79**(23):14622–31.
52. Leopold PL, Wendland RL, Vincent T, Crystal RG. Neutralized adenovirus-immune complexes can mediate effective gene transfer via an Fc receptor-dependent infection pathway. *J Virol* 2006;**80**(20):10237–47.
53. Zaiss AK, Vilaysane A, Cotter MJ, Clark SA, Meijndert HC, Colarusso P, et al. Antiviral antibodies target adenovirus to phagolysosomes and amplify the innate immune response. *J Immunol* 2009;**182**(11):7058–68.
54. Cichon G, Boeckh-Herwig S, Schmidt HH, Wehnes E, Muller T, Pring-Akerblom P, et al. Complement activation by recombinant adenoviruses. *Gene Ther* 2001;**8**(23):1794–800.

55. Tian J, Xu Z, Smith JS, Hofherr SE, Barry MA, Byrnes AP. Adenovirus activates complement by distinctly different mechanisms in vitro and in vivo: indirect complement activation by virions in vivo. *J Virol* 2009;**83**:5648–58.
56. Xu Z, Qiu Q, Tian J, Smith JS, Conenello GM, Morita T, et al. Coagulation factor X shields adenovirus type 5 from attack by natural antibodies and complement. *Nat Med* 2013;**19**(4):452–7.
57. Carlisle RC, Di Y, Cerny AM, Sonnen AF, Sim RB, Green NK, et al. Human erythrocytes bind and inactivate type 5 adenovirus by presenting Coxsackie virus-adenovirus receptor and complement receptor 1. *Blood* 2009;**113**(9):1909–18.
58. Tam JC, Bidgood SR, McEwan WA, James LC. Intracellular sensing of complement C3 activates cell autonomous immunity. *Science* 2014;**345**(6201):1256070.
59. Tsai V, Varghese R, Ravindran S, Ralston R, Vellekamp G. Complement component C1q and anti-hexon antibody mediate adenovirus infection of a CAR-negative cell line. *Viral Immunol* 2008;**21**(4):469–76.
60. Xu Z, Tian J, Smith JS, Byrnes AP. Clearance of adenovirus by Kupffer cells is mediated by scavenger receptors, natural antibodies and complement. *J Virol* 2008;**82**:11705–13.
61. He JQ, Katschke Jr KJ, Gribling P, Suto E, Lee WP, Diehl L, et al. CRIg mediates early Kupffer cell responses to adenovirus. *J Leukoc Biol* 2013;**93**:301–6.
62. McEwan WA, Tam JC, Watkinson RE, Bidgood SR, Mallery DL, James LC. Intracellular antibody-bound pathogens stimulate immune signaling via the Fc receptor TRIM21. *Nat Immunol* 2013;**14**(4):327–36.
63. Moore ML, McKissic EL, Brown CC, Wilkinson JE, Spindler KR. Fatal disseminated mouse adenovirus type 1 infection in mice lacking B cells or Bruton's tyrosine kinase. *J Virol* 2004;**78**(11):5584–90.
64. Sumida SM, Truitt DM, Kishko MG, Arthur JC, Jackson SS, Gorgone DA, et al. Neutralizing antibodies and CD8+ T lymphocytes both contribute to immunity to adenovirus serotype 5 vaccine vectors. *J Virol* 2004;**78**(6):2666–73.
65. Parker AL, Waddington SN, Buckley SM, Custers J, Havenga MJ, van Rooijen N, et al. Effect of neutralizing sera on factor X-mediated adenovirus serotype 5 gene transfer. *J Virol* 2009;**83**(1):479–83.
66. Small JC, Haut LH, Bian A, Ertl HC. The effect of adenovirus-specific antibodies on adenoviral vector-induced, transgene product-specific T cell responses. *J Leukoc Biol* 2014;**96**(5):821–31.
67. Pichla-Gollon SL, Drinker M, Zhou X, Xue F, Rux JJ, Gao GP, et al. Structure-based identification of a major neutralizing site in an adenovirus hexon. *J Virol* 2007;**81**(4):1680–9.
68. Pichla-Gollon SL, Lin SW, Hensley SE, Lasaro MO, Herkenhoff-Haut L, Drinker M, et al. Effect of preexisting immunity on an adenovirus vaccine vector: in vitro neutralization assays fail to predict inhibition by antiviral antibody in vivo. *J Virol* 2009;**83**(11):5567–73.
69. Khare R, Hillestad ML, Xu Z, Byrnes AP, Barry MA. Circulating antibodies and macrophages as modulators of adenovirus pharmacology. *J Virol* 2013;**87**(7):3678–86.
70. Unzu C, Melero I, Morales-Kastresana A, Sampedro A, Serrano-Mendioroz I, Azpilikueta A, et al. Innate functions of immunoglobulin M lessen liver gene transfer with helper-dependent adenovirus. *PLoS One* 2014;**9**(1):e85432.
71. Appledorn DM, Kiang A, McBride A, Jiang H, Seregin S, Scott JM, et al. Wild-type adenoviruses from groups A-F evoke unique innate immune responses, of which HAd3 and SAd23 are partially complement dependent. *Gene Ther* 2008;**15**(12):885–901.
72. Appledorn DM, McBride A, Seregin S, Scott JM, Schuldt N, Kiang A, et al. Complex interactions with several arms of the complement system dictate innate and humoral immunity to adenoviral vectors. *Gene Ther* 2008;**15**:1606–17.

73. Vogels R, Zuijdgeest D, van RR, Hartkoorn E, Damen I, de Bethune MP, et al. Replication-deficient human adenovirus type 35 vectors for gene transfer and vaccination: efficient human cell infection and bypass of preexisting adenovirus immunity. *J Virol* 2003;**77**(15): 8263–71.
74. Yang Y, Nunes FA, Berencsi K, Furth EE, Gonczol E, Wilson JM. Cellular immunity to viral antigens limits E1-deleted adenoviruses for gene therapy. *Proc Natl Acad Sci USA* 1994;**91**(10):4407–11.
75. Smith CA, Woodruff LS, Rooney C, Kitchingman GR. Extensive cross-reactivity of adenovirus-specific cytotoxic T cells. *Hum Gene Ther* 1998;**9**(10):1419–27.
76. Leen AM, Christin A, Khalil M, Weiss H, Gee AP, Brenner MK, et al. Identification of hexon-specific CD4 and CD8 T-cell epitopes for vaccine and immunotherapy. *J Virol* 2008;**82**(1):546–54.
77. Flomenberg P, Piaskowski V, Truitt RL, Casper JT. Characterization of human proliferative T cell responses to adenovirus. *J Infect Dis* 1995;**171**(5):1090–6.
78. Chen H, Xiang ZQ, Li Y, Kurupati RK, Jia B, Bian A, et al. Adenovirus-based vaccines: comparison of vectors from three species of adenoviridae. *J Virol* 2010;**84**(20):10522–32.
79. Lenaerts L, Kelchtermans H, Geboes L, Matthys P, Verbeken E, De CE, et al. Recovery of humoral immunity is critical for successful antiviral therapy in disseminated mouse adenovirus type 1 infection. *Antimicrob Agents Chemother* 2008;**52**(4):1462–71.
80. Vaysburd M, Watkinson RE, Cooper H, Reed M, O'Connell K, Smith J, et al. Intracellular antibody receptor TRIM21 prevents fatal viral infection. *Proc Natl Acad Sci USA* 2013;**110**(30):12397–401.
81. Kainulainen L, Nikoskelainen J, Vuorinen T, Tevola K, Liippo K, Ruuskanen O. Viruses and bacteria in bronchial samples from patients with primary hypogammaglobulinemia. *Am J Respir Crit Care Med* 1999;**159**(4 Pt 1):1199–204.
82. Winkelstein JA, Marino MC, Lederman HM, Jones SM, Sullivan K, Burks AW, et al. X-linked agammaglobulinemia: report on a United States registry of 201 patients. *Med Baltim* 2006;**85**(4):193–202.
83. Leen AM, Rooney CM. Adenovirus as an emerging pathogen in immunocompromised patients. *Br J Haematol* 2005;**128**(2):135–44.
84. Breuer S, Rauch M, Matthes-Martin S, Lion T. Molecular diagnosis and management of viral infections in hematopoietic stem cell transplant recipients. *Mol Diagn Ther* 2012;**16**(2):63–77.
85. Leen AM, Bollard CM, Mendizabal AM, Shpall EJ, Szabolcs P, Antin JH, et al. Multicenter study of banked third-party virus-specific T cells to treat severe viral infections after hematopoietic stem cell transplantation. *Blood* 2013;**121**(26):5113–23.
86. Ochsenbein AF, Fehr T, Lutz C, Suter M, Brombacher F, Hengartner H, et al. Control of early viral and bacterial distribution and disease by natural antibodies. *Science* 1999;**286**(5447):2156–9.
87. Notkins AL. Polyreactivity of antibody molecules. *Trends Immunol* 2004;**25**(4):174–9.
88. Racine R, Winslow GM. IgM in microbial infections: taken for granted? *Immunol Lett* 2009;**125**(2):79–85.
89. Khare R, May SM, Vetrini F, Weaver EA, Palmer D, Rosewell A, et al. Generation of a Kupffer cell-evading adenovirus for systemic and liver-directed gene transfer. *Mol Ther* 2011;**19**(7):1254–62.
90. Ma J, Duffy MR, Deng L, Dakin RS, Uil T, Custers J, et al. Manipulating adenovirus hexon hypervariable loops dictates immune neutralisation and coagulation factor X-dependent cell interaction in vitro and in vivo. *PLoS Pathog* 2015;**11**(2):e1004673.
91. Bastian A, Bewig B. Inhibition of adenovirus-mediated gene transfer by bronchoalveolar lavage fluid. *Gene Ther* 1999;**6**(4):637–42.

92. Blackwell JL, Li H, Gomez-Navarro J, Dmitriev I, Krasnykh V, Richter CA, et al. Using a tropism-modified adenoviral vector to circumvent inhibitory factors in ascites fluid. *Hum Gene Ther* 2000;**11**(12):1657–69.
93. Goossens PH, Vogels R, Pieterman E, Havenga MJ, Bout A, Breedveld FC, et al. The influence of synovial fluid on adenovirus-mediated gene transfer to the synovial tissue. *Arthritis Rheum* 2001;**44**(1):48–52.
94. Tsai V, Johnson DE, Rahman A, Wen SF, LaFace D, Philopena J, et al. Impact of human neutralizing antibodies on antitumor efficacy of an oncolytic adenovirus in a murine model. *Clin Cancer Res* 2004;**10**(21):7199–206.
95. Morrissey RE, Horvath C, Snyder EA, Patrick J, Collins N, Evans E, et al. Porcine toxicology studies of SCH 58500, an adenoviral vector for the p53 gene. *Toxicol Sci* 2002;**65**(2):256–65.
96. Machemer T, Engler H, Tsai V, Lee S, Cannon-Carlson S, Voloch M, et al. Characterization of hemodynamic events following intravascular infusion of recombinant adenovirus reveals possible solutions for mitigating cardiovascular responses. *Mol Ther* 2005;**12**(2):254–63.
97. Xu Z, Smith JS, Tian J, Byrnes AP. Induction of shock after intravenous injection of adenovirus vectors: a critical role for platelet-activating factor. *Mol Ther* 2010;**18**(3):609–16.
98. Varnavski AN, Zhang Y, Schnell M, Tazelaar J, Louboutin JP, Yu QC, et al. Preexisting immunity to adenovirus in rhesus monkeys fails to prevent vector-induced toxicity. *J Virol* 2002;**76**(11):5711–9.
99. Vlachaki MT, Hernandez-Garcia A, Ittmann M, Chhikara M, Aguilar LK, Zhu X, et al. Impact of preimmunization on adenoviral vector expression and toxicity in a subcutaneous mouse cancer model. *Mol Ther* 2002;**6**(3):342–8.
100. Varnavski AN, Calcedo R, Bove M, Gao G, Wilson JM. Evaluation of toxicity from high-dose systemic administration of recombinant adenovirus vector in vector-naive and pre-immunized mice. *Gene Ther* 2005;**12**(5):427–36.
101. Chen Y, Yu DC, Charlton D, Henderson DR. Pre-existent adenovirus antibody inhibits systemic toxicity and antitumor activity of CN706 in the nude mouse LNCaP xenograft model: implications and proposals for human therapy. *Hum Gene Ther* 2000;**11**(11):1553–67.
102. Dhar D, Spencer JF, Toth K, Wold WS. Effect of preexisting immunity on oncolytic adenovirus vector INGN 007 antitumor efficacy in immunocompetent and immunosuppressed Syrian hamsters. *J Virol* 2009;**83**(5):2130–9.
103. O'Brien KL, Liu J, King SL, Sun YH, Schmitz JE, Lifton MA, et al. Adenovirus-specific immunity after immunization with an Ad5 HIV-1 vaccine candidate in humans. *Nat Med* 2009;**15**(8):873–5.
104. McCoy K, Tatsis N, Korioth-Schmitz B, Lasaro MO, Hensley SE, Lin SW, et al. Effect of preexisting immunity to adenovirus human serotype 5 antigens on the immune responses of nonhuman primates to vaccine regimens based on human- or chimpanzee-derived adenovirus vectors. *J Virol* 2007;**81**(12):6594–604.
105. Tamminga C, Sedegah M, Regis D, Chuang I, Epstein JE, Spring M, et al. Adenovirus-5-vectored *P. falciparum* vaccine expressing CSP and AMA1. Part B: safety, immunogenicity and protective efficacy of the CSP component. *PLoS One* 2011;**6**(10):e25868.
106. Smaill F, Jeyanathan M, Smieja M, Medina MF, Thanthrige-Don N, Zganiacz A, et al. A human type 5 adenovirus-based tuberculosis vaccine induces robust T cell responses in humans despite preexisting anti-adenovirus immunity. *Sci Transl Med* 2013;**5**(205):205ra134.
107. Harro CD, Robertson MN, Lally MA, O'Neill LD, Edupuganti S, Goepfert PA, et al. Safety and immunogenicity of adenovirus-vectored near-consensus HIV type 1 clade B gag vaccines in healthy adults. *AIDS Res Hum Retroviruses* 2009;**25**(1):103–14.

108. Ledgerwood JE, Costner P, Desai N, Holman L, Enama ME, Yamshchikov G, et al. A replication defective recombinant Ad5 vaccine expressing Ebola virus GP is safe and immunogenic in healthy adults. *Vaccine* 2010;**29**(2):304–13.
109. Buchbinder SP, Mehrotra DV, Duerr A, Fitzgerald DW, Mogg R, Li D, et al. Efficacy assessment of a cell-mediated immunity HIV-1 vaccine (the Step Study): a double-blind, randomised, placebo-controlled, test-of-concept trial. *Lancet* 2008;**372**(9653):1881–93.
110. Duerr A, Huang Y, Buchbinder S, Coombs RW, Sanchez J, del RC, et al. Extended follow-up confirms early vaccine-enhanced risk of HIV acquisition and demonstrates waning effect over time among participants in a randomized trial of recombinant adenovirus HIV vaccine (Step study). *J Infect Dis* 2012;**206**(2):258–66.
111. Perreau M, Pantaleo G, Kremer EJ. Activation of a dendritic cell-T cell axis by Ad5 immune complexes creates an improved environment for replication of HIV in T cells. *J Exp Med* 2008;**205**(12):2717–25.
112. Benlahrech A, Harris J, Meiser A, Papagatsias T, Hornig J, Hayes P, et al. Adenovirus vector vaccination induces expansion of memory CD4 T cells with a mucosal homing phenotype that are readily susceptible to HIV-1. *Proc Natl Acad Sci USA* 2009;**106**(47):19940–5.
113. Zak DE, Andersen-Nissen E, Peterson ER, Sato A, Hamilton MK, Borgerding J, et al. Merck Ad5/HIV induces broad innate immune activation that predicts CD8(+) T-cell responses but is attenuated by preexisting Ad5 immunity. *Proc Natl Acad Sci USA* 2012;**109**(50):E3503–12.
114. Stephenson KE, Hural J, Buchbinder SP, Sinangil F, Barouch DH. Preexisting adenovirus seropositivity is not associated with increased HIV-1 acquisition in three HIV-1 vaccine efficacy trials. *J Infect Dis* 2012;**205**(12):1806–10.
115. Thorner AR, Vogels R, Kaspers J, Weverling GJ, Holterman L, Lemckert AA, et al. Age dependence of adenovirus-specific neutralizing antibody titers in individuals from sub-Saharan Africa. *J Clin Microbiol* 2006;**44**(10):3781–3.
116. Appaiahgari MB, Pandey RM, Vrati S. Seroprevalence of neutralizing antibodies to adenovirus type 5 among children in India: implications for recombinant adenovirus-based vaccines. *Clin Vaccine Immunol* 2007;**14**(8):1053–5.
117. Barouch DH, Pau MG, Custers JH, Koudstaal W, Kostense S, Havenga MJ, et al. Immunogenicity of recombinant adenovirus serotype 35 vaccine in the presence of pre-existing anti-Ad5 immunity. *J Immunol* 2004;**172**(10):6290–7.
118. Schmitz H, Wigand R, Heinrich W. Worldwide epidemiology of human adenovirus infections. *Am J Epidemiol* 1983;**117**(4):455–66.
119. Mast TC, Kierstead L, Gupta SB, Nikas AA, Kallas EG, Novitsky V, et al. International epidemiology of human pre-existing adenovirus (Ad) type-5, type-6, type-26 and type-36 neutralizing antibodies: correlates of high Ad5 titers and implications for potential HIV vaccine trials. *Vaccine* 2010;**28**(4):950–7.
120. Barouch DH, Kik SV, Weverling GJ, Dilan R, King SL, Maxfield LF, et al. International seroepidemiology of adenovirus serotypes 5, 26, 35, and 48 in pediatric and adult populations. *Vaccine* 2011;**29**(32):5203–9.
121. Lopez-Gordo E, Podgorski II , Downes N, Alemany R. Circumventing antivector immunity: potential use of nonhuman adenoviral vectors. *Hum Gene Ther* 2014;**25**(4):285–300.
122. Bangari DS, Mittal SK. Porcine adenoviral vectors evade preexisting humoral immunity to adenoviruses and efficiently infect both human and murine cells in culture. *Virus Res* 2004;**105**(2):127–36.
123. Patel A, Tikoo S, Kobinger G. A porcine adenovirus with low human seroprevalence is a promising alternative vaccine vector to human adenovirus 5 in an H5N1 virus disease model. *PLoS One* 2010;**5**(12):e15301.

124. Bru T, Salinas S, Kremer EJ. An update on canine adenovirus type 2 and its vectors. *Viruses* 2010;**2**(9):2134–53.
125. Singh N, Pandey A, Jayashankar L, Mittal SK. Bovine adenoviral vector-based H5N1 influenza vaccine overcomes exceptionally high levels of pre-existing immunity against human adenovirus. *Mol Ther* 2008;**16**(5):965–71.
126. Hofmann C, Loser P, Cichon G, Arnold W, Both GW, Strauss M. Ovine adenovirus vectors overcome preexisting humoral immunity against human adenoviruses in vivo. *J Virol* 1999;**73**(8):6930–6.
127. Farina SF, Gao GP, Xiang ZQ, Rux JJ, Burnett RM, Alvira MR, et al. Replication-defective vector based on a chimpanzee adenovirus. *J Virol* 2001;**75**(23):11603–13.
128. Roy S, Gao G, Lu Y, Zhou X, Lock M, Calcedo R, et al. Characterization of a family of chimpanzee adenoviruses and development of molecular clones for gene transfer vectors. *Hum Gene Ther* 2004;**15**(5):519–30.
129. Colloca S, Barnes E, Folgori A, Ammendola V, Capone S, Cirillo A, et al. Vaccine vectors derived from a large collection of simian adenoviruses induce potent cellular immunity across multiple species. *Sci Transl Med* 2012;**4**(115):115ra2.
130. Xiang Z, Li Y, Cun A, Yang W, Ellenberg S, Switzer WM, et al. Chimpanzee adenovirus antibodies in humans, sub-Saharan Africa. *Emerg Infect Dis* 2006;**12**(10):1596–9.
131. Dudareva M, Andrews L, Gilbert SC, Bejon P, Marsh K, Mwacharo J, et al. Prevalence of serum neutralizing antibodies against chimpanzee adenovirus 63 and human adenovirus 5 in Kenyan children, in the context of vaccine vector efficacy. *Vaccine* 2009;**27**(27):3501–4.
132. Ersching J, Hernandez MI, Cezarotto FS, Ferreira JD, Martins AB, Switzer WM, et al. Neutralizing antibodies to human and simian adenoviruses in humans and New-World monkeys. *Virology* 2010;**407**(1):1–6.
133. Sharma A, Tandon M, Ahi YS, Bangari DS, Vemulapalli R, Mittal SK. Evaluation of cross-reactive cell-mediated immune responses among human, bovine and porcine adenoviruses. *Gene Ther* 2010;**17**(5):634–42.
134. Hutnick NA, Carnathan D, Demers K, Makedonas G, Ertl HC, Betts MR. Adenovirus-specific human T cells are pervasive, polyfunctional, and cross-reactive. *Vaccine* 2010,**28**(8):1932–41.
135. Abbink P, Lemckert AA, Ewald BA, Lynch DM, Denholtz M, Smits S, et al. Comparative seroprevalence and immunogenicity of six rare serotype recombinant adenovirus vaccine vectors from subgroups B and D. *J Virol* 2007;**81**(9):4654–63.
136. Tan WG, Jin HT, West EE, Penaloza-MacMaster P, Wieland A, Zilliox MJ, et al. Comparative analysis of simian immunodeficiency virus gag-specific effector and memory CD8+ T cells induced by different adenovirus vectors. *J Virol* 2013;**87**(3):1359–72.
137. Stone D, Liu Y, Li ZY, Tuve S, Strauss R, Lieber A. Comparison of adenoviruses from species B, C, E, and F after intravenous delivery. *Mol Ther* 2007;**15**(12):2146–53.
138. Arnberg N. Adenovirus receptors: implications for targeting of viral vectors. *Trends Pharmacol Sci* 2012;**33**(8):442–8.
139. Coughlan L, Bradshaw AC, Parker AL, Robinson H, White K, Custers J, et al. Ad5:Ad48 hexon hypervariable region substitutions lead to toxicity and increased inflammatory responses following intravenous delivery. *Mol Ther* 2012;**20**(12):2268–81.
140. Chirmule N, Propert K, Magosin S, Qian Y, Qian R, Wilson J. Immune responses to adenovirus and adeno-associated virus in humans. *Gene Ther* 1999;**6**(9):1574–83.
141. Molnar-Kimber KL, Sterman DH, Chang M, Kang EH, ElBash M, Lanuti M, et al. Impact of preexisting and induced humoral and cellular immune responses in an adenovirus-based gene therapy phase I clinical trial for localized mesothelioma. *Hum Gene Ther* 1998;**9**(14):2121–33.

142. Liu J, O'Brien KL, Lynch DM, Simmons NL, La PA, Riggs AM, et al. Immune control of an SIV challenge by a T-cell-based vaccine in rhesus monkeys. *Nature* 2009;**457**(7225): 87–91.
143. Hill AV, Reyes-Sandoval A, O'Hara G, Ewer K, Lawrie A, Goodman A, et al. Prime-boost vectored malaria vaccines: progress and prospects. *Hum Vaccin* 2010;**6**(1):78–83.
144. Weaver EA, Nehete PN, Buchl SS, Senac JS, Palmer D, Ng P, et al. Comparison of replication-competent, first generation, and helper-dependent adenoviral vaccines. *PLoS One* 2009;**4**(3):e5059.
145. Kass-Eisler A, Leinwand L, Gall J, Bloom B, Falck-Pedersen E. Circumventing the immune response to adenovirus-mediated gene therapy. *Gene Ther* 1996;**3**(2):154–62.
146. Mastrangeli A, Harvey BG, Yao J, Wolff G, Kovesdi I, Crystal RG, et al. "Sero-switch" adenovirus-mediated in vivo gene transfer: circumvention of anti-adenovirus humoral immune defenses against repeat adenovirus vector administration by changing the adenovirus serotype. *Hum Gene Ther* 1996;**7**(1):79–87.
147. Mack CA, Song WR, Carpenter H, Wickham TJ, Kovesdi I, Harvey BG, et al. Circumvention of anti-adenovirus neutralizing immunity by administration of an adenoviral vector of an alternate serotype. *Hum Gene Ther* 1997;**8**(1):99–109.
148. Morral N, O'Neal W, Rice K, Leland M, Kaplan J, Piedra PA, et al. Administration of helper-dependent adenoviral vectors and sequential delivery of different vector serotype for long-term liver-directed gene transfer in baboons. *Proc Natl Acad Sci USA* 1999;**96**(22):12816–21.
149. Parks R, Evelegh C, Graham F. Use of helper-dependent adenoviral vectors of alternative serotypes permits repeat vector administration. *Gene Ther* 1999;**6**(9):1565–73.
150. Moffatt S, Hays J, Hogenesch H, Mittal SK. Circumvention of vector-specific neutralizing antibody response by alternating use of human and non-human adenoviruses: implications in gene therapy. *Virology* 2000;**272**(1):159–67.
151. Tandon M, Sharma A, Vemula SV, Bangari DS, Mittal SK. Sequential administration of bovine and human adenovirus vectors to overcome vector immunity in an immunocompetent mouse model of breast cancer. *Virus Res* 2012;**163**(1):202–11.
152. Thorner AR, Lemckert AA, Goudsmit J, Lynch DM, Ewald BA, Denholtz M, et al. Immunogenicity of heterologous recombinant adenovirus prime-boost vaccine regimens is enhanced by circumventing vector cross-reactivity. *J Virol* 2006;**80**(24):12009–16.
153. Paris R, Kuschner RA, Binn L, Thomas SJ, Colloca S, Nicosia A, et al. Adenovirus type 4 and 7 vaccination or adenovirus type 4 respiratory infection elicits minimal cross-reactive antibody responses to nonhuman adenovirus vaccine vectors. *Clin Vaccine Immunol* 2014;**21**(5):783–6.
154. Teigler JE, Penaloza-MacMaster P, Obeng R, Provine NM, Larocca RA, Borducchi EN, et al. Hexon hypervariable region-modified adenovirus type 5 (Ad5) vectors display reduced hepatotoxicity but induce T lymphocyte phenotypes similar to Ad5 vectors. *Clin Vaccine Immunol* 2014;**21**(8):1137–44.
155. Wu H, Dmitriev I, Kashentseva E, Seki T, Wang M, Curiel DT. Construction and characterization of adenovirus serotype 5 packaged by serotype 3 hexon. *J Virol* 2002;**76**(24):12775–82.
156. Roberts DM, Nanda A, Havenga MJ, Abbink P, Lynch DM, Ewald BA, et al. Hexon-chimaeric adenovirus serotype 5 vectors circumvent pre-existing anti-vector immunity. *Nature* 2006;**441**(7090):239–43.
157. Milla P, Dosio F, Cattel L. PEGylation of proteins and liposomes: a powerful and flexible strategy to improve the drug delivery. *Curr Drug Metab* 2012;**13**(1):105–19.
158. O'Riordan CR, Lachapelle A, Delgado C, Parkes V, Wadsworth SC, Smith AE, et al. PEGylation of adenovirus with retention of infectivity and protection from neutralizing antibody in vitro and in vivo. *Hum Gene Ther* 1999;**10**(8):1349–58.

159. Kreppel F, Gackowski J, Schmidt E, Kochanek S. Combined genetic and chemical capsid modifications enable flexible and efficient de- and retargeting of adenovirus vectors. *Mol Ther* 2005;**12**(1):107–17.
160. Kreppel F, Kochanek S. Modification of adenovirus gene transfer vectors with synthetic polymers: a scientific review and technical guide. *Mol Ther* 2008;**16**(1):16–29.
161. Fisher KD, Seymour LW. HPMA copolymers for masking and retargeting of therapeutic viruses. *Adv Drug Deliv Rev* 2010;**62**(2):240–5.
162. Croyle MA, Chirmule N, Zhang Y, Wilson JM. "Stealth" adenoviruses blunt cell-mediated and humoral immune responses against the virus and allow for significant gene expression upon readministration in the lung. *J Virol* 2001;**75**(10):4792–801.
163. Fisher KD, Stallwood Y, Green NK, Ulbrich K, Mautner V, Seymour LW. Polymer-coated adenovirus permits efficient retargeting and evades neutralising antibodies. *Gene Ther* 2001;**8**(5):341–8.
164. Eto Y, Gao JQ, Sekiguchi F, Kurachi S, Katayama K, Mizuguchi H, et al. Neutralizing antibody evasion ability of adenovirus vector induced by the bioconjugation of methoxypolyethylene glycol succinimidyl propionate (MPEG-SPA). *Biol Pharm Bull* 2004;**27**(6):936–8.
165. Wortmann A, Vohringer S, Engler T, Corjon S, Schirmbeck R, Reimann J, et al. Fully detargeted polyethylene glycol-coated adenovirus vectors are potent genetic vaccines and escape from pre-existing anti-adenovirus antibodies. *Mol Ther* 2008;**16**(1):154–62.
166. Croyle MA, Chirmule N, Zhang Y, Wilson JM. PEGylation of E1-deleted adenovirus vectors allows significant gene expression on readministration to liver. *Hum Gene Ther* 2002;**13**(15):1887–900.
167. Weaver EA, Barry MA. Effects of shielding adenoviral vectors with polyethylene glycol on vector-specific and vaccine-mediated immune responses. *Hum Gene Ther* 2008;**19**(12):1369–82.
168. Yotnda P, Chen DH, Chiu W, Piedra PA, Davis A, Templeton NS, et al. Bilamellar cationic liposomes protect adenovectors from preexisting humoral immune responses. *Mol Ther* 2002;**5**(3):233–41.
169. Steel JC, Cavanagh HM, Burton MA, Kalle WH. Microsphere-liposome complexes protect adenoviral vectors from neutralising antibody without losses in transfection efficiency, in-vitro. *J Pharm Pharmacol* 2004;**56**(11):1371–8.
170. Beer SJ, Matthews CB, Stein CS, Ross BD, Hilfinger JM, Davidson BL. Poly (lactic-glycolic) acid copolymer encapsulation of recombinant adenovirus reduces immunogenicity in vivo. *Gene Ther* 1998;**5**(6):740–6.
171. Sun C, Feng L, Zhang Y, Xiao L, Pan W, Li C, et al. Circumventing antivector immunity by using adenovirus-infected blood cells for repeated application of adenovirus-vectored vaccines: proof of concept in rhesus macaques. *J Virol* 2012;**86**(20):11031–42.
172. Butterfield LH, Vujanovic L. New approaches to the development of adenoviral dendritic cell vaccines in melanoma. *Curr Opin Investig Drugs* 2010;**11**(12):1399–408.
173. Xiang ZQ, Gao GP, Reyes-Sandoval A, Li Y, Wilson JM, Ertl HC. Oral vaccination of mice with adenoviral vectors is not impaired by preexisting immunity to the vaccine carrier. *J Virol* 2003;**77**(20):10780–9.
174. Croyle MA, Patel A, Tran KN, Gray M, Zhang Y, Strong JE, et al. Nasal delivery of an adenovirus-based vaccine bypasses pre-existing immunity to the vaccine carrier and improves the immune response in mice. *PLoS One* 2008;**3**(10):e3548.
175. Pandey A, Singh N, Vemula SV, Couetil L, Katz JM, Donis R, et al. Impact of preexisting adenovirus vector immunity on immunogenicity and protection conferred with an adenovirus-based H5N1 influenza vaccine. *PLoS One* 2012;**7**(3):e33428.

176. Yei S, Mittereder N, Tang K, O'Sullivan C, Trapnell BC. Adenovirus-mediated gene transfer for cystic fibrosis: quantitative evaluation of repeated in vivo vector administration to the lung. *Gene Ther* 1994;**1**(3):192–200.
177. Dong JY, Wang D, van Ginkel FW, Pascual DW, Frizzell RA. Systematic analysis of repeated gene delivery into animal lungs with a recombinant adenovirus vector. *Hum Gene Ther* 1996;**7**(3):319–31.
178. Koehler DR, Martin B, Corey M, Palmer D, Ng P, Tanswell AK, et al. Readministration of helper-dependent adenovirus to mouse lung. *Gene Ther* 2006;**13**(9):773–80.
179. Richardson JS, Abou MC, Tran KN, Kumar A, Sahai BM, Kobinger GP. Impact of systemic or mucosal immunity to adenovirus on Ad-based Ebola virus vaccine efficacy in guinea pigs. *J Infect Dis* 2011;**204**(Suppl. 3):S1032–42.
180. Fontanellas A, Hervas-Stubbs S, Mauleon I, Dubrot J, Mancheno U, Collantes M, et al. Intensive pharmacological immunosuppression allows for repetitive liver gene transfer with recombinant adenovirus in nonhuman primates. *Mol Ther* 2010;**18**(4):754–65.
181. Smith TA, White BD, Gardner JM, Kaleko M, McClelland A. Transient immunosuppression permits successful repetitive intravenous administration of an adenovirus vector. *Gene Ther* 1996;**3**(6):496–502.
182. Kuriyama S, Tominaga K, Mitoro A, Tsujinoue H, Nakatani T, Yamazaki M, et al. Immunomodulation with FK506 around the time of intravenous re-administration of an adenoviral vector facilitates gene transfer into primed rat liver. *Int J Cancer* 2000;**85**(6):839–44.
183. Ye X, Robinson MB, Pabin C, Batshaw ML, Wilson JM. Transient depletion of CD4 lymphocyte improves efficacy of repeated administration of recombinant adenovirus in the ornithine transcarbamylase deficient sparse fur mouse. *Gene Ther* 2000;**7**(20):1761–7.
184. Alzuguren P, Hervas-Stubbs S, Gonzalez-Aseguinolaza G, Poutou J, Fortes P, Mancheno U, et al. Transient depletion of specific immune cell populations to improve adenovirus-mediated transgene expression in the liver. *Liver Int* 2014;**35**(4):1274–89.
185. Stein CS, Pemberton JL, van RN, Davidson BL. Effects of macrophage depletion and anti-CD40 ligand on transgene expression and redosing with recombinant adenovirus. *Gene Ther* 1998;**5**(4):431–9.
186. Kushwah R, Oliver JR, Duan R, Zhang L, Keshavjee S, Hu J. Induction of immunological tolerance to adenoviral vectors by using a novel dendritic cell-based strategy. *J Virol* 2012;**86**(7):3422–35.
187. Kalyuzhniy O, Di Paolo NC, Silvestry M, Hofherr SE, Barry MA, Stewart PL, et al. Adenovirus serotype 5 hexon is critical for virus infection of hepatocytes in vivo. *Proc Natl Acad Sci USA* 2008;**105**(14):5483–8.
188. Vigant F, Descamps D, Jullienne B, Esselin S, Connault E, Opolon P, et al. Substitution of hexon hypervariable region 5 of adenovirus serotype 5 abrogates blood factor binding and limits gene transfer to liver. *Mol Ther* 2008;**16**(8):1474–80.
189. Waddington SN, McVey JH, Bhella D, Parker AL, Barker K, Atoda H, et al. Adenovirus serotype 5 hexon mediates liver gene transfer. *Cell* 2008;**132**(3):397–409.

# Methods to Mitigate Immune Responses to Adenoviral Vectors

16

*Yasser A. Aldhamen*[1], *Andrea Amalfitano*[1,2]
[1]Department of Microbiology and Molecular Genetics, Michigan State University, East Lansing, MI, USA; [2]College of Osteopathic Medicine, Michigan State University, East Lansing, MI, USA

## 1. Introduction

Recombinant adenovirus (Ad)-based vectors have been the focus of considerable interest from 1994 to 2014 for their potential applications as gene delivery vectors for use in human and animal diseases.[1–7] As a result, Ad vectors became the most widely utilized gene transfer vectors in human gene therapy clinical trials worldwide in clinical studies of cancer and infectious (e.g., vaccine development), neurologic, and metabolic disorders.[8] More to the point, as of 2014, over 463 human clinical trials have administered Ad-based vectors using various administration routes both to normal human volunteers and to patients suffering from different types of diseases potentially treatable by an Ad-based gene transfer approach (http://www.wiley.com/legacy/wileychi/genmed/clinical/). There are no current FDA-approved Ad-based vectors. In China, however, since 2003, Ad vectors expressing the p53 tumor suppressor gene (Gendicine or Advexin) have been routinely utilized in patients affected by several forms of cancer. To understand and then overcome the multiple immunological problems inherent in the use of viruses as gene transfer vectors, a keen understanding of virus–host biological interactions, both generally and specifically, is required, as summarized below.

Adenoviruses were initially isolated from patients suffering from acute respiratory infection in 1954 by Hilleman and Werner.[9] Ads are nonenveloped, with an icosahederal capsid containing a linear, double-stranded DNA (dsDNA) genome of about 36 kb.[10,11] Wild-type Ads usually cause relatively mild, self-limiting respiratory illnesses in immunocompetent individuals. However, Ad infection represents a major health issue in immunocompromised hosts such as patients with primary immune deficiencies, HIV-infected patients, and hematopoietic stem cell and organ transplant recipients.[12–15]

Currently, at least 52 serotypes of human Ad have been identified. They are classified into seven subgroups (A–G), primarily based on erythrocyte agglutination capabilities and DNA sequence similarity of the various subgroups.[16] Importantly, of the immunologically distinct human Ad serotypes, none are associated with any neoplastic disease.[17] The most extensively utilized and characterized members (both in relation to general Ad biology and in regard to utilization as a gene transfer vector) of the Ad family are the Ad serotypes 2 (Ad2) and Ad5, which belong to Ad subclass C. Recently, alternative serotypes of human Ads (e.g., Ad4, Ad7, Ad9, Ad11, Ad26, Ad35,

**Adenoviral Vectors for Gene Therapy.**

Ad48, and Ad50) and chimpanzee-derived adenoviruses (ChAds) (e.g., ChAd3 and ChAd63) have also been genetically engineered (primarily as replication-incompetent vectors) for use in multiple gene therapy and vaccine development trials.[18–23] The use of these alternative Ad serotypes allowed for improved and prolonged transgene expression in individuals harboring preexisting immunity (both neutralizing antibodies (NAbs) and $CD8^+$ T cells) to the commonly utilized Ads, such as Ad5.[24] However, their use will also have several limitations, including altered biodistribution and biosafety profiles and shared T-cell epitopes; they are also subject to neutralization on their readministration.[25–28] The use of an alternative Ad serotype strategy will be discussed in more detail later in the chapter.

Structurally, the human Ad capsid is composed of 11 different structural proteins.[29] These proteins are divided into major and minor capsid proteins. The major capsid proteins include the trimeric hexon protein (also called protein II), penton base (protein III), and fiber (protein IV).[10] Hexon is a homotrimeric protein composed of 720 copies, thus representing the major structural protein in Ad capsid. Hexon is highly conserved among Ad serotypes. However, the sequence variation on the nine surface-exposed hypervariable regions (HVRs) of the hexon protein distinguishes Ad serotypes.[30] The minor (cement) capsid proteins include proteins pIIIa, pVI, pVIII, and pIX.[16] In addition, Ad contains another four proteins that are packaged in the core and are tightly associated with the viral DNA. They are known as protein V, protein VII, μ, and the terminal protein TP.[16] The capsid proteins are the most commonly targeted proteins for viral capsid protein manipulation, such as insertion of foreign peptides. In addition, hexon, penton, and fiber represent the major host immune system-targeted proteins that result in Ad neutralization, and therefore are targets for immune evasion strategies, as discussed later in this chapter.

The human Ad genome is generally composed of five early (E1A, E1B, E2, E3, and E4) and late (L1–L5) gene segments. First-generation Ad vectors were rendered "replication incompetent" such that a transgene replaces only the E1 region of genes ([E1-] Ad) or E1 and a portion of E3 ([E1-, E3-] Ad); thus, 90% of the wild-type (WT) Ad genome is retained in the vector.[31,32] Human cells lines, engineered to express the E1 gene products, such as human HEK-293 cells, are used to propagate the first generation Ad vectors.[32,33] Newer generation Ad vectors, with additional deletions in their genome in the E2A, E2B, and E4, or the entire Ad genome, are produced via the use of newer generation, trans-complementing packaging cell lines and/or helper viruses.[34–38] These next generation Ad vectors have several advantages over the first generation Ad vectors including accommodation of larger segments of foreign DNA, reduction in cytotoxic effects in transduced host cells, and reductions in vector-specific B- and T-cell-adaptive immune responses, thought to be a result of reductions in the expression of numerous Ad protein products.[34,39–42]

Compared to other gene transfer vectors, Ad vectors possess many important characteristics that make them ideal platforms for gene therapy. This includes the ability to easily and routinely scale the vectors to high titers in a good manufacturing practice (GMP) compliant fashion (up to $1 \times 10^{13}$ vp/ml). Ad vectors do not depend on transfection-based packaging systems for vector production and thus a likely a major reason for their efficiency. This feature has likely limited the widespread clinical use of other vector systems that have also been available over the past decades. Additionally, Ad vectors allow for efficient transduction of various dividing and quiescent cell types in vitro and in vivo, and can allow for transgene expression in an episomal fashion, thereby mitigating risks associated with vectors that depend on genome integration to function optimally.[43]

Despite the several advantages of Ad vectors, it is also clear that administration of large concentrations of Ad vectors that express high levels of their respective transgene payload Ad-based vectors rapidly activates innate immune responses (that have evolved to detect minute amounts of pathogens generally) and also robustly induces T- and B-cell-adaptive immune responses against both the vector and the transgene product being expressed.[44] Moreover, due to the susceptibility of humans to infection by WT Ad serotype infections (both in childhood or in adulthood), it is now estimated that more than 60% of the worldwide human population have B- and T-cell memory immune responses to the most common Ad serotypes.

## 2. Activation of Innate and Adaptive Immunity by Adenovirus Vectors

The innate immune system is the first line of defense against invading pathogens.[45,46] It is composed not only of physical barriers (e.g., the skin, the tight endothelial cell junctions of the vascular system, the blood–brain barrier) but also of a network of different cell types including dendritic cells (DCs), natural killer (NK) cells, monocytes/macrophages, NKT cells, gamma delta (γδ) T cells, and neutrophils.[47] Each cell of the innate immune system expresses a variety of germline-encoded innate immune receptors that recognize and respond to molecular structures commonly present in a variety of pathogens, collectively known as "pathogen-associated molecular patterns" (PAMP).[46,48–50] These PAMPs essentially identify a molecule, virus, or cell as "foreign" to the host even on first encounter, and are detected by a wide range of secreted (extracellular), cell surface, and cytosolic receptors, commonly known as "pattern recognition receptors" (PRRs).[46,47,51] For example, on binding to PAMPs, secreted PRRs can activate the complement arm of the innate immune system, facilitating phagocytosis by macrophages and neutrophils.[52]

The cell surface PRRs, such as TLRs (TLR1/2, TLR2/6, TLR4, and TLR5) and the C-type lectins, recognize extracellular PAMPs.[53] In contrast, the intracellular PPRs, such as TLR3, TLR7, TLR9, the retinoic acid-inducible gene-1-like receptors, and the nucleotide-binding oligomerization domain/leucine-rich repeat receptors (NLRs), recognize intracellular PAMPs.[54] PRRs can also recognize intracellular stress and/or damage signals released during cell death or inflammation, commonly known as "damage-associated molecular patterns".[55] Recognition of a specific ligand by these various PRRs results in the activation of several innate immune signaling pathways, such as the mitogen-activated protein kinase (MAPK), PI3K, NFkB, interferon regulatory factors (IRFs) 3 and 7, and/or AP-1.[56] Activation of these innate molecules and transcription factors orchestrates the transcription of several innate immune response genes, including those for proinflammatory cytokines and chemokines and type I interferons (IFNα and IFNβ). Production of these innate response proteins controls pathogen infection and regulates the development of pathogen-specific B- and T-cell-adaptive immune responses.

Clearly, any gene transfer vector, based on chemicals, viruses, or cells, will encounter identification by the highly evolved, human innate immune system. Ad vectors induce innate immune responses and this occurs at early stages during cellular transduction[57] (Figure 1). In addition, these innate immune responses are usually exaggerated due to the need to utilize higher doses of Ad vectors in order to achieve relevant transgene expression levels and therapeutic efficacy. This is also true for many other gene transfer vectors, both viral and nonviral based. As a result, Ad vectors serve as useful tools for further understanding gene transfer-triggered innate immune responses. The production of proinflammatory cytokines and chemokines and type I interferons occurs as early as 1–6 h post Ad vector injection (hpi).[58,59] Specifically, our lab previously found that systemic (i.v.) administration of high doses of Ad vectors induced higher production levels of several proinflammatory cytokines and chemokines including IL-1α, IL-1β, IL-6, TNFα, IL12, KC, RANTES, MCP-1, MIP-1α, IFNα, IFNβ, IFNγ, and IP-10.[56,60–62] In addition, it has been shown previously that Ad vectors can induce potent type I interferon responses, responses that result in the elimination and clearance of Ad vectors by NK cells, in a perforin- and granzyme B-dependent manner.[63] Several immune and nonimmune

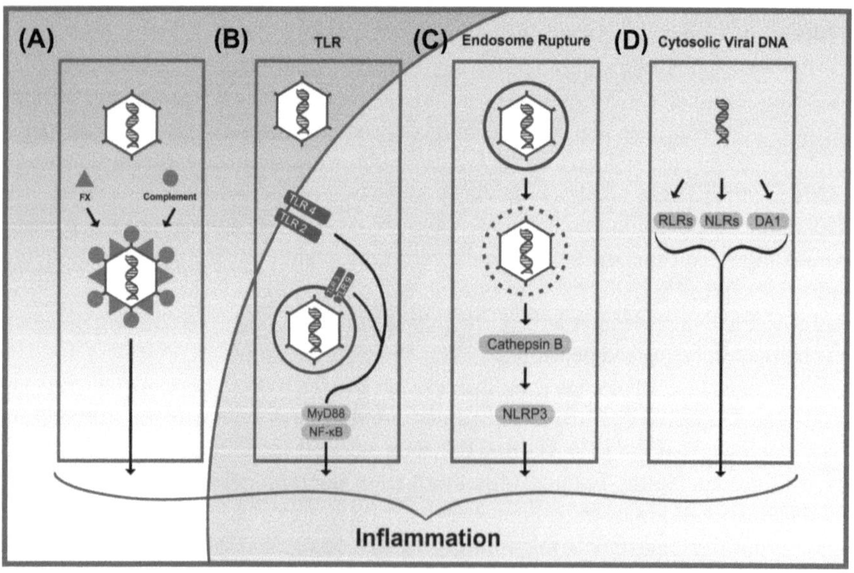


**Figure 1 Adenoviral vector innate immune sensing mechanisms.** Recognition of adenovirus capsid proteins or nucleic acids by the extracellular (A), cell surface (B), or intracellular pattern recognition receptors (PRRs) (B–D) activates several downstream signaling pathways that regulate the expression and production of several innate immune genes and inflammatory mediators.

cells contribute to the production of these proinflammatory cytokines and chemokines. These include Kupffer cells, DCs, NK cells, neutrophils, macrophages, and endothelial cells.[64–67]

At the molecular level, activation of the innate immune responses by Ad vectors is mediated by both TLR-dependent and TLR-independent mechanisms.[18,56,60,63,68–72] Studies in our laboratory previously showed that i.v. administration of Ad vectors activates MAPK and NFkB signaling pathways in MyD88- and TLR2/TLR9-dependent mechanisms.[56] We also identified a critical role for TLR2, TLR3, and TLR4 in regulating the expression of several innate immune genes following Ad vector administration in vivo, a mechanism that requires a functional MyD88 and TRIF (TIR domain-containing adaptor-inducing interferon-beta) signaling pathways.[56,73] Additionally, we found that production of several other proinflammatory cytokines and chemokines, such as KC and MCP-1, was also dependent on MyD88, but not the TLR9 pathway, indicating that Ad vectors induce several innate pathways, independent of its DNA recognition by the TLR9 receptor.[7] In line with this observation, we and others have found that several varieties of Ad vector, such as Ad4 and Ad5 serotype-based vectors, capsid activate innate immune responses via a mechanism that involves the complement system.[61,74]

It has been shown previously that cellular penetration by Ad vectors also activates innate immune responses. Interactions between the Ad capsid fiber knob and its cellular receptor, coxsackie-adenovirus receptor (CAR), promote the clustering of junctional adhesion molecule-like protein and activate PI3K-dependent innate immune responses.[75,76] Moreover, the interaction between the Ad penton base RGD (Arg–Gly–Asp) motif and the cellular vitronectin-binding integrins, mainly $\alpha_v$-β3 and $\alpha_v$-β5, activates innate immune responses via a mechanism that involves PI3K signaling and a TNFα autocrine-dependent mechanism, a mechanism that results in enhanced DC maturation and function.[77,78]

Besides TLRs, Ad vectors can also activate innate immunity by several other innate immune sensing mechanisms. For example, Ad vectors can induce caspase-1-dependent IL-1β release from macrophages via a mechanism that involves the NLRP3/ASC inflammasome complex.[70] In addition, it has been shown previously that the nucleic acid sensors, absent in melanoma-2 (AIM-2), DNA-dependent activator of IRFs, and MDA5, are all essential for innate immune response activation following Ad DNA recognition.[79–81] Moreover, it has been shown recently that the DNA sensor cyclic GMP-AMP synthase (cGAS) is also involved in activating innate immunity following Ad vector administration, a sensing mechanism that requires a functional STING–TBK1signaling pathway.[82] Moreover, disruption of lysosomes by Ad vectors during lysosomal escape triggers the release of lysosomal cathepsin B into the cytoplasm, a danger signal that induces NLRP3 inflammasome activation and IL-1β and IL-18 processing and release.[71] Finally, activation of NLRP3 by Ad vectors induces necrotic cell death and the release of the nuclear protein, high-mobility group box 1 protein (HGMB1), a critical danger/damage signal that mediates potent innate immune responses.[71] Together, these data indicate that Ad vectors induce innate immune responses by activating a complex network of innate immune sensing and signaling pathways. In some respects, a sobering reality may be that the more efficient a vector may be at escaping the lysosome (and allowing for efficacious gene transduction and subsequent expression) the more likely it will trigger these and other innate immune responses.

The innate immune system subsequently delivers essential signals that regulate and orchestrate the initiation and the progression of the adaptive immune response.[60,83,84] Preexisting B- and T-cell immunity to Ads is serotype dependent and usually targeted toward the Ad capsid proteins.[85] These responses are generated following infection with WT Ads or, in the case of gene therapy trials, subsequent to administration of first-generation Ad vectors (both commonly utilized or alternative Ad serotypes[86]), which still express Ad viral early and late genes.[87–89] It is important to note that these adaptive immune responses will also be generated against Ad vector-encoded transgene products as well, especially if these products are recognized as immunologically foreign. The Ad-specific adaptive immune responses reduce the efficacy of Ad vectors in both gene therapy (i.e., blunted transgene expression of therapeutic transgene)[90–92] and vaccine development (i.e., diminished antigen-specific adaptive immune response to vaccine antigens)[93] applications.

Neutralizing antibodies to the commonly utilized Ad vectors, Ad2 and Ad5, are primarily directed toward the surface loops of the major viral capsid protein, hexon, especially against the seven hexon hypervariable regions (HVRs1–7) on the Ad capsid surface.[94,95] In addition, NAbs to penton base have also been described.[96] Moreover, NAbs against the fiber knob protein were also described and recently found to be present in over 90% of Ad5-positive serum samples.[86] Characterization of the neutralizing antibody responses to Ad capsid proteins fosters the development of novel Ad vectors that had the immunogenic portions of the Ad capsid replaced with homologous regions from alternative, rare, Ad serotypes.[97,98] It, however, must also be understood that clinical applications that utilize delivery of high doses of Ad vectors in some tissues (such as an intramuscular (i.m.) or intradermal injection for vaccine applications, or into a viscous such as the bladder, or intestine) may not be as susceptible to the presence of preexisting neutralizing antibodies as other clinical applications that target other systems (such as intravascular administration so as to target specific organs).

Ad vectors have also been shown to induce potent cellular (mainly CD8$^+$ T cells) immune responses to the Ad-derived protein antigens, not only the capsid proteins but also nonstructural Ad proteins, such as the early region 2 proteins DNA polymerase (Pol) and DNA-binding protein (DBP).[99] These Ad-specific CD8$^+$ T-cell responses can eliminate Ad transduced cells that happen to express these viral products, but not capsid proteins.[99] As a result, diminished transgene expression and reduced Ad efficacy will occur. Besides Ad-specific cellular immunity, Ad vectors also induce CD8$^+$ T-cell responses to the expressed transgenes, especially if they are perceived as foreign by the host.[100] As a result of this activity, Ad vectors have been utilized in numerous preclinical and clinical vaccine-based applications.[101–104] For example, the E1-deleted Ad5 vectors expressing HIV *gag*, *pol*, and *nef* antigens have been utilized in human clinical trial subjects, the so-called "STEP trial," and shown to induce potent HIV antigen-specific CD8$^+$ T-cell responses.[105,106] Moreover, advanced generation Ad vectors, such as the E1- and E2b-deleted Ad5 vectors (E1-, E2b-_Ad5) expressing the tumor-associated antigen carcinoembryonic antigen, have also been utilized in human clinical trials and allowed for breakage of immunological tolerance to tumor-associated self-antigens.[107]

# 3. Therapeutic Strategies for Overcoming Immune Barriers to Adenovirus Vectors

Despite the multiple immunological hurdles the human innate and adaptive immune systems present to gene transfer therapeutics generally, and Ad-based therapeutics specifically, several elegant strategies have been developed to improve the clinical application of Ad vectors. These strategies aim to evade and/or harness Ad vector and transgene-induced innate and adaptive immune responses, thereby enhancing target cell transduction, increasing the persistence of Ad vectors, reducing Ad-induced toxicity, and improving the overall efficacy of Ad-mediated gene transfer. These approaches can be categorized into two major categories: preemptive transient immune modulation of the host and the selective modification of Ad vectors themselves (Figure 2).

## 3.1 *Suppression of Adenovirus-Induced Immune Responses in the Host*

### 3.1.1 *Global Immunosuppression of the Host Immune System*

This strategy involves the use of currently approved, noninvasive, immune-suppressive drugs or specific compounds to transiently suppress important immune pathways in the host, pathways that were previously identified to be induced by Ad vectors. For example, studies in our laboratory previously showed that systemic (i.v.) administration of Ad vectors in dexamethasone (DEX) pretreated mice (10 mg/kg, 15 and 2 h prior to Ad injection) significantly reduced Ad-triggered innate immune responses including (1) reduced proinflammatory cytokine and chemokine gene expression in liver and spleen tissues; (2) reduced production of several proinflammatory cytokines and chemokines such as IL-6, IL-12, G-CSF, MIP-1β, KC, RANTES, and MCP-1; (3) reduced thrombocytopenia; (4) reduced leukocyte infiltration in the liver; and (5) reduced endothelial cell activation.[108] These responses were positively correlated with reduced Ad capsid-specific humoral immune responses, including capsid-specific NAbs.[108] Intraperitoneal (i.p.) administration of DEX 3 days prior to systemic administration of β galactosidase-expressing Ad vectors reduced Ad-induced innate immune responses (including reduced IL-1β, IFNγ, IL-6, TNFα, MIP-1α, MCP-1, and leukocyte infiltration in the lung) and prolonged transgene expression.[109]

Several other immune-suppressive agents such as cyclosporine,[110] cyclophosphamide,[111,112] deoxyspergualin,[113–115] and FK506[116] have also been utilized in multiple Ad gene therapy trials (using mouse and nonhuman primate preclinical models) and shown to reduce immune responses and prolong transgene expression as well as the ability to readminister the homologous Ad vectors in previously Ad-treated animals. For example, cyclophosphamide administration along with oncolytic Ads reduced induction of Ad-specific humoral immune responses, thereby improving Ad vector-induced antitumor activity on readministration in several Ad-based cancer immune therapy systems.[117,118] In addition, combined treatment with cyclophosphamide and innate modulating cytokines, such as IL-12 (IL-12-expressing Ad vectors), in a subcutaneous

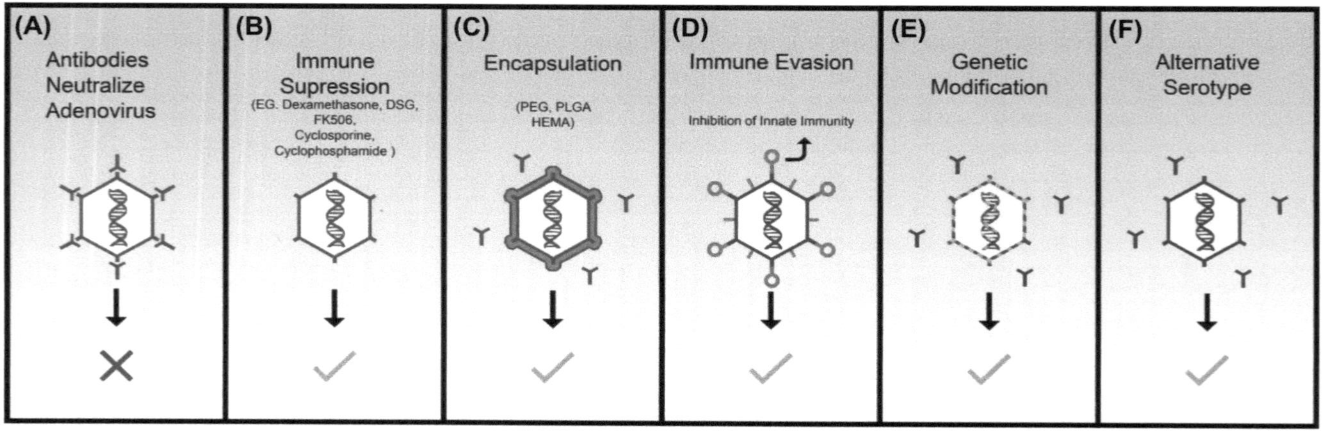


**Figure 2 Methods to evade the host immune response to Ad vectors.** (A) Clearance of adenoviral vector by the preexisting Ad-specific neutralizing antibodies is shown. (B–F) Various immune evasion strategies that were utilized to prevent Ad vector neutralization and clearance by the host innate and adaptive immune systems.

colorectal carcinoma mouse model modulated regulatory T-cell (Tregs) functions (i.e., reduced intratumoral IL-10 and TGB-β concentration) and decreased the number of myeloid-derived suppressor cells, as compared to control mice.[119] Interestingly, these reduced innate immune responses were also correlated with a strong antitumor $CD4^+$ T-cell response.[119]

### 3.1.2 Selective Immunosuppression of the Host Immune System

#### 3.1.2.1 Selective Immunosuppression of the Host Innate Immune System

Modulation of a specific arm of the host immune system has also been studied in the context of Ad vector-mediated gene transfer therapies. The primary goal of these approaches is to increase the safety and efficacy of Ad-based gene transfer systems while modulating a specific innate immune pathway known to be activated by Ads.[120–123] For example, the use of Ad vectors expressing a soluble TNFα receptor, which neutralizes the circulatory TNFα protein, significantly reduced Ad vector-associated innate immune responses (reduced proinflammatory cytokine/chemokine and leukocyte infiltration in the liver) and prolonged transgene expression.[124] Similarly, systemic administration of Ad vectors expressing a fusion protein encoding the extracellular domain of human TNF receptor 1 (p55) and the Fc portion of mouse IgG1, Ad-TNFR1-IgG, increased transgene expression, reduced Ad-specific Nabs, and diminished Ad vector-induced innate immune responses, including recruitment of macrophages and NK cells in the liver.[125] Moreover, preadministration of the TNFα pharmacological inhibitor, etanercept (Enbrel), or the use of TNFα-specific monoclonal antibodies modulated Ad vector-induced innate and adaptive immune responses in several preclinical mouse models, such as the bronchioloalveolar cancer mouse model.[122] Strategies targeting other Ad-induced inflammatory pathways, such as IL-6 or IL-1, have also been developed and shown to reduce Ad-triggered proinflammatory immune responses and enhance transgene expression.[125–127]

Approaches that target specific innate immune pathways that recognize/sense Ad vector components and initiate a downstream innate immune response have also been developed and utilized in preclinical Ad-based gene transfer applications. For example, inhibition of the dsDNA sensor, TLR9 (which recognized Ad dsDNA and induced systemic proinflammatory cytokine release (mainly IL-6, TNFα, and MCP-1) via the NFkB pathway), with the blocking oligonucleotide, ODN-2088, significantly reduced Ad vector-associated innate immune toxicities.[56,128] Similarly, IP injection of the MEK1/2 inhibitor, U0126, along with systemic Ad administration reduced Ad-triggered innate immune responses, including the release of the inflammatory chemokine IP-10 (CXCL10).[129] Moreover, Ad vectors expressing the suppressor of cytokine signaling-1 (SOCS-1) gene (Ad-SOCS1) have also been utilized and shown to mitigate Ad-induced innate immune responses, such as liver toxicities and cytokine (e.g., TNFα, RANTES, MCP-1, and IL-6) production.[130] SOCS1 gene transfer by Ad vectors has also been utilized in Ad vector-mediated anticancer therapies and shown to induce a potent antitumor activity in vivo in mice.[131] The use of these innate modulatory agents in Ad vector gene transfer has yet to be tested in large animals and in human clinical trials.

Another approach for modulating innate immune responses during Ad gene transfer therapy is via depletion of specific innate immune cells.[66] For example, injection of an anti-asialo GM1 antibody, to deplete NK cells prior to Ad vector administration, increased transgene expression and enhanced the persistence of Ad vectors in the liver in mice.[63,132] Similarly, depletion of splenic DCs (via anti-PDCA-1 or anti-CD11c antibodies[65,66]) and/or tissue macrophages (via gadolinium chloride,[133] liposome-encapsulated clodronate,[134] or dichloromethylene bisphosphate[135]) enhanced Ad vector efficacy by reducing the production levels of several proinflammatory cytokines (e.g., IFNα/β, IL-6, TNFα, and IL-12), increasing transgene expression, and diminishing cellular and humoral immune responses to Ad vectors and the expressed antigens, thus, improving Ad vector persistence in immune-competent hosts. Moreover, Kupffer cell depletion also improved Ad vector efficacy in a similar manner.[134,136] Finally, preventing Ad vector uptake by hepatocytes via warfarin (which reduces the plasma concentration levels of the coagulation factors IX and X, which interact with Ad capsid and facilitate hepatocytes transduction) injection reduced Ad-mediated hepatotoxicity.[137] Similarly, pretreatment with polyinosinic acid, a scavenger receptor A ligand, transiently increased (10-fold) peripheral Ad vector concentration, and enhanced (up to 15-fold) transgene expression in the liver along with preventing necrosis of Kupffer cells.[138] Depletion of platelets prior to Ad vector administration also reduced Ad-triggered proinflammatory cytokines and Ad sequestration in the liver sinusoids.[139]

#### 3.1.2.2 Selective Immunosuppression of the Host-Adaptive Immune System

Modulation of adaptive immune responses to Ad vectors by selective inhibition of a specific arm of the adaptive immune system has also been studied in Ad-mediated gene transfer applications. For example, the use of CTLA4Ig recombinant protein (which inhibits T-cell activation by blocking the interaction between B7 molecules (B7.1 (CD80) and B7.2 (CD86)) on antigen presenting cells (APCs) and CD28 on T cells) along with im administration of a β-galactosidase-expressing Ad vector (Ad-LacZ) reduced $CD4^+$ and $CD8^+$ T-cell infiltration into the injected muscle;[140] however, this approach failed to repress the subsequent induction of Ad-specific humoral immune responses. A similar group also showed that a combinatorial therapy approach using Ad-LacZ and CTLA4Ig combined with antibodies directed toward CD4 (YTS 191), CD8 (YTS 169), and CD11a (TIB 213) is more efficacious in blocking both cellular and humoral immune responses and prolonging transgene expression.[140,141] Moreover, providing an anti-CD4 antibody along with clodronated liposomes reduced Ad-triggered innate immunity as well as Ad-specific humoral and cellular immune responses, responses that were associated with increased liver transduction (1000-fold) on Ad vector readministration.[132] Pretreatment with an anti-CD8 T-cell receptor antibody prolonged transgene expression and reduced lung inflammation after Ad-mediated gene transfer in mice.[142] Additionally, administration of Ad vectors and the humanized anti-CD40 antibody, hu5C8 (to block CD40–CD40L interactions), has been shown to suppress Ad vector-induced lymphoproliferation, reduce cytokine responses (IL-2, IL-4, IFNγ, and IL-10), reduce Ad vector-specific Nabs, and allowed for Ad vector readministration in nonhuman primates.[143] Finally, combined use of anti-CD86 and anti-CD40 antibodies to block costimulation by APCs during Ad administrations has also been utilized and shown to increase transgene expression and allowed for repeated systemic administration of Ad vector in rhesus monkeys.[144] It is important to note that only marginal increases in the duration of transgene expression were observed in nonhuman primate studies using such approaches, indicating that these approaches might not be beneficial in Ad-based gene transfer therapies in human.

## 3.2 Modulation Strategies of Adenovirus Vectors to Evade the Host Immune System

In addition to modifying the host immune response to mitigate immune responses against Ad vectors, a number of novel strategies have been developed to modify the Ad vector itself to further these goals. These include several innovative strategies such as Ad capsid display of certain ligands or inhibitory molecules, covalent modifications of Ad vector capsid, construction of chimeric Ad vectors, Ad vector genome modification, and the use of Ad alternative/rare serotypes, as discussed below.

### 3.2.1 Modification of Adenovirus Capsid

#### 3.2.1.1 Ad Capsid Display of Foreign Peptides and Immune Evasion Proteins

As noted earlier, the Ad capsid is composed of nine proteins, and of these protein IX, fiber, hexon, and penton proteins have all been targeted for genetic insertion of foreign peptides, either as in-frame insertions within the proteins or as in-frame C-terminal fusions.[145] The goal of some of these modifications has included reducing Ad-associated immune toxicity. To achieve this goal, several innovative strategies of Ad capsid modification (including modification of the HVR5 of hexon protein, the HI loop or C-terminus of fiber knob, and protein IX C-terminus) have been developed and tested in vivo in mice,[97] nonhuman primates,[146] and human clinical trials.[147] Studies in our laboratory previously showed that Ad capsid display of the natural complement inhibitor, human decay-accelerating factor (DAF) as a fusion protein from the C-terminus of the Ad capsid protein IX (Ad-DAF), improved the efficacy and the safety of Ad-based gene transfer in mice.[148,149] Moreover, systemic administration of DAF-displaying Ad vectors significantly reduced Ad-triggered innate immune responses (including reduced inflammatory cytokines and chemokines, activation of innate (NK, macrophage, and DCs) and adaptive ($CD8^+$ and $CD4^+$ T cells) immune cells, thrombocytopenia, liver toxicity, and endothelial cell activation), as compared to unmodified Ad vectors.[148,149] Importantly, the reduced innate immune responses following Ad-DAF vector administration significantly correlated with reduced Ad-specific NAbs responses as well as reduced transgene (HIV-Gag and GFP)-specific T-cell immune responses.[148] Similarly, our laboratory also constructed Ad vectors displaying a 13 amino acid long peptide (COMPinh),[150] known to inhibit complement activation in human and nonhuman primates.[151] COMPinh peptide was inserted in-frame as a C-terminal fusion in protein IX or embedded within the HI loop of the Ad fiber protein.[150] We found that, COMPinh-displaying Ad vectors minimized Ad-dependent activation of human and nonhuman primate complement systems in vitro,[150] indicating that capsid display strategies of immune modulation proteins/peptides might be beneficial in humans.

#### 3.2.1.2 Chemical Modification of Ad Capsid

The rationale for some Ad capsid modifications is to completely mask the viral capsid antigenic epitopes in order to prevent Ad vector interactions with various blood components, including complement proteins, coagulation factors, and preexisting NAbs. Achieving this goal should, in principle, reduce Ad-induced toxicity at the least, and possibly prolong transgene expression, and/or allow for repeated Ad administrations to be more efficacious. Hexon and fiber proteins are the most frequently modified Ad capsid proteins in these examples.[145] For example, the Ad capsid hexon protein has been complexed with several biodegradable polymers to shield the vector from the deleterious effect of the innate and adaptive immune responses.[152,153] One of the commonly utilized polymers is the FDA-approved compound, polyethylene glycol (PEG), a capsid modification methodology known as "PEGylation."[154] Importantly, up to 18,000 PEG molecules can be complexed to a single Ad capsid. Several studies have shown that PEGylation of the Ad capsid diminished the activation of cytotoxic $CD8^+$ T cells, reduced the skewing of $CD4^+$ T helper cells toward Th1-type response, and decreased the development NAbs against native Ad capsid antigens, thus increasing the persistence of Ad vectors.[153] These diminished responses to Ad vector capsid proteins positively correlated with prolonged transgene expression and allowed for partial readministration of Ad vectors in homologously treated animals.[154]

Similarly, the use of polylactic glycolic acid,[155,156] lipids (liposomes, both cationic and anionic),[157,158] and the *N*-(2-hydroxypropyl) methacrylamide (HPMA) copolymer significantly improved the efficiency of gene transfer by capsid-modified Ad vectors. For example, the use of anionic liposomes to encapsulate Ad vectors enhanced Ad vector transduction (in CAR-deficient cells), evaded Ad vector neutralization by Ad-specific NAbs, and induced lower cytotoxicity, as compared to uncoated Ad vectors.[159] Similarly, the use of HPMA to coat Ad vectors resulted in lower complement and NAb binding, increased persistence, and decreased interaction with the complement receptor on the human erythrocytes.[160,161] As a result, enhanced cellular transduction was observed when HPMA-coated Ad vectors were utilized in vivo in mice, even in the presence of Ad-specific NAbs.[160] Furthermore, modification of Ad vectors by arginine-grafted bioreducible polymers (ABPs) (Ad-ΔE1/GFP-ABP) has been shown to increase cellular transduction and to reduce Ad-triggered liver toxicity and IL-6 production both in vitro and in vivo, as compared to unmodified Ad-ΔE1/GFP vector.[162] Moreover, ABP-modified Ad vector persisted 45-fold more in the blood circulation and evaded neutralization by anti-Ad-specific NAbs.[162]

Besides diminishing adaptive immune responses to Ad capsid proteins, modification of Ad capsid, such as via PEGylation, has been shown to decrease Ad-triggered innate immune responses and liver toxicity including decreased production of proinflammatory cytokines and chemokines (mainly IL-6, TNFα, IL-12, MCP-1, and IP-10), reduced thrombocytopenia, and decreased leukocyte infiltration in various host tissues after systemic Ad vector administration.[163] In addition to direct PEGylation of hexon protein, a group recently developed a strategy to improve Ad vector gene transfer efficacy by mixing Ad vectors with PEGylated blood coagulation factor X.[164] Systemic administration of PEG-FX-associated Ad vectors improved the efficacy of Ad vector gene transfer, including increased Ad persistence and prolonged transgene expression.[164]

Despite various advantages of PEGylation, several studies have also suggested that PEGylation of Ad capsid proteins may interfere with Ad vector interactions with the CAR receptor, thus significantly reducing their transductional efficiencies, especially when the PEGylation ratio is increased.[165,166] Moreover, several studies showed that i.v. administration of PEGylated Ad vectors into previously immunized mice failed to induce transgene expression.[154] Therefore, Ad PEGylation strategies that improve their therapeutic efficacy without decreasing cellular transduction and transgene expression are needed. It is also unclear if these strategies can be scaled to human use, for example, if batch-to-batch variations in the overall PEGylation state of the Ad capsid can be controlled to permit eventual FDA or other appropriate governmental approvals.

#### 3.2.1.3 Genetic Modification of Ad Capsid Proteins

Several strategies were also utilized to mutate several regions of Ad capsid proteins in order to improve their gene transfer efficacy, but may also allow for improvements in evading anti-Ad capsid immune responses. For example, systemic administration of Ad vectors containing fiber protein mutants (deleted of the fiber shaft KKTK motif) and/or penton protein mutants (RGD deleted Ad (Ad-βGal-ΔRGD)) resulted in significantly reduced expression of several inflammatory genes as well as reduced production of several proinflammatory cytokines and chemokines (e.g., MIP-1α, MIP-1β, RANTES, TNFα, IP-10, and IFNβ).[167,168] However, these fiber- and penton base-modified Ad vectors still induced high levels of Ad-specific NAbs.[94] Moreover, these modified Ad vectors also produce increased transgene-specific humoral immune responses, as compared to unmodified Ad vectors,[167] suggesting that this modification strategy may be beneficial in Ad-based vaccine development approaches. In addition to modulating innate and humoral immune responses, these capsid-modified Ad vectors have reduced tropism/transduction efficiency in vitro and in vivo.[167,168]

Innate immune responses by Ad vectors are induced following fiber knob interactions with coagulation factor IX and complement factor C4BP.[127] Decoupling these interactions, preventing liver and Kupffer cell transduction, may allow for improved safety profiles of systemically delivered Ads. For example, administration of fiber knob mutated Ad5 vectors, Ad5ΔF(AB)ΔPS35-L2, resulted in about 15,000-fold (after i.v. injection) and 500 fold (after IP injection) lower mouse liver transduction, as compared to unmodified Ad5 vectors.[169] In addition, reduced innate inflammatory responses (such as serum IL-6) and liver toxicity (reduced AST and ALT levels) were also observed following Ad5ΔF(AB)ΔPS35-L2 administration, as compared to unmodified Ad5 control vectors.[169]

### *3.2.2 Chimeric Adenovirus Vectors in Gene Therapy and Immune Modulation Strategies*

This strategy aims to improve Ad vector efficacy by enhancing transgene expression in the presence of Ad-specific Nabs by utilizing Ad vectors derived in part or wholly from alternative serotype Ads. This strategy also aims to change the tissue tropism of Ad vectors and to evade the host's innate and adaptive immune responses. Therefore, several elegant approaches have been utilized to swap/replace Ad capsid proteins with capsid proteins from alternative/rare Ad serotypes. To construct a chimeric capsid-modified Ad vector, a subportion of the genomic elements of Ad is replaced with a fragment from the alternative Ad serotype, or, in some cases, the entire Ad vector genome is composed of proteins exclusively derived from the alternative serotype Ad vector genome.[170–173] For these modification, both hexon and fiber proteins of the commonly utilized Ad vectors, such as Ad5, were routinely modified with proteins from the rare alternative Ad serotypes. For example, in a side-by-side comparison of IFNα production from human peripheral mononuclear cells (PBMCs) following infection with unmodified Ad5 vectors or with Ad5 vectors containing fiber proteins from Ad16, Ad35, or Ad37,[174] it has been shown that Ad5 vectors expressing the Ad35 fiber (which transduced CD46, instead of CRA-expressing cells) produce significantly increased levels of IFNα from human PBMCs.[174] These data indicate that binding of Ad35 fiber knob to CD46 receptor induces more potent innate immune responses, as compared to Ad5 interaction with the CAR receptor. Since Ad5-specific NAbs are primarily targeted toward the hexon protein,[94] several studies showed that systemic administration of Ad5 vectors expressing hexon proteins from alternative Ad serotypes, such as Ad3, Ad6, or Ad12 into Ad5-immune mice, significantly evaded Ad5-specific NAb responses, and thereby, increased transgene expressions were observed.[171,175,176] Despite these exciting results, it has also been shown that NAbs against Ad3, Ad26, and Ad12 hexons already exist in human subjects,[177] indicating that this approach might not fully mitigate preexisting anti-Ad immunity present in certain human populations.

A hexon-modified Ad5 vector containing seven hexon HVRs (HVR1–7) from the relatively rare Ad serotype, Ad48 (Ad5HVR48), has also been intensively studied.[147] Administration of Ad5HVR48 into Ad5-immune mice and/or nonhuman primates significantly evaded Ad5-specific preexisting immunity against the hexon HVRs and improved Ad5-based gene transfer therapy.[97,146] In addition, a single administration of Ad5HVR48-modified vector expressing portions of the SIV Env/Gag/Nef/and Pol proteins resulted in improved immune responses to these same proteins, and allowed improved resistance of these vaccinated rhesus macaques to subsequent SIV challenge.[178] It has also been shown that vaccination with $10^9$–$10^{11}$ viral particles (vps) doses of Ad5HVR48 expressing the HIV1 EnvA antigen is safe and well tolerated and induced significantly higher levels of EnvA-specific humoral immune responses, despite the presence of Ad5-specific Nabs in human clinical trials.[147]

### *3.2.3 Advanced Generations of Adenovirus Vectors*

It is now widely accepted that E1 and E3 deleted (e.g., first generation) the replication incompetent and continue expressing Ad viral genes "leaky expression," including genes for Ad capsid proteins.[41,42,179,180] Therefore, transduction of cells with first generation Ad vectors can result in processing and display of capsid protein-derived immunogenic epitopes on MHC class I (MHC-I) molecules. Consequently, Ad-specific T- and B-cell responses will be developed, a phenomenon that is associated with continued inflammation, transient transgene expression due to viral clearance, and development of long-lived, Ad-specific cytotoxic T cells and NAb memory responses.

To prevent leaky expression from first-generation Ad vectors, several next generation Ad vectors, harboring additional deletions in the E2A, E2b, and E4 regions, have been developed and shown to prolong transgene expression while reducing various forms of toxicity.[34,40,42,181–184] Fully deleted, helper virus-dependent Ad vectors (HD-Ads, also called "gutless" or third generation adenoviruses) have been developed in order to further prolong transgene expression by removing all Ad genes from the fully deleted HD-Ad vector.[185] These vectors have the entire viral genome deleted, except the inverted terminal repeats and the packaging signals; therefore, no viral genes will be transcribed.[186] In addition, they can accommodate up to 37 kb of foreign DNA, which allow for delivery of large segments of therapeutic genes.[187] Despite this unique feature and due to the need to use a large amount of the therapeutic HD-Ads in order to achieve acceptable therapeutic efficacy, HD-Ad vectors have been shown to induce some innate immune responses in a manner similar to that of first generation Ad vectors. For example, systemic administration of HD-Ad vectors into mice[188] or nonhuman primates[189] induces potent innate immune responses and liver toxicities, in a dose-dependent manner. It is important to note that, unlike first-generation Ad vectors that induce a second wave of innate immunity (originated primarily from Ad-specific cytolytic T cells) at day 7 after administration,[188] HD-Ad vectors do not induce innate immunity beyond 24 h post injection.[188] For example, intracranial injection of an HSV-1-derived thymidine kinase and fms-like tyrosine kinase ligand 3 (Flt3L)-expressing HD-Ad vectors induced tumor regression and long-term survival in a mouse[190] and rat[191] model of intracranial glioblastoma, despite the presence of high levels of Ad-specific immunity.

The induction of innate toxicity by HD-Ad vectors has been significantly reduced by improving the vector delivery methods themselves such as the use of the "balloon occlusion catheter-based method" to isolate the nonhuman primate liver, to minimize systemic distribution of the vectors during liver-targeting gene therapy.[192,193] For example, the use of a balloon occlusion catheter-based method in nonhuman primates increased HD-Ad vector hepatic transduction and reduced proinflammatory cytokines responses as well as prolonged transgene expression for at least 413 days, as compared to conventional administration methods.[192] Furthermore, HD-Ad vectors have also been utilized in neurological gene therapy trials and achieved a long-term transgene expression in the brain.

### *3.2.4 Alternative Adenovirus Serotypes*

#### 3.2.4.1 Human Adenoviruses

The presence of preexisting immunity to the commonly utilized Ad vectors, Ad2 and Ad5, represents a major limitation for their use in human gene therapy and vaccine applications.[90] As a result, several strategies have been developed to overcome this problem by constructing alternative/rare Ad serotype-based vectors.[194] Several human Ad serotypes have been constructed and used as a gene transfer vector in both gene therapy and vaccine applications. For example, subgroup B (such as Ad3, Ad7, Ad11, Ad14, Ad35, and Ad50), subgroup D (such as Ad26, Ad48, Ad24, and Ad49), subgroup E (Ad4), and subgroup F adenoviruses (such as Ad41) have all been constructed and used in various preclinical and clinical applications. The seroprevalence of these alternative Ads is very low, as compared to Ad2 and Ad5 vectors. For example, screening for the presence of NAbs to Ad5, Ad26, Ad35, and Ad48 in 4381 individuals from North America, South America, Sub-Sahara Africa, and Southeast Asia revealed that NAbs against Ad35 are infrequent and very low in all regions studied.[24] In contrast, NAbs to Ad48 vectors were only detected in individuals from East Africa.[24] Similarly, Ad26-specific NAbs were only moderately present in individuals from Sub-Sahara Africa and Southeast Asia, but significantly lower than Ad5-specific NAbs in all areas.[24] Very high levels of NAbs against Ad5 were detected in all regions. These data indicate that alternative Ad serotype-based vectors can be used as gene transfer delivery vectors in the developing world.

These alternative Ad serotype vectors are different from each other in terms of cellular receptor utilization and immunogenicity. The commonly utilized subgroup C of human Ad (HAd) vectors use the CAR receptor for cellular transduction. In contrast, subgroup B HAd vectors utilize desmoglein-2 and the cellular coreceptor CD46,[195,196] which make them better gene delivery vectors for CAR-deficient cells. Moreover, it has been shown that HAd26 and HAd48 can also utilize CD46 as their primary cell entry receptor.[146,197] These studies make clear that use of alternative serotype Ads to avoid preexisting Ad5 immunity also entails new obstacles, not the least of which is the unique "innate immune response" signature that each Ad serotype likely generates after administration in vivo.

With regard to immunogenicity, our lab previously characterized and studied innate immune responses after systemic administration of various Ad serotypes HAd31, HAd3, HAd5, Simian-derived Ad23, HAd37, and HAd41, which represent subgroups A–F.[58] We observed significant differences in vector biodistribution between these Ad serotypes, with high levels of HAd3 genomes found in the liver and lung, and HAd37 genomes found in the spleen. Human Ad3 and simian-derived Ad23 induced higher innate immune responses (including higher plasma levels of several proinflammatory cytokines and chemokines, and higher expression levels of various innate immune genes) as compared to other viruses.[58] We also constructed an alternative serotype Ad vector that belongs to subgroup E, HAd4, and evaluated innate and adaptive immune responses following systemic, i.v., and local, i.m., administration. Compared to HAd5 vectors, administration of HAd4-based vectors induced significant increases in innate immune responses both in vitro and in vivo,[74] in a complement-dependent manner. Homologous and/or heterologous prime-boost vaccination with Ad4 vectors expressing the *Plasmodium falciparum*-derived antigen, circumsporozoite protein (CSP), significantly induced higher CSP-specific cellular and humoral immune responses in HAd5 preimmune mice, as compared to Ad5-CSP-treated mice.[198] Similarly, it has been shown previously that vaccination of rhesus monkeys with Ad26 and Ad48 vectors induced higher levels of IFN$\gamma$, IL-1RA, IL-6, and IP-10 cytokines, as compared to Ad5 vaccinated monkeys,[146] indicating that CD46-induced pathways are more efficient in activating innate immunity than the CAR-induced pathways. The enhanced innate immune responses triggered following Ad35, Ad26, and Ad48 vector vaccination resulted in increased adaptive immune responses, as compared to Ad5-treated animals.[146]

In addition to subgroup E, subgroup B-derived vectors, such as Ad35 and Ad11, have been repeatedly used in gene therapy and vaccine applications.[199] For example, utilization of SIV-Gag-expressing Ad11 and Ad35 vectors in Ad5 immune mice induced high-frequency immune responses both in the presence and in the absence of anti-Ad5 immunity,[200] indicating that these rare serotypes can evade Ad5-specific immunity. As a result of these preclinical studies, a phase I randomized, double-blind, placebo-controlled trial to assess the safety and immunogenicity of escalating doses of two Ad35 vectors expressing the HIV1 Gag, reverse transcriptase, integrase and nef (Ad35-GRIN), and env (Ad35-ENV) was conducted in 56 healthy HIV-uninfected individuals.[201] Ad35 vectors were found to be safe and tolerated and were able to induce HIV antigen-specific humoral and cellular (polyfunctional and broad $CD4^+$ and $CD8^+$ T cells) immune responses, in a dose-dependent manner. Similarly, Ad35 vectors have also been utilized as a vaccine platform for malaria in human clinical trials.[21,202] CSP-expressing Ad35 vectors (Ad35.CS.01) were im administered in healthy adult subjects (up to $10^{11}$ vps) and found to be safe and were able to induce CSP-specific humoral and $CD8^+$ T-cell immune responses, in a dose-dependent manner. However, Ad35-specific NAbs were also induced in these human subjects following Ad35. CS.01 administration,[21] suggesting that this platform will also face neutralization on readministration.

The other promising alternative Ad serotype platform that was recently used in human clinical trials is the subgroup D-derived vector, Ad26.[203,204] Initial studies utilizing Ad26 as a vaccine vector in mice[205] and nonhuman primates[28,206] indicated that Ad26-based vaccines are able to induce potent transgene-specific adaptive immune responses, in Ad5-preimmunized animals. As a result of these promising studies, a human clinical trial using Ad26 as a vaccine platform for HIV (Ad26-EnvA) was recently conducted in 60 Ad26-seronegative healthy, HIV-uninfected subjects.[203] The Ad26-EnvA vaccine platform was shown to be safe and immunogenic (Ad26-EnvA was able to induce durable (up to 52 weeks) EnvA specific humoral and cellular immune responses in all vaccinated subjects),[203] indicating that Ad26 might be a beneficial platform for HIV1 vaccine. Due to the complication of Ad5-specific immunity in HIV vaccine trials,[207,208] it is necessary to evaluate the ability of Ad26 to induce transgene-specific immune responses in Ad26 preimmune subjects. Also, it is necessary to evaluate the possible complication of Ad26 preexisting immunity in the vaccinated human subjects, especially in the case of HIV high-risk individuals.

#### 3.2.4.2 Nonhuman Ad-Derived Ad Vectors

Besides human Ads, adenoviruses derived from nonhuman species were also constructed and utilized in several preclinical and clinical applications.[209] The goal of this approach is to evade preexisting humoral and cellular immunity against human Ad serotypes, due to the fact that diseases caused by these nonhuman Ads are species specific and these viruses do not infect humans. Several vaccine and gene therapy vectors derived from chimpanzee Ads (such as ChAd3, ChAd6, ChAd7, ChAd63, and ChAd68),[210–213] canine (such as canine serotype 2 Ad (CAd2)),[214] bovine (such as bovine serotype 3 Ad (BAd3)),[215,216] fowl Ads (such as FAd1, FAd4, FAd8, FAd9, and FAd10),[217,218] ovine (such as the ovine serotype 7 Ad (OAd7)),[219] and porcine[220,221] were constructed and tested in various experimental animal models and humans. Results utilizing these nonhuman Ad vectors in mice, nonhuman primates, and human clinical trials revealed that these vectors can evade HAd-specific preexisting immunity, suggesting that nonhuman Ad vectors may be utilized as a gene transfer vectors in humans.

Some of the best characterized nonhuman Ad vectors are the ones that are derived from canine serotype 2 Ad (CAd2) vectors.[222] One of the unique features of CAd2 is the structure of their capsid, which allowed them to evade preexisting NAbs against human adenoviruses. For example, a side-by-side comparison for the three-dimensional structure of the capsid protein of CAd2 and HAd5 vectors revealed significant differences between these two platforms.[223] The CAd2 capsid contains a smooth surface with shorter external loops, as compared to the longer loops that are commonly present in HAd5 penton base and hexon proteins, which are the common targets for neutralization by HAd5-specific NAbs.[223] In addition, significant differences in the fiber shaft and pIX structures were also observed between CAd2 and HAd5.[223] Together, these differences allowed CAd2 to evade HAd5-specific NAb responses.[224] Another unique feature for CAd2 vectors is their ability to transduce neuronal cells and their high level of retrograde axonal transport in the central and peripheral nervous system,[225] suggesting that these vectors can be used as a gene transfer delivery system for neurological disorders.

Another promising type of nonhuman Ads is the chimpanzees-derived Ad vectors.[211] Initial studies using the ChAd68 vectors (ChAd68/Pan9/SAdV25; species HAd E) confirmed the ability of these vectors to grow in HEK293 cells, transduce a number of CAR receptor-expressing human and murine cell lines, and to avoid neutralization by HAd-specific NAbs.[226,227] Structurally, ChAd68 viruses were also found to be similar to human adenoviruses.[226] Additionally, they contain a genome (about 36 kb) that is closely related to subgroup E of human adenoviruses, with 90% sequence homology to human Ad4 virus.[226] Importantly, significant differences were observed between ChAd68 and human adenoviruses in the hexon HVRs, which enabled them to evade neutralization by human and rhesus monkeys sera, as well as sera from human Ad2-, Ad4-, Ad5-, Ad7-, and Ad12-immunized mice.[226] These features make ChAds a good candidate for gene transfer applications in humans. Results from these studies confirmed the ability of ChAd vectors to infect human cells and induce potent antigen-specific immune responses in humans.[228–233] ChAd3 vectors expressing the glycoprotein of Zaire and Sudan species of Ebola virus (cAd3-EBO) have been tested in a phase I human clinical trial (note: the trial was still ongoing at the time of the preparation of this chapter) and found to be safe and tolerated.[234] In addition, vaccination with cAd3-EBO induced Ebola glycoprotein-specific humoral and cellular responses in all vaccinees (20 participants), in a dose-dependent manner.[234] Similarly, ChAd63 vectors were also utilized as a vaccine candidate for several infectious diseases, including malaria and HIV.[229–233] For example, a recent phase I human clinical trial utilizing a ChAd63 vector that expresses the *P. falciparum*-derived CSP antigen confirms the ability of this platform to induce high levels of T-cell responses to CSP. ChAd63-CSP vectors also induced CSP-specific antibody responses up to 140 days.[235] These results suggest that ChAd vectors are safe and immunogenic in humans, and do not pose increased toxicity risks in vaccine applications.

## 4. Concluding Remarks

Induction of potent innate and adaptive immune responses by the commonly utilized human Ad5 vectors is a major limitation for their utilization in human applications. In some respects, the more efficient a gene transfer vector is in its ability to transduce a gene into a host tissue, the more likely it is to induce significant innate and adaptive immune responses. The human innate system has evolved to detect small amounts of pathogens, pathogens that attempt to invade the human body by a number of evolving strategies as well. Therefore, it is no wonder that one of the most efficient gene transfer vectors of all, Ad-based vectors, has also led the way in elucidating the many aspects of innate and adaptive immune responses faced by any gene transfer vector. Collectively, these responses induce unwanted toxicity, reduce cell transduction efficiencies, and limit the duration of therapeutic transgene expression. To overcome these significant problems, we have summarized some of the various strategies developed to improve Ad vector-mediated gene transfer in a variety of clinical applications. Global transient immune suppression of the host prior to Ad vector administration is a beneficial strategy in certain settings; however, this strategy requires careful attention to the host, especially in immune-compromised patients. Modulation of Ad vector-induced innate and adaptive immune responses, encapsulation of Ad vectors, modification of Ad capsid structure, and the use of alternative Ad serotypes are all promising innovative strategies; however, these strategies require further evaluation in order to improve the outcome of Ad vector-based therapies. Further testing of these novel strategies and gene transfer vectors in nonhuman primates and human clinical trials will shed light on their safety and molecular mechanisms of action as well as their clinical utility in humans.

## Acknowledgments

Dr Andrea Amalfitano was supported by the Michigan State University Foundation, as well the Osteopathic Heritage Foundation. Dr Yasser A. Aldhamen was supported by Michigan State University Clinical and Translational Sciences Institute (MSU-CTSI).

## References

1. Abe K. Gene therapy for ischemic stroke. *Brain Nerve* 2008;**60**:1373–81.
2. Seregin SS, Amalfitano A. Gene therapy for lysosomal storage diseases: progress, challenges and future prospects. *Curr Pharm Des* 2011;**17**:2558–74.
3. Appledorn DM, Seregin S, Amalfitano A. Adenovirus vectors for renal-targeted gene delivery. *Contrib Nephrol* 2008;**159**:47–62.
4. Kiang A, Amalfitano A. Progress and problems when considering gene therapy for GSD-II. *Acta Myol* 2007;**26**:49–52.
5. Amalfitano A. Utilization of adenovirus vectors for multiple gene transfer applications. *Methods* 2004;**33**:173–8.

6. Huang PI, Chang JF, Kirn DH, Liu TC. Targeted genetic and viral therapy for advanced head and neck cancers. *Drug Discov Today* 2009;**14**:570–8.
7. Pearson S, Jia H, Kandachi K. China approves first gene therapy. *Nat Biotechnol* 2004;**22**:3–4.
8. Ginn SL, Alexander IE, Edelstein ML, Abedi MR, Wixon J. Gene therapy clinical trials worldwide to 2012-an update. *J Gene Med* 2013;**15**:65–77.
9. Hilleman MR, Werner JH. Recovery of new agent from patients with acute respiratory illness. *Proc Soc Exp Biol Med Soc Exp Biol Med* 1954;**85**:183–8.
10. Russell WC. Adenoviruses: update on structure and function. *J Gen Virol* 2009;**90**:1–20.
11. Nemerow GR, Pache L, Reddy V, Stewart PL. Insights into adenovirus host cell interactions from structural studies. *Virology* 2009;**384**:380–8.
12. Hoffman JA. Adenoviral disease in pediatric solid organ transplant recipients. *Pediatr Transplant* 2006;**10**:17–25.
13. Rynans S, Dzieciatkowski T, Mlynarczyk G. Adenovirus infection in immunocompromised patients. *Postepy Hig Med Dosw* 2013;**67**:964–72.
14. Florescu DF, Kwon JY, Dumitru I. Adenovirus infections in heart transplantation. *Cardiol Rev* 2013;**21**:203–6.
15. Florescu MC, Miles CD, Florescu DF. What do we know about adenovirus in renal transplantation? *Nephrol Dial Transplant* 2013;**28**:2003–10.
16. Crawford-Miksza L, Schnurr DP. Analysis of 15 adenovirus hexon proteins reveals the location and structure of seven hypervariable regions containing serotype-specific residues. *J Virol* 1996;**70**:1836–44.
17. Mackey JK, Rigden PM, Green M. Do highly oncogenic group A human adenoviruses cause human cancer? Analysis of human tumors for adenovirus 12 transforming DNA sequences. *Proc Natl Acad Sci USA* 1976;**73**:4657–61.
18. Colloca S, Barnes E, Folgori A, Ammendola V, Capone S, Cirillo A, et al. Vaccine vectors derived from a large collection of simian adenoviruses induce potent cellular immunity across multiple species. *Sci Transl Med* 2012;**4**:115ra2.
19. Stanley DA, Honko AN, Asiedu C, Trefry JC, Lau-Kilby AW, Johnson JC, et al. Chimpanzee adenovirus vaccine generates acute and durable protective immunity against ebolavirus challenge. *Nat Med* 2014;**20**:1126–9.
20. Baden LR, Liu J, Li H, Johnson JA, Walsh SR, Kleinjan JA, et al. Induction of HIV-1-specific mucosal immune responses following intramuscular recombinant adenovirus serotype 26 HIV-1 vaccination of humans. *J Infect Dis* 2014.
21. Ouedraogo A, Tiono AB, Kargougou D, Yaro JB, Ouedraogo E, Kabore Y, et al. A phase 1b randomized, controlled, double-blinded dosage-escalation trial to evaluate the safety, reactogenicity and immunogenicity of an adenovirus type 35 based circumsporozoite malaria vaccine in Burkinabe healthy adults 18 to 45 years of age. *PLoS One* 2013;**8**:e78679.
22. Hodgson SH, Choudhary P, Elias SC, Milne KH, Rampling TW, Biswas S, et al. Combining viral vectored and protein-in-adjuvant vaccines against the blood-stage malaria antigen AMA1: report on a phase 1a clinical trial. *Mol Ther J Am Soc Gene Ther* 2014;**22**:2142–54.
23. Abbink P, Lemckert AA, Ewald BA, Lynch DM, Denholtz M, Smits S, et al. Comparative seroprevalence and immunogenicity of six rare serotype recombinant adenovirus vaccine vectors from subgroups B and D. *J Virol* 2007;**81**:4654–63.
24. Barouch DH, Kik SV, Weverling GJ, Dilan R, King SL, Maxfield LF, et al. International seroepidemiology of adenovirus serotypes 5, 26, 35, and 48 in pediatric and adult populations. *Vaccine* 2011;**29**:5203–9.

25. Appledorn DM, Kiang A, McBride A, Jiang H, Seregin S, Scott JM, et al. Wild-type adenoviruses from groups A–F evoke unique innate immune responses, of which HAd3 and SAd23 are partially complement dependent. *Gene Ther* 2008.
26. Thorner AR, Lemckert AA, Goudsmit J, Lynch DM, Ewald BA, Denholtz M, et al. Immunogenicity of heterologous recombinant adenovirus prime-boost vaccine regimens is enhanced by circumventing vector cross-reactivity. *J Virol* 2006;**80**:12009–16.
27. Barouch DH, Pau MG, Custers JH, Koudstaal W, Kostense S, Havenga MJ, et al. Immunogenicity of recombinant adenovirus serotype 35 vaccine in the presence of pre-existing anti-Ad5 immunity. *J Immunol* 2004;**172**:6290–7.
28. Liu J, O'Brien KL, Lynch DM, Simmons NL, La Porte A, Riggs AM, et al. Immune control of an SIV challenge by a T-cell-based vaccine in rhesus monkeys. *Nature* 2009;**457**:87–91.
29. Nemerow GR, Stewart PL, Reddy VS. Structure of human adenovirus. *Curr Opin Virol* 2012;**2**:115–21.
30. Burnett RM. The structure of the adenovirus capsid. II. The packing symmetry of hexon and its implications for viral architecture. *J Mol Biol* 1985;**185**:125–43.
31. Luo J, Deng ZL, Luo X, Tang N, Song WX, Chen J, et al. A protocol for rapid generation of recombinant adenoviruses using the AdEasy system. *Nat Protoc* 2007;**2**:1236–47.
32. Jager L, Hausl MA, Rauschhuber C, Wolf NM, Kay MA, Ehrhardt A. A rapid protocol for construction and production of high-capacity adenoviral vectors. *Nat Protoc* 2009;**4**:547–64.
33. Graham FL, Smiley J, Russell WC, Nairn R. Characteristics of a human cell line transformed by DNA from human adenovirus type 5. *J Gen Virol* 1977;**36**:59–74.
34. Engelhardt JF, Ye X, Doranz B, Wilson JM. Ablation of E2A in recombinant adenoviruses improves transgene persistence and decreases inflammatory response in mouse liver. *Proc Natl Acad Sci USA* 1994;**91**:6196–200.
35. Amalfitano A, Parks RJ. Separating fact from fiction: assessing the potential of modified adenovirus vectors for use in human gene therapy. *Curr Gene Ther* 2002;**2**:111–33.
36. Raper SE, Haskal ZJ, Ye X, Pugh C, Furth EE, Gao GP, et al. Selective gene transfer into the liver of non-human primates with E1-deleted, E2A-defective, or E1-E4 deleted recombinant adenoviruses. *Hum Gene Ther* 1998;**9**:671–9.
37. Ehrhardt A, Kay MA. Gutted adenovirus: a rising star on the horizon? *Gene Ther* 2005;**12**:1540–1.
38. Brunetti-Pierri N, Ng T, Iannitti D, Cioffi W, Stapleton G, Law M, et al. Transgene expression up to 7 years in nonhuman primates following hepatic transduction with helper-dependent adenoviral vectors. *Hum Gene Ther* 2013;**24**:761–5.
39. Ding EY, Hodges BL, Hu H, McVie-Wylie AJ, Serra D, Migone FK, et al. Long-term efficacy after [E1-, polymerase-] adenovirus-mediated transfer of human acid-α-glucosidase gene into glycogen storage disease type II knockout mice. *Hum Gene Ther* 2001;**12**:955–65.
40. Amalfitano A. Next-generation adenoviral vectors: new and improved. *Gene Ther* 1999;**6**:1643–5.
41. Hodges BL, Serra D, Hu H, Begy CA, Chamberlain JS, Amalfitano A. Multiply deleted [E1, polymerase-, and pTP-] adenovirus vector persists despite deletion of the preterminal protein. *J Gene Med* 2000;**2**:250–9.
42. Hodges BL, Evans HK, Everett RS, Ding EY, Serra D, Amalfitano A. Adenovirus vectors with the 100K gene deleted and their potential for multiple gene therapy applications. *J Virol* 2001;**75**:5913–20.

43. Howe SJ, Mansour MR, Schwarzwaelder K, Bartholomae C, Hubank M, Kempski H, et al. Insertional mutagenesis combined with acquired somatic mutations causes leukemogenesis following gene therapy of SCID-X1 patients. *J Clin Invest* 2008;**118**:3143–50.
44. Hartman ZC, Appledorn DM, Amalfitano A. Adenovirus vector induced innate immune responses: impact upon efficacy and toxicity in gene therapy and vaccine applications. *Virus Res* 2008;**132**:1–14.
45. Hoffmann JA, Kafatos FC, Janeway CA, Ezekowitz RA. Phylogenetic perspectives in innate immunity. *Science* 1999;**284**:1313–8.
46. Kawai T, Akira S. The role of pattern-recognition receptors in innate immunity: update on Toll-like receptors. *Nat Immunol* 2010;**11**:373–84.
47. Beutler BA. TLRs and innate immunity. *Blood* 2009;**113**:1399–407.
48. Aoshi T, Koyama S, Kobiyama K, Akira S, Ishii KJ. Innate and adaptive immune responses to viral infection and vaccination. *Curr Opin Virol* 2011;**1**:226–32.
49. Kawai T, Akira S. Toll-like receptors and their crosstalk with other innate receptors in infection and immunity. *Immunity* 2011;**34**:637–50.
50. Takeuchi O, Akira S. Pattern recognition receptors and inflammation. *Cell* 2010;**140**:805–20.
51. Girardin SE, Sansonetti PJ, Philpott DJ. Intracellular vs extracellular recognition of pathogens–common concepts in mammals and flies. *Trends Microbiol* 2002;**10**:193–9.
52. Medzhitov R. Recognition of microorganisms and activation of the immune response. *Nature* 2007;**449**:819–26.
53. Bortoluci KR, Medzhitov R. Control of infection by pyroptosis and autophagy: role of TLR and NLR. *Cell Mol Life Sci* 2010;**67**:1643–51.
54. Chen G, Shaw MH, Kim YG, Nunez G. NOD-like receptors: role in innate immunity and inflammatory disease. *Annu Rev Pathol* 2009;**4**:365–98.
55. Hou W, Zhang Q, Yan Z, Chen R, Zeh Iii HJ, Kang R, et al. Strange attractors: DAMPs and autophagy link tumor cell death and immunity. *Cell Death Dis* 2013;**4**:e966.
56. Appledorn DM, Patial S, McBride A, Godbehere S, Van Rooijen N, Parameswaran N, et al. Adenovirus vector-induced innate inflammatory mediators, MAPK signaling, as well as adaptive immune responses are dependent upon both TLR2 and TLR9 in vivo. *J Immunol* 2008;**181**:2134–44.
57. Rhee EG, Blattman JN, Kasturi SP, Kelley RP, Kaufman DR, Lynch DM, et al. Multiple innate immune pathways contribute to the immunogenicity of recombinant adenovirus vaccine vectors. *J Virol* 2011;**85**:315–23.
58. Appledorn DM, Kiang A, McBride A, Jiang H, Seregin S, Scott JM, et al. Wild-type adenoviruses from groups A–F evoke unique innate immune responses, of which HAd3 and SAd23 are partially complement dependent. *Gene Ther* 2008;**15**:885–901.
59. Basner-Tschakarjan E, Gaffal E, O'Keeffe M, Tormo D, Limmer A, Wagner H, et al. Adenovirus efficiently transduces plasmacytoid dendritic cells resulting in TLR9-dependent maturation and IFN-α production. *J Gene Med* 2006;**8**:1300–6.
60. Hartman ZC, Kiang A, Everett RS, Serra D, Yang XY, Clay TM, et al. Adenovirus infection triggers a rapid, MyD88-regulated transcriptome response critical to acute-phase and adaptive immune responses in vivo. *J Virol* 2007;**81**:1796–812.
61. Seregin SS, Aldhamen YA, Appledorn DM, Schuldt NJ, McBride AJ, Bujold M, et al. CR1/2 is an important suppressor of Adenovirus-induced innate immune responses and is required for induction of neutralizing antibodies. *Gene Ther* 2009;**16**:1245–59.
62. Aldhamen YA, Seregin SS, Aylsworth CF, Godbehere S, Amalfitano A. Manipulation of EAT-2 expression promotes induction of multiple beneficial regulatory and effector functions of the human innate immune system as a novel immunomodulatory strategy. *Int Immunol* 2014;**26**:291–303.

63. Zhu J, Huang X, Yang Y. A critical role for type I IFN-dependent NK cell activation in innate immune elimination of adenoviral vectors in vivo. *Mol Ther* 2008;**16**:1300–7.
64. Lindsay RW, Darrah PA, Quinn KM, Wille-Reece U, Mattei LM, Iwasaki A, et al. $CD8^+$ T cell responses following replication-defective adenovirus serotype 5 immunization are dependent on $CD11c^+$ dendritic cells but show redundancy in their requirement of TLR and nucleotide-binding oligomerization domain-like receptor signaling. *J Immunol* 2010;**185**:1513–21.
65. Fejer G, Drechsel L, Liese J, Schleicher U, Ruzsics Z, Imelli N, et al. Key role of splenic myeloid DCs in the IFN-αβ response to adenoviruses in vivo. *PLoS Pathog* 2008;**4**:e1000208.
66. Zhang Y, Chirmule N, Gao GP, Qian R, Croyle M, Joshi B, et al. Acute cytokine response to systemic adenoviral vectors in mice is mediated by dendritic cells and macrophages. *Mol Ther* 2001;**3**:697–707.
67. Smith JS, Xu Z, Tian J, Stevenson SC, Byrnes AP. Interaction of systemically delivered adenovirus vectors with Kupffer cells in mouse liver. *Hum Gene Ther* 2008;**19**:547–54.
68. Zhu J, Huang X, Yang Y. Innate immune response to adenoviral vectors is mediated by both Toll-like receptor-dependent and -independent pathways. *J Virol* 2007;**81**:3170–80.
69. Yamaguchi T, Kawabata K, Koizumi N, Sakurai F, Nakashima K, Sakurai H, et al. Role of MyD88 and TLR9 in the innate immune response elicited by serotype 5 adenoviral vectors. *Hum Gene Ther* 2007;**18**:753–62.
70. Muruve DA, Petrilli V, Zaiss AK, White LR, Clark SA, Ross PJ, et al. The inflammasome recognizes cytosolic microbial and host DNA and triggers an innate immune response. *Nature* 2008;**452**:103–7.
71. Barlan AU, Griffin TM, McGuire KA, Wiethoff CM. Adenovirus membrane penetration activates the NLRP3 inflammasome. *J Virol* 2011;**85**:146–55.
72. Hensley SE, Amalfitano A. Toll-like receptors impact on safety and efficacy of gene transfer vectors. *Mol Ther* 2007;**15**:1417–22.
73. Appledorn DM, Patial S, Godbehere S, Parameswaran N, Amalfitano A. TRIF, and TRIF-interacting TLRs differentially modulate several adenovirus vector-induced immune responses. *J Innate Immun* 2009;**1**:376–88.
74. Hartman ZC, Appledorn DM, Serra D, Glass O, Mendelson TB, Clay TM, et al. Replication-attenuated Human Adenoviral Type 4 vectors elicit capsid dependent enhanced innate immune responses that are partially dependent upon interactions with the complement system. *Virology* 2008;**374**:453–67.
75. Verdino P, Witherden DA, Havran WL, Wilson IA. The molecular interaction of CAR and JAML recruits the central cell signal transducer PI3K. *Science* 2010;**329**:1210–4.
76. Verdino P, Wilson IA. JAML and CAR: two more players in T-cell activation. *Cell Cycle* 2011;**10**:1341–2.
77. Philpott NJ, Nociari M, Elkon KB, Falck-Pedersen E. Adenovirus-induced maturation of dendritic cells through a PI3 kinase-mediated TNF-α induction pathway. *Proc Natl Acad Sci USA* 2004;**101**:6200–5.
78. Li E, Stupack D, Klemke R, Cheresh DA, Nemerow GR. Adenovirus endocytosis via α(v) integrins requires phosphoinositide-3-OH kinase. *J Virol* 1998;**72**:2055–61.
79. Schulte M, Sorkin M, Al-Benna S, Stupka J, Hirsch T, Daigeler A, et al. Innate immune response after adenoviral gene delivery into skin is mediated by AIM2, NALP3, DAI and mda5. *SpringerPlus* 2013;**2**:234.
80. Nociari M, Ocheretina O, Schoggins JW, Falck-Pedersen E. Sensing infection by adenovirus: Toll-like receptor-independent viral DNA recognition signals activation of the interferon regulatory factor 3 master regulator. *J Virol* 2007;**81**:4145–57.

81. Takaoka A, Wang Z, Choi MK, Yanai H, Negishi H, Ban T, et al. DAI (DLM-1/ZBP1) is a cytosolic DNA sensor and an activator of innate immune response. *Nature* 2007; **448**:501–5.
82. Lam E, Stein S, Falck-Pedersen E. Adenovirus detection by the cGAS/STING/TBK1 DNA sensing cascade. *J Virol* 2014;**88**:974–81.
83. Schenten D, Medzhitov R. The control of adaptive immune responses by the innate immune system. *Adv Immunol* 2011;**109**:87–124.
84. Iwasaki A, Medzhitov R. Regulation of adaptive immunity by the innate immune system. *Science* 2010;**327**:291–5.
85. Dai Y, Schwarz EM, Gu D, Zhang WW, Sarvetnick N, Verma IM. Cellular and humoral immune responses to adenoviral vectors containing factor IX gene: tolerization of factor IX and vector antigens allows for long-term expression. *Proc Natl Acad Sci USA* 1995;**92**:1401–5.
86. Yu B, Dong J, Wang C, Zhan Y, Zhang H, Wu J, et al. Characteristics of neutralizing antibodies to adenovirus capsid proteins in human and animal sera. *Virology* 2013;**437**:118–23.
87. Kafri T, Morgan D, Krahl T, Sarvetnick N, Sherman L, Verma I. Cellular immune response to adenoviral vector infected cells does not require de novo viral gene expression: implications for gene therapy. *Proc Natl Acad Sci USA* 1998;**95**:11377–82.
88. Yang Y, Nunes FA, Berencsi K, Furth EE, Gonczol E, Wilson JM. Cellular immunity to viral antigens limits E1-deleted adenoviruses for gene therapy. *Proc Natl Acad Sci USA* 1994;**91**:4407–11.
89. Yang Y, Li Q, Ertl HC, Wilson JM. Cellular and humoral immune responses to viral antigens create barriers to lung-directed gene therapy with recombinant adenoviruses. *J Virol* 1995;**69**:2004–15.
90. Seregin SS, Amalfitano A. Overcoming pre-existing adenovirus immunity by genetic engineering of adenovirus-based vectors. *Expert Opin Biol Ther* 2009;**9**:1521–31.
91. Aldhamen YA, Seregin SS, Amalfitano A. Immune recognition of gene transfer vectors: focus on adenovirus as a paradigm. *Front Immunol* 2011;**2**:40.
92. Ahi YS, Bangari DS, Mittal SK. Adenoviral vector immunity: its implications and circumvention strategies. *Curr Gene Ther* 2011;**11**:307–20.
93. Small JC, Haut LH, Bian A, Ertl HC. The effect of adenovirus-specific antibodies on adenoviral vector-induced, transgene product-specific T cell responses. *J Leukoc Biol* 2014;**96**:821–31.
94. Sumida SM, Truitt DM, Lemckert AA, Vogels R, Custers JH, Addo MM, et al. Neutralizing antibodies to adenovirus serotype 5 vaccine vectors are directed primarily against the adenovirus hexon protein. *J Immunol* 2005;**174**:7179–85.
95. Pichla-Gollon SL, Drinker M, Zhou X, Xue F, Rux JJ, Gao GP, et al. Structure-based identification of a major neutralizing site in an adenovirus hexon. *J Virol* 2007; **81**:1680–9.
96. Hong SS, Habib NA, Franqueville L, Jensen S, Boulanger PA. Identification of adenovirus (ad) penton base neutralizing epitopes by use of sera from patients who had received conditionally replicative ad (addl1520) for treatment of liver tumors. *J Virol* 2003;**77**:10366–75.
97. Roberts DM, Nanda A, Havenga MJ, Abbink P, Lynch DM, Ewald BA, et al. Hexon-chimaeric adenovirus serotype 5 vectors circumvent pre-existing anti-vector immunity. *Nature* 2006;**441**:239–43.
98. Bruder JT, Chen P, Semenova E, Thomas CA, Konovalova S, Ekberg G, et al. Identification of a suppressor mutation that improves the yields of hexon-modified adenovirus vectors. *J Virol* 2013;**87**:9661–71.

99. Joshi A, Tang J, Kuzma M, Wagner J, Mookerjee B, Filicko J, et al. Adenovirus DNA polymerase is recognized by human CD8+ T cells. *J Gen Virol* 2009;**90**:84–94.
100. Bassett JD, Swift SL, Bramson JL. Optimizing vaccine-induced CD8(+) T-cell immunity: focus on recombinant adenovirus vectors. *Expert Rev Vaccines* 2011;**10**:1307–19.
101. Johnson JA, Barouch DH, Baden LR. Nonreplicating vectors in HIV vaccines. *Curr Opin HIV AIDS* 2013;**8**:412–20.
102. Barouch DH. Novel adenovirus vector-based vaccines for HIV-1. *Curr Opin HIV AIDS* 2010;**5**:386–90.
103. Lasaro MO, Ertl HC. New insights on adenovirus as vaccine vectors. *Mol Ther* 2009;**17**:1333–9.
104. Schuldt NJ, Amalfitano A. Malaria vaccines: focus on adenovirus based vectors. *Vaccine* 2012;**30**:5191–8.
105. Gray GE, Moodie Z, Metch B, Gilbert PB, Bekker LG, Churchyard G, et al. Recombinant adenovirus type 5 HIV gag/pol/nef vaccine in South Africa: unblinded, long-term follow-up of the phase 2b HVTN 503/Phambili study. *Lancet Infect Dis* 2014;**14**:388–96.
106. Hanke T. STEP trial and HIV-1 vaccines inducing T-cell responses. *Expert Rev Vaccines* 2008;**7**:303–9.
107. Morse MA, Chaudhry A, Gabitzsch ES, Hobeika AC, Osada T, Clay TM, et al. Novel adenoviral vector induces T-cell responses despite anti-adenoviral neutralizing antibodies in colorectal cancer patients. *Cancer Immunol Immunother* 2013;**62**:1293–301.
108. Seregin SS, Appledorn DM, McBride AJ, Schuldt NJ, Aldhamen YA, Voss T, et al. Transient pretreatment with glucocorticoid ablates innate toxicity of systemically delivered adenoviral vectors without reducing efficacy. *Mol Ther* 2009;**17**:685–96.
109. Otake K, Ennist DL, Harrod K, Trapnell BC. Nonspecific inflammation inhibits adenovirus-mediated pulmonary gene transfer and expression independent of specific acquired immune responses. *Hum Gene Ther* 1998;**9**:2207–22.
110. Shen WY, Lai MC, Beilby J, Barnett NL, Liu J, Constable IJ, et al. Combined effect of cyclosporine and sirolimus on improving the longevity of recombinant adenovirus-mediated transgene expression in the retina. *Archives Ophthalmol* 2001;**119**:1033–43.
111. Thomas MA, Spencer JF, Toth K, Sagartz JE, Phillips NJ, Wold WS. Immunosuppression enhances oncolytic adenovirus replication and antitumor efficacy in the Syrian hamster model. *Mol Ther* 2008;**16**:1665–73.
112. Hamada K, Sakaue M, Sarkar A, Buchl S, Satterfield W, Keeling M, et al. Immune responses to repetitive adenovirus-mediated gene transfer and restoration of gene expression by cyclophosphamide or etoposide. *Gynecol Oncol* 2005;**99**:S177–86.
113. Smith TA, White BD, Gardner JM, Kaleko M, McClelland A. Transient immunosuppression permits successful repetitive intravenous administration of an adenovirus vector. *Gene Ther* 1996;**3**:496–502.
114. Cichon G, Strauss M. Transient immunosuppression with 15-deoxyspergualin prolongs reporter gene expression and reduces humoral immune response after adenoviral gene transfer. *Gene Ther* 1998;**5**:85–90.
115. Kaplan JM, Smith AE. Transient immunosuppression with deoxyspergualin improves longevity of transgene expression and ability to readminister adenoviral vector to the mouse lung. *Hum Gene Ther* 1997;**8**:1095–104.
116. Lochmuller H, Petrof BJ, Pari G, Larochelle N, Dodelet V, Wang Q, et al. Transient immunosuppression by FK506 permits a sustained high-level dystrophin expression after adenovirus-mediated dystrophin minigene transfer to skeletal muscles of adult dystrophic (mdx) mice. *Gene Ther* 1996;**3**:706–16.

117. Dhar D, Toth K, Wold WS. Cycles of transient high-dose cyclophosphamide administration and intratumoral oncolytic adenovirus vector injection for long-term tumor suppression in Syrian hamsters. *Cancer Gene Ther* 2014;**21**:171–8.
118. Young BA, Spencer JF, Ying B, Tollefson AE, Toth K, Wold WS. The role of cyclophosphamide in enhancing antitumor efficacy of an adenovirus oncolytic vector in subcutaneous Syrian hamster tumors. *Cancer Gene Ther* 2013;**20**:521–30.
119. Malvicini M, Ingolotti M, Piccioni F, Garcia M, Bayo J, Atorrasagasti C, et al. Reversal of gastrointestinal carcinoma-induced immunosuppression and induction of antitumoural immunity by a combination of cyclophosphamide and gene transfer of IL-12. *Mol Oncol* 2011;**5**:242–55.
120. Elkon KB, Liu CC, Gall JG, Trevejo J, Marino MW, Abrahamsen KA, et al. Tumor necrosis factor α plays a central role in immune-mediated clearance of adenoviral vectors. *Proc Natl Acad Sci USA* 1997;**94**:9814–9.
121. Sung RS, Qin L, Bromberg JS. TNFα and IFNγ induced by innate anti-adenoviral immune responses inhibit adenovirus-mediated transgene expression. *Mol Ther* 2001;**3**:757–67.
122. Wilderman MJ, Kim S, Gillespie CT, Sun J, Kapoor V, Vachani A, et al. Blockade of TNF-α decreases both inflammation and efficacy of intrapulmonary Ad.IFNβ immunotherapy in an orthotopic model of bronchogenic lung cancer. *Mol Ther* 2006;**13**:910–7.
123. Salako MA, Kulbe H, Ingemarsdotter CK, Pirlo KJ, Williams SL, Lockley M, et al. Inhibition of the inflammatory cytokine TNF-α increases adenovirus activity in ovarian cancer via modulation of cIAP1/2 expression. *Mol Ther* 2011;**19**:490–9.
124. Peng Y, Trevejo J, Zhou J, Marino MW, Crystal RG, Falck-Pedersen E, et al. Inhibition of tumor necrosis factor α by an adenovirus-encoded soluble fusion protein extends transgene expression in the liver and lung. *J Virol* 1999;**73**:5098–109.
125. Benihoud K, Esselin S, Descamps D, Jullienne B, Salone B, Bobe P, et al. Respective roles of TNF-α and IL-6 in the immune response-elicited by adenovirus-mediated gene transfer in mice. *Gene Ther* 2007;**14**:533–44.
126. Saggio I, Ciapponi L, Savino R, Ciliberto G, Perricaudet M. Adenovirus-mediated gene transfer of a human IL-6 antagonist. *Gene Ther* 1997;**4**:839–45.
127. Shayakhmetov DM, Gaggar A, Ni S, Li ZY, Lieber A. Adenovirus binding to blood factors results in liver cell infection and hepatotoxicity. *J Virol* 2005;**79**:7478–91.
128. Cerullo V, Seiler MP, Mane V, Brunetti-Pierri N, Clarke C, Bertin TK, et al. Toll-like receptor 9 triggers an innate immune response to helper-dependent adenoviral vectors. *Mol Ther* 2007;**15**:378–85.
129. Tibbles LA, Spurrell JC, Bowen GP, Liu Q, Lam M, Zaiss AK, et al. Activation of p38 and ERK signaling during adenovirus vector cell entry lead to expression of the C-X-C chemokine IP-10. *J Virol* 2002;**76**:1559–68.
130. Sakurai H, Tashiro K, Kawabata K, Yamaguchi T, Sakurai F, Nakagawa S, et al. Adenoviral expression of suppressor of cytokine signaling-1 reduces adenovirus vector-induced innate immune responses. *J Immunol* 2008;**180**:4931–8.
131. Liu L, Li W, Wei X, Cui Q, Lou W, Wang G, et al. Potent antitumor activity of oncolytic adenovirus-mediated SOCS1 for hepatocellular carcinoma. *Gene Ther* 2013;**20**:84–92.
132. Alzuguren P, Hervas-Stubbs S, Gonzalez-Aseguinolaza G, Poutou J, Fortes P, Mancheno U, et al. Transient depletion of specific immune cell populations to improve adenovirus-mediated transgene expression in the liver. *Liver Int* 2014.
133. Lieber A, He CY, Meuse L, Schowalter D, Kirillova I, Winther B, et al. The role of Kupffer cell activation and viral gene expression in early liver toxicity after infusion of recombinant adenovirus vectors. *J Virol* 1997;**71**:8798–807.

134. Kuzmin AI, Finegold MJ, Eisensmith RC. Macrophage depletion increases the safety, efficacy and persistence of adenovirus-mediated gene transfer in vivo. *Gene Ther* 1997;**4**:309–16.
135. Worgall S, Leopold PL, Wolff G, Ferris B, Van Roijen N, Crystal RG. Role of alveolar macrophages in rapid elimination of adenovirus vectors administered to the epithelial surface of the respiratory tract. *Hum Gene Ther* 1997;**8**:1675–84.
136. Schiedner G, Hertel S, Johnston M, Dries V, van Rooijen N, Kochanek S. Selective depletion or blockade of Kupffer cells leads to enhanced and prolonged hepatic transgene expression using high-capacity adenoviral vectors. *Mol Ther* 2003;**7**:35–43.
137. Koski A, Rajecki M, Guse K, Kanerva A, Ristimaki A, Pesonen S, et al. Systemic adenoviral gene delivery to orthotopic murine breast tumors with ablation of coagulation factors, thrombocytes and Kupffer cells. *J Gene Med* 2009;**11**:966–77.
138. Haisma HJ, Kamps JA, Kamps GK, Plantinga JA, Rots MG, Bellu AR. Polyinosinic acid enhances delivery of adenovirus vectors in vivo by preventing sequestration in liver macrophages. *J Gen Virol* 2008;**89**:1097–105.
139. Stone D, Liu Y, Shayakhmetov D, Li ZY, Ni S, Lieber A. Adenovirus-platelet interaction in blood causes virus sequestration to the reticuloendothelial system of the liver. *J Virol* 2007;**81**:4866–71.
140. Guerette B, Vilquin JT, Gingras M, Gravel C, Wood KJ, Tremblay JP. Prevention of immune reactions triggered by first-generation adenoviral vectors by monoclonal antibodies and CTLA4Ig. *Hum Gene Ther* 1996;**7**:1455–63.
141. Ye X, Robinson MB, Pabin C, Batshaw ML, Wilson JM. Transient depletion of CD4 lymphocyte improves efficacy of repeated administration of recombinant adenovirus in the ornithine transcarbamylase deficient sparse fur mouse. *Gene Ther* 2000;**7**:1761–7.
142. Zsengeller ZK, Boivin GP, Sawchuk SS, Trapnell BC, Whitsett JA, Hirsch R. Anti-T cell receptor antibody prolongs transgene expression and reduces lung inflammation after adenovirus-mediated gene transfer. *Hum Gene Ther* 1997;**8**:935–41.
143. Chirmule N, Raper SE, Burkly L, Thomas D, Tazelaar J, Hughes JV, et al. Readministration of adenovirus vector in nonhuman primate lungs by blockade of CD40–CD40 ligand interactions. *J Virol* 2000;**74**:3345–52.
144. Haegel-Kronenberger H, Haanstra K, Ziller-Remy C, Ortiz Buijsse AP, Vermeiren J, Stoeckel F, et al. Inhibition of costimulation allows for repeated systemic administration of adenoviral vector in rhesus monkeys. *Gene Ther* 2004;**11**:241–52.
145. Kurachi S, Koizumi N, Sakurai F, Kawabata K, Sakurai H, Nakagawa S, et al. Characterization of capsid-modified adenovirus vectors containing heterologous peptides in the fiber knob, protein IX, or hexon. *Gene Ther* 2007;**14**:266–74.
146. Teigler JE, Iampietro MJ, Barouch DH. Vaccination with adenovirus serotypes 35, 26, and 48 elicits higher levels of innate cytokine responses than adenovirus serotype 5 in rhesus monkeys. *J Virol* 2012;**86**:9590–8.
147. Baden LR, Walsh SR, Seaman MS, Johnson JA, Tucker RP, Kleinjan JA, et al. First-in-human evaluation of a hexon chimeric adenovirus vector expressing HIV-1 Env (IPCAVD 002). *J Infect Dis* 2014;**210**:1052–61.
148. Seregin SS, Aldhamen YA, Appledorn DM, Zehnder J, Voss T, Godbehere S, et al. Use of DAF-displaying adenovirus vectors reduces induction of transgene- and vector-specific adaptive immune responses in mice. *Hum Gene Ther* 2011;**22**:1083–94.
149. Seregin SS, Aldhamen YA, Appledorn DM, Hartman ZC, Schuldt NJ, Scott J, et al. Adenovirus capsid-display of the retro-oriented human complement inhibitor DAF reduces Ad vector-triggered immune responses in vitro and in vivo. *Blood* 2010;**116**:1669–77.

150. Seregin SS, Hartman ZC, Appledorn DM, Godbehere S, Jiang H, Frank MM, et al. Novel adenovirus vectors 'capsid-displaying' a human complement inhibitor. *J Innate Immun* 2010;**2**:353–9.
151. Sahu A, Morikis D, Lambris JD. Compstatin, a peptide inhibitor of complement, exhibits species-specific binding to complement component C3. *Mol Immunol* 2003;**39**:557–66.
152. Kim J, Kim PH, Kim SW, Yun CO. Enhancing the therapeutic efficacy of adenovirus in combination with biomaterials. *Biomaterials* 2012;**33**:1838–50.
153. O'Riordan CR, Lachapelle A, Delgado C, Parkes V, Wadsworth SC, Smith AE, et al. PEGylation of adenovirus with retention of infectivity and protection from neutralizing antibody in vitro and in vivo. *Hum Gene Ther* 1999;**10**:1349–58.
154. Croyle MA, Chirmule N, Zhang Y, Wilson JM. PEGylation of E1-deleted adenovirus vectors allows significant gene expression on readministration to liver. *Hum Gene Ther* 2002;**13**:1887–900.
155. Matthews C, Jenkins G, Hilfinger J, Davidson B. Poly-L-lysine improves gene transfer with adenovirus formulated in PLGA microspheres. *Gene Ther* 1999;**6**:1558–64.
156. Mok H, Park JW, Park TG. Microencapsulation of PEGylated adenovirus within PLGA microspheres for enhanced stability and gene transfection efficiency. *Pharm Res* 2007;**24**: 2263–9.
157. Lee SG, Yoon SJ, Kim CD, Kim K, Lim DS, Yeom YI, et al. Enhancement of adenoviral transduction with polycationic liposomes in vivo. *Cancer Gene Ther* 2000;**7**:1329–35.
158. Mendez N, Herrera V, Zhang L, Hedjran F, Feuer R, Blair SL, et al. Encapsulation of adenovirus serotype 5 in anionic lecithin liposomes using a bead-based immunoprecipitation technique enhances transfection efficiency. *Biomaterials* 2014;**35**:9554–61.
159. Zhong Z, Shi S, Han J, Zhang Z, Sun X. Anionic liposomes increase the efficiency of adenovirus-mediated gene transfer to coxsackie-adenovirus receptor deficient cells. *Mol Pharm* 2010;**7**:105–15.
160. Wang CH, Chan LW, Johnson RN, Chu DS, Shi J, Schellinger JG, et al. The transduction of Coxsackie and Adenovirus Receptor-negative cells and protection against neutralizing antibodies by HPMA-co-oligolysine copolymer-coated adenovirus. *Biomaterials* 2011;**32**:9536–45.
161. Fisher KD, Seymour LW. HPMA copolymers for masking and retargeting of therapeutic viruses. *Adv Drug Deliv Rev* 2010;**62**:240–5.
162. Kim PH, Kim J, Kim TI, Nam HY, Yockman JW, Kim M, et al. Bioreducible polymer-conjugated oncolytic adenovirus for hepatoma-specific therapy via systemic administration. *Biomaterials* 2011;**32**:9328–42.
163. Croyle MA, Le HT, Linse KD, Cerullo V, Toietta G, Beaudet A, et al. PEGylated helper-dependent adenoviral vectors: highly efficient vectors with an enhanced safety profile. *Gene Ther* 2005;**12**:579–87.
164. Matsui H, Sakurai F, Katayama K, Yamaguchi T, Okamoto S, Takahira K, et al. A hexon-specific PEGylated adenovirus vector utilizing blood coagulation factor X. *Biomaterials* 2012;**33**:3743–55.
165. Mok H, Palmer DJ, Ng P, Barry MA. Evaluation of polyethylene glycol modification of first-generation and helper-dependent adenoviral vectors to reduce innate immune responses. *Mol Ther* 2005;**11**:66–79.
166. Gao JQ, Eto Y, Yoshioka Y, Sekiguchi F, Kurachi S, Morishige T, et al. Effective tumor targeted gene transfer using PEGylated adenovirus vector via systemic administration. *J Control Release* 2007;**122**:102–10.
167. Schoggins JW, Falck-Pedersen E. Fiber and penton base capsid modifications yield diminished adenovirus type 5 transduction and proinflammatory gene expression with retention of antigen-specific humoral immunity. *J Virol* 2006;**80**:10634–44.

168. Schoggins JW, Nociari M, Philpott N, Falck-Pedersen E. Influence of fiber detargeting on adenovirus-mediated innate and adaptive immune activation. *J Virol* 2005;**79**:11627–37.
169. Koizumi N, Kawabata K, Sakurai F, Watanabe Y, Hayakawa T, Mizuguchi H. Modified adenoviral vectors ablated for coxsackievirus-adenovirus receptor, $\alpha_v$ integrin, and heparan sulfate binding reduce in vivo tissue transduction and toxicity. *Hum Gene Ther* 2006;**17**:264–79.
170. Wu H, Curiel DT. Fiber-modified adenoviruses for targeted gene therapy. *Methods Mol Biol* 2008;**434**:113–32.
171. Youil R, Toner TJ, Su Q, Chen M, Tang A, Bett AJ, et al. Hexon gene switch strategy for the generation of chimeric recombinant adenovirus. *Hum Gene Ther* 2002;**13**:311–20.
172. Parker AL, White KM, Lavery CA, Custers J, Waddington SN, Baker AH. Pseudotyping the adenovirus serotype 5 capsid with both the fibre and penton of serotype 35 enhances vascular smooth muscle cell transduction. *Gene Ther* 2013;**20**:1158–64.
173. Koski A, Kangasniemi L, Escutenaire S, Pesonen S, Cerullo V, Diaconu I, et al. Treatment of cancer patients with a serotype 5/3 chimeric oncolytic adenovirus expressing GMCSF. *Mol Ther* 2010;**18**:1874–84.
174. Nepomuceno RR, Pache L, Nemerow GR. Enhancement of gene transfer to human myeloid cells by adenovirus-fiber complexes. *Mol Ther* 2007;**15**:571–8.
175. Roy S, Shirley PS, McClelland A, Kaleko M. Circumvention of immunity to the adenovirus major coat protein hexon. *J Virol* 1998;**72**:6875–9.
176. Wu H, Dmitriev I, Kashentseva E, Seki T, Wang M, Curiel DT. Construction and characterization of adenovirus serotype 5 packaged by serotype 3 hexon. *J Virol* 2002;**76**:12775–82.
177. Vogels R, Zuijdgeest D, van Rijnsoever R, Hartkoorn E, Damen I, de Bethune MP, et al. Replication-deficient human adenovirus type 35 vectors for gene transfer and vaccination: efficient human cell infection and bypass of preexisting adenovirus immunity. *J Virol* 2003;**77**:8263–71.
178. Barouch DH, Liu J, Lynch DM, O'Brien KL, La Porte A, Simmons NL, et al. Protective efficacy of a single immunization of a chimeric adenovirus vector-based vaccine against simian immunodeficiency virus challenge in rhesus monkeys. *J Virol* 2009;**83**:9584–90.
179. Amalfitano A, Hauser MA, Hu H, Serra D, Begy CR, Chamberlain JS. Production and characterization of improved adenovirus vectors with the E1, E2b, and E3 genes deleted. *J Virol* 1998;**72**:926–33.
180. Hu H, Serra D, Amalfitano A. Persistence of an [E1-, polymerase-] adenovirus vector despite transduction of a neoantigen into immune-competent mice. *Hum Gene Ther* 1999;**10**:355–64.
181. Everett RS, Hodges BL, Ding EY, Xu F, Serra D, Amalfitano A. Liver toxicities typically induced by first-generation adenoviral vectors can be reduced by use of E1, E2b-deleted adenoviral vectors. *Hum Gene Ther* 2003;**14**:1715–26.
182. Wang Q, Greenburg G, Bunch D, Farson D, Finer MH. Persistent transgene expression in mouse liver following in vivo gene transfer with a ΔE1/ΔE4 adenovirus vector. *Gene Ther* 1997;**4**:393–400.
183. Chirmule N, Hughes JV, Gao GP, Raper SE, Wilson JM. Role of E4 in eliciting CD4 T-cell and B-cell responses to adenovirus vectors delivered to murine and nonhuman primate lungs. *J Virol* 1998;**72**:6138–45.
184. Do Thi NA, Saillour P, Ferrero L, Dedieu JF, Mallet J, Paunio T. Delivery of GDNF by an E1,E3/E4 deleted adenoviral vector and driven by a GFAP promoter prevents dopaminergic neuron degeneration in a rat model of Parkinson's disease. *Gene Ther* 2004;**11**:746–56.
185. Kreppel F. Production of high-capacity adenovirus vectors. *Methods Mol Biol* 2014;**1089**:211–29.

186. Fisher KJ, Choi H, Burda J, Chen SJ, Wilson JM. Recombinant adenovirus deleted of all viral genes for gene therapy of cystic fibrosis. *Virology* 1996;**217**:11–22.
187. Pastore N, Nusco E, Piccolo P, Castaldo S, Vanikova J, Vetrini F, et al. Improved efficacy and reduced toxicity by ultrasound-guided intrahepatic injections of helper-dependent adenoviral vector in Gunn rats. *Hum Gene Ther Methods* 2013;**24**:321–7.
188. Muruve DA, Cotter MJ, Zaiss AK, White LR, Liu Q, Chan T, et al. Helper-dependent adenovirus vectors elicit intact innate but attenuated adaptive host immune responses in vivo. *J Virology* 2004;**78**:5966–72.
189. Brunetti-Pierri N, Palmer DJ, Beaudet AL, Carey KD, Finegold M, Ng P. Acute toxicity after high-dose systemic injection of helper-dependent adenoviral vectors into nonhuman primates. *Hum Gene Ther* 2004;**15**:35–46.
190. King GD, Muhammad AK, Xiong W, Kroeger KM, Puntel M, Larocque D, et al. High-capacity adenovirus vector-mediated anti-glioma gene therapy in the presence of systemic antiadenovirus immunity. *J Virol* 2008;**82**:4680–4.
191. Muhammad AK, Puntel M, Candolfi M, Salem A, Yagiz K, Farrokhi C, et al. Study of the efficacy, biodistribution, and safety profile of therapeutic gutless adenovirus vectors as a prelude to a phase I clinical trial for glioblastoma. *Clin Pharmacol Ther* 2010;**88**:204–13.
192. Brunetti-Pierri N, Stapleton GE, Palmer DJ, Zuo Y, Mane VP, Finegold MJ, et al. Pseudo-hydrodynamic delivery of helper-dependent adenoviral vectors into non-human primates for liver-directed gene therapy. *Mol Ther* 2007;**15**:732–40.
193. Brunetti-Pierri N, Palmer DJ, Mane V, Finegold M, Beaudet AL, Ng P. Increased hepatic transduction with reduced systemic dissemination and proinflammatory cytokines following hydrodynamic injection of helper-dependent adenoviral vectors. *Mol Ther* 2005;**12**:99–106.
194. Hall K, Blair Zajdel ME, Blair GE. Unity and diversity in the human adenoviruses: exploiting alternative entry pathways for gene therapy. *Biochem J* 2010;**431**:321–36.
195. Gaggar A, Shayakhmetov DM, Lieber A. CD46 is a cellular receptor for group B adenoviruses. *Nat Med* 2003;**9**:1408–12.
196. Wang H, Li ZY, Liu Y, Persson J, Beyer I, Moller T, et al. Desmoglein 2 is a receptor for adenovirus serotypes 3, 7, 11 and 14. *Nat Med* 2011;**17**:96–104.
197. Li H, Rhee EG, Masek-Hammerman K, Teigler JE, Abbink P, Barouch DH. Adenovirus serotype 26 utilizes CD46 as a primary cellular receptor and only transiently activates T lymphocytes following vaccination of rhesus monkeys. *J Virol* 2012;**86**:10862–5.
198. Schuldt NJ, Aldhamen YA, Godbehere-Roosa S, Seregin SS, Kousa YA, Amalfitano A. Immunogenicity when utilizing adenovirus serotype 4 and 5 vaccines expressing circumsporozoite protein in naive and adenovirus (Ad5) immune mice. *Malar J* 2012;**11**:209.
199. Stone D, di Paolo NC, Lieber A. Development of group B adenoviruses as gene transfer vectors. *Biotechnol Genet Eng Rev* 2006;**22**:101–23.
200. Lemckert AA, Sumida SM, Holterman L, Vogels R, Truitt DM, Lynch DM, et al. Immunogenicity of heterologous prime-boost regimens involving recombinant adenovirus serotype 11 (Ad11) and Ad35 vaccine vectors in the presence of anti-ad5 immunity. *J Virol* 2005;**79**:9694–701.
201. Keefer MC, Gilmour J, Hayes P, Gill D, Kopycinski J, Cheeseman H, et al. A phase I double blind, placebo-controlled, randomized study of a multigenic HIV-1 adenovirus subtype 35 vector vaccine in healthy uninfected adults. *PLoS One* 2012;**7**:e41936.
202. Creech CB, Dekker CL, Ho D, Phillips S, Mackey S, Murray-Krezan C, et al. Randomized, placebo-controlled trial to assess the safety and immunogenicity of an adenovirus type 35-based circumsporozoite malaria vaccine in healthy adults. *Hum Vaccines Immunother* 2013;**9**:2548–57.

203. Baden LR, Walsh SR, Seaman MS, Tucker RP, Krause KH, Patel A, et al. First-in-human evaluation of the safety and immunogenicity of a recombinant adenovirus serotype 26 HIV-1 Env vaccine (IPCAVD 001). *J Infect Dis* 2013;**207**:240–7.
204. Barouch DH, Liu J, Peter L, Abbink P, Iampietro MJ, Cheung A, et al. Characterization of humoral and cellular immune responses elicited by a recombinant adenovirus serotype 26 HIV-1 Env vaccine in healthy adults (IPCAVD 001). *J Infect Dis* 2013;**207**:248–56.
205. Penaloza-MacMaster P, Provine NM, Ra J, Borducchi EN, McNally A, Simmons NL, et al. Alternative serotype adenovirus vaccine vectors elicit memory T cells with enhanced anamnestic capacity compared to Ad5 vectors. *J Virol* 2013;**87**:1373–84.
206. Li H, Liu J, Carville A, Mansfield KG, Lynch D, Barouch DH. Durable mucosal simian immunodeficiency virus-specific effector memory T lymphocyte responses elicited by recombinant adenovirus vectors in rhesus monkeys. *J Virol* 2011;**85**:11007–15.
207. Duerr A, Huang Y, Buchbinder S, Coombs RW, Sanchez J, del Rio C, et al. Extended follow-up confirms early vaccine-enhanced risk of HIV acquisition and demonstrates waning effect over time among participants in a randomized trial of recombinant adenovirus HIV vaccine (Step Study). *J Infect Dis* 2012;**206**:258–66.
208. Buchbinder SP, Mehrotra DV, Duerr A, Fitzgerald DW, Mogg R, Li D, et al. Efficacy assessment of a cell-mediated immunity HIV-1 vaccine (the Step Study): a double-blind, randomised, placebo-controlled, test-of-concept trial. *Lancet* 2008;**372**:1881–93.
209. Lopez-Gordo E, Podgorski II, Downes N, Alemany R. Circumventing antivector immunity: potential use of nonhuman adenoviral vectors. *Hum Gene Ther* 2014;**25**:285–300.
210. Roshorm Y, Cottingham MG, Potash MJ, Volsky DJ, Hanke T. T cells induced by recombinant chimpanzee adenovirus alone and in prime-boost regimens decrease chimeric EcoHIV/NDK challenge virus load. *Eur J Immunol* 2012;**42**:3243–55.
211. Capone S, D'Alise AM, Ammendola V, Colloca S, Cortese R, Nicosia A, et al. Development of chimpanzee adenoviruses as vaccine vectors: challenges and successes emerging from clinical trials. *Expert Rev Vaccines* 2013;**12**:379–93.
212. Zhou D, Zhou X, Bian A, Li H, Chen H, Small JC, et al. An efficient method of directly cloning chimpanzee adenovirus as a vaccine vector. *Nat Protoc* 2010;**5**:1775–85.
213. Cervasi B, Carnathan DG, Sheehan KM, Micci L, Paiardini M, Kurupati R, et al. Immunological and virological analyses of rhesus macaques immunized with chimpanzee adenoviruses expressing the simian immunodeficiency virus Gag/Tat fusion protein and challenged intrarectally with repeated low doses of SIVmac. *J Virol* 2013;**87**:9420–30.
214. Tordo N, Foumier A, Jallet C, Szelechowski M, Klonjkowski B, Eloit M. Canine adenovirus based rabies vaccines. *Dev Biol* 2008;**131**:467–76.
215. Ayalew LE, Kumar P, Gaba A, Makadiya N, Tikoo SK. Bovine adenovirus-3 as a vaccine delivery vehicle. *Vaccine* 2015;**33**:493–9.
216. Zhu YM, Yu Z, Cai H, Gao YR, Dong XM, Li ZL, et al. Isolation, identification, and complete genome sequence of a bovine adenovirus type 3 from cattle in China. *Virol J* 2011;**8**:557.
217. Corredor JC, Nagy E. The non-essential left end region of the fowl adenovirus 9 genome is suitable for foreign gene insertion/replacement. *Virus Res* 2010;**149**:167–74.
218. Kim MS, Lim TH, Lee DH, Youn HN, Yuk SS, Kim BY, et al. An inactivated oil-emulsion fowl Adenovirus serotype 4 vaccine provides broad cross-protection against various serotypes of fowl Adenovirus. *Vaccine* 2014;**32**:3564–8.
219. Bridgeman A, Roshorm Y, Lockett LJ, Xu ZZ, Hopkins R, Shaw J, et al. Ovine atadenovirus, a novel and highly immunogenic vector in prime-boost studies of a candidate HIV-1 vaccine. *Vaccine* 2009;**28**:474–83.

220. Hammond JM, Johnson MA. Porcine adenovirus as a delivery system for swine vaccines and immunotherapeutics. *Vet J* 2005;**169**:17–27.
221. Patel A, Tikoo S, Kobinger G. A porcine adenovirus with low human seroprevalence is a promising alternative vaccine vector to human adenovirus 5 in an H5N1 virus disease model. *PLoS One* 2010;**5**:e15301.
222. Kremer EJ. CAR chasing: canine adenovirus vectors-all bite and no bark? *J Gene Med* 2004;**6**(Suppl. 1):S139–51.
223. Schoehn G, El Bakkouri M, Fabry CM, Billet O, Estrozi LF, Le L, et al. Three-dimensional structure of canine adenovirus serotype 2 capsid. *J Virol* 2008;**82**:3192–203.
224. Bru T, Salinas S, Kremer EJ. An update on canine adenovirus type 2 and its vectors. *Viruses* 2010;**2**:2134–53.
225. Soudais C, Laplace-Builhe C, Kissa K, Kremer EJ. Preferential transduction of neurons by canine adenovirus vectors and their efficient retrograde transport in vivo. *FASEB J* 2001;**15**:2283–5.
226. Farina SF, Gao GP, Xiang ZQ, Rux JJ, Burnett RM, Alvira MR, et al. Replication-defective vector based on a chimpanzee adenovirus. *J Virol* 2001;**75**:11603–13.
227. Cohen CJ, Xiang ZQ, Gao GP, Ertl HC, Wilson JM, Bergelson JM. Chimpanzee adenovirus CV-68 adapted as a gene delivery vector interacts with the coxsackievirus and adenovirus receptor. *J Gen Virol* 2002;**83**:151–5.
228. Ewer KJ, O'Hara GA, Duncan CJ, Collins KA, Sheehy SH, Reyes-Sandoval A, et al. Protective $CD8^+$ T-cell immunity to human malaria induced by chimpanzee adenovirus-MVA immunisation. *Nat Commun* 2013;**4**:2836.
229. Ogwang C, Afolabi M, Kimani D, Jagne YJ, Sheehy SH, Bliss CM, et al. Safety and immunogenicity of heterologous prime-boost immunisation with *Plasmodium falciparum* malaria candidate vaccines, ChAd63 ME-TRAP and MVA ME-TRAP, in healthy Gambian and Kenyan adults. *PLoS One* 2013;**8**:e57726.
230. Sheehy SH, Duncan CJ, Elias SC, Choudhary P, Biswas S, Halstead FD, et al. ChAd63-MVA-vectored blood-stage malaria vaccines targeting MSP1 and AMA1: assessment of efficacy against mosquito bite challenge in humans. *Mol Ther* 2012;**20**:2355–68.
231. O'Hara GA, Duncan CJ, Ewer KJ, Collins KA, Elias SC, Halstead FD, et al. Clinical assessment of a recombinant simian adenovirus ChAd63: a potent new vaccine vector. *J Infect Dis* 2012;**205**:772–81.
232. Sheehy SH, Duncan CJ, Elias SC, Collins KA, Ewer KJ, Spencer AJ, et al. Phase Ia clinical evaluation of the *Plasmodium falciparum* blood-stage antigen MSP1 in ChAd63 and MVA vaccine vectors. *Mol Ther* 2011;**19**:2269–76.
233. Sheehy SH, Duncan CJ, Elias SC, Biswas S, Collins KA, O'Hara GA, et al. Phase Ia clinical evaluation of the safety and immunogenicity of the *Plasmodium falciparum* blood-stage antigen AMA1 in ChAd63 and MVA vaccine vectors. *PLoS One* 2012;**7**:e31208.
234. Ledgerwood JE, DeZure AD, Stanley DA, Novik L, Enama ME, Berkowitz NM, et al. Chimpanzee adenovirus vector ebola vaccine – preliminary report. *N Engl J Med* 2014.
235. de Barra E, Hodgson SH, Ewer KJ, Bliss CM, Hennigan K, Collins A, et al. A phase Ia study to assess the safety and immunogenicity of new malaria vaccine candidates ChAd63 CS administered alone and with MVA CS. *PLoS One* 2014;**9**:e115161.

# Helper-Dependent Adenoviral Vectors

17

*Nicola Brunetti-Pierri*[1,2], *Philip Ng*[3]
[1]Telethon Institute of Genetics and Medicine, Pozzuoli, Italy; [2]Department of Translational Medicine, Federico II University, Naples, Italy; [3]Department of Molecular and Human Genetics, Baylor College of Medicine, Houston, TX, USA

## 1. Introduction

Helper-dependent adenoviral vectors (HDAds) (also referred to as gutless, gutted, mini, fully deleted, high capacity, Δ, pseudo, encapsidated adenovirus minichromosome) are deleted of all viral genes. Like early-generation adenoviral (Ad) vectors, HDAds can efficiently transduce a wide variety of dividing and nondividing cells to mediate high-level transgene expression. However, unlike early-generation Ad vectors, the absence of viral genes enables HDAds to mediate long-term transgene expression without chronic toxicity. Moreover, deletion of viral genes permits a large cloning capacity of 37 kb. Because HDAd genomes exist episomally in transduced cells, the risks of germline transmission and insertional mutagenesis are negligible. Several strategies for producing HDAds, and numerous in vivo, in vitro, and ex vivo applications of HDAd-mediated gene transfer for gene and cell therapy have been published. The purpose of this chapter is not to provide a comprehensive review of all of these studies, but rather to describe examples of particular significance or interest.

## 2. Production of Helper-Dependent Adenoviral Vectors

Currently, the most efficient and widely used method for producing HDAds is based on the Cre/loxP system[1] (Figure 1). In this system the HDAd genome is first constructed in a bacterial plasmid. Minimally, the HDAd genome includes the expression cassette of interest and ~500 bp of *cis*-acting Ad sequences required for vector DNA replication (ITRs) and packaging (ψ). In addition, a small segment of noncoding Ad sequence from the E4 region adjacent to the right ITR can be included to increase vector yield, possibly by enhancing packaging of the HDAd DNA.[2] Stuffer DNA is often required to bring the size of the HDAd genome up to the packaging requirements of the viral capsid, which is between 27.7 and 37.8 kb.[3,4]

To convert the plasmid form of the HDAd genome into the viral form, the plasmid is first digested with the appropriate restriction enzyme to liberate the HDAd genome from the bacterial plasmid sequences. The 293 cells expressing Cre recombinase are then transfected with the linearized HDAd genome and subsequently infected with a

**Adenoviral Vectors for Gene Therapy.**

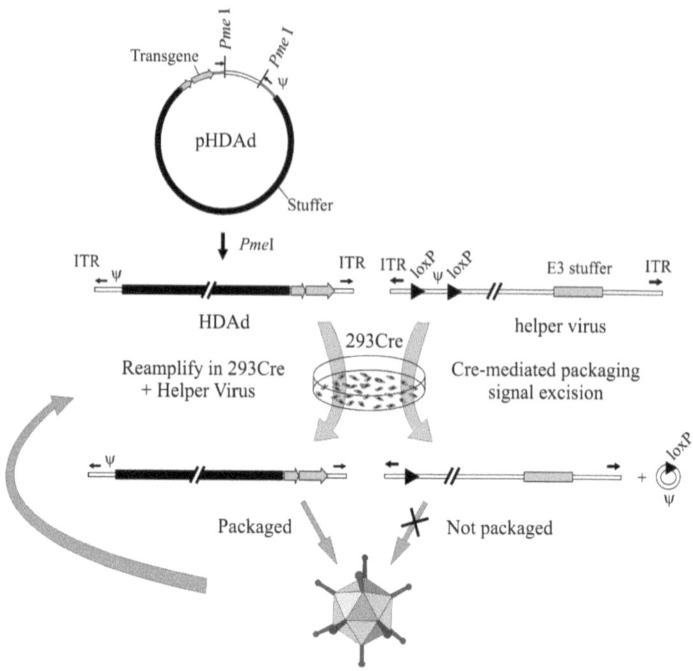


**Figure 1** The Cre/loxP system for generating HDAds. The HDAd contains only ~500 bp of *cis*-acting Ad sequences required for DNA replication (ITRs) and packaging (ψ); the remainder of the genome consists of the desired transgene and non-Ad "stuffer" sequences. The HDAd genome is constructed as a bacterial plasmid (pHDAd) and is liberated by restriction enzyme digestion (e.g., *Pme*I). To rescue the HDAd, the liberated genome is transfected into 293 cells expressing Cre and infected with an HV bearing a packaging signal (ψ) flanked by loxP sites. The HV genome also contains a stuffer sequence in E3 to prevent the formation of RCA in 293-derived cells. Cre-mediated excision of ψ renders the HV genome unpackagable, but still able to replicate and provide all of the necessary *trans*-acting factors for propagation of the HDAd. The titer of the HDAd vector is increased by serial coinfections of 293Cre cells with the HDAd and the HV.

helper virus (HV). The HV is an E1-deleted Ad that bears a packaging signal flanked by loxP sites and thus following infection of 293Cre cells, the packaging signal is excised from the helper viral genome by Cre-mediated site-specific recombination between the loxP sites (Figure 1). This renders the helper viral genome unpackagable but still able to replicate to *trans*-complement the replication and encapsidation of the HDAd genome. The titer of the HDAd is increased by serial coinfection of 293Cre cells with the HDAd and the HV. Improved reagents and optimized methods have permitted rapid and robust large-scale production of high-quality HDAds with very low HV contamination levels.[5] Detailed methodologies for producing HDAds are described elsewhere.[6,7] Other alternative strategies for producing HDAds have also been developed.[8–12] However, due to space constraints, these will not be discussed because the Cre/loxP system remains the method of choice for HDAd production.

Because the HV is an E1-deleted Ad, the generation of replication-competent Ad (RCA; $E1^+$) as a consequence of homologous recombination between the HV and the Ad sequences present in 293-derived cells can occur.[1,13] To prevent the formation of RCA, a "stuffer" sequence was inserted into the E3 region to render any $E1^+$ recombinants too large to be packaged.[1,9,14] The length of the E3 stuffer is such that the total size of the HV genome is <105% of wild type but >105% following homologous recombination with Ad sequences in the 293 cells, and thus unpackagable.

Since the HDAd genome exists episomally, integrating into the host chromosomes only at very low frequencies, it is likely that transgene expression will not be permanent, especially if the target cells are dividing. Should transgene expression fade over time, it would be desirable to simply readminister the vector. Unfortunately, this approach is not possible using the same vector because the initial administration elicits neutralizing anti-Ad antibody that renders subsequent readministrations ineffective. One strategy, known as "serotype switching", may help to overcome this problem. In addition to the Ad serotype 5-based HV, other HV based on serotypes 1, 2, and 6 have been generated.[15–17] Therefore, genetically identical HDAds of different serotypes can be generated simply by changing the serotype of the HV used for vector production. There are ~50 human serotypes of Ad. Therefore, it may be possible to create a panel of different serotype HVs for producing different serotypes but genetically identical HDAds. These HDAds could then be given sequentially when transgene expression wanes from the previous vector administration, as shown by several in vivo studies.[15–18] Because of the large number of Ad serotypes, this could be theoretically repeated for the lifetime of the patient.

A variety of serotype five HVs with genetic elements from other serotype Ads have been described. These chimeric HVs permit the production of chimeric HDAds with novel and useful properties. For example, cells that do not express CAR are inefficiently transduced by the most commonly used serotype 5 Ad-based vectors. To overcome this, the fiber gene of a serotype 5 HV was replaced with the fiber gene from serotype 35 and this chimeric HV was used to produce a chimeric HDAd that utilized CD46 as the cellular receptor instead of CAR.[19] Likewise, serotype 5 HV bearing the fiber knob domain from serotype 3 was used to produce a chimeric HDAd that could more efficiently transduce adult muscle in mice.[20] As discussed below, Kupffer cells of the liver trap intravenously injected Ad vectors. To avoid this, the hypervariable region (HVR) of the capsid hexon protein in the serotype 5 HV was replaced with the HVR from serotype 6, and this was used to produce a hexon chimeric HDAd better able to evade Kupffer cell uptake.[21]

Although current systems cannot produce HDAds free of HV, it should be noted that the HV is an E1-deleted Ad at low contaminating amounts, which are far below the much higher quantities of E1-deleted Ad that have been given to numerous patients in clinical trials without adverse events. It should also be noted that in mouse models, intravenous[13] or intramuscular[22] injections of HDAds with up to 10% HV contamination did not reduce the duration of transgene expression or result in significantly higher toxicity compared to preparations with only 0.1–0.5% contamination.

The large cloning capacity of HDAds permits accommodation of transgenes in their native genomic context. This is important because the level and duration of expression of transgenes in their native genomic context are superior compared to their cDNA counterparts.[18,23,24] This is likely due to a more physiological regulation of gene expression. HDAds offer the advantage of transferring many genes in their genomic context whereas other vectors [e.g., early-generation Ads, retroviral, lentiviral, or adeno-associated viral (AAV) vectors] cannot, due to their limited cloning capacity.

## 3. Intracellular Status of Helper-Dependent Adenoviral Vectors

HDAd genomes appear to exist in the nucleus as replication-deficient linear monomers both in cell culture and in mouse livers.[25] However, one study found, at least in cell culture, that approximately 1–3% of HDAd genomes have circularized and contain end-to-end joining of the Ad termini.[26] The intracellular HDAd genome is assembled into chromatin through association with cellular histones, promoting efficient transgene expression.[27,28] Because the vector genome is episomal, it is lost during cell division. However, Ehrhardt et al. discovered that the episomal persistence of the HDAd genome in murine liver is greater than plasmid DNA by a mechanism that has not been yet identified.[29]

Although Ad and Ad vectors are episomal, it is reasonable to assume that integration of vector DNA into the host genome occurs sporadically. Several studies in cell culture have investigated the frequency of HDAd genome integration and found random integration frequencies to be $10^{-3}$ to $10^{-5}$ per cell, depending on the experimental conditions.[30–34] Further, these studies revealed that vector genomes appear to integrate as intact monomers with little or no loss of sequences at the vector ends, and an apparent preference for integration into genes.[31,32] However, artificial conditions of cell culture and inherent genetic instability of cultured cells have led one group to investigate HDAd genomic integration in mouse livers.[35] The investigators found a lower in vivo integration frequency of $6.72 \times 10^{-5}$ per hepatocyte compared to culture cells, and the vector appeared to integrate through its termini.[35]

## 4. Helper-Dependent Adenoviral Vectors as a Platform for Hybrid Vectors

In nondividing cells, the HDAd genome exists as a stable episome, thus conferring stable transgene expression. However, in dividing cells, the HDAd genome is lost because it integrates rarely, and it does not possess any replicative or nuclear retention mechanisms. Thus, HDAd are not useful for long-term expression in actively dividing cells. However, because Ad-based vectors remain one of the most efficient at gene transfer, numerous hybrid vectors using the HDAd platform have been devised to permit high efficiency transduction and long-term transgene expression in dividing cells. One class of hybrids consists of integrating vectors encoded by HDAd; the HDAd provides high-efficiency transduction of the target cell to express the integrating vector, which then mediates integration of the transgene into the target cell chromosomes. Integrating systems encoded by HDAds included retroviral vectors, AAV, sleeping beauty transposon, bacteriophage phiC31, and retrotransposons. In these systems, integration requires expression of an integrase or transposase, which may have off-target effects and could be immunogenic. Also, there is always a concern with genotoxicity with any integrating vector. The reader is referred to an excellent review by Müther et al. for more details on these HDAd-based hybrid vectors.[36] To overcome potential genotoxicity of integrating vectors, HDAds encoding episomal replicons have been developed. Instead of integrating, these HDAd-encoded episomal replicons can replicate and segregate into daughter cells. Toward this end, HDAds encoding Epstein–Barr virus replication and retention mechanisms have been developed.[26,37–39] However, these systems require expression of the foreign EBNA1 protein in transduced cells to maintain episomal persistence and this might be immunogenic. To address this, Voigtlander et al. recently developed a hybrid HDAd that utilized only a 2 kb S/MAR (scaffold/matrix attachment region) sequence derived from the 5′-region of the human β-interferon gene that mediates episomal persistence and replication. This vector resulted in persistent transgene expression following transduction of actively dividing cells in vitro and in vivo.[40] It should be noted that all of the above systems require the episomal replicon to be circular. Because the HDAd genome is linear, all the above methods require excision of the episomal replicon from the HDAd as a circular genome and this is achieved in all cases by site-specific recombination (Cre or FLP). Consequently, a potentially immunogenic recombinase must be expressed in the transduced cells. However, expression of recombinase (delivered by another Ad vector) is only needed transiently, thus possibly minimizing this risk.

## 5. Liver Gene Therapy

The liver is an attractive target for gene therapy because it is the affected organ in many acquired and genetic diseases. The fenestrated structure of its endothelium permits exposure of hepatocytes to intravenously delivered vector, and permits secretion of vector-encoded therapeutic proteins into the circulation for systemic delivery. To date, numerous examples of in vivo liver-directed gene therapy for disease models using HDAds have been reported. In general, all these studies demonstrate that HDAds can lead to long-term phenotypic correction without chronic toxicity. Corrections of hypercholesterolemia in apolipoprotein E-deficient mice and hyperbilirubinemia in the rat model of Crigler–Najjar syndrome are paradigmatic examples; these studies demonstrated that a single intravenous injection of HDAd resulted in lifelong expression of the therapeutic transgene and permanent phenotypic correction of a genetic disease with negligible toxicity.[18,41] Several other rodent studies have illustrated the potential of HDAd for long-term phenotypic correction of several disease models.[42] Importantly, long-term expression by HDAds has also been recapitulated in clinically relevant large animal models.[43–48]

Until recently, it was believed that all Ad5 infection was dependent on receptors for cellular attachment (the coxsackie and adenovirus receptor, CAR) and entry

($\alpha_v$ integrins).[49–51] However, this may not be the case for hepatocytes.[52,53] In the bloodstream, coagulation factor X (FX) (among other plasma proteins) binds to Ad particles.[54–58] Binding to FX occurs with extremely high affinity and appears to function as a bridge facilitating the attachment of Ad5 to cells: the γ-carboxyglutamic acid domain of FX binds to Ad5 hexon protein and the serine protease domain of FX binds cell-surface heparan sulfate proteoglycans.[59] FX is required for Ad5 vectors to transduce hepatocytes in vivo: mutations in the hexon protein or FX ablation by warfarin treatment[60–62] both resulted in hepatocyte detargeting and negligible hepatocyte transduction.[54] FX also appears to protect Ad vector particles; within the blood, natural IgM binds Ad and activates the classical complement pathway resulting in virus neutralization. Binding of FX to the virus shields it from IgM binding, thereby permitting hepatocyte transduction.[63,64] This study contends that in vivo hepatocyte transduction is primarily due to FX's protective role rather than acting as a bridge between the virus and the hepatocytes.

## 5.1 Threshold Effect to Hepatocyte Transduction

Relatively high vector doses were required to achieve efficient hepatic transduction following systemic intravascular delivery. Following intravenous Ad injection, there is a nonlinear dose response to hepatic transduction, with low doses yielding very low to undetectable levels of transgene expression, but with higher doses resulting in disproportionately high levels of transgene expression. Kupffer cells are responsible for this nonlinear dose response by avidly sequestering bloodborne Ad particles.[65,66] Natural or preexisting neutralizing antibodies are involved in vector clearance by Kupffer cells[63,67] through opsonization of vector particles that enhance Fc-receptor-mediated vector uptake. Antibodies can also bind indirectly to viral particles through binding to complement factor C3.[68] Antibody–virus complexes activate the classical complement proteins C1, C2, and C4.[63,69] Complement activation results in covalent binding of C3 fragments to viral capsid and Ad particle uptake by Kupffer cells via the complement receptor Ig superfamily (CRIg) that regulates death of these cells in the liver.[70] Following uptake of Ad, Kupffer cells undergo rapid proinflammatory necrotic death that is controlled by interferon-regulatory factor 3 (IRF3).[71–74] Ad uptake by Kupffer cells and their necrosis appear to play a protective role and may represent a defensive suicide strategy, preventing disseminated virus infection.[75] Consequently, for the host that lacks this macrophage population, even a sublethal virus infection may lead to compromised resistance and be detrimental for survival. Mice depleted of tissue macrophages by clodronate liposomes showed indeed higher virus DNA burden, greater hepatotoxicity, and increased lethality.[75] Nevertheless, intravenously injected Ad5 causes a rapid hemodynamic response presenting with hypotension, hemoconcentration, tissue edema, and vasocongestion[73,76] that is dependent, at least in part, on upregulation in macrophages of platelet-activating factor, a known shock inducer lipid signaling molecule.[76] The observation that administration of polyinosine, as well as other polyanionic ligands, into mice prior to intravenous Ad injection drastically reduces Ad accumulation in Kupffer cells and increases hepatocyte gene transfer[68,77] has led to the recognition of scavenger receptor-A (SR-A) and scavenger receptor expressed on endothelial cell type I (SREC-I) on both Kupffer cells and endothelial cells as mediators of Ad vector uptake.[78–81]

Several studies suggest that the diameter of liver endothelial fenestrations plays an important role in the efficiency of Ad-mediated hepatocyte transduction.[82–84] Ad5 particles have a diameter of 93 nm with protruding fibers of 30 nm[83] whereas the diameter of human liver fenestration is ~107 nm and thus, the relative smaller size of liver fenestrations may be an obstacle for hepatocyte transduction in humans.[84]

Systemic administration of high doses of Ad vectors likely results in widespread transduction of a large number of various extrahepatic cell types (e.g., blood cells, endothelium, spleen, and lung), which may also be important barriers to efficient hepatocyte transduction. Over 90% of Ad vectors bind to human erythrocytes ex vivo[85] through the CAR that is expressed on erythrocytes from humans but not from mice or rhesus macaques.[86] Furthermore, erythrocytes from humans but not from mice bear the complement receptor 1, which binds Ad5 in the presence of antibodies and complement.[86] Although the liver takes up more vector than any other organ following systemic injection, the total amount of vector, on a vector genome copy number per microgram DNA basis, is abundantly distributed throughout the body in mice,[87] nonhuman primates,[88,89] and human patients.[90]

## 5.2 Acute Toxicity

Systemic intravascular administration of Ad-based vectors, including HDAds, can result in acute toxicity. This acute response occurs immediately after vector administration; its severity is dose dependent and is characterized by elevations in serum proinflammatory cytokines as a consequence of activation of the innate inflammatory immune response against the viral capsid.[87,88,91,92] Indeed, dose-dependent acute toxicity, consistent with activation of the innate inflammatory immune response, was observed following systemic administration of HDAds into nonhuman primates.[93] The role of the viral capsid in causing the acute toxicity was confirmed by Muruve et al.[94] showing that intravenous injection of either FGAd or HDAd into mice induced an acute expression of several inflammatory cytokine and chemokine genes in the liver. However, FGAd, but not HDAd, also induced a second phase of liver inflammation from 7 days postinjection. These results demonstrate that HDAds induce intact innate but attenuated adaptive immune responses in vivo.

The mechanism(s) responsible for Ad-mediated activation of the acute inflammatory response is not entirely known, and several have been postulated. Vector uptake and activation to secrete proinflammatory cytokines by reticuloendothelial system cells including macrophages and dendritic cells in the spleen have been involved.[87,88] Vector transduction of endothelial cells,[95,96] peripheral mononuclear cells,[97] and Ad-mediated complement activation[98–100] have all been implicated to play a role in acute toxicity. Several studies also suggest that antibodies may play a role. Both neutralizing anti-Ad antibodies[89,98] and nonneutralizing or naturally occurring (nonspecific cross-reacting) antibodies[65,66,98] may contribute to acute toxicity, perhaps by opsonizing the viral particles and rendering them more susceptible to Fc-mediated uptake by macrophages, which in turn may become activated to secrete proinflammatory cytokines. For example, systemic vector administration into nonhuman primates resulted in significantly higher IL-6 levels in animals with neutralizing anti-Ad antibodies compared to naïve animals.[89] As described above, FX binds Ad within the blood and appears to play a role in the innate response to Ad; following macrophage internalization of FX-decorated Ad particles, intracellular FX triggers activation of innate immunity via the TLR4/NFKB pathway and thus plays a role in activation of the acute toxicity.[75] Besides TLR4,[75] other Toll-like receptors (TLRs) are involved in Ad-induced innate responses, including TLR9 that senses vector dsDNA,[101,102] TLR3,[103] and TLR2.[104,105]

Ultimately, the precise mechanism responsible for activation of the acute inflammatory response by systemic Ad is multifactorial and complex and remains to be fully elucidated. However, regardless of the precise mechanism, it is likely that a threshold of innate immune activation must first be attained, as a consequence of high doses and systemic exposure of the vector to many cell types and bloodborne components, before severe and lethal acute toxicity is manifested. Evidence of robust activation of the acute inflammatory response is observed in both rodents and nonhuman primates given comparable systemic high-dose Ad (on a per kg basis). However, it is important to emphasize that unlike primates, lethal systemic inflammatory response syndrome does not develop in rodents and, in fact, such high doses are well tolerated. This may reflect species-to-species differences in the quality of the innate immune response or sensitivities of the end organs to pathologic sequelae. This likely accounts for the plethora of studies reporting negligible toxicity in mice given high-dose HDAds and underscores the importance of safety and toxicity evaluations in larger animals.

## 5.3 Overcoming the Threshold to Hepatic Transduction and Acute Toxicity

Several groups have investigated different strategies to overcome the threshold to hepatocyte transduction and the obstacle of acute toxicity. Because the severity of the acute response is dose dependent and appears to correlate with extrahepatic systemic vector dissemination, one of these approaches aimed at preferential targeting of the vector to the liver, thereby allowing the use of lower vector doses. One strategy involved injection of HDAd directly into the surgically isolated liver in nonhuman primates and was shown to achieve higher efficiency hepatic transduction with reduced systemic vector dissemination compared to systemic injection.[44] However, this approach is invasive and consequently, minimally invasive percutaneous balloon occlusion catheter-based methods were developed to achieve preferential hepatocyte transduction. One such strategy mimics hydrodynamic injection but without rapid, large-volume injection[43] and was based on the observation that hydrodynamic injection of HDAd into mice resulted in increased hepatocyte transduction with reduced markers of acute toxicity and reduced systemic vector dissemination.[106] This so-called pseudohydrodynamic injection involved transient occlusion of hepatic venous outflow using two balloon occlusion catheters percutaneously placed in the inferior vena cava (IVC), above and below the hepatic veins. Because blood entering the liver from the hepatic artery (HA) and portal vein remains unobstructed, an increase in intrahepatic

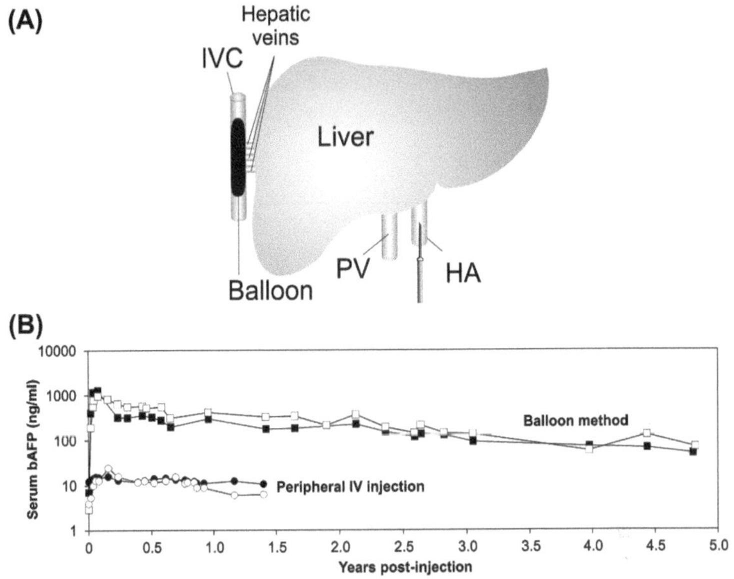


**Figure 2** (A) A sausage-shaped balloon catheter is positioned in the inferior vena cava (IVC) under fluoroscopic guidance in nonhuman primates (baboons). Inflation of the balloon results in hepatic venous outflow occlusion from the hepatic veins. The HDAd is administered by injection through a percutaneously positioned hepatic artery (HA) catheter. (B) Serum bAFP levels in baboons following administration of HDAd expressing the baboon α-fetoprotein (bAFP) as reporter gene either by the balloon method or by peripherial intravenous (IV) injection. Adapted from Brunetti-Pierri et al.[107]

pressure is achieved, mimicking the high pressures achieved by systemic hydrodynamic injection in mice. Following balloon deflation, the vector is then administered by peripheral intravenous injection in a small volume, which resulted in high efficiency hepatic transduction with minimal toxicity.[43] In a subsequent refined method, a balloon occlusion catheter was percutaneously positioned in the IVC to occlude hepatic venous outflow and HDAd was injected directly into the occluded liver via a percutaneously placed HA catheter (Figure 2(A)). This resulted in up to 80-fold improvement in hepatic transduction compared to systemic vector injection with negligible toxicity (Figure 2(B)).[107] This balloon catheter method was used to deliver a low dose of HDAd expressing human factor IX (hFIX) into rhesus macaques, which resulted in plasma hFIX levels within the therapeutic range for up to 2.8 years postinjection.[108] A follow-up of the three methods above reported that transgene expression had persisted for up to 7 years (which is more than half the life span of most captive baboons[109]) for all injected nonhuman primates without long-term adverse effects. However, in all cases, transgene expression levels slowly declined over time to less than 10% of peak values by the end of the observation period but remained 2.3- to 111-fold above baseline values (Figure 3).[46] The slow, steady decline in transgene expression over time is likely dependent on the gradual loss of transduced hepatocytes due to physiologic hepatocyte turnover, loss of the extrachromosomal vector genome, or a combination of both.

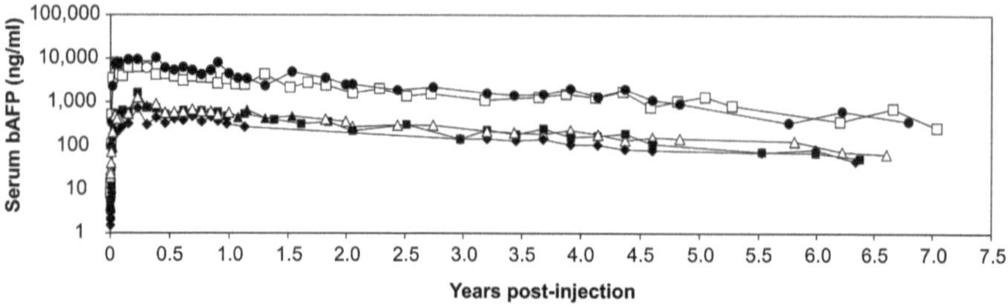


**Figure 3** Long-term transgene expression in nonhuman primates following injection of HDAd directly into the isolated liver. Duration of transgene expression following administration of an HDAd expressing the baboon α-fetoprotein (bAFP) in baboons by the method described previously.[44]
Adapted from Brunetti-Pierri et al.[46]

Physical or chemical methods to enlarge fenestration diameter could increase hepatocyte transduction with lower vector doses, thereby reducing the acute toxicity. For example, *N*-acetylcysteine combined with transient liver ischemia[83] and Na-decanoate[82] have been shown to increase the size of sinusoidal fenestrae and augment Ad-mediated hepatocyte transduction. As well, pretreatment with vasoactive intestinal peptide increases hepatic transduction and reduces the innate immune response following administration of HDAd.[110] Recently, Prill et al. showed that specific coupling of 5K PEG or transferrin to the hexon capsid protein of FGAd and HDAd can improve liver transduction by 11-fold and 18-fold, respectively.[111] The mechanism for this improvement appears to be evasion of Kupffer cells. Another recent study has shown that avoiding Kupffer cell uptake using a chimeric HDAd, in which the serotype 5 hypervariable region was replaced with that of serotype 6, increases liver transduction approximately 10-fold in BALB/c mice. Additionally, ALT levels were significantly lower in mice given the HDAd5/6 chimeric vector than mice receiving the HDAd5 vector.[21] Scavenger receptors on Kupffer cells bind Ad particles and remove them from the circulation, thus preventing hepatocyte transduction.[68,77] Piccolo et al. found that HDAd particles interact in vitro and in vivo with SR-A and with SREC-I. Interestingly, this knowledge was exploited to increase the efficiency of hepatocyte transduction and improve the HDAd therapeutic index in vivo through blocking of SR-A and SREC-I with specific antigen-binding fragments or small peptides.[80,81]

"Masking" the viral capsid has also been reported to attenuate the severity of the innate inflammatory response. Yotanda et al.[112] demonstrated that systemic injection of Ad encapsidated within bilamellar cationic liposomes resulted in a 70–80% decrease in serum IL-6 compared to unencapsidated virions without compromising hepatic transduction efficiency. Likewise, two independent groups were able to demonstrate that systemic administration of PEGylated Ad into mice resulted in a 50–70% reduction in serum IL-6 compared to unPEGylated vector without compromising hepatic transduction efficiency.[113–115] In another study the combination of methylprednisolone, an anti-inflammatory glucocorticoid, and PEGylated Ad potently inhibited IL-6 elevation.[116] Similarly, a single administration of dexamethasone, another anti-inflammatory glucocorticoid, prior to Ad administration significantly reduced both innate and adaptive immune responses.[117] These approaches, individually or in combination, may improve the therapeutic index of HDAds. Although the above studies showed that hepatic transduction in mice was not compromised by PEGylated HDAd, this was not the case in nonhuman primates where hepatic transduction was reduced,[118] emphasizing that caution should be taken in extrapolating results from rodents to larger animals and humans.

## 6. Brain and Eye Gene Therapy

Because of their ability to infect postmitotic cells, such as cells of the brain,[119] and to mediate long-term transgene expression, Ad-based vectors are particularly attractive for brain gene therapy. Intravascular delivery of Ad vectors is ineffective because the blood–brain barrier prevents access of bloodborne vector to the brain. Therefore Ad vectors for brain gene therapy have been investigated with direct intracranial injection.

Unlike the rapid decline observed in transgene expression in peripheral organs following intravenous administration, FGAd-mediated transduction of adult brain cells leads to stable transgene expression[120,121] because the brain is relatively protected from the effects of the immune response.[122] However, if a peripheral immune response against Ad is elicited after natural infection or vector readministration, loss of transgene expression and chronic inflammation are observed.[123] Interestingly, these negative effects are not seen with HDAds[123,124] that mediate significantly higher transgene expression levels and induce a substantially reduced inflammatory and immune response.[125]

For brain cancer applications, FGAd vectors have been used in clinical trials for glioblastoma multiforme and intratumoral injection has been associated with increased survival in two different trials.[126,127] Given the high risk that FGAd treatment of glioblastoma multiforme can be compromised by exposure to natural Ad infection, HDAd vectors encoding regulated therapeutic genes could offer a significantly safer and more effective treatment for patients with this type as well as other forms of brain cancer.[128–130]

For brain-directed gene therapy, helper-dependent canine adenovirus (CAV-2) vectors have also been developed and investigated particularly for therapy of lysosomal storage disorders, such as MPSIIIA and MPSVII.[131,132] These vectors preferentially transduce neurons, resulting in stable, high-level expression, and efficiently traffic via axonal retrograde transport.[133] The main advantages of helper-dependent CAV-2 vectors over serotype 5 HDAd vectors include the lower prevalence of preexisting humoral immunity because 98% of subjects are negative for neutralizing antibodies against CAV-2,[134] reduced activation of the innate response[135] and dendritic cells,[136] and neuronal retrograde transport.[137]

Eye gene therapy is attractive because affected patients are only exposed to low vector doses that remain confined to an immunologically privileged site, thus enhancing safety. Intravitreal injections of HDAd expressing antiangiogenic factors have been investigated to counteract ocular neovascularization occurring in disorders such as diabetic retinopathy and age-related macular degeneration.[138,139] These studies showed long-term therapeutic transgene expression, inhibition of retinal neovascularization, and substantially reduced inflammatory response with HDAd compared to FGAd. Subretinal injection of HDAd for ocular gene therapy has also been investigated and these studies show that HDAds transduce and rescue cells from the neurosensory retina in a mouse model of retinal degeneration.[140] Moreover, HDAds are able to mediate long-term expression of therapeutic genes in the retinal pigment epithelium (RPE) in mice following subretinal injection without evidence of adverse immune reactions or significant toxicity.[141] Ad-based vectors preferentially target the RPE following subretinal injections. However, long-term transduction of murine photoreceptors can be achieved by increasing the vector doses without adverse effects.[142] HDAd, because of its large cloning capacity, is particularly attractive for several inherited retinopathies due to large therapeutic genes that are beyond the cargo capacity of other vectors commonly used for eye gene therapy, such as the AAV. For example, Stargardt disease, the most common form of juvenile onset macular degeneration, is caused by mutations in the *ABCA4* gene, whose cDNA is 6.8 kb, and Leber's congenital amaurosis (LCA10) is due to mutations in *CEP290*, whose cDNA is 7.4 kb. Unfortunately, none of the Ad serotypes investigated so far has shown efficiency greater than AAV8 vectors for transduction of photoreceptors that are the target cells for treatment of inherited retinopathies.[143]

## 7. Lung Gene Therapy

Ad vectors have been extensively used for pulmonary gene transfer with the goal of treating cystic fibrosis (CF) due to mutations in the cystic fibrosis conductance regulator (CFTR) gene. However, there are a number of shortcomings that limit the efficacy of Ad vectors for lung-directed gene transfer. First, the pulmonary delivery of Ad in small animals, large animals, and humans is inefficient[144–148] because the cellular receptor for Ad (and other viral vectors) resides on the basolateral surface of the airway epithelial cells and tight junctions prevent vector–receptor interactions required for transduction.[149] However, transient disruption of the tight junctions can significantly increase the efficiency of transduction, thus dramatically decreasing the vector dose required to achieve therapeutic levels of transduction. Various strategies have been proposed to relax the tight junctions and improve Ad entry into airway epithelia; they include calcium phosphate coprecipitates,[150] EGTA,[151] EDTA,[152] polycations,[153] sodium caprate,[154] L-α-lysophosphatidylcholine (LPC),[155,156] and other agents. Second, pulmonary delivery of FGAd resulted in dose-dependent inflammation and pneumonia.[157–160] This has been attributed to expression of the viral genes present in the vector backbone of FGAds, which is directly cytotoxic and also provokes an adaptive cellular immune response against the transduced cells, consequently resulting in transient transgene expression and toxicity.[161–166]

In contrast to FGAd, lungs of mice receiving HDAd following EGTA to disrupt tight junctions by nasal instillations were free of inflammation and indistinguishable from mice instilled with saline.[167] Moreover, HDAd resulted in extensive transduction of proximal and distal airways; transduction was detected in the trachea, submucosal glands, and bronchiolar epithelium with little transduction of alveolar cells. Using the relatively large (4.1 kb) human cytokeratin 18 (K18) control elements, which can be easily accommodated into HDAds given their large cloning capacity, it was also possible to obtain a transgene expression pattern that is similar to that of CFTR[168] because

K18, like CFTR, is expressed in the epithelium of large airways and bronchioles and in submucosal glands, but has little expression in alveoli. Using an HDAd expressing the human α-fetoprotein (AFP) reporter gene[169] from the K18 promoter, pulmonary transgene expression was detected long term, for at least 15 weeks postadministration. These outcomes indicated that intranasal administration of HDAd following disruption of tight junctions resulted in high-efficiency pulmonary transduction, and use of the K18 promoter greatly restricts expression to the desired target cell types relevant to CF gene therapy. Moreover, HDAd can express properly localized CFTR in the appropriate target cell types for CF gene therapy in vivo and protects the airways from infections by opportunistic pathogens, such as *Burkholderia cepacia*.[170]

Despite the encouraging results in mice described above, the requirement for two separate administrations, one to deliver EGTA to disrupt the tight junction and then a second, 30 min later, to deliver the vector, is suboptimal in terms of safety and efficacy as this increases the procedure time and compromises transduction efficiency because both the EGTA and the vector must be applied to the same location, which is not guaranteed in the case of separate, independent administrations. This obstacle was addressed by Koehler et al. who demonstrated that high-efficiency pulmonary transduction by HDAds can be obtained by formulating the vector in 0.1% LPC, thus permitting a single administration containing the vector and the tight junction opening agent.[156] The intranasal delivery as performed in mice (spontaneous liquid inhalation) is not applicable to larger animals, and a clinically relevant method of vector delivery was therefore developed by Koehler et al.[156] In this study, an intracorporeal nebulizing catheter called the AeroProbe (Trudell Medical International) was developed to aerosolize material directly into the trachea and lungs. Using the AeroProbe, Koehler et al. aerosolized 1.25 ml of 0.1% LPC containing $5 \times 10^{11}$ vp HDAd-K18LacZ into the lungs of 2.8 kg rabbits.[156] The X-gal staining revealed extensive transduction of the trachea. Although the intratracheal aerosolization was intended to deliver vector to the entire lung, X-gal staining revealed high interlobular variation. Nevertheless, in those lobes that received vector, exceedingly high and unprecedented levels of transduction were achieved in all cell types of the proximal and distal airway epithelium, from the trachea to terminal bronchioles. All rabbits, including those given LPC only as controls, exhibited a transient decrease in dynamic lung compliance immediately following aerosol delivery. Fever and mild-to-moderate patchy pneumonia without edema were also observed, leading the investigators to speculate that LPC may have been a contributing factor and that these may be eliminated/minimized by further optimizing the dose of LPC and/or vector. Nevertheless, this study is significant because it is the first to demonstrate high efficiency transduction of the airway epithelium in a large animal, which had been a major obstacle to CF gene therapy.

## 8. Muscle Gene Therapy

The muscle is an attractive tissue for transduction because (1) it is the target tissue for therapy of Duchenne muscular dystrophy (DMD) due to mutations in the dystrophin gene, one of the most common genetic diseases without effective treatments, (2) it constitutes as much as 40% of the total body mass and much of it is readily accessible, (3) skeletal myocytes can be transduced in vivo, (4) skeletal myofibers have a relatively long half-life and therefore represent a stable platform for transgene expression, (5) muscle is highly vascularized and skeletal muscle can efficiently secrete recombinant proteins into the circulation for systemic delivery, and (6) the high seroprevalence in the adult human population of preexisting anti-Ad neutralizing antibodies, an obstacle for intravenous vector delivery, may be minimized through localized delivery into the muscle. The length of the dystrophin cDNA (14 kb) precluded its inclusion into most gene therapy viral vectors and HDAds with their large cloning capacity can accommodate not only one full-length dystrophin cDNA but also two copies of the gene.[171,172] HDAd vectors expressing full-length dystrophin have been shown to restore in the skeletal muscle the full dystrophin–glycoprotein complex, resulting in a reduced level of muscle degeneration and amelioration of the physiological and pathological indices of muscle disease.[171–175] Compared to dystrophin-deficient *mdx* mice treated as neonates, higher doses of vector are needed to treat adult *mdx* mice because of inefficient infection of mature muscle by Ad.[174] Moreover, direct intramuscular injection of HDAd encoding full-length murine dystrophin resulted in only transient expression in adult-injected mice[175] because of a humoral immune response against the dystrophin protein. Importantly, such a response has not been observed in immunodeficient SCID mice, suggesting that sustained expression could be achieved in the absence of an immune response to the transgene product.[175,176] A similar outcome could be predicted to occur in humans as well, since many DMD patients have large dystrophin gene deletions preventing the expression of epitopes encoded by deleted exons. In these cases the dystrophin produced by HDAd-transduced cells could be perceived by the immune system as a neoantigen, thus preventing long-term expression.[177]

An important hurdle that hampers muscle-directed gene therapy with Ad vectors is inefficient transduction of mature muscle presumably due to low CAR receptor levels. Incorporation of polylysine into the H-I loop of the Ad fiber protein can improve HDAd transduction of mature muscle cells, resulting in up to 21-fold increase in transduction compared to the unmodified counterpart.[178] In addition, HDAd bearing the fiber knob domain from Ad serotype 3 significantly improved skeletal muscle transduction following intramuscular injections in adult mice and mediated stable transgene expression for at least 1 year.[20]

## 9. Helper-Dependent Adenoviral Vectors as Genetic Vaccines

FGAds have been developed to express antigens and have proven to be valuable genetic vaccines. However, a number of recent studies have demonstrated that HDAd induces higher transgene-specific cytotoxic T-lymphocytes (CTL) and greater anti-transgene antibody response than FGAd by intraperitoneal,[179] intramuscular,[16,180,181] intravenous,[16] and intranasal[182] administrations. In addition, HDAd-based vaccines resulted in lower tissue damage and anti-Ad T cell responses than FGAd.[16] Importantly, Kron et al. showed that HDAd generates a potentially more potent multi-specific T cell response against the antigen transgene compared to FGAd, which stimulates a response primarily against an immunodominant epitope because expression of viral genes from the FGAd backbone induces CTLs against viral epitopes and, thus, restricts the specificity of the response against the transgene product.[178] Weaver et al. demonstrated that rhesus macaques vaccinated with HDAd expressing HIV-1 envelope were protected from subsequent mucosal simian HIV challenge.[17] Because most humans are seropositive for Ad serotype 5, HDAd vaccines based on serotype 5 may be minimally, if at all, effective. To overcome this, Weaver et al. used HDAd vaccines based on serotypes 1, 2, and 6 and showed that preexisting immunity to Ad serotype 5 in both mice and rhesus macaques did not prevent successful vaccination.[16,17] These studies demonstrate the potential utility of HDAd as a genetic vaccine. Furthermore, the large cloning capacity of HDAd can accommodate multiple antigen genes and thus may allow simultaneous immunization against multiple antigens from the same or from different pathogens.

## 10. Helper-Dependent Adenoviral Vectors and Stem Cells

Embryonic stem (ES) cells and induced pluripotent stem cells (iPSCs) have recently attracted much attention because of their self-renewal capacity and pluripotency that make them promising for cell-based therapies and regenerative medicine. Targeted gene repair/editing of stem cells offer the potential for autologous cell therapy for the treatment of a wide spectrum of human diseases. As a gene-targeting vector for stem cells, HDAds offers numerous advantages. HDAds can transduce stem cells with high efficiency and low cytotoxicity. The large cloning capacity of HDAds can accommodate very long regions of homology to the target chromosome locus to enhance homologous recombination and permit a wide range of choices for promoters and selectable markers. Moreover, HDAd vector may result in correction of multiple mutations at the target locus. Furthermore, highly efficient transduction by HDAd contributes to high targeting efficiency, thus allowing for smaller numbers of starting cells. Introduction of artificial double-strand breaks at the target loci, which are potentially mutagenic and may result in unpredictable and undetectable off-target effects, is not required. Indeed, high efficiency HDAd-mediated gene targeting has been accomplished in mouse ES cells,[33] monkey ES cells,[34] and into a wide variety of loci in different human ES cells and iPS cells.[183–186] Collectively, these studies showed that HDAd could mediate efficient knockin and knockout at transcriptionally active or inactive loci in human ES cells and iPS cells with no effect on undifferentiated state and pluripotency, no ectopic random integration/off-target effects, and/or introduction of additional mutations, while maintaining genetic and epigenetic integrity. Indeed, targeted gene correction of human iPS cells by HDAd minimally impacts whole-genome mutational load as determined by whole-genome sequencing at single-base resolution.[187] Importantly, the disease phenotype was also reversed in patient-derived cells after HDAd-mediated gene targeting.[183,184]

## 11. Human Gene Therapy with Helper-Dependent Adenoviral Vectors

There has been a single case of intravascular administration of HDAd into a human patient. In this clinical trial, $4.3 \times 10^{11}$ vp/kg of a HDAd expressing factor VIII (FVIII) was intravenously injected into a hemophilia A patient.[188] This subject developed grade 3 liver toxicity, marked increase in IL-6, thrombocytopenia, and laboratory signs of disseminated intravascular coagulopathy, but all these values returned to baseline by day 19 postinfusion. Unfortunately, no evidence of FVIII expression was detected.[188] Also unfortunate is that this study has yet to be published in a peer-reviewed format so that much of the details remain unknown.

HDAd has recently been used in an ex vivo clinical trial to treat patients with anemia secondary to chronic kidney failure.[189] In this phase I–II study, a small number of dermal fibroblasts were removed from the skin of patients with anemia secondary to chronic renal failure, transduced ex vivo with an HDAd expressing erythropoietin (EPO), and implanted autologously in the subcutaneous tissue under local anesthesia. Following HDAd transduction, the amount of EPO produced by the transduced cells was measured so that the precise number of transduced cells was implanted to achieve the predetermined blood levels of EPO. No adverse events were reported in this trial and, importantly, elevated hemoglobin levels were sustained for up to 1 year after a single treatment with the HDAd-transduced cells.[189]

## 12. Concluding Remarks

Dose-dependent activation of the innate inflammatory response by the viral capsid following intravascular delivery remains an important concern, particularly for liver-directed applications requiring high-vector doses to achieve clinically relevant phenotypic improvements. Although important knowledge has been recently gained about Ad–host interactions occurring following systemic intravascular administrations, further studies are clearly needed to completely elucidate the mechanism(s) of Ad-mediated activation of the innate inflammatory response. Perhaps with a better understanding of this phenomenon, more effective strategies can be developed to minimize, if not eliminate, this innate inflammatory response. In the meantime, in vivo applications that require very low and/or localized vector doses, or ex vivo gene and cell-based strategies that do not provoke an innate inflammatory response may hold immediate potential for clinical translation.

In all animal models studied, HDAd-transduced hepatocytes (as well as all other target cell types examined) are not destroyed by an adaptive cellular immune response, thus leading to multiyear transgene expression. However, whether this holds true for humans is not known, especially considering the outcomes of recent liver-directed clinical trials for FIX deficiency with AAV vectors.[190,191] AAV vectors, like HDAd, do not contain any viral genes and mediates long-term transgene expression following hepatocyte transduction in all animal models investigated. However, in humans,

AAV-mediated transgene expression from transduced hepatocytes is subject to killing by AAV-specific CTLs. The source of immunogen has been attributed to AAV capsid peptides derived directly from the injected vector particles.[190,192,193] Similar to AAV, HDAd capsid proteins derived from the administered particles may be a source of immunogen.[194,195] In this regard, Roth et al.[196] showed that HDAd transduction of dendritic cells in vitro can stimulate activation of anti-Ad CD8+ T cells. Indeed, Muruve et al.[94] showed that Ad-specific CTLs were generated following intravascular administration of HDAd into mice. Similarly, Kushwah et al.[197] showed that intranasal administration of HDAd resulted in Ad-specific CD8+ T cells. Collectively, these studies show that following administration into mice, HDAd can indeed provoke a CTL response directed against viral proteins derived from the capsid, independent of de novo viral protein synthesis. However, whether these Ad-specific CTLs will eliminate HDAd-transduced cells in vivo remains to be shown, and animal modeling may not be useful for addressing this important issue as it was not in the case of AAV vectors.[198–200]

## Acknowledgments

P.N. is supported by the National Institutes of Health (R01 DK067324 and P51OD011133). N.B.-P. is supported by the Italian Telethon Foundation (TCBMT3TELD), the European Research Council (IEMTx), and the Italian Ministry of Health (GR-2009-1594913).

## References

1. Parks RJ, Chen L, Anton M, Sankar U, Rudnicki MA, Graham FL. A helper-dependent adenovirus vector system: removal of helper virus by Cre-mediated excision of the viral packaging signal. *Proc Natl Acad Sci USA* 1996;**93**:13565–70.
2. Sandig V, Youil R, Bett AJ, Franlin LL, Oshima M, Maione D, et al. Optimization of the helper-dependent adenovirus system for production and potency in vivo. *Proc Natl Acad Sci USA* 2000;**97**:1002–7.
3. Parks RJ, Graham FL. A helper-dependent system for adenovirus vector production helps define a lower limit for efficient DNA packaging. *J Virol* 1997;**71**:3293–8.
4. Bett AJ, Prevec L, Graham FL. Packaging capacity and stability of human adenovirus type 5 vectors. *J Virol* 1993;**67**:5911–21.
5. Palmer D, Ng P. Improved system for helper-dependent adenoviral vector production. *Mol Ther* 2003;**8**:846–52.
6. Palmer D, Ng P. *Methods for the production and characterization of helper-dependent adenoviral vectors*. Cold Spring Harbor Press; 2007.
7. Palmer D, Ng P. Methods for the preparation of helper-dependent adenoviral vectors. *Methods Mol Med* 2008;**433**:33–53.
8. Ng P, Beauchamp C, Evelegh C, Parks R, Graham FL. Development of a FLP/frt system for generating helper-dependent adenoviral vectors. *Mol Ther* 2001;**3**:809–15.
9. Umana P, Gerdes CA, Stone D, Davis JR, Ward D, Castro MG, et al. Efficient FLPe recombinase enables scalable production of helper-dependent adenoviral vectors with negligible helper-virus contamination. *Nat Biotechnol* 2001;**19**:582–5.

10. Sargent KL, Ng P, Evelegh C, Graham FL, Parks RJ. Development of a size-restricted pIX-deleted helper virus for amplification of helper-dependent adenovirus vectors. *Gene Ther* 2004;**11**:504–11.
11. Parks RJ. Adenovirus protein IX: a new look at an old protein. *Mol Ther* 2005;**11**:19–25.
12. Alba R, Hearing P, Bosch A, Chillon M. Differential amplification of adenovirus vectors by flanking the packaging signal with attB/attP-ΦC31 sequences: implications for helper-dependent adenovirus production. *Virology* 2007;**367**:51–8.
13. Reddy PS, Sakhuja K, Ganesh S, Yang L, Kayda D, Brann T, et al. Sustained human factor VIII expression in hemophilia A mice following systemic delivery of a gutless adenoviral vector. *Mol Ther* 2002;**5**:63–73.
14. Suzuki T, Sasaki T, Yano K, Sakurai F, Kawabata K, Kondoh M, et al. Development of a recombinant adenovirus vector production system free of replication-competent adenovirus by utilizing a packaging size limit of the viral genome. *Virus Res* 2011;**158**:154–60.
15. Parks R, Evelegh C, Graham F. Use of helper-dependent adenoviral vectors of alternative serotypes permits repeat vector administration. *Gene Ther* 1999;**6**:1565–73.
16. Weaver EA, Nehete PN, Buchl SS, Senac JS, Palmer D, Ng P, et al. Comparison of replication-competent, first generation, and helper-dependent adenoviral vaccines. *PLoS One* 2009;**4**:e5059.
17. Weaver EA, Nehete PN, Nehete BP, Buchl SJ, Palmer D, Montefiori DC, et al. Protection against mucosal SHIV challenge by peptide and helper-dependent adenovirus vaccines. *Viruses* 2009;**1**:920.
18. Kim IH, Jozkowicz A, Piedra PA, Oka K, Chan L. Lifetime correction of genetic deficiency in mice with a single injection of helper-dependent adenoviral vector. *Proc Natl Acad Sci USA* 2001;**98**:13282–7.
19. Shayakhmetov DM, Li ZY, Gaggar A, Gharwan H, Ternovoi V, Sandig V, et al. Genome size and structure determine efficiency of postinternalization steps and gene transfer of capsid-modified adenovirus vectors in a cell-type-specific manner. *J Virol* 2004;**78**:10009–22.
20. Guse K, Suzuki M, Sule G, Bertin TK, Tyynismaa H, Ahola-Erkkila S, et al. Capsid-modified adenoviral vectors for improved muscle-directed gene therapy. *Hum Gene Ther* 2012;**23**:1065–70.
21. Khare R, May SM, Vetrini F, Weaver EA, Palmer D, Rosewell A, et al. Generation of a Kupffer cell-evading adenovirus for systemic and liver-directed gene transfer. *Mol Ther* 2011;**19**:1254–62.
22. Maione D, Della Rocca C, Giannetti P, D'Arrigo R, Liberatoscioli L, Franlin LL, et al. An improved helper-dependent adenoviral vector allows persistent gene expression after intramuscular delivery and overcomes preexisting immunity to adenovirus. *Proc Natl Acad Sci USA* 2001;**98**:5986–91.
23. Schiedner G, Hertel S, Johnston M, Biermann V, Dries V, Kochanek S. Variables affecting in vivo performance of high-capacity adenovirus vectors. *J Virol* 2002;**76**:1600–9.
24. Schiedner G, Morral N, Parks RJ, Wu Y, Koopmans SC, Langston C, et al. Genomic DNA transfer with a high-capacity adenovirus vector results in improved in vivo gene expression and decreased toxicity. *Nat Genet* 1998;**18**:180–3.
25. Jager L, Ehrhardt A. Persistence of high-capacity adenoviral vectors as replication-defective monomeric genomes in vitro and in murine liver. *Hum Gene Ther* 2009;**20**:883–96.
26. Kreppel F, Kochanek S. Long-term transgene expression in proliferating cells mediated by episomally maintained high-capacity adenovirus vectors. *J Virol* 2004;**78**:9–22.
27. Ross PJ, Kennedy MA, Parks RJ. Host cell detection of noncoding stuffer DNA contained in helper-dependent adenovirus vectors leads to epigenetic repression of transgene expression. *J Virol* 2009;**83**:8409–17.

28. Ross PJ, Kennedy MA, Christou C, Risco Quiroz M, Poulin KL, Parks RJ. Assembly of helper-dependent adenovirus DNA into chromatin promotes efficient gene expression. *J Virol* 2011;**85**:3950–8.
29. Ehrhardt A, Xu H, Kay MA. Episomal persistence of recombinant adenoviral vector genomes during the cell cycle in vivo. *J Virol* 2003;**77**:7689–95.
30. Harui A, Suzuki S, Kochanek S, Mitani K. Frequency and stability of chromosomal integration of adenovirus vectors. *J Virol* 1999;**73**:6141–6.
31. Stephen SL, Sivanandam VG, Kochanek S. Homologous and heterologous recombination between adenovirus vector DNA and chromosomal DNA. *J Gene Med* 2008;**10**:1176–89.
32. Hillgenberg M, Tonnies H, Strauss M. Chromosomal integration pattern of a helper-dependent minimal adenovirus vector with a selectable marker inserted into a 27.4-kilobase genomic stuffer. *J Virol* 2001;**75**:9896–908.
33. Ohbayashi F, Balamotis MA, Kishimoto A, Aizawa E, Diaz A, Hasty P, et al. Correction of chromosomal mutation and random integration in embryonic stem cells with helper-dependent adenoviral vectors. *Proc Natl Acad Sci USA* 2005;**102**:13628–33.
34. Suzuki K, Mitsui K, Aizawa E, Hasegawa K, Kawase E, Yamagishi T, et al. Highly efficient transient gene expression and gene targeting in primate embryonic stem cells with helper-dependent adenoviral vectors. *Proc Natl Acad Sci USA* 2008;**105**:13781–6.
35. Stephen SL, Montini E, Sivanandam VG, Al-Dhalimy M, Kestler HA, Finegold M, et al. Chromosomal integration of adenoviral vector DNA in vivo. *J Virol* 2010;**84**:9987–94.
36. Muther N, Noske N, Ehrhardt A. Viral hybrid vectors for somatic integration – are they the better solution? *Viruses* 2009;**1**:1295–324.
37. Gallaher SD, Gil JS, Dorigo O, Berk AJ. Robust in vivo transduction of a genetically stable Epstein–Barr virus episome to hepatocytes in mice by a hybrid viral vector. *J Virol* 2009;**83**:3249–57.
38. Dorigo O, Gil JS, Gallaher SD, Tan BT, Castro MG, Lowenstein PR, et al. Development of a novel helper-dependent adenovirus-Epstein-Barr virus hybrid system for the stable transformation of mammalian cells. *J Virol* 2004;**78**:6556–66.
39. Gil JS, Gallaher SD, Berk AJ. Delivery of an EBV episome by a self-circularizing helper-dependent adenovirus: long-term transgene expression in immunocompetent mice. *Gene Ther* 2010;**17**:1288–93.
40. Voigtlander R, Haase R, Muck-Hausl M, Zhang W, Boehme P, Lipps HJ, et al. A novel adenoviral hybrid-vector system carrying a plasmid replicon for safe and efficient cell and gene therapeutic applications. *Mol Ther Nucleic Acids* 2013;**2**:e83.
41. Toietta G, Mane VP, Norona WS, Finegold MJ, Ng P, McDonagh AF, et al. Lifelong elimination of hyperbilirubinemia in the Gunn rat with a single injection of helper-dependent adenoviral vector. *Proc Natl Acad Sci USA* 2005;**102**:3930–5.
42. Brunetti-Pierri N, Ng P. Helper-dependent adenoviral vectors for liver-directed gene therapy. *Hum Mol Genet* 2011;**20**:R7–13.
43. Brunetti-Pierri N, Stapleton GE, Palmer DJ, Zuo Y, Mane VP, Finegold MJ, et al. Pseudo-hydrodynamic delivery of helper-dependent adenoviral vectors into non-human primates for liver-directed gene therapy. *Mol Ther* 2007;**15**:732–40.
44. Brunetti-Pierri N, Ng T, Iannitti DA, Palmer DJ, Beaudet AL, Finegold MJ, et al. Improved hepatic transduction, reduced systemic vector dissemination, and long-term transgene expression by delivering helper-dependent adenoviral vectors into the surgically isolated liver of nonhuman primates. *Hum Gene Ther* 2006;**17**:391–404.
45. Morral N, O'Neal W, Rice K, Leland M, Kaplan J, Piedra PA, et al. Administration of helper-dependent adenoviral vectors and sequential delivery of different vector serotype for long-term liver-directed gene transfer in baboons. *Proc Natl Acad Sci USA* 1999;**96**:12816–21.

46. Brunetti-Pierri N, Ng T, Iannitti D, Cioffi W, Stapleton G, Law M, et al. Transgene expression up to 7 years in nonhuman primates following hepatic transduction with helper-dependent adenoviral vectors. *Hum Gene Ther* 2013;**24**:761–5.
47. Brunetti-Pierri N, Nichols TC, McCorquodale S, Merricks E, Palmer DJ, Beaudet AL, et al. Sustained phenotypic correction of canine hemophilia B after systemic administration of helper-dependent adenoviral vector. *Hum Gene Ther* 2005;**16**:811–20.
48. McCormack Jr WM, Seiler MP, Bertin TK, Ubhayakar K, Palmer DJ, Ng P, et al. Helper-dependent adenoviral gene therapy mediates long-term correction of the clotting defect in the canine hemophilia A model. *J Thromb Haemost* 2006;**4**:1218–25.
49. Bergelson JM, Cunningham JA, Droguett G, Kurt-Jones EA, Krithivas A, Hong JS, et al. Isolation of a common receptor for Coxsackie B viruses and adenoviruses 2 and 5. *Science* 1997;**275**:1320–3.
50. Tomko RP, Xu R, Philipson L. HCAR and MCAR: the human and mouse cellular receptors for subgroup C adenoviruses and group B coxsackieviruses. *Proc Natl Acad Sci USA* 1997;**94**:3352–6.
51. Wickham TJ, Mathias P, Cheresh DA, Nemerow GR. Integrins $\alpha_v\beta_3$ and $\alpha_v\beta_5$ promote adenovirus internalization but not virus attachment. *Cell* 1993;**73**:309–19.
52. Martin K, Brie A, Saulnier P, Perricaudet M, Yeh P, Vigne E. Simultaneous CAR- and alpha V integrin-binding ablation fails to reduce Ad5 liver tropism. *Mol Ther* 2003;**8**:485–94.
53. Alemany R, Curiel DT. CAR-binding ablation does not change biodistribution and toxicity of adenoviral vectors. *Gene Ther* 2001;**8**:1347–53.
54. Parker AL, Waddington SN, Nicol CG, Shayakhmetov DM, Buckley SM, Denby L, et al. Multiple vitamin K-dependent coagulation zymogens promote adenovirus-mediated gene delivery to hepatocytes. *Blood* 2006;**108**:2554–61.
55. Parker AL, McVey JH, Doctor JH, Lopez-Franco O, Waddington SN, Havenga MJ, et al. Influence of coagulation factor zymogens on the infectivity of adenoviruses pseudotyped with fibers from subgroup D. *J Virol* 2007;**81**:3627–31.
56. Shayakhmetov DM, Gaggar A, Ni S, Li ZY, Lieber A. Adenovirus binding to blood factors results in liver cell infection and hepatotoxicity. *J Virol* 2005;**79**:7478–91.
57. Waddington SN, Parker AL, Havenga M, Nicklin SA, Buckley SM, McVey JH, et al. Targeting of adenovirus serotype 5 (Ad5) and 5/47 pseudotyped vectors in vivo: fundamental involvement of coagulation factors and redundancy of CAR binding by Ad5. *J Virol* 2007;**81**:9568–71.
58. Waddington SN, McVey JH, Bhella D, Parker AL, Barker K, Atoda H, Pink R, et al. Adenovirus serotype 5 hexon mediates liver gene transfer. *Cell* 2008;**132**:397–409.
59. Duffy MR, Bradshaw AC, Parker AL, McVey JH, Baker AH. A cluster of basic amino acids in the factor X serine protease mediates surface attachment of adenovirus/FX complexes. *J Virol* 2011;**85**:10914–9.
60. Di Paolo NC, van Rooijen N, Shayakhmetov DM. Redundant and synergistic mechanisms control the sequestration of blood-born adenovirus in the liver. *Mol Ther* 2009;**17**:675–84.
61. Bradshaw AC, Coughlan L, Miller AM, Alba R, van Rooijen N, Nicklin SA, et al. Biodistribution and inflammatory profiles of novel penton and hexon double-mutant serotype 5 adenoviruses. *J Control Release* 2012;**164**:394–402.
62. Alba R, Bradshaw AC, Coughlan L, Denby L, McDonald RA, Waddington SN, et al. Biodistribution and retargeting of FX-binding ablated adenovirus serotype 5 vectors. *Blood* 2010;**116**:2656–64.
63. Xu Z, Qiu Q, Tian J, Smith JS, Conenello GM, Morita T, et al. Coagulation factor X shields adenovirus type 5 from attack by natural antibodies and complement. *Nat Med* 2013;**19**:452–7.
64. Qiu Q, Xu Z, Tian J, Moitra R, Gunti S, Notkins AL, et al. Impact of natural IgM concentration on gene therapy with adenovirus type 5 vectors. *J Virol* 2015;**89**:3412–6.

65. Tao N, Gao GP, Parr M, Johnston J, Baradet T, Wilson JM, et al. Sequestration of adenoviral vector by Kupffer cells leads to a nonlinear dose response of transduction in liver. *Mol Ther* 2001;**3**:28–35.
66. Schiedner G, Hertel S, Johnston M, Dries V, van Rooijen N, Kochanek S. Selective depletion or blockade of Kupffer cells leads to enhanced and prolonged hepatic transgene expression using high-capacity adenoviral vectors. *Mol Ther* 2003;**7**:35–43.
67. Khare R, Hillestad ML, Xu Z, Byrnes AP, Barry MA. Circulating antibodies and macrophages as modulators of adenovirus pharmacology. *J Virol* 2013;**87**:3678–86.
68. Xu Z, Tian J, Smith JS, Byrnes AP. Clearance of adenovirus by Kupffer cells is mediated by scavenger receptors, natural antibodies, and complement. *J Virol* 2008;**82**:11705–13.
69. Tian J, Xu Z, Smith JS, Hofherr SE, Barry MA, Byrnes AP. Adenovirus activates complement by distinctly different mechanisms in vitro and in vivo: indirect complement activation by virions in vivo. *J Virol* 2009;**83**:5648–58.
70. He JQ, Katschke Jr KJ, Gribling P, Suto E, Lee WP, Diehl L, et al. CRIg mediates early Kupffer cell responses to adenovirus. *J Leukoc Biol* 2013;**93**:301–6.
71. Di Paolo NC, Doronin K, Baldwin LK, Papayannopoulou T, Shayakhmetov DM. The transcription factor IRF3 triggers "defensive suicide" necrosis in response to viral and bacterial pathogens. *Cell Rep* 2013;**3**:1840–6.
72. Smith JS, Xu Z, Tian J, Stevenson SC, Byrnes AP. Interaction of systemically delivered adenovirus vectors with Kupffer cells in mouse liver. *Hum Gene Ther* 2008;**19**:547–54.
73. Schiedner G, Bloch W, Hertel S, Johnston M, Molojavyi A, Dries V, et al. A hemodynamic response to intravenous adenovirus vector particles is caused by systemic Kupffer cell-mediated activation of endothelial cells. *Hum Gene Ther* 2003;**14**:1631–41.
74. Nociari M, Ocheretina O, Schoggins JW, Falck-Pedersen E. Sensing infection by adenovirus: Toll-like receptor independent viral DNA recognition signals activation of the interferon regulatory factor 3 master regulator. *J Virol* 2007;**81**:4145–57.
75. Doronin K, Flatt JW, Di Paolo NC, Khare R, Kalyuzhniy O, Acchione M, et al. Coagulation factor X activates innate immunity to human species C adenovirus. *Science* 2012;**338**:795–8.
76. Xu Z, Smith JS, Tian J, Byrnes AP. Induction of shock after intravenous injection of adenovirus vectors: a critical role for platelet-activating factor. *Mol Ther* 2010;**18**:609–16.
77. Haisma HJ, Kamps JA, Kamps GK, Plantinga JA, Rots MG, Bellu AR. Polyinosinic acid enhances delivery of adenovirus vectors in vivo by preventing sequestration in liver macrophages. *J Gen Virol* 2008;**89**:1097–105.
78. Haisma HJ, Boesjes M, Beerens AM, van der Strate BW, Curiel DT, Pluddemann A, et al. Scavenger receptor A: a new route for adenovirus 5. *Mol Pharm* 2009;**6**:366–74.
79. Khare R, Reddy VS, Nemerow GR, Barry MA. Identification of adenovirus serotype 5 hexon regions that interact with scavenger receptors. *J Virol* 2012;**86**:2293–301.
80. Piccolo P, Vetrini F, Mithbaokar P, Grove NC, Bertin T, Palmer D, et al. SR-A and SREC-I are Kupffer and endothelial cell receptors for helper-dependent adenoviral vectors. *Mol Ther* 2013;**21**:767–74.
81. Piccolo P, Annunziata P, Mithbaokar P, Brunetti-Pierri N. SR-A and SREC-I binding peptides increase HDAd-mediated liver transduction. *Gene Ther* 2014;**21**:950–7.
82. Lievens J, Snoeys J, Vekemans K, Van Linthout S, de Zanger R, Collen D, et al. The size of sinusoidal fenestrae is a critical determinant of hepatocyte transduction after adenoviral gene transfer. *Gene Ther* 2004;**11**:1523–31.
83. Snoeys J, Lievens J, Wisse E, Jacobs F, Duimel H, Collen D, et al. Species differences in transgene DNA uptake in hepatocytes after adenoviral transfer correlate with the size of endothelial fenestrae. *Gene Ther* 2007;**14**:604–12.

84. Wisse E, Jacobs F, Topal B, Frederik P, De Geest B. The size of endothelial fenestrae in human liver sinusoids: implications for hepatocyte-directed gene transfer. *Gene Ther* 2008;**15**:1193–9.
85. Lyons M, Onion D, Green NK, Aslan K, Rajaratnam R, Bazan-Peregrino M, et al. Adenovirus type 5 interactions with human blood cells may compromise systemic delivery. *Mol Ther* 2006;**14**:118–28.
86. Carlisle RC, Di Y, Cerny AM, Sonnen AF, Sim RB, Green NK, et al. Human erythrocytes bind and inactivate type 5 adenovirus by presenting Coxsackievirus-adenovirus receptor and complement receptor 1. *Blood* 2009;**113**:1909–18.
87. Zhang Y, Chirmule N, Gao GP, Qian R, Croyle M, Joshi B, et al. Acute cytokine response to systemic adenoviral vectors in mice is mediated by dendritic cells and macrophages. *Mol Ther* 2001;**3**:697–707.
88. Schnell MA, Zhang Y, Tazelaar J, Gao GP, Yu QC, Qian R, et al. Activation of innate immunity in nonhuman primates following intraportal administration of adenoviral vectors. *Mol Ther* 2001;**3**:708–22.
89. Varnavski AN, Zhang Y, Schnell M, Tazelaar J, Louboutin JP, Yu QC, et al. Preexisting immunity to adenovirus in rhesus monkeys fails to prevent vector-induced toxicity. *J Virol* 2002;**76**:5711–9.
90. Raper SE, Chirmule N, Lee FS, Wivel NA, Bagg A, Gao GP, et al. Fatal systemic inflammatory response syndrome in a ornithine transcarbamylase deficient patient following adenoviral gene transfer. *Mol Genet Metab* 2003;**80**:148–58.
91. Muruve DA, Barnes MJ, Stillman IE, Libermann TA. Adenoviral gene therapy leads to rapid induction of multiple chemokines and acute neutrophil-dependent hepatic injury in vivo. *Hum Gene Ther* 1999;**10**:965–76.
92. Liu Q, Muruve DA. Molecular basis of the inflammatory response to adenovirus vectors. *Gene Ther* 2003;**10**:935–40.
93. Brunetti-Pierri N, Palmer DJ, Beaudet AL, Carey KD, Finegold M, Ng P. Acute toxicity after high-dose systemic injection of helper-dependent adenoviral vectors into nonhuman primates. *Hum Gene Ther* 2004;**15**:35–46.
94. Muruve DA, Cotter MJ, Zaiss AK, White LR, Liu Q, Chan T, et al. Helper-dependent adenovirus vectors elicit intact innate but attenuated adaptive host immune responses in vivo. *J Virol* 2004;**78**:5966–72.
95. Ramalingam R, Rafii S, Worgall S, Hackett NR, Crystal RG. Induction of endogenous genes following infection of human endothelial cells with an E1(−) E4(+) adenovirus gene transfer vector. *J Virol* 1999;**73**:10183–90.
96. Morral N, O'Neal WK, Rice K, Leland MM, Piedra PA, Aguilar-Cordova E, et al. Lethal toxicity, severe endothelial injury, and a threshold effect with high doses of an adenoviral vector in baboons. *Hum Gene Ther* 2002;**13**:143–54.
97. Higginbotham JN, Seth P, Blaese RM, Ramsey WJ. The release of inflammatory cytokines from human peripheral blood mononuclear cells in vitro following exposure to adenovirus variants and capsid. *Hum Gene Ther* 2002;**13**:129–41.
98. Cichon G, Boeckh-Herwig S, Schmidt HH, Wehnes E, Muller T, Pring-Akerblom P, et al. Complement activation by recombinant adenoviruses. *Gene Ther* 2001;**8**:1794–800.
99. Jiang H, Wang Z, Serra D, Frank MM, Amalfitano A. Recombinant adenovirus vectors activate the alternative complement pathway, leading to the binding of human complement protein C3 independent of anti-ad antibodies. *Mol Ther* 2004;**10**:1140–2.
100. Kiang A, Hartman ZC, Everett RS, Serra D, Jiang H, Frank MM, et al. Multiple innate inflammatory responses induced after systemic adenovirus vector delivery depend on a functional complement system. *Mol Ther* 2006;**14**:588–98.

101. Cerullo V, Seiler MP, Mane V, Brunetti-Pierri N, Clarke C, Bertin TK, et al. Toll-like receptor 9 triggers an innate immune response to helper-dependent adenoviral vectors. *Mol Ther* 2007;**15**:378–85.
102. Basner-Tschakarjan E, Gaffal E, O'Keeffe M, Tormo D, Limmer A, Wagner H, et al. Adenovirus efficiently transduces plasmacytoid dendritic cells resulting in TLR9-dependent maturation and IFN-α production. *J Gene Med* 2006;**8**:1300–6.
103. Appledorn DM, Patial S, Godbehere S, Parameswaran N, Amalfitano A. TRIF, and TRIF-interacting TLRs differentially modulate several adenovirus vector-induced immune responses. *J Innate Immun* 2009;**1**:376–88.
104. Appledorn DM, Patial S, McBride A, Godbehere S, Van Rooijen N, Parameswaran N, et al. Adenovirus vector-induced innate inflammatory mediators, MAPK signaling, as well as adaptive immune responses are dependent upon both TLR2 and TLR9 in vivo. *J Immunol* 2008;**181**:2134–44.
105. Suzuki M, Cerullo V, Bertin TK, Cela R, Clarke C, Guenther M, et al. MyD88-dependent silencing of transgene expression during the innate and adaptive immune response to helper-dependent adenovirus. *Hum Gene Ther* 2010;**21**:325–36.
106. Brunetti-Pierri N, Palmer DJ, Mane V, Finegold M, Beaudet AL, Ng P. Increased hepatic transduction with reduced systemic dissemination and proinflammatory cytokines following hydrodynamic injection of helper-dependent adenoviral vectors. *Mol Ther* 2005;**12**: 99–106.
107. Brunetti-Pierri N, Stapleton GE, Law M, Breinholt J, Palmer DJ, Zuo Y, et al. Efficient, long-term hepatic gene transfer using clinically relevant HDAd doses by balloon occlusion catheter delivery in nonhuman primates. *Mol Ther* 2009;**17**:327–33.
108. Brunetti-Pierri N, Liou A, Patel P, Palmer D, Grove N, Finegold M, et al. Balloon catheter delivery of helper-dependent adenoviral vector results in sustained, therapeutic hFIX expression in rhesus macaques. *Mol Ther* 2012;**20**:1863–70.
109. Martin LJ, Mahaney MC, Bronikowski AM, Dee Carey K, Dyke B, Comuzzie AG. Lifespan in captive baboons is heritable. *Mech Ageing Dev* 2002;**123**:1461–7.
110. Vetrini F, Brunetti-Pierri N, Palmer DJ, Bertin T, Grove NC, Finegold MJ, et al. Vasoactive intestinal peptide increases hepatic transduction and reduces innate immune response following administration of helper-dependent Ad. *Mol Ther* 2010;**18**:1339–45.
111. Prill JM, Espenlaub S, Samen U, Engler T, Schmidt E, Vetrini F, et al. Modifications of adenovirus hexon allow for either hepatocyte detargeting or targeting with potential evasion from Kupffer cells. *Mol Ther* 2011;**19**:83–92.
112. Yotnda P, Chen DH, Chiu W, Piedra PA, Davis A, Templeton NS, et al. Bilamellar cationic liposomes protect adenovectors from preexisting humoral immune responses. *Mol Ther* 2002;**5**:233–41.
113. Mok H, Palmer DJ, Ng P, Barry MA. Evaluation of polyethylene glycol modification of first-generation and helper-dependent adenoviral vectors to reduce innate immune responses. *Mol Ther* 2005;**11**:66–79.
114. Croyle MA, Le HT, Linse KD, Cerullo V, Toietta G, Beaudet A, et al. PEGylated helper-dependent adenoviral vectors: highly efficient vectors with an enhanced safety profile. *Gene Ther* 2005;**12**:579–87.
115. Leggiero E, Astone D, Cerullo V, Lombardo B, Mazzaccara C, Labruna G, et al. PEGylated helper-dependent adenoviral vector expressing human Apo A-I for gene therapy in LDLR-deficient mice. *Gene Ther* 2013;**20**:1124–30.
116. De Geest B, Snoeys J, Van Linthout S, Lievens J, Collen D. Elimination of innate immune responses and liver inflammation by PEGylation of adenoviral vectors and methylprednisolone. *Hum Gene Ther* 2005;**16**:1439–51.

117. Seregin SS, Appledorn DM, McBride AJ, Schuldt NJ, Aldhamen YA, Voss T, et al. Transient pretreatment with glucocorticoid ablates innate toxicity of systemically delivered adenoviral vectors without reducing efficacy. *Mol Ther* 2009;**17**:685–96.
118. Wonganan P, Clemens CC, Brasky K, Pastore L, Croyle MA. Species differences in the pharmacology and toxicology of PEGylated helper-dependent adenovirus. *Mol Pharm* 2011;**8**:78–92.
119. Persson A, Fan X, Widegren B, Englund E. Cell type- and region-dependent coxsackie adenovirus receptor expression in the central nervous system. *J Neurooncol* 2006;**78**:1–6.
120. Le Gal La Salle G, Robert JJ, Berrard S, Ridoux V, Stratford-Perricaudet LD, Perricaudet M, et al. An adenovirus vector for gene transfer into neurons and glia in the brain. *Science* 1993;**259**:988–90.
121. Davidson BL, Allen ED, Kozarsky KF, Wilson JM, Roessler BJ. A model system for in vivo gene transfer into the central nervous system using an adenoviral vector. *Nat Genet* 1993;**3**:219–23.
122. Byrnes AP, Wood MJ, Charlton HM. Role of T cells in inflammation caused by adenovirus vectors in the brain. *Gene Ther* 1996;**3**:644–51.
123. Thomas CE, Schiedner G, Kochanek S, Castro MG, Lowenstein PR. Peripheral infection with adenovirus causes unexpected long-term brain inflammation in animals injected intracranially with first-generation, but not with high-capacity, adenovirus vectors: toward realistic long-term neurological gene therapy for chronic diseases. *Proc Natl Acad Sci USA* 2000;**97**:7482–7.
124. Xiong W, Goverdhana S, Sciascia SA, Candolfi M, Zirger JM, Barcia C, et al. Regulatable gutless adenovirus vectors sustain inducible transgene expression in the brain in the presence of an immune response against adenoviruses. *J Virol* 2006;**80**:27–37.
125. Zou L, Yuan X, Zhou H, Lu H, Yang K. Helper-dependent adenoviral vector-mediated gene transfer in aged rat brain. *Hum Gene Ther* 2001;**12**:181–91.
126. Germano IM, Fable J, Gultekin SH, Silvers A. Adenovirus/herpes simplex-thymidine kinase/ganciclovir complex: preliminary results of a phase I trial in patients with recurrent malignant gliomas. *J Neurooncol* 2003;**65**:279–89.
127. Immonen A, Vapalahti M, Tyynela K, Hurskainen H, Sandmair A, Vanninen R, et al. AdvHSV-tk gene therapy with intravenous ganciclovir improves survival in human malignant glioma: a randomised, controlled study. *Mol Ther* 2004;**10**:967–72.
128. Candolfi M, Curtin JF, Xiong WD, Kroeger KM, Liu C, Rentsendorj A, et al. Effective high-capacity gutless adenoviral vectors mediate transgene expression in human glioma cells. *Mol Ther* 2006;**14**:371–81.
129. Candolfi M, Pluhar GE, Kroeger K, Puntel M, Curtin J, Barcia C, et al. Optimization of adenoviral vector-mediated transgene expression in the canine brain in vivo, and in canine glioma cells in vitro. *Neuro Oncol* 2007;**9**:245–58.
130. Muhammad AK, Xiong W, Puntel M, Farrokhi C, Kroeger KM, Salem A, et al. Safety profile of gutless adenovirus vectors delivered into the normal brain parenchyma: implications for a glioma phase 1 clinical trial. *Hum Gene Ther Methods* 2012;**23**:271–84.
131. Lau AA, Rozaklis T, Ibanes S, Luck AJ, Beard H, Hassiotis S, et al. Helper-dependent canine adenovirus vector-mediated transgene expression in a neurodegenerative lysosomal storage disorder. *Gene* 2012;**491**:53–7.
132. Ariza L, Gimenez-Llort L, Cubizolle A, Pages G, Garcia-Lareu B, Serratrice N, et al. Central nervous system delivery of helper-dependent canine adenovirus corrects neuropathology and behavior in mucopolysaccharidosis type VII mice. *Hum Gene Ther* 2014;**25**:199–211.
133. Bru T, Salinas S, Kremer EJ. An update on canine adenovirus type 2 and its vectors. *Viruses* 2010;**2**:2134–53.

134. Kremer EJ, Boutin S, Chillon M, Danos O. Canine adenovirus vectors: an alternative for adenovirus-mediated gene transfer. *J Virol* 2000;**74**:505–12.
135. Keriel A, Rene C, Galer C, Zabner J, Kremer EJ. Canine adenovirus vectors for lung-directed gene transfer: efficacy, immune response, and duration of transgene expression using helper-dependent vectors. *J Virol* 2006;**80**:1487–96.
136. Perreau M, Mennechet F, Serratrice N, Glasgow JN, Curiel DT, Wodrich H, et al. Contrasting effects of human, canine, and hybrid adenovirus vectors on the phenotypical and functional maturation of human dendritic cells: implications for clinical efficacy. *J Virol* 2007;**81**:3272–84.
137. Hnasko TS, Perez FA, Scouras AD, Stoll EA, Gale SD, Luquet S, et al. Cre recombinase-mediated restoration of nigrostriatal dopamine in dopamine-deficient mice reverses hypophagia and bradykinesia. *Proc Natl Acad Sci USA* 2006;**103**:8858–63.
138. Lamartina S, Cimino M, Roscilli G, Dammassa E, Lazzaro D, Rota R, et al. Helper-dependent adenovirus for the gene therapy of proliferative retinopathies: stable gene transfer, regulated gene expression and therapeutic efficacy. *J Gene Med* 2007;**9**:862–74.
139. Chen P, Hamilton M, Thomas CA, Kroeger K, Carrion M, Macgill RS, et al. Persistent expression of PEDF in the eye using high-capacity adenovectors. *Mol Ther* 2008;**16**:1986–94.
140. Kumar-Singh R, Farber DB. Encapsidated adenovirus mini-chromosome-mediated delivery of genes to the retina: application to the rescue of photoreceptor degeneration. *Hum Mol Genet* 1998;**7**:1893–900.
141. Kreppel F, Luther TT, Semkova I, Schraermeyer U, Kochanek S. Long-term transgene expression in the RPE after gene transfer with a high-capacity adenoviral vector. *Invest Ophthalmol Vis Sci* 2002;**43**:1965–70.
142. Wu L, Lam S, Cao H, Guan R, Duan R, van der Kooy D, et al. Subretinal gene delivery using helper-dependent adenoviral vectors. *Cell Biosci* 2011;**1**:15.
143. Puppo A, Cesi G, Marrocco E, Piccolo P, Jacca S, Shayakhmetov DM, et al. Retinal transduction profiles by high-capacity viral vectors. *Gene Ther* 2014;**21**:855–65.
144. Joseph PM, O'Sullivan BP, Lapey A, Dorkin H, Oren J, Balfour R, et al. Aerosol and lobar administration of a recombinant adenovirus to individuals with cystic fibrosis. I. Methods, safety, and clinical implications. *Hum Gene Ther* 2001;**12**:1369–82.
145. Grubb BR, Pickles RJ, Ye H, Yankaskas JR, Vick RN, Engelhardt JF, et al. Inefficient gene transfer by adenovirus vector to cystic fibrosis airway epithelia of mice and humans. *Nature* 1994;**371**:802–6.
146. Harvey BG, Leopold PL, Hackett NR, Grasso TM, Williams PM, Tucker AL, et al. Airway epithelial CFTR mRNA expression in cystic fibrosis patients after repetitive administration of a recombinant adenovirus. *J Clin Invest* 1999;**104**:1245–55.
147. Zuckerman JB, Robinson CB, McCoy KS, Shell R, Sferra TJ, Chirmule N, et al. A phase I study of adenovirus-mediated transfer of the human cystic fibrosis transmembrane conductance regulator gene to a lung segment of individuals with cystic fibrosis. *Hum Gene Ther* 1999;**10**:2973–85.
148. Perricone MA, Morris JE, Pavelka K, Plog MS, O'Sullivan BP, Joseph PM, et al. Aerosol and lobar administration of a recombinant adenovirus to individuals with cystic fibrosis. II. Transfection efficiency in airway epithelium. *Hum Gene Ther* 2001;**12**:1383–94.
149. Pickles RJ, Fahrner JA, Petrella JM, Boucher RC, Bergelson JM. Retargeting the coxsackievirus and adenovirus receptor to the apical surface of polarized epithelial cells reveals the glycocalyx as a barrier to adenovirus-mediated gene transfer. *J Virol* 2000;**74**:6050–7.
150. Lee JH, Zabner J, Welsh MJ. Delivery of an adenovirus vector in a calcium phosphate coprecipitate enhances the therapeutic index of gene transfer to airway epithelia. *Hum Gene Ther* 1999;**10**:603–13.

151. Chu Q, St George JA, Lukason M, Cheng SH, Scheule RK, Eastman SJ. EGTA enhancement of adenovirus-mediated gene transfer to mouse tracheal epithelium in vivo. *Hum Gene Ther* 2001;**12**:455–67.
152. Wang G, Zabner J, Deering C, Launspach J, Shao J, Bodner M, et al. Increasing epithelial junction permeability enhances gene transfer to airway epithelia in vivo. *Am J Respir Cell Mol Biol* 2000;**22**:129–38.
153. Kaplan JM, Pennington SE, St George JA, Woodworth LA, Fasbender A, Marshall J, et al. Potentiation of gene transfer to the mouse lung by complexes of adenovirus vector and polycations improves therapeutic potential. *Hum Gene Ther* 1998;**9**:1469–79.
154. Johnson LG, Vanhook MK, Coyne CB, Haykal-Coates N, Gavett SH. Safety and efficiency of modulating paracellular permeability to enhance airway epithelial gene transfer in vivo. *Hum Gene Ther* 2003;**14**:729–47.
155. Limberis M, Anson DS, Fuller M, Parsons DW. Recovery of airway cystic fibrosis transmembrane conductance regulator function in mice with cystic fibrosis after single-dose lentivirus-mediated gene transfer. *Hum Gene Ther* 2002;**13**:1961–70.
156. Koehler DR, Frndova H, Leung K, Louca E, Palmer D, Ng P, et al. Aerosol delivery of an enhanced helper-dependent adenovirus formulation to rabbit lung using an intratracheal catheter. *J Gene Med* 2005;**7**:1409–20.
157. Simon RH, Engelhardt JF, Yang Y, Zepeda M, Weber-Pendleton S, Grossman M, et al. Adenovirus-mediated transfer of the CFTR gene to lung of nonhuman primates: toxicity study. *Hum Gene Ther* 1993;**4**:771–80.
158. Yei S, Mittereder N, Wert S, Whitsett JA, Wilmott RW, Trapnell BC. In vivo evaluation of the safety of adenovirus-mediated transfer of the human cystic fibrosis transmembrane conductance regulator cDNA to the lung. *Hum Gene Ther* 1994;**5**:731–44.
159. Wilmott RW, Amin RS, Perez CR, Wert SE, Keller G, Boivin GP, et al. Safety of adenovirus-mediated transfer of the human cystic fibrosis transmembrane conductance regulator cDNA to the lungs of nonhuman primates. *Hum Gene Ther* 1996;**7**:301–18.
160. Harvey BG, Maroni J, O'Donoghue KA, Chu KW, Muscat JC, Pippo AL, et al. Safety of local delivery of low- and intermediate-dose adenovirus gene transfer vectors to individuals with a spectrum of morbid conditions. *Hum Gene Ther* 2002;**13**:15–63.
161. Yang Y, Nunes FA, Berencsi K, Furth EE, Gonczol E, Wilson JM. Cellular immunity to viral antigens limits E1-deleted adenoviruses for gene therapy. *Proc Natl Acad Sci USA* 1994;**91**:4407–11.
162. Dai Y, Schwarz EM, Gu D, Zhang WW, Sarvetnick N, Verma IM. Cellular and humoral immune responses to adenoviral vectors containing factor IX gene: tolerization of factor IX and vector antigens allows for long-term expression. *Proc Natl Acad Sci USA* 1995;**92**:1401–5.
163. Morral N, O'Neal W, Zhou H, Langston C, Beaudet A. Immune responses to reporter proteins and high viral dose limit duration of expression with adenoviral vectors: comparison of E2a wild type and E2a deleted vectors. *Hum Gene Ther* 1997;**8**:1275–86.
164. O'Neal WK, Zhou H, Morral N, Aguilar-Cordova E, Pestaner J, Langston C, et al. Toxicological comparison of E2a-deleted and first-generation adenoviral vectors expressing $\alpha_1$-antitrypsin after systemic delivery. *Hum Gene Ther* 1998;**9**:1587–98.
165. Yang Y, Li Q, Ertl HC, Wilson JM. Cellular and humoral immune responses to viral antigens create barriers to lung-directed gene therapy with recombinant adenoviruses. *J Virol* 1995;**69**:2004–15.
166. Yang Y, Xiang Z, Ertl HC, Wilson JM. Upregulation of class I major histocompatibility complex antigens by interferon gamma is necessary for T-cell-mediated elimination of recombinant adenovirus-infected hepatocytes in vivo. *Proc Natl Acad Sci USA* 1995;**92**:7257–61.

167. Toietta G, Koehler DR, Finegold MJ, Lee B, Hu J, Beaudet AL. Reduced inflammation and improved airway expression using helper-dependent adenoviral vectors with a K18 promoter. *Mol Ther* 2003;**7**:649–58.
168. Koehler DR, Hannam V, Belcastro R, Steer B, Wen Y, Post M, et al. Targeting transgene expression for cystic fibrosis gene therapy. *Mol Ther* 2001;**4**:58–65.
169. O'Neal WK, Rose E, Zhou H, Langston C, Rice K, Carey D, et al. Multiple advantages of α-fetoprotein as a marker for in vivo gene transfer. *Mol Ther* 2000;**2**:640–8.
170. Koehler DR, Sajjan U, Chow YH, Martin B, Kent G, Tanswell AK, et al. Protection of Cftr knockout mice from acute lung infection by a helper-dependent adenoviral vector expressing Cftr in airway epithelia. *Proc Natl Acad Sci USA* 2003;**100**:15364–9.
171. Dudley RW, Lu Y, Gilbert R, Matecki S, Nalbantoglu J, Petrof BJ, et al. Sustained improvement of muscle function one year after full-length dystrophin gene transfer into mdx mice by a gutted helper-dependent adenoviral vector. *Hum Gene Ther* 2004;**15**:145–56.
172. Gilbert R, Dudley RW, Liu AB, Petrof BJ, Nalbantoglu J, Karpati G. Prolonged dystrophin expression and functional correction of mdx mouse muscle following gene transfer with a helper-dependent (gutted) adenovirus-encoding murine dystrophin. *Hum Mol Genet* 2003;**12**:1287–99.
173. Clemens PR, Kochanek S, Sunada Y, Chan S, Chen HH, Campbell KP, et al. In vivo muscle gene transfer of full-length dystrophin with an adenoviral vector that lacks all viral genes. *Gene Ther* 1996;**3**:965–72.
174. DelloRusso C, Scott JM, Hartigan-O'Connor D, Salvatori G, Barjot C, Robinson AS, et al. Functional correction of adult mdx mouse muscle using gutted adenoviral vectors expressing full-length dystrophin. *Proc Natl Acad Sci USA* 2002;**99**:12979–84.
175. Gilchrist SC, Ontell MP, Kochanek S, Clemens PR. Immune response to full-length dystrophin delivered to Dmd muscle by a high-capacity adenoviral vector. *Mol Ther* 2002;**6**:359–68.
176. Gilbert R, Liu A, Petrof B, Nalbantoglu J, Karpati G. Improved performance of a fully gutted adenovirus vector containing two full-length dystrophin cDNAs regulated by a strong promoter. *Mol Ther* 2002;**6**:501–9.
177. Ohtsuka Y, Udaka K, Yamashiro Y, Yagita H, Okumura K. Dystrophin acts as a transplantation rejection antigen in dystrophin-deficient mice: implication for gene therapy. *J Immunol* 1998;**160**:4635–40.
178. Bramson JL, Grinshtein N, Meulenbroek RA, Lunde J, Kottachchi D, Lorimer IA, et al. Helper-dependent adenoviral vectors containing modified fiber for improved transduction of developing and mature muscle cells. *Hum Gene Ther* 2004;**15**:179–88.
179. Harui A, Roth MD, Kiertscher SM, Mitani K, Basak SK. Vaccination with helper-dependent adenovirus enhances the generation of transgene-specific CTL. *Gene Ther* 2004;**11**:1617–26.
180. Kron MW, Engler T, Schmidt E, Schirmbeck R, Kochanek S, Kreppel F. High-capacity adenoviral vectors circumvent the limitations of ΔE1 and ΔE1/ΔE3 adenovirus vectors to induce multispecific transgene product-directed CD8 T-cell responses. *J Gene Med* 2011;**13**:648–57.
181. Zong S, Kron MW, Epp C, Engler T, Bujard H, Kochanek S, et al. ΔE1 and high-capacity adenoviral vectors expressing full-length codon-optimized merozoite surface protein 1 for vaccination against *Plasmodium falciparum*. *J Gene Med* 2011;**13**:670–9.
182. Fu YH, He JS, Zheng XX, Wang XB, Xie C, Shi CX, et al. Intranasal vaccination with a helper-dependent adenoviral vector enhances transgene-specific immune responses in BALB/c mice. *Biochem Biophys Res Commun* 2010;**391**:857–61.
183. Liu GH, Suzuki K, Qu J, Sancho-Martinez I, Yi F, Li M, et al. Targeted gene correction of laminopathy-associated LMNA mutations in patient-specific IPSCs. *Cell Stem Cell* 2011;**8**:688–94.

184. Li M, Suzuki K, Qu J, Saini P, Dubova I, Yi F, et al. Efficient correction of hemoglobinopathy-causing mutations by homologous recombination in integration-free patient IPSCs. *Cell Res* 2011;**21**:1740–4.
185. Aizawa E, Hirabayashi Y, Iwanaga Y, Suzuki K, Sakurai K, Shimoji M, et al. Efficient and accurate homologous recombination in hESCs and hiPSCs using helper-dependent adenoviral vectors. *Mol Ther* 2012;**20**:424–31.
186. Umeda K, Suzuki K, Yamazoe T, Shiraki N, Higuchi Y, Tokieda K, et al. Albumin gene targeting in human embryonic stem cells and induced pluripotent stem cells with helper-dependent adenoviral vector to monitor hepatic differentiation. *Stem Cell Res* 2013;**10**:179–94.
187. Suzuki K, Yu C, Qu J, Li M, Yao X, Yuan T, et al. Targeted gene correction minimally impacts whole-genome mutational load in human-disease-specific induced pluripotent stem cell clones. *Cell Stem Cell* 2014;**15**:31–6.
188. White GI, Monahan PE. *Gene therapy for hemophilia A*. Oxford, (UK): Blackwell Publishing Ltd.; 2007.
189. Mitrani E, Pearlman A, Stern B, Miari R, Goltsman H, Kunicher N, et al. Biopump: Autologous skin-derived micro-organ genetically engineered to provide sustained continuous secretion of therapeutic proteins. *Dermatol Ther* 2011;**24**:489–97.
190. Manno CS, Pierce GF, Arruda VR, Glader B, Ragni M, Rasko JJ, et al. Successful transduction of liver in hemophilia by AAV-Factor IX and limitations imposed by the host immune response. *Nat Med* 2006;**12**:342–7.
191. Nathwani AC, Tuddenham EG, Rangarajan S, Rosales C, McIntosh J, Linch DC, et al. Adenovirus-associated virus vector-mediated gene transfer in hemophilia B. *N Engl J Med* 2011;**365**:2357–65.
192. Mingozzi F, Maus MV, Hui DJ, Sabatino DE, Murphy SL, Rasko JE, et al. $CD8^+$ T-cell responses to adeno-associated virus capsid in humans. *Nat Med* 2007;**13**:419–22.
193. Pien GC, Basner-Tschakarjan E, Hui DJ, Mentlik AN, Finn JD, Hasbrouck NC, et al. Capsid antigen presentation flags human hepatocytes for destruction after transduction by adeno-associated viral vectors. *J Clin Invest* 2009;**119**:1688–95.
194. Smith CA, Woodruff LS, Kitchingman GR, Rooney CM. Adenovirus-pulsed dendritic cells stimulate human virus-specific T-cell responses in vitro. *J Virol* 1996;**70**:6733–40.
195. Molinier-Frenkel V, Gahery-Segard H, Mehtali M, Le Boulaire C, Ribault S, Boulanger P, et al. Immune response to recombinant adenovirus in humans: capsid components from viral input are targets for vector-specific cytotoxic T lymphocytes. *J Virol* 2000;**74**:7678–82.
196. Roth MD, Cheng Q, Harui A, Basak SK, Mitani K, Low TA, et al. Helper-dependent adenoviral vectors efficiently express transgenes in human dendritic cells but still stimulate antiviral immune responses. *J Immunol* 2002;**169**:4651–6.
197. Kushwah R, Cao H, Hu J. Characterization of pulmonary T cell response to helper-dependent adenoviral vectors following intranasal delivery. *J Immunol* 2008;**180**:4098–108.
198. Wang L, Figueredo J, Calcedo R, Lin J, Wilson JM. Cross-presentation of adeno-associated virus serotype 2 capsids activates cytotoxic T cells but does not render hepatocytes effective cytolytic targets. *Hum Gene Ther* 2007;**18**:185–94.
199. Li H, Murphy SL, Giles-Davis W, Edmonson S, Xiang Z, Li Y, et al. Pre-existing AAV capsid-specific $CD8^+$ T cells are unable to eliminate AAV-transduced hepatocytes. *Mol Ther* 2007;**15**:792–800.
200. Li H, Lin SW, Giles-Davis W, Li Y, Zhou D, Xiang ZQ, et al. A preclinical animal model to assess the effect of pre-existing immunity on AAV-mediated gene transfer. *Mol Ther* 2009;**17**:1215–24.

# Hybrid Adenoviral Vectors

**18**

*Stephen J. Murphy, Richard G. Vile*
Molecular Medicine Program, Mayo Clinic, Rochester, MN, USA

## 1. Introduction

The characterization of disease at the genetic level facilitates potential genotypic and/or phenotypic correction by gene therapy. Although the concept of gene therapy has been extensively established over the past two decades, the development of effective clinical protocols to facilitate efficacious reversal of disease has proven highly problematic. The development of an effective gene delivery system to the site of therapeutic significance has proven to be the major hurdle to the advancement of gene therapies. Many questions currently remain unanswered and these raise major debates over the best vector systems to treat a specific clinical disorder, and at a more fundamental level the choice of gene to be applied. The ultimate goal of a gene therapy protocol is the efficient targeted delivery of a therapeutic transgene, whose expression can be sufficiently regulated, in a defective tissue. Vector delivery would ideally involve a single, lifetime treatment by a simple, non-invasive, and safe protocol, which can be incorporated into clinical practice. The vast array of clinical diseases, for which gene therapy presents clinical promise, demands a multitude of different requirements for a vector system to meet.

Ideologies for gene therapy vectors will differ considerably among different disorders. The treatment of severely disabling genetic disorders such as Duchenne muscular dystrophy would require lifelong genetic complementation of the defective genes in an immense amount of both skeletal and smooth muscular tissues, as well as brain tissues to correct cognitive functions. Whereas somatic gene therapy for hemophilia B holds out greater potential for treatment, only a few percent of normal reversed phenotype cells would be sufficient to provide a constant level of factor IX in plasma, offering patients significant clinical improvements. In contrast to the aim of preservation of host physiology for inherited disorders, gene therapy for cancer focuses on efficient cell killing (Table 1). Hence genetic cancer therapies require different vector functions, requiring initial high local transduction of primary tumor masses to effect clinical removal, followed by subsequent systemic vector surveillance to eliminate metastatic disease. In essence, ideological concepts are rarely fully achieved, and the current minimal aim of gene therapy is reversal of clinical phenotypes to an extent of easy maintenance, facilitating improvements in standards of life for patients.

Despite the development of increasingly complex nonviral gene delivery systems, it is virally derived vector systems that still offer most promise to the clinic. Viruses throughout the evolution have developed highly skilled methods of entering cells, evading the host immune defense, and delivering their viral payloads.

**Adenoviral Vectors for Gene Therapy.**

## Table 1 Comparison of the Ideologies of a Gene Therapy Vector for Genetic Disorders and Cancer

| |
|---|
| ***Genetic Disorder*** |
| **Aim:** Cell preservation<br>Targeting diseased tissues<br>Efficient transduction of affected cells<br>Therapeutic levels of transgene expression<br>Adequate maintenance of gene expression levels<br>Long-term, stable transgene expression<br>Minimal vector toxicity |
| ***Cancer*** |
| **Aim:** Cell eradication<br>Targeting diseased tissues<br>Efficient transduction of tumor cells<br>Therapeutic levels of transgene expression<br>Transient vector expression for tumor clearance<br>Vector toxicity—danger signals attack tumor cells |

Hence phenomenal amounts of research have been directed at harnessing the finely tuned transduction functions and obligate parasite lifestyles of viruses. A plethora of genetically modified viral vector systems have now been reported, all ingeniously subverting the parasitic viral life cycles for the presentation of therapeutic transgenes aimed at reversal of disease phenotype. The development of viruses as clinical vectors will revolutionize the medical world, providing an invaluable new tool for the treatment of disease. Our present understanding of the molecular genetics of many viruses renders possible their manipulation as cloning vectors for gene transfer both in cell culture and directly in patients. As the major objective is usually long-lasting gene transfer, a deletion of the key regulatory viral genes was deemed essential to manipulate the genetic program of the virus and to ensure that infection of the target cell does not lead to cell death. Conversely, for the treatment of cancer, more recent strategies have reversed this thinking and selectively retain the replicative functions of the virus to enhance tumor cell killing. Viruses have thus been designed with predictable biological properties, retaining the beneficial targeting/infectivity properties, while dissociating them from the major virulent determinants of pathology in normal tissues.

Currently four classes of viral vector have presented most promise as gene delivery vehicles: retrovirus (RV), adenovirus (Ad), adeno-associated virus (AAV) and herpes simplex-based virus (HSV). Although retroviruses embodied the pioneering vector when the concept of gene therapy began to emerge as a reality in the early 1980s, adenoviruses have since become the major vector choice in the clinic. More recent advances in the production technologies of HSV-and AAV-based vectors have greatly increased their clinical potentials. Additionally, the lentiviral (LV) subclass of retroviral vectors, with distinct biological properties, has emerged with great potential and has gained individual acclaim from the rest of the group. The major properties of each viral vector are presented in Table 2, as well as being briefly discussed below.

| | Adenovirus Ad2/Ad5 | Retrovirus: MoMuLV | AAV AAV2 | HSV HSV-1 | Lentivirus HSV-1 |
|---|---|---|---|---|---|
| Nuclear status | Episomal | Integrating | Episomal/integrating | Episomal | Integrating |
| Genome | dsDNA | RNA | ssDNA | dsDNA | RNA |
| Structure | Encapsidated | Enveloped | Encapsidated | Enveloped/encapsidated | Enveloped |
| Genome size | 36 kb | 10 kb | 4.7 kb | 152 kb | 10 kb |
| Nuclear targeting | Yes | No | Yes | Yes | Yes |
| Titers | $10^{11}$ | $10^8$–$10^9$ | $10^{11}$ | $10^8$–$10^9$ | $10^8$–$10^9$ |
| Insert capacity | 7–8 kb (E1/3-) to 35 kb HD | 9–10 kb | 4.5–5 kb | 25–150 kb HD | 9–10 kb |
| Immunogenicity | High | Low | Low | Low (HD) | Low |
| Genetics | Early: E1–E4-transduction<br>Late: multiple structural | *Gag*: Capsid<br>*Pol*: RT, INT, PR<br>*Env*: envelope proteins | *Rep* replication/integration cap; capsid | Multiple | *Gag*, *Pol*, and *Env* as retrovirus plus accessory genes. |
| Pros | • High transduction efficiencies.<br>• Stable particles<br>• High titers<br>• Broad host range<br>• Infect dividing/non-dividing cells<br>• Large insert capacity: HD | • Low IR<br>• Integrating<br>• Potential long term expression | • High transduction efficiencies.<br>• Stable particles<br>• High titers<br>• Targeted integration (+*Rep*)<br>• Broad host range<br>• Infect dividing/non-dividing cells<br>• Large insert capacity: HD<br>• Limited IR | • Wide tropism<br>• Large insert capacities<br>• Minimal IR to virion particles<br>• High tropism for neuronal tissues<br>• Large genome copy number in amplicon concatomers | • Integrating<br>• Nuclear localization<br>• Easily pseudotyped with MoMuLV |
| Cons | • Transient<br>• High IR to viral proteins<br>• Humoral IR<br>• Cytotoxic IR<br>• Innate IR<br>• Leaky expression of viral genes (first gen.)<br>• Infect dividing/non-dividing cells<br>• RCA | • Relatively low titers<br>• Low transduction efficiencies in vivo<br>• Virions relatively unstable<br>• Complement-inactivation<br>• No nuclear targeting<br>• Transduction dep. On cell division<br>• Random integration<br>• Endogenous RVs recombination | • Low insert capacity<br>• *Rep* toxicity<br>• Helper contamination | • Transient<br>• Low titers<br>• No stable maintenance mechanism<br>• Helper contamination-<br>• High IR<br>• Unpredictable biology of concatomers in amplicon | • Transient<br>• Comparatively low titers<br>• Significant pathology of native virus<br>• RCL<br>• Limited host range<br>• Virions relatively unstable<br>• Endogenous RV recombination |

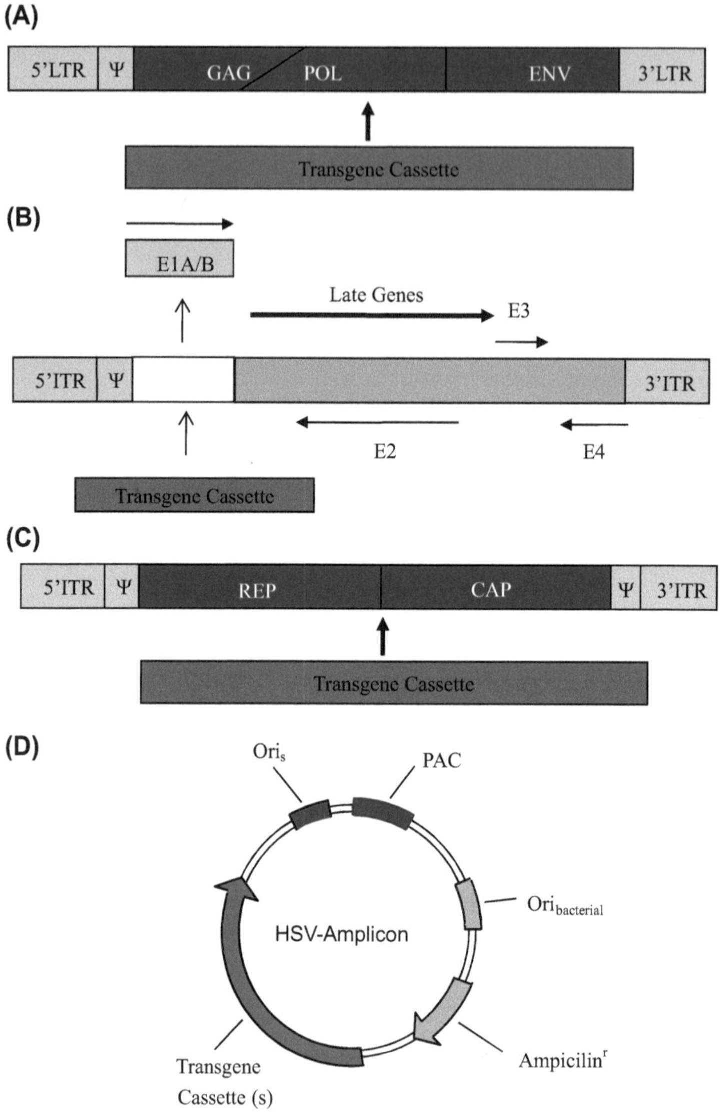


**Fig. 1 Vector genome structures.** The wild type viral genomes and the strategy of transgene substitution are presented for (A) retrovirus, (B) Adenovirus and (C) Adeno-associated virus. (D) represents the minimal structure of the HSV-1 based amplicon vector.

## 1.1 Retroviral Vectors

Retroviruses are enveloped RNA viruses, whose genomes consist of three core genetic units: *gag*, *pol*, and *env* (Fig. 1(A)).[1] Retroviruses stably transduce cells by integrating their genomes into the host cell chromosomes and subsequently release progeny virus by continuously budding viral particles from the cell membrane. The *gag* gene encodes proteins which form the viral core, while the *pol* gene encodes reverse transcriptase (RT), the viral integrase (INT), and a viral protease which acts on the *gag* gene products. The *env* gene encodes the glycosylated envelope proteins that determine the tropism of the virus. These genetic elements are flanked by the long-terminal repeat (LTR) sequences and a packaging signal (ψ) which directs the assembly of the genome into the viral particles (Fig. 1(A)).[1] The LTR sequences contain the *cis*-acting elements required to regulate viral genome replication and transcription and to mediate stable integration into the host genome.[1] Retroviral vectors have been principally based on the well-studied Moloney murine leukemia virus (MoMuLV). Recombinant MoMuLV vectors are engineered by replacing the *gag*, *pol*, and *env* coding units with a transgene of interest, while retaining the LTRs and packaging *cis*-acting sequences. Producer cell lines stably transformed with independent gag/pol and *env* expression cassettes are used to fully complement the viral polypeptides for packaging of the vector proviruses.[2,3] Hence by transfecting these packaging cell lines with plasmid-based LTR-flanked retroviral cassettes, retroviral particles efficiently bud from the host cells containing the recombinant retroviral genome. These retroviral particles are capable of infecting cells and directing the expression of the transgene of interest, but cannot replicate or generate progeny virus.

## 1.2 Adenoviral Vectors

Adenoviral particles consist of lipid-free "spiked" regular icosahedra of 60–90 nm in diameter, consisting of three main structural proteins: hexon, penton base, and fiber.[4] The genome consists of a double-stranded linear DNA molecule of approximately 36 kb in length, functionally divided into two major noncontiguous overlapping regions, early and late, defined by the onset of transcription after infection (Fig. 1(B)).[5] There are five distinct early regions (E1A, E1B, E2, E3, and E4) and one major late region (MLR) with five principal coding units (L1 to L5), plus several minor intermediate and/or late regions. At the extremities of the viral chromosome are the inverted terminal repeats (ITRs) and the encapsidation signals (ψ), encompassing the *cis*-elements necessary for viral DNA replication and packaging.[5]

Recombinant adenoviral (Ad) vectors are constructed by deleting the essential early genes *E1A* and *E1B*, whose expression enables transformation of the host cell and *trans*-activates expression of the other early viral genes, as well as some host factors.[6] Transgenes are inserted into this deleted region (Fig. 1(B)) and can be assembled into infectious adenoviral particles in cell lines which *trans*-complement the E1A/B functions.[7] Additional deletions in the nonessential E3 region are also often performed to increase cloning capacities.[8] Thus infection of cells with the Ad vector enables expression of the transgene in the absence of expression of viral proteins. Further incapacitation of the Ad vector genomes, limiting leaky expression of viral proteins by further deletions in the E2 or E4, has also proved advantageous, further enhancing the cloning capacities, but requiring further complementation functions in packaging cell lines.[9–12] The development of so-called "gutless" or helper-dependent (HD) adenoviral vectors has also greatly expanded the potential of Ad vectors. These vectors retain just the terminal ITRs and ψ, required for replication and packaging of adenoviral genomes, thus greatly increasing the cloning capacity.[13]

## 1.3 Adeno-Associated Viruses

Adeno-associated virus (AAV) has recently become attractive candidates for gene transfer. AAV belong to the family *Parvoviridae* and consist of nonenveloped icosahedral virions of 18–26 nm diameter, with linear single-stranded DNA genomes of 4680 nucleotides for the most characterized AAV2 strain.[14,15] The genome consists of two coding regions, *cap* and *rep*, which are flanked by ITRs and encapsidation signals ($\psi$), at either end of the genome (Fig. 1(C)). The *cap* gene encodes the capsid (coat) proteins, and *rep* encodes proteins involved in replication and integration functions.[15] After infection, AAV genomes can persist extrachromosomally in an episomal form[16,17] or integrate into the cellular genome.[18,19] AAV have been demonstrated to preferentially integrate into human chromosome 19 at site q13.4 (AAVS1), directed by the *rep* genes, facilitating latent infection for the life of the cell.[20] AAV is, however, naturally replication-incompetent and requires additional genes from a helper virus infection, which in nature is generally complemented by Ad or HSV coinfection.[21] AAV-based vectors generally involve replacement of the *rep* and *cap* genes with a transgene of interest (Fig. 1(C)), retaining the terminal repeats and packaging sequences essential to direct replication and packaging of the genome.[15] These AAV vectors can be packaged into infectious AAV particles upon complementation of the *rep/cap* genes and Ad/HSV helper functions in *trans*. Deletion of the *rep* genes, however, eliminates targeted integration of the AAV cassettes at AAVS1. The nonpathogenic nature of AAV, having not been associated with any disease or tumor in humans, makes them potentially powerful clinical vectors.

## 1.4 Herpes Simplex Viruses

Herpes simplex virus (HSV) belongs to the herpesvirus family, a diverse family of large DNA viruses, all of which have the potential to establish lifelong latent infection.[22,23] HSV consists of 110 nm diameter particles comprising an icosahedral nucleocapsid, surrounded by a protein matrix, the tegument, which in turn is surrounded by a glycolipid-containing envelope.[24] The HSV-1 genome consists of a giant linear double-stranded DNA molecule of 152 kb encoding 81 known genes, 38 of which are essential for virus production in vitro.[24] First generation HSV-based vectors involve replacement of one or more of the seven immediate early genes whose functions are *trans*-complemented by packaging cell lines.[24] Second-generation HSV-amplicon vectors consist of plasmids containing just the HSV-1 origin of replication ($Ori_s$) for replication in packaging cell lines by the rolling circle mechanism and the cleavage/packaging signal (pac) (Fig. 1(D)). These amplicon vectors can accommodate inserts of up to 15 kb, enabling the assembly of concatomer structures of up to 10 genomes, reconstituting the packaging size of 150 kb.[25] The future construction of "full size" gutless HSV vectors could accommodate up to 150 kb of insert DNA.[26]

## 1.5 Lentiviral Vectors

Lentiviruses are a subclass of retroviral vectors which have become infamous in world affairs by the HIV family members. LV vectors are characterized by the presence of additional accessory genes to the *gag/pol/env*-based genomes.[27] These accessory genes extend the functions of the viruses, with the major gene therapy focus being on their ability to infect nondividing as well dividing cells, in distinct contrast to other retroviral family members such as MoMuLV. These karyotropic properties of lentiviruses provide a promising tool to direct retrovirus-mediated gene therapies to nondividing cells.[28] LV vectors are constructed by an analogous mechanism to conventional MoMuLV vectors.

## 1.6 The Choice of Gene Therapy Vector

No single vector system can presently provide the necessary flexibility for all the possible clinical applications of gene therapy. Vast variabilities exist in vector host range and uptake potentials for the many tissues of the human body, which together with the many biological barriers to reaching the target tissues make a universal vector unlikely. Thus disease-specific gene targeting strategies are likely to be required, involving the development of multiple gene delivery systems. Hence the technology of gene therapy stands to benefit from the vast range of clinical vectors being designed, each system having distinct properties which can complement each other in the clinic.

Extensive research has focused on the potential of adenoviruses as transducing viruses for use in gene therapy. The translation of laboratory-derived viral vectors as practical pharmaceutical tools is a major determinant of gene therapy interest in the clinic. In essence, the ease of generating Ad vectors, the efficiency of purification, and the superior titers, which can be obtained ($>10^{11}$ pfu/ml), have made Ad the vector of choice for many applications of in vivo gene therapy.[4] The rapid technical advances in the construction and purification of alternative viral vector systems have, however, expanded clinical interests. The vastly improved techniques of helper-free AAV production have significantly increased the potential of these vectors. Titers of AAV vectors equivalent to those of Ad vectors are now routinely achievable, which are free of the once-problematic helper-virus contamination.[29] The production procedures, however, are still relatively laborious and problematic. The comparatively low titers of the MoMuLV-, HSV-, and LV based vectors, generally greater than two logs lower stable titers, limit the effectiveness of these vector systems especially upon translation to the clinic. However, current immune system barriers preclude the beneficial attributes of administration of Ad vectors at their maximal titers, with significant safety concerns apparent with the maximal doses of Ad vectors in the clinic.[30]

The generation of large-scale, high-titer vector preparations with stable shelf lives is essential for clinical applications. The stable pharmaceutical properties of Ad virions, as well as the similarly encapsidated AAV and HSV virions, present significant advantages over the much less stable enveloped retrovirus-based vectors. The integrative functions of retroviral vectors, however, confer on them the potential of long-term stable expression, fulfilling an additional highly desirable vector property. These integrative functions together with rapidly advancing methods of enhancing viral titers using concentration procedures,[31] maintains major clinical interest in retroviral vectors. The integrative functions of AAV vectors are also highly desirable, specifically the chromosomal targeting mechanism in the presence of the Rep protein.[32]

The absence of any cellular retention mechanisms for Ad and HSV vectors presents a distinct disadvantage to many gene therapy applications. In the context of tumor eradication, however, high-titer vector transduction is unlikely to require long-term maintenance of vectors.

In deciding the most appropriate vector for treatment of a clinical disorder, the main selection criterion for vector choice comes in the ability of a specific vector to efficiently transduce the target tissue. Ad vectors have a wide distribution of their target receptors dispersed throughout the body tissues. AAV and HSV have similar diverse tropism to most cells in the human body, with HSV-1 vectors having a major selective tropism for neuronal tissues. MoMuLV viruses are, however, severely limited by their dependence on host cell mitosis to enable stable transduction of a cell, limiting their efficacy in quiescent cell populations.[33] These cell-cycle restrictions are not apparent with the LV subclass of retroviral vectors, which possess the additional nuclear targeting functions.[34] The additional nuclear targeting property of LV vectors, together with the integration functions, has significantly raised the clinical interest with respect to gene therapy. Additionally the ability of MoMuLV vectors to only infect dividing cells can be deemed an advantage in targeting actively dividing tumor cells which are surrounded by nondividing normal tissues. The ideal vector system is thus very much dependent on the diseased tissue to be treated.

The extent of genetic material that is required to be delivered to a specific tissue is also a major influence on the vector system. Ad vectors offer a wide range of insert potentials from 7 to 8 kb insert capacities for first generation vectors and up to 36 kb inserts in the "gutless" HD vector system.[6,35] Whereas the relatively small packageable genome sizes of retroviral vectors (~8 kb), but more significantly for AAV vectors (~4.5 kb), severely limit their applications to some gene therapy protocols,[15] specifically where the delivery of multiple genes or the insertion of large regulatory elements is deemed essential. It is, however, the HSV-1-based vectors that offer the superior transgene delivery potentials with inserts of up to 150 kb feasible in a "gutless" vector rationale.[36] Additionally, in the alternative HSV-1 amplicon vector system, as well as providing an insert capacity of up to 15 kb, the assembly of concatomers vastly increases the copy number of transgene cassettes being delivered to target cells.[25]

The immune system is a perpetual barrier to viral transduction. The compromised state of many diseases would be severely stressed by further immunological effects/inflammation induced by a "therapeutic" vector challenge. The exception again is cancer gene therapy where activation of local immune responses can be advantageous in tumor recognition and possibly aid in breaking immune tolerance.[37] Viral vectors are designed to exploit specific biological properties of viruses, such as recognition of cell receptors for entry and mechanisms of host genome integration, which have evolved over time in relationship with the host. The natural response of the host has, however, also developed to eliminate disease inducing viral pathogens. Current strategies of viral vector design are working to engineer viruses with predictable biological properties, maintaining the biological advantages of the virus that have been selected by nature while reducing the immunogenicity of the viral components. The majority of Ad vector-derived immunogenicity was deemed to be due to the leaky expression of retained viral transcripts in the vector genome.[30,38] For AAV and HSV amplicon vectors, contaminating helper virus was also deemed highly immunogenic. The more recent improvements in Ad vector design[9–12] and generation of "helper-free" packaging systems for AAV and HSV amplicon vectors[29,39] have stunted this immunogenicity to some extent. However, the immune system still stands as a major barrier to gene therapy efficacy. The mere physical presence of the virus can induce significant cytopathology. The current requirement of repeated administration to boost expression levels further augments the immune memory responses to the presence of the virus, until eventual complete immunity is developed to the applied vector.[40] The power of the immune system is emphasized by practically eliminating 95% of Ad virions by the natural nonspecific innate immune response on each administration.[41]

## 1.7 How to Maintain Stable Transgene Expression?

The transient natures of Ad and HSV-1 vectors, as well as the rapid loss of transgene expression upon stable integration of AAV and RV vectors due to nuclear effects on the transgene cassettes, have dramatically limited the efficacy of each vector system. Hence the question remains: how do we maintain stable transgene expression following recombinant viral vector transduction? One solution may come from looking closer at the wild-type mechanisms of preservation evolved by the parental viruses.

Viruses have developed diverse mechanisms of self-preservation and maintenance to enable them to infect cells and direct self-replication and propagation. Mechanisms of maintenance vary according to the life cycle of the virus. Viruses such as retroviruses have developed life cycles that live in harmony with the host cell. They utilize the host cellular machinery to enable continuous shedding of the virus and thus require stable preservation of the viral genetic material. Retroviruses facilitate this function by stable integration into the host genome, permitting continuous replication/maintenance of the viral genome in the context of host cell replication.[1] Conversely lytic viruses such as adenoviruses subvert the host's cellular functions solely for their own preservation. Infected cells become short-term factories of virus production, amassing viral particles until host cell saturation is achieved, and cell lysis occurs in less than 36 h.[5] The short-term association of virus and host does not therefore necessitate mechanisms for long-term persistence of the viral genome. The Ad genome is thus maintained extrachromosomally with a very efficient mechanism of replication to enable large-scale genome packaging into the vast numbers of viral particles generated. Herpes viruses, such as HSV, Epstein–Barr virus (EBV) and cytomegalovirus (CMV), have developed more complex mechanisms of self-preservation.[42] Upon infection a lysogenic life cycle enables the virus to live in harmony with the cell, casually maintaining the genome in an extrachromosomal state, where methylation and histone binding to the viral genome keep viral gene expression essentially quiescent.[22] The switch of the life cycle from the quiescent latent state to the major virulent lytic phase, upon signals of cell stress, rapidly reveals the viral presence. This terminal lytic stage of rapid viral genome reproduction and mass assembly of virions enables the virus to rapidly multiply and abandon the host. The AAV life cycle is a further intriguing evolutionary mechanism, being naturally dependent on helper Ad or HSV coinfection to effect lytic AAV virion assembly and viral progeny release. In the absence of such helper functions, AAV remains lysogenic by either stable integration into the host genome or independent episomal replication in the infected cell.[14]

# 2. Hybrid Viral Vectors

The inadequacies of each viral vector system are illustrated in Table 2. The negative attributes of one vector, however, generally emphasize the positive attributes of another. Thus most of the criteria defined for a hypothetical perfect gene therapy might actually be met by considering defined properties of the currently available vectors defined in Table 2. Hence although at present no individual virus system alone can meet all the criteria, current research is focusing on combining individual viral properties into single vector constructs, termed "Hybrid" or "Chimeric" vectors.

Adenoviral vectors are currently the major vector choice for a variety of clinical disorders, despite the limited efficacy due to the transient nature of the vector. Mechanisms of enhancing the pharmaceutical properties of Ad vectors are thus highly desirable. The incorporation of other viral vector functions that could enhance the duration of Ad-directed transgene expression and/or target the vectors to a specific disease tissue would be extremely beneficial. In essence, whether the aim is to kill or cure the target cell, a vector encompassing the advantageous properties of high titer, broad host range, and infectivity of an Ad vector, together with the low immunogenicity and potential of long-term stable expression of a retrovirus, AAV or EBV vector would be extremely useful for gene therapy for a wide range of genetic and acquired disorders. Hence the main focus of this chapter is to review the properties of other viral vectors which have been utilized to generate "hybrid" adenoviral vectors in the aim of enhancing vector efficacy in the clinic.

## *2.1 Are Hybrid Vectors Truly New Technology?*

The formation of hybrid adenoviruses is not a new technology and has been extensively reported to occur naturally in nature. Adenoviral/Simian virus 40 (SV40) hybrids have been documented to occur in nature.[43,44] Although human adenoviruses do not normally replicate in primate cells, upon coinfection with SV40, Ad genomes acquired sequences from the SV40 genomes (large T antigen) which permitted replication and assembly of hybrid genomes into wild-type Ad capsid particles.[43] Additionally it may be that the helper-dependent AAV genome represents a segment of an extinct or undiscovered virus that was selected upon coinfection of an Ad or an HSV. Perhaps the parental virus was too virulent to coexist in a human host, therefore explaining the nonpathogenic nature of the dependovirus.

The development of hybrid viral vectors is fundamentally not a new technology in gene therapy. Since the dawn of gene therapy, scientists have utilized alternative *cis*-acting sequences from other viruses, specifically promoters and enhancers, to drive transgene expression. Most significantly the CMV immediate early promoter and enhancer has been utilized in almost every viral vector reported to date and is well characterized as an extremely strong constitutive promoter in most tissues.[45,46] Other well-utilized viral promoters have included the Rous sarcoma virus LTR promoter, the SV40 early promoter, hepatitis B virus (HBV), and the EBV promoter.[45,46] Additionally, application of the picornaviral functions of "cap-independent" initiation of translation has also been extensively applied in viral vectors. These translational regulatory elements, termed internal ribosomal entry site sequences, enable bicistronic expression from a single mRNA transcript.[47] The application of these elements greatly complemented the limited insert capacities of viral vectors, thereby negating the need for separate promoters to drive two transgene cassettes.

Retroviral vectors have been studied in hybrid vector systems since the early 1980s, "pseudotyping" them with functions from other retroviral vectors. Specifically heterotrophic viral glycoproteins from other retroviral *env* genes have been stably incorporated into MoMuLV vector particles. The incorporation of vesicular somatic virus G (VSV-G) glycoprotein,[48] gibbon ape leukemia virus (GALV) and HIV-1 glycoproteins[49] into murine leukemia virus (MLV) particles has been reported. These "hybrid" MoMuLV virions attain the tropism of the pseudotyped env proteins, retargeting or broadening the host range of the MoMuLV vector. Additionally, incorporation of VSV-G env has been demonstrated to increase the stability of the virions, enabling higher titer-yielding purification techniques to be applied.[50,51] Hybrid retroviral vectors have also been constructed incorporating different *cis* acting elements contained in the U3 region of the LTR, which direct the transcriptional activity of the virus. Replacement of these U3 regulatory elements can impart tissue-specific transcriptional activity on the RV vector.[52,53] Hence the concept of hybrid vectors is not a new technology, but the new strategies proposed could vastly expand the repertoire of viral vectors available to the clinic.

## 3. Hybrid Adenoviral Vector Systems

A number of hybrid adenoviral vector systems have been reported in the literature, combining the properties of RV, AAV, and EBV vectors, as well as elements of other Ad serotypes, to enhance the therapeutic efficacy of Ad vectors in vivo. The principal aim of these new hybrid vectors is to overcome the limitations of transient Ad vector retention in infected cells. In addition to the well-documented limitations of Ad vectors (Table 2), some initially perceived advantageous properties of Ad vectors do actually limit their effectiveness toward therapy for some diseases. The broad host range of Ad vectors induces significant disadvantages when tissue targeting is required and compromises systemic administration. Additionally, the low pathogenicity of adenoviruses in humans has resulted in many serotypes, including the conventional vector strains of Ad2 and Ad5, being endemic. Hence a potent natural anti-adenoviral immunity is fashioned generally at a very early age. The highly immunogenic nature of the proteinous Ad virion further confounds the system, with a rapid and highly effective host humoral response being developed to the Ad vector. Research is thus being channeled into both retargeting Ad vectors to specific tissues and silencing the structural immune stimuli to facilitate enhanced Ad vector transduction.

### 3.1 Pseudotyping and Retargeting Adenoviral Vectors

As targeting and humoral immunity are connected in essence to the same surface moieties of the Ad particles, both disciplines are fundamentally interlinked. Methods applied to limit the humoral responses have focused on two main strategies: application of alternative "immune-silent" Ad serotypes or display of alternative ligands on the surface of the virions, which is also the major strategy for retargeting the vector.

The use of alternative serotypes enables the consecutive application of immunologically distinct Ad particles, enabling avoidance of specific humoral responses to previously applied vectors.[54,55] This system has presented some success in vivo[56] although the presence of cross-reacting antibodies is problematic due to the evolutionary similarities of Ad serotypes. The application of alternative Ad serotypes with different surface markers also provides a mechanism of alternative targeting, as different serotypes posses tropism for different tissues in the human body. For instance, the conventional gene therapy subtypes Ad2 and Ad5 have natural tropism for the gut epithelial layer. Hence in terms of gene therapy for cystic fibrosis, initial vectors proved disappointing due to their low infectivity of the airway epithelia. To overcome this restriction, Zabner and colleagues investigated other Ad serotypes for airway epithelia tropism.[57] A number of other Ad serotypes, specifically Ad17, were found to infect the airway epithelia with increased efficiency to wtAd2.[57] They therefore proceeded to generate Ad2 hybrid vectors pseudotyped with the Ad17 fiber, where the endogenous Ad2 fiber gene was replaced with the Ad17 fiber gene. The resultant chimeric vector displayed increased efficiency of binding and gene transfer to well-differentiated human epithelial cells. A similar study by Croyle and colleagues demonstrated that binding of wild-type Ad41 had enhanced transduction properties in intestines compared to Ad5.[58] These studies emphasize the potential of alternative Ad serotypes with tropism for different tissues in the human body. Pseudotyping also provides an invaluable mechanism of integrating alternative serotype fiber (and/or penton base) genes from other Ad serotypes into the currently well-researched Ad vectors, without having to reconstruct the vector backbones. The use of nonhuman adenoviruses as vectors for gene therapy is also under investigation, with bovine, ovine, canine, feline, and avian adenoviruses being researched.[59–62] As well as being potentially unexposed to the immune system, they may also have specific tropism for selective tissues in humans. The potential of pseudotyping nonhuman Ad vector components with conventional human Ad vectors is therefore of interest.

The use of targeted viral vectors to localize gene therapy to specific cell types introduces significant advances over vectors with conventional natural tropism. As well as the safety aspects of reduced immunogenicity and toxicity, the reduced uptake by nontargeted cell types may enable application of systemic delivery with feasible viral titers and loads. In order to retarget Ad vectors, firstly, the natural tropism of the virus must be removed, and secondly, novel, tissue-specific ligands introduced.[63] Two main mechanisms have been used to retarget Ad vectors. Firstly, the use of external molecules with affinities for both the Ad surface structural moieties as well as a cell-type specific surface ligand. These bispecific molecules act as bridges between the virions and the cell. A neutralizing antibody or high-affinity peptide for the fiber or penton base can act as the Ad-binding moiety, which can be covalently linked to a high-affinity ligand for a tissue-specific receptor.[63] A drawback of the bridging molecule approach is that native receptor binding is never 100% blocked. To truly block native Ad binding to its cognitive receptor requires removal of the intrinsic receptor-binding domains.

A second approach involves creation of hybrid Ad vectors, pseudotyped with novel receptor recognition functions. Genetic modification of the Ad genome by incorporating targeting ligands inside the genome, while deleting or ablating sequences of the penton and fiber involved in receptor recognition, has been reported. High-affinity peptide motifs have been subsequently demonstrated to be functionally incorporated into Ad particles. These "proof-of-concept" studies focused on the incorporation of ligands without ablating natural receptor interactions and resulted in expanding the vector tropism, which have proved beneficial in vivo in transducing both vascular smooth muscle and some tumor types.[64–66] Future studies will focus on honing the targeting functions to specific cell types. High-affinity ligands have been stably inserted into the HI loop or on the C-terminus of the fiber or into the integrin-binding RGD domain of the penton base.[63] However, the size, location, and type of ligand to be inserted are currently under debate and remain to be determined. Wickham and colleagues demonstrated 10- to 1000-fold reductions in transduction of cells expressing the coxsackievirus and adenovirus receptor (CAR) with CAR-ablated vectors.[63] The residual transduction is being penton-base-mediated, emphasizing the requirement of additional ablation of penton-base binding.[63] The further requirement of novel packaging cell lines to facilitate infection and propagation of the CAR-/integrin-binding ablated particles also remains an issue.

## 3.2 Adenoviral/Retroviral Hybrid Vector Technologies

A hybrid vector system incorporating the advantageous long-term stable integrative functions of retroviral vectors into adenoviral vectors could provide a major clinical advancement to gene therapy. Hybrid vector systems are thus being investigated incorporating retroviral components into the backbones of adenoviral vectors. Initial studies have focused on utilizing adenoviral vectors as directors of retroviral vector production, delivering the *gag*, *pol*, and *env* genes as well as retroviral LTR cassettes to cell populations both in vitro and in vivo.

Conventional retroviral packaging cell lines are stably transformed with *gag*, *pol*, and *env* functions and release retroviral particles upon plasmid transfection of a retroviral LTR transgene cassette.[2] High-titer retroviral stocks of greater than $10^7$ infectious units (iu)/ml can now be obtained from conventional stable producer cell lines.[3] To achieve the highest vector titer, it is necessary to select clones of vector-transduced cells individually due to the varying titers of producer cell clones.[67] Direct injection of retroviral vectors in vivo has, however, yielded limited efficiencies due to the limited transducing titers and poor infectivity. Application of retroviral vectors in the clinic has thus focused on ex vivo protocols. This involves the removal of patient tissues, which can be cultured for a brief period in the laboratory, transduction with the RV vector, and re-implanting back into the patient. The ex vivo approach has yielded some success though the procedure is cumbersome and costly, and in most cases, it can only transduce a small fraction of the target cells.[68,69] The establishment of retroviral producer cells in situ provides a further mechanism of enhancing the efficacy of retroviral gene therapy. Transient transfection of target cells in vivo with the retroviral vector and packaging plasmids, previously used to generate producer cell lines in vitro, by direct DNA injection has been reported.[70] Although stable integration of subsequently generated retroviral particle genomes could be detected, the efficiency was very low. The implantation of retroviral producer cell lines into patients has presented a far greater potential for the in situ production of retroviral vectors. Gene therapy using MoMuLV-based producer cells to treat brain tumors[71] has been carried out in a clinical trial, but no clear clinical benefit has been reported to date.

The infectivity of Ad vectors both in vitro and in vivo provides great potential in increasing the efficiencies of retroviral production technology. The group of David Curiel pioneered the development of hybrid retroviral/adenoviral vectors by using the infectivity of adenoviral vectors to efficiently deliver the requisite retroviral packaging and vector functions to target cells in vivo, thereby rendering them retroviral producer cells in situ (Fig. 2). The subsequent release of high local concentrations of

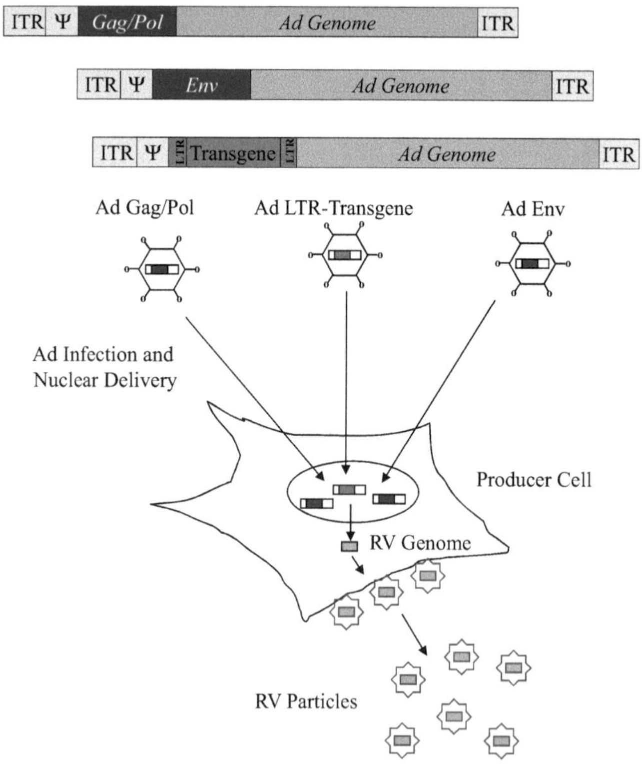


**Fig. 2 Hybrid-Ad/RV vector mediated production of RV particles.** Hybrid adenoviral vectors expressing the gag/pol and env RV genes either together or on split adenoviral constructs (as shown) are coinfected with the Ad-LTR transgene vector into cells in vitro or in vivo. Subsequent expression of the gag/pol and env genes in the cells establishes in situ retroviral producer cells which direct the packaging of the expressed retroviral genomes. RV particles expressing the transgene cassettes subsequently bud from the cells and are released into the surrounding environment.

retroviral particles in situ would enable stable transduction of neighboring tissues, for the transient period of adenovirus transduction. The Ad/RV hybrid system reported by Feng and colleagues utilized a two-adenovirus delivery strategy.[72] The first adenovirus contained an LTR-flanked retroviral vector cassette encompassing the green fluorescent protein (GFP) marker and neomycin resistance genes, Ad/RV-vector. The second adenovirus contained the replication-defective retroviral helper machinery, encoding the *gag*, *pol*, and *env* genes of MoMuLV, Ad-*gag/pol/env*. High-titer adenoviral vectors could be generated containing the RV cassettes, which could efficiently direct the in vitro packaging of RV particles at titers similar to convention packaging cell lines.[67,72] These studies clearly demonstrate the compatibility of both the adenoviral and retroviral life cycles in the context of a hybrid vector configuration.

Upon infection of cells in vitro with the Ad/RV-vector alone, high initial levels of GFP expression were observed, but gradual loss of expression was documented over a period of 60 days, as the non-integrated adenovirus was lost from dividing target cells. Conversely, upon application of both adenoviruses to cells in vitro, GFP expression was persistent for extended periods of time. However, the persistent level of gene expression was reduced beyond the time at which expression could be solely attributed to the Ad/RV vector. The stable integration of the retroviral cassette in surrounding cells was believed to be responsible for this extended expression. The longer-term GFP-expressing cells in cultures transduced with both Ad vectors were present in clustered outgrowths suggesting local retroviral spreading and/or clonal origin. Subsequent demonstration of proviral integration was confirmed by the presence of retroviral transgene sequences in high molecular weight cellular DNA.[72] No replication-competent retroviruses (RCR) were detected with the Ad/RV chimera, despite the large genome copy numbers associated with adenovirus production in vitro. [72]

The ex vivo efficacy of the Ad/RV hybrid vector system was investigated by transducing the ovarian carcinoma cell line SKOV3 in vitro with the Ad/RV vector alone or in combination with the Ad-*gag/pol/env* vector at a multiplicity of infection (MOI) of 50 pfu/cell. The infected cells were then mixed at a ratio of 3:1 with uninfected SKOV3 cells and subcutaneously implanted in athymic nude mice to allow tumor formation. Tumors were assessed 20 days post transplantation for GFP expression. Large expansive clusters of GFP-expressing cells were only observed in tumors treated with both vectors. Further in vivo studies involved direct intraperitoneal injection of 5-day-old established SKOV3 tumors in nude mice ($1 \times 10^7$ cells) with the single or two adenoviral strategy (1 × 109 pfu/mouse). Sixteen days post Ad infection, the two-virus-treated tumors were observed to have islands of GFP-positive cells (10–15% transduction), consistent with secondary retroviral transduction. In contrast, single-virus-treated tumors revealed very limited (<1%) GFP-positive cells. This pioneering study thus established the great potential of hybrid Ad/RV vectors, whose pros and cons are presented in Table 3 and discussed further in the concluding remarks of this chapter.

Following the initial proof of concept, a number of other laboratories have further investigated the concept of Ad-mediated establishment of retroviral producer cell lines in situ. Duisit and colleagues, in collaboration with Francois-Loic Cosset, reported on an extension of the hybrid vector system.[67] These studies further restricted the

Table 3 **Pros and Cons of Hybrid Adeno-/Retroviral Vectors for Gene Therapy**

| *Pros* |
|---|
| • Exploit high-efficiency adenoviral infection to deliver retroviral assembly machinery<br>• Utilize stable high titer adenoviral vectors<br>• Increase the duration of biological activity of delivered transgene<br>• Avoid initial limitations of retroviral infectivity to nondividing tissues<br>• Utilize adenoviral vector tropism<br>• Therapeutic gene expressed by retroviral cassette will still be expressed in context of sole delivery of the Ad/RV cassette vector<br>• Initial burst of transgene expression can be converted to a stable lower level expression<br>• Delivery deep into cell layers |
| *Cons* |
| • Progeny retroviral vector can still only infect dividing cells<br>• In situ released retroviral vector limited according to RV infectivity and tropism<br>• Requires codelivery of two or more adenoviral vectors<br>• Risk of RCR<br>• Safety<br>• Rescue of endogenous retroviral elements<br>• Interactions with host cell functions<br>• Diffusion may still be very limited around the initial needle tract<br>• In situ titers of retroviral particles maybe limited<br>• Different cells have different intrinsic potentials for retrovirus production |

potential of RCR by separating the gag/pol core particle-expressing elements from the *env* surface glycoprotein gene, which they supply on a separate Ad vector (Ad-*gag/pol* and Ad-*env*; Fig. 2), to minimize retroviral sequence overlaps. Additionally, in the context of pseudotyping retroviral vectors they replaced the natural MoMuLV *env* gene with the GALV *env* gene. In small-scale pilot experiments, TE671 cells simultaneously infected with the three Ad vectors efficiently released helper-free retroviral particles at titers of up to $5 \times 10^6$ iu/ml for at least 3 days following infection.[67]

The further separation of the key retroviral elements facilitated the individual characterization of each retroviral function in terms of variable copy load, on complementing retroviral cell lines. The results helped to shed light on the factors currently limiting retroviral vector production and investigating particular cell-type-specific features of the producer cells. The availability of packageable RNAs of the retroviral genome itself was not found to be rate limiting, with Ad-mediated overexpression resulting in increasing, non-saturatable retroviral titers.[67] The results indicated that high expression of Gag-Pol and Env proteins through the introduction of high copy numbers of their genes was not required to achieve an efficient retroviral production, and that there is probably a limit to the number of particles that a given cell may release. Increased GALV *env* copy number resulted in augmented glycoprotein synthesis, with RV-particle production plateauing between MOIs of 10 and 50. At higher MOIs, titers decreased, possibly by a break of tolerance by the cell to efficient RV particle assembly or budding. It was observed that Pr65 gag precursor expression saturated with Ad-Gag/Pol MOIs of greater than 5.[67] At higher titers, premature maturation of Pr65 transcripts became apparent, which normally occurs at maturation of the retroviral particles.[73] However, despite these observations reported, titers were equivalent to those generated on high titer generating stable packaging cells.[67] A detailed analysis of the release of non-infectious/incorrectly processed budding particles may be of interest. Additionally, despite the expression of similar levels of Gag precursors and premature forms, a wide variability was observed in the capacity of different cell types examined to assemble and release retroviral particles.[67] In the context of the hybrid adenoviral/retroviral vector system, when applied at optimal MOIs, a critical limiting factor(s) for the production of retrovirus is the ability to avoid premature activation and convert the bulk of Gag and Gag-Pol precursors in nascent viral infectious particles.[67] Despite the high copy number of all three retroviral units introduced into cells, no RCR could be detected. These Ad/RV hybrid vector studies go some way in aiding the elucidation of the limiting factors involved in retroviral production in packaging cell lines and they indicate that the careful selection of packaging cell type is crucial. This observation is highly significant to therapeutic applications of the hybrid vector system where different tissues of the body will be more suited to RV complementation than others. The hybrid Ad/RV system can also facilitate the rapid screening of various primate cells for their retroviral production potentials and allows simple substitution/pseudotyping of components in the system.

### 3.2.1 Tetracycline-Inducible Env Pseudotyping of Ad/RV Hybrid Vectors

Pseudotyping the VSV-G retroviral envelope in the MoMuLV back bone, as discussed earlier, enhances the stability and tropism of the native virus.[31] This enhanced stability enables higher titer preparation to be prepared by centrifugation. Generating Ad vectors expressing the VSV-G envelope glycoprotein has, however, proven technically difficult due to the cytotoxic nature of the protein product. Yang and colleagues demonstrated that the VSV-G gene could be effectively controlled under the tetracycline-inducible system[74] in packaging cell lines obtaining unconcentrated titers of $10^5$ to $10^6$ iu/ml.[75] Yoshida and colleagues extended these studies by applying the tetracycline-inducible system in the context of an Ad vector.[76] Ad vectors were generated encoding VSV-G and MoMuLV *gag/pol* genes, both under the control of the tetracycline-controllable promoter. Hence only upon the supply of doxycycline (a tetracycline homolog) efficient expression would proceed from the *gag*, *pol,* and *env* genes. Minimum "leaky" expression of cytotoxic VSV-G under the control of the inducible promoter remained low enough to allow Ad propagation to titers of $4 \times 10^9$ pfu/ml. The drawback of this system is the necessity to provide a further Ad construct containing the tetracycline transcriptional regulator (Ad-rtTA), expanding the system to a four-adenovirus transduction strategy, together with Ad-TetGag/Pol, Ad-TetEnv, and the Ad/RV-vector expressing neomycin resistance. Application of the four viruses in vitro generated retroviral transgene titers of up to $5 \times 10^5$ iu/ml, which were further purified to titers of $>10^7$ iu/ml following simple centrifuge concentration of the virus from culture fluids at 50–80% recovery efficiency.[76] Caplen and colleagues extended these studies in two tumor model systems in vivo by subcutaneous injection of (1) 9L glioma tumors in rat or (2) human A735 melanoma xenografts in nude mice.[77] Only upon application of all four viruses in the 9L rat model were neomycin-resistant cell cultures established from harvested tumor tissues. Molecular analysis of genomic DNA extracted from neomycin-expressing 9L rat cultures, derived both in vitro and in vivo, showed the appropriate integration of the retroviral transgene cassette.[77]

The human-xenograft nude mouse model system meant that Ad was not cleared in the time frame examined (4 weeks), hence efficacy was assessed as increases in G418R cells compared to single hybrid Ad/RV-vector transgene transduction.[77] In the human-xenograft mouse model system, tumors harvested at 1, 3, and 4 weeks post transduction displayed increased numbers of neomycin-resistant colonies with time only upon transduction of the full complement of adeno–retroviral constructs. At 4 weeks up to 7.2% of xenografted cells were retrovirally transduced. Transduction of tumors with Ad/RV vector alone yielded no increase in the number of neomycin-resistant clones. DNA extracted from the xenograft tumors, as for the rat model, only showed the presence of integrated proviral sequences when transduced with the full complement of adeno–retroviruses.

Titers of retrovirus particles generated from the 9L rat glioma cells in vitro were dependent on the input MOI of the adenoviruses, with maximum titers of up to $1 \times 10^5$ iu/ml generated at MOIs in the range of 200–300 for each virus.[77] Under these optimal conditions the presence of doxycycline (1 μM) enhanced the titers by a factor of 2000-fold.[77] Interestingly, in vivo similar numbers of clones were observed after the four-adenovirus transduction strategy in the presence or absence of doxycycline: 30 and 20 colonies per $10^6$ cells plated. Less than 1 colony per $10^6$ plated cells was observed with the Ad/RV vector alone. These low transduction titers do, however, indicate the current inefficiencies of the system, which are reduced compared to other reports.[72] But the inefficiencies can to some extent be explained by the application of four separate Ad vectors for the system to function, significantly increasing the kinetic complexity of the generating retroviral vector producer cells in vivo. Additionally the poor efficiency of transduction of the rodent cells by Ad is emphasized by the required MOIs applied (>200) to generate optimum titers.[77]

### 3.2.2 Cooperative Adenoviral/Retroviral Vector Delivery

Other mechanisms of combining the advantageous properties of adenoviral and retroviral vectors have involved combinatory application of the separate vectors. Delivery of the retroviral genome in the context of a retroviral particle (RV vector) coinfected with an adenovirus expressing the *gag*, *pol*, and *env* genes (Ad-*gag*/*pol*/*env*) has been reported.[78] Coinfection of the vectors into NIH 3T3 cells generated retroviral titers $>10^5$ iu/ml. The advantages of this system over the hybrid Ad/RV-vector delivery of the transgene cassette are questionable, specifically from a cell-targeting aspect.

Several groups have also recently demonstrated stable ecotropic retrovirus-mediated gene transduction of human cells using preinfection of Ad or AAV vectors expressing an ecotropic receptor.[79–81] In order to target retroviruses specifically to malignant hepatic tissues, an adenovirus expressing the ecotropic receptor under the control of a hepatoma-specific promoter[82,83] was constructed. Although tissue-specific expression of the retroviral ecotropic receptor and subsequent tissue-specific targeting of ecotropic enveloped-retroviral vectors were demonstrated in vitro,[84] the clinical benefits of this system are limited. In essence, direct application of the tissue-specific promoter to expression of a therapeutic transgene in the adenoviral vector would be more beneficial.

### 3.2.3 Non-Specific Integration of RV LTR Cassettes in the Context of Ad Vectors

The studies presented above all emphasize the requirement of all the gag/pol and env components to derive stable integration of an RV LTR-flanked transgene cassette. However, a controversial report printed in *Nature Biotechnology* challenged that doctrine. Zheng and colleagues reported that an RV LTR-flanked cassette contained in an Ad vector (Ad/RV vector) could integrate efficiently in the absence of the retroviral enzymatic proteins.[85] The group studied a conventional MoMuLV LTR-flanked luciferase reporter gene cloned in the E1-deleted region of a first generation Ad vector (AdLTR-Luc), analogous to previous hybrid Ad/RV-vectors. A variety of cells and tissues permissive to Ad infection (epithelial cells, macrophages, and hippocampal cells) were transduced in vitro and in vivo by the hybrid AdLTR-Luc, and compared with AdCMV-Luc transduction, containing the CMV promoter in place of the LTRs. The AdLTR-Luc vector was demonstrated to direct sustained luciferase expression compared to the CMV promoter driven vector. Despite probable well-documented CMV promoter inactivation events, the authors present evidence for stable integration of the LTR-Luc cassette at sites within the LTR elements, by a mechanism independent of classical retroviral integration. Fluorescence in situ hybridization (FISH) analysis using probes for the 5′LTR and the luciferase gene revealing integration of the AdLTR-Luc vector with an apparent frequency of 10–15% in vitro and 5% in vivo. Southern blot analysis also implied integration of the 5′LTR of the hybrid vector, which was subsequently supported by sequencing of the region adjacent to the 5′LTR integration site. No integration of the AdCMV-Luc was reported. The frequency of spontaneous Ad integration has previously been reported at much lower frequencies ($10^{-3}$ to $10^{-5}$),[86] suggesting the presence of the retroviral LTR elements in the AdLTR-Luc somehow potentiates integration. The major question is whether an endogenous retrovirus is present in these cells; however, the authors reported negative results for RT activity. Additionally the integration events reported are not classic retroviral integration, as integration does not proceed at a conserved terminal position and results in the random loss of substantial terminal LTR sequence. *In vivo* studies involved injection of rat submandibular glands by retrograde ductal instillation of $1 \times 10^9$ pfu/gland. After initial high-luciferase expression, the levels plunged to near zero for AdCMV-Luc after 9 weeks but stabilized with AdLTR-Luc after 2 weeks although at significantly reduced levels. Although these findings are consistent with low-level integration of the LTR cassette, no specific mechanism of integration has been proposed, and alternative interpretations of nonspecific LTR-independent mechanisms are probable. As no drug selection gene was present in the vectors, clonal populations were derived on the basis of sustained luciferase expression. Hence, considering firstly the well-documented in situ inactivation of CMV promoters, specifically in the context of integration, luciferase expression would be absent in long-term cultures transduced with AdCMV-Luc. Therefore, the studies with AdLTR-Luc may have inadvertently selected for random integration events within the 5′LTR that maintained luciferase expression. The probability of such an event is fairly significant considering the limited extent of genetic material upstream of the LTR cassette in the Ad vector (ITR and ψ). Additionally with the high MOIs applied, selection of high copy number-transduced cells is highly probable, under which conditions spontaneous integration of the Ad vector is more probable. The sequencing data presented also demonstrate integration occurring at sites within the U3 region of the 5′LTR for several clones that would ablate LTR promoter activity. The selection of these luciferase-expressing cells would more probably be due to multiple integration events, which were clearly demonstrated by FISH analysis,[85] rather than integration site promoter effects.

Integrated proviral sequences were not reported in animals that received the Ad/RV vectors alone in other similar studies.[72,77] However these reports were looking specifically for retrovirus-mediated integration events and not the proposed alternative mechanisms reported by Zheng and colleagues.[85] The report by Caplen and colleagues[77] investigated the integration event based on the retroviral mechanism of reproducing the 3′LTR sequences to the 5′LTR structures.[1] The specific duplication of a nucleotide restriction site upon retroviral replication was used as a marker of integration in southern blot analyses. Analysis of a pooled population of neomycin-resistant colonies revealed efficient band size switching indicative of the duplication event, and thus retroviral replication. However, consistent with the observation of Zheng,[85] randomly integrated Ad–RV transgene cassette fragments could be seen in context of generalized hybridization of the probe to high-molecular weight DNA from single-vector transduced animals.[77] These bands were, however, weak and consistent with random integration. Further studies are therefore merited to evaluate the efficacy of integration of LTR-flanked cassettes in the context of an Ad vector to determine whether a specific mechanism does exist and whether it could be further refined for vector use.

### 3.2.4 Integration of Closed Circle Retroviral Cassettes Delivered by Adenoviral Vectors

Following retroviral infection, reverse transcribed proviral DNA serves as a substrate for an integration reaction catalyzed by the retroviral INT protein, which, along with viral GAG proteins, forms the pre-integration complex.[87] This complex brings the 5′ end of the 5′LTR (the U3 region) into close juxtaposition with the 3′ end of the 3′LTR (the U5 region).[87] The direct substrate for INT is most likely a linear, double-stranded molecule with blunt ends.[88] INT-mediated integration then occurs by a very precise mechanism in which the terminal-two base pairs of each LTR are lost prior to integration into the target cell genome.[87,89] However, closed circular molecules have also been detected in the nuclei of retrovirally infected cells, which contain 2LTRs joined covalently together at the so-called circle junction.[87,90,91] Although there is considerable evidence that MLV probably does not use a 2LTR circle as the principal integration intermediate,[87,89,92] it was hypothesized that it may still be possible for INT to use such a molecule as a template for integration if it were the only, or predominant, species delivered into the nucleus.[87] This hypothesis is supported by the existence of the 2LTR circles in MLV-infected cells[90] and evidence from the spleen necrosis virus (SNV) system that the LTR junction fragment can be an effective substrate for integration.[91]

We investigated whether a 5′LTR–3′LTR junction fragment, in a closed circular DNA molecule excised from an incoming plasmid by Cre recombinase and in the absence of the preferred, linear viral DNA molecules, could be recognized by the retroviral integration machinery (Fig. 3). A fused LTR junction fragment was thus cloned, containing the entire 3′LTR and just 28 bp of the U3 region of the 5′LTR (Fig. 4(A)). This LTR junction together with the puromycin resistance gene was flanked by LoxP sites and was demonstrated to efficiently excise a circular proviral intermediate in vitro upon supply of Cre recombinase in *trans*.[93] Further studies in cell lines *trans*-complementing Gag/Pol gene functions, together with Cre recombinase generated long-term neomycin-resistant clones. Genomic DNA extracted from stable clones was used to investigate the proviral integration structures by utilizing a panel of diagnostic PCR primers. The PCR demonstrated that integration following plasmid transfection, Cre excision, and puromycin selection for >1 month can produce a very specific molecular structure which is distinct from that produced by random plasmid integration. PCR results demonstrated that the 5′ and 3′ LTRs, which are adjoining in the plasmid backbone, become separated by the intervening sequences of the retroviral vector genome (between the loxP sites).

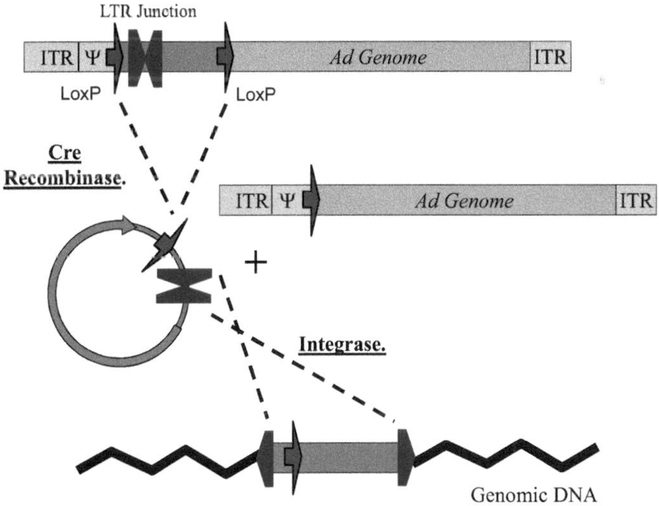


**Fig. 3 Genomic integration of an adenovirally delivered retroviral circular provirus cassette.** A LoxP flanked cassette containing a fused terminal LTR junction and transgenes of interest were inserted into an adenoviral vector. Upon infection of cells expressing cre recombinase, this cassette is efficiently excised as a closed circular molecule. The fused LTR junction contained in this circular proviral molecule are subsequently recognized by retroviral integrase directing integration into the host chromosome.

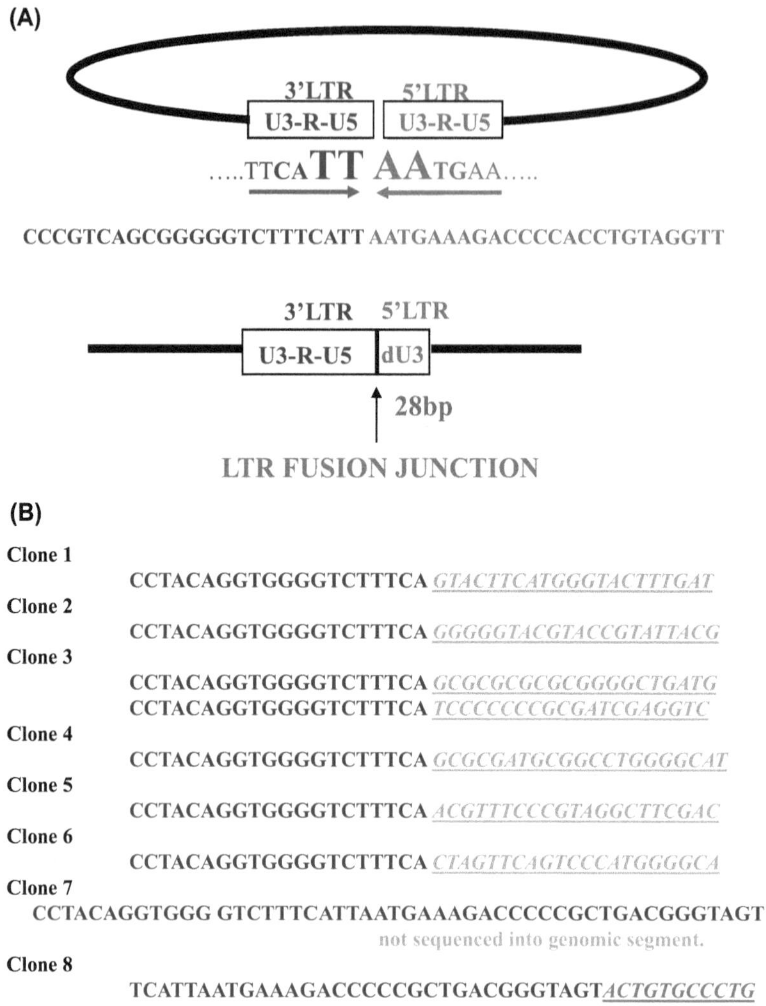


**Fig. 4 Sequencing of the integration junctions of the circular RV proviral cassette.** (A) Schematic representation of the RV genome conformation in the non-covalently linked circular pre-integration complex and the subsequently cloned fused LTR junction. (B) A human cell line expressing the retroviral gag/pol genes and cre recombinase (TelCre) were infected with the Ad/RV hybrid vector expressing puromycin resistance. Colonies were selected which had stably integrated the RV proviral cassettes and the genomic DNA extracted. The integration junctions were subsequently cloned by PCR amplification of re-ligated restriction digested fragments containing the integration site,[93] which were subsequently sequenced through the integration site.

A molecule is thus generated in which the proviral genome is now bounded by the LTRs in a manner typical of INT-mediated integration (Fig. 3). Critically, the terminal two base pairs of both the 5′LTR (U3 region) and the 3′LTR (U5 region) were lost (Fig. 4(B)). Hence these studies confirm that a circular retroviral genome with terminally fused LTR structures can indeed serve as a substrate for the retroviral machinery.

From this initial proof of concept, the LoxP cassette was subsequently assembled into an E1-deleted Ad vector. The Ad virus is used to deliver the LTR junction fragment into the nuclei of cells; the proviral-like intermediate can then be excised from the Ad genome by the Cre/lox system and forms a template for INT-mediated integration. This hybrid Ad/RV system thus has the high transient titer of Ad vectors, does not depend upon cell division for infection, and leads to long-term gene expression via integration of a proviral transgene cassette. Delivery of the Lox-Puro-Junc-Lox cassette in an Ad vector, in the presence of Cre and GAG and POL allowed cloning of cells which are resistant to puromycin for long periods in culture. Without Cre, such clones were impossible to obtain. Moreover, these clones contain a molecular structure consistent with proviral integration by PCR and contain integration sites which, for the majority of the clones (7 of 9), are typical of INT-mediated, rather than random integration processes (Fig. 4(B)). Codelivery of three separate Ad vectors, Ad-GAG/POL, AdCre, and Ad.LTR.Junc, was also ablc to produce long-term integrants. Therefore, we are currently optimizing the design and use of this novel hybrid vector system into a single, or double, Ad delivery system. Recent experiments have shown that Pol-expressed INT alone is sufficient to drive the integration of the Cre-excised proviral form in vitro without the need for additional Pol or Gag proteins. An Ad vector was thus cloned incorporating the INT gene in the same cassette as the transgene cassette to enable a two-vector transduction strategy, which is currently under investigation in our laboratory. This novel hybrid vector system presents great potential in enabling the stable transduction of all cells primarily infected by the Ad vectors.

## 3.3 Ad/EBV Hybrid Vectors

An alternative application of the Cre/LoxP recombinase system of excising a circular proviral molecule from an Ad vector[93] has replaced the retroviral component with the genetic stability of the EBV replicon system.[94,95] This hybrid Ad/EBV vector system utilizes Ad-mediated nuclear delivery of a Cre-excisable EBV replicon which can be stably maintained as an EBV episome.[96] EBV episomes contain the EBV latent origin of replication ($Ori_p$) and the EBV nuclear antigen-1 (EBNA-1) which acts on $Ori_p$, driving episomal replication (Fig. 5(A)). Previous studies have demonstrated that EBV nuclear episomes are stably maintained through multiple cell divisions in primate and canine cells, replicating once during S phase and segregating to both daughter cells with approximately 95% efficiency.[97] Tan and colleagues flanked $Ori_p$ and EBNA-1, together with the puromycin resistance gene, with LoxP sites and cloned them into an E1-deleted Ad vector.[94] However, multiple attempts to make an adenovirus failed due to suspected inhibition of Ad replication upon binding of ENBA-1 to $Ori_p$. Hence a vector was assembled which only brought EBNA-1 upstream of its promoter following Cre excision of the proviral cassette (Fig. 5(B)), thus silencing its expression in the absence of Cre recombinase.[94] The resultant Ad/EBV hybrid vector stably transformed 37% of surviving canine D-17 cells to puromycin resistance following cotransfection with Ad-Cre. The circular EBV replicons were maintained in daughter cells for 14 weeks, ~110 cell generations. Surprisingly the puromycin resistance gene was also discovered in an integrated form, in the cellular chromosomal DNA.[94] Integration of EBV episomes has not been reported previously in human cells, although a differential function in the canine

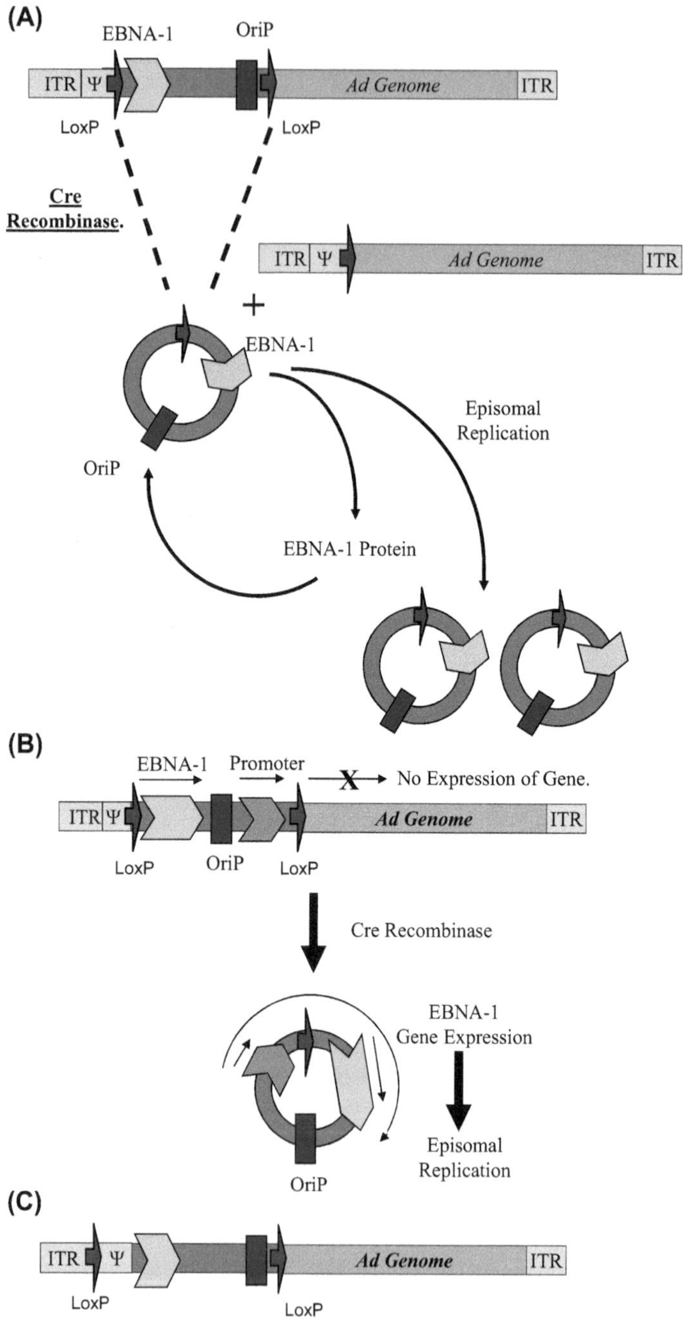


**Fig. 5 Episomal replication and maintenance of a cre-excised EBV replicon from an adenoviral vector.** (A) A LoxP flanked cassette containing the EBV origin of replication ($Ori_P$) nuclear antigen (EBNA-1) and transgenes of interest were inserted into an adenoviral vector. Upon infection of cells expressing cre recombinase, this cassette is efficiently excised as a closed

cells may be involved. One major limitation of the hybrid system was, however, that a large cell fatality was observed upon transduction with the Ad vector into the D-17 cells at the optimum transduction conditions (MOI 30). This was discussed by the authors as a function of leaky Ad gene expression from the first-generation vector in the canine cells and not related to EBNA-1 toxicity.[94] Reports have previously shown that EBNA-1 does not elicit a cytotoxic T-cell response, due to the presence of a series of glycine–alanine repeats.[98] These repeats act in *cis* to prevent major histocompatibility complex class I presentation by inhibiting antigen processing by the ubiquitous processing pathway.[98]

A very similar Ad/EBV replicon vector has also been described in the context of an E4-deleted, second-generation Ad vector.[95] Coinfection of human cells with this vector, together with AdCre (also E4-deleted), resulted in efficient delivery and excision of the replicon in the absence of vector-induced toxicity. The replicons were maintained following successive cell divisions both in vitro and in vivo, suggesting efficient extrachromosomal replication as well as nuclear retention of the episome. The residual Ad backbones were, however, progressively lost by a dilution mechanism occurring in the absence of DNA replication.[95]

As for all gene therapy vector systems, incorporation of all the components into a single vector would simplify delivery and therapeutic efficacy. As for the previously described Cre-excisable RV provirus strategy,[93] combination of the vector elements for the Ad/EBV has its limitations. Any expression of Cre recombinase would result in premature excision of the EBV replicon, specifically upon initial Ad propagation. Wang and colleagues, however, enabled incorporation of all the components into a gutless HD Ad vector by use of a tissue-specific promoter to control Cre expression.[99] Placing Cre under the control of a synthetic promoter (HCR12), consisting of hepatic locus control elements from the human ApoE/C locus fused to the first intron of the human EF1α gene allowed adequate suppression of expression in 293 cells while permitting recombination and subsequent gene expression in the target tissue. However, promoter activity was not completely extinguished in all nonhepatic cells. In order to limit the effects of leaky Cre excision of the LoxP cassette, the Ad packaging signal was included in the excisable cassette (Fig. 5(C)). The placing of a LoxP site between the LTR and ψ has been demonstrated not to inhibit Ad expression extensively.[13] Thus leaky excision would remove ψ from the Ad vector backbone, rendering the Ad genome non-packagable and hence preventing contamination in the final viral stocks. Additionally, removal of ψ from the cassette also eliminates the E1 enhancer elements, which are inter-linked with the packaging elements, which have been reported to additionally limit leaky viral expression events from the adenoviral tripartite leader sequence (TPL).[99]

◀ circular molecule. Upon action of EBNA-1 on $Ori_P$, the circular cassette is efficiently replicated by the rolling circle mechanism facilitating maintenance of the casstte in the infected cells. (B) In order to control the expression of the EBNA-1 genes the promoter and gene are separated in the Adenoviral cassette and only become productively in line in the excised replicon form following cre-mediated excision. (C) By inserting the left LoxP site between the Ad LTR and Ψ, leaky excision of cre recombinase and subsequent premature excision of the loxP cassette would render the resulting Ad vector non-packagable due to elimination of Ψ. This strategy eliminates contamination of the excised adenoviral form upon propogation of the hybrid Ad/EBV vector.

## 3.4 Hybrid Retroviruses Trafficking to the Nucleus

While previously described strategies have focused on combining the advantageous properties of retroviruses into adenoviral vectors, research has also investigated the reverse scenario. An interesting study by Lieber and colleagues investigated the potential of inserting the nuclear localization functions of an adenovirus into a retrovirus.[100] The failure of MoMuLV to cross the nuclear membrane in the absence of cell division has limited retroviral vectors. Large proteins or complexes (>40–60 kDa), such as the retroviral pre-integration complex are too large to pass into the nuclear membrane by simple diffusion and require nuclear localization signals (NLS). NLS interact with cytoplasmic receptors initiating an energy-dependent multistep translocation into the nucleus. The efficient nuclear-targeting properties of Ad vectors have made them ideal gene delivery vehicles. It is generally believed that NLS in the preterminal protein (pTP) and the core protein V play a crucial role in directing the Ad genome complex to the nucleus. The Ad preterminal protein (pTP) binds alone or in a complex with the Ad polymerase to specific sequences at the termini of the adenoviral ITRs. Lieber and colleagues investigated whether coexpression of pTP with retroviral DNA carrying pTP-binding sites would facilitate nuclear import of the pre-integration complex and transduction of quiescent cells. Preliminary experiments demonstrated successful nuclear import of plasmid DNA via the karyotypic pTP (in the presence or absence of Ad-polymerase) into the nuclei of growth-arrested cells.[100]

The pTP-binding motif was initially established by engineering two head-to-head adenoviral ITRs, but was later reduced to an 18-bp terminal fragment of the ITR, deemed the minimum required unit.[100] Interestingly attempts to introduce the full Ad ITR fragment into retrovirus vectors resulted in viruses with very low titers ($<10^2$ iu/ml), indicating adverse effects on retroviral replication. The minimal ITR 18-mer oligonucleotide, however, allowed high-titer retrovirus production. The pTP-binding site was placed in the center of the recombinant vector between hAAT and neo in order to avoid potential interference of pTP binding on pre-integration complex stoichiometry. Results demonstrated that the incorporation of the pTP karyotypic machinery in the context of the retroviral backbone could indeed efficiently translocate the RV genome across the nuclear divide. pTP-mediated transduction was, however, always less than in proliferating cells, possibly indicating weak binding to the viral DNA, which is supported by Ad-Pol increasing nuclear import and transduction. Alternatively the nuclear matrix-binding properties of pTP could interfere with the retroviral transduction functions. Disappointingly, however, pTP nuclear import of MoMuLV DNA in nondividing cells was found not to be sufficient for stable transduction. Undetermined additional cellular factors activating during S phase and/or DNA repair are required for efficient retroviral integration.[100]

## 3.5 Hybrid Ad/AAV Vectors

Incorporation of AAV nuclear retention functions into hybrid Ad vectors has also become a great interest in gene therapy. AAV vectors have emerged strongly as candidates for gene therapy, being nonpathogenic and presenting a mechanism of stable integration into a specific locus of the human host chromosome. The terminal ITR structures contain all the *cis*-acting elements required to drive episomal replication, host genome integration, and packaging into infectious AAV particles.[101–104] The *rep* gene products mediate the amplification of the AAV genome and facilitate site-specific integration into the human chromosome 19q13.3, termed AAVS1.[32,105] In the context of double-stranded circular DNA plasmid vectors, the presence of the two AAV ITRs was demonstrated to be sufficient to rescue AAV genomes from the plasmid backbone and to mediate its integration into host DNA.[101,106] These findings paved the way to the development of AAV vectors and initiated the application of AAV genomes in hybrid vector systems. The ITR-flanked AAV cassettes were subsequently demonstrated to also be efficiently rescued from the backbones of other viral vectors. In cultured cells, AAV integrates into the host chromosome with a relatively low frequency of $1 \times 10^{-4}$ to $3 \times 10^{-4}$ genomes per cell, with alternative episomal replication of its genome permitting long-term persistent expression in cells.[107] However, the integration efficiency can be enhanced by stimulation of the host DNA repair machinery by gamma irradiation or topoisomerase inhibitors.[108,109] The only requirement for AAV integration and episomal concatemerization appears to be the presence of AAV ITRs and as yet undetermined cellular factors.[101,102,106]

Following the hybrid Ad/RV studies, AAV ITR-flanked transgene cassettes have been similarly applied in the context of an Ad backbone. AAV ITR cassettes can be efficiently rescued from Ad genomes and assembled into AAV particles upon the supply of rep and cap functions in *trans*.[15] In the absence of the *cap* genes, Ad-mediated delivery of the AAV ITR cassettes can result in its stable integration into the host genome, in the presence or absence of the *rep* genes (Fig. 6). However, studies on the relationship between Ad and AAV demonstrate a strong interference of AAV on the Ad life cycle.[110] Although the precise mechanism is undetermined, Rep expression is sufficient to suppress the maturation of Ad replication centers.[111] Hence the major complication in the union of the Ad/AAV hybrid vector system has been strategies to facilitate rep expression in the context of an adenoviral vector.

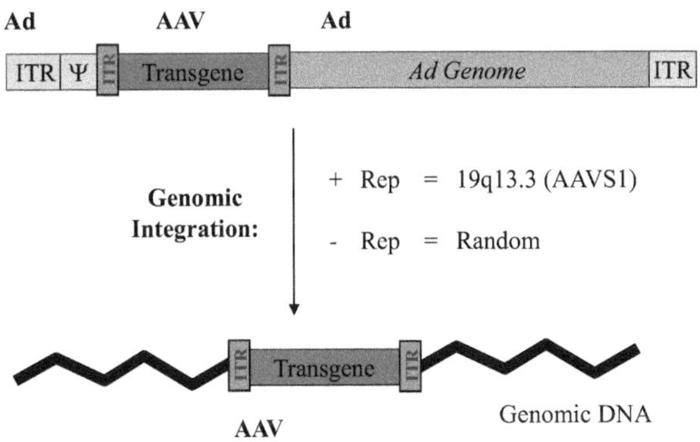


**Fig. 6 Hybrid-Ad/AAV vector mediated integration of the AAV ITR cassettes.** Infection of cells with the hybrid Ad/AAV vector enables precise excision of the ITR-flanked cassette from the adenoviral genome, which can subsequent be integrated into the host genome. This mechanism can occur in the absence or presence of the AAV Rep proteins. In the presence of Rep the cassette is predictably integrated into the AAVS1 loci on human chromosome 19.

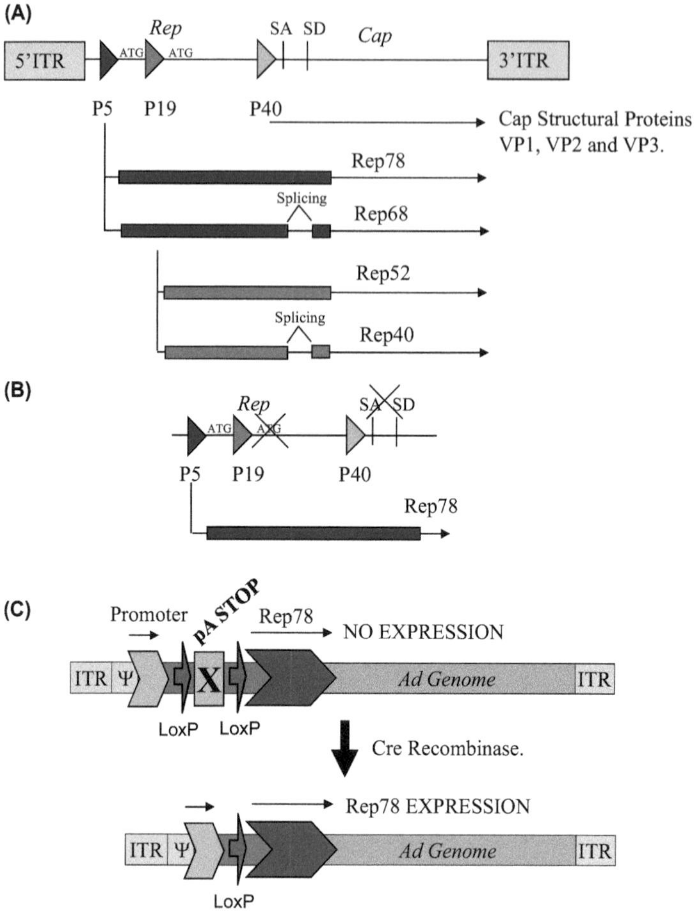


**Fig. 7 Control of the expression of the AAV Rep proteins in the context of their inhibitory effects on Ad production.** (A). Schematic representation of the AAV genome. Three promoter are contained in the AAV genome; P5 and p19 control expression of the alternatively spliced Rep68/78 and Rep40/52 transcripts, respectively, while P40 controls expression of the Cap gene products. (B) In order to restrict expression from the Rep cassette to just the Rep78 form, a point mutation was introduced into the P19 promoters ATG start site, preventing Rep40/52 expression and a similar mutation in the P5 transcripts splice site eliminated Rep68 expression. (C) In order to further restrict Rep78 expression from the Ad vector a Loxp cassette flanking a polyA stop site was cloned between the Rep78 gene and its promoter. This cassette completely silences Rep78 by preventing translation by premature termination at the introduced polyA site. Cre-mediated excision, removes this cassette permiting Rep78 expression to proceed.

The AAV *rep* gene encodes four proteins that are expressed from independent promoters (Fig. 7(A)). The differentially spliced Rep68 and Rep78 products are expressed from the P5 promoter and individually are capable of catalyzing AAV genome integration.[112,113] The poorly characterized Rep40 and Rep52 proteins are expressed from the P19 promoter and, although having similar catalytic properties to the other proteins, their function is undetermined and believed to be distinct from Rep68/78.[114] Recchia and colleagues investigated amplification conditions that were likely to minimize Rep inhibition of vector production.[115] Specifically, P19 promoter expression of Rep52 and Rep40 was reported to impose significant inhibitory functions to Ad replication, although Rep68 and Rep78 functions were also apparent. To minimize the complications of Rep-mediated interference of Ad production, the expression of Rep proteins was restricted to just the Rep78 isotype, by inactivating the Rep52 and Rep40 transcripts by ATG mutation and preventing Rep68 splicing by similar point mutations at the splice site (Fig. 7(B)). Rep79 was placed under the control of either the T7 promoter, a promoter previously applied to the production of adenoviruses expressing toxic genes, or the α1-antitrypsin (α1at) liver specific promoter to additionally minimize any interference. The AdT7–Rep78 shuttle vector was successfully recombined in 293 cells to generate the Ad vector, whereas shuttle vectors containing the wild-type Rep or only expressing the Rep52 and Rep40 isotypes, did not yield any viral plaques. The functionality of the AdT7–Rep78 vector was demonstrated in an AAV rescue study.[115]

As expression from the α1at promoter restricts expression to hepatic tissues, Ueno and colleagues applied an alternative Cre/LoxP bacteriophage P1 system as a switch to regulate Rep expression from an Ad vector.[116] A LoxP-flanked cassette containing a transcriptional silencing sequence (SV40 polyA) was cloned between the *Rep78* gene and its CAG promoter (Fig. 7(C)). Hence upon Cre recombinase expression the LoxP cassette is excised, uniting promoter and transgene and allowing transcription to proceed. The authors failed to yield any virus with *Rep78* driven by the CAG promoter in the absence of the Lox-stop cassette. The vector system thus required a third Ad expressing Cre (AdCre) to be coexpressed with AdLoxP–Rep78 and the Ad/AAV hybrid vector. Only upon application of all three vectors to HeLa cells (MOIs of 10–20) site-specific integration into AAVS1 was detected by PCR analysis of genomic DNA.[116] As with the previous systems, incorporation of all the vector components into a single gutless vector is a major aim. The application of Cre recombinase would thus, as with the previous Ad/EBV replicon system,[94] require tightly controlled expression to be incorporated into the same HD vector as the LoxP cassette.

### 3.5.1 Helper-Dependent Ad/AAV

Recchia and colleagues furthered the studies of Ad/AAV hybridology by incorporating the system into HD Ad vectors.[115] The system applied a similar two-vector strategy with the AAV LTR transgene cassette and *Rep78* genes on separate gutless vectors, HD-AAV and HDα1at-Rep78, respectively. The gutless Ad constructs consisted of the terminal *cis*-acting regions of the Ad genome (ITRs and ψ) together with the transgene cassette(s), as well as additional inert stuffer sequences, to bring the vector genome size above the efficient packaging size threshold (>27 kb).[35] Large-scale production of HDα1at-Rep78 generated titers of $3\times10^9$ iu from $5\times10^7$ cells, indicating 50–100 Rep-expressing viruses per cell could be produced. This HDα1at-Rep78 virus expressed Rep78 selectively in hepatic cells (Hep3B). Rescue of the AAV-LTR transgene cassette from HD-AAV into infectious AAV particles was observed upon coinfection of Hep3B cells with HDα1at-Rep78 and wild-type Ad2 helper. No AAV rescue was detected upon elimination of any of the three vector components, demonstrating the functionality of each component. Coinfection of HD-AAV with HDα1at-Rep78 into a number of cell lines of hepatic origin showed stable integration of the AAV transgene cassettes into the host cell genome specifically at AAVS1, by nested PCR analysis, southern blotting and integration site junction sequencing.[115] FISH studies on HepG2 cells infected with both vectors demonstrated targeted integration to AAVS1 in 14 of 39 (35%) metaphases analyzed. In the absence of the Rep78 vector only one integration in chromosome 19 was observed in 34 metaphases analyzed (3%). Hence, the study by Recchia and colleagues demonstrates that Rep78 expression increases the targeted integration of AAV-ITR-flanked DNA without affecting the overall integration frequency in cells of hepatic origin.[115] In contrast to other studies on 293 cells,[112,117,118] however, Rep78 did not increase the stable transduction efficiency on the hepatic cell lines investigated, which was believed to be due to cell type effects.[115] The next advance in this study will be incorporation of both cassettes into a single HD Ad vector. This will be much more complex than originally perceived considering the action of Rep78 on the AAV cassette, especially in the high copy number context of adenovirus production. Additionally, considering the Rep independent processing of the AAV ITR cassette, the fate of the cassette at high copy number in the producer cells would be of great interest.

### 3.5.2 Generation of Mini-Ad/AAV Hybrid Vectors by in vitro Hybridization

Inverted repeat (IR) sequences inserted into first-generation Ad vector genomes were recently reported to mediate precise genomic rearrangements resulting in vector genomes devoid of all viral genes but which were efficiently packaged into functional Ad virions.[119] These genomes were generated by a *trans*-recombination between two Ad genomes exchanging sequences either side of the IR regions. Hence two species are generated, firstly, a small genome containing only the transgene cassette flanked on both sides by precisely duplicated IRs, Ad packaging signals (ψ) and Ad ITRs (Fig. 8). Secondly, a larger genome is generated containing the transgene cassette flanked by the IRs and also the rest of the Ad genome (Fig. 8). The presence of the Ad packaging signal only in the mini-genome product meant that only this form could be packaged, whereas the larger genome just facilitated helper functions. Application of this precise recombination mechanism to generate mini-Ad genomes deleted of all viral genes could minimize the immunogenicity apparent with first generation vectors. By modifying the IR regions to increase the efficiency of recombination, further selection for the recombinant mini-genomes could be achieved.[119] The mini-Ad virions could be efficiently separated on CsCl gradients by buoyant density, with great resolution from the larger helper viral genomes enabling efficient purification.

The generation of the recombinant mini-Ad genomes was very efficient ($\sim 5 \times 10^4$ genomes per cell) and did not depend on the sequences within or adjacent to the IRs.[119] The mini-Ad vectors efficiently infected cultured cells with the same efficiency as first generation vectors. However, in the absence of any vector selection in

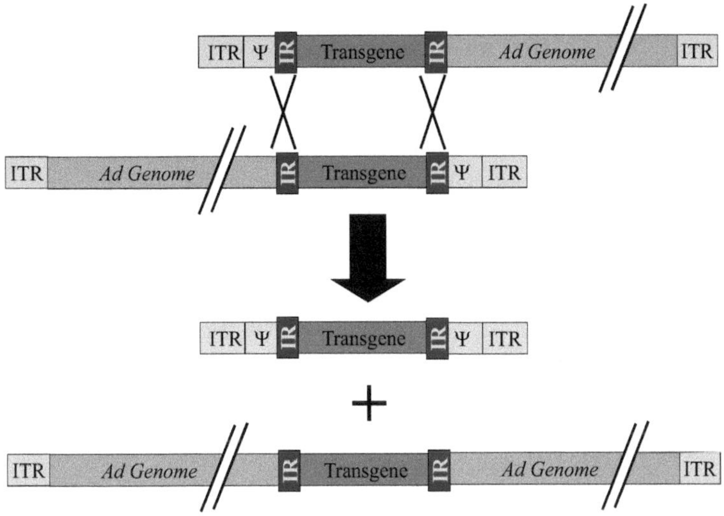


**Fig. 8 Generation of mini-Ad genomes by recombination between inverted repeat (IR) regions.** The presence of IR regions in adenoviral cassettes enables precise recombinations between different Ad genomes within the IR regions. This recombination generates mini-Ad genomes with the precisely replicated IR cassettes being flanked at either end by Ad ITR and Ψ sequences. A second, much larger, recombinant species is also generated which also contains a precisely replicated IR cassettes but is flanked on either side by the rest of the adenoviral genome. This larger recombinant, as well as being of a size non-packable into an adenoviral virion, also lacks the Ad Ψ and hence is not packaged. Whereas the smaller mini-Ad genomes can be efficiently assembled into adenoviral particles assembled by the larger genomes helper functions.

the cell (episomal replication or integration), transgene expression was only transient (~7 days) due to the instability of the deleted genomes within transduced cells.

Lieber and colleagues further developed the system to incorporate AAV cassettes into a hybrid vector system.[119] The AAV ITRs flanking the transgene cassette were used as the IRs to mediate the recombination event, as well as stimulating transgene integration into the host genome of transduced cells. The Ad-AAV vectors efficiently generated mini-genomes by IR recombination as by-products of first-generation Ad-AAV vector amplification. The mini-genomes containing only the transgene flanked by AAV ITRs, Adψs, and Ad ITRs could be efficiently assembled in Ad capsids and purified to high titers and purity. The mini-Ad-AAV hybrid vectors transduced cells with efficiencies comparable to AAV, but were less efficient than conventional Ad vectors due to elevated particle to infectious unit ratios.[119] Since the hybrid mini-vectors contained no cytotoxic viral genes, the hybrid virus could be applied at very high MOIs to increase transduction rates. The AAV transgene cassettes randomly integrated into the host cell genomes as head-to-tail concatemers, as shown by southern blot analysis and pulsed-field gel electrophoresis.

Amplification of Ad-AAV hybrid vectors in 293 cells routinely yielded final mini-Ad-AAV genome titers of $5\times10^{12}$ genomes per ml, or $\sim10^{4}$ packaged genomes per

293 cell, comprising 10% of the total number of adenoviral virions.[119] The 5.5kd mini-Ad-AAV hybrid vectors which contain two Ad packaging signals were, however, packaged approximately fivefold less efficiently than the corresponding full-length genomes.[119] These results are compatible with the published observations that Ad vector genomes of less than 27 kb package with much reduced potentials compared to full-length genomes.[35] Additionally, contamination of mini-Ad-AAV hybrid vector preparations with the parental Ad-AAV hybrid vector was less than 0.1%, consistent with conventional gutless Ad purification.[13] The efficiency of vector production measured on a genome-per-cell basis was reported to be comparable or higher than the labor-intensive techniques for AAV production. The transducing titers expressed as neomycin-resistant colonies per milliliter were $9 \times 10^5$ for AAV and $2.5 \times 10^8$ for the mini-Ad-AAV hybrid vector.[119] These results present significant clinical promise for mini-Ad/AAV hybrid vectors in the clinic.

## 4. Conclusion

The establishment of hybrid Ad vectors incorporating the advantageous properties of other viruses greatly expands their therapeutic potential. In the early 1990s, after the initial decade of proof-of-concept for Ad-mediated gene therapy, the main focus was on limiting the immunogenicity of the vectors to enhance transgene expression. Further restricting the expression of the highly immunogenic late viral transcripts by E4 deletions or by complete deletion of all viral genes in the gutless vectors notably enhanced transgene expression.[11–13,35] However, complete ablation of immunogenicity is restricted by the highly inflammatory nature of the Ad particles themselves, in the absence of viral expression. While suppression of specific immune responses can counter these effects to some extent, in many disease states where the immune system is already compromised, this rationale is not ideal. Pseudotyping the Ad vectors with alternative surface moieties does, however, offers great potential. Firstly, novel surface structures can be introduced which can reduce the immunogenicity of the viral particles by either shielding or replacing the highly immunogenic wild type structures. Future research may permit the complete replacement of the viral external domains with immune-tolerated surface structures. Secondly, the introduction of new targeting ligands will enable selected infection of desired tissue populations, limiting the required vector doses. Additionally the avoidance of infection of non-targeted tissues, specifically cells of the immune system, will negate potentially immunogenic signaling which the vectors can initialize upon receptor docking.

It is now well accepted that the immune system is not the major limiting factor in the transient expression attained from Ad vectors. The absence of a specific mechanism of long-term retention of the viral genomes in infected cells is critical. As presented at the start of this chapter, the rapid lytic life cycle of wild type adenoviruses does not require long-term persistence of the genomes. Adenovirus infection, genome replication, virion packaging and lysis of the host cell are generally completed within 48 h. While these properties have proved highly beneficial in the area of vector production, they do not aid in vivo stability. By combining the long-term stable persistence mechanisms of other viral systems into the Ad vectors, the efficacy of Ad-mediated gene therapy has been significantly enhanced.

A number of mechanisms are presented in this chapter for the combination of adenoviral and retroviral vectors. Initial applications utilized Ad vectors to deliver RV packaging functions to producer cells in vitro, in attempts to increase the efficiency of RV production. From these initial studies the potential of the hybrid Ad/RV vectors for the establishment of RV producer cell in vivo was realized. As producers of RV packaging cells in vitro the Ad/RV hybrid vector system has a number of advantages over conventional packaging cell lines. RV titers generated from transient plasmid-transfected producer cells are generally several orders of magnitude lower than the best stable clones.[3] Therefore, as the Ad/RV hybrid system has been demonstrated to generate RV titers of the same orders of magnitude as conventional producer cells, it bypasses the need to isolate clonal populations and makes scaling up production more manageable. The requirement of GMP screening for replication-competent adenoviruses (RCA), as well as RCR, would however be a concern. The separation of the different retroviral components on separate viruses, as well as limiting the potential of RCR, also makes pseudotyping very simple. For instance, in the treatment of specific tissues, an envelope gene best suited for tropism to that tissue can be easily incorporated into the vector system. The separation also enables characterization of individual RV components. Application of the individual Ad/RV hybrid vectors at varying MOIs enables study of the RV production at varying copy load. Additionally, as different target tissues have been well documented to have different potentials as RV producer cells, the hybrid Ad/RV system enables the rapid screening of tissues for their suitability as RV producer cells. However, this measure of suitability is also influenced by the susceptibility of the cells to Ad transduction. Understanding the effects of saturating RV components could allow us to determine what factors need to be further regulated in future hybrid vectors to enable enhanced RV production both in vitro and in vivo. Elucidation of host factors vital to efficient assembly of RV particles would be very valuable. In future vectors, these host factors could be codelivered or upregulated to enhance RV titers.

One major question is: what advantages does the hybrid Ad/RV-vector system have over conventional Ad or RV delivery? The ability to establish RV producer cells in vivo following Ad infection is a major step forward in gene therapy. Previous methods of ex vivo transduction with retroviral vectors and re-implantation are laborious and inefficient. Vector spread is limited by the restricted migratory properties of the re-implanted cells. Application of the hybrid Ad/RV vector enables a non-invasive therapy with the enhanced distribution and infectivity in target tissues. The subsequent local release of retroviral particles following adenoviral transduction also tackles the problem of inserting high levels of vector deep into the middle of tissue or tumor masses, rather than to just the peripheral layers. The major advantage over conventional Ad vectors is the establishment of a stable population of transgene-expressing cells in the surrounding tissues, through RV integration, following the initial transient Ad transduction. This permanency of therapeutic transgene has major implications in the clinic, specifically for the treatment of inherited diseases. The separation of the RV genes, as well as the introduction of additional regulatory elements carried on separate rAds, instils a multiple-vector transduction strategy. Vector systems involving more than one vector are limited by codelivery kinetics. The greater the number of individual vectors, the lower the probability that a cell will receive the full vector repertoire to allow retroviral production to occur. Therapeutic transgene expression can, however, proceed from the adenoviral vector itself, initiating an initial boost of gene expression, followed by a secondary level of sustained expression in RV transduced cells. Currently, however, the secondary phase of RV expression is much reduced compared to the initial Ad-mediated expression. This would minimize the sustained therapeutic effect of the vector system. Nevertheless this hybrid Ad/RV-vector system has great potential for the treatment of genetic disorders.

The dual transduction properties of the Ad/RV hybrid vector also present the possibility of combinatory gene therapy where the Ad and RV portions of the vector provide different therapeutic effects. This mechanism could have specific advantages to the treatment of cancer. The initial Ad transduction could act to initially immunostimulate the tumor mass, aiding a break in tolerance by drawing in immune effector cells and initiating "danger signal." The secondary RV transduction could deliver a cytoreductive transgene aimed at tumor cell killing, to eliminate tumor tissue and further immunostimulate the tumor environment. A major limitation of the system is the requirement of active cell division in neighboring cell populations to enable RV transduction. Hence, inserting a gene in the Ad vector, separate from the retroviral cassette, could trigger cell division of neighboring tissues so that they become fully receptive to the subsequently available retrovirus. An alternative strategic context of application could be applied to tumors, where the actively dividing tumor tissue is generally surrounded by virtually quiescent normal tissue. Utilizing a highly regulated cytotoxic gene, under the control of an inducible or tissue-specific promoter system, the primary Ad infection would enable production of retroviral particles which in theory would selectively infect dividing tumor cells. Subsequent cytotoxic gene expression could, to some extent, restrict cell killing to tumor cells.

Ad-mediated delivery of an excisable closed circular RV cassette that can subsequently be integrated into the host genome would be of great value to gene therapy in the clinic. The system provides the potential to direct stable transgene expression in each primarily Ad-infected cell. This would be a significant advance on the previous Ad/RV hybrid system, where secondary RV transduction is extensively reduced compared to the primary Ad transduction. Although the closed circular form is not the primary substrate for retrovirus integration, in the absence of the wild-type substrate, INT has been demonstrated to integrate such structures into the host genome. While the efficiency of such a system has still to be addressed, further elaboration of the integration mechanism could enable increased affinity of the RV machinery for closed circle LTR proviral forms. The system would also have the potential of combinatory gene therapy by the inclusion of transgene cassettes within or outside the integrating RV cassette. Transgenes outside the excisable cassette would provide transient expression for the duration of the Ad genome retention in infected cells. The integrated cassette could provide stable expression for the lifetime of the cell. The major advance of this system will come from the development of highly regulated expression systems that can completely silence Cre expression. Silencing of Cre recombinase expression would enable assembly of all the vector components into a single gutless HD Ad vector. Although the system proposed by Wang and colleagues[99] goes some way to prevent expression, the system is still leaky and restricts therapeutic application to hepatic tissue.

The alternative strategy of maintaining a Cre-excised circular molecule by utilizing the EBV episomal replication system provides another potentially powerful gene therapy vector, providing many of the advantageous properties detailed above. The Ad/EBV hybrid system would again require absolute control of Cre expression to combine all the components into a single vector. One limitation of this system is that EBNA-1 gene expression would have to be permanently maintained in the host cell, which could involve long-term cell regulatory or immunological problems in vivo. Conversely, the integration mechanism only requires transient expression of INT to facilitate integration, and the transgene cassette is then maintained in the context of host cell chromosome replication. While the EBV replicon has the advantage of avoiding integration-related shutdown of transgene expression, other cellular factors are believed to be involved in the eventual loss transgene expression. EBV retention in human cells has been deemed limited and lost with time.[120,121] Without drug selection, plasmids carrying the EBV elements are lost from human cells at rates of between 1% and 5% per generation.[122]

The Ad/AAV hybrid vector system provides a powerful mechanism of maintained transgene expression by integration or episomal replication. The system also provides the potential of predictable integration at a specific locus in the human genome in the presence of Rep78. The establishment of targeted integration strategies introduces valuable safety features into a gene therapy protocol. This advantageous property of integration also carries with it the potential hazard of insertional mutagenesis and the risk of activating cancer oncogenes in vivo. Although there are limited literature reports on the impact of such phenomena in a gene therapy protocol, as vector technology increases and the efficiencies of integration in human tissues are potentiated these effects could become more significant. However, even in the context of the targeted integration of AAV, the exact phenotype of integration at chromosome 19q13.4, as well as the activity of genes integrated at such a loci are still to be determined.

The generation of the mini-Ad/AAV hybrid vectors enables the high titer purification of adenoviral particles deleted of all the Ad genes, analogous to the HD rationale. The mechanism of preparation and purification, however, appear to be simpler. The Ad/AAV hybrid vector is applied to the producer cells as an Ad, which also supplies the helper functions. This bypasses the necessity of HD plasmid transfection and subsequent serial passage to enhance titers to enable purification from the contaminating helper virus. The extensive size difference of the derived mini-Ad/AAV genomes, from the parental Ad/AAV genomes, also enables more efficient purification by buoyant density on CsCl gradients. Additionally, any contaminating parental vector will be a functional Ad/AAV hybrid vector. The biological stability of these mini-adenoviruses, both in terms of particle stability outside the cell and genome stability within the cell, still needs to be addressed. Nevertheless, considering the integration of the ITR AAV cassettes, the mini-Ad genome stability is not as important. Additionally although the transduction efficiency of the mini-Ad particles is similar to AAV, the ratio of total particles to infectious virions is enhanced, limiting their efficiency compared to conventional Ad vectors. The vector is also limited in terms of codelivery of the *rep* gene, which unlike the HD vectors cannot be easily incorporated into the same vector.

HSV-based hybrid vectors have also been well reported in literature, presenting a number of advantageous properties over the described Ad-based systems. The development of the HSV-1 amplicon technology and helper-free packaging systems has made HSV-based vectors a very promising clinical tool for gene therapy.[39] The large insert capacity of the amplicons (15kb) and the concatemer-styled packaging, with up to 10 genome copies per virion introduces powerful features to gene therapy vectorology.[25] HSV vectors, like Ads, have tropism for most cells in the human body, but have particular affinity neuronal tissues. The HSV-1 based amplicons do not, however, retain the episomal maintenance functions of the parental herpes viruses and thus, as with Ad vectors, the genomes are rapidly lost in dividing cells. Hence, hybrid technology has been investigated to enhance the expression from HSV amplicon vectors. As with the Ad/AAV hybrid system, the presence of an AAV ITR-flanked cassette in the HSV amplicon vector can promote both extrachromosomal amplification and integration of the transgene cassette into the host genome. The HSV/AAV hybrid vector system has been demonstrated to stably transform dividing cells for over 25 passages in culture.[123] Hepatic transduction in vivo with an HSV/AAV hybrid vector supported gene expression in vivo for considerably longer periods than traditional HSV-1 amplicons, with minimal toxicity and immunogenicity.[117] An additional feature of the HSV/AAV studies was the placement of the *Rep* gene under the control of its own promoter, as literature has reported potential downregulation feedback inhibition of transcription when rep levels increase.[124] Thus the natural expression machinery of rep is utilized to regulate its expression. Compared to Ad vectors, the HSV amplicon vector titers are limited ($10^7$ to $10^8$ tu/ml). Increased copy number can compensate for reduced titers in some fields of gene therapy, although for many corrective genetic therapies higher transduction efficiencies from higher titer viral applications may prove more efficacious. The reduced immunogenicity of the HSV-1 amplicons is, however, a major advantage over Ad vectors. Other hybrid vectors have also combined RV and EBV functions within the HSV-1 amplicons, which also have great potential as gene delivery vectors.[120,121,125]

The development of hybrid viral vector systems has thus revolutionized the way gene therapy vectors are conceived. The combination of the advantageous properties of different vectors goes some way to establishing a vector system approaching the ideologies of a perfect gene transfer vehicle. The technologies are, however, in their infancy and many factors need to be elucidated before the full potentials of the vectors can be achieved. In essence, further detailed elucidation of the viral life cycles and their interactions with host cell factors is necessary. Understanding these factors will allow vectors to be developed which can interact with the host cellular machinery to facilitate long-term stable gene expression. Future "hybrid" vectors will be developed quite distinct from the currently perceived parental vectors. Virtually synthetic viral vectors will be established with predictable biological properties, which can effect desired clinical functions. The vast array of clinical phenotypes and biological properties of target tissues involved in human disease will, however, require a wide spectrum of clinical vectors, fashioned to specific disorders. Nevertheless, the current advances in the development of hybrid viral vector technology.

## References


1. Coffin JM. Retroviridae: the viruses and their replication. In: Fields BN, et al., editor. *Fundamental virology*. 3rd ed. New York: Raven Press; 1996. p. 771–813.
2. Vile RG, Russell SJ. Retroviruses as vectors. *Br Med Bull* 1995;**51**:139–58.
3. Cossett F-L, Takeuchi Y, Battini J-L, Weiss RA, Collins MKL. High titer packaging cells producing recombinant retroviruses resistant to human complement. *J Virol* 1995;**69**:7430–6.
4. Zhang WW. Review: adenovirus vectors: development and application. *Exp Opin Invest Drugs* 1997;**6**:1419–57.
5. Shenk T. Adenoviridae: the viruses and their replication. In: Fields BN, et al., editor. *Fields virology*, 3rd ed. vol. 2. New York: Lippincott-Raven; 1996. p. 2111–48.
6. Bett AJ, Prevec L, Graham FL. Packaging capacity and stability of human adenovirus type 5 vectors. *J Virol* 1993;**67**:5911–21.
7. Graham FL, Smiley J, Russell WC, Nairn R. Characterisation of a human cell line transformed by DNA from human adenovirus type 5. *J Gen Virol* 1977;**36**:59–72.
8. Bett AJ, Haddara W, Prevec L, Graham FL. An efficient and flexible system for construction of adenovirus vectors with inserts or deletions in early regions 1 and 3. *Proc Natl Acad Sci USA* 1994;**91**:8802–6.
9. Gorziglia MI, Kadan MJ, Yei S, Lim J, Lee GM, Luthra R, et al. Elimination of both E1 and E2a from adenovirus vectors further improves prospects for *in vivo* human gene therapy. *J Virol* 1996;**70**:4173–8.
10. Fang B, Wang H, Gordon G, Bellinger DA, Read MS, Brinkhous KM, et al. Lack of persistence of E1- recombinant adenoviral vectors containing a temperature-sensitive E2A mutation in immunocompetent mice and hemophilia B dogs. *Gene Ther* 1996;**3**:217–22.
11. Armentano D, Zabner J, Sacks C, Sookdeo CC, Smith MP, St George JA, et al. Effect of the E4 region on the persistence of transgene expression from adenovirus vectors. *J Virol* 1997;**71**:2408–16.
12. Wang Q, Greenburg G, Bunch D, Farson D, Finer MH. Persistent transgene expression in mouse liver following *in vivo* gene transfer with a δE1/δE4 adenovirus vector. *Gene Ther* 1997;**4**:393–400.
13. Parks RJ, Chen L, Anton M, Sankar U, Rudnicki A, Graham FL. A helper-dependent adenovirus vector system: removal of helper virus by Cre-mediated excision of the viral packaging signal. *Proc Natl Acad Sci USA* 1996;**93**:13565–70.
14. Berns KI. Parvoviridae and their replication. In: Fields BN, et al., editor. *Fundamental virology*. 3rd ed. New York: Raven Press; 1991. p. 817–37.
15. Kremer EJ, Perricaudet M. Adenovirus and adeno-associated virus-mediated gene-transfer. *Br Med Bull* 1995;**51**:31–44.
16. Flotte TR, Afione SA, Zeitlin PL. Adeno-associated virus vector gene expression occurs in nondividing cells in the absence of vector DNA integration. *Am J Resp Cell Mol Biol* 1994;**11**:517–21.
17. Russell DW, Miller AD, Alexander IE. Adeno-associated virus vectors preferentially transduce cells in S phase. *Proc Natl Acad Sci USA* 1994;**91**:8915–9.

18. Walsh CE, Liu JM, Xiao X, Young NS, Nienhuis AW, Samulski RJ. Regulated high level expression of a human gamma-globin gene introduced into erythroid cells by an adeno-associated virus vector. *Proc Natl Acad Sci USA* 1992;**89**:7257–61.
19. Linden RM, Ward P, Giraud C, Winocour E, Berns KI. Site-specific integration by adeno-associated virus. *Proc Natl Acad Sci USA* 1996;**93**:11288–94.
20. Muzyczka N. Use of AAV as a general transduction vector for mammalian cells. *Curr Top Micro Immunol* 1992;**158**:97–129.
21. Verma IM, Somia N. Gene therapy - promises, problems and prospects. *Nature* 1997;**389**:239–42.
22. Roizman B, Sears AE. Herpes simplex viruses and their replication. In: Fields BN, et al., editor. *Fundamental virology*. 3rd ed. New York: Raven Press; 1991. p. 849–95.
23. Efstathiou S, Minson AC. Herpes virus-based vectors. *Br Med Bull* 1995;**51**:45–55.
24. Huard J, Akkaraju G, Watkins SC, Pikecavalcoli M, Glorioso JC. LacZ gene transfer to skeletal muscle using a replication-defective herpes simplex virus type 1 mutant vector. *Hum Gene Ther* 1997;**8**:439–52.
25. Kwong AD, Frenkel N. Biology of herpes-simplex virus (HSV) defective viruses and development of the amplicon system. In: Kaplitt MG, Loewy AD, editors. *Viral vectors*. New York: Academic Press; 1995. p. 25–42.
26. Frenkel N, Singer O, Kwong AD. The herpes simplex virus amplicon - a versatile defective virus vector. *Gene Ther* 1994;**1**(Suppl. 1):S40–6.
27. Wong-Staal F. Human immunodeficiency viruses and their replication. In: Fields BN, et al., editor. *Fundamental virology*. 3rd ed. New York: Raven Press; 1991. p. 709–23.
28. Zufferey R, Nagy D, Mandel RJ, Naldini L, Trono D. Multiply attenuated lentiviral vector achieves efficient gene delivery *in vivo*. *Nat Biotech* 1997;**15**:871–5.
29. Xiao X, Li J, Samulski RJ. Production of high-titer recombinant adeno-associated virus vectors in the absence of helper adenovirus. *J Virol* 1998;**72**:2224–32.
30. Yang Y, Jooss KU, Su Q, Ertl HCJ, Wilson JM. Immune response to viral antigens versus transgene product in the elimination of recombinant adenovirus-infected hepatocytes *in vivo*. *Gene Ther* 1996;**3**:137–44.
31. Burns JC, Friedmann T, Driever W, Burrascano M, Yee JK. Vesicular Stomatitis virus G glycoprotein pseudotyped retroviral vectors: concentration to very high titers and efficient transfer to mammalian and non-mammalian cells. *Proc Natl Acad Sci USA* 1993;**90**:8033–7.
32. Weitzman MD, Kyostio SR, Kotin RM, Owens RA. Adeno-associated virus (AAV) Rep proteins mediate complex formation between AAV DNA and its integration site in human DNA. *Proc Natl Acad Sci USA* 1994;**91**:5808–12.
33. Miller DB, Adam MA, Miller AD. Gene transfer by retrovirus vectors occurs only in *Cells* that are actively replicating at the time of infection. *Mol Cell Biol* 1990;**10**:4239–42.
34. Poeschla EM, Wong-Staal F, Looney DJ. Development of HIV vectors for anti-HIV gene therapy. *Proc Natl Acad Sci USA* 1996;**93**:11396–9.
35. Parks RJ, Graham FL. A helper-dependent system for adenovirus vector production helps define a lower limit for efficient DNA packaging. *J Virol* 1997;**71**:3293–8.
36. Vos J-MH. Herpesviruses as genetic vectors. In: *Viruses in human gene therapy*. Durham, NC, and London: Carolina Academic Press and Chapman and Hall; 1995.
37. Gough MJ, Vile RG. Different approaches in the gene therapy of cancer. *Forum* 1999;**9**:225–36.
38. Yang Y, Li Q, Ertl CJH, Wilson J. Cellular and humoral immune responses to viral antigens create barriers to lung-directed gene therapy with recombinant adenoviruses. *J Virol* 1995;**69**:2004–15.

39. Fraefel C, Song S, Lim F, Lang P, Yu L, Wang Y, et al. Helper virus-free transfer of herpes simplex virus type 1 plasmid vectors into neural cells. *J Virol* 1996;**70**:7190–7.
40. Yei S, Mittereder N, Te K, O'Sullivan C, Trapnell BC. Adenovirus-mediated gene transfer for cystic fibrosis: quantitative evaluation of repeated *in vivo* vector administration to the lung. *Gene Ther* 1994;**1**:192–200.
41. Worgall S, Wolff G, Falck-Pedersen E, Crystal RG. Innate immune mechanisms dominate elimination of adenoviral vectors following *in vivo* administration. *Hum Gene Ther* 1997;**8**:37–44.
42. Kieff E, Liebowitz D. Epstein-Barr virus and its replication. In: Fields BN, et al., editor. *Fundamental virology*. 2nd ed. New York: Raven Press; 1991. p. 897–928.
43. Rapp F, Melnick JL, Brutel JS, Kitahara T. The incorporation of SV40 genetic material into adenovirus 7 as measured by intranuclear synthesis of SV40 tumor antigen. *Proc Natl Acad Sci USA* 1964;**52**:1348–52.
44. Lewis BA, Tullis G, Seto E, Horikoshi N, Weinmann R, Shenk T. Adenovirus E1A proteins interact with the cellular YY1 transcription factor. *J Virol* 1995;**69**:1628–36.
45. Freund CT, Tong XW, Block A, Contant CF, Kieback DG, Rowley DR, et al. Adenovirus-mediated suicide gene therapy for bladder cancer: comparison of the cytomegalovirus- and Rous sarcoma virus-promoter. *Anticancer Res* 2000;**20**:2811–6.
46. Chen L, Perlick H, Morgan RA. Comparison of retroviral and adeno-associated viral vectors designed to express human clotting factor IX. *Hum Gene Ther* 1997;**8**:125–35.
47. Jackson RJ, Kaminski A. Internal initiation of translation in eukaryotes: the picornavirus paradigm and beyond. *RNA* 1995;**1**:985–1000.
48. Yee JK, Friedmann T, Burns JC. Generation of high-titer pseudotyped retroviral vectors with very broad host range. *Methods Cell Biol* 1994;**43**(Pt A):99–112.
49. Schnierle BS, Stitz J, Bosch V, Nocken F, Merget-Millitzer H, Engelstadter M, et al. Pseudotyping of murine leukemia virus with the envelope glycoproteins of HIV generates a retroviral vector with specificity of infection for CD4-expressing cells. *Proc Natl Acad Sci USA* 1997;**94**:8640–5.
50. Emi N, Friedmann T, Yee JK. Pseudotype formation of murine leukemia virus with the G protein of vesicular stomatitis virus. *J Virol* 1991;**65**:1202–7.
51. Pensiero MN, Wysocki CA, Nader K, Kikuchi GE. Development of amphotropic murine retrovirus vectors resistant to inactivation by human serum. *Hum Gene Ther* 1996;**7**:1095–101.
52. Diaz RM, Eisen T, Hart IR, Vile RG. Exchange of viral promoter/enhancer elements with heterologous regulatory sequences generates targeted hybrid vectors to enhance both local and systemic antitumor effects of HSVtk or cytokine expression in a murine melanoma model. *J Virol* 1998;**72**:789–95.
53. Danno S, Itoh K, Baum C, Ostertag W, Ohnishi N, Kido T, et al. Efficient gene transfer by hybrid retroviral vectors to murine spermatogenic cells. *Hum Gene Ther* 1999;**10**:1819–31.
54. Kass-Eisler A, Leinwand L, Gall J, Bloom B, Falck-Pedersen E. Circumventing the immune-response to adenovirus-mediated gene-therapy. *Gene Ther* 1996;**3**:154–62.
55. Mack CA, Song WR, Carpenter H, Wickham TJ, Kovesdi I, Harvey BG, et al. Circumvention of anti-adenovirus neutralizing immunity by administration of an adenoviral vector of an alternate serotype. *Hum Gene Ther* 1997;**8**:99–109.
56. Mastrangeli A, Harvey BG, Yao J, Wolff G, Kovesdi I, Crystal RG, et al. Sero-switch adenovirus-mediated *in-vivo* gene-transfer - circumvention of anti-adenovirus humoral immune defenses against repeat adenovirus vector administration by changing the adenovirus serotype. *Hum Gene Ther* 1996;**7**:79–87.
57. Zabner J, Chillon M, Grunst T, Moninger TO, Davidson BL, Gregory R, et al. A chimeric type 2 adenovirus vector with a type 17 fiber enhances gene transfer to human airway epithelia. *J Virol* 1999;**73**:8689–95.

58. Croyle MA, Stone M, Amidon GL, Roessler BJ. *In vitro* and *in vivo* assessment of adenovirus 41 as a vector for gene delivery to the intestine. *Gene Ther* 1998;**5**:645–54.
59. Kelleher ZT, Vos JM. Long-term episomal gene delivery in human lymphoid cells using human and avian adenoviral-assisted transfection. *Biotechniques* 1994;**17**:1110–7.
60. Paillard F. Advantages of non-human adenoviruses versus human adenoviruses. *Hum Gene Ther* 1997;**8**:2007–9.
61. Hofmann C, Loser P, Cichon G, Arnold W, Both GW, Strauss M. Ovine adenovirus vectors overcome preexisting humoral immunity against human adenoviruses *in vivo*. *J Virol* 1999;**73**:6930–6.
62. Moffatt S, Hays J, HogenEsch H, Mittal SK. Circumvention of vector-specific neutralizing antibody response by alternating use of human and non-human adenoviruses: implications in gene therapy. *Virology* 2000;**272**:159–67.
63. Wickham TJ. Targeting adenovirus. *Gene Ther* 2000;**7**:110–4.
64. Wickham TJ, Segal DM, Roelvink PW, Carrion ME, Lizonova A, Lee GM, et al. Targeted adenovirus gene transfer to endothelial and smooth muscle cells by using bispecific antibodies. *J Virol* 1996;**70**:6831–8.
65. Wickham TJ, Tzeng E, Shears 2nd LL, Roelvink PW, Li Y, Lee GM, et al. Increased *in vitro* and *in vivo* gene transfer by adenovirus vectors containing chimeric fiber proteins. *J Virol* 1997;**71**:8221–9.
66. Shinoura N, Yoshida Y, Tsunoda R, Ohashi M, Zhang W, Asai A, et al. Highly augmented cytopathic effect of a fiber-mutant E1B-defective adenovirus for gene therapy of gliomas. *Cancer Res* 1999;**59**:3411–6.
67. Duisit G, Salvetti A, Moullier P, Cosset FL. Functional characterization of adenoviral/retroviral chimeric vectors and their use for efficient screening of retroviral producer cell lines. *Hum Gene Ther* 1999;**10**:189–200.
68. Rettinger SD, Kennedy SC, Wu X, Saylors RL, Hafenrichter DG, Flye MW, et al. Liver-directed gene therapy: quantitative evaluation of promoter elements by using *in vivo* retroviral transduction. *Proc Natl Acad Sci USA* 1994;**91**:1460–4.
69. Salmons B, Saller RM, Baumann JG, Gunzburg WH. Construction of retroviral vectors for targeted delivery and expression of therapeutic genes. *Leukemia* 1995;**9**(Suppl. 1):S53–60.
70. Noguiez-Hellin P, Meur MR, Salzmann JL, Klatzmann D. Plasmoviruses: nonviral/viral vectors for gene therapy. *Proc Natl Acad Sci USA* 1996;**93**:4175–80.
71. Culver KW, Ram Z, Wallbridge S, Ishii H, Oldfield EH, Blaese RM. In vivo gene transfer with retroviral vector-producer cells for treatment of experimental brain tumors. *Science* 1992;**256**:1550–2.
72. Feng M, Jackson Jr WH, Goldman CK, Rancourt C, Wang M, Dusing SK, et al. Stable *in vivo* gene transduction via a novel adenoviral/retroviral chimeric vector. *Nat Biotech* 1997;**15**:866–70.
73. Kaplan AH, Manchester M, Smith T, Yang YL, Swanstrom R. Conditional human immunodeficiency virus type 1 protease mutants show no role for the viral protease early in virus replication. *J Virol* 1996;**70**:5840–4.
74. Gossen M, Bujard H. Anhydrotetracycline, a novel effector for tetracycline controlled gene expression systems in eukaryotic cells. *Nuc Acid Res* 1993;**21**:4411–2.
75. Yang Y, Vanin EF, Whitt MA, Fornerod M, Zwart R, Schneiderman RD, et al. Inducible, high-level production of infectious murine leukemia retroviral vector particles pseudotyped with vesicular stomatitis virus G envelope protein. *Hum Gene Ther* 1997;**6**:1203–13.
76. Yoshida Y, Emi N, Hamada H. VSV-G-pseudotyped retroviral packaging through adenovirus-mediated inducible gene expression. *Biochem Biophys Res Commun* 1997;**232**:379–82.

77. Caplen NJ, Higginbotham JN, Scheel JR, Vahanian N, Yoshida Y, Hamada H, et al. Adeno-retroviral chimeric viruses as *in vivo* transducing agents. *Gene Ther* 1999;**6**:454–9.
78. Lin X. Construction of new retroviral producer cells from adenoviral and retroviral vectors. *Gene Ther* 1998;**5**:1251–8.
79. Bertran J, Miller JL, Yang Y, Fenimore-Justman A, Rueda F, Vanin EF, et al. Recombinant adeno-associated virus-mediated high-efficiency, transient expression of the murine cationic amino acid transporter (ecotropic retroviral receptor) permits stable transduction of human HeLa cells by ecotropic retroviral vectors. *J Virol* 1996;**70**:6759–66.
80. Qing K, Bachelot T, Mukherjee P, Wang XS, Peng L, Yoder MC, et al. Adeno-associated virus type 2-mediated transfer of ecotropic retrovirus receptor cDNA allows ecotropic retroviral transduction of established and primary human cells. *J Virol* 1997;**71**:5663–7.
81. Scott-Taylor TH, Gallardo HF, Gansbacher B, Sadelain M. Adenovirus facilitated infection of human cells with ecotropic retrovirus. *Gene Ther* 1998;**5**:621–9.
82. Huber BE, Richards CA, Krenitsky TA. Retroviral-mediated gene therapy for the treatment of hepatocellular carcinoma: an innovative approach for cancer therapy. *Proc Natl Acad Sci USA* 1991;**88**:8039–43.
83. Bui LA, Butterfield LH, Kim JY, Ribas A, Seu P, Lau R, et al. *In vivo* therapy of hepatocellular carcinoma with a tumor-specific adenoviral vector expressing interleukin-2. *Hum Gene Ther* 1997;**8**:2173–82.
84. Uto H, Ido A, Hori T, Hirono S, Hayashi K, Tamaoki T, et al. Hepatoma-specific gene therapy through retrovirus-mediated and targeted gene transfer using an adenovirus carrying the ecotropic receptor gene. *Biochem Biophys Res Commun* 1999;**265**:550–5.
85. Zheng C, Baum BJ, Iadarola MJ, O'Connell BC. Genomic integration and gene expression by a modified adenoviral vector. *Nat Biotechnol* 2000;**18**:176–80.
86. Harui A, Suzuki S, Kochanek S, Mitani K. Frequency and stability of chromosomal integration of adenovirus vectors. *J Virol* 1999;**73**:6141–6.
87. Brown PO. Integration of retroviral DNA. In: Swanstrom R, Vogt PK, editors. *Retroviruses. strategies of replication*. Springer Verlag; 1990. p. 19–48.
88. Varmus HE, Swanstrom R. Replication of retroviruses. In: Weiss R, Teich N, Varmus H, Coffin J, editors. *RNA tumor viruses*. Cold Spring Harbor, New York: Cold Spring Harbor Laboratory Press; 1985. p. 75–134.
89. Brown PO, Boweman P, Varmus HE, Bishop JM. Retroviral integration: structure of the initial covalent product and its precursor, and a role for the viral IN protein. *Proc Natl Acad Sci USA* 1989;**86**:2525–9.
90. Panganiban AT, Temin HM. Circles with two tandem LTRs are precursors to integrated retrovirus DNA. *Cell* 1984;**36**:673–9.
91. Shoemaker C, Goff S, Gilboa E, Pasking M, Mitra SW, Baltimore D. Structure of a cloned circular Moloney murine leukemia virus molecule containing an inverted segment: implications for retrovirus integration. *Proc Natl Acad Sci USA* 1980;**77**:3932–6.
92. Lobel LI, Murphy J, Goff SP. The palindromic LTR-LTR junction is not an efficient substrate for proviral integration. *J Virol* 1989;**63**:2629–37.
93. Murphy SJ, Chong H, Bell S, Diaz RM, Vile RG. A novel integrating adenoviral/retroviral hybrid vector for gene therapy, submitted for publication.
94. Tan BT, Wu L, Berk AJ. An adenovirus-Epstein-Barr virus hybrid vector that stably transforms cultured cells with high efficiency. *J Virol* 1999;**73**:7582–9.
95. Leblois H, Roche C, Di Falco N, Orsini C, Yeh P, Perricaudet M. Stable transduction of actively dividing cells via a novel adenoviral/episomal vector. *Mol Ther* 2000;**1**:314–22.
96. Sugden B, Marsh K, Yates J. A vector that replicates as a plasmid and can be efficiently selected in B-lymphoblasts transformed by Epstein-Barr virus. *Mol Cell Biol* 1985;**5**:410–3.

97. Yates JL, Guan N. Epstein-Barr virus-derived plasmids replicate only once per cell cycle and are not amplified after entry into cells. *J Virol* 1991;**65**:483–8.
98. Levitskaya J, Sharipo A, Leonchiks A, Ciechanover A, Masucci MG. Inhibition of ubiquitin/proteasome-dependent protein degradation by the Gly-Ala repeat domain of the Epstein-Barr virus nuclear antigen 1. *Proc Natl Acad Sci USA* 1997;**94**:12616–21.
99. Wang X, Zeng W, Murakawa M, Freeman MW, Seed B. Episomal segregation of the adenovirus enhancer sequence by conditional genome rearrangement abrogates late viral gene expression. *J Virol* 2000;**74**:11296–303.
100. Lieber A, Kay MA, Li ZY. Nuclear import of Moloney murine leukemia virus DNA mediated by adenovirus preterminal protein is not sufficient for efficient retroviral transduction in nondividing cells. *J Virol* 2000;**74**:721–34.
101. Xiao X, Xiao W, Li J, Samulski RJ. A novel 165-base-pair terminal repeat sequence is the sole cis requirement for the adeno-associated virus life cycle. *J Virol* 1997;**71**:941–8.
102. Yang CC, Xiao X, Zhu X, Ansardi DC, Epstein ND, Frey MR, et al. Cellular recombination pathways and viral terminal repeat hairpin structures are sufficient for adeno-associated virus integration *in vivo* and *in vitro*. *J Virol* 1997;**71**:9231–47.
103. Duan D, Sharma P, Yang J, Yue Y, Dudus L, Zhang Y, et al. Circular intermediates of recombinant adeno-associated virus have defined structural characteristics responsible for long-term episomal persistence in muscle tissue. *J Virol* 1998;**72**:8568–77.
104. Wu P, Phillips MI, Bui J, Terwilliger EF. Adeno-associated virus vector-mediated transgene integration into neurons and other nondividing cell targets. *J Virol* 1998;**72**:5919–26.
105. Walker SL, Wonderling RS, Owens RA. Mutational analysis of the adeno-associated virustype 2 Rep68 protein helicase motifs. *J Virol* 1997;**71**:6996–7004.
106. Balague C, Kalla M, Zhang WW. Adeno-associated virus Rep78 protein and terminal repeats enhance integration of DNA sequences into the cellular genome. *J Virol* 1997;**71**:3299–306.
107. Rutledge EA, Russell DW. Adeno-associated virus vector integration junctions. *J Virol* 1997;**71**:8429–36.
108. Alexander IE, Russell DW, Spence AM, Miller AD. Effects of gamma irradiation on the transduction of dividing and nondividing cells in brain and muscle of rats by adeno-associated virus vectors. *Hum Gene Ther* 1996;**7**:841–50.
109. Russell DW, Alexander IE, Miller AD. DNA synthesis and topoisomerase inhibitors increase transduction by adeno-associated virus vectors. *Proc Natl Acad Sci USA* 1995;**92**:5719–23.
110. Carter BJ, Laughlin CA, de la Maza LM, Myers M. Adeno-associated virus autointerference. *Virology* 1979;**92**:449–62.
111. Weitzman MD, Fisher KJ, Wilson JM. Recruitment of wild-type and recombinant adeno-associated virus into adenovirus replication centres. *J Virol* 1996;**70**:1845–54.
112. Surosky RT, Urabe M, Godwin SG, McQuiston SA, Kurtzman GJ, Ozawa K, et al. Adeno-associated virus Rep proteins target DNA sequences to a unique locus in the human genome. *J Virol* 1997;**71**:7951–9.
113. Lamartina S, Roscilli G, Rinaudo D, Delmastro P, Toniatti C. Lipofection of purified adeno-associated virus Rep68 protein: toward a chromosome-targeting nonviral particle. *J Virol* 1998;**72**:7653–8.
114. Berns KI, Linden RM. The cryptic life style of adeno-associated virus. *Bioessays* 1995;**17**:237–45.
115. Recchia A, Parks RJ, Lamartina S, Toniatti C, Pieroni L, Palombo F, et al. Site-specific integration mediated by a hybrid adenovirus/adeno-associated virus vector. *Proc Natl Acad Sci USA* 1999;**96**:2615–20.

116. Ueno T, Matsumura H, Tanaka K, Iwasaki T, Ueno M, Fujinaga K, et al. Site-specific integration of a transgene mediated by a hybrid adenovirus/adeno-associated virus vector using the Cre/loxP-expression-switching system. *Biochem Biophys Res Commun* 2000;**273**:473–8.
117. Fraefel C, Jacoby DR, Lage C, Hilderbrand H, Chou JY, Alt FW, et al. Gene transfer into hepatocytes mediated by helper virus-free HSV/AAV hybrid vectors. *Mol Med* 1997;**3**:813–25.
118. Shelling AN, Smith MG. Targeted integration of transfected and infected adeno-associated virus vectors containing the neomycin resistance gene. *Gene Ther* 1994;**1**:165–9.
119. Lieber A, Steinwaerder DS, Carlson CA, Kay MA. Integrating adenovirus-adeno-associated virus hybrid vectors devoid of all viral genes. *J Virol* 1999;**73**:9314–24.
120. Wang S, Vos JM. A hybrid herpesvirus infectious vector based on Epstein-Barr virus and herpes simplex virus type 1 for gene transfer into human cells *in vitro* and *in vivo*. *J Virol* 1996;**70**:8422–30.
121. Sena-Esteves M, Saeki Y, Camp SM, Chiocca EA, Breakefield XO. Single-step conversion of cells to retrovirus vector producers with herpes simplex virus-Epstein-Barr virus hybrid amplicons. *J Virol* 1999;**73**:10426–39.
122. Yates JL, Warren N, Sugden B. Stable replication of plasmids derived from Epstein-Barr virus in various mammalian cells. *Nature* 1985;**313**:812–5.
123. Jacoby DR, Fraefel C, Breakefield XO. Hybrid vectors: a new generation of virus-based vectors designed to control the cellular fate of delivered genes. *Gene Ther* 1997;**4**:1281–3.
124. Labow MA, Hermonat PL, Berns KI. Positive and negative autoregulation of the adeno-associated virus type 2 genome. *J Virol* 1986;**60**:251–8.
125. Savard N, Cosset FL, Epstein AL. Defective herpes simplex virus type 1 vectors harboring *gag, pol,* and *env* genes can be used to rescue defective retrovirus vectors. *J Virol* 1997;**71**:4111–7.

# Xenogenic Adenoviral Vectors

19

*Suresh K. Mittal*[1], *Yadvinder S. Ahi*[2], *Sai V. Vemula*[3]
[1]Department of Comparative Pathobiology, College of Veterinary Medicine and Purdue University Center for Cancer Research, Purdue University, West Lafayette, IN, USA; [2]HIV Drug Resistance Program, National Cancer Institute, Frederick National Laboratory for Cancer Research, Frederick, MD, USA; [3]Laboratory of Molecular Virology, Center for Biologics Evaluation and Research, Food and Drug Administration, Silver Spring, MD, USA

## 1. Introduction

Invention of novel vectors for gene delivery into a host has garnered great interest among the scientific community. The foremost applications for gene delivery include the development of vaccines and therapeutics for several infectious or genetic diseases and cancers. For the success of a gene delivery platform, the two most important criteria are safety and efficiency. A tremendous increase in the knowledge of molecular biology, genomics, proteomics, and immunology from 1980 to 2014 has allowed remarkable progress in recombinant DNA technology, which has greatly influenced the development of novel vector systems both viral and nonviral.

While clinical trials for gene therapy and vaccine applications have seen utilization of both viral and nonviral vectors, a significant majority of clinical trials have relied on viral vectors primarily due to the following reasons: (1) Viral vectors are naturally well adapted to infect the host due to the coevolution of viruses and host, thus enabling efficient delivery of the target gene into host cells; (2) Expression of the antigen in host cells allows for the presentation of antigenic peptides on MHC class I resulting in stimulation of cellular immune response; (3) Some viral vectors can infect antigen-presenting cells such as dendritic cells and macrophages allowing for direct presentation of antigenic peptides on both MHC class I and MHC class II and (4) Viral proteins themselves can act as immune stimulants providing a strong adjuvant effect which can boost the immune response to the antigen.[1–4] The most commonly used viral vectors in clinical trials are based on adenoviruses (AdVs), retroviruses, poxviruses, and herpesviruses. Of all gene therapy clinical trials, AdV vectors are currently the most commonly used gene delivery platform (http://www.abedia.com/wiley/vectors.php).

AdVs were first isolated more than 60 years ago from human adenoid tissues undergoing spontaneous degeneration in tissue culture.[5] Although the first evaluation of AdVs as an antitumor agent dates back to 1956,[6] it was not until the advent of recombinant DNA technology during the early 1980s that the therapeutic potential of AdVs was recognized, and AdV vectors were first developed. Since then, AdV biology has been studied thoroughly, and AdV vectors have been greatly optimized for various applications including vaccine vectors, gene therapy, oncolytic cancer therapy, and cancer immunotherapy.

AdVs are nonenveloped viruses with icosahedral capsids varying in size from 70 to 90nm with the dsDNA genome ranging from 26 to 46kb.[7] AdVs are classified under the family *Adenoviridae,* which is further divided into five genera: *Mastadenovirus*, *Aviadenovirus*, *Atadenovirus*, *Ichtadenovirus*, and *Siadenovirus*.[8] The currently known 57 serotypes of human AdVs (HAdVs) encompass species A through G based on genome sequence.[9] AdV capsids are primarily composed of the major capsid proteins, the hexon, the fiber, and the penton base which determines the tropism and immunogenicity of the virus. The relatively less abundant proteins in AdV capsids are called minor capsid proteins and include the core proteins (Tp, V, VII, and mu) and the cement proteins (IIIa, VI, VIII, and IX). Depending on the species type, AdVs use a variety of receptors for cell entry and include the coxsackievirus-adenovirus receptor (CAR), CD46, VCAM-1, CD80, CD86, heparin sulfate proteoglycans, desmoglein-2, and others.[10,11]

## 2. Advantages of Adenovirus Vectors

AdV-derived vectors fulfill many of the desirable features of a gene delivery platform. Although AdVs are known to infect a wide range of species, they rarely cause a serious disease in an immune-competent host. Most HAdVs are associated with self-limiting disease in humans and may manifest as mild upper respiratory tract infections, conjunctivitis, gastroenteritis, or ear infections, depending on the HAdV species involved. In most instances, the severity of the disease is a function of the host's immune function.[7] Their relatively innocuous nature makes AdVs particularly suitable for developing into a safe vector system. AdV vectors offer several other advantages including (1) the ability to infect a wide range of dividing and nondividing mammalian cells; (2) a robust transgene expression that can be restricted to a specific tissue by employing a tissue specific promoter; (3) the fact that they are easy to generate and can be grown to high titers; (5) their inability to integrate into host genome with a minimal risk of insertional mutagenesis; (6) the induction of strong humoral and cellular adaptive immune response to the transgene; and (7) strong activation of the innate immune response to provide an adjuvant effect that can potentiate booster immunity and may be desirable in vaccine applications.[12]

## 3. Preexisting Adenovirus Immunity

Among the known HAdV serotypes, the biology of the subgroup C serotype HAdV-5 is the most thoroughly studied; therefore, HAdV-5-derived vectors were the first to be evaluated in preclinical studies. The HAdV-5 vectors show great promise for gene delivery due to extremely strong transgene expression, a highly optimized system for production and preparation on a large scale, and their ability to induce strong humoral and cellular immune responses in nonhuman primates as well as in rodents.[13,14] As a result, the majority of the initial clinical studies involving AdV vectors rely on

HAdV-5 vectors.[15] The high prevalence of preexisting AdV immunity in humans has been noted as a major concern in the success of HAdV-5 vectors in clinical studies.

The development of preexisting AdV immunity is mainly due to exposure of a majority of the human population worldwide to one or more HAdV serotypes from natural infections resulting in strong anti-HAdV humoral and cellular immune responses.[16–19] Given that the natural exposure to HAdVs can be highly variable, the level of preexisting AdV immunity is expected to vary among individuals and human populations making the efficacy of HAdV vectors in clinical trials unpredictable. The high immunogenicity of AdV vectors implies that a first inoculation with an AdV vector will result in AdV serotype-specific immune responses that will preclude subsequent use of the same AdV serotype. Thus, high efficacy of any single AdV serotype is not expected even in individuals lacking preexisting AdV immunity.

The humoral immune response to HAdVs consists of both neutralizing and nonneutralizing antibodies.[20] Although the anti-HAdV-neutralizing antibodies (nAbs) targeting the three major capsid proteins, hexon, penton, and fiber, have been confirmed,[21,22] the anti-hexon nAbs are the most significant in vector neutralization. The major epitopes for anti-hexon nAbs are located in the exposed loops, also known as hypervariable regions, located on the surface of the virus particle.[23] The nAbs bind to the AdV vector in the blood, preventing its binding to the cell surface and subsequent internalization, thus resulting in rapid clearance from the host. The end result is a drastic reduction in transgene expression and an induction of transgene-specific immune responses. Such blunting of the efficacy of HAdV-5 vectors in the presence of preexisting immunity has been demonstrated by generating artificial serotype-specific immunity in animals preinoculated with HAdV-5.[24–26] The nonneutralizing, but cross-reactive, anti-HAdV antibodies are generated after repeated administration or infections with the same AdV serotype.[26] The nonneutralizing antibodies can also significantly reduce the transgene expression through Fc receptor-dependent and -independent mechanisms resulting in poor efficacy of the AdV vector.[27]

The cellular component of preexisting AdV immunity consists of both $CD4^+$ and $CD8^+$ T cells that are highly cross-reactive and widely prevalent in human populations.[28] The epitopes recognized by anti-AdV T cells are located primarily in the hexon and are highly conserved among various HAdV serotypes,[29–33] implying that exposure to a single HAdV serotype may hamper the efficacy of other HAdV serotypes. Studies have revealed that HAdV-5-specific T cells are prevalent in 80–100% of human subjects from Europe and the United States.[32,34,35] More importantly, HAdV-specific T cells may be prevalent even in the absence of nAbs.[20,35] The presence of vector-specific CD8 T cells causes cytolysis and elimination of the vector-transduced cells, while the $CD4^+$ T cells will potentiate the humoral immune response to the vector.

In addition to the development of adaptive immune responses, AdV vectors can also induce an immune activation characterized by overproduction of proinflammatory cytokines, chemokines,[36–38] high levels of type I interferons (IFNs),[39] complement activation, and phagocytosis.[40–42] The pattern recognition receptors (PRR) involved in the recognition of AdV vectors receptors include Toll-like receptors (TLR) 4 and 9, nucleotide-binding oligomerization domain-like receptors, Factor X, and retinoic acid-inducible gene 1.[28,43] Binding of AdVs to PRR results in the activation of downstream signaling pathways such as NF-κB culminating in the production of proinflammatory mediators including tumor necrosis factor alpha (TNFα), IL-1, IL-6, IL-8, IL-10, IL-12, IFN-γ, granulocyte colony-stimulating factor, macrophage inflammatory protein (MIP)-1α/β, IP-10, RANTES, and monocyte chemoattractant protein-1.[28,41] The AdV-induced innate immune response may be beneficial for targeted cancer gene therapy and vaccine applications where the goal is to eliminate the AdV vector-transduced cells and generate a robust immune response. However, activation of innate immune responses can result in rapid destruction of AdV-transduced cells, thus reducing the transgene expression. Furthermore, intravenous inoculations of large doses of AdV vectors can also have life-threatening consequences mediated by strong activation of innate immune reactions.[44]

The tremendous potential of AdV vectors is hampered by the immune responses to AdVs. For the success of these vectors, it is critical to understand thoroughly the underlying mechanisms. Several strategies have been developed to circumvent the preexisting immunity and minimize the innate immune reactions to AdVs. They can be broadly classified into either genetic or chemical modifications. The genetic modifications include hexon pseudotyping, hexon HVR swapping, fiber pseudotyping, fiber truncation, use of alternative trimerization motifs for fiber, and the use of vectors derived from rare HAdV serotypes or nonhuman AdVs.[12,45–48] The chemical modifications include derivation of AdV vectors with polyethylene glycol or cationic polymers such as poly-L-lysine, *N*-[2-hydroxypropyl] methacrylamide, arginine-grafted bioreducible polymers, coating of AdV vectors with cationic liposomes, and lipidic envelopes.[12,46,47] This review will focus on AdV vectors derived from nonhuman AdVs.

## 4. Nonhuman Adenovirus Vectors

The idea of developing vectors derived from nonhuman AdVs is based on the premise that nonhuman AdVs are less prevalent and consequently will have lower seroprevalence among human populations. Similar to the HAdVs, the nonhuman AdVs do not cause serious disease in their natural host. Preexisting immunity to nonhuman AdVs can be expected only in their natural hosts. In addition, the overall genome organization and structural features of nonhuman AdVs are not vastly different from the HAdVs; vector design strategies and techniques used in working with HAdV vectors could, in principle, be applied to the nonhuman AdV vectors. The initial impetus for developing nonhuman AdV vectors was seen in the early 1990s when the limitations of HAdV vectors were starting to become evident. Since then gene expression vectors based on various species of nonhuman AdVs have been constructed and extensively characterized. These include simian AdV serotypes (SAdVs), bovine AdV serotype 3 (BAdV-3), porcine AdV serotypes 3 and 5 (PAdV-3 and PAdV-5), ovine AdV serotype 7 (OAdV-7), canine AdV serotype 2 (CAdV-2), murine AdV serotype 1 (MAdV-1), and fowl AdV serotypes (FAdVs). In this review, various aspects of these nonhuman AdV vectors including strategies for construction, types of vector/s, genome organization (Figure 1), insertion sites, foreign gene insertion capacity, receptors, tropism, impact of preexisting immunity, and preclinical and clinical studies will be discussed. Examples of nonhuman AdV-based vectors as gene delivery vehicles for recombinant vaccines or gene therapy applications are shown (Table 1).

### 4.1 Simian Adenovirus Vectors

The first SAdV vector developed was a chimpanzee AdV (chAdV)-based vector derived from chAdV-68 (SAdV-25) by homologous recombination of a shuttle vector and vector genome in the 293 cell line expressing HAdV-5 early region 1 (E1) proteins. It appears that the E1 of chAdV-68 is closely related to the E1 of HAdV-5 and thus allows for efficient replication of the E1-deleted chAdV-68.[49] This similarity offers two advantages: (1) there is no need to create cell lines expressing E1 from a chAdV and (2) there is less chance of vector contamination with a replication-competent AdV due to variation in the common sequences of the chAdV-68 vector and HAdV-5 E1. In addition, there are substantial variations in the hypervariable region within the hexon sequence of chAdV-68 and several HAdVs. Given that the hypervariable regions of the hexon carry the type-specific epitopes for HAdV-neutralizing antibodies,[50] it is not surprising that chAdV-68 is not cross-neutralized by HAdV preexisting immunity.

The AB loop within the knob domain of chAdV-68 fiber, which mediates binding of CAR-binding HAdV serotypes, is identical to HAdV-4 and is very similar to HAdV-2, HAdV-5, and HAdV-12, thus explaining the observation that CAR serves as a receptor for chAdV-68.[51] Other chAdVs such as chAdV-6 (SAdV-23) and chAdV-7 (SAdV-22) are closely related to subgroup E HAdVs; chAdV-68[52] may possibly use CAR as a receptor for cell entry. The chAdV-1 (SAdV-21), which is more closely related to subgroup B HAdVs, utilizes CD46 as a receptor for cell entry.[53] It is conceivable that the tropism of chAdV vectors utilizing CAR or CD46 as receptors would be similar to the CAR or CD46 binding HAdV serotypes.

Depending on the E1 deletion with or without early region 3 (E3) deletion, the chAdV-68 vectors allow for transgene lengths of up to 5000 or 1600 bp respectively, without significant decreases in virus yields or virus particle-to-PFU ratios compared to smaller size transgenes.[54] The E1- and E1/E3-deleted chAdV-68 vectors are comparable in terms of the level of transgene expression and induction of transgene-specific $CD8^+$ T cell response. However, incorporation of an E4 deletion results in decreased transgene expression and $CD8^+$ T cell response. Interestingly, the level of protein expression by the chAdV-68 vector seems to be a function of transgene length. A chAdV-68 vector expressing simian immunodeficiency virus (SIV) gag alone resulted in much higher protein expression than a vector expressing SIV gag–pol or SIV gag–pol–nef fusion proteins, and, consequently, resulted in a higher gag-specific $CD8^+$ T cell response.[54] It is likely that the competition of peptides for the MHC class I peptide-binding site decreased the overall magnitude of immune response to gag. Thus, even though the chAdV-68 vectors are capable of incorporating large inserts, the choice of transgene length should be carefully evaluated.

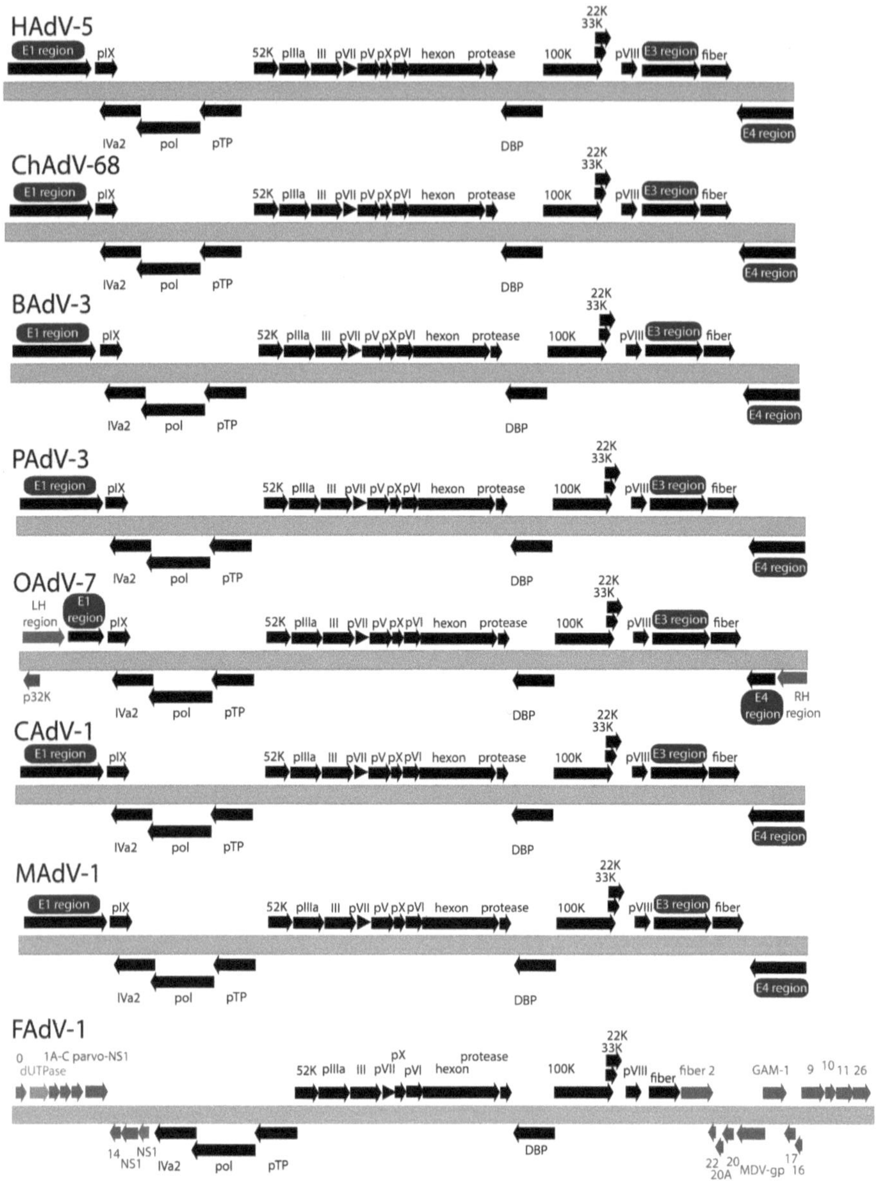


**Figure 1** Simplified depiction of gene arrangements of simian adenovirus (AdV) (chAdV-68),[a] bovine AdV (BAdV-3),[105] porcine AdV (PAdV-3),[201] canine AdV (CAdV-1),[154] ovine AdV (OAdV-7),[b] murine AdV (MAdV-1),[c] and avian AdV (FAdV-1)[d,196] compared to human AdV (HAdV-5).[e] The conserved genes are indicated with black arrows. The early regions (E) 1 (E1), E3, and E4 are depicted with black boxes, respectively; however, the variable features of E1, E3, and E4 regions in various AdVs are not shown. The unique genes are shown with dark or light gray arrows. The maps are not at scale. [a]Roy S, Gao G, Clawson DS, Vandenberghe LH, Farina SF, Wilson JM. Complete nucleotide sequences and genome organization of four chimpanzee adenoviruses. *Virology* 2004; 324:361–72.

**Preexisting immunity**. The chAdVs show great potential as gene therapy and vaccine vectors and are the only nonhuman AdV vectors that have progressed to clinical trials to date. The chAdVs have little to no seroprevalence in the human population worldwide which makes them suitable vector candidates for use in humans.[55,56] In the hexons of chAdV and HAdVs,[57] there is very little sequence similarity in the hypervariable region, which is the location of the major neutralization determinant, suggesting that the chAdV can circumvent preexisting immunity against HAdV vectors. A low frequency of nAbs against chAdV, ranging from 0 to 4% has been observed in humans in the United States, Europe, Thailand,[58] and China,[59,60] whereas nAbs to HAdV serotypes may reach up to 30–45% in these countries. Up to ~20% of sera from sub-Saharan countries have been found to be positive for nAb against chAdV, whereas ~60–80% sera are positive for nAb against HAdV serotypes.[58,61,62] In Brazil, the seroprevalence rates of chAdV nAbs are between 20 and 23%; HAdV nAbs were between 40 and 70% depending on the serotype.[16,56] Therefore, even though the seroreactivity to chAdVs is higher in sub-Saharan countries and Brazil than the rest of the world, it is still lower than that of HAdVs.

Neutralizing antibody titers of >200 are considered to hamper the efficacy of AdV vectors[63]; therefore, it is important to consider the titers of chAdV nAbs. As anticipated, 10% or less of human sera with chAdV nAb had titers of >200, whereas about 40% of the human sera positive for HAdV nAbs had titers of >200. The lower titers of chAdV nAbs, in combination with the lower overall seroreactivity compared to HAdV serotypes, suggest that chAdV serotypes are less likely to suffer from preexisting immunity than HAdV serotypes.

**Immunological potential**. The chAdV vectors belonging to group C of HAdV species have the highest immunogenic potential among all chAdV vectors.[57] In sub-Saharan Africa where the target population for endemic diseases such as human immunodeficiency virus (HIV), malaria, and tuberculosis exists, it is particularly important for a vaccine to have the ability to induce a $CD8^+$ T cell response.[64] The most effective HAdV serotypes (HAdV-5 and HAdV-26) in generating a $CD8^+$ T cell response would be inefficient in these areas owing to modest to high levels of preexisting immunity. The other HAdV serotype, HAdV-35, which has the least preexisting immunity worldwide, is not as effective as HAdV-5 and HAdV-26 in generating a $CD8^+$ T cell response in particular and overall immunogenicity in general.[19,65,66]

[b]Kumin D, Hofmann C, Rudolph M, Both GW, Loser P. Biology of ovine adenovirus infection of nonpermissive cells. *J Virol* 2002; 76:10,882–93. [c]Hemmi S, Vidovszky MZ, Ruminska J, Ramelli S, Decurtins W, Greber UF, et al. Genomic and phylogenetic analyses of murine adenovirus 2. *Virus Res* 2011; 160:128–35. [d]Marek A, Kosiol C, Harrach B, Kaján GL, Schlötterer C, Hess M. *Vet Microbiol* 2013; 166:250–56; and Harrach B, Benko M, Both GW, Brown M, Davison AJ, Echavarria M, et al. The Double Stranded DNA Viruses: Adenoviridae. In: King AMQ, Adams MJ, Carstens EB, Lefkowitz EJ, editors. *Virus taxonomy: ninth report of the international committee on taxonomy of viruses*. Waltham (MA): Academic Press: Elsevier; 2011. p. 125–41. [e]Chroboczek J, Bieber F, Jacrot B. The sequence of the genome of adenovirus type 5 and its comparison with the genome of adenovirus type 2. *Virology* 1992; 186:280–85.

**Table 1** Examples of Nonhuman Adenovirus (AdV)-Based Vectors as Gene Delivery Vehicles for Recombinant Vaccines or Gene Therapy Applications

| Adenovirus (AdV) | Vector Type | Insertion Site | Pathogen/Cancer/Gene Therapy | Antigen | References |
|---|---|---|---|---|---|
| **Simian AdVs** | | | | | |
| chAdV-3, chAdV-63 | E1, E3 deleted | E1 | Simian immunodeficiency virus | Gag | 67 |
| chAdV-6, chAdV-68 | E1 deleted | E1 | Human immunodeficiency virus | Gag | 68 |
| chAdV-7 | E1, E3 deleted | E1 | Ebola | Env GP | 72 |
| chAdV-7 | E1, E3 deleted | E1 | Respiratory syncytial virus | Fusion protein | 75 |
| chAdV-68 | E1 deleted | E1 | Rabies | Glycoprotein | 76,77 |
| chAdV-7 | E1 deleted | E1 | H5N1 influenza | NP | 78 |
| PanAdV-3 | E1, E3 deleted | E1 | H5N1 influenza | NP-M1 fusion | 79 |
| chAdV-7 | E1 deleted | E1 | Severe acute respiratory syndrome (SARS) | Spike protein | 83 |
| chAdV-63 | E1, E3 deleted | E1 | Human immunodeficiency virus | HIV*cons* | 89[a] |
| chAdV-63 | E1, E3 deleted | E1 | *P. falciparum* | ME-TRAP | 96 |
| chAdV-63 | E1, E3 deleted | E1 | *P. falciparum* | AMA1 | 97 |
| chAdV-63 | E1, E3 deleted | E1 | *P. falciparum* | MSP1 | 98 |
| chAdV-3 | E1, E3 deleted | E1 | Hepatitis C virus | NS | 104 |
| chAdV-3 | E1, E3 deleted | E1 | Colon carcinoma | CEA | 69 |
| **Bovine AdVs** | | | | | |
| BAdV-3 | E3 deleted | E3 | Bovine herpes virus-1 | Glycoprotein D (gDt) | 126 |
| BAdV-3 | E3 deleted | E3 | Bovine herpes virus-1 | gDt | 127 |
| BAdV-3 | E3 deleted | E3 | Bovine viral diarrhea virus | Glycoprotein E2 (gE2) | 128 |
| BAdV-3 | E3 deleted | E3 | Bovine respiratory syncytial virus | Glycoprotein G (gG) and IL-6 | 129 |
| BAdV-3 | E1, E3 deleted | E1 | A/Hong Kong/156/97 (H5N1) | Hemagglutinin (HA) | 131 |
| BAdV-3 | E3 deleted | E3 | Bovine herpes virus –1 and bovine respiratory syncytial virus | gG and gDt | 130 |

| | | | | | |
|---|---|---|---|---|---|
| **Porcine AdVs** | | | | | |
| PAdV-3 | E3 deleted | E3 | Classical swine fever virus | gp55/E2 gene | 136 |
| PAdV-3 | E3 deleted | E3 | Classical swine fever virus | gp55/E2 gene (DNA/PAdV-3) | 137 |
| PAdV-3 | E3 deleted | E3 | Pseudorabies virus (PRV) | Glycoprotein D (gD) | 138 |
| PAdV-3 | E1, E3 deleted | E1 | A/Hanoi/30408/2005 (H5N1) | HA | 139 |
| PAdV-5 | E3 deleted | E3 | Transmissible gastroenteritis virus | Spike protein | 141 |
| **Ovine AdVs** | | | | | |
| OAdV-7 | NA | Site III | Hepatitis C virus | Nonstructural protein 3 (NS3) | 146 |
| OAdV-7 | NA | Site II | *Taenia ovis* | 45W | 147 |
| OAdV-7 | NA | Site III | Human immunodeficiency virus 1 | Gag | 148 |
| OAdV-7 | NA | Site II | Skeletal muscle gene therapy | Human alpha-1 anti-trypsin (hAAT) | 150 |
| OAdV-7 | NA | Site III | Prostate cancer | *E. coli* purine fludarabine phosphorylase (PNP) | 152 |
| **Canine AdVs** | | | | | |
| CAdV-1 | E3-deleted | E3 | Canine parvovirus | Capsid | 156 |
| CAdV-2 | cRAd | | Canine osteosarcoma | | 161 |
| CAdV-2 | Gutless | | Canine mucopolysaccharidosis VII | Human GUSB | 172–174 |
| **Murine AdVs** | | | | | |
| MAdV-1 | cRAd | | Murine colon carcinoma | mGM-CSF | 189 |
| **Avian AdVs** | | | | | |
| FAdV-10 | NA | Near right ITR | Infectious bronchitis virus | VP2 | 197 |
| FAdV-8 | NA | Near right ITR | Infectious bronchitis virus | Spike peplomer S1 subunit (S1) | 199 |

[a]Ondondo B, Brennan C, Nicosia A, Crome SJ, Hanke T. Absence of systemic toxicity changes following intramuscular administration of novel pSG2.HIV*consv* DNA, ChAdV63.HIV*consv* and MVA.HIV*consv* vaccines to BALB/c mice. *Vaccine* 2013; 31:5594–601.

In HIV, both chAdV-derived vectors have markedly lower seroreactivity titers than HAdV-5, HAdV-26, and HAV-35 across populations in southern Africa, eastern Africa, central Africa, India, the United States, South America, and the Caribbean.[67] In addition, chAdV vectors result in the gag-specific $CD8^+$ T cell response and expression levels of IFN-γ, IL-2, and TNFα that are comparable to or better than those of the HAdV serotypes at a wide range of doses.[67] The chAdV vectors are also found to induce the SIV gag-specific $CD4^+$ response better than HAdV-5.[67]

In addition, chAdV-3 is found to be as effective as HAdV-5 in boosting a DNA or a HAdV-28 primed SIV gag-specific $CD8^+$ T cell response.[67] A heterologous prime–boost regimen with chAdV vectors carrying the HIV-1 gag gene results in strong and robust $CD8^+$ T cells in various systemic compartments including the genital tract with intramuscular immunization within 2 weeks.[68] Moreover, the $CD8^+$ T cell response seems to be long lasting with the gag-specific $CD8^+$ T cell being detected in all systemic compartments for up to one year after immunization.[68] These observations suggest that chAdV-based vectors can be used for developing vaccines against infections requiring the $CD8^+$ T cell-mediated immune (CMI) response.

A single immunization with a chAdV vector expressing HIV-1 gag elicited a potent antigen-specific cellular immune response [~2000 IFN-γ spot-forming cells (SFCs) per million PBMCs] that contracted eventually but persisted beyond 5 years (~400 IFN-γ SFCs per million PBMCs). A booster at this 5 year stage with a chAdV vector carrying the same antigen resulted in a rapid increase in the number of the gag-specific IFN-γ producing T cells. Additionally, gag-specific antibodies could be detected before the booster and increased by a factor of 10 after the booster,[57] implying that chAdV vectors are capable of inducing long-lasting T and B cell memory.

A recent report compared the immunological potency of chAdV-3 with that of HAdV-5 as an anticancer vaccine vector.[69] The chAdV-3-based vector, expressing transgene at levels similar to that of HAdV-5, was able to induce higher levels of a $CD8^+$ IFN-γ-positive T cell response at significantly lower doses compared to a HAdV-5-based vector. Additionally, a stronger immune response was observed in a chAdV-3/chAdV-3 prime–boost regimen than either a HAdV-5/HAdV-5 or a HAdV-5/chAdV-3 prime–boost regimen.[69] Most importantly, the chAdV-3-derived vector was able to confer antitumor protection in mice with anti-HAdV-5 immunity.[69]

Another study[70] reported that a chAdV-23 (also known as SAdV-22 or Pan-5)-derived vector was equally or more efficient than an HAdV-5-derived vector in transducing low-passage brain tumor cells, $CD133^+$ and $CD133^-$ glioma tumor stem cells derived from human patients. Given that the $CD133^+$ cells are difficult to treat with traditional chemotherapy, the aforementioned observations warrant testing of SAdV-23-based vectors as a treatment for human brain tumors.

The chAdV-based vectors are not only capable of inducing robust humoral and cellular adaptive immune responses but also have been shown to induce protective immunity against a variety of infections including Ebola virus in mice and nonhuman primates,[71–74] respiratory syncytial virus in neonates,[75] rabies virus,[76,77] H5N1 influenza virus,[78,79] malaria,[64,80,81] Rift Valley fever virus,[82] and severe acute respiratory syndrome virus.[83] The ability to induce protective immunity in human populations in the presence of preexisting immunity to HAdV serotypes is a highly desirable trait of chAdV-based vectors. These vectors have the potential to develop into a common platform for vaccine or gene therapy purposes.

### 4.1.1 Chimpanzee Adenovirus Vectors in Clinical Trials

**HIV-1**. In a heterologous prime–boost regimen, a chAdV-63 vector in combination with a plasmid DNA or modified vaccinia vector Ankara (MVA) as a vaccine for delivering HIV*consv* was evaluated in uninfected human subjects.[84] The HIV*consv* is based on the 14 functionally most conserved subprotein domains of HIV-1 and is common to most virus variants worldwide.[85–88] The combinatorial regimen that included the chAd63.HIV-*consv* resulted in an HIV*consv*-specific IFN-γ-producing T cells with remarkably high frequencies. Furthermore, both CD4+ and CD8+ T cells with broad specificity and multiple intercellular signaling molecules were induced. In addition, the effector T cells which were found to be specific for gag and pol were efficacious in inhibiting HIV-1 replication in cultured autologous CD4+ T cells by a factor of 5.79 log. No adverse side effects were reported in any of the subjects receiving the vaccine formulation.[84] A follow-up Phase I clinical trial HIV-CORE 002 with the same combinatorial regimens discussed above did not report significant safety or tolerability concerns.[89] Only mild-to-moderate local reactogenicity and systemic effects such as mild pain, erythema, fever, and headache lasting for fewer than 3 days postadministration were described in the majority of vaccines receiving chAd63.HIV*consv*. Following these encouraging results, a Phase I clinical trial using chAd63.HIV*consv* and MVA.HIV*consv* is underway (ClinicalTrials.gov; NCT01712425).

**Malaria**. Malaria is caused by *Plasmodium falciparum* and is a major pathogen responsible for childhood morbidity and mortality in Africa and other countries. The primary objective of an effective malaria vaccine will be to control the disease transmission. Multiple preclinical studies have demonstrated a strong correlation between CD8+ T cells and protective efficacy.[90–94] However, the existing subunit vaccines are not good enough to induce high levels of malaria antigen-specific CD8+ T cells.[95]

Several studies have shown that heterologous prime–boost immunization with chAdV-63 and MVA expressing ME-TRAP, a liver stage antigen of *P. falciparum*, are capable of inducing robust CD8+ T cell response and protection.[64,96–98] The chAd63.ME TRAP and MVA.ME-TRAP did not cause any significant local or systemic adverse effects.[96] The chAd63.ME-TRAP demonstrated a good safety profile in individuals in Kenya, Gambia, and United Kingdom [96,99] with a dose of up to $2\times10^{11}$.[99] The chAd63.ME-TRAP and MVA.ME-TRAP prime–boost regimen induced high frequencies of antigen-specific IFN-γ secreting T cells, reaching beyond 2000 SFCs/million PBMCs in two independent Phase Ib clinical trials.[96] The induced cellular immune response included cytokine secreting, antigen-specific CD4+ and CD8+ T cell populations.[64] In addition, the response seemed to be long lasting since it had not significantly decreased at 9 months postvaccination.[96] The high levels of antigen-specific IFN-γ secreting T cells are highly remarkable as it is difficult to induce more than 1000 SFCs/million PBMCs.[64,96] ChAd63.ME-TRAP priming followed by MVA.ME-TRAP boosting induced high levels of IgG response to the TRAP antigen. Importantly, the low levels of anti-chAd63 nAb titers detected pre- and postadministration of chAd63.ME-TRAP did not attenuate the immunogenicity of the vaccines. Finally, the chAd63.ME-TRAP and

MVA.ME-TRAP prime–boost also provided sterile to partial protection against controlled human malaria infection with a heterologous malaria strain, thereby demonstrating strong correlation between immunogenicity and protective efficacy.

Another heterologous prime–boost vaccine regimen employed the chAd63 and MVA encoding for either one or both of the blood stage malaria antigens AMAI and MSPI, alone or in combination with the liver stage antigen ME-TRAP.[97] The chAdV-63 vectors displayed an excellent safety profile and induced significantly higher T cell responses compared to other delivery platforms tested (DNA/MVA or Fowl Pox virus/MVA). The cellular immune response was broadly specific to the AMA1 and MSP1, persisted at high levels, and comprised of relevant cytokine secreting $CD8^+$ and $CD4^+$ T cell populations. The chAdV-63 vectors also induced an antigen-specific IgG response.[97,98] The chAdV-63-MVA prime–boost combination is now a commonly employed and versatile vaccine delivery system capable of inducing strong cellular and humoral immune responses.

**Hepatitis C**. Hepatitis C virus (HCV) is another pathogen of immense public health significance against which the chAdV-derived vectors have been used for vaccine delivery. HCV is capable of establishing persistent, stealthy infection in immunocompetent hosts in which cirrhosis has affected a significant portion of the liver. The current treatments are becoming increasingly effective but are costly and have serious side effects.[100]

Induction of high levels of virus-specific $CD4^+$ and $CD8^+$ T cells is important for protection against HCV,[101–103] implying that a successful HCV vaccine will be able to mount a strong and long-lasting T cell response. A chAdV-3-based vector, alongside the rare HAdV serotype of HAdV-6, expressing the NS proteins of HCV genotype 1B was recently tested in a Phase I clinical trial.[104] Both vectors were found to be overall safe and well tolerated, resulting in only mild local and systemic side effects at various doses ranging from $5 \times 10^8$ to $7.5 \times 10^{10}$ vp. The chAdV-3 vector used at $2.5 \times 10^{10}$ vp dose induced more than 1000 IFN-γ SFCs/million PBMCs—a level similar to or higher than the HAdV-6 vector. Furthermore, both viral vectors were able to induce broad T cell responses, targeting peptides encompassing multiple viral gene products including NS3, NS4A/B, NS5A, and NS5B. The induced T cell responses for each vector included both $CD4^+$ and $CD8^+$ T cells with the $CD8^+$ response being the more dominant of the two. T cell populations secreting IFN-γ/TNFα, IFN-γ/IL-2, or all three cytokines were detected with a minimal IL17A response. In addition, the induced T cells responses were found to be cross-reactive with other HCV genotypes 1A, 3A, and 1B with genotype 1B representing the immunogen. Finally, detectable and functional memory T cell could be observed up to one year postboost, suggesting that the vaccines induced a long-lasting T cell response.[104]

In the same clinical trial, chAdV-3 and HAdV-6 were also tested in a heterologous prime–boost regimen. The chAdV-3 prime/HAdV-6 boost regimen, in contrast to the HAdV-6 prime/chAdV-3 boost regimen, resulted in a stronger boost response that was also consistent among different vaccines. However, the boost response was still not as high as that observed for the boost in chAdV-63/MVA malaria trials. The poorer chAdV-3 prime/HAdV-6 boost response is surprising given the strong and functional responses at priming. The authors proposed that the nAbs induced on priming were possibly interfering with the heterologous vector used at priming and resulted in attenuation of AdV vector-specific and NS antigen-specific T cell responses. Nonetheless, the overall findings of this clinical trial of a T cell-based HCV vaccine have again demonstrated the utility of chAdV vectors as a vaccine delivery platform.

## 4.2 Bovine Adenovirus-Based Vectors

AdVs infecting cattle, termed as bovine AdVs (BAdVs), have been classified under the two genera, *Mastadenovirus* and *Atadenovirus*. Ten serotypes of BAdVs (BAdV-1–10) have been isolated thus far, and most of these are responsible for mild diseases of the gastrointestinal or respiratory tracts in bovines. Of the 10 BAdV serotypes, BAdV-3 is the best characterized. The complete nucleotide sequence and genome map of BAdV-3 is available.[105] The BAdV-3 genome shares a high level of similarity with HAdV-5 with certain differences. The E3 region in BAdV-3 is relatively smaller and less complex compared to the corresponding E3 region from HAdV-5.[106] BAdV-3 E1 proteins (E1A, E1B-157R, and E1B-420R) show functional or amino acid sequence homologies with their counterpart E1 proteins of HAdV-5 and BAdV-3. E1A complements HAdV-5 E1A functions[107,108] suggesting functional similarities between BAdV-3 E1 and HAdV-5 E1. Structurally, the BAdV-3 fiber is exceptionally long and bent at several sites[109] indicating that the bending may be necessary to allow the penton base to make contact with the secondary receptors on the cell surface.

Cotton rats (*Sigmodon hispidus*) can serve as a replication-competent small-animal model for evaluating the pathogenesis and vaccine efficacy of BAdV-3 vectors.[110,111] In the presence of circulating BAdV-3-neutralizing antibodies, intranasal inoculation of cattle with BAdV-3 results in inapparent infection[112] suggesting that a BAdV-3-based vaccine would be successful.

BAdV-3-based vectors have been created using homologous recombination in both mammalian cells[113] and bacteria[114] in addition to I-*Sce*I-based approaches.[105,115,116] Several bovine cell lines have been used to rescue BAdV-based vectors. These include MDBK, BHH3 (bovine human hybrid cell lines expressing HAdV-5 E1),[117] FBRT-HE1 (fetal bovine retinal cells expressing HAdV-5 E1[118] and VIDO R2 (fetal bovine retinal cells expressing HAdV-5 E1).[115] Initially BAdV-3 as a replication-competent vector containing the firefly luciferase gene in the E3 region was constructed in 1995,[113] and the BAdV-3 vector-infected 293 cells efficiently expressed luciferase, suggesting the suitability of BAdV-3 vectors for gene transfer into human cells.

There are a number of findings that further explore the potential of BAdV-3 as a gene delivery platform. The anti-HAdV-5 or anti-BAdV-3 antibodies raised in rabbits or mice are not cross-virus neutralizing, and the reporter gene expression with HAdV-5-LacZ in BAdV-3-primed mice has been significantly higher ($P>0.05$) than that obtained in HAdV-5-primed animals.[26] BAdV-3 internalization is independent of the HAdV-5 receptors (CAR and $\alpha v\beta 3$- or $\alpha v\beta 5$-integrin)[119] but utilizes $\alpha(2,3)$-linked as well as $\alpha(2,6)$-linked sialic acid as a major receptor for internalization.[120] Preexisting HAdV-neutralizing antibodies in humans do not cross-neutralize BAdV-3.[108] HAdV-specific CMI response does not cross react with BAdV-3.[121] BAdV-3 has tropism distinct from that of HAdV-5 and efficiently transduces diverse human and nonhuman cells in culture.[108,122] Unlike HAdV-5, BAdV-3 is a strong inducer of TLR4. There is an absence of Kupffer cell depletion with BAdV-3 in mice[123] while Kupffer cell depletion with HAdV-5 is the main reason for a faster vector depletion from the host. Intravenous inoculation with a BAdV-3 vector efficiently transduces the heart, kidney, lung, liver, and spleen. The vector persists for a longer duration compared to a HAdV-5 vector especially in the heart, kidney, and lung in a mouse model.[122] Sequential administration of HAdV-5 and BAdV-3 vectors overcomes vector immunity in an immunocompetent mouse model of breast cancer.[124] Persistence of the BAdV-3 genome in human and nonhuman cell lines is similar to that of the HAdV-5 vectors.[125] These various findings underscore why BAdV-3 vectors offer an attractive alternative to HAdV vectors for effectively immunizing individuals with high levels of preexisting HAdV immunity with safety aspects similar to those of HAdV-5 vectors.

BAdV-3-based vectors have been successfully used in gene delivery for vaccination purposes in experimental animals. Intranasal vaccination of cotton rats with a replication-competent BAdV-3 vector, BAV3.E3gD, expressing bovine herpesvirus-1 (BHV-1) gDt glycoprotein induced strong gD-specific IgA and IgG immune responses.[126] In a subsequent study, vaccination of calves with BAV3.E3gD induced gD-specific antibody responses in serum and nasal secretions that conferred protection against BHV-1 challenge.[127] A BAdV-3 vector expressing bovine viral diarrhea virus glycoprotein E2 induced E2-specific IgA and IgG in nasal secretions and serum, respectively, in cotton rats following intranasal immunization.[128] A BAdV-3 vector (BAV327) coexpressing bovine respiratory syncytial virus glycoprotein G and bovine IL-6 from the E3 region was developed.[129] A BAdV-3 vector (BAV851) coexpressing bovine respiratory syncytial virus G and BHV-1gDt proteins was generated, and vaccination of cotton rats with this BAV851 induced strong antigen-specific immune responses.[130]

In order to evaluate whether a BAdV vector can effectively elude high levels of preexisting HAdV-5 vector immunity, naïve and HAdV-5-primed mice were immunized with BAd-H5HA (a BAdV vector expressing HA of a H5N1 influenza virus).[131] Even in the presence of very high levels of HAdV-5-specific neutralizing antibody titer (2133 ± 660), no reductions in HA-specific humoral and CMI responses were observed in mice immunized with BAd-H5HA. In the presence of exceptionally high levels of preexisting vector immunity, mice immunized with BAd-H5HA resulted in approximately 2.8-fold higher hemagglutination inhibition (HI) titers or a 2.3-fold higher percentage of HA-specific $CD8^+$ T cells compared to the levels in naïve mice inoculated with HAd-H5HA. The immunization of naïve or HAdV-primed mice with BAd-H5HA resulted in full protection from morbidity and mortality following a potentially lethal challenge with A/Hong Kong/483/97 (H5N1). Furthermore, a heterologous prime–boost regimen comprised of HAd-H5HA priming and boosting with BAd-H5HA elicited a significantly higher HI response compared with HAd-H5HA or BAd-H5HA alone. These results strongly suggest the importance of using BAdV-based vectors as an alternate to HAdV-based vectors for eluding preexisting vector immunity and in a heterologous prime–boost strategy for enhanced immune responses.

## 4.3 Porcine Adenovirus-Based Vectors

AdVs have been isolated from pigs and termed as porcine AdVs (PAdVs). There are five serotypes of PAdVs currently known to infect swine (PAdV-1–5). Similar to the other

AdVs, PAdVs are prevalent primarily in swine species and are responsible for gastrointestinal disease and multifactorial porcine respiratory disease complexes. Among the five PAdVs, PAdV-3 is the most prevalent and well-characterized serotype. Initially isolated from a healthy young piglet, PAdV-3 is associated with subclinical infection. The complete nucleotide sequence and transcription map of PAdV-3 is available.[105] PAdV-3 shares genomic and structural similarities with HAdV-5. Like HAdV-5, PAdV-3 belongs to the genus *Mastadenovirus*, and its genome consists of five early transcription units (E1A, E1B, E2, E3, and E4).[105] Further, the PAdV-3 E1 transaction unit has been shown to complement the functions of the HAdV-5 E1 transcription unit.

However, unlike HAdV-5, PAdV-3 internalization into cells is independent of CAR and $\alpha_v\beta_3$- or $\alpha_v\beta_5$-integrin receptors; the primary receptors for HAdV-5 and PAdV-3 are distinct.[132] Anti-HAdV-5 or anti-PAdV-3 antibodies raised in rabbits or in mice do not cross-neutralize, and the reporter gene expression with HAdV-5-LacZ in PAdV-3-primed mice show significantly higher *P* values ($P > 0.05$) than those obtained in HAdV-5-primed animals.[26] Preexisting HAdV-specific neutralizing antibodies in humans do not cross-neutralize PAdV-3.[133] A PAdV-3 vector efficiently transduces a number of human, murine, porcine, and bovine cells in culture,[108,133] suggesting that PAdV-3 vectors are a promising supplement to HAdV vectors.

In studies, the biodistribution of a PAdV-3 vector was comparable to that of a HAdV-5 vector in the mouse model but showed more rapid vector clearance. Only linear episomal forms of PAdV-3 vector genomes were detected in inoculated mice.[10] In addition, PAdV-3-specific T cell responses did not show significant cross-reactivity with HAdV-5 or BAdV-3.[121] Compared to the HAdV-5 vector, the PAdV-3 vector induced higher levels of innate immune responses, including TLRs and proinflammatory chemokines and cytokines.[123] The persistence of a PAdV-3 vector in a number of cell lines was comparable to that of the HAdV-5 or BAdV-3 vectors, and only the linear episomal form of the vector genome was observed.[125] These findings suggest the uniqueness of receptor usage by PAdV-3 and highlights its potential to elude HAdV specific humoral and CMI responses with a safety similar to that of HAdV-5 vectors.

Construction of recombinant PAdV-3-based vectors has been pursued using homologous recombination in *Escherichia coli* to generate a full-length infectious clone followed by transfection of fetal porcine retinal cell lines transformed with HAdV-5 E1 (FPRT-HE1-5 and VIDO R1).[133,134] Both replication-competent (containing deletion in E3 region) and replication-defective (containing deletions in E1 and/or E3 regions) PAdV vectors have been developed and evaluated as delivery tools for vaccine or gene therapy purposes.[135] A PAdV-3-based classical swine fever virus (CSFV) vaccine (rPAV-gp55) containing the gp55 (E2) gene from the CSFV 'Weybridge' strain into the right-hand end of the PAdV-3 genome has been developed.[136] Transgene expression was driven from the major late promoter and tripartite leader sequences of PAdV-3. Subcutaneous vaccination of outbred pigs with a single dose of the rPAV-gp55 vaccine induced high levels of gp55-specific antibodies and conferred complete protection from lethal challenge with CSFV. Furthermore, all the vaccinated animals showed no adverse clinical signs of CSFV. In a subsequent study, 6-week-old weaned pigs and 7-day-old preweaned piglets were vaccinated with a DNA vaccine expressing the gp55/E2 gene from CSFV and then boosted with rPAV-gp55.[137] This prime–boost vaccine approach induced high levels of gp55 antibody titers. Following challenge with CFSV, 100% of the weaned pigs and 75% of preweaned piglets were protected.

Similarly, a PAdV-3-based pseudorabies virus (PRV) vaccine encoding the glycoprotein D gene from a PRV strain was developed.[138] Vaccination of 5-week-old pigs with a single or two doses of the PAdV-3-PRV vaccine induced high levels of serum-neutralizing antibodies to PRV and conferred protection against challenge with a PRV virus. Pigs vaccinated with two doses of the PAdV-3-PRV vaccine had relatively higher PRV antibody titers and demonstrated better protection efficacy compared to those receiving a single vaccine dose.

A PAdV-3-based influenza vaccine (PAV3-HA) expressing HA of A/Hanoi/30408/2005 (H5N1) virus was evaluated for its potential to induce protective immunity in mice.[139] Vaccination of BALB/c mice with PAV3-HA induced high levels of HA-specific humoral and cellular responses. Vaccinated mice were protected against lethal challenge with a highly pathogenic avian H5N1 influenza virus. Interestingly, compared to a HAdV-5-based H5N1 vaccine, the immunity induced by PAV3-HA was present even twelve months postvaccination.

In addition to PAdV-3, PAdV-5 is also being investigated as a gene delivery tool. Tuboly and coworkers developed a recombinant PAdV-5 vector encoding the spike gene of transmissible gastroenteritis virus. Oral vaccination of pigs induced high levels of spike-specific antibodies.[140,141]

Like other AdV vectors, the prevalence of PAdV-neutralizing antibodies in the swine population is thought to hinder the use of PAdV vectors for vaccination purposes. A survey for PAdV-3 immunity in pigs from Australia demonstrated up to 90% prevalence of virus-neutralizing antibodies.[140] However, a study evaluating the performance of rPAV-gp55 in the presence of high levels of PAdV-3 antibodies demonstrated that the effect of rPAV-gp55 was not inhibited by the presence of elevated levels of PAdV-3-neutralizing antibodies.[135] Since preexisting HAdV-specific neutralizing antibodies in humans do not cross-neutralize PAdV-3,[133] PAdV-3 vectors would be a promising supplement to HAdV vectors as a delivery vehicle for recombinant vaccines and gene therapy applications in humans.

## 4.4 Ovine Adenovirus-Based Vectors

AdVs isolated from sheep are termed as ovine AdVs (OAdVs). Since 1969 seven serotypes of OAdVs have been isolated and classified (OAdV-1–7). OAdV-1 to 6 belong to the genus *Mastadenovirus*, while OAdV-7 belongs to the genus *Atadenovirus*. Most of these viruses are associated with either respiratory tract or intestinal tract infections. Among these serotypes, OAdV-7 has been well characterized and evaluated as a gene delivery vector. The nucleotide sequences of all seven OAdV serotypes are currently available. The genome organization of OAdV-7 is distinct from the other AdVs in the genus *Mastadenovirus* since its genome is AT rich.[142] Moreover, it lacks a clear distinguishable E1 region. The OAdV-7 entry into cells is independent of CAR and has in vivo tissue tropism distinct from HAdV-5.[143]

OAdV-7-based vectors were created using homologous recombination in *E. coli* and cosmid-based approaches followed by virus rescue in the ovine fetal lung cell line (CSL503) or ovine fetal skin fibroblastic cell line.[142,144,145] Transgenes have been expressed from three unique sites, referred to as sites I, II, and III, in the OAdV-7 genome. Site I is located between the PVIII and fiber genes, site II is present within the RH2 gene approximately 902 bp from the 3′ end of the viral genome, and site III is located within a short noncoding region present between the E4 and the right-end transcription regions. Site III is considered the most stable among the three insertion sites, resulting in high levels of transgene expression independent of its orientation. Moreover, OAdV-7 vectors with site III insertions are easy to rescue and grow to high titers.[144]

An OAdV-7-based HCV vaccine (OAdV-NS3) encoding the nonstructural protein 3 (NS3) of the HCV BK strain has been developed.[146] The transgene was inserted at site III and was driven by the RSV 3′ LTR. Intramuscular immunization of BALB/c mice induced high levels of NS3-specific IFN-γ-secreting T-lymphocytes as measured by ELISpot assay. Interestingly, the NS3-specific T cell response persisted for up to 10 weeks postvaccination. Moreover, the OAdV-NS3-induced T cell response was not altered in the presence of immunity to HAdV-5.

In another study, an OAdV-based vector encoding protective recombinant antigen (45W) of *Taenia ovis* was generated and evaluated for its immunogenicity and protective efficacy in sheep when used alone or in combination with a DNA-based *T. ovis* vaccine.[147] Immunization of sheep with two doses of either OAdV or DNA vaccine induced low levels of 45W-specific antibody responses. However, immunization with the DNA vaccine followed by boosting with OAdV-based vaccine induced antibody responses >65-fold higher than those vaccinated with either the DNA or the OAdV vaccine alone, conferring protection from challenge with *T. ovis*.

An OAdV-7-based vaccine for HIV-1 (OAdV.HIVA) encoding the HIV-1 clade A consensus gag-derived protein coupled to a T cell polyepitope was developed.[148] Vaccination of mice with OAdV.HIVA either alone or in combination with HAdV-5 or MVA-vectored vaccines induced high levels of the HIV-1-specific T cell responses necessary to confer protection against HIV-1. This study demonstrated the potential of OAdV-7 as a delivery vehicle for HIV vaccines. Furthermore, the feasibility of using OAdV.HIVA in combination with BCG.HIVA(401) (a *Mycobacterium bovis* bacillus Calmette–Guérin (BCG)-based HIV-1) and MVA.HIVA was evaluated.[149] Unfortunately, vaccination with the BCG.HIVA(401) alone induced undetectable and weak CD8 T-cell responses in BALB/c mice and rhesus macaques, respectively. However, priming with the BCG.HIVA(401) followed by boosting with MVA.HIVA and OAdV.HIVA induced robust HIV-1-specific T-cell responses.

A recombinant OAdV-7 vector expressing the human alpha-1 antitrypsin gene (OAVhAAT) was generated and evaluated for its utility in human gene therapy in the skeletal muscle.[150] Injection of low doses of $3 \times 10^7$ infectious particles of OAVhAAT resulted in high serum levels of hAAT (>100 ng/ml), which was accompanied by a weak immune response to the vector. OAdV-7 infection was restricted to the smooth muscle with the level of hAAT expression comparable to that of a HAdV-5-based vector expressing the hAAT gene.[151]

Systemic administration of a single dose of an OAdV-7 vector encoding the *E. coli* purine fludarabine phosphorylase gene followed by prodrug fludarabine phosphate significantly inhibited the progression of prostate cancer in an immunocompetent mouse model. The role of an OAdV-7-based vector expressing ovalbumin (OVA) was evaluated for inducing antitumor immunity in a mouse model.[152] Incubation of bone marrow-derived dendritic cells with the OAdV-7 vectors expressing OVA resulted in upregulation of costimulatory markers and production of IL-12. Splenocytes collected from the immunized animals actively responded to in vitro antigen stimulation. Furthermore, the in vivo cytotoxicity assays demonstrated efficient killing (up to 75%) of antigenic peptide-pulsed target cells. In mice inoculated with B16-OVA tumor cells, immunization with the OAdV7-OVA significantly suppressed tumor growth.

These studies highlight the potential of OAdV as a potential vehicle for gene delivery for vaccination and gene therapy purposes. Neutralizing antibodies against OAdV are not prevalent in humans, making OAdV vectors as promising tools for gene delivery in humans.[153]

## 4.5 Canine Adenovirus-Based Vectors

Canine AdV (CAdV) serotypes 1 and 2 have been well characterized. In dogs, CAdV-1 is responsible for infectious canine hepatitis, whereas CAdV-2 causes only mild upper respiratory tract infection. Despite the long history of cohabitation of dogs and humans, CAdVs are not able to cross the specific barriers and have not been associated with any human disease. The complete genome sequences of both CAdV-1[154] and CAdV-2[155] are available. Genome sizes of CAdV-1 and CAdV-2 are about 30.5 and 31.3 kb, respectively. Based on sequence analysis and genome organization, these CAdVs are classified under the genus *Mastadenovirus*.

An E3-deleted vector system based on CAdV-1 has been developed by homologous recombination in bacteria.[156] The E1-deleted CAdV-2 vectors were developed in the late 1990s and have an insertion capacity of ~4 kb.[157,158] The strategy employed for constructing CAdV-2 vectors involves homologous recombination between the viral genome and a shuttle vector carrying a transgene expression cassette along with other necessary sequences, and is similar to that used for generating HAdV vectors.[158,159] The E1-deleted CAdV-2 vectors cannot be *trans*-complemented by cell lines expressing human E1, necessitating the development of canine kidney cell lines expressing the CAdV-2 E1.[157,158] The E1-deleted CAdV-2 vectors can be grown to high titers ($10^{13}$ vp/ml) and seem to have an excellent infectious particles-to-virus particles ratio.[158] However, the initial system for CAdV-2 vector development was cumbersome and very inefficient, often resulting in a CAdV-2 vector titer that was $10^4$- to $10^5$-fold lower than that of a HAdV-5 vector. Apparently this was because the canine kidney cells are difficult to transfect with linear CAdV-2 vector genomes of >30 kb.[157] Recently, an improved system for CAdV-2 vector generation has been described that uses a canine kidney cell line *trans*-complementing the CAdV-2 E1 and expressing I-*Sce*I fused to estrogen receptor.[160] This system allows highly efficient transfection of the supercoiled CAdV-2 vector genome into the canine kidney cells followed by an intracellular release of the vector genome and results in a 1000-fold increase in CAdV-2 vector titers.

A conditionally replicative AdV (CrAd) vector based on CAdV-2 has been developed and shows efficient replication and oncolytic potential in canine cell lines and a mouse xenograft model.[161] The subsequent study has demonstrated enhanced binding and internalization of the CAdV-2 CrAd vector into canine osteosarcoma cells on vector modification by incorporation of polylysine into the C-terminus of fiber knobs.[162]

Tropism of CAdV-2 is distinct but overlaps with HAdV-5. Cells transduced by CAdV-2 can also be transduced by HAdV-5 but not vice versa.[163] CAdV-2 utilizes CAR as receptor but does not depend on $\alpha_M\beta_2$-integrins or the heavy chain of the MHC-I for entry. Interestingly, the CAdV-2 capsid lacks the RGD motif in the penton base required for interaction with integrins.[164–166] CAdV-2 also does not interact with other AdV receptors such as lactoferrin and CD46.[163] CAdV-2 is replication-defective for human cells, although it can infect human cells—a highly desirable trait for preventing any complication caused by a replication-competent AdV contamination.[157]

CAdV-2 vectors preferentially transduced rodent olfactory neurons and central nervous system (CNS) neurons in vitro and in vivo and demonstrated more efficient retrograde axonal transport than HAdV-5 following intramuscular and intrastriatum injections allowing for transgene expression throughout the substantia nigra.[167] Injection of a CAdV-2 vector into multiple sites into the striatum instead of a single site resulted in five times more dopaminergic neurons in the substantia nigra, suggesting that a multiple site injection strategy could prove more effective if a large area of the CNS is to be targeted.[168] CAdV-2 vectors are capable of transducing young mouse neurons without affecting the functional maturation of the neurons.[169] In addition, CAdV-2 vectors did not induce significant cellular infiltration in the rat brain, reflecting their poor immunogenicity in the CNS.[168] Collectively, these observations suggest that CAdV-2 vectors could prove to be a very effective tool for therapy of neurodegenerative diseases requiring widespread expression of transgene in the CNS without the complications arising from cellular infiltrations as a result of unintended immune activation. In addition to the CNS, the CAdV-2 vectors efficiently transduce mouse airway epithelial cells in vitro, ex vivo, and in vivo.[170]

Human sera containing high levels of HAdV-5-neutralizing antibodies demonstrated little to no detectable neutralization of CAdV-2.[157,158] In addition, sera from mice containing exceptionally high levels (titer 3360) of anti-HAdV-5-neutralizing antibodies did not affect transduction by CAdV-2 but caused greater than 97% inhibition of HAdV-5 in vitro. Similarly, the preexisting HAdV immunity did not significantly affect transduction of the murine respiratory tract by CAdV-2.[170] Collectively, these observations imply that the preexisting HAdV immunity does not affect CAdV-2 transduction and that CAdV-2 vectors are an effective tool for circumventing preexisting AdV immunity.

In addition to the E1-deleted, E3-deleted, and CrAdV vectors, helper-dependent (HD) vectors have also been developed based on CAdV-2.[168] The strategy for construction of HD CAdV-2 vectors involves homologous recombination in *E. coli* BJ5183 between the plasmid pEJK25 containing a CAdV-2 ITR (without the packaging sequences and~25 kb stuffer sequence) and the shuttle plasmid (pGut containing a transgene expression cassette, CAdV-2 ITR, with the packaging sequences, and a 2 kb overlap region with pEJK25).[168] The HD AdV vectors are generated by transfection of E1 complementing cell lines with the linearized HD vector genome followed by infection with the helper virus. Amplification of the HD vectors also requires coinfection with the helper virus. After several rounds of amplification, the HD vector is purified by a cesium chloride density gradient.[171] Flanking the packaging domain of the helper virus with loxP sites does not prevent a significant level of contamination of the CAdV-2 vectors with the helper virus as the Cre-mediated excision at the loxP sites is not very efficient. However, mutations in *cis*-acting sequences in the packaging domain of the helper virus reduced the contamination to ~1% in HD CAdV-2 vector preparation with a final titer of $2.5 \times 10^{10}$ infectious units/ml and $\sim 2 \times 10^{11}$ particles/ml. In addition, no replication-competent AdV could be detected in $10^{11}$ particles after multiple rounds of amplifications.

Importantly, the HD CAdV-2 vectors are capable of inducing long-term, sustained transgene expression in the rat brain lasting for at least one year postinjection.[168] An HD CAdV-2 vector also has induced transgene expression lasting up to 3 months postinstillation in the mouse upper respiratory tract. At 3 months, some decrease was detected compared to earlier time points, likely due to natural turnover of pulmonary epithelium. Additionally, the HD CAdV-2 vectors induced some degree of innate immune response in the mouse lung, but the level of induction was lower than that induced by HAdV-5 vectors.[170]

HD CAdV-2 vectors have also been tested for use in therapy of mucopolysaccharidosis (MPS) type VII which results from the deficiency of β-glucuronidase (β-glu) and is manifested as corneal clouding due to the accumulation of glycosaminoglycans (GAG).[172] The HD CAdV-2 vector encoding for human β-glu (HD-RIGIE) efficiently transduced CAR-positive keratocytes in mice and nonhuman primates following direct intrastromal injection. Apart from a temporary corneal edema that lasted for ~24 h, no major complication occurred. The CAdV-2 vector transduction was as efficient, if not more, as that of the HAdV-5 vector. It covered the entire cornea but declined after 1 week, suggesting that further improvements are necessary to obtain long-term expression. Interestingly, the CAdV-2 vector induced a histological correction of GAG and cell morphology in the canine cornea, possibly due to the CAR expression by keratocytes and a high-level expression of β-glu by the vector.[172] In addition, the HD-RIGIE could also reverse neuropathological changes associated with MPS VII in the dog's brain.[173] This HD-RIGIE has the potential to induce long-term β-glu expression in mice, resulting in decreased GAG levels, lysosomal enzyme activity, and most importantly, a dramatic improvement in cognitive function.[174] For these reasons, HD CAdV-2 vectors can potentially provide tools for therapy of MPS VII and other lysosomal storage diseases.

## 4.6 Murine Adenovirus-Based Vectors

The murine AdVs are classified under the genus *Mastadenovirus*—species A, B, and C. The murine AdV serotype 1 (MAdV-1) was first isolated and characterized in 1960[175] and has been used as a model system for exploring virus–host interactions, AdV pathogenesis, and antiviral therapies.[176] Infection with MAdV-1 can cause serious disease in both newborn and adult mice. MAdV-1 infections, even at low doses, in newborn mice can cause serious mortality. However, akin to HAdVs, MAdV-1 infections in immunocompetent adult mice cause only mild infections with low mortality.[177,178] Unlike HAdVs, which initially infect the respiratory epithelium, MAdV-1 replicates in the primary endothelial cells of various body systems and thus results in widespread systemic infection affecting the liver, spleen, kidneys, intestines, adrenals, heart, brain, and spinal cord.[179,180]

The genome of MAdV-1 is ~30.9 kbp and is similar to HAdV-5 in organization except that it does not encode virus-associated RNAs.[181,182] Vectors based on MAdV-1 were first described in the late 1990s.[181,183,184] E1A-deleted MAdV-1 vectors are constructed either by replacing an initiation codon with a stop codon or by deleting each of the three conserved regions CR1, CR2, or CR3 within E1A. These vectors grew to titers only one log lower than those of wild-type virus in mouse fibroblasts, implying that MAdV-1 E1A is not essential for efficient virus replication.[184] Deletion of E1A did not alter the gene expression levels of the other early region genes, suggesting that, unlike HAdVs, the MAdV-1 E1A is not required for transactivation of other early region genes.[184] However, the pathogenicity of E1A null mutants of MAdV-1 is significantly lower than wild-type virus, suggesting the requirement of E1A in the host but not in the cell culture.[179]

The E3 region of MAdV-1 encodes for three proteins with a common N-terminal sequence but unique C-terminal sequences.[183] E3-deleted MAdV-1 vectors were constructed by mutagenesis to block expression of each one of the E3 proteins.[181,185] The E3 mutants of MAdV-1 induced significantly less endothelial cell damage and inflammatory response in the brain and spinal cord than the wild-type virus, indicating that there is a role for E3 proteins in these areas.[181] The MAdV-1 does not utilize CAR as a primary receptor[176] and does not contain the RGD motif in the penton base.[186] Instead, the MAdV-1 fiber knob domain carries an RGD motif that plays an important role in MAdV-1 entry mediated by αV-integrins. In addition, cell surface glycosaminoglycan heparan sulfate is also involved in MAdV-1 infection.[186,187] MAdV-1 binds to primary human smooth muscles with significantly higher affinity than HAdV-5. The biodistribution of MAdV-1 is not altered by the presence of physiological concentrations of coagulation factor XI or the vitamin K-dependent factors that play a role in the targeting of HAdV-5 to the liver. Although MAdV-1 does bind to the factor XI, contrary to HAdV-5, the binding does not result in cell attachment. Consequently, the targeting of MAdV-1 to the liver is significantly lower than for HAdV-5.[188]

A MAdV-1 vector carrying a deletion in the CR2 region of E1A (dlE102) has been shown to be an excellent oncolytic AdV system that can help in understanding the mechanism of the action of oncolytic AdVs in an immunocompetent host.[189] The ability to replicate in an immunocompetent host is particularly noteworthy as most of the oncolytic AdV vectors based on HAdV-5 cannot replicate in mouse tissue, and, therefore, their safety and efficacy cannot be adequately evaluated. The dlE102 virus replicates efficiently in murine tumor cells but its replication is attenuated in nontransformed cells. In addition, it demonstrates potent antitumor activity in an immunocompetent xenograft tumor model.[189] A feasible approach to arm the dlE102 virus with an immunomodulatory molecule has further enhanced its antitumor efficacy.[189] MAdV-1 has also been used as a model for understanding the pathogenic mechanisms of pediatric myocarditis[190,191] and acute respiratory infection[192] caused by AdVs.

### 4.7 Avian Adenovirus-Based Vectors

Several AdVs have been isolated from birds including fowl (FAdV), falcon (FaAdV), goose (GoAdV), ducks (DAdV), and turkey (TAdV). Most of these viruses are included under the genus *Aviadenovirus* and grouped in to four serotypes (FaAdVs A–E, GoAdV A, DAdV A, and TAdV B). Duck AdV A, AdV 127, and egg drop syndrome 76 virus, which is also known as duck AdV 1 (DAdV-1), are included under the genus *Atadenovirus* based on their unique genomic and structural characteristics. These viruses are serologically distinct from other AdV genera and infect only birds.

The complete nucleotide sequences of five *Aviadenovirus* species (FAdV-1, FAdV-4, FAdV-8, FAdV-9, and TAdV-1) have been determined.[193,194] The genomes of aviadenoviruses are 20–45% larger than other AdVs and range in size between 43,804 and 45,667 bp. Furthermore, the genome organization of aviadenoviruses is distinct compared with other AdVs and contains a high G + C content (ranging from 50% to 67%). The virions from FAdV-1, FAdV-4, and TAdV-1 contain two fibers per penton base[195] while genomes of aviadenoviruses lack the genes encoding E1, E3, V, and IX proteins. The aviadenovirus genomes contain several uncharacterized transcription units at the right end that are unique to this genus.[196]

FAdV-based vectors devoid of GAM-1 have been created using homologous recombination in *E. coli* with cosmid-based approaches followed by transfection in permissive Leghorn male hepatoma cells. Three aviadenoviruses (FAdV-1, FAdV-8, and FAdV-10) have been evaluated as gene delivery vectors for vaccination. A FAdV-10 vector containing the VP2 gene from the Australian classical strain 002/73 of infectious bursal disease virus (IBDV) has been generated.[197] Vaccination of specific pathogen-free chickens with this FAdV10-based vaccine induced VP2-specific antibodies that protected chickens against challenge with the IBDV V877 strain. A FAdV-1-based vector has been evaluated for cancer gene therapy and demonstrated efficient transduction of several human cells including HepG2, A549, and primary human dermal fibroblasts.[198] In another study, a FAdV-8-based vaccine was developed against infectious bronchitis virus (IBV) after incorporating the spike peplomer S1 subunit from the IBV Vic S strain under the control of the FAdV major late promoter.[199] Using this FAdV-8 vector-based vaccine, immunization of commercial broiler chickens induced S1-specific antibody responses and conferred protection against challenge with either Vic S (serotype B) or N1/62 (serotype C) strains of IBV. A FAdV-8 vector expressing cytokines including chicken IFN-γ has been developed and evaluated to enhance the immunogenicity of vaccines in chickens.[200]

## 5. Concluding Remarks

A number of nonhuman AdVs including simian (SAdV), bovine (BAdV), porcine (PAdV), canine (CAdV), ovine (OAdV), murine (MAdV, and fowl (FAdV) are at various stages of development as gene delivery systems for recombinant vaccines and gene therapy applications. In addition to effectively circumventing preexisting HAdV immunity, these nonhuman AdV vectors can utilize a number of other receptors in addition to CAR for vector internalization, thereby expanding the range of cell types that can be targeted. The safety aspects of these vectors appear to be similar to or better than HAdV vectors. In addition to their utility for human applications, nonhuman AdV vectors also provide excellent platforms for veterinary vaccines. A specific nonhuman AdV vector when used in its species of origin could provide an excellent animal model for evaluating the efficacy and pathogenesis of these vectors.

The mechanism/s of activation of innate immunity including TLR expression by nonhuman AdVs needs to be determined to fully explore their potential as gene delivery systems. These vectors will be useful in prime–boost approaches with other AdV vectors or with other gene delivery systems including DNA immunization or other viral or bacterial vectors. In situations where multiple vector inoculations are required for a desired effect, nonhuman AdV vectors could supplement HAdV or other viral vectors. Only SAdV vectors can be grown in certified human cell lines that are used for HAdV replication and purification; therefore, there is a need to certify a number of other cell lines that are suitable for growing and purifying other nonhuman AdV vectors. To fully exploit the desired impact of using nonhuman AdVs, further changes in nonhuman AdV vector design will be necessary.

## Acknowledgments

This work was supported by Hatch Fund and we are thankful to Jane Kovach for her excellent secretarial assistance.

## References

1. Bonnet MC, Tartaglia J, Verdier F, Kourilsky P, Lindberg A, Klein M, et al. Recombinant viruses as a tool for therapeutic vaccination against human cancers. *Immunol Lett* 2000;**74**:11–25.
2. Souza AP, Haut L, Reyes-Sandoval A, Pinto AR. Recombinant viruses as vaccines against viral diseases. *Braz J Med Biol Res* 2005;**38**:509–22.
3. Rocha CD, Caetano BC, Machado AV, Bruna-Romero O. Recombinant viruses as tools to induce protective cellular immunity against infectious diseases. *Int Microbiol* 2004;**7**:83–94.
4. Barouch DH. Rational design of gene-based vaccines. *J Pathol* 2006;**208**:283–9.
5. Rowe WP, Huebner RJ, Gilmore LK, Parrot RH, Ward TG. Isolation of a cytopathogenic agent from human adenoids undergoing spontaneous degeneration in tissue culture. *Proc Soc Exp Biol Med* 1953;**84**:570–3.
6. Huebner RJ, Rowe WP, Schatten WE, Smith RR, Thomas LB. Studies on the use of viruses in the treatment of carcinoma of the cervix. *Cancer* 1956;**9**:1211–8.
7. Russell WC. Adenoviruses: update on structure and function. *J Gen Virol* 2009;**90**:1–20.
8. Harrach B, Benko M, Both GW, Brown M, Davison AJ, Echavarria M, et al. The double stranded DNA viruses: adenoviridae. In: King AMQ, Adams MJ, Carstens EB, Lefkowitz EJ, editors. *Virus taxonomy: ninth report of the international committee on taxonomy of viruses*. 1st ed Waltham (MA): Academic Press: Elsevier; 2011. p. 125–41.

9. Davison AJ, Benko M, Harrach B. Genetic content and evolution of adenoviruses. *J Gen Virol* 2003;**84**:2895–908.
10. Sharma A, Li X, Bangari DS, Mittal SK. Adenovirus receptors and their implications in gene delivery. *Virus Res* 2009;**143**:184–94.
11. Arnberg N. Adenovirus receptors: implications for targeting of viral vectors. *Trends Pharmacol Sci* 2012;**33**:442–8.
12. Bangari DS, Mittal SK. Current strategies and future directions for eluding adenoviral vector immunity. *Curr Gene Ther* 2006;**6**:215–26.
13. Sakurai F, Kawabata K, Mizuguchi H. Adenovirus vectors composed of subgroup B adenoviruses. *Curr Gene Ther* 2007;**7**:229–38.
14. Russell WC. Update on adenovirus and its vectors. *J Gen Virol* 2000;**81**:2573–604.
15. Edelstein ML, Abedi MR, Wixon J, Edelstein RM. Gene therapy clinical trials worldwide 1989–2004-an overview. *J Gene Med* 2004;**6**:597–602.
16. Mast TC, Kierstead L, Gupta SB, Nikas AA, Kallas EG, Novitsky V, et al. International epidemiology of human pre-existing adenovirus (Ad) type-5, type-6, type-26 and type-36 neutralizing antibodies: correlates of high Ad5 titers and implications for potential HIV vaccine trials. *Vaccine* 2010;**28**:950–7.
17. Yu B, Zhou Y, Wu H, Wang Z, Zhan Y, Feng X, et al. Seroprevalence of neutralizing antibodies to human adenovirus type 5 in healthy adults in China. *J Med Virol* 2012;**84**:1408–14.
18. Sumida SM, Truitt DM, Lemckert AA, Vogels R, Custers JH, Addo MM, et al. Neutralizing antibodies to adenovirus serotype 5 vaccine vectors are directed primarily against the adenovirus hexon protein. *J Immunol* 2005;**174**:7179–85.
19. Barouch DH, Kik SV, Weverling GJ, Dilan R, King SL, Maxfield LF, et al. International seroepidemiology of adenovirus serotypes 5, 26, 35, and 48 in pediatric and adult populations. *Vaccine* 2011;**29**:5203–9.
20. Chirmule N, Propert K, Magosin S, Qian Y, Qian R, Wilson J. Immune responses to adenovirus and adeno-associated virus in humans. *Gene Ther* 1999;**6**:1574–83.
21. Gahery-Segard H, Juillard V, Gaston J, Lengagne R, Pavirani A, Boulanger P, et al. Humoral immune response to the capsid components of recombinant adenoviruses: routes of immunization modulate virus-induced Ig subclass shifts. *Eur J Immunol* 1997;**27**:653–9.
22. Gahery-Segard H, Farace F, Godfrin D, Gaston J, Lengagne R, Tursz T, et al. Immune response to recombinant capsid proteins of adenovirus in humans: antifiber and anti-penton base antibodies have a synergistic effect on neutralizing activity. *J Virol* 1998;**72**:2388–97.
23. Bruder JT, Semenova E, Chen P, Limbach K, Patterson NB, Stefaniak ME, et al. Modification of Ad5 hexon hypervariable regions circumvents pre-existing Ad5 neutralizing antibodies and induces protective immune responses. *PLoS One* 2012;**7**:e33920.
24. Mittal SK, McDermott MR, Johnson DC, Prevec L, Graham FL. Monitoring foreign gene expression by a human adenovirus-based vector using the firefly luciferase gene as a reporter. *Virus Res* 1993;**28**:67–90.
25. Chirmule N, Raper SE, Burkly L, Thomas D, Tazelaar J, Hughes JV, et al. Readministration of adenovirus vector in nonhuman primate lungs by blockade of CD40-CD40 ligand interactions. *J Virol* 2000;**74**:3345–52.
26. Moffatt S, Hays J, HogenEsch H, Mittal SK. Circumvention of vector-specific neutralizing antibody response by alternating use of human and non-human adenoviruses: implications in gene therapy. *Virology* 2000;**272**:159–67.
27. Pichla-Gollon SL, Lin SW, Hensley SE, Lasaro MO, Herkenhoff-Haut L, Drinker M, et al. Effect of preexisting immunity on an adenovirus vaccine vector: in vitro neutralization assays fail to predict inhibition by antiviral antibody in vivo. *J Virol* 2009;**83**:5567–73.

28. Fausther-Bovendo H, Kobinger GP. Pre-existing immunity against Ad vectors: humoral, cellular and innate response, what's important? *Hum Vaccin Immunother* 2014;**10**:2875–84.
29. Olive M, Eisenlohr L, Flomenberg N, Hsu S, Flomenberg P. The adenovirus capsid protein hexon contains a highly conserved human CD4+ T-cell epitope. *Hum Gene Ther* 2002;**13**:1167–78.
30. Leen AM, Sili U, Vanin EF, Jewell AM, Xie W, Vignali D, et al. Conserved CTL epitopes on the adenovirus hexon protein expand subgroup cross-reactive and subgroup-specific CD8+ T cells. *Blood* 2004;**104**:2432–40.
31. Tang J, Olive M, Pulmanausahakul R, Schnell M, Flomenberg N, Eisenlohr L, et al. Human CD8+ cytotoxic T cell responses to adenovirus capsid proteins. *Virology* 2006;**350**:312–22.
32. Heemskerk B, Veltrop-Duits LA, van VT, ten Dam MM, Heidt S, Toes RE, et al. Extensive cross-reactivity of CD4+ adenovirus-specific T cells: implications for immunotherapy and gene therapy. *J Virol* 2003;**77**:6562–6.
33. Hutnick NA, Carnathan D, Demers K, Makedonas G, Ertl HC, Betts MR. Adenovirus-specific human T cells are pervasive, polyfunctional, and cross-reactive. *Vaccine* 2010;**28**:1932–41.
34. Veltrop-Duits LA, Heemskerk B, Sombroek CC, van VT, Gubbels S, Toes RE, et al. Human CD4+ T cells stimulated by conserved adenovirus 5 hexon peptides recognize cells infected with different species of human adenovirus. *Eur J Immunol* 2006;**36**:2410–23.
35. Calcedo R, Vandenberghe LH, Roy S, Somanathan S, Wang L, Wilson JM. Host immune responses to chronic adenovirus infections in human and nonhuman primates. *J Virol* 2009;**83**:2623–31.
36. Schnell MA, Zhang Y, Tazelaar J, Gao GP, Yu QC, Qian R, et al. Activation of innate immunity in nonhuman primates following intraportal administration of adenoviral vectors. *Mol Ther* 2001;**3**:708–22.
37. Zhang Y, Chirmule N, Gao GP, Qian R, Croyle M, Joshi B, et al. Acute cytokine response to systemic adenoviral vectors in mice is mediated by dendritic cells and macrophages. *Mol Ther* 2001;**3**:697–707.
38. Raper SE, Chirmule N, Lee FS, Wivel NA, Bagg A, Gao GP, et al. Fatal systemic inflammatory response syndrome in a ornithine transcarbamylase deficient patient following adenoviral gene transfer. *Mol Genet Metab* 2003;**80**:148–58.
39. Basner-Tschakarjan E, Gaffal E, O'Keeffe M, Tormo D, Limmer A, Wagner H, et al. Adenovirus efficiently transduces plasmacytoid dendritic cells resulting in TLR9-dependent maturation and IFN-alpha production. *J Gene Med* 2006;**8**:1300–6.
40. Thaci B, Ulasov IV, Wainwright DA, Lesniak MS. The challenge for gene therapy: innate immune response to adenoviruses. *Oncotarget* 2011;**2**:113–21.
41. Chen RF, Lee CY. Adenoviruses types, cell receptors and local innate cytokines in adenovirus infection. *Int Rev Immunol* 2014;**33**:45–53.
42. Hendrickx R, Stichling N, Koelen J, Kuryk L, Lipiec A, Greber UF. Innate immunity to adenovirus. *Hum Gene Ther* 2014;**25**:265–84.
43. Huang X, Yang Y. Innate immune recognition of viruses and viral vectors. *Hum Gene Ther* 2009;**20**:293–301.
44. Fejer G, Freudenberg M, Greber UF, Gyory I. Adenovirus-triggered innate signalling pathways. *Eur J Microbiol Immunol (Bp)* 2011;**1**:279–88.
45. Lopez-Gordo E, Podgorski II, Downes N, Alemany R. Circumventing antivector immunity: potential use of nonhuman adenoviral vectors. *Hum Gene Ther* 2014;**25**:285–300.
46. Capasso C, Garofalo M, Hirvinen M, Cerullo V. The evolution of adenoviral vectors through genetic and chemical surface modifications. *Viruses* 2014;**6**:832–55.

47. Ahi YS, Bangari DS, Mittal SK. Adenoviral vector immunity: its implications and circumvention strategies. *Curr Gene Ther* 2011;**11**:307–20.
48. Bangari DS, Mittal SK. Development of nonhuman adenoviruses as vaccine vectors. *Vaccine* 2006;**24**:849–62.
49. Farina SF, Gao GP, Xiang ZQ, Rux JJ, Burnett RM, Alvira MR, et al. Replication-defective vector based on a chimpanzee adenovirus. *J Virol* 2001;**75**:11603–13.
50. Crawford-Miksza L, Schnurr DP. Analysis of 15 adenovirus hexon proteins reveals the location and structure of seven hypervariable regions containing serotype-specific residues. *J Virol* 1996;**70**:1836–44.
51. Cohen CJ, Xiang ZQ, Gao GP, Ertl HC, Wilson JM, Bergelson JM. Chimpanzee adenovirus CV-68 adapted as a gene delivery vector interacts with the coxsackievirus and adenovirus receptor. *J Gen Virol* 2002;**83**:151–5.
52. Roy S, Gao G, Clawson DS, Vandenberghe LH, Farina SF, Wilson JM. Complete nucleotide sequences and genome organization of four chimpanzee adenoviruses. *Virology* 2004;**324**:361–72.
53. Tatsis N, Blejer A, Lasaro MO, Hensley SE, Cun A, Tesema L, et al. A CD46-binding chimpanzee adenovirus vector as a vaccine carrier. *Mol Ther* 2007;**15**:608–17.
54. Tatsis N, Tesema L, Robinson ER, Giles-Davis W, McCoy K, Gao GP, et al. Chimpanzee-origin adenovirus vectors as vaccine carriers. *Gene Ther* 2006;**13**:421–9.
55. Chen H, Smith GJ, Li KS, Wang J, Fan XH, Rayner JM, et al. Establishment of multiple sublineages of H5N1 influenza virus in Asia: implications for pandemic control. *Proc Natl Acad Sci USA* 2006;**103**:2845–50.
56. Ersching J, Hernandez MI, Cezarotto FS, Ferreira JD, Martins AB, Switzer WM, et al. Neutralizing antibodies to human and simian adenoviruses in humans and New-World monkeys. *Virology* 2010;**407**:1–6.
57. Colloca S, Barnes E, Folgori A, Ammendola V, Capone S, Cirillo A, et al. Vaccine vectors derived from a large collection of simian adenoviruses induce potent cellular immunity across multiple species. *Sci Transl Med* 2012;**4**:115ra2.
58. Xiang Z, Li Y, Cun A, Yang W, Ellenberg S, Switzer WM, et al. Chimpanzee adenovirus antibodies in humans, sub-Saharan Africa. *Emerg Infect Dis* 2006;**12**:1596–9.
59. Zhang S, Huang W, Zhou X, Zhao Q, Wang Q, Jia B. Seroprevalence of neutralizing antibodies to human adenoviruses type-5 and type-26 and chimpanzee adenovirus type-68 in healthy Chinese adults. *J Med Virol* 2013;**85**:1077–84.
60. Jian L, Zhao Q, Zhang S, Huang W, Xiong Y, Zhou X, et al. The prevalence of neutralising antibodies to chimpanzee adenovirus type 6 and type 7 in healthy adult volunteers, patients with chronic hepatitis B and patients with primary hepatocellular carcinoma in China. *Arch Virol* 2014;**159**:465–70.
61. Dudareva M, Andrews L, Gilbert SC, Bejon P, Marsh K, Mwacharo J, et al. Prevalence of serum neutralizing antibodies against chimpanzee adenovirus 63 and human adenovirus 5 in Kenyan children, in the context of vaccine vector efficacy. *Vaccine* 2009;**27**:3501–4.
62. Chen H, Xiang ZQ, Li Y, Kurupati RK, Jia B, Bian A, et al. Adenovirus-based vaccines: comparison of vectors from three species of adenoviridae. *J Virol* 2010;**84**:10522–32.
63. Buchbinder SP, Mehrotra DV, Duerr A, Fitzgerald DW, Mogg R, Li D, et al. Efficacy assessment of a cell-mediated immunity HIV-1 vaccine (the Step Study): a double-blind, randomised, placebo-controlled, test-of-concept trial. *Lancet* 2008;**372**:1881–93.
64. Ewer KJ, O'Hara GA, Duncan CJ, Collins KA, Sheehy SH, Reyes-Sandoval A, et al. Protective $CD8^+$ T-cell immunity to human malaria induced by chimpanzee adenovirus-MVA immunisation. *Nat Commun* 2013;**4**:2836.

65. Abbink P, Lemckert AA, Ewald BA, Lynch DM, Denholtz M, Smits S, et al. Comparative seroprevalence and immunogenicity of six rare serotype recombinant adenovirus vaccine vectors from subgroups B and D. *J Virol* 2007;**81**:4654–63.
66. Kahl CA, Bonnell J, Hiriyanna S, Fultz M, Nyberg-Hoffman C, Chen P, et al. Potent immune responses and in vitro pro-inflammatory cytokine suppression by a novel adenovirus vaccine vector based on rare human serotype 28. *Vaccine* 2010;**28**:5691–702.
67. Quinn KM, Da CA, Yamamoto A, Berry D, Lindsay RW, Darrah PA, et al. Comparative analysis of the magnitude, quality, phenotype, and protective capacity of simian immunodeficiency virus gag-specific $CD8^+$ T cells following human-, simian-, and chimpanzee-derived recombinant adenoviral vector immunization. *J Immunol* 2013;**190**:2720–35.
68. Haut LH, Lin SW, Tatsis N, DiMenna LJ, Giles-Davis W, Pinto AR, et al. Robust genital gag-specific $CD8^+$ T-cell responses in mice upon intramuscular immunization with simian adenoviral vectors expressing HIV-1-gag. *Eur J Immunol* 2010;**40**:3426–38.
69. Peruzzi D, Dharmapuri S, Cirillo A, Bruni BE, Nicosia A, Cortese R, et al. A novel chimpanzee serotype-based adenoviral vector as delivery tool for cancer vaccines. *Vaccine* 2009;**27**:1293–300.
70. Skog J, Edlund K, Bergenheim AT, Wadell G. Adenoviruses 16 and CV23 efficiently transduce human low-passage brain tumor and cancer stem cells. *Mol Ther* 2007;**15**:2140–5.
71. Wang D, Raja NU, Trubey CM, Juompan LY, Luo M, Woraratanadharm J, et al. Development of a cAdVax-based bivalent ebola virus vaccine that induces immune responses against both the Sudan and Zaire species of Ebola virus. *J Virol* 2006;**80**:2738–46.
72. Kobinger GP, Feldmann H, Zhi Y, Schumer G, Gao G, Feldmann F, et al. Chimpanzee adenovirus vaccine protects against Zaire Ebola virus. *Virology* 2006;**346**:394–401.
73. Roy S, Zhi Y, Kobinger GP, Figueredo J, Calcedo R, Miller JR, et al. Generation of an adenoviral vaccine vector based on simian adenovirus 21. *J Gen Virol* 2006;**87**:2477–85.
74. Stanley DA, Honko AN, Asiedu C, Trefry JC, Lau-Kilby AW, Johnson JC, et al. Chimpanzee adenovirus vaccine generates acute and durable protective immunity against ebolavirus challenge. *Nat Med* 2014;**20**:1126–9.
75. Sharma A, Wendland R, Sung B, Wu W, Grunwald T, Worgall S. Maternal immunization with chimpanzee adenovirus expressing RSV fusion protein protects against neonatal RSV pulmonary infection. *Vaccine* 2014;**32**:5761–8.
76. Xiang ZQ, Greenberg L, Ertl HC, Rupprecht CE. Protection of non-human primates against rabies with an adenovirus recombinant vaccine. *Virology* 2014;**450-451**:243–9.
77. Zhou D, Cun A, Li Y, Xiang Z, Ertl HC. A chimpanzee-origin adenovirus vector expressing the rabies virus glycoprotein as an oral vaccine against inhalation infection with rabies virus. *Mol Ther* 2006;**14**:662–72.
78. Roy S, Kobinger GP, Lin J, Figueredo J, Calcedo R, Kobasa D, et al. Partial protection against H5N1 influenza in mice with a single dose of a chimpanzee adenovirus vector expressing nucleoprotein. *Vaccine* 2007;**25**:6845–51.
79. Vitelli A, Quirion MR, Lo CY, Misplon JA, Grabowska AK, Pierantoni A, et al. Vaccination to conserved influenza antigens in mice using a novel simian adenovirus vector, PanAd3, derived from the bonobo *Pan paniscus*. *PLoS One* 2013;**8**:e55435.
80. Reyes-Sandoval A, Sridhar S, Berthoud T, Moore AC, Harty JT, Gilbert SC, et al. Single-dose immunogenicity and protective efficacy of simian adenoviral vectors against *Plasmodium berghei*. *Eur J Immunol* 2008;**38**:732–41.
81. Sridhar S, Reyes-Sandoval A, Draper SJ, Moore AC, Gilbert SC, Gao GP, et al. Single-dose protection against *Plasmodium berghei* by a simian adenovirus vector using a human cytomegalovirus promoter containing intron A. *J Virol* 2008;**82**:3822–33.

82. Warimwe GM, Lorenzo G, Lopez-Gil E, Reyes-Sandoval A, Cottingham MG, Spencer AJ, et al. Immunogenicity and efficacy of a chimpanzee adenovirus-vectored Rift Valley fever vaccine in mice. *Virol J* 2013;**10**:349.
83. Kobinger GP, Figueredo JM, Rowe T, Zhi Y, Gao G, Sanmiguel JC, et al. Adenovirus-based vaccine prevents pneumonia in ferrets challenged with the SARS coronavirus and stimulates robust immune responses in macaques. *Vaccine* 2007;**25**:5220–31.
84. Borthwick N, Ahmed T, Ondondo B, Hayes P, Rose A, Ebrahimsa U, et al. Vaccine-elicited human T cells recognizing conserved protein regions inhibit HIV-1. *Mol Ther* 2014;**22**:464–75.
85. Altfeld M, Addo MM, Rosenberg ES, Hecht FM, Lee PK, Vogel M, et al. Influence of HLA-B57 on clinical presentation and viral control during acute HIV-1 infection. *Aids* 2003;**17**:2581–91.
86. Ferguson AL, Mann JK, Omarjee S, Ndung'u T, Walker BD, Chakraborty AK. Translating HIV sequences into quantitative fitness landscapes predicts viral vulnerabilities for rational immunogen design. *Immunity* 2013;**38**:606–17.
87. Kelleher AD, Long C, Holmes EC, Allen RL, Wilson J, Conlon C, et al. Clustered mutations in HIV-1 gag are consistently required for escape from HLA-B27-restricted cytotoxic T lymphocyte responses. *J Exp Med* 2001;**193**:375–86.
88. Leslie AJ, Pfafferott KJ, Chetty P, Draenert R, Addo MM, Feeney M, et al. HIV evolution: CTL escape mutation and reversion after transmission. *Nat Med* 2004;**10**:282–9.
89. Hayton EJ, Rose A, Ibrahimsa U, Del SM, Capone S, Crook A, et al. Safety and tolerability of conserved region vaccines vectored by plasmid DNA, simian adenovirus and modified vaccinia virus ankara administered to human immunodeficiency virus type 1-uninfected adults in a randomized, single-blind phase I trial. *PLoS One* 2014;**9**:e101591.
90. Hoffman SL, Isenbarger D, Long GW, Sedegah M, Szarfman A, Waters L, et al. Sporozoite vaccine induces genetically restricted T cell elimination of malaria from hepatocytes. *Science* 1989;**244**:1078–81.
91. Khusmith S, Sedegah M, Hoffman SL. Complete protection against *Plasmodium yoelii* by adoptive transfer of a CD8+ cytotoxic T-cell clone recognizing sporozoite surface protein 2. *Infect Immun* 1994;**62**:2979–83.
92. Romero P, Maryanski JL, Corradin G, Nussenzweig RS, Nussenzweig V, Zavala F. Cloned cytotoxic T cells recognize an epitope in the circumsporozoite protein and protect against malaria. *Nature* 1989;**341**:323–6.
93. Schofield L, Villaquiran J, Ferreira A, Schellekens H, Nussenzweig R, Nussenzweig V. Gamma interferon, CD8+ T cells and antibodies required for immunity to malaria sporozoites. *Nature* 1987;**330**:664–6.
94. Weiss WR, Mellouk S, Houghten RA, Sedegah M, Kumar S, Good MF, et al. Cytotoxic T cells recognize a peptide from the circumsporozoite protein on malaria-infected hepatocytes. *J Exp Med* 1990;**171**:763–73.
95. Hill AV, Reyes-Sandoval A, O'Hara G, Ewer K, Lawrie A, Goodman A, et al. Prime-boost vectored malaria vaccines: progress and prospects. *Hum Vaccin* 2010;**6**:78–83.
96. Ogwang C, Afolabi M, Kimani D, Jagne YJ, Sheehy SH, Bliss CM, et al. Safety and immunogenicity of heterologous prime-boost immunisation with *Plasmodium falciparum* malaria candidate vaccines, ChAd63 ME-TRAP and MVA ME-TRAP, in healthy Gambian and Kenyan adults. *PLoS One* 2013;**8**:e57726.
97. Sheehy SH, Duncan CJ, Elias SC, Biswas S, Collins KA, O'Hara GA, et al. Phase Ia clinical evaluation of the safety and immunogenicity of the *Plasmodium falciparum* blood-stage antigen AMA1 in ChAd63 and MVA vaccine vectors. *PLoS One* 2012;**7**:e31208.

98. Sheehy SH, Duncan CJ, Elias SC, Collins KA, Ewer KJ, Spencer AJ, et al. Phase Ia clinical evaluation of the *Plasmodium falciparum* blood-stage antigen MSP1 in ChAd63 and MVA vaccine vectors. *Mol Ther* 2011;**19**:2269–76.
99. O'Hara GA, Duncan CJ, Ewer KJ, Collins KA, Elias SC, Halstead FD, et al. Clinical assessment of a recombinant simian adenovirus ChAd63: a potent new vaccine vector. *J Infect Dis* 2012;**205**:772–81.
100. Hayes CN, Chayama K. Emerging treatments for chronic hepatitis C. *J Formos Med Assoc* 2015;**114**:204–15.
101. Shoukry NH, Grakoui A, Houghton M, Chien DY, Ghrayeb J, Reimann KA, et al. Memory $CD8^+$ T cells are required for protection from persistent hepatitis C virus infection. *J Exp Med* 2003;**197**:1645–55.
102. Urbani S, Amadei B, Fisicaro P, Tola D, Orlandini A, Sacchelli L, et al. Outcome of acute hepatitis C is related to virus-specific CD4 function and maturation of antiviral memory CD8 responses. *Hepatology* 2006;**44**:126–39.
103. Folgori A, Capone S, Ruggeri L, Meola A, Sporeno E, Ercole BB, et al. A T-cell HCV vaccine eliciting effective immunity against heterologous virus challenge in chimpanzees. *Nat Med* 2006;**12**:190–7.
104. Barnes E, Folgori A, Capone S, Swadling L, Aston S, Kurioka A, et al. Novel adenovirus-based vaccines induce broad and sustained T cell responses to HCV in man. *Sci Transl Med* 2012;**4**:115ra1.
105. Reddy PS, Idamakanti N, Zakhartchouk AN, Baxi MK, Lee JB, Pyne C, et al. Nucleotide sequence, genome organization, and transcription map of bovine adenovirus type 3. *J Virol* 1998;**72**:1394–402.
106. Mittal SK, Prevec L, Babiuk LA, Graham FL. Sequence analysis of bovine adenovirus type 3 early region 3 and fibre protein genes. *J Gen Virol* 1992;**73**:3295–300.
107. Zheng B, Mittal SK, Graham FL, Prevec L. The E1 sequence of bovine adenovirus type 3 and complementation of human adenovirus type 5 E1A function in bovine cells. *Virus Res* 1994;**31**:163–86.
108. Bangari DS, Shukla S, Mittal SK. Comparative transduction efficiencies of human and nonhuman adenoviral vectors in human, murine, bovine, and porcine cells in culture. *Biochem Biophys Res Commun* 2005;**327**:960–6.
109. Ruigrok RW, Barge A, Mittal SK, Jacrot B. The fibre of bovine adenovirus type 3 is very long but bent. *J Gen Virol* 1994;**75**:2069–73.
110. Mittal SK, Middleton DM, Tikoo SK, Babiuk LA. Pathogenesis and immunogenicity of bovine adenovirus type 3 in cotton rats (*Sigmodon hispidus*). *Virology* 1995;**213**:131–9.
111. Mittal SK, Middleton DM, Tikoo SK, Prevec L, Graham FL, Babiuk LA. Pathology and immunogenicity in the cotton rat (*Sigmodon hispidus*) model after infection with a bovine adenovirus type 3 recombinant virus expressing the firefly luciferase gene. *J Gen Virol* 1996;**77**:1–9.
112. Mittal SK, Tikoo SK, Van Donkersgoed J, Beskorwayne T, Godson DL, Babiuk LA. Experimental inoculation of heifers with bovine adenovirus type 3. *Can J Vet Res* 1999;**63**:153–6.
113. Mittal SK, Prevec L, Graham FL, Babiuk LA. Development of a bovine adenovirus type 3-based expression vector. *J Gen Virol* 1995;**76**:93–102.
114. van Olphen AL, Mittal SK. Generation of infectious genome of bovine adenovirus type 3 by homologous recombination in bacteria. *J Virol Methods* 1999;**77**:125–9.
115. Reddy PS, Idamakanti N, Chen Y, Whale T, Babiuk LA, Mehtali M, et al. Replication-defective bovine adenovirus type 3 as an expression vector. *J Virol* 1999;**73**:9137–44.

116. Du E, Tikoo SK. Efficient replication and generation of recombinant bovine adenovirus-3 in nonbovine cotton rat lung cells expressing I-SceI endonuclease. *J Gene Med* 2010;**12**:840–7.
117. van Olphen AL, Mittal SK. Development and characterization of bovine x human hybrid cell lines that efficiently support the replication of both wild-type bovine and human adenoviruses and those with E1 deleted. *J Virol* 2002;**76**:5882–92.
118. van Olphen AL, Tikoo SK, Mittal SK. Characterization of bovine adenovirus type 3 E1 proteins and isolation of E1-expressing cell lines. *Virology* 2002;**295**:108–18.
119. Bangari DS, Sharma A, Mittal SK. Bovine adenovirus type 3 internalization is independent of primary receptors of human adenovirus type 5 and porcine adenovirus type 3. *Biochem Biophys Res Commun* 2005;**331**:1478–84.
120. Li X, Bangari DS, Sharma A, Mittal SK. Bovine adenovirus serotype 3 utilizes sialic acid as a cellular receptor for virus entry. *Virology* 2009;**392**:162–8.
121. Sharma A, Tandon M, Ahi YS, Bangari DS, Vemulapalli R, Mittal SK. Evaluation of cross-reactive cell-mediated immune responses among human, bovine and porcine adenoviruses. *Gene Ther* 2010;**17**:634–42.
122. Sharma A, Bangari DS, Tandon M, Pandey A, HogenEsch H, Mittal SK. Comparative analysis of vector biodistribution, persistence and gene expression following intravenous delivery of bovine, porcine and human adenoviral vectors in a mouse model. *Virology* 2009;**386**:44–54.
123. Sharma A, Bangari DS, Tandon M, HogenEsch H, Mittal SK. Evaluation of innate immunity and vector toxicity following inoculation of bovine, porcine or human adenoviral vectors in a mouse model. *Virus Res* 2010;**153**:134–42.
124. Tandon M, Sharma A, Vemula SV, Bangari DS, Mittal SK. Sequential administration of bovine and human adenovirus vectors to overcome vector immunity in an immunocompetent mouse model of breast cancer. *Virus Res* 2012;**163**:202–11.
125. Sharma A, Bangari DS, Vemula SV, Mittal SK. Persistence and the state of bovine and porcine adenoviral vector genomes in human and nonhuman cell lines. *Virus Res* 2011;**161**:181–7.
126. Zakhartchouk AN, Reddy PS, Baxi M, Baca-Estrada ME, Mehtali M, Babiuk LA, et al. Construction and characterization of E3-deleted bovine adenovirus type 3 expressing full-length and truncated form of bovine herpesvirus type 1 glycoprotein gD. *Virology* 1998;**250**:220–9.
127. Zakhartchouk AN, Pyne C, Mutwiri GK, Papp Z, Baca-Estrada ME, Griebel P, et al. Mucosal immunization of calves with recombinant bovine adenovirus-3: induction of protective immunity to bovine herpesvirus-1. *J Gen Virol* 1999;**80**(Pt 5):1263–9.
128. Baxi MK, Deregt D, Robertson J, Babiuk LA, Schlapp T, Tikoo SK. Recombinant bovine adenovirus type 3 expressing bovine viral diarrhea virus glycoprotein E2 induces an immune response in cotton rats. *Virology* 2000;**278**:234–43.
129. Kumar P, Ayalew LE, Godson DL, Gaba A, Babiuk LA, Tikoo SK. Mucosal immunization of calves with recombinant bovine adenovirus-3 coexpressing truncated form of bovine herpesvirus-1 gD and bovine IL-6. *Vaccine* 2014;**32**:3300–6.
130. Brownlie R, Kumar P, Babiuk LA, Tikoo SK. Recombinant bovine adenovirus-3 co-expressing bovine respiratory syncytial virus glycoprotein G and truncated glycoprotein gD of bovine herpesvirus-1 induce immune responses in cotton rats. *Mol Biotechnol* 2015;**57**:58–64.
131. Singh N, Pandey A, Jayashankar L, Mittal SK. Bovine adenoviral vector-based H5N1 influenza vaccine overcomes exceptionally high levels of pre-existing immunity against human adenovirus. *Mol Ther* 2008;**16**:965–71.

132. Bangari DS, Mittal SK. Porcine adenovirus serotype 3 internalization is independent of CAR and alpha(v)beta(3) or alpha(v)beta(5) integrin. *Virology* 2005;**332**:157–66.
133. Bangari DS, Mittal SK. Porcine adenoviral vectors evade preexisting humoral immunity to adenoviruses and efficiently infect both human and murine cells in culture. *Virus Res* 2004;**105**:127–36.
134. Zakhartchouk A, Zhou Y, Tikoo SK. A recombinant E1-deleted porcine adenovirus-3 as an expression vector. *Virology* 2003;**313**:377–86.
135. Hammond JM, Johnson MA. Porcine adenovirus as a delivery system for swine vaccines and immunotherapeutics. *Vet J* 2005;**169**:17–27.
136. Hammond JM, McCoy RJ, Jansen ES, Morrissy CJ, Hodgson AL, Johnson MA. Vaccination with a single dose of a recombinant porcine adenovirus expressing the classical swine fever virus gp55 (E2) gene protects pigs against classical swine fever. *Vaccine* 2000;**18**:1040–50.
137. Hammond JM, Jansen ES, Morrissy CJ, Goff WV, Meehan GC, Williamson MM, et al. A prime-boost vaccination strategy using naked DNA followed by recombinant porcine adenovirus protects pigs from classical swine fever. *Vet Microbiol* 2001;**80**:101–19.
138. Hammond JM, Jansen ES, Morrissy CJ, van der HB, Goff WV, Williamson MM, et al. Vaccination of pigs with a recombinant porcine adenovirus expressing the gD gene from pseudorabies virus. *Vaccine* 2001;**19**:3752–8.
139. Patel A, Tikoo S, Kobinger G. A porcine adenovirus with low human seroprevalence is a promising alternative vaccine vector to human adenovirus 5 in an H5N1 virus disease model. *PLoS One* 2010;**5**:e15301.
140. Nagy M, Tuboly T. Porcine adenoviruses: an update on genome analysis and vector development. *Acta Vet Hung* 2000;**48**:491–9.
141. Tuboly T, Nagy E. Construction and characterization of recombinant porcine adenovirus serotype 5 expressing the transmissible gastroenteritis virus spike gene. *J Gen Virol* 2001;**82**:183–90.
142. Loser P, Hofmann C, Both GW, Uckert W, Hillgenberg M. Construction, rescue, and characterization of vectors derived from ovine atadenovirus. *J Virol* 2003;**77**:11941–51.
143. Kumin D, Hofmann C, Rudolph M, Both GW, Loser P. Biology of ovine adenovirus infection of nonpermissive cells. *J Virol* 2002;**76**:10882–93.
144. Loser P, Kumin D, Hillgenberg M, Both GW, Hofmann C. Preparation of ovine adenovirus vectors. *Methods Mol Med* 2002;**69**:415–26.
145. Xu ZZ, Hyatt A, Boyle DB, Both GW. Construction of ovine adenovirus recombinants by gene insertion or deletion of related terminal region sequences. *Virology* 1997;**230**:62–71.
146. Wuest T, Both GW, Prince AM, Hofmann C, Loser P. Recombinant ovine atadenovirus induces a strong and sustained T cell response against the hepatitis C virus NS3 antigen in mice. *Vaccine* 2004;**22**:2717–21.
147. Rothel JS, Boyle DB, Both GW, Pye AD, Waterkeyn JG, Wood PR, et al. Sequential nucleic acid and recombinant adenovirus vaccination induces host-protective immune responses against *Taenia ovis* infection in sheep. *Parasite Immunol* 1997;**19**:221–7.
148. Bridgeman A, Roshorm Y, Lockett LJ, Xu ZZ, Hopkins R, Shaw J, et al. Ovine atadenovirus, a novel and highly immunogenic vector in prime-boost studies of a candidate HIV-1 vaccine. *Vaccine* 2009;**28**:474–83.
149. Rosario M, Hopkins R, Fulkerson J, Borthwick N, Quigley MF, Joseph J, et al. Novel recombinant *Mycobacterium bovis* BCG, ovine atadenovirus, and modified vaccinia virus Ankara vaccines combine to induce robust human immunodeficiency virus-specific CD4 and CD8 T-cell responses in rhesus macaques. *J Virol* 2010;**84**:5898–908.

150. Loser P, Hillgenberg M, Arnold W, Both GW, Hofmann C. Ovine adenovirus vectors mediate efficient gene transfer to skeletal muscle. *Gene Ther* 2000;**7**:1491–8.
151. Martiniello-Wilks R, Dane A, Mortensen E, Jeyakumar G, Wang XY, Russell PJ. Application of the transgenic adenocarcinoma mouse prostate (TRAMP) model for pre-clinical therapeutic studies. *Anticancer Res* 2003;**23**:2633–42.
152. Tang R, Li K, Wilson M, Both GW, Taylor JA, Young SL. Potent antietumor immunity in mice induced by vaccination with an ovine atadenovirus vector. *J Immunother* 2012;**35**:32–41.
153. Hofmann C, Loser P, Cichon G, Arnold W, Both GW, Strauss M. Ovine adenovirus vectors overcome preexisting humoral immunity against human adenoviruses in vivo. *J Virol* 1999;**73**:6930–6.
154. Morrison MD, Onions DE, Nicolson L. Complete DNA sequence of canine adenovirus type 1. *J Gen Virol* 1997;**78**(Pt 4):873–8.
155. Spibey N, McClory RS, Cavanagh HM. Identification and nucleotide sequence of the early region 1 from canine adenovirus types 1 and 2. *Virus Res* 1989;**14**:241–55.
156. Morrison MD, Reid D, Onions D, Spibey N, Nicolson L. Generation of E3-deleted canine adenoviruses expressing canine parvovirus capsid by homologous recombination in bacteria. *Virology* 2002;**293**:26–30.
157. Klonjkowski B, Gilardi-Hebenstreit P, Hadchouel J, Randrianarison V, Boutin S, Yeh P, et al. A recombinant E1-deleted canine adenoviral vector capable of transduction and expression of a transgene in human-derived cells and in vivo. *Hum Gene Ther* 1997;**8**:2103–15.
158. Kremer EJ, Boutin S, Chillon M, Danos O. Canine adenovirus vectors: an alternative for adenovirus-mediated gene transfer. *J Virol* 2000;**74**:505–12.
159. Chartier C, Degryse E, Gantzer M, Dieterle A, Pavirani A, Mehtali M. Efficient generation of recombinant adenovirus vectors by homologous recombination in *Escherichia coli*. *J Virol* 1996;**70**:4805–10.
160. Ibanes S, Kremer EJ. Canine adenovirus type 2 vector generation via I-Sce1-mediated intracellular genome release. *PLoS One* 2013;**8**:e71032.
161. Hemminki A, Kanerva A, Kremer EJ, Bauerschmitz GJ, Smith BF, Liu B, et al. A canine conditionally replicating adenovirus for evaluating oncolytic virotherapy in a syngeneic animal model. *Mol Ther* 2003;**7**:163–73.
162. Le LP, Rivera AA, Glasgow JN, Ternovoi VV, Wu H, Wang M, et al. Infectivity enhancement for adenoviral transduction of canine osteosarcoma cells. *Gene Ther* 2006;**13**:389–99.
163. Bru T, Salinas S, Kremer EJ. An update on canine adenovirus type 2 and its vectors. *Viruses-Basel* 2010;**2**:2134–53.
164. Soudais C, Boutin S, Hong SS, Chillon M, Danos O, Bergelson JM, et al. Canine adenovirus type 2 attachment and internalization: coxsackievirus-adenovirus receptor, alternative receptors, and an RGD-independent pathway. *J Virol* 2000;**74**:10639–49.
165. Chillon M, Kremer EJ. Trafficking and propagation of canine adenovirus vectors lacking a known integrin-interacting motif. *Hum Gene Ther* 2001;**12**:1815–23.
166. Zussy C, Salinas S. Study of adenovirus and CAR axonal transport in primary neurons. *Methods Mol Biol* 2014;**1089**:71–8.
167. Soudais C, Laplace-Builhe C, Kissa K, Kremer EJ. Preferential transduction of neurons by canine adenovirus vectors and their efficient retrograde transport in vivo. *FASEB J* 2001;**15**:2283–5.
168. Soudais C, Skander N, Kremer EJ. Long-term in vivo transduction of neurons throughout the rat CNS using novel helper-dependent CAV-2 vectors. *FASEB J* 2004;**18**:391–3.

169. Morante-Oria J, Carleton A, Ortino B, Kremer EJ, Fairen A, Lledo PM. Subpallial origin of a population of projecting pioneer neurons during corticogenesis. *Proc Natl Acad Sci USA* 2003;**100**:12468–73.
170. Keriel A, Rene C, Galer C, Zabner J, Kremer EJ. Canine adenovirus vectors for lung-directed gene transfer: efficacy, immune response, and duration of transgene expression using helper-dependent vectors. *J Virol* 2006;**80**:1487–96.
171. Mitani K, Graham FL, Caskey CT, Kochanek S. Rescue, propagation, and partial purification of a helper virus-dependent adenovirus vector. *Proc Natl Acad Sci USA* 1995;**92**:3854–8.
172. Serratrice N, Cubizolle A, Ibanes S, Mestre-Frances N, Bayo-Puxan N, Creyssels S, et al. Corrective GUSB transfer to the canine mucopolysaccharidosis VII cornea using a helper-dependent canine adenovirus vector. *J Control Release* 2014;**181**:22–31.
173. Cubizolle A, Serratrice N, Skander N, Colle MA, Ibanes S, Gennetier A, et al. Corrective GUSB transfer to the canine mucopolysaccharidosis VII brain. *Mol Ther* 2014;**22**:762–73.
174. Ariza L, Gimenez-Llort L, Cubizolle A, Pages G, Garcia-Lareu B, Serratrice N, et al. Central nervous system delivery of helper-dependent canine adenovirus corrects neuropathology and behavior in mucopolysaccharidosis type VII mice. *Hum Gene Ther* 2014;**25**:199–211.
175. Hartley JW, Rowe WP. A new mouse virus apparently related to the adenovirus group. *Virology* 1960;**11**:645–7.
176. Lenaerts L, Verbeken E, De CE, Naesens L. Mouse adenovirus type 1 infection in SCID mice: an experimental model for antiviral therapy of systemic adenovirus infections. *Antimicrob Agents Chemother* 2005;**49**:4689–99.
177. Wigand R. Age and susceptibility of Swiss mice for mouse adenovirus, strain FL. *Arch Virol* 1980;**64**:349–57.
178. van der Veen J, Mes A. Experimental infection with mouse adenovirus in adult mice. *Arch Gesamte Virusforsch* 1973;**42**:235–41.
179. Smith K, Brown CC, Spindler KR. The role of mouse adenovirus type 1 early region 1A in acute and persistent infections in mice. *J Virol* 1998;**72**:5699–706.
180. Kajon AE, Brown CC, Spindler KR. Distribution of mouse adenovirus type 1 in intraperitoneally and intranasally infected adult outbred mice. *J Virol* 1998;**72**:1219–23.
181. Cauthen AN, Brown CC, Spindler KR. In vitro and in vivo characterization of a mouse adenovirus type 1 early region 3 null mutant. *J Virol* 1999;**73**:8640–6.
182. Meissner JD, Hirsch GN, LaRue EA, Fulcher RA, Spindler KR. Completion of the DNA sequence of mouse adenovirus type 1: sequence of E2B, L1, and L2 (18-51 map units). *Virus Res* 1997;**51**:53–64.
183. Beard CW, Spindler KR. Characterization of an 11K protein produced by early region 3 of mouse adenovirus type 1. *Virology* 1995;**208**:457–66.
184. Ying B, Smith K, Spindler KR. Mouse adenovirus type 1 early region 1A is dispensable for growth in cultured fibroblasts. *J Virol* 1998;**72**:6325–31.
185. Beard CW, Spindler KR. Analysis of early region 3 mutants of mouse adenovirus type 1. *J Virol* 1996;**70**:5867–74.
186. Raman S, Hsu TH, Ashley SL, Spindler KR. Usage of integrin and heparan sulfate as receptors for mouse adenovirus type 1. *J Virol* 2009;**83**:2831–8.
187. Lenaerts L, van DW, Persoons L, Naesens L. Interaction between mouse adenovirus type 1 and cell surface heparan sulfate proteoglycans. *PLoS One* 2012;**7**:e31454.
188. Lenaerts L, McVey JH, Baker AH, Denby L, Nicklin S, Verbeken E, et al. Mouse adenovirus type 1 and human adenovirus type 5 differ in endothelial cell tropism and liver targeting. *J Gene Med* 2009;**11**:119–27.

189. Robinson M, Li B, Ge Y, Ko D, Yendluri S, Harding T, et al. Novel immunocompetent murine tumor model for evaluation of conditionally replication-competent (oncolytic) murine adenoviral vectors. *J Virol* 2009;**83**:3450–62.
190. McCarthy MK, Procario MC, Twisselmann N, Wilkinson JE, Archambeau AJ, Michele DE, et al. Proinflammatory effects of interferon gamma in mouse adenovirus 1 myocarditis. *J Virol* 2015;**89**:468–79.
191. McCarthy MK, Levine RE, Procario MC, McDonnell PJ, Zhu L, Mancuso P, et al. Prostaglandin E2 induction during mouse adenovirus type 1 respiratory infection regulates inflammatory mediator generation but does not affect viral pathogenesis. *PLoS One* 2013;**8**:e77628.
192. Weinberg JB, Stempfle GS, Wilkinson JE, Younger JG, Spindler KR. Acute respiratory infection with mouse adenovirus type 1. *Virology* 2005;**340**:245–54.
193. Ojkic D, Krell PJ, Tuboly T, Nagy E. Characterization of fowl adenoviruses isolated in Ontario and Quebec, Canada. *Can J Vet Res* 2008;**72**:236–41.
194. Ojkic D, Nagy E. The complete nucleotide sequence of fowl adenovirus type 8. *J Gen Virol* 2000;**81**:1833–7.
195. Marek A, Nolte V, Schachner A, Berger E, Schlotterer C, Hess M. Two fiber genes of nearly equal lengths are a common and distinctive feature of Fowl adenovirus C members. *Vet Microbiol* 2012;**156**:411–7.
196. Corredor JC, Garceac A, Krell PJ, Nagy E. Sequence comparison of the right end of fowl adenovirus genomes. *Virus Genes* 2008;**36**:331–44.
197. Sheppard M, Werner W, Tsatas E, McCoy R, Prowse S, Johnson M. Fowl adenovirus recombinant expressing VP2 of infectious bursal disease virus induces protective immunity against bursal disease. *Arch Virol* 1998;**143**:915–30.
198. Renaut L, Colin M, Leite JP, Benko M, D'Halluin JC. Abolition of hCAR-dependent cell tropism using fiber knobs of Atadenovirus serotypes. *Virology* 2004;**321**:189–204.
199. Johnson MA, Pooley C, Ignjatovic J, Tyack SG. A recombinant fowl adenovirus expressing the S1 gene of infectious bronchitis virus protects against challenge with infectious bronchitis virus. *Vaccine* 2003;**21**:2730–6.
200. Johnson MA, Pooley C, Lowenthal JW. Delivery of avian cytokines by adenovirus vectors. *Dev Comp Immunol* 2000;**24**:343–54.
201. Reddy PS, Idamakanti N, Song JY, Lee JB, Hyun BH, Park JH, et al. Nucleotide sequence and transcription map of porcine adenovirus type 3. *Virology* 1998;**251**:414–26.

# Engineering Chimeric Adenoviruses: Exploiting Virus Diversity for Improved Vectors, Vaccines, and Oncolytics

20

*Johanna K. Kaufmann*, Dirk M. Nettelbeck*
German Cancer Research Center (DKFZ), Heidelberg, Germany

## 1. Bedside to Bench: Viral Chimerism as a Tool for Addressing Challenges of Adenovirus-Based Therapeutics and Vaccines

Recombinant adenoviruses (Ads) represent a drug platform that offers various opportunities for applications in prevention or treatment of major medical problems, including infectious diseases and cancer. Ads have been developed and clinically explored as vectors for gene therapy and genetic vaccination or as oncolytic viruses.[1–5] These diverse medical applications of a single virus family are owing to outstanding knowledge of the Ad structure, genome organization, and replication cycle and to the development of advanced technologies for engineering of recombinant Ads. In fact, using Ad as a model organism in molecular biology research has revealed fundamental genetic processes of eukaryotes, such as gene splicing as well as mechanisms of DNA replication and transcription.[6] Also, recombinant Ad vector technology is a widely used laboratory tool for gene transfer and recombinant protein expression. For these purposes, the most widely used Ads have been species C Ads, especially Ad serotype 5 (HAdV-5). Therefore, it is no surprise that also previous medical applications have nearly exclusively exploited HAdV-5.[1–5,7] Comprehensive strategies of HAdV-5 engineering have been pursued: the replacement of essential viral genes with heterologous genes for gene therapy and genetic vaccination[1,2,5]; mutations of viral genes establishing tumor-targeted replication competency for viral oncolysis[3]; genetic insertion of peptide ligands into the virus capsid for improved cell entry[8]; and the insertion of promoter elements or other regulatory elements for targeted expression of therapeutic or viral genes in gene therapy or viral oncolysis, respectively.[9]

Extensive clinical studies of Ad-based gene therapy, vaccination, and oncolysis have demonstrated that therapeutic Ads are well tolerated by patients, and also revealed key roadblocks that need to be overcome to achieve clinical efficacy: For potent and safe medical applications, therapeutic Ads should be efficiently and possibly specifically

* Current affiliation: Genocea Biosciences Inc., Cambridge, MA, USA

**Adenoviral Vectors for Gene Therapy.**

delivered to a certain target tissue or cell. However, blood components (e.g., neutralizing antibodies and blood coagulation factors) as well as anatomical (e.g., extracellular matrix) and physiological barriers (e.g., high tissue pressure) cause unfavorable viral biodistribution after systemic applications.[1–5] Moreover, the HAdV-5 receptor is expressed on healthy tissues, resulting in virus sequestration and toxic side effects. In contrast, the receptor is absent or marginally expressed on several tissues that therapeutic Ads should target, including hematopoietic and tumor cells, limiting therapeutic efficacy. Finally, replication and lysis of tumor cells should be improved for oncolytic Ads.

Educated by results from clinical studies, researchers back at the bench can once more benefit from the noted exceptional opportunities of Ad engineering to develop new generations of improved therapeutic viruses. In this context, Ads offer an additional opportunity: The availability of a high and growing number of serotypes infecting humans or animals,[6] not available for most other viruses used in medical applications. These serotypes possess similar structures and genome organizations, but offer distinct features with respect to cellular receptors, immunogenicity, and replication, hence representing a reservoir of structural and functional modules for therapeutic applications. Importantly, the availability of a large number of natural Ad serotypes enables the generation of chimeric Ads.

Virus chimeras contain genetic and/or structural components originating from different virus serotypes, species, or families. They are engineered by genetically exchanging regulatory sequences, whole coding sequences or parts thereof; by complementing proteins; or by genetically or physically inserting heterologous virus genomes into other virus genomes or particles. Chimeric viruses have been previously used in basic research to reveal mechanisms of viral replication, virulence factors, and transformation mechanisms (see Bowser et al. for discussion and references).[10] Virus chimerism is an attractive strategy for developing improved therapeutic viruses[11] as documented by a remarkable variety of chimeric Ads that have been engineered and characterized in preclinical and clinical studies.

The following sections of this chapter describe different strategies for engineering of chimeric Ads and how they offer solutions for major challenges in Ad gene therapy, vaccination, and oncolysis. After explaining general strategies of capsid engineering (Section 2), we follow the virus' application route through the bloodstream to and into the target cell (Sections 3–6). We then discuss chimeric Ads that deliver virus genomes (Section 7) and address how Ad chimerism can be combined with further Ad engineering strategies (Section 8). Finally, we conclude with an outlook (Section 9).

## 2. General Strategies for Engineering of Chimeric Ad Capsids

Ads possess a protein capsid built of the major proteins hexon, penton base, and fiber, and of further stabilizing minor or "cement" proteins.[6] The 240 hexon trimers build the icosahedric protein shell; 12 trimeric fibers protrude from the capsids vertices linked to the hexons via a pentameric penton base (Figure 1). The hexon, fiber, and

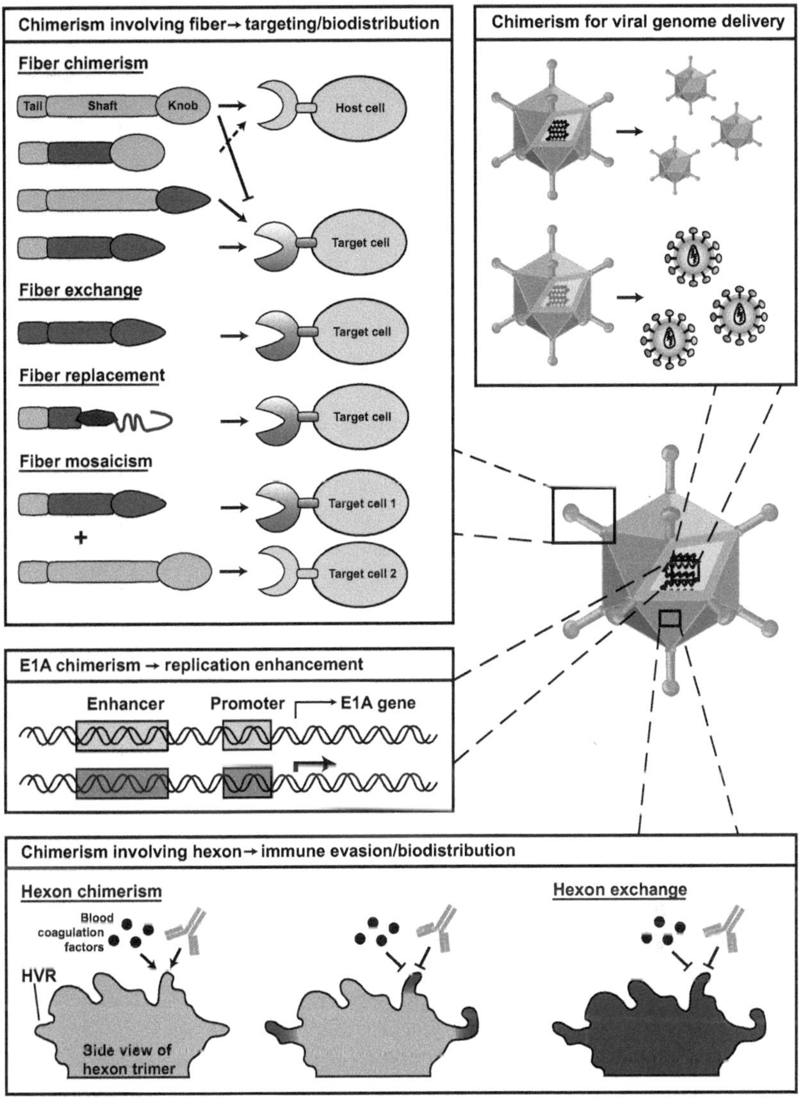


**Figure 1 Adenovirus chimerism.** Different shades of gray symbolize the different origins of the respective virus protein, protein domain, or genetic element (with permission, modified after Kaufmann and Nettelbeck).[11] Top left panel: For targeting and biodistribution purposes, the Ad fiber is modified. The knob domain, responsible for primary receptor recognition, and the shaft can be exchanged individually or in combination (fiber chimerism). Alternatively, the entire fiber molecule can be exchanged. Furthermore, the knob domain can be swapped with heterologous sequences (fiber replacement). Finally, Ads can contain two different fibers (mosaicism), and in this scenario any combination of modified or wild-type fibers is possible. Top right panel: Ads can be used for delivery of virus genomes or replicons. In this strategy, viral DNA encoding for a different virus or replicon is incorporated into the Ad vector genome. Middle panel: Chimerism on the level of noncapsid proteins is shown for promoter and enhancer chimerism to regulate E1A expression, the driving force of Ad replication. Bottom panel: The second major capsid protein, hexon (an outline of the upper part of the side view of the hexon trimer is shown), is modified to evade preexisting immunity or blood coagulation factors that influence biodistribution in vivo. Either the complete hexon or only the protruding parts, also called hypervariable regions (HVRs), are exchanged.

penton base proteins, or certain domains of these proteins, represent candidate capsid modules feasible to be switched between Ad serotypes.

The fiber trimer is responsible for binding the attachment receptor on the host cell. Each fiber monomer consists of an N-terminal tail domain, a central shaft domain, and a C-terminal globular knob domain (Figure 1, top left panel). The tail domain interacts with the penton base, the trimeric knob is responsible for receptor binding, and the shaft domain determines accessibility of the knob domain for cell binding as a result of varying length and rigidity.[6,12,13] Besides this variation, the fiber is structurally conserved between Ad serotypes, even if they use different receptors. In consequence, structurally stable chimeric Ads possessing heterologous fibers or fiber domains can be engineered in order to redirect cell binding (see Section 5) and also to avoid virus sequestration triggered by interactions with blood components (see Sections 3 and 4).

Exchange of the complete fiber protein is feasible (Figure 1, top left panel; Sections 3–5) because the tail domain is highly conserved between Ad serotypes. Technically, fiber exchange can be achieved by protein complementation or exchange of the fiber gene in its entirety. More frequently, however, individual fiber domains, that is, the knob and/or shaft, have been switched (Figure 1, top left panel; Sections 3–5), generating stable chimeric fiber proteins. For some serotype combinations, this strategy might be superior, since it was shown that conserving the natural fiber–hexon interaction (by retaining the fiber tail domain) can result in superior fiber incorporation into virus particles.[14] A similar approach takes capsid chimerism beyond the Ad world by replacing the knob and shaft domains with structurally similar domains of the capsid protein σ1 of reovirus, a nonrelated RNA virus (Chapter 5). Notably, chimeric Ads that possess two distinct fibers, termed "mosaic Ads" (Figure 1, top left panel), have been engineered by genetic and coinfection approaches (Section 5,[15] and references therein).

The penton bases structurally link the fiber proteins to the hexons. They trigger virus cell entry and disassembly via an interaction between their flexible RGD peptide-containing loops and cellular integrins, which function as secondary receptors. Exchange of the penton base together with the fiber was explored for improving Ad entry (see Section 5). Furthermore, capsid chimerism by penton base exchange was pursued to evade inactivation of the virus by innate host factors (Section 4).

Capsid chimeric Ads have also been generated by hexon exchange or by hexon chimerism via replacement of hypervariable regions (HVRs) (Figure 1, bottom panel). Representing the major capsid protein, hexon has been shown to be the main antigenic component of Ad particles[16–18] and to interact with further blood components (see Section 4). Thus, hexon exchange or chimerism is being pursued to evade neutralizing antibodies that would otherwise recognize the parental vector (Section 3) and to avoid virus sequestration mediated by blood components (Section 4).

Another application of Ad capsid chimerism is the insertion of viral immunogenic epitopes into the Ad capsid for vaccination purposes, as shown for epitopes of HIV, enterovirus 71, or heterologous Ads.[19–22] Although we will not discuss this further in this chapter, we would like to point out the conceptual difference of this approach to genetic vaccination, for which Ad vectors are engineered to encode the antigenic proteins that are produced as transgenes in infected cells.

## 3. Chimeric Ad Capsids for Evasion of Neutralizing Antibodies

Most therapeutic modalities using Ads in vivo, whether aiming at gene therapy, vaccination, or oncolysis, rely on the injection of the virus into patients followed by homing of the virus to a certain target cell or tissue. Especially after systemic injection, this viral biodistribution is not solely determined by receptor usage. Before the virus reaches the target cell, it must overcome various barriers, the most prominent of which are blood components, such as coagulation factors, and preexisting antibodies. Both were shown to sequester the virus and trigger toxic side effects.

A major hurdle for systemically injected HAdV-5-based therapeutic Ads is the pre-existing humoral immunity in a high percentage of the human population resulting from natural infections. Moreover, many applications depend on multiple sequential applications of therapeutic Ads, where readministered viruses will definitely face neutralizing antibodies induced by the initially applied viral dose(s). In both scenarios, neutralizing antibodies bind the therapeutic Ads, strongly reducing their active dose and causing harmful immune effects. In addition to pharmacological immunosuppression and sequential use of Ad vectors or viruses built from different serotypes Ad chimerism is an attractive strategy for avoiding neutralizing antibodies. Exchanging immunodominant epitopes of a single parental virus with those of several non-cross-reactive Ad serotypes can generate a panel of viruses that feature the identical therapeutic mode of action for individualized or sequential therapy while evading the patients' neutralizing antibody responses.

As noted above, the major capsid protein hexon represents the main antigenic component of Ads with the HVRs being the immunodominant epitopes.[16–18] Correspondingly, exchange of the whole hexon protein or hexon chimerism by switching HVRs has been explored, indeed demonstrating evasion of HAdV-5 neutralizing antibodies.[17,23,24] However, exchange of the whole hexon protein may result in reduced virus particle titers, most likely due to tempering with the structural integrity of the mixed-serotype capsid.[17] Thus, the number of serotype combinations that can build functional virus chimeras is limited. Another issue to be considered in the context of hexon chimerism is the number of discontinuous HVRs (seven for HAdV-5), because neutralizing antibodies were shown to target several if not all of them.[18] Consequently, each of the HVRs should be switched to realize maximal immunoevasion. Hexon chimeras have been generated using HVRs of low seroprevalence subgroup D Ad serotypes HAdV-48[25–28] or HAdV-43,[29] demonstrating reduced inactivation by neutralizing antibodies in vitro and in vivo. In fact, a clinical study evaluating a HAdV-5/48 HIV vaccine vector demonstrated that the chimera is well tolerated, triggers anti-HIV antibody and T cell responses, and also induces neutralizing antibodies specific for HAdV-5 and even more so for HAdV-48[30] (NCT00695877).

In addition to the hexon protein, the fiber protein contributes to Ad immunogenicity.[27,31,32] Of interest, fiber chimeras were reported to facilitate immunoevasion of Ads, even though the fiber is less immunodominant than the hexon. One study demonstrated that a fiber chimera generated by switching the knob of an HAdV-5 vector with the HAdV-3 knob (Ad5/3) shows superior transduction of lung tumors in mice preimmunized with HAdV-5.[32]

For immunoevasion purposes, chimerism technology exploiting Ads that naturally infect animal hosts is especially appealing because the majority of humans should be naïve to these viruses.[33] This strategy is called "xenotype switching" and has been investigated using fiber chimerism as an example. To this end, an HAdV-5 vector has been equipped with the fiber shaft and knob domain of the bovine Ad BAdV-4 (named Ad5FB4), successfully resulting in evasion of neutralizing antibodies.[34]

Note that strategies exploring fiber chimerism for evasion of neutralizing immune responses need to consider receptor binding of the resulting chimeric Ad (see Section 5). Of interest, for both examples noted above, Ad5/3 and Ad5FB4, receptors are overexpressed on tumor cells (see Section 5). Therefore, these chimeras address two challenges by one engineering strategy: neutralizing antibodies and unfavorable virus receptor expression.

Considering that neutralizing antibodies target both hexon and fiber proteins, optimal immunoevasion of therapeutic Ads is expected from viruses combining modifications to both proteins. In fact a double hexon/fiber chimera has been generated on the basis of this hypothesis: Near-complete evasion of neutralizing antibodies was shown in vitro and in an animal model for an HAdV-5-based vector built of a hexon chimera containing HAdV-48 HVRs plus a fiber "xeno-" chimera with a knob domain derived from the chimpanzee SAdV-25.[27]

## 4. Chimeric Ad Capsids for Improved Viral Biodistribution: Evasion of Blood Coagulation and Innate Immune Factors

Preclinical in vivo studies with HAdV-5-based therapeutic Ads revealed that systemically applied viruses are rapidly cleared from the bloodstream as the majority of viruses are sequestered in the liver.[35] Since natural Ad infections are not bloodborne, studies with Ad vectors needed to reveal the underlying mechanism. It turned out that HAdV-5-based vectors are captured by hepatocytes, resulting in their transduction,[36] and by resident macrophages (Kupffer cells), leading to virus phagocytosis and degradation.[37] Thus, liver-targeted gene transfer by Ad vectors is efficient, while effective virus doses for extrahepatic target sites are substantially reduced after systemic injection. At the same time, liver toxicity turned out to be the major clinical side effect of systemic Ad therapies.[3,5] Both hepatocyte transduction and phagocytosis by macrophages were revealed to be mostly independent of binding to the HAdV-5 attachment receptor CAR.[38–42] A hallmark discovery that established the foundation for manipulating Ad biodistribution was the identification that blood coagulation factors mediate hepatocyte transduction. Initial studies indicated that the blood coagulation factor IX (FIX) and the complement factor C4BP bridge the HAdV-5 knob to hepatocyte heparin sulfate proteoglycans.[43] A subsequent study revealed a role for the hexon protein in hepatocyte transduction with coagulation factor FX as bridging factor.[25] This came as a surprise, as hexons were previously not reported to possess functions in virus attachment.

Identification of the mechanism of hepatocyte transduction by HAdV-5, together with the discovery that hexons of several Ad serotypes do not bind to FX,[25] was the rationale used for the engineering of therapeutic Ads with exchanged or chimeric hexon, aiming at detargeting of viruses from the liver and thereby increasing effective virus doses for extrahepatic target tissues. The original study[25] already showed that HAdV-5-based vectors with hexon chimeras containing HVRs of HAdV-48 that does not bind FX possess reduced hepatocyte transduction capacity in vitro and in vivo. Subsequently, the amino acids responsible for the HAdV-5 hexon–FX interaction were more precisely specified[44] and several chimeric Ads with HVR or hexon exchange demonstrated reduced liver transduction and toxicity.[26,28,44–46] However, when engineering liver-detargeted chimera one must bear in mind that increased virus doses are made available not only for target tissues, such as tumors, but also for other healthy organs. Indeed, one study showed increased off-target transduction of the spleen, triggering unfavorable side effects.[46] Moreover, intravenous injection of the HAdV-5/48 chimera was reported to trigger increased serum transaminases and liver infiltration by immune cells despite reduced hepatocyte transduction and diminished uptake by Kupffer cells.[47] Another concern was that ablating FX binding would not only reduce binding to the liver but also the transduction of target tissues, as shown for tumors in one report.[48] Whether this is a concern for applications is still being discussed, because other studies with hexon-chimeric or hexon-exchanged oncolytic Ads confirmed in different tumor models that reduced liver toxicity resulted in an increased maximum tolerated dose as well as retained or increased antitumor activity.[26,45]

Unproductive uptake of Ads by Kupffer cells is mediated by scavenger receptors.[37,49] First studies indicate that hexon HVRs are responsible for the interaction of Ads with scavenger receptors.[37,49] Of importance for engineering therapeutic Ads is the observation that Ad serotypes differ also with respect to their scavenger receptor-binding properties.[49] Thus, by engineering hexon-chimeric viruses yet another roadblock to systemic Ad application might be overcome.

Recently, it was demonstrated that defensins, innate immune factors that inactivate bacteria and viruses by various diverse mechanisms,[50] can inhibit virus uncoating after uptake into target cells by stabilization of the penton base–fiber interaction.[51,52] Importantly, defensin-insensitive Ad species, such as HAdV-19c, can be exploited for preventing inactivation of sensitive Ads, like HAdV-5: Complete defensin resistance was obtained by replacing both the penton base and the defensin-binding tetrapeptide of the HAdV-5 fiber with the corresponding sections of HAdV-19c.[51,53] Such chimeric Ads might offer advantages for therapeutic applications by circumventing inactivation by defensins produced by blood or intestinal cells.

# 5. Chimeric Ad Capsids for Improved Viral Cell Binding and Entry

Therapeutic Ads that have overcome inactivation and sequestration by soluble blood factors and homed to target tissues face yet another barrier to therapeutic efficacy, which is at the cellular level: effective entry into target cells is strongly determined by cellular receptor expression. Indeed, expression of the HAdV-5 receptor in humans represents a major limitation for several applications of HAdV-5-based vectors. Widespread expression of the receptor facilitates transduction of many different target tissues, but also causes virus sequestration and undesirable side effects.[54] On the other hand, receptor expression turned out to be weak in many therapeutic target cells, including hematopoietic cells and most cancer cells, whether in situ or freshly purified.[55–57] Accordingly, virus tropism has been engineered to achieve improved efficiency and/or specificity of therapeutic Ads at the level of direct cell binding. Strategies that have been pursued to this end include genetic and physical ablation of native tropism, genetic insertion, or physical addition of cell-binding ligands[8] and capsid chimerism further discussed here.

Cell binding and entry of chimeric Ads, as described in this section, has been explored early on due to the relative ease of investigation in established cell culture models. However, at this point it must be noted that a higher complexity was observed in in vivo models. This includes the activity of soluble blood factors as attachment factors described in the previous section and the revelation that tumor cells in situ or freshly purified, in contrast to established tumor cell lines, lack expression of or weakly express the HAdV-5 receptor.[55–57]

The cell-binding fiber protein of Ads is structurally conserved between serotypes, yet different attachment receptors are used. This feature was supported by competition experiments in the first study using a chimeric fiber protein containing the HAdV-5 tail and shaft genetically linked to the HAdV-3 knob.[58] A consecutive study generated the corresponding fiber chimeric virus HAdV-5/3 and showed that receptor specificity was switched. Thus, Ad transduction can indeed be altered by fiber chimerism.[59] While these studies explored basic mechanisms of Ad biology, the fact that the chimeric virus possessed a markedly increased cell entry potency for primary cells pointed at the impact of capsid chimerism for therapeutic applications of Ads. In fact, a later study demonstrated an impressive three orders of magnitude increase in transduction of primary melanoma cells by HAdV-5/3 fiber chimerism.[60] Following the development of HAdV-5/3 chimeras, fiber chimeras with the knob domain of HAdV-35[61] and of several other serotypes have been explored for transduction of many target cell types.

A slightly different approach to fiber chimerism is the replacement of the shaft domain in addition to the knob domain (first developed by Shayakhmetov and Lieber).[62] As the shaft domain differs in length and flexibility between Ad serotypes, this strategy is not only technically but also functionally distinct. In fact, Ad transduction can be modified by changing the fiber shaft and thereby the accessibility of the knob domain for receptor binding.[13,62,63] For example, this was convincingly demonstrated by a strong increase of the transduction ratio tumor/liver, when the HAdV-5 fiber shaft was replaced by the considerably shorter shaft domain of the HAdV-3 fiber.[64]

As noted in Section 2, exchanging the complete fiber protein is an alternative to engineering chimeric fibers in cases where the interaction of penton base with the heterologous fiber tail allows incorporation of the fiber into virus particles.[16] By this approach, enhanced transduction efficiency and/or therapeutic activity has been shown for HAdV-5-based vectors with fibers from HAdV-16, -35, or -50.[65,66]

Exchange of both fiber and penton proteins of HAdV-5 has been investigated in an effort to increase smooth muscle cell (SMC) transduction via CD46, the receptor for HAdV-35.[67] Pseudotyping HAdV-5 with both proteins of HAdV-35 resulted in increased gene transfer into SMCs in cell culture and in human saphenous veins ex vivo, when compared to parental HAdV-5 or HAdV-35 vectors or to a chimera containing only the HAdV-35 fiber. It remains to be determined whether this approach, being dependent on a heterologous hexon–penton interaction, is superior with respect to capsid stability to the strategies described above that depend on either heterologous penton base–fiber interactions or chimeric fiber proteins.

Opportunities for tropism modification of therapeutic Ads are extended by exploitation of Ad serotypes that naturally infect animal hosts. In addition to avoiding fiber-directed neutralizing antibodies (see Section 3), this "xenotype" switching can also redirect Ads to novel receptors. For example, fiber chimeric Ads that contain the shaft and knob domain of bovine BAdV-4 are attractive agents especially for cancer therapy. This is because the BAdV-4 fiber components, beside their immunoevasion activity,[34] mediate cell entry via binding to proteins of the B7 family, which are overexpressed in dormant cancer cells.[68] Further examples for improving transduction properties of Ads by "xenotype switching" are chimeric Ads with canine CAdV-2 knob[69] or ovine OAdV-7 fiber.[70]

Taking this concept even further, receptor-binding capsid proteins from distant viruses can be displayed by Ad-based virus chimeras. Reoviruses are distant nonenveloped RNA viruses that contain the capsid protein σ1, which is structurally similar to the Ad fiber. Fusion of the near-complete σ1 protein to the fiber tail was feasible and the resulting virus HAdV5/σ1 could be successfully rescued.[71,72] This chimera preferentially infects mucosal and dendritic cells (which express the reovirus receptor) and is therefore an attractive tool for vaccination strategies.

With a few exceptions (see Section 8), human Ad particles contain 12 copies of identical fiber trimers. Chimeric viruses have been generated to contain two distinct fiber genes (Ref. 15 and references therein) in order to enhance infectivity and expand tropism, aiming at better transduction of tumors composed of heterogeneous cell types. These "mosaic" Ads were obtained in two ways: (i) by engineering of the virus genome to contain two fiber genes and (ii) by coinfection of cell cultures with two viruses containing different fiber genes. However, at this point it is difficult to predict and control the ratio of the different fiber trimers in the resulting mosaic virus particle, although several factors have been proposed to influence the incorporation stoichiometry, for example, the choice of fiber tail domains, the relative expression efficacy of the two fiber proteins, or their stability.[15]

Having multiple strategies of virus engineering established, Ads have the compelling advantage as therapeutic agents that several of these strategies can be combined in a modular way to overcome different roadblocks to Ad efficiency in patients (see also Section 8). An example is a multichimeric oncolytic Ad that combines enhanced infectivity by HAdV-5/3 fiber chimerism with liver detargeting by exchange of the HAdV-5 with the HAdV-3 hexon.[45] This double chimera was shown to possess potent antitumor activity in an ovarian cancer model. Importantly, compared with the control virus, liver tropism was reduced by approximately 3 orders of magnitude while tumor infection was strongly increased.

Engineering receptor interactions of therapeutic Ads by capsid chimerism has mostly focused on enhancing binding and entry into target cells. However, it must be noted that manipulating blood cell binding might also improve biodistribution of Ads after systemic application. Especially, erythrocyte binding is of concern, as erythrocytes bind large fractions of injected HAdV-5.[73] HAdV-5 binds to erythrocytes directly via fiber:receptor interactions or indirectly via antibodies (see Section 3) and complement (see Section 4) as bridging factors.[74] As Ad serotypes differ in their erythrocyte-binding properties, as shown for HAdV-5, HAdV-37, and CAdV-2,[75] virus chimerism offers an opportunity to ablate blood cell binding.

The technology for manipulating Ad tropism by fiber exchange or chimerism has been applied for the design of several therapeutic viruses that are in preclinical and clinical development. The great majority of clinical studies exploring therapeutic Ads have used HAdV-5-based vectors that were attenuated by deletion of essential genes (gene therapy, genetic vaccination) and by mutation or selective expression of viral genes (oncolysis). With respect to Ad safety in patients, capsid chimerism constitutes an added level of complexity, because the resulting viruses possibly bind to or even infect cells that parental viruses would not affect. Therefore, it is encouraging that several different oncolytic Ads with chimeric HAdV-5/3 fiber have been well tolerated by patients suffering from various cancers after systemic or local injection.[76–82] Although these studies were not powered to assess therapeutic efficacy, indications of virus replication and tumor marker responses were observed in individual patients. Based on these case studies, more extensive clinical trials of capsid chimeric oncolytic Ads have recently been initiated.

As we discuss in the next section, virus chimerism can also be exploited to improve replication of oncolytic Ads after tumor cell entry. Of note, postentry processes must be considered also when engineering the virus capsid. This is because switching receptor binding may alter intracellular virus trafficking. For example, exchange of the penton base was speculated to improve gene transfer into SMCs because of improved intracellular trafficking.[67] Virus rerouting to destructive cell compartments or exocytosis following enhanced binding and entry was demonstrated for both the HAdV-5/7 chimera with exchanged fiber and the fiber chimeric HAdV-5/35 virus (containing the shaft and knob domain of the HAdV-35 fiber).[61,83–85] Interestingly, aberrant intracellular trafficking of the HAdV-5/35 chimera was also revealed as a mechanism responsible for a strongly reduced liver transduction even in the presence of blood coagulation factors.[85] Another important consequence of redirected intracellular trafficking, resulting from modified recognition by cellular pattern recognition receptors, is that capsid chimeric Ads can trigger an altered immune response.[86,87]

# 6. Beyond the Capsid: Chimerism of Regulatory Proteins and Genomic Elements by Rational Engineering or Directed Evolution

Having discussed strategies to overcome roadblocks for therapeutic Ads during biodistribution and cell binding by switching structural modules, we now discuss virus chimerism to improve therapeutic Ads at the postentry level (Figure 1, middle panel). This is relevant for oncolytic viruses that should achieve optimal replication in tumor cells, which can considerably differ from natural Ad host cells with respect to metabolism, cell signaling, and gene expression profile. Thus, there is a window of opportunity for improving the efficiency of tumor cell lysis. Again, the panel of Ad serotypes provides a natural library of virus functions, determined by gene or genome modules that can be switched to improve oncolytic Ad replication.

Two technological approaches have been pursued and are described in the following: (i) rational engineering and (ii) directed evolution by bioselection. In contrast to all viruses noted in previous sections, which were based on HAdV-5, both viruses discussed in this section are based on HAdV-11p. Like HAdV-3 and HAdV-35, this virus belongs to species B Ads, which use different receptors than species C HAdV-5.

One study compared HAdV-11p with HAdV-5, because HAdV-11p uses receptors upregulated on many tumor cells, CD46 and desmoglein-2, and has low seroprevalence in the human population.[88] The authors observed superior entry of HAdV-11p into tumor cells. Strikingly, however, this did not result in improved cell killing for several cell lines. For these cells, cell killing by HAdV-11p is restricted by expression levels of E1A, which were much lower than those of HAdV-5. By switching the E1A promoter and enhancer of HAdV-11p with the corresponding sequences of HAdV-5 (Figure 1, middle panel), E1A expression in tumor cells was indeed strongly increased. As a consequence, the HAdV-11p/5 chimera possessed enhanced oncolytic activity in cell cultures and animal models. This study demonstrates that several favorable properties of different Ad serotypes, here reduced immune neutralization and efficient cell entry of HAdV-11p and potent expression of E1A by HAdV-5, can be combined in one therapeutic agent by linking the corresponding functional genomic modules.[88]

Directed evolution of oncolytic viruses with improved properties is an attractive alternative to rational virus design. This has been established for Ads by random chemical or biological mutagenesis and bioselection in relevant cell cultures or animal tumor models.[89–93] An alternative fascinating strategy, which has already provided a chimera that progressed to clinical applications, exploits Ad serotype diversity: A complex library of chimeric Ad genomes was generated by random natural recombination in cells after high titer coinfection.[94] This library was used for bioselection of efficiently replicating chimeras in colon cancer cell cultures. One selected candidate virus, termed ColoAd1, was further characterized, demonstrating increased potency and selectivity than the parental viruses in vitro and in vivo. Analysis of the chimeric genome revealed that the virus is based on serotype HAdV-11p, but possesses deletions in the early genes E3 and E4 and a chimeric E2B region containing sequences of HAdV-3. Such a combination would most likely not have been identified by rational design. Correspondingly, the mechanisms underlying the improved features of ColoAd1 remain obscure. E3 and E4 genes encode proteins of various functions in virus replication and modulation of host cell activities, one example being evasion of host defense mechanisms. The E2B region encodes the viral DNA polymerase and the preterminal protein pTP, responsible for priming of DNA replication and association with the nuclear matrix. Therefore, one can speculate that ColoAd1 implements molecular virus–host interactions optimized for tumor cells. Alternatively, as discussed by the authors of the study, the chimeric pTP–DNA polymerase complex might have an increased affinity toward the viral origin of replication.[94] Of note, ColoAd1 also possesses the beneficial features of the HAdV-11p capsid described above. Indeed, a recent study demonstrated evasion of inactivation by human serum and a dramatically reduced inactivation by whole human blood in comparison with HAdV-5.[95] A clinical study of systemically injected ColoAd1 for treatment of disseminated cancer has been initiated.[96]

Both HAdV-11p/5 and ColoAd1 chimeras represent promising pioneers for engineering of potency-enhanced oncolytic viruses for two reasons: (i) they are based on a different serotype than the "standard" HAdV-5 and (ii) they result from the exchange of nonstructural sequences.

Furthermore, regulatory nucleotide sequences from distant heterologous viruses have been inserted into Ad vectors to enable the expression of multiple proteins from a single promoter, including internal ribosome entry sites, derived from encephalomyocarditis virus and others, and 2A sequences for skipping peptide bonds during protein translation, as present, for example, in the insect virus *Thosea asigna* (Ref. 97 and refs therein). These elements allow for economical usage of genomic vector space and temporal or spatial regulation of the expression of two proteins from the same promoter.

## 7. Chimeras of Ads and Viruses of Other Families for Delivery of Viral Genomes

Several properties of Ads, such as particle stability, efficient transduction mechanism, capacity for insertion of large heterologous DNA sequences, and ease of genetic manipulation, have boosted their application as vectors in therapeutic gene transfer and genetic vaccination. The development of Ad-based vectors has been extended to the delivery of complete viruses/vectors genomes (resembling a genetic Matryoshka doll, Figure 1, top right panel). Here, virus chimerism is advanced from individual elements or proteins to whole genomes, thereby linking advantages of Ads for delivery of genetic material with advantageous features of other viral vectors for several therapeutic approaches as described in the following.

First, stable gene transfer using Ads, whose episomal genomes are otherwise lost in proliferative tissues over time, was established by insertion of an episomally stable

Epstein Barr virus vector.[98] A second application is the delivery of replicons that trigger improved vaccination or increased gene transfer efficiency, as shown using an alphavirus replicon encoding a swine fever vaccine antigen or GFP.[99–101] Similarly, a chimeric Ad-hepatitis B virus (HBV) vector was engineered for enhanced and specific delivery of a matrix protease to cirrhotic liver.[102] The HBV vector genome, shuttled by Ad into the liver, circumvents superinfection exclusion, yet allows for liver-specific *trans*-complementation by endogenous HBV. This strategy facilitated the amelioration of liver cirrhosis in a rodent model. Fourth, improved molecular chemotherapy was reported by Ad delivery of a replication-competent retroviral vector genome encoding a prodrug convertase.[103] Viral oncolysis is yet another variation of this genome delivery approach. In this context, Ads were established for genetic delivery of the oncolytic parvovirus H-1PV.[104] Rodent H-1PV preferentially replicates in tumor cells and is currently under clinical investigation for treatment of recurrent glioblastoma.[105] Parvoviruses are considerably smaller than Ad, thus presumably allowing for better perfusion of tumor tissues. However, their compact structure and small genome size limit engineering opportunities for improved efficiency and selectivity of virus delivery. The Ad-parvovirus approach had to circumvent inhibition of Ad replication by the inserted virus genome. This was achieved by suppressing parvoviral gene expression during production of the chimeric virus. Importantly, infectious parvoviruses were produced after infection of cancer cells with the Ad-parvovirus chimera, which showed superior and targeted therapeutic activity in vitro. Further Ad chimeras for virus genome delivery are reviewed in Lam and Breakefield.[106]

## 8. Combining Ad Chimerism with Other Virus Engineering Strategies

In Section 1 we noted that the availability of diverse virus engineering strategies is a key advantage of Ad-based therapeutics. Virus chimerism is one of these strategies, broadening Ad application possibilities as described above. In this section we give examples for strategies that successfully combine virus chimerism with other engineering technologies.

The first example addressed the question of how to restrict replication of Ad capsid chimeras that enable enhanced, but not targeted infectivity (see Section 5) to tumors. This was achieved by engineering a postentry spatial control mechanism: transcriptional control of the essential Ad gene E1A using an optimized cellular tyrosinase promoter provided a potent melanoma-targeted HAdV-5/3-based oncolytic Ad.[107] Transcriptional targeting of capsid-chimeric Ads has since been widely pursued, for example, using the mesothelin promoter for targeting ovarian cancer[108] or using the COX-2 promoter for targeting pancreatic cancer and gallbladder carcinoma.[109,110]

Fiber exchange or chimerism can be exploited to modify Ad tropism in a more indirect way by combining fibers that lack cell-binding properties (yet ensure particle stability) with the insertion of peptides that mediate targeted cell binding. This strategy was established exploiting two human Ad serotypes, HAdV-40 and -41, which have a unique feature: each possesses two different fibers, a long and a short fiber. The long fiber mediates cell binding, whereas no receptor interaction has been reported for the short fiber. Accordingly, chimerism based on the short fiber of HAdV-40 or -41 achieved detargeting of HAdV-5 Ads from healthy cells and tissues.[42,111] The HAdV-5/41s chimera was used as a base virus in subsequent studies to target Ad infection to novel receptors by inserting peptide ligands into suitable locales of the "receptor-blind" scaffold.[14,112,113] In one of these applications, this technology could redirect Ad infection to tumor cells by insertion of a peptide ligand that binds to the receptor tyrosine kinase EphA2, which is strongly overexpressed on several advanced tumors.[113]

An alternative approach for Ad detargeting, allowing for subsequent retargeting by genetic ligand insertion, is to fully delete the fiber knob and parts of the fiber shaft (Figure 1, top left panel). This strategy is hampered by the fact that the knob domain contains not only the receptor-binding moiety but also the fiber trimerization signal. Thus, knobless fibers need to be complemented by heterologous trimerization domains to ensure virus stability and enable ligand insertion (Figure 1, top left panel). To this end, different virus-derived protein trimerization domains were used, including those of the structurally related reovirus fiber,[114] of fibritin of T4 bacteriophage,[115] and of the envelope glycoprotein of Moloney murine leukemia virus.[116] Future studies will need to improve the incorporation of knobless fibers into virus particles and virus manufacturing. One approach is the generation of fiber mosaic viruses (see Section 2), combining knobless with conventional fibers.[114] However, the resulting viruses possess an expanded, rather than targeted tropism.

## 9. Future Perspectives

To date, chimeric Ads have been nearly exclusively developed on the basis of HAdV-5. Recent advances in Ad engineering technologies facilitate the rapid cloning of Ad genomes from purified virus DNA and rapid traceless modification of Ad genomes.[117,118] This paves the way for future efforts to realize the full potential of Ad chimerism by building therapeutic viruses based on any serotype. The two Ad11p-based virus chimeras described in Section 6 clearly highlight the opportunities this approach offers.

In addition to progress in Ad chimerism technology, which will further increase the pipeline of therapeutic Ad candidates, advances in the clinical implementation of chimeric Ads can be expected, as preclinical studies have revealed the molecular mode of action and clearly demonstrated optimized features for a panel of quite different Ad chimeras. First clinical trials with capsid-chimeric and bioselected mixed-serotype Ad oncolytics have already been initiated[96] and it will be exciting to reveal whether the preclinical promise of chimeric Ads translates into effective gene therapies, vaccines, and virotherapies to improve patients' lives.

## References


1. McConnell MJ, Imperiale MJ. Biology of adenovirus and its use as a vector for gene therapy. *Hum Gene Ther* 2004;**15**:1022–33.
2. Lasaro MO, Ertl HC. New insights on adenovirus as vaccine vectors. *Mol Ther* 2009;**17**:1333–9.
3. Yamamoto M, Curiel DT. Current issues and future directions of oncolytic adenoviruses. *Mol Ther* 2010;**18**:243–50.
4. Majhen D, Calderon H, Chandra N, Fajardo CA, Rajan A, Alemany R, et al. Adenovirus-based vaccines for fighting infectious diseases and cancer: progress in the field. *Hum Gene Ther* 2014;**25**:301–17.
5. Crystal RG. Adenovirus: the first effective in vivo gene delivery vector. *Hum Gene Ther* 2014;**25**:3–11.
6. Knipe DM, Howley P. *Fields virology*. 6th ed. Philadelphia: Lippincott Williams & Wilkins; 2013.
7. Stone D, Liu Y, Li ZY, Tuve S, Strauss R, Lieber A. Comparison of adenoviruses from species b, c, e, and f after intravenous delivery. *Mol Ther* 2007;**15**:2146–53.
8. Glasgow JN, Everts M, Curiel DT. Transductional targeting of adenovirus vectors for gene therapy. *Cancer Gene Ther* 2006;**13**:830–44.
9. Dorer DE, Nettelbeck DM. Targeting cancer by transcriptional control in cancer gene therapy and viral oncolysis. *Adv Drug Deliv Rev* 2009;**61**:554–71.
10. Bowser BS, Chen HS, Conway MJ, Christensen ND, Meyers C. Human papillomavirus type 18 chimeras containing the l2/l1 capsid genes from evolutionarily diverse papillomavirus types generate infectious virus. *Virus Res* 2011;**160**:246–55.
11. Kaufmann JK, Nettelbeck DM. Virus chimeras for gene therapy, vaccination, and oncolysis: adenoviruses and beyond. *Trends Mol Med* 2012;**18**:365–76.
12. Wu E, Fernandez J, Fleck SK, Von Seggern DJ, Huang S, Nemerow GR. A 50-kDa membrane protein mediates sialic acid-independent binding and infection of conjunctival cells by adenovirus type 37. *Virology* 2001;**279**:78–89.
13. Wu E, Pache L, Von Seggern DJ, Mullen TM, Mikyas Y, Stewart PL, et al. Flexibility of the adenovirus fiber is required for efficient receptor interaction. *J Virol* 2003;**77**:7225–35.
14. Hesse A, Kosmides D, Kontermann RE, Nettelbeck DM. Tropism modification of adenovirus vectors by peptide ligand insertion into various positions of the adenovirus serotype 41 short-fiber knob domain. *J Virol* 2007;**81**:2688–99.
15. Murakami M, Ugai H, Wang M, Belousova N, Dent P, Fisher PB, et al. An adenoviral vector expressing human adenovirus 5 and 3 fiber proteins for targeting heterogeneous cell populations. *Virology* 2010;**407**:196–205.
16. Gall J, Kass-Eisler A, Leinwand L, Falck-Pedersen E. Adenovirus type 5 and 7 capsid chimera: fiber replacement alters receptor tropism without affecting primary immune neutralization epitopes. *J Virol* 1996;**70**:2116–23.
17. Wu H, Dmitriev I, Kashentseva E, Seki T, Wang M, Curiel DT. Construction and characterization of adenovirus serotype 5 packaged by serotype 3 hexon. *J Virol* 2002;**76**:12775–82.
18. Bradley RR, Maxfield LF, Lynch DM, Iampietro MJ, Borducchi EN, Barouch DH. Adenovirus serotype 5-specific neutralizing antibodies target multiple hexon hypervariable regions. *J Virol* 2012;**86**:1267–72.
19. Matthews QL. Capsid-incorporation of antigens into adenovirus capsid proteins for a vaccine approach. *Mol Pharm* 2011;**8**:3–11.

20. Matthews QL, Fatima A, Tang Y, Perry BA, Tsuruta Y, Komarova S, et al. HIV antigen incorporation within adenovirus hexon hypervariable 2 for a novel HIV vaccine approach. *PLoS One* 2010;**5**:e11815.
21. Tian X, Su X, Li X, Li H, Li T, Zhou Z, et al. Protection against enterovirus 71 with neutralizing epitope incorporation within adenovirus type 3 hexon. *PLoS One* 2012;**7**:e41381.
22. Liu M, Tian X, Li X, Zhou Z, Li C, Zhou R. Generation of neutralizing monoclonal antibodies against a conformational epitope of human adenovirus type 7 (hadv-7) incorporated in capsid encoded in a HAdv-3-based vector. *PLoS One* 2014;**9**:e103058.
23. Roberts DM, Nanda A, Havenga MJ, Abbink P, Lynch DM, Ewald BA, et al. Hexon-chimaeric adenovirus serotype 5 vectors circumvent pre-existing anti-vector immunity. *Nature* 2006;**441**:239–43.
24. Tian X, Su X, Li H, Li X, Zhou Z, Liu W, et al. Construction and characterization of human adenovirus serotype 3 packaged by serotype 7 hexon. *Virus Res* 2011;**160**:214–20.
25. Waddington SN, McVey JH, Bhella D, Parker AL, Barker K, Atoda H, et al. Adenovirus serotype 5 hexon mediates liver gene transfer. *Cell* 2008;**132**:397–409.
26. Zhang Z, Krimmel J, Hu Z, Seth P. Systemic delivery of a novel liver-detargeted oncolytic adenovirus causes reduced liver toxicity but maintains the antitumor response in a breast cancer bone metastasis model. *Hum Gene Ther* 2011;**22**:1137–42.
27. Bradley RR, Lynch DM, Iampietro MJ, Borducchi EN, Barouch DH. Adenovirus serotype 5 neutralizing antibodies target both hexon and fiber following vaccination and natural infection. *J Virol* 2012;**86**:625–9.
28. Teigler JE, Penaloza-MacMaster P, Obeng R, Provine NM, Larocca RA, Borducchi EN, et al. Hexon hypervariable region-modified adenovirus type 5 (Ad5) vectors display reduced hepatotoxicity but induce T lymphocyte phenotypes similar to Ad5 vectors. *Clin Vaccine Immunol* 2014;**21**:1137–44.
29. Bruder JT, Semenova E, Chen P, Limbach K, Patterson NB, Stefaniak ME, et al. Modification of Ad5 hexon hypervariable regions circumvents pre-existing Ad5 neutralizing antibodies and induces protective immune responses. *PLoS One* 2012;**7**:e33920.
30. Baden LR, Walsh SR, Seaman MS, Johnson JA, Tucker RP, Kleinjan JA, et al. First-in-human evaluation of a hexon chimeric adenovirus vector expressing HIV-1 Env (IPCAVD 002). *J Infect Dis* 2014;**210**:1052–61.
31. Nanda A, Lynch DM, Goudsmit J, Lemckert AA, Ewald BA, Sumida SM, et al. Immunogenicity of recombinant fiber-chimeric adenovirus serotype 35 vector-based vaccines in mice and rhesus monkeys. *J Virol* 2005;**79**:14161–8.
32. Sarkioja M, Pesonen S, Raki M, Hakkarainen T, Salo J, Ahonen MT, et al. Changing the adenovirus fiber for retaining gene delivery efficacy in the presence of neutralizing antibodies. *Gene Ther* 2008;**15**:921–9.
33. Sharma A, Tandon M, Ahi YS, Bangari DS, Vemulapalli R, Mittal SK. Evaluation of cross-reactive cell-mediated immune responses among human, bovine and porcine adenoviruses. *Gene Ther* 2010;**17**:634–42.
34. Rogee S, Grellier E, Bernard C, Jouy N, Loyens A, Beauvillain JC, et al. Influence of chimeric human-bovine fibers on adenoviral uptake by liver cells and the antiviral immune response. *Gene Ther* 2010;**17**:880–91.
35. Baker AH, McVey JH, Waddington SN, Di Paolo NC, Shayakhmetov DM. The influence of blood on in vivo adenovirus bio-distribution and transduction. *Mol Ther* 2007;**15**:1410–6.
36. Huard J, Lochmuller H, Acsadi G, Jani A, Massie B, Karpati G. The route of administration is a major determinant of the transduction efficiency of rat tissues by adenoviral recombinants. *Gene Ther* 1995;**2**:107–15.

37. Alemany R, Suzuki K, Curiel DT. Blood clearance rates of adenovirus type 5 in mice. *J Gen Virol* 2000;**81**:2605–9.
38. Alemany R, Curiel DT. CAR-binding ablation does not change biodistribution and toxicity of adenoviral vectors. *Gene Ther* 2001;**8**:1347–53.
39. Leissner P, Legrand V, Schlesinger Y, Hadji DA, van Raaij M, Cusack S, et al. Influence of adenoviral fiber mutations on viral encapsidation, infectivity and in vivo tropism. *Gene Ther* 2001;**8**:49–57.
40. Smith T, Idamakanti N, Kylefjord H, Rollence M, King L, Kaloss M, et al. In vivo hepatic adenoviral gene delivery occurs independently of the coxsackievirus-adenovirus receptor. *Mol Ther* 2002;**5**:770–9.
41. Smith TA, Idamakanti N, Rollence ML, Marshall-Neff J, Kim J, Mulgrew K, et al. Adenovirus serotype 5 fiber shaft influences in vivo gene transfer in mice. *Hum Gene Ther* 2003;**14**:777–87.
42. Nicol CG, Graham D, Miller WH, White SJ, Smith TA, Nicklin SA, et al. Effect of adenovirus serotype 5 fiber and penton modifications on in vivo tropism in rats. *Mol Ther* 2004;**10**:344–54.
43. Shayakhmetov DM, Gaggar A, Ni S, Li ZY, Lieber A. Adenovirus binding to blood factors results in liver cell infection and hepatotoxicity. *J Virol* 2005;**79**:7478–91.
44. Alba R, Bradshaw AC, Parker AL, Bhella D, Waddington SN, Nicklin SA, et al. Identification of coagulation factor (F)X binding sites on the adenovirus serotype 5 hexon: effect of mutagenesis on FX interactions and gene transfer. *Blood* 2009;**114**:965–71.
45. Short JJ, Rivera AA, Wu H, Walter MR, Yamamoto M, Mathis JM, et al. Substitution of adenovirus serotype 3 hexon onto a serotype 5 oncolytic adenovirus reduces factor X binding, decreases liver tropism, and improves antitumor efficacy. *Mol Cancer Ther* 2010;**9**:2536–44.
46. Alba R, Bradshaw AC, Coughlan L, Denby L, McDonald RA, Waddington SN, et al. Biodistribution and retargeting of FX-binding ablated adenovirus serotype 5 vectors. *Blood* 2010;**116**:2656–64.
47. Coughlan L, Bradshaw AC, Parker AL, Robinson H, White K, Custers J, et al. Ad5:Ad48 hexon hypervariable region substitutions lead to toxicity and increased inflammatory responses following intravenous delivery. *Mol Ther* 2012;**20**:2268–81.
48. Gimenez-Alejandre M, Cascallo M, Bayo-Puxan N, Alemany R. Coagulation factors determine tumor transduction in vivo. *Hum Gene Ther* 2008;**19**:1415–9.
49. Khare R, Reddy VS, Nemerow GR, Barry MA. Identification of adenovirus serotype 5 hexon regions that interact with scavenger receptors. *J Virol* 2012;**86**:2293–301.
50. Wilson SS, Wiens ME, Smith JG. Antiviral mechanisms of human defensins. *J Mol Biol* 2013;**425**:4965–80.
51. Smith JG, Silvestry M, Lindert S, Lu W, Nemerow GR, Stewart PL. Insight into the mechanisms of adenovirus capsid disassembly from studies of defensin neutralization. *PLoS Pathog* 2010;**6**:e1000959.
52. Snijder J, Reddy VS, May ER, Roos WH, Nemerow GR, Wuite GJ. Integrin and defensin modulate the mechanical properties of adenovirus. *J Virol* 2013;**87**:2756–66.
53. Flatt JW, Kim R, Smith JG, Nemerow GR, Stewart PL. An intrinsically disordered region of the adenovirus capsid is implicated in neutralization by human alpha defensin 5. *PLoS One* 2013;**8**:e61571.
54. Nicklin SA, Wu E, Nemerow GR, Baker AH. The influence of adenovirus fiber structure and function on vector development for gene therapy. *Mol Ther* 2005;**12**:384–93.
55. Miller CR, Buchsbaum DJ, Reynolds PN, Douglas JT, Gillespie GY, Mayo MS, et al. Differential susceptibility of primary and established human glioma cells to adenovirus infection: targeting via the epidermal growth factor receptor achieves fiber receptor-independent gene transfer. *Cancer Res* 1998;**58**:5738–48.

56. Hemmi S, Geertsen R, Mezzacasa A, Peter I, Dummer R. The presence of human coxsackievirus and adenovirus receptor is associated with efficient adenovirus-mediated transgene expression in human melanoma cell cultures. *Hum Gene Ther* 1998;**9**:2363–73.
57. Nettelbeck DM, Rivera AA, Kupsch J, Dieckmann D, Douglas JT, Kontermann RE, et al. Retargeting of adenoviral infection to melanoma: combining genetic ablation of native tropism with a recombinant bispecific single-chain diabody (scDb) adapter that binds to fiber knob and HMWMAA. *Int J Cancer* 2004;**108**:136–45.
58. Stevenson SC, Rollence M, White B, Weaver L, McClelland A. Human adenovirus serotypes 3 and 5 bind to two different cellular receptors via the fiber head domain. *J Virol* 1995;**69**:2850–7.
59. Krasnykh VN, Mikheeva GV, Douglas JT, Curiel DT. Generation of recombinant adenovirus vectors with modified fibers for altering viral tropism. *J Virol* 1996;**70**:6839–46.
60. Volk AL, Rivera AA, Kanerva A, Bauerschmitz G, Dmitriev I, Nettelbeck DM, et al. Enhanced adenovirus infection of melanoma cells by fiber-modification: incorporation of rgd peptide or Ad5/3 chimerism. *Cancer Biol Ther* 2003;**2**:511–5.
61. Shayakhmetov DM, Li ZY, Ternovoi V, Gaggar A, Gharwan H, Lieber A. The interaction between the fiber knob domain and the cellular attachment receptor determines the intracellular trafficking route of adenoviruses. *J Virol* 2003;**77**:3712–23.
62. Shayakhmetov DM, Lieber A. Dependence of adenovirus infectivity on length of the fiber shaft domain. *J Virol* 2000;**74**:10274–86.
63. Seki T, Dmitriev I, Suzuki K, Kashentseva E, Takayama K, Rots M, et al. Fiber shaft extension in combination with HI loop ligands augments infectivity for CAR-negative tumor targets but does not enhance hepatotropism in vivo. *Gene Ther* 2002;**9**:1101–8.
64. Breidenbach M, Rein DT, Wang M, Nettelbeck DM, Hemminki A, Ulasov I, et al. Genetic replacement of the adenovirus shaft fiber reduces liver tropism in ovarian cancer gene therapy. *Hum Gene Ther* 2004;**15**:509–18.
65. Ni S, Gaggar A, Di Paolo N, Li ZY, Liu Y, Strauss R, et al. Evaluation of adenovirus vectors containing serotype 35 fibers for tumor targeting. *Cancer Gene Ther* 2006;**13**:1072–81.
66. Kuhlmann KF, van Geer MA, Bakker CT, Dekker JE, Havenga MJ, Elferink RP, et al. Fiber-chimeric adenoviruses expressing fibers from serotype 16 and 50 improve gene transfer to human pancreatic adenocarcinoma. *Cancer Gene Ther* 2009;**16**:585–97.
67. Parker AL, White KM, Lavery CA, Custers J, Waddington SN, Baker AH. Pseudotyping the adenovirus serotype 5 capsid with both the fibre and penton of serotype 35 enhances vascular smooth muscle cell transduction. *Gene Ther* 2013;**20**:1158–64.
68. Grellier E, Lecolle K, Rogee S, Couturier C, D'Halluin JC, Hong SS, et al. A fiber-modified adenoviral vector interacts with immunoevasion molecules of the B7 family at the surface of murine leukemia cells derived from dormant tumors. *Mol Cancer* 2011;**10**:105.
69. Glasgow JN, Kremer EJ, Hemminki A, Siegal GP, Douglas JT, Curiel DT. An adenovirus vector with a chimeric fiber derived from canine adenovirus type 2 displays novel tropism. *Virology* 2004;**324**:103–16.
70. Nakayama M, Both GW, Banizs B, Tsuruta Y, Yamamoto S, Kawakami Y, et al. An adenovirus serotype 5 vector with fibers derived from ovine atadenovirus demonstrates CAR-independent tropism and unique biodistribution in mice. *Virology* 2006;**350**:103–15.
71. Mercier GT, Campbell JA, Chappell JD, Stehle T, Dermody TS, Barry MA. A chimeric adenovirus vector encoding reovirus attachment protein sigma1 targets cells expressing junctional adhesion molecule 1. *Proc Natl Acad Sci USA* 2004;**101**:6188–93.
72. Weaver EA, Mercier GT, Gottschalk S, Barry MA. T-cell-biased immune responses generated by a mucosally targeted adenovirus-sigma1 vaccine. *Mucosal Immunol* 2012;**5**:311–9.

73. Lyons M, Onion D, Green NK, Aslan K, Rajaratnam R, Bazan-Peregrino M, et al. Adenovirus type 5 interactions with human blood cells may compromise systemic delivery. *Mol Ther* 2006;**14**:118–28.
74. Carlisle RC, Di Y, Cerny AM, Sonnen AF, Sim RB, Green NK, et al. Human erythrocytes bind and inactivate type 5 adenovirus by presenting coxsackie virus-adenovirus receptor and complement receptor 1. *Blood* 2009;**113**:1909–18.
75. Seiradake E, Henaff D, Wodrich H, Billet O, Perreau M, Hippert C, et al. The cell adhesion molecule "CAR" and sialic acid on human erythrocytes influence adenovirus in vivo biodistribution. *PLoS Pathog* 2009;**5**:e1000277.
76. Koski A, Kangasniemi L, Escutenaire S, Pesonen S, Cerullo V, Diaconu I, et al. Treatment of cancer patients with a serotype 5/3 chimeric oncolytic adenovirus expressing GMCSF. *Mol Ther* 2010;**18**:1874–84.
77. Pesonen S, Helin H, Nokisalmi P, Escutenaire S, Ribacka C, Sarkioja M, et al. Oncolytic adenovirus treatment of a patient with refractory neuroblastoma. *Acta Oncol* 2010;**49**:117–9.
78. Pesonen S, Nokisalmi P, Escutenaire S, Sarkioja M, Raki M, Cerullo V, et al. Prolonged systemic circulation of chimeric oncolytic adenovirus Ad5/3-Cox2l-D24 in patients with metastatic and refractory solid tumors. *Gene Ther* 2010;**17**:892–904.
79. Pesonen S, Diaconu I, Kangasniemi L, Ranki T, Kanerva A, Pesonen SK, et al. Oncolytic immunotherapy of advanced solid tumors with a CD40l-expressing replicating adenovirus: assessment of safety and immunologic responses in patients. *Cancer Res* 2012;**72**:1621–31.
80. Kim KH, Dmitriev IP, Saddekni S, Kashentseva EA, Harris RD, Aurigemma R, et al. A phase I clinical trial of Ad5/3-δ24, a novel serotype-chimeric, infectivity-enhanced, conditionally-replicative adenovirus (CRAd), in patients with recurrent ovarian cancer. *Gynecol Oncol* 2013;**130**:518–24.
81. Kanerva A, Nokisalmi P, Diaconu I, Koski A, Cerullo V, Liikanen I, et al. Antiviral and antitumor t-cell immunity in patients treated with GM-CSF-coding oncolytic adenovirus. *Clin Cancer Res* 2013;**19**:2734–44.
82. Bramante S, Koski A, Kipar A, Diaconu I, Liikanen I, Hemminki O, et al. Serotype chimeric oncolytic adenovirus coding for GM-CSF for treatment of sarcoma in rodents and humans. *Int J Cancer* 2014;**135**:720–30.
83. Miyazawa N, Leopold PL, Hackett NR, Ferris B, Worgall S, Falck-Pedersen E, et al. Fiber swap between adenovirus subgroups B and C alters intracellular trafficking of adenovirus gene transfer vectors. *J Virol* 1999;**73**:6056–65.
84. Miyazawa N, Crystal RG, Leopold PL. Adenovirus serotype 7 retention in a late endosomal compartment prior to cytosol escape is modulated by fiber protein. *J Virol* 2001;**75**:1387–400.
85. Corjon S, Gonzalez G, Henning P, Grichine A, Lindholm L, Boulanger P, et al. Cell entry and trafficking of human adenovirus bound to blood factor X is determined by the fiber serotype and not hexon:heparan sulfate interaction. *PLoS One* 2011;**6**:e18205.
86. Schoggins JW, Falck-Pedersen E. Fiber and penton base capsid modifications yield diminished adenovirus type 5 transduction and proinflammatory gene expression with retention of antigen-specific humoral immunity. *J Virol* 2006;**80**:10634–44.
87. Sakurai F, Nakashima K, Yamaguchi T, Ichinose T, Kawabata K, Hayakawa T, et al. Adenovirus serotype 35 vector-induced innate immune responses in dendritic cells derived from wild-type and human CD46-transgenic mice: comparison with a fiber-substituted ad vector containing fiber proteins of ad serotype 35. *J Control Release* 2010;**148**:212–8.
88. Wong HH, Jiang G, Gangeswaran R, Wang P, Wang J, Yuan M, et al. Modification of the early gene enhancer-promoter improves the oncolytic potency of adenovirus 11. *Mol Ther* 2012;**20**:306–16.

89. Yan W, Kitzes G, Dormishian F, Hawkins L, Sampson-Johannes A, Watanabe J, et al. Developing novel oncolytic adenoviruses through bioselection. *J Virol* 2003;**77**:2640–50.
90. Subramanian T, Vijayalingam S, Chinnadurai G. Genetic identification of adenovirus type 5 genes that influence viral spread. *J Virol* 2006;**80**:2000–12.
91. Gros A, Martinez-Quintanilla J, Puig C, Guedan S, Mollevi DG, Alemany R, et al. Bioselection of a gain of function mutation that enhances adenovirus 5 release and improves its antitumoral potency. *Cancer Res* 2008;**68**:8928–37.
92. Uil TG, Vellinga J, de Vrij J, van den Hengel SK, Rabelink MJ, Cramer SJ, et al. Directed adenovirus evolution using engineered mutator viral polymerases. *Nucleic Acids Res* 2011;**39**:e30.
93. Puig-Saus C, Gros A, Alemany R, Cascallo M. Adenovirus i-leader truncation bioselected against cancer-associated fibroblasts to overcome tumor stromal barriers. *Mol Ther* 2012;**20**:54–62.
94. Kuhn I, Harden P, Bauzon M, Chartier C, Nye J, Thorne S, et al. Directed evolution generates a novel oncolytic virus for the treatment of colon cancer. *PLoS One* 2008;**3**:e2409.
95. Di Y, Seymour L, Fisher K. Activity of a group B oncolytic adenovirus (ColoAd1) in whole human blood. *Gene Ther* 2014;**21**:440–3.
96. Pol J, Bloy N, Obrist F, Eggermont A, Galon J, Cremer I, et al. Trial watch: oncolytic viruses for cancer therapy. *Oncoimmunology* 2014;**3**:e28694.
97. Quirin C, Rohmer S, Fernandez-Ulibarri I, Behr M, Hesse A, Engelhardt S, et al. Selectivity and efficiency of late transgene expression by transcriptionally targeted oncolytic adenoviruses are dependent on the transgene insertion strategy. *Hum Gene Ther* 2011;**22**:389–404.
98. Gallaher SD, Gil JS, Dorigo O, Berk AJ. Robust in vivo transduction of a genetically stable Epstein-Barr virus episome to hepatocytes in mice by a hybrid viral vector. *J Virol* 2009;**83**:3249–57.
99. Sun Y, Li HY, Tian DY, Han QY, Zhang X, Li N, et al. A novel alphavirus replicon-vectored vaccine delivered by adenovirus induces sterile immunity against classical swine fever. *Vaccine* 2011;**29**:8364–72.
100. Sun Y, Tian DY, Li S, Meng QL, Zhao BB, Li Y, et al. Comprehensive evaluation of the adenovirus/alphavirus-replicon chimeric vector-based vaccine rAdV-SFV-E2 against classical swine fever. *Vaccine* 2013;**31**:538–44.
101. Yang Y, Xiao F, Lu Z, Li Z, Zuo H, Zhang Q, et al. Development of a novel adenovirus-alphavirus hybrid vector with RNA replicon features for malignant hematopoietic cell transduction. *Cancer Gene Ther* 2013;**20**:429–36.
102. Liu J, Cheng X, Guo Z, Wang Z, Li D, Kang F, et al. Truncated active human matrix metalloproteinase-8 delivered by a chimeric adenovirus-hepatitis B virus vector ameliorates rat liver cirrhosis. *PLoS One* 2013;**8**:e53392.
103. Kubo S, Haga K, Tamamoto A, Palmer DJ, Ng P, Okamura H, et al. Adenovirus-retrovirus hybrid vectors achieve highly enhanced tumor transduction and antitumor efficacy in vivo. *Mol Ther* 2011;**19**:76–82.
104. El-Andaloussi N, Bonifati S, Kaufmann JK, Mailly L, Daeffler L, Deryckere F, et al. Generation of an adenovirus–parvovirus chimera with enhanced oncolytic potential. *J Virol.* 2012;**86**(19):10418–31.
105. Vacchelli E, Eggermont A, Sautes-Fridman C, Galon J, Zitvogel L, Kroemer G, et al. Trial watch: oncolytic viruses for cancer therapy. *Oncoimmunology* 2013;**2**:e24612.
106. Lam PY, Breakefield XO. Hybrid vector designs to control the delivery, fate and expression of transgenes. *J Gene Med* 2000;**2**:395–408.

107. Rivera AA, Davydova J, Schierer S, Wang M, Krasnykh V, Yamamoto M, et al. Combining high selectivity of replication with fiber chimerism for effective adenoviral oncolysis of CAR-negative melanoma cells. *Gene Ther* 2004;**11**:1694–702.
108. Tsuruta Y, Pereboeva L, Breidenbach M, Rein DT, Wang M, Alvarez RD, et al. A fiber-modified mesothelin promoter-based conditionally replicating adenovirus for treatment of ovarian cancer. *Clin Cancer Res* 2008;**14**:3582–8.
109. Tekant Y, Davydova J, Ramirez PJ, Curiel DT, Vickers SM, Yamamoto M. Oncolytic adenoviral therapy in gallbladder carcinoma. *Surgery* 2005;**137**:527–35.
110. Ramirez PJ, Vickers SM, Ono HA, Davydova J, Takayama K, Thompson TC, et al. Optimization of conditionally replicative adenovirus for pancreatic cancer and its evaluation in an orthotopic murine xenograft model. *Am J Surg* 2008;**195**:481–90.
111. Schoggins JW, Gall JG, Falck-Pedersen E. Subgroup B and F fiber chimeras eliminate normal adenovirus type 5 vector transduction in vitro and in vivo. *J Virol* 2003;**77**:1039–48.
112. Kashentseva EA, Douglas JT, Zinn KR, Curiel DT, Dmitriev IP. Targeting of adenovirus serotype 5 pseudotyped with short fiber from serotype 41 to c-erbB2-positive cells using bispecific single-chain diabody. *J Mol Biol* 2009;**388**:443–61.
113. Behr M, Kaufmann JK, Ketzer P, Engelhardt S, Muck-Hausl M, Okun PM, et al. Adenoviruses using the cancer marker EphA2 as a receptor in vitro and in vivo by genetic ligand insertion into different capsid scaffolds. *PLoS One* 2014;**9**:e95723.
114. Tsuruta Y, Pereboeva L, Glasgow JN, Rein DT, Kawakami Y, Alvarez RD, et al. A mosaic fiber adenovirus serotype 5 vector containing reovirus sigma 1 and adenovirus serotype 3 knob fibers increases transduction in an ovarian cancer ex vivo system via a coxsackie and adenovirus receptor-independent pathway. *Clin Cancer Res* 2007;**13**:2777–83.
115. Krasnykh V, Belousova N, Korokhov N, Mikheeva G, Curiel DT. Genetic targeting of an adenovirus vector via replacement of the fiber protein with the phage T4 fibritin. *J Virol* 2001;**75**:4176–83.
116. van Beusechem VW, van Rijswijk AL, van Es HH, Haisma HJ, Pinedo HM, Gerritsen WR. Recombinant adenovirus vectors with knobless fibers for targeted gene transfer. *Gene Ther* 2000;**7**:1940–6.
117. Ruzsics Z, Lemnitzer F, Thirion C. Engineering adenovirus genome by bacterial artificial chromosome (BAC) technology. *Methods Mol Biol* 2014;**1089**:143–58.
118. Muck-Hausl M, Solanki M, Zhang W, Ruzsics Z, Ehrhardt A. Ad 2.0: a novel recombineering platform for high-throughput generation of tailored adenoviruses. *Nucleic Acids Res* 2015;**43**(8):e50.

# Adenoviral Vector Vaccines Antigen Transgene

21

*Hildegund C.J. Ertl*
Wistar Institute, Philadelphia, PA, USA

## 1. Introduction

Vaccines save millions of human lives. They increase the lifespan of our pets and reduce the loss of livestock. Traditionally, vaccines were based on attenuated or inactivated pathogens, or in some cases bacterial toxins. Advances in molecular biology now allow for the generation of subunit vaccines that express one or a few of the pathogens' antigens that induce protective immune responses. Most viral and bacterial infections can be prevented by sufficiently high titers of neutralizing antibodies, which are elicited by viral surface antigens, such as the glycoprotein of rabies virus, the hemagglutinin of influenza virus, or the envelope protein of human immunodeficiency virus (HIV). For other pathogens such as those that are highly variable and thereby escape neutralization, cellular immune responses in form of $CD8^+$ T cells may need to be elicited to prevent clinically symptomatic infections. Although unlike neutralizing antibodies $CD8^+$ T cells fail to provide sterilizing immunity, they can rapidly clear infected cells and thereby block the spread of pathogens and ensuing disease.

$CD8^+$ T cells come in different varieties.[1] Upon activation, $CD8^+$ T cells expand and assume effector functions. Most effector cells, which in general produce the antiviral cytokine interferon (IFN)-gamma and lytic enzymes such as perforin and granzyme B that allow for lysis of infected cells, die once the infection is cleared. A fraction differentiates into effector or central memory cells. Effector memory cells, which are longer-lived than effector cells although their numbers gradually decline, circulate through the periphery. They can rapidly assume effector functions although their proliferative capacity is limited. Central memory cells reside in lymphatic tissues. They require reactivation and expansion before they commence effector functions. Unlike effector or effector memory cells, their proliferative capacity is high and they persist through cytokine-driven cell renewal for the lifespan of an individual. Central memory cells in general produce interleukin (IL)-2 but not IFN-gamma or lytic enzymes. Which type of $CD8^+$ T cell response is most suited to protect against an infection remains debated and most likely depends on the type of invading pathogen. For example, infections of mice with lymphocytic choriomeningitis virus can be prevented by central memory $CD8^+$ T cells,[2] whereas chronic infections of rhesus macaques with simian immunodeficiency virus (SIV) can be averted by effector memory $CD8^+$ T cells.[3] The nature of vaccine-induced immune responses is largely dictated by the type of vaccine carrier. Protein vaccines and most viral vector vaccines, with the exception of vectors based on cytomegaloviruses (CMV), can induce antibodies including

**Adenoviral Vectors for Gene Therapy.**

neutralizing antibodies to foreign antigens. Viral vectors that are rapidly cleared, such as those based on poxviruses or most ribonucleic acid (RNA) viruses, for example, recombinant influenza or rabies virus vectors, induce sustained central memory $CD8^+$ T cell responses. Vectors that persist, such as those derived from CMV or adenoviruses, maintain more activated effector or effector memory $CD8^+$ T cell responses.[4,5] Adenoviral vectors persist at low levels and therefore, unlike CMV vectors, induce both sustained effector-like and central memory T cell responses.

This chapter described some basic characteristics of adenoviruses and vectors based on adenoviruses and their performance as vaccine carriers. Vaccine carriers based on adenoviruses can be designed to retain the ability to replicate or they can be rendered replication-defective upon deletion of gene segments needed for viral reproduction. Numerous serotypes of adenoviruses have been isolated from humans, nonhuman primates, and other species, which allow for generation of vectors that can be used sequentially for prime-boosting or for immunizations against different pathogens. Preexisting neutralizing antibodies to adenoviruses, which are commonly found in humans to many of the human serotypes, can impair the immunogenicity of adenoviral vectors; this can be circumvented by using vaccines based on rare human serotypes or adenoviruses isolated from other species such as chimpanzees. Unless used at excessively high doses, adenoviral vectors are well tolerated. They induce exceptionally potent and sustained B and $CD8^+$ T cell responses to foreign antigens encoded by a transgene. Adenoviral vectors thus provide a highly suitable platform as vaccine carriers for numerous pathogens.

## 2. Characteristics of Adenoviruses

Adenoviruses are members of the Adenoviridae family. They are nonenveloped viruses with a double-stranded genome ranging from 25 to 48 kilobases (kb). Adenoviruses infect numerous species, including mammals, birds, and even frogs. They are subdivided into five genera depending on their target species.[6] Human adenoviruses, which belong to the genus Mastadenovirus, are divided into seven families (A–G) and then further into 57 distinct serotypes (HAdV-1–57). Chimpanzee adenoviruses, which have been vectored for gene transfer, are grouped within human adenoviruses.

The adenovirus genome encodes from both deoxyribonucleic acid (DNA) strands a number of early and late gene products. Products of three of the four early domains, that is, E1, E2, and E4, are essential for viral replication, whereas products of E3, which are antiapoptotic or allow the virus to escape immunosurveillance,[7] are nonessential. The late genes L1–L5 encode the icosahedral viral capsid composed of the major proteins hexon, penton, and fiber and minor proteins IIIa, VI, VIII, and IX. Hexon, the most abundant viral protein, has a conserved stalk and a head domain with several highly variable loops, which serve as targets of virus-neutralizing antibodies.[8,9] Different serotypes of adenoviruses of the same family mainly differ in the sequence of these variable loops.[10,11] Fiber binds the viral receptor, which for most adenoviruses is the Coxsackie adenovirus receptor (CAR[12,13]), whereas others bind to CD46.[14] Coxsackie adenovirus receptor, a type I membrane protein, is expressed on endothelial and epithelial cells; it is not expressed on antigen-presenting cells such as dendritic cells or macrophages. CD46, an inhibitory complement receptor, is ubiquitously expressed. It also serves as receptor for measles virus[15] and human herpes virus-6.[16] Penton, which anchors fiber into the viral capsid, contains an arginyl-glycyl-aspartic acid (RGD) sequence, which binds to integrins and together with CAR facilitates the virus' entry into its target cells.[17]

Adenoviruses spread mainly by aerosols. Most cause upper respiratory diseases, conjunctivitis, tonsillitis, and ear infections. Adenovirus 40 and 41 infections are associated with gastroenteritis. In general, adenoviruses cause mild disease although severe and even fatal infections can occur in immunocompromised individuals, and on rare occasions in healthy individuals. After infection, humans shed adenoviruses in the feces for a few days whereas nonhuman primates such as chimpanzees shed viruses for very long periods of time. Adenoviruses persist at low levels in activated T cells; they can be recovered for years after infection from lymphatic tissues.[18]

A live attenuated vaccine is available for human adenoviruses 4 and 7; it is used by the United States (US) Army for immunization of recruits.[19]

## 3. Characteristics and Construction of Adenovirus Vectors

Adenoviruses have been constructed as replication-competent or replication-defective vectors. Replication-competent vectors used in a nontarget species are in fact replication-defective whereas E1-deleted replication-defective vectors can replicate to some degree in cells with transcription machineries that substitute for the deleted E1 gene products.[20] Foreign sequences of up to 1.8 kb can be incorporated into the adenovirus genome with no deletions. Incorporation of longer sequences interferes with packaging and viral rescue. The packaging size can be increased by deletion of the E3 domain. E3 only–deleted adenovirus vectors remain replication-competent and such vectors are used as vaccine carriers. Deletions of E1, E2, or E4 render the vector replication-defective and the deleted gene products have to be provided in trans during vector production. Most vaccine vectors used to date are deleted in E1, which allows for insertion of approximately 4 kb of foreign sequences. Some are also deleted in E3, which increases the packaging capacity to about 7.5 kb. Further deletions of E4, which encodes polypeptides that affect host cell proliferation and survival and provide for nuclear export of RNA, have been explored. Additional deletion of E4 reduces the synthesis of adenoviral proteins; this in turn may reduce the stimulation of T cell response to adenoviral antigens.[21] Originally, E1-deleted adenovirus vectors were constructed by homologous recombination in a packaging cell line that provides E1 in trans.[22] Currently, adenoviruses are constructed from viral molecular clones by ligating a transgene expression cassette directly into the deleted E1 domain.[23] The bacterial clone upon linearization is then transfected into the packaging cell line for viral rescue. Several packaging cell lines are available. HEK 293 cells carry the 5′ gene region of HAdV-5 virus.[24] They are suited for rescue of E1-deleted HAdV-5 vectors as well as for some of the family E simian adenoviruses (SAdV) that are being explored as vaccine carriers. Homologous recombination between the adenovirus sequences within HEK 293 cells and the E1-deleted vector can lead to outgrowth of replication-competent viruses[25,26]; this is not a problem for SAdV vectors grown on HEK 293 cells because sequence differences between the E1-flanking regions of HAdV-5 and SAdV viruses prevent homologous recombination. PER.C6 cells offer an alternative[27]; they only carry the E1 domain of HAdV-5, which reduces the chance of insertion of E1 into the genome of HAdV-5 vectors.[28] For other serotypes, packaging cells have to be constructed typically by stable E1 transfection of a cell line that is readily infected with adenoviruses.

Upon transfection of packaging cells with a recombinant molecular clone of adenovirus, viral plaques typically become visible after 7–10 days. Virus once expanded in packaging cells is then purified. For small-scale production, vectors are purified by CsCl gradient centrifugation, whereas for large-scale production chromatographic separation is more suited.[29–31] The virus particle (vp) content of purified vectors is determined by spectrophotometry. Content of infectious virus particles (IU) is measured by plaque assays by staining for hexon or by reverse transcription polymerase chain reaction methods. Although the level of transgene product expression depends on IU, adenovirus vectors are dosed according to vp, which determines the toxicity of a preparation. The vp to IU ratios vary and are typically 5:1–400:1. Vectors with higher vp to IU ratios tend to perform poorly.

Depending on the diluent, adenovirus vectors are stable for several days at room temperature and for several months if kept on ice. Highly concentrated adenovirus vectors can be lyophilized with minimal loss in titers. Adenovirus vectors become unstable at low pH, which may pose problems for shipment in dry ice because $CO_2$ seeping into the vials may lower the pH, causing loss of viral titers.[32]

Transgene product expression by adenoviral vectors is influenced by a number of parameters. In our hands, HAdV-5 vectors express higher levels of their transgene compared with HAdV-26 or family E chimpanzee-origin adenovirus vectors.[23,33,34] A chimeric vector based on a CD46 binding chimpanzee vector was shown to express lower levels of transgene products compared with HAdV-5 vectors or a CAR-binding chimpanzee origin adenovirus vector.[35] E1- versus E1- and E3-deleted HAdV-5 or chimpanzee-origin vectors had comparable expression whereas further deletions in E4 reduced expression. The length of the transgene product also affected levels of protein expression; longer transgenes resulted in reduced expression.[35] The promoter regulating transgene expression was shown to influence both the magnitude and the kinetics of protein expression.[36,37] The orientation of the transgene expression cassette within E1 or E3 significantly influences protein expression.[37,38] We attempted to produce dual-expression adenoviral vectors, which carried one cassette in the deleted E1 domain and the other in the deleted E3 domain. Although we were able to generate stable vectors that expressed both transgene products, vectors that shared regulatory sequences within both expression cassettes were unstable, presumably owing to excision of large fragments of the genome upon homologous recombination.[38]

## 4. Preexisting Immunity to Antigen of Adenoviruses

Infections with adenoviruses are common and most humans carry T cells and binding and neutralizing antibodies to adenoviruses. Neutralizing antibodies directed mainly against the adenovirus hexon are serotype-specific whereas other antibodies or T cells, which are directed to multiple antigens, including those that are highly conserved between adenovirus serotypes and even families, are highly cross-reactive.[39] Seroconversion happens early in life. A study in Northern China showed that newborns compared with toddlers commonly have higher titers of antibodies to HAdV-2 and -5, which presumably reflects transmission of antibodies from their mothers. Infants aged 6–12 months tend to have low titers, which then steadily increase.[40] A study conducted in India showed that neutralizing antibodies to HAdV-5 virus are slightly higher in infants aged 1–6 months than aged 7–12 months. Titers then steadily increase until adulthood.[41] A study conducted in North America, South America, Sub-Saharan Africa, and Southeast Asia showed low seroprevalence rates in infants (up to age 6 months) to HAdV-26 and HAdV-35 virus. Children had high prevalence rates of antibodies to HAdV-5 and HAdV-26, although titers to HAdV-26 tended to be lower than those to HAdV-5.[42] Prevalence of neutralizing antibodies in human adults varies depending on the serotype and geographic region. For example, neutralizing antibodies to common serotypes such as HAdV-5 are found in about 40% of individuals residing in the US and in up to 90% of those living in Sub-Saharan Africa.[43,44] The prevalence of neutralizing antibodies to HAdV-26, which was initially described as a rare serotype, is low in the US and Europe but high in most African countries.[36] Seroprevalence rates to HAdV-48 were shown to be high only in Africa but low in the US and Asia.[42] HAdV-35 appeared to be a truly rare serotype; less than 20% of sera from children or adults independent of their geographic localization were able to neutralize this virus.[42] Because adenoviruses are species-specific, most humans lack antibodies to those that infect other species, although this depends on the serotype as well as potential contact with the infected species. For examples, we described antibodies to three different chimpanzee adenoviruses (SAdV-23, SAdV-24, and SAdV-25) that were virtually absent in sera from humans residing in the US but that could be detected in up to 20% of individuals from Sub-Saharan Africa.[44,45] We assumed that this reflected close contact between humans and chimpanzees in Sub-Saharan Africa, where hunting and eating monkeys is common. We subsequently conducted a study in Brazil and again found slightly enhanced prevalence rates of neutralizing antibodies to chimpanzee adenoviruses in humans residing in Brazil; this was especially pronounced in cohorts from Amazonia. Testing of New World monkeys from Brazil showed the presence of neutralizing antibodies to chimpanzee adenoviruses in nearly all sera of common marmosets, which suggested spillover of these viruses from monkeys into the human population.[46] The same viruses tested against sera from different regions in China showed low prevalence rates for high-titer antibodies, although sera with low titers (≤1:40) were found in up to 20% of adults.[47] Another study measured the prevalence rate of neutralizing antibodies in Caucasians residing in the US or Europe against a large panel of chimpanzee adenoviruses of family B, C, and E. The prevalence rate for the chimpanzee-origin adenoviruses with titers greater than 1:200 was in 5% of sera with the exception of SAdV-3-neutralizing antibodies, which was found in about 10% of sera. One of the viruses, CAdV-63, was tested for neutralization by sera from Kenyan children. Sera with high titers of CAdV-63 neutralizing antibodies were rare (approximately 0.5%) whereas low titers were common (17.5%).[48]

Preexisting neutralizing antibodies to adenovirus cause rapid uptake of the virions by cells of the reticuloendothelial system. They also lead to activation of complement.[49] Specific neutralizing antibodies prevent cell transduction by adenoviruses and transcription and translation of the transgene product. This results in a reduction of the vaccine antigen and diminishes the vectors' ability to stimulate adaptive immune responses[33,34,50] while paradoxically increasing innate responses.[51,52] The latter presumably reflects that neutralizing antibodies target the vectors to Fc-receptor–positive cells. Binding antibodies or preexisting T cell responses to antigens of adenovirus do not appear to have a major effect on transgene product-specific B or T cell responses. The effect of preexisting neutralizing antibodies can be circumvented by increasing the vaccine dose; this is problematic in humans because, depending on titers of neutralizing antibodies, increases by 100- to 1000-fold are required,[53] which would likely result in unacceptable toxicity. Alternatively, adenovirus vectors could be combined with another vaccine modality such as a DNA vaccine or a poxvirus vector in a so-called prime-boost regimen.[54–58] Although this is known to increase transgene product-specific immune responses markedly, it also increases the cost and complexity of a vaccine. Repeated use of the same adenovirus vector has been tested to augment immune responses; this is relatively ineffective because neutralizing antibodies induced by the first vaccine dose impair uptake of repeat doses.[59] Loss of transduction in the presence of preexisting neutralizing antibodies can be circumvented by coating viruses with polyethylene glycol[60] or hydrophilic polymers,[61] or encapsidating them into microspheres.[62] It is not yet known whether these approaches also rescue adaptive immune responses to the transgene product. It is also possible to swap variable loops of the hexon of common human serotypes with those of rare human serotypes.[63] We used a straightforward approach by developing vectors that typically fail to infect humans, such as those derived from nonhuman primates.[33,34] By now a large number of vectors have been isolated from chimpanzees.[48,64] They are phylogenetically grouped within human serotypes[65]; their molecular organization, receptor usage, and growth characteristics are similar to those of human serotypes. Antibodies to human serotypes of adenoviruses fail to cross-react with chimpanzee-derived adenoviruses. Most important, as already mentioned, neutralizing antibodies to chimpanzee adenoviruses are only rarely found in humans. Prevalence rates of such antibodies are slightly increased in Africa, presumably owing to closer contact with infected chimpanzees. Chimpanzee adenovirus vectors are thus highly suited as vaccine carriers for use in humans.

## 5. Innate Immune Responses to Adenovirus Vectors

Adenoviruses and adenovirus vectors stimulate potent innate immune responses that cause dose-limiting toxicity. Innate responses, which lead to the release of proinflammatory cytokines and chemokines such as type I IFN, IL-6, IL-12, Regulated on Activation, Normal T cell Expressed and Secreted, macrophage inflammatory protein-1β, IFN-gamma–induced protein (IP)-10, and others, are typically triggered through recognition of pathogen-associated molecular patterns by cellular pathogen recognition receptors located on the cell surface, on endosomes, or within the cytoplasm. Several motifs present on adenovirus vectors are recognized by innate sensor. The RGD motif on penton activates the nuclear factor-kappa light chain enhancer of activated B cells pathway, which triggers an inflammatory reaction.[66] The double-stranded adenovirus genome is recognized by Toll-like receptor (TLR)-9, one of the endosomal sensors.[67] Adenovirus vectors appear to be recognized by additional TLRs because innate responses are diminished in cells lacking TIR domain–containing adapter-inducing IFN-beta, which serves as an adaptor to TLR-4 and -3.[68,69] The adenovirus genome further activates retinoic acid-inducible gene-1, another intracellular sensor of double-stranded DNA.[70]

Adenoviruses applied to the airways interact with surfactant,[71] which appears to block inflammatory responses, presumably by causing opsonization of the virus particles. High doses of adenovirus vectors further activate the complement system through interactions between fiber knob with blood Factor IX and C4 binding protein. These interactions increase inflammatory responses.[72,73] Adenovirus vectors are sequestered by preexisting neutralizing antibodies; for unknown reasons, this interaction increases inflammatory responses to the vectors.[51,52]

The potency of innate immune responses partly depends on the vector serotype. For example, in mice, some of the chimpanzee-derived adenovirus vectors were shown to induce more potent type I IFN responses compared with HAdV-5 vectors.[74,75] This may relate to the frequency of CpG motifs, which tend to be lower in genomes of HAdV-5 (2168 CG motifs) and HAdV-26 (2206 CG motifs) viruses than in those of chimpanzee-origin adenoviruses (approximately 2400–2700 CG motifs). Whether more intense inflammatory reactions in mice translate into heightened toxicity in humans remains to be tested.

## 6. Humoral Immune Responses to Adenoviral Vectors

Adenoviral vectors induce potent B cell responses to the capsid antigens of the vector[46,44] and to foreign transgene products.[45,76] The latter allow for the use of adenovirus vectors as prophylactic vaccines to infections, which can be prevented by neutralizing antibodies. The advantages of adenovirus vectors for delivery of antigens from other pathogens for induction of protective antibody responses are numerous because they interact with innate sensor drive maturation of antigen-presenting cells and do not require addition of adjuvants. E1-deleted adenoviral vectors are safe at immunogenic doses. They can be given through multiple routes including intranasal, intramuscular, and oral, provided vectors are encapsidated. Similar to wild-type adenoviruses, E1-deleted adenovirus vectors persist in activated T cells[5] and induce sustained antibody responses. Adenoviruses use the host cell machinery for posttranslational modifications of antigens so that viral antigens are presented in their native form unless their structure depends on co-interactions with other antigens. We compared the magnitude of transgene product-specific antibody responses to two chimpanzee-derived adenovirus vectors, both of which are family E members, with those induced by vectors derived from HAdV5, a family C member, and HAdV26, a family D member.[45] Vectors expressed the rabies virus glycoprotein as the transgene product. The HAdV-5 vector induced markedly higher rabies virus-specific antibody responses compared with the other vectors.

E3-deleted HAdV-5 vectors expressing the rabies virus glycoprotein have been licensed for immunization of wildlife such as foxes or skunks.[77] Neither of these species supports replication of HAdV-5 virus, so that in its intended target species the vaccine is replication-defective. E1-deleted adenovirus vectors based on simian serotypes also expressing the rabies virus glycoprotein are under development for human immunization.[33] These vectors are highly immunogenic and induce protective titers of rabies virus–neutralizing antibodies in nonhuman primates after a single low dose ($10^9$ vp) given intramuscularly before challenge.[78] Antibody titers and protective immunity were shown to be sustained for at least 2 years. Ebola virus in the winter and spring of 2014 caused an outbreak in Guinea, which by fall of the same year had spread to Liberia and Sierra Leone with isolated cases imported to Europe and the US. The World Health Organization estimated that without improved interventions, the virus would continue to spread and potentially infect up to a million individuals by 2015.[79] Others voiced concern about viral mutation that may allow Ebola virus to be transmitted by aerosols, even further increasing infection rates. One of the first Ebola vaccines that underwent testing in humans was based on an E1-deleted HAdV-5 vector expressing the glycoproteins of Zaire and Sudan Ebola species.[80] The vaccine, which had been shown to protect nonhuman primates against Ebola virus infection,[81] was found to be safe at $2 \times 10^{10}$ vp per dose and human subjects developed specific T and B cell responses. Because of concerns about preexisting HAdV-5 neutralizing antibodies,[82] which tend to be prevalent and robust in African human populations, a second set of vectors based on chimpanzee serotype 3 was constructed.[83] This vaccine is currently undergoing testing in human volunteers. Provided the vaccine is immunogenic and safe, it may then be used under an investigational new drug application in consenting individuals at high risk of contracting the virus. One potential setback of this Ebola vaccine is that although humoral responses to the viral glycoprotein developed rapidly in nonhuman primates, antibody titers were not sustained, but rather declined to baseline within less than a year. This could potentially be addressed by a booster immunization with an adenovirus vector based on a different serotype or an unrelated vaccine prototype such as a poxvirus vector.

A plethora of adenovirus vectors based on human and simian serotypes have been tested preclinically for induction of protective antibody responses to numerous pathogens including hepatitis B virus,[84] Dengue virus,[85] *Severe acute respiratory syndrome* coronavirus,[86] rotavirus,[87] respiratory syncytial virus,[88] rabies virus,[33,77] herpes simplex virus type 2,[89] Hantaan virus,[90] influenza virus,[91] and plasmodium vivax.[92] Results consistently showed that the vectors induced potent antibody responses dominated by isotypes linked to type 1 T helper cell responses upon systemic immunization whereas mucosal injections also led to local immunoglobulin A production.[93]

# 7. Cellular Immune Responses to Adenoviral Vectors

Although neutralizing antibodies are the primary correlate of protection against most pathogens for some of the more complex viruses, intracellular bacteria, or protozoa, protection can be provided by cellular immune responses that achieve accelerated clearance of infected cells. By the same token, chronic viral infections or cancer cells are best combated with vaccine-induced $CD8^+$ T cell responses. Adenoviruses acquired by natural infections induce $CD4^+$ and $CD8^+$ T cell responses to a number of viral proteins. Such T cells can be found in most human adults.[94] They cross-react between different adenovirus serotypes including those derived from chimpanzee-origin adenoviruses.[39] Adenovirus-specific $CD4^+$ T cells were found to be monofunctional; they largely belong to the memory subset. In contrast, adenovirus-specific $CD8^+$ T cells are more polyfunctional. They are highly activated and are phenotypically mainly grouped into effector/effector memory subsets. This may reflect repeated exposures to different serotypes of adenoviruses or internal reactivations of $CD8^+$ T cells by persisting viruses. Adenoviral vectors induce remarkably high transgene product-specific $CD8^+$ T cell responses and, at least in mice, only modest $CD4^+$ T cell responses.[34,45] Induction of T cells is triggered by cross-presentation of antigen.[95] Transgene product-specific $CD8^+$ T cells similar to those induced by natural infections to the adenovirus antigens are polyfunctional, and because of the vectors' persistence, a large proportion remains activated. Nevertheless, because levels of persistence are low, adenovirus-induced T cells in part transition into the memory pool, which allows for increased responses after booster immunizations.[88,49]

We compared the magnitude of transgene product-specific $CD8^+$ T cell responses to HAdV-5, HAdV-26, and chimpanzee-derived adenovirus vectors SAdV-24 and SAdV-25.[45] In mice, responses were largely comparable. Others compared T cell responses to different human and chimpanzee serotypes in mice and rhesus macaques. In both species, human serotypes adenoviruses such as HAdV-5 and -6 based on family C induced higher $CD8^+$ T cell responses compared with those from family D, such as HAdV-26 or HAdV-24. Adenovirus vectors based on family B, such as HAdV-35 and HAdV34, both of which use CD46 rather than CAR for cell entry, were the least immunogenic.[48] Similar trends were seen for chimpanzee-origin adenovirus vectors; those from group C, such as SAdV-3, SAdV-20, or SAdV-11, tended to be more immunogenic than those of group E, such as SAdV-4 or -5, whereas the one group B virus, SAdV-30, was the least immunogenic.[48] Nevertheless, some of the family E–derived vectors were as immunogenic as or even more so than some of the family C vectors.

Large numbers of different adenovirus vectors have undergone preclinical testing as T cell–inducing vaccines for a variety of pathogens, including HIV-1/SIV,[34,50,68] hepatitis C virus (HCV),[58] *Trypanosome cruzii*,[96] *Mycobacterium tuberculosis*,[97,98] dengue virus,[99] human CMV,[100] influenza A virus,[101,102] Rift valley fever virus,[103] and Epstein Barr virus.[104] Results showed that adenovirus vectors induced exceptionally potent and sustained $CD8^+$ T cell responses in animals that were higher than those induced by other recombinant vaccines such as DNA vaccines or poxvirus vectors. T cell responses induced by an adenovirus vector can be enhanced by prime boost-regimens using serologically distinct adenovirus vectors or other vaccine platforms for priming or boosting.

## 8. Clinical Experience with Vaccines Based on Adenoviral Vector

Early-stage safety studies have been conducted with several human- as well as chimpanzee-derived adenovirus vectors expressing antigens of HIV-1,[58] *M. tuberculosis*,[105] *Plasmodium falciparum*,[106] Ebola virus,[80] influenza A virus,[107] and HCV.[108] Results showed that toxicity was dose-related and unaffected by preexisting neutralizing antibodies to the vectors.[109,110] At high doses of $10^{11}$ vp, individuals mainly experienced mild to moderate flulike symptoms with fever, myalgia, fatigue, and headache. Injection-site reactions such as erythema and local pain were seen regardless of vector dose. Repeated injections of the same vector resulted in decreased systemic reactions upon sequential immunizations.[111] Vaccinated individuals did not develop significant changes in blood values. Overall, the vaccines were well tolerated. Analyses of vaccine-induced immune responses showed that in a dose-dependent manner, vaccine recipients developed T and B cells to the transgene product. They also had increases in immune responses to antigens of the adenoviral vector.

The first large-scale phase IIb trial for an adenoviral vector vaccine, the STEP trial, was conducted by Merck with HAdV-5 vectors expressing Gag, Pol, and Nef of HIV-1 clade B for induction of T cells. Individuals from North or South America, the Caribbean, or Australia at high risk for HIV-1 acquisition received three injections of $5 \times 10^{10}$ vp of the vaccine or placebo on day 1 and weeks 4 and 26. The trial was designed to enroll 1500 individuals with HAdV-5–specific neutralizing antibody titers less than 1:200 at baseline and 1500 individuals with titers greater than 1:200. As expected, the vaccine elicited potent $CD8^+$ T cell responses that were slightly attenuated in individuals with high titers of preexisting neutralizing antibodies to HAdV-5. $CD4^+$ T cell responses to HIV-1 were observed in only about a third of vaccine recipients.[112] The trial was stopped prematurely after an interim analysis by the Safety Monitoring Board showed that the vaccine neither prevented HIV-1 infections nor reduced viral loads in individuals who became infected.[113] A trial conducted in parallel in South Africa, the Phambili trial,[114] was stopped shortly thereafter and in both trials participants were unblinded. In the STEP trial, male vaccine recipients, who were mainly homosexual and engaged in high-risk anal intercourse had increased rates of HIV-1 acquisition (49 of 941 male vaccine recipients) compared with placebo controls (33 of 922 participants). Increased acquisition was linked to high titers of HAdV-5–neutralizing antibodies at baseline as well as the lack of circumcision.[115] There was no link to differences in risk behavior.[115] This trend for increased HIV-1 acquisition, which was transient and waned after about 18 months,[116] was not seen in the Phambili trial, which enrolled mainly heterosexual men and women. Additional studies comparing HIV-1 acquisition rates of participants in other HIV-1 trials based on vaccines other than adenoviral vectors showed that titers of neutralizing antibodies to several adenoviral serotypes did not increase HIV-1 acquisition.[117] By the same token, the army failed to observe increases in HIV-1 infection rates in recruits who received the live attenuated adenovirus vaccines compared with those who did not.[118] High neutralizing antibody titers to adenovirus per se or vaccination with an adenovirus thus do not appear to increase the risk of HIV-1 infection. The mechanisms underlying the increased acquisition of HIV-1 in HAdV-5–seropositive individuals of the STEP trial have been studied extensively but remain unexplained. One possibility that has been explored is that CD4+ T cells induced by adenovirus vectors become highly susceptible to infection with HIV-1.[119] Such T cells also express homing markers for gut mucosa. The increased presence of HIV-1 target cells at the port of viral entry could explain increased HIV-1 acquisition after vaccination with an adenoviral vector; nevertheless, it does not explain why this increased acquisition was seen only in homosexual males with high baseline titers of HAdV-5–specific neutralizing antibodies.

The next large-scale efficacy trial was again based on an HAdV-5 vector combined with DNA vaccine priming to prevent HIV-1. The DNA vaccine expressed Gag, Pol, Nef, and Env from clade A, B, and C. The HAdV-5 vectors given as mixtures expressed a gag–pol fusion protein and Env of clades A, B, and C.[120] The HAdV-5 vector of HVTN505 differed from the vector used in the STEP trial by the additional deletion of E4 and by the inclusion of Env as a vaccine target antigen. The DNA vaccines were given three times on days 0 and weeks 4 and 8; the HAdV-5 vectors were given once on week 24. The study, conducted in the US, enrolled circumcised males and transgender women with preexisting neutralizing antibody titers to HAdV-5 below 1:18, who were at high risk for HIV-1 acquisition. A total of 1251 participants were enrolled into the vaccine arm; 1245 participants were enrolled into the placebo arm. End points were the prevention of HIV-1 acquisition or lowering of viral loads in infected vaccine recipients. An interim analysis showed that neither of these end points would be met and the trial was halted and unblinded. The vaccines did not increase HIV-1 acquisition rates. The vaccine induced HIV-1–specific CD4+ and/or CD8+ T cells in about 60% of recipients, antibodies to gp41 in all recipients, and antibodies to gp120 in about 50% of vaccine recipients. Antibodies to the V1–V2 loop, which in another, more successful HIV-1 vaccine efficacy trial[121,122] had been identified as correlates of protection against viral acquisition, were detected in only about 20% of vaccine recipients. Responses rates for neutralizing antibodies were low and only tier 1 HIV-1 virus could be neutralized.

The futility of three HAdV-5 vector-based HIV-1 vaccine efficacy trials combined with the increased risk for HIV-1 acquisition in a subcohort of the STEP trial raised questions about the use of adenoviral vectors as a vaccine platform for HIV-1 in general. The initial early-phase immunogenicity trials with the STEP and HVTN505 vaccines had shown that the breadth of HIV-1–specific T cell responses was limited because T cells from most vaccinated individuals only recognized one to four epitopes of HIV-1.[123] This limited epitope specificity was recapitulated in a small phase I trial with an HAdV-26 vaccine expressing Env. We tested chimpanzee-derived adenovirus vectors expressing HIV-1 clade B Gag in rhesus macaques, and after a single immunization the breadth of the response was also limited and most animals responded to only one or two epitopes. A boost with a serologically distinct adenovirus vectors markedly increased the breadth of the response and most animals responded after the boost to 15–30 different Gag epitopes.[124] The restricted epitope specificity observed in the clinical trials, which carries the risk that HIV-1 may escape cellular immunosurveillance, can thus potentially be addressed by prime-boosting with two different serotypes of adenovirus vector vaccines. Also, the efficacy of the STEP and HVTN505 vaccines was limited in preclinical studies. The STEP vaccine protected rhesus macaques against challenge with the SIV/HIV chimera SHIV89.6P[125] but provided no protection against a more stringent challenge with SIVmac239.[126] By the same token, the HVTN505 vaccine failed to protect rhesus macaques from SIV-mac251 challenges but conferred MHC class I allele-dependent protection against the less stringent SIVsmE660 challenge.[127] Others described induction of protective immunity in rhesus macaques immunized with adenovirus vectors from other human as well as chimpanzee serotypes[15,128] against SIVmac239 or SIVmac251, and the use of these alternative adenovirus vector for HIV-1 prophylaxis continues to be explored.

Other clinical efficacy trials explored HAdV-5 or CAdV-63 as carriers for malaria vaccines. An HAdV-5 vector expressing the *P. falciparum* circumsporozoite protein and apical membrane antigen-1 failed to protect against a controlled malaria infection,[129] whereas some protection (about 27%) was achieved when this vaccine was combined with a DNA vaccine prime.[130] A similar level of efficacy was achieved with a CAdV-63 expressing T cell–inducing antigens of *P. falciparum* followed by a modified vaccinia Ankara boost.[131]

Adenovirus vector-based vaccines for *M. tuberculosis* have thus far undergone only early-stage safety and immunogenicity trials. An HAdV-5 vector expressing Ag85A was shown to induce T cells in individuals who were naive to antigens of this pathogen; responses were more robust in previously BCG vaccinated trial volunteers. Immunogenicity was not affected by preexisting HAdV-5–neutralizing antibodies.[105]

One phase I trial tested HAdV-6 and CAdV-3 vectors expressing the NS protein of HCV. Both vectors induced HCV-specific T cell responses that cross-reacted between heterologous HCV strains. Responses to either vaccine were polyfunctional and sustained. Responses were boosted when the two vectors were used sequentially.[108]

## References


1. Wherry EJ, Ahmed R. Memory CD8 T-cell differentiation during viral infection. *J Virol* 2004;**78**:5535–45.
2. Bachmann MF, Wolint P, Schwarz K, Oxenius A. Recall proliferation potential of memory CD8+ T cells and antiviral protection. *J Immunol* 2005;**175**:4677–85.
3. Hansen SG, Ford JC, Lewis MS, Ventura AB, Hughes CM, Coyne-Johnson L, et al. Profound early control of highly pathogenic SIV by an effector memory T-cell Vaccine. *Nature* 2011;**473**:523–7.
4. Hansen SG, Piatak Jr M, Ventura AB, Hughes CM, Gilbride RM, Ford JC, et al. Immune clearance of highly pathogenic SIV infection. *Nature* 2013;**502**:100–4.
5. Tatsis N, Fitzgerald JC, Reyes-Sandoval A, Harris-McCoy KC, Hensley SE, Zhou D, et al. Adenoviral vectors persist in vivo and maintain activated CD8+ T cells: implications for their use as vaccines. *Blood* 2007;**110**:1916–23.
6. http://www.vmri.hu/~harrach/AdVtaxlong.htm.
7. Wold WS, Gooding LR. Region E3 of adenovirus: a cassette of genes involved in host immunosurveillance and virus-cell interactions. *Virology* 1991;**184**:1–8.

8. Roberts MM, White JL, Grütter MG, Burnett RM. Three-dimensional structure of the adenovirus major coat protein hexon. *Science* 1986;**232**:1148–51.
9. Pichla-Gollon SL, Drinker M, Zhou X, Xue F, Rux JJ, Gao GP, et al. Structure-based identification of a major neutralizing site in an adenovirus hexon. *J Virol* 2007;**81**:1680–9.
10. Crawford-Miksza L, Schnurr DP. Analysis of 15 adenovirus hexon proteins reveals the location and structure of seven hypervariable regions containing serotype-specific residues. *J Virol* 1996;**70**:1836–44.
11. Sumida SM, Truitt DM, Lemckert AA, Vogels R, Custers JH, Addo MM, et al. Neutralizing antibodies to adenovirus serotype 5 vaccine vectors are directed primarily against the adenovirus hexon protein. *J Immunol* 2005;**174**:7179–85.
12. Bergelson JM, Cunningham JA, Droguett G, Kurt-Jones EA, Krithivas A, Hong JS, et al. Isolation of a common receptor for Coxsackie B viruses and adenoviruses 2 and 5. *Science* 1997;**275**:1320–3.
13. Segerman A, Atkinson JP, Marttila M, Dennerquist V, Wadell G, Arnberg N. Adenovirus type 11 uses CD46 as a cellular receptor. *J Virol* 2003;**77**:9183–91.
14. Gaggar A, Shayakhmetov DM, Lieber A. CD46 is a cellular receptor for group B adenoviruses. *Nat Med* 2003;**9**:1408–12.
15. Dörig RE, Marcil A, Chopra A, Richardson CD. The human CD46 molecule is a receptor for measles virus (Edmonston strain). *Cell* 1993;**75**:295–305.
16. Santoro F, Kennedy PE, Locatelli G, Malnati MS, Berger EA, Lusso P. CD46 is a cellular receptor for human herpesvirus 6. *Cell* 1999;**99**:817–27.
17. Mathias P, Wickham T, Moore M, Nemerow G. Multiple adenovirus serotypes use alpha v integrins for infection. *J Virol* 1994;**68**:6811–4.
18. Pereira HG. Persistent infection by adenoviruses. *J Clin Pathol Suppl* 1972;**6**:39–42.
19. Hoke Jr CH, Hawksworth A, Snyder Jr CE. Initial assessment of impact of adenovirus type 4 and type 7 vaccine on febrile respiratory illness and virus transmission in military basic trainees, March 2012. *MSMR* 2012;**19**:2–4.
20. Marienfeld U, Haack A, Thalheimer P, Schneider-Rasp S, Brackmann HH, Poller W. 'Autoreplication' of the vector genome in recombinant adenoviral vectors with different E1 region deletions and transgenes. *Gene Ther* 1999;**6**:1101–13.
21. Koup RA, Lamoreaux L, Zarkowsky D, Bailer RT, King CR, Gall JG, et al. Replication-defective adenovirus vectors with multiple deletions do not induce measurable vector-specific T cells in human trials. *J Virol* 2009;**83**:6318–22.
22. Thummel C, Tjian R, Grodzicker T. Construction of adenovirus expression vectors by site-directed in vivo recombination. *J Mol Appl Genet* 1982;**1**:435–46.
23. Zhou D, Zhou X, Bian A, Li H, Chen H, Small JC, et al. An efficient method of directly cloning chimpanzee adenovirus as a vaccine vector. *Nat Protoc* 2010;**5**:1775–85.
24. Shaw G, Morse S, Ararat M, Graham FL. Preferential transformation of human neuronal cells by human adenoviruses and the origin of HEK 293 cells. *FASEB J* 2002;**16**:869–71.
25. Zhu J, Grace M, Casale J, Chang AR, Musco ML, Bordens L. Characterization of replication-competent adenovirus isolates from large-scale production of a recombinant adenoviral vector. *Hum Gene Ther* 1999;**10**:113–21.
26. Lochmuller H, Jani A, Huard J, Prescott S, Simoneau M, Massie B, et al. Emergence of early region 1-containing replication-competent adenovirus in stocks of replication-defective adenovirus recombinants (delta E1 + delta E3) during multiple passages in 293 cells. *Hum Gene Ther* 1994;**5**:1485–91.
27. Fallaux FJ, Bout A, van der Velde I, van den Wollenberg DJ, Hehir KM, Keegan J, et al. New helper cells and matched early region 1-deleted adenovirus vectors prevent generation of replication-competent adenoviruses. *Hum Gene Ther* 1998;**9**:1909–17.

28. Murakami P, Havenga M, Fawaz F, Vogels R, Marzio G, Pungor E, et al. Common structure of rare replication-deficient E1-positive particles in adenoviral vector batches. *J Virol* 2004;**78**:6200–8.
29. Eglon MN, Duffy AM, O'Brien T, Strappe PM. Purification of adenoviral vectors by combined anion exchange and gel filtration chromatography. *J Gene Med* 2009;**11**:978–89.
30. Lee DS, Kim BM, Seol DW. Improved purification of recombinant adenoviral vector by metal affinity membrane chromatography. *Biochem Biophys Res Commun* 2009;**37**:640–4.
31. Dormond E, Kamen AA. Manufacturing of adenovirus vectors: production and purification of helper dependent adenovirus. *Methods Mol Biol* 2011;**737**:139–56.
32. Cassandra Nyberg-Hoffman C, Aguilar-Cordo E. Instability of adenoviral vectors during transport and its implication for clinical studies. *Nat Med* 1999;**5**:955–7.
33. Xiang Z, Gao G, Reyes-Sandoval A, Cohen CJ, Li Y, Bergelson JM, et al. Novel, chimpanzee serotype 68-based adenoviral vaccine carrier for induction of antibodies to a transgene product. *J Virol* 2002;**76**:2667–75.
34. Fitzgerald JC, Gao GP, Reyes-Sandoval A, Pavlakis GN, Xiang ZQ, Wlazlo AP, et al. A simian replication-defective adenoviral recombinant vaccine to HIV-1 gag. *J Immunol* 2003;**170**:1416–22.
35. Tatsis N, Blejer A, Lasaro MO, Hensley SE, Cun A, Tesema L, et al. A CD46-binding chimpanzee adenovirus vector as a vaccine carrier. *Mol Ther* 2007;**15**:608–17.
36. Chen P, Tian J, Kovesdi I, Bruder JT. Promoters influence the kinetics of transgene expression following adenovector gene delivery. *J Gene Med* 2008;**10**:123–31.
37. Addison CL, Hitt M, Kunsken D, Graham FL. Comparison of the human versus murine cytomegalovirus immediate early gene promoters for transgene expression by adenoviral vectors. *J Gen Virol* 1997;**78**(Pt 7):1653–61.
38. Small JC, Kurupati RK, Zhou X, Bian A, Chi E, Li Y, et al. Construction and characterization of E1- and E3-deleted adenovirus vectors expressing two antigens from two separate expression cassettes. *Hum Gene Ther* 2014;**25**:328–38.
39. Hutnick NA, Carnathan D, Demers K, Makedonas G, Ertl HC, Betts MR. Adenovirus-specific human T cells are pervasive, polyfunctional, and cross-reactive. *Vaccine* 2010;**28**:1932–41.
40. Yu B, Wang Z, Dong J, Wang C, Gu L, Sun C, et al. A serological survey of human adenovirus serotype 2 and 5 circulating pediatric populations in Changchun, China, 2011. *Virol J* 2012;**9**:287.
41. Appaiahgari MB, Pandey RM, Vrati S. Seroprevalence of neutralizing antibodies to adenovirus type 5 among children in India: implications for recombinant adenovirus-based vaccines. *Clin Vaccine Immunol* 2007;**14**:1053–5.
42. Barouch DH, Kik SV, Weverling GJ, Dilan R, King SL, Maxfield LF, et al. International seroepidemiology of adenovirus serotypes 5, 26, 35, and 48 in pediatric and adult populations. *Vaccine* 2011;**29**:5203–9.
43. Farina SF1, Gao GP, Xiang ZQ, Rux JJ, Burnett RM, Alvira MR, et al. Replication-defective vector based on a chimpanzee adenovirus. *J Virol* 2001;**75**:11603–13.
44. Xiang Z1, Li Y, Cun A, Yang W, Ellenberg S, Switzer WM, et al. Chimpanzee adenovirus antibodies in humans, Sub-Saharan Africa. *Emerg Infect Dis* 2006;**12**:1596–9.
45. Chen H, Xiang ZQ, Li Y, Kurupati RK, Jia B, Bian A, et al. Adenovirus-based vaccines: comparison of vectors from three species of adenoviridae. *J Virol* 2010;**84**:10522–32.
46. Ersching J, Hernandez MI, Cezarotto FS, Ferreira JD, Martins AB, Switzer WM, et al. Neutralizing antibodies to human and simian adenoviruses in humans and New-World monkeys. *Virology* 2010;**407**:1–6.

47. Jian L, Zhao Q, Zhang S, Huang W, Xiong Y, Zhou X, et al. The prevalence of neutralising antibodies to chimpanzee adenovirus type 6 and type 7 in healthy adult volunteers, patients with chronic hepatitis B and patients with primary hepatocellular carcinoma in China. *Arch Virol* 2014;**159**:465–70.
48. Colloca S, Barnes E, Folgori A, Ammendola V, Capone S, Cirillo A, et al. Vaccine vectors derived from a large collection of simian adenoviruses induce potent cellular immunity across multiple species. *Sci Transl Med* 2012;**4**:115ra2.
49. Cichon G, Boeckh-Herwig S, Schmidt HH, Wehnes E, Muller T, Pring-Akerblom P, et al. Complement activation by recombinant adenoviruses. *Gene Ther* 2001;**8**:1794–800.
50. McCoy K, Tatsis N, Korioth-Schmitz B, Lasaro MO, Hensley SE, Lin SW, et al. Effect of preexisting immunity to adenovirus human serotype 5 antigens on the immune responses of nonhuman primates to vaccine regimens based on human- or chimpanzee-derived adenovirus vectors. *J Virol* 2007;**81**:6594–604.
51. Varnavski AN, Calcedo R, Bove M, Gao G, Wilson JM. Evaluation of toxicity from high-dose systemic administration of recombinant adenovirus vector in vector-naive and pre-immunized mice. *Gene Ther* 2005;**12**:427–36.
52. Varnavski AN, Zhang Y, Schnell M, Tazelaar J, Louboutin JP, Yu QC, et al. Preexisting immunity to adenovirus in rhesus monkeys fails to prevent vector-induced toxicity. *J Virol* 2002;**76**:5711–9.
53. Casimiro DR, Chen L, Fu TM, Evans RK, Caulfield MJ, Davies ME, et al. Comparative immunogenicity in rhesus monkeys of DNA plasmid, recombinant vaccinia virus, and replication-defective adenovirus vectors expressing a human immunodeficiency virus type 1 gag gene. *J Virol* 2003;**77**:6305–13.
54. Pinto AR, Fitzgerald JC, Giles-Davis W, Gao GP, Wilson JM, Ertl HC. Induction of CD8+ T cells to an HIV-1 antigen through a prime boost regimen with heterologous E1 deleted adenoviral vaccine carriers. *J Immunol* 2003;**171**:6774–9.
55. Tatsis N, Lin SW, Harris-McCoy K, Garber DA, Feinberg MB, Ertl HC. Multiple immunizations with adenovirus and MVA vectors improve CD8+ T cell functionality and mucosal homing. *Virology* 2007;**367**:156–67.
56. Graham BS, Enama ME, Nason MC, Gordon IJ, Peel SA, Ledgerwood JE, et al. DNA vaccine delivered by a needle-free injection device improves potency of priming for antibody and CD8+ T-cell responses after rAd5 boost in a randomized clinical trial. *PLoS One* 2013;**8**:e59340.
57. Bart PA, Huang Y, Karuna ST, Chappuis S, Gaillard J, Kochar N, et al. HIV-specific humoral responses benefit from stronger prime in phase Ib clinical trial. *J Clin Invest* 2014;**124**:4843–56.
58. Asmuth DM, Brown EL, DiNubile MJ, Sun X, del Rio C, Harro C, et al. Comparative cell-mediated immunogenicity of DNA/DNA, DNA/adenovirus type 5 (Ad5), or Ad5/Ad5 HIV-1 clade B gag vaccine prime-boost regimens. *J Infect Dis* 2010;**201**:132–41.
59. Casimiro DR, et al. Vaccine-induced immunity in baboons by using DNA and replication-incompetent adenovirus type 5 vectors expressing a human immunodeficiency virus type 1 gag gene. *J Virol* 2003;**77**:7663–8.
60. O'Riordan CR, et al. PEGylation of adenovirus with retention of infectivity and protection from neutralizing antibody in vitro and in vivo. *Hum Gene Ther* 1999;**10**:1349–58.
61. Fisher KD, Stallwood W, Ulbrich K, Mautner V, Seymour LW. Protection and retargeting of adenovirus using a multifunctional hydrophilic polymer. *Mol Ther* 2000;**1S**:57.
62. Sailaja G, HogenEsch H, North A, Hays J, Mittal SK. Encapsulation of recombinant adenovirus into alginate microspheres circumvents vector-specific immune response. *Gene Ther* 2002;**9**:1722–9.

63. Baden LR, Walsh SR, Seaman MS, Johnson JA, Tucker RP, Kleinjan JA, et al. First-in-human evaluation of a hexon chimeric adenovirus vector expressing HIV-1 Env (IPCAVD 002). *J Infect Dis* 2014;**210**:1052–61.
64. Roy S, Sandhu A, Medina A, Clawson DS, Wilson JM. Adenoviruses in fecal samples from asymptomatic rhesus macaques, United States. *Emerg Infect Dis* 2012;**18**:1081–8.
65. Rux JJ, Kuser PR, Burnett RM. Structural and phylogenetic analysis of adenovirus hexons by use of high-resolution x-ray crystallographic, molecular modeling, and sequence-based methods. *J Virol* 2003;**77**:9553–66.
66. Hirschowitz EA, Weaver JD, Hidalgo GE, Doherty DE. Murine dendritic cells infected with adenovirus vectors show signs of activation. *Gene Ther* 2000;**7**:1112–20.
67. Cerullo V, Seiler MP, Mane V, Brunetti-Pierri N, Clarke C, Bertin TK, et al. Toll-like receptor 9 triggers an innate immune response to helper-dependent adenoviral vectors. *Mol Ther* 2007;**15**:378–85.
68. Hartman ZC, Black EP, Amalfitano A. Adenoviral infection induces a multi-faceted innate cellular immune response that is mediated by the toll-like receptor pathway in A549 cells. *Virology* 2007;**358**:357–72.
69. Hartman ZC, Kiang A, Everett RS, Serra D, Yang XY, Clay TM, et al. Adenovirus infection triggers a rapid, MyD88-regulated transcriptome response critical to acute-phase and adaptive immune responses in vivo. *J Virol* 2007;**81**:1796–812.
70. Cheng G, Zhong J, Chung J, Chisari FV. Double-stranded DNA and double-stranded RNA induce a common antiviral signaling pathway in human cells. *Proc Natl Acad Sci USA* 2007;**104**:9035–40.
71. Harrod KS, Trapnell BC, Otake K, Korfhagen TR, Whitsett JA. SP-A enhances viral clearance and inhibits inflammation after pulmonary adenoviral infection. *Am J Physiol* 1999;**277**(3 Pt 1):L580–8.
72. Jiang H, Wang Z, Serra D, Frank MM, Amalfitano A. Recombinant adenovirus vectors activate the alternative complement pathway, leading to the binding of human complement protein C3 independent of anti-ad antibodies. *Mol Ther* 2004;**10**:1140–2.
73. Shayakhmetov DM, Gaggar A, Ni S, Li ZY, Lieber A. Adenovirus binding to blood factors results in liver cell infection and hepatotoxicity. *J Virol* 2005;**79**:7478–91.
74. Hensley SE, Cun AS, Giles-Davis W, Li Y, Xiang Z, Lasaro MO, et al. Type I interferon inhibits antibody responses induced by a chimpanzee adenovirus vector. *Mol Ther* 2007;**15**:393–403.
75. Hensley SE, Giles-Davis W, McCoy KC, Weninger W, Ertl HC. Dendritic cell maturation, but not CD8+ T cell induction, is dependent on type I IFN signaling during vaccination with adenovirus vectors. *J Immunol* 2005;**175**:6032–41.
76. Xiang ZQ, Yang Y, Wilson JM, Ertl HC. A replication-defective human adenovirus recombinant serves as a highly efficacious vaccine carrier. *Virology* May 1, 1996;**219**:220–7.
77. Lutze-Wallace C, Wandeler A, Prevec L, Sidhu M, Sapp T, Armstrong J. Characterization of a human adenovirus 5: rabies glycoprotein recombinant vaccine reisolated from orally vaccinated skunks. *Biologicals* 1995;**23**:271–7.
78. Xiang ZQ, Greenberg L, Ertl HC, Rupprecht CE. Protection of non-human primates against rabies with an adenovirus recombinant vaccine. *Virology* 2014;**450–451**:243–9.
79. Meltzer MI, Atkins CY, Santibanez S, Knust B, Petersen BW, Ervin ED, et al. Estimating the future number of cases in the Ebola epidemic – Liberia and Sierra Leone, 2014–2015. *MMWR Surveill Summ* 2014;**63**:1–14.
80. Ledgerwood JE, Costner P, Desai N, Holman L, Enama ME, Yamshchikov G, et al. VRC 205 Study Team. A replication defective recombinant Ad5 vaccine expressing Ebola virus GP is safe and immunogenic in healthy adults. *Vaccine* 2010;**29**:304–13.

81. Sullivan NJ, Geisbert TW, Geisbert JB, Shedlock DJ, Xu L, Lamoreaux L, et al. Immune protection of nonhuman primates against Ebola virus with single low-dose adenovirus vectors encoding modified GPs. *PLoS Med* 2006;**3**:e177.
82. Geisbert TW, Bailey M, Hensley L, Asiedu C, Geisbert J, Stanley D, et al. Recombinant adenovirus serotype 26 (Ad26) and Ad35 vaccine vectors bypass immunity to Ad5 and protect nonhuman primates against ebolavirus challenge. *J Virol* 2011;**85**:4222–33.
83. http://www.fiercevaccines.com/story/nih-ebola-vaccine-trial-will-begin-september/2014-08-01.
84. Hung PP, Morin JE, Lubeck MD, Barton JE, Molnar-Kimber KL, Mason BB, et al. Expression of HBV surface antigen or HIV envelope protein using recombinant adenovirus vectors. *Nat Immun Cell Growth Regul* 1988;**7**:135–43.
85. Khanam S, Pilankatta R, Khanna N, Swaminathan S. An adenovirus type 5 (AdV5) vector encoding an envelope domain III-based tetravalent antigen elicits immune responses against all four dengue viruses in the presence of prior AdV5 immunity. *Vaccine* 2009;**27**:6011–21.
86. Shim BS, Stadler K, Nguyen HH, Yun CH, Kim DW, Chang J, et al. Sublingual immunization with recombinant adenovirus encoding SARS-CoV spike protein induces systemic and mucosal immunity without redirection of the virus to the brain. *Virol J* 2012;**9**:215.
87. Zhou H, Guo L, Wang M, Qu J, Zhao Z, Wang J, et al. Prime immunization with rotavirus VLP 2/6 followed by boosting with an adenovirus expressing VP6 induces protective immunization against rotavirus in mice. *Virol J* 2011;**8**:3.
88. Sharma A, Wendland R, Sung B, Wu W, Grunwald T, Worgall S. Maternal immunization with chimpanzee adenovirus expressing RSV fusion protein protects against neonatal RSV pulmonary infection. *Vaccine* 2014;**32**:5761–8.
89. Gallichan WS, Rosenthal KL. Specific secretory immune responses in the female genital tract following intranasal immunization with a recombinant adenovirus expressing glycoprotein B of herpes simplex virus. *Vaccine* 1995;**13**:1589–95.
90. Li K, Li PY, Wu XA, Zhang L, Liu ZY, Yu L, et al. Induction of Hantaan virus-specific immune responses in C57BL/6 mice by immunization with a modified recombinant adenovirus containing the chimeric gene, GcS0.7. *Int J Mol Med* 2013;**32**:709–16.
91. Kim EH, Park HJ, Han GY, Song MK, Pereboev A, Hong JS, et al. Intranasal adenovirus-vectored vaccine for induction of long-lasting humoral immunity-mediated broad protection against influenza in mice. *J Virol* 2014;**88**:9693–703.
92. Teixeira LH, Tararam CA, Lasaro MO, Camacho AG, Ersching J, Leal MT, et al. Immunogenicity of a prime-boost vaccine containing the circumsporozoite proteins of *Plasmodium vivax* in rodents. *Infect Immun* 2014;**82**:793–807.
93. Xiang Z, Ertl HC. Induction of mucosal immunity with a replication-defective adenoviral recombinant. *Vaccine* 1999;**17**:2003–8.
94. Calcedo R, Vandenberghe LH, Roy S, Somanathan S, Wang L, Wilson JM. Host immune responses to chronic adenovirus infections in human and nonhuman primates. *J Virol* 2009;**83**:2623–31.
95. Prasad SA, Norbury CC, Chen W, Bennink JR, Yewdell JW. Cutting edge: recombinant adenoviruses induce CD8 T cell responses to an inserted protein whose expression is limited to nonimmune cells. *J Immunol* 2001;**166**:4809–12.
96. Miyahira Y, Takashima Y, Kobayashi S, Matsumoto Y, Takeuchi T, Ohyanagi-Hara M, et al. Immune responses against a single CD8+-T-cell epitope induced by virus vector vaccination can successfully control *Trypanosoma cruzi* infection. *Infect Immun* 2005;**73**:7356–65.

97. Santosuosso M, McCormick S, Zhang X, Zganiacz A, Xing Z. Intranasal boosting with an adenovirus-vectored vaccine markedly enhances protection by parenteral *Mycobacterium* bovis BCG immunization against pulmonary tuberculosis. *Infect Immun* 2006;**74**:4634–43.
98. Radosevic K, Wieland CW, Rodriguez A, Weverling GJ, Mintardjo R, Gillissen G, et al. Protective immune responses to a recombinant adenovirus type 35 tuberculosis vaccine in two mouse strains: CD4 and CD8 T-cell epitope mapping and role of gamma interferon. *Infect Immun* 2007;**75**:4105–15.
99. Gao G, Wang Q, Dai Z, Calcedo R, Sun X, Li G, et al. Adenovirus-based vaccines generate cytotoxic T lymphocytes to epitopes of NS1 from dengue virus that are present in all major serotypes. *Hum Gene Ther* 2008;**19**:927–36.
100. Zhong J, Khanna R. Ad-gBCMVpoly: a novel chimeric vaccine strategy for human cytomegalovirus-associated diseases. *J Clin Virol* 2009;**46**(Suppl 4):S68–72.
101. DiMenna L, Latimer B, Parzych E, Haut LH, Töpfer K, Abdulla S, et al. Augmentation of primary influenza A virus-specific CD8+ T cell responses in aged mice through blockade of an immunoinhibitory pathway. *J Immunol* 2010;**184**:5475–84.
102. Lambe T, Carey JB, Li Y, Spencer AJ, van Laarhoven A, Mullarkey CE, et al. Immunity against heterosubtypic influenza virus induced by adenovirus and MVA expressing nucleoprotein and matrix protein-1. *Sci Rep* 2013;**3**:1443.
103. Warimwe GM, Lorenzo G, Lopez-Gil E, Reyes-Sandoval A, Cottingham MG, Spencer AJ, et al. Immunogenicity and efficacy of a chimpanzee adenovirus-vectored Rift Valley fever vaccine in mice. *Virol J* 2013;**10**:349.
104. Leskowitz R, Fogg MH, Zhou XY, Kaur A, Silveira EL, Villinger F, et al. Adenovirus-based vaccines against rhesus lymphocryptovirus EBNA-1 induce expansion of specific CD8+ and CD4+ T cells in persistently infected rhesus macaques. *J Virol* 2014;**88**:4721–35.
105. Smaill F, Jeyanathan M, Smieja M, Medina MF, Thanthrige-Don N, Zganiacz A, et al. A human type 5 adenovirus-based tuberculosis vaccine induces robust T cell responses in humans despite preexisting anti-adenovirus immunity. *Sci Transl Med* 2013;**5**:205ra134.
106. Ogwang C, Afolabi M, Kimani D, Jagne YJ, Sheehy SH, Bliss CM, et al. Safety and immunogenicity of heterologous prime-boost immunisation with *Plasmodium falciparum* malaria candidate vaccines, ChAd63 ME-TRAP and MVA ME-TRAP, in healthy Gambian and Kenyan adults. *PLoS One* 2013;**8**:e57726.
107. Van Kampen KR, Shi Z, Gao P, Zhang J, Foster KW, Chen DT, et al. Safety and immunogenicity of adenovirus-vectored nasal and epicutaneous influenza vaccines in humans. *Vaccine* 2005;**23**:1029–36.
108. Barnes E, Folgori A, Capone S, Swadling L, Aston S, Kurioka A, et al. Novel adenovirus-based vaccines induce broad and sustained T cell responses to HCV in man. *Sci Transl Med* 2012;**4**:115ra1.
109. Nicholson O, Dicandilo F, Kublin J, Sun X, Quirk E, Miller M, et al. For the Merck V520-018/HIV Vaccine Trials Network 050 Study Team. Safety and immunogenicity of the MRKAd5 gag HIV type 1 vaccine in a Worldwide phase 1 study of healthy adults. *AIDS Res Hum Retroviruses* 2011;**27**:557–67.
110. Hayton EJ, Rose A, Ibrahimsa U, Del Sorbo M, Capone S, Crook A, et al. Safety and tolerability of conserved region vaccines vectored by plasmid DNA, simian adenovirus and modified vaccinia virus ankara administered to human immunodeficiency virus type 1-uninfected adults in a randomized, single-blind phase I trial. *PLoS One* 2014;**9**:e101591.
111. Harro CD, Robertson MN, Lally MA, O'Neill LD, Edupuganti S, Goepfert PA, et al. Safety and immunogenicity of adenovirus-vectored near-consensus HIV type 1 clade B gag vaccines in healthy adults. *AIDS Res Hum Retroviruses* 2009;**25**:103–14.

112. McElrath MJ, De Rosa SC, Moodie Z, Dubey S, Kierstead L, Janes H, et al. HIV-1 vaccine-induced immunity in the test-of-concept Step Study: a case-cohort analysis. *Lancet* 2008;**372**:1894–905.
113. Buchbinder SP, Mehrotra DV, Duerr A, Fitzgerald DW, Mogg R, Li D, et al. Efficacy assessment of a cell-mediated immunity HIV-1 vaccine (the Step Study): a double-blind, randomised, placebo-controlled, test-of-concept trial. *Lancet* 2008;**372**:1881–93.
114. Gray GE, Moodie Z, Metch B, Gilbert PB, Bekker LG, Churchyard G, et al. HVTN 503/Phambili study team. Recombinant adenovirus type 5 HIV gag/pol/nef vaccine in South Africa: unblinded, long-term follow-up of the phase 2b HVTN 503/Phambili study. *Lancet Infect Dis* 2014;**14**:388–96.
115. Koblin BA, Mayer KH, Noonan E, Wang CY, Marmor M, Sanchez J, et al. Sexual risk behaviors, circumcision status, and preexisting immunity to adenovirus type 5 among men who have sex with men participating in a randomized HIV-1 vaccine efficacy trial: step study. *J Acquir Immune Defic Syndr* 2012;**60**:405–13.
116. Duerr A, Huang Y, Buchbinder S, Coombs RW, Sanchez J, del Rio C, et al. Step/HVTN 504 Study Team. Extended follow-up confirms early vaccine-enhanced risk of HIV acquisition and demonstrates waning effect over time among participants in a randomized trial of recombinant adenovirus HIV vaccine (Step Study). *J Infect Dis* 2012;**206**:258–66.
117. Stephenson KE, Hural J, Buchbinder SP, Sinangil F, Barouch DH. Preexisting adenovirus seropositivity is not associated with increased HIV-1 acquisition in three HIV-1 vaccine efficacy trials. *J Infect Dis* 2012;**205**:1806–10.
118. http://www.niaid.nih.gov/topics/HIVAIDS/Research/vaccines/Pages/adenovirusPlatforms.aspx.
119. Hu H, Eller MA, Zafar S, Zhou Y, Gu M, Wei Z, et al. Preferential infection of human Ad5-specific CD4 T cells by HIV in Ad5 naturally exposed and recombinant Ad5-HIV vaccinated individuals. *Proc Natl Acad Sci USA* 2014;**111**:13439–44.
120. http://www.niaid.nih.gov/news/newsreleases/2013/Pages/HVTN505April2013.aspx.
121. Rerks-Ngarm S, Pitisuttithum P, Nitayaphan S, Kaewkungwal J, Chiu J, Paris R, et al. MOPH-TAVEG Investigators. Vaccination with ALVAC and AIDSVAX to prevent HIV-1 infection in Thailand. *N Engl J Med* 2009;**361**:2209–20.
122. Yates NL, Liao HX, Fong Y, Decamp A, Vandergrift NA, Williams WT, et al. Vaccine-induced env V1-V2 IgG3 correlates with lower HIV-1 infection risk and declines soon after vaccination. *Sci Transl Med* 2014;19;**6**(228):228ra39.
123. Koup RA, Roederer M, Lamoreaux L, Fischer J, Novik L, Nason MC, et al. Priming immunization with DNA augments immunogenicity of recombinant adenoviral vectors for both HIV-1 specific antibody and T-cell responses. *PLoS One* 2010;**5**:e9015.
124. Lasaro MO, Haut LH, Zhou X, Xiang Z, Zhou D, Li Y, et al. Vaccine-induced T cells provide partial protection against high-dose rectal SIVmac239 challenge of rhesus macaques. *Mol Ther* 2011;**19**:417–26.
125. Shiver JW, Fu TM, Chen L, Casimiro DR, Davies ME, Evans RK, et al. Replication-incompetent adenoviral vaccine vector elicits effective anti-immunodeficiency-virus immunity. *Nature* 2002;**415**:331–5.
126. Reynolds MR, Weiler AM, Piaskowski SM, Piatak Jr M, Robertson HT, Allison DB, et al. A trivalent recombinant Ad5 gag/pol/nef vaccine fails to protect rhesus macaques from infection or control virus replication after a limiting-dose heterologous SIV challenge. *Vaccine* 2012;**30**:4465–75.
127. Letvin NL, Rao SS, Montefiori DC, Seaman MS, Sun Y, Lim SY, et al. Immune and genetic correlates of vaccine protection against mucosal infection by SIV in monkeys. *Sci Transl Med* 2011;**3**:81ra36.

128. Liu J, O'Brien KL, Lynch DM, Simmons NL, La Porte A, Riggs AM, et al. Immune control of an SIV challenge by a T-cell-based vaccine in rhesus monkeys. *Nature* 2009;**457**:87–91.
129. Tamminga C, Sedegah M, Maiolatesi S, Fedders C, Reyes S, Reyes A, et al. Human adenovirus 5-vectored Plasmodium falciparum NMRC-M3V-Ad-PfCA vaccine encoding CSP and AMA1 is safe, well-tolerated and immunogenic but does not protect against controlled human malaria infection. *Hum Vaccin Immunother* 2013;**9**:2165–77.
130. Chuang I, Sedegah M, Cicatelli S, Spring M, Polhemus M, Tamminga C, et al. DNA prime/Adenovirus boost malaria vaccine encoding *P. falciparum* CSP and AMA1 induces sterile protection associated with cell-mediated immunity. *PLoS One* 2013;**8**:e55571.
131. Ewer KJ, O'Hara GA, Duncan CJ, Collins KA, Sheehy SH, Reyes-Sandoval A, et al. Protective CD8+ T-cell immunity to human malaria induced by chimpanzee adenovirus-MVA immunisation. *Nat Commun* 2013;**4**:2836.

# Adenoviral Vectors Vaccine: Capsid Incorporation of Antigen

22

*Anurag Sharma*[1], *Stefan Worgall*[1,2]
[1]Department of Pediatrics, Weill Cornell Medical College, New York, NY, USA; [2]Department of Genetic Medicine, Weill Cornell Medical College, New York, NY, USA

## 1. Introduction

Initially developed for gene therapy, recombinant adenovirus (Ad)-based vectors have been extensively investigated as vaccine platforms for a variety of pathogens and tumors.[1,2] Several inherent features of Ad vectors make them attractive as a vaccine delivery vehicle. This includes their low pathogenicity, the ability to infect a wide repertoire of cell types, large cloning capacities, well-established good manufacturing practices (cGMP) to generate high titers, the lack of genomic integration in host cells, strong immune-adjuvant properties, a well-understood biology, and a genome that is relatively easy to manipulate. In addition, there are an abundance of safety and efficacy data on the use of Ad vectors in humans based on hundreds of preclinical[1,3] and clinical trials (http://www.wiley.co.uk/genmed/clinical/). Because of these valuable attributes, it is not surprising that Ad vectors are at the forefront of viral vectors for vaccine development. Most conventional Ad vaccine candidates are transgene expression vectors that express a gene coding for an antigen protein of the vaccine target pathogen. This transgene is usually inserted in place of the early region 1 (E1) of the Ad genome. This strategy allows the expression of the vaccine antigen in host cells that then induces humoral and cellular immunity against the target pathogen. Insertion of the transgene also renders the Ad vector replication-defective because E1 is critical for Ad replication and thus further improves the safety profile of Ad vectors. However, despite the optimism with Ad vectors, performance of conventional Ad vector-based vaccines has so far remained below expectations in clinical trials.[4,5] One major limitation for the use of Ad vaccines is the high prevalence of anti-Ad immunity in the general population.[6,7] Most adults develop varying levels of immunity to a number of common Ad serotypes after exposure to wild-type Ads during childhood. Anti-Ad immunity is particularly prevalent against serotype 5, traditionally the most commonly used Ad vector for vaccine development. It is estimated that 30–90% of the population, depending on geographic location, has medium to high levels of Ad5 antibodies.[6,7] Infection of host cells, an essential step for expression of the transgene antigen, is hampered by neutralizing anti-Ad immunity. Even in naive recipients, a potent antivector cellular and humoral immunity is induced on first administration that compromises the efficiency of subsequent administrations of the same vector for boosting the immune response. Several strategies have been suggested to overcome this limitation.[8–10] Chimeric Ads with components of Ad capsid proteins, such as

**Adenoviral Vectors for Gene Therapy.**

fiber or hexon, switched from other, less prevalent serotypes have been developed. Novel Ad vectors have been generated from non-Ad5 serotypes such as Ad35 or Ad48 or from those derived from nonhuman species such as chimpanzee,[11] bovine,[12] or canine.[13] However, repeat administration will not be facilitated unless varying serotypes are used for boosting.

Native Ad capsid proteins are highly immunogenic and induce potent innate and adaptive immune responses.[14,15] This has provided the basis for various interesting strategies to display the epitopes of pathogens on Ad capsids to induce robust pathogen-specific immunity. This approach presents numerous possible advantages over the approach of conventional transgene expression. The incorporation of immunogenic epitopes into Ad capsid may circumvent preexisting anti-Ad immunity.[16–18] This also allows for these capsid-modified vectors to be administered repeatedly to boost epitope-specific immunity. Adenoviruses efficiently infect antigen-presenting cells; once internalized, the capsid proteins are processed through the exogenous pathway, resulting in presentation via major histocompatibility complex II molecules to $CD4^+$ T cells. These $CD4^+$ T cells in turn stimulate the differentiation of B cells into antibody-producing plasma cells.[19] Furthermore, the densely packed and highly repetitive structure of Ad capsid can efficiently activate B cells in the absence of T cell stimulation.[20] In particular, the capsid-incorporation strategy appears to be promising against infections that require the strong induction of humoral immunity.

Tremendous flexibility offered by Ad to accommodate heterologous peptides into its capsid, together with a detailed understanding of the capsid structure made this approach feasible. The Ad icosahedral capsid shell contains three major (hexon, penton base, and fiber) and four minor (IIIa, VI, VIII, and IX) proteins.[21] The principal component is the homotrimeric hexon, and each virion contains 240 hexons on the faces and edges of the capsid and 12 penton base pentamers (penton base) at the vertices of icosahedrons, each of which is bound with one fiber trimer (fiber). Hexons and pentons are joined together by three minor proteins: IX, IIIa, and VIII. Minor protein IX (pIX) (240 copies) acts as a capsid cement to bind hexons together are located at the outer surface of the capsid. The Ad capsid shell is formed by two kinds of building blocks: group-of-nine (GON) hexons and group of six capsomers (a penton base and its five surrounding hexons).[21]

Numerous locations within the various Ad capsid proteins have been identified to accommodate foreign epitopes without affecting Ad vector viability.[22] The outer capsid proteins of Ad, which have been targets for genetic modification, include the three major capsid proteins hexon, fiber knob, and penton base, as well as pIX, a minor capsid protein (Figure 1(A)). Each has distinct characteristics and can be readily engineered to present heterologous peptides (Table 1).

## 2. Fiber

Adenovirus fiber protrudes from the 12 vertices of the virion and has a central role in host cell attachment and entry.[21] The fiber consists of a conserved N-terminal tail that noncovalently binds to the penton base, a central rod-like shaft with repeating

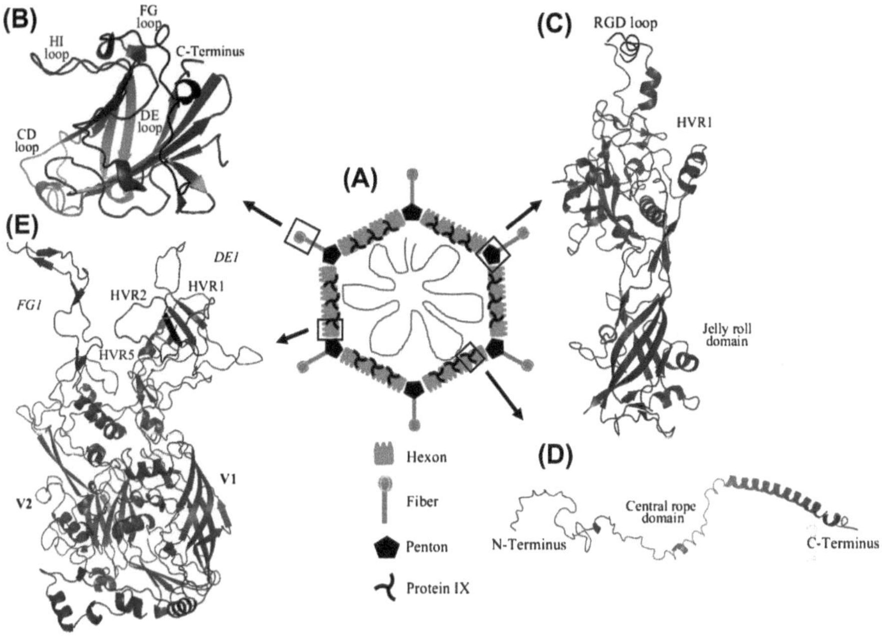


**Figure 1** Antigen incorporation sites within adenovirus capsid. (A) Model of adenovirus structure depicting capsid proteins that can be modified to display antigenic epitopes: fiber, penton base, protein IX, and hexon. (B E) Ribbon models of adenovirus capsid protein monomers illustrating epitope insertion sites. (B) Insertion sites in fiber knob monomer: loops HI (blue (gray in print versions)), FG (red (very dark gray in print versions)), CD (green (light gray in print versions)), DE (brown (dark gray in print versions)), and C-terminus (black); (PDB 1KNB). (C) Insertion site in penton base: RGD loop (red (very dark gray in print versions)); (PDB 1X9P). (D) Insertion site in protein IX: C-terminus (red (very dark gray in print versions)); (PDB 3IYN). (E) Insertion sites in hexon monomer: HVR1 (green (gray in print versions)), HVR2 (red (very dark gray in print versions)), and HVR5 (blue (dark gray in print versions)); (PDB 3IYN).

motifs and a C-terminal globular knob that is the ligand for direct interaction with the Ad receptor.[23] The length of the fiber is variable and depends on the number of repeats (specific to the particular serotype) of 15–20 aa connected by a β-turn in the fiber shaft. Adenovirus 5 fiber contains 582 aa and 22 repeats in its shaft, and is 35–40 nm long. The trimerization of fiber is controlled by sequences in both the knob and shaft regions. The terminal knob has the topology of an eight-stranded antiparallel β-sandwich with interspersing loop regions. These loop regions vary from 8 to 55 aa residues and are generally designated as AB, CD, DG, GH, HI, and IJ loops[127,128]. Some of these loops are surface exposed and flexible and allow the insertion of foreign peptides[17] (Figure 1(B)). The insertion of foreign ligands into the fiber knob to ablate or alter the natural tropism of the Ad vector to reduce vector-associated toxicity or specifically target the transduction of particular cell types has been well described. The HI loop of the fiber knob has been the most common insertion site.[16,24,25] The incorporation of the FLAG octapeptide into the

**Table 1** **Adenoviral Capsid Proteins for Incorporation of Antigenic Epitopes to Induce Epitope-Specific Immune Response**

| Protein | Structure | Copy Number per Virion | Location on Virion | Function | Locations for Antigen Incorporation | Advantages for Capsid Incorporation | Limitations in Capsid Incorporation |
|---|---|---|---|---|---|---|---|
| Fiber | Trimer | 12×3=36 | Protrude from capsid at each vertex | Fiber knob domain interacts with primary receptor for virus attachment | Loops: CD, DE, FG, HI, C-terminus | Efficient Ag presentation to immune system; potent boosting on re-administration | Affect virus transduction; can affect virus stability; usually incorporate small peptides only; low copy number per virion |
| Hexon | Trimer | 240×3=720 | On the 20 facets of Ad capsid | Major capsid building block; harbors major virus neutralization sites | HVR1, 2, 5 | High copy number per virion; Ag incorporation can mask neutralizing epitopes; reduce hepatotropism | Only modest boost on vector re-administration |
| Penton base | Pentamer | 12×5=60 | At 12 vertices of Ad icosahedron | Binds to cellular integrins and mediates cellular internalization | RGD loop; PPxY motif | Pentons from some serotypes self-assemble to form highly stable and immunogenic Dd VLPs | Disrupt RGD motif and potential infectivity to dendritic cells; low copy number per virion |
| PIX | Trimer | 80×3=240 | On outer capsid; forms network between hexons | Cement hexons together and stabilizes virion | C-terminus | Readily incorporate large full-length functional proteins | Poor accessibility because of steric hindrance from fiber |

HI loop did not ablate fiber trimerization or disturb receptor-binding site localized in the knob and was accessible to anti-FLAG antibodies, indicating surface exposure.[26] Subsequently, CD, IJ, and C terminus have also been shown to accommodate RGD peptide successfully to enhance vector interaction with cellular αv integrins.[27] In principle, the same locations can also display epitopes to introduce pathogen-derived antigens into a host to generate immune responses to the respective antigen. Our group demonstrated that an epitope from influenza hemagglutinin (HA) displayed on the HI loop induced high levels of HA-specific immunity.[25] In fact, this study compared the immune response generated by incorporating the HA epitope into different Ad capsid proteins (hexon, fiber, penton base, and pIX) and the fiber-modified Ads induced the strongest anti-HA humoral and CD4 cellular immunity. This effect was independent of administering the vectors at equal Ad particle numbers or equal HA epitope doses. These results were surprising considering that fiber is the least abundant of the capsid proteins. We subsequently explored additional epitope-insertion sites within the Ad fiber knob to elicit epitope-specific immunity.[17] Adenovirus vectors expressing the 14-mer *Pseudomonas aeruginosa* immune-dominant outer membrane protein F (OprF) epitope 8 (Epi8) in five distinct sites of the Ad5 fiber (loops CD, DE, FG, HI, and C-terminus) were compared in mouse model of *Pseudomonas* infection.[17] Insertions in FG or HI loops induced strongest humoral, cellular, and protective immunity against *Pseudomonas* infections. Repeat administration of the fiber-modified Ad vectors effectively boosted the antigen-specific humoral immune response to levels that exceeded those generated by immunization with an Ad vector expressing the full-length OprF transgene. These fiber-modified vectors also performed well even in the presence of anti-Ad5 immunity.[17] Similar observations were also reported by Lanzi et al. using ovalbumin-derived epitopes inserted into Ad hexon or fiber proteins.[16] After initial inoculation, anti-ovalbumin antibody titers were higher with hexon-insertion of the epitope. However, after re-administration, fiber-modified Ad showed stronger amplification of epitope-specific immunity. The ovalbumin epitope was more accessible on the naturally protruding fiber protein compared with insertion into hexon. This study also demonstrated that anti-Ad antibodies strongly shape the anti-epitope antibody responses. Interestingly, mice with serum anti-Ad antibodies injected with fiber-modified Ad carrying ovalbumin epitope had higher anti-ovalbumin titer compared with naive mice injected with the same vector. It is likely that the binding of anti-Ad antibodies may facilitate vector uptake by antigen-presenting cells through the Fcγ receptor resulting in subsequent induction of $CD4^+$ T cells to provide help for ovalbumin epitope-specific B cells.[28]

Overall, Ad fiber knob presents an excellent choice for incorporating heterologous epitopes for vaccine applications, particularly because of the excellent presentation of antigens to the host immune system and exceptional boosting potential. However, because foreign insertions can potentially affect the trimerization and stability of fiber protein or alter Ad infectivity, caution is advised when considering the location on the fiber knob or the nature of epitope. Considering the use of incorporation of foreign peptides in fiber knob, the optimal locales in fiber knob region of additional Ad serotypes have been investigated.[29,30]

## 3. Penton Base

The penton capsomere is composed of a homotrimeric fiber and homopentameric penton base at the 12 vertices of the icosahedron.[21] Together with fiber, penton base has a major role in Ad cellular internalization.[31,32] After initial attachment of the virus by binding the distal C-terminal knob domain of the fiber to primary cellular receptors (currently known ones are CAR, CD46, CD86, sialic acid, and desmoglein 2 for various Ad serotypes), cell entry is gained by the interaction of penton base with secondary receptors followed by endocytosis. For most serotypes, RGD loops extending from penton base bind to cellular αvβ3 or αvβ5 integrins.[32]

Adenovirus 2 and 5 penton bases are composed of 571 aa and consist of two domains: the lower one is a typical jelly roll of two four-stranded antiparallel β-sheets forming a β-barrel and the upper one has irregular folds formed by two insertions arising from the lower jelly roll strands[33] (Figure 1(C)). The first insertion includes two external loops: The first is the highly flexible integrin-binding RGD domain (RGD loop) that varies among serotypes in sequence and length. The other hypervariable region 1 (HVR1) loop is shorter and projects from the outer surface of the penton base. The insertion of an HA epitope in the RGD loop of Ad5 penton was well tolerated and immunogenic.[25] However, the vector elicited only modest levels of epitope-specific immunity compared with fiber- or hexon-incorporated epitope. It is likely that insertional disruption of the RGD motif impaired the vector's capacity to transduce the host cells, particularly antigen-presenting cells. Furthermore, the location of the epitope at the base of the fiber might pose some steric hindrance in its immune presentation. Additional studies on the structural and immunogenicity aspects of penton-modified Ad vectors are required to assess the value of this site for epitope insertion.

During its life cycle, Ad3 synthesizes excess amounts of free pentons that spontaneously assemble into dodecahedral virus-like nanoparticles containing 12 pentons.[34] Beside this natural expression, the Ad dodecahedron (Dd) can be expressed in the heterologous baculovirus system. Dodecahedron particles composed solely of 12 penton base pentamers and without fibers can also be generated. Interestingly, these Dd virus-like particles (VLPs) have remarkably efficient intracellular penetration and can be engineered to deliver several million foreign cargo molecules to a single target cell.[35] Dodecahedrons are efficiently taken up by human dendritic cells and induce their maturation.[36] The crystal structure of Ad3 penton base Dd revealed that the N-terminal strand-swapping between neighboring penton base molecules imparts remarkable stability to its structure.[37] Other advantages of these VLPs as potential vaccine platforms are that they do not carry genetic information, can be easily produced on a large scale, seem safe, and induce potent B and T cell responses. The fact that two VLP-based vaccines against human papillomavirus and hepatitis B has already been approved for human use has heightened interest in Ad3 penton Dd as a vaccine platform.[38] The distal N-terminus residues of penton base in Dd are externally exposed and available for attaching cargo.[37] To attach a protein to the vector, an adaptor containing WW domains derived from Nedd4 is used.[35,39] The WW domain, a fragment of 23–35 amino acids flanked by two tryptophans (W), is a partner of the proline-proline-x-tyrosine (PPxY) motif present at the N-terminal extremity of the penton base protein. A tandem of three WW domains, when cloned in fusion with the protein of interest, acts as an adaptor attaching this protein to Dd without impairing its endocytosis efficiency. A penton Dd-based vaccine complexed with influenza M1 epitopes and generated through the baculovirus expression system[40] was efficiently internalized, processed, and presented via HLA class II and cross-presented via class I molecules triggering $CD4^+$ and $CD8^+$ T cell responses. Upon vaccination of chicken with Dd carrying M1 epitopes, both cellular and humoral immune responses were elicited in the absence of an adjuvant.[40] Similarly, a soluble complex of influenza HA with Dd has been engineered.[41] Its immunogenicity remains to be investigated. Besides targeting infectious diseases, in vivo delivery of antigens by Ad Dd can elicit antitumor immunity.[42] Adenovirus Dd also improved intracellular delivery of an oncogene inhibitor (mRNA cap analog) to hepatocellular carcinoma and resulted in the significant inhibition of tumor growth.[43]

## 4. Protein IX

Protein IX is a 14.3-kDa minor structural protein of 140 residues on the exterior surface of the Ad capsid that enhances its structural integrity by stabilizing hexon–hexon interactions[44] (Figure 1(D)). Protein IX is dispensable for virus assembly because pIX-deletion mutants assembled to form viral particles but had lower thermostability.[45] Protein IX forms a network that lines the boundaries between hexons and cements the GON hexons together on each facet. Each pIX monomer has an N-terminal domain, a central rope domain, and a conserved C-terminal coiled coil (CC). The N-terminal domains of three pIX monomers form a triskelion with four triskelions per 20 facets (240 molecules per virion). In Ad5, four CC domains from different triskelions associate into a unique four-helix bundle. Three C-terminal domains associate together in parallel structure, whereas the fourth domain originates from a triskelion of the neighboring facet, and associates to the timer in an antiparallel manner.[46–49] The structural model of Ad cement proteins is currently being debated.[50–52] Protein IX-to-pIX interaction is not required for inclusion in the capsid or thermostability of the particles and N-terminal domain of pIX is enough to confer capsid thermostability.[53] In addition to providing stability to the Ad particle, pIX acts as transcriptional activator of Ad genes, reorganizes host cell nuclear domains,[49] modulates viral tropism, and inhibits antiviral immune responses.[54]

Immunoaccessibility studies and cryo-electron microscopy have demonstrated that the N-terminus lies hidden between the hexon capsomers whereas the C-terminal domain is exposed on the outer surface and accessible to immunoglobulins.[46,55] The remarkable characteristic of the C-terminus of pIX to tolerate the insertion of small or large ligands has made it a popular platform for Ad vector modification for retargeting, imaging, or immunization applications (Figure 1(D)). The addition of lysine octapeptide to C-terminus of pIX redirected the Ad virus infectivity to CAR-deficient cell lines containing heparin sulfate.[56] Similarly, incorporation of RGD motif to pIX enhanced Ad binding to cells expressing $\alpha v\beta$ integrins.[57] Biotinylated Ad vectors can be generated through insertion of biotin acceptor peptide to pIX that can be conjugated to a wide variety of avidin-tagged targeting ligands.[58] Alternatively, targeting ligands can be coupled to pIX via the genetic inclusion of cysteine residues and subsequent chemical coupling of ligands to the reactive thiol groups.[59] Protein IX is an efficient platform to present larger targeting ligands such as single-chain[60] or single-domain[61] antibodies, single-chain T-cell receptor targeted against MAGE-A1 antigen[62] that have high antigen-binding affinity, and exquisite specificity to redirect Ad infectivity.

Protein IX has also been used to anchor imaging ligands on the Ad surface, providing a valuable tool for virus tracking in vitro and in vivo. Fusion of pIX with green fluorescent protein was efficiently incorporated into Ad5 capsids, showing normal growth characteristics.[63,64] Protein IX of canine[65] and bovine[66] Ads have been efficiently tagged with fluorescent proteins to generate viable viral particles. Adenovirus pIX has been labeled with functional HSV thymidine kinase,[67,68] luciferase,[68] and metallothione[69] proteins. The versatility of pIX was demonstrated in a proof-of-principle study to generate a triple-mosaic Ad that simultaneously displayed a FLAG tag, a hexahistidine tag, and a monomeric red fluorescent protein, thus combining targeting, therapeutic, and imaging modalities on a single virion.[70]

Few studies have demonstrated that pathogen-specific antigens fused to pIX can stimulate robust protective immune responses in animals. Because of its capacity to accept large proteins, pIX provides an interesting option for the incorporation of whole antigenic protein, thus presenting multiple epitopes to host immune system for the induction of a broader and more diverse immune response. Furthermore, pIX may present the proteins in their natural trimeric form, resulting in the induction of conformation-dependent antibodies. Boyer et al. developed vaccines against pneumonic plague by incorporating *Yersinia pestis* entire V antigen or the F1 capsular antigen to pIX C-terminus.[71] These Ad vector vaccines allowed boosting by repeat administration and stimulated more robust protective immune responses compared with equivalent recombinant protein-based subunit vaccines administered with conventional adjuvant. In an innovative combination approach, Bayer et al. developed an Ad vector that expressed as transgene and displayed entire Friend virus (FV) antigen on pIX of Ad capsid.[72] This vaccine elicited increased neutralizing antibody titers, strong $CD4^+$ T cell responses, and protection against systemic FV challenge.

Despite the paucity of studies investigating incorporation of heterologous proteins on pIX for immunizations, pIX is an attractive location particularly because of its ability to display large, full-length, conformational proteins. However, pIX incorporation might not be as effective as fiber knob incorporation, in part because of its relatively low accessibility by virtue of steric hindrance from fibers.[25] It has been suggested that addition of α-helical spacers to raise the antigenic protein above the surface of the virion might improve its presentation to the immune cells.[57] This has been disputed by others.[72,73] The C-terminal fusion of exogenous proteins disorganizes the 4-helix bundle; however, this does not preclude the incorporation of pIX to the Ad capsid or affect thermostability.[53] In contrast to the fiber, for which trimerization is essential for its association with the penton base, it may not be essential for pIX to retain its capacity to form trimers.

## 5. Hexon

Hexon is the largest (>900 residues) and most abundant protein of the Ad capsid, with 720 copies per virion. The hexons exist as homotrimeric capsomeres distributed symmetrically with 12 copies forming each of the 20 capsid facets.[21] The hexons are classified as H1, H2, H3, and H4, according to their location within each facet of the icosahedral capsid.[74] Each hexon trimer has a pseudohexonal base and three tower domains that are presented to the exterior of capsid. The base of each hexon monomer contains two eight-stranded jelly roll domains—V1 and V2—stabilized by an internal loop[75] (Figure 1(E)). Three long loops (DG1, FG1, and FG2) extend from the base structure to form the tower region. These loops from adjacent monomers interact, providing stability to the capsomere. Analysis of the protein sequences of different hexons has revealed high variability restricted to the nine (reassigned from previous seven) HVRs confined to the hexon towers.[76,77] The DE1 loop is the largest and most variable and flexible region, which contains HVR 1–6, whereas the FG1 loop contains HVR 7–9.[77]

Most Ad5-specific neutralizing antibodies are directed against hexon, although neutralizing antibodies against fiber have also been reported.[78–83] Efforts have been made to replace the Ad5 hexon with alternate serotypes to circumvent the problems raised by preexisting immunity. Switching the entire hexon is complicated because hexon proteins interact extensively with other capsid proteins that impart stability to the Ad structure.[83,84] Instead, it is more feasible to exchange the exposed HVRs against which the serotype-specific immune responses are directed.[82] Although partial swap of HVRs has been investigated to escape anti-Ad5 preexisting immunity,[18,85] it may be necessary to replace all HVRs to achieve complete evasion of vector-specific neutralizing antibodies.[79] The Ad5-based hexon-modified vectors with all nine HVRs replaced with those from Ad43 or Ad48 were not affected by Ad5-specific neutralizing antibodies and elicited robust transgene-specific immunity in murine and nonhuman primate vaccine models.[86,87] Chimeric hexon-modified Ad5 vector with HVRs from Ad48 expressing simian immunodeficiency virus/human immunodeficiency virus (HIV) antigens evaded most preexisting Ad5 immunity and induced protective immunity in preclinical studies in mice and rhesus monkeys.[86,88] A similar vector encoding HIV-1 EnvA protein was safe, well-tolerated, and immunogenic in a first assessment of hexon chimeric Ad in humans.[89]

Based on the poor conservation of HVRs and their lack of involvement in maintaining structural integrity of hexons, it was hypothesized that HVRs could tolerate incorporation of heterologous peptides without impairing virus viability. Hexon was the first capsid protein that was genetically modified to display vaccine epitopes.[90] An eight–amino acid epitope from the VP1 capsid protein of poliovirus type 3 was incorporated into the exposed loops of Ad2 hexon and elicited antibodies against VP1.[90] Various HVRs were investigated to incorporate $His_6$ peptides into the hexon. The modified viruses were viable and thermostable and retained the infectivity and growth rate of an unmodified virus. $His_6$ epitopes in HVR2 and HVR5 bound to the anti-His tag antibody, suggesting that the epitopes were exposed to the surface when inserted at these locations.[91] The capacities and flexibility of Ad5 HVR2 and HVR5 were compared by genetically incorporating identical epitopes of increasing size (33–83 amino acids). Hypervariable region 5 was more permissive, allowing incorporation of 65–amino acid–long peptide, whereas the maximum length accommodated in HVR2 was 45 amino acids.[92]

Our group applied capsid-incorporation strategies to develop a vaccine against *P. aeruginosa*.[17,93–95] An Ad5 vector was modified to include an immunodominant 14–amino acid epitope (Epi8) of *P. aeruginosa* outer membrane protein F (OprF) at HVR5 of the hexon (AdZ.Epi8).[93] Immunization of mice with AdZ.Epi8 induced robust Epi8-specific humoral and IFN-gamma–positive CD4 and CD8 T cell responses and resulted in protection against a lethal pulmonary challenge with *P. aeruginosa*. Importantly, repeated administration of hexon-modified vector resulted in boosting of the anti-OprF immune responses.[93] Another multimodality vaccine candidate targeting *P. aeruginosa* (AdOprF.RGD.Epi8) expressed OprF as a transgene, displayed Epi8 epitope at HVR5 of hexon, and carried integrin-binding RGD sequence at fiber HI loop.[94] The RGD motif enhances the transduction of dendritic cells and increases the transgene-specific immune response.[96] Immunization of mice with AdOprF.RGD. Epi8 induced humoral immunity comparable to AdOprF (non-capsid-modified vector expressing OprF) but increased OprF-specific IFN-gamma–positive CD4 and CD8 T cell responses as well as improved protection against *P. aeruginosa* challenge. Moreover, in contrast to AdOprF, repeat administration of AdOprF.RGD.Epi8 resulted in boosting of the humoral anti-OprF response as well as increased protection. Because humoral, mucosal, or systemic opsonizing immunity is most effective to prevent *P. aeruginosa* colonization and infection, capsid antigen-incorporation is a promising strategy to develop anti-*P. aeruginosa* vaccines.

A B-cell epitope from *Bacillus anthracis* protective antigen incorporated to HVR5 of Ad5 hexon elicited IgG1 and IgG2a antibodies in mice that failed to protect mice against challenge with anthrax lethal toxin.[97] The inability of vaccine to protect might be because of the selection of epitope or the requirement of high titers to neutralize secreted bacterial toxin.

In recent years, capsid-modified Ad-based platforms have been investigated for malaria vaccine.[98–100] A malaria vaccine was generated by inserting a B cell epitope derived from a *Plasmodium yoelii* circumsporozoite (CS) protein (PyCS-B epitope) into the capsid proteins (hexon HVR1, HVR5, or fiber HI loop) of Ad5 vector that also expresses CS protein and GFP as its transgene.[98] Repeated immunizations of mice with the capsid-modified Ad induced a substantially increased level of protection against *P. yoelii* that correlated with increased anti-PyCS-B antibodies. Hypervariable region 1 modification circumvented vector neutralization by preexisting Ad-specific antibodies and maintained immunogenicity of PyCS-B epitope in the HVR1 and CS transgene. Most of the capsid-modified Ad vaccines are replication-defective because of safety concerns. The exemplary safety and efficacy of orally given lyophilized live vaccine to protect United States military personals from serious respiratory disease caused by Ad4 and Ad7 could encourage the exploration of vaccines based on replicating Ad vectors. An orally delivered, affordable, replicating Ad-based malaria vaccine would be tremendously beneficial particularly for the developing world where malaria immunization presents its greatest challenges. A replication-competent Ad displaying B-cell epitopes from *Plasmodium falciparum* CS protein in HVR1 of hexon induced high anti-CS antibodies in mice that recognized and neutralized *P. falciparum* sporozoites in vitro.[99] However, because human Ads do not replicate in mice, their safety and efficacy in humans cannot be reliably evaluated in mice. An hr404 mutation in Ad DNA binding protein gene allows Ad5 to replicate in monkey cells and macaques.[101,102] *Aotus nancymaae* monkeys intratracheally immunized with similar capsid-modified replicating Ad displaying *P. falciparum* CS protein and harboring mutation to allow replication in monkeys elicited sporozoite-specific antibodies.[100] Although enteric immunization elicited only low immunoglobulin (Ig) M and no IgG, it efficiently primed the immune response to subsequent intratracheal inoculation. The authors speculated that the low viral dose administered in small capsules, failure of the hand-coated capsules to survive through stomach, or the difference in susceptibility of Ad5 infection of intestinal cells of *Aotus* and humans could be potential explanations for suboptimal efficacy for enteric immunization. The enteric route for such vaccines needs to be reinvestigated. A capsid-modified Ad vaccine was developed against *Trypanosoma cruzi*, the parasite causing Chagas disease. An epitope from the gp83 protein, a ligand used by *T. cruzi* to attach to host cells, was incorporated at HVR1 of Ad5 that induced neutralizing immunity and protection in mice.[103]

Adenovirus 3, member of species B Ad, has been developed as a gene delivery vector.[104] Adenovirus 3 has a distinct tropism and is considered safe compared with Ad5.[105] Foreign peptides were incorporated at various HVRs of Ad3 hexon to elicit effective immune responses.[85,106,107] Enterovirus 71 (EV71) causes hand, foot, and mouth disease responsible for high mortality in children. An Ad3-based capsid-modified EV71 vaccine incorporating a neutralizing epitope SP70 into HVR1, HVR2, or HVR7 induced sufficient immunity to protect mice against a lethal EV71 challenge.[107] This vaccine was subsequently modified by incorporation of two EV71 neutralizing epitopes in HVR1 and HVR2 without affecting stability or growth characteristics of the vectors.[85] The capsid-incorporation strategy was used to develop Ad7-neutralizing monoclonal antibodies by replacing HVR5 of Ad3 with HVR5 or Ad7.[108] Besides the development of a vaccine, this may be a useful strategy for development of therapeutic neutralizing monoclonal antibodies and may aid in the structural analysis of Ads.

Most humans lack antibodies to chimpanzee-derived replication-defective Ad (AdC68); therefore, AdC68-based vectors can surpass preexisting anti-Ad5 immunity. As determined by X-ray crystallography, AdC68 contains five variable regions (VR1–5) at the top of the molecule; VR1 is the primary target of AdC68-neutralizing antibodies. Influenza A virus vaccines based on AdC68, modified to express a linear B cell epitope of the ectodomain of matrix (M2e) incorporated in VR1 or VR4 regions of hexon were generated.[109] Variable region 1–modified vectors provided significant protection against influenza A challenge, which could be further improved with simultaneous expression of M2e-NP transgene.

A safe and effective vaccine against HIV has eluded researchers for decades. Adenovirus vectors have been extensively investigated for HIV vaccines; however, clinical trials with Ad5-based vectors have been disappointing.[110] Alternate Ad serotypes or heterologous prime-boost strategies have been suggested to improve efficacy.[111,112] Capsid-modified Ad vectors present a promising platform for development of HIV vaccines.[18,72,88,89,113–116] A multivalent vaccine incorporated the 24–amino acid region of the HIV membrane proximal ectodomain region (MPER) derived from HIV gp41 at hexon HVR2 combined with expression of HIV gag as transgene.[113] Robust anti-HIV cellular and humoral responses were elicited in immunized mice. The MPER epitope was surface-exposed and elicited epitope-specific humoral response that could be boosted. In a follow-up cryo-electron microscopy structural study, the structure, flexibility, and accessibility of MPER epitope in context with Ad capsid and its effect on host immune response suggested potential ways to optimize epitope presentation on Ad capsids.[114] Gu et al. generated proof-of-concept multivalent Ad5 vectors displaying HIV glycoprotein 41 epitope at HVR1 and $His_6$ at HVR2 or HVR5. These vectors successfully elicited HIV and $His_6$ epitope-specific humoral immune response.[116] The V3 loop of HIV-1 glycoprotein 120 is an important target for broadly neutralizing antibodies and vaccine development. Capsid-incorporation of V3 at HVR1 presented V3 similar to native conformation and successfully induced HIV-1–specific humoral immune responses.[115]

A novel vaccine platform for addictive drugs has been developed by covalently linking cocaine or nicotine analogs to capsid proteins of noninfectious disrupted Ad vector. The disrupted Ad capsid proteins retain the immunologic adjuvant properties of an infectious Ad and induced long-lasting high titers to cocaine in mice, rats, or nonhuman primates.[117–120] Administration of this vaccine to mice or rats inhibited cocaine-induced hyperlocomotor activity and blocked cocaine-seeking behavior. In nonhuman primates, a significant reduction in cocaine occupancy of dopamine transporter in the brain was observed.[121] Although the vaccine effectively blocked the systemically administered cocaine from reaching the brain by sequestration in blood, no cocaine-mediated toxicity was observed in the peripheral organs of nonhuman primates.[118] In a similar approach, disrupted Ad5 coupled with nicotine analog induced high anti-nicotine antibody titers that inhibited nicotine-induced behavior.[122] In fact, purified hexon protein alone conjugated with nicotine analog was sufficient to evoke high anti-nicotine titers.[123] This is not surprising because hexon is the most immunogenic component of Ad capsid.[14,15] Interestingly, the disrupted Ad5 platform maintains immunopotency even in the presence of preexisting anti-Ad5 immunity.[122]

In summary, hexon is the most extensively used protein for capsid-incorporation strategies, primarily because of its high copy numbers within the virion and its natural role in eliciting anti-Ad immunity. Adenovirus 5 HVRs also interact with coagulation factors such as Factor X, which is responsible for hepatocyte transduction, and with scavenger receptors on Kupffer cells.[24,124–126] Modification of hexon HVRs can potentially suppress such interactions and may improve the bioavailability of vector to the host. In addition, the placement of epitopes on hexon can mask natural hexon-neutralizing epitopes that might otherwise curtail vector efficacy.[18,85] Comparative studies have indicated that hexon modification interferes minimally with virus entry to the cells in vitro compared with fiber, penton base, or pIX modification.[25] However, fiber incorporation of epitope was superior to hexon modification in eliciting epitope-specific immunity, particularly after repeat vector administrations.[16,17,25] Better presentation of fiber-incorporated epitopes to the immune cells could be responsible for this observation. The nature of epitope and site of incorporation could also affect the differential induction of epitope-specific immune responses. It is also suggested that anti-Ad antibodies can reduce the accessibility of the epitope when inserted into hexon protein and not fiber.[16] Additional conformational studies are required to explain this discrepancy and to further improve the hexon-incorporation approach.

## 6. Conclusion

Incorporation of antigenic epitopes on distinct Ad capsid proteins emerged as a promising vaccine platform. Antigen-displaying Ads combine the benefits of epitope-based vaccines and VLPs while retaining the immunogenicity and adjuvant effects of Ad vectors. In particular, the ability of capsid-incorporated vectors to bypass vector immunity, boost immune response on repeat administration, and induce exceptionally high humoral immunity are the most attractive features of capsid-modified Ad vectors. Certain other limitations such as Ad-mediated hepatotoxicity and vector uptake by liver Kupffer cells can also be addressed. The tremendous flexibility of the Ad platform and the ease of introducing elegant genetic modifications to Ad capsid together with knowledge of intricate Ad structure had an important role in realizing this strategy. To date, numerous preclinical studies targeting relevant infectious diseases and agents such as HIV, malaria, *P. aeruginosa*, *B. anthracis*, *T. cruzi*, enterovirus 71, poliovirus, and Ad7 have investigated Ad capsid incorporation strategy with encouraging results. Further studies need to address issues relevant for use in humans, particularly viability, stability, and growth to high titers in cGMP-compliant fashion.

Different loci within Ad fiber, hexon, penton base, and pIX proteins have been used to present heterologous epitopes to the host immune system. Investigation of additional sites within these proteins or in additional capsid proteins of different Ad serotypes will further broaden the repertoire of locales and provide extra flexibility to improve epitope-specific immunity.

It is critical to exercise caution when selecting peptides and the location for incorporation. The same peptides might vary in configuration and may not perform identically in distinct capsid environments.[114] The structure, size, and electric charge of peptides may affect virion stability and viability. Introduction of certain mutations can potentially improve the growth rate of capsid-modified Ad vectors.[87] Similarly, the use of appropriate spacers or adapters may improve presentation and immunological recognition of incorporated peptides.[57,91,114,115] A major hindrance in the rational design of capsid-modified vectors is an inadequate understanding of the principles that regulate the relationship between epitope presentation and the induction of host immunity. It is expected that future studies will include structural analyses to provide precise details of epitope structure and conformational presentation in context with Ad capsid.

One of the most cited constraints to capsid-incorporation strategy is that the size restriction of incorporated peptides can limit the breadth of the immune response. However, most peptide epitopes are 6–14 amino acids in length and most capsid locations may be able to accommodate larger peptides with multiple epitopes, thus generating polyvalent vaccines. Adenovirus pIX has the exceptional capacity to accommodate full-length large conformational antigenic proteins. More than one location can be modified to insert multiple peptides concomitantly to further expand the epitope count. Furthermore, repeat administration can boost capsid-incorporated epitope-specific antibody titers to levels even higher than the full-length protein expressed as transgene.[16,17] Numerous studies have consistently reported that simultaneous expression of antigenic protein as transgene combined with capsid display of epitopes generates strongest cellular and humoral immunity compared with transgene expression or capsid display alone.[72,94,109,113] The outstanding performance of these multivalent vaccines is the most encouraging attribute and should be advanced further. Motivated by the success of Ad4 and Ad7 vaccines, application of replication-competent capsid-displaying Ad vectors that can be delivered orally is another fresh approach worth exploring, particularly for endemic infections in the developing world.

## References

1. Majhen D, et al. Adenovirus-based vaccines for fighting infectious diseases and cancer: progress in the field. *Hum Gene Ther* 2014;**25**:301–17.
2. Sharma A, Tandon M, Bangari DS, Mittal SK. Adenoviral vector-based strategies for cancer therapy. *Curr Drug Ther* 2009;**4**:117–38.
3. Tatsis N, Ertl HC. Adenoviruses as vaccine vectors. *Mol Ther* 2004;**10**:616–29.
4. McElrath MJ, et al. HIV-1 vaccine-induced immunity in the test-of-concept step study: a case-cohort analysis. *Lancet* 2008;**372**:1894–905.
5. Stanley DA, et al. Chimpanzee adenovirus vaccine generates acute and durable protective immunity against Ebolavirus challenge. *Nat Med* 2014;**20**:1126–9.
6. Barouch DH, et al. International seroepidemiology of adenovirus serotypes 5, 26, 35, and 48 in pediatric and adult populations. *Vaccine* 2011;**29**:5203–9.
7. Mast TC, et al. International epidemiology of human pre-existing adenovirus (Ad) type-5, type-6, type-26 and type-36 neutralizing antibodies: correlates of high Ad5 titers and implications for potential HIV vaccine trials. *Vaccine* 2010;**28**:950–7.
8. Ahi YS, Bangari DS, Mittal SK. Adenoviral vector immunity: its implications and circumvention strategies. *Curr Gene Ther* 2011;**11**:307–20.
9. Thacker EE, Timares L, Matthews QL. Strategies to overcome host immunity to adenovirus vectors in vaccine development. *Expert Rev Vaccines* 2009;**8**:761–77.
10. Seregin SS, Amalfitano A. Overcoming pre-existing adenovirus immunity by genetic engineering of adenovirus-based vectors. *Expert Opin Biol Ther* 2009;**9**:1521–31.
11. Capone S, et al. Development of chimpanzee adenoviruses as vaccine vectors: challenges and successes emerging from clinical trials. *Expert Rev Vaccines* 2013;**12**:379–93.
12. Sharma A, et al. Comparative analysis of vector biodistribution, persistence and gene expression following intravenous delivery of bovine, porcine and human adenoviral vectors in a mouse model. *Virology* 2009;**386**:44–54.
13. Hemminki A, et al. A canine conditionally replicating adenovirus for evaluating oncolytic virotherapy in a syngeneic animal model. *Mol Ther* 2003;**7**:163–73.
14. Haase AT, Mautner V, Pereira HG. The immunogenicity of adenovirus type 5 structural proteins. *J Immunol* 1972;**108**:483–5.
15. Molinier-Frenkel V, et al. Adenovirus hexon protein is a potent adjuvant for activation of a cellular immune response. *J Virol* 2002;**76**:127–35.

16. Lanzi A, Ben Youssef G, Perricaudet M, Benihoud K. Anti-adenovirus humoral responses influence on the efficacy of vaccines based on epitope display on adenovirus capsid. *Vaccine* 2011;**29**:1463–71.
17. Sharma A, et al. Adenovirus-based vaccine with epitopes incorporated in novel fiber sites to induce protective immunity against *Pseudomonas aeruginosa*. *PLoS One* 2013;**8**:e56996.
18. Abe S, et al. Adenovirus type 5 with modified hexons induces robust transgene-specific immune responses in mice with pre-existing immunity against adenovirus type 5. *J Gene Med* 2009;**11**:570–9.
19. Germain RN. MHC-dependent antigen processing and peptide presentation: providing ligands for T lymphocyte activation. *Cell* 1994;**76**:287–99.
20. Bachmann MF, et al. The influence of antigen organization on B cell responsiveness. *Science* 1993;**262**:1448–51.
21. Rux JJ, Burnett RM. Adenovirus structure. *Hum Gene Ther* 2004;**15**:1167–76.
22. Matthews QL. Capsid-incorporation of antigens into adenovirus capsid proteins for a vaccine approach. *Mol Pharm* 2011;**8**:3–11.
23. Bergelson JM, et al. Isolation of a common receptor for Coxsackie B viruses and adenoviruses 2 and 5. *Science* 1997;**275**:1320–3.
24. Kalyuzhniy O, et al. Adenovirus serotype 5 hexon is critical for virus infection of hepatocytes in vivo. *Proc Natl Acad Sci USA* 2008;**105**:5483–8.
25. Krause A, et al. Epitopes expressed in different adenovirus capsid proteins induce different levels of epitope-specific immunity. *J Virol* 2006;**80**:5523–30.
26. Krasnykh V, et al. Characterization of an adenovirus vector containing a heterologous peptide epitope in the HI loop of the fiber knob. *J Virol* 1998;**72**:1844–52.
27. Vanderkwaak TJ, et al. An advanced generation of adenoviral vectors selectively enhances gene transfer for ovarian cancer gene therapy approaches. *Gynecol Oncol* 1999;**74**:227–34.
28. Leopold PL, Wendland RL, Vincent T, Crystal RG. Neutralized adenovirus-immune complexes can mediate effective gene transfer via an Fc receptor-dependent infection pathway. *J Virol* 2006;**80**:10237–47.
29. Coughlan L, et al. Retargeting adenovirus serotype 48 fiber knob domain by peptide incorporation. *Hum Gene Ther* 2014;**25**:385–94.
30. Matsui H, et al. Development of fiber-substituted adenovirus vectors containing foreign peptides in the adenovirus serotype 35 fiber knob. *Gene Ther* 2009;**16**:1050–7.
31. Wickham TJ, Mathias P, Cheresh DA, Nemerow GR. Integrins alpha v beta 3 and alpha v beta 5 promote adenovirus internalization but not virus attachment. *Cell* 1993;**73**:309–19.
32. Mathias P, Wickham T, Moore M, Nemerow G. Multiple adenovirus serotypes use alpha v integrins for infection. *J Virol* 1994;**68**:6811–4.
33. Zubieta C, Schoehn G, Chroboczek J, Cusack S. The structure of the human adenovirus 2 penton. *Mol Cell* 2005;**17**:121–35.
34. Fender P, Ruigrok RW, Gout E, Buffet S, Chroboczek J. Adenovirus dodecahedron, a new vector for human gene transfer. *Nat Biotechnol* 1997;**15**:52–6.
35. Garcel A, Gout E, Timmins J, Chroboczek J, Fender P. Protein transduction into human cells by adenovirus dodecahedron using WW domains as universal adaptors. *J Gene Med* 2006;**8**:524–31.
36. Naskalska A, et al. Influenza recombinant vaccine: matrix protein M1 on the platform of the adenovirus dodecahedron. *Vaccine* 2009;**27**:7385–93.
37. Szolajska E, et al. The structural basis for the integrity of adenovirus Ad3 dodecahedron. *PLoS One* 2012;**7**:e46075.
38. Fender P. Use of dodecahedron "VLPs" as an alternative to the whole adenovirus. *Methods Mol Biol* 2014;**1089**:61–70.

39. Villegas-Mendez A, et al. Functional characterisation of the WW minimal domain for delivering therapeutic proteins by adenovirus dodecahedron. *PLoS One* 2012;**7**:e45416.
40. Szurgot I, et al. Self-adjuvanting influenza candidate vaccine presenting epitopes for cell-mediated immunity on a proteinaceous multivalent nanoplatform. *Vaccine* 2013;**31**:4338–46.
41. Naskalska A, Szolajska E, Andreev I, Podsiadla M, Chroboczek J. Towards a novel influenza vaccine: engineering of hemagglutinin on a platform of adenovirus dodecahedron. *BMC Biotechnol* 2013;**13**:50.
42. Villegas-Mendez A, et al. In vivo delivery of antigens by adenovirus dodecahedron induces cellular and humoral immune responses to elicit antitumor immunity. *Mol Ther* 2010;**18**:1046–53.
43. Zochowska M, et al. Virus-like particle-mediated intracellular delivery of mRNA cap analog with in vivo activity against hepatocellular carcinoma. *Nanomedicine* 2015;**11**:67–76.
44. Parks RJ. Adenovirus protein IX: a new look at an old protein. *Mol Ther* 2005;**11**:19–25.
45. Colby WW, Shenk T. Adenovirus type 5 virions can be assembled in vivo in the absence of detectable polypeptide IX. *J Virol* 1981;**39**:977–80.
46. Fabry CM, et al. The C-terminal domains of adenovirus serotype 5 protein IX assemble into an antiparallel structure on the facets of the capsid. *J Virol* 2009;**83**:1135–9.
47. Marsh MP, et al. Cryoelectron microscopy of protein IX-modified adenoviruses suggests a new position for the C terminus of protein IX. *J Virol* 2006;**80**:11881–6.
48. Liu H, et al. Atomic structure of human adenovirus by cryo-EM reveals interactions among protein networks. *Science* 2010;**329**:1038–43.
49. Rosa-Calatrava M, Grave L, Puvion-Dutilleul F, Chatton B, Kedinger C. Functional analysis of adenovirus protein IX identifies domains involved in capsid stability, transcriptional activity, and nuclear reorganization. *J Virol* 2001;**75**:7131–41.
50. Campos SK. New structural model of adenoviral cement proteins is not yet concrete. *Proc Natl Acad Sci USA* 2014;**111**:E4542–3.
51. Reddy VS, Nemerow GR. Reply to Campos: revised structures of adenovirus cement proteins represent a consensus model for understanding virus assembly and disassembly. *Proc Natl Acad Sci USA* 2014;**111**:E4544–5.
52. Reddy VS, Nemerow GR. Structures and organization of adenovirus cement proteins provide insights into the role of capsid maturation in virus entry and infection. *Proc Natl Acad Sci USA* 2014;**111**:11715–20.
53. Vellinga J, van den Wollenberg DJ, van der Heijdt S, Rabelink MJ, Hoeben RC. The coiled-coil domain of the adenovirus type 5 protein IX is dispensable for capsid incorporation and thermostability. *J Virol* 2005;**79**:3206–10.
54. de Vrij J, et al. Enhanced transduction of CAR-negative cells by protein IX-gene deleted adenovirus 5 vectors. *Virology* 2011;**410**:192–200.
55. Akalu A, Liebermann H, Bauer U, Granzow H, Seidel W. The subgenus-specific C-terminal region of protein IX is located on the surface of the adenovirus capsid. *J Virol* 1999;**73**:6182–7.
56. Dmitriev IP, Kashentseva EA, Curiel DT. Engineering of adenovirus vectors containing heterologous peptide sequences in the C terminus of capsid protein IX. *J Virol* 2002;**76**:6893–9.
57. Vellinga J, et al. Spacers increase the accessibility of peptide ligands linked to the carboxyl terminus of adenovirus minor capsid protein IX. *J Virol* 2004;**78**:3470–9.
58. Campos SK, Parrott MB, Barry MA. Avidin-based targeting and purification of a protein IX-modified, metabolically biotinylated adenoviral vector. *Mol Ther* 2004;**9**:942–54.

59. Corjon S, et al. Targeting of adenovirus vectors to the LRP receptor family with the high-affinity ligand RAP via combined genetic and chemical modification of the pIX capsomere. *Mol Ther* 2008;**16**:1813–24.
60. Vellinga J, et al. Efficient incorporation of a functional hyper-stable single-chain antibody fragment protein-IX fusion in the adenovirus capsid. *Gene Ther* 2007;**14**:664–70.
61. Poulin KL, et al. Retargeting of adenovirus vectors through genetic fusion of a single-chain or single-domain antibody to capsid protein IX. *J Virol* 2010;**84**:10074–86.
62. de Vrij J, et al. Adenovirus targeting to HLA-A1/MAGE-A1-positive tumor cells by fusing a single-chain T-cell receptor with minor capsid protein IX. *Gene Ther* 2008;**15**:978–89.
63. Meulenbroek RA, Sargent KL, Lunde J, Jasmin BJ, Parks RJ. Use of adenovirus protein IX (pIX) to display large polypeptides on the virion–generation of fluorescent virus through the incorporation of pIX-GFP. *Mol Ther* 2004;**9**:617–24.
64. Le LP, et al. Fluorescently labeled adenovirus with pIX-EGFP for vector detection. *Mol Imaging* 2004;**3**:105–16.
65. Le LP, Li J, Ternovoi VV, Siegal GP, Curiel DT. Fluorescently tagged canine adenovirus via modification with protein IX-enhanced green fluorescent protein. *J Gen Virol* 2005;**86**:3201–8.
66. Zakhartchouk A, Connors W, van Kessel A, Tikoo SK. Bovine adenovirus type 3 containing heterologous protein in the C-terminus of minor capsid protein IX. *Virology* 2004;**320**:291–300.
67. Li J, Le L, Sibley DA, Mathis JM, Curiel DT. Genetic incorporation of HSV-1 thymidine kinase into the adenovirus protein IX for functional display on the virion. *Virology* 2005;**338**:247–58.
68. Matthews QL, et al. Genetic incorporation of a herpes simplex virus type 1 thymidine kinase and firefly luciferase fusion into the adenovirus protein IX for functional display on the virion. *Mol Imaging* 2006;**5**:510–9.
69. Liu L, et al. Construction and radiolabeling of adenovirus variants that incorporate human metallothionein into protein IX for analysis of biodistribution. *Mol Imaging* 2014;**13**.
70. Tang Y, et al. Derivation of a triple mosaic adenovirus based on modification of the minor capsid protein IX. *Virology* 2008;**377**:391–400.
71. Boyer JL, et al. Protective immunity against a lethal respiratory *Yersinia pestis* challenge induced by V antigen or the F1 capsular antigen incorporated into adenovirus capsid. *Hum Gene Ther* 2010;**21**:891–901.
72. Bayer W, et al. Vaccination with an adenoviral vector that encodes and displays a retroviral antigen induces improved neutralizing antibody and $CD4^+$ T-cell responses and confers enhanced protection. *J Virol* 2010;**84**:1967–76.
73. Kurachi S, et al. Characterization of capsid-modified adenovirus vectors containing heterologous peptides in the fiber knob, protein IX, or hexon. *Gene Ther* 2007;**14**:266–74.
74. Burnett RM. The structure of the adenovirus capsid. II. The packing symmetry of hexon and its implications for viral architecture. *J Mol Biol* 1985;**185**:125–43.
75. Rux JJ, Burnett RM. Type-specific epitope locations revealed by X-ray crystallographic study of adenovirus type 5 hexon. *Mol Ther* 2000;**1**:18–30.
76. Crawford-Miksza L, Schnurr DP. Analysis of 15 adenovirus hexon proteins reveals the location and structure of seven hypervariable regions containing serotype-specific residues. *J Virol* 1996;**70**:1836–44.
77. Rux JJ, Kuser PR, Burnett RM. Structural and phylogenetic analysis of adenovirus hexons by use of high-resolution X-ray crystallographic, molecular modeling, and sequence-based methods. *J Virol* 2003;**77**:9553–66.

78. Bradley RR, Lynch DM, Iampietro MJ, Borducchi EN, Barouch DH. Adenovirus serotype 5 neutralizing antibodies target both hexon and fiber following vaccination and natural infection. *J Virol* 2012;**86**:625–9.
79. Bradley RR, et al. Adenovirus serotype 5-specific neutralizing antibodies target multiple hexon hypervariable regions. *J Virol* 2012;**86**:1267–72.
80. Gahery-Segard H, et al. Immune response to recombinant capsid proteins of adenovirus in humans: antifiber and anti-penton base antibodies have a synergistic effect on neutralizing activity. *J Virol* 1998;**72**:2388–97.
81. Hong SS, Habib NA, Franqueville L, Jensen S, Boulanger PA. Identification of adenovirus (ad) penton base neutralizing epitopes by use of sera from patients who had received conditionally replicative ad (addl1520) for treatment of liver tumors. *J Virol* 2003;**77**:10366–75.
82. Sumida SM, et al. Neutralizing antibodies to adenovirus serotype 5 vaccine vectors are directed primarily against the adenovirus hexon protein. *J Immunol* 2005;**174**:7179–85.
83. Youil R, et al. Hexon gene switch strategy for the generation of chimeric recombinant adenovirus. *Hum Gene Ther* 2002;**13**:311–20.
84. Gall JG, Crystal RG, Falck-Pedersen E. Construction and characterization of hexon-chimeric adenoviruses: specification of adenovirus serotype. *J Virol* 1998;**72**:10260–4.
85. Xue C, et al. Construction and characterization of a recombinant human adenovirus type 3 vector containing two foreign neutralizing epitopes in hexon. *Virus Res* 2014;**183**:67–74.
86. Roberts DM, et al. Hexon-chimaeric adenovirus serotype 5 vectors circumvent pre-existing anti-vector immunity. *Nature* 2006;**441**:239–43.
87. Bruder JT, et al. Modification of Ad5 hexon hypervariable regions circumvents pre-existing Ad5 neutralizing antibodies and induces protective immune responses. *PLoS One* 2012;**7**:e33920.
88. Barouch DH, et al. Protective efficacy of a single immunization of a chimeric adenovirus vector-based vaccine against simian immunodeficiency virus challenge in rhesus monkeys. *J Virol* 2009;**83**:9584–90.
89. Baden LR, et al. First-in-human evaluation of a hexon chimeric adenovirus vector expressing HIV-1 Env (IPCAVD 002). *J Infect Dis* 2014;**210**:1052–61.
90. Crompton J, Toogood CI, Wallis N, Hay RT. Expression of a foreign epitope on the surface of the adenovirus hexon. *J Gen Virol* 1994;**75**(Pt 1):133–9.
91. Wu H, et al. Identification of sites in adenovirus hexon for foreign peptide incorporation. *J Virol* 2005;**79**:3382–90.
92. Matthews QL, et al. Optimization of capsid-incorporated antigens for a novel adenovirus vaccine approach. *Virol J* 2008;**5**:98.
93. Worgall S, et al. Protection against *P. aeruginosa* with an adenovirus vector containing an OprF epitope in the capsid. *J Clin Invest* 2005;**115**:1281–9.
94. Worgall S, et al. Protective immunity to pseudomonas aeruginosa induced with a capsid-modified adenovirus expressing *P. aeruginosa* OprF. *J Virol* 2007;**81**:13801–8.
95. Krause A, et al. RGD capsid modification enhances mucosal protective immunity of a non-human primate adenovirus vector expressing *Pseudomonas aeruginosa* OprF. *Clin Exp Immunol* 2013;**173**:230–41.
96. Worgall S, et al. Modification to the capsid of the adenovirus vector that enhances dendritic cell infection and transgene-specific cellular immune responses. *J Virol* 2004;**78**:2572–80.
97. McConnell MJ, Danthinne X, Imperiale MJ. Characterization of a permissive epitope insertion site in adenovirus hexon. *J Virol* 2006;**80**:5361–70.
98. Shiratsuchi T, Rai U, Krause A, Worgall S, Tsuji M. Replacing adenoviral vector HVR1 with a malaria B cell epitope improves immunogenicity and circumvents preexisting immunity to adenovirus in mice. *J Clin Invest* 2010;**120**:3688–701.

99. Palma C, et al. Adenovirus particles that display the *Plasmodium falciparum* circumsporozoite protein NANP repeat induce sporozoite-neutralizing antibodies in mice. *Vaccine* 2011;**29**:1683–9.
100. Karen KA, et al. A replicating adenovirus capsid display recombinant elicits antibodies against *Plasmodium falciparum* sporozoites in *Aotus nancymaae* monkeys. *Infect Immun* 2015;**83**:268–75.
101. Buge SL, et al. An adenovirus-simian immunodeficiency virus env vaccine elicits humoral, cellular, and mucosal immune responses in rhesus macaques and decreases viral burden following vaginal challenge. *J Virol* 1997;**71**:8531–41.
102. Klessig DF, Grodzicker T. Mutations that allow human Ad2 and Ad5 to express late genes in monkey cells map in the viral gene encoding the 72K DNA binding protein. *Cell* 1979;**17**:957–66.
103. Farrow AL, et al. Immunization with hexon modified adenoviral vectors integrated with gp83 epitope provides protection against *Trypanosoma cruzi* infection. *PLoS Negl Trop Dis* 2014;**8**:e3089.
104. Zhang Q, et al. Construction and characterization of a replication-competent human adenovirus type 3-based vector as a live-vaccine candidate and a viral delivery vector. *Vaccine* 2009;**27**:1145–53.
105. Stone D, et al. Comparison of adenoviruses from species B, C, E, and F after intravenous delivery. *Mol Ther* 2007;**15**:2146–53.
106. Zhong T, et al. Characterization of malleability and immunological properties of human adenovirus type 3 hexon hypervariable region 1. *Arch Virol* 2012;**157**:1709–18.
107. Tian X, et al. Protection against enterovirus 71 with neutralizing epitope incorporation within adenovirus type 3 hexon. *PLoS One* 2012;**7**:e41381.
108. Liu M, et al. Generation of neutralizing monoclonal antibodies against a conformational epitope of human adenovirus type 7 (HAdv-7) incorporated in capsid encoded in a HAdv-3-based vector. *PLoS One* 2014;**9**:e103058.
109. Zhou D, et al. Hexon-modified recombinant E1-deleted adenovirus vectors as dual specificity vaccine carriers for influenza virus. *Mol Ther* 2013;**21**:696–706.
110. Kim JH, Rerks-Ngarm S, Excler JL, Michael NL. HIV vaccines: lessons learned and the way forward. *Curr Opin HIV AIDS* 2010;**5**:428–34.
111. Barouch DH. Novel adenovirus vector-based vaccines for HIV-1. *Curr Opin HIV AIDS* 2010;**5**:386–90.
112. Michael NL. Rare serotype adenoviral vectors for HIV vaccine development. *J Clin Invest* 2012;**122**:25–7.
113. Matthews QL, et al. HIV antigen incorporation within adenovirus hexon hypervariable 2 for a novel HIV vaccine approach. *PLoS One* 2010;**5**:e11815.
114. Flatt JW, et al. CryoEM visualization of an adenovirus capsid-incorporated HIV antigen. *PLoS One* 2012;**7**:e49607.
115. Gu L, et al. A recombinant adenovirus-based vector elicits a specific humoral immune response against the V3 loop of HIV-1 gp120 in mice through the "Antigen Capsid-Incorporation" strategy. *Virol J* 2014;**11**:112.
116. Gu L, et al. Using multivalent adenoviral vectors for HIV vaccination. *PLoS One* 2013;**8**:e60347.
117. Hicks MJ, et al. Cocaine analog coupled to disrupted adenovirus: a vaccine strategy to evoke high-titer immunity against addictive drugs. *Mol Ther* 2011;**19**:612–9.
118. Hicks MJ, et al. Fate of systemically administered cocaine in nonhuman primates treated with the dAd5GNE anticocaine vaccine. *Hum Gene Ther Clin Dev* 2014;**25**:40–9.
119. Koob G, et al. Anti-cocaine vaccine based on coupling a cocaine analog to a disrupted adenovirus. *CNS Neurol Disord Drug Targets* 2011;**10**:899–904.

120. Wee S, et al. Novel cocaine vaccine linked to a disrupted adenovirus gene transfer vector blocks cocaine psychostimulant and reinforcing effects. *Neuropsychopharmacology* 2012;**37**:1083–91.
121. Maoz A, et al. Adenovirus capsid-based anti-cocaine vaccine prevents cocaine from binding to the nonhuman primate CNS dopamine transporter. *Neuropsychopharmacology* 2013;**38**:2170–8.
122. De BP, et al. Disrupted adenovirus-based vaccines against small addictive molecules circumvent anti-adenovirus immunity. *Hum Gene Ther* 2013;**24**:58–66.
123. Rosenberg JB, et al. Suppression of nicotine-induced pathophysiology by an adenovirus hexon-based antinicotine vaccine. *Hum Gene Ther* 2013;**24**:595–603.
124. Waddington SN, et al. Adenovirus serotype 5 hexon mediates liver gene transfer. *Cell* 2008;**132**:397–409.
125. Khare R, Reddy VS, Nemerow GR, Barry MA. Identification of adenovirus serotype 5 hexon regions that interact with scavenger receptors. *J Virol* 2012;**86**:2293–301.
126. Khare R, et al. Generation of a Kupffer cell-evading adenovirus for systemic and liver-directed gene transfer. *Mol Ther* 2011;**19**:1254–62.
127. Santis G, et al. Molecular determinants of adenovirus serotype 5 fibre binding to its cellular receptor CAR. *J Gen Virol* 1999;**80**:1519–27.
128. Xia D, Henry LJ, Gerard RD, Deisenhofer J. Crystal structure of the receptor-binding domain of adenovirus type 5 fiber protein at 1.7 A resolution. *Structure* 1994;**2**:1259–70.

# Utility of Adenoviral Vectors in Animal Models of Human Disease I: Cancer

**23**

*Raj K. Batra*[1,2], *Sherven Sharma*[3], *Lily Wu*[2,4,5]
[1]UCLA School of Medicine, Division of Pulmonary and Critical Care Medicine, GLA-VAHCS, Los Angeles, CA, USA; [2]Jonsson Comprehensive Cancer Center, UCLA, Los Angeles, CA, USA; [3]UCLA/Wadsworth Pulmonary Immunology Laboratory, Division of Pulmonary and Critical Care Medicine, GLA-VAHCS, Los Angeles, CA, USA; [4]Department of Urology, UCLA School of Medicine, Los Angeles, CA, USA; [5]Department of Pediatrics, UCLA School of Medicine, Los Angeles, CA, USA

## 1. Introduction

The development of molecular therapeutics for the treatment of human disease has a rational and predictable course. Because these therapeutics are generally derived from an understanding of molecular mechanisms underlying a disease process, the treatment strategies are hypothesis-driven and specifically targeted toward a pathway underlying the molecular and cellular pathogenesis. Accordingly, in addition to establishing therapeutic efficacy, the evaluation of a molecular therapeutic also confirms the importance of a specific genetic or biological pathway in the *pathogenesis* of a disease process. The evaluation of a molecular therapeutic typically begins by providing a molecular/cellular proof-of-concept in vitro, followed by an expansion of therapeutic principles and toxicological analyses of the intervention in animal models, and finally a systematic sequence of safety and clinical efficacy trials in human subjects.

Logically, gene therapy paradigms using adenoviral (Ad) vectors can be expected to proceed along this course in order to be considered for the treatment of human disease. This chapter will focus on the use of animal models in the process of evaluating Ad gene transfer strategies for the treatment of human cancer. In this respect, we will offer a personal perspective, concentrated on outlining principles rather than cataloging individual examples. Because the focus is on principles, the review will not be an inventory of the various experimental therapeutic strategies for cancer that utilize the Ad vector, although specific examples may be cited. Rather, we will use our background and experience to illustrate the problems inherent in testing experimental hypotheses in animal models of cancer, with the confidence that themes particular to our research may have broader applicability. Lastly, because the authors have an interest in utilizing Ad gene transfer techniques for the treatment of lung and prostrate

**Adenoviral Vectors for Gene Therapy.**

cancer, respectively, this chapter will emphasize experimental designs relevant to those clinical entities.

A primary goal of in vivo/animal experimentation is to build on an in vitro proof of concept and to strengthen the rationale for clinical testing of an experimental therapeutic intervention. To justify animal studies, there should already be an existent pathophysiological rationale and/or in vitro experimental data suggesting that a strategy is likely to be effective. At this juncture, the investigator is faced with the formidable challenge of approximating a human disease in an animal model. Although animal models cannot be exact replicas of the human disease, they should, at the very least, provide useful molecular and cellular similarities to the pathogenesis and clinical manifestations of the target disease.[1] For a variety of reasons (low expense for breeding and maintenance, susceptibility to tumorigenesis, well-defined immunosuppressive states, feasible duration of experimental studies, etc.) mice are considered the prototypic animal model for experimentation. Ideally, a mouse model would mimic the target human disease in its etiology, genetics, clinical presentation, and progression. To model human lung cancer, for example, the ideal mouse model would systematically (in defined pathological stages) develop lung cancer from exposure to cigarette smoke, and the disease could be characterized by sequential gene defects that culminate in the clinical progression that typifies the human disease.

Secondly, in designing the experimental approach, the investigator must also take into consideration the "pharmacological intervention or drug" (the Ad vector here) that is being tested. Because the ideal drug should have reliable delivery, specific targeted distribution and mechanism of action, and predictable elimination, the challenge to test an Ad vector based-therapeutic in an apt animal model becomes particularly daunting for a gene therapist. Thus, to test whether an Ad-based therapeutic will have efficacy for the treatment of human cancer, we must (1) model the complex human disease in vivo and then (2) test a multi-faceted biological compound with ill-defined pharmacokinetic and pharmacodynamic properties that are likely unique to the host and/or the disease state. In order to overcome the inherent complexity of the problem, we have adopted an approach that uses a combination of models to overcome specific deficiencies that accompany each individually. Consequently, we utilize xenogeneic models (engraftment of heterologous tissue derived from donors of a different species, typically into an immunodeficient host) to study the therapeutic gene effects and Ad vector–target cell interactions. Syngeneic (engraftment of tissue from genetically identical donors) and allogeneic (engraftment of tissues from a genetically dissimilar member of the same species) models are used to study host–tumor interactions in terms of immunological parameters and metastasis. To further discern the specific immunologic parameters important for tumor rejection in mice, specific knockout (targeted gene disruption) mice are utilized. Lastly, we extensively utilize transgenic (in this context, referring to the tissue-specific expression of a transforming oncogene) models to study the effects of a molecular therapeutic in the setting of established orthotopic (referring to organ- or site-specific) malignancy. We believe that integrating the results of these individual approaches will enable us to meet the goals of in vivo experimentation for advancing Ad gene therapy.

## 2. Animal Models of Lung Cancer

### 2.1 Human Lung Cancer

Lung cancer is the leading cause of cancer-associated mortality in ***both*** men and women. Although susceptibility to environmental carcinogens may be predetermined and follow a pattern of autosomal dominant Mendelian inheritance,[2,3] lung cancer results from an accumulation of acquired genetic mutations.[4–6] In fact, it is suggested that 10–20 genetic mutations may be necessary for the development of lung cancer,[7] although the discrete steps for the progression of a hyperplastic bronchial lesion to metaplasia and anaplasia have not been uncovered. Tobacco use is the strongest epidemiologic risk for the development of lung cancer and it is anticipated that approximately 10% of all smokers will develop lung cancer over their lifetime.[8] Current paradigms predict that lung cancer results from the widespread exposure of the carcinogen, leading to a process of "field cancerization" whereby the entire aerodigestive track is exposed to the offending agents and leads to the occurrence of synchronous and metachronous tumors.[9] The tobacco carcinogens apparently invoke the multiple clonal chromosomal abnormalities found throughout the airways and alveoli of smokers.[10,11] Following, the series of genetic mutations likely results in patterned aberrancies in signal transduction and cell cycle pathways, eventuating in malignant and metastatic phenotypes.[12] The general pattern of genetic changes is characteristic but not specific for pathologic subtypes of lung cancer (see below). Overall, *K-ras* mutations are observed in 20–50%,[13] *p53* mutations are present in 50%,[14] 60% exhibit reduced expression of *p16-ink4a*,[15,16] and 30% show deletion of *Rb*. Small cell lung cancers (SCLCs) display a greater proclivity to *c-myc* amplification and a greater degree of *p53* (80%) and *Rb* mutations (90%). Chromosome 3p deletions, occurring at a chromosomal fragile site that includes the *FHIT* locus, are found in 50% of NSCLC and in 90% of SCLC primary tumors.[17] Overexpression of the tyrosine growth factor receptor *erbB2-neu* is seen in 10–30% and overexpression of *bcl-2*[18] in 10–25%of NSCLC tumors.[19]

Clinically, lung cancer is discriminated into SCLC and non-small cell (NSCLC) categories by histopathology or cytopathology and by their characteristic clinical presentations and divergent responses to conventional cytoreductive therapies. NSCLC may be further subclassified pathologically into squamous cell, adenocarcinoma, bronchoalveolar cell carcinoma (BAC), adenosquamous (mixed pathology), or large cell carcinoma. As noted above, the progression of lung cancer from a premalignant state to the clinical/pathological entity that is diagnosed in the vast majority of patients is unknown. This is because although the disease is very prevalent, it is typically diagnosed when it has already spread outside the lungs and is pathologically advanced. Not surprisingly, because of the late stage of diagnosis, progressive genetic instability confers ***marked genetic and phenotypic heterogeneity*** within lung cancers, even in individual patients. The late stage of diagnosis also results in an absolute lack of premalignant material, making it difficult to assign specific roles for the genetic mutations in the systematic progression of lung cancer. Recently, however, some of the characteristic genetic mutations of lung cancer (e.g., loss of heterozygosity (LOH) at chromosome 3p, *p53* mutations) are being identified in microdissected dysplastic epithelium.[20] Similar observations are implicating the characteristic *K-ras* abnormalities in lung cancer as a correlate of mucinous differentiation.[21] A precursor to lung adenocarcinoma, a lesion pathologically termed alveolar atypical hyperplasia (AAH), is being advanced. AAH is described by increased cellular proliferation when compared to adjacent normal parenchyma, and by immunohistochemical evidence of *p53* stabilization, *k-ras* mutations, and c-*erb-B2* overexpression.[22–24] The presence of these mutations in AAH may explain why such mutations may be detectable in sputum cytology specimens that predate the onset of clinical lung cancer.[25] Identification of these early events is a particular focus of study because they may serve as genetic markers for malignant progression, or as targets of specific genetic or chemopreventive approaches. More relevant to this discussion, perhaps, these early events may be better modeled in murine models than late stage lung cancer (see below). Thus, there exists an inherent complexity in human lung cancer and to precisely recapitulate the disease process in animals is not possible.

## 2.2 Animal Models of Human Lung Cancer

### 2.2.1 Murine Lung Cancer and Transplantable Allografts

Due to time of model development, ease of experimentation, and cost restraints, murine models of disease are the accepted standards. However, there are generic shortcomings in this approach. For example, cigarette smoke, which is a strong epidemiological risk for the development of human lung cancer and is proximally responsible for approximately 85–90% of lung cancer cases in humans,[26] is only weakly carcinogenic in mice.[27,28] In addition, although both mouse and human lung adenocarcinomas may share common molecular defects,[27] the histopathological repertoire of spontaneous or induced tumors in mice is very limited,[29,30] and morphologically, nearly all mouse lung tumors bear structural similarities only to BAC or well-differentiated adenocarcinomas. Consequently, whereas humans typically die from lung cancer of "late stage" metastatic disease, mice succumb to respiratory failure following the diffuse involvement of their lungs by "early stage" carcinoma in situ.[1]

Spontaneous lung cancer develops in 3% of wild mice[31,32] with strain-dependent sensitivity. Clones have been isolated from spontaneously arising tumors and established as cultures in vitro. These cultures now serve as a readily available source for the generation of transplantable allografts. Many investigators, including our group,[33–38] have extensively utilized line 1 alveolar carcinoma (L1C2), a murine lung cancer cell line that is syngeneic to BALB/c, and 3LL (Lewis lung cancer) that is syngeneic to C57Bl/6. Usually, these cell lines are utilized to generate transplantable heterotopic (referring to a location outside of the organ of origin, typically subcutaneous) tumors in syngeneic mice. Our group has utilized these models to investigate, in general, the interplay between the immune system and the host. Both L1C2 and 3LL tumors are relatively "nonimmunogenic," as is human lung cancer, and immunogenetic strategies that modulate the immune system to generate an antitumor immune response can be systematically investigated in these models. However, other lung tumor allografts, especially when cells are selected to express "marker antigens" to enable their easy detection in culture systems, may indeed become immunogenic. Notably, the transplantable allograft system is artificial, and all recipient hosts have a "stress" response to the implanted tumor that cannot be recapitulated in control animals. In addition, extrapolating antitumor responses in mice to humans is not a straightforward proposition, and many therapies that reliably "cure" tumors in inbred strains of mice are not as effective in humans. In part, these differences may be attributable to differences in immune responses in the two hosts. For example, cluster determinants (CD antigens) in murine strains may not have homologous or functional cellular analogs in the human host.

Laboratory animals used for medical experimentation are genetically inbred strains with reliable phenotypic characteristics. Although this feature imposes a generic limitation on the extrapolation of results in lab animal studies to outbred populations, and thus, human disease, there are significant advantages that need to be considered. The inbred nature of laboratory animals enables investigators to reliably establish disease in an animal host and subsequently to study that disease process in controlled subsets. With respect to tumorigenesis, murine-A/J and SWR strains are the most sensitive, BALB/c is of intermediate sensitivity, and DBA and C57BL/6 are the most resistant. Crosses between susceptible and resistant inbred mouse strains may allow for the mapping of modifier loci for the development of lung cancer.[39] For example, it is reported that the propensity of strains to develop lung tumors correlates with a polymorphism in the second intron of *K-ras*.[40] Practical experience suggests that there are common genetic alterations affecting known tumor suppressor genes and proto oncogenes occur during mouse lung carcinogenesis. Molecular abnormalities may also be shared with human lung cancer, and *K-ras* activation is a conspicuous example.[41] Human adenocarcinomas commonly carry *K-ras* mutations; most of these mutations are in codon 12 and are transversions of GGT to either TGT or GTT. It is postulated that these mutations occur early in lung cancer pathogenesis since they can be detected in sputum samples of smokers prior to the clinical diagnosis of lung cancer. Analogously, 80–90% of both spontaneous and chemically induced murine lung tumors contain *K-ras* mutations. Moreover, *K-ras* mutations also occur early in murine lung tumorigenesis, and remarkably, codon 12 is the site of genetic change induced by many chemical carcinogens.[1] Furthermore, a consistent loss of mouse chromosome region 4, an area that contains the mouse homolog of the human *p16-ink4a*,[42,43] has been described to result in an allelic loss of the *p16-ink4a* seen in 50% of mouse adenocarcinomas. Similarly, *p53* mutations are found, albeit infrequently,[44] although mouse chromosomal regions containing *p53* and *Rb* more commonly exhibit LOH.[43] Reduced expression of *Rb* and *p16* and increased *c-myc* expression[39] have also been reported. These commonalties have suggested some to conclude that mouse and human lung carcinomas are sufficiently similar for the murine model to be informative[1] and have formed the rationale for the testing of chemopreventive strategies[39] in mice. Analogously, these commonalties may be advanced to form the basis for the testing of genetic therapies in murine tumors as well, perhaps with questionable legitimacy.

Mice strains also vary with respect to inducible tumorigenesis. Generally, mice that are sensitive to the development of spontaneous lung tumors are also at the highest risk for chemically induced tumors[31] and form the basis for the quantitative carcinogenicity bioassays. Although a variety of agents, including urethane, metals, concentrated components of tobacco smoke such as polyaromatic hydrocarbons and nitrosamines,[45,46] can induce lung cancer in mice, tobacco smoke per se is only weakly carcinogenic.[28] Murine lung tumors histologically resemble early lesions that originate peripherally (from type 2 alveolar cells or Clara cells) and simulate papillary or BAC. In contrast, the bulk of human tumors are bronchogenic (arise in the airways) and, as described above, display a broad histopathologic variation. In fact, individual human lung cancers may be histologically heterogeneous, i.e., they often displaying mixed morphologies within the same tumor specimen. So how does one reconcile these differences between murine lung cancer and human lung cancer? and moreover, can one generalize observations and results from one species to another, or even from one human being to another? When considered in the context of adenoviral gene delivery, there is a limiting paucity of in vivo data to generate any broad conclusion. On the contrary, our observations in vitro suggest that gene transfer into subtypes of human lung cancer is highly variable, and strategies directed toward achieving intratumoral gene transfer may require patient or disease-specific vector formulations.[47]

The biological heterogeneity of human lung cancer drives our investigations along specified pathways, utilizing many different models and strategies to come up with viable treatment approaches. For instance, we believe that a systematic assessment of the efficiency and optimal route of Ad gene delivery in vivo into murine lung tumors and transplanted human xenografts needs to be performed. Researchers are beginning to identify the Ad cellular attachment receptor (termed the coxsackievirus–adenovirus receptor, CAR[48]) as a major determinant underlying efficient transduction.[49] Along these lines, the scope and "polarity" of CAR expression in tumors in vivo needs to be defined. Thus, one focus of our program is to systematically evaluate gene transfer into these model systems using conventional and retargeted Ad vectors with the aim of optimizing a vector system and a mode of delivery. This focus evolves from the premise supported by our in vitro data that the histological heterogeneity of lung tumors may be a harbinger of variable responsiveness to both Ad entry and/or the efficacy of Ad gene therapy.[47] Because targeting of tumor in vivo may be unattainable, we have also generated protocols in which the Ad vector is used in precisely controlled ex vivo "dosing" approaches to genetically modify antigen-presenting cells (APCs) or tumor cells to vaccinate the host against their tumor.[37]

### 2.2.2 Murine Models that Spontaneously Develop Lung Cancer

Murine models of lung cancer include strains susceptible to chemically induced tumors and transgenic strains that express viral and cellular oncogenes. The simian virus-40 large TAg (SV40-TAg) has been commonly used to produce tumors in transgenic mice.[50,51] SV40-TAg binds and incapacitates the cell cycle checkpoint and DNA-binding capabilities of the *p53* and *Rb* gene products, resulting in uncontrolled cellular proliferation.[52] To develop a murine model of lung cancer, Wikenheiser and colleagues chose to express the SV40-TAg under the transcriptional control of the lung-specific human surfactant protein C (SP-C) promoter in transgenic mice.[53,54] They demonstrated that these mice consistently developed multifocal lung adenocarcinomas that had pathological features similar to some human lung adenocarcinomas, and that the mice succumbed to respiratory distress by age 4–5 months. As expected, the transgenic animals developed no tumors in any other organ systems, although some nonmalignant tissue also expressed the transgene.[53] Within the lungs, tumors consistently involved the bronchiolar and alveolar regions of the lung while sparing the large airways. The tumors of these mice also varied with respect to the expression of the large TAg, suggesting perhaps that SV-40-TAg may contribute to transformation, but continued expression may not be necessary for tumor progression. Likewise, organ-specific expression of SV40-TAg using the regulatory regions of uteroglobin[55] and the Clara cell-specific $M_r$ 10,000 protein (CC-10) also results in the induction of lung tumors.[56] Uteroglobin is a marker protein for the non-ciliated epithelial Clara cells, the source of xenobiotic metabolism in the lung, lining the respiratory and terminal bronchioli of the lungs. In animals expressing SV40-TAg under the uteroglobin promoter control, the pulmonary epithelium was morphologically normal at 2 months, dysplastic by 4 months, and transgenic animals were described as developing multifocal pulmonary adenocarcinoma present in various stages of differentiation by 5 months of age. In situ hybridization studies suggested that tumors did not contain the transcripts of the uteroglobin gene, and again, late stage tumors lost expression of the large TAg. Tumors also formed in the urogenital tract where uteroglobin is also expressed.

Transgenic mice were also generated using the CC-10 kDa promoter driving SV-40 large TAg,[56] and it is in this model that we have chosen to test the immunomodulatory capacity of secondary lymphoid chemokine or *slc*.[36] In the 7736-mouse line, CC-10TAg-transgenic mice develop multifocal pulmonary adenocarcinomas and succumb to respiratory failure at 16–20 weeks of age. Pathology is localized to the lungs, and the tumors express the large TAg in normal Clara cells and in transformed tumor cells. Pathological progression is similar to that described above, with the lungs appearing morphologically normal at 2 months of age, a number of tumor foci are grossly discernable by 3 months, and the majority of the lung is replaced by coalesced nodules by 4 months of age. As tumor progresses, the expression of endogenous CC-10 expression diminishes, and there is an increased nuclear p53 expression, suggesting binding and stabilization of the protein by the large TAg.[56] From our standpoint, we have found that the reliable progression of lethal tumors in these transgenic mice enables us to test a number of hypotheses, dosing schemes, and dosing routes. Importantly, the effects of immunomodulation by the gene transfer of specific cytokines and chemokines into tumor cells in vivo can be determined. Moreover, one can compare this direct-delivery strategy with alternative approaches, including ex vivo modification of autologous APCs using recombinant Ad vectors. The subsequent reintroduction of gene-modified APCs back into the tumor environment overcomes the inability of dendritic cells (DCs) to maturate in the presence of tumor in vivo[57] by providing functional APCs that are capable of processing and presenting tumor antigens to cytolytic T cells in vivo.[6]

### 2.2.3 Murine Models with Transplantable Xenografts

Xenotransplantation of human tumors into immunocompromised mice began in the late 1960s[58] following the discovery of the nude mouse in 1962 and its characterization as an athymic mutant in 1968.[59] The morphologic and karyotypic stability of tumors serially passaged in nude mice was described,[60] and it was established that xenotransplanted tumors in nude mice often retained distinctive phenotypic and functional characteristics found in the human host.[61] However, the "tumor-take" rate for nude mouse xenotransplants is tumor-specific, and generally, carcinomas are more difficult to establish than melanomas or sarcomas.[62] Thus, progressive tumor growth from inoculated primary tumors (i.e., cultured directly from the patient) is observed in only 33% for lung cancers[61,63] and is virtually nil for primary breast or prostate cancers. In addition to properties inherent to the tumor, nude mouse-related factors also impact on tumor-take. For example, mice infected with the mouse hepatitis virus do not accept xenotransplants, presumably because of enhanced NK cell activity.[64] In this regard, it is important to recall that although nude mice lack functionally mature T cells, they are capable of mounting normal humoral responses to T-cell-independent antigens[65] and they exhibit high NK cell activity,[66] and these properties probably impact negatively on the tumor-take rate of xenotransplants. The high NK cell activity also abrogates the metastatic potential of implanted tumors, and the incidence of metastasis is higher in mice with lower NK cell activity, e.g., young (3-week-old) syngeneic mice or the beige ($bg^j/bg^j$) mutants derived from the C57BL/6 mice.[67,68]

The discovery of a severe-combined immunodeficiency in mice[69] offered yet another option for hosting human tumor xenografts. The *scid/scid* mice are characterized by the virtual absence of functional T and B lymphocytes due to aberrancies in the rearrangement of antigen receptor genes.[70] The first successful engraftment of human solid organ tumors into scid mice began with the subcutaneous inoculation using the A549 lung adenocarcinoma cell line.[71] Since that time, a variety of human solid organ cancers, both from cell lines and primary tumor specimens, have been successfully engrafted.[72] The higher rates of successful engraftment, presumably because of the lack of residual B-cell function in *scid* mice, have led many investigators to prefer *scid/scid* mice over *nu/nu* mice as the host recipients of human xenograft tumors. Xenografts are still impacted upon by the *scid* host's innate immunity, and NK and monocyte/macrophage activities can be upregulated in these hosts. For specific needs, selective breeding of other available mutants (beige mutants with reduced NK cell activity and osteopetrosis with altered macrophage differentiation) enables the generation of strains that harbor overlapping defects in immune function.[73] Furthermore, genetic engineering and gene-targeting technology has helped create murine mutants with exquisitely specific immune defects, including mice in which CD4 or CD8 T cells are deleted[74] and mice which lack β2 microglobulin and thus do not express transplantation antigens.[75]

Xenotransplants have many advantages, the primary being that they provide a replenishable source of human tumor. This enables the genetic characterization and gene discovery of tumor-specific phenotypes, and in rare occasions, the progression toward an advanced or metastatic phenotype of the tumor (e.g., from an androgen-dependent prostate tumor to one that is androgen-independent, see below). Xenografts incorporating human tumor cells in immune deficient mice are plentiful. For example, we have developed a novel animal model mimicking intrapleural malignancy that allows for a controlled, focal dosing of reagents, and evaluation of therapeutic benefit.[76] The model is comprised of 2.5 cm segments of rat intestine that is denuded and then everted so that the serosal surface is converted into the lumenal surface of a tube. Lung cancer cells are instilled into the lumen via a polyethylene cannula on day −1, allowed to adhere to the serosal surface overnight, and this tubular xenograft is implanted into the interscapular subcutaneous tissue of a nude mouse on day 0.[76] The graft simulates metastatic tumor growth on the pleural surface *basal lamina* both grossly and histopathologically and enables robust quantitation of tumor kinetics.[76] The appearance of tumor on this surface is nodular, and these nodules coalesce over time with intervening fibrous stroma. Neovascularization is evident on histological exam of the graft, and tumor growth is continuous with a variety of NSCLC cell lines. We have found this model to offer certain tangible advantages. For example, with respect to the transduction characteristics of tumor, the value of this model is evidenced by (1) the cells are representative of human lung cancer; (2) the location of the tumor is precisely known and tumor is directly accessible; (3) the vast majority of cells that repopulate the graft are derived from those instilled (host leukocytes and fibrocytes comprise the remaining minority); (4) the mode of delivery of reagents (fluid inoculation rather than intratumoral injection) is designed so as to be clinically applicable for installation into pleural space; (5) the size of the xenograft enables quantitative assessments of transgene expression and morphometry simultaneously containing human tumor into nude mice.[76]

## 2.3 Gene Therapy of Lung Cancer Using Ad Vectors

### 2.3.1 Gene-Based Therapies Targeting Molecular Transformation

Abnormalities at the cell surface (e.g., *erbB2*), signal transduction (e.g., *ras* oncogene), gene regulation and cell cycle control (e.g., *p53*, *Rb*, *c-myc* oncogene), or apoptosis (e.g., *p53*, *BCL-2*) are all implicated in the process of transformation and can serve as targets for rational therapeutic intervention. For example, to overcome the deficits due to mutated *p53*, one strategy for lung cancer gene therapy has opted to replace the mutated *p53* gene with a normal copy.[77] Restoring *p53* function in these cells has led to decreased tumorigenicity of human cancer cells in vitro and in animal models.[78,79] Based on these preliminary studies, the first clinical gene therapy trial for human NSCLC also utilized a *p53* gene transfer strategy.[77] In this study, nine patients with advanced NSCLC were treated with either bronchoscopic or percutaneous CT-guided injections with a retroviral *p53* expression vector (a genetically reengineered retrovirus that is designed to integrate into the cell genome and express the normal *p53* protein). Of the seven patients evaluated, three showed evidence of tumor regression at the treatment site and six showed increased apoptosis of tumor cells on posttreatment biopsies. Importantly, there was no significant toxicity associated with the therapy, and in situ gene transfer was achieved. However, limited therapeutic efficacy was observed and the mechanisms responsible for the antitumor effects are still under study. For example, although it was originally believed that mutated *p53* function would have to be compensated in each and every cell for restoration of the normal apoptosis program, the results suggested otherwise. Because there was substantive tumor regression despite poor in situ gene transfer, mechanisms for the observed "bystander effect" were hypothesized.[80] The term "bystander effect" refers to the ability of gene-modified tumor cells to mediate killing of neighboring nontransfected cells. One plausible explanation is that wild-type *p53* induces release of angiostatic factors, thus undermining the blood supply to the tumor.[81] In addition, the expression of *p53* may also contribute to an immune-mediated response.[82,83] These issues have led to more mechanism-based bench and animal studies, as well as other Phase 1 clinical trials using Ad vectors encoding the *p53* gene for a variety of cancers, including lung tumors.[84]

Because of the high frequency of *p53* mutations, another strategy that uses *replication-competent viruses* has been hypothesized to be ideally suited for lung cancer. This approach employs adenoviruses (mutant dl1520 or ONYX-015) that are suggested to selectively replicate in *p53*-mutated (therefore, selectively in cancer) cells.[85,86] Consequently, these mutant viruses are promoted as "magic bullets" that kill tumor cells and leave normal tissues intact. This particular approach has generated considerable controversy both in terms of its reputed efficacy as well as its proposed mechanism of action.[87–89] In brief, its effectiveness in both in vitro and in vivo models of lung cancer needs to be confirmed. Nevertheless, the approach represents a prime example of a novel hypothesis-driven strategy that attempts to exploit the biology of a mutant virus to clinical advantage.

### 2.3.2 Immunogene Therapy

Effective immunotherapy has the potential for *systemic* eradication of disease, a payoff that is especially enticing for the treatment of lung cancer. Previous, largely unsuccessful immunomodulatory campaigns utilized nonspecific immune strategies (e.g., BCG adjuvants). Increasingly, the interest now is in developing *specific* immune interventions for lung cancer. The major obstacle for effective immunotherapy of lung cancer has been a meager understanding of the immunobiology of this disease. However, a better understanding of the reciprocal interaction between the tumor and the immune system is starting to emerge, lending itself to plausible hypotheses for intervention. We realize that an effective antitumor response may either provoke the immune system to recognize and attack the tumor, or conversely, it may serve to reduce the immunosuppression encumbered upon the host by the tumor.

Specific and effective antitumor immunity requires both ***adequate tumor antigen presentation*** and the subsequent ***generation of effector lymphocytes***. A variety of cytokines have been investigated to implement such a program in situ,[90–97] and many of the studies have utilized the Ad vector for gene delivery. For several reasons, our efforts have focused on IL-7, IL-12, and more recently on the chemokine *slc*, for the treatment of lung cancer. The rationale underlying the use of these particular cytokines and chemokines is that they all optimize conditions for tumor antigen processing and presentation by the host's APCs, and they help appropriately localize and sustain the effector lymphocytes response.[36,37,97,98]

Although the cellular infiltrates differ depending on the cytokine and model used, many studies indicate that tumor cells that have been transfected with cytokine genes can generate specific and systemic antitumor immunity in vivo. Based on these promising animal studies, what prevents these strategies from being translated into successful and curative human clinical trials? One major problem in human cancer patients may be that although lung cancers express tumor antigens,[99] they are ineffective as APCs.[100] Tumor cells cannot function as APCs because (1) they lack costimulatory molecules, (2) they are unable to adequately process Ag, and (3) they secrete a variety of inhibitory peptides which promote a state of specific T cell anergy. Thus, even for highly immunogenic tumors, professional APCs are required for antigen presentation.[101] As described above, local augmentation of IL-7 and IL-12 may help to overcome some of these defects.[37] In addition, one may bring into the tumor environment professional APCs to orchestrate a satisfactory immune response against the tumor. In this regard, DCs are potent APCs that are ideal for interacting with and activating naïve T cells to generate antigen-specific immunity.[102,103] Recent advances in the isolation and in vitro propagation of DC have stimulated great interest in the use of these cells for clinical cancer therapy.[104,105] In such approaches, DC may be envisioned to serve as vehicles for genes expressing antigens[106] or expressing cytokines in lung cancer gene therapy.[33] In addition, DC-based immunogenetic therapies may be used in combination with other strategies that have been optimized for Ag presentation.[34,37] Importantly, of the various approaches tested to gene modify the DCs, our colleagues at UCLA have determined that the Ad vector is best suited for DC transduction.[107]

### 2.3.3 Targeting Tumor Invasion and Angiogenesis

Overcoming metastatic disease is paramount for effective lung cancer therapy, and the biology underlying metastasis is gaining clarity. Metastasis is a process involving several complementary yet distinct elements, including the capacity for tumor cells to invade and traverse the basement membrane and to reestablish viable tumor foci in distant organs. Each step in this process may serve as a point for therapeutic intervention in lung cancer. As the molecular biology becomes better understood, the opportunity to incorporate specific genes into vector systems invariably materializes. The initial step, tumor invasion, requires proteolysis which has been suggested to be mediated by an overexpression and secretion of matrix metalloproteinases (MMPs) by lung cancer cells.[108–111] Therapeutically, gene transfer strategies have incorporated tissue inhibitors of metalloproteinases to inhibit invasion and metastasis[112] or have utilized antisense abrogation of MMPs to inhibit tumorigenicity.[113]

Similarly, angiogenesis (induced growth of blood vessels) is suspected to be critical for tumor survival and progression at each stage of metastasis.[114] Angiogenic progression in lung cancer is felt to be due to an imbalance of angiogenic and angiostatic factors, and the risk of metastasis in NSCLC directly correlates with the extent of tumor-derived angiogenesis.[114] Thus, strategies that inhibit angiogenic mediators or restore angiostatic factors have potential utility for all stages of lung cancer.[115–119] The important mediators implicated in promoting or inhibiting angiogenesis lend themselves favorably for inclusion into gene therapy strategies. For example, recent studies indicate that vascular endothelial growth factor (VEGF) is an important angiogenic factor produced by a variety of tumors, including lung cancer. Lymph nodes with

NSCLC metastases express significantly higher levels of VEGF than do normal, uninvolved nodes,[120] consistent with the speculation that VEGF plays an important role in the metastasis of lung cancer. In addition to VEGF, recent studies have also implicated CXC chemokines in the abnormal angiogenic/angiostatic balance in NSCLC.[121] Members of this family containing the ELR motif (e.g., IL-8) are angiogenic, whereas those that lack this motif (e.g., interferon-inducible protein 10, IP-10) are angiostatic. Accordingly, neutralizing antibodies to IL-8 reduces angiogenesis and consequently the growth of human lung tumors in SCID mice.[122]

Other molecular strategies to specifically target angiogenic vessels are also being developed. For example, the adhesion protein $\alpha_v\beta_3$ is relatively specific for angiogenic vessels where it mediates endothelial cells interaction with extracellular matrix components[123] and enables cell motility.[124] Importantly, its blockade can promote tumor regression in vivo in lung cancer models by inducing apoptosis of tumor-associated blood vessels.[125] More recently, phage display peptide libraries, which are used to screen the specific binding of a massive array of peptides, have isolated small peptides which selectively bind to receptors (including $\alpha_v\beta_3$) on angiogenic vessels. Conjugating these peptides to chemotherapeutic agents has enabled investigators to specifically target tumor vasculature and abrogate tumor growth.[126]

### 2.3.4 Adjuvants to Conventional Therapeutic Approaches for Lung Cancer

Conventional multimodality therapy for lung cancer incorporates surgery, radiation, and chemotherapy using a variety of clinical protocols dictated by the subtype and extent of disease. Theoretically, gene therapies may play important synergistic roles in augmenting the effectiveness of conventional approaches. For many such strategies, there already exists a scientific rationale to test them in combination with conventional multimodality therapy. For example, one may enhance the radiation sensitivity or chemosensitivity of tumor cells (e.g., *p53* or IκBα gene therapy)[127,128] or modify normal tissue susceptibility to cytoablative therapy (e.g., mucosal/tissue protection: by virtue of *MDR-1* or *bFGF* gene transfer). Examples of synergism with the suicide gene therapy approaches have also been studied. The herpes simplex virus (HSV) thymidine kinase gene/ganciclovir system induces radiation sensitivity into transduced tumor cells[129] suggesting that these two forms of therapy can be combined to potentiate antitumor responses.[130] Similarly, tumor cells transduced with the cytosine deaminase transgene exhibit enhanced radiation sensitivity following pretreatment with 5-fluorocytosine.[131] Because the loss of *p53* function can result in tumor resistance to ionizing radiation,[132] restoring *p53* function may restore apoptotic pathways, and promote effective radiation or chemotherapy. In fact, gene transfer of wild-type *p53* has been shown to enhance radiation sensitivity[133] and can act synergistically with cis-platinum-based chemotherapy to augment cytotoxicity.[134]

Many of the approaches outlined above as being strategies for gene therapy of "lung cancer" are generic; these approaches can be generalized to a variety of malignancy since transformed cells have in common the same aberrant growth regulatory and signal transduction pathways. The molecular and cellular pathogenesis of tumor invasion and immune evasion is also similar between tumors originating in diverse organ systems. Unfortunately, this commonality may not confer a broad-based advantage when gene therapy strategies are advanced clinically. In this respect, vectors need to provide both efficient gene delivery as well as tumor specificity, and as a result, the gene transfer strategies have to become "disease specific." Targeted vectors (as discussed elsewhere in this compilation) have to incorporate features rendering them capable of selective cell surface adherence or entry, or alternatively, have to express their therapeutic transgenes under tumor-specific regulation. Unfortunately, a lung cancer-specific cell surface target (for transductional targeting) has not been identified, and one is left trying to use targets that are generally overexpressed in tumor cells or tumor-induced endothelium.[135,136] Similarly, lung cancer also does not express a specific tumor marker. Thus, transcriptional targeting approaches largely utilize elements that are "tissue specific" rather than "cancer specific." Accordingly, constructs where transgene expression is regulated by tissue-specific promoters (e.g., SLPI, SP-A, CC-10) are being actively developed and tested.

# 3. Animal Models of Human Prostate Cancer

## 3.1 Human Prostate Cancer

After lung cancer, cancer of the prostate (CaP) is the second most common cause of cancer death in American males. A latent disease, many men have prostate cancer cells long before overt signs of the disease are apparent. The annual incidence of CaP is over 100,000 in the United States, of which over 40,000 will die of the disease. Nearly one-third of patients present with locally advanced or metastatic disease, and androgen deprivation therapy forms the basis of conventional therapy for the majority of these patients. However, currently available approaches for advanced CaP are not curative,[137] primarily because the cells lose their dependence on androgenic stimulation. The mechanisms of progression of CaP cells to hormone independence under androgen ablation therapy remain unclear. To investigate the factors and mechanisms that underlie the development of androgen resistance and metastasis, reliable in vivo models that mimic human CaP progression are essential. Moreover, it is critical that tumor models mirror the pathology, cellular, and molecular characteristics of human CaP if it is to serve as a useful tool for basic research, drug screening, or for the evaluation of new therapeutic strategies.

## 3.2 Spontaneous and Transgenic Models of Human Prostate Cancer

Currently, a single animal model cannot epitomize the multifaceted aspects of CaP pathogenesis and progression. Rodent models of prostate carcinoma have been developed by hormone treatment,[138] spontaneous development,[139] transgenic prostate-specific oncogene expression,[140] and knockout of CaP tumor suppressor genes.[141] However, these models are largely inadequate in recapitulating the progression of human disease as bone metastasis,[142] the major cause of clinical morbidity attributable to CaP. Despite pitfalls, the mouse transgenic TRAMP model has been useful for studying the development and progression of prostatic adenocarcinoma. TRAMP mice, generated by expressing SV40-TAg specifically in prostatic epithelium,[140] develop prostatic intraepithelial neoplasia (PIN) by 10–12 weeks of age, and eventually progress to adenocarcinoma with metastasis to lymph nodes and lungs.[143] As in human disease, androgen ablation therapy in these mice contributes to the emergence of androgen-independent disease with a poorly differentiated phenotype.[144]

## 3.3 Xenograft Models of Human Prostate Cancer

As for lung cancer, a number of investigators have chosen to utilize xenograft models of CaP. Unfortunately, CaP xenografts are far more fastidious than lung cancer xenografts, and the generation of models that are representative of typical human disease has only recently been accomplished. Until recently, the majority of research conducted for CaP relied on the cell lines PC-3, DU145, and LNCaP. Among these, only LNCaP cells exhibit androgen responsiveness and express the prostate-specific antigen (PSA) and androgen receptor (AR). Thus, the relevance of DU145 and PC-3 cells to clinical CaP has been questioned. To overcome the shortage of representative models of human CaP, a number of investigators began establishing xenografts in immune-deficient *scid/scid* mice using samples obtained directly from patients.[145–149] These xenografts offered the following advantages: (1) the expansion of small amounts of starting clinical material, (2) the enrichment of relatively homogeneous cell populations from heterogeneous tumor cell populations, (3) the ability to investigate progression to metastasis and androgen independence,[145,146,148] and (4) representative diversity that provided a more realistic picture of the heterogeneous nature of this disease. Investigators at UCLA established six distinct CaP xenografts from patients with locally advanced or metastatic diseases into *scid/scid* mice. Two of these xenografts, LAPC-4 and LAPC-9, have been maintained continuously for more than 2 years by serial passage in *scid/scid* mice,[145,146] and LAPC-4 has also been successfully established as a cell line in tissue culture to enable correlation with investigations performed in vitro.[145] LAPC-4 and LAPC-9 offer several advantages over previous models; both express the wild-type AR, both xenografts have intact AR signal transduction pathways, and both secrete high levels of the androgen-dependent protein, PSA. Accordingly, they grow as androgen-dependent cancers in male SCID mice and respond to androgen ablation treatment, but interestingly, they eventually progress to a hormone-refractory, androgen-independent state.[145,146] LAPC-4 and LAPC-9 can be implanted subcutaneously, orthotopically into the mouse prostate, or intratibially. Orthotopic tumors metastasize reproducibly to regional lymph nodes and lung, providing an opportunity to study prostate cancer metastasis. Intratibial injection results in the formation of osteoblastic tumors typical of human CaP where bony metastasis is the major cause of morbidity.

From a research standpoint, the generation of these xenografts has provided significant dividends. Given the inability to culture CaP by other means, the xenografts have been used to identify chromosomal abnormalities and to pinpoint the genes important to the pathogenesis of CaP. For example, loss of chromosome 10q was a frequently observed genetic defect in prostate cancer. Recently, the PTEN/MMAC tumor suppressor gene was identified and mapped to chromosome 10q23.3.[150,151] PTEN encodes a protein/lipid phosphatase which has been clearly established to function as a negative regulator of the PI3 kinase/Akt signaling pathway.[152–158] Loss of PTEN leads to constitutive activation of PI3-kinase, and in turn the Akt signaling pathway.[158] PI3 kinase is also a downstream target of several growth factors implicated in CaP pathogenesis including epidermal growth factor receptor, insulin-like growth factor receptor, and Her-2/neu, and it is possible that deregulation of this pathway in PTEN-deficient cells may indeed be responsible for the cancer phenotype. Of note, knockout mice lacking PTEN as a consequence of targeted deletion develop multiple cancers, including prostatic hyperplasia and PIN.[141,159] Correspondingly, 50–60% of all prostate cancer xenografts established contain deletions, mutations, or absent expression of PTEN,[160,161] making the xenografts a relevant and valuable source for biological and therapeutic discovery. Prostate cancer gene therapy approaches that specifically target this pathway are now underway in these models.

In addition to modeling the abnormalities of the PTEN/MMAC pathway, xenografts are important in delineating the role of androgens and AR signaling in CaP. Prostate epithelial cells utilize androgen as a growth and differentiation factor and are dependent on androgen for survival. Once transformed, androgen deprivation is associated with a transition of CaP cells through a range of diminishing androgen dependence and ultimately androgen independence. Although not well understood, this process likely involves perturbations in AR signaling of cellular growth control. Potential AR related perturbations may involve (1) AR mutation or gene amplification, (2) cross talk between AR and other signal pathways, and/or (3) alterations in transcriptional coregulators. Greater than 80% of clinical CaP specimens have confirmed AR expression, even in advanced androgen-independent diseases.[162,163] Among these, AR gene mutation or amplification has been documented in 20–40% of CaP cases[164–166]. Both LNCaP and the CWR22 xenografts bear AR mutations that enable the receptor to be activated by nonandrogenic steroid hormones such as progesterone and estrogen. In addition, in a patient who had failed androgen ablation, it was recently demonstrated that his or her CaP cells possessed a mutated AR with altered ligand affinity. Essentially, the mutant AR functioned as a high-affinity cortisol receptor, enabling the CaP cells to circumvent the androgen requirement for growth.[167] Another emergent theme is that some hormone refractory cancers have activated the AR signaling pathway through a ligand-independent mechanism. For example, in LAPC-4 cells expressing wild-type AR, the overexpression of Her-2/neu has been shown to activate AR.[168] Not surprisingly, the LAPC-4 xenograft progresses to androgen independence after androgen ablation and differential gene expression studies reveal a consistent increase in Her-2/neu protein expression in androgen-independent tumors. Furthermore, forced overexpression of Her-2/neu in androgen-dependent CaP cells is sufficient to confer androgen-independent growth in vitro and to accelerate androgen-independent growth in castrated animals. Thus, Her-2/neu overexpression activates the AR signaling pathway in the absence of ligand and enhances the magnitude of AR response in the presence of low levels of androgen. Lastly, reconstitution experiments in a heterologous cell type expressing low levels of endogenous AR suggest that these effects of Her-2/neu on the AR pathway require AR expression.[168] Although the point where Her-2/neu and AR pathway intersects is still undefined, nuclear receptor coactivators might be potential targets since amplification of steroid receptor coactivator, AIB1, is documented in breast and ovarian cancer.[169] Cross talk between Her-2/neu and AR signaling pathways should provide a novel mechanistic insight into the development of androgen independence.

### 3.4 Gene Therapy Approaches with Adenovectors in Prostate Cancer

Recombinant adenovirus vectors (Ad) are most commonly used for CaP because they have demonstrated the capacity to deliver genes intraprostatically in animal models.[170] Hence, several ongoing human CaP clinical gene therapy trials are using Ad.[171,172] With respect to these applications, several groups are developing transcriptionally targeted prostate-specific Ad.[172–175] These strategies are beneficial in gene therapy applications in that they potentially restrict the expression of cytotoxic therapeutic genes to the malignant cells. Most commonly, the kallikrein protease PSA gene regulatory regions have been used to direct prostate-specific expression because prostate epithelia, normal or malignant, specifically express the PSA.[176] Unfortunately, the transcriptional output from the native PSA enhancer and promoter (as from most highly regulated tissue-specific promoters) is much lower than from strong constitutive viral promoters such as cytomegalovirus (CMV). For example, our studies suggest that the native PSA enhancer and promoter inserted into Ad can direct tissue-specific and androgen-inducible expression in LNCaP cells, but the transcriptional activity is 50-fold lower than the constitutive CMV promoter.

By exploiting the known properties of the native PSA control regions, we have improved the activity and specificity of the prostate-specific PSA enhancer. Previous studies had established that AR molecules bound cooperatively to AREs in the PSA enhancer core (−4326 to −3935) act synergistically with AR bound to the proximal promoter to regulate transcriptional output.[177,178] To exploit the synergistic nature of AR action, we generated chimeric enhancer constructs by (1) insertion of a synthetic element containing four tandem copies of the proximal PSA promoter AREI (ARE4) element or (2) duplication of enhancer core, and (3) removal of intervening sequences (−3744 to −2875) between the enhancer and promoter. Each of these three strategies augments activity, androgen inducibility, and retained a high degree of tissue discriminatory ability. As a result of these combined approaches, two most active constructs are termed PSE-BC (duplication of core) and PSE-BAC (insertion of core and ARE4) are approximately 20-fold higher in activity than native PSA enhancer/promoter construct, PSE, composed of the PSA enhancer (−5322 to −2855) fused to the proximal promoter (−541 to +12). Most importantly, the enhanced activity and specificity of the new PSA enhancer/promoter constructs are retained in an Ad vector. The recently developed human CaP xenografts should be excellent models to refine and evaluate this novel prostate-targeted gene therapy because their AR pathways are intact and their growth regulatory pathways bear close resemblance to clinical disease.

## 4. Summary and Discussion

We have presented for discussion a broad-based review of the utility of Ad vectors in animal models of lung cancer. Since this entire compilation is devoted to Ad gene therapy, we have particularly embellished the sections on "animal models" of disease, especially as they pertain to lung and prostate cancer. These examples illustrate that the development of our approaches may need to be disease specific, especially with respect to targeting and mode of delivery. From this review, it is evident that to realize the full potential of cancer gene therapy, advances need to be made on a number of fronts. Not only do we need to construct better Ad vectors or more relevant animal models, we also need to incorporate emerging technologies to a useful purpose within the experimental design. For example, the pathway to human clinical trials may be better paved by an improved ability to gather interim surrogate measures of gene transfer and expression in animal models. The implementation of a quantitative and noninvasive method capable of monitoring transgene expression in living animals repetitively would be useful toward validating the efficacy of any gene therapy strategy. In this respect, a number of investigators, including those at UCLA, are developing sensitive technologies for imaging transgene expression using positron emission tomography (PET) and optical measurements. PET is a noninvasive, tomographic imaging modality that already has clinical applications for the diagnosis and management of several diseases including cancer. Newer high-resolution animal microPET technology developed at UCLA is allowing for the study of smaller animal systems (mice, rats, and small primates) previously difficult to image with a resolution approaching 2 mm.[179] With relevance to gene therapy for cancer, the herpes simplex virus 1 thymidine kinase (HSV1-tk) gene has been demonstrated to be an excellent "PET reporter gene" by virtue of trapping positron-emitting 8-[18F] fluoroganciclovir (FGCV) specifically only in cells expressing HSV1-tk.[180] Using FGCV, repetitive PET imaging of adenovirus-directed hepatic expression of the HSV1-tk reporter gene in living mice has been achieved.[180–182] More importantly direct correlation between the retained PET reporter probe and the levels of HSV1-tk gene expression in the targeted organ have also been demonstrated.[180–182] Thus, PET is a sensitive and quantitative modality to image the location and magnitude of Ad vector-mediated gene expression in living animals which could be translated to clinical gene therapy application. Similarly, a charge-coupled device (CCD) camera is a highly sensitive camera for measuring photons. Advances in CCD technology can now enable investigators to quantitatively and reliably image low levels of luminescence (from the heterologous expression of the firefly luciferase gene) arising from within living animals.[183] Although tomographic images are not possible, and the signal is dependent on the depth of tissue from which the light source emanates, it is possible to get reproducible and semiquantitative images. The simplicity and minimal background signal of optical CCD luciferase approach may complement the detailed tomographic imaging of microPET and the newer confocal microscopy techniques and ultimately, be more predictive of gene transfer strategies in the treatment of human disease.

## Acknowledgments

We wish to thank Dr Steven M Dubinett, Dr Charles L. Sawyers in the Department of Medicine, Dr Robert E. Reiter in the Department of Urology and Dr Sanjiv Gambhir in the Department of Molecular and Medical Pharmacology at the UCLA School of Medicine for thoughtful advice. This project has been supported by the Veterans Administration-Career Development Award and Medical Research Funds (rkb), NIH-R01-CA78654 (rkb), California Cancer Research Program (lw), Department of Army (lw), the UCLA-Jonsson Comprehensive Cancer Center and the UCLA-Gene Medicine Program.

## References

1. Malkinson AM. Molecular comparison of human and mouse pulmonary adenocarcinomas. *Exp Lung Res* 1998;**24**:541.
2. Sellers TA, Bailey-Wilson JE, Elston RC, Wilson AF, Elston GZ, Ooi WL, et al. Evidence for mendelian inheritance in the pathogenesis of lung cancer. *J Natl Cancer Inst* 1990;**82**:1272.
3. Schwartz AG, Yang P, Swanson GM. Familial risk of lung cancer among nonsmokers and their relatives. *Am J Epidemiol* 1996;**144**:554.
4. Carbone D. The biology of lung cancer. *Semin Oncol* 1997;**24**:388.
5. Salgia R, Skarin AT. Molecular abnormalities in lung cancer. *J Clin Oncol* 1998;**16**:1207.
6. Dubinett SM, Miller PW, Sharma S, Batra RK. Gene therapy for lung cancer. *Hematol Oncol Clin North Am* 1998;**12**:569.
7. Sethi T. Science, medicine, and the future. Lung cancer. *Br Med J* 1997;**314**:652.
8. Shopland DR, Eyre HJ, Pechacek TF. Smoking-attributable cancer mortality in 1991: is lung cancer now the leading cause of death among smokers in the United States? *J Natl Cancer Inst* 1991;**83**:1142.
9. Sozzi G, Miozzo M, Pastorino U, Pilotti S, Donghi R, Giarola M, et al. Genetic evidence for an independent origin of multiple preneoplastic and neoplastic lung lesions. *Cancer Res* 1995;**55**:135.
10. Mao L, Lee JS, Kurie JM, Fan YH, Lippman SM, Lee JJ, et al. Clonal genetic alterations in the lungs of current and former smokers. *J Natl Cancer Inst* 1997;**89**:857.
11. Wistuba II, Lam S, Behrens C, Virmani AK, Fong KM, LeRiche J, et al. Molecular damage in the bronchial epithelium of current and former smokers. *J Natl Cancer Inst* 1997;**89**:1366.
12. Batra R, Sharma S, Dubinett S. New gene and cell-based therapies for lung cancer. *Semin Respir Med* 2000;**21**:463.
13. Slebos R, Kibbelaar R, Dalesio O, Kooistra A, Stam J, Meijer C, et al. K-ras oncogene activation as a prognostic marker in adenocarcinoma of the lung. *N Engl J Med* 1990;**323**:561.
14. Mitsudomi T, Steinberg SM, Nau MM, Carbone D, D'Amico D, Bodner S, et al. p53 gene mutations in non-small-cell lung cancer cell lines and their correlation with the presence of ras mutations and clinical features. *Oncogene* 1992;**7**:171.
15. Kamb A, Shattuck-Eidens D, Eeles R, Liu Q, Gruis NA, Ding W, et al. Analysis of the p16 gene (CDKN2) as a candidate for the chromosome 9p melanoma susceptibility locus. *Nat Genet* 1994;**8**:23.
16. Shapiro GI, Edwards CD, Kobzik L, Godleski J, Richards W, Sugarbaker DJ, et al. Reciprocal Rb inactivation and p16INK4 expression in primary lung cancers and cell lines. *Cancer Res* 1995;**55**:505.

17. Sozzi G, Veronese ML, Negrini M, Baffa R, Cotticelli MG, Inoue H, et al. The FHIT gene 3p14.2 is abnormal in lung cancer. *Cell* 1996;**85**:17.
18. Pezzella F, Turley H, Kuzu I, Tungekar MF, Dunnill MS, Pierce CB, et al. bcl-2 protein in non-small-cell lung carcinoma. *N Engl J Med* 1993;**329**:690.
19. Kern JA, Schwartz DA, Nordberg JE, Weiner DB, Greene MI, Torney L, et al. p185neu expression in human lung adenocarcinomas predicts shortened survival. *Cancer Res* 1990;**50**:5184.
20. Sundaresan V, Ganly P, Hasleton P, Rudd R, Sinha G, Bleehen NM, et al. p53 and chromosome 3 abnormalities, characteristic of malignant lung tumours, are detectable in preinvasive lesions of the bronchus. *Oncogene* 1992;**7**:1989.
21. Marchetti A, Pellegrini S, Bertacca G, Buttitta F, Gaeta P, Carnicelli V, et al. FHIT and p53 gene abnormalities in bronchioloalveolar carcinomas. Correlations with clinicopathological data and K-ras mutations. *J Pathol* 1998;**184**:240.
22. Kerr KM, Carey FA, King G, Lamb D. Atypical alveolar hyperplasia: relationship with pulmonary adenocarcinoma, p53, and c-erbB-2 expression. *J Pathol* 1994;**174**:249.
23. Westra WH, Baas IO, Hruban RH, Askin FB, Wilson K, Offerhaus GJ, et al. K-ras oncogene activation in atypical alveolar hyperplasias of the human lung. *Cancer Res* 1996;**56**:2224.
24. Cooper CA, Carby FA, Bubb VJ, Lamb D, Kerr KM, Wyllie AH. The pattern of K-ras mutation in pulmonary adenocarcinoma defines a new pathway of tumour development in the human lung. *J Pathol* 1997;**181**:401.
25. Mao L, Hruban RH, Boyle JO, Tockman M, Sidransky D. Detection of oncogene mutations in sputum precedes diagnosis of lung cancer. *Cancer Res* 1994;**54**:1634.
26. Yesner R. Pathogenesis and pathology. *Clin Chest Med* 1993;**14**:17.
27. Tuveson DA, Jacks T. Modeling human lung cancer in mice: similarities and shortcomings. *Oncogene* 1999;**18**:5318.
28. Witschi H. Tobacco smoke as a mouse lung carcinogen. *Exp Lung Res* 1998;**24**:385.
29. Malkinson A, Belinsky S. The use of animal models in preclinical studies. In: Pass H, Mitchell J, Johnson D, Turrisi A, editors. *Lung cancer: principles and practice*. Philadelphia: Lippincott-Raven Publishers; 1996.
30. Stoner GD. Introduction to mouse lung tumorigenesis. *Exp Lung Res* 1998;**24**:375.
31. Shimkin MB, Stoner GD. Lung tumors in mice: application to carcinogenesis bioassay. *Adv Cancer Res* 1975;**21**:1.
32. Malkinson AM. Primary lung tumors in mice: an experimentally manipulable model of human adenocarcinoma. *Cancer Res* 1992;**52**:2670s.
33. Sharma S, Miller P, Stolina M, Zhu L, Huang M, Paul R, et al. Multi-component gene therapy vaccines for lung cancer: effective eradication of established murine tumors in vivo with interleukin 7/herpes simplex thymidine kinase-transduced autologous tumor and ex vivo-activated dendritic cells. *Gene Ther* 1997;**4**:1361.
34. Miller PW, Sharma S, Stolina M, Chen K, Zhu L, Paul RW, et al. Dendritic cells augment granulocyte-macrophage colony-stimulating factor (GM-CSF)/herpes simplex virus thymidine kinase-mediated gene therapy of lung cancer. *Cancer Gene Ther* 1998;**5**:380.
35. Sharma S, Stolina M, Lin Y, Gardner B, Miller PW, Kronenberg M, et al. T cell-derived IL-10 promotes lung cancer growth by suppressing both T cell and APC function. *J Immunol* 1999;**163**:5020.
36. Sharma S, Stolina M, Luo J, Strieter RM, Burdick M, Zhu LX, et al. Secondary lymphoid tissue chemokine mediates T cell-dependent antitumor responses in vivo. *J Immunol* 2000;**164**:4558.
37. Miller PW, Sharma S, Stolina M, Butterfield LH, Luo J, Lin Y, et al. Intratumoral administration of adenoviral interleukin 7 gene-modified dendritic cells augments specific antitumor immunity and achieves tumor eradication. *Hum Gene Ther* 2000;**11**:53.

38. Stolina M, Sharma S, Lin Y, Dohadwala M, Gardner B, Luo J, et al. Specific inhibition of cyclooxygenase 2 restores antitumor immunity by altering the balance of IL-10 and IL-12 synthesis. *J Immunol* 2000;**164**:361.
39. Herzog CR, Lubet RA, You M. Genetic alterations in mouse lung tumors: implications for cancer chemoprevention. *J Cell Biochem Suppl* 1997;**28–29**:49.
40. Chen B, Johanson L, Wiest JS, Anderson MW, You M. The second intron of the K-ras gene contains regulatory elements associated with mouse lung tumor susceptibility. *Proc Natl Acad Sci USA* 1994;**91**:1589.
41. You M, Candrian U, Maronpot RR, Stoner GD, Anderson MW. Activation of the Ki-ras protooncogene in spontaneously occurring and chemically induced lung tumors of the strain A mouse. *Proc Natl Acad Sci USA* 1989;**86**:3070.
42. Herzog CR, Wiseman RW, You M. Deletion mapping of a putative tumor suppressor gene on chromosome 4 in mouse lung tumors. *Cancer Res* 1994;**54**:4007.
43. Wiseman RW, Cochran C, Dietrich W, Lander ES, Soderkvist P. Allelotyping of butadiene-induced lung and mammary adenocarcinomas of B6C3F1 mice: frequent losses of heterozygosity in regions homologous to human tumor-suppressor genes. *Proc Natl Acad Sci USA* 1994;**91**:3759.
44. Horio Y, Chen A, Rice P, Roth JA, Malkinson AM, Schrump DS. Ki-ras and p53 mutations are early and late events, respectively, in urethane-induced pulmonary carcinogenesis in A/J mice. *Mol Carcinog* 1996;**17**:217.
45. Hecht SS, Morse MA, Amin S, Stoner GD, Jordan KG, Choi CI, et al. Rapid single-dose model for lung tumor induction in A/J mice by 4-(methylnitrosamino)-1-(3-pyridyl)-1-butanone and the effect of diet. *Carcinogenesis* 1989;**10**:1901.
46. Kim SH, Lee CS. Induction of benign and malignant pulmonary tumours in mice with benzo(a)pyrene. *Anticancer Res* 1996;**16**:465.
47. Batra R, Olsen J, Pickles R, Hoganson S, Boucher R. Transduction of non-small cell lung cancer cells by adenoviral and retroviral vectors. *Am J Respir Cell Mol Biol* 1998;**18**:402.
48. Bergelson JM, Cunningham JA, Droguett G, Kurt-Jones EA, Krithivas A, Hong JS, et al. Isolation of a common receptor for Coxsackie B viruses and adenoviruses 2 and 5. *Science* 1997;**275**:1320.
49. Hutchin ME, Pickles RJ, Yarbrough WG. Efficiency of adenovirus-mediated gene transfer to oropharyngeal epithelial cells correlates with cellular differentiation and human coxsackie and adenovirus receptor expression. *Hum Gene Ther* 2000;**11**:2365.
50. Compere SJ, Baldacci P, Jaenisch R. Oncogenes in transgenic mice. *Biochim Biophys Acta* 1988;**948**:129.
51. Kao C, Huang J, Wu SQ, Hauser P, Reznikoff CA. Role of SV40 T antigen binding to pRB and p53 in multistep transformation in vitro of human uroepithelial cells. *Carcinogenesis* 1993;**14**:2297.
52. Levine AJ, Momand J. Tumor suppressor genes: the p53 and retinoblastoma sensitivity genes and gene products. *Biochim Biophys Acta* 1990;**1032**:119.
53. Wikenheiser K, Clark J, Linnoila R, Stahlman M, Whitsett J. Simian virus 40 large T antigen directed by transcriptional elements of the human surfactant protein C gene produces pulmonary adenocarcinomas in transgenic mice. *Cancer Res* 1992;**52**:5342.
54. Wikenheiser K, Whitsett J. Tumor progression and cellular differentiation of pulmonary adenocarcinomas in SV40 large T antigen transgenic mice. *Am J Respir Cell Mol Biol* 1997;**16**:713.
55. Sandmoller A, Halter R, Suske G, Paul D, Beato M. A transgenic mouse model for lung adenocarcinoma. *Cell Growth Differ* 1995;**6**:97.

56. Magdaleno S, Wang G, Mireles V, Ray M, Finegold M, Demayo F. Cyclin-dependent kinase inhibitor expression in pulmonary clara cells transformed with SV40 large T antigen in transgenic mice. *Cell Growth Differ* 1997;**8**:145.
57. Gabrilovich DI, Chen HL, Girgis KR, Cunningham HT, Meny GM, Nadaf S, et al. Production of vascular endothelial growth factor by human tumors inhibits the functional maturation of dendritic cells. *Nat Med* 1996;**2**:1096.
58. Rygaard J, Povlsen CO. Heterotransplantation of a human malignant tumour to "Nude" mice. *Acta Pathol Microbiol Scand* 1969;**77**:758.
59. Pantelouris EM. Absence of thymus in a mouse mutant. *Nature* 1968;**217**:370.
60. Povlsen CO, Visfeldt J, Rygaard J, Jensen G. Growth patterns and chromosome constitutions of human malignant tumours after long-term serial transplantation in nude mice. *Acta Pathol Microbiol Scand A* 1975;**83**:709.
61. Shimosato Y, Kameya T, Hirohashi S. Growth, morphology, and function of xenotransplanted human tumors. *Pathol Annu* 1979;**14 Pt 2**:215.
62. Fidler IJ. Rationale and methods for the use of nude mice to study the biology and therapy of human cancer metastasis. *Cancer Metastasis Rev* 1986;**5**:29.
63. Bepler G, Neumann K. Nude mouse xenografts as in vivo models for lung carcinomas. *In Vivo* 1990;**4**:309.
64. Kyriazis AP, DiPersio L, Michael JG, Pesce AJ. Influence of the mouse hepatitis virus (MHV) infection on the growth of human tumors in the athymic mouse. *Int J Cancer* 1979;**23**:402.
65. Reed ND, Manning JK, Baker PJ, Ulrich JT. Analysis of 'thymus-independent' immune responses using nude mice. In: Rygaard J, Povlsen CO, editors. *Proceedings of the first international workshop on nude mice.* Stuttgart, Verlag; 1974. p. 95–103. 19:19.
66. Hanna N. Role of natural killer cells in control of cancer metastasis. *Cancer Metastasis Rev* 1982;**1**:45.
67. Hanna N, Burton RC. Definitive evidence that natural killer (NK) cells inhibit experimental tumor metastases in vivo. *J Immunol* 1981;**127**:1754.
68. Talmadge JE, Meyers KM, Prieur DJ, Starkey JR. Role of NK cells in tumour growth and metastasis in beige mice. *Nature* 1980;**284**:622.
69. Bosma GC, Custer RP, Bosma MJ. A severe combined immunodeficiency mutation in the mouse. *Nature* 1983;**301**:527.
70. Bosma MJ, Carroll AM. The SCID mouse mutant: definition, characterization, and potential uses. *Annu Rev Immunol* 1991;**9**:323.
71. Reddy S, Piccione D, Takita H, Bankert RB. Human lung tumor growth established in the lung and subcutaneous tissue of mice with severe combined immunodeficiency. *Cancer Res* 1987;**47**:2456.
72. Williams SS, Alosco TR, Croy BA, Bankert RB. The study of human neoplastic disease in severe combined immunodeficient mice. *Lab Anim Sci* 1993;**43**:139.
73. Croy BA, Percy DH, Smith AL. What are scid mice and why is it timely to devote a special topic issue to them? *Lab Anim Sci* 1993;**43**:120.
74. Mak TW, Rahemtulla A, Schilham M, Koh DR, Fung-Leung WP. Generation of mutant mice lacking surface expression of CD4 or CD8 by gene targeting. *J Autoimmun* 1992;**5**(Suppl. A):55.
75. Koller BH, Smithies O. Altering genes in animals by gene targeting. *Annu Rev Immunol* 1992;**10**:705.
76. Hoganson D, Matsui H, Batra R, Boucher R. Toxin gene-mediated growth inhibition of lung adenocarcinoma in an animal model of pleural malignancy. *Hum Gene Ther* 1998;**9**:1143.

77. Roth JA, Nguyen D, Lawrence DD, Kemp BL, Carrasco CH, Ferson DZ, et al. Retrovirus-mediated wild-type p53 gene transfer to tumors of patients with lung cancer. *Nat Med* 1996;**2**:985.
78. Qazilbash M, Xiao X, Cowan K, Walsh C. Cancer gene therapy using a novel adeno-associated virus vector expressing human wild-type p53. *Gene Ther* 1997;**4**:675.
79. Takahashi T, Carbone D, Takahashi T, Nau M, Hida T, Linnoila I, et al. Wild-type but not mutant p53 suppresses the growth of human lung cancer cells bearing multiple genetic lesions. *Cancer Res* 1992;**52**:2340.
80. Freeman SM, Abboud CN, Whartenby KA, Packman CH, Koeplin DS, Moolten FL, et al. The "bystander effect": tumor regression when a fraction of the tumor mass is genetically modified. *Cancer Res* 1993;**53**:5274.
81. Nishizaki M, Fujiwara T, Tanida T, Hizuta A, Nishimori H, Tokino T, et al. Recombinant adenovirus expressing wild-type p53 is antiangiogenic: a proposed mechanism for bystander effect. *Clin Cancer Res* 1999;**5**:1015.
82. Chen H, Carbone D. p53 as a target for anti-cancer immunotherapy. *Mol Med Today* 1997;**3**:160.
83. Vierboom MP, Nijman HW, Offringa R, van der Voort EI, van Hall T, van den Broek L, et al. Tumor eradication by wild-type p53-specific cytotoxic T lymphocytes. *J Exp Med* 1997;**186**:695.
84. Swisher SG, Roth JA, Nemunaitis J, Lawrence DD, Kemp BL, Carrasco CH, et al. Adenovirus-mediated p53 gene transfer in advanced non-small-cell lung cancer. *J Natl Cancer Inst* 1999;**91**:763.
85. Bischoff JR, Kirn DH, Williams A, Heise C, Horn S, Muna M, et al. An adenovirus mutant that replicates selectively in p53-deficient human tumor cells. *Science* 1996;**274**:373.
86. Heise C, Sampson-Johannes A, Williams A, McCormick F, Von Hoff D, Kirn D. ONYX-015, an E1B gene-attenuated adenovirus, causes tumor-specific cytolysis and antitumoral efficacy that can be augmented by standard chemotherapeutic agents. *Nat Med* 1997;**3**:639.
87. Hall AR, Dix BR, O'Carroll SJ, Braithwaite AW. p53-dependent cell death/apoptosis is required for a productive adenovirus infection. *Nat Med* 1998;**4**:1068.
88. Harada JN, Berk AJ. p53-Independent and -dependent requirements for E1B-55K in adenovirus type 5 replication. *J Virol* 1999;**73**:5333.
89. Rothmann T, Hengstermann A, Whitaker NJ, Scheffner M, zur Hausen H. Replication of ONYX-015, a potential anticancer adenovirus, is independent of p53 status in tumor cells. *J Virol* 1998;**72**:9470.
90. Fearon E, Pardoll D, Itaya T, Golumbek P, Levitsky H, Simons J, et al. Interleukin-2 production by tumor cells bypasses T helper function in the generation of an antitumor response. *Cell* 1990;**60**:397.
91. Allione A, Consalvo M, Nanni P, Lollini PL, Cavallo F, Giovarelli M, et al. Immunizing and curative potential of replicating and nonreplicating murine mammary adenocarcinoma cells engineered with interleukin (IL)-2, IL-4, IL-6, IL-7, IL-10, tumor necrosis factor alpha, granulocyte-macrophage colony-stimulating factor, and gamma-interferon gene or admixed with conventional adjuvants. *Cancer Res* 1994;**54**:6022.
92. Bottazzi B, Walter S, Govoni D, Colotta F, Mantovani A. Monocyte chemotactic cytokine gene transfer modulates macrophage infiltration, growth, and susceptibility to IL-2 therapy of a murine melanoma. *J Immunol* 1992;**148**:1280.
93. Colombo MP, Ferrari G, Stoppacciaro A, Parenza M, Rodolfo M, Mavilio F, et al. Granulocyte colony-stimulation factor (G-CSF) gene transfer suppress tumorigenicity of a murine adenocarcinoma in vivo. *J Exp Med* 1991;**173**:889.

94. Heike Y, Takahashi M, Kanegae Y, Sato Y, Saito I, Saijo N. Interleukin-2 gene transduction into freshly isolated lung adenocarcinoma cells with adenoviral vectors. *Hum Gene Ther* 1997;**8**:1.
95. Zitvogel L, Tahara H, Robbins P, et al. Cancer immunotherapy of established tumors with IL-12: effective delivery by genetically engineered fibroblasts. *J Immunol* 1995;**155**:1393.
96. Dranoff G, Jaffee E, Lazenby A, Golumbek P, Levitsky H, Brose K, et al. Vaccination with irradiated tumor cells engineered to secrete murine granulocyte-macrophage colony-stimulating factor stimulates potent, specific, and long-lasting anti-tumor immunity. *Proc Natl Acad Sci USA* 1993;**90**:3539.
97. Sharma S, Wang J, Huang M, Paul R, Lee P, McBride W, et al. Interleukin-7 gene transfer in non-small cell lung cancer decreases tumor proliferation, modifies cell surface molecule expression, and enhances antitumor reactivity. *Cancer Gene Ther* 1996;**3**:302.
98. Sica D, Rayman P, Stanley J, Edinger M, Tubbs RR, Klein E, et al. Interleukin 7 enhances the proliferation and effector function of tumor-infiltrating lymphocytes from renal-cell carcinoma. *Int J Cancer* 1993;**53**:941.
99. Yoshino I, Goedegebuure PS, Peoples GE, Parikh AS, DiMaio JM, Lyerly HK, et al. HER2/neu-derived peptides are shared antigens among human non-small cell lung cancer and ovarian cancer. *Cancer Res* 1994;**54**:3387.
100. Restifo NP, Esquivel F, Kawakami Y, Yewdell JW, Mule JJ, Rosenberg SA, et al. Identification of human cancers deficient in antigen processing. *J Exp Med* 1993;**177**:265.
101. Huang AYC, Golumbek P, Ahmadzadeh M, Jaffee E, Pardoll D, Levitsky H. Role of bone marrow-derived cells in presenting MHC class I-restricted tumor antigens. *Science* 1994;**264**:961.
102. Caux C, Liu Y, Banchereau J. Recent advances in the study of dendritic cells and follicular dendritic cells. *Immunol Today* 1995;**16**:2.
103. Steinman RM. The dendritic cell system and its role in immunogenicity. *Annu Rev Immunol* 1991;**9**:271.
104. Hsu F, Benike C, Fagnoni F, Liles T, Czerwinski D, Taidi B, et al. Vaccination of patients with B-cell lymphoma using autologous antigen-pulsed dendritic cells. *Nat Med* 1996;**2**:52.
105. Nestle F, Alijagic S, Gilliet M, Sun Y, Grabbe S, Dummer R, et al. Vaccination of melanoma patients with peptide- or tumor lysate-pulsed dendritic cells. *Nat Med* 1998;**4**:328.
106. Ribas A, Butterfield L, McBride W, Jilani S, Bui L, Vollmer C, et al. Genetic immunization for the melanoma antigen MART-1/Melan-A using recombinant adenovirus-transduced murine dendritic cells. *Cancer Res* 1997;**57**:2865.
107. Arthur J, Butterfield L, Roth M, Bui L, Kiertscher S, Lau R, et al. A comparison of gene transfer methods in human dendritic cells. *Cancer Gene Ther* 1997;**4**:17.
108. Bolon I, Devouassoux M, Robert C, Moro D, Brambilla C, Brambilla E. Expression of urokinase-type plasminogen activator, stromelysin 1, stromelysin 3, and matrilysin genes in lung carcinomas. *Am J Pathol* 1997;**150**:1619.
109. Garbisa S, Scagliotti G, Masiero L, Di Francesco C, Caenazzo C, Onisto M, et al. Correlation of serum metalloproteinase levels with lung cancer metastasis and response to therapy. *Cancer Res* 1992;**52**:4548.
110. Kawano N, Osawa H, Ito T, Nagashima Y, Hirahara F, Inayama Y, et al. Expression of gelatinase A, tissue inhibitor of metalloproteinases-2, matrilysin, and trypsin(ogen) in lung neoplasms: an immunohistochemical study. *Hum Pathol* 1997;**28**:613.
111. Mari B, Anderson I, Mari S, Ning Y, Lutz Y, Kobzik L, et al. Stromelysin-3 is induced in tumor/stroma cocultures and inactivated via a tumor-specific and basic fibroblast growth factor-dependent mechanism. *J Biol Chem* 1998;**273**:618.

112. DeClerck Y, Perez N, Shimada H, Boone T, Langley K, Taylor S. Inhibition of invasion and metastasis in cells transfected with an inhibitor of metalloproteinases. *Cancer Res* 1992;**52**:701.
113. Noel A, Lefebvre O, Maquoi E, VanHoorde L, Chenard M, Mareel M, et al. Stromelysin-3 expression promotes tumor take in nude mice. *J Clin Invest* 1996;**97**:1924.
114. Skobe M, Rockwell P, Goldstein N, Vosseler S, Fusenig N. Halting angiogenesis suppresses carcinoma cell invasion. *Nat Med* 1997;**3**:1222.
115. Folkman J. Angiogenesis in cancer, vascular, rheumatoid and other disease. *Nat Med* 1995;**1**:27.
116. Fontanini G, Vignati S, Lucchi M, Mussi A, Calcinai A, Boldrini L, et al. Neoangiogenesis and p53 protein in lung cancer: their prognostic role and their relation with vascular endothelial growth factor (VEGF) expression. *Br J Cancer* 1997;**75**:1295.
117. O'Reilly MS, Holmgren L, Shing Y, Chen C, Rosenthal RA, Moses M, et al. Angiostatin: a novel angiogenesis inhibitor that mediates the suppression of metastases by a Lewis lung carcinoma. *Cell* 1994;**79**:315.
118. O'Reilly MS, Boehm T, Shing Y, Fukai N, Vasios G, Lane WS, et al. Endostatin: an endogenous inhibitor of angiogenesis and tumor growth. *Cell* 1997;**88**:277.
119. Cyster J. Chemokines and cell migration in secondary lymphoid organs. *Science* 1999;**286**:2098.
120. Ohta Y, Watanabe Y, Murakami S, Oda M, Hayashi Y, Nonomura A, et al. Vascular endothelial growth factor and lymph node metastasis in primary lung cancer. *Br J Cancer* 1997;**76**:1041.
121. Arenberg D, Polverini P, Kunkel S, Shanafelt A, Hesselgesser J, Horuk R, et al. The role of CXC chemokines in the regulation of angiogenesis in non-small cell lung cancer. *J Leukoc Biol* 1997;**62**:554.
122. Arenberg D, Kunkel S, Polverini P, Glass M, Burdick M, Strieter R. Inhibition of Interleukin-8 reduces tumorigenesis of human non-small cell lung cancer in SCID mice. *J Clin Invest* 1996;**97**:2792.
123. Hynes R. Integrins: versatility, modulation, and signaling in cell adhesion. *Cell* 1992;**69**:11.
124. Leavesley P, Schwartz M, Rosenfeld M, Cheresh D. Integrin b1- and b3-mediated endothelial cell migration is triggered through distinct signaling mechanisms. *J Cell Biol* 1993;**121**:163.
125. Brooks P, Montgomery A, Rosenfeld M, Reisfeld R, Hu T, Klier G, et al. Integrin $\alpha_v\beta_3$ antagonists promote tumor regression by inducing apoptosis of angiogenic blood vessels. *Cell* 1994;**79**:1157.
126. Arap W, Pasqualini R, Ruoslahti E. Cancer treatment by targeted drug delivery to tumor vasculature in a mouse model. *Science* 1998;**279**:377.
127. Wang C-Y, Mayo MW, Baldwin ASJ. TNF- and cancer therapy-induced apoptosis: potentiation by inhibition of NF-κB. *Science* 1996;**274**:784.
128. Batra RK, Guttridge DC, Brenner DA, Dubinett SM, Baldwin AS, Boucher RC. Ikappa-Balpha gene transfer is cytotoxic to squamous-cell lung cancer cells and sensitizes them to tumor necrosis factor-alpha-mediated cell death. *Am J Respir Cell Mol Biol* 1999;**21**:238.
129. Kim JH, Kim SH, Brown SL, Freytag SO. Selective enhancement be an antiviral agent of the radiation-induced cell killing of human glioma cells transduced with *HSV-tk* gene. *Cancer Res* 1994;**54**:6053.
130. McBride W, Dougherty G. Radiotherapy for genes that cause cancer. *Nat Med* 1995;**1**:1215.
131. Hanna N, Mauceri H, Wayne J, Hallahan D, Kufe D, Weichselbaum R. Virally directed cytosine deaminase/5-fluorocytosine gene therapy enhances radiation response in human cancer xenografts. *Cancer Res* 1997;**57**:4205.

132. McIlwrath A, Vasey P, Ross G, Brown R. Cell cycle arrests and radiosensitivity of human tumor cell lines: dependence on wild-type p53 for radiosensitivity. *Cancer Res* 1994;**54**:3718.
133. Gallardo D, Drazen ZE, McBride WH. Adenovirus-based transfer of wild-type p53 gene increases ovarian tumor radiosensitivity. *Cancer Res* 1996;**56**:4891.
134. Nguyen D, Spitz F, Yen N, Cristiano R, Roth J. Gene therapy for lung cancer: enhancement of tumor suppression by a combination of sequential systemic cisplatin and adenovirus-mediated p53 gene transfer. *J Thorac Cardiovasc Surg* 1996;**112**:1372.
135. Douglas JT, Rogers BE, Rosenfeld ME, Michael SI, Feng M, Curiel DT. Targeted gene delivery by tropism-modified adenoviral vectors. *Nat Biotechnol* 1996;**14**:1574.
136. Wickham T, Roelvink P, Brough D, Kovesdi I. Adenovirus targeted to heparan-containing receptors increases its gene delivery efficiency to multiple cell types. *Nat Biotechnol* 1996;**14**:1570.
137. Jones GW, Mettlin C, Murphy GP, Guinan P, Herr HW, Hussey DH, et al. Patterns of care for carcinoma of the prostate gland: results of a national survey of 1984 and 1990. *J Am Coll Surg* 1995;**180**:545.
138. Noble RL. The development of prostatic adenocarcinoma in Nb rats following prolonged sex hormone administration. *Cancer Res* 1977;**37**:1929.
139. Voigt W, Dunning WF. In vivo metabolism of testosterone-3H in R-3327, an androgen-sensitive rat prostatic adenocarcinoma. *Cancer Res* 1974;**34**:1447.
140. Greenberg NM, DeMayo F, Finegold MJ, Medina D, Tilley WD, Aspinall JO, et al. Prostate cancer in a transgenic mouse. *Proc Natl Acad Sci USA* 1995;**92**:3439.
141. Podsypanina K, Ellenson LH, Nemes A, Gu J, Tamura M, Yamada KM, et al. Mutation of Pten/Mmac1 in mice causes neoplasia in multiple organ systems. *Proc Natl Acad Sci USA* 1999;**96**:1563.
142. Zhau HE, Li CL, Chung LW. Establishment of human prostate carcinoma skeletal metastasis models. *Cancer* 2000;**88**:2995.
143. Gingrich JR, Barrios RJ, Morton RA, Boyce BF, DeMayo FJ, Finegold MJ, et al. Metastatic prostate cancer in a transgenic mouse. *Cancer Res* 1996;**56**:4096.
144. Gingrich JR, Barrios RJ, Kattan MW, Nahm HS, Finegold MJ, Greenberg NM. Androgen-independent prostate cancer progression in the TRAMP model. *Cancer Res* 1997;**57**:4687.
145. Klein KA, Reiter RE, Redula J, Moradi H, Zhu XL, Brothman AR, et al. Progression of metastatic human prostate cancer to androgen independence in immunodeficient SCID mice. *Nat Med* 1997;**3**:402.
146. Craft N, Chhor C, Tran C, Belldegrun A, DeKernion J, Witte ON, et al. Evidence for clonal outgrowth of androgen-independent prostate cancer cells from androgen-dependent tumors through a two-step process. *Cancer Res* 1999;**59**:5030.
147. Ellis WJ, Vessella RL, Buhler KR, Bladou F, True LD, Bigler SA, et al. Characterization of a novel androgen-sensitive, prostate-specific antigen-producing prostatic carcinoma xenograft: LuCaP 23. *Clin Cancer Res* 1996;**2**:1039.
148. Wainstein MA, He F, Robinson D, Kung HJ, Schwartz S, Giaconia JM, et al. CWR22: androgen-dependent xenograft model derived from a primary human prostatic carcinoma. *Cancer Res* 1994;**54**:6049.
149. van Weerden WM, de Ridder CM, Verdaasdonk CL, Romijn JC, van der Kwast TH, Schroder FH, et al. Development of seven new human prostate tumor xenograft models and their histopathological characterization. *Am J Pathol* 1996;**149**:1055.
150. Li J, Yen C, Liaw D, Podsypanina K, Bose S, Wang SI, et al. PTEN, a putative protein tyrosine phosphatase gene mutated in human brain, breast, and prostate cancer. *Science* 1997;**275**:1943.

151. Steck PA, Pershouse MA, Jasser SA, Yung WK, Lin H, Ligon AH, et al. Identification of a candidate tumour suppressor gene, MMAC1, at chromosome 10q23.3 that is mutated in multiple advanced cancers. *Nat Genet* 1997;**15**:356.
152. Davies MA, Lu Y, Sano T, Fang X, Tang P, LaPushin R, et al. Adenoviral transgene expression of MMAC/PTEN in human glioma cells inhibits Akt activation and induces anoikis. *Cancer Res* 1998;**58**:5285.
153. Furnari FB, Huang HJ, Cavenee WK. The phosphoinositol phosphatase activity of PTEN mediates a serum-sensitive G1 growth arrest in glioma cells. *Cancer Res* 1998;**58**:5002.
154. Haas-Kogan D, Shalev N, Wong M, Mills G, Yount G, Stokoe D. Protein kinase B (PKB/Akt) activity is elevated in glioblastoma cells due to mutation of the tumor suppressor PTEN/MMAC. *Curr Biol* 1998;**8**:1195.
155. Myers MP, Pass I, Batty IH, Van der Kaay J, Stolarov JP, Hemmings BA, et al. The lipid phosphatase activity of PTEN is critical for its tumor suppressor function. *Proc Natl Acad Sci USA* 1998;**95**:13513.
156. Stambolic V, Suzuki A, de la Pompa JL, Brothers GM, Mirtsos C, Sasaki T, et al. Negative regulation of PKB/Akt-dependent cell survival by the tumor suppressor PTEN. *Cell* 1998;**95**:29.
157. Sun H, Lesche R, Li DM, Liliental J, Zhang H, Gao J, et al. PTEN modulates cell cycle progression and cell survival by regulating phosphatidylinositol 3,4,5,-trisphosphate and Akt/protein kinase B signaling pathway. *Proc Natl Acad Sci USA* 1999;**96**:6199.
158. Wu X, Senechal K, Neshat MS, Whang YE, Sawyers CL. The PTEN/MMAC1 tumor suppressor phosphatase functions as a negative regulator of the phosphoinositide 3-kinase/Akt pathway. *Proc Natl Acad Sci USA* 1998;**95**:15587.
159. Di Cristofano A, Pesce B, Cordon-Cardo C, Pandolfi PP. Pten is essential for embryonic development and tumour suppression. *Nat Genet* 1998;**19**:348.
160. Vlietstra RJ, van Alewijk DC, Hermans KG, van Steenbrugge GJ, Trapman J. Frequent inactivation of PTEN in prostate cancer cell lines and xenografts. *Cancer Res* 1998;**58**:2720.
161. Whang YE, Wu X, Suzuki H, Reiter RE, Tran C, Vessella RL, et al. Inactivation of the tumor suppressor PTEN/MMAC1 in advanced human prostate cancer through loss of expression. *Proc Natl Acad Sci USA* 1998;**95**:5246.
162. Trapman J, Brinkmann AO. The androgen receptor in prostate cancer. *Pathol Res Pract* 1996;**192**:752.
163. Hobisch A, Culig Z, Radmayr C, Bartsch G, Klocker H, Hittmair A. Distant metastases from prostatic carcinoma express androgen receptor protein. *Cancer Res* 1995;**55**:3068.
164. Gaddipati JP, McLeod DG, Heidenberg HB, Sesterhenn IA, Finger MJ, Moul JW, et al. Frequent detection of codon 877 mutation in the androgen receptor gene in advanced prostate cancers. *Cancer Res* 1994;**54**:2861.
165. Taplin ME, Bubley GJ, Shuster TD, Frantz ME, Spooner AE, Ogata GK, et al. Mutation of the androgen-receptor gene in metastatic androgen-independent prostate cancer. *N Engl J Med* 1995;**332**:1393.
166. Visakorpi T, Hyytinen E, Koivisto P, Tanner M, Keinanen R, Palmberg C, et al. In vivo amplification of the androgen receptor gene and progression of human prostate cancer. *Nat Genet* 1995;**9**:401.
167. Zhao XY, Malloy PJ, Krishnan AV, Swami S, Navone NM, Peehl DM, et al. Glucocorticoids can promote androgen-independent growth of prostate cancer cells through a mutated androgen receptor. *Nat Med* 2000;**6**:703.
168. Craft N, Shostak Y, Carey M, Sawyers CL. A mechanism for hormone-independent prostate cancer through modulation of androgen receptor signaling by the HER-2/neu tyrosine kinase. *Nat Med* 1999;**5**:280.

169. Anzick SL, Kononen J, Walker RL, Azorsa DO, Tanner MM, Guan XY, et al. AIB1, a steroid receptor coactivator amplified in breast and ovarian cancer. *Science* 1997;**277**:965.
170. Steiner MS, Zhang Y, Carraher J, Lu Y. In vivo expression of prostate-specific adenoviral vectors in a canine model. *Cancer Gene Ther* 1999;**6**:456.
171. Herman JR, Adler HL, Aguilar-Cordova E, Rojas-Martinez A, Woo S, Timme TL, et al. In situ gene therapy for adenocarcinoma of the prostate: a phase I clinical trial. *Hum Gene Ther* 1999;**10**:1239.
172. Rodriguez R, Schuur ER, Lim HY, Henderson GA, Simons JW, Henderson DR. Prostate attenuated replication competent adenovirus (ARCA) CN706: a selective cytotoxic for prostate-specific antigen-positive prostate cancer cells. *Cancer Res* 1997;**57**:2559.
173. Gotoh A, Ko SC, Shirakawa T, Cheon J, Kao C, Miyamoto T, et al. Development of prostate-specific antigen promoter-based gene therapy for androgen-independent human prostate cancer. *J Urol* 1998;**160**:220.
174. Yu DC, Sakamoto GT, Henderson DR. Identification of the transcriptional regulatory sequences of human kallikrein 2 and their use in the construction of calydon virus 764, an attenuated replication competent adenovirus for prostate cancer therapy. *Cancer Res* 1999;**59**:1498.
175. Latham JP, Searle PF, Mautner V, James ND. Prostate-specific antigen promoter/enhancer driven gene therapy for prostate cancer: construction and testing of a tissue-specific adenovirus vector. *Cancer Res* 2000;**60**:334.
176. Aumuller G, Seitz J, Lilja H, Abrahamsson PA, von der Kammer H, Scheit KH. Species- and organ-specificity of secretory proteins derived from human prostate and seminal vesicles. *Prostate* 1990;**17**:31.
177. Huang W, Shostak Y, Tarr P, Sawyers C, Carey M. Cooperative assembly of androgen receptor into a nucleoprotein complex that regulates the prostate-specific antigen enhancer. *J Biol Chem* 1999;**274**:25756.
178. Reid KJ, Hendy SC, Saito JL, Sorensen P, Nelson CC. Two classes of androgen receptor elements mediate cooperativity through allosteric interactions. *J Biol Chem* 2000;**60**:24.
179. Shao Y, Cherry SR, Farahani K, Meadors K, Siegel S, Silverman RW, et al. Simultaneous PET and MR imaging. *Phys Med Biol* 1997;**42**:1965.
180. Gambhir SS, Barrio JR, Wu L, Iyer M, Namavari M, Satyamurthy N, et al. Imaging of adenoviral-directed herpes simplex virus type 1 thymidine kinase reporter gene expression in mice with radiolabeled ganciclovir. *J Nucl Med* 1998;**39**:2003.
181. MacLaren DC, Gambhir SS, Satyamurthy N, Barrio JR, Sharfstein S, Toyokuni T, et al. Repetitive, non-invasive imaging of the dopamine D2 receptor as a reporter gene in living animals. *Gene Ther* 1999;**6**:785.
182. Gambhir SS, Barrio JR, Phelps ME, Iyer M, Namavari M, Satyamurthy N, et al. Imaging adenoviral-directed reporter gene expression in living animals with positron emission tomography. *Proc Natl Acad Sci USA* 1999;**96**:2333.
183. Contag PR, Olomu IN, Stevenson DK, Contag CH. Bioluminescent indicators in living mammals. *Nat Med* 1998;**4**:245.

# In Situ Vaccination with Adenoviral Vectors to Treat Cancer

24

*Steven M. Albelda, Edmund Moon*
Thoracic Oncology Research Group, Pulmonary, Allergy, and Critical Care Division, Department of Medicine, Perelman School of Medicine, University of Pennsylvania, Philadelphia, PA, USA

## 1. Vaccination Strategies and the Advantages of In Situ Vaccination with Adenovirus

Therapeutic vaccination for cancer involves sensitizing a patient to tumor-specific or tumor-selective antigens that enable the immune system to recognize and eliminate the tumor cells.[1] Because, by definition, existing tumors have not been eliminated, they have escaped detection by the immune system (ignorance), expressed antigens to which tolerance exists, or developed immunosuppressive mechanisms that prevent existing immune cells from being effective. Some human tumors are considered "nonimmunogenic" (i.e., have avoided activating the immune system); others (i.e., melanoma) appear to have stimulated an antitumor immune response that has been effectively suppressed by tolerance mechanisms such as upregulation of T cell inhibitory ligands (i.e., *programmed death-ligand 1*), T-regulatory cells, myeloid suppressor cells, or secretion of immune-inhibitory agents such as transforming growth factor β, prostaglandin E2, adenosine, arginase, or IL-10.[2–5]

The first step in any successful vaccination strategy requires that a tumor antigen (or antigens) be presented to the immune system in an "immunogenic" context.[1] In addition to the antigen, the immune system also requires the co-administration of a "danger signal".[6] In the context of vaccines, this is called an adjuvant. The classically used adjuvants have been alum or toll-like receptor stimulants (modified Freund's adjuvant/montenide or, more recently, agents such as Poly-IC). Traditionally, the most effective way to vaccinate has been to inject antigen or adjuvant into the skin.

Viruses (i.e., adenovirus [Ad]) or bacteria (i.e., modified *Listeria*) vectors can also be used to present antigens in an immunogenic context.[7,8] The vector infects the tumor or surrounding tissue and induces expression of the antigen. The antigen is then taken up by dendritic cells (DCs), which are also activated by the adjuvant or danger signal.[9] In some cases, the DCs themselves may be infected by the vector where the antigen is expressed intracellularly. In either case, these activated DCs then travel to the lymph nodes and come in close contact with CD4 and CD8 T cells, where they present peptides on human leukocyte antigen (HLA) molecules along with the proper costimulatory molecules. This results in activated T cells that can

**Adenoviral Vectors for Gene Therapy.**

then leave the lymph node and traffic to areas of inflammation where antigen can activate these cells and induce effector function.[10] Thus, most antigen presentation and T cell activation occurs in lymph nodes. B cell activation, with resultant production of antibodies, also occurs. Interestingly, the site of vaccination often imprints the T cells with adhesion molecules and chemokine receptors that tend to bring them back to the site.

Which antigens should be used for tumor vaccination? One approach has been to identify tumor-specific or tumor-selective antigens and then inject whole proteins or specific peptides along with adjuvants. One limitation of peptide vaccines is that they are HLA-type specific, and thus their use is limited to patients with the matching HLA type. A partial list of commonly targeted tumor antigens includes cell surface–associated mucin-1, carcinoembryonic antigen, Wilms tumor-1 protein, melanoma-specific proteins (i.e., MART-1), cancer testis antigens (i.e., NY-ESO), telomerase, prostate-specific antigen (PSA), or viral-tumor antigens such as human papillomavirus-E7. Instead of injecting the protein or peptides, investigators have also injected deoxyribonucleic acid (DNA). Alternatively, one can generate DC ex vivo by differentiating blood monocytes, directly load the DCs with proteins, DNA, or ribonucleic acid (RNA) in culture, and inject the "loaded and activated DCs" back into the patient.[11]

A major advantage of using a specific target protein or peptide is the ability of the vaccine to generate antigen-specific T cells in the blood can be easily tracked (although, interestingly, this has not been found to correlate with clinical effectiveness). A major disadvantage to this approach is that only a single or small number of tumor antigens are targeted. Because of tumor heterogeneity, this could allow some tumor cells to escape immunity and thus be unaffected. Probably more important, specific targeting allows for tumor editing, resulting in loss of the antigen. Another potential problem is that the number of known tumor-specific antigens is limited.

An alternative approach has been not to select one specific antigen, but to use polyclonal activation strategies using whole tumor cells, tumor cell extracts, or even tumor cell RNA or DNA libraries.[12] An advantage of this approach is that it presents a broad range of tumor antigens, allowing each individual's immune system to select the most potent and important tumor antigens. This broad, polyclonal antitumor response has a better chance of avoiding rapid tumor editing. A disadvantage is that it is difficult to monitor and quantify this response. Another issue is that it can be difficult, time-consuming, and expensive to prepare tumor cell-based vaccines, especially if it is based on a patient's own tumor. The most successful of these approaches clinically has probably been the use of "GVAX" in pancreatic tumor.[14] In this instance, a set of pancreatic tumor cell lines have been transduced to express granulocyte macrophage–colony-stimulating factor (GM-CSF), which attracts and activates DC.

One limitation inherent in both of these approaches is that they do nothing to alter the tumor microenvironment. This leaves problems with regard to effective trafficking of antitumor T cells into the tumors, and more important, does nothing to counter the formidable immunosuppressive mechanisms within the tumor.[4,5]

A third approach that can potentially overcome key disadvantages of either previous approach is "in situ vaccination".[12,13] In this strategy, the tumor site itself is used as a target and source of antigen. One way this can be accomplished is by intratumorally injecting activating cytokines and/or chemokines to stimulate endogenous DCs. Although the tumor microenvironment is altered, disadvantages of this approach are that the effects are short-lived and there may be a limited amount of "immunogenic" cell death occurring (apoptotic cell death may not trigger optimal antigen-presentation). Another idea has been direct intratumoral injection of autologous DCs that have been activated or enhanced (i.e., to secrete CCL21) in some way. A disadvantage of this approach is that the tumor microenvironment is still tolerogenic and may not support induction of an immune response.

As a way to overcome these limitations, many investigators are now combining agents that will both induce immunogenic cell death and make the tumor microenviroment less tolerogenic. One strategy has been to combine local tumor radiation (to induce cell death) with intratumoral injection of a Toll-like receptor agonist (i.e., TLR9).[15,16] A second strategy is to use the strong immune-activating activity of an adenoviral vector combined with a transgene that induces immunogenic cell death and/or even more powerfully activates the tumor microenvironment toward an immunostimulatory state. This second strategy is the one to be discussed in detail here.

Advantages and disadvantages of each vaccine approach are summarized in Table 1.

**Table 1 Advantages and Disadvantages of Vaccination Approaches**

| | Specific Protein or Peptide Vaccine | Polyclonal Activation | In situ Vaccination |
|---|---|---|---|
| Requirement to identify target antigens | Yes | No | No |
| Likelihood of immune escape | High | Low | Low |
| Ability to treat "scale up" treatments | Depends on platform (low for DC vaccines) | Moderate as depends on platform (lower for cells or libraries) | High |
| Can track ability of vaccine to induce immune response | Simple | Difficult | Difficult |
| Has favorable effects on tumor microenvironment | No | No | Yes |
| Requires access to tumor | No | No | Yes |
| Imprints T cells to tumors | No | No | Yes |
| Avoids tolerogenic micro-environment during T cell activation stage | Yes | Yes | No |

## 2. Adenovirus as an Immune Stimulant

Adenovirus is well suited for the role of an in situ immune stimulant. The ability of Ad to activate the immune system was well studied in the early days of gene therapy, when Ad was proposed as a vector for use in genetic diseases such as cystic fibrosis. It soon became apparent that Ad was highly immunogenic and able to induce both innate and adaptive immunity.[7–9] The presence of adenovirus activates pathogen recognition receptors, leading to powerful stimulation of the nuclear factor-κB pathway inducing cytokines such as tumor necrosis factor, IL-6, IL-1, IL-12, and type I interferons, as well as chemokines such as MIP1, RANTES, IL-8, and monocyte chemotactic protein-1. Adenoviral DNA can be recognized by TLRs, especially TLR9, which can signal through MyD88. Adenoviruses can also activate NLRP3 and the inflammasome causing IL-1 release. More recently, the cGAS/STING/TBK1 pathway has been implicated as a DNA-sensing cascade. These multiple pathways are effective in stimulating DC and result in the generation of strong anti-Ad immune responses. However, this reaction is relatively nonspecific, and if other antigens are present at the site of Ad instillation (i.e., infectious or tumor antigens), immune stimulation against these antigen will also occur. Thus, a vaccine effect can be generated.

A large number of Ad-encoded transgenes that could potentially exert in situ vaccination activity have been studied in preclinical models and, in a number of cases, small, phase I studies.[17–30] Table 2 provides some examples of these approaches.

**Table 2 Examples of In Situ Adenoviral Immunogene Therapy Studies**

| Transgene | Tumor Model | Comments | Author | Reference |
|---|---|---|---|---|
| IL-2 | Breast | Preclinical | Addison et al., 1995 | 17 |
| GM-CSF | Lung | Preclinical | Lee et al., 1997 | 18 |
| CD40L | Colon | Preclinical | Kikuchi et al., 2000 | 19 |
| CD40L | Bladder | Human trial | Malmstrom et al., 2010 | 20 |
| CD40L | Advanced cancer patients | Human trial | Pesonen et al., 2012 | 21 |
| IL-12 | Colon/breast | Preclinical | Bramson et al., 1996<br>Caruso et al., 1996 | 22<br>23 |
| IL-12 | Gastrointestinal malignancies | Human trial | Sangro et al., 2004 | 24 |
| MDA-7/IL-24 | Fibrosarcoma | Preclinical | Miyahara et al., 2006 | 25 |
| MDA-7/IL-24 | Advanced cancer patients | Human trial | Tong et al., 2005; Cunningham et al., 2005 | 26<br>27 |
| IFN-gamma | Lymphomas | Human trial | Drummer et al., 2004; Drummer et al., 2010; Dreno et al., 2014 | 28<br>29<br>30 |

Although some tumor cell death likely occurs after injection of any Ad into a tumor (providing some tumor antigens for presentation), it is likely that the antitumor vaccine effect can be amplified by including transgenes that can activate the immune system and induce additional cell death in an immunogenic manner. Two transgenes that seem to accomplish this in an especially potent way are the herpes simplex virus thymidine kinase (HSV.tk) suicide gene and the type I interferon gene.

## 3. Adenoviral (HSV.tk)

Gene therapy using a suicide gene was one of the earliest applications of cancer gene therapy.[31] The most commonly used suicide gene has been the HSV.tk gene, which converts the normally nontoxic drug ganciclovir (GCV) into a monophosphorylated form that can then be metabolized by mammalian thymidine kinases into a triphosphorylated moiety that is incorporated in DNA and causes cell death. The vector is thus injected intratumorally followed by administration of GCV or a GCV-like compound (i.e., valacyclovir). Many preclinical studies using HSV.tk inserted into an adenoviral vector (Ad.tk) were conducted and showed good antitumor activity (reviewed in Ref. 31). Initially, efficacy was thought to be primarily supported by a bystander effect mediated by gap-junctional transport of GCV-triphosphate. However, it was observed that local tumor injection of Ad.TK could cause systemic antitumor effects and efficacy was lost in immunodeficient animals, leading to the hypothesis that a much more powerful bystander mechanism was "immunogenic cell death" (perhaps related to release of heat-shock proteins,[32] production of danger signals, and a "superantigen-like" effect of the TK molecule), leading to the induction of CD8 T cell–dependent antitumor immune response. Natural killer (NK) cell activation may also be involved.[33] The reader is referred to a complete discussion of these effects in a review article by Aguilar et al.[31]

The Ad.tk approach has been shown to be safe and has shown hints of potential efficacy in phase I and phase II clinical trials in a variety of tumors (see Table 1 in Ref. 31). Tumor types most well studied include mesothelioma, prostate cancer, and glioma.

### 3.1 Ad.tk in Mesothelioma

Malignant pleura mesothelioma (MPM) is caused by asbestos exposure. It is a rapidly progressive neoplasm of the lining of the lungs that carries a high mortality rate and is poorly responsive to standard medical regimens.[34] Our group has used Ad-based in situ immunogene therapy for the treatment of MPM for a number of reasons. First, the median survival for patients with mesothelioma from the time of diagnosis ranges between 1 and 2 years, depending on comorbid disease, stage at presentation, and histological subtype. Long-term survival (>5 years) with any treatment modality is exceedingly rare in MPM. Even the best combination chemotherapy using cisplatin and a new multitargeted antifolate agent, pemetrexed (Alimta®), has been shown to improve the median survival of mesothelioma patients by only a few months. Second, MPM's location within the thoracic cavity makes the tumor uniquely accessible, facilitating directed administration of novel agents and subsequent analysis of treatment effects. Third, local persistence or recurrence of disease, rather than the development of widespread distant metastases, is responsible for most of the morbidity and mortality associated with this neoplasm. Eradication of local disease conceivably could lead to significant improvement in palliation or survival.

Between 1995 and 1999, our group treated 34 subjects with malignant mesothelioma with a single dose of a nonreplicative adenoviral vector encoding the HSV.tk "suicide gene" administered through a pleural catheter in combination with systemic ganciclovir.[35–38] We found evidence of tumor gene transfer in the 21 patients receiving "high-dose" therapy, defined by a dose of vector ($\geq 1.6 \times 10^{13}$ particles) in which transgene-encoded protein was reliably seen by immunohistochemical staining. In 13 patients, the vector was deleted in the E1 and E3 regions of the Ad; in the other 8 patients the vector had deletions in the early Ad genes *E1* and *E4*.

Our conclusions were that intrapleural administration of Ad.HSV.tk/GCV was safe and well tolerated. We did not reach a maximally tolerated dose. These data fit well with a large clinical experience showing the relative safety of Ad vectors.[20,21,24,26–30,39–47] We also noted that although intrapleural gene transfer was detectable, we saw it at the surface of the pleural tumors. Despite this localized expression of transgene, several patients had clear reductions in tumor size and two patients had slow but durable objective responses that lasted for more than 6.5 years in one case and 15 years in a second one. This was most consistent with an immune-mediated effect. The detection of antibodies against mesothelioma proteins that were not present in the pretreatment serum was also evidence of the generation of an immune response. Although these studies were stopped because of issues related to the availability of clinical-grade vector, we have opened a similar trial for mesothelioma and metastatic pleural effusions in collaboration with Advantagene.

## 3.2 Ad.tk in Other Cancers

A number of relatively small phase I and II trials of Ad.tk in prostate cancer have been conducted.[42,45] A phase II trial treated 36 patients with local recurrence after radiotherapy with intraprostate injections.[45] The therapy was well tolerated with slowing in the PSA doubling time noted and an actual reduction in PSA levels in 78% of subjects. No long-term recurrence or survival data were reported.

Ad.tk has also been examined in malignant glioma. The most recent trials include a phase IB trial enrolling 13 patients.[46] Toxicity was tolerable and survival at 2 and 3 years was longer than reported in historical controls. The largest Ad.tk trial ever conducted was a randomized, open-label phase III trial for 250 glioblastoma multiforme subjects.[47] Patients were randomized to receive either surgical resection or surgical resection plus perilesional injections of Ad.tk followed by ganciclovir. The results of the trial were controversial and complicated by the use of temozolomide, which became available in the middle of the study. The median time to death or reintervention was slightly longer in the Ad.tk group. However, the overall survival differences did not reach significance. In any case, Kaplan–Meier curves showed only small differences. There may have been a slightly stronger benefit in patients who were resistant to temozolomide.

# 4. Ad.IFN

Type 1 interferons (IFNs) have a number of characteristics that would make them ideal transgenes for in situ immunotherapy.[48,49] Interferons have immunoregulatory effects upon antibody production, NK and T cell activation, macrophage function, cross-presentation and delayed-type hypersensitivity, and major histocompatibility complex antigen expression.[50–52] They also have antiangiogenic properties as well as direct antiproliferative and cytotoxic effects.[53]

## 4.1 Ad.IFN in Mesothelioma

On the basis of preclinical studies in mice showing that Ad.IFN had better efficacy than we saw with Ad.tk,[54] we began a series of phase I trials using an Ad expressing type 1 IFN. Our first trial used intrapleural instillation of an E1-deleted Ad containing the human *IFN-β* gene in MPM (seven patients) and metastatic pleural malignancies (three patients).[55] Gene transfer was detected in 7 of the 10 patients by measurement of pleural fluid IFN-β mRNA or protein. Antitumor immune responses, including humoral responses to known tumor antigens (e.g., SV40 virus T-antigen, mesothelin) and unknown tumor antigens were elicited in 7 of 10 patients. Four patients demonstrated meaningful clinical responses, defined as disease stability and/or partial regression on 18-fluoro-deoxyglucose positron emission tomography ($^{18}$FDG-PET) and computed tomography (CT) imaging 2 months after vector administration.

In light of these encouraging results, a second study was performed with the aim of augmenting these immunologic and clinical responses.[56] Based on preclinical studies showing enhanced effects after two doses, a second phase I trial involving two administrations of Ad.IFN-β (levels ranging from $3\times10^{11}$ to $3\times10^{12}$ viral particles [vp]) via an indwelling pleural catheter separated by 1 to 2 weeks was conducted in 17 patients (10 with MPM and 7 with malignant pleural effusions). Again, overall treatment was well tolerated and antitumor humoral immune responses similar to those seen in the initial trials were induced. Several patients had meaningful clinical responses (mixed and/or partial responses) as determined by pre- and postvector delivery PET/CT scans. However, high antiadenoviral neutralizing antibody (NAb) titers were detected, even with a dose interval as short as 7 days, inhibiting effective gene transfer of the second dose.

Combined survival data from the MPM patients in the one-dose[55] and two-dose[56] Ad.IFN-β trials showed a median survival of 22 months, with three patients surviving more than 2 years.

**Two doses of Ad.IFN-α.** Unfortunately, Ad.IFN-β became unavailable for use in clinical trials. However, we were able to form a partnership with Schering-Plough, which had an ongoing intravesical trial for bladder cancer using a similar nonreplicating, type 5 Ad vector expressing another type 1 IFN, interferon α-2b (IFN-α). Preclinical studies showed comparable efficacy of Ad.IFN-α compared with Ad.IFN-β. A "bridging trial" using Ad.IFN-α was initiated in February 2009.[57] We made two additional changes: (1) exclusion of patients with baseline antiadenovirus Nab titers greater than 1:1000, and (2) reduction in vector dosing interval to 3 days to avoid inactivation owing to rapid increases in NAb. The NAb exclusion was justified based on analysis of our previous trials showing poor pleural gene transfer in patients with titers greater than 1:1000.

Nine subjects were given intrapleural instillation of two Ad.IFN-α2b doses separated by 3 days. Three patients treated with $1\times10^{12}$ vp had extremely high levels of pleural fluid IFN-α (75–2000 ng/ml) and serum IFN-α (3000–7700 pg/ml) associated with classic "IFN-type symptoms" necessitating a dose reduction. At $3\times10^{11}$ vp, pleural fluid IFN-α levels were still high (12 ng/ml), serum levels ranged from 370 to 524 pg/ml, but this dose was much better tolerated. An increase in pleural fluid IFN-α levels after the second dose was seen in five of seven subjects. All patients demonstrated significant serum anti-Ad-Nab levels by day 8 but not by day 3. No dose-limiting toxicities were observed. Seven of eight patients who could be evaluated developed anti-MPM antibodies.

No long-term clinical responses were seen in the four subjects with advanced disease. However, evidence of disease stability or tumor regression was seen in the remaining five patients, including one dramatic example of partial tumor regression at sites not in contiguity with vector infusion (see below). Using modified *Response Evaluation Criteria in Solid Tumors (*RECIST) criteria, at day 64, four of nine patients had stable disease and two patients had partial responses. At 6 months (end of the trial), three patients remained alive and well. Two had minimal residual disease and were able to undergo successful radical pleurectomies.

One of the patients (patient 309) exhibited a striking response. Figure 1 shows the $^{18}$FDG-PET scans pretherapy and at 2 and 6 months after intrapleural Ad.IFN-α instillation. The patient received no other treatment over this time. Note the relatively slow but dramatic decrease in tumor size and FDG uptake.

## 4.2 Ad.IFN in Bladder Cancer

Ad.IFN-α has also been used to treat subjects with recurrent nonmuscle invasive bladder cancer.[58] Seventeen patients in a phase I trial received a single intravesicular injection of Ad.IFN-α formulated in an excipient that maximized vector transduction.

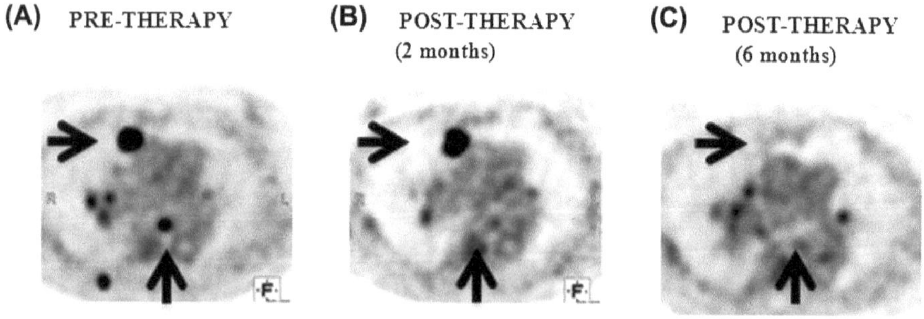


**Figure 1** Antitumor responses after intrapleural instillation of Ad.IFN. The chest images are taken from $^{18}$FDG-PET scans (axial images) at three time points. Tumors are shown by intense (black) uptake of tracer. (A) The baseline scan shows multiple lesions in the chest, chest wall, and mediastinum (horizontal arrow shows a large anteromedial mass; vertical arrow shows a malignant lymph node). (B) Two months after Ad.IFN vector instillation, a number of lesions are smaller. (C) Six months after Ad.IFN instillation (with no other therapy), most of the lesions have disappeared, including the two masses marked by arrows.

The treatment was well tolerated, with high urine levels of IFN-α detected. Of the 14 patients treated with higher doses, 43% experienced a complete response. A phase III trial is being initiated.

# 5. Use of Combination Therapies to Augment In Situ Vaccination with Ad

A relatively small subset of patients treated with Ad.tk or Ad.IFN seemed to have impressive, long-lived responses, indicating activity of the approach but also suggesting that additional improvements were needed to broaden the response. Although it would be attractive to give repeated administration of vector, unfortunately injection of Ad vectors induces a strong anti-Ad neutralizing antibody response that limits repeat administration of vector for months or even years. As an alternative, investigators have explored a second approach combining intratumoral Ad vectors with either radiation therapy or chemotherapy.

## 5.1 *Ad.tk Plus Radiation*

There is preclinical evidence that intratumoral Ad.tk followed by radiation therapy can augment antitumor efficacy.[59] The Freytag group has been exploring a strategy of prostatic radiation preceded by intraprostatic injection of a replicating Ad encoding HSV.tk plus a second suicide gene (cytosine deaminase) followed by valacyclovir and f-fluorocytosine versus giving radiation with no preceding Ad therapy.[39,60] Their most recent randomized phase II trial has shown encouraging trends, with a 42% reduction in biopsy positivity at 2 years in the investigational group.[60]

## 5.2 *Ad.IFN Plus Chemotherapy*

In the laboratory setting, we were able to augment the efficacy of Ad immunogene therapy by the administration of cyclo-oxygenase-2 inhibition[61] and by subsequent administration of chemotherapy.[62] This latter approach fits well with the growing body of information showing that immune stimulation by certain forms of chemotherapy is common and important in efficacy.[63–68] Accordingly, we designed a phase II trial in MPM patients in which two doses of intrapleural administration of a replication-defective recombinant Ad vector containing the human interferon-α (*hIFN-α2b*) gene at a dose of $3 \times 10^{11}$ vp were given along with a 14-day course of high-dose COX-2 inhibitor (celecoxib) to reduce side effects and modify the tumor microenvironment by decreasing prostaglandin-E2 levels. This was followed by standard first-line or second-line chemotherapy agents. Forty patients were treated in the study. Treatment was well tolerated and adverse events were comparable to historical controls. Follow-up chest CT scans demonstrated an overall response rate of 20% by modified RECIST criteria and a disease control rate of 85% (partial and complete responses plus stable disease) at initial follow-up scan after the first two cycles of chemotherapy.

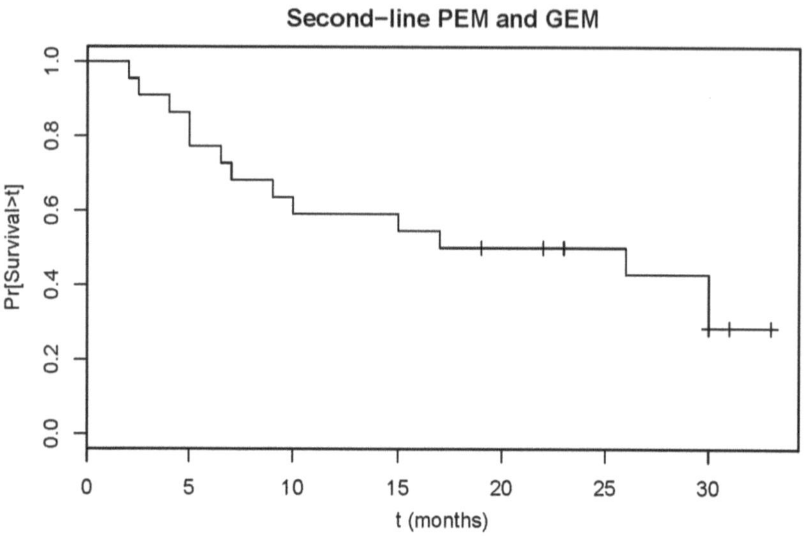


**Figure 2** Kaplan–Meier survival curve from the Ad.IFN/celecoxib/chemotherapy trial. This curves shows survival in the patients who received Ad.IFN/celecoxib followed by second-line chemotherapy (either pemetrexed or gemcitabine). Median overall survival is 22 months with a significant survival tail. The historical control MOS for a similar group of patients is 6–9 months.

Encouragingly, median overall survival (MOS) for all patients with epithelial histology (including both first- and second-line) was 26 months. Historical MOS with first-line chemotherapy alone is 13.3 months. We saw especially impressive results in the second-line patients (Figure 2). Our MOS (18 versus 9 months) was approximately twice that seen in similar second-line chemotherapy trials reported in the literature.[69] Importantly, not reflected in these MOS numbers was a substantial "survival tail" on the Kaplan–Meier plots that is rarely if ever seen in reported second-line therapy trials. Of course, these data need to be validated in a larger, randomized trial.

## 6. Conclusions and Future Directions

In situ vaccination with agents that both stimulate antitumor T cell responses and alter the tumor microenvironment to be more supportive of immune responses have the potential for effective immunotherapy. The ability of Ad vectors to effectively transduce tumor cells in vivo, their proven safety, their ability to express transgenes effectively, and their strong induction of innate inflammation and activation of DCs make them ideal agents for this purpose. One limitation may be difficulties in giving repeated doses because of the presence of Nabs. Thus, they are best used as a way to prime antitumor immune responses and will likely need some sort of alternative boosting mechanisms. This could be accomplished with a different serotype of Ad, other types of vectors, or other nanoparticles. An alternative that fits well into current treatment paradigms is to follow Ad-mediated immune-gene therapy with chemotherapy or radiation therapy as a way to stimulate the release of tumor antigen and favorably alter the tumor microenvironment.

Adenoviral vectors using the HSV.tk suicide gene or expression of type 1 IFNs coupled with radiation therapy or chemotherapy have shown promising results in early stage trials and are ready to be evaluated in larger randomized phase II or III trials.

# References


1. Palena C, Schlom J. Vaccines against human carcinoma: strategies to improve antitumor immune responses. *J Biomed Biotechnol* 2010;**2010**:380697.
2. Whiteside TL. The tumor microenvironment and its role in promoting tumor growth. *Oncogene* 2008;**27**:5904–12.
3. Vasievich EA, Huang L. The suppressive tumor microenvironment: a challenge in cancer therapy. *Mol Pharm* 2011;**8**:635–41.
4. Motz GT, Coukos G. Deciphering and reversing tumor immune suppression. *Immunity* 2013;**39**:61–73.
5. Hanahan D, Coussens LM. Accessories to the crime: functions of cells recruited to the tumor microenvironment. *Cancer Cell* 2012;**21**:309–22.
6. Melero I, Gaudernack G, Gerritsen W, Huber C, Parmiani, et al. Therapeutic vaccines for cancer: an overview of clinical trials. *Nat Rev Clin Oncol* 2014;**11**:509–24.
7. Gallo P, Dharmapuri S, Cipriani B, Monaci P. Adenovirus as vehicle for anticancer genetic immunotherapy. *Gene Ther* 2005;**12**:S84–91.
8. Majhen D, Calderon H, Chandra N, Gajardo CA, Rajan A, Alemany R, et al. Adenovirus-based vaccines for fighting infectious diseases and cancer: progress in the field. *Hum Gene Ther* 2014;**25**:301–17.
9. Vujanovic L, Whiteside TL, Potter DM, Chu J, Ferrone S, Butterfield LH. Regulation of antigen presentation machinery in human dendritic cells by recombination adenovirus. *Cancer Immunol Immunother* 2009;**58**:121–33.
10. Fehres CM, Unger WWJ, Garcia-Vallejo JJ, van Kooyk Y. Understanding the biology of antigen cross-presentation for the design of vaccines against cancer. *Front Immunol* 2014;**5**:1–10.
11. Cohn L, Delamarre L. Dendritic cell-targeted vaccines. *Front Immunol* 2014;**5**:255–66.
12. van den Boorn JG, Hartmann G. Turning tumors into vaccines: co-opting the innate immune system. *Immunity* 2013;**39**:27–37.
13. Crittenden MR, Thanarajasingam U, Vile RG. Intratumoral immunotherapy; using the thumor against itself. *Immunology* 2005;**114**:11–22.
14. Lutz ER, Wu A, Bigelow E, Sharma R, Mo G, Soares K, et al. Immunotherapy converts nonimmunogenic pancreatic tumors into immunogenic foci of immune regulation. *Cancer Immunol Res* 2014;**2**:1–16.
15. Brody JD, Ai WZ, Czerwinski DK, Torchia JA, Levy M, Advani RH, et al. In situ vaccination with a TLR9 agonist induces systemic lymphoma regression: a phase I/II study. *J Clin Oncol* 2010;**28**:4324–32.
16. Kim YH, Gratzinger D, Harrison C, Brody JD, Czerwinski DK, Ai WZ, et al. In situ vaccination against mycosis fungoides by intratumoral injection of a TLR9 agonist combined with radiation: a phase 1/2 study. *Blood* 2012;**119**:355–63.
17. Addison CL, Braciak T, Ralston R, Muller WJ, Gaulidie J, Graham FL. Intratumoral injection of an adenovirus expressing interleukin 2 induces regression and immunity in a murine breast cancer model. *Proc Natl Acad Sci* 1995;**92**:8522–6.

18. Lee C, Wu S, Ciernik IF, Chen H, Nadaf-Rahrov S, Gabrilovich D, et al. Genetic immunotherapy of established tumors with adenovirus-murine granulocyte-macrophage colony-stimulating factors. *Hum Gene Ther* 1997;**8**:187–93.
19. Kikuchi T, Miyazawa N, Moor MAS, Crystal RG. Tumor regression induced by intratumor administration of adenovirus vector expressing CD40 ligand and naïve dendritic cells. *Cancer Res* 2002;**60**:6391–5.
20. Malmstrom P, Loskog ASI, Lindqvist CA, Magsbo SM, Fransson M, Wanders A, et al. AdCD40L immunogene therapy for bladder carcinoma – the first phase I/IIa trial. *Clin Cancer Res* 2010;**16**:3279–87.
21. Pesonen S, Diaconu, Kangasniemi, Ranki T, Kanerva A, Pesonen SK, et al. Oncolytic immunotherapy of advanced solid tumors with a CD40L-expressing replicating adenovirus: assessment of safety and immunologic responses in patients. *Cancer Res* 2012;**72**:1621–31.
22. Bramson JL, Hitt M, Addison CL, Muller WJ, Gauldie J, Graham FL. Direct intratumoral injection of an adenovirus expressing interleukin-12 induces regression and long-lasting immunity that is associated with highly localized expression of interleukin-12. *Hum Gene Ther* 1996;**7**:1995–2002.
23. Caruso M, Pham-Nguyen K, Kwong YL, Xu B, Kosai KL, Finegold M, et al. Adenovirus-mediated interleukin-12 gene therapy for metastatic colon carcinoma. *Proc Natl Acad Sci* 1996;**93**:11302–6.
24. Sangro B, Mazzolini G, Ruiz M, Quiroga J, Herrero I, Benito A, et al. Phase I trial of intratumoral injection of an adenovirus encoding interleukin-12 for advanced digestive tumors. *J Clin Oncol* 2004;**22**:1389–97.
25. Miyahara R, Banerjee S, Kawano K, Efferson C, Tsuda N, Miyahara Y, et al. Melanoma differentiation-associated gene-7 (Mda-7) interleukin (IL)-24 induces anticancer immunity in a syngeneic murine model. *Cancer Gene Ther* 2006;**13**:753–61.
26. Tong AW, Nemunaitis J, Su D, Xhang Y, Cunningham C, Senzer N, et al. Intratumoral injection of INGN 241, a nonreplicating adenovector expressing the melanoma-differentiation associated gene-7 (mda-7/IL24): biologic outcome in advanced cancer patients. *Mol Ther* 2005;**11**:160–72.
27. Cunningham CC, Chada S, Merritt JA, Tong A, Senzer N, Zhang Y, et al. Clinical and local biological effects of an intratumoral injection of mda-7 (IL24; INGN 241) in patients with advanced carcinoma: a phase I study. *Mol Ther* 2005;**11**:149–59.
28. Drummer R, Hassel JC, Fellenberg F, Eichmuller S, Maier R, Slos P, et al. Adenovirus-mediated intralesional interferon-γ gene transfer induces tumor regressions in cutaneous lymphomas. *Blood* 2004;**104**:1631–8.
29. Drummer R, Eichmuller S, Gellrich S, Assaf C, Dreno B, Schiller M, et al. Phase II clinical trial of intratumoral application of TG1042 (adenovirus-interferon-gamma) in patients with advanced cutaneous T-cell lymphomas and multilesional cutaneous B-cell lymphomas. *Mol Ther* 2010;**18**:1244–7.
30. Dreno B, Uroseciv-Maiwald M, Kim Y, Guitart J, Duvic M, Dereure O, et al. TG1042 (Adenovirus-interferon-γ) in primary cutaneous B-cell lymphomas: a phase II clinical trial. *PLoS One* 2014;**9**:e83670.
31. Aguilar LK, Guzik BW, Aguilar-Cordova E. Cytotoxic immunotherapy strategies for cancer: mechanisms and clinical development. *J Cell Biochem* 2011;**112**:1969–77.
32. Vile RG, Castleden S, Marshall J, Camplejohn R, Upton C, Chong H. Generation of an anti-tumor immune response in a non-immunolgenic tumor: HSVtk killing in vivo stimulates a mononuclear cell infiltrate and a Th1-like profile of intratumoural cytokine expression. *Int J Cancer* 1997;**71**:267–74.

33. Sanchez-Perez L, Gough M, Qiao J, Thanarajasingam U, Kottke A, Ahmed A, et al. Synergy of adoptive T-cell therapy and intratumoral suicide gene therapy is mediated by host NK cells. *Gene Ther* 2007;**14**:998–1009.
34. Sterman DH, Albelda SM. Advances in the diagnosis, evaluation, and management of malignant pleural mesothelioma. *Respirology* 2005;**10**:266–83.
35. Sterman DH, Treat J, Litzky LA, Amin KM, Coonrod L, Molnar-Kimber K, et al. Adenovirus-mediated herpes simplex virus thymidine kinase/ganciclovir gene therapy in patients with localized malignancy: results of a phase I clinical trial in malignant mesothelioma. *Hum Gene Ther* 1998;**9**:1083–92.
36. Molnar-Kimber KL, Sterman DH, Chang M, Kang EH, ElBash M, Lanuti M, et al. Impact of preexisting and induced humoral and cellular immune responses in an adenovirus-based gene therapy phase I clinical trial for localized mesothelioma. *Hum Gene Ther* 1998;**9**:2121–33.
37. Sterman DH, Molnar-Kimber K, Iyengar T, Chang M, Lanuti M, Amin KM, et al. A pilot study of systemic corticosteroid administration in conjunction with intrapleural adenoviral vector administration in patients with malignant pleural mesothelioma. *Cancer Gene Ther* 2000;**7**:1511–8.
38. Sterman DH, Recio A, Vachani A, Sun J, Cheung L, DeLong P, et al. Long-term follow-up of patients with malignant pleural mesothelioma receiving high-dose adenovirus herpes simplex thymidine kinase/ganciclovir suicide gene therapy. *Clin Cancer Res* 2005;**11**:7444–53.
39. Freytag SO, Stricker H, Pegg J, Paielli D, Pradhan DG, Peabody J, et al. Phase I study of replication-competent adenovirus-mediated double-suicide gene therapy in combination with conventional-dose three-dimensional conformal radiation therapy for the treatment of newly diagnosed, intermediate- to high-risk prostate cancer. *Cancer Res* 2003;**63**:7497–506.
40. Hasenburg A, Tong XW, Rojas-Martinez A, Nyberg-Hoffman C, Kieback CC, Kaplan A, et al. Thymidine kinase gene therapy with concomitant topotecan chemotherapy for recurrent ovarian cancer. *Mol Ther* 2001;**4**:182–91.
41. Sung MW, Yeh HC, Thung SN, Schwartz ME, Mandeli JP, Chen SH, et al. Intratumoral adenovirus-mediated suicide gene transfer for hepatic metastases from colorectal adenocarcinoma: results of a phase I clinical trial. *Mol Ther* 2001;**4**:182–91.
42. Herman JR, Adler HL, Aguilar-Cordova E, Rojas-Martinez A, Woo S, Timme TL, et al. In situ gene therapy for adenocarcinoma of the prostate: a phase I clinical trial. *Hum Gene Ther* 1999;**10**:1239–49.
43. Trask TW, Trask RP, Aguilar-Cordova E, Shine HD, Wyde PR, Goodman JC, et al. Phase I study of adenoviral delivery of the HSV-tk gene and ganciclovir administration in patients with current malignant brain tumors. *Mol Ther* 2000;**1**:195–203.
44. Alvarez RD, Gomez-Navarro J, Wang M, Barnes MN, Strong TV, Arani RB, et al. Adenoviral-mediated suicide gene therapy for ovarian cancer. *Mol Ther* 2000;**2**:524–30.
45. Miles BJ, Shalev M, Aguilar-Cordova E, Timme TL, Lee H, Yang G, et al. Prostate-specific antigen response and systemic T cell activation after in situ gene therapy in prostate cancer patients filing radiotherapy. *Hum Gene Ther* 2001;**12**:1955–67.
46. Chiocca EA, Aguilar LK, Bell SD, Kaur B, Hardcastle J, Cavaliere R, et al. Phase IB study of gene-mediated cytotoxic immunotherapy adjuvant to up-front surgery and intensive timing radiation for malignant glioma. *J Clin Oncol* 2011;**29**:3611–9.
47. Westphal M, Yla-Herttuala, Martin J, Warnke P, Menei D, Kinley J, et al. Adenovirus-mediated gene therapy with sitimagene ceradenovec followed by intravenous ganciclovir for patients with operable high-grade glioma (ASPECT): a randomized, open-label, phase 3 trial. *Lancet Oncol* 2013;**14**:823–33.

48. Ferrantini M, Capanoe I, Belardelli F. Interferon-α and cancer: mechanisms of action and new perspectives of clinical use. *Biochimie* 2007;**89**:884–93.
49. Trinchieri G. Type I interferon: friend or foe? *J Exp Med* 2010;**207**:2053–63.
50. Yang X, Zhang X, Fu ML, Weichselbaum RR, Gajewski TF, Guo Y, et al. Targeting the tumor microenvironment with interferon-α bridges innate and adaptive immune responses. *Cancer Cell* 2014;**25**:37–48.
51. Fuertes MB, Kacha AK, Kline J, Woo S, Kranz DM, Murphy KM, et al. Host type I IFN signals are required for antitumor CD8+ T cell responses through CD8α+ dendritic cells. *J Exp Med* 2011;**208**:2005–16.
52. Le Bon A, Etchart N, Rossman C, Ashton M, Hou S, Gewert D, et al. Cross-priming of CD8+ T cells stimulated by virus-induced type I interferon. *Nat Immunol* 2003;**4**:1009–15.
53. Thyrell L, Erickson S, Zhivotovsky B, Pokrovskaja K, Sangfelt O, Castro J, et al. Mechanisms of interferon-alpha induced apoptosis in malignant cells. *Oncogene* 2002;**21**:1251–62.
54. Odaka M, Sterman DH, Wiewridt R, Zhang Y, Keifer M, Amin KM, et al. Eradiation of intraperitoneal and distant tumor by adenovirus-mediated interferon-β gene therapy is attributable to induction of systemic immunity. *Cancer Res* 2001;**61**:6201–12.
55. Sterman DH, Reico A, Carroll RG, Gillespie CT, Haas AR, Vachani A, et al. A phase I clinical trial of single-dose intrapleural INF-β gene transfer for malignant pleural mesothelioma and metastatic pleural effusions: high rate of antitumor immune responses. *Clin Cancer Res* 2007;**13**:4456–66.
56. Sterman DH, Reico A, Haas AR, Vachani A, Katz SI, Gillespie CT, et al. A phase I trial of repeated intrapleural adenoviral-mediated interferon-β gene transfer for mesothelioma and metastatic pleural effusion. *Mol Ther* 2010;**18**:852–60.
57. Sterman DH, Haas A, Moon E, Recio A, Schwed D, Vachani A, et al. A trial of intrapleural adenoviral-mediated interferon-α2b gene transfer for malignant pleural mesothelioma. *Am J Respir Crit Care Med* 2011;**184**:1395–9.
58. Dinney CPN, Fisher MB, Navai N, O'Donnell MA, Cutler D, Abraham A, et al. Phase I trial of intravesical recombinant adenovirus mediated interferon-α2b formulated in Syn3 for Bacillus Calmette-Guerin failures in nonmuscle invasive bladder cancer. *J Urol* 2013;**190**:850–6.
59. Vlachaki MT, Chhikara M, Aguilar L, Zhu X, Chiu KJ, Woo S, et al. Enhanced therapeutic effect of multiple injections of HSV-TK+GCV gene therapy in combination with ionizing radiation in a mouse mammary tumor model. *Int J Radiat Oncol Phys* 2001;**51**:1008–17.
60. Freytag S, Stricker H, Lu M, Elshaikh M, Aref I, Pradhan D, et al. Prospective randomized phase 2 trial of intensity modulated radiation therapy with or without oncolytic adenovirus-mediated cytotoxic gene therapy in intermediate-risk prostate cancer. *Int J Radiat Oncol Biol Phys* 2014;**89**:268–76.
61. Haas A, Sun J, Vachani A, Wallace AF, Silverberg M, Kapoor V, et al. Cyclooxygenase-2 inhibition augments efficacy of a cancer vaccine. *Clin Cancer Res* 2006;**12**:214–22.
62. Fridlender ZG, Sun J, Singhal S, Kapoor V, Cheng G, Suzuki E, et al. Chemotherapy delivered after viral immune-gene therapy augments anti-tumor efficacy via multiple immune-mediated mechanisms. *Mol Ther* 2010;**18**:1947–59.
63. Nowak AK, Robinson BWS, Lake RA. Synergy between chemotherapy and immunotherapy in the treatment of established murine solid tumors. *Cancer Res* 2003;**63**:4490–6.
64. Lake RA, Robinson BWS. Immunotherapy and chemotherapy – a practical partnership. *Nat Rev Cancer* 2005;**5**:397–405.
65. McCoy MJ, Nowak AK, Lake RA. Chemoimmunotherapy: an emerging strategy for the treatment of malignant mesothelioma. *Tissue Antigens* 2009;**74**:1–10.

66. Ramakrishnan R, Gabrilovich DI. Mechanism of synergistic effect of chemotherapy and immunotherapy of cancer. *Cancer Immunol Immunother* 2011;**60**:419–23.
67. Zitvogel L, Apetoh L, Ghiringhelli, Kroemer G. Immunological aspects of cancer chemotherapy. *Nat Rev* 2008;**8**:59–73.
68. Zitvogel L, Galluzi L, Smyth MJ, Kroemer G. Mechanism of action of conventional and targeted anticancer therapies: reinstating immunosurveillance. *Immunity* 2013;**39**:74–88.
69. Zucali PA, Simonelli M, Michetti G, Tiseo M, Ceresoli GL, Collova E, et al. Second-line therapy chemotherapy in malignant pleural mesothelioma; results of a retrospective multicenter survey. *Lung Cancer* 2012;**75**:360–7.

# Utility of Adenoviral Vectors in Animal Models of Human Disease II: Genetic Disease

25

*Raymond John Pickles*
Cystic Fibrosis/Pulmonary Research and Treatment Center, University of North Carolina at Chapel Hill, Chapel Hill, NC, USA

## 1. Introduction

A disease at the forefront of gene therapy research over the last decade is cystic fibrosis (CF). This hereditary, single gene defect disease although affecting epithelial cells of multiple organs of the body results most often in mortality due to complications associated with the lung. CF lung disease has been considered as a prototypic disease state for "proof-of-concept" gene-therapy strategies. The lack of an alternative long-term treatment for the pulmonary manifestations of this disease, the accessibility of the lung via the airway lumen, and the fact that viruses known to infect the lung were being developed into nonreplicating gene transfer vectors led investigators to believe that administration of gene transfer vectors to the lung could potentially result in an effective treatment of this disease.

Shortly after the cloning of the gene responsible for CF pathophysiology, two groundbreaking observations made gene therapy for CF lung disease appear imminent. First, isolated epithelial cells cultured from the airway epithelium of CF patients could be phenotypically "corrected" by transferring into the cells the cDNA corresponding to the CF gene.[1–5] Second, adenoviral (Ad) vectors engineered to express the CF gene were administered to the airways of experimental animals and transgene expression observed in cells that were considered to require "correction."[6] These initial observations produced a flurry of scientific activity and excitement in both the gene therapy and CF scientific communities and within 3 years of these observations the first clinical trials describing successful Ad-mediated gene transfer to the airway epithelium of CF patients in vivo were reported.[7]

These promising early observations have unfortunately not withstood further investigation. After approximately 20 gene therapy clinical trials for CF lung disease (of which greater than 70% utilized Ad) it has become apparent that gene transfer to airway epithelium in vivo is not a simple procedure. The difficulty lies in the evolution of the respiratory epithelium as an effective barrier to invading pathogens entering the lung (e.g., viruses). The epithelium achieves this "barrier function" by presentation of a host of innate and cell-mediated immune systems, which for gene transfer vectors culminate in reduced uptake and expression of the transgene. In this chapter, I will describe the evidence that led investigators to believe that Ad would be useful in CF

**Adenoviral Vectors for Gene Therapy.**

lung disease, why subsequently this simplistic approach failed and how increasing knowledge of lung biology and viral bioengineering has and will allow novel strategies to be tested. In light of this emphasis on basic research new strategies and models will need to be tested and successful demonstration of efficiency and safety will be required before we once again enter the clinic with Ad for CF lung disease gene therapy.

## 2. Pathophysiology of CF Lung Disease

CF is a multifaceted disease with major morbidity and mortality resulting from chronic decline of lung function. This disease is the most common fatal inherited disease in Caucasians with 1 in 2500 live births affected.[8] Although CF is most devastating to the lung (accounting for 90% of mortality), resulting in chronic repetitive infections, chronic obstructive pulmonary disease, and respiratory failure, other tissues are also affected including the liver, pancreas, the gastrointestinal tract, and the sweat glands. The abnormal CF gene (250 kb) encodes an mRNA of 6.5 kb which translates into an 180kD protein that has been extensively characterized as a cAMP-activated chloride ion channel, named the CF transmembrane conductance regulator (CFTR).[3,9] In the lung, CFTR is normally expressed in the respiratory epithelium and, although the specific functions of CFTR are complex, is predominately involved in maintenance of ionic homeostasis in this tissue. Over 900 different mutations of CFTR have now been reported resulting in a range of clinical manifestations and differing severity of the disease. However, 70% of these mutations are due to a three base-pair deletion leading to the absence of phenylalanine (F) at position 508 (ΔF508).[10] This particular mutation leads to misfolded CFTR being retained within the endoplasmic reticulum of cells so reducing CFTR function at the plasma membrane.[11] Currently, although the specific localization and functional capacity of ΔF508 CFTR in the different affected organs is a matter of controversial debate,[12] and other mutations can display partial CFTR function, for CF patients, expression of abnormal CFTR in the airway epithelium generally results in reduced chloride ion secretion, hyperabsorption of sodium ions, increased viscosity of airway secretions, impaired mucociliary clearance, chronic bacterial infection, bronchiectasis, and premature death.[8,13] Given that all of these effects are likely primary or secondary to loss of CFTR function, the most efficacious way to treat the broad range of effects would be to replace the defective CFTR gene with a normal copy. Gene therapy for CF lung disease therefore seeks to replace normal CFTR in the airway epithelial cells to hopefully "correct" lung epithelium function.

## 3. Trials and Tribulations with Ad Vectors for CF Lung Disease

Clinical gene transfer trials with CF patients investigating the safety and efficacy of gene transfer vectors (predominately adenoviral and liposomal vectors) have been performed in both the US and UK. Details of these trials and the background preclinical studies have been comprehensively reviewed in a recent review.[14] Although preclinical data have been largely promising for lung-directed gene transfer, the trials performed to date have shown, at best, only partial "correction" (<20%) of the CF bioelectrical defect.[7,15–18] This relatively low degree of correction is most likely due to inefficient transfer of the CFTR cDNA to the airway epithelial cells, i.e., a low efficiency of gene transfer, and is most likely not sufficient to be of benefit to CF patients although long-term reversal of disease symptoms were not monitored in this studies.

The gene transfer efficiency required for physiological correction of CF lung disease has been a matter of recent debate. While Johnson and colleagues have shown that "correction" of ~10% of CF cells restores normal chloride secretory function to an epithelium, this degree of "correction" was insufficient to correct the hyperabsorption of sodium.[19] Since "correction" of the sodium defect is likely to be necessary for resolving CF lung disease, then transduction of a higher proportion of epithelial cells will be required.[1,20] Indeed, it has been suggested that greater than 80% of epithelial cells will have to express CFTR to restore the normal sodium transporting capabilities of the epithelium.[20] With regard to efficiency of gene expression on a per cell basis, it appears that CFTR is normally expressed at levels as low as 10 copies per cell and heterozygotes for the CF gene although only expressing 50% of normal CFTR show no disease symptoms. This suggests that the level of expression per cell does not need to be high in order to correct function. On the other hand, overexpression of CFTR has been shown to have deleterious effects on cell function although the effects on polarized airway epithelial cells are not documented.[21]

Issues of safety have arisen due to elicitation of inflammatory responses after Ad instillation in both animal and human experiments.[22–29] These effects have often been due to the large "loads" of vector that has been administered. A current hypothesis is that improvements in gene transfer efficiency may allow smaller quantities of Ad to be administered possibly circumventing much of the inflammatory response.

## 4. The Airway Epithelium: Cellular Targets for CF Gene Therapy

Airway epithelial cells are present throughout the conducting airways of the lung including the nasal, tracheal, bronchial, and bronchiolar regions. In the upper airway, the surface epithelium lines these structures and is continuous with the tubuloacinar submucosal mucus-secreting glands that invaginate from the airway surface. Airway epithelial cell-type composition is dependent both on the regional location and on the particular species studied and the reader is referred to comprehensive reviews that describe species-specific epithelial cell distribution in more detail.[30,31] The epithelial cell types present in the lung are numerous and include ciliated cells, mucus-secreting cells (goblet), serous cells, Clara cells, and basal cells. The cell types of the alveolar structures of the lung (alveolar Type I and II cells) are not thought to participate in the pathophysiology of CF lung disease. In human airways, the upper airway regions (nasal, tracheal, bronchial) are composed of a pseudostratified mucociliary epithelium in which ciliated cells predominate with interspersed mucus-secreting goblet cells. The columnar cells overlie intermediary differentiated cells and basal cell layers which interface with the basement membrane. In addition, the human upper airways contain numerous submucosal glands. In the human lower airways, the bronchioles are lined with a simple cuboidal ciliated epithelium containing few mucus-secreting cells, no basal cells, and an absence of submucosal glands. An important morphological difference between the upper airways of human and mice, the most common animal model for investigating airway administration of gene transfer vectors, is that for the mouse upper airway (excluding the nasal cavity epithelium) the columnar cells are roughly an equal distribution of ciliated and Clara cells, compared to the predominance of ciliated cells in the human upper airway.[32] Clara cells are a nonciliated bronchiolar mucus-secreting cell type with distinct properties from ciliated cells. Clara cells although present in human airway are located only in the distal airways and only account for a fraction of the cells present in that region.[32]

The airway basal cells, or at least a subpopulation, are considered to be stem cell precursors for all other airway epithelial cells in the upper airway regions. Basal cells can differentiate into mucus or ciliated cell phenotypes.[33] Whereas mucus cells may also be able to differentiate into ciliated cells, the ciliated cell is considered as a terminally differentiated cell type.

An important observation with regard to experimental models of human airway epithelial cells is that isolation of upper airway epithelial cells for tissue culture purposes results initially in a predominately basal cell-like culture since isolated basal cells proliferate at a greater rate than isolated ciliated and mucus cells. Furthermore, for cells isolated from CF airways, the rate of proliferation of basal cells is even greater than that in normal airway probably reflecting responses to ongoing inflammatory processes.[34] Therefore, morphological differences need to be considered when designing models to study the interactions of gene transfer vectors with airway cells that are presumed to represent the cells in the lung that are exposed to lumenally delivered vectors.

Although CF is a disease of the respiratory epithelium, the exact airway region where CF lung disease initiates is still a matter of debate. It does appear that the first signs of pathology occur in the distal airways with findings of bronchiolitis and mucus plugging in the small airways and although the exact nature of how the CFTR defect initiates the disease is not totally resolved, it does appear that hydration of the periciliary fluid layer in these regions may be a major cause.[35,36] Currently, both the airway surface columnar cells lining the lumen of the small bronchiolar airways and the serous cells of the submucosal glands are candidates for the location for the onset of the disease. The cell type that is believed to be predominately involved in the onset of disease and therefore the specific target for gene transfer is the ciliated cell since these cells exhibit all of the ion- and fluid-transporting functions of CFTR and display abnormal function in patients with CF.[37] However, the submucosal gland serous cell is the highest CFTR-expressing cell type in the lung, suggesting that these cells may also be an important target for gene replacement.[38]

Ultimately it will be important to determine the location of disease initiation since it is likely that for a lumenal gene therapy to be successful, administration of vector will have to occur early in the life of a CF patient. Later in life, when the airways posses overwhelming mucus plugging and associated bacterial colonization and inflammation, delivery of genes to the target cells will likely become restricted. The current thrust for CF gene therapy strategies is to deliver transgenes to target cell types before such other barriers to treatment are present.

# 5. Ad Vectors as Gene Transfer Vectors in the Lung

## *5.1 Animal Models for CF Airway Gene Transfer Studies*

The generation of CF mouse models was an important step for understanding the physiology of CF disease. There have now been over 10 different mouse models produced displaying a range of CF-associated genetic mutations.[39] Although most of the models reflect the most common human mutation, either a complete gene knockout or a ΔF508 mutation of the mouse CFTR, other models with less common human mutations (e.g., G551D) have also been reported. The multiorgan pathophysiology associated with the different models has been recently reviewed.[39] Interestingly, although the gastrointestinal phenotype of CF mice is similar to that observed in CF patients, there is no CF-like pathology associated with the CF mouse lung. A comparison of bioelectrical measurements between CF human and CF murine airways has revealed that both species exhibit, relative to normals, hyperabsorption of sodium and an absent or reduced cAMP-induced chloride secretory response. However, it has been deduced that the ion transport defects in the CF mouse airway do not lead to CF-like lung pathology because CF murine airways compensate for the loss of CFTR activity by upregulating an alternative chloride secretory channel that is regulated by changes in intracellular calcium.[40] However, from a practical standpoint, the ability to measure the "bioelectrical defect" in CF mouse airways makes the model useful in terms of monitoring "bioelectrical correction" with gene transfer strategies, but the ability to monitor inhibition or reversal of CF-like pathology induced by transfer of normal CFTR is not possible in these current models. Therefore, the current gold standard for success in CF gene transfer to mouse lung in vivo is correction of the chloride (and sodium) ion transport defects.

Most gene delivery strategies to murine airways have focused on the epithelium of the nasal mucosa and trachea mainly because of accessibility to these regions but also because these regions are similar to those targeted for human CF gene therapy trials. Unfortunately, baseline bioelectric measurements of murine trachea indicate that these tissues do not display sodium hyperabsorption,[41] a key indicator for the human disease, and one that will likely need to be corrected for a treatment to be successful. In contrast, the epithelium of the CF mouse nasal cavity and freshly isolated CF murine nasal mucosa both display sodium hyperabsorption and reduced cAMP-induced chloride secretory activity providing an ideal model for study.[42] A further difficulty with murine airways (excluding nasal epithelium) is the large proportion of Clara cells that are present throughout the upper airway. The distribution of this cell type in the mouse may be misleading when comparing gene transfer efficiency between mouse and human upper airways (see later). The murine nasal mucosa however has few Clara cells and exists as a pseudostratified mucociliary epithelium with a cell-type distribution similar to human nasal mucosa, again demonstrating the usefulness of this tissue for gene transfer studies.

Therefore, in conclusion, the CF murine models do not display spontaneous or induced pathological signs of human CF lung disease. However, CF murine airways do display bioelectric abnormalities associated with human CF and correction of these parameters by gene transfer can be measured both in vitro and in vivo. Given these considerations, since most clinical trials have focused on studying gene transfer to the nasal mucosa, the CF mouse nose is considered a good model for studying these strategies. In addition, since the epithelial cell-type distribution in human nose is similar to that of the human trachea and bronchus the nasal epithelium would appear a good model for a large proportion of the human airway epithelium.

## 5.2 Success and Limitations of Ad

### 5.2.1 Efficiency of Gene Transfer

#### 5.2.1.1 Cell Types

The major cell types that support wild-type Ad infection in the lung are the epithelial cells of the respiratory mucosa lining the airway passages. The tropism of Ad to the respiratory epithelium established this vector as an obvious candidate for delivering transgenes to the lung. Indeed, Ad-mediated gene transfer to airway epithelial cells grown under standard culture conditions in vitro is highly efficient,[43,44] with cellular transduction efficiencies of 90–100% and when the transgene is CFTR, full correction of the spectrum of CF bioelectrical defects is obtained.[1] In contrast, observations from in vivo epithelial cell models derived from cartilaginous (upper airway) regions of the airways of rodents, nonhuman and human primates, show that transgenes are expressed after in vivo dosing in less than 20% of the surface epithelial cells, an efficiency unlikely to benefit to the defective physiology of a CF airway.[43,45] Although the efficiency of gene transfer can be enhanced by prolonging the contact time of Ad with the epithelium for 12–24 h, it is difficult to envision this strategy as being practical in a clinical scenario.[46,47] In the case of intralumenal delivery of Ad to the lower airways of rodents, gene transfer to 10–80% of the airway epithelial cells has been reported with apparently no cell-type-specific selectivity,[48,49] although, in a detailed study of Ad administration to murine airways, only the nonciliated bronchiolar epithelial cells (i.e., Clara cells) were observed to express transgenes.[50] Clara cells are not thought to require correction in the CF lung and this observation casts a shadow on the use of murine airway epithelium as a model for Ad-mediated gene transfer to the human airway epithelium where Clara cells are less common. Therefore, it appears that lumenal facing well-differentiated (WD) airway epithelial cells in vivo, at least in the upper airway regions, are resistant to efficient Ad-mediated gene transfer.

How can we envision that the airway epithelial cells facing the lumen of the airway are not transduced by Ad given the large body of clinical data that show that these cells are targeted in wild-type infections? In a series of studies using human tracheal epithelium ex vivo and murine trachea in vivo it was discovered that injury to the epithelium by physical abrasion of the columnar cells revealed epithelial cell types that are susceptible to efficient Ad transduction, as depicted in Figure 1.[43,51,52]

This cell-type-specific variable efficiency led to the finding that underlying basal cell-like cells were efficiently transduced by Ad. These cells, as precursors to columnar cells could once transduced, over time proliferate and differentiate into transgene-expressing columnar epithelial cells. Since the epithelial basal cells are probably stimulated to proliferate and differentiate upon injury these susceptible cells were described as "basal cell-like cells" or the poorly differentiated (PD) airway epithelial cells, i.e., injured or regenerating cells, and this cellular phenotype is similar to that displayed by airway epithelial cells grown on plastic that are also highly transducible by Ad.[43,44]

One consideration when comparing wild-type Ad infection to Ad vectors is that the latter rely on delivering many virus particles to a target tissue whereas wild-type Ad needs only access to a small number of cells from which Ad replication and spread can then occur. Therefore, wild-type Ad may be able to take advantage of regions of the airway in which epithelium integrity is compromised or injured. Initiation of wild-type infection in injured regions would then be able to spread as a "basal cellitus" effectively beneath the resistant superficial columnar cells.

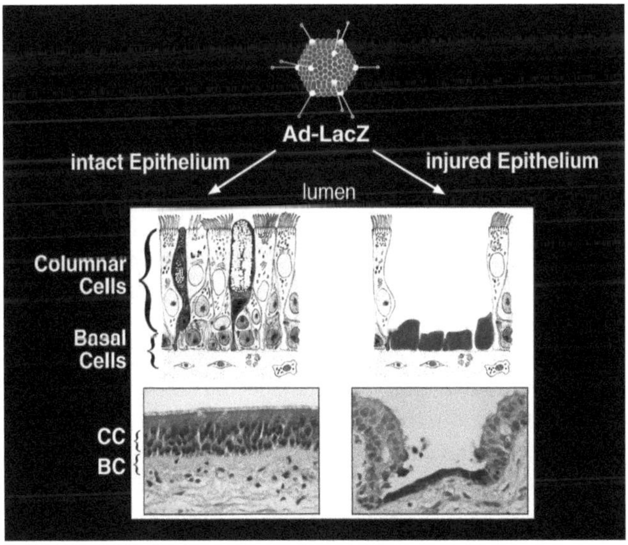


**Figure 1 Increased susceptibility of injured epithelium to Ad-mediated gene transfer.** Exposure of Ad vectors to intact pseudostratified columnar cells (CC) results in low gene transfer efficiency. Physical abrasion of columnar cells before Ad exposure results in efficient gene transfer to the underlying basal cells (BC). Upper figures show schematic of intact and injured pseudostratified columnar respiratory epithelium and lower figures are intact and abraded human tracheal epithelium exposed to Ad-LacZ ex vivo.
Reprinted from Ref. 20 with permission from Elsevier Science.

#### 5.2.1.2 Receptors

The differences in the gene transfer efficiencies for the two cellular phenotypes of airway epithelial cells, the PD and WD columnar cells, suggest that an early step in the virus–cell interaction is deficient for the WD cells. Ad enters epithelial cells by a two-step process: (1) initial attachment of the viral fiber-knob protein to a high affinity receptor, the human coxsackie B and Ad 2 and 5 receptor (hCAR)[53,54] and (2) translocation of the virus into the cell cytoplasm via clathrin-coated pit internalization processes, in part mediated by an interaction of the viral penton base with $\alpha_v\beta_{3/5}$ integrins.[55]

Since quantitative studies of the interactions of Ad with the airway epithelium in vivo are difficult and prone to considerable variation, specialized cell culture models have been generated to aid characterization of the interaction of Ad with both PD and WD cell types. These models have been shown by a number of groups to reproduce[1]: the WD (ciliated) and PD cellular phenotypes and[2] the relative resistance of WD and permissiveness of PD cells to Ad-mediated gene transfer as observed in vivo.[56,57] In addition, although these models were originally generated to ask specific questions regarding gene transfer strategies, they have subsequently become valuable in a whole series of studies were quantitative and qualitative measurement of events in the airway epithelium are difficult to perform in vivo.[35,36,58–63]

Using these models of human airway epithelium, immunofluorescent and functional analyses of the interactions of Ad with human airway epithelial cells have shown that decreased gene transfer efficiency to WD compared to PD cultures is due to limited entry (penetration) of Ad across the apical membrane of WD cultures which reflects a reduced specific Ad attachment due to the absence of hCAR and $\alpha_v\beta_{3/5}$ integrins from the apical surface. Interestingly, columnar cells and basal cell-like cells express all the necessary receptors to efficiently allow Ad entry but for columnar cells these processes are segregated and limited to the basolateral membranes as depicted in Figure 2. In these culture model systems, Ad has been shown to efficiently transduce epithelial cells when applied to the basolateral epithelial surfaces.[56,57,64,65]

It appears that the most significant Ad–cell interaction in determining efficiency is that of the Ad–hCAR interaction. Many cell types usually resistant to Ad infection have been shown to be efficiently transduced after heterologous expression of hCAR, although the status of integrin expression in these cell types is not always clear.[56,66] Earlier observations had suggested that inefficient Ad-mediated gene transfer to a bronchial xenograft model of human *in vivo*-like ciliated airway epithelial cells reflected the absence of $\alpha_v\beta_{3/5}$ integrins from the lumenal membrane of the epithelium.[65] However, $\alpha_v\beta_{3/5}$ integrins may not alone account for decrements in gene transfer efficiency. In support of this hypothesis, Ad mutants lacking penton base RGD sequences (normally required for Ad–$\alpha_v\beta_{3/5}$ integrin interactions) are able to efficiently transduce human epithelial cells although *the rate of* internalization is reduced.[67] In addition, in a $\beta_5$ integrin knockout mouse model, airway epithelial cells were equally susceptible to Ad-mediated gene transfer as were wild-type airway cells,[68] again suggesting that $\alpha_v\beta_{3/5}$ integrins may be facilitative rather than necessary for efficient vector entry into the cell.

These observations are important for the design of targeted vectors that attempt to increase gene transfer efficiency to normally unsusceptible cell types.[69,70] Retargeted vectors attached via nonspecific interactions or to noninternalizing receptors

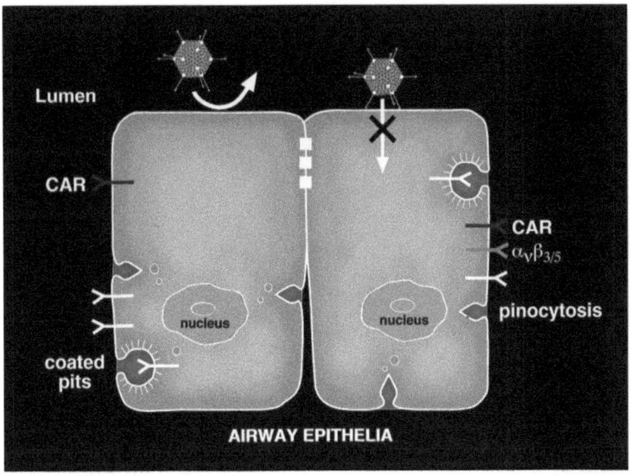


**Figure 2 Schematic of polarized epithelial cells displaying resistance of the lumenal surface to adenovirus attachment and entry.** The receptors required for Ad entry are located on basolateral membranes and excluded from the apical membrane by the tight junctional complexes.
Republished with permission from Boucher R. Current status of CF gene therapy. *J Clin Invest* 1999.

will probably depend on nonspecific uptake pathways to enter cells and while this approach is useful for PD cells in vitro, increasing Ad attachment to WD cultures that do not exhibit these pathways is unlikely to improve gene transfer efficiency.[56]

#### 5.2.1.3 The Innate Immune System of the Lung

Despite the progress on the cell biological aspects of vector–cell interactions, surprisingly little attention has been devoted to another fundamental component of innate airway defense that will almost certainly impact on the efficiency of lumenally delivered vectors, the barrier/shielding function of epithelial surfaces by the carbohydrate-rich cell surface glycocalyx. Expression of hCAR, engineered to be expressed at the apical surface of polarized epithelia by incorporation of a glycosylphosphatidylinositol linker (GPI-CAR), identified glycocalyx components as barriers for lumenally applied Ad accessing these receptors as depicted in Figure 3.[71] Electron micrographs demonstrate a "fuzzy coat" on the cell surface,[72,73] termed the glycocalyx, and on epithelial cell apical surfaces it is comprised of several families of carbohydrate-rich molecules, including glycoproteins (most notably the mucins), proteoglycans, and glycolipids. Glycoconjugates are variably modified by sialic acid and sulfate that impart a strong anionic charge to the cell surface. A major component of the airway glycocalyx will likely be the "tethered" mucins and the molecular biologic advances in the mucin field have revealed that the MUC1 and MUC4 are highly expressed in airway epithelium and have transmembrane anchoring (tethering) domains.[74–82] With respect to airway gene transfer, sialoglycoconjugates (including MUC1) comprising the glycocalyx on MDCK cells appear to inhibit Ad gene transfer, presumably due, in part, to

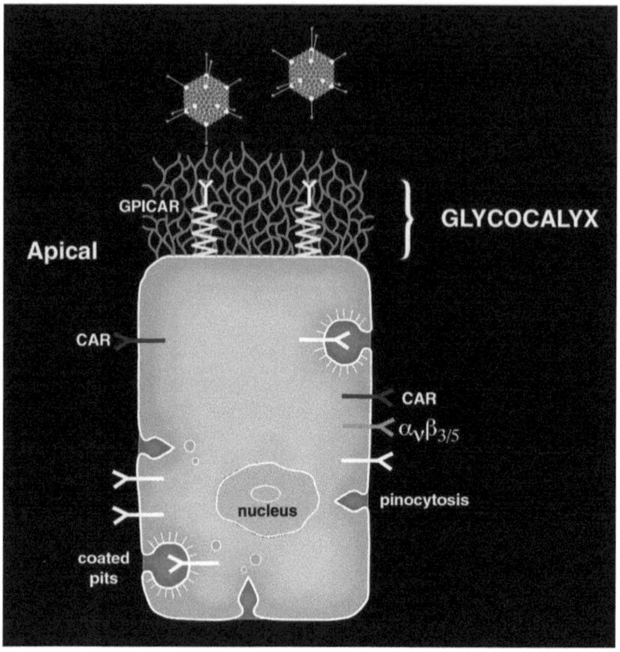


**Figure 3 Schematic of polarized epithelial cell expressing reengineered Ad receptors at the apical surface.** These studies revealed that the apical surface glycocalyx was an effective barrier to Ad-accessing receptors located on the apical surface.

their negative charge since neuraminidase treatment to selectively remove sialic acid can circumvent the glycocalyx barrier in these cell types.[83,84] Although apical surface mucins expressed on WD cells are also restrictive to Ad, neuraminidase alone is not sufficient to allow Ad permeation through the glycocalyx, and more stringent proteolytic treatments are required.[115,116] Presumably, the mucins, including both tethered and secreted mucins, may also be present in the mucus layer in the airway and may act as false attachment sites for Ad thus effectively reducing the amount of Ad that ultimately reaches the epithelial surface. The reported rheological properties of CF mucus producing a more viscous, more dehydrated, and immobile barrier suggest that this obstacle to gene transfer will be even more pronounced in the CF lung.

Other components of the innate immune system, not studied in specialized cell culture models, may also have barrier effects on gene transfer efficiency. Ordinarily, such barriers occur in the lung as primary defense mechanisms and may be aggravated in the CF lung where airway lumens are inflamed. For example, alveolar macrophages have been reported to sequester up to 70% of Ad genomes within 24 h following tracheal administration to mouse airways.[85] In a mouse nasal model of CF lung bacterial colonization, Pseudomonas infection (PA01 strain) was shown to inhibit Ad gene transfer by 10-fold relative to noninfected control nasal airways.[86]

In conclusion, there appears to be numerous potential barriers to Ad gene transfer in the lung especially in the CF lung that exhibits an overactive inflammatory milieu, and strategies to circumvent these barriers will likely need to be designed. However, even if all of these barriers are circumvented, the major cause of low-efficiency gene transfer is the lack of entry of Ad into the target cells. Strategies to improve the transduction efficiency will therefore be crucial to proving that the concept of gene transfer into the airway may actually be a feasible one.

In summary, human WD cultures are resistant to Ad-mediated gene transfer because of decreased specific attachment sites and reduced nonspecific entry paths that can internalize a fraction of a large vector load typical of CF gene therapy protocols using Ad. To circumvent the inefficiency of Ad-mediated gene transfer to the respiratory epithelium, either alterations of the host will be required, i.e., ability to access Ad receptors expressed on basolateral cell surfaces or Ad will require retargeting to receptor types that are present in sufficient number on the airway epithelial lumenal surface which allow for efficient uptake of Ad into the cell.

### 5.2.2 Safety

Initial attempts to improve efficiency of Ad gene transfer to the airway epithelium in vivo have mostly involved delivery of greater doses of Ad to the lung. These doses can represent a relatively large protein load and the subsequent gene expression (even in nonepithelial cells) can produce an unusually high level of transgene in an organ that is designed for monitoring invading pathogen assaults. It is therefore not surprising that inflammatory and immune responses are observed when Ad is delivered to the lung and numerous studies have reported Ad induced lung inflammation. In general, Ad induces an acute nonspecific mixed cellular inflammatory response and a late-specific, dose-related, lymphocyte-predominant, cell-mediated immune response in all species so far studied.[22,23,25–27,29,87–90] The acute response is nonspecific and likely induced by cytokine production in response to the protein load. It has also been suggested that neurogenic inflammation results after administration of Ad in rat airways, an effect shown to be partially due to vector gene expression but also to the viral proteins of the capsid coat.[91] The later, specific immune response to Ad is mediated by major histocompatibility complex class I-restricted cytotoxic (CD8) T lymphocytes directed against viral gene products and transgene proteins in expressing cells. The subsequent destruction of these cells leads to loss of persistence of transgene expression and so reduces efficiency of gene transfer.[28,92,93] The use of second-generation and high-capacity "gutless" vectors aims to limit the amount of viral gene expression to decrease the effects of this late immune response and these approaches are the topics of other chapters.[94,95]

In addition to cellular immune responses, Ad also elicits humoral immune responses with the production of mucosal and neutralizing antibodies.[25,87,90,96–98] These responses have been shown to be against the viral capsid proteins and are secondary to a helper (CD4+) T lymphocyte response. The production of such an antibody response results in neutralization of subsequent readministration of Ad, resulting in loss of gene transfer assuming the same Ad serotype is used (see later).

Therefore, in addition to the innate immunity of the lung (receptor localization, glycocalyx, macrophages, and mucus) reducing the efficiency of gene transfer, the cellular and humoral immune systems also respond to Ad delivery into the airway and as a result reduce the efficiency of gene transfer and the persistence of expression in the target epithelial cells.

## 5.3 Overcoming the Limitations of Ad

### 5.3.1 Efficiency

The localization of entry pathways for Ad to the basolateral surfaces of airway epithelial cells suggests that a delivery strategy to access these regions would be beneficial to improving gene transfer efficiency. This approach may also allow targeting of the epithelial stem cells (basal cells) resulting in transgene expression in the lung for the lifetime of the individual. This is an important consideration for gene transfer to the airway epithelium since fully differentiated lumenal facing cells (e.g., ciliated cells) have a relatively short lifetime in the order of 40–100 days and targeting these cell types specifically will require regular readministration of vectors.

Access to basal cells/basolateral surfaces may possibly be achieved by intravenous administration of vectors if penetration of the blood vessel wall, the connective tissue, and the basal lamina of the basement membrane were achievable. Unfortunately, studies that have attempted intravenous delivery strategies have not been successful since vectors do not appear to gain access to sufficient lung epithelial cells to make this approach feasible.[99–102] Barrier functions provided by the blood vessel endothelial cells and connective tissue surrounding the airway passages seem impenetrable by Ad. Indeed, the particle permeability of the basal lamina alone is thought to exclude inert particles of greater than 10 nm, which would certainly be restrictive to particles the size of Ad (100 nm). In an in vivo experimental mouse model where Ad was externally administered directly to the tracheal basement membrane, efficient gene transfer to the connective tissue fibroblasts adjacent to the basement membrane was observed without gene transfer to the epithelial cells of the juxtaposed epithelium.[51]

To date, two main strategies to improve intralumenal delivery of Ad vectors have been focused on. One approach is to access the basolateral surfaces of the epithelial cells by disruption of the epithelial "tight" junctions, and the other is to retarget Ad vectors to nonviral receptors that are present on the apical surface of lumenal epithelial cells that allow for entry of Ad into these cell types. Retargeting has so far been achieved by chemically, immunologically, or genetically modifying the Ad capsid coat by incorporating new receptor ligands that can target candidate receptors.

#### 5.3.1.1 Modification of the Host by Opening Tight Junctions

Epithelial cell "tight" junctions (zonula occludens) are collar-like structures composed of a diverse number of proteins that separate the apical and basolateral domains of the lumenal columnar epithelial cells. As well as functioning as a restrictive barrier to mixing of apical and basolateral membrane components, these intercellular junctions limit the transepithelial transport of solutes across the epithelium. A number of disease states have been shown to alter tight junction permeability (e.g., asthma), and reagents to increase the permeability of the junction are available. The key to successful disruption of tight junctions to allow Ad access to basolateral epithelial cell surfaces will be to use a reagent that open tight junctions sufficiently for Ad to pass through but that is rapidly reversible to limit the passage of other lumenal contents (e.g., bacteria) or serosal fluid into the airway lumen.

A property exploited for this purpose is the calcium ion dependency of the structural integrity of the junction. Walters et al., have successfully shown that treatment of the apical surface of human WD airway cells with the calcium chelator EGTA or hypotonic solutions (e.g., water) allow for improvements in Ad-mediated gene transfer presumably by allowing Ad access to basolateral receptors.[64,103] The slow reversibility of this effect however is problematic, tight junction reformation takes a least a couple of hours, a time period that would be unacceptable in a clinical setting. In vivo studies in mouse airways have confirmed that these treatments improve gene transfer efficiency although parameters of safety were not assessed fully.[62,104]

More specific reagents are available for studying tight junction permeability and the effect on Ad gene transfer. Parsons et al. used a detergent, polidocanol, in murine airways in vivo to enhance Ad-mediated gene transfer, an effect shown to be due to the ability of this reagent to transiently open tight junctions.[86] The short-chain fatty acid, sodium caprate, has also been used to increase Ad-mediated gene transfer to human WD cultures and results in full correction of CF cultures when AdCFTR is subsequently applied to the apical surface. This result is exciting since the effect is rapidly reversible effect and has previously been used clinically for enhancing pharmaceutics absorption across the GI tract, again presumably by an effect on tight junctional permeability.

These studies although fraught with inherent safety issues are beginning to establish that this strategy for delivering transgenes to the lung may be a viable option. The possibility of targeting the basal stem cells by this procedure is reason enough to continue pursuing the usefulness of these strategies.

#### 5.3.1.2 Targeted Ad to Increase Gene Transfer Efficiency

Targeted Ad directed against specific receptors has been used to successfully transduce cell types that are usually refractory to Ad infection. The epidermal growth factor receptor, stem cell factor receptor, fibroblast growth factor receptor, $\alpha_V$ integrins, and T cell receptors (CD3) have all been used as surrogate receptors for Ad entry in a variety of cell types.[105–108] Given the lack of Ad receptors at the apical surface of lumenal airway epithelial cells, a retargeting strategy to receptors known to present on the airway lumen may allow for gene transfer efficiency to be improved. However, a successful targeting strategy to the lung epithelium will require the identification of target molecules that allow for attachment and internalization of AdV across the apical membrane of columnar airway epithelial cells.

The identification of target receptors to which to redirect Ad tropism on the lumen of airway epithelium is difficult because most receptors and entry mechanisms occur on the basolateral surfaces of the cells. Certain members of a specific 7-transmembrane-spanning G-protein-coupled receptor family (i.e., P2Y2 purinoceptors, B2 kinin receptors, and adenosine type 2b receptors) have been identified as putative utile target receptors for redirecting Ad tropism to the surface epithelium of the lung. These receptors have been shown to be present on the lumenal surface of human airway epithelium and internalized into clathrin-coated pits when activated by their respective agonists.[109] The utilization of clathrin-coated pit internalization pathways for native Ad receptors suggests that the G-protein-coupled receptors may provide an ideal surrogate entry pathway for Ad. The high potency of P2Y2 agonists (e.g., ATP, UTP) combined with the low affinity of these agonists for the receptor suggests that the P2Y2 purinoceptors are abundant in number on the lumenal surface of the human respiratory epithelium.[110] Since pharmacological activation of airway epithelial P2Y2 receptors does not result in untoward effects in human airways, this receptor is an ideal target receptor to redirect Ad tropism. However, since the only available ligands for this receptor are low-affinity, small organic molecules, certain technical difficulties are associated with conjugating these molecules to Ad. Other receptor types suitable for Ad retargeting exist on the airway although specific retargeting data for Ad are lacking. The urokinase plasminogen activator receptor, uPA-R and the SEC-2 receptor have also been proposed as target receptors for Ad and adeno-associated virus (AAV), respectively.[111,112]

**5.3.1.2.1 Immunologically Modified Targeted Vectors** One immunological approach for targeting gene transfer vectors is using bispecific antibodies linking Ad directly to non-Ad receptor types present on the cell surface.[107,113] For example, chemically conjugated antibodies, one of which is directed against an epitope-tagged Ad coat protein and the other against $\alpha_V$ integrin membrane proteins have been reported to increase gene transfer efficiency by seven- to ninefold compared to that of nonmodified Ad, indicating that increased Ad attachment results in increased gene transfer efficiency.[113] In a similar approach, Ad was retargeted to nonviral receptor types in conjunction with ablation of the natural Ad tropism using an anti-fiber-knob protein antibody conjugated to folate.[114] Folate-conjugated antibody was the ligand of choice since the folate receptor is reported to be upregulated on the surface of malignant cells, thus providing a targeted vector for a variety of cancers. Retargeting Ad to cells expressing folate receptors was shown to be specific and successful with significant increases in gene transfer efficiency.

As "proof-of-concept" studies, an hemagglutin (HA) epitope-tagged P2Y2 receptor expressed at the apical surface of human WD cultures and targeted with bispecific antibodies consisting of antibodies to Ad fiber-knob protein/HA tag has been shown to facilitate Ad entry into these cell types, shown schematically in Figure 4.[115,116] This effect is enhanced by coadministration of exogenous ATP to activate the receptor, an effect that can be reduced by desensitization of the P2Y2 receptors prior to addition of targeted Ad. Importantly, the apical surfaces of the HA-tagged P2Y2-expressing cultures required a brief exposure to specific proteases before targeting was effective suggesting that the apical surface glycocalyx hindered access of the targeted vector to the target receptors.[115] This approach also relied on the expression of an HA-tagged receptor that may be overexpressed relative to the endogenously expressed P2Y2 receptors in the culture system. The number of target receptors and the affinity of the targeting ligand are both likely to be critical parameters for the success of such a targeting strategy.

**5.3.1.2.2 Chemically Modified Targeted Vectors** Since antibodies to the external domains of P2Y2 receptors are not currently available, a strategy to target Ad to the endogenous P2Y2 receptor was to chemically conjugate small molecule agonists (UTP)

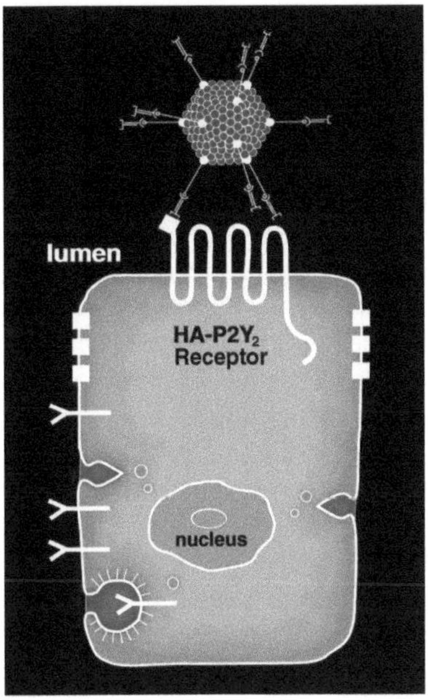


**Figure 4 Schematic of targeting strategy used to redirect Ad tropism to P2Y2 receptors on the apical surface of human airway epithelial cells.** Bispecific antibodies against the virus and the receptor were used as a targeting link and activation of the receptor results in receptor internalization and entry of Ad with subsequent gene transfer.

to the proteins of the Ad capsid coat. Using chemically reactive biotin derivatives, biotin was coupled to the Ad capsid coat predominately via hexon protein. This strategy is reported to couple 2–300 biotins to a single Ad particle and does not significantly alter the fiber-knob–hCAR interaction. By using commercially available biotin-linked UTP in combination with streptavidin as a "bridge" linking biotin-Ad to biotin-UTP, these molecular conjugates were shown to mediate gene transfer by an interaction specifically with endogenous P2Y2 receptors on the apical surface of WD cultures.[109] Again, the effectiveness of this approach was reduced by the presence of apical surface glycocalyx since gene transfer was only observed in cultures pretreated with agents that degrade this barrier. Regardless, gene transfer efficiency using these conjugates was still inefficient, probably due to the clumsiness of the "streptavidin bridge" and the low affinity of UTP for this receptor. Future experiments using this targeting strategy will require the identification of receptor agonists with higher affinity in addition to improved methods to directly couple the agonist ligands to the Ad capsid coat.

Another method for chemically conjugating receptor ligands to Ad is by the use of polyethylene glycol (PEG) that can be covalently linked directly to the Ad capsid coat. A number of groups have now shown that PEG-conjugated viruses can be used to target Ad.[111,117] For example, Ad conjugated to a 12-amino acid peptide, identified from phage display assays on the apical surfaces of human WD cultures, resulted in a 10-fold increase in gene transfer efficiency to these cell types.[117] Similarly, Ad conjugated via PEG to a peptide that binds to uPA-R has been shown to target Ad to this receptor type and enhance gene transfer to polarized airway epithelia.[111] An additional bonus of using PEG-conjugated Ad is that these vectors appear to be less immunogenic that non-PEG-conjugated Ad. This effect is due to the masking of antigenic Ad capsid proteins (mainly hexon) from neutralizing antibodies.[118]

**5.3.1.2.3 Genetically Modified Targeted Vectors** The ideal targeted vector would be one in which the target ligand could be incorporated into the capsid coat with minimal disruption of the physical and biological properties of Ad. For targeting strategies in which a peptide ligand is used, the most desirable method would be to generate an Ad vector genetically modified to express a functional peptide ligand on the viral surface. Such an approach for targeting vectors has been reported, where the Ad viral coat has been genetically modified to express multiple polylysine groups on the C-terminus of the Ad fiber-knob protein.[70] This redirects Ad tropism to heparan sulfate moieties that are present on the surfaces of most mammalian cells. With certain nonepithelial cell types, which lack hCAR, this modified vector has been shown to increase gene transfer efficiency from 10- to 300-fold in comparison to nonmodified Ad. However, the modified vector will likely not be useful for gene transfer to the airway epithelium since heparan sulfate is not expressed at the apical surface of airway epithelial cells.[119] Targeted Ad in which the fiber-knob protein (responsible for Ad attachment to the hCAR) has been modified to express novel ligands that can interact with other receptor types are being developed and the feasibility of this approach has now been reported by a number of groups.[106,120,121] A recent development in this type of approach was reported by Krasnykh et al.,[122] who hypothesized that the HI loop region of the fiber-knob structure can withstand the insertion of heterologous peptide sequences without significantly compromising the tertiary structure of the fiber-knob protein nor the production and infectivity of the modified Ad. These authors incorporated the FLAG octapeptide marker sequence into the HI loop region and were able to produce functional Ad. Importantly, they also showed that the sequence contained within intact virions was accessible to a FLAG-specific antibody suggesting that sequences inserted into this region are capable of interacting with other target substrates such as cell surface receptors.

A significant technical advance in Ad-targeting strategies evolved from studies that deduced the viral sequences in fiber-knob protein that interact with hCAR. Genetic ablation of these sequences from Ad vectors led to the generation of Ad that no longer binds to hCAR and no longer transduces cells that are permissive for normal Ad transduction.[123] The broad cellular tropism of Ad vectors can now be reduced and by the addition of targeting moieties to these Ad vectors specific cell-type targeting is possible. Reduced Ad interactions with nontarget cells will lessen the potential for adverse effects with these vectors. In the lung, however, the significance of natural tropism ablation is unclear since most of the epithelial cells targeted with delivery strategies do not express Ad receptors at the lumenal surface. However, the loss of transduction to other cell types that may interact with Ad delivered to the lung (e.g., macrophages, dendritic cells) may benefit from the hCAR-binding ablation mutant.

Recent developments in immunologically, chemically, and genetically modified targeted Ad suggest that "designer" gene transfer vectors will one day be available. Although Ad vectors, in their present form, may not be ideal for a number of gene transfer target tissues, notably the lung epithelium, this vector clearly remains at the forefront of gene therapy research since it is still one of the most efficacious gene transfer vectors available and will continue to be useful at least in "proof-of-concept" studies.

**5.3.1.2.4 Screening with Other Adenoviral Subtypes** Although over 51 different serotypes of wild-type Ad exist, the predominant serotypes used for gene transfer experiments are serotypes 2 and 5. The reason for this is largely historical since these two serotypes have been extensively studied over the last 30 years and understanding of the viral genome has allowed the manipulations necessary to evolve these viruses into gene transfer vectors. With regard to the airway epithelium, other serotypes have been suggested to be efficacious at delivering transgenes to human WD cultures. Serotypes 17 and 12 have been shown to bind/deliver transgenes 10-fold over Ad2 vectors.[124] However as of yet no conclusive results have been presented that suggest that the improvements warrant future investigations with these vectors. One approach to determining if any of the other serotypes may be more efficacious in the lung epithelium could be envisioned using a recently reported system of generating an Ad5 capsid expressing fiber proteins from the other serotypes.[125] This system was used to screen vascular endothelial and smooth muscle cells and the efficiency of gene transfer compared against the efficiency of gene transfer with Ad5. This screening procedure identified Ad5 with Ad16 fibers as being significantly more efficient at gene transfer than Ad5 in these particular cell types. It will be of interest to screen these serotypes on human WD cultures relative to Ad5 to determine whether other Ad serotypes may be of benefit to airway epithelial cell gene transfer. A serotype which may be of particular interest is Ad37, since it has been reported that Ad37 utilizes sialic acid residues that are present on the extracellular surfaces of most cells.[126] An abundance of sialic acid residues on the lumenal surface of airway epithelial cells as components of glycoconjugates may allow for improved gene transfer. Whether attachment of Ad37 to sialic acid residues located on the airway lumen leads to efficient entry and gene transfer awaits further study.

**5.3.1.2.5 Other Methods to Increase Gene Transfer Efficiency** Nonspecific methods to enhance Ad-mediated gene transfer to airway epithelial cells have been reported.[127,128] Calcium phosphate coprecipitation has been used to precipitate aggregates of Ad and other vectors to increase gene transfer to airway epithelia both in vitro and in vivo. It has been suggested that in in vivo these aggregates increase the rate of nonspecific endocytosis of Ad across the apical membrane of polarized epithelial cells. The possible effects of this technique on cellular and paracellular permeability have not been investigated.

Another method to improve both the delivery and efficiency of Ad to the lung epithelium in vivo is using the inert perfluorochemicals (PFCs). These compounds are liquid in nature but due to high-oxygen saturation capacities can be instilled into the lung for periods of time with maintenance of passive oxygen diffusion. Several studies have now shown that administration of gene transfer vectors (including Ad) with PFC results in increased gene transfer to rodent and nonhuman primate lungs.[129–131] The improvements in gene transfer are predominately localized to the alveolar regions with only modest improvements in the efficiency of gene transfer to the respiratory epithelium. The exact mechanism by which PFCs produce these effects remains to be determined but may be due to prolonged contact time for the vector on the cells and reduced ingestion of Ad by macrophages and/or due to some nonspecific effect on the paracellular permeability. Nonetheless, this method provides an example of a new strategy to deliver transgenes to the lung without the need for direct instillation or aerosolization, which are both inefficient methods for airway epithelium delivery.

### 5.3.2 Safety

Strategies that improve gene transfer efficiency, as described above, will allow for lower doses of Ad to be administered to the lung. This achievement alone will be beneficial in reducing the inflammatory responses seen with Ad administration. However, attempts are also been made to reduce the inflammation produced by expression of viral genes that produce the cell-mediated immune responses described above. The identification of specific viral genes that initiate or amplify the immune response has led to the reengineering of Ad vectors to ablate the specific gene expression. For example, vectors deleted of E2a and E4 have been reported to display reduced immune responses and improve persistence of transgene expression.[92,93] The ultimate vector is one that contains no viral genes and the high-capacity "gutless" vectors have been generated and appear to blunt the immune response considerably.[132–134] In contrast, several viral genes have been identified that have evolved to subvert the immune response and the inclusion of these genes into new vectors may be desirable (e.g., E3).[135]

Strategies to circumvent the humoral immune response are also been considered. Since this arm of the immune system results in the inability of readministration of specific Ad serotypes, serotype switching has been proposed as a method to allow repeat administration. Indeed, Ad5 administration but not Ad4 or Ad30 has been reported to prevent the gene transfer obtained with subsequent Ad5 administration to the lung.[97] However, in addition to this being a somewhat limited procedure, it is not yet clear whether these different serotypes are as inefficient for gene transfer to the airway epithelium as Ad5. Transient immunosuppression has also been suggested to reduce the inhibitory effects of neutralizing antibodies. Intratracheal administration of immunosuppressive factors (IL-12, interferon gamma, antibodies to CD40, corticosteroids, and cyclophosphamide) at the time of vector administration have all shown a reduction in generation of neutralizing antibodies.[136–140] The longer-term effects of administering these factors to lung have not been reported. Finally, covalent conjugation of PEG to the Ad capsid coat that permits addition of targeting moieties is also a strategy for the virus to elude neutralizing antibodies by masking capsid coat proteins, especially hexon protein. Although PEGylation of Ad leads to some loss of viral titer and aggregation the ability of this procedure to develop targeted vectors combined with reduction in immune response makes this a promising method for future study.[118]

## 6. Other Vectors

The focus of this review has been on Ad vectors for use in CF lung disease. However, a number of other vectors have been suggested as candidates for CF lung gene transfer vectors. AAV, retrovirus, lentivirus, and liposomal vectors have all shown promise in preclinical studies in the lung and some have been tested in clinical trials. The general observation is that all of these vectors, like Ad, do not appear to display the efficiency of gene transfer in WD airway epithelial cells as they do in nonpolarized cells suggesting that these vectors confront similar barriers in the airways as do Ad vectors. Strategies to improve gene transfer efficiency for these other vectors have followed the progression of experiments with Ad, i.e., tight junction modulation, targeting, serotype switching, and immune response reduction, and all have been shown as for Ad to improve efficiency to some degree. Whether efficiency can ever be improved to a point that shows efficacy in the lungs of CF patients remains to be determined. Meanwhile, other viruses (Sendai virus[141] and lentiviruses pseudotyped with filovirus coat proteins[142]) may show promise for gene delivery to the airway and preliminary reports suggest that these viruses or components thereof may one day provide us with a method to deliver transgenes to the lung in an efficient and safe manner.

## 7. Conclusion

It is clear that the evolution of gene therapy has been aided by many different aspects of basic biological and medical research efforts and the possibility of a gene therapy for CF lung disease will only take time and a continuation of these efforts. These findings will not only be beneficial to the treatment of CF lung disease but also other disease states, which are continually being brought closer to a treatment and perhaps a cure by this new and exciting biomedical technology.

## Acknowledgments

The author would especially like to thank Dr Ashley Henderson of the UNC Pulmonary Division for useful discussions and critical review of the manuscript and Jennifer Nank for help with both manuscript and figure production.

## References

1. Johnson LG, Boyles SE, Wilson J, Boucher RC. Normalization of raised sodium absorption and raised calcium-mediated chloride secretion by adenovirus-mediated expression of cystic fibrosis transmembrane conductance regulator in primary human cystic fibrosis airway epithelial cells. *J Clin Invest* 1995;**95**:1377–82.
2. Drumm ML, Pope HA, Cliff WH, Rommens JM, Marvin SA, Tsui LC, et al. Correction of the cystic fibrosis defect in vitro by retrovirus-mediated gene transfer. *Cell* 1990;**62**:1227–33.
3. Rommens JM, Iannuzzi MC, Kerem B, Drumm ML, Melmer G, Dean M, et al. Identification of the cystic fibrosis gene: chromosome walking and jumping. *Science* 1989;**245**:1059–65.

4. Stutts MJ, Canessa CM, Olsen JC, Hamrick M, Cohn JA, Rossier BC, et al. CFTR as a cAMP-dependent regulator of sodium channels. *Science* 1995;**269**:847–50.
5. Boucher RC, Stutts MJ, Knowles MR, Cantley L, Gatzy JT. Na+ transport in cystic fibrosis respiratory epithelia. Abnormal basal rate and response to adenylate cyclase activation. *J Clin Invest* 1986;**78**:1245–52.
6. Rosenfeld MA, Yoshimura K, Trapnell BC, Yoneyama K, Rosenthal ER, Dalemans W, et al. In vivo transfer of the human cystic fibrosis transmembrane conductance regulator gene to the airway epithelium. *Cell* 1992;**68**:143–55.
7. Zabner J, Couture LA, Gregory RJ, Graham SM, Smith AE, Welsh MJ. Adenovirus-mediated gene transfer transiently corrects the chloride transport defect in nasal epithelia of patients with cystic fibrosis. *Cell* 1993;**75**:207–16.
8. Boat T, Welsh MJ, Beaudet AL. Cystic fibrosis. In: Scriver ER, Beaudet AL, Sly WS, Valle D, editors. *The metabolic basis of inherited disease*. New York: McGraw-Hill; 1989. p. 2649–80.
9. Riordan JR, Rommens JM, Kerem B, Alon N, Rozmahel R, Grzelczak Z, et al. Identification of the cystic fibrosis gene: cloning and characterization of complementary DNA. *Science* 1989;**245**:1066–73.
10. Welsh MJ, Smith AE. Molecular mechanisms of CFTR chloride channel dysfunction in cystic fibrosis. *Cell* 1993;**73**:1251–4.
11. Cheng SH, Gregory RJ, Marshall J, Paul S, Souza DW, White GA, et al. Defective intracellular transport and processing of CFTR is the molecular basis of most cystic fibrosis. *Cell* 1990;**63**:827–34.
12. Kalin N, Claass A, Sommer M, Puchelle E, Tummler B. DeltaF508 CFTR protein expression in tissues from patients with cystic fibrosis. *J Clin Invest* 1999;**103**:1379–89.
13. Quinton PM. Cystic fibrosis: a disease in electrolyte transport. *Faseb J* 1990;**4**:2709–17.
14. Johnson L, RC B. Towards correction of the genetic defect in cystic fibrosis. In: Brigham KL, editor. *Gene therapy for diseases of the lung*, vol. 104. New York: Marcel Dekker; 1997. p. 239–65.
15. Crystal RG, McElvaney NG, Rosenfeld MA, Chu CS, Mastrangeli A, Hay JG, et al. Administration of an adenovirus containing the human CFTR cDNA to the respiratory tract of individuals with cystic fibrosis. *Nat Genet* 1994;**8**:42–51.
16. Knowles MR, Hohneker KW, Zhou Z, Olsen JC, Noah TL, Hu PC, et al. A controlled study of adenoviral-vector-mediated gene transfer in the nasal epithelium of patients with cystic fibrosis. *N Engl J Med* 1995;**333**:823–31.
17. Caplen NJ, Kinrade E, Sorgi F, Gao X, Gruenert D, Geddes D, et al. In vitro liposome-mediated DNA transfection of epithelial cell lines using the cationic liposome DC-Chol/DOPE. *Gene Ther* 1995;**2**:603–13.
18. Gill DR, Southern KW, Mofford KA, Seddon T, Huang L, Sorgi F, et al. A placebo-controlled study of liposome-mediated gene transfer to the nasal epithelium of patients with cystic fibrosis. *Gene Ther* 1997;**4**:199–209.
19. Johnson LG, Olsen JC, Sarkadi B, Moore KL, Swanstrom R, Boucher RC. Efficiency of gene transfer for restoration of normal airway epithelial function in cystic fibrosis. *Nat Genet* 1992;**2**:21–5.
20. Boucher RC. Current status of CF gene therapy. *Trends Genet* 1996;**12**:81–4.
21. Teramoto S, Johnson LG, Huang W, Leigh MW, Boucher RC. Effect of adenoviral vector infection on cell proliferation in cultured primary human airway epithelial cells. *Hum Gene Ther* 1995;**6**:1045–53.
22. Dong JY, Wang D, Van Ginkel FW, Pascual DW, Frizzell RA. Systematic analysis of repeated gene delivery into animal lungs with a recombinant adenovirus vector. *Hum Gene Ther* 1996;**7**:319–31.

23. Ginsberg HS, Lundholm-Beauchamp U, Horswood RL, Pernis B, Wold WS, Chanock RM, et al. Role of early region 3 (E3) in pathogenesis of adenovirus disease. *Proc Natl Acad Sci USA* 1989;**86**:3823–7.
24. Ginsberg HS, Prince GA. The molecular basis of adenovirus pathogenesis. *Infect Agents Dis* 1994;**3**:1–8.
25. Kaplan JM, St George JA, Pennington SE, Keyes LD, Johnson RP, Wadsworth SC, et al. Humoral and cellular immune responses of nonhuman primates to long-term repeated lung exposure to Ad2/CFTR-2. *Gene Ther* 1996;**3**:117–27.
26. Look DC, Brody SL. Engineering viral vectors to subvert the airway defense response. *Am J Respir Cell Mol Biol* 1999;**20**:1103–6.
27. St George JA, Pennington SE, Kaplan JM, Peterson PA, Kleine LJ, Smith AE, et al. Biological response of nonhuman primates to long-term repeated lung exposure to Ad2/CFTR-2. *Gene Ther* 1996;**3**:103–16.
28. Yang Y, Nunes FA, Berencsi K, Gonczol E, Engelhardt JF, Wilson JM. Inactivation of E2a in recombinant adenoviruses improves the prospect for gene therapy in cystic fibrosis. *Nat Genet* 1994;**7**:362–9.
29. Yei S, Mittereder N, Wert S, Whitsett JA, Wilmott RW, Trapnell BC. In vivo evaluation of the safety of adenovirus-mediated transfer of the human cystic fibrosis transmembrane conductance regulator cDNA to the lung. *Hum Gene Ther* 1994;**5**:731–44.
30. Jeffery P. Form and function of airway epithelium. In: Jones CJ, editor. *Epithelia: advances in cell physiology and cell culture*. London: Kluwer Academic Publishers; 1990. p. 195–220.
31. Harkema J, Mariassy A, St George J, Hyde DM, Plopper CG. Epithelial cells of the conducting airways: a species comparison. In: Farmer SG, Hay DWP, editors. *The airway epithelium*, vol. 55. New York: Marcel Dekker; 1994. p. 3–39.
32. Mercer RR, Russell ML, Roggli VL, Crapo JD. Cell number and distribution in human and rat airways. *Am J Respir Cell Mol Biol* 1994;**10**:613–24.
33. Randell SH. Progenitor-progeny relationships in airway epithelium. *Chest* 1992;**101**:11S–6S.
34. Leigh MW, Kylander JE, Yankaskas JR, Boucher RC. Cell proliferation in bronchial epithelium and submucosal glands of cystic fibrosis patients. *Am J Respir Cell Mol Biol* 1995;**12**:605–12.
35. Matsui H, Randell SH, Peretti SW, Davis CW, Boucher RC. Coordinated clearance of periciliary liquid and mucus from airway surfaces. *J Clin Invest* 1998;**102**:1125–31.
36. Matsui H, Grubb BR, Tarran R, Randell SH, Gatzy JT, Davis CW, et al. Evidence for periciliary liquid layer depletion, not abnormal ion composition, in the pathogenesis of cystic fibrosis airways disease. *Cell* 1998;**95**:1005–15.
37. Cotton CU, Stutts MJ, Knowles MR, Gatzy JT, Boucher RC. Abnormal apical cell membrane in cystic fibrosis respiratory epithelium. An in vitro electrophysiologic analysis. *J Clin Invest* 1987;**79**:80–5.
38. Engelhardt JF, Yankaskas JR, Ernst SA, Yang Y, Marino CR, Boucher RC, et al. Submucosal glands are the predominant site of CFTR expression in the human bronchus. *Nat Genet* 1992;**2**:240–8.
39. Grubb BR, Boucher RC. Pathophysiology of gene-targeted mouse models for cystic fibrosis. *Physiol Rev* 1999;**79**:S193–214.
40. Clarke LL, Grubb BR, Yankaskas JR, Cotton CU, McKenzie A, Boucher RC. Relationship of a non-cystic fibrosis transmembrane conductance regulator-mediated chloride conductance to organ-level disease in Cftr(-/-) mice. *Proc Natl Acad Sci USA* 1994;**91**:479–83.
41. Grubb BR, Paradiso AM, Boucher RC. Anomalies in ion transport in CF mouse tracheal epithelium. *Am J Physiol* 1994;**267**:C293–300.

42. Grubb BR, Vick RN, Boucher RC. Hyperabsorption of Na+ and raised Ca(2+)-mediated Cl- secretion in nasal epithelia of CF mice. *Am J Physiol* 1994;**266**:C1478–83.
43. Grubb BR, Pickles RJ, Ye H, Yankaskas JR, Vick RN, Engelhardt JF, et al. Inefficient gene transfer by adenovirus vector to cystic fibrosis airway epithelia of mice and humans. *Nature* 1994;**371**:802–6.
44. Mittereder N, Yei S, Bachurski C, Cuppoletti J, Whitsett JA, Tolstoshev P, et al. Evaluation of the efficacy and safety of in vitro, adenovirus-mediated transfer of the human cystic fibrosis transmembrane conductance regulator cDNA. *Hum Gene Ther* 1994;**5**: 717–29.
45. Engelhardt JF, Yang Y, Stratford-Perricaudet LD, Allen ED, Kozarsky K, Perricaudet M, et al. Direct gene transfer of human CFTR into human bronchial epithelia of xenografts with E1-deleted adenoviruses. *Nat Genet* 1993;**4**:27–34.
46. Jiang C, Akita GY, Colledge WH, Ratcliff RA, Evans MJ, Hehir KM, et al. Increased contact time improves adenovirus-mediated CFTR gene transfer to nasal epithelium of CF mice. *Hum Gene Ther* 1997;**8**:671–80.
47. Zabner J, Zeiher BG, Friedman E, Welsh MJ. Adenovirus-mediated gene transfer to ciliated airway epithelia requires prolonged incubation time. *J Virol* 1996;**70**:6994–7003.
48. Mastrangeli A, Danel C, Rosenfeld MA, Stratford-Perricaudet L, Perricaudet M, Pavirani A, et al. Diversity of airway epithelial cell targets for in vivo recombinant adenovirus-mediated gene transfer. *J Clin Invest* 1993;**91**:225–34.
49. Hansen SH, Sandvig K, van Deurs B. Internalization efficiency of the transferrin receptor. *Exp Cell Res* 1992;**199**:19–28.
50. St. George J, Sacks CR, Lukason MJ, Nichols M, Peterson PA, Vaccaro C, et al. Efficacy of adenoviral vectors in airway epithelium. *Pediatr Pulmonol* 1995;**12**(Suppl.):151.
51. Pickles RJ, Barker PM, Ye H, Boucher RC. Efficient adenovirus-mediated gene transfer to basal but not columnar cells of cartilaginous airway epithelia. *Hum Gene Ther* 1996;**7**:921–31.
52. Dupuit F, Zahm JM, Pierrot D, Brezillon S, Bonnet N, Imler JL, et al. Regenerating cells in human airway surface epithelium represent preferential targets for recombinant adenovirus. *Hum Gene Ther* 1995;**6**:1185–93.
53. Bergelson JM, Cunningham JA, Droguett G, Kurt-Jones EA, Krithivas A, Hong JS, et al. Isolation of a common receptor for Coxsackie B viruses and adenoviruses 2 and 5. *Science* 1997;**275**:1320–3.
54. Tomko RP, Xu R, Philipson L. HCAR and MCAR: the human and mouse cellular receptors for subgroup C adenoviruses and group B coxsackie viruses. *Proc Natl Acad Sci USA* 1997;**94**:3352–6.
55. Wickham TJ, Mathias P, Cheresh DA, Nemerow GR. Integrins alpha v beta 3 and alpha v beta 5 promote adenovirus internalization but not virus attachment. *Cell* 1993;**73**:309–19.
56. Pickles RJ, McCarty D, Matsui H, Hart PJ, Randell SH, Boucher RC. Limited entry of adenovirus vectors into well-differentiated airway epithelium is responsible for inefficient gene transfer. *J Virol* 1998;**72**:6014–23.
57. Zabner J, Freimuth P, Puga A, Fabrega A, Welsh MJ. Lack of high affinity fiber receptor activity explains the resistance of ciliated airway epithelia to adenovirus infection. *J Clin Invest* 1997;**100**:1144–9.
58. Matsui H, Davis CW, Tarran R, Boucher RC. Osmotic water permeabilities of cultured, well-differentiated normal and cystic fibrosis airway epithelia. *J Clin Invest* 2000;**105**:1419–27.
59. Walters RW, Yi S, Keshavjee S, Brown KE, Welsh MJ, Chiorini JA, et al. Binding of adeno-associated virus type 5 to 2,3-linked sialic acid is required for gene transfer. *J Biol Chem* 2001;**21**:21.

60. Jepsen M, Graham S, Karp PH, Zabner J. Effect of topical nasal pharmaceuticals on sodium and chloride transport by human airway epithelia. *Am J Rhinol* 2000;**14**:405–9.
61. Zabner J, Seiler MP, Launspach JL, Karp PH, Kearney WR, Look DC, et al. The osmolyte xylitol reduces the salt concentration of airway surface liquid and may enhance bacterial killing. *Proc Natl Acad Sci USA* 2000;**97**:11614–9.
62. Wang G, Zabner J, Deering C, Launspach J, Shao J, Bodner M, et al. Increasing epithelial junction permeability enhances gene transfer to airway epithelia in vivo. *Am J Respir Cell Mol Biol* 2000;**22**:129–38.
63. Wang G, Davidson BL, Melchert P, Slepushkin VA, van Es HH, Bodner M, et al. Influence of cell polarity on retrovirus-mediated gene transfer to differentiated human airway epithelia. *J Virol* 1998;**72**:9818–26.
64. Walters RW, Grunst T, Bergelson JM, Finberg RW, Welsh MJ, Zabner J. Basolateral localization of fiber receptors limits adenovirus infection from the apical surface of airway epithelia. *J Biol Chem* 1999;**274**:10219–26.
65. Goldman MJ, Wilson JM. Expression of alpha v beta 5 integrin is necessary for efficient adenovirus-mediated gene transfer in the human airway. *J Virol* 1995;**69**:5951–8.
66. Hidaka C, Milano E, Leopold PL, Bergelson JM, Hackett NR, Finberg RW, et al. CAR-dependent and CAR-independent pathways of adenovirus vector-mediated gene transfer and expression in human fibroblasts. *J Clin Invest* 1999;**103**:579–87.
67. Freimuth P. A human cell line selected for resistance to adenovirus infection has reduced levels of the virus receptor. *J Virol* 1996;**70**:4081–5.
68. Griffiths M, Huang XZ, Wu JF, Driscoll R, Sheppard D. Inactivation of the beta5 integrin subunit gene does not prevent expression of adenovirus genes in mouse airway epithelium. *Respir Crit Care Med* 1997;**155**:A549.
69. Fasbender A, Zabner J, Chillon M, Moninger TO, Puga AP, Davidson BL, et al. Complexes of adenovirus with polycationic polymers and cationic lipids increase the efficiency of gene transfer in vitro and in vivo. *J Biol Chem* 1997;**272**:6479–89.
70. Wickham TJ, Roelvink PW, Brough DE, Kovesdi I. Adenovirus targeted to heparan-containing receptors increases its gene delivery efficiency to multiple cell types. *Nat Biotechnol* 1996;**14**:1570–3.
71. Pickles R, Fahrner J, Petrella J, Boucher R, Bergelson J. Retargeting the coxsackievirus and adenovirus receptor to the apical surface of polarised epithelial cells reveals the glycocalyx as a barrier to adenovirus mediated gene transfer. *J Virol* 2000;**74**:6050–7.
72. Rambourg A, Neutra M, Leblond CP. Presence of a "cell coat" rich in carbohydrate at the surface of cells in the rat. *Anat Rec* 1966;**154**:41–71.
73. Bennett HS. Morphological aspects of extracellular polysaccharides. *J Histochem Cytochem* 1963;**11**:23.
74. Bernacki SH, Nelson AL, Abdullah L, Sheehan JK, Harris A, Davis CW, et al. Mucin gene expression during differentiation of human airway epithelia in vitro. *Am J Respir Cell Mol Biol* 1999;**20**:595–604.
75. Buisine MP, Devisme L, Copin MC, Durand-Reville M, Gosselin B, Aubert JP, et al. Developmental mucin gene expression in the human respiratory tract. *Am J Respir Cell Mol Biol* 1999;**20**:209–18.
76. Chambers JA, Hollingsworth MA, Trezise AE, Harris A. Developmental expression of mucin genes MUC1 and MUC2. *J Cell Sci* 1994;**107**(Pt 2):413–24.
77. Braga VM, Pemberton LF, Duhig T, Gendler SJ. Spatial and temporal expression of an epithelial mucin, Muc-1, during mouse development. *Development* 1992;**115**:427–37.
78. Pemberton L, Taylor-Papadimitriou J, Gendler SJ. Antibodies to the cytoplasmic domain of the MUC1 mucin show conservation throughout mammals. *Biochem Biophys Res Commun* 1992;**185**:167–75.

79. Porchet N, Nguyen VC, Dufosse J, Audie JP, Guyonnet-Duperat V, Gross MS, et al. Molecular cloning and chromosomal localization of a novel human tracheo-bronchial mucin cDNA containing tandemly repeated sequences of 48 base pairs. *Biochem Biophys Res Commun* 1991;**175**:414–22.
80. McNeer RR, Huang D, Fregien NL, Carraway KL. Sialomucin complex in the rat respiratory tract: a model for its role in epithelial protection. *Biochem J* 1998;**330**(Pt 2):737–44.
81. Sheng Z, Wu K, Carraway KL, Fregien N. Molecular cloning of the transmembrane component of the 13762 mammary adenocarcinoma sialomucin complex. A new member of the epidermal growth factor superfamily. *J Biol Chem* 1992;**267**:16341–6.
82. Wu K, Fregien N, Carraway KL. Molecular cloning and sequencing of the mucin subunit of a heterodimeric, bifunctional cell surface glycoprotein complex of ascites rat mammary adenocarcinoma cells. *J Biol Chem* 1994;**269**:11950–5.
83. Arcasoy SM, Latoche J, Gondor M, Watkins SC, Henderson RA, Hughey R, et al. MUC1 and other sialoglycoconjugates inhibit adenovirus-mediated gene transfer to epithelial cells. *Am J Respir Cell Mol Biol* 1997;**17**:422–35.
84. Arcasoy SM, Latoche JD, Gondor M, Pitt BR, Pilewski JM. Polycations increase the efficiency of adenovirus-mediated gene transfer to epithelial and endothelial cells in vitro. *Gene Ther* 1997;**4**:32–8.
85. Worgall S, Leopold PL, Wolff G, Ferris B, Van Roijen N, Crystal RG. Role of alveolar macrophages in rapid elimination of adenovirus vectors administered to the epithelial surface of the respiratory tract. *Hum Gene Ther* 1997;**8**:1675–84.
86. Parsons DW, Grubb BR, Johnson LG, Boucher RC. Enhanced in vivo airway gene transfer via transient modification of host barrier properties with a surface-active agent. *Hum Gene Ther* 1998;**9**:2661–72.
87. Otake K, Ennist DL, Harrod K, Trapnell BC. Nonspecific inflammation inhibits adenovirus-mediated pulmonary gene transfer and expression independent of specific acquired immune responses. *Hum Gene Ther* 1998;**9**:2207–22.
88. Van Ginkel FW, Liu C, Simecka JW, Dong JY, Greenway T, Frizzell RA, et al. Intratracheal gene delivery with adenoviral vector induces elevated systemic IgG and mucosal IgA antibodies to adenovirus and beta-galactosidase. *Hum Gene Ther* 1995;**6**: 895–903.
89. Simon RH, Engelhardt JF, Yang Y, Zepeda M, Weber-Pendleton S, Grossman M, et al. Adenovirus-mediated transfer of the CFTR gene to lung of nonhuman primates: toxicity study. *Hum Gene Ther* 1993;**4**:771–80.
90. Yei S, Mittereder N, Tang K, O'Sullivan C, Trapnell BC. Adenovirus-mediated gene transfer for cystic fibrosis: quantitative evaluation of repeated in vivo vector administration to the lung. *Gene Ther* 1994;**1**:192–200.
91. Piedimonte G, Pickles RJ, Lehmann JR, McCarty D, Costa DL, Boucher RC. Replication-deficient adenoviral vector for gene transfer potentiates airway neurogenic inflammation. *Am J Respir Cell Mol Biol* 1997;**16**:250–8.
92. Goldman MJ, Litzky LA, Engelhardt JF, Wilson JM. Transfer of the CFTR gene to the lung of nonhuman primates with E1- deleted, E2a-defective recombinant adenoviruses: a preclinical toxicology study. *Hum Gene Ther* 1995;**6**:839–51.
93. Engelhardt JF, Litzky L, Wilson JM. Prolonged transgene expression in cotton rat lung with recombinant adenoviruses defective in E2a. *Hum Gene Ther* 1994;**5**:1217–29.
94. Lieber A, He CY, Kirillova I, Kay MA. Recombinant adenoviruses with large deletions generated by Cre-mediated excision exhibit different biological properties compared with first-generation vectors in vitro and in vivo. *J Virol* 1996;**70**:8944–60.

95. Morsy MA, Gu M, Motzel S, Zhao J, Lin J, Su Q, et al. An adenoviral vector deleted for all viral coding sequences results in enhanced safety and extended expression of a leptin transgene. *Proc Natl Acad Sci USA* 1998;**95**:7866–71.
96. Mack CA, Song WR, Carpenter H, Wickham TJ, Kovesdi I, Harvey BG, et al. Circumvention of anti-adenovirus neutralizing immunity by administration of an adenoviral vector of an alternate serotype. *Hum Gene Ther* 1997;**8**:99–109.
97. Mastrangeli A, Harvey BG, Yao J, Wolff G, Kovesdi I, Crystal RG, et al. "Sero-switch" adenovirus-mediated in vivo gene transfer: circumvention of anti-adenovirus humoral immune defenses against repeat adenovirus vector administration by changing the adenovirus serotype. *Hum Gene Ther* 1996;**7**:79–87.
98. Scaria A, St George JA, Gregory RJ, Noelle RJ, Wadsworth SC, Smith AE, et al. Antibody to CD40 ligand inhibits both humoral and cellular immune responses to adenoviral vectors and facilitates repeated administration to mouse airway. *Gene Ther* 1997;**4**:611–7.
99. Griesenbach U, Chonn A, Cassady R, Hannam V, Ackerley C, Post M, et al. Comparison between intratracheal and intravenous administration of liposome-DNA complexes for cystic fibrosis lung gene therapy. *Gene Ther* 1998;**5**:181–8.
100. Liu F, Qi H, Huang L, Liu D. Factors controlling the efficiency of cationic lipid-mediated transfection in vivo via intravenous administration. *Gene Ther* 1997;**4**:517–23.
101. Lemarchand P, Jaffe HA, Danel C, Cid MC, Kleinman HK, Stratford-Perricaudet LD, et al. Adenovirus-mediated transfer of a recombinant human alpha 1-antitrypsin cDNA to human endothelial cells. *Proc Natl Acad Sci USA* 1992;**89**:6482–6.
102. Lemarchand P, Jones M, Danel C, Yamada I, Mastrangeli A, Crystal RG. In vivo adenovirus-mediated gene transfer to lungs via pulmonary artery. *J Appl Physiol* 1994;**76**:2840–5.
103. Coyne CB, Kelly MM, Boucher RC, Johnson LG. Enhanced epithelial gene transfer by modulation of tight junctions with sodium caprate. *Am J Respir Cell Mol Biol* 2000;**23**:602–9.
104. Chu Q, St George JA, Lukason M, Cheng SH, Scheule RK, Eastman SJ. Egta enhancement of adenovirus-mediated gene transfer to mouse tracheal epithelium in vivo. *Hum Gene Ther* 2001;**12**:455–67.
105. Watkins SJ, Mesyanzhinov VV, Kurochkina LP, Hawkins RE. The 'adenobody' approach to viral targeting: specific and enhanced adenoviral gene delivery. *Gene Ther* 1997;**4**:1004–12.
106. Wickham TJ, Carrion ME, Kovesdi I. Targeting of adenovirus penton base to new receptors through replacement of its RGD motif with other receptor-specific peptide motifs. *Gene Ther* 1995;**2**:750–6.
107. Wickham TJ, Lee GM, Titus JA, Sconocchia G, Bakacs T, Kovesdi I, et al. Targeted adenovirus-mediated gene delivery to T cells via CD3. *J Virol* 1997;**71**:7663–9.
108. Hoganson DK, Sosnowski BA, Pierce GF, Doukas J. Uptake of adenoviral vectors via fibroblast growth factor receptors involves intracellular pathways that differ from the targeting ligand. *Mol Ther* 2001;**3**:105–12.
109. Kreda SM, Pickles RJ, Lazarowski ER, Boucher RC. G-protein-coupled receptors as targets for gene transfer vectors using natural small-molecule ligands. *Nat Biotechnol* 2000;**18**:635–40.
110. Mason SJ, Paradiso AM, Boucher RC. Regulation of transepithelial ion transport and intracellular calcium by extracellular ATP in human normal and cystic fibrosis airway epithelium. *Br J Pharmacol* 1991;**103**:1649–56.

111. Drapkin PT, O'Riordan CR, Yi SM, Chiorini JA, Cardella J, Zabner J, et al. Targeting the urokinase plasminogen activator receptor enhances gene transfer to human airway epithelia. *J Clin Invest* 2000;**105**:589–96.
112. Ziady A, R K, T F, Davies P. Serpin enzyme complex receptor targeted DNA complexes deliver genes to airway epithelia. *Pediatr Pulmonol* 1999;**19**:233.
113. Wickham TJ, Segal DM, Roelvink PW, Carrion ME, Lizonova A, Lee GM, et al. Targeted adenovirus gene transfer to endothelial and smooth muscle cells by using bispecific antibodies. *J Virol* 1996;**70**:6831–8.
114. Douglas JT, Rogers BE, Rosenfeld ME, Michael SI, Feng M, Curiel DT. Targeted gene delivery by tropism-modified adenoviral vectors. *Nat Biotechnol* 1996;**14**:1574–8.
115. Pickles R, Johnson LG, Olsen JC, Gerard R, Segal D, Boucher RC. Correction of the CF bioelectric defect in human CF well-differentiated airway epithelial cells by retargeting adenoviral vectors to luminal P2Y2 purinoceptors. *Pediatr Pulmonol* 1999;**19**:222.
116. Pickles R, Kreda S, Olsen J, Johnson L, Gerard R, Segal D, et al. High efficiency gene transfer to polarised epithelial cells by retargeting adenoviral vectors to P2Y2 purinoceptors with bispecific antibodies. *Pediatr Pulmonol* 1998;**17**:261.
117. Romanczuk H, Galer C, Zabner J, Barsomian G, Wadsworth S, O'Riordan C. Modification of an adenoviral vector with biologically selected peptides: a novel strategy for gene delivery to cells of choice. *Hum Gene Ther* 1999;**10**:2615–26.
118. O'Riordan C, Lachapelle A, Delgado C, Parkes V, Wadsworth S, Smith A, et al. PEGylation of adenovirus with retention of infectivity and protection from neutralizing antibody in vitro and in vivo. *Hum Gene Ther* 1999;**10**:1349–58.
119. Summerford C, Samulski RJ. Membrane-associated heparan sulfate proteoglycan is a receptor for adeno-associated virus type 2 virions. *J Virol* 1998;**72**:1438–45.
120. Krasnykh VN, Mikheeva GV, Douglas JT, Curiel DT. Generation of recombinant adenovirus vectors with modified fibers for altering viral tropism. *J Virol* 1996;**70**:6839–46.
121. Michael S, Hong J, Curiel D, Engler J. Addition of a short peptide ligand to the adenovirus fibre protein. *Gene Ther* 1995;**2**:660–8.
122. Krasnykh V, Dmitriev I, Mikheeva G, Miller CR, Belousova N, Curiel DT. Characterization of an adenovirus vector containing a heterologous peptide epitope in the HI loop of the fiber knob. *J Virol* 1998;**72**:1844–52.
123. Roelvink PW, Mi Lee G, Einfeld DA, Kovesdi I, Wickham TJ. Identification of a conserved receptor-binding site on the fiber proteins of CAR-recognizing adenoviridae. *Science* 1999;**286**:1568–71.
124. Zabner J, Chillon M, Grunst T, Moninger TO, Davidson BL, Gregory R, et al. A chimeric type 2 adenovirus vector with a type 17 fiber enhances gene transfer to human airway epithelia. *J Virol* 1999;**73**:8689–95.
125. Havenga MJ, Lemckert AA, Grimbergen JM, Vogels R, Huisman LG, Valerio D, et al. Improved adenovirus vectors for infection of cardiovascular tissues. *J Virol* 2001;**75**:3335–42.
126. Arnberg N, Edlund K, Kidd AH, Wadell G. Adenovirus type 37 uses sialic acid as a cellular receptor. *J Virol* 2000;**74**:42–8.
127. Fasbender A, Lee JH, Walters RW, Moninger TO, Zabner J, Welsh MJ. Incorporation of adenovirus in calcium phosphate precipitates enhances gene transfer to airway epithelia in vitro and in vivo. *J Clin Invest* 1998;**102**:184–93.
128. Lee JH, Zabner J, Welsh MJ. Delivery of an adenovirus vector in a calcium phosphate coprecipitate enhances the therapeutic index of gene transfer to airway epithelia. *Hum Gene Ther* 1999;**10**:603–13.
129. Weiss DJ, Strandjord TP, Liggitt D, Clark JG. Perflubron enhances adenovirus-mediated gene expression in lungs of transgenic mice with chronic alveolar filling. *Hum Gene Ther* 1999;**10**:2287–93.

130. Weiss DJ, Strandjord TP, Jackson JC, Clark JG, Liggitt D. Perfluorochemical liquid-enhanced adenoviral vector distribution and expression in lungs of spontaneously breathing rodents. *Exp Lung Res* 1999;**25**:317–33.
131. Weiss DJ, Bonneau L, Allen JM, Miller AD, Halbert CL. Perfluorochemical liquid enhances adeno-associated virus-mediated transgene expression in lungs. *Mol Ther* 2000;**2**:624–30.
132. Kochanek S, Clemens PR, Mitani K, Chen HH, Chan S, Caskey CT. A new adenoviral vector: replacement of all viral coding sequences with 28 kb of DNA independently expressing both full-length dystrophin and beta-galactosidase. *Proc Natl Acad Sci USA* 1996;**93**:5731–6.
133. Mitani K, Graham FL, Caskey CT, Kochanek S. Rescue, propagation, and partial purification of a helper virus-dependent adenovirus vector. *Proc Natl Acad Sci USA* 1995;**92**:3854–8.
134. Clemens PR, Kochanek S, Sunada Y, Chan S, Chen HH, Campbell KP, et al. In vivo muscle gene transfer of full-length dystrophin with an adenoviral vector that lacks all viral genes. *Gene Ther* 1996;**3**:965–72.
135. Bruder JT, Jie T, McVey DL, Kovesdi I. Expression of gp19K increases the persistence of transgene expression from an adenovirus vector in the mouse lung and liver. *J Virol* 1997;**71**:7623–8.
136. Wilson CB, Embree LJ, Schowalter D, Albert R, Aruffo A, Hollenbaugh D, et al. Transient inhibition of CD28 and CD40 ligand interactions prolongs adenovirus-mediated transgene expression in the lung and facilitates expression after secondary vector administration. *J Virol* 1998;**72**:7542–50.
137. Yang Y, Su Q, Grewal IS, Schilz R, Flavell RA, Wilson JM. Transient subversion of CD40 ligand function diminishes immune responses to adenovirus vectors in mouse liver and lung tissues. *J Virol* 1996;**70**:6370–7.
138. Jooss K, Yang Y, Wilson JM. Cyclophosphamide diminishes inflammation and prolongs transgene expression following delivery of adenoviral vectors to mouse liver and lung. *Hum Gene Ther* 1996;**7**:1555–66.
139. Yang Y, Greenough K, Wilson JM. Transient immune blockade prevents formation of neutralizing antibody to recombinant adenovirus and allows repeated gene transfer to mouse liver. *Gene Ther* 1996;**3**:412–20.
140. Yang Y, Trinchieri G, Wilson JM. Recombinant IL-12 prevents formation of blocking IgA antibodies to recombinant adenovirus and allows repeated gene therapy to mouse lung. *Nat Med* 1995;**1**:890–3.
141. Yonemitsu Y, Kitson C, Ferrari S, Farley R, Griesenbach U, Judd D, et al. Efficient gene transfer to airway epithelium using recombinant Sendai virus. *Nat Biotechnol* 2000;**18**:970–3.
142. Kobinger GP, Weiner DJ, Yu QC, Wilson JM. Filovirus-pseudotyped lentiviral vector can efficiently and stably transduce airway epithelia in vivo. *Nat Biotechnol* 2001;**19**:225–30.

# Adenoviral Vectors for Pulmonary Disease (Pulmonary Vascular Disease)

# 26

*Paul N. Reynolds*
Department of Thoracic Medicine and Lung Research Laboratory, Royal Adelaide Hospital, Adelaide

Adenoviral vectors have a number of features that make them a suitable platform for pulmonary disease, including high transfection efficiency and stability after systemic administration. Early studies investigated the potential for direct airway administration with the aim of achieving airway epithelial transduction, for example, in the context of cystic fibrosis (CF) gene therapy. Unfortunately, transfection efficiency in this setting was poor due to difficulties in traversing airway epithelial mucus and glycocalyx, as well as the fact that the coxsackie and adenoviral receptor (CAR) is located at the basolateral surface of epithelial cells, below the intracellular tight junctions. These difficulties, coupled with the issues of the proinflammatory effects of first generation Ad vectors in Phase I CF trials, have led to a move away from Ad as an airway gene delivery vehicle. However, advances in helper-dependent Ads (discussed elsewhere), as well as strategies to bypass the glycocalyx barrier, may see renewed interest in Ad for this setting.

An alternate strategy for pulmonary gene therapy is systemic vascular delivery with the aim of transducing pulmonary vascular endothelium. Such an approach may lead to direct correction of genetic aberrations in endothelial cells, or deliver genes that code for proteins that are secreted in the local milieu for a more widespread effect. For this application, the systemic stability of Ad is a clear advantage, but the recognized shortcomings, including hepatic sequestration and relatively poor pulmonary transduction after intravenous administration, must be overcome. This chapter will primarily discuss the translation of targeting principles to a gene-based approach to therapy for pulmonary vascular disease.

In the clinical setting, pulmonary arterial hypertension (PAH) is a condition for which new treatments are needed and gene-based approaches may provide an important new option for therapy.[1] PAH is defined physiologically as a mean pulmonary artery pressure of greater than or equal to 25 mm Hg, with a normal pulmonary capillary wedge pressure, thereby excluding secondary pulmonary hypertension due to left heart disease.[2] Pathologically, PAH is characterized by abnormal arteriolar endothelial proliferation with the development of plexiform lesions composed of monoclonal expansions of endothelial cells and associated smooth muscle hypertrophy. There have been substantial improvements in our understanding of the pathological basis of PAH over the past 15 years. A major milestone was the discovery by two independent groups of the role of mutations in the gene for bone morphogenetic protein receptor

**Adenoviral Vectors for Gene Therapy.**

type 2 (BMPR2) in the pathogenesis of PAH.[3,4] The causal link between BMPR2 mutations and PAH is now clearly established based on studies of familial PAH and confirmed by a number of transgenic mouse studies.[5] Patients with BMPR2-related PAH are heterozygous for a mutant allele—homozygous mutations in mice are congenitally lethal due to the key role BMPR2 plays in organogenesis. In animal models (such as hypoxia and monocrotaline-induced PAH in rats) in which mutations are not present, there is an acquired deficiency of BMPR2 characterized by reduced protein expression.[6,7] This deficiency has now also been shown in some patients, including those with scleroderma-associated PAH. In this latter context, BMPR2 downregulation in skin microvessels is due to promoter methylation, and it is likely that the same basis exists for scleroderma-associated pulmonary vascular disease, although this is yet to be proven.[8] On balance, this body of evidence suggested that correction of BMPR2 deficiency by gene delivery could be a novel approach to therapy.

To achieve BMPR2 upregulation, McMurtry et al. first attempted aerosol gene delivery in a rat model using an Ad vector carrying the BMPR2 gene.[9] Detailed immunohistochemistry indicated that some transduction of pulmonary vascular smooth muscle was achieved; however, in the monocrotaline PAH model that was used, no therapeutic impact was achieved. Immunohistochemistry assessment of BMPR2 expression in the lungs reveals predominant expression in endothelial cells, with lesser staining seen in smooth muscle.[10] We therefore rationalized that direct gene delivery to endothelial cells may be necessary to achieve a therapeutic impact with BMPR2.

Ad vectors with native tropism for CAR are relatively inefficient in achieving pulmonary endothelial transduction after vascular administration due to the vast bulk of vector being sequestered by the liver. To achieve gene delivery to pulmonary vascular endothelium it is necessary to alter Ad tropism. These modifications require altering the Ad capsid, either by direct genetic manipulation of Ad proteins or by using adapter molecules.

Achieving efficient, selective targeting to pulmonary vascular endothelium requires the selection of a target protein expressed on pulmonary endothelial cells. Candidate molecules include platelet—endothelial cell adhesion molecule (PECAM) and angiotensin converting enzyme (ACE), among others.[11] With respect to ACE targeting, a monoclonal antibody, Mab9B9, has good intravascular targeting properties and has been used to target proteins to pulmonary endothelium.[12] In the context of PAH, ACE levels are increased in the endothelium in the remodeled pulmonary vessels, suggesting that this target may have particular utility in this disease.[13]

To redirect Ad vectors to pulmonary endothelium it was necessary to alter the tropism to achieve binding of the virus to cell surface ACE. Initial attachment of virus to cells is generally achieved by binding of the knob domain at the tip of the Ad fiber protein to CAR. Internalization is then mediated by an interaction between the cell surface integrins and an RGD motif located in the penton base. An alternative entry pathway for hepatic cell transduction is mediated by an interaction between plasma coagulation Factor X and protein motifs in the Ad hexon.[14]

To establish the proof of principle that Ad vectors could be retargeted after systemic in vivo injection, an antibody conjugate approach was used.[15] A monoclonal antibody was derived against the Ad knob domain, and then the Fab fragment of this antibody was chemically crosslinked to Mab9B9, thus forming the bispecific conjugate "Fab–9B9."[16]

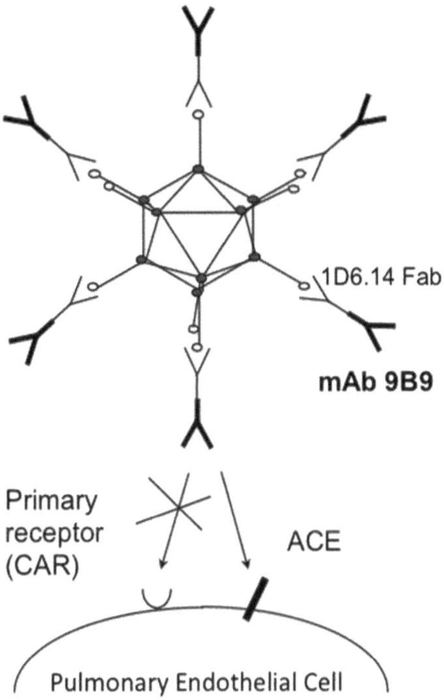


**Figure 1** Retargeting schema using Fab–9B9 to redirect Ad binding away from CAR and toward pulmonary endothelial ACE.

The concept was that this would block the native fiber knob—CAR binding and redirect the Ad to bind to ACE (Figure 1). It must be noted, however, that this strategy by itself does not significantly affect hexon—Factor X binding and thus liver sequestration of vector remains a limiting factor. Nevertheless, significant increases in pulmonary transduction were achieved with this approach. Immunohistochemistry and electron microscopy with immunogold staining confirmed endothelial gene delivery (Figure 2).[16]

To determine whether retargeted Ad could be used as a therapeutic platform, a vector carrying the BMPR2 gene (incorporating a myc tag to facilitate detection) was constructed. The therapeutic potential was then evaluated in two widely used animal models of pulmonary hypertension: the rat hypoxia and monocrotaline (MCT) models.[7,17] Rats were placed in 10% oxygen atmosphere for 3 weeks, administered AdBMPR2myc + Fab–9B9 or control vector + Fab–9B9, returned to hypoxia, and then assessed 3 weeks later. Detection of BMPR2 expression by the Western blot of lung lysate confirmed that levels were reduced by hypoxia and restored by AdBMPR2myc gene delivery (Figure 3). Physiological measurements showed that hypoxia led to an increase in right ventricular systolic pressure (RVSP) and mean pulmonary artery pressure (mPAP) that was substantially ameliorated by targeted AdBMPR2myc (Figure 4). In conjunction, right ventricular hypertrophy (determined by the Fulton index, RV weight/LV + septum weight) was significantly reduced and cardiac output (CO, measured by thermodilution) was significantly increased.

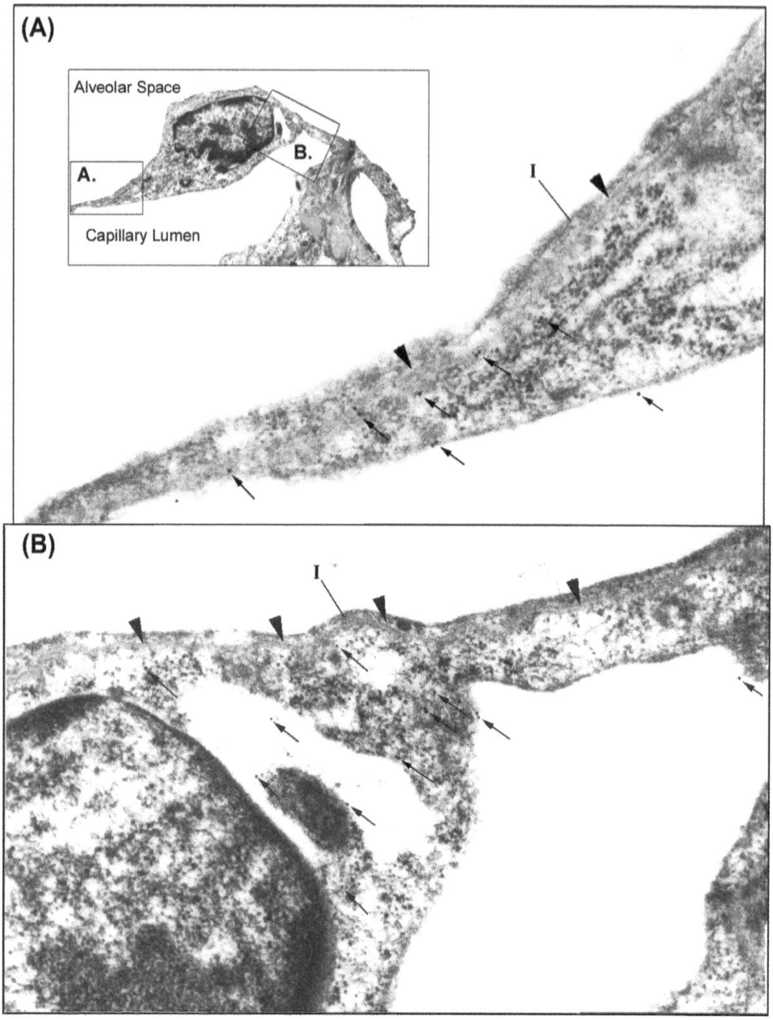


**Figure 2** Electron microscopy detection of transgene expression. Portion of alveolar wall from rat that received ACE-targeted AdCMVCEA, immunogold staining for CEA transgene expression. An endothelial cell nucleus can be seen. Beads are localized to endothelial cell aspect of the alveolar wall. Arrowheads indicate basement membranes. Arrows indicate gold beads. I indicates type I alveolar epithelial cell. Original magnification: (inset) 3000×, (A and B) 20,000×. Taken from Reynolds et al.[16]

To further investigate the potential for Ad-mediated therapy of PAH, the MCT model was then used.[18] In this setting PAH develops due to an inflammatory response in the pulmonary vasculature induced by a subcutaneous injection of MCT. Thus, rats were administered MCT, and then 10 days later they were administered AdBMPR2myc + Fab–9B9 or control and assessed 10 days after that. In this setting reduced BMPR2 was again seen with the PAH stimulus and was increased with gene

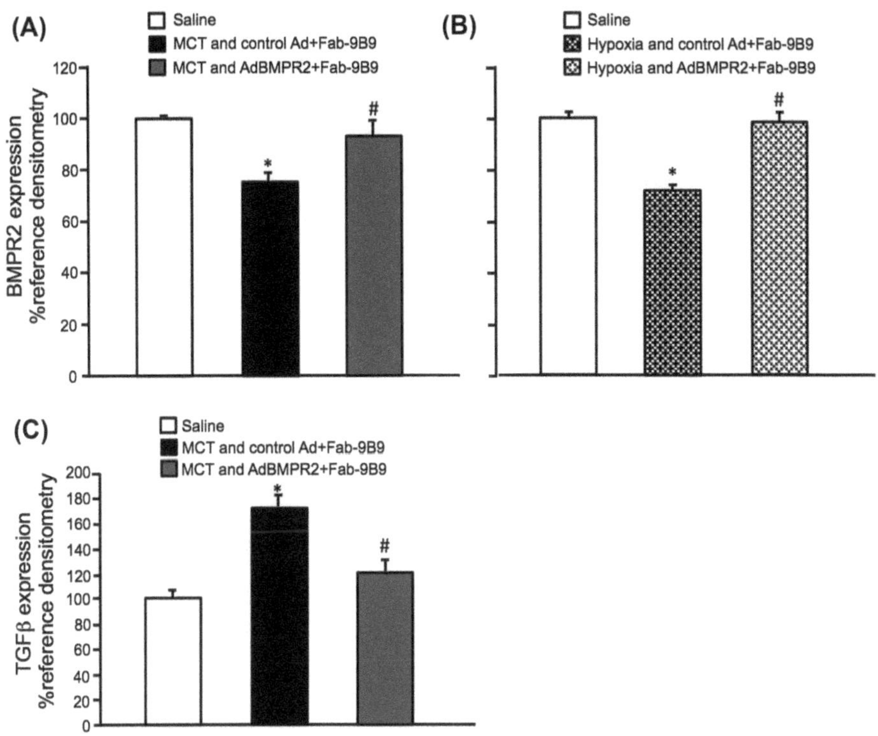


**Figure 3** Effect of PAH and BMPR2 gene delivery on BMPR2 protein expression in (A) MCT PAH model and (B) hypoxia model. Effect of PAH and BMPR2 gene delivery on TGFβ levels in MCT model. Data are presented as mean ± SEM. $n=5$–6 animals in each treatment group. *$P$, 0.05 versus saline; $^{\#}P$, 0.05 treatment and control Ad + Fab–9B9 versus treatment and AdBMPR2 + Fab–9B9. Taken from Reynolds et al.[7]

delivery. In the MCT-induced PAH a significant increase in TGFβ was noted, reflecting the inflammatory basis of the model. This was reduced with BMPR2 gene delivery (Figure 3). As had been seen for the hypoxia model, AdBMPR2 significantly reduced the RVSP, mPAP, and Fulton index and improved CO (Figure 4).[18]

In both models, immunohistochemical analysis of lung sections revealed reduced small vessel muscularization and reduced cellular proliferation in the AdBMPR2myc-treated groups (not shown).

Some of the downstream signaling mechanisms involved in the therapeutic effect of BMPR2 gene delivery have now been evaluated. The predominant finding has been a shift from TGFβ-induced Smad 2/3 signaling in PAH, toward Smad 1/5/8 signaling with BMPR2 treatment (unpublished observations).

These studies served to establish the rationale for further development of gene delivery, particularly with BMPR2, as a therapy for PAH. To enable this approach to achieve clinical translation, however, much more work is needed to optimize the gene delivery system. First generation vectors of the type used in these studies are not ideal in view of the proinflammatory effects they may generate. That

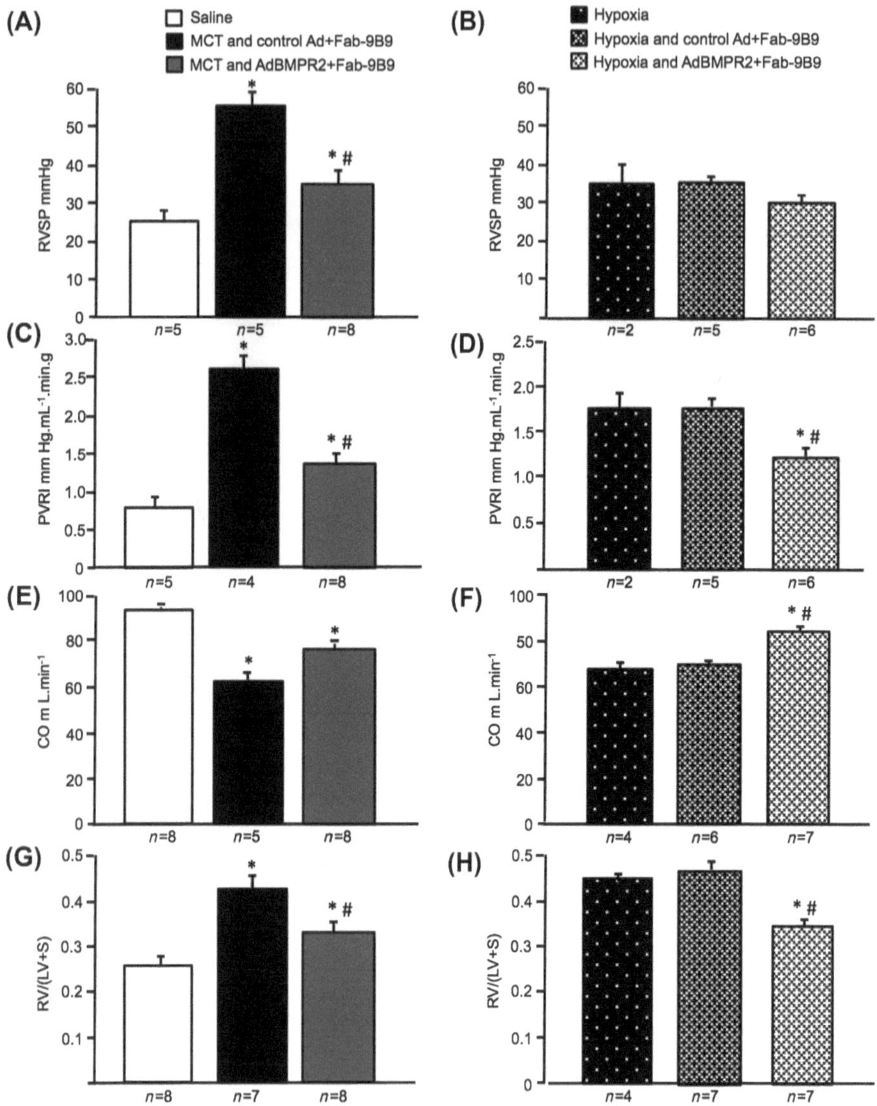


**Figure 4** Effect of AdBMPR2 administration on the development of pulmonary hypertension in response to MCT and chronic hypoxia. Right ventricular systolic pressure (RVSP) in MCT (A) and hypoxia (B); pulmonary vascular resistance index in MCT (C) and hypoxia (D); cardiac output in MCT (E) and hypoxia (F); and Fulton index—ratio of right ventricle to left ventricle + septum weight (RV/[LV + S]) in MCT (G) and hypoxia (H). Values are means ± SEM. $^{*}P<0.05$ versus control; $^{\dagger}P<0.05$ treatment + control Ad + Fab–9B9 versus treatment + AdBMPR2 + Fab–9B9. Number in parentheses refers to the number of measurements per group. Taken from Reynolds et al.[7]

having been said, any such effect in this model clearly did not compromise the therapeutic benefit seen with BMPR2, even in the "inflammatory" MCT model. This finding suggests that later generation vectors, particularly helper-dependent Ads, may have even greater potential benefits, especially as they have been shown to achieve much longer duration of transgene expression than first generation Ads.

In addition to redirecting Ad tropism, improved selectivity of gene expression can be achieved through the use of cell-specific promoters. Such an approach can reduce off-target effects as well as potentially enable greater duration of transgene expression due to the avoidance of promoter silencing that can occur with nonspecific promoters such as CMV. In the context of pulmonary vasculature a number of endothelial-specific promoters may be considered. The VEGF receptor promoter, flt-1 and Robo4 are two such candidates.

The potential utility of the flt-1 promoter was assessed in an in vitro an ex vivo context (human saphenous vein) and found to have good endothelial selectivity and to be capable of driving reporter gene expression with similar strength to the CMV promoter.[19] The potential to combine transductional targeting using Fab–9B9 with transcriptional targeting using flt-1 was then assessed in vivo. A substantial improvement in overall selectivity for the pulmonary endothelial target was achieved with the combined approach (Figure 5).[20]

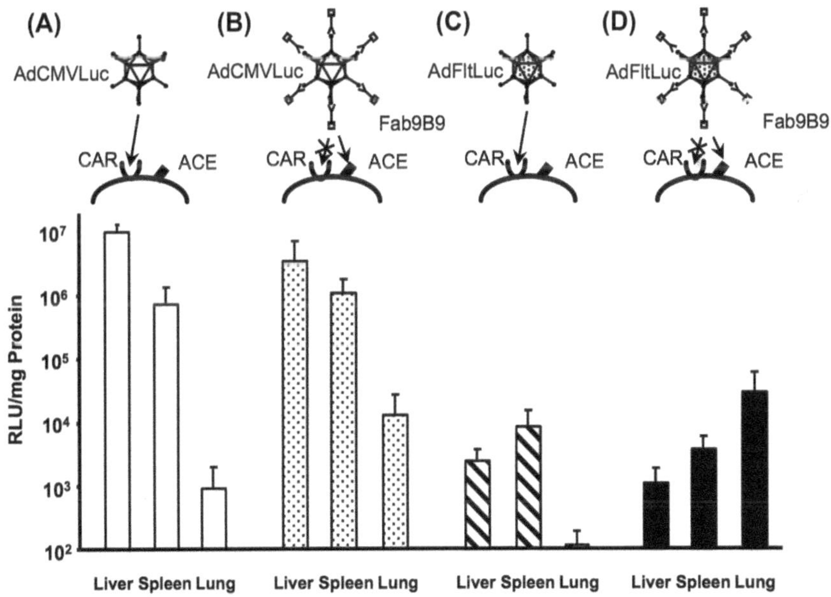


**Figure 5** Combined transductional and transcriptional targeting improves the specificity of transgene expression in vivo. Rats were injected (tail vein) with $5 \times 10^{10}$ viral particles of AdCMVLuc or AdfltLuc, either alone (A,C) or in combination with the pulmonary endothelial targeting conjugate Fab–9B9 (B,D), and sacrificed 3 days later, and luciferase activity was quantified. Data are means ± SD of 8–10 rats per group. Taken from Reynolds et al.[20]

The above studies with early generation Ads provide a sound rationale to further develop and improve the vector technology. While the Fab–9B9 approach has established important principles, the heterogeneous nature of the chemically crosslinked conjugate produces a product that is not suitable for clinical translation. To address this limitation, a great deal of effort has been devoted to producing improved bispecific targeting conjugates of a defined molecular structure. One such approach is to produce recombinant bispecific molecules comprising soluble CAR (sCAR) as an Ad-binding motif linked to targeting moieties (such as recombinant ligands or single chain antibodies).[21] This general approach is discussed in other chapters. As a proof of principle that this approach could work in vivo for pulmonary endothelial gene delivery, a novel model system was developed. A recombinant molecule was developed consisting of a fusion protein between sCAR and an anticarcinoembryonic antigen (CEA) single chain antibody (MFE). This targeting conjugate was shown to enhance Ad gene delivery to CEA-expressing cells in vitro. To assess in vivo utility, a transgenic mouse expressing human CAR (hCAR mouse[22]) was first injected intravenously with AdCMVCEA to induce expression of CEA in the pulmonary vasculature. This was then followed by an injection of a luciferase reporter gene carrying Ad with either the sCAR–MFE fusion protein or a control fusion protein. A 10-fold increase in absolute luciferase expression was seen in the lungs of animals given sCAR–MFE versus control, along with a 10-fold increase in lung:liver ratio of luciferase expression.[23] These studies established the important principle that sCAR fusion proteins retain their binding to Ad after systemic injection and thus are a potential platform for improved pulmonary endothelial gene delivery. A challenge remains to construct a suitable single chain antibody fusion protein for direct targeting to a native endothelial target. While attempts have been made using an anti-ACE single chain antibody as a logical extension of the Mab–9B9 work, a construct with adequate affinity for the target has not yet been achieved.

While adapter molecules and conjugates have been established as having in vivo targeting utility, the complexities of the "two component" approach will complicate clinical translation. A single particle targeted vector is thus more appealing. To this end, a number of genetic modifications have been made to Ad coat proteins to directly incorporate targeting moieties, several of which have been assessed for pulmonary vascular gene delivery.

A strategy to broaden Ad tropism, especially in the context of relatively "CAR-deficient" tumor cells, is to insert peptide targeting motifs into the knob domain, with the c-terminus and HI-loop being two sites that have been evaluated. In this regard, an RGD integrin binding motif was inserted into the HI loop, and CAR-independent gene transfer to a variety of cells was achieved.[24] On systemic vascular administration, some differences were seen in the biodistribution of transgene expression in comparison to native vector, with significant absolute increases in gene expression seen in the lungs, as well as the spleen, kidney, and liver.[25] There was some increase in the relative expression in lung versus liver, but this construct cannot be considered to have significant pulmonary selectivity.

To achieve a vector with greater pulmonary transduction efficiency, more extensive engineering of Ad has been performed. Owing to structural constraints, only relatively short peptides can be successfully inserted into the native Ad knob domain, thus limiting the options available for high affinity targeting moieties such as single chain antibodies. To overcome this limitation, a strategy was devised to completely remove the Ad fiber protein and replace it with an alternative which would still form fibers but would be much more amenable to ligand insertion. This approach also ensured complete ablation of CAR binding, thereby helping to remove off-target transduction. To this end, the fibritin protein from phage T4 was used, and Ad vectors bearing these fibers were successfully constructed.[26] Using this platform an Ad vector was constructed incorporating CD40 ligand at the end of the fibritin fiber (Ad5Luc.FF/CD40L), for the purposes of improving transduction of dendritic cells in the context of genetic vaccines.[27] To establish whether this type of construct might have in vivo targeting potential, an approach using the hCAR mouse was again employed. In this case an Ad vector carrying the gene for CD40 was first used to achieve pulmonary endothelial CD40 expression. This was followed by injection of Ad5luc.FF/CD40. A substantial increase in luciferase transgene expression was seen in the lungs of mice expressing endothelial CD40, thereby indicating the systemic targeting capacity of the fibritin construct (Figure 6).[28]

To develop a vector of utility for direct pulmonary transduction the fibritin platform was then used as a basis for a construct incorporating a novel targeting peptide that proved to have excellent pulmonary targeting potential. In the first instance a strategy was employed to define a peptide motif that had good binding ability for myeloid cells.[29] First, a loop-constrained random heptapeptide phage display library was panned over murine bone marrow cells. Eventually, a peptide, WTLDRGY, (myelin binding peptide, MBP) was defined. This peptide was then genetically incorporated into the fiber–fibritin platform, and a stable vector was produced, Ad.MBP. In vitro, good transduction to myeloid cells was achieved.

When the gene delivery properties of Ad.MBP were assessed after systemic vascular administration a remarkable degree of pulmonary selectivity was noted. Reporter gene expression levels in the lung (per mg protein) were 3–4 logs higher than those in liver, spleen, heart, or kidney. Detailed analysis revealed that the vector was actually binding to myeloid cells in the circulation, but then pulmonary endothelial gene delivery (confirmed by immunohistochemistry) was achieved by a "handover" phenomenon.[30] The remarkable degree of pulmonary endothelial selectivity achieved may prove to be an ideal platform for the further development of BMPR2 gene therapy for PAH, as well as other possible conditions.

In summary, it has now been clearly shown that adenoviral vectors can achieve therapeutic outcomes in models of pulmonary vascular disease. New advances in vector technology as discussed herein provide the potential to achieve very high levels of pulmonary-specific gene expression. These platforms will be useful not only for PAH but also for other conditions where loco-regional high-level expression can be predicted to be advantageous, for example, in the context of passive antibody protection against respiratory pathogens, and for expression of other extracellar therapeutic proteins such as alpha-1 anti-trypsin in emphysema.

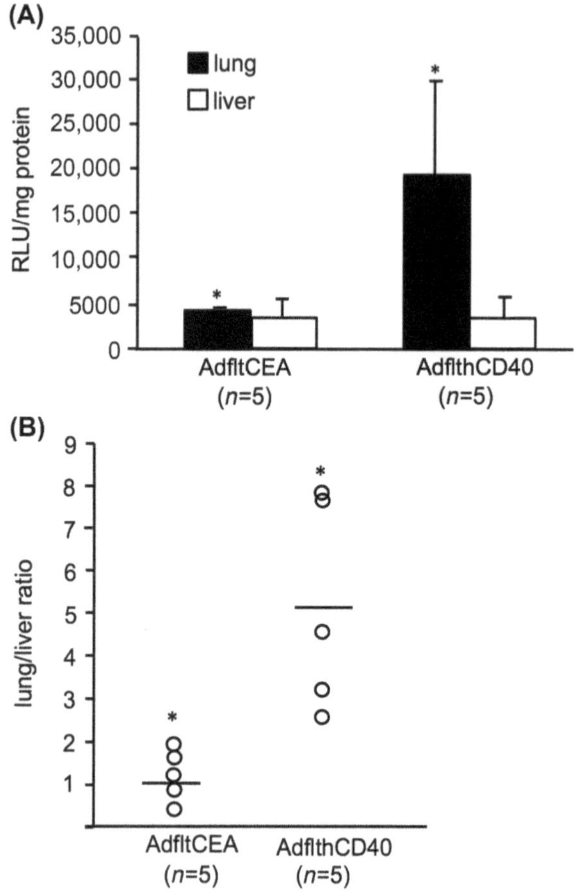


**Figure 6** In vivo function of Ad5Luc.FF/CD40L in CD40 transient transgenic hCAR mice. hCAR mice were injected with $1.0 \times 10^{11}$ vp of AdflthCD40 or AdfltCEA via the tail vein, followed by injection with $7.0 \times 10^{10}$ vp of Ad5Luc.FF/CD40L via the tail vein 48 h later. Seventy-two hours after injection with Ad5Luc.FF/CD40L, the mice were sacrificed. (A) Lung and liver luciferase activities 72 h after intravenous administration of Ad5Luc.FF/CD40L into AdflthCD40- and AdfltCEA-injected mice (means ± standard deviations, RLU/mg protein, $n=5$). Data are presented as means ± standard deviation of five mice per group (*$P<0.02$). (B) Lung-to-liver ratios of luciferase activities in each mouse. Bars indicate the mean in each group (*$P<0.01$). Taken from Izumi et al.[28]

## References

1. Reynolds PN. Gene therapy for pulmonary hypertension: prospects and challenges. *Expert Opin Biol Ther* 2011;**11**(2):133–43.
2. Simonneau G, Gatzoulis MA, Adatia I, Celermajer D, Denton C, Ghofrani A, et al. Updated clinical classification of pulmonary hypertension. *J Am Coll Cardiol* 2013;**62**(Suppl. 25): D34–41.

3. Lane KB, Machado RD, Pauciulo MW, Thomson JR, Phillips 3rd JA, Loyd JE, et al. Heterozygous germline mutations in BMPR2, encoding a TGF-beta receptor, cause familial primary pulmonary hypertension. The International PPH Consortium. *Nat Genet* 2000;**26**(1):81–4.
4. Deng Z, Morse JH, Slager SL, Cuervo N, Moore KJ, Venetos G, et al. Familial primary pulmonary hypertension (Gene PPH1) is caused by mutations in the bone morphogenetic protein receptor-II gene. *Am J Hum Genet* 2000;**67**(3):737–44.
5. West J, Fagan K, Steudel W, Fouty B, Lane K, Harral J, et al. Pulmonary hypertension in transgenic mice expressing a dominant-negative BMPRII gene in smooth muscle. *Circ Res* 2004;**94**(8):1109–14.
6. Long L, Crosby A, Yang X, Southwood M, Upton PD, Kim DK, et al. Altered bone morphogenetic protein and transforming growth factor-beta signaling in rat models of pulmonary hypertension: potential for activin receptor-like kinase-5 inhibition in prevention and progression of disease. *Circulation* 2009;**119**(4):566–76.
7. Reynolds AM, Holmes MD, Danilov SM, Reynolds PN. Targeted gene delivery of BMPR2 attenuates pulmonary hypertension. *Eur Respir J* 2012;**39**(2):329–43.
8. Wang Y, Kahaleh B. Epigenetic repression of bone morphogenetic protein receptor II expression in scleroderma. *J Cell Mol Med* 2013;**17**(10):1291–9.
9. McMurtry MS, Moudgil R, Hashimoto K, Bonnet S, Michelakis ED, Archer SL. Overexpression of human bone morphogenetic protein receptor II does not ameliorate monocrotaline pulmonary arterial hypertension. *Am J Physiol Lung Cell Mol Physiol* 2006.
10. Atkinson C, Stewart S, Upton PD, Machado R, Thomson JR, Trembath RC, et al. Primary pulmonary hypertension is associated with reduced pulmonary vascular expression of type II bone morphogenetic protein receptor. *Circulation* 2002;**105**(14):1672–8.
11. Danilov SM, Gavrilyuk VD, Franke FE, Pauls K, Harshaw DW, McDonald TD, et al. Lung uptake of antibodies to endothelial antigens: key determinants of vascular immunotargeting. *Am J Physiol Lung Cell Mol Physiol* 2001;**280**(6):L1335–47.
12. Danilov SM, Muzykantov VR, Martynov AV, Atochina EN, Sakharov I, Trakht IN, et al. Lung is the target organ for a monoclonal antibody to angiotensin- converting enzyme. *Lab Invest* 1991;**64**(1):118–24.
13. Orte C, Polak JM, Haworth SG, Yacoub MH, Morrell NW. Expression of pulmonary vascular angiotensin-converting enzyme in primary and secondary plexiform pulmonary hypertension. *J Pathol* 2000;**192**(3):379–84.
14. Lopez-Gordo E, Denby L, Nicklin SA, Baker AH. The importance of coagulation factors binding to adenovirus: historical perspectives and implications for gene delivery. *Expert Opin Drug Deliv* 2014;**11**(11):1795–813.
15. Douglas JT, Rogers BE, Rosenfeld ME, Michael SI, Feng M, Curiel DT. Targeted gene delivery by tropism-modified adenoviral vectors. *Nat Biotechnol* 1996;**14**:1574–8.
16. Reynolds PN, Zinn KR, Gavrilyuk VD, Balyasnikova IV, Rogers BE, Buchsbaum DJ, et al. A targetable, injectable adenoviral vector for selective gene delivery to pulmonary endothelium in vivo. *Mol Ther* 2000;**2**(6):562–78.
17. Reynolds AM, Xia W, Holmes MD, Hodge SJ, Danilov S, Curiel DT, et al. Bone morphogenetic protein type 2 receptor gene therapy attenuates hypoxic pulmonary hypertension. *Am J Physiol Lung Cell Mol Physiol* 2007;**292**(5):L1182–92.
18. Reynolds AM, Holmes M, Morrell NW, Danilov S, Reynolds PN. Gene delivery of bone morphogenetic protein receptor type-2 attenuates established hypoxic and monocrotaline-induced pulmonary hypertension. *Am J Respir Crit Care Med* 2010;**181**:A6333.
19. Nicklin SA, Reynolds PN, Brosnan MJ, White SJ, Curiel DT, Dominiczak AF, et al. Analysis of cell-specific promoters for viral gene therapy targeted at the vascular endothelium. *Hypertension* 2001;**38**(1):65–70.

20. Reynolds PN, Nicklin SA, Kaliberova L, Boatman BG, Grizzle WE, Balyasnikova IV, et al. Combined transductional and transcriptional targeting improves the specificity of transgene expression in vivo. *Nat Biotechnol* 2001;**19**(9):838–42.
21. Pereboev AV, Asiedu CK, Kawakami Y, Dong SS, Blackwell JL, Kashentseva EA, et al. Coxsackievirus-adenovirus receptor genetically fused to anti-human CD40 scFv enhances adenoviral transduction of dendritic cells. *Gene Ther* 2002;**9**(17):1189–93.
22. Tallone T, Malin S, Samuelsson A, Wilbertz J, Miyahara M, Okamoto K, et al. A mouse model for adenovirus gene delivery. *Proc Natl Acad Sci USA* 2001;**98**(14):7910–5.
23. Everts M, Kim-Park SA, Preuss MA, Passineau MJ, Glasgow JN, Pereboev AV, et al. Selective induction of tumor-associated antigens in murine pulmonary vasculature using double-targeted adenoviral vectors. *Gene Ther* 2005;**12**(13):1042–8.
24. Dmitriev I, Krasnykh V, Miller CR, Wang M, Kashentseva E, Mikheeva G, et al. An adenovirus vector with genetically modified fibers demonstrates expanded tropism via utilization of a Coxsackievirus and adenovirus receptor-independent cell entry mechanism. *J Virol* 1998;**72**(12):9706–13.
25. Reynolds P, Dmitriev I, Curiel D. Insertion of an RGD motif into the HI loop of adenovirus fiber protein alters the distribution of transgene expression of the systemically administered vector. *Gene Ther* 1999;**6**(7):1336–9.
26. Krasnykh V, Belousova N, Korokhov N, Mikheeva G, Curiel DT. Genetic targeting of an adenovirus vector via replacement of the fiber protein with the phage T4 fibritin. *J Virol* 2001;**75**(9):4176–83.
27. Belousova N, Korokhov N, Krendelshchikova V, Simonenko V, Mikheeva G, Triozzi PL, et al. Genetically targeted adenovirus vector directed to CD40-expressing cells. *J Virol* 2003;**77**(21):11367–77.
28. Izumi M, Kawakami Y, Glasgow JN, Belousova N, Everts M, Kim-Park S, et al. In vivo analysis of a genetically modified adenoviral vector targeted to human CD40 using a novel transient transgenic model. *J Gene Med* 2005;**7**(12):1517–25.
29. Alberti MO, Roth JC, Ismail M, Tsuruta Y, Abraham E, Pereboeva L, et al. Derivation of a myeloid cell-binding adenovirus for gene therapy of inflammation. *PLoS One* 2012;**7**(5):e37812.
30. Alberti MO, Deshane JS, Chaplin DD, Pereboeva L, Curiel DT, Roth JC. A myeloid cell-binding adenovirus efficiently targets gene transfer to the lung and escapes liver tropism. *Gene Ther* 2013;**20**(7):733–41.

# Utility of Adenoviral Vectors in Animal Models of Human Disease III: Acquired Diseases

27

*Erik Lubberts*[1,2], *Jay K. Kolls*[3,4]
[1]Department of Immunology, Erasmus MC, University Medical Center, Rotterdam, The Netherlands; [2]Department of Rheumatology, Erasmus MC, University Medical Center, Rotterdam, The Netherlands; [3]Richard King Mellon Foundation Institute for Pediatric Research, Children's Hospital of Pittsburgh of UPMC, Pittsburgh, PA, USA; [4]Department of Pediatrics, School of Medicine, University of Pittsburgh, Pittsburgh, PA, USA

## 1. Adenoviral Vectors for Infectious Disease

Recombinant adenoviral vectors for infectious diseases can generally be categorized into three general approaches. The first is the use of a vector-based vaccine where the vector encodes for proteins to achieve an immune response. In fact adenoviruses have been used in the US military for vaccines.[1] The second approach is to use adenoviral vectors, which encode immunostimulatory genes to achieve in vivo immunotherapy. Lastly, these vectors can be used to provide critical accessory molecules for T- or B-cell activation for patients who are deficient in these molecules, or theoretically direct anti-infectious genes such as antibacterial peptides. These general paradigms hold true for most gene therapy approaches with adenoviral-based vector systems regardless if the targets are infectious disease, an inherited deficiency state, or cancer. In this chapter we will focus on these paradigms in the context of specific disease entities that may be candidates for treatment with adenovirus-based vector systems.

### 1.1 Tuberculosis

*Mycobacterium tuberculosis*, the etiologic agent of tuberculosis, is a facultative intracellular pathogen which remains the foremost cause of death from a single infectious agent among adults.[2] It has been estimated that approximately one-third of the world's population in 1990 (1.7 billion individuals) were infected with *M. tuberculosis* affecting mostly people living in developing countries, and that with the global control measures in place at that time, 30 million people were expected to die due to tuberculosis by the year 2000.[2] The most effective vaccine against tuberculosis in man is the BCG vaccine, an attenuated substrain of *Mycobacterium bovis*, which has been used for more than 50 years worldwide. However, this vaccine is very erratic in conferring protection, varying as much as 0–80% in separate clinical trials.[3] In countries with a lower incidence of tuberculosis, such as the United States, the emergence of multidrug-

Adenoviral Vectors for Gene Therapy.

resistant strains threatens control measures with antimycobacterial drugs. It is apparent that current immunotherapeutic and chemotherapeutic approaches for the control of tuberculosis need to be improved.

After inhalation, the organism replicates within the lung macrophage. The protective response to infection with intracellular bacteria is cell mediated.[4] Protective immunity in the mouse model of tuberculosis is mediated by T lymphocytes that secrete interferon-γ (IFN-γ), which activates infected macrophages to control intracellular bacilli in a manner believed to be similar to the protective response in man. Several subpopulations of T lymphocytes contribute to the protective response in the lungs of infected mice.[5] Most of this protection is conferred by a short-lived population of rapidly dividing, IFN-γ-secreting $CD4^+$ T lymphocytes[6] which peak within 3 weeks of infection, a time which correlates with the control of further Mycobacterial growth in the host.[7]

The pivotal role of IFN-γ in protective immunity to *M. tuberculosis* was unequivocally demonstrated using IFN-γ KO mice. The single gene encoding IFN-γ was disrupted, and these mice were originally shown to (1) be incapable of IFN-γ production, (2) poorly express class II MHC, (3) be deficient in the production of reactive oxygen and reactive nitrogen radicals, and (4) be very susceptible to *Mycobacterium bovis* BCG.[8] IFN-γ KO mice succumbed to infection with *M. tuberculosis* fairly rapidly whether the virulent bacilli were delivered intravenously at moderate[9] to high doses[10] or via a low-dose aerosol.[9] There is also an absolute requirement for interleukin (IL)-12 in the protective response against TB. This has been demonstrated using IL-12p40 KO mice. These mice do not produce the heavy chain of the IL-12 heterodimer and therefore do not make the bioactive p70 form of IL-12, which results in a poor cell-mediated response to antigen.[11] Recently, it was shown that IL-12p40 KO succumbed to an intravenous infection with *M. tuberculosis* within 50 days.[12] Whereas wild-type controls contained the infection and strongly expressed genes encoding IFN-γ, tumor necrosis factor (TNF), and inducible nitric oxide synthase (iNOS) in infected tissues, these KO mice produced no IFN-γ message and delayed TNF and iNOS message.

As protective cytokines, which play a pivotal role in protection against tuberculosis, IFN-γ and IL-12 represent attractive targets for cytokine-based therapy approaches designed to enhance protective cell-mediated immunity.[13] Recently, a replication-deficient adenoviral vector designed to deliver IFN-γ (AdIFN) was delivered intratracheally into the lungs of BALB/c mice, which were subsequently challenged, with a sublethal aerosol of *M. tuberculosis*. Prior pharmacokinetic analyses of adenoviral-mediated expression of IFN-γ in BALB/c mice had indicated that transfected mice expressed increased IFN-γ in the lungs for as long as 21 days following delivery of the vector. Other mice were transfected with a control virus-expressing lacZ (AdLacZ) shortly before the low-dose aerosol exposure to *M. tuberculosis*. AdIFN-treated mice initially contained the infection in the lungs much better than the control nontransfected mice or AdLacZ-treated mice (Fig. 1). The protective effect in the lungs paralleled the local production of IFN-γ by the vector and thus, was relatively short-lived, such that the load of viable bacilli in AdIFN-γ-treated lungs reached levels similar to the controls by 30 days of infection. There was no protective effect on the control of Mycobacterial dissemination or growth in other primary target organs. Similar AdIFN-γ-mediated control of bacterial growth in the lungs was not seen in mice, which already had established chronic *M. tuberculosis* infection.

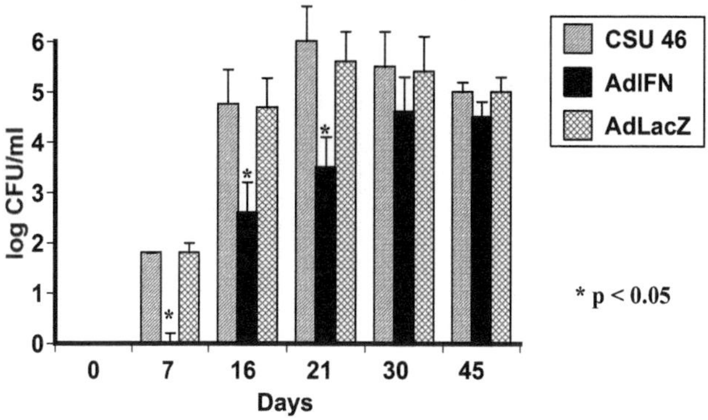


**Figure 1** AdIFN reduces growth of *Mycobacterium tuberculosis* in the lung. Mice were pretreated with AdIFN, AdLacZ, or vehicle and then challenged with CSU 46, a clinical isolate of *M. tuberculosis* and lung organism burden was quantified by quantitative organ culture serially after aerosol challenge.
Data provided by Dr Elizabeth Rhoades and Dr Ian Orme, Colorado State University.

Based on these preclinical data, several clinical trials have been initiated for both multidrug-resistant *M. tuberculosis* and persistent Mycobacteria avium complex (MAC) infection in non-HIV-infected hosts. Williams and colleagues, from our group, recently reported on a Phase I trial of aerosolized IFN-γ to patients with persistent MAC infection.[14] All patients tolerated the aerosol well and 3 of 8 had sputum acid-fast bacilli (AFB) smears convert to negative. Condos and colleagues have recently reported on five patients in New York City with multidrug-resistant tuberculosis who received 500 μg of IFN-γ aerosolized three times a week for 1 month.[15] Again the aerosol form of the drug was well-tolerated and all patients had sputum smears for AFB convert to negative and the time to positive culture increased (from 17 to 24 days, not significant) suggesting a reduction in organism burden. Moreover, patient's weight increased or stabilized, and there were objective decreases in the size of cavitary lesions in all patients, 2 months after treatment had ended. It is important to note that data to date suggest that IFN whether in protein or vector form needs to be provided for a relatively long time to control *M. tuberculosis* growth. Thus, it is possible that newer generations of adenoviral-based or other vector systems may achieve longer-term control of infection.

## 1.2 Pneumonia

Pneumonia and influenza infection remains the sixth leading cause of death in the United States.[16] Drug-resistant organisms are increasingly isolated from infected patients presumably due to the broad use of antibiotics. As mentioned above several biological response modifiers such as granulocyte colony-stimulating factor (G-CSF) and IFN-γ have been investigated in patients as protein-based therapies. However, due to pharmacological advantages of adenovirus and other gene-based vector systems, gene therapy may provide an alternative approach for in vivo immunomodulation.

Adenoviral-mediated gene transfer of the murine IFN gene (AdIFN) results in dose-dependent increases in IFN in bronchoalveolar lavage fluid (BALF) in both

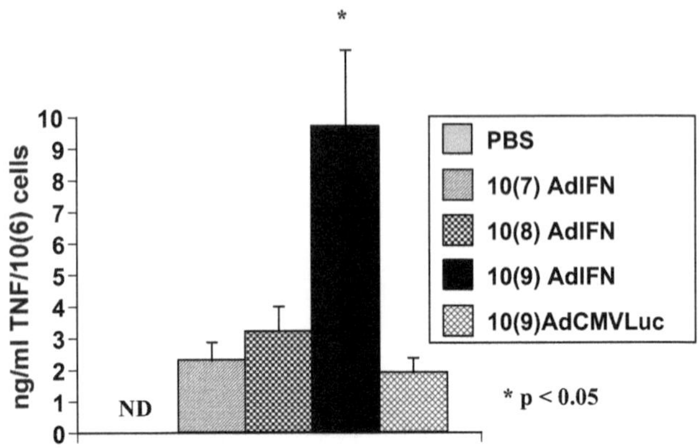


**Figure 2** Dose-dependent increase of LPS-induced lung TNF production by AdIFN. 6–8-week-old BALB/c mice were pretreated with AdIFN, AdCMVLuc, or PBS 3 days prior to administration of intratracheal LPS. TNF was measured in bronchoalveolar lavage (BAL) fluid 3 h after LPS administration by ELISA and corrected for macrophage cell number in the BAL fluid.

mice and rats.[17] Recombinant protein expression occurs up to 28 days in Sprague–Dawley rats and up to 21 days in Balb/c mice[17] Expression of IFN has a biological effect for at least 14 days in the lung as class II MHC is significantly upregulated in lavaged alveolar macrophages, over this time.[17] Moreover, although AdIFN does not result in spontaneous release of TNF in the lung, a subsequent challenge with intratracheal endotoxin results in a greater than fivefold increase in peak TNF levels in BALF in AdIFN-transduced animals compared to control animals (Fig. 2(A)). This enhanced TNF response is associated with increased neutrophil recruitment[17] and increased clearance of *Pseudomonas aeruginosa* (PA) up to 14 days after gene transfer.[17] Although the high levels of both IFN and TNF in the BALF were quite high, these cytokines were confined to the lung and remained essentially undetectable in the plasma (data not shown). Thus, these compartmentalized effects may offer cytokine gene transfer and advantage over systemically delivered proteins-based therapies.

Alcohol abuse is a risk factor for bacterial pneumonia[18] as well as acute lung injury.[19] Alcohol intoxication increases the risk of aspiration and suppresses macrophage free-radical protection and bacterial killing.[20,21] Moreover, alcohol can suppress the elaboration of alarm cytokines such as TNF.[22,23] Alcohol-induced suppression of TNF production by macrophages can be reversed by IFN in vitro. To investigate whether IFN gene therapy could augment TNF and bacterial host defense in vivo, we administered AdIFN intratracheally to rats followed 3 days later with an acutely intoxicating dose of ethanol (5.5 gm/kg intraperitoneal). Thirty minutes later animals were challenged intratracheally with endotoxin (LPS) or live *Klebsiella pneumoniae* to measure LPS-induced TNF response and lung neutrophil recruitment or bacterial clearance of *K. pneumoniae* respectively. This dose of alcohol has previously been shown by our group to suppress LPS-induced lung macrophage production of TNF. AdIFN pretreatment prevented alcohol-induced TNF suppression as well as lung neutrophil recruitment (Figs 3(A) and (B)). Moreover, we observed a significant increase in lung bacterial clearance of *K. pneumoniae* (Fig. 3(C)).[24]

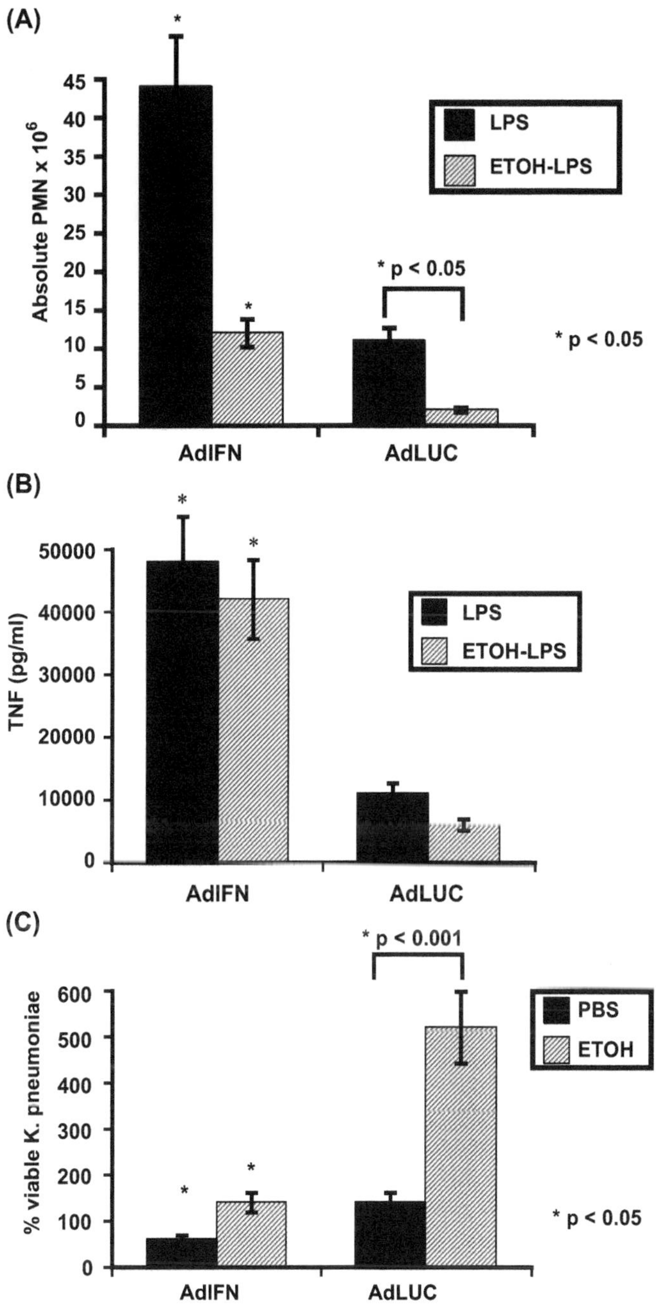


**Figure 3** Enhancement of pulmonary host defense in an acute model of ethanol (ETOH) intoxication. Male Sprague–Dawley rats were treated with $10^9$ pfu of AdIFN or AdLUC as a control. Three days later rats were treated with intratracheal LPS or live *Klebsiella pneumoniae*. Animals receiving LPS were sacrificed 3 h later to determine cell migration into or TNF concentration in the BAL fluid. Rats receiving *K. pneumoniae* were sacrificed immediately or 4 h after bacterial inoculation to determine bacterial clearance. Panel A: AdIFN reverses ETOH-induced suppression of lung neutrophil migration after LPS. Panel B: AdIFN enhances lung TNF release into BAL fluid after LPS in control and ETOH-treated animals. Panel C: AdIFN improves lung clearance of *K. pneumoniae* in a ETOH-treated rat model.

Standiford and colleagues have shown that adenoviral gene transfer of functional IL-12[25] produces the p70 heterodimer of IL-12 in the lung lavage fluid in a dose-dependent fashion for up to 7 days.[26] Mice pretreated with this vector, then challenged with $3 \times 10^2$ *K. pneumoniae*, had significantly improved survival, compared to AdCMVLacZ-treated or untreated controls.[26] The beneficial effect of IL-12 overexpression was mediated by both TNF-α and IFN-γ, as survival in Ad5IL-12-treated mice was attenuated by concomitant neutralization of endogenous TNF or IFN-γ.[26] This same group demonstrated the feasibility of using TNF-α, a critical proinflammatory cytokine in lung host defenses,[27,28] for in vivo immunomodulation of pulmonary host defense. To overexpress TNF compartmentally in the lung a recombinant adenovirus expressing TNF (AdmTNF) has been reported.[29] Concomitant bacterial challenge with *K. pneumoniae* and low-dose AdmTNF ($10^8$ pfu) resulted in improved host defenses against the organism. However, a higher dose of vector ($5 \times 10^8$ pfu) was not beneficial in terms of bacterial clearance. Thus, understanding dose–response relationships in gene-based immunotherapies will be critical for this form of treatment to have an impact in clinical infections.

Crystal and colleagues have used an adenoviral vector encoding CD40 ligand (AdCD40L) in a vaccine approach to protect against PA lung infection.[30] CD40L is expressed on activated T cells and allows dendritic cells, which are specialized antigen-presenting cells, to interact directly with either $CD8^+$ cytotoxic T cells[31,32] and B cells.[33] By transfecting DC, with AdCD40L, Kikuchi and colleagues demonstrated that gene-modified DC pulsed with PA could stimulate naïve B cells to produce anti-PA antibodies. Moreover, if these pulsed, gene-modified DCs were administered in vivo, that an in vivo anti-PA response was achieved which protected the vaccinated mice against a subsequent challenge with PA.[30] To support the fact that B cells were critical to this response, both passive transfer of serum or B cells form vaccinated mice conferred protection to naïve mice subsequently challenged with PA.

## 1.3 Opportunistic Infections

In addition to its known effects on upregulating macrophage function and innate host defenses, IFN-γ is also the prototypic TH1 cytokine that facilitates TH0 $CD4^+$ T-cell differentiation into TH1-expressing $CD4^+$ T cells.[34] Moreover, IFN can also modulate the cytokine expression of $CD8^+$ T cells to a Tc1 phenotype.[35,36] As IFN is produced by activated $CD4^+$ T cells, a lack of IFN secretion could partly explain the pulmonary host defense defect associated with HIV infection. Among HIV-associated opportunistic infections, *Pneumocystis carinii* pneumonia remains a persistent complication of HIV infection. There is an inverse relationship between $CD4^+$ T-cell count and acquisition of this infection. Furthermore, IFN-γ, in the form of recombinant protein given as an aerosol, has been shown to reduce the intensity of *P. carinii* infection in a mouse model.[37] Based on these data, our laboratory investigated whether adenoviral-mediated gene transfer of IFN-γ to the lung would have a therapeutic effect in a mouse model of *P. carinii* pneumonia. To test this concept with gene delivery, we used the AdIFN model, which results in prolonged expression of IFN in the lungs of mice depleted of $CD4^+$ T cells.[38] AdIFN-transduced or control (AdLuc) animals were challenged with $2 \times 10^5$ *P. carinii* cysts and sacrificed at serial time points. There was similar growth of *P. carinii* in both AdIFN and control animals for the first 2 weeks of the infection. However, after this time point AdIFN-treated mice showed resolution of the infection over 4–6 weeks in spite of continued depletion of $CD4^+$ T cells (Fig. 4(A)). AdIFN-treated mice recruited greater numbers of T cells, which were largely $CD8^+$ cells.[38] There was also a significant increase in recruited NK cells in the AdIFN-treated mice.[38] AdIFN was ineffective in improving *P. carinii* infection in both *scid* mice (which have intact macrophages and NK cells) or in mice depleted of both $CD4^+$ and $CD8^+$ T cells, suggesting that $CD8^+$ T cells are required for the clearance effect imparted by AdIFN treatment. In further support of $CD8^+$ T cells having effector function is the fact that there is a greater precursor frequency of IFN-producing $CD8^+$ T-cell clones in AdIFN-treated mice as measured by Elispot (Fig. 4(B)).

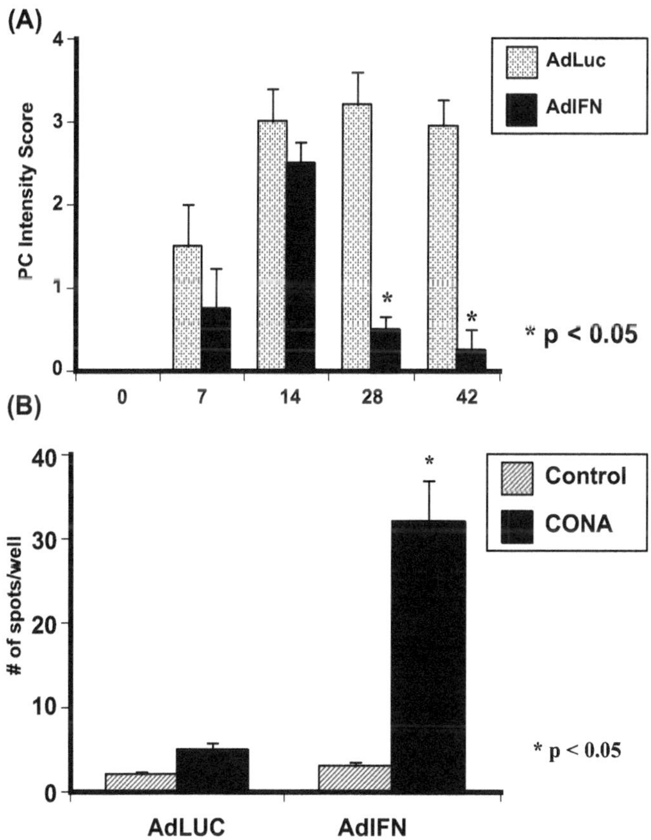


**Figure 4** Interferron (IFN)-mediated clearance of *Pneumocystis carinii* in CD4 T-cell-depleted mice. Panel A: Pretreatment with AdIFN resulted in significant clearance of *P. carinii* by 28 weeks. Panel B: Specific modulation of CD8 phenotype by AdIFN. Lung CD8 cells form AdIFN-treated mice show a significant higher precursor frequency of IFN-producing clones, as measured by Elispot, compared to AdLUC controls.

Understanding effector function of $CD8^+$ T cells in the context of *P. carinii* infection may have a significant impact in future therapies designed to support HIV-infected individuals against opportunistic infections.

### 1.4 Viral Hepatitis

Both hepatitis B and hepatitis C are important causes of chronic hepatitis and hepatitis B has been linked to hepatocellular cancer. Hepatitis C virus (HCV), is a positive-strand RNA virus, and is the major infectious agent responsible for causing chronic hepatitis. Presently, there is no vaccine for HCV infection. There have been recent advances in drug therapy for this disease using a combination of ribavirin with IFN-α,[39] however, there is still need for improved sustained therapy. Lieber and colleagues have demonstrated that adenoviral-mediated gene transfer of hammerhead ribozymes directed against a conserved region of the plus strand and minus strand of the HCV genome was efficient at reducing or eliminating the respective plus- or minus-strand HCV RNAs expressed in cultured cells and from primary human hepatocytes obtained from chronic HCV-infected patients.

Another therapeutic approach has been locally to upregulate innate antiviral immunity. Toward this end Aurisicchio and colleagues demonstrated that adenoviral-mediated gene transfer of the IFN-α2 under the control of a liver-specific promoter protected mice for a challenge with mouse hepatitis virus type 3.[40] Lastly, another approach has been to construct dominant negative core mutants of hepatitis B and when these are expressed in hepatocytes cell lines in the context of a recombinant adenoviral vector, these molecules were capable of significantly suppressing viral replication.[41]

## 2. Chronic Inflammatory Diseases

### 2.1 Inflammatory Bowel Disease

The gut has been proposed as a target for gene delivery for a variety of diseases including both metabolic diseases and primary diseases affecting the intestine including the inflammatory bowel diseases, Chron's disease, and ulcerative colitis.[42,43] Toward this end Hogaboam and colleagues have shown that intraperitoneal delivery of adenovirus-encoding IL-4 (AdIL-4), a prototypic TH2 cytokine, attenuates colitis induced by trinitrobenzene sulfonic acid (TNB).[44] TNB-induced colitis is associated with an acute phase followed by an immunologically mediated phase, which is thought to be hapten-induced.[45] The attenuation as a result of AdIL-4 in colitis was associated with a reduction in colonic IFN levels and less induction of iNOS.[44] The same group has shown similar data for another TH2 cytokine, IL-10 in a similar model of colitis.[46] Adenovirus IL-10 treatment was again done by the intraperitoneal route and associated with a significant reduction in colonic myeloperoxidase activity and leukotriene four levels, both markers of acute inflammation. What remains unclear from these studies is whether T-cell activation is modified and whether there is protection against a second bout of colitis. Lastly, the intraperitoneal approach is essentially a systemic form of therapy since IL-4 and IL-10 can be detected in the serum of these mice. Since the gut can be transduced directly with adenovirus vectors, this raises the possibility that local administration of vectors to inflamed intestine could be used to compartmentally upregulate an immunomodulatory gene that would prevent or attenuate existing colitis.

Toward this end, Wirtz and colleagues have investigated adenovirus-mediated gene transfer to the inflamed colon using intrarectal administration of Ad5-based vectors. These investigators observed significant gene transfer to colonic epithelium, whereas no colonic gene transfer was observed then the vector was given systemically (intravenously or intraperitoneally). Moreover, gene transfer was enhanced in the setting of TNB-induced inflammation. Lastly, the investigators investigated an Ad5-based vector with a lysine repeat engineered in the fiber gene, the protein responsible for initial interactions with the coxsackie–adenovirus receptor. With this genetically modified vector, the investigators observed enhanced gene transfer to cells in the lamina propria and spleen, suggesting that antigen-specific T cells could be modified with this vector approach.

## 2.2 Arthritis

Like inflammatory bowel disease, rheumatoid arthritis (RA) is thought to be dominated by TH1-like inflammation (Fig. 5).[47,48] Among chronic inflammatory diseases, more has been published on gene therapy for arthritis than any other disease. This is likely due to the fact that (1) it is a common disease entity, (2) current treatment,

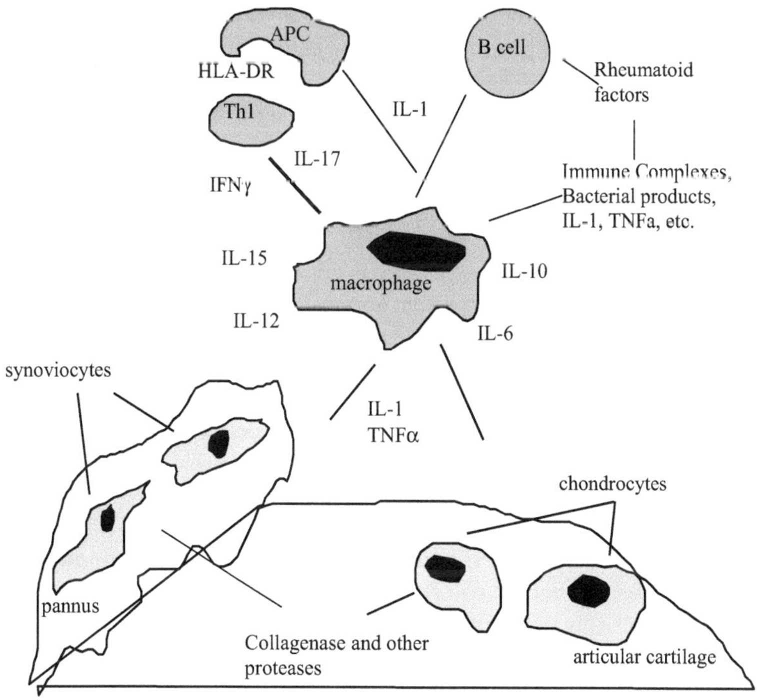


**Figure 5** Schematic scheme of the pathogenesis of rheumatoid arthritis.

although effective in many cases, can be improved upon, (3) readily accessible site for gene transfer, and (4) relevant clinical models of the disease, particularly RA, and (5) gene transfer can be accomplished locally to the synovial lining cells using adenovirus-based vectors.[49] The pathogenesis of RA is complex but data to date suggest there exist alloreactive T cells that secrete TH1-like cytokines such as TNF-α, TNF-β, IL-2, and IFN, which drives inflammation. Accessory cells can also secrete TNF and IL-1β, and IL-18 which are also proinflammatory and can drive TH1 inflammation. This leads to an inflammatory synovial pannus, which mediates destruction of cartilage and joint erosion, which results in loss of joint function over time. A novel T-cell-derived cytokine, IL-17, has also been implicated in the pathogenesis of RA.[50]

Since TH1 inflammation can be downregulated by TH2 cytokines, such as IL-4 or IL-10, these cytokines have been investigated as candidate genes to modify RA inflammation. Woods and colleagues investigated adenoviral-mediated gene transfer of the human IL-4 gene into synovial explants from RA patients and demonstrated a significant reduction in IL-1β, TNF-α, and IL-8 elaboration in the explant cultures treated with AdIL-4.[51] In follow-up to this work, the same group demonstrated in vivo efficacy of intra-articular AdIL-4 treatment in adjuvant-induced arthritis in a rat model.[52] Of note was that AdIL-4 was effective in both a pretreatment and post-treatment paradigm.[52] Similar to the in vitro findings in human explants, the in vivo treatment with AdIL-4 in the rat model was associated with lower TNF-α and IL-1β levels.[52] Lubberts and colleagues have also shown efficacy of AdIL-4 in a murine model of collagen-induced arthritis (CIA).[53,54] Interestingly in these studies, IL-4 had less effect on the joint inflammation than it appeared to have on preservation of cartilage and in preventing bone erosion.[53] These latter effects were associated with a reduction of mRNAs for IL-17, TNF, and IL-1β, as well as a decrease in metalloproteinase activity.[53,54] These investigators also demonstrated that IL-4 can increase type I procollagen synthesis and thus this may explain the joint sparing/repair effect of IL-4.[53] Lastly, Kim and colleagues demonstrated that both periarticular and systemic AdIL-4 was effective in a model of CIA.[55]

Whalen and colleagues have investigated another TH2 cytokine, viral IL-10, encoded by a adenoviral vector (AdvIL-10) given by periarticular injection in the same model of CIA and found significant benefit in terms of development of arthritis and arthritis score. Moreover, the investigators showed that the injection of AdvIL-10 into one joint prevented arthritis in a second joint.[56] This may be due to in vivo T-cell immunomodulation by viral IL-10. In further support of a role for TH2 cytokine gene therapy in RA, Woods and colleagues have recently demonstrated that adenovirus-mediated gene transfer of IL-13, another TH2 cytokine, also suppresses TNF and IL-1β production in RA explant cultures.[57]

In addition to the TH2 cytokine approach, the other approach of adenoviral gene transfer for arthritis has largely focused on the proinflammatory cytokines TNF-α and IL-1β. Toward this end, our laboratory has created soluble type-1 receptors for both IL-1[58] and TNF-α[59] (Fig. 6). Both these molecules are dimerized by the addition of murine IgG Fc fragment and in the case of the TNF inhibitor, this molecule has been found to be more potent in TNF inhibition than monoclonal antibodies that only bind to one epitope.[60] Moreover, the proteins have longer half-lives in vivo compared to the

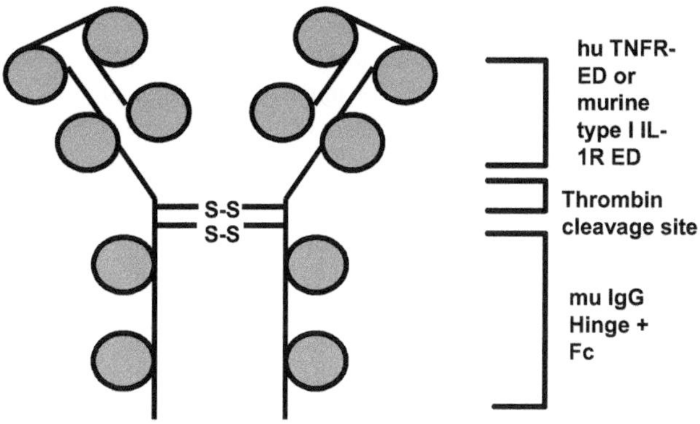


**Figure 6** Schematic diagram of tumor necrosis factor and interlukin-1 (IL-1) receptor fusion proteins utilized in arthritis gene therapy.

monomeric soluble receptors.[59] Adenoviral-mediated gene transfer of either of these constructs into the joint space in a rabbit model of arthritis showed less white blood cell infiltration as well as less joint swelling. However, the IL-1 inhibitor showed a better effect in preventing a reduction in cartilage matrix degradation. Moreover, the two vectors together appeared to have an additive effect on white blood cell infiltration into the joint space. There was also an effect observed on contralateral joints in this study.[58] Le and colleagues also demonstrated efficacy of the TNF inhibitor gene in a rat model of CIA.[61] Interestingly, Quattrocchi have reported in a mouse model of CIA that there is an acute beneficial effect of the TNF inhibitor fusion protein, however there is a subsequent rebound with enhanced inflammation despite continued circulating levels of the TNF inhibitor. The investigators speculated that this may be due to antibody formation against the extracellular domain of the receptor that the cross-linked endogenous TNF receptors in the joint.[62] It is important to note that these studies were performed with a chimeric fusion protein (mouse Fc/human p55 TNF receptor) and thus whether the exacerbation of arthritis would be seen with the mouse p55 TNF receptor remains to be seen. Lastly, Zhang and colleagues have shown that adenoviral-mediated gene transfer of a dominant negative form of inhibitory kappa-B which facilitates nuclear translocation of nuclear factor-kappa-B enhanced TNF-mediated apoptosis in synovial tissue.[63]

## 2.3 Fibrotic Lung Disease

Idiopathic pulmonary fibrosis (IPF) is an insidious disorder that results in the deposition of collagen and fibrous tissue in the lung. The etiology of this disorder is unknown but several groups have reported decreased fibrinolytic activity[64,65] and elevated tissue growth factor-beta-1 expression[66,67] in the lungs of patients with IPF. Moreover, IFN-$\gamma$ has been shown in a pilot study to improve lung function in patients with IPF and this impairment was associated with a decreased levels of messenger RNA for transforming growth factor-beta-1 (TGF-$\beta$1) and connective tissue growth factor, the main growth factor product of TGF-$\beta$ stimulation.[67] To date this is the first compound to show improvement in lung function. Many trials have been performed with corticosteroids (prednisone) alone or in combination with cyclophosphamide.[68] However, these agents have not been shown to be effective in preserving lung function in randomized clinical trials, and moreover, their use is associated with significant side effects. Since IPF is associated with dysregulated growth factor gene expression, and a lack of definitive therapy, there is a rationale for gene therapy.

Simon and colleagues recently reported on enhancing fibrinolytic activity in the lung in an effort to ameliorate lung fibrotic injury in response to bleomycin, a chemotherapeutic agent that can cause lung fibrosis.[69] These investigators constructed a recombinant adenovirus-encoding urokinase-type plasminogen activator (AduPA), a fibrinolytic activator protein. When expressed in the lung, AduPA resulted in a significant attenuation of bleomycin-induced increases in hydroxyproline content, a measure of collagen deposition.[69] Furthermore, Nakao and colleagues have shown that adenovirus-mediated gene transfer of smad7, a downstream inhibitor of TGF-β signaling could also block bleomycin-induced fibrosis.[70] This finding was specific to *smad7* and not to *smad6*, which does not inhibit TGF-β signaling and thus the data suggest that the effect is through the downregulation of TGF-β signaling. Thus, other molecules such as dominant negatives or soluble receptors for TGF-β may also be good candidate constructs for pulmonary fibrosis.

In addition to TGF-β, other proinflammatory cytokines such as IL-1 and TNF have been implicated in pulmonary fibrosis.[71] Our lab reported several years ago on a soluble inhibitor of TNF that consists of the extracellular domain of the human p55 TNF receptor coupled to the murine CH2 and CH3 domains of mouse IgG1 (Fig. 6).[59] This molecule forms as a dimer and is a potent inhibitor of TNF.[60] When expressed in the context of a recombinant adenovirus, after intravenous administration, the construct results in high circulating levels of TNF inhibitory activity.[59] In fact these mice provide a phenocopy of p55 TNF receptor knockout mice, in that they are resistant to mortality induced by endotoxin and D-galactosamine administration, however, they are susceptible to the intracellular pathogen *Listeria monocytogenes*.[59] However, this molecule also readily crosses into the lung[72] and inhibits TNF activity in this compartment. Moreover, this construct, by virtue of it's ability to inhibit TNF in the lung (after systemic vector administration), attenuates the fibrotic response to intratracheal silica.[73]

## 3. Conclusions

There are numerous acquired diseases in which adenoviral-mediated gene transfer has shown in proof-of-principle experiments a therapeutic benefit. The challenges for researchers in the field are to take these data and try to develop safe and effective therapies for these diseases. Toward this end, there will need to be advances in targeted vector therapy and regulated gene expression. One area, which may yield promising results in the near future, is in adenovirus-based vaccines either into somatic cells or professional antigen-presenting cells such as dendritic cells or in compartmentalized chronic inflammation such as arthritis. In this case, precise gene expression is less likely and thus, there are less technological hurdles to overcome.

## References


1. Gurwith MJ, Horwith GS, Impellizzeri CA, Davis AR, Lubeck MD, Hung PP. Current use and future directions of adenovirus vaccine. *Semin Respir Infect* 1989;**4**:299–303.
2. Dolin PJ, Raviglione MC, Kochi A. Global tuberculosis incidence and mortality during 1990–2000. *Bull World Health Organ* 1994;**72**:213–20.
3. Sudre P, ten DG, Kochi A. Tuberculosis: a global overview of the situation today. *Bull World Health Organ* 1992;**70**:149–59.
4. Mackaness GB. Resistance to intracellular infection. [Review] [31 refs] *J Infect Dis* 1971;**123**:439–45.
5. Orme IM, Andersen P, Boom WH. T cell response to *Mycobacterium tuberculosis*. [Review] [142 refs] *J Infect Dis* 1993;**167**:1481–97.
6. Orme IM. Characteristics and specificity of acquired immunologic memory to *Mycobacterium tuberculosis* infection. *J Immunol* 1988;**140**:3589–93.
7. Orme IM, Miller ES, Roberts AD, Furney SK, Griffin JP, Dobos KM, et al. T lymphocytes mediating protection and cellular cytolysis during the course of *Mycobacterium tuberculosis* infection. Evidence for different kinetics and recognition of a wide spectrum of protein antigens. *J Immunol* 1992;**148**:189–96.
8. Dalton DK, Pitts Meek S, Keshav S, Figari IS, Bradley A, Stewart TA. Multiple defects of immune cell function in mice with disrupted interferon-gamma genes. *Science* 1993;**259**: 1739–42.
9. Cooper AM, Dalton DK, Stewart TA, Griffin JP, Russell DG, Orme IM. Disseminated tuberculosis in interferon gamma gene-disrupted mice. *J Exp Med* 1993;**178**:2243–7.
10. Flynn JL, Chan J, Triebold KJ, Dalton DK, Stewart TA, Bloom BR. An essential role for interferon gamma in resistance to *Mycobacterium tuberculosis* infection. *J Exp Med* 1993;**178**: 2249–54.
11. Magram J, Connaughton SE, Warrier RR, Carvajal DM, Wu CY, Ferrante J, et al. IL-12-deficient mice are defective in IFN gamma production and type 1 cytokine responses. *Immunity* 1996;**4**:471–81.
12. Cooper AM, Magram J, Ferrante J, Orme IM. Interleukin 12 (IL-12) is crucial to the development of protective immunity in mice intravenously infected with *Mycobacterium tuberculosis*. *J Exp Med* 1997;**186**:39–45.
13. Nathan CF, Kaplan G, Levis WR, Nusrat A, Witmer MD, Sherwin SA, et al. Local and systemic effects of intradermal recombinant interferon-gamma in patients with lepromatous leprosy. *N Engl J Med* 1986;**315**:6–15.
14. Williams LM, Snyder DC, Deblieux P, Ali J, Kuebel D, deBoisblanc BP, et al. Safety and feasibility of combined aerosolized and subcutaneous interferon-gamma as adjuvant treatment of *Mycobacterium avium* complex pulmonary infection in non-HIV infected hosts. *Am J Respir Crit Care Med* 1994:A110.
15. Condos R, Rom WN, Schluger NW. Treatment of multidrug-resistant pulmonary tuberculosis with interferon-gamma via aerosol. [see comments] *Lancet* 1997;**349**:1513–5.
16. Ewig S. Community-acquired pneumonia: definition, epidemiology, and outcome. *Semin Respir Infect* 1999;**14**:94–102.
17. Lei D, Lancaster Jr JR, Joshi MS, Nelson S, Stoltz D, Bagby GJ, et al. Activation of alveolar macrophages and lung host defenses using transfer of the interferon-gamma gene. *Am J Physiol Lung Cell Mol Physiol* 1997;**272**:L852–9.
18. MacGregor RR. Alcohol and immune defense. *J Am Med Assoc* 1986;**256**:1474–9.
19. Moss M, Bucher B, Moore FA, Moore EE, Parsons PE. The role of chronic alcohol abuse in the development of acute respiratory distress syndrome in adults. *J Am Med Assoc* 1996;**275**:50–4.

20. Tamura DY, Moore EE, Partrick DA, Johnson JL, Offner PJ, Harbeck RJ, et al. Clinically relevant concentrations of ethanol attenuate primed neutrophil bactericidal activity. *J Trauma* 1998;**44**:320–4.
21. Dorio RJ, Forman HJ. Ethanol inhibition of signal transduction in superoxide production by rat alveolar macrophages. A proposed mechanism for ethanol related pneumonia. *Ann Clin Lab Sci* 1988;**18**:190–4.
22. Nelson S, Bagby G, Summer WR. Alcohol suppresses lipopolysaccharide-induced tumor necrosis factor activity in serum and lung. *Life Sci* 1989;**44**:673–6.
23. D'Souza NB, Bagby GJ, Nelson S, Lang CH, Spitzer JJ. Acute alcohol infusion suppresses endotoxin-induced tumor necrosis factor production. *Alcohol Clin Exp Res* 1989;**13**:295–8.
24. Kolls JK, Lei D, Stoltz D, Zhang P, Schwarzenberger PO, Ye P, et al. Adenoviral-mediated interferon-gamma gene therapy augments pulmonary host defense of ethanol-treated rats. *Alcohol Clin Exp Res* 1998;**22**:157–62.
25. Bramson J, Hitt M, Gallichan WS, Rosenthal KL, Gauldie J, Graham FL. Construction of a double recombinant adenovirus vector expressing a heterodimeric cytokine: In vitro and in vivo production of biologically active interleukin-12. *Hum Gene Ther* 1996;**7**:333–42.
26. Greenberger MJ, Kunkel SL, Strieter RM, Lukacs NW, Bramson J, Gauldie J, et al. IL-12 gene therapy protects mice in lethal *Klebsiella pneumonia*. *J Immunol* 1996;**157**:3006–12.
27. Standiford TJ, Huffnagle GB. Cytokines in host defense against pneumonia. *J Invest Med* 1997;**45**:335–45.
28. Nelson S, Bagby G, Andresen J, Nakamura C, Shellito J, Summer W. The effects of ethanol, tumor necrosis factor, and granulocyte colony-stimulating factor on lung antibacterial defenses. *Adv Exp Med Biol* 1991;**288**:245–53.
29. Standiford TJ, Wilkowski JM, Sisson TH, Hattori N, Mehrad B, Bucknell KA, et al. Intrapulmonary tumor necrosis factor gene therapy increases bacterial clearance and survival in murine gram-negative pneumonia. *Hum Gene Ther* 1999;**10**:899–909.
30. Kikuchi T, Worgall S, Singh R, Moore MA, Crystal RG. Dendritic cells genetically modified to express CD40 ligand and pulsed with antigen can initiate antigen-specific humoral immunity independent of CD4+ T cells. *Nat Med* 2000;**6**:1154–9.
31. Banchereau J, Steinman RM. Dendritic cells and the control of immunity. *Nature* 1998;**392**:245–52.
32. Grewal IS, Flavell RA. CD40 and CD154 in cell-mediated immunity. *Annu Rev Immunol* 1998;**16**:111–35.
33. Clark EA, Ledbetter JA. How B and T cells talk to each other. *Nature* 1994;**367**:425–8.
34. Mosmann TR, Sad S. The expanding universe of T-cell subsets: Th1, Th2 and more. [Review] [94 refs] *Immunol Today* 1996;**17**:138–46.
35. Li L, Sad S, Kagi D, Mosmann TR. CD8Tc1 and Tc2 cells secrete distinct cytokine patterns in vitro and in vivo but induce similar inflammatory reactions. *J Immunol* 1997;**158**:4152–61.
36. Mosmann TR, Li L, Sad S. Functions of CD8 T-cell subsets secreting different cytokine patterns. [Review] [26 refs] *Semin Immunol* 1997;**9**:87–92.
37. Beck JM, Liggit HD, Brunette EN, Fuchs HJ, Shellito JE, Debs RJ. Reduction in intensity of *Pneumocystis carinii* pneumonia in mice by aerosol administration of interferon-gamma. *Infect Immun* 1991;**59**:3859–62.
38. Kolls JK, Habetz S, Shean MK, Vazquez C, Brown JA, Lei D, et al. IFN-gamma and CD8+ T cells restore host defenses against *pneumocystis carinii* in mice depleted of CD4+ T cells. *J Immunol* 1999;**162**:2890–4.
39. Cummings KJ, Lee SM, West ES, Cid-Ruzafa J, Fein SG, Aoki Y, et al. Interferon and ribavirin vs interferon alone in the re-treatment of chronic hepatitis C previously nonresponsive to interferon: a meta-analysis of randomized trials. *J Am Med Assoc* 2001;**285**:193–9.

40. Aurisicchio L, Delmastro P, Salucci V, Paz OG, Rovere P, Ciliberto G, et al. Liver-specific alpha 2 interferon gene expression results in protection from induced hepatitis. *J Virol* 2000;**74**:4816–23.
41. Scaglioni P, Melegari M, Takahashi M, Chowdhury JR, Wands J. Use of dominant negative mutants of the hepadnaviral core protein as antiviral agents. *Hepatology* 1996;**24**: 1010–7.
42. Noel RA, Shukla P, Henning SJ. Optimization of gene transfer into intestinal cells using a retroviral vector. *J Pediatr Gastroenterol Nutr* 1994;**19**:43–9.
43. Cheng DY, Kolls JK, Lei D, Noel RA. In vivo and in vitro gene transfer and expression in rat intestinal epithelial cells by E1-deleted adenoviral vector. *Hum Gene Ther* 1997;**8**: 755–64.
44. Hogaboam CM, Vallance BA, Kumar A, Addison CL, Graham FL, Gauldie J, et al. Therapeutic effects of interleukin-4 gene transfer in experimental inflammatory bowel disease. *J Clin Invest* 1997;**100**:2766–76.
45. Macdonald TT. Viral vectors expressing immunoregulatory cytokines to treat inflammatory bowel disease. *Gut* 1998;**42**:460–1.
46. Barbara G, Xing Z, Hogaboam CM, Gauldie J, Collins SM. Interleukin 10 gene transfer prevents experimental colitis in rats. *Gut* 2000;**46**:344–9.
47. Miossec P. Are T cells in rheumatoid synovium aggressors or bystanders? *Curr Opin Rheumatol* 2000;**12**:181–5.
48. Muller B, Gimsa U, Mitchison NA, Radbruch A, Sieper J, Yin Z. Modulating the Th1/Th2 balance in inflammatory arthritis. *Springer Semin Immunopathol* 1998;**20**:181–96.
49. Roessler BJ, Allen ED, Wilson JM, Hartman JW, Davidson BL. Adenoviral-mediated gene transfer to rabbit synovium in vivo. *J Clin Invest* 1993;**92**:1085–92.
50. Chabaud M, Fossiez F, Taupin JL, Miossec P. Enhancing effect of IL-17 on IL-1-induced IL-6 and leukemia inhibitory factor production by rheumatoid arthritis synoviocytes and its regulation by Th2 cytokines. *J Immunol* 1998;**161**:409–14.
51. Woods JM, Tokuhira M, Berry JC, Katschke KJ, Kurata H, Damergis JA, et al. Interleukin-4 adenoviral gene therapy reduces production of inflammatory cytokines and prostaglandin E2 by rheumatoid arthritis synovium ex vivo. *J Invest Med* 1999;**47**:285–92.
52. Woods JM, Katschke KJ, Volin MV, Ruth JH, Woodruff DC, Amin MA, et al. IL-4 adenoviral gene therapy reduces inflammation, proinflammatory cytokines, vascularization, and bony destruction in rat adjuvant- induced arthritis. *J Immunol* 2001;**166**:1214–22.
53. Lubberts E, Joosten LA, Chabaud M, Van Den BL, Oppers B, Coenen-De Roo CJ, et al. IL-4 gene therapy for collagen arthritis suppresses synovial IL-17 and osteoprotegerin ligand and prevents bone erosion. *J Clin Invest* 2000;**105**:1697–710.
54. Lubberts E, Joosten LA, Van Den BL, Helsen MM, Bakker AC, van Meurs JB, et al. Adenoviral vector-mediated overexpression of IL-4 in the knee joint of mice with collagen-induced arthritis prevents cartilage destruction. *J Immunol* 1999;**163**:4546–56.
55. Kim SH, Evans CH, Kim S, Oligino T, Ghivizzani SC, Robbins PD. Gene therapy for established murine collagen-induced arthritis by local and systemic adenovirus-mediated delivery of interleukin-4. *Arthritis Res* 2000;**2**:293–302.
56. Whalen JD, Lechman EL, Carlos CA, Weiss K, Kovesdi I, Glorioso JC, et al. Adenoviral transfer of the viral IL-10 gene periarticularly to mouse paws suppresses development of collagen-induced arthritis in both injected and uninjected paws. *J Immunol* 1999;**162**: 3625–32.
57. Woods JM, Katschke KJ, Tokuhira M, Kurata H, Arai KI, Campbell PL, et al. Reduction of inflammatory cytokines and prostaglandin E2 by IL-13 gene therapy in rheumatoid arthritis synovium. *J Immunol* 2000;**165**:2755–63.

58. Ghivizzani SC, Lechman ER, Kang R, Tio C, Kolls J, Evans CH, et al. Direct adenovirus-mediated gene transfer of interleukin 1 and tumor necrosis factor alpha soluble receptors to rabbit knees with experimental arthritis has local and distal anti-arthritic effects. *Proc Natl Acad Sci USA* 1998;**95**:4613–8.
59. Kolls J, Peppel K, Silva M, Beutler B. Prolonged and effective blockade of tumor necrosis factor activity through adenovirus-mediated gene transfer. *Proc Natl Acad Sci USA* 1994;**91**:215–9.
60. Peppel K, Crawford D, Beutler B. A tumor necrosis factor (TNF) receptor-IgG heavy chain chimeric protein as a bivalent antagonist of TNF activity. *J Exp Med* 1991;**171**:1483–9.
61. Le CH, Nicolson AG, Morales A, Sewell KL. Suppression of collagen-induced arthritis through adenovirus-mediated transfer of a modified tumor necrosis factor alpha receptor gene. *Arthritis Rheum* 1997;**40**:1662–9.
62. Quattrocchi E, Walmsley M, Browne K, Williams RO, Marinova-Mutafchieva L, Buurman W, et al. Paradoxical effects of adenovirus-mediated blockade of TNF activity in murine collagen-induced arthritis. *J Immunol* 1999;**163**:1000–9.
63. Zhang HG, Huang N, Liu D, Bilbao L, Zhang X, Yang P, et al. Gene therapy that inhibits nuclear translocation of nuclear factor kappaB results in tumor necrosis factor alpha-induced apoptosis of human synovial fibroblasts. *Arthritis Rheum* 2000;**43**:1094–105.
64. Hattori N, Degen JL, Sisson TH, Liu H, Moore BB, Pandrangi RG, et al. Bleomycin-induced pulmonary fibrosis in fibrinogen-null mice. *J Clin Invest* 2000;**106**:1341–50.
65. Olman MA, Mackman N, Gladson CL, Moser KM, Loskutoff DJ. Changes in procoagulant and fibrinolytic gene expression during bleomycin-induced lung injury in the mouse. *J Clin Invest* 1995;**96**:1621–30.
66. Gauldie J, Jordana M, Cox G. Cytokines and pulmonary fibrosis. *Thorax* 1993;**48**:931–5.
67. Ziesche R, Hofbauer E, Wittmann K, Petkov V, Block LH. A preliminary study of long-term treatment with interferon gamma-1b and low-dose prednisolone in patients with idiopathic pulmonary fibrosis. *N Engl J Med* 1999;**341**:1264–9.
68. Selman M, King TE, Pardo A. Idiopathic pulmonary fibrosis: prevailing and evolving hypotheses about its pathogenesis and implications for therapy. *Ann Intern Med* 2001;**134**:136–51.
69. Sisson TH, Hattori N, Xu Y, Simon RH. Treatment of bleomycin-induced pulmonary fibrosis by transfer of urokinase-type plasminogen activator genes. *Hum Gene Ther* 1999;**10**:2315–23.
70. Nakao A, Fujii M, Matsumura R, Kumano K, Saito Y, Miyazono K, et al. Transient gene transfer and expression of Smad7 prevents bleomycin-induced lung fibrosis in mice. *J Clin Invest* 1999;**104**:5–11.
71. Piguet PF. Cytokines involved in pulmonary fibrosis. *Int Rev Exp Pathol* 1993;**34**(Pt B):173–81.
72. Kolls JK, Lei D, Greenberg S, Nelson S, Beutler B. Adenovirus-mediated blockade of tumor necrosis factor in mice protects against endotoxic shock yet impairs pulmonary host defense. *J Infect Dis* 1995;**171**:570–5.
73. Curiel DT, Pilewski JM, Albelda SM. Gene therapy approaches for inherited and acquired lung diseases. *Am J Respir Cell Mol Biol* 1996;**14**:1–18.

# Animal Models of Gene Therapy for Cardiovascular Disease

**28**

*Johanna P. Laakkonen*[1], *Seppo Ylä-Herttuala*[1,2,3]
[1]Department of Biotechnology and Molecular Medicine, A.I. Virtanen Institute for Molecular Sciences, University of Eastern Finland, Kuopio, Finland; [2]Department of Medicine, University of Eastern Finland, Kuopio, Finland; [3]Gene Therapy Unit, Kuopio University Hospital, Kuopio, Finland

## 1. Introduction

Cardiovascular gene therapy can be used to grow neovessels into ischemic tissues, reduce neointimal hyperplasia, improve cardiomyocyte function, and treat hypertension and dyslipidemias. Human adenovirus (Ad) serotypes 2 and 5 are commonly used vectors in clinical trials because of their high gene transfer efficiency, ease of production, and well-known virus–cell biology. Adenovirus vectors can be administrated to tissues via intramyocardial and intramuscular injections, coronary infusion, and pericardial or systemic delivery. Delivery method, serotype, and vector dosage are critical determinants of the therapeutic outcome. Animal models for cardiovascular diseases (CVDs) are essential to develop new therapeutic strategies. Increased understanding of the pathogenesis of CVDs enables identification of novel molecular targets for therapy. In the following chapter, methods and animal models used for gene therapy of CVDs are described in the context of Ad vector–mediated transgene delivery. Administration of vectors, barriers of Ad transduction, and vector retargeting are discussed.

## 2. Adenoviral Vectors for Cardiovascular Gene Therapy

### *2.1 Barriers in Vascular Gene Transfer*

Human Ads are the most commonly used vectors in gene therapy clinical trials (y. 2015, 22.5%; n=496) (http://www.abedia.com/wiley/vectors.php) owing to their high transduction efficiency, ease of production in good manufacturing practice quality, and well-known virus–cell interactions. To date, mainly human Ad serotypes 2 and 5, belonging to the subgroup C of *Adenoviridae*, have been used in clinical trials. Although systemic delivery restricts Ad-mediated gene transfer owing to various interactions of the vector with cellular factors/circulating blood components, localized Ad vector delivery, e.g., via intramyocardial and intramuscular injections, has shown great promise in treating CVDs.

**Adenoviral Vectors for Gene Therapy.**

Human Ads are nonenveloped, icosahedral (90 nm), double-stranded deoxyribonucleic acid (dsDNA) viruses that have more than 60 identified serotypes. Wild-type infections of Ads cause mild respiratory tract infections, but depending on the serotype, they can also result in conjunctivitis, gastroenteritis, and myocarditis. Coxsackievirus and adenovirus receptor (CAR), the primary receptor for subgroup C Ads, including Ad2 and Ad5 serotypes, is expressed in heart, pancreas, and intestine, and in lower levels in lung and liver.[1] Among cardiovascular cell types, low levels of CAR expression can be found from endothelial cells, smooth muscle cells (SMCs), mature myofibers, and intact human vessels, whereas high expression levels have been found from cardiomyocytes.[2–4] Adenoviruses have shown to transduce endothelial cells after administration of the vectors to vessel segments of peripheral or carotid artery and jugular vein.[5–8] Medial SMCs can be transduced after endothelial denudation.[6] High gene transfer efficiencies have also been obtained in cardiomyocytes after coronary perfusion or intramyocardial injection.[9,10]

CAR is involved in the formation of cell–cell junctions in epithelial cells and belongs to the immunoglobulin superfamily. CAR binds to fiber knob of Ad vector, followed by interaction of RGD motif in penton base and $\alpha v\beta 3$ and $\alpha v\beta 5$ integrins.[11] Also, heparan sulfate proteoglycans (HSPGs) and $\alpha v\beta 1$ integrins may act as co-receptors in some cell types.[12,13] As with many other viruses, respiratory, digestive, and ocular tracts are the main portals for Ad entry. In respiratory tract, macrophages are suggested to engulf Ad, triggering release of chemokines and cytokines that further remodel cell–cell junctions and enable passage of macrophages across the airway epithelium. This enables transport of CAR and integrins from basolateral to apical cell side and Ad infection via the CAR-mediated pathway.[14] Adhesion of neutrophils to airway epithelium can also facilitate Ad entry.[15] In vitro, Ad is taken up into cells by clathrin-mediated endocytosis. Other uptake mechanism, macropinocytosis also have been suggested.[16,17] Uncoating of Ad capsid occurs as it is transported from the cell membrane to the nucleus. Initially uncoating was suggested to occur pH-dependently in the endosomes. However, data suggest that Ad fiber shedding and exposure of the lytic viral protein initiate at the cell membrane induced by CAR drifting motions and binding of Ad to immobile integrins.[18] After endosomal escape, Ad is transported to nuclear pores by microtubules via interaction of Ad hexon protein and dynein intermediate light chain.[19] pH-induced conformational changes in the hypervariable region (HVR1) of the hexon protein lead to initial recruitment of dynein.[20,21] After docking to the nuclear pore, kinesin-1 mediates disassembly of the virus capsid by interacting with nucleoporins. This increases nuclear permeability and enables transport of the virus genome into the nucleus.[22,23] Virus entry from the cell surface to nucleus occurs within 30–60 min.[16]

A major limitation for the use of Ad vectors in gene therapy trials is the development of strong innate and adaptive immune responses against the Ad capsid. Adenovirus–host interactions with circulating blood components and cellular factors restrict Ad5-mediated gene transfer efficiency and the length of the transgene expression, and affect biodistribution of the vectors. Briefly, after intravascular delivery, Ad capsids induce the innate immune response by activating macrophages and dendritic cells. This results in the release of various chemokines and proinflammatory cytokines.[24,25] Uptake and sequestration of Ad vectors from blood by Kupffer cells and liver sinusoidal endothelial cells occur via scavenger receptors.[26–28] Activation of the complement system further limits gene transfer efficiency.[29,30] Binding of platelets to Ad via CAR or integrins leads to activation of endothelial cells and thrombocytopenia.[31,32] Although CAR acts as the primary receptor for Ads, binding to blood coagulation factors redirects the virus to alternative cellular receptors and facilitates high transduction efficiency in liver. Coagulation Factor X (FX) binds to HVR of hexon proteins and enables binding to HSPGs in hepatocytes.[33–36] Factor X has also been suggested to shield Ad vectors from the classical complement pathway.[37] Besides FX, Ad transduction of hepatocytes is suggested to be enhanced by coagulation Factor IX, complement C4-binding protein, protein C, and plasma zymogens including Factor VII.[25,38]

Adenovirus vectors induce biphasic toxicity, acute and late. Late toxicity occurs as a result of adaptive immune responses against Ad vectors. Preexisting humoral anti-Ad immunity restricts Ad-vector-mediated gene transfer further. Neutralizing antibodies are mainly against Ad hexon protein, but antibodies against fiber and penton base may also exist.[39,40] Besides platelets, erythrocytes also bind to Ad5 in humans via CAR and complement receptor CR1 and sequester the viruses.[41] Interestingly, similar expression of CAR is not present in murine erythrocytes.[42,43] Although systemic delivery restricts Ad-mediated gene transfer and requires high vector doses, localized Ad vector delivery has shown great promise in treating CVDs by having reduced immune responses against the vector and high gene transfer efficiency. Genetic and/or chemical modification of Ad vectors, pseudotyping of Ad5, or use of other Ad serotypes may further broaden Ad virus vector usability in cardiovascular gene therapy.

## 2.2 Adenovirus Vectors and Other Virus Vectors in Vascular Gene Transfer

Adenovirus vectors that are used in cardiovascular gene therapy are mainly nonreplicative first-generation vectors that have their early gene regions E1 and/or E3 deleted. To reduce immunogenicity in second-generation vectors, gene regions E2 and E4 can be removed. In third-generation Ad vectors (i.e., gutless Ad, helper-dependent Ad, or high-capacity Ad), all viral coding sequences except the inverted terminal repeats flanking the Ad genome and packaging signal are deleted. The insertion capacity of first- and second-generation vectors is 8–12 kb, whereas helper-dependent Ad allows an insertion size up to 36 kb. Adenovirus 5 vectors have been shown to transduce cardiomyocytes, endothelial cells, and SMCs after local administration or denudation of endothelial cells.[5–9,44] In rabbit carotid arteries, second-generation Ad5 vectors did not improve gene transfer efficiency compared with first-generation vectors.[45] Instead, third-generation vectors were able to induce stable transgene expression for 8 weeks in rabbit carotids, whereas first-generation Ad5 vectors expressed transgene for 2 weeks.[46] In murine models, helper-dependent Ads have shown reduced inflammatory response in myocardium compared with first-generation vectors,[47] as well as reduced late toxicity after intravenous administration.[48] In hyperlipidemic rabbits, helper-dependent Ad vectors caused less lesion growth and reduced macrophage and lipid accumulation to plaques compared with first-generation Ad vectors.[49]

Despite promising results, most research is currently performed using first-generation Ad vectors because of their ease of large-scale manufacturing with high quantities. Limitations of helper-dependent Ad vectors are their less efficient virus production compared with first-generation Ad vectors and helper Ad contamination in the produced virus stocks.[50] Helper-dependent Ad5 vectors have also been suggested to induce similar acute toxicity as first-generation Ad5 vectors.[51] Generating helper-dependent Ad vectors from serotypes other than Ad5 could be beneficial for vascular gene transfer owing to their reduced late toxicity. Adenoviruses vectors from *Adenoviridae* subgroups B and D (Ad35 and Ad49) showed increased gene transfer efficiency of vascular cells compared with Ad5 after ex vivo delivery to coronary artery bypass graft tissues.[52] Reduced immunogenicity has also been observed in mice after intravenous administration of nonhuman Ad vectors.[53]

Because of the immunogenic nature of the Ad vectors and their transient gene expression, other viral gene transfer vectors have been used in gene therapy trials for CVDs. Briefly, adeno-associated virus (AAV) allows longer transgene expression in target tissues and is considered nonpathogenic and less immunogenic than first-generation Ad5 vectors. AAV vectors have been used especially in myocardial gene transfer, and have shown high transduction efficiencies of cardiomyocytes.[54] Disadvantages of AAV vectors compared with Ad vectors are their less efficient vector production and limited cloning capability (<5 kb). Besides AAV and Ad vectors, lentivirus vectors (cloning capacity <8 kb) have been used in cardiovascular gene therapy because of their ability to transduce nondividing cells and their ability to induce long-term transgene expression. Higher transgene expression efficiency of primary SMCs has been detected by lentivirus vectors in comparison to Ad5 or AAV serotypes 2–6.[55] Other than their integration capability, and therefore safety concerns, a restriction in the use of lentivirus vectors have been in production of adequate amounts of the vectors for studies in large-animal models. For this reason, most applications have used intramyocardial injections instead of e.g., coronary perfusion.[56]

## 2.3 Adenovirus Vectors and Vascular Targeting Strategies

To enhance the gene transfer efficiency of Ad vectors, multiple targeting strategies have been applied. In transductional targeting, the vector is modified genetically and/or chemically. The aims are to inhibit interaction of the Ad capsid with its receptors (CAR, integrins, and HSPGs), retarget the vectors to other cellular receptors, and reduce immune responses as well as dose-dependent toxicity. Examples of vascular targeting strategies by Ad5 vectors are presented in Table 1. By pseudotyping, fiber protein of Ad5 can be changed to those of other Ad serotypes. Substitution of Ad5 fiber knob to alternative serotypes results in altered vascular tropism. With Ad16 chimera, increased transduction efficiency is observed in SMCs and endothelial cells and in human saphenous vein segments ex vivo.[57] Pseudotyping of Ad5 with Ad19p and Ad37 results in reduced transduction of hepatocytes and increased vascular tropism.[58] Transduction of SMCs was shown to be enhanced by substituting Ad5 fiber and penton base with that of Ad35.[59] Adenovirus capsids can also be modified by selective mutation or inserting ligands, e.g., targeting peptides. Mutation of Ad5 knob was shown to reduce acute toxicity compared with unmodified Ad5.[60] Mutation of both CAR and HSPG binding sites of Ad5 also reduced

**Table 1 Examples of Vascular Retargeting Strategies of Ad5 Vectors**

| Targeting Strategy | Benefit in Vascular Gene Transfer |
|---|---|
| A. Pseudotyping<br>e.g., Ad5/16, Ad5/19p, Ad5/35, Ad5/37 | Enhanced vascular tropism |
| B. Genetic modification | |
| i. Insertion of EC or SMC-binding peptides | i. Enhanced EC or SMC transduction efficiency |
| ii. Selective mutation of CAR, integrin, or HSPG binding sites | ii. Reduced immunogenicity and liver targeting after systemic delivery |
| C. Chemical modification | |
| i. Coating: e.g., PEG, HPMA, PAMAM, lipids, pDMAEMA, chitosan | i. Reduced immunogenicity and liver targeting after systemic delivery |
| ii. Retargeting: Ligands, e.g., EGF, FGF2, transferrin, magnetic particles, cyclic RGD, CGKRK, EGF-mimetic peptide GE11. Antibodies, e.g., selectins, ACE, VEGFR, integrins. | ii. Enhanced EC and/or SMC transduction efficiency |

ACE, Angiotensin-converting enzyme; EC, endothelial cell; EGF, epidermal growth factor; FGF2, fibroblast growth factor 2; PAMAM, polyamidoamine; pDMAEMA, poly(2-(dimethylamino)ethylmethacrylate; PEG, polyethylene glycol; pHPMA, poly-*N*-hydroxy-propylmethacrylamide; SMC, smooth muscle cell; VEGFR, vascular endothelial growth factor.

liver transduction. Insertion of targeting peptides to Ad5 fiber, such as RGD and endothelial cell or SMC-targeting peptides, has been shown to enhance vascular tropism.[61,62] Similarly, peptide-targeted Ad5/19p chimera vectors were able to target cardiomyocytes.[63]

Polymers, bispecific molecules, magnetic particles, microbeads, and lipids have been used to retarget Ad vectors. In murine models, antibodies against the Ad fiber knob and endothelial targeting peptides (Tie2, vascular endothelial growth factor-2 [VEGFR2], and integrin) were shown to retarget Ad vectors to tumors.[64] Angiotensin-converting enzyme antibody instead was shown to target Ad vectors to pulmonary vascular endothelium.[65] Enhanced transduction efficiencies of endothelial cells and SMCs were detected after coating Ad5 vectors with CAR protein and RGD peptides.[3] Besides bifunctional molecules, coating Ad vectors with polymers, e.g., polyethylene glycol (PEG), altered tropism and decreased the immunogenicity of Ad5 vectors after intravenous delivery.[66–68] In murine models, coating Ad5 vector with PEG or poly-*N*-hydroxy-propylmethacrylamide (pHPMA) moieties and further targeting with E-selectin antibody increased vascular tropism after intravenous delivery.[69] Conjugation of murine epidermal growth factor (EGF) to HPMA polymer[70] or EGF-mimetic peptide to polyamidoamine (PAMAM) dendrimer[71] retargeted polymer-coated Ad5 vectors to EGFR-positive cells. Similarly, fibroblast growth factor 2 (FGF2) was used to target polymer-coated Ad5 vectors to FGFR-positive cells.[72,73] Microbubbles and magnetic particles have also been used to retarget Ad vectors. Adenovirus 5/3 vector coated with microbubbles and VEGFR2, integrins, or selectins showed increased tumor targeting after intravenous administration.[74] Successful targeting of Ad vectors with magnetic particles/epicardial magnet was also shown in an acute myocardial infarction model in rats.[75] Besides total shielding of Ad vectors, thiol-based chemistry has been used for site-specific attachment of ligands to Ad5 capsid.[68,76–78] Despite significant advancements in Ad retargeting, most modified Ad vectors have not yet made the transition from small- to large-animal models or clinical trials.

## 3. Animal Models for Cardiovascular Gene Transfer

Genetically modified mice using various transgenic and knockout models have provided researchers with several powerful experimental settings for studying the effects of single genes on CVD. However, for preclinical efficacy, safety, and toxicology studies, larger animal models are usually needed before entering clinical testing. Animal models for CVD are listed in Table 2.

### 3.1 Mouse Models

A large number of transgenic and knockout mouse models have been developed for CVD.[79] Because mice are relatively cheap and easy to handle, most early-phase animal studies are conducted in murine models. Apolipoprotein E and low-density lipoprotein (LDL) receptor knockout mice have become standard models for hyperlipidemia and atherosclerosis studies.[80] Various other knockout models in which specific apolipoproteins, growth factors, cytokines, or extracellular matrix components are deleted have also been developed for CVD research. Despite the small size, several surgical applications are possible, including coronary artery ligation for myocardial infarction models and reperfusion injury studies.[81] Carotid artery denudation and collar models have been used for restenosis and atherosclerosis studies.[82] Peripheral skeletal muscle ischemia using ligation of femoral arteries is a common screening model for proangiogenic factors.[83] Heart failure induced by transaortic banding or angiotensin infusion is commonly used in heart failure studies.[84] More recently, so-called vulnerable plaques that lead to acute plaque rupture and thrombosis have been modeled in mice as well.[82]

In general, mouse models allow adequate numbers of experimental animals per study group and various interventions based on systemic gene delivery via tail vein, intraperitoneal, or periorbital delivery. Also, intramuscular applications in peripheral skeletal muscles, direct intramyocardial injections, and intraretinal injections are feasible. However, because of the small size of tissues and organs, significant damage can be caused by local injection techniques. Systemic delivery of Ad usually leads to transduction of liver, which can modify treatment effects and cause immune responses. Nude mice can be used to avoid immune reactions and prolong Ad gene expression in target organs. However, the relevance of these models to human clinical situation remains unclear.

### 3.2 Rat Models

Like mice, rats are commonly used in CVD gene transfer studies. They are especially useful for carotid artery restenosis studies and myocardial ischemia and infarction studies using coronary ligation models.[85] Also, transaortic banding and hypertensive models are commonly used for heart failure studies.[86] Femoral artery ligation is commonly

**Table 2 Animal Models for Cardiovascular Gene Therapy**

| Mouse | Rat | Rabbit | Dog | Pig |
|---|---|---|---|---|
| Genetically modified models for hyperlipidemia and atherosclerosis<br>Surgical models<br>• Myocardial ischemia, reperfusion injury<br>• Peripheral skeletal muscle ischemia<br>• Restenosis; denudation and collar model<br>• Atherosclerosis and vulnerable plaques<br>• Heart failure<br>• Vein graft stenosis | Some genetically modified models available<br>Surgical models<br>• Myocardial infarction; reperfusion injury<br>• Restenosis<br>• Peripheral skeletal muscle ischemia<br>• Heart failure<br>• Vein graft stenosis | Naturally occurring genetic mutations for hyperlipidemia<br>Surgical models<br>• Myocardial ischemia and reperfusion injury<br>• Restenosis, in-stent restenosis<br>• Peripheral skeletal muscle ischemia<br>• Vein graft stenosis | Naturally occurring genetic mutation for hemophilia<br>Surgical and catheter interventions<br>• Myocardial infarction, reperfusion injury<br>• Ameroid models<br>• Catheter-mediated coil models for coronary arteries for ischemia<br>• Restenosis; in-stent restenosis<br>• Heart failure | Naturally occurring genetic mutations for hyperlipidemia<br>Surgical and catheter interventions<br>• Myocardial infarction; reperfusion injury<br>• Ameroid models<br>• Catheter-mediated coil models for coronary arteries for ischemia<br>• Vein graft stenosis<br>• Modification of lymphatic vasculature |

used for peripheral skeletal muscle ischemia studies. Because of the larger size compared with mice, rats provide an easier model for targeted gene transfer: for example, in heart, with less local damage caused by needle injections and other mechanical manipulations. However, rats cannot be made hyperlipidemic without extensive hormonal and dietary modifications. This limits the usefulness of rat models in studies related to atherosclerosis and vascular pathologies other than restenosis and in-stent restenosis. Currently, some genetically modified rat models have become available for CVD research that should improve the value of rat models in CVD gene transfer studies.[87]

## 3.3 Rabbit Models

Rabbits have been most commonly used for atherosclerosis and hyperlipidemia studies owing to the easy induction of high cholesterol levels with dietary means.[6] Rabbits also have been useful in carotid and femoral artery restenosis and in-stent restenosis studies, vein graft studies,[88] and myocardial and peripheral skeletal muscle ischemia studies after local vascular manipulations.[89,90] Because of the larger size, rabbits allow the use of several similar imaging modalities as used in the clinics. Ultrasound, magnetic resonance imaging, and positron emission tomography studies are useful in characterizing tissue conditions after local or systemic gene transfers. Also, some naturally occurring gene defects such as Watanabe heritable hyperlipidemic (WHHL) rabbits with LDL receptor mutations have been very useful in atherosclerosis, restenosis, and vulnerable plaque studies.[91] Also, gene therapy for LDL receptor deficiency has been efficiently tested in a WHHL rabbit model.[92] Rabbits have also been used for CVD gene therapy safety, biodistribution, and toxicology studies because the dosing and delivery of the vectors better resemble those used in human clinical studies.[93]

## 3.4 Dog Models

Dogs have been used less frequently in CVD gene transfer studies for ethical, practical, and logistic reasons. However, especially in cardiac ischemia reperfusion and infarction modeling, dogs are useful because the dog heart closely resembles the human heart and is large enough to allow intracoronary and intracardiac manipulations using the same equipment as that used in the clinics.[94] Most commonly used ischemia models in the heart are the ameroid constrictor and coronary ligation techniques, which lead to predictable ischemia, compromised heart function, and pathologies that resemble human clinical conditions.[94,95] Also, dogs have been used in hemophilia studies because a naturally occurring clotting factor gene defect is available and offers an excellent model for preclinical safety, efficacy, and toxicity studies.

## 3.5 Pig Models

Pigs have commonly been used for myocardial infarction, ischemia reperfusion, and in-stent restenosis studies. Pig heart is similar to human heart and the size allows the use of the same catheter-based interventions as those used in the clinics. All imaging modalities are available for pigs. Commonly used models are the ameroid constrictor and coils that are inserted in coronary arteries using surgical or catheter techniques.[10,96] Also, open transthoracic surgical models have been commonly used in pigs. A limitation of pig model is the difficulty in making pigs hyperlipidemic and atherosclerotic because of their efficient lipid and lipoprotein metabolism. Pig myocardium is also sensitive to ventricular fibrillation, which leads to increased mortality rates in cardiac ischemia models. Currently, efficient anti-arrhythmic drug cocktails have significantly improved the situation. Pigs allow easy, affordable, long-term follow-up, which is important in CVD gene therapy for myocardial infarction and heart failure. Preclinical safety, biodistribution, and toxicology studies are commonly done in pigs before entering clinical trials because gene vector dosage and delivery routes are similar to those used in human CVD gene transfer applications.[97] A new pig model has also been developed for gene therapy of lymphatic vessel disorders.[98]

## 4. Conclusions

Adenoviruses are useful vectors for efficient transient gene transfer applications in CVD. These include myocardial infarction, proangiogenic studies, restenosis, in-stent restenosis, vein graft stenosis, peripheral ischemia, and heart failure studies. Transductional and transcriptional targeting of Ad vectors to specific cardiovascular tissues via systemic delivery has not yet been successful and requires the use of high vector doses. Most applications currently rely on direct applications into treated tissue with catheters, needle injections, or equivalent techniques. Mouse models are useful for proof-of-principle and large-scale screening studies. Availability of several transgenic and knockout models greatly enhances possibilities of obtaining physiologically significant results using gene transfer applications. Final preclinical efficacy, safety, biodistribution, and toxicology studies are usually performed in larger animals because they produce a more realistic outcome regarding the usefulness of CVD gene transfer in terms of human clinical applications.

## Acknowledgments

This work was supported by grants from the Finnish Academy Center of Excellence, Academy of Finland, Kuopio University Hospital, an ERC Advanced Grant, the Finnish Cultural Foundation, and the Ella and Georg Ehrnrooth Foundation.

## References

1. Fechner H, Haack A, Wang H, et al. Expression of coxsackie adenovirus receptor and alphav-integrin does not correlate with adenovector targeting in vivo indicating anatomical vector barriers. *Gene Ther* 1999;**6**(9):1520–35.
2. Nalbantoglu J, Pari G, Karpati G, Holland PC. Expression of the primary coxsackie and adenovirus receptor is downregulated during skeletal muscle maturation and limits the efficacy of adenovirus-mediated gene delivery to muscle cells. *Hum Gene Ther* 1999;**10**(6): 1009–19.

3. Krom YD, Gras JC, Frants RR, et al. Efficient targeting of adenoviral vectors to integrin positive vascular cells utilizing a CAR-cyclic RGD linker protein. *Biochem Biophys Res Commun* 2005;**338**(2):847–54.
4. Noutsias M, Fechner H, de Jonge H, et al. Human coxsackie-adenovirus receptor is colocalized with integrins alpha(v)beta(3) and alpha(v)beta(5) on the cardiomyocyte sarcolemma and upregulated in dilated cardiomyopathy: implications for cardiotropic viral infections. *Circulation* 2001;**104**(3):275–80.
5. Lemarchand P, Jones M, Yamada I, Crystal RG. In vivo gene transfer and expression in normal uninjured blood vessels using replication-deficient recombinant adenovirus vectors. *Circ Res* 1993;**72**(5):1132–8.
6. Gruchala M, Bhardwaj S, Pajusola K, et al. Gene transfer into rabbit arteries with adeno-associated virus and adenovirus vectors. *J Gene Med* 2004;**6**(5):545–54.
7. Qian HS, Channon K, Neplioueva V, et al. Improved adenoviral vector for vascular gene therapy: beneficial effects on vascular function and inflammation. *Circ Res* 2001;**88**(9):911–7.
8. Laitinen M, Makinen K, Manninen H, et al. Adenovirus-mediated gene transfer to lower limb artery of patients with chronic critical leg ischemia. *Hum Gene Ther* 1998;**9**(10):1481–6.
9. Sasano T, Kikuchi K, McDonald AD, Lai S, Donahue JK. Targeted high-efficiency, homogeneous myocardial gene transfer. *J Mol Cell Cardiol* 2007;**42**(5):954–61.
10. Rutanen J, Rissanen TT, Markkanen JE, et al. Adenoviral catheter-mediated intramyocardial gene transfer using the mature form of vascular endothelial growth factor-D induces transmural angiogenesis in porcine heart. *Circulation* 2004;**109**(8):1029–35.
11. Wickham TJ, Mathias P, Cheresh DA, Nemerow GR. Integrins alpha v beta 3 and alpha v beta 5 promote adenovirus internalization but not virus attachment. *Cell* 1993;**73**(2):309–19.
12. Dechecchi MC, Melotti P, Bonizzato A, Santacatterina M, Chilosi M, Cabrini G. Heparan sulfate glycosaminoglycans are receptors sufficient to mediate the initial binding of adenovirus types 2 and 5. *J Virol* 2001;**75**(18):8772–80.
13. Li E, Brown SL, Stupack DG, Puente XS, Cheresh DA, Nemerow GR. Integrin alpha(v) beta1 is an adenovirus coreceptor. *J Virol* 2001;**75**(11):5405–9.
14. Lutschg V, Boucke K, Hemmi S, Greber UF. Chemotactic antiviral cytokines promote infectious apical entry of human adenovirus into polarized epithelial cells. *Nat Commun* 2011;**2**:391.
15. Kotha PL, Sharma P, Kolawole AO, et al. Adenovirus entry from the apical surface of polarized epithelia is facilitated by the host innate immune response. *PLoS Pathog* 2015;**11**(3):e1004696.
16. Meier O, Greber UF. Adenovirus endocytosis. *J Gene Med* 2004;**6**(Suppl. 1):S152–63.
17. Meier O, Boucke K, Hammer SV, et al. Adenovirus triggers macropinocytosis and endosomal leakage together with its clathrin-mediated uptake. *J Cell Biol* 2002;**158**(6):1119–31.
18. Burckhardt CJ, Suomalainen M, Schoenenberger P, Boucke K, Hemmi S, Greber UF. Drifting motions of the adenovirus receptor CAR and immobile integrins initiate virus uncoating and membrane lytic protein exposure. *Cell Host Microbe* 2011;**10**(2):105–17.
19. Bremner KH, Scherer J, Yi J, Vershinin M, Gross SP, Vallee RB. Adenovirus transport via direct interaction of cytoplasmic dynein with the viral capsid hexon subunit. *Cell Host Microbe* 2009;**6**(6):523–35.
20. Suomalainen M, Nakano MY, Keller S, Boucke K, Stidwill RP, Greber UF. Microtubule-dependent plus- and minus end-directed motilities are competing processes for nuclear targeting of adenovirus. *J Cell Biol* 1999;**144**(4):657–72.
21. Scherer J, Vallee RB. Conformational changes in the adenovirus hexon subunit responsible for regulating cytoplasmic dynein recruitment. *J Virol* 2015;**89**(2):1013–23.

22. Strunze S, Engelke MF, Wang IH, et al. Kinesin-1-mediated capsid disassembly and disruption of the nuclear pore complex promote virus infection. *Cell Host Microbe* 2011;**10**(3):210–23.
23. Greber UF, Suomalainen M, Stidwill RP, Boucke K, Ebersold MW, Helenius A. The role of the nuclear pore complex in adenovirus DNA entry. *EMBO J* 1997;**16**(19): 5998–6007.
24. Zhang Y, Chirmule N, Gao GP, et al. Acute cytokine response to systemic adenoviral vectors in mice is mediated by dendritic cells and macrophages. *Mol Ther* 2001; **3**(5 Pt. 1):697–707.
25. Shayakhmetov DM, Gaggar A, Ni S, Li ZY, Lieber A. Adenovirus binding to blood factors results in liver cell infection and hepatotoxicity. *J Virol* 2005;**79**(12):7478–91.
26. Lieber A, He CY, Meuse L, et al. The role of kupffer cell activation and viral gene expression in early liver toxicity after infusion of recombinant adenovirus vectors. *J Virol* 1997;**71**(11):8798–807.
27. Piccolo P, Annunziata P, Mithbaokar P, Brunetti-Pierri N. SR-A and SREC-I binding peptides increase HDAd-mediated liver transduction. *Gene Ther* 2014;**21**(11):950–7.
28. Tao N, Gao GP, Parr M, et al. Sequestration of adenoviral vector by kupffer cells leads to a nonlinear dose response of transduction in liver. *Mol Ther* 2001;**3**(1):28–35.
29. Cichon G, Boeckh-Herwig S, Schmidt HH, et al. Complement activation by recombinant adenoviruses. *Gene Ther* 2001;**8**(23):1794–800.
30. Tian J, Xu Z, Smith JS, Hofherr SE, Barry MA, Byrnes AP. Adenovirus activates complement by distinctly different mechanisms in vitro and in vivo: indirect complement activation by virions in vivo. *J Virol* 2009;**83**(11):5648–58.
31. Stone D, Liu Y, Shayakhmetov D, Li ZY, Ni S, Lieber A. Adenovirus-platelet interaction in blood causes virus sequestration to the reticuloendothelial system of the liver. *J Virol* 2007;**81**(9):4866–71.
32. Othman M, Labelle A, Mazzetti I, Elbatarny HS, Lillicrap D. Adenovirus-induced thrombocytopenia: the role of von Willebrand factor and P-selectin in mediating accelerated platelet clearance. *Blood* 2007;**109**(7):2832–9.
33. Kalyuzhniy O, Di Paolo NC, Silvestry M, et al. Adenovirus serotype 5 hexon is critical for virus infection of hepatocytes in vivo. *Proc Natl Acad Sci USA* 2008;**105**(14):5483–8.
34. Waddington SN, McVey JH, Bhella D, et al. Adenovirus serotype 5 hexon mediates liver gene transfer. *Cell* 2008;**132**(3):397–409.
35. Vigant F, Descamps D, Jullienne B, et al. Substitution of hexon hypervariable region 5 of adenovirus serotype 5 abrogates blood factor binding and limits gene transfer to liver. *Mol Ther* 2008;**16**(8):1474–80.
36. Ma J, Duffy MR, Deng L, et al. Manipulating adenovirus hexon hypervariable loops dictates immune neutralisation and coagulation factor X-dependent cell interaction in vitro and in vivo. *PLoS Pathog* 2015;**11**(2):e1004673.
37. Xu Z, Qiu Q, Tian J, et al. Coagulation factor X shields adenovirus type 5 from attack by natural antibodies and complement. *Nat Med* 2013;**19**(4):452–7.
38. Parker AL, Waddington SN, Nicol CG, et al. Multiple vitamin K-dependent coagulation zymogens promote adenovirus-mediated gene delivery to hepatocytes. *Blood* 2006;**108**(8):2554–61.
39. Sumida SM, Truitt DM, Lemckert AA, et al. Neutralizing antibodies to adenovirus serotype 5 vaccine vectors are directed primarily against the adenovirus hexon protein. *J Immunol* 2005;**174**(11):7179–85.
40. Gahery-Segard H, Farace F, Godfrin D, et al. Immune response to recombinant capsid proteins of adenovirus in humans: antifiber and anti-penton base antibodies have a synergistic effect on neutralizing activity. *J Virol* 1998;**72**(3):2388–97.

41. Carlisle RC, Di Y, Cerny AM, et al. Human erythrocytes bind and inactivate type 5 adenovirus by presenting coxsackie virus-adenovirus receptor and complement receptor 1. *Blood* 2009;**113**(9):1909–18.
42. Seiradake E, Henaff D, Wodrich H, et al. The cell adhesion molecule "CAR" and sialic acid on human erythrocytes influence adenovirus in vivo biodistribution. *PLoS Pathog* 2009;**5**(1):e1000277.
43. Cichon G, Boeckh-Herwig S, Kuemin D, et al. Titer determination of Ad5 in blood: a cautionary note. *Gene Ther* 2003;**10**(12):1012–7.
44. Puumalainen AM, Vapalahti M, Agrawal RS, et al. Beta-galactosidase gene transfer to human malignant glioma in vivo using replication-deficient retroviruses and adenoviruses. *Hum Gene Ther* 1998;**9**(12):1769–74.
45. Wen S, Schneider DB, Driscoll RM, Vassalli G, Sassani AB, Dichek DA. Second-generation adenoviral vectors do not prevent rapid loss of transgene expression and vector DNA from the arterial wall. *Arterioscler Thromb Vasc Biol* 2000;**20**(6):1452–8.
46. Wen S, Graf S, Massey PG, Dichek DA. Improved vascular gene transfer with a helper-dependent adenoviral vector. *Circulation* 2004;**110**(11):1484–91.
47. Fleury S, Driscoll R, Simeoni E, et al. Helper-dependent adenovirus vectors devoid of all viral genes cause less myocardial inflammation compared with first-generation adenovirus vectors. *Basic Res Cardiol* 2004;**99**(4):247–56.
48. Schiedner G, Morral N, Parks RJ, et al. Genomic DNA transfer with a high-capacity adenovirus vector results in improved in vivo gene expression and decreased toxicity. *Nat Genet* 1998;**18**(2):180–3.
49. Jiang B, Qian K, Du L, Luttrell I, Chitaley K, Dichek DA. Helper-dependent adenovirus is superior to first-generation adenovirus for expressing transgenes in atherosclerosis-prone arteries. *Arterioscler Thromb Vasc Biol* 2011;**31**(6):1317–25.
50. Alba R, Bosch A, Chillon M. Gutless adenovirus: last-generation adenovirus for gene therapy. *Gene Ther* 2005;**12**(Suppl. 1):S18–27.
51. Muruve DA, Cotter MJ, Zaiss AK, et al. Helper-dependent adenovirus vectors elicit intact innate but attenuated adaptive host immune responses in vivo. *J Virol* 2004;**78**(11):5966–72.
52. Dakin RS, Parker AL, Delles C, Nicklin SA, Baker AH. Efficient transduction of primary vascular cells by the rare adenovirus serotype 49 vector. *Hum Gene Ther* 2015;**26**(5):312–9.
53. Lopez-Gordo E, Podgorski II, Downes N, Alemany R. Circumventing antivector immunity: potential use of nonhuman adenoviral vectors. *Hum Gene Ther* 2014;**25**(4):285–300.
54. Bish LT, Morine K, Sleeper MM, et al. Adeno-associated virus (AAV) serotype 9 provides global cardiac gene transfer superior to AAV1, AAV6, AAV7, and AAV8 in the mouse and rat. *Hum Gene Ther* 2008;**19**(12):1359–68.
55. Dishart KL, Denby L, George SJ, et al. Third-generation lentivirus vectors efficiently transduce and phenotypically modify vascular cells: implications for gene therapy. *J Mol Cell Cardiol* 2003;**35**(7):739–48.
56. Wolfram JA, Donahue JK. Gene therapy to treat cardiovascular disease. *J Am Heart Assoc* 2013;**2**(4):e000119.
57. Havenga MJ, Lemckert AA, Grimbergen JM, et al. Improved adenovirus vectors for infection of cardiovascular tissues. *J Virol* 2001;**75**(7):3335–42.
58. Denby L, Work LM, Graham D, et al. Adenoviral serotype 5 vectors pseudotyped with fibers from subgroup D show modified tropism in vitro and in vivo. *Hum Gene Ther* 2004;**15**(11):1054–64.
59. Parker AL, White KM, Lavery CA, Custers J, Waddington SN, Baker AH. Pseudotyping the adenovirus serotype 5 capsid with both the fibre and penton of serotype 35 enhances vascular smooth muscle cell transduction. *Gene Ther* 2013;**20**(12):1158–64.

60. Coughlan L, Vallath S, Gros A, et al. Combined fiber modifications both to target alpha(v) beta(6) and detarget the coxsackievirus-adenovirus receptor improve virus toxicity profiles in vivo but fail to improve antitumoral efficacy relative to adenovirus serotype 5. *Hum Gene Ther* 2012;**23**(9):960–79.
61. Nicklin SA, Von Seggern DJ, Work LM, et al. Ablating adenovirus type 5 fiber-CAR binding and HI loop insertion of the SIGYPLP peptide generate an endothelial cell-selective adenovirus. *Mol Ther* 2001;**4**(6):534–42.
62. Work LM, Nicklin SA, Brain NJ, et al. Development of efficient viral vectors selective for vascular smooth muscle cells. *Mol Ther* 2004;**9**(2):198–208.
63. Nicol CG, Denby L, Lopez-Franco O, et al. Use of in vivo phage display to engineer novel adenoviruses for targeted delivery to the cardiac vasculature. *FEBS Lett* 2009;**583**(12): 2100–7.
64. Haisma HJ, Kamps GK, Bouma A, et al. Selective targeting of adenovirus to alphavbeta3 integrins, VEGFR2 and Tie2 endothelial receptors by angio-adenobodies. *Int J Pharm* 2010;**391**(1–2):155–61.
65. Reynolds PN, Nicklin SA, Kaliberova L, et al. Combined transductional and transcriptional targeting improves the specificity of transgene expression in vivo. *Nat Biotechnol* 2001;**19**(9):838–42.
66. Croyle MA, Chirmule N, Zhang Y, Wilson JM. PEGylation of E1-deleted adenovirus vectors allows significant gene expression on readministration to liver. *Hum Gene Ther* 2002;**13**(15):1887–900.
67. Fisher KD, Stallwood Y, Green NK, Ulbrich K, Mautner V, Seymour LW. Polymer-coated adenovirus permits efficient retargeting and evades neutralising antibodies. *Gene Ther* 2001;**8**(5):341–8.
68. Prill JM, Espenlaub S, Samen U, et al. Modifications of adenovirus hexon allow for either hepatocyte detargeting or targeting with potential evasion from kupffer cells. *Mol Ther* 2011;**19**(1):83–92.
69. Bachtarzi H, Stevenson M, Subr V, Ulbrich K, Seymour LW, Fisher KD. Targeting adenovirus gene delivery to activated tumour-associated vasculature via endothelial selectins. *J Control Release* 2011;**150**(2):196–203.
70. Morrison J, Briggs SS, Green N, et al. Virotherapy of ovarian cancer with polymer-cloaked adenovirus retargeted to the epidermal growth factor receptor. *Mol Ther* 2008;**16**(2):244–51.
71. Vetter A, Virdi KS, Espenlaub S, et al. Adenoviral vectors coated with PAMAM dendrimer conjugates allow CAR independent virus uptake and targeting to the EGF receptor. *Mol Pharm* 2013;**10**(2):606–18.
72. Lanciotti J, Song A, Doukas J, et al. Targeting adenoviral vectors using heterofunctional polyethylene glycol FGF2 conjugates. *Mol Ther* 2003;**8**(1):99–107.
73. Green NK, Morrison J, Hale S, et al. Retargeting polymer-coated adenovirus to the FGF receptor allows productive infection and mediates efficacy in a peritoneal model of human ovarian cancer. *J Gene Med* 2008;**10**(3):280–9.
74. Warram JM, Sorace AG, Saini R, Borovjagin AV, Hoyt K, Zinn KR. Systemic delivery of a breast cancer-detecting adenovirus using targeted microbubbles. *Cancer Gene Ther* 2012;**19**(8):545–52.
75. Zhang Y, Li W, Ou L, et al. Targeted delivery of human VEGF gene via complexes of magnetic nanoparticle-adenoviral vectors enhanced cardiac regeneration. *PLoS One* 2012;**7**(7):e39490.
76. Corjon S, Wortmann A, Engler T, van Rooijen N, Kochanek S, Kreppel F. Targeting of adenovirus vectors to the LRP receptor family with the high-affinity ligand RAP via combined genetic and chemical modification of the pIX capsomere. *Mol Ther* 2008;**16**(11):1813–24.

77. Kreppel F, Gackowski J, Schmidt E, Kochanek S. Combined genetic and chemical capsid modifications enable flexible and efficient de- and retargeting of adenovirus vectors. *Mol Ther* 2005;**12**(1):107–17.
78. Laakkonen JP, Engler T, Romero IA, et al. Transcellular targeting of fiber- and hexon-modified adenovirus vectors across the brain microvascular endothelial cells in vitro. *PLoS One* 2012;**7**(9):e45977.
79. Getz GS, Reardon CA. Animal models of atherosclerosis. *Arterioscler Thromb Vasc Biol* 2012;**32**(5):1104–15.
80. Lucerna M, Zernecke A, de Nooijer R, et al. Vascular endothelial growth factor-A induces plaque expansion in ApoE knock-out mice by promoting de novo leukocyte recruitment. *Blood* 2007;**109**(1):122–9.
81. Turunen MP, Husso T, Musthafa H, et al. Epigenetic upregulation of endogenous VEGF-A reduces myocardial infarct size in mice. *PLoS One* 2014;**9**(2):e89979.
82. Van der Donckt C, Van Herck JL, Schrijvers DM, et al. Elastin fragmentation in atherosclerotic mice leads to intraplaque neovascularization, plaque rupture, myocardial infarction, stroke, and sudden death. *Eur Heart J* 2014;**36**(17):1049–58.
83. Koponen JK, Kekarainen T, Heinonen SE, et al. Umbilical cord blood-derived progenitor cells enhance muscle regeneration in mouse hindlimb ischemia model. *Mol Ther* 2007;**15**(12):2172–7.
84. Huusko J, Lottonen L, Merentie M, et al. AAV9-mediated VEGF-B gene transfer improves systolic function in progressive left ventricular hypertrophy. *Mol Ther* 2012;**20**(12):2212–21.
85. Hiltunen MO, Turunen MP, Yla-Herttuala S. Gene therapy methods in cardiovascular diseases. *Methods Enzymol* 2002;**346**:311–20.
86. Serpi R, Tolonen AM, Huusko J, et al. Vascular endothelial growth factor-B gene transfer prevents angiotensin II-induced diastolic dysfunction via proliferation and capillary dilatation in rats. *Cardiovasc Res* 2011;**89**(1):204–13.
87. Bry M, Kivela R, Holopainen T, et al. Vascular endothelial growth factor-B acts as a coronary growth factor in transgenic rats without inducing angiogenesis, vascular leak, or inflammation. *Circulation* 2010;**122**(17):1725–33.
88. Puhakka HL, Turunen P, Gruchala M, et al. Effects of vaccinia virus anti-inflammatory protein 35K and TIMP-1 gene transfers on vein graft stenosis in rabbits. *In Vivo* 2005;**19**(3):515–21.
89. Laukkanen MO, Kivela A, Rissanen T, et al. Adenovirus-mediated extracellular superoxide dismutase gene therapy reduces neointima formation in balloon-denuded rabbit aorta. *Circulation* 2002;**106**(15):1999–2003.
90. Korpisalo P, Hytonen JP, Laitinen JT, et al. Capillary enlargement, not sprouting angiogenesis, determines beneficial therapeutic effects and side effects of angiogenic gene therapy. *Eur Heart J* 2011;**32**(13):1664–72.
91. Brasen JH, Leppanen O, Inkala M, et al. Extracellular superoxide dismutase accelerates endothelial recovery and inhibits in-stent restenosis in stented atherosclerotic Watanabe heritable hyperlipidemic rabbit aorta. *J Am Coll Cardiol* 2007;**50**(23):2249–53.
92. Kankkonen HM, Vahakangas E, Marr RA, et al. Long-term lowering of plasma cholesterol levels in LDL-receptor-deficient WHHL rabbits by gene therapy. *Mol Ther* 2004;**9**(4):548–56.
93. Hiltunen MO, Turunen MP, Turunen AM, et al. Biodistribution of adenoviral vector to nontarget tissues after local in vivo gene transfer to arterial wall using intravascular and periadventitial gene delivery methods. *FASEB J* 2000;**14**(14):2230–6.
94. Schaper W, Scholz D. Factors regulating arteriogenesis. *Arterioscler Thromb Vasc Biol* 2003;**23**(7):1143–51.

95. Cai W, Vosschulte R, Afsah-Hedjri A, et al. Altered balance between extracellular proteolysis and antiproteolysis is associated with adaptive coronary arteriogenesis. *J Mol Cell Cardiol* 2000;**32**(6):997–1011.
96. Lahteenvuo JE, Lahteenvuo MT, Kivela A, et al. Vascular endothelial growth factor-B induces myocardium-specific angiogenesis and arteriogenesis via vascular endothelial growth factor receptor-1- and neuropilin receptor-1-dependent mechanisms. *Circulation* 2009;**119**(6):845–56.
97. Rissanen TT, Korpisalo P, Markkanen JE, et al. Blood flow remodels growing vasculature during vascular endothelial growth factor gene therapy and determines between capillary arterialization and sprouting angiogenesis. *Circulation* 2005;**112**(25):3937–46.
98. Lahteenvuo M, Honkonen K, Tervala T, et al. Growth factor therapy and autologous lymph node transfer in lymphedema. *Circulation* 2011;**123**(6):613–20.

# Polymer-Anchored Adenovirus as a Therapeutic Agent for Cancer Gene Therapy

29

*Dayananda Kasala, Chae-Ok Yun*
Department of Bioengineering, College of Engineering, Hanyang University, Seongdong-gu, Seoul, Republic of Korea

## 1. Introduction

Cancer is one of the most serious diseases; it causes large numbers of deaths worldwide every year. Despite the development of sophisticated diagnostic protocols and treatments, the overall survival rate is significantly low, and new strategies are needed to provide better treatment for cancer patients. Over the past two decades, various viral gene delivery systems have been established. Adenovirus (Ad) is a promising candidate for the treatment of various maligned cancers. Specific attention has been paid to increasing Ad's ability to target solid tumors and selectively kill tumor cells without harming normal cells.[1–3] Adenovirus is a nonenveloped virus with a capsid size of 100nm that can carry therapeutic genes. Adenovirus is internalized into cells by binding to the primary Ad receptor, the coxsackie-adenoviral receptor (CAR), and interacting with the cell surface integrins $\alpha_v\beta_3$ and $\alpha_v\beta_5$, which induces Ad endocytosis.[4] Heparin sulfate proteoglycans on the cell surface interact with protein epitopes on the fiber shaft of the virus, acting as additional Ad receptors.[5,6] Adenovirus has several benefits, including high gene transfer efficiency in both dividing and nondividing cells, easy production of high-titer Ad stocks, no risk of insertional mutagenesis, and induction of oncolysis by viral replication.[7–9]

To date, over 2000 gene therapy clinical trials have been performed using viral vectors. Among these, since 1993, Ad vectors have been used in more than 488 protocols and have become the most common gene delivery method for clinical gene therapy.[10] Adenovirus has been modified by deletion of the E1B protein, which inhibits viral replication in normal cells while permitting it in tumor cells. This strategy has been demonstrated as a clinical proof-of-concept for oncolytic virotherapy using ONYX-015. ONYX-015 (dl1520) was one of the first-generation oncolytic Ads used to treat human cancer. This Ad is genetically engineered by deleting the 55-kD gene in the E1B region, resulting in a complete lack of E1B expression that allows it to specifically infect and kill p53-deficient cancer cells without harming normal cells.[11] Numerous phase I and II clinical trials have been performed using ONYX-015 delivered by intratumoral, intravenous, intra-arterial, and intraperitoneal routes in malignant gliomas, head and neck, hepatocellular, pancreatic, ovarian, and metastatic colorectal carcinomas. The results from these trials have demonstrated that ONYX-015 is well tolerated, but it has shown only moderate antitumor efficacy.[1,12]

**Adenoviral Vectors for Gene Therapy.**

Several genetic engineering strategies have been developed to improve the cancer-killing potency of the Ad vector.[13] The various strategies are summarized as follows:[14] (1) insertion of cytotoxic genes such as the suicide gene (herpes simplex virus-thymidine kinase),[15] cytosine deaminase (CD),[16] the tumor-suppressor gene (p53),[17] and tumor necrosis factor (TNF)-related apoptosis-inducing ligand (TRAIL)[18]; (2) overexpression of immune-stimulatory cytokine genes, including granulocyte macrophage colony-stimulating factor, interleukin-2 (IL-2), IL-12, IL-18, and IL-23, interferon-α, etc.[19–27]; (3) downregulation of sequence-specific gene expression by small interfering ribonucleic acid (siRNA) against messenger RNA related to angiogenesis[28,29]; (4) acceleration of cancer cell lysis induced by viral replication using adenosine diphosphate and deletion of the Ad E1B 19-kD gene[30,31]; (5) elimination of the extracellular matrix barrier within the tumor tissue for efficient viral spread (relaxin and decorin).[30,32,33] Despite these efforts, most of the Ad delivered systemically through blood circulation is eliminated within 2 min.[34] Recent human clinical trials using Ad have therefore been strictly limited to local injection against exposed tumors such as head and neck, prostate, and melanoma cancer.[35]

There are many challenges to overcome before the systemic delivery of therapeutic Ad particles can be used to treat metastatic cancer. First, the complexes are frequently neutralized by preexisting Ad-specific neutralizing antibodies (NAbs) in the bloodstream; and thus, a new technology is needed to shield therapeutic Ad from blood-mediated inactivation. Second, Ad must be modified with appropriate material to avoid infecting nonspecific cells, particularly in the liver where therapeutic viruses may be captured by phagocytes.[36,37] Third, selective tumor targeting and complete eradication of tumors are highly desirable. The ability to avoid interactions through the use of inert biomaterials that protect the virus capsid may minimize the adverse side effects. In addition, if the inert materials deshield after the target cells are internalized, the virus can function normally and transfer the genetic materials to the nucleus with high efficacy.[38] This chapter focuses predominantly on the recent developments made in polymer-modified Ad systems to overcome the challenges of long-term and systemic administration, which will ultimately enhance the efficacy and safety of this anticancer therapeutic strategy. After a brief summary of the commonly used polymers, the polymer-modified Ad complex is described in detail with emphasis on the various strategies used and the in vitro and in vivo efficacy of each study.

## 2. Polymer Coating on Adenovirus Surface

The major purposes of the polymer coating on the Ad surface are to increase transgene expression, provide protection from the immune system, and target the tumor site. Moreover, conjugating the surface chemistry of the viral capsid without changing the viral genome is a prerequisite for maintaining the biological activity of therapeutic Ad. Efficient and safe gene transfer carriers are also critically important for gene delivery. Therefore, techniques for the surface modification of Ad have received significant attention and various procedures have been successfully developed to overcome the barriers to systemic administration.

Surface modification approaches are classified in terms of how the polymers interact with the Ad capsid proteins. There are two approaches: "physical engineering through electrostatic interactions," and "chemical modification", in which polymers are covalently anchored to the amine of Ad surface proteins (Figures 1 and 2). Various cationic polymers have been investigated for coating the Ad surface using the engineering through the electrostatic interaction approach. The backbones of cationic polymers contain several amine groups which can induce interactions between the positively charged cationic polymers and the negatively charged Ad surface, leading to the formation of Ad–polymer complex in an aqueous solution.[39] This polymer-coated Ad has a positive surface charge that increases cellular uptake and results in enhanced transgene expression. In addition, these cationic polymers possess tertiary and secondary amines with high buffering capacity which facilitate the therapeutic agents to effectively escape from the endosomes and release the virus into the cytosol via the "proton sponge effect".[40,41] This strategy has several benefits, such as the easy manipulation of

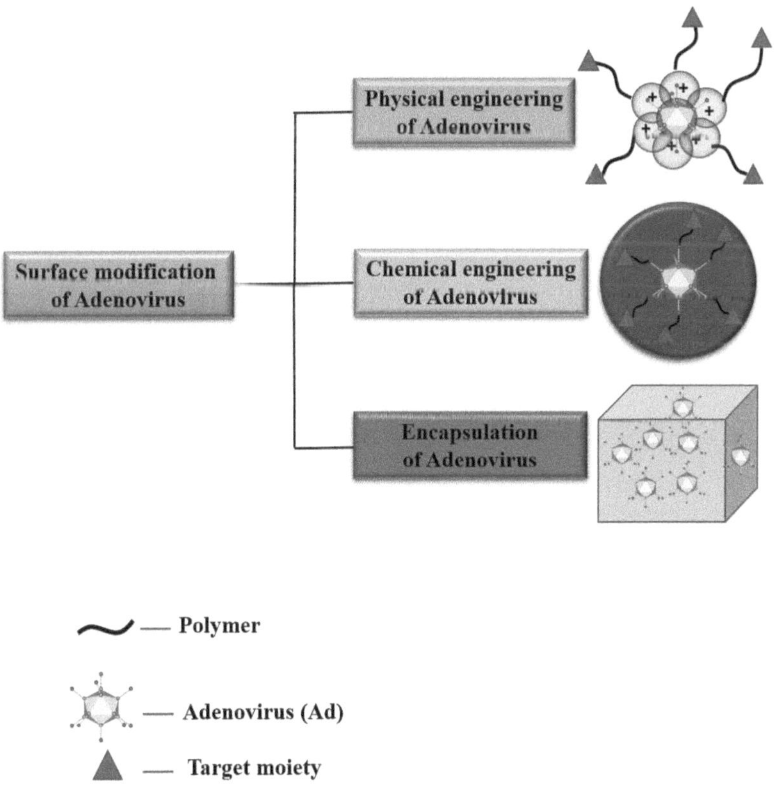


**Figure 1** Approaches to the surface modification of Ad for overcoming the challenge of Ad immunogenicity.

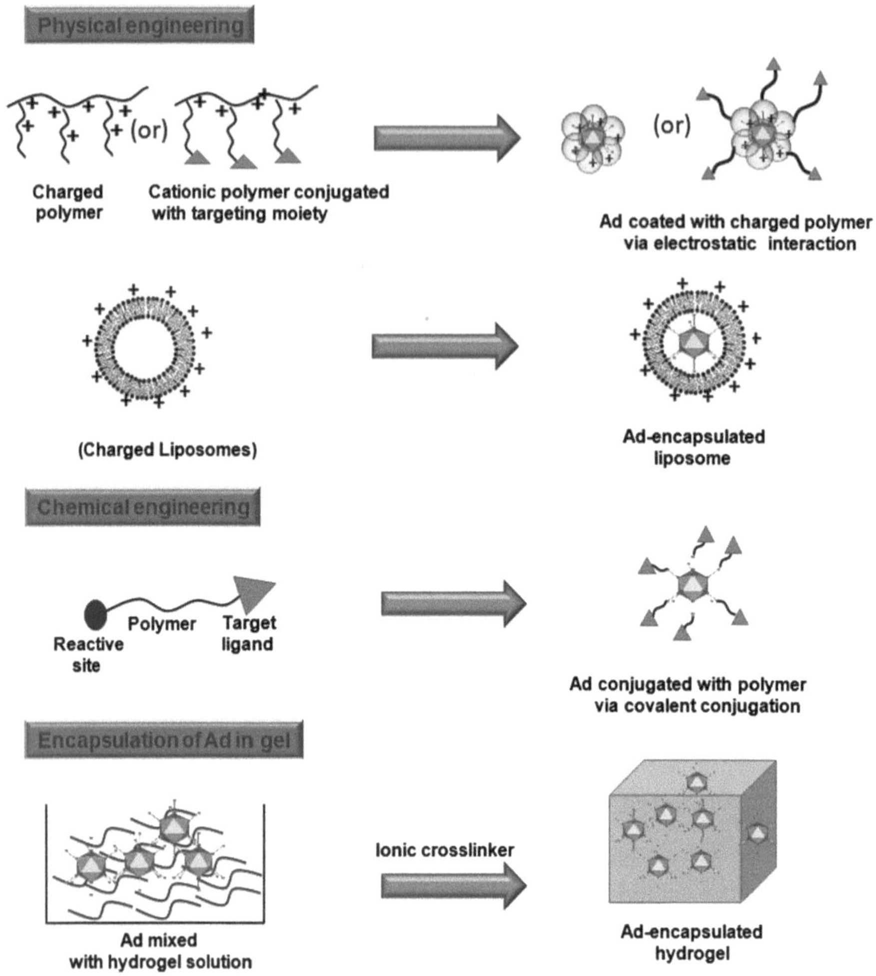


**Figure 2** Overview of various strategies used for surface modification of Ad.

molecular weight and facilitated conjugation of ligands to polymers. It is also easy and straightforward to generate and handle the complexes without the need for stringent conditions. Importantly, polymer-modified Ad maintains its viral activity and intracellular trafficking, whereas transduction efficiency is enhanced through increased cellular binding affinity. However, the main disadvantage of this technique is its nonspecific uptake in cells because of the highly positive surface charge and cytotoxicity of the cationic polymers (e.g., polyethyleneimine (PEI) 25 kDa). It can rapidly dissociate in the bloodstream after intravenous injection owing to the negatively charged serum protein, which has a strong affinity with the positively charged polymer encapsulating Ad. Certain serum proteins can be adsorbed on the Ad–polymer surface, triggering interactions between the positively charged polymer and the membrane receptors of macrophages and monocytes.[42]

Covalent conjugation of polymers to Ad is a good alternative strategy for physical engineering. In this approach, polymers are chemically cross-linked with the Ad surface using coupling agents or cross-linkers. Adenoviruses contain large numbers of free lysines located on the hexon, penton, and fiber proteins, which are used for surface modification via covalent conjugation with bifunctional or heterofunctional polymers.[14,43,44] Compared with the genetic approach, the major advantage of the covalent conjunction approach is that several hundred amino acids can be modified on the capsid in a single step, which is difficult to achieve genetically. Amino reactive functional polymers, such as polyethylene glycol (PEG) and *N*-(2-hydroxypropyl) methacrylamide (HPMA), have been covalently conjugated with Ad. The results have shown high serum stability and enhanced blood circulation time. This approach also prevents antibody neutralization and ultimately reduces unwanted interactions with blood components after systemic administration. However, these techniques have some limitations compared with the noncovalent coating of Ad: (1) The polymer monolayers are permanently linked to the viral vectors, which can interfere with the intracellular mechanisms, resulting in reduced transduction efficacy. Enzymatically cleavable linkers are required to overcome this obstacle. (2) Targeting ligands are required to increase the therapeutic efficacy of Ad because CAR receptor binding site is often masked by the polymer layer. To this end, various cationic polymers, such as poly(L-lysine) (PLL), PEI, poly(amidoamine) (PAMAM) dendrimers, chitosan, and other biodegradable polymers such as arginine-grafted bioreducible polymer (ABP) and methoxy poly(ethylene glycol)-b-poly{*N*-[*N*-(2-aminoethyl)-2-aminoethyl]-L-glutamate} (PNLG), have been used to deliver Ad. In the following sections we discuss the techniques developed to modify Ad to overcome these hurdles.

## 2.1 Poly(ethyleneimine)-Coated Ad

Among the various sizes of PEI that have been developed, PEI (25 kDa) is used as the standard for newly developed cationic polymers because of its high transgene expression.[39] However, its clinical application is limited by its systemic cytotoxicity and low serum stability as a result of its strong binding affinity with serum proteins in vivo. Therefore, nontoxic polymers or moieties have been used to modify PEI. Hans et al. developed a low-molecular-weight PEI cross-linked with bioreducible disulfide bonds (PEI-DEG-*bis*-NPC) and physically engineered it with Ad. Adenovirus complexed with PEI-DEG-*bis*-NPC showed higher transduction efficiency than naked Ad (Table 1).[45] Ad–PEI-DEG-*bis*-NPC complexes promoted the inhibition of tumor growth more efficiently than Ad–PEI25 kDa in vivo. Lee et al. demonstrated that Ad complexed with bile acid-conjugated PEI (DA3) had increased transduction efficiency in both CAR-high and CAR-negative cancer cells.[46] The Ad–DA3 complex was more rapidly internalized into cells than the unmodified Ad. The cellular uptake mechanism revealed that the endocytosis of Ad–DA3 was not mediated primarily by CAR but involved clathrin-, caveolae-, and macropinocytosis-mediated endocytosis. In an aggressive model of HT1080 fibrosarcoma, the oncolytic Ad–DA3 complex showed significantly improved therapeutic effects compared with naked oncolytic Ad.

## Table 1 Overview of Polymers Used for Physical Modification of Ad Surface

| Polymers | Adenovirus | Application | References |
|---|---|---|---|
| Polyethyleneimine cross-linked diethylene glycol *bis*-4-nitrophenyl chloroformate (PEI-DEG-*bis*-NPC) (PDN) | Replication-incompetent Ad encoding β- galactosidase or SDF-1α (Ad-LacZ and Ad-SDF1α) | In vitro and in vivo | 45 |
| Bile acid-conjugated PEI (DA3) | Oncolytic Ad encoding ZFP that is targeted to VEGF (RdB-KOX) | In vitro and in vivo | 46 |
| PEG-grafted -poly-L-lysine (PLL) | Replication-incompetent Ad encoding GFP (ADV) | In vitro | 49 |
| Citraconylated biotin-PEG-PLL | Replication-incompetent Ad encoding GFP (ADV) | In vitro | 50 |
| Poly(diethyleneglycol)-modified L-lysine($K_{60}K^{p}{}_{370}$) | Replication-incompetent Ad encoding firefly luciferase or GFP (AdCMV-fl, AdCMV-GFP) | In vitro and in vivo | 52 |
| Poly(2-hydroxypropyl) methylacrylate conjugated oligo-L-lysine (pHK) | Replication-incompetent Ad encoding GFP (Ad5-GFP) | In vitro | 54 |
| Arginine-grafted bioreducible polymer (ABP) | Replication-incompetent Ad encoding GFP (dl324-ΔE1/GFP) | In vitro | 57 |
| Methoxy poly(ethylene glycol)-b-poly(*N*-[*N*-(2-aminoethyl)-2-aminoethyl]-L-glutamate)(PNLG) | Replication-incompetent Ad encoding GFP or oncolytic Ad encoding short hairpin RNA (shRNA) against interleukin-8 (IL-8) (dE1/GFP or Ad-ΔB7-U6ShIL8) | In vitro and in vivo | 59 |
| Poly(amidoamine) (PAMAM) dendrimers | Replication-incompetent Ad encoding luciferase fusion protein, Alexa Fluor488, enhanced green fluorescent protein, or enhanced green fluorescent and luciferase fusion protein (Ad-Luc, Ad-Alexa488, Ad-EGFP, Ad-EGFPLuc) | In vitro | 61 |

**Table 1 Overview of Polymers Used for Physical Modification of Ad Surface—cont'd**

| Polymers | Adenovirus | Application | References |
|---|---|---|---|
| Poly(amidoamine) (PAMAM) dendrimers | Replication-incompetent Ad encoding hNIS (Ad5-CMV/NI0053) Oncolytic Ad encoding hNIS gene (Ad5-E1/AFP-E3/NIS) | In vitro and in vivo | 62 |
| Liposome; 1,2-dioleoyl-sn-glycero-3-phosphoethanolamine/cholesterol semisuccinate (DOPE:CHEMS) | Replication-incompetent Ad encoding $\beta$- galactosidase (Ad $\beta$-gal) | In vitro and in vivo | 64 |
| Anionic liposome (PPC-AL; PEG-peptide-cholesterol (PPC-AL/PEG-peptide-Chol) | Replication-incompetent Ad encoding firefly luciferase (Ad/FL) | In vitro and in vivo | 66 |
| Cationic liposome; 1,2-dioleoyl-3-trimethylammonium propane/1,2-dioleoyl-sn-glycero-3-phosphoethanolamine (DOTAP/DOPE) | Oncolytic Ad expressing soluble TRAIL gene (pmT-d19/stTR) | In vitro and in vivo | 67 |
| PEG-grafted bioreducible polymer (ABP-g-PEG) | Replication-incompetent Ad encoding GFP (Ad-ΔE1/GFP) | In vitro and in vivo | 68 |
| Chitosan | Replication-incompetent Ad encoding $\beta$- galactosidase (Ad-$\beta$-gal) | In vitro | 84 |

The potent antitumor efficacy of the oncolytic Ad–DA3 complex may be attributable to the consecutive amplification of oncolytic Ad by active replication in tumor tissues after the rapid and effective infection of cancer cells.

## 2.2 Poly(L-lysine)-Coated Ad

Poly(L-lysine) is a linear polypeptide bearing free amine side arms that is able to interact with negatively charged nucleic acids or Ad, leading to the formation of a complex with a high net positive charge via electrostatic interaction. Complexes with a high positive charge often induce aggregation in physiological media, and cationic polymers at high concentrations elicit undesirable cytotoxicity.[47,48] To overcome these limitations, Park et al. studied the transduction efficacy and in vitro serum stability of

Ad coated with PEG grafted and blocked with PLL (PLL-g-PEG and PLL-b-PEG) via electrostatic interaction.[49] The Ad-PLL-g-PEG and Ad-PLL-b-PEG complexes exhibited gradually increased gene expression efficiency, followed by saturated or slightly decreased levels above a certain PLL concentration. Although the Ad–PLL complex showed marginally enhanced gene expression, it was much lower than that of Ad coated with PLL–PEG copolymers. Gene expression in the PLL-b-PEG–coated Ad was about six times higher than in the PLL-g-PEG–coated Ad in the presence of serum. It is possible that PLL-b-PEG is more effective in protecting Ad from spatial access by serum proteins because it produces a more densely PEGylated shielding layer around the Ad surface.

Mok et al.[50] developed a pH-sensitive poly(lysine)-modified Ad (Ad–citraconylated biotin-PEG-PLL) to target the hypoxic tumor microenvironment. The extracellular environment of a tumor has a pH of 6.5 to 7.2,[51] which triggers the signal for the tumor-targeted drug delivery of nanoparticles. In this study, PLL modified with citraconic anhydride caused the amide to link with the negative surface charge of the polymer. This amide bond can be cleaved under acidic conditions resulting in a conversion from negative to positive charge. This charge conversion can enhance the cellular uptake of the nanocomplex and facilitate the release of its cargo which leads to high gene transduction. The surface charge reversal of Ad–citraconylated biotin-PEG-PLL at acidic pH 5.8 enhanced nonspecific cellular uptake compared with pH 7.4. Because the surface charge value of Ad–citraconylated biotin-PEG-PLL at pH 5.8 was still negative at −2.3 mV, a significant change in the surface charge value of Ad complex toward a positive direction could lead to enhanced cellular uptake and gene expression. These results demonstrate the feasibility and importance of modifying the Ad surface to enable Ad–pH-sensitive polymers to be widely applied to solid tumors that are known to have an acidic tumor microenvironment.

In another study, Ads coated with $PLL_{60}$ and poly(diethyleneglycol)-modified L-lysine($K_{60}K^{p}_{370}$) were monitored to measure in vivo transduction efficacy and stability.[52] Exposure to the Ad5-NAb did not ablate the transductional activity of polymer-coated Ad (Ad-$K_{60}K^{p}_{370}$), which suggests that the PEG of $K_{60}K^{p}_{370}$ protected the viral surface, prevented access by the NAb, and enhanced the blood circulation time. Biodistribution studies have shown that the accumulation of $K_{60}K^{p}_{370}$-coated Ad in the lungs is significantly higher than that of naked Ad, despite similar accumulations within the liver, and suggest that $K_{60}K^{p}_{370}$-coated Ad acts differently from naked Ad. The HPMA copolymer has also been used to coat the surface of Ad vectors via electrostatic interaction. *N*-(2-hydroxypropyl) methacrylamide copolymer-modified Ad vectors have been shown to enhance the blood circulation time by decreasing the binding affinity with NAbs, and to reduce the interaction with complement receptor 1 on human erythrocytes.[53] Wang et al.[54] complexed poly(lysine)s conjugated with the HMPA copolymer with Ad5 vectors. The polymer-modified Ad showed higher transduction efficiency than naked Ad in CAR-negative RAW 264.7 cells. The polymer-coated Ad also retained significant transduction efficiency in the presence of NAbs, which suggests that the Ad was well protected and the polymer prevented access by the immune cells. It also has the potential to target CAR receptor-deficient cells and to overcome the systemic delivery problem. Taken together, PLL and its derivatives have the potential to enhance the transduction efficacy of Ad in CAR receptor-deficient cells and confer protection against the immune response.

## 2.3 Bioreducible/Biodegradable Polymer-Coated Ad

To decrease the cytotoxicity of cationic polymers, several bioreducible/biodegradable polymers have been developed and their complexation ability was examined.[55] These bioreducible polymers, which carry a disulfide moiety in the block copolymer, can be degraded in the cytoplasm to release the viral vector. The bioreducible block can be degraded into small fragments of monomer units in the cytoplasm. These fragments are easily excreted from the body, resulting in low systemic cytotoxicity. Arginine, a frequently used amino acid for cell penetrating peptides, has been exploited for grafting onto bioreducible polymers. Arginine contains a highly water-soluble guanidine side-chain functional group, is nonnucleophilic, and has comparatively strong basic characteristics ($pKa \approx 12.5$).[56] Guanidine has been used as a soft cationic site which facilitates the modulation of and interaction with the cell membrane to generate a membrane translocation or nuclear translocation. These reducing conditions can facilitate the efficient release of Ad from the complex and minimize polymer-induced cytotoxicity. Kim et al.[57] generated Ad complexed with an ABP via ionic interaction, which showed sixfold enhanced gene transduction efficacy over naked Ad in cells expressing high or low levels of CAR. These results suggest that the cationic ABP polymer complex mediates a CAR-independent cell entry and can facilitate gene transfer to cells that are resistant to Ad infection or express low levels of CAR protein. Hepatoma-specific oncolytic Ad (YKL-1001) chemically conjugated with ABP enhanced the replication of Ad-mediated hepatoma cell killing compared with naked YKL-1001, indicating that YKL-1001 retains the ability to induce hepatoma-specific cytotoxicity after conjugation with ABP.[58] Arginine-grafted bioreducible polymer-conjugated YKL-1001 reduced the immune response against Ad, increased the blood circulation time by 45-fold, decreased accumulation in the liver, and exhibited strong antitumor effects with minimal liver toxicity after systemic administration. Bioreducible polymers such as ABP, which involve disulfide bonds, are dissociated with glutathione and thioredoxin reductases into nontoxic small molecules in cytosolic reducing environments. Adenovirus conjugated with ABP polymers induces no obvious cytotoxic effects because the ABP is efficiently broken down in the reductive environment of cytosol. Bioreducible ABP polymers support the stability of Ad complexes in the bloodstream because the total concentration of reducing agents in the extracellular environment is low, whereas the concentration in cytosol is high.[55] These characteristics help ABP-conjugated Ad complexes to circulate in the blood for longer. The prolonged circulation time gives the polymer-conjugated Ad complexes a greater opportunity to reach the target cells through enhanced permeability and retention (EPR) effects.

A different approach examined the systemic administration of biodegradable PNLG polymer-coated oncolytic Ad (Ad-ΔB7-U6shIL8).[59] The PNLG-coated Ad (Ad–PNLG) complex showed increased transgene expression in both high and low CAR-expressing cells compared with naked Ad, Ad–ABP, and Ad-PEI. The cells treated with the Ad–PNLG complex also showed a significant killing effect compared with naked Ad, Ad–ABP, and Ad-PEI. Stability results showed that more than 10% of the biological activity of the PNLG-coated Ad complex was preserved over a 13-day period under physiological conditions. In contrast, the activity of naked Ad and Ad–ABP declined to less than 5% after 5 days of incubation. A biodistribution study in HT1080 tumor-bearing mice revealed that the accumulation of Ad–PNLG within tumors was enhanced by 164- and 750-fold compared with naked Ad or Ad–ABP, respectively, after systemic administration. The tumor-to-liver ratio of the mice injected with Ad–ABP (130-fold increase) or Ad–PNLG (1229-fold increase) was significantly higher than the mice injected with naked Ad. These results indicate that the PNLG polymer-complexed Ad remarkably increased transduction efficiency, cancer-killing ability, stability, and reduced nonspecific interactions with cells and anticancer therapeutic efficacy in malignant tumor cells with low or deficient CAR expression. Together, these results demonstrate that bioreducible/biodegradable polymer-coated Ad shows a synergism between the cancer cell-specific cytolysis of oncolytic Ad therapy and the systemic delivery of polymers, thus enabling effective treatment of disseminated metastatic tumors.

## 2.4 Poly(amidoamine) Dendrimer-Coated Ad

Recently, dendrimers, synthetic macromolecules that are highly symmetric and highly branched and have a flexible size range, have been investigated as a nonviral vector. Dendrimers have distinct structural features, including tunable structures, molecular size, easily accessible terminal amine groups, and the ability to encapsulate cargos, which increase their potential as drug carriers.[60] Vetter and colleagues developed a targeting peptide-conjugated PAMAM dendrimer-coated Ad complex for selective treatment of epidermal growth factor receptor (EGFR)-overexpressing tumor cells.[61] A synthetic peptide GE11 (CYHWYGYTPQNVI) is an EGFR-targeting ligand with high receptor affinity without activating the receptor tyrosine kinase. The results showed that PAMAM-coated Ad significantly enhanced transductional efficacy even in the presence of Ad-NAbs. The net positive charge of PAMAM-PEG-GE11–coated Ad showed higher cellular binding and uptake, leading to a 2.3-fold increase in transduction efficiency, specifically in highly EGFR-positive A549 lung cancer cells compared with Ad-PAMAM-PEG-cys. These findings imply that noncovalent charge-based coating of Ad vectors with ligand-equipped dendrimers is a viable strategy for efficient transduction of cells that are otherwise refractory to Ad transduction.

In another study, sodium iodide symporter (NIS)-expressing Ad coated with PAMAM-G5 demonstrated partial protection from NAbs and enhanced transduction efficacy in CAR-negative cells in vitro.[62] The in vivo $^{123}$I scintigraphy of nude mice revealed significantly reduced levels of hepatic transgene expression after intravenous injection of dendrimer-coated Ad5-cytomegalovirus (CMV)/NIS. The low accumulation in the liver resulted in significantly reduced liver toxicity and increased the transduction efficiency of dcAd5-CMV/NIS in hepatoma xenografts. After PAMAM-G5 coating of the replication-selective Ad5-E1/AFP-E3/NIS, a significantly enhanced oncolytic effect was observed after intravenous application (virotherapy). The effect was further increased by additional treatment with a therapeutic dose of $^{131}I$ (radiovirotherapy) and was associated with markedly improved survival. These results demonstrate efficient detargeting of the liver and retargeting to the tumor by Ad vectors after coating with synthetic dendrimers, which exploits the synergies between oncolytic virotherapy and NIS-mediated radionuclide therapy.

## 2.5 Lipid-Modified Ad

Liposomes are widely used as biomimicking carriers for therapeutic agents. They can form vesicular structures from single and multiple phospholipid bilayers, which are capable of encapsulating water insoluble and soluble drugs within the bilayer membrane. For more than two decades, cationic lipids have been used as carriers to transfect cells with plasmid deoxyribonucleic acid (DNA), oligonucleotides, and siRNA. It is possible to further engineer liposomes with cell- or tissue-specific signals by conjugation with targeting moieties such as peptides, antibodies, or small molecules.[63] It is well known that liposomes undergo endocytosis, and the use of pH-sensitive liposomes has been extensively explored in the design of vectors for drug and gene delivery.[64,65] One of the simplest pH-sensitive liposome systems is based on the construction of lipid bilayers consisting of the fusogenic lipid dioleoyl-phosphatidylethanolamine (DOPE) with protonable amphiphilcs such as cholesteryl hemisuccinate (CHEMS).[64] The mechanism underlying the pH-dependent destabilization of this bilayer system is that head group protonation of CHEMS occurs within an acidic environment, causing DOPE molecules to revert from a bilayer (lamellar) to an inverted hexagonal II phase (nonlamellar). To overcome the problem of endosomal virion release from artificially enveloped Ad, Van den Bossche et al. modified Ad by building pH-sensitive lipid bilayer envelopes around virions made of DOPE:CHEMS. This study showed that a high lipid concentration of DOPE:CHEMS (6:4 mM) was required to offer full envelopment of the Ad ($1 \times 10^{10}$ particles), and pH-sensitive enveloped Ad showed similar levels of gene expression to those of naked Ad in many different cell lines. Intracellular trafficking studies showed that both Ad and pH-sensitive enveloped Ad successfully escaped the endosomes and trafficked to the perinuclear region within 3 h of interaction with cells.

Another approach formulated Ad vectors using enzyme-responsive liposome.[66] The enzymatically cleavable PEG-lipids (PPC-Al) consisting of polyethylene glycol–matrix metalloproteinase (PEG/MMP) substrate are sensitive to the protease enzyme type IV collagenases (MMP-2 and MMP-9) and are easily degradable in the tumor microenvironment. The MMP-specific cleavable formulation of PPC-Al-Ad showed higher transduction than naked Ad and noncleavable PEG-lipid-modified Ad vectors in tumor cells. PPC-Al-Ad reduced the innate and adaptive immune response against Ad and resulted in lower liver toxicity than naked Ad. Taken together, these approaches imply that cationic liposome-based Ad complexes can improve antitumor efficacy without toxicity.

## 2.6 Encapsulation of Viral Genomes Using Polymers

Adenovirus genome DNA has been delivered to tumor sites using nonviral vectors to minimize the time and cost of viral mass production. Adenovirus genome DNA delivery systems can effectively synergize oncolytic Ad-mediated high therapeutic efficacy and nonviral vector-mediated systemic delivery. Oncolytic viral genomes first arrive at the tumor foci via systemic blood circulation with the aid of a nonviral carrier, where they translocate into the nucleus and subsequently replicate, generating infectious oncolytic Ad progenies within the tumor cells. These newly generated oncolytic Ads subsequently lyse the cancer cells and infect neighboring cancer cells, augmenting their cancer-cell-killing efficacy. This approach also has advantages over nonviral-vector-encapsulated conventional oncolytic Ad because the viral genome DNA and nonviral vector complex contain no viral capsid proteins, whereas Ad encapsulated by nonviral vectors still contain viral proteins that can induce nondesirable cytotoxic and immunogenic effects. This attractive combination of features makes this approach safer because of the reduced propensity for inducing humoral and cellular adaptive immune responses to Ad capsid proteins. In addition, the robust efficacy of the approach lies in its ability to induce active viral replication in local tumor tissues via systemic delivery, thereby eliciting the full potential of systemic antitumor therapeutic effects.

In this context, Kwon et al. investigated the systemic delivery of TNF-related apoptosis-inducing ligand-expressing oncolytic Ad genome DNA (pmT-d19/stTR) via lipid envelopment in an orthotopic lung cancer model.[67] The pmT-d19/stTR+DOTAP/DOPE lipoplexes-transfected cancer cells induced predominant cell apoptosis, which suggests that the Ad genome DNA in the lipoplexes effectively produced the active oncolytic Ad progenies once viral replication had occurred. The systemic delivery of pmT-d19/stTR+DOTAP/DOPE elicited more potent antitumor responses compared with naked oncolytic Ad, mT-d19/stTR, in orthotopic lung cancer-bearing mice. Innate immune responses and Ad-specific NAbs were significantly decreased in the pmT-d19/stTR+DOTAP/DOPE-treated mice compared with those in the mT-d19/stTR-treated group. The biodistribution profile analyzed by quantitative polymerase chain reaction (PCR) and immunohistochemical analysis demonstrated that viral replication occurred preferentially in tumor tissues. Moreover, the viral genome tumor-to-liver ratio was significantly elevated in the pmT-d19/stTR+DOTAP/DOPE-treated mice, which was 934- and 27-fold greater than in the mT-d19/stTR- and pmT-d19/stTR-treated mice, respectively. These results demonstrate that systemic delivery of oncolytic viral genome DNA with liposomes is a powerful alternative to naked Ad, overcoming the limited clinical applicability of conventional Ad and enabling effective treatment of disseminated metastatic tumors.

Kim et al.[68] studied the transduction efficacy and in vivo therapeutic ability of oncolytic Ad plasmid DNA complexed with two bioreducible polymers (ABP and PEG5K grafted with ABP [ABP5K]) via electrostatic interaction. The replication of Ad DNA occurs by a strand displacement mechanism initiated at both ends of the virus genome.[69] The presence of inverted terminal repeated (ITR) sequences on the single-stranded DNA molecules permits the formation of a double-stranded region (called a panhandle) that is used as the template for the initiation of DNA replication.[70,71] Thus, the initiation of Ad replication takes place at either end of the Ad genome, which indicates that circular Ad plasmid DNA does not produce infectious

Ad particles.[72] With reference to these strategies, circular and linear plasmid DNA was complexed with polymers and the viral production efficiency was assessed. Only the 293 cells transfected with the linearized oncolytic Ad DNA–polymers (ABP or ABP5k) induced active viral replication, indicating that to produce progeny infectious Ad particles, the Ad plasmid should be linearized to expose both ITR regions. Adenovirus DNA delivered by ABP or ABP5k markedly reduced the innate immune response mediated by the secretion of IL-6 cytokines from the splenocytes and the adaptive immune response related to existing Ad-specific NAb. The biodistribution results revealed that the systemic administration of Ad DNA–ABP5k decreased liver tropism by 99% compared with the naked Ad. In addition, the liver-to-tumor ratio of the Ad genome DNA complexed with ABP or ABP5k was 229- and 419-fold greater than that of naked Ad. Together, these results demonstrate that the PEG-conjugated bioreducible polymer delivery of linearized Ad DNA can overcome the limitations of systemic administration of oncolytic Ad and increase the accumulation of polyplex at the tumor site via passive tumor targeting by the EPR effect.

## 2.7 Polymer-Conjugated Ad

Polyethylene glycol is a neutral polymer that is hydrophilic and has negligible cytotoxicity. It has been used frequently as a delivery vehicle in biomedical applications and in food and cosmetics, and is approved by the United States Food and Drug Administration. Anchoring of PEG to the surface of Ad through covalent bonds is referred to as PEGylation of Ad. This is a promising approach for overcoming the various obstacles of Ad for systemic administration. PEGylation of Ad can enhance the plasma half-life and alter the biodistribution.[73] PEGylation also minimizes nonspecific uptake by macrophages because of the steric hindrance of PEG molecules, which protects the Ad from preexisting antibodies and retains its infectivity both in vitro and in vivo.[74] In particular, according to the pharmacokinetics of murine models, the half-life of systemically administered naked Ad is less than 2 min, whereas PEGylation of Ad significantly reduces blood clearance by a factor of 4. Various PEGylation protocols have been reported in the literature.[75]

A wide range of commercially available functional polymers, such as mono- and bifunctional PEG derivatives, have been used for PEGylation of Ad.[75,76] O'Riordan et al.[77] pioneered the covalent conjugation of PEG with Ad using monofunctional PEGs, such as methoxy PEG activated by tresyl chloride, cyanuric chloride, and succinimidyl succinate, which react with the free amine of lysine on the viral capsid protein. PEGylation of Ad significantly reduced the innate immune response and protected the Ad from preexisting NAbs. However, the transduction was significantly decreased because the polymer monolayers masked the binding site of vectors with integrins of the cell membrane.

To overcome this issue, Kreppel et al.[78] developed a classic PEGylation approach using bioreducible disulfide bonds between the viral vector and the polymer, which can be cleaved in a reducing environment such as cytosol, allowing the detachment of the polymer inside the endosomes. Mok et al.[79] (Table 2) demonstrated that varying the percentage of PEGylation to Ad affected the transduction efficacy in vitro and in vivo. Increasing PEGylation ablated in vitro transduction, showing that heavy PEGylation

**Table 2 Overview of Polymers Which Were Used for Chemical Modification of Ad Surface**

| Polymers | Adenovirus | Application | References |
|---|---|---|---|
| Arginine-grafted bioreducible polymer (ABP) | Replication-incompetent Ad encoding GFP (Ad-$\Delta$E1/GFP)<br>Replication-incompetent Ad encoding GFP on pIX of viral capsid (Ad-$\Delta$E1/IX-GFP)<br>Hepatoma-specific oncolytic Ad (YKL-1001) | In vitro and in vivo | 58 |
| Polyethylene glycol (PEG) | Helper-dependent Ad5 encoding LacZ transgene (HD$\Delta$28E4LacZ) | In vitro and in vivo | 79 |
| PEG | Replication-incompetent Ad encoding firefly luciferase or HSV-tk under control of CMV or TERT promoter (Ad-CMV/Luc, Ad-CMV/HSV-tk, Ad-TERT/Luc, and Ad-TERT/HSV-tk) | In vitro and in vivo | 80 |
| PEG | Replication-incompetent Ad encoding firefly luciferase or TNF-$\alpha$ (Adv-CMV/Luc or Adv-CMV/TNF-$\alpha$) | In vivo | 74 |
| Poly(2-hydroxy propyl) methacrylamide-activatable cell penetrated peptide (HPMA-ACPP) | Replication-incompetent Ad encoding GFP (Ad-eGFP) | In vitro | 81 |
| Chitosan | Replication-incompetent Ad encoding GFP (Ad-CMV-gfp) | In vitro and in vivo | 85 |

of Ad (where 85% of the primary amines on the viral capsid had reacted) caused a 3-log decrease in transduction in vitro. However, viral function that was inactivated in vitro by PEG saturation was largely irrelevant in the in vivo system, suggesting that PEGylated Ad retained its ability to interact with cellular heparin sulfate proteoglycans and integrins. This may explain why the vector remains functional in vivo. Other mechanisms of viral transduction, such as blood factor-mediated uptake and the pharmacodynamic environment of the liver, as well as a longer circulation time, may offer additional explanations for the heavily PEGylated vector's ability to produce the same amount of transduction as unmodified vectors in vivo.

Yao et al.[80] examined the characteristics of PEGylated Ad using 5-kDa PEG or 20-kDa PEG. The transgene expression from PEGylated Ad decreased as the PEG modification ratio and molecular weight of the PEG increased. PEG[20K/45%]-Ad showed fivefold higher transgene expression in the tumor and 185-fold lower expression in the liver than naked Ad. In a separate experiment, the researchers showed that PEG[5K/90%]-Ad had 40-fold higher transgene expression in the tumor and twofold higher in the liver than naked Ad. Thus, PEG[20K/45%]-Ad produced lower transgene expression in the liver than PEG[5K/90%]-Ad, although the transgene expression of PEG[20K/45%]-Ad in the tumor was about eightfold lower than that of PEG[5K/90%]-Ad. Various serum proteins bind to the surface of Ad and may serve as a bridge to the receptors on hepatocytes. The large 20-kDa PEG is likely to cover a larger area of the Ad surface than smaller forms of PEG, preventing the binding of blood factors that would otherwise mediate liver transduction. The blood circulation time of Ad PEGylated with 20 kDa was also enhanced compared with that of 5-kDa PEG after systemic injection. These results indicate that the molecular size of PEG has a vital role in increasing the blood circulation of Ad vectors.

Eto et al.[74] evaluated the effect of the modification rate of PEGylated Ad (30%, 50%, and 90%) on the production of anti-Ad antibodies in vivo. Rats treated with PEGylated Ad (90% modification rate) had lower levels of anti-Ad immunoglobulin (Ig)G and anti-Ad IgM than those treated with naked Ad, whereas rats treated with PEGylated Ad (30% or 45% modification rate) had levels of anti-Ad IgG and IgM similar to those treated with naked Ad. These results suggest that PEGylated Ad (90%) is the optimal PEGylated condition for reducing the immune reaction against Ad. Importantly, 90% PEGylated Ad showed 17-fold higher accumulation and 35-fold greater transgene expression in the primary tumor tissue as a result of the EPR effect, and 17-fold lower accumulation and sixfold lower transgene expression in the liver compared with naked Ad. Systemic treatment with PEGylated Ad (90%) significantly reduced the number of metastatic colonies in metastasis-model mice, compared with treatment with PBS or naked Ad.

Poly(*N*-(2-hydroxypropyl)) methacrylamide (pHPMA) has been used as an alternative to PEG. Poly(*N*-(2-hydroxypropyl)) methacrylamide is a hydrophilic and nontoxic polymer that was invented to deliver therapeutic agents. Seymour et al. developed multifunctional polymers consisting of amino reactive 4-nitrophenoxy groups on pendent diglycyl side chains of pHPMA conjugate with Ad vectors.[53] These modified vectors also significantly reduced toxicity, enhanced the plasma circulation time, and augmented the evasion of NAbs. Li et al. demonstrated a new strategy for selectively delivering HPMA polymer-coated Ad particles into MMP-overexpressing tumor cells[81] by attaching an activatable cell-penetrating peptide (ACPP) to the reactive 4-nitrophenoxy groups of HPMA polymers by the C-terminal amino acid (asparagine, N). Activatable cell-penetrating peptides are polycationic peptides (polyarginine) that are neutralized by polyanionic (polyglutamic acid) sequences through the fusion of cleavable linkers. Activatable cell-penetrating peptides are released from the polycationic peptides only in the immediate vicinity of extracellular proteases (MMPs) in tumors, allowing their cargo to attach to and enter cells. Incorporation of these targeting ACPPs into the HPMA polymer-coated virus enabled ACPP-mediated and CAR-independent binding, selective infection of the MMP-overexpressing tumor, and increased resistance toward NAbs. These results demonstrate the use of a polymer-based system for targeted delivery into MMP-overexpressing solid tumors.

The use of chitosan, a naturally occurring linear polysaccharide composed of β(1,4)-linked glucosamine with some *N*-acetyl-glucosamine, has been extensively studied for the delivery of drug and nucleic acids.[82,83] Kawamata et al.[84] developed chitosan-coated Ad to enhance the transduction efficiency of Ad into mammalian cells. They observed an enhanced effect of chitosan on the infectivity of Ad in Chinese hamster ovary cells that did not express the receptor for Ad, thus indicating the receptor-independent mechanism(s) for this enhancement effect. The pH of the culture medium and the molecular mass and concentration of chitosan were also shown to be critical factors. The highest effect was obtained with 0.1 to 1 μg/ml chitosan with a molecular mass of 19 and 40 kDa in the culture medium with a pH of 6.4, whereas the effect was negligible with higher chitosan concentrations (10 μg/ml or more), a lower or higher molecular mass (11 and 110 kDa) of chitosan, or with a pH of 7.4. Because chitosan is biocompatible and inexpensive, these data indicate that it may be a potential candidate for use as a nonviral vector to safely increase Ad infectivity in mammalian cells, particularly those with poor susceptibility to Ad infection.

Wang et al. demonstrated the covalent conjugation of chitosan to Ad through a thioether link between chitosan modified with 2-iminothiolane and Ad cross-linked with *N*-[C-maleimidobutyryloxy]succinimide ester (GMBS) as an alternative method of vector coating, which allowed transduction into corneal epithelial cells in vitro.[85] Chitosan modification did not significantly change the particle size of the Ad, but its surface charge increased significantly from −24.3 mV to nearly neutral. The transduction efficiency was attenuated gradually with increasing amounts of GMBS. However, the incorporation of chitosan partially restored the infection ability of Ad and induced resistance to antibody neutralization. Compared with PEGylated Ad, chitosan-modified Ad at a ratio of 1:10:700 (Ad:GMBS:ChiSH) protected Ad from antibody neutralization at a low concentration. Unlike PEG, chitosan has a rigid hexose structure that allows chitosan modification to protect the virus effectively with a lower concentration of NAb.

Taken together, these studies imply that Ad covalently conjugated with a variety of polymers significantly enhances stability, increases blood circulation time, and prevents access to NAbs, which emphasizes the potential for polymers to overcome the obstacles of Ad application in clinical practice.

## 3. Active Targeting-Mediated Smart Ad Nanohybrid Systems

Active targeting approaches rely on the binding affinity between ligand-conjugated polymers and specific receptors on the surface of the cell membrane. The surface decoration of polymeric nanoparticles by a specific tumor-homing moiety such as an antibody, antibody fragment, peptide, aptamer, polysaccharide, saccharide, folic acid, and so forth can considerably increase the retention and accumulation of nanoparticles in the tumor vasculature, together with selective and efficient internalization by the target tumor cells, which is termed active tumor targeting. In addition to the type of targeting moiety, the size, shape, and stability of the nanoparticles, and the density and affinity of the targeting moiety have an important role in active cancer targeting. With its potential to maximize therapeutic efficacy and minimize systemic side effects, the targeting moiety-mediated active tumor targeting strategy has become an emerging and indispensable platform for safe and efficient cancer treatment. Various strategies have been proposed for generating active targeting-mediated smart Ad nanohybrid systems using polymer macromolecules, a target moiety, and a spacer or cross-linker.[86,87] To attach targeting ligands or therapeutic agents, the polymeric nanoparticles need a functional moiety. The targeting ligands can be anchored directly or through a degradable or nondegradable cross-linker to the polymer backbone, which controls the release of the therapeutic agent.

Yao and coworkers[73] developed an Ad nanohybrid system that targets both tumor tissues and tumor vasculatures after systemic administration, by conjugating a CGKRK tumor vasculature homing peptide to the end of a 20-kDa PEG chain (Ad-$PEG_{CGKRK}$) (Table 3). In a primary tumor model, systemic administration of Ad-$PEG_{CGKRK}$ resulted in ~500- and 100-fold higher transgene expression in the tumor than administration of naked Ad and Ad-PEG, respectively. In contrast, the transgene expression of Ad-$PEG_{CGKRK}$ in the liver was about 400-fold lower than that of naked Ad, and was almost the same as that of Ad-PEG. These results indicate that administration of Ad-$PEG_{CGKRK}$ results in significant tumor-specific transgene expression compared with Ad-PEG and naked Ad. Furthermore, in mice with established B16BL6 pulmonary metastasis, transgene expression in lungs with Ad-$PEG_{CGKRK}$ was 27-fold and 12-fold higher than with naked Ad and Ad-PEG, respectively. This suggests that Ad-$PEG_{CGKRK}$ not only reduces transgene expression in the liver, it also actively targets both the primary tumor and metastases. Intravenous administration of Ad-$PEG_{CGKRK}$ resulted in strong transgene expression in endothelial cells and colocalization with $CD31^+$ endothelial cells, which indicates that the transgene expression of Ad-$PEG_{CGKRK}$ is localized in the tumor endothelial cells.

The expression of the adhesion molecule integrin αvβ3 on sprouting capillary cells and their interaction with specific matrix ligands have been shown to have a key role in tumor angiogenesis and metastasis. Xiong et al.[88] studied luciferase-expressing Ad modified with cRGD-PEG to specifically target αvβ3 integrins, which are highly expressed in tumor cells. The addition of monomeric cyclic RGD peptides showed enhanced transduction efficiency in cells expressing integrin αvβ3 and the effect was observed in both in vitro and in vivo studies. This enhancement in transduction depended on the binding of the coupled RGD peptide to integrin and was independent of CAR viral receptors. Nontarget tissues such as liver showed a marked decrease in transduction after intravenous delivery of the cRGD-PEG–modified Ad, which suggests that PEGylation may reduce the in vivo sequestration of the vector. The expression of α6β1 and α6β4 integrins is frequently altered in tumor cells with reduced β4 subunit expression, leading to increased α6β1 integrin heterodimerization and consequently increased migratory and invasive properties. Therefore, α6β1 integrin provides an intriguing target for selective gene delivery to metastatic cancer.

Table 3 **Overview of Active Target-Mediated Delivery Systems for Systemic Injection**

| Polymer Materials (Modification Type) | Targeting Ligand | Adenovirus | Target Cell Type | Application | References |
|---|---|---|---|---|---|
| PEG-20kDa (Covalent conjugation) | CGKRK | Replication-incompetent Ad encoding firefly luciferase, GFP, or HSV-tk (Ad5/FL, Ad5/GFP, Ad5/HSV-tk) | Primary and metastatic tumors | In vitro and in vivo | 73 |
| PEG (Covalent conjugation) | cRGD | Replication-incompetent Ad encoding HSV1-sr39tk (Adtk) | CAR-positive and CAR-negative cells | In vitro and in vivo | 88 |
| pHPMA (Covalent conjugation) | YESIKVAVS | Replication-incompetent Ad encoding firefly luciferase (Adluc) | $\alpha_6\beta_1$ integrin-positive prostate cancer cells | In vitro and in vivo | 89 |
| Bioreducible polymer(PEG-CBA-DAH) (Electrostatic interaction) | cRGD | Oncolytic Ad encoding short hairpin RNA against interleukin-8 (Ad-ΔB7-U6shIL8) | CAR-positive and CAR-negative cells | In vitro | 90 |
| PEG (Covalent conjugation) | E-selectin–specific antibody | Replication-incompetent Ad encoding TL (AdTL) | TNF-α activated HUVECs | In vitro and in vivo | 93 |
| Biotin-PEG (Electrostatic interaction) | EGF | Replication- incompetent Ad encoding firefly luciferase (AdCMV-LucΔE1) | EGFR-positive and EGFR-negative cells | In vitro and in vivo | 94 |
| pHPMA (Covalent conjugation) | FGF-2 | Replication-incompetent Ad encoding firefly luciferase (Adluc)<br>Ad5 wild-type adenovirus (Adwt) | FGFR-positive and FGFR-negative cells | In vitro and in vivo | 96 |
| PEG (Covalent conjugation) | Herceptin | Oncolytic Ad expressing relaxin (DWP418) | EGFR-positive and EGFR-negative cells | In vitro and in vivo | 100 |
| Chitosan-PEG (Electrostatic interaction) | Folic acid | Oncolytic Ad (Hmt) | FR-positive cells | In vitro and in vivo | 101 |
| PEG (Covalent conjugation) | Transferrin | Replication-incompetent Ad encoding GFP (Ad1stGFP) | CAR-positive and CAR-negative cells | In vitro | 78 |

Stevenson et al.[89] explored the use of α6β1 integrin-targeted peptide (YESIKVAVS)-conjugated PEG to modify the surface of Ad for tumor-specific targeting. Competition studies confirmed that uptake was mediated by the incorporated ligand and was CAR-independent. The application of retargeted Ad to a panel of prostate cancer cell lines demonstrated enhanced infection of cells derived from metastatic cancers, and the effect was related to the expression of α6 integrins. Intravenous administration of YESIKVAVS-retargeted polymer-coated Ad to tumor-bearing mice showed highly efficient detargeting of the liver and hence greatly reduced toxicity, while maintaining tumor tropism.

Bioreducible poly(cystaminebisacrylamine-diaminohexane) (poly[CBA-DAH] [CD]) has been investigated as a polymer carrier for tumor-targeted oncolytic Ad delivery.[90] These polymers have low cytotoxicity compared with 25 kDa PEI because they are biodegraded to nontoxic small molecules upon exposure to the reductive environment of the cytoplasm through the cleavage of disulfide bonds by glutathione and are no longer harmful. A cyclic RGD peptide has been widely investigated as an active targeting moiety in anti-angiogenic gene therapy for cancer.[91] This ligand can specifically recognize and bind with αvβ3 and αvβ5 integrin receptors, which is an important biomarker overexpressed in sprouting tumor vessels and most tumor cells. In an attempt to take advantage of each system's strengths, actively targeting RGD peptide was conjugated to the bioreducible CD polymer connected with polyethylene glycol (PEG) (CD-PEG-RGD). Compared with unmodified Ad, oncolytic Ad complexed with cRGD-conjugated polymers via physical engineering showed enhanced transduction and greater cancer cell-killing efficacy in a dose-dependent manner, particularly toward selective tumor cells. Furthermore, a competition assay using anti-CAR or anti-integrin antibodies revealed that both CAR and integrins were required for naked Ad to infect target cells, but only integrins were needed for effective infection by Ad–CD-PEG-cRGD. Cells treated with Ad–CD-PEG-cRGD also showed high levels of apoptosis and decreased IL-8 and vascular epithelial growth factor (VEGF) overexpression. Together, these results suggest that the infection pathway of CD-PEG-cRGD–complexed Ad is not a CAR-mediated interaction; rather, it is exclusively regulated by the interaction between integrins and tumor-homing peptides on the virus surface.

E-selectin, known as endothelial leukocyte adhesion molecule-1 and CD62E, is the only endothelial-specific cell adhesion molecule within the selectin family. The recognition that E-selectin is a key molecular marker in inflammation and cancer[92] has led to several independent investigations targeting therapeutic agents to diseased sites. Indeed, E-selectin represents an attractive vascular target because of its physiological roles in binding circulating blood cells under high shear blood flow conditions and its restricted pattern of expression at sites of inflammation, including tumor-associated vasculature. Ogawara et al.[93] investigated the use of E-selectin–specific antibody-conjugated Ad for retargeting the Ad to activated endothelial cells. First, Ad was covalently modified with bifunctional PEG, and E-selectin–specific antibody was subsequently introduced to the other PEG molecule functional group through a coupling reaction. Compared with its unmodified counterpart, systematic administration of anti–E-selectin antibody-modified PEGylated Ad demonstrated improved effects in vivo. It improved blood circulation time and selectively homed to activated endothelium in the skin of mice with a delayed-type hypersensitivity skin inflammation, resulting in local expression of the reporter transgene. Taken together, these results suggest that tumor/endothelial cell-targeting moiety–PEG conjugation is an effective way to modify Ad tropism for improved systemic gene delivery.

A retargeting strategy to ablate native tropism and redirect Ad infection via a receptor that is highly expressed on the target cell surface has been shown to enhance transgene expression significantly. Epidermal growth factor receptor (EGFR), a family of receptor tyrosine kinase proteins that includes EGFR, HER2/erbB2, and HER3/erbB3, is activated in various tumor types of epithelial origin and non–small cell lung cancer. Epidermal growth factor receptor activation in cancer cells results in signal cascades, cell growth, drug and radiation sensitivity, and ligand-independent activation. Epidermal growth factor has been used as an attractive targeting ligand for various diagnostic and therapeutic nanoparticulates because of its high affinity binding to EGFR and subsequent cellular internalization by receptor-mediated endocytosis. Park et al. conjugated biotin–PEG–EGF to avidin-modified Ad via biotin–avidin interaction, generating an Ad-Avi/biotin–PEG–EGF complex,[94] The EGF tethered and PEGylated Ad complex exhibited significantly increased green fluorescent protein expression for EGFR high-expressing A431 cells by EGFR-mediated endocytosis but not for MCF-7 cells (an EGFR-deficient cell line). These results suggest that retargeting of Ad to specific cells can be achieved by tethering a cell-specific targeting ligand to the distal end of a PEG chain anchored onto the Ad surface.

The fibroblastic growth factor (FGF) family has been shown to be highly effective in retargeting Ad because of its cognate high-affinity FGF receptors (FGFRs). Fibroblastic growth factor ligand was conjugated either directly or through polymers such as PEG or HPMA.[95–97] Adenoviruses modified by the addition of FGF2 showed enhanced transduction efficiency in various cancer cells. This enhanced transduction depended on binding of the coupled FGF2 to its high-affinity receptor and was independent of CAR expression. In an intraperitoneal model of ovarian cancer, Ad–PEG–FGF2 elicited greater transgene expression in tumor tissue than naked Ad. Polymer modification of Ad resulted in reduced localization of Ad to nontarget tissues and a marked decrease in thymidine (Th)1 and Th2 T cell responses. Factor VIII staining for endothelial cells demonstrated no increase in angiogenesis in tumors transduced with FGF2-retargeted Ad. Overall, these findings indicate that gene delivery to tumors by FGF2-retargeted Ad is feasible in vivo without inducing unwanted angiogenesis.

Her2/neu is a human EGF2 receptor that is known to be overexpressed in 20–30% of breast cancer patients. Her2/neu has a crucial role as an oncogene in these cancers, and drugs that target Her2/neu, such as trastuzumab and lapatinib, are in clinical use.[98,99] Trastuzumab (Herceptin), a Her2/neu-specific monoclonal antibody, is also widely used to treat both metastatic and early breast cancer. Kim et al. covalently conjugated heterobifunctional PEG to relaxin (matrix-degrading hormone)-expressing oncolytic Ad (DWP418), and subsequently conjugated the Her2/neu-specific Herceptin antibody to the end terminals of PEG for Her2/neu-targeted cancer gene therapy.[100] Adenovirus-PEG-HER increased the blood circulation time 16-fold compared with the naked Ad. Her2/neu-positive SKOV3 and MDA-MB435 xenograft tumor models treated with DWP418-PEG-HER elicited greater antitumor activity than that of naked Ad. However, DWP418-PEG-HER-treated Her2/neu-negative MCF7-mot xenograft tumors had similar antitumor activity to naked Ad. Therefore, the enhanced antitumor activity of DWP418-PEG-HER in Her2/neu-positive tumors depended on targeting the Ad–nanocomplex to the tumor through the specific interaction between the Herceptin and Her2/neu on the cell surface. Specific targeting of DWP418-PEG-HER to tumor cells with Herceptin led to 58,000-fold higher virus accumulation than in the tumors of mice treated with naked Ad. The enhanced accumulation was probably a result of the efficient secondary infection of tumor cells by progeny DWP418. Overall, the liver-to-tumor biodistribution ratio for DWP418-PEG-HER was $10^{10}$-fold greater than for naked DWP418, demonstrating that PEGylation and Herceptin directed the accumulation in the tumor beds and tumor-specific Ad replication in cancer cells, whereas the expression of relaxin promoted the secondary spread of the virus in the tumor bed. These results suggest that an Ad-targeting platform based on the conjugation of a polymer and targeting moiety onto Ad may lead to the development of a gene therapy vector capable of targeting a therapeutic gene to diseased cells, while minimizing toxicity and expression in other tissues.

Chitosan has been used for DNA gene delivery because it contains amine functional groups that are subject to convenient chemical modification and condensed DNA packing via electrostatic interaction.[82,83] These chitosan–DNA complexes have been demonstrated to enhance DNA transduction because of charge interactions with the negatively charged cellular membrane. Similarly, chemically conjugated Ad–chitosan complexes enhance transduction as the result of a similar mechanism involving positive charges that improve attachment to the cell membrane.[85] This is an important feature because it does not negatively affect the infectiousness of the Ad–chitosan complex. Green fluorescent protein-expressing replication-incompetent Ad cross-linked with chitosan–PEG–folic acid (FA) was generated through an electrospinning process, demonstrating the advantages of electrospinning for large-scale production of Ad nanocomplexes.[101] The electrospinning of Ad–chitosan nanocomplexes and ionic cross-linking efficiently coated Ad without reducing its biological activity or infectivity. The ionically cross-linked chitosan layer on the Ad surface provided chemical conjugation sites for PEG and further for FA, as a targeting moiety at the end of heterofunctional PEG. The transduction efficiency of the Ad–chitosan–PEG–FA increased as a function of the FA ratio in FA receptor-expressing cancer cells, but not in FA-receptor negative cancer cells, thus demonstrating FA-receptor–targeted viral transduction. The transduction efficiency of Ad–chitosan–PEG–FA was 57.2% higher than that of Ad–chitosan, which showed the superiority of FA-receptor–mediated endocytosis for viral transduction. The blood clearance assay results revealed a 48.9-fold enhanced blood retention time for the Ad–chitosan-PEG–FA nanocomplex compared with naked Ad. Interestingly, Ad–chitosan–PEG–FA showed a significant increase in blood circulation, 5.44-fold higher than nontargeted Ad–chitosan–PEG, 24 h after injection. This result suggests that endowment of the targeting moiety, FA, onto the surface of the PEGylated Ad nanocomplex further increases the blood circulation time, probably by shielding the surface of Ad more tightly. The immune response was markedly reduced by PEG conjugation, and the tumor-to-liver ratio of the FR-targeted oncolytic Ad nanocomplex was significantly increased. The FR-targeted oncolytic Ad nanocomplex elicited a more potent antitumor efficacy than the nontargeted oncolytic Ad nanocomplex, demonstrating the efficacy of active tumor targeting in addition to EPR-mediated passive targeting.

A unique combination of genetic and chemical vector particle modifications has been investigated in an attempt to overcome the typical restrictions in virus vector targeting. The reactive thiol cysteine group was genetically engineered on the viral capsid, which was conjugated to the target ligand using heterofunctional PEG. First, target transferrin (Tf) was reacted with maleimide-PEG-*N*-hydroxysuccinimide, resulting in maleimide-PEG-Tf, and then coupled to the cysteine thiol of the viral capsid to provide Tf-PEG-Ad.[78] Transferrin was used to enhance the Ad binding affinity to tumor cells expressing transferrin receptors (TfR). After coupling of the high-affinity ligand transferrin to the vector particle surface, the transferrin-modified particles were specifically and efficiently taken up by the Tf/TfR pathway, demonstrating successful receptor targeting.

In summary, these results demonstrate that shielding the oncolytic Ad surface with a biocompatible polymer and a targeting moiety may overcome the limitations of conventional oncolytic Ad vectors, such as hepatotoxicity, immunogenicity, and short blood circulation time, and allow tumor-selective targeting, which is needed for safe and effective systemic cancer treatment.

## 4. Biocompatible Hydrogels for Ad Depot System

The development of optimized Ad and an effective Ad delivery system would further advance Ad vector therapy by maximizing the safety, efficacy, and duration of transgene expression. In particular, the controlled release of viral vectors into the disease-affected site to increase the local vector concentration and therapeutic index are key factors for long-term therapy. Hydrogels are ideal candidates because of their high hydration ability. Injectable hydrogels formed in situ can deliver therapeutic drugs to specific sites in a controlled manner without surgery and implantation procedures.[102] Hydrogels have been used in various biomedical applications such as gene, drug, and cell delivery, and in regenerative medicine.[103,104] The release rate of therapeutic molecules is governed by various parameters such as the degradation of scaffold/gels, the microenvironment, and gel strength.

Hydrogels are generated from naturally occurring polymers such as fibrin, gelatin, alginate, chitosan, and silk elastin-like protein (SELP), as well as from synthetic polymers such as poly(lactic-*co*-glycolic)acid and Pluronic F-127. Natural polymers may be safer than synthetic polymers because of the biocompatibility inherent to biological activities that affect the attachment, migration, and differentiation of cells. Importantly, the biological activity of the encapsulated vectors is well maintained within the hydrogel system. A locally injectable virus delivery system has been used for bone morphogenic protein-2–expressing Ad gene therapy of osteogenesis using a collagen carrier, an antibody complexation with Ad in a collagen carrier, lentivirus entrapment in a hyaluronic acid–collagen matrix, and a mixture of SELP polymers–Ad for breast, head, and neck cancer xenograft tumor models.[105–108] The use of a matrix-based Ad transduction system assisted widespread and uniform virus transduction. The local injection of a matrix-based Ad delivery system prevented rapid viral dissemination and infusion from the tumor to normal organs during intratumoral injection resulting in low normal cell transduction efficiency.

To maximize the therapeutic potential of Ad-mediated gene therapeutics, Choi et al. investigated the efficacy of locally sustained Ad delivery using an injectable alginate gel matrix system. A biodegradable alginate gel, a natural polymer, is frequently used as an injectable gel carrier for protein and cells because of its biocompatibility, low toxicity, relatively low cost, and mild gelation behavior with divalent cations (Table 4).[109,110] Adenovirus loaded in alginate gel had prolonged biological activity compared with naked Ad over an extended period, which suggests that alginate gel–encapsulated Ad may provide a biocompatible environment for maintaining the viral activity of Ad. Considering that long-term transduction is needed for in vivo applications, the use of Ad–alginate gel as a depot system may be useful because it acts as a reservoir that releases Ad in a sustained manner while maintaining the Ad's biological activity. Oncolytic Ad encapsulated in alginate gel elicited significantly greater antitumor activity than naked Ad in human tumor xenograft models, which indicates that local administration of Ad in 5% alginate gel by intratumoral injection enhanced oncolytic Ad-mediated antitumor efficacy. Histological analysis confirmed that oncolytic Ad–gel treatment resulted in wider and denser dissemination of the oncolytic Ad across the tumor bed compared with naked oncolytic Ad treatment. Importantly, the accumulation and spreading of Ad through the tumor tissue was sustained over time. Quantitative polymerase chain reaction analysis showed that the oncolytic Ad–alginate gel matrix system significantly increased the preferential replication and dissemination of oncolytic Ad in a larger area of tumor tissue in vivo. Taken together, these results show that local, sustained delivery of oncolytic Ad in alginate gel augments the

**Table 4 Overview of Hydrogels Used to Sustain Release of Ad**

| Hydrogels | Nature of Gelation | Adenovirus | Tumor Model | Application | References |
|---|---|---|---|---|---|
| Alginate | Ionic cross-linking (i.e., divalent cation $Ca^{2+}$) | Oncolytic Ad expressing relaxin (DWP418) | Glioma xenografts | In vitro and in vivo | 109 |
| Silk elastin-like protein polymers (SELP) | Physical cross-linking | Oncolytic Ad-expressing c-met–specific shRNA (Ad-C-Met) | Head and neck cancer xenografts | In vitro and in vivo | 113 |

therapeutic effect through the selective infection of tumor cells, sustained release, and prolonged maintenance of Ad activity.

Biomimetic SELP hydrogels have been used to control the release of Ad into solid tumors via intratumoral injection.[111,112] Recombinant SELP polymers were designed to insert typical Ala–Gly amino acid sequences from silk and Gyl-Val-Gyl-Val-Pro amino acid sequences from elastin. Some of the positively charged moieties of SELP polymers can interact with viral vectors. The SELP polymers have several unique properties that make them suitable for use as a controlled release matrix, including control of the polymer length and sequence, tunable release kinetics, and an irreversible sol-to-gel transition when elevated to body temperature. This transition is useful in allowing an injectable formulation that becomes a depot for the sustained delivery of Ads to solid tumors.

A study showed that c-met–specific shRNA-expressing oncolytic Ad encapsulated SELP hydrogels for the matrix-mediated delivery of oncolytic Ad to tumors.[113] The Ad released from the SELP hydrogel had much greater biological activity than the naked Ad from days 7 to 21 (1.50- to 3.03-fold increase), which suggests that the SELP matrix prevented the biological inactivation of the Ad induced by incubation at 37 C. c-Met–specific shRNA-expressing oncolytic Ad in an SELP matrix had 1.5-fold greater antitumor efficacy than naked Ad in human xenograft tumor models. A histological analysis demonstrated that treatment with Ad in an SELP matrix resulted in apoptosis over a wider area of tumor tissue and a higher density of Ad infection compared with Ad administered alone. c-Met activation is associated with tumor survival, growth, and metastasis. c-Met also induces VEGF expression through a signaling cascade. In this regard, c-Met–specific shRNA-expressing oncolytic Ad in an SELP matrix efficiently inhibited c-Met–mediated VEGF expression, which suggests that SELP-mediated delivery of oncolytic Ad containing shRNA can be used to treat various types of cancer.

Taken together, these studies support the significance of local Ad delivery using biocompatible hydrogels in a controlled manner for long-term treatment. Importantly, the activity of the gelated Ad was maintained over an extended period, thus maximizing the full potential of Ad-mediated cancer gene therapy by spatiotemporal control.

## 5. Conclusion

We have highlighted recent developments in the modification of Ad by a variety of polymers for use in cancer gene therapy. Clinical trials of oncolytic Ad have demonstrated that it is safe for human cancer treatment via systemic injection, but its therapeutic efficacy is insufficient to ensure the complete eradication of tumors. To overcome the current shortcomings of viral vector-based gene delivery, a variety of polymer-based hybrid systems have been investigated. The conjugation of polymers to Ad enhances stability and protects the Ad from the immune system by reducing innate and adaptive immunogenicity. The polymers coating Ad through ionic charge interactions also increase the transduction efficiency via passive targeting promoted by the EPR effect. Importantly, these polymers can also be used as linkers to conjugate the targeting ligands to the Ad surface for target-specific delivery of Ad. Further developments in multifunctional biomaterials and the fusion of bioengineering and biopharmaceutical technologies are expected to improve the safety and therapeutic efficacy of Ad for human cancer gene therapy.

## References


1. Kirn D. Oncolytic virotherapy for cancer with the adenovirus dl1520 (Onyx-015): results of phase I and II trials. *Expert Opin Biol Ther* 2001;**1**:525–38.
2. Liu TC, Galanis E, Kirn D. Clinical trial results with oncolytic virotherapy: a century of promise, a decade of progress. *Nat Clin Pract Oncol* 2007;**4**:101–17.
3. Donnelly OG, Errington-Mais F, Prestwich R, Harrington K, Pandha H, Vile R, et al. Recent clinical experience with oncolytic viruses. *Curr Pharm Biotechnol* 2012;**13**:1834–41.
4. Einfeld DA, Brough DE, Roelvink PW, Kovesdi I, Wickham TJ. Construction of a pseudoreceptor that mediates transduction by adenoviruses expressing a ligand in fiber or penton base. *J Virol* 1999;**73**:9130–6.
5. Coughlan L. Genetically engineering adenoviral vectors for gene therapy. *Methods Mol Biol* 2014;**1108**:23–40.
6. Douglas JT. Adenoviral vectors for gene therapy. *Mol Biotechnol* 2007;**36**:71–80.
7. Russell SJ, Peng KW, Bell JC. Oncolytic virotherapy. *Nat Biotechnol* 2012;**30**:658–70.
8. Green NK, Hale A, Cawood R, Illingworth S, Herbert C, Hermiston T, et al. Tropism ablation and stealthing of oncolytic adenovirus enhances systemic delivery to tumors and improves virotherapy of cancer. *Nanomedicine (Lond)* 2012;**7**:1683–95.
9. Bachtarzi H, Stevenson M, Fisher K. Cancer gene therapy with targeted adenoviruses. *Expert Opin Drug Deliv* 2008;**5**:1231–40.
10. Ginn SL, Alexander IE, Edelstein ML, Abedi MR, Wixon J. Gene therapy clinical trials worldwide to 2012-an update. *J Gene Med* 2013;**15**:65–77.
11. Capasso C, Garofalo M, Hirvinen M, Cerullo V. The evolution of adenoviral vectors through genetic and chemical surface modifications. *Viruses* 2014;**6**:832–55.
12. Nemunaitis J, Ganly I, Khuri F, Arseneau J, Kuhn J, McCarty T, et al. Selective replication and oncolysis in p53 mutant tumors with ONYX-015, an E1B-55kD gene-deleted adenovirus, in patients with advanced head and neck cancer: a phase II trial. *Cancer Res* 2000;**60**:6359–66.
13. Crompton AM, Kirn DH. From ONYX-015 to armed vaccinia viruses: the education and evolution of oncolytic virus development. *Curr Cancer Drug Target* 2007;**7**:133–9.
14. Choi JW, Lee JS, Kim SW, Yun CO. Evolution of oncolytic adenovirus for cancer treatment. *Adv Drug Deliv Rev* 2012;**64**:720–9.
15. Kim J, Kim PH, Kim SW, Yun CO. Enhancing the therapeutic efficacy of adenovirus in combination with biomaterials. *Biomaterials* 2012;**33**:1838–50.
16. Shewach DS, Zerbe LK, Hughes TL, Roessler BJ, Breakefield XO, Davidson BL. Enhanced cytotoxicity of antiviral drugs mediated by adenovirus directed transfer of the herpes simplex virus thymidine kinase gene in rat glioma cells. *Cancer Gene Ther* 1994;**1**:107–12.
17. Hirschowitz EA, Ohwada A, Pascal WR, Russi TJ, Crystal RG. In vivo adenovirus-mediated gene transfer of the *Escherichia coli* cytosine deaminase gene to human colon carcinoma-derived tumors induces chemosensitivity to 5-fluorocytosine. *Hum Gene Ther* 1995;**6**:1055–63.

18. Zhang WW, Alemany R, Wang J, Koch PE, Ordonez NG, Roth JA. Safety evaluation of Ad5CMV-p53 in vitro and in vivo. *Hum Gene Ther* 1995;**6**:155–64.
19. Griffith TS, Anderson RD, Davidson BL, Williams RD, Ratliff TL. Adenoviral-mediated transfer of the TNF-related apoptosis-inducing ligand/Apo-2 ligand gene induces tumor cell apoptosis. *J Immunol* 2000;**165**:2886–94.
20. Choi KJ, Kim JH, Lee YS, Kim J, Suh BS, Kim H, et al. Concurrent delivery of GM-CSF and B7-1 using an oncolytic adenovirus elicits potent antitumor effect. *Gene Ther* 2006;**13**:1010–20.
21. Lee YS, Kim JH, Choi KJ, Choi IK, Kim H, Cho S, et al. Enhanced antitumor effect of oncolytic adenovirus expressing interleukin-12 and B7-1 in an immunocompetent murine model. *Clin Cancer Res* 2006;**12**:5859–68.
22. Choi IK, Lee JS, Zhang SN, Park J, Sonn CH, Lee KM, et al. Oncolytic adenovirus co-expressing IL-12 and IL-18 improves tumor-specific immunity via differentiation of T cells expressing IL-12Rbeta2 or IL-18Ralpha. *Gene Ther* 2011;**18**:898–909.
23. Zhang SN, Choi IK, Huang JH, Yoo JY, Choi KJ, Yun CO. Optimizing DC vaccination by combination with oncolytic adenovirus coexpressing IL-12 and GM-CSF. *Mol Ther* 2011;**19**:1558–68.
24. Bai FL, Yu YH, Tian H, Ren GP, Wang H, Zhou B, et al. Genetically engineered Newcastle disease virus expressing interleukin-2 and TNF-related apoptosis-inducing ligand for cancer therapy. *Cancer Biol Ther* 2014;**15**:1226–38.
25. Choi SH, Kwon OJ, Park JY, Kim do Y, Ahn SH, Kim SU, et al. Inhibition of tumour angiogenesis and growth by small hairpin HIF-1alpha and IL-8 in hepatocellular carcinoma. *Liver Int* 2014;**34**:632–42.
26. Choi KJ, Zhang SN, Choi IK, Kim JS, Yun CO. Strengthening of antitumor immune memory and prevention of thymic atrophy mediated by adenovirus expressing IL-12 and GM-CSF. *Gene Ther* 2012;**19**:711–23.
27. Kim W, Seong J, Oh HJ, Koom WS, Choi KJ, Yun CO. A novel combination treatment of armed oncolytic adenovirus expressing IL-12 and GM-CSF with radiotherapy in murine hepatocarcinoma. *J Radiat Res* 2011;**52**:646–54.
28. Choi IK, Li Y, Oh E, Kim J, Yun CO. Oncolytic adenovirus expressing IL-23 and p35 elicits IFN-gamma- and TNF-alpha-co-producing T cell-mediated antitumor immunity. *PLoS One* 2013;**8**:e67512.
29. Yoo JY, Kim JH, Kwon YG, Kim EC, Kim NK, Choi HJ, et al. VEGF-specific short hairpin RNA-expressing oncolytic adenovirus elicits potent inhibition of angiogenesis and tumor growth. *Mol Ther* 2007;**15**:295–302.
30. Yoo JY, Kim JH, Kim J, Huang JH, Zhang SN, Kang YA, et al. Short hairpin RNA-expressing oncolytic adenovirus-mediated inhibition of IL-8: effects on antiangiogenesis and tumor growth inhibition. *Gene Ther* 2008;**15**:635–51.
31. Yun CO, Kim E, Koo T, Kim H, Lee YS, Kim JH. ADP-overexpressing adenovirus elicits enhanced cytopathic effect by induction of apoptosis. *Cancer Gene Ther* 2005;**12**:61–71.
32. Yoon AR, Kim JH, Lee YS, Kim H, Yoo JY, Sohn JH, et al. Markedly enhanced cytolysis by E1B-19kD-deleted oncolytic adenovirus in combination with cisplatin. *Hum Gene Ther* 2006;**17**:379–90.
33. Kim JH, Lee YS, Kim H, Huang JH, Yoon AR, Yun CO. Relaxin expression from tumor-targeting adenoviruses and its intratumoral spread, apoptosis induction, and efficacy. *J Natl Cancer Inst* 2006;**98**:1482–93.
34. Choi IK, Lee YS, Yoo JY, Yoon AR, Kim H, Kim DS, et al. Effect of decorin on overcoming the extracellular matrix barrier for oncolytic virotherapy. *Gene Ther* 2010;**17**:190–201.

35. Alemany R, Suzuki K, Curiel DT. Blood clearance rates of adenovirus type 5 in mice. *J Gen Virol* 2000;**81**:2605–9.
36. Nemunaitis J, O'Brien J. Head and neck cancer: gene therapy approaches. Part II: genes delivered. *Expert Opin Biol Ther* 2002;**2**:311–24.
37. Xu Z, Tian J, Smith JS, Byrnes AP. Clearance of adenovirus by Kupffer cells is mediated by scavenger receptors, natural antibodies, and complement. *J Virol* 2008;**82**:11705–13.
38. Smith JS, Xu Z, Byrnes AP. A quantitative assay for measuring clearance of adenovirus vectors by Kupffer cells. *J Virol Methods* 2008;**147**:54–60.
39. Morimoto K, Nishikawa M, Kawakami S, Nakano T, Hattori Y, Fumoto S, et al. Molecular weight-dependent gene transfection activity of unmodified and galactosylated polyethyleneimine on hepatoma cells and mouse liver. *Mol Ther* 2003;**7**:254–61.
40. Varkouhi AK, Scholte M, Storm G, Haisma HJ. Endosomal escape pathways for delivery of biologicals. *J Control Release* 2011;**151**:220–8.
41. Pack DW, Hoffman AS, Pun S, Stayton PS. Design and development of polymers for gene delivery. *Nat Rev Drug Discov* 2005;**4**:581–93.
42. Moghimi SM, Hunter AC, Murray JC. Long-circulating and target-specific nanoparticles: theory to practice. *Pharmacol Rev* 2001;**53**:283–318.
43. Wortmann A, Vohringer S, Engler T, Corjon S, Schirmbeck R, Reimann J, et al. Fully detargeted polyethylene glycol-coated adenovirus vectors are potent genetic vaccines and escape from pre-existing anti-adenovirus antibodies. *Mol Ther* 2008;**16**:154–62.
44. Kasala D, Choi JW, Kim SW, Yun CO. Utilizing adenovirus vectors for gene delivery in cancer. *Expert Opin Drug Deliv* 2014;**11**:379–92.
45. Han J, Zhao D, Zhong Z, Zhang Z, Gong T, Sun X. Combination of adenovirus and cross-linked low molecular weight PEI improves efficiency of gene transduction. *Nanotechnology* 2010;**21**:105106.
46. Lee CH, Kasala D, Na Y, Lee MS, Kim SW, Jeong JH, et al. Enhanced therapeutic efficacy of an adenovirus-PEI-bile-acid complex in tumors with low Coxsackie and adenovirus receptor expression. *Biomaterials* 2014;**35**:5505–16.
47. Ward CM, Read ML, Seymour LW. Systemic circulation of poly(L-lysine)/DNA vectors is influenced by polycation molecular weight and type of DNA: differential circulation in mice and rats and the implications for human gene therapy. *Blood* 2001;**97**:2221–9.
48. Fasbender A, Zabner J, Chillon M, Moninger TO, Puga AP, Davidson BL, et al. Complexes of adenovirus with polycationic polymers and cationic lipids increase the efficiency of gene transfer in vitro and in vivo. *J Biol Chem* 1997;**272**:6479–89.
49. Park JW, Mok H, Park TG. Physical adsorption of PEG grafted and blocked poly-L-lysine copolymers on adenovirus surface for enhanced gene transduction. *J Control Release* 2010;**142**:238–44.
50. Mok H, Park JW, Park TG. Enhanced intracellular delivery of quantum dot and adenovirus nanoparticles triggered by acidic pH via surface charge reversal. *Bioconjug Chem* 2008;**19**:797–801.
51. Lee ES, Gao Z, Bae YH. Recent progress in tumor pH targeting nanotechnology. *J Control Release* 2008;**132**:164–70.
52. Jiang ZK, Koh SB, Sato M, Atanasov IC, Johnson M, Zhou ZH, et al. Engineering polypeptide coatings to augment gene transduction and in vivo stability of adenoviruses. *J Control Release* 2013;**166**:75–85.
53. Fisher KD, Seymour LW. HPMA copolymers for masking and retargeting of therapeutic viruses. *Adv Drug Deliv Rev* 2010;**62**:240–5.

54. Wang CH, Chan LW, Johnson RN, Chu DS, Shi J, Schellinger JG, et al. The transduction of Coxsackie and adenovirus receptor-negative cells and protection against neutralizing antibodies by HPMA-co-oligolysine copolymer-coated adenovirus. *Biomaterials* 2011;**32**:9536–45.
55. Lee YS, Kim SW. Bioreducible polymers for therapeutic gene delivery. *J Control Release* 2014;**190**:424–39.
56. Börner HG. Strategies exploiting functions and self-assembly properties of bioconjugates for polymer and materials sciences. *Prog Polym Sci* 2009;**34**:811–51.
57. Kim PH, Kim TI, Yockman JW, Kim SW, Yun CO. The effect of surface modification of adenovirus with an arginine-grafted bioreducible polymer on transduction efficiency and immunogenicity in cancer gene therapy. *Biomaterials* 2010;**31**:1865–74.
58. Kim PH, Kim J, Kim TI, Nam HY, Yockman JW, Kim M, et al. Bioreducible polymer-conjugated oncolytic adenovirus for hepatoma-specific therapy via systemic administration. *Biomaterials* 2011;**32**:9328–42.
59. Kim J, Li Y, Kim SW, Lee DS, Yun CO. Therapeutic efficacy of a systemically delivered oncolytic adenovirus – biodegradable polymer complex. *Biomaterials* 2013;**34**:4622–31.
60. Shcharbin D, Shakhbazau A, Bryszewska M. Poly(amidoamine) dendrimer complexes as a platform for gene delivery. *Expert Opin Drug Deliv* 2013;**10**:1687–98.
61. Vetter A, Virdi KS, Espenlaub S, Rodl W, Wagner E, Holm PS, et al. Adenoviral vectors coated with PAMAM dendrimer conjugates allow CAR independent virus uptake and targeting to the EGF receptor. *Mol Pharm* 2013;**10**:606–18.
62. Grunwald GK, Vetter A, Klutz K, Willhauck MJ, Schwenk N, Senekowitsch-Schmidtke R, et al. Systemic image-guided liver cancer radiovirotherapy using dendrimer-coated adenovirus encoding the sodium iodide symporter as theranostic gene. *J Nucl Med* 2013;**54**:1450–7.
63. Ashizawa AT, Cortes J. Liposomal delivery of nucleic acid-based anticancer therapeutics: BP-100-1.01. *Expert Opin Drug Deliv* 2014:1–14.
64. Van den Bossche J, Al-Jamal WT, Yilmazer A, Bizzarri E, Tian B, Kostarelos K. Intracellular trafficking and gene expression of pH-sensitive, artificially enveloped adenoviruses in vitro and in vivo. *Biomaterials* 2011;**32**:3085–93.
65. Futaki S, Masui Y, Nakase I, Sugiura Y, Nakamura T, Kogure K, et al. Unique features of a pH-sensitive fusogenic peptide that improves the transfection efficiency of cationic liposomes. *J Gene Med* 2005;**7**:1450–8.
66. Wan Y, Han J, Fan G, Zhang Z, Gong T, Sun X. Enzyme-responsive liposomes modified adenoviral vectors for enhanced tumor cell transduction and reduced immunogenicity. *Biomaterials* 2013;**34**:3020–30.
67. Kwon OJ, Kang E, Kim S, Yun CO. Viral genome DNA/lipoplexes elicit in situ oncolytic viral replication and potent antitumor efficacy via systemic delivery. *J Control Release* 2011;**155**:317–25.
68. Kim J, Kim PH, Nam HY, Lee JS, Yun CO, Kim SW. Linearized oncolytic adenoviral plasmid DNA delivered by bioreducible polymers. *J Control Release* 2012;**158**:451–60.
69. Sussenbach JS, Ellens DJ, Jansz HS. Studies on the mechanism of replication of adenovirus DNA II. The nature of single-stranded DNA in replicative intermediates. *J Virol* 1973;**12**:1131–8.
70. Hay RT. The origin of adenovirus DNA replication: minimal DNA sequence requirement in vivo. *EMBO J* 1985;**4**:421–6.
71. van Bergen BG, van der Ley PA, van Driel W, van Mansfeld AD, van der Vliet PC. Replication of origin containing adenovirus DNA fragments that do not carry the terminal protein. *Nucleic Acids Res* 1983;**11**:1975–89.

72. Berkner KL, Sharp PA. Generation of adenovirus by transfection of plasmids. *Nucleic Acids Res* 1983;**11**:6003–20.
73. Yao X, Yoshioka Y, Morishige T, Eto Y, Narimatsu S, Kawai Y, et al. Tumor vascular targeted delivery of polymer-conjugated adenovirus vector for cancer gene therapy. *Mol Ther* 2011;**19**:1619–25.
74. Eto Y, Yoshioka Y, Ishida T, Yao X, Morishige T, Narimatsu S, et al. Optimized PEGylated adenovirus vector reduces the anti-vector humoral immune response against adenovirus and induces a therapeutic effect against metastatic lung cancer. *Biol Pharm Bull* 2010;**33**:1540–4.
75. Eto Y, Yoshioka Y, Mukai Y, Okada N, Nakagawa S. Development of PEGylated adenovirus vector with targeting ligand. *Int J Pharm* 2008;**354**:3–8.
76. Wonganan P, Croyle MA. PEGylated adenoviruses: from mice to monkeys. *Viruses* 2010;**2**:468–502.
77. O'Riordan CR, Lachapelle A, Delgado C, Parkes V, Wadsworth SC, Smith AE, et al. PEGylation of adenovirus with retention of infectivity and protection from neutralizing antibody in vitro and in vivo. *Hum Gene Ther* 1999;**10**:1349–58.
78. Kreppel F, Gackowski J, Schmidt E, Kochanek S. Combined genetic and chemical capsid modifications enable flexible and efficient de- and retargeting of adenovirus vectors. *Mol Ther* 2005;**12**:107–17.
79. Mok H, Palmer DJ, Ng P, Barry MA. Evaluation of polyethylene glycol modification of first-generation and helper-dependent adenoviral vectors to reduce innate immune responses. *Mol Ther* 2005;**11**:66–79.
80. Yao X, Yoshioka Y, Morishige T, Eto Y, Watanabe H, Okada Y, et al. Systemic administration of a PEGylated adenovirus vector with a cancer-specific promoter is effective in a mouse model of metastasis. *Gene Ther* 2009;**16**:1395–404.
81. Li S, Chen J, Xu H, Long J, Xie X, Zhang Y. The targeted transduction of MMP-overexpressing tumor cells by ACPP-HPMA copolymer-coated adenovirus conjugates. *PloS One* 2014;**9**:e100670.
82. Bozkir A, Saka OM. Chitosan-DNA nanoparticles: effect on DNA integrity, bacterial transformation and transfection efficiency. *J Drug Target* 2004;**12**:281–8.
83. Mao HQ, Roy K, Troung-Le VL, Janes KA, Lin KY, Wang Y, et al. Chitosan-DNA nanoparticles as gene carriers: synthesis, characterization and transfection efficiency. *J Control Release* 2001;**70**:399–421.
84. Kawamata Y, Nagayama Y, Nakao K, Mizuguchi H, Hayakawa T, Sato T, et al. Receptor-independent augmentation of adenovirus-mediated gene transfer with chitosan in vitro. *Biomaterials* 2002;**23**:4573–9.
85. Wang IJ, Jhuang MC, Chen YH, Yeh LK, Liu CY, Young TH. Chitosan modification of adenovirus to modify transfection efficiency in bovine corneal epithelial cells. *PloS One* 2010;**5**:e12085.
86. Jung Y, Park HJ, Kim PH, Lee J, Hyung W, Yang J, et al. Retargeting of adenoviral gene delivery via Herceptin-PEG-adenovirus conjugates to breast cancer cells. *J Control Release* 2007;**123**:164–71.
87. Reetz J, Herchenroder O, Putzer BM. Peptide-based technologies to alter adenoviral vector tropism: ways and means for systemic treatment of cancer. *Viruses* 2014;**6**:1540–63.
88. Xiong Z, Cheng Z, Zhang X, Patel M, Wu JC, Gambhir SS, et al. Imaging chemically modified adenovirus for targeting tumors expressing integrin alphavbeta3 in living mice with mutant herpes simplex virus type 1 thymidine kinase PET reporter gene. *J Nucl Med* 2006;**47**:130–9.
89. Stevenson M, Hale AB, Hale SJ, Green NK, Black G, Fisher KD, et al. Incorporation of a laminin-derived peptide (SIKVAV) on polymer-modified adenovirus permits tumor-specific targeting via alpha6-integrins. *Cancer Gene Ther* 2007;**14**:335–45.

90. Kim J, Nam HY, Kim TI, Kim PH, Ryu J, Yun CO, et al. Active targeting of RGD-conjugated bioreducible polymer for delivery of oncolytic adenovirus expressing shRNA against IL-8 mRNA. *Biomaterials* 2011;**32**:5158–66.
91. Ge Z, Chen Q, Osada K, Liu X, Tockary TA, Uchida S, et al. Targeted gene delivery by polyplex micelles with crowded PEG palisade and cRGD moiety for systemic treatment of pancreatic tumors. *Biomaterials* 2014;**35**:3416–26.
92. Reynolds PR, Larkman DJ, Haskard DO, Hajnal JV, Kennea NL, George AJ, et al. Detection of vascular expression of E-selectin in vivo with MR imaging. *Radiology* 2006;**241**:469–76.
93. Ogawara K, Rots MG, Kok RJ, Moorlag HE, Van Loenen AM, Meijer DK, et al. A novel strategy to modify adenovirus tropism and enhance transgene delivery to activated vascular endothelial cells in vitro and in vivo. *Hum Gene Ther* 2004;**15**:433–43.
94. Park JW, Mok H, Park TG. Epidermal growth factor (EGF) receptor targeted delivery of PEGylated adenovirus. *Biochem Biophys Res Commun* 2008;**366**:769–74.
95. Morrison J, Briggs SS, Green N, Fisher K, Subr V, Ulbrich K, et al. Virotherapy of ovarian cancer with polymer-cloaked adenovirus retargeted to the epidermal growth factor receptor. *Mol Ther* 2008;**16**:244–51.
96. Green NK, Morrison J, Hale S, Briggs SS, Stevenson M, Subr V, et al. Retargeting polymer-coated adenovirus to the FGF receptor allows productive infection and mediates efficacy in a peritoneal model of human ovarian cancer. *J Gene Med* 2008;**10**:280–9.
97. Wang W, Zhu NL, Chua J, Swenson S, Costa FK, Schmitmeier S, et al. Retargeting of adenoviral vector using basic fibroblast growth factor ligand for malignant glioma gene therapy. *J Neurosurg* 2005;**103**:1058–66.
98. Gajria D, Chandarlapaty S. HER2-amplified breast cancer: mechanisms of trastuzumab resistance and novel targeted therapies. *Expert Rev Anticancer Ther* 2011;**11**:263–75.
99. Alba E, Albanell J, de la Haba J, Barnadas A, Calvo L, Sanchez-Rovira P, et al. Trastuzumab or lapatinib with standard chemotherapy for HER2-positive breast cancer: results from the GEICAM/2006-14 trial. *Br J Cancer* 2014;**110**:1139–47.
100. Kim PH, Sohn JH, Choi JW, Jung Y, Kim SW, Haam S, et al. Active targeting and safety profile of PEG-modified adenovirus conjugated with herceptin. *Biomaterials* 2011;**32**:2314–26.
101. Kwon OJ, Kang E, Choi JW, Kim SW, Yun CO. Therapeutic targeting of chitosan-PEG-folate-complexed oncolytic adenovirus for active and systemic cancer gene therapy. *J Control Release* 2013;**169**:257–65.
102. He C, Kim SW, Lee DS. In situ gelling stimuli-sensitive block copolymer hydrogels for drug delivery. *J Control Release* 2008;**127**:189–207.
103. Seliktar D. Designing cell-compatible hydrogels for biomedical applications. *Science* 2012;**336**:1124–8.
104. Nicodemus GD, Bryant SJ. Cell encapsulation in biodegradable hydrogels for tissue engineering applications. *Tissue Eng Part B Rev* 2008;**14**:149–65.
105. Gugala Z, Davis AR, Fouletier-Dilling CM, Gannon FH, Lindsey RW, Olmsted-Davis EA. Adenovirus BMP2-induced osteogenesis in combination with collagen carriers. *Biomaterials* 2007;**28**:4469–79.
106. Levy RJ, Song C, Tallapragada S, DeFelice S, Hinson JT, Vyavahare N, et al. Localized adenovirus gene delivery using antiviral IgG complexation. *Gene Ther* 2001;**8**:659–67.
107. Shin S, Shea LD. Lentivirus immobilization to nanoparticles for enhanced and localized delivery from hydrogels. *Mol Ther* 2010;**18**:700–6.
108. Greish K, Araki K, Li D, O'Malley Jr BW, Dandu R, Frandsen J, et al. Silk-elastinlike protein polymer hydrogels for localized adenoviral gene therapy of head and neck tumors. *Biomacromolecules* 2009;**10**:2183–8.

109. Choi JW, Kang E, Kwon OJ, Yun TJ, Park HK, Kim PH, et al. Local sustained delivery of oncolytic adenovirus with injectable alginate gel for cancer virotherapy. *Gene Ther* 2013;**20**:880–92.
110. Park H, Kim PH, Hwang T, Kwon OJ, Park TJ, Choi SW, et al. Fabrication of cross-linked alginate beads using electrospraying for adenovirus delivery. *Int J Pharm* 2012;**427**:417–25.
111. Gustafson JA, Ghandehari H. Silk-elastinlike protein polymers for matrix-mediated cancer gene therapy. *Adv Drug Deliv Rev* 2010;**62**:1509–23.
112. Greish K, Frandsen J, Scharff S, Gustafson J, Cappello J, Li D, et al. Silk-elastinlike protein polymers improve the efficacy of adenovirus thymidine kinase enzyme prodrug therapy of head and neck tumors. Kirn D. Oncolytic virotherapy for cancer with the adenovirus dl1520 (Onyx-015): results of phase I and II trials. *Expert Opin Biol Ther* 2001;**1**:525–38.
113. Jung SH, Choi JW, Yun CO, Yhee JY, Price R, Kim SH, et al. Sustained local delivery of oncolytic short hairpin RNA adenoviruses for treatment of head and neck cancer. *J Gene Med* 2014;**16**:143–52.

# Adenoviral Vectors for RNAi Delivery

# 30

*Hideyo Ugai*
Cancer Biology Division, Department of Radiation Oncology, School of Medicine, Washington University, St. Louis, MO, USA

## 1. Introduction

In the 1980s, antisense RNAs were shown to inhibit gene expression efficiently in bacteria,[1] *Dictyostelium*,[2] *Xenopus oocytes*,[3] *Drosophila*,[4] plant cells,[5] and mammalian cells.[6] In other work, the introduction of the chalcone synthase gene, the key enzyme in flavonoid biosynthesis, was attempted in petunias to increase or alter pigmentation of flower color.[7] Unexpectedly, it resulted in less pigmentation, yielding fully or partially white flowers, indicating silencing of flower colors.[7] Thus, the downregulation of target genes was observed by using antisense RNAs or a transgene cassette through unknown mechanism(s). After these observations, researchers investigated this phenomenon in other organisms. In 1995, analysis of asymmetric cell division of *Caenorhabditis elegans* by injecting antisense ribonucleic acid (RNA) for *par-1* mRNA induced *per-1* messenger RNA (mRNA) degradation and mediated the disruption of asymmetric cell division.[8] In addition, injection of both sense and antisense RNAs dictated *par-1* phenocopies.[8] Interestingly, injection of in vitro–synthesized sense RNA for *par-1* mRNA induced *par-1* phenotypes at a high frequency. Antisense and sense RNAs seemed to function independently during the induction of *par-1* phenotypes. In 1998, Drs. Andrew Z. Fire and Craig C. Mello reported a potent gene-silencing effect by injecting double-stranded RNA (dsRNA) into *C. elegans*,[9] demonstrating that dsRNA successfully silenced the targeted gene expression. They initially and systematically demonstrated RNA interference (RNAi) in the nematode *C. elegans*. The Nobel Prize in Physiology or Medicine 2006 was jointly awarded to Drs. Fire and Mello for their discovery of RNAi-gene silencing by dsRNA.

Ribonucleic acid interference is a regulatory mechanism conserved in most eukaryotic cells. The events of RNAi are initiated through a dsRNA of approximately 21–23 nucleotides in length, such as short interfering RNAs (siRNAs), short hairpin RNAs (shRNAs), or microRNAs (miRNAs).[10–14] Ribonucleic acid interference leads to sequence-dependent degradation or translation repression of target mRNAs. Ribonucleic acid interference is a biological process observed in cells and occurs without triggering a toxic response. Therefore, it is used as the standard tool for the sequence-specific knockdown of gene expression in molecular biology. In this chapter, we provide an overview of RNAi in adenoviral biology and adenoviral vectors for delivery of RNAi-mediated gene silencing.

**Adenoviral Vectors for Gene Therapy.**

## 2. MicroRNAs and Human Diseases

Different types of noncoding RNAs (ncRNAs) such as microRNAs (miRNAs), transcribed ultraconserved regions (T-UCRs), small nucleolar RNAs (snoRNAs), PIWI-interacting RNAs (piRNAs), large intergenic ncRNAs (lincRNAs), and the heterogeneous group of long ncRNAs (lncRNAs) are transcribed in human cells.[15] They might contribute to the development of many different human disorders.[16] Of ncRNAs, the biogenesis of miRNAs and their functions have been well characterized in mammalian cells and have been the most widely investigated in human disease. The human genome contains approximately 20,000 to 25,000 protein-coding genes[17] and over 2000 miRNAs.[18] It has been predicted that miRNAs regulate the translation rate of more than 60% of protein-coding genes,[19] and over 45,000 miRNA target sites within human 3′-untranslated regions (UTRs) are conserved.[19] MicroRNAs are associated with gene regulation of protein-coding genes and control developmental and oncogenic processes.[20] Notably, miRNA processing defects enhance tumorigenesis[21] and result in unique gene expression profiles of different miRNAs observed in a variety of normal tissue and cancers.[22–26] MicroRNAs are often differentially expressed or downregulated in various human cancers[20–26] as a result of the deletion of gene regions and epigenetic changes, including aberrant deoxyribonucleic acid (DNA) methylation or histone modification.[15,27–30] Therefore, miRNAs can function as oncogenes or tumor suppressor genes and be used as novel biomarkers in cancer disease diagnostics.[31–33] MicroRNAs are classified into several categories—angiomiRs, apoptomiRs, hypoxamiRs, metastamiRs, and oncomiRs—based on functional actions.[31–35] In addition to cancer diseases, aberrant expression of miRNAs is reported in nontumoral disorders, including neurological and cardiovascular disorders.[15]

## 3. MicroRNA Biogenesis and Gene Silencing

The biogenesis of miRNAs has been well characterized in mammalian cells to examine the mechanism of RNAi-induced gene silencing. MicroRNAs are transcribed as long primary transcripts named pri-miRNAs by RNA polymerase II.[36,37] Pri-miRNAs can contain one or more immature miRNAs, possess a 7-methyl guanosine cap at the 5′-terminus, and are polyadenylated.[38] In the nucleoplasm, they are processed by Drosha, which is a nuclear RNase III involved in the Microprocessor complex,[39] generating imperfectly pairing stem-loop molecules of approximately 70 nucleotides known as pre-miRNAs that have a 5′ phosphate and a 2-nucleotide 3′ overhang. Drosha-containing complex was examined to identify the components of the miRNA-processing machinery.[39] Approximately 20 kinds of Drosha-associating proteins were identified by affinity chromatography and mass spectrometry using 293 cells stably expressing FLAG-Drosha.[39] FLAG-Drosha is expressed as approximately 145 and 160 kDa recombinant proteins in 293 cells. However, whether the 160-kDa form of Drosha (Drosha-a) is caused by a posttranslational modification of the 145-kDa form (Drosha-b) or is proteolytically cleaved to the 145-kDa form remains unknown. Drosha-containing complexes were isolated at three different molecular masses (>2 MDa, approximately 600 kDa, and approximately 400 kDa) in 293 cells stably expressing FLAG-Drosha. Drosha-a is involved in both the >2-MDa and approximately 600-kDa larger complexes, whereas Drosha-b is associated with both the 600-kDa and the 400-kDa complexes. The 2-MDa Drosha complex contains 19 classes of RNA-associated proteins including the DEAD-box and DEAH-box family of RNA helicases, dsRNA-binding proteins including nuclear factor 90 (NF90) and NF45, novel heterogeneous nuclear ribonucleoproteins, and the Ewing sarcoma family of proteins.[39] Although the 600-kDa and 400-kDa Drosha complexes contain the gene product of DiGeorge syndrome critical region gene 8 (DGCR8), which is a dsRNA-binding protein, the 2-MDa Drosha complex does not include DGCR8.[39] Whereas the 2-MDa Drosha complex has weak activity to process pri-miRNAs because of nonspecific RNase activity, the 600-kDa Drosha complex containing DGCR8 exhibits stronger pri-miRNA processing activity compared with the 2-MDa Drosha complex.[39] The knockdown analysis of DGCR8 and in vitro reconstitution analysis using recombinant Drosha and DGCR8 proteins demonstrated that both Drosha and DGCR8 are responsible for the processing of pri-miRNAs into pre-miRNAs.[39] Thus, the 600-kDa Drosha complex containing Drosha-a, Drosha-b, and DGCR8 is defined as the Microprocessor complex processing pri-miRNAs into pre-miRNAs.[39] One of the proteins is NF90/Nuclear Factor Associated with double-stranded RNA (NFAR), which is a dsRNA-binding protein similar to DGCR8. NF90 forms a heterodimer with NF45, which is also involved in the 2-MDa Drosha complex, and is called nuclear factor of activated T cells (NFAT).[40] In 293T cells transiently overexpressing both NF90 and NF45 proteins, the NF90 and NF45 complex binds to pri-miRNAs more strongly than DGCR8 owing to its overexpression and inhibits the processing of pri-miRNAs into pre-miRNAs.[41] These data suggest that the NF90 and NF45 complex may function as a negative regulator of the Microprocessor complex.[41] Nevertheless, how the NF90 and NF45 complex involved in the 2-MDa Drosha complex negatively regulates the function of the Microprocessor complex (the 600-kDa Drosha complex) in normal and cancer cells remains unclear. After the processing of pri-miRNAs into pre-miRNAs, pre miRNAs are transferred from the nucleoplasm to the cytoplasm by Exportin-5 (Exp-5/XPO5) in the presence of Ran-GTP cofactor.[42] Subsequently, cytoplasmic pre-miRNAs are processed by a complex consisting of Dicer, which is a cytoplasmic RNase III, TAR RNA-binding protein (TRBP), and protein activator of protein kinase PKR (PACT) into a mature double-stranded miRNA of 21 to 25 nucleotides, which has 3′ overhangs of 2-nucleotides and 5′-monophosphate groups.[43–45] Also, Exp-5 exports Dicer mRNA from the nucleus to the cytoplasm and regulates its protein expression levels.[46] TRBP and PACT proteins contain dsRNA-binding sites and are associated with the processing of miRNA and shRNA into siRNAs by Dicer in a different manner.[45] Whereas PACT activates PKR without dsRNA,[47,48] TRBP is an inhibitor for PKR.[49] Recent reports demonstrate that PKR is involved in the RNA-induced silencing complex (RISC) containing PACT and TRBP.[43,50] However, the function of PKR in the RNAi machinery remains largely unknown and thus is an area requiring further investigation. The siRNA guide strands from mature miRNAs recognize imperfect sequences of mRNAs, which are usually located in 3′-UTRs, and are incorporated into the RISC containing an endonuclease, Argonaute 2 (AGO2). Argonaute 2 is the only member of the Argonaute subfamily of proteins, which contain a PAZ domain required for miRNA/siRNA binding and have RNaseH endonuclease activity,[51] which catalyzes direct mRNA cleavage in the RISC complex and mediates translational inhibition.[10,11] In addition, miRNAs with perfect complementary sequences to target sites in mRNAs are believed to induce mRNA degradation, similar to siRNAs.[52,53]

## 4. Posttranscriptional Gene Silencing by ncRNAs

Short interfering RNAs are chemically synthesized RNA duplexes of 21 to 23 nucleotides with 3′ overhangs of 2 nucleotides that are transported as dsRNAs into cells by direct lipofection procedure or delivered into target cells via vector systems. In particular, a simple procedure to introduce siRNAs into mammalian cells is to use cationic lipid transfection agents, generating approximately 500 nm lipoplexes; the complexes are composed of DNA and liposomes.[54,55] The lipoplexes containing large amounts of siRNAs (in the range of $10^7$ to $10^9$ molecules per cell) are transfected cells by standard transfection protocols, targeting approximately $10^1$ to $10^4$ molecules of mRNAs.[56] A fluorescent resonance energy transfer-based visualization method shows that siRNAs are accumulated into the nuclei within 4 h.[56] Exogenous siRNAs target complementary mRNAs and mediate posttranscriptional gene silencing for cleavage and the degradation of target mRNAs. To maximize the effect of posttranscriptional gene silencing by siRNAs, the guide strand of the siRNA needs to perfectly or nearly perfectly form to the target mRNA by Watson-Crick base-pairing, resulting in the cleavage of the mRNA by the RISC.[57]

Short hairpin RNAs are the precursor of siRNAs, defined by base-paired stems and a loop region.[58] The nucleotide sequence for an shRNA is cloned into plasmid DNA under the U6 or H1 promoter activated by RNA polymerase III[58,59] and is delivered into target cells by lipoplexes using vector systems. To analyze gene silencing and protein expression, cells are transduced with plasmid DNAs in the range of $10^6$ to $10^9$ molecules per cell to transduce cells, according to the manufacturer's instructions of lipofection reagents. The cellular uptake of plasmid DNAs into cells is approximately 45% of the transfected dose, which is equivalent to $10^6$ molecules of plasmid DNAs.[60] Of transfected plasmid DNAs, approximately 1–5% of plasmid DNAs are transferred into the nuclei.[60,61] Thus, larger amounts of shRNA-expressing plasmid DNA are required for gene silencing analysis when using plasmid DNA. Expressed shRNAs such as pre-miRNAs are transported by Exp-5 from the nuclei to the cytoplasm and are processed by Dicer into siRNAs. AGO2 is responsible for cleaving the antiguide strand of siRNA during RISC activation,[62] leading to subsequent degradation of the cleaved mRNA transcript by cellular exonucleases and mediating posttranscriptional gene silencing. Thus, the RNAi pathway mediated by siRNAs and shRNAs are related with the pathway of cellular miRNA biogenesis.

A prudent approach for the application of RNAi is to identify potent shRNAs that are capable of downregulating target mRNAs. In addition, a major goal is to properly express potent shRNAs at low levels and sustain gene silencing in target cells.[63] Moreover, efficient gene delivery tools are required for RNAi studies.[63] When ncRNAs are properly designed and delivered into target cells, RNAi will be a powerful approach for the analysis of gene function and the treatment of a variety of acute or chronic diseases.[63,64] Although the phenotypes of gene silencing are observed by transfection of siRNAs or plasmid vector-derived siRNAs, higher amounts of chemically synthesized siRNAs or shRNA-plasmid DNAs are required for in vivo analysis. For example, intravenous administration of shRNA-expressed adeno-associated virus vectors at $10^{12}$ particle doses resulted in liver toxicity (36 of 49 mice; 73%) and caused death within 2 months (23 of 49 mice; 47%) as a result of saturation of Exp-5 and subsequent inhibition of endogenous pre-miRNA nuclear export.[65] Thus, gene delivery vectors using much lower amounts of ncRNAs will be needed for effective RNAi delivery without toxicity to cells in RNAi-mediated gene therapy.

## 5. Adenovirus Vectors for ncRNA Gene Delivery

Human adenoviruses (HAdVs) are nonenveloped viruses containing double-stranded DNA and the biology of species C HAdV serotype 5 (HAdV-C5) is relatively well characterized.[66] The advantages of this virus as vector include efficient transduction of target cells at a low multiplicity of infection (MOI), well-established methods for manipulation and propagation,[67] and relative safety because the viral genome is not integrated into the host genome.[68,69] Human adenovirus C5 is amplified in infected cells at much higher titers (approximately $10^{11}$ plaque-forming units [PFU]/ml), which is critical for in vivo gene delivery and clinical applications, compared with retroviruses (approximately $10^6$ to $10^7$ PFU/ml). Also, they are stable during the process of concentration and purification by cesium-chloride ultracentrifugation.[70] They can deliver large therapeutic genes,[71,72] up to approximately 37 kb.[73] The HAdV-C5 vectors more effectively accomplish transient expression of transgenes compared with other gene delivery vehicles.[74] Therefore, HAdV-C5 vectors have been widely used as transgene delivery vectors in the research fields of gene therapy and basic science. Human adenovirus C5 vectors are one feasible vehicle to deliver ncRNAs as well as protein-coding genes in mammalian cells.

Initially, the first-generation HAdV vector was tested to examine the gene-silencing effect of shRNA for enhanced green fluorescent protein (eGFP) (siGFP) using the cytomegalovirus (CMV) promoter, controlled by RNA polymerase II, and a full-length simian virus 40 (SV40) polyadenylation signal.[75] However, there was no effect on gene silencing of eGFP.[75] On the other hand, siGFP, which is inserted within six base pairs from the transcription start site of the CMV promoter followed by a synthetic minimal polyadenylation signal, reduced the target mRNA and protein expression in cells. By analyzing the gene silencing effects at different lengths of spacer (a spacer of 9, 12, or 21 nucleotides), it was found that the spacer was critically important for functional gene silencing by the CMV promoter controlled by RNA polymerase

II, along with a synthetic minimal polyadenylation signal.[75] In addition to cell lines, first-generation HAdV vector expressing siGFP and shRNA for α-glucuronidase mediated gene silencing in the brain striatal region of eGFP-transgenic mice and in the mouse liver, respectively.[75]

The use of RNA polymerase II-based promoters can be a promising strategy for ncRNA delivery. However, optimization will be required to transcribe a short length of nucleotide sequences. The human H1 promoter controlled by RNA polymerase III, which efficiently transcribes ncRNA, was tested in an HAdV vector, pAdEasy-1 (Agilent Technologies, Santa Clara, CA).[76] Human adenovirus-mediated delivery of shRNA for tumor suppressor p53 downregulated its protein expression in different cell lines.[76] In addition to human H1 promoter, mouse U6 and human U6 promoters were tested to examine the RNAi effect of an shRNA for glyceraldehyde 3-phosphate dehydrogenase (GAPDH) (shGAPDH).[77] The human H1 promoter-mediated shGAPDH effectively reduced approximately 88% of GAPDH expression compared with the human U6 and the mouse U6 promoters in primary murine culture cell line, whereas the mouse U6 promoter was more effective in downregulating its target mRNA (approximately 76%) in mouse NIH3T3 cells compared with the human H1 and the human U6 promoters.[77] Currently, efficient HAdV expression kits to insert ncRNAs have been developed and are available from some companies. A list of HAdV vector systems for noncoding gene delivery is presented in Table 1.

**Table 1 Academically and Commercially Available HAdV Vector Systems**

| pShuttle Vector | Promoter | Adenoviral Plasmid | Method | Reference or Source |
|---|---|---|---|---|
| pShuttle-H1 | Human H1 | pAdEasy-1 | Homologous recombination in *Escherichia coli* strain BJ5183 | 74 |
| pAd shRNA/H1, pAd shRNA/hU6, pAd shRNA/mU6, | Human H1, Human U6, Mouse U6 | pAdEasy-1 | Homologous recombination in BJ5183 | 75 |
| pSIREN | Human U6 | pAdeno-X | In vitro ligation | Clontech Laboratories, Inc. |
| pENTR/U6 | Human H1 | pAd/BLOCK-iT-DEST | Gateway system | Life Technologies |
| pRNA-H1.1/Adeno, pRNAT-H1.1/Adeno, pRNATin-H1.2/Adeno | Human H1 | pAdEasy-1 | Homologous recombination in BJ5183 | Genscript USA, Inc. |
| pRNA-U6/Shuttle | Human U6 | pAdeno-X | In vitro ligation | Genscript USA, Inc. |

# 6. In Vivo Delivery of ncRNAs Using Adenoviral Vectors

A first-generation HAdV vector was tested to examine in vivo processing of shRNA, accumulation, functional kinetics, and side effects related to shRNA saturation of the cellular gene-silencing machinery.[78] Murine adenosine triphosphate (ATP)-binding cassette multidrug resistance protein 2 (Abcc2/MRP2/cMOAT) is a liver export pump located in the canalicular membrane of hepatocytes and is responsible for secretion of conjugated bilirubin to bile.[79–81] Intravenous injection of an HAdV vector expressing shRNA for Abcc2 (shAbcc2) at $2.0 \times 10^9$ infective units showed that functional gene silencing of Abcc2 was detectable for up to 3 weeks in C57BL/6 mice and is related with increased bilirubin levels in serum.[78] Abcc2 mRNA were detected at normal levels in the liver at 24 days postadministration, resulting in bilirubin clearance observed by 24 days postadministration.[78] Although processed siAbcc2 in the mouse liver was accumulated and detected for up to 150 days postadministration,[78] it was rendered nonfunctional by 24 days postadministration in the liver. Thus, the bilirubin clearance resulted from an alternation of siAbcc2 function rather than disappearance of siAbcc2 from the liver.[78] Also, overexpression of exogenous shAbcc2 in vivo did not interfere with processing or accumulation of endogenous miRNAs.[78] Thus, these experiments concluded that overexpression of shAbcc2 mediated by HAdV vector did not alter miRNA biogenesis, accumulation, or the functionality of RNAi machinery. On the other hand, these experiments presented the following points to RNAi therapy: (1) Gene-silencing effects of shAbcc2 were not detected in the liver when using lower amounts than $2.0 \times 10^9$ infectious units of the HAdV vector; therefore, side effects by HAdV vector-mediated RNAi therapy will be undetectable in unwanted tissues. (2) There is a threshold amount of siRNAs needed to achieve gene silencing. Also, oversaturation of therapeutic ncRNAs may inactivate RNAi per se. Whether a threshold model is fitted into endogenous and exogenous miRNAs remains unclear. Even if endogenous miRNAs are expressed in cells, and their miRNAs are expressed at lower levels than the threshold amounts, they may be nonfunctional. In addition, the threshold amount of each ncRNA may be determined by the amounts of target mRNAs. (3) Because amounts of siAbcc2 were decreased in the liver by 24 days postinjection, human H1 promoter functions in the liver for short time periods followed by inactivation of the promoter. Therefore, human promoter(s) may not be suitable for long-term in vivo murine analysis. In summary, more effective promoters will be required for long-term RNAi therapy.

In addition to a first-generation HAdV vector, helper-dependent HAdV vectors carrying shRNAs have been tested to examine gene-silencing effects in vivo.[82,83] Intravenous administration of helper-dependent HAdV vector carrying shRNA for fatty acid-binding protein 5 (FABP5) at $1.0 \times 10^{11}$ viral particles resulted in reduction of its target protein by approximately 75% for 1 week in the liver of C57BL/6J mice without inflammatory infiltration and toxicity.[82] However, the injection of $2.0 \times 10^{11}$ viral particles caused harmful side effects, toxicity in the liver, and activated cellular antiviral defense pathways as a result of the high expression level of shRNA for FABP5 (shFABP5).[82] Thus, these experiments indicated that optimization of vector dose is necessary for maximizing knockdown efficacy without cytotoxicity mediated by the vector[82,84] and that the shRNA design is important to avoid activation of interferon response through Toll-like receptor 7.[82] Moreover, miR-122 and *lethal*-7a (*let*-7a) were examined by Northern blot analysis to investigate whether overexpression of shFABP5 in the liver inhibits expression of cellular miRNAs.[82] miR-122 is highly expressed in the adult liver,[85–87] where it accounts for more than 70% of all miRNAs[85] and functions as a regulator of fatty-acid metabolism in mouse studies.[87–89] On the other hand, *let*-7a is a member of a larger class of miRNAs,[85,90] a developmental key regulator,[91] and is highly overexpressed in the adult lung tissue.[92] Overexpression of shFABP5 mediated by helper-dependent HAdV vector administrated at $2.0 \times 10^{11}$ viral particles did not interfere with gene expression of these miRNAs in the liver.[82] Because mature miRNAs were expressed in the liver, whether cellular miRNA biogenesis is inhibited by exogenous shRNAs remains to be elucidated. Another example is when a helper-dependent HAdV vector carrying an shRNA for sterol regulatory element-binding protein-1 (SREBP-1) was tested to examine the functionality of the vector in vivo.[83] Overexpression of shRNA for SREBP-1 reduced over 90% of target protein expression in the liver of C57BL/6J mice for approximately 1 week.[83] Also, it decreased more than 90% of target protein expression for 3 weeks in the liver of an animal model of obesity and type 2 diabetes, the *db*/*db* mouse (leptin receptor deficient). Liver toxicity was not observed in both kinds of mice after the vector administration at $1.0 \times 10^{11}$ viral particles.[83] Together, gene-silencing effects for endogenous gene mRNAs and/or downregulation of endogenous proteins mediated by HAdV vectors were observed in mice. However, RNAi in vivo was detected at limited ranges of viral particles.[78,82,83] Thus, the use of conventional HAdV vectors seems to be restricted in obtaining effective silencing of endogenous mRNAs in vivo, which potentially limits their effectiveness in certain therapeutic settings.

Ribonucleic acid interference in vivo was observed for limited periods.[78,82,83] On the other hand, human Factor IX expression in the serum obtained from C57BL/6 mice intravenously injected first-generation HAdV vector ($2.0 \times 10^9$ transducing particles) as well as helper-dependent HAdV vector continued to be detected over a period of 1 year.[93] In comparison, posttranscriptional gene silencing effects mediated by conventional HAdV-C5 vectors (first-generation and helper-dependent HAdV vectors) were observed in vivo for only short time periods as described above. Indeed, a major limitation of the use of HAdV vectors is the host immune response.[94–97] First-generation HAdV vectors induce an innate immune response that results in inflammation of infected tissues and efficient clearance of administered vectors.[97] Although helper-dependent HAdV vectors show lower side effects in the liver and vector-directed T cell responses,[98] they also induce innate and adaptive immune responses in DBA/2 and Balb/c mice.[84,98] In contrast to adaptive immunity, the innate immune response results from the HAdV particles per se and does not need viral gene expression.[97] A study suggested that transfection of viral-associated (VA) RNAs that were synthesized in vitro induced a signal pathway for innate immune responses in a cell culture model.[99] In addition, comparative analysis using a VA-deletion mutant (VAI– and VAII–) suggested that VA

RNAs are capable of inducing type I interferon in infected cells.[100] Because VAI RNA is also considered to be an RNAi inhibitor in infected cells, its elimination will be needed for long-term RNAi function.

# 7. MicroRNA-Mediated Regulation of Adenovirus Vector Tropism

The development of replication-deficient and replication-competent HAdV vectors for transgene expression and oncolysis, respectively, has been under way for many years to customize those vectors for experimental and therapeutic purposes. We essentially focus on controlling gene expression to specific target cells, reducing viral immunogenicity, increasing efficacy, and decreasing cytotoxicity to normal tissues. For the best therapeutic benefit, HAdV vectors should be delivered into specific target cells while avoiding sequestration in other organs or toxicity from infection of unwanted cells. Many strategies have been tested to target the tissue tropism for gene therapy using transcriptional targeting and transductional targeting.[101,102] Studies have shown that tissue-specific or tumor-specific miRNAs offer promising tools to control the tropisms of transgene expression,[103] leaky virus gene expression,[104] and replication-competent adenoviruses.[105–108]

To examine whether endogenous miRNAs related to tissue lineage and differentiation state mediate gene silencing through multiple artificial target elements, recombinant lentiviral vectors containing multiple target elements inserted into the 3′-UTR of the reporter gene were developed.[109] This article provided us a few important lessons in developing miRNA-regulated HAdV vectors.[103,105,106,108] (1) The silencing of the reporter gene expression depends on a threshold value of miRNA expression in cells.[109] Even if miRNA is detectable in cells, the gene-silencing effect will not function effectively if at all in cells expressing miRNA at low levels. Therefore, validation of miRNA target elements is required in a variety of cells including target cells. (2) Although a conventional lentivirus vector expressing eGFP transduced Kupffer cells and hepatocytes in the liver through intravenous administration, eGFP signal was eliminated in most Kupffer cells and hepatocytes infected with a recombinant lentivirus vector expressing combinatorial miRNA target elements for liver-specific miR-122a and hematopoietic cell-specific miR-142-3p.[109] Thus, a strategy using combinatorial miRNA target elements can more effectively and nearly completely block leaky expression of therapeutic genes in unwanted multiple tissue types. Based on this evidence, miR-122a-target sequences were examined in HAdV vectors.

Intratumorally injected HAdV vectors are disseminated into systemic circulation, resulting in severe hepatotoxicity owing to effective transduction to the liver.[110–114] To examine whether miR-122a-target elements can mediate reduction of a reporter protein expression in the liver, an HAdV vector expressing firefly luciferase with inserted miR-122 target elements (Ad-L-122aT) was intratumorally administrated at $3.0 \times 10^9$ or $3.0 \times 10^{10}$ viral particles into murine melanoma B16 tumors of mice.[103] Ad-L-122aT reduced luciferase expression in the liver by 130- to 1500-fold compared with that of the control virus. Next, an HAdV vector carrying the herpes simplex virus thymidine kinase (HSV-tk) gene and miR-122a-target elements (Ad-tk-122 aT) was tested to examine therapeutic efficacy in B16 cells and hepatotoxicity.[103] The suicide gene therapy strategy against B16 cells using Ad-tk-122 aT and ganciclovir showed antitumor effect to the B16 cells with reduced hepatotoxicity.[103] Thus, the use of miR-122a-target elements provided enhanced safety to hepatocytes and increased the specificity of suicide gene therapy. In addition to the first-generation HAdV vector, miR-122a–regulated conditionally replicating adenoviruses (CRAds) were developed by inserting the miR-122a target elements into the 3′-UTR of the E1a gene[105,107] or the E1a/firefly luciferase fusion gene.[108] Initially, Dr. Ylösmäki et al. adapted miR-122a target elements to regulate gene expression of E1a-Δ24 protein and developed the CRAd Ad5/3-Δ24-122.[105] The insertion of miR-122a target elements resulted in reduction of E1A mRNA and protein in the hepatocellular carcinoma cell line Huh7 cells, highly expressing miR-122a, infected at an MOI of 0.05 PFU/cell but not human lung carcinoma A549 cells. Although 95% of the cells are uninfected at an MOI of 0.05 PFU/cell, according to Poisson distribution, the insertion of miR-122a target elements significantly reduced E1A mRNA and protein in Ad5/3-Δ24-122-infected Huh7 cells compared with Ad5/3-Δ24 (control)–infected Huh7 cells.[105] However, the insertion of miR-122a target elements did not inhibit late viral protein expression in Ad5/3-Δ24-122–infected Huh7 cells as well as Ad5/3-Δ24–infected cells.[105] These data suggested that E1A protein may be expressed at undetectable levels in infected cells and further attenuation is required for preventing CPE induction in Huh7 cells.[105] In addition to the in vitro study,[105] a replication-competent adenovirus carrying the E1a and firefly luciferase fusion gene controlled by miR-122a target elements (Ad5 E1A-Luc-mir122) was tested in vitro and in vivo.[108] The insertion of miR-122a–binding sites achieved dramatic attenuation of E1A-luciferase fusion protein expression in Huh7 cells and the liver of Balb/c mice. To examine hepatotoxicity, $5.0 \times 10^{10}$ viral particles, which are a potentially lethal dose, were administrated to Balb/c mice. The intravenous administration of Ad5 E1A-Luc-mir122 decreased E1A-luciferase expression up to approximately 80-fold in the liver at 72 h postinjection compared with Ad5 E1A-Luc.[108] Also, miR-122a target elements reduced genome replication in the liver by approximately 50-fold and clearly abrogated liver toxicity was shown by reduced levels of alanine aminotransferase and aspartate aminotransferase in serum. These data suggested that posttranscriptional gene silencing strategy is useful for control adenoviral E1a gene expression in vitro and in vivo. Also, miR-122a target elements are a feasible tool to reduce E1A protein-mediated liver toxicity.

Tumor-specific promoters such as the human telomerase reverse transcriptase (hTERT) promoter,[106] the cyclooxygenase-2 gene promoter,[115] and the midkine gene promoter[116] are promising tools for transcriptional targeting to achieve the selective replication between cancer and normal cells. On the other hand, a study demonstrated that a CRAd controlled by the hTERT promoter replicates in normal cells.[106] Therefore, target sequences for several tumor-specific miRNAs, miR-143, miR-145, miR-199a, and *let*-7a, were tested along with the hTERT promoter in CRAds to enhance the safety of the CRAd strategy.[106] Although replication of these viruses was not examined in infected normal cells, the combination of miRNA-target elements and the hTERT promoter resulted in reduction of viral genome replication in normal cells compared with the control.[106] Thus, the data suggest that posttranscriptional gene silencing along with transcriptional targeting can improve the safety of CRAds. Collectively, miRNA strategy is a potential approach to achieve selective gene expression in vitro and in vivo. In particular, miR-122a target elements are suitable for eliminating hepatotoxicity in vivo.[103,108]

## 8. Adenoviral Virus-Associated RNAs and Their Biogenesis

In 1966, the presence of a low-molecular-weight RNA was shown in HAdV-C2–infected KB cells but not uninfected cells.[117] This RNA was designated VA RNA, readily detected in infected cells and expressed in the cytoplasm.[118] The nucleotide sequence and predicted secondary structure of a VA RNA (5.5S), which is equivalent to VAI RNA, from HAdV-C2 were reported.[119] In addition to VAI RNA, VAII RNA (5.2S) was isolated from HAdV-C2–infected cells and the gene encoding VAII RNA was identified as downstream of the VAI RNA gene.[120,121] Both VA RNAs are detected at 2 h postinfection; thus, their synthesis starts before viral DNA replication.[120] At 2 h postinfection, VAI and VAII RNAs account for approximately $6 \times 10^6$ and approximately $2 \times 10^6$ molecules, respectively, per infected HeLa cell.[120,122] Whereas the synthesis of VAI RNA is increased after onset of viral DNA replication, that of VAII RNA is already saturated.[120] At 14 h postinfection, VAI and VAII RNAs account for approximately $10^8$ and approximately $10^7$ molecules, respectively, per infected HeLa cell.[120,122–124] Transcription analysis of HAdV-C2 VAI RNA shows that it contains two initiation sites for transcription at nucleotide (nt) position 10,607 and at nt position 10,610 producing VAI(A) and VAI(G) RNAs, respectively.[125] The latter is the major product of VAI RNA and accounts for approximately 75% of the total VAI RNA amounts.[125] On the other hand, the transcription VAII RNA starts at nt position 10,866. Virus-associated I RNAs of HAdV-C2 and C5 are well characterized and are about 160 nucleotides in length as well as GC-rich. The secondary structures of VA RNAs, consisting of three kinds of domains—the terminal stem, apical stem loop, and central domain—are well conserved in VA RNAs of all HAV serotypes.[126] Virus-associated RNAs are transcribed by RNA polymerase III through their own promoters.[120,127,128] Using a cell-free in vitro transcription assay for HAdV-C2 and HAdV-C5 VAI RNA deletion constructs, an element (+9 to +72) required for its transcription was identified within the VAI RNA-encoding gene.[129] Also, this experiment revealed a few important characteristics of VA RNAs. The 5′-flanking sequences of the VAI RNA gene are more effective as a transcriptional element than the equivalent VAII RNA gene sequences.[129] In addition, the VAI RNA gene reduces expression of the VAII RNA gene.[129] This phenomenon may result from a competition for the limited factor(s) required for their transcriptions, resulting in further reduction of the VAII RNA gene expression level.[129] Therefore, the nucleotide sequence for the

VAI RNA gene is considered to be a strong competitor of VAII RNA transcription.[129] Subsequently, two elements, element A (+10 to +18) and element B (+54 to +69), required for the transcription of the VAI RNA gene, was determined in the intragenic promoter sequences.[130] Virus-associated RNAs are mainly localized in the cytoplasm of infected cells, where they accumulate at high levels. Exportin-5 binds to HAdV-C5 VAI RNA and transports it from the nuclei to the cytoplasm.[131,132]

## 9. Inactivation of Protein Kinase R by VAI RNA

The first hint as to the function of VA RNAs came from the analysis of a VAI-deleted HAdV, dl331, which contains a deletion of 28 base pairs in the Box B element of the VAI RNA gene promoter. The lack of VAI RNA resulted in dramatic suppression of viral replication in infected cells.[133] The defect mediated inefficient synthesis of viral proteins during the late phase of infection,[134,135] resulting from phosphorylation of eukaryotic initiation factor-2 (eIF-2) through protein kinase R (PKR) (interferon-induced, dsRNA-activated protein kinase, or eukaryotic translation initiation factor 2-α kinase 2 [EIF2AK2]). Eukaryotic initiation factor-2 is composed of three subunits, α, β, and γ, and functions as a translational control.[136,137] At the early stage in the initiation process of translation, eIF-2 binds the initiator transfer RNA (Met-tRNA) in a guanosine triphosphate (GTP)-dependent manner to form the ternary complex, which interacts with the 40S ribosomal subunit. Subsequently, the complexes bind mRNA and the 60S ribosomal subunit for translation. Once the initiation process of translation is completed, eIF-2 is released from the ribosome, lacking Met-tRNA, binding guanosine diphosphate (GDP) instead of GTP. The eIF-2–GDP complex is an inactive form and is converted to an active form (the eIF-2–GTP complex) by phosphorylating GDP to GTP via guanosine nucleotide exchange factor (eIF-2B) for another round of translation initiation. Thus, eIF-2 is required for the initiation of translation. In wild-type adenoviral-infected cells, PKR is thought to be activated by dsRNAs, which is produced as a result of the symmetric transcription of the viral genome.[138,139] However, VAI RNA directly binds to PKR and inhibits not only its autophosphorylation but also its kinase activity.[139–141] Therefore, the VAI RNA-mediated inactivation of PKR allows the synthesis of viral proteins in wild-type infected cells. In contrast to wild-type infected cells, PKR is activated in dl331-infected cells evidenced by its autophosphorylation and subsequent phosphorylation of the eIF-2α subunit, resulting in inhibition of viral protein synthesis. Collectively, VAI RNA is necessary for effective synthesis of viral proteins and blocks the antiviral defense of the host via the PKR pathway. On the other hand, VAII RNA does not block PKR activation as effectively as VAI RNA or may not impair it.[139] Therefore, VAII RNA is considered to exhibit either only limited ability to block PKR activation or not to have an intrinsic function to inhibit RKR activation.

## 10. Virus-Associated I RNA as an RNAi Inhibitor

In 2004, Lu et al. reported that transient expression of HAdV-C5 VAI RNA is capable of partially blocking the function of Exp-5 to transfer pre-miRNAs from the nucleus to the cytoplasm and competitively inhibits RNAi in transiently transfected cells.[142] In addition, HAdV-C2 and C5 VAI RNAs bind to Dicer,[142,143] blocking its function in processing cellular pre-miRNAs to mature miRNAs in vitro, and show an intrinsic capacity to inhibit RNAi in vitro.[142] Thus, in vitro studies suggested that VAI RNAs function as an RNAi inhibitor owing to the hijacking of Exp-5 and Dicer, key components of the RNAi machinery.[142–144] Human adenovirus C2 VAI RNA as well as HAdV-C5 VAI RNA are capable of inhibiting RNAi mediated by an shRNA or an miRNA in transiently transfected cells.[142,143] Whereas HAdV-C2 VAII RNA did not inhibit RNAi in transiently transfected cells, the H1 promoter-driven HAdV-C2 VAII RNA blocked RNAi in transiently transfected cells.[143] Therefore, the result obtained using the H1-promoter suggested that VAII RNA has an intrinsic capacity to inhibit RNAi.[143] Of note, Dicer processed VA RNAs in vitro and in vivo, generating different lengths of viral microRNAs (mivaRNAs).[142,143,145,146] Interestingly, the most prominent mivaRNAs are made from the 3′-strands of VAI and VAII RNAs.[142,145] The 3′-strands of VAI and VAII RNAs are more efficiently incorporated into the RISC compared to cellular miRNAs[145] and account for approximately 80% of RISC-bound miRNAs at the late phase of infection in 293-Ago2 cells.[145] In particular, the 3′-strands of VAII RNA are preferentially detected in the RISC compared with those of VAI RNA.[145] Together, full-length VA RNAs and mivaRNAs are considered to block the key molecules involved in the RNA pathway and inhibit cellular miRNA biogenesis as competitor(s), respectively.

## 11. Virus-Associated II RNA

To investigate the function of VAII RNA, VAII RNA-binding proteins were identified by Northwestern blot analysis.[147] Virus-associated II RNA directly binds to RNA helicase A and NF90, a component of the heterodimeric NFAT.[147] Although NF45 was isolated along with RNA helicase A and NF90, it did not directly bind to VAII RNA. Therefore, NF45 was involved in the complex through NF90. Interestingly, Exp-5 was not isolated as a VAII RNA-binding protein by Northwestern blot analysis.[147] Although Exp-5 transports NF90 along with VAI RNA from the nuclei to the cytoplasm,[132] the mechanism of transportation of VAII RNA from the nucleus to the cytoplasm remains to be elucidated. Whereas RNA helicase A interacts with siRNA, AGO2, TRBP, and Dicer and functions in the RNAi pathway,[148] it is not required for RISC activity.[149] NF90/NFAR is activated by PKR[150] and is incorporated into the Drosha-a complex in the nuclei, but is not associated with the Microprocessor complex.[39] Thus, VAII RNA interconnects the PKR and the RNAi pathways through NF90. However, the functional mechanism of VAII RNA in the PKR and the RNAi pathways remains to be elucidated as described above.

# 12. Virus-Associated–Deleted HAdV Vectors for RNAi Analysis

Studies of viral biology including HAdV have reported that viral genomes encode functional genes to inhibit RNAi and that the gene products disturb key molecules in the RNAi machinery during productive infection.[142–145,151–155] The HAdV-C5 genome encodes two kinds of VA RNAs: VAI RNAI and VAII RNAs.[121] Of these, VAI RNA is considered to be an RNAi inhibitor of exogenous ncRNAs.[142] Because full utility of the current vector to analyze the function of ncRNAs would be limited, it is clear that new types of HAdV vectors containing deletion of the VA RNA genes are required for ncRNA and gene-silencing analyses. Two kinds of methods to generate VA-deleted vectors have been reported.[156,157] One method is to eliminate gene expression of VAI and VAII RNAs by deleting the Box B elements in their own promoters.[157] However, inactivation of both VAI and VAII impairs viral growth in host cells.[158] Therefore, a VA-deletion vector (AdΔVR) is not generated by transfecting its plasmid DNA in 293 cells, which suggests that the AdΔVR genome is noninfectious genome. Although the AdΔVR vector was generated and propagated in a VAI RNA-expressing 293 cell line,[157] it was not purified from VA-expressing 293 cells by cesium chloride ultracentrifugation.[157] Subsequently, the function of the AdΔVR vector was compared with first-generation and helper-dependent HAdV vectors in cells stably expressing luciferase by comparing knockdown efficiency of shRNA for firefly luciferase (shFluc).[159] The AdΔVR vector as well as helper-dependent HAdV vector downregulated the expression of firefly luciferase and improved knockdown efficiency by approximately 10% compared with the first-generation HAdV vector. Interestingly, siFluc processed from shFluc as well as mivaRNAs was efficiently incorporated into the RISC in first-generation HAdV-infected cells.[159]

Another strategy was to use the FLP-FRT homologous recombination system to excise the gene regions for VAI and VAII RNAs.[156] A first-generation AdV vector called Pre-AdV vector contains two FRT sequences upstream and downstream of the VA RNA genes. Pre-AdV vector is initially propagated in 293 cells and then amplified in 293hde12 cells[160] that stably express humanized FLPe (hFLPe) recombinase.[156] In 293hde12 cells, the VA RNA genes are excised from the Pre-vector genomes by hFLPe recombinase, resulting in its circular form. Because VA RNAs seem to be expressed from circular VA RNA genes in AxdV-infected 293hde12 cells, VA-deleted AdV (AxdV) vector can be effectively produced in 293hde12 cells. In this vector system, a second infection in 293hde12 cells is required to obtain sufficient amounts of AxdV for purification by ultracentrifugation and minimize the contamination of Pre-vectors.[156] In fact, 1–3% of VA RNAs were detected in the RNA samples extracted from AxdV vector-infected HuH7 cells by Northern blot analysis compared with VA RNA expression levels detected in Pre-AdV vector-infected cells.[156] Transduction efficiency, using EGFP as a reporter protein, was reduced approximately 90% because of the deletion of the VA RNA genes.[156] However, RNAi mediated by AxdV vectors carrying shRNAs for hepatitis C virus RNA was effective at a low MOI compared with its VA-intact vector.[161] In contrast to low MOIs, the use of AxdV vectors at higher MOI may interfere with gene silencing by VA RNAs originating from the

Pre-vector. Also, the analysis obtained from the AxdV vector system suggested that helper-dependent HAdV vectors may have a potential problem in RNAi analysis when used at higher doses.

## 13. Adenovirus Full-Length VA RNAs and VA RNA-Derived miRNAs Suppress Cellular Gene Expression

In 2005, Andersson et al. reported that both VAI and VAII RNAs are processed by Dicer in vitro and in vivo, generating viral small RNAs which are identical with 5′- and 3′-strands.[143] Also, artificial experiments using the transcripts containing the complementary sequences for these strands demonstrated that both the 5′- and 3′-strands of VAI RNA are incorporated into the RISC, whereas the 3′-strand of VAII RNA is selectively incorporated during a lytic cycle of infection.[143] These results demonstrated that viral small RNAs generated from VA RNAs function as miRNAs, if the 3′-UTR of cellular mRNAs contain complementary sequences. In addition, Aparicio et al. identified small RNAs derived from VA RNAs (svaRNAs) in HAdV-infected cells.[162] They showed that svaRNAs interacted with AGO-2 proteins and function as siRNAs under an artificial reporter assay.[162] Of VAI RNA produced in infected 293 cells, only an approximate 2–5% of VAI RNA were processed into svaRNAs.[162] On the other hand, Xu et al. showed that Dicer processes VAI and VAII RNAs into different lengths of mivaRNAs; major products for VAI RNA are mivaRI-137 and -138 cleaved at the position 137 and 138, respectively, whereas a major product for VAII RNA is mivaRII-138 cleaved at the position 138.[145] Also, Xu et al. showed that Dicer preferentially processed VAII RNA compared with VAI RNA. Both mivaRNAs processed from VAI and VAII RNAs, along with an artificial reporter with their target sequence, were isolated from Ago-2–containing complex.[145] Moreover, they examined small RNAs bound in the RISC in HAdV-infected 293-Ago2 cells by analyzing a small RNA cDNA library prepared from the Ago2 protein-containing complexes.[145] mivaRIIs were present in approximately twofold more molecules than total small RNAs in the cytoplasm of infected cells and accounted for approximately 60% of total small RNAs incorporated into the RISC.[145] In their estimation, approximately 75,000 molecules of mivaRII-138, which was cleaved from approximately 1.5% of VAII RNA, were produced at the late phase of infection.[145] On the other hand, mivaRIs accounts for approximately 20% of total small RNAs incorporated into the RISC.[145] Approximately 80% of total small RNAs incorporated into the RISC were derived from VA RNAs at the late phase of infection.[145] These data suggested that the mivaRNAs of VAI and VAII RNAs are efficiently incorporated the RISC compared with cellular miRNAs.[145] Because mivaR-138 is associated with polyribosomes,[145] Xu et al. also suggested that mivaRNAs may function as miRNAs to downregulate cellular gene expression during productive infection.

To investigate the target genes for mivaRNAs, microarray analysis was performed using a human cervix cancer cell line, HeLa cells, transiently expressing HAdV-C2 VAI and VAII RNAs for 3 days.[163] Using Affymetrix HG-U133-Plus2 microarrays containing 54,675 probes, transient expression of VA RNAs significantly altered 1099 genes: 637 upregulated and 462 downregulated genes. TargetScan and Motif Locator analyses suggested that the T cell intracellular antigen 1 (TIA-1) gene is a direct candidate for mivaRI-138.[163] TIA-1 is a factor that activates apoptosis by regulating the RNA metabolism of some mRNAs that encode for proapoptotic molecules.[164] Comparative experiments using mivaRI-138–expressing plasmid and a ΔVA mutant (dl331 is a VAI-deleted mutant), and wild-type adenovirus also suggested that TIA-1 is a direct target for mivaRI-138 in infected 293 and HeLa cells. Also, the downregulation of TIA-1 was observed by 48 h postinfection in wild-type (AdWT)-infected 293 cells compared with dl331-infected 293 cells.[163] Therefore, TIA-1 is considered to be a factor reduced after viral replication. In addition to TIA-1, VA RNAs and mivaRNAs modulated many cellular genes expression involved in cell signaling, cell growth, apoptosis, transcription, DNA repair, and RNA metabolism.[163]

Using a first-generation HAdV-C5 vector deleting the region of the VA RNA genes (VA-deleted AdV[156]) and the VA-intact HAdV-C5 vector, the target genes of VA RNAs were examined by microarray analysis.[165] Hepatoma-derived growth factor (HDGF) was identified as a candidate for VA RNAs in A549 cells infected with AdV (VA+) or VA-deleted AdV at an MOI of 0.5 relative vector titer (rVT)/cell[166] for 1 day postinfection, but TIA-1 was not.[165] Also, significant suppression of HDGF mRNA was observed in AdV-infected HeLa and a human hepatocellular carcinoma cell line, HuH-7 cells, compared with VA-deleted AdV. Although Kondo et al. found a putative site for mvaRI-138 in the 3′-UTR of the HDGF gene, they did not observe the suppression of the HDGF mRNA in 293 cells transiently expressing VA RNAs and a reporter plasmid carrying the 3′-UTR of the HDGF gene. Therefore, Kondo et al. suggested that its suppression may be mediated by full-length VA RNAs and does not result from gene silencing via mivaRNAs. Because the suppression of HDGF mRNA was detected at an early time point in AdV-infected 293 cells compared with VA-deleted AdV-infected 293 cells, they also suggested that full-length VA RNAs suppressed the HDGF gene expression per se by an unknown mechanism.[165] Furthermore, VA-intact and VA-deleted AdVs carrying the HDGF cDNA gene controlled by the EF1$\alpha$ promoter and artificial polyadenylation signal were generated in 293 cells or 293hde12 cells, respectively. This experiment's data indicated that overexpression of HDGF protein did not essentially affect VA-deleted AdV viral production in infected 293hde12 cells. However, 293 cells overexpressing HDGF restricted the VA-deleted AdV DNA production by 1 day postinfection in 293 cells, but not wild-type DNA replication.[165] Therefore, HDGF suppression may be needed to effectively maximize the VA-deleted HAdV DNA replication and to produce the viruses in 293 cells. Collectively, microarray analysis using first-generation HAdV vectors suggested that the full length of VA RNAs as well as mivaRI-138[163] functions as suppressors of cellular gene expression.[165]

## 14. The PKR Pathway Rather than the RNAi Pathway Is Critical for Productive Infection

One question regarding adenoviral replication is whether inactivation of the PKR pathway or the RNAi pathway is critical to achieve productive infection. Cellular and viral protein expression was drastically reduced in dl331-infected cells owing to blocked translation initiation by active PKR-mediated eIF-2α phosphorylation.[124,133,139,167] Thus, VAI RNA per se is essential for viral protein expression in infected cells. The base substitution mutations in the 5′- or 3′-strand of VAI RNA were nondefective for viral growth and virus protein expression.[168] Also, these substitution mutants of VAI RNA retained the function of inactivating PKR, indicating no effect on viral protein synthesis.[168] In 2010, Bennasser et al. examined the relationship between adenovirus infection and the RNAi pathway in cells with the knockdown of Exp-5 and Dicer.[46] dl-sub720 is a double mutant deleting the genes for VAI and VAII RNAs and shows a phenotype of translational defects in viral proteins.[169–171] Therefore, dl-sub720 viral replication was dramatically suppressed in infected cells.[169,170] Since the knockdown of Exp-5 and Dicer restored the defect of dl-sub720 DNA replication in cells with the knockdown of Exp-5 or Dicer,[46] the RNAi pathway seems to be associated with productive infection of adenovirus. In 2013, Kamel et al. constructed and characterized recombinant HAdVs with substitution mutations in the seed sequences of the 5′- or 3′-strands of the VAI RNA gene.[172] They also investigated whether mivaRIs are critical for achieving productive infection. These mutants, as well as wild-type and an isogenic VAI-expressed mutant, grew similarly in 293 cells.[172] Viral proteins were sufficiently expressed in the substitution mutants-infected cells.[172] Therefore, they also concluded that mivaRIs are not essential for viral replication and viral protein expression.[172] Interestingly, these base-substituted VAI RNAs were processed into small RNAs, resulting in incorporation into the RISC.[172] This data implied that these mutants still have the function of inhibiting exogenous shRNA and cellular miRNA biogenesis.[172] Therefore, the experiments using the base substitution mutants did not answer whether inhibition of cellular miRNA biogenesis is necessary for productive infection. Moreover, Kamel et al. examined whether the PKR or the RNAi pathway is essential to achieve productive infection. Although Dicer knockdown compensated for the defect in dl-sub720 DNA replication,[46] it did not restore viral protein synthesis of dl-sub720.[172] In contrast, gene silencing of PKR compensated for the defect of viral protein synthesis of dl-sub720.[172] Therefore, they concluded that a suppressive effect of VAI RNA on the PKR pathway, rather than gene-silencing effects by VAI RNA and mivaRIs on the RNAi pathway, is critical to achieve productive infection.

## 15. Conclusions and Future Directions

There are numerous publications on functional analysis of ncRNAs or target genes for ncRNAs using conventional HAdV vectors, both first-generation and helper-dependent HAdV vectors. Also, CRAds controlled by miRNA-target sequences were reported to increase safety in the research field of cancer gene therapy. On the other hand, studies demonstrated that VA RNAs inhibit gene silencing or cellular gene expression in transiently expressed human cells. Microarray analysis showed that VAI RNA and miRI-138 suppress cellular gene expression in adenovirus infected cells. A major problem is whether VA RNAs are critical when examining the function of exogenous ncRNAs or endogenous miRNAs. Based on this evidence, HAdV vectors lacking VA RNAs were developed and validated for RNAi analysis. However, there are many technical problems in these processes: generation, propagation, purification, and transgene expression. To circumvent these limitations, we will need to develop novel VA-deleted vectors lacking RNAi inhibitory activity. Moreover, conventional HAdV vectors mediate the gene-silencing effect only for short periods in vivo. The gene expression of HAdV VA RNAs is considered to be dramatically suppressed in mouse cells compared with human cells.[173] When we use conventional HAdV vectors, we may overestimate the gene-silencing potency of therapeutic ncRNAs in in vitro and in vivo animal models. In addition, transcription and gene expression of HAdV VA RNAs in rodent animals remains unknown. Thus, the validation of ncRNA-expressed HAdV vectors is challenging and difficult because of a lack of characterization in critical animal models. There are some technological barriers to developing novel HAdV vectors lacking not only RNAi inhibitory activity but also PKR activity. However, the novel HAdV vectors will be necessary for advances in RNAi analysis, basic science, cancer gene therapy, and ncRNA-mediated gene therapy.

## Acknowledgments

I thank Mr. William Everett (Washington University in St. Louis) and Mr. Calvin Stephens (Washington University in St. Louis) for the language editing of this book chapter and for useful advice and fruitful discussions. This work was supported by grants from Susan G. Komen for the Cure KG100194 (Drs. David T. Curiel and Hideyo Ugai).

## References

1. Light J, Molin S. The sites of action of the two copy number control functions of plasmid R1. *Mol Gen Genet* 1982;**187**:486–93.
2. Crowley TE, Nellen W, Gomer RH, Firtel RA. Phenocopy of discoidin I-minus mutants by antisense transformation in Dictyostelium. *Cell* 1985;**43**:633–41.
3. Melton DA. Injected anti-sense RNAs specifically block messenger RNA translation in vivo. *Proc Natl Acad Sci USA* 1985;**82**:144–8.
4. Rosenberg UB, Preiss A, Seifert E, Jackle H, Knipple DC. Production of phenocopies by Kruppel antisense RNA injection into Drosophila embryos. *Nature* 1985;**313**:703–6.
5. Ecker JR, Davis RW. Inhibition of gene expression in plant cells by expression of antisense RNA. *Proc Natl Acad Sci USA* 1986;**83**:5372–6.
6. Izant JG, Weintraub H. Inhibition of thymidine kinase gene expression by anti-sense RNA: a molecular approach to genetic analysis. *Cell* 1984;**36**:1007–15.
7. Napoli C, Lemieux C, Jorgensen R. Introduction of a chimeric chalcone synthase gene into petunia results in reversible co-suppression of homologous genes in trans. *Plant Cell* 1990;**2**:279–89.
8. Guo S, Kemphues KJ. par-1, a gene required for establishing polarity in *C. elegans* embryos, encodes a putative Ser/Thr kinase that is asymmetrically distributed. *Cell* 1995;**81**:611–20.
9. Fire A, Xu S, Montgomery MK, Kostas SA, Driver SE, Mello CC. Potent and specific genetic interference by double-stranded RNA in *Caenorhabditis elegans*. *Nature* 1998;**391**:806–11.
10. Bartel DP. MicroRNAs: genomics, biogenesis, mechanism, and function. *Cell* 2004;**116**:281–97.

11. van den Berg A, Mols J, Han J. RISC-target interaction: cleavage and translational suppression. *Biochim Biophys Acta* 2008;**1779**:668–77.
12. Ambros V. MicroRNA pathways in flies and worms: growth, death, fat, stress, and timing. *Cell* 2003;**113**:673–6.
13. Kim VN, Nam JW. Genomics of microRNA. *Trends Genet* 2006;**22**:165–73.
14. Matranga C, Zamore PD. Small silencing RNAs. *Curr Biol* 2007;**17**:R789–93.
15. Esteller M. Non-coding RNAs in human disease. *Nat Rev Genet* 2011;**12**:861–74.
16. Mercer TR, Dinger ME, Mattick JS. Long non-coding RNAs: insights into functions. *Nat Rev Genet* 2009;**10**:155–9.
17. International Human Genome Sequencing Consortium. Finishing the euchromatic sequence of the human genome. *Nature* 2004;**431**:931–45.
18. Kozomara A, Griffiths-Jones S. miRBase: integrating microRNA annotation and deep-sequencing data. *Nucleic Acids Res* 2011;**39**:D152–7.
19. Friedman RC, Farh KK, Burge CB, Bartel DP. Most mammalian mRNAs are conserved targets of microRNAs. *Genome Res* 2009;**19**:92–105.
20. Schickel R, Boyerinas B, Park SM, Peter ME. MicroRNAs: key players in the immune system, differentiation, tumorigenesis and cell death. *Oncogene* 2008;**27**:5959–74.
21. O'Rourke JR, Swanson MS, Harfe BD. MicroRNAs in mammalian development and tumorigenesis. *Birth Defects Res C Embryo Today* 2006;**78**:172–9.
22. Calin GA, Croce CM. MicroRNA signatures in human cancers. *Nat Rev Cancer* 2006;**6**:857–66.
23. Jay C, Nemunaitis J, Chen P, Fulgham P, Tong AW. miRNA profiling for diagnosis and prognosis of human cancer. *DNA Cell Biol* 2007;**26**:293–300.
24. Volinia S, Calin GA, Liu CG, Ambs S, Cimmino A, Petrocca F, et al. A microRNA expression signature of human solid tumors defines cancer gene targets. *Proc Natl Acad Sci USA* 2006;**103**:2257–61.
25. Yu SL, Chen HY, Yang PC, Chen JJ. Unique MicroRNA signature and clinical outcome of cancers. *DNA Cell Biol* 2007;**26**:283–92.
26. Lu J, Getz G, Miska EA, Alvarez-Saavedra E, Lamb J, Peck D, et al. MicroRNA expression profiles classify human cancers. *Nature* 2005;**435**:834–8.
27. Lopez-Serra P, Esteller M. DNA methylation-associated silencing of tumor suppressor microRNAs in cancer. *Oncogene* 2012;**31**:1609–22.
28. Blandino G, Fazi F, Donzelli S, Kedmi M, Sas-Chen A, Muti P, et al. Tumor suppressor microRNAs: a novel non-coding alliance against cancer. *FEBS Lett* 2014;**588**:2639–52.
29. Saito Y, Liang G, Egger G, Friedman JM, Chuang JC, Coetzee GA, et al. Specific activation of microRNA-127 with downregulation of the proto-oncogene BCL6 by chromatin-modifying drugs in human cancer cells. *Cancer Cell* 2006;**9**:435–43.
30. Lujambio A, Ropero S, Ballestar E, Fraga MF, Cerrato C, Setien F, et al. Genetic unmasking of an epigenetically silenced microRNA in human cancer cells. *Cancer Res* 2007;**67**:1424–9.
31. White NM, Fatoohi E, Metias M, Jung K, Stephan C, Yousef GM. Metastamirs: a stepping stone towards improved cancer management. *Nat Rev Clin Oncol* 2011;**8**:75–84.
32. Krutovskikh VA, Herceg Z. Oncogenic microRNAs (OncomiRs) as a new class of cancer biomarkers. *Bioessays* 2010;**32**:894–904.
33. Esquela-Kerscher A, Slack FJ. Oncomirs – microRNAs with a role in cancer. *Nat Rev Cancer* 2006;**6**:259–69.
34. Manikandan J, Aarthi JJ, Kumar SD, Pushparaj PN. Oncomirs: the potential role of non-coding microRNAs in understanding cancer. *Bioinformation* 2008;**2**:330–4.
35. Cho WC. OncomiRs: the discovery and progress of microRNAs in cancers. *Mol Cancer* 2007;**6**:60.

36. Lee Y, Kim M, Han J, Yeom KH, Lee S, Baek SH, et al. MicroRNA genes are transcribed by RNA polymerase II. *EMBO J* 2004;**23**:4051–60.
37. Dieci G, Fiorino G, Castelnuovo M, Teichmann M, Pagano A. The expanding RNA polymerase III transcriptome. *Trends Genet* 2007;**23**:614–22.
38. Cai X, Hagedorn CH, Cullen BR. Human microRNAs are processed from capped, polyadenylated transcripts that can also function as mRNAs. *RNA* 2004;**10**:1957–66.
39. Gregory RI, Yan KP, Amuthan G, Chendrimada T, Doratotaj B, Cooch N, et al. The microprocessor complex mediates the genesis of microRNAs. *Nature* 2004;**432**:235–40.
40. Kao PN, Chen L, Brock G, Ng J, Kenny J, Smith AJ, et al. Cloning and expression of cyclosporin A- and FK506-sensitive nuclear factor of activated T-cells: NF45 and NF90. *J Biol Chem* 1994;**269**:20691–9.
41. Sakamoto S, Aoki K, Higuchi T, Todaka H, Morisawa K, Tamaki N, et al. The NF90-NF45 complex functions as a negative regulator in the microRNA processing pathway. *Mol Cell Biol* 2009;**29**:3754–69.
42. Yi R, Qin Y, Macara IG, Cullen BR. Exportin-5 mediates the nuclear export of pre-microRNAs and short hairpin RNAs. *Genes Dev* 2003;**17**:3011–6.
43. Kok KH, Ng MH, Ching YP, Jin DY. Human TRBP and PACT directly interact with each other and associate with dicer to facilitate the production of small interfering RNA. *J Biol Chem* 2007;**282**:17649–57.
44. Lee Y, Hur I, Park SY, Kim YK, Suh MR, Kim VN. The role of PACT in the RNA silencing pathway. *EMBO J* 2006;**25**:522–32.
45. Lee HY, Zhou K, Smith AM, Noland CL, Doudna JA. Differential roles of human Dicer-binding proteins TRBP and PACT in small RNA processing. *Nucleic Acids Res* 2013;**41**:6568–76.
46. Bennasser Y, Chable-Bessia C, Triboulet R, Gibbings D, Gwizdek C, Dargemont C, et al. Competition for XPO5 binding between Dicer mRNA, pre-miRNA and viral RNA regulates human Dicer levels. *Nat Struct Mol Biol* 2011;**18**:323–7.
47. Patel RC, Sen GC. PACT, a protein activator of the interferon-induced protein kinase, PKR. *EMBO J* 1998;**17**:4379–90.
48. Li S, Peters GA, Ding K, Zhang X, Qin J, Sen GC. Molecular basis for PKR activation by PACT or dsRNA. *Proc Natl Acad Sci USA* 2006;**103**:10005–10.
49. Daher A, Longuet M, Dorin D, Bois F, Segeral E, Bannwarth S, et al. Two dimerization domains in the trans-activation response RNA-binding protein (TRBP) individually reverse the protein kinase R inhibition of HIV-1 long terminal repeat expression. *J Biol Chem* 2001;**276**:33899–905.
50. Redfern AD, Colley SM, Beveridge DJ, Ikeda N, Epis MR, Li X, et al. RNA-induced silencing complex (RISC) proteins PACT, TRBP, and Dicer are SRA binding nuclear receptor coregulators. *Proc Natl Acad Sci USA* 2013;**110**:6536–41.
51. Lingel A, Sattler M. Novel modes of protein-RNA recognition in the RNAi pathway. *Curr Opin Struct Biol* 2005;**15**:107–15.
52. He L, Hannon GJ. MicroRNAs: small RNAs with a big role in gene regulation. *Nat Rev Genet* 2004;**5**:522–31.
53. Meister G, Tuschl T. Mechanisms of gene silencing by double-stranded RNA. *Nature* 2004;**431**:343–9.
54. Dodds E, Dunckley MG, Naujoks K, Michaelis U, Dickson G. Lipofection of cultured mouse muscle cells: a direct comparison of Lipofectamine and DOSPER. *Gene Ther* 1998;**5**:542–51.
55. Regelin AE, Fankhaenel S, Gurtesch L, Prinz C, von Kiedrowski G, Massing U. Biophysical and lipofection studies of DOTAP analogs. *Biochim Biophys Acta* 2000;**1464**:151–64.

56. Jarve A, Muller J, Kim IH, Rohr K, MacLean C, Fricker G, et al. Surveillance of siRNA integrity by FRET imaging. *Nucleic Acids Res* 2007;**35**:e124.
57. Martinez J, Patkaniowska A, Urlaub H, Luhrmann R, Tuschl T. Single-stranded antisense siRNAs guide target RNA cleavage in RNAi. *Cell* 2002;**110**:563–74.
58. Brummelkamp TR, Bernards R, Agami R. A system for stable expression of short interfering RNAs in mammalian cells. *Science* 2002;**296**:550–3.
59. Miyagishi M, Taira K. U6 promoter-driven siRNAs with four uridine 3' overhangs efficiently suppress targeted gene expression in mammalian cells. *Nat Biotechnol* 2002; **20**:497–500.
60. Hama S, Akita H, Ito R, Mizuguchi H, Hayakawa T, Harashima H. Quantitative comparison of intracellular trafficking and nuclear transcription between adenoviral and lipoplex systems. *Mol Ther* 2006;**13**:786–94.
61. Cohen RN, van der Aa MA, Macaraeg N, Lee AP, Szoka Jr FC. Quantification of plasmid DNA copies in the nucleus after lipoplex and polyplex transfection. *J Control Release* 2009;**135**:166–74.
62. Rand TA, Petersen S, Du F, Wang X. Argonaute2 cleaves the anti-guide strand of siRNA during RISC activation. *Cell* 2005;**123**:621–9.
63. An DS, Qin FX, Auyeung VC, Mao SH, Kung SK, Baltimore D, et al. Optimization and functional effects of stable short hairpin RNA expression in primary human lymphocytes via lentiviral vectors. *Mol Ther* 2006;**14**:494–504.
64. Kim DH, Rossi JJ. Strategies for silencing human disease using RNA interference. *Nat Rev Genet* 2007;**8**:173–84.
65. Grimm D, Streetz KL, Jopling CL, Storm TA, Pandey K, Davis CR, et al. Fatality in mice due to oversaturation of cellular microRNA/short hairpin RNA pathways. *Nature* 2006;**441**:537–41.
66. Shenk TE. *Adenoviridae: the viruses and their replication*. Philadelphia: Lippincott Williams & Wilkins; 2001.
67. Wivel NA, Gao G, Wilson JM. Adenovirus vectors. In: Friedmann T, editor. *The development of human gene therapy*. New York: Cold Spring Harbor Laboratory Press; 1999. p. 87–110.
68. Thomas CE, Ehrhardt A, Kay MA. Progress and problems with the use of viral vectors for gene therapy. *Nat Rev Genet* 2003;**4**:346–58.
69. Verma IM, Weitzman MD. Gene therapy: twenty-first century medicine. *Annu Rev Biochem* 2005;**74**:711–38.
70. Ugai H, Yamasaki T, Hirose M, Inabe K, Kujime Y, Terashima M, et al. Purification of infectious adenovirus in two hours by ultracentrifugation and tangential flow filtration. *Biochem Biophys Res Commun* 2005;**331**:1053–60.
71. Fisher KJ, Choi H, Burda J, Chen SJ, Wilson JM. Recombinant adenovirus deleted of all viral genes for gene therapy of cystic fibrosis. *Virology* 1996;**217**:11–22.
72. Kochanek S, Clemens PR, Mitani K, Chen HH, Chan S, Caskey CT. A new adenoviral vector: replacement of all viral coding sequences with 28 kb of DNA independently expressing both full-length dystrophin and beta-galactosidase. *Proc Natl Acad Sci USA* 1996;**93**:5731–6.
73. Palmer DJ, Ng P. Helper-dependent adenoviral vectors for gene therapy. *Hum Gene Ther* 2005;**16**:1–16.
74. Waehler R, Russell SJ, Curiel DT. Engineering targeted viral vectors for gene therapy. *Nat Rev Genet* 2007;**8**:573–87.
75. Xia H, Mao Q, Paulson HL, Davidson BL. siRNA-mediated gene silencing in vitro and in vivo. *Nat Biotechnol* 2002;**20**:1006–10.

76. Shen C, Buck AK, Liu X, Winkler M, Reske SN. Gene silencing by adenovirus-delivered siRNA. *FEBS Lett* 2003;**539**:111–4.
77. Ro S, Hwang SJ, Ordog T, Sanders KM. Adenovirus-based short hairpin RNA vectors containing an EGFP marker and mouse U6, human H1, or human U6 promoter. *Biotechniques* 2005;**38**:625–7.
78. Narvaiza I, Aparicio O, Vera M, Razquin N, Bortolanza S, Prieto J, et al. Effect of adenovirus-mediated RNA interference on endogenous microRNAs in a mouse model of multidrug resistance protein 2 gene silencing. *J Virol* 2006;**80**:12236–47.
79. Buchler M, Konig J, Brom M, Kartenbeck J, Spring H, Horie T, et al. cDNA cloning of the hepatocyte canalicular isoform of the multidrug resistance protein, cMrp, reveals a novel conjugate export pump deficient in hyperbilirubinemic mutant rats. *J Biol Chem* 1996;**271**:15091–8.
80. Keppler D, Konig J. Hepatic canalicular membrane 5: expression and localization of the conjugate export pump encoded by the MRP2 (cMRP/cMOAT) gene in liver. *FASEB J* 1997;**11**:509–16.
81. Keppler D, Konig J, Buchler M. The canalicular multidrug resistance protein, cMRP/MRP2, a novel conjugate export pump expressed in the apical membrane of hepatocytes. *Adv Enzyme Regul* 1997;**37**:321–33.
82. Witting SR, Brown M, Saxena R, Nabinger S, Morral N. Helper-dependent adenovirus-mediated short hairpin RNA expression in the liver activates the interferon response. *J Biol Chem* 2008;**283**:2120–8.
83. Ruiz R, Witting SR, Saxena R, Morral N. Robust hepatic gene silencing for functional studies using helper-dependent adenoviral vectors. *Hum Gene Ther* 2009;**20**:87–94.
84. Muruve DA, Cotter MJ, Zaiss AK, White LR, Liu Q, Chan T, et al. Helper-dependent adenovirus vectors elicit intact innate but attenuated adaptive host immune responses in vivo. *J Virol* 2004;**78**:5966–72.
85. Lagos-Quintana M, Rauhut R, Yalcin A, Meyer J, Lendeckel W, Tuschl T. Identification of tissue-specific microRNAs from mouse. *Curr Biol* 2002;**12**:735–9.
86. Landgraf P, Rusu M, Sheridan R, Sewer A, Iovino N, Aravin A, et al. A mammalian microRNA expression atlas based on small RNA library sequencing. *Cell* 2007;**129**:1401–14.
87. Castoldi M, Vujic Spasic M, Altamura S, Elmen J, Lindow M, Kiss J, et al. The liver-specific microRNA miR-122 controls systemic iron homeostasis in mice. *J Clin Invest* 2011;**121**:1386–96.
88. Moore KJ, Rayner KJ, Suarez Y, Fernandez-Hernando C. microRNAs and cholesterol metabolism. *Trends Endocrinol Metab* 2010;**21**:699–706.
89. Fernandez-Hernando C, Suarez Y, Rayner KJ, Moore KJ. MicroRNAs in lipid metabolism. *Curr Opin Lipidol* 2011;**22**:86–92.
90. Pasquinelli AE, Reinhart BJ, Slack F, Martindale MQ, Kuroda MI, Maller B, et al. Conservation of the sequence and temporal expression of let-7 heterochronic regulatory RNA. *Nature* 2000;**408**:86–9.
91. Reinhart BJ, Slack FJ, Basson M, Pasquinelli AE, Bettinger JC, Rougvie AE, et al. The 21-nucleotide let-7 RNA regulates developmental timing in *Caenorhabditis elegans*. *Nature* 2000;**403**:901–6.
92. Johnson CD, Esquela-Kerscher A, Stefani G, Byrom M, Kelnar K, Ovcharenko D, et al. The let-7 microRNA represses cell proliferation pathways in human cells. *Cancer Res* 2007;**67**:7713–22.
93. Ehrhardt A, Kay MA. A new adenoviral helper-dependent vector results in long-term therapeutic levels of human coagulation factor IX at low doses in vivo. *Blood* 2002;**99**:3923–30.

94. Yamaguchi T, Kawabata K, Koizumi N, Sakurai F, Nakashima K, Sasaki T, et al. Adenovirus vectors stimulate innate immunity via MYD88/TLR9-dependent and-independent pathways. *J Gene Med* 2008;**10**:453–4.
95. Sakurai H, Kawabata K, Sakurai F, Nakagawa S, Mizuguchi H. Innate immune response induced by gene delivery vectors. *Int J Pharm* 2008;**354**:9–15.
96. Machitani M, Yamaguchi T, Shimizu K, Sakurai F, Katayama K, Kawabata K, et al. Adenovirus vector-derived VA-RNA-mediated innate immune responses. *Pharmaceutics* 2011;**3**:338–53.
97. Muruve DA. The innate immune response to adenovirus vectors. *Hum Gene Ther* 2004; **15**:1157–66.
98. Weaver EA, Nehete PN, Buchl SS, Senac JS, Palmer D, Ng P, et al. Comparison of replication-competent, first generation, and helper-dependent adenoviral vaccines. *PLoS ONE* 2009;**4**:e5059.
99. Minamitani T, Iwakiri D, Takada K. Adenovirus virus-associated RNAs induce type I interferon expression through a RIG-I-mediated pathway. *J Virol* 2011;**85**:4035–40.
100. Yamaguchi T, Kawabata K, Kouyama E, Ishii KJ, Katayama K, Suzuki T, et al. Induction of type I interferon by adenovirus-encoded small RNAs. *P Natl Acad Sci USA* 2010;**107**:17286–91.
101. Glasgow JN, Everts M, Curiel DT. Transductional targeting of adenovirus vectors for gene therapy. *Cancer Gene Ther* 2006;**13**:830–44.
102. Glasgow JN, Bauerschmitz GJ, Curiel DT, Hemminki A. Transductional and transcriptional targeting of adenovirus for clinical applications. *Curr Gene Ther* 2004;**4**:1–14.
103. Suzuki T, Sakurai F, Nakamura S, Kouyama E, Kawabata K, Kondoh M, et al. miR-122a-regulated expression of a suicide gene prevents hepatotoxicity without altering antitumor effects in suicide gene therapy. *Mol Ther* 2008;**16**:1719–26.
104. Shimizu K, Sakurai F, Machitani M, Katayama K, Mizuguchi H. Quantitative analysis of the leaky expression of adenovirus genes in cells transduced with a replication incompetent adenovirus vector. *Mol Pharm* 2011;**8**:1430–5.
105. Ylösmäki E, Hakkarainen T, Hemminki A, Visakorpi T, Andino R, Saksela K. Generation of a conditionally replicating adenovirus based on targeted destruction of E1A mRNA by a cell type-specific MicroRNA. *J Virol* 2008;**82**:11009–15.
106. Sugio K, Sakurai F, Katayama K, Tashiro K, Matsui H, Kawabata K, et al. Enhanced safety profiles of the telomerase-specific replication-competent adenovirus by incorporation of Normal cell-specific microRNA-targeted sequences. *Clin Cancer Res* 2011; **17**:2807–18.
107. Cawood R, Wong SL, Di Y, Baban DF, Seymour LW. MicroRNA controlled adenovirus mediates anti-cancer efficacy without affecting endogenous microRNA activity. *PLoS One* 2011;**6**:e16152.
108. Cawood R, Chen HH, Carroll F, Bazan-Peregrino M, van Rooijen N, Seymour LW. Use of tissue-specific microRNA to control pathology of wild-type adenovirus without attenuation of its ability to kill cancer cells. *PLoS Pathog* 2009;**5**:e1000440.
109. Brown BD, Gentner B, Cantore A, Colleoni S, Amendola M, Zingale A, et al. Endogenous microRNA can be broadly exploited to regulate transgene expression according to tissue, lineage and differentiation state. *Nat Biotechnol* 2007;**25**:1457–67.
110. Wang Y, Hu JK, Krol A, Li YP, Li CY, Yuan F. Systemic dissemination of viral vectors during intratumoral injection. *Mol Cancer Ther* 2003;**2**:1233–42.
111. Lohr F, Huang Q, Hu K, Dewhirst MW, Li CY. Systemic vector leakage and transgene expression by intratumorally injected recombinant adenovirus vectors. *Clin Cancer Res* 2001;**7**:3625–8.

112. Toloza EM, Hunt K, Swisher S, McBride W, Lau R, Pang S, et al. In vivo cancer gene therapy with a recombinant interleukin-2 adenovirus vector. *Cancer Gene Ther* 1996;**3**:11–7.
113. Tjuvajev JG, Chen SH, Joshi A, Joshi R, Guo ZS, Balatoni J, et al. Imaging adenoviral-mediated herpes virus thymidine kinase gene transfer and expression in vivo. *Cancer Res* 1999;**59**:5186–93.
114. Bramson JL, Hitt M, Gauldie J, Graham FL. Pre-existing immunity to adenovirus does not prevent tumor regression following intratumoral administration of a vector expressing IL-12 but inhibits virus dissemination. *Gene Ther* 1997;**4**:1069–76.
115. Yamamoto M, Davydova J, Wang M, Siegal GP, Krasnykh V, Vickers SM, et al. Infectivity enhanced, cyclooxygenase-2 promoter-based conditionally replicative adenovirus for pancreatic cancer. *Gastroenterology* 2003;**125**:1203–18.
116. Adachi Y, Reynolds PN, Yamamoto M, Wang M, Takayama K, Matsubara S, et al. A midkine promoter-based conditionally replicative adenovirus for treatment of pediatric solid tumors and bone marrow tumor purging. *Cancer Res* 2001;**61**:7882–8.
117. Rose JA, Reich PR, Weissman SM. RNA production in adenovirus-infected KB cells. *Virology* 1965;**27**:571–9.
118. Reich PR, Forget BG, Weissman SM. RNA of low molecular weight in KB cells infected with adenovirus type 2. *J Mol Biol* 1966;**17**:428–39.
119. Ohe K, Weissman SM. The nucleotide sequence of a low molecular weight ribonucleic acid from cells infected with adenovirus 2. *J Biol Chem* 1971;**246**:6991–7009.
120. Soderlund H, Pettersson U, Vennstrom B, Philipson L, Mathews MB. A new species of virus-coded low molecular weight RNA from cells infected with adenovirus type 2. *Cell* 1976;**7**:585–93.
121. Mathews MB. Genes for VA-RNA in adenovirus 2. *Cell* 1975;**6**:223–9.
122. Quincey RV, Wilson SH. The utilization of genes for ribosomal RNA, 5S RNA, and transfer RNA in liver cells of adult rats. *Proc Natl Acad Sci USA* 1969;**64**:981–8.
123. Mathews MB. Structure, function, and evolution of adenovirus virus-associated RNAs. *Curr Top Microbiol Immunol* 1995;**199**(Pt 2):173–87.
124. Mathews MB, Shenk T. Adenovirus virus-associated RNA and translation control. *J Virol* 1991;**65**:5657–62.
125. Vennstrom B, Pettersson U, Philipson L. Two initiation sites for adenovirus 5.5S RNA. *Nucleic Acids Res* 1978;**5**:195–204.
126. Ma Y, Mathews MB. Structure, function, and evolution of adenovirus-associated RNA: a phylogenetic approach. *J Virol* 1996;**70**:5083–99.
127. Thimmappaya B, Jones N, Shenk T. A mutation which alters initiation of transcription by RNA polymerase III on the Ad5 chromosome. *Cell* 1979;**18**:947–54.
128. Weinmann R, Raskas HJ, Roeder RG. Role of DNA-dependent RNA polymerases II and III in transcription of the adenovirus genome late in productive infection. *Proc Natl Acad Sci USA* 1974;**71**:3426–39.
129. Fowlkes DM, Shenk T. Transcriptional control regions of the adenovirus VAI RNA gene. *Cell* 1980;**22**:405–13.
130. Bhat RA, Metz B, Thimmappaya B. Organization of the noncontiguous promoter components of adenovirus VAI RNA gene is strikingly similar to that of eukaryotic tRNA genes. *Mol Cell Biol* 1983;**3**:1996–2005.
131. Gwizdek C, Ossareh-Nazari B, Brownawell AM, Doglio A, Bertrand E, Macara IG, et al. Exportin-5 mediates nuclear export of minihelix-containing RNAs. *J Biol Chem* 2003;**278**:5505–8.
132. Gwizdek C, Ossareh-Nazari B, Brownawell AM, Evers S, Macara IG, Dargemont C. Minihelix-containing RNAs mediate exportin-5-dependent nuclear export of the double-stranded RNA-binding protein ILF3. *J Biol Chem* 2004;**279**:884–91.

133. Thimmappaya B, Weinberger C, Schneider RJ, Shenk T. Adenovirus VAI RNA is required for efficient translation of viral mRNAs at late times after infection. *Cell* 1982; **31**:543–51.
134. Reichel PA, Merrick WC, Siekierka J, Mathews MB. Regulation of a protein synthesis initiation factor by adenovirus virus-associated RNA. *Nature* 1985;**313**:196–200.
135. Schneider RJ, Weinberger C, Shenk T. Adenovirus VAI RNA facilitates the initiation of translation in virus-infected cells. *Cell* 1984;**37**:291–8.
136. Sadler AJ, Williams BR. Interferon-inducible antiviral effectors. *Nat Rev Immunol* 2008;**8**:559–68.
137. Williams BR. PKR; a sentinel kinase for cellular stress. *Oncogene* 1999;**18**:6112–20.
138. Maran A, Mathews MB. Characterization of the double-stranded RNA implicated in the inhibition of protein synthesis in cells infected with a mutant adenovirus defective for VA RNA. *Virology* 1988;**164**:106–13.
139. O'Malley RP, Mariano TM, Siekierka J, Mathews MB. A mechanism for the control of protein synthesis by adenovirus VA RNAI. *Cell* 1986;**44**:391–400.
140. Ghadge GD, Swaminathan S, Katze MG, Thimmapaya B. Binding of the adenovirus VAI RNA to the interferon-induced 68-kDa protein kinase correlates with function. *Proc Natl Acad Sci USA* 1991;**88**:7140–4.
141. Galabru J, Katze MG, Robert N, Hovanessian AG. The binding of double-stranded RNA and adenovirus VAI RNA to the interferon-induced protein kinase. *Eur J Biochem* 1989;**178**:581–9.
142. Lu S, Cullen BR. Adenovirus VA1 noncoding RNA can inhibit small interfering RNA and MicroRNA biogenesis. *J Virol* 2004;**78**:12868–76.
143. Andersson MG, Haasnoot PC, Xu N, Berenjian S, Berkhout B, Akusjarvi G. Suppression of RNA interference by adenovirus virus-associated RNA. *J Virol* 2005;**79**:9556–65.
144. de Vries W, Berkhout B. RNAi suppressors encoded by pathogenic human viruses. *Int J Biochem Cell Biol* 2008;**40**:2007–12.
145. Xu N, Segerman B, Zhou X, Akusjarvi G. Adenovirus virus-associated RNAII-derived small RNAs are efficiently incorporated into the rna-induced silencing complex and associate with polyribosomes. *J Virol* 2007;**81**:10540–9.
146. Carnero E, Sutherland JD, Fortes P. Adenovirus and miRNAs. *Biochim Biophys Acta* 2011;**1809**:660–7.
147. Liao HJ, Kobayashi R, Mathews MB. Activities of adenovirus virus-associated RNAs: purification and characterization of RNA binding proteins. *Proc Natl Acad Sci USA* 1998;**95**:8514–9.
148. Robb GB, Rana TM. RNA helicase A interacts with RISC in human cells and functions in RISC loading. *Mol Cell* 2007;**26**:523–37.
149. Liang XH, Crooke ST. RNA helicase A is not required for RISC activity. *Biochim Biophys Acta* 2013;**1829**:1092–101.
150. Parker LM, Fierro-Monti I, Mathews MB. Nuclear factor 90 is a substrate and regulator of the eukaryotic initiation factor 2 kinase double-stranded RNA-activated protein kinase. *J Biol Chem* 2001;**276**:32522–30.
151. Li WX, Li H, Lu R, Li F, Dus M, Atkinson P, et al. Interferon antagonist proteins of influenza and vaccinia viruses are suppressors of RNA silencing. *Proc Natl Acad Sci USA* 2004;**101**:1350–5.
152. Pantaleo V, Szittya G, Burgyan J. Molecular bases of viral RNA targeting by viral small interfering RNA-programmed RISC. *J Virol* 2007;**81**:3797–806.
153. Bucher E, Hemmes H, de Haan P, Goldbach R, Prins M. The influenza A virus NS1 protein binds small interfering RNAs and suppresses RNA silencing in plants. *J Gen Virol* 2004;**85**:983–91.

154. Haasnoot J, de Vries W, Geutjes EJ, Prins M, de Haan P, Berkhout B. The Ebola virus VP35 protein is a suppressor of RNA silencing. *PLoS Pathog* 2007;**3**:e86.
155. Leonard JN, Shah PS, Burnett JC, Schaffer DV. HIV evades RNA interference directed at TAR by an indirect compensatory mechanism. *Cell Host Microbe* 2008;**4**:484–94.
156. Maekawa A, Pei Z, Suzuki M, Fukuda H, Ono Y, Kondo S, et al. Efficient production of adenovirus vector lacking genes of virus-associated RNAs that disturb cellular RNAi machinery. *Sci Rep* 2013;**3**:1136.
157. Machitani M, Katayama K, Sakurai F, Matsui H, Yamaguchi T, Suzuki T, et al. Development of an adenovirus vector lacking the expression of virus-associated RNAs. *J Control Release* 2011;**154**:285–9.
158. Bhat RA, Thimmappaya B. Adenovirus mutants with DNA sequence perturbations in the intragenic promoter of VAI RNA gene allow the enhanced transcription of VAII RNA gene in HeLa cells. *Nucleic Acids Res* 1984;**12**:7377–88.
159. Machitani M, Sakurai F, Katayama K, Tachibana M, Suzuki T, Matsui H, et al. Improving adenovirus vector-mediated RNAi efficiency by lacking the expression of virus-associated RNAs. *Virus Res* 2013;**178**:357–63.
160. Takata Y, Kondo S, Goda N, Kanegae Y, Saito I. Comparison of efficiency between FLPe and Cre for recombinase-mediated cassette exchange in vitro and in adenovirus vector production. *Genes Cells* 2011;**16**:765–77.
161. Pei Z, Shi G, Kondo S, Ito M, Maekawa A, Suzuki M, et al. Adenovirus vectors lacking virus-associated RNA expression enhance shRNA activity to suppress hepatitis C virus replication. *Sci Rep* 2013;**3**:3575.
162. Aparicio O, Razquin N, Zaratiegui M, Narvaiza I, Fortes P. Adenovirus virus-associated RNA is processed to functional interfering RNAs involved in virus production. *J Virol* 2006;**80**:1376–84.
163. Aparicio O, Carnero E, Abad X, Razquin N, Guruceaga E, Segura V, et al. Adenovirus VA RNA-derived miRNAs target cellular genes involved in cell growth, gene expression and DNA repair. *Nucleic Acids Res* 2010;**38**:750–63.
164. Forch P, Valcarcel J. Molecular mechanisms of gene expression regulation by the apoptosis-promoting protein TIA-1. *Apoptosis* 2001;**6**:463–8.
165. Kondo S, Yoshida K, Suzuki M, Saito I, Kanegae Y. Adenovirus-encoding virus-associated RNAs suppress HDGF gene expression to support efficient viral replication. *PLoS One* 2014;**9**:e108627.
166. Pei Z, Kondo S, Kanegae Y, Saito I. Copy number of adenoviral vector genome transduced into target cells can be measured using quantitative PCR: application to vector titration. *Biochem Biophys Res Commun* 2012;**417**:945–50.
167. Kitajewski J, Schneider RJ, Safer B, Shenk T. An adenovirus mutant unable to express VAI RNA displays different growth responses and sensitivity to interferon in various host cell lines. *Mol Cell Biol* 1986;**6**:4493–8.
168. Furtado MR, Subramanian S, Bhat RA, Fowlkes DM, Safer B, Thimmappaya B. Functional dissection of adenovirus VAI RNA. *J Virol* 1989;**63**:3423–34.
169. Rajan P, Swaminathan S, Zhu J, Cole CN, Barber G, Tevethia MJ, et al. A novel translational regulation function for the simian virus 40 large-T antigen gene. *J Virol* 1995;**69**:785–95.
170. Subramanian S, Bhat RA, Rundell MK, Thimmappaya B. Suppression of the translation defect phenotype specific for a virus-associated RNA-deficient adenovirus mutant in monkey cells by simian virus 40. *J Virol* 1986;**60**:363–8.

171. Davies MV, Furtado M, Hershey JW, Thimmappaya B, Kaufman RJ. Complementation of adenovirus virus-associated RNA I gene deletion by expression of a mutant eukaryotic translation initiation factor. *Proc Natl Acad Sci USA* 1989;**86**:9163–7.
172. Kamel W, Segerman B, Oberg D, Punga T, Akusjarvi G. The adenovirus VA RNA-derived miRNAs are not essential for lytic virus growth in tissue culture cells. *Nucleic Acids Res* 2013;**41**:4802–12.
173. White RJ, Trouche D, Martin K, Jackson SP, Kouzarides T. Repression of RNA polymerase III transcription by the retinoblastoma protein. *Nature* 1996;**382**:88–90.

# Imaging and Adenoviral Gene Therapy

# 31

*Jillian R. Richter, Benjamin B. Kasten, Kurt R. Zinn*
Department of Radiology, The University of Alabama at Birmingham, Birmingham, AL, USA

## 1. Introduction

Imaging instrumentation and software have experienced tremendous growth between 2002 and 2015 since publication of the first edition of this book, and therefore imaging now offers even more potential for gene therapy applications. Faster instruments with higher resolution and ease of operation are widely available. There are new modalities, multimodality instruments, and many well-established methods for detecting the efficacy of gene therapy as well as to measure where transgenes are expressed in living subjects. The purpose of this chapter is to provide an overview of current and emerging imaging technologies, review how imaging has been applied to assess adenoviral-based gene therapy, and discuss genetic-based adenoviral imaging reporter systems with a focus on human applications. Selected preclinical examples of relevant reporter systems introduced since 2002 will also be discussed, although an in-depth overview of all adenoviral vector imaging applications is beyond the scope of this chapter.

Imaging technologies all require electromagnetic energy; each modality uses different parts of the energy spectrum. The spectrum includes gamma rays, X-rays, visible and near-infrared light, ultrasonic (sound) waves, and radiowaves. These various photons of energy differ in their wavelength and, therefore, energy. Each imaging instrument is designed to detect a particular range of electromagnetic energy. For most imaging techniques, the instrument also generates the requisite energy for the imaging application, and the detection occurs after the photons interact with the imaging subject. In bioluminescence and nuclear or gamma-ray imaging, the requisite energy is provided by photons emitted during an enzymatic reaction or during radioactive decay, respectively, rather than by the instrument.

For imaging to be useful there must be differences in the brightness of the image, or contrast, such that normal anatomy, function, or pathology can be appreciated. Each imaging modality achieves the contrast by a different mechanism. For gamma-ray imaging, the contrast is due to localized accumulation of radioactivity. Bone, soft tissues, and contrast-enhancing agents absorb the X-rays to a different degree, which leads to contrast for radiography and computed tomography (CT). Dual energy CT uses two different X-ray energies, with subtle differences in attenuation of the two energies for tissues and contrast-enhancing agents to improve image quality. Contrast is achieved in magnetic resonance imaging (MRI) because the local environment of the proton is different in fat, water, and various soft tissues. With ultrasound (US), contrast is achieved because the reflectance of the ultrasonic wave is dependent on the tissue

**Adenoviral Vectors for Gene Therapy.**

architecture or blood flow. Contrast for optical imaging is provided by localized light emission, or fluorescence. Contrast for radiography, CT, MRI, and US can be increased by administration of an exogenous contrast-enhancing agent. Angiography is always better with a contrast agent, whether done by fluoroscopy, CT, MRI, or US.

# 2. Review of Imaging Modalities

Table 1 presents a list of imaging modalities, including those that are routinely applied in clinical imaging (in italics) and emerging technologies. Images are broadly classified as two dimensional (2D) or three dimensional (3D). For CT, single photon emission computed tomography (SPECT), and many optical modalities, the 3D presentation of images is accomplished by acquiring a series of 2D images of different views of the imaging subject, and then reconstructing the images into a 3D model provided by defined algorithms and transformations. Positron emission tomography (PET) uses coincidence algorithms to determine 3D locations of positron emitters, whereas US detects differences in the reflected sound waves of tissues and structures to determine depth and/or 3D location. MRI has data that are registered in 3D space as the signal is specifically generated in that way.

## *2.1 Nuclear (Gamma-Ray) Imaging*

Gamma rays are the highest energy photons (shortest wavelength, highest frequency), arising out of nuclear events during radioactive decay. For in vivo applications, the best gamma rays are of low energy (100–511 keV) because they can penetrate tissues. Gamma rays in this energy range can also be efficiently stopped, and therefore measured by external detectors. Most human imaging procedures with radioactivity are accomplished using $^{99m}$Tc, which emits a 140 keV gamma ray during decay. $^{99m}$Tc has a 6 h half-life and is continuously available from generators at hospitals or at regional nuclear pharmacies. It is the decay product of $^{99}$Mo (half-life = 66 h) and is eluted daily from the $^{99}$Mo/$^{99m}$Tc generator system, and therefore available at very high specific activity and low cost. $^{99m}$Tc can be chelated (complexed) with various compounds that have different biological characteristics, or it can be attached to proteins.

$^{99m}$Tc is typically imaged with a gamma camera that includes a collimator, a lead gamma-ray attenuator that is placed between the imaging subject and the gamma-ray detector. There are various types of collimators, some specific for low energy gamma rays, whereas others are thicker and designed for higher energy gamma rays. Examples of collimators are the parallel-hole collimator and the pinhole collimator. The parallel-hole collimator allows passage of gamma rays that are perpendicular to the plane of the collimator. In contrast, the pinhole collimator has a small round hole at the end that allows projection of the gamma rays onto the detector crystal, thus forming an image like a pinhole camera. Figure 1 presents images of an Ad5 vector encoding luciferase that was radiolabeled with $^{99m}$Tc and injected intravenously in four mice. Each mouse was positioned below the pinhole collimator at 10 min after injection, and a static gamma camera image was collected. The gamma rays emitted from the animal were stopped

| Modality (Abbreviation)[a] | 3D | Comments | Advantages | Disadvantages |
|---|---|---|---|---|
| *Ultrasound (US)* | + | Doppler, color Doppler, elasticity | Fast, high frame rate | Requires great skill and experience |
| *Magnetic resonance imaging (MRI)* | + | There are varied magnetic field strengths, in Tesla (T), ranging from 0.5 to 9.4 T; BOLD, DCE-MRI, DWI | Robust, high resolution, many techniques | Long imaging time, analyses complicated |
| *Magnetic resonance spectroscopy (MRS)* | + | Hyperpolarized imaging agents | Multiple metabolites | Expensive |
| *Planar X-ray radiography* | – | often part of multimodality instrument | Simple, fast | Ionizing radiation |
| *X-ray computed tomography (CT)* | + | Varied voxel size and speed | High resolution | Ionizing radiation |
| *Planar gamma camera imaging* | – | Variable collimators, multipinhole, single pinhole, parallel-hole | Fast | 2D imaging, ionizing radiation |
| *Single photon emission computed tomography (SPECT, SPECT/CT)* | + | Variable voxel size | High resolution | Lower sensitivity, ionizing radiation |
| *Positron emission tomography (PET/CT; PET/MR)* | + | Variable voxel size | High sensitivity | Ionizing radiation |
| Bioluminescence (BLI) | – | +/– planar X-ray radiography | Fast, inexpensive multiple animals | 2D |
| Bioluminescence (BLI) | + | +/– CT, +/– fluorescence | 3D | Slow |
| Fluorescence[b] | – | +/– CT, +/– spectral unmixing | Fast | Limited tissue depth, 2D imaging |
| Fluorescence molecular tomography (FLT) | + | | 3D | Slow |
| Intravital imaging, with fluorescence | + | Requires special animal models | Highest resolution | |
| Endoscopic fluorescence imaging[c] | – | | Fast | |
| Endoscopic confocal microscopy | + | | Highest resolution | Limited field of view |
| Photoacoustic imaging | + | | Rapid | |
| Raman spectroscopy | + | Multiple labels simultaneously | Multiple labels | |

[a]*Italics* under modality indicates widespread applications in clinical imaging (human) and well as preclinical imaging in animal models.
[b]Commercially approved systems include Spy system (Novadaq), Firefly in de Vinci Robotic system (Intuitive); both use human-approved indocyanine green (ICG) fluorophore; and a fluorescent stereomicroscope (Leica).
[c]There is one commercial system currently approved for human imaging (Novadaq's Pinpoint) using the ICG fluorophore.

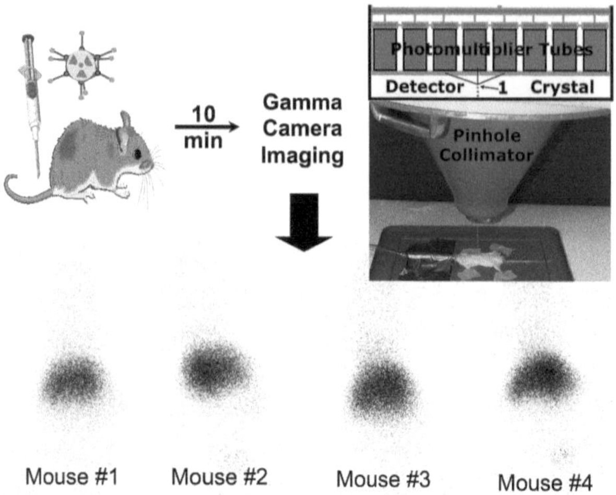


**Figure 1** Gamma camera imaging of $^{99m}$Tc-labeled Ad5 encoding luciferase following intravenous injection in 4 mice.

by the detector crystal and visible light photons were emitted. These photons were captured by the photomultiplier tubes adjacent to the crystal, and converted to a voltage pulse. The X,Y location of the interaction event was recorded, as well as the magnitude of the voltage pulse (Z, pulse height), which was proportional to the energy of the gamma ray that was stopped. The 2D gamma camera image in this example showed the expected liver pattern of the Ad5 vector distribution.

Proteins can be radiolabeled with $^{99m}$Tc for imaging applications. Most often, the $^{99m}$Tc is attached to proteins with a bifunctional chelator. With this system the chelator is first attached to the protein, then the $^{99m}$Tc in complexed to the chelator in a separate step. An example was the $^{99m}$Tc-labeling of Ad5 knob.[1] Besides $^{99m}$Tc, other radionuclides that are used for imaging include $^{67}$Ga, $^{111}$In, $^{123}$I, $^{125}$I, and $^{131}$I (see Table 2). These radionuclides have different gamma-ray emissions; therefore, simultaneous imaging with $^{99m}$Tc is possible.

The image presented in Figure 1 is a planar image that represents a 2D projection of the $^{99m}$Tc-Ad5 at 10 min after intravenous injection. SPECT is also possible with specialized gamma cameras that are routinely available in small animal imaging cores and nuclear medicine departments. SPECT is accomplished by collecting multiple images (or projections) at various angles around the subject; the detectors move while the subject remains static. A tomographic image of the distribution of the radioactivity is produced following reconstruction of these projections.

Gamma camera imaging is differentiated from PET as PET can image only 511 keV gamma rays that arise from positron decay. PET is a 3D imaging technique for the indirect detection of positrons. Positrons are positively charged electrons that are emitted from a proton-rich nucleus during radioactive decay. The lifetime of positrons is relatively short since they undergo annihilation by combining with an electron, giving rise to two 511 keV gamma rays at opposite (180°) orientations. The 511 keV gamma rays are actually detected in PET, not the positrons. PET scanners have a circular array of detectors that are designed to detect photons in coincidence, and the exact time of detection can be recorded. This means that during analyses it can be precisely known when two detectors at opposite orientations simultaneously detect the 511 keV gamma rays, arising from the positron annihilation event. Since various 511 keV pairs of photons strike different opposite pairs of detectors, the location of the actual decay events can be determined when the image is reconstructed. PET scanners do not require collimators, since the coincidence detection method accomplishes the same objective. There have been many recent advances in SPECT and PET hardware.[2]

Radionuclides that are used in PET imaging are proton rich and produced at cyclotrons using charged-particle reactions. A list of common PET radionuclides is included in Table 2. Most PET radionuclides have short half-lives; therefore, production must be in close proximity to where imaging will be done. In addition, PET radionuclides such as $^{11}C$, $^{13}N$, and $^{15}O$ are suitable as intrinsic labels for many molecules, thereby enabling imaging studies of the actual molecule of interest. For example, fatty acid metabolism could be imaged with the $^{11}C$-labeled fatty acid, where the $^{11}C$ replaced the normal $^{12}C$ in the molecular structure. Intrinsic labeling of this type cannot be accomplished with $^{99m}Tc$, since the radionuclide is not part of the molecule. A bifunctional chelate would be required for the $^{99m}Tc$ to attach it to the fatty acid, and due to the size of the chelator the $^{99m}Tc$-labeled fatty acid might have different in vivo uptake and elimination characteristics than the natural fatty acid.

## 2.2 Optical Imaging

Bioluminescence imaging (BLI) can be accomplished in 2D or 3D modes, and is based on detecting the location of expressed enzymes that emit light when reacting with their substrates. Most commonly bioluminescence is accomplished with luciferase and this technique is very efficient, sensitive, and low cost in animal models. Figure 2 shows the liver expression of luciferase after 7 days in the same mice that were injected with $^{99m}Tc$-Ad5 encoding luciferase (Figure 1). There is a great utility of BLI for preclinical evaluation of gene therapy vectors.[3,4] Unfortunately, BLI cannot be accomplished in humans.

Fluorescence imaging is an optical technique that can be easily accomplished in humans, currently in a 2D manner. In animal models it has 2D and 3D modes. The first fluorescent genetic reporter was green fluorescence protein (GFP). Now there are many different genetic reporters that span the visible spectrum (400–700 nm) and out to the near-infrared (NIR) range, for improved deep in vivo imaging.[5,6] In vivo fluorescence imaging can be accomplished with fluorescent stereomicroscopes, fluorescent endoscopes/laparoscopes, intraoperative cameras (Novadaq SPY system), and by light-tight boxes with charge-coupled device (CCD) cameras and appropriate filters (many commercial instruments). Spectral fluorescence imaging is now widely applied to improve sensitivity and specificity as autofluorescence signal can be removed when the specific fluorescence signal is "unmixed" from the overall fluorescence signal. Spectral imaging can be done with multiple filters or a tunable-wavelength filter system to cover the emission spectrum. Figure 3 presents an example of imaging mRFP in a conditionally replication-competent Ad vector targeting to HER2 in tumors.[7]

## Table 2 Common PET and SPECT Radionuclides Used in Imaging

| PET Isotopes | Half-life | Decay Modes (% Abundance) | Energy of Emissions Relevant for Imaging, keV (% Abundance) | Source |
|---|---|---|---|---|
| $^{11}C$ | 20.4 min | $\beta^+$ (100%) | 385.7 $\beta^+$ (99.8%)<br>511.0 $\gamma$ (199.5%) | Cyclotron |
| $^{13}N$ | 10.0 min | $\beta^+$ (100%) | 41.8 $\beta^+$ (99.8%)<br>511.0 $\gamma$ (199.6%) | Cyclotron |
| $^{15}O$ | 2.0 min | $\beta^+$ (100%) | 735.3 $\beta^+$ (99.9%)<br>511.0 $\gamma$ (198.8%) | Cyclotron |
| $^{18}F$ | 1.8 h | $\beta^+$ (100%) | 249.8 $\beta^+$ (96.7%)<br>511.0 $\gamma$ (193.75%) | Cyclotron |
| $^{64}Cu$ | 12.7 h | $\beta^-$ (38.5%)<br>$\varepsilon$ (61.5%) | 278.2 $\beta^+$ (17.6%)<br>511.0 $\gamma$ (35.2%) | Cyclotron |
| $^{68}Ga$ | 1.1 h | $\varepsilon$ (100%) | 836.0 $\beta^+$ (87.9%)<br>511.0 $\gamma$ (178.3%) | Generator |
| $^{86}Y$ | 14.7 h | $\varepsilon$ (100%) | 394.1 $\beta^+$ (1.1%)<br>454.2 $\beta^+$ (1.9%)<br>509.4 $\beta^+$ (1.3%)<br>535.4 $\beta^+$ (11.9%)<br>681.1 $\beta^+$ (5.6%)<br>767.8 $\beta^+$ (1.7%)<br>883.3 $\beta^+$ (3.6%)<br>1436.8 $\beta^+$ (2.0%)<br>307.0 $\gamma$ (3.5%)<br>382.9 $\gamma$ (3.6%)<br>443.1 $\gamma$ (16.9%)<br>511.0 $\gamma$ (64%) | Cyclotron |
| $^{89}Zr$ | 3.3 days | $\varepsilon$ (100%) | 395.5 $\beta^+$ (22.7%)<br>511.0 $\gamma$ (45.5%) | Cyclotron |
| $^{124}I$ | 4.2 days | $\varepsilon$ (100%) | 366.8 $\beta^+$ (0.3%)<br>687.0 $\beta^+$ (11.7%)<br>974.7 $\beta^+$ (10.7%)<br>511.0 $\gamma$ (45.0%) | Cyclotron |
| **SPECT isotopes** | | | | |
| $^{67}Ga$ | 3.3 days | $\varepsilon$ (100%) | 91.3 $\gamma$ (3.1%)<br>93.3 $\gamma$ (38.8%)<br>184.6 $\gamma$ (21.4%)<br>209.0 $\gamma$ (2.5%)<br>300.2 $\gamma$ (16.6%)<br>393.5 $\gamma$ (4.6%) | Cyclotron |
| $^{99m}Tc$ | 6.0 h | IT (99.99%) | 140.5 $\gamma$ (89.1%) | Generator |
| $^{111}In$ | 2.8 days | $\varepsilon$ (100%) | 171.3 $\gamma$ (90.7%)<br>245.4 $\gamma$ (94.1%) | Cyclotron |
| $^{123}I$ | 13.2 h | $\varepsilon$ (100%) | 159.0 $\gamma$ (83.3%) | Cyclotron |
| $^{131}I$ | 8.0 days | $\beta^-$ (100%) | 284.3 $\gamma$ (6.1%)<br>364.5 $\gamma$ (81.5%) | Reactor |

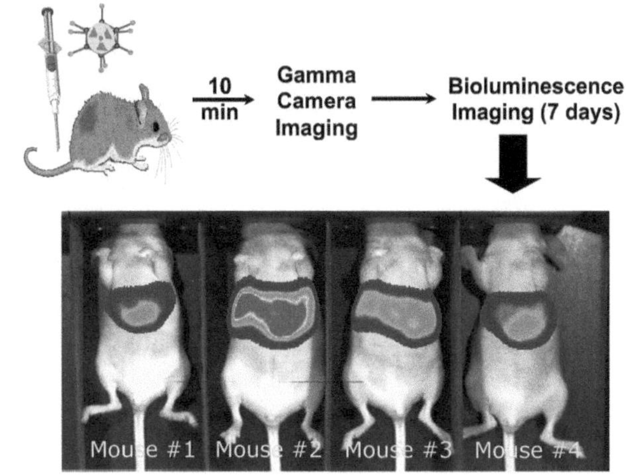


**Figure 2** Bioluminescence imaging of luciferase expression in liver of 4 mice at 7 days following intravenous injection of $^{99m}$Tc-labeled Ad5 encoding luciferase.

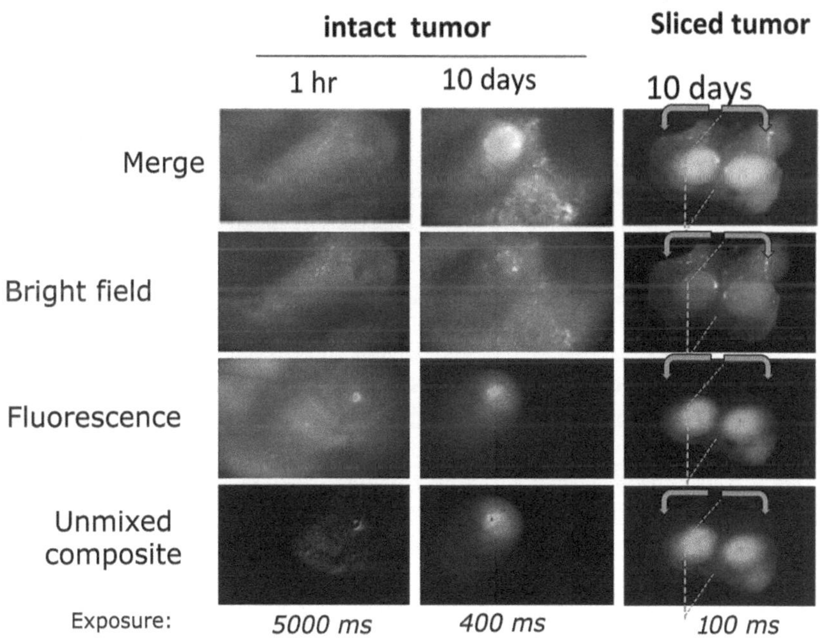


**Figure 3** Fluorescence imaging (including spectral imaging) of mRFP following intratumor injection of a conditionally replicative Ad5 vector with mRFP fused to viral capsid protein IX.

## 2.3 *Magnetic Resonance Imaging and Magnetic Resonance Spectroscopy*

The signal of proton MRI is based on the unpaired proton, most abundant in water for in vivo applications. The physics of MRI was recently reviewed.[8] Typical clinical MRI instruments have a magnetic field strength of 0.5–3 T, but 7–9 T MRI instruments are now more widely used. A low-field MRI with a permanent magnet for animal imaging was reported,[9] and is available at relatively lower cost. MRI can achieve contrast in many ways, including T1-, T2-, and T2*-weighted images; additional techniques include dynamic contrast-enhanced (DCE) MRI, diffusion-weighted imaging (DWI), and diffusion tensor imaging (DTI).

Magnetic resonance spectroscopy (MRS) is a specialized acquisition on the same instruments that measures the chemical shifts associated with molecules that incorporate $^{1}H$, $^{13}C$, $^{19}F$, and $^{31}P$. Each particular molecule has a different signature (chemical shift), allowing for assignment of individual metabolites. This information can provide fundamental insights concerning normal biochemical pathways, disease processes, or responses during therapy.[10] Also, brain phospholipid metabolites (such as free phosphate, phosphocholine, etc.) can be studied using $^{31}P$ NMR by following the unique resonances associated with each compound. MRS often requires magnets of high-field strength in order to separate the overlapping signals of individual metabolites. An additional disadvantage of MRS is the relatively low sensitivity for detecting the metabolites, which requires either very long imaging times or large voxel sizes (volume area). MRS methods and applications are detailed in a recent review.[11]

## 3. What Information Is Provided by Imaging?

Noninvasive imaging technologies have become increasingly important from 1980 to 2015 in the management of human diseases. Diagnostic radiology is the medical specialty that is responsible for imaging, providing information impacting clinical care in the general areas of (i) anatomy and perfusion, (ii) function and metabolism, and more recently (iii) molecular imaging. Anatomy and perfusion are the most widely applied in terms of the number of studies. Thus, imaging detects abnormalities, since many conditions result in the disruption of normal anatomy, function, or blood flow. One example is the detection of a mass in an abnormal location on a chest radiograph, which with further tests leads to diagnosis of cancer. Another example is the identification of fractures following traumatic injury, or decreased bone density resulting from osteoporosis. These basic radiology techniques remain an important component of disease management. They are routinely accomplished by radiography and angiography, CT, MRI, and US. PET studies can also assess blood flow.

The second general area that can be evaluated by imaging is metabolism and function, including organ functions. Examples include noninvasive imaging to assess heart perfusion under stress, gastric emptying, ventilation/perfusion of the lung, and renal and liver function. Imaging metabolism includes many modalities. MRS techniques can detect altered metabolites in disease processes. Another aspect of metabolism that can be assessed is energy utilization. The increased metabolic rate of cancerous tissue relative to normal tissue can be imaged using PET tracers that accumulate in areas of higher metabolic activity. These studies are accomplished by administration of a radioactive drug; the increased uptake of the radioactive drug in the cancerous lesion is imaged with gamma-ray detection instruments. In a similar manner, the glucose or fatty acid metabolism in myocardium can be evaluated following ischemic injury.

Molecular imaging is the latest evolution. A Society of Nuclear Medicine task force defined molecular imaging in 2007 as "the visualization, characterization, and measurement of biological processes at the molecular and cellular levels in humans and other living systems. Molecular imaging typically includes 2D or 3D imaging as well as quantification over time. The techniques used include radiotracer imaging/nuclear medicine, MR imaging, MR spectroscopy, optical imaging, ultrasound, and others."[12] Several examples of molecular imaging are provided later in this chapter.

# 4. Imaging to Monitor Gene Therapy

## 4.1 Computed Tomography Imaging

Traditionally, noninvasive imaging to evaluate therapeutic efficacy has relied primarily on anatomical assessments using CT and/or MRI. Guidelines for characterizing treatment outcome in cancer patients, such as those established by the World Health Organization (WHO) in 1979, the Response Evaluation Criteria in Solid Tumors (RECIST) group in 2000,[13] and RECIST version 1.1 in 2009,[14] have helped standardize radiological evaluations of cytotoxic therapies. These criteria use early tumor growth and/or the development of new lesions as indicators of progressive disease (PD) and treatment inefficacy. RECIST and WHO criteria have been implemented in a number of clinical trials to evaluate the antitumor response of adenoviral gene therapy.[15–22]

Immunologically active treatments such as oncolytic viruses and immunotherapies often initiate inflammatory swelling that results in a temporary increase in tumor size. Thus, response evaluations that rely solely on volumetric measurements may not be suitable for assessing the treatment efficacy of adenoviral therapies. The paradoxical increase in tumor size followed by shrinkage has been demonstrated in several clinical trials involving oncolytic viral treatments.[17–19] Some immunotherapies have also been shown to lead to complete remission, partial remission, or stable disease (SD) only after the appearance of new lesions or an initial increase in total tumor burden. Other therapies result in tissue necrosis and/or cavitation without necessary changes in tumor size.

Between 2004 and 2009, systematic criteria to enhance the characterization response of immunotherapeutic agents were established, called the immune response-related criteria (irRC). These guidelines were formed based on the premise that, in comparison to cytotoxic therapies, treatments that initiate an immune response may take longer to generate an antitumor response and/or may demonstrate antitumor effects after conventionally diagnosed PD. The apparent increase in tumor burden that often precedes antitumor activity can be attributed to continued tumor growth prior to a sufficient immune response or transient inflammation caused by immune cell infiltration. Thus, the irRC account for clinically insignificant PD (e.g., small new lesions in the presence of responsive tumors) or durable SD in assessments of treatment efficacy.[23–25]

The utility of using CT and irRC guidelines for evaluating adenoviral therapeutic response is promising, given the inflammatory and delayed treatment response that is similar to that of other immunotherapies (e.g., monoclonal antibodies, immune adjuvants, viral vaccines) by which the irRC were established and validated. In addition, the ability of CT in combination with contrast enhancement and perfusion to assess functional changes has clinical value for assessing early response to therapy before volumetric changes occur.[26] The high spatial resolution of contrast-enhanced CT, for example, allows changes in microvasculature to be detected in response to treatment (e.g., changes in vessel size, relative blood volume, 3D vessel distribution, and vascular branching).[27]

Other imaging modalities such as PET, US, and MR-based techniques are also being implemented to supplement radiologic anatomical information with functional and metabolic insight. These noninvasive tools hold great value for detecting early response to therapy and can, therefore, be used to effectively tailor and optimize treatment regimens much sooner than traditional assessments that rely on changes in tumor size (or lack thereof).

## 4.2 Positron Emission Tomography Imaging

Fluorodeoxyglucose ($^{18}$F-FDG)-PET imaging is capable of providing information on the metabolic state of tissues and is, therefore, useful in many cancer types for diagnosing, staging, and monitoring treatment response. PET Response Criteria in Solid Tumors (PERCIST, version 1.0) was established in 2009 to standardize assessments of early treatment response using $^{18}$F-FDG.[28] For some patient populations, PERCIST has been shown to be a stronger independent predictor of chemotherapeutic outcome than RECIST.[29]

The utility of PERCIST guidelines in assessing the antitumor response of adenoviral gene therapy has yet to be evaluated. Often, oncolytic viral treatments cause an influx of metabolically active inflammatory cells that can be mistakenly interpreted as increased tumor metabolism in FDG-PET imaging. In addition, falsely positive signals in local lymph nodes can be misinterpreted as metabolic progression. Utilization of PERCIST criteria for interpreting the metabolic information provided by FDG-PET imaging offers a structured approach that may enable standardization among quantitative clinical reporting.

A study by Koski et al.[30] in 2013 evaluated the utility of CT and PET for evaluating the response to oncolytic adenoviral treatment in patients with advanced cancer. A retrospective study of 17 individual case reports of cancer patients imaged with both contrast-enhanced CT and FDG-PET before and after adenovirus treatment demonstrated that response evaluations correlated moderately well and that the two imaging modalities were equally reliable as prognostic markers for long-term survival following treatment. Studies in hamsters suggested that FDG-PET is a more sensitive method for detecting antitumor effects since reductions in tumor metabolism occurred prior to reductions in tumor size as assessed by CT. These data were corroborated by the patient series, in which a favorable trend for disease control was detected more often by FDG-PET than CT. Importantly, both the hamster and the human studies also demonstrated the development of FDG-avid lymph nodes following adenoviral treatment, most likely the result of an inflammatory response and not treatment failure.[30]

This same group also used FDG-PET and CT imaging to evaluate patient response to treatment with a double-targeted chimeric oncolytic adenovirus (CGCT-401) controlled by the human telomerase reverse transcriptase (hTERT) promoter for tumor selectivity and expressing CD40 ligand (CD40L) for antitumor effects.[21] In preclinical studies, CD40L transgene expression correlated with several antitumor effects including oncolysis, apoptosis, induction of T-cell responses, and upregulation of $T_H1$ cytokines.[31–33] Consistent with these findings, in human trials, evidence was provided to support the antitumor immune response and therapeutic efficacy of CGCT-401 treatment; five out of six patients displayed disease control as determined by CT or PET-CT.[21]

A common limitation of FDG-PET that is cited in these clinical reports of adenoviral therapy involves the difficulty of quantifying therapeutic effects due to false-positive signals that arise from tumor swelling and/or recruitment of metabolically active immune cells. Furthermore, the type of viral vector used and the presence of immunostimulatory transgenes likely influence the utility of PET imaging, as evidenced by a clinical trial using an unarmed oncolytic herpes virus that did not find diagnostic or prognostic utility in FDG-PET evaluations.[19] Understanding the role of PET imaging in monitoring responses to adenoviral gene therapy is dependent on more clinical evaluations and standardization of PET response criteria for adenoviral vectors.

## 4.3 Single Photon Emission Computed Tomography Imaging

SPECT is a standard method for monitoring myocardial perfusion and has been utilized to assess adenoviral gene therapy efficacy in cardiovascular disease. Adenoviral vectors encoding angiogenic proteins that stimulate blood vessel growth have been proposed to restore perfusion to ischemic regions. Grines et al. reported a trial aimed at evaluating the efficacy of replication-defective adenoviral particles encoding fibroblast growth factor 4 (FGF-4) for reducing ischemia defects in patients with stable angina.[34] SPECT imaging of myocardial perfusion with $^{99m}$Tc-sestamibi was used to assess therapeutic outcome. A slight reduction (<5%) in reversible perfusion defect size was observed in the treatment group compared to the placebo control group at 4 and 8 weeks following viral administration in this study when one outlier patient was removed from data analysis. However, the small sample size and minimal improvement observed did not allow definitive establishment of gene therapy efficacy in this trial. A phase II study of patients with coronary heart disease assessed the efficacy of vascular endothelial growth factor-165 ($VEGF_{165}$) gene therapy for improving vascular perfusion.[35] Six months after treatment with either replication-defective adenoviral particles or liposomes encoding the gene, SPECT imaging of myocardial perfusion using $^{99m}$Tc-sestamibi indicated a significant improvement in cardiac perfusion compared to baseline values for the adenoviral experimental group but not for the liposome-treated or control groups.

## 4.4 Magnetic Resonance-Based Imaging

MR-based imaging techniques are capable of providing functional, metabolic, and structural information about tissues,[36] which can be useful for tumor localization, phenotyping, and treatment monitoring.[37] The lack of ionizing radiation makes these approaches attractive for human use, and the high endogenous contrast and spatial resolution of MRI enable macroscopic tumor heterogeneity to be addressed.

T2-weighted MRI is useful for identifying coagulative necrotic regions within a tumor, which appear as hypointense areas with shortened T2 relaxation times. This technique was implemented in an immunocompetent hamster model to determine the antitumor response of a granulocyte-macrophage colony-stimulating factor (GM-CSF)-coding adenoviral therapy.[38] In this study, tumors treated with an adenovirus armed with GM-CSF exhibited a hypointense core region consistent with coagulative necrosis that persisted throughout the 7-day follow-up; these tumors were undetectable by day 28. In contrast, tumors treated with an unarmed adenovirus or phosphate buffered saline alone did not maintain a homogeneous necrotic core and demonstrated continuous growth. Similar observations were observed in a single neuroblastoma patient treated with oncolytic adenovirus. As identified using T2-weighted MRI, initial posttreatment tumor regression was accompanied by a hypointense core region indicative of coagulative necrosis. However, 4 months after a second treatment, tumor progression resumed and the tumors appeared moderately hyperintense, suggesting regrowth of viable tumor tissues.[38]

MRS of tumor metabolites such as unsaturated fatty acids, choline, inositol, and taurine is another noninvasive method for monitoring the antitumor response of gene therapy treatments. The spectral peaks provided by these biochemical compounds can be distinguished by the larger peaks that arise from tumor water content and used to gather information on the viability and aggressiveness of tumor cells. Choline, for example, produces a peak at 3.2 parts per million, and an increase in the resonance of choline-containing compounds has been shown to correlate with cell membrane synthesis. Increased tumor levels of choline are, thus, predictive of tumor aggressiveness.[39] Taurine is another cancer biomarker whose presence in tumors is associated with malignant growth,[40–43] and spectroscopic detection of cell membrane unsaturated fatty acids has been used to indicate tumor cell apoptosis[44] or autophagy.[45]

In the 2014 study by Hemminki et al.,[38] tumors that responded to treatment with the GM-CSF armed adenovirus showed decreased levels of choline and taurine metabolites compared to untreated tumors, even at early time points before volumetric differences emerged. Thus, MRS identification of these metabolites may prove useful as predictive or prognostic biomarkers for evaluating the efficacy of oncolytic adenoviral therapies.

Diffusion-weighted MRI (DW-MRI) detects the movement of free water molecules via the use of magnetic field gradients. This technique is sensitive to tissue microstructure and cellular-level function, and can be used to assess early response to therapy by measuring changes in the perfusion and diffusion of water that occur following exposure to drugs. The apparent diffusion coefficient (ADC) is a quantification of the decay of the diffusion-weighted signal and is indicative of cellular density and membrane permeability. Highly cellularized tissues, such as tumors, typically have lower ADC values due to the impedance of water, whereas necrotic regions or areas with compromised cell permeability are associated with increased ADC values.

Clinically, DW-MRI has been used extensively to detect early changes in the tumor microenvironment in response to a variety of different treatment regimens (reviewed in

2012 by Bains et al.).[46] Although many cancer therapies result in an increase in ADC value, treatments that act via certain mechanisms of action (e.g., swelling, vascular restriction) may lead to decreased ADC values. Furthermore, certain types of cancer (e.g., rectal carcinoma[47,48] and bone metastasis[49]) have also demonstrated posttreatment reductions in ADC values, which could be attributed, at least in part, to tissue inflammation and/or fibrosis.[50] Clinically, the predictive value of ADC, therefore, is likely dependent on tailoring DW-MRI evaluations and protocols to individual tissue types and therapeutic strategies. Specific to adenoviral-based gene therapy strategies, DW-MRI has been used to assess the response of antiangiogenic and antilymphangiogenic adenoviral treatments in a human ovarian cancer xenograft model. In this study, however, ADC values did not correlate with tumor necrosis. Instead, increases in ADC values were observed in tumors following treatment that led to histologically confirmed fibrosis.[51]

DCE-MRI is a noninvasive tool that is optimal for measuring tumor responses to antiangiogenic therapies given its ability to measure relative blood volume, perfusion, and permeability.[27,52] In a preclinical model, DCE-MRI has also been used in conjunction with T2-weighted MRI coregistered with electron paramagnetic resonance to correlate tissue perfusion with quantitative absolute oxygen measurements. This multimodality approach is aimed at characterizing the "signature" of tumor response to therapy to be used as an adaptive image-guided scheme for identifying heterogeneous tumor regions and for optimizing the scheduling of antiangiogenic and cytotoxic agents.[53]

A prime example of imaging adenoviral gene therapy is documented by Menendez et al.[54] in a preclinical study to assess cartilage repair following injury. A pony model of healing in an osteochondral defect used quantitative CT and MRI (T2 mapping and DCE-MRI using Gd-DTPA) to noninvasively assess the therapeutic efficacy of adenoviral vectors encoding bone morphogenetic proteins (BMP)-2 or BMP-6 genes. Serial in vivo imaging enabled the investigators to study osteochondral regeneration within individual subjects across time, which avoided the need for more invasive evaluations relying on biopsy. In particular, the T1 values of Gd-DTPA observed using delayed DCE-MRI provided useful information regarding the biochemical composition of the osteochondral lesion (e.g., glycosaminoglycan content), and live pony CT was used to calculate the bone mineral density. The imaging results, which indicated signs of chondrogenesis and osteogenesis in response to BMP-2/6 adenoviral treatment, were in close agreement with those of postmortem histological evaluations.[54]

Other preclinical studies have also employed MRI for assessing therapeutic growth factor gene efficacy with adenoviral vectors. Examples include recovery from ischemia in the brain,[55–59] cardiovascular reconstruction,[60–63] and blood flow to extremities,[64] among other applications.[65,66]

## 4.5 Ultrasound Imaging

Contrast-enhanced ultrasound (CEUS) using targeted, site-specific microbubbles has been used to study angiogenesis in cancer and cardiovascular models and to preclinically evaluate antiangiogenic therapies (reviewed in Ehling et al. and Leong-Poi).[27,67] For example, molecular imaging of the vascular phenotype can be achieved using RGD-labeled microbubbles to characterize vascular expression of alpha(V)-beta-3 integrin. This approach was used by Sirsi et al. to evaluate vascular response following treatment with the anti-VEGF antibody bevacizumab (BV), and CEUS molecular imaging was shown to be an early prognostic indicator of BV efficiency.[68] CEUS using hard-shell microbubbles has also been used to depict a significant decrease in tumor vascularity as early as 1 day after antiangiogenic therapy.[69] Given the low cost, sensitivity, and ease of use, CEUS techniques will likely have significant clinical utility for assessing tissue response to adenoviral gene therapies.

## 5. Imaging and Gene Therapy Vectors

In addition to evaluating gene therapy efficacy, noninvasive imaging is widely utilized in other aspects of adenoviral vector applications. Imaging has been employed to assess viral accumulation and retention as well as transgene expression both regionally and temporally following vector administration. While preliminary trials have demonstrated the potential of noninvasive imaging techniques with adenoviral vectors in a variety of settings, each of these specific applications has unique challenges that must be considered prior to their widespread clinical implementation.[70–74]

Viral particle delivery, accumulation, and retention over time can be visualized by attaching a reporter probe to the viral capsid prior to administration to the subject. This is typically accomplished by covalent attachment of the reporter of interest to the viral particles, although noncovalent interactions (e.g., electrostatics) can be employed as well. For instance, superparamagnetic iron oxide particles (microparticles or nanoparticles), which are suitable reporters for detection by MRI due to their T2 and T2* contrast-enhancing effects, have been used to label viral capsids[75,76] as well as transduced cells[77,78] prior to in vivo delivery. While these strategies may be beneficial for short-term assessment of viral localization, they typically do not provide information on long-term viral behavior, replication, or gene expression. Alternatively, the viral construct can be genetically altered to express a reporter gene during its replication and assembly. Vectors encoding fusion constructs that combine optical reporter proteins[79] or that have affinity for exogenous reporting metals (e.g., $^{99m}$Tc for SPECT imaging[80,81]) are examples from this category. Such constructs would be favorable for monitoring viral replication over time in addition to visualizing their localization within organisms. Although preclinical studies have demonstrated encouraging potential for noninvasively monitoring vector accumulation, these techniques have not yet been widely employed in clinical adenoviral applications.

As opposed to imaging the virus particles themselves, imaging transgene expression is an attractive means for monitoring successful transfection and proper gene translation to functional protein products over time. Reporter genes, which allow this aspect of imaging, have been commonly employed in adenoviral vectors. Two main classes of reporter genes used in human studies can be identified: receptor/transporter genes and enzyme/activator genes. The former encode extracellular protein receptors, transporters, or symporters, which increase the local concentration of the reporter probes within the transduced tissues. The latter encode intracellular enzymes that convert cell-permeable reporter probe substrates into products that become trapped within the cells or that provide unique spectroscopic signals compared to the substrates. The high local concentration of the probe relative to nontransduced tissues allows visualization of gene expression. The proportional relationship between the number of reporter protein molecules present and the subsequent reporter probe accumulation allows quantitative assessment of transgene expression, albeit in an indirect manner. Direct assessment of transgene expression is possible in cases where the reporter molecule is the product encoded by the reporter gene, such as with fluorescent proteins. While these intrinsic reporter systems have been commonly explored in vitro and in animal models, they have not been applied in human imaging applications.

## 5.1 Nuclear Imaging Techniques

Nuclear imaging has been utilized more than any other modality for noninvasively detecting adenoviral transgene expression in clinical settings. The following discussion will focus on the relevant reporter genes/reporter probe systems incorporating PET and SPECT radionuclides that have been utilized in human studies. Additional systems that have yet to be validated in human trials will be summarized briefly at the close of this section. For further reading, several excellent reviews concerning imaging adenoviral gene delivery and expression in preclinical reports are available.[70–73,82–94]

### *HSV1-tk* Reporter Gene

The herpes simplex virus type 1 thymidine kinase (HSV1-*tk*) gene has been extensively utilized in human gene therapy due to the ability of viral TK to phosphorylate pyrimidine analogs (e.g., 2′-fluoro-2′-deoxy-1-β-D-arabinofuranosyl-5-iodouracil [FIAU], 2′-fluoro-2′-deoxyarabinofuranosyl-5-ethyluracil [FEAU], 2′-deoxy-2′-fluoro-5-methyl-1-β-L-arabinofuranosyluracil [FMAU]), or acycloguanosine analogs (e.g., 9-[4-fluoro-3-(hydroxymethyl)butyl]guanine [FHBG]), which are not substrates for human TK. On phosphorylation by HSV1-TK, the cell-permeable precursors become trapped within the cells and are ultimately incorporated into the host cell DNA, resulting in termination of DNA synthesis and subsequent cell death. This promiscuity of HSV1-TK has been harnessed for imaging applications by utilizing radiolabeled substrates as reporter probes for detecting HSV1-*tk* expression.

Prior to the first reported clinical trial for noninvasively imaging viral transgene expression, the pharmacokinetics and dosimetry of several PET reporter probes intended for use with the HSV1-*tk* reporter gene were examined in human subjects who had not been administered the viral gene.[95,96] These preliminary studies showed tissues where the probes were nonspecifically retained (e.g., intestinal region for $^{18}$F-FHBG) as well as where they had negligible accumulation (e.g., brain for $^{124}$I-FIAU and $^{18}$F-FHBG) following intravenous injection, highlighting areas where these probes would not be useful for detecting transgene expression in subsequent studies. Additional probes (e.g., $^{18}$F-FMAU[97]) have since been developed and used for preliminary studies in healthy human volunteers.

The first study utilizing a radioactive reporter molecule to detect transgene expression in humans was reported in 2001.[98] Following HSV1-*tk* gene delivery via a nonviral vector to five patients with glioblastoma, PET imaging utilizing $^{124}$I-FIAU demonstrated enhanced probe uptake in the tumor of one patient relative to baseline levels prior to transfection. MRI, $^{18}$F-FDG, and $^{11}$C-MET (methionine) were used to assess the response to ganciclovir therapy in this individual. Imaging analysis showed necrotic tissue in the transfected region of the tumor, thus signifying successful therapeutic outcome. While transgene expression was not observed in the other four individuals, this study showed the potential of imaging both gene expression and viral gene therapeutic efficacy via noninvasive nuclear imaging probes. Similar results were observed in a related phase I/II trial with eight patients suffering from glioblastoma multiforme, where $^{124}$I-FIAU accumulation was detectable in only one of the patients following vector administration.[99]

Noninvasive imaging of adenoviral-delivered gene expression in human subjects was first reported in 2005.[100] In this study, four of seven patients with hepatocellular carcinoma showed accumulation of $^{18}$F-FHBG within tumor lesions 2 days after intratumoral injection of viral vectors bearing the HSV1-*tk* gene. The results showed that PET probe accumulation within tumoral lesions was not directly correlated with the initial viral load administered, and malignant areas were not visible by $^{18}$F-FHBG in patients receiving fewer than $10^{12}$ viral particles. The patients who received doses above this threshold and showed transgene expression via PET imaging exhibited stable disease 30 days following transfection, suggesting a favorable response to gene therapy in combination with valganciclovir. Subsequent $^{18}$F-FHBG imaging 1 week after transfection showed no probe accumulation in the previously detected lesion. This result confirmed the expected transient expression of the viral gene. Furthermore, one patient who received a second viral dose 1 month after the initial dose failed to exhibit transgene expression via $^{18}$F-FHBG imaging following the second viral administration. An explanation suggested for this result was adenoviral neutralization by the patient's immune system following the initial dose. Since host immune responses against viral vectors are common, subsequent adenoviral administrations would likely not pose significant clinical benefits to patients. Therefore, repeated imaging of the particular replication-defective adenoviral vector and therapeutic gene strategy employed in this study are limited to a several-day window following initial administration.

A related study employing 10 patients with hepatocellular carcinoma was reported in 2010.[101] Only those patients (four out of 10) who received $10^{12}$ or more viral particles showed $^{18}$F-FHBG uptake within the lesions of interest. Two of the patients exhibited tumor necrosis following chemotherapy, although no partial responses were established. Dempsey et al. attempted to image HSV-TK expression with $^{123}$I-FIAU in eight patients with malignant glioma following intratumoral administration of $10^6$ pfu of a herpes simplex virus mutant (HSV1716) rather than an adenoviral vector.[102] No tumoral uptake of the probe was apparent in any subjects up to 5 days following viral administration relative to pretreatment levels. This finding is likely attributable to the low viral dose employed and the inability of $^{123}$I-FIAU to cross the blood–brain barrier.

## *hNIS* Reporter Gene

The human sodium iodide symporter (*hNIS*) gene is naturally present in humans with fairly restricted tissue expression (thyroid, stomach, salivary, and lactating mammary glands).

Since these tissues are often removed from areas associated with cancer metastases, utilizing adenoviral vectors to transfect tissues with *hNIS* for imaging and therapy with suitable radioactive anions has been of great interest in cancer applications. The symporter efficiently transports iodide and pertechnetate ($[TcO_4]^-$) anions within cells; although gradual efflux of the anions is unavoidable, the continual activity of hNIS allows sufficient radionuclide accumulation for nuclear imaging applications.

The first study to successfully image adenoviral-mediated *hNIS* gene expression noninvasively in human patients was reported in 2008.[103] In this phase I trial of 12 individuals with prostate cancer, a replication-competent adenoviral vector bearing the *hNIS* gene was injected into the diseased portions of the prostate within the patients. External intensity-modified radiation in addition to 5-fluorocytosine (5-FC) and valganciclovir prodrugs (activated by expression of viral-encoded yeast cytosine deaminase [yCD]/HSV1-TK) were used as therapeutics in the study. SPECT imaging with $[^{99m}TcO_4]^-$ demonstrated maximal hNIS expression 1–2 days following viral delivery, with detectable levels remaining in one patient 1 week after viral injection. As was observed by Penuelas et al. with HSV1-*tk* gene delivery, a threshold value of viral particles ($>10^{11}$) was required for adequate detection by the SPECT probe with hNIS in this study. Overall, seven of the subjects exhibited $[^{99m}TcO_4]^-$ accumulation attributed to transgene expression.

In a subsequent dosimetry study intended to estimate the potential for utilizing $^{131}I$ as a radiotherapeutic nuclide following *hNIS* gene delivery, six individuals with prostate cancer were administered $5 \times 10^{12}$ particles of the same adenoviral agent and imaged with $[^{99m}TcO_4]^-$ via SPECT on days 1–3 and 7 thereafter unless imaging showed no uptake of the radionuclide in the prostate.[104] Transgene expression was apparent in noncancerous prostatic tissue in this study, and several individuals did not show gene expression in tissues where virus was administered. While the viral dose and method of application were intended to deliver the virus to the entire prostate, the results indicated that ~45% of the prostate showed adequate hNIS expression for SPECT imaging. Dosimetry calculations suggested that this would correlate to ~7 Gy dose to the prostate had 200 mCi $^{131}I$ been employed, indicating that the level of hNIS expression with the vector utilized in this study would be too low for successful $\beta^-$ radiotherapy. Two of the subjects from the study that showed hNIS expression and completed the therapy regimen had negative (e.g., cancer-free) biopsy results 2 years after completion of the study. While not definitive, this result suggests that the viral gene/prodrug regimen combined with external radiation therapy may be effective for combating localized prostate cancer.

Another example, in which an adenoviral vector containing the *hNIS* gene was delivered, was reported for a single patient with advanced and metastatic cervical carcinoma.[105] A dose ($3 \times 10^{11}$ particles) of the virus, which was designed to express hNIS under the viral E3 promoter, was administered to pelvic and liver cancer lesions. However, no viral-mediated hNIS expression was apparent on imaging with $^{123}I$ at 20–44 h or with $[^{99m}TcO_4]^-$ 3 days after viral administration. These negative findings could have been due to the different gene promoter, injection technique, or lower viral dose used as compared to the other reports employing adenoviral agents.

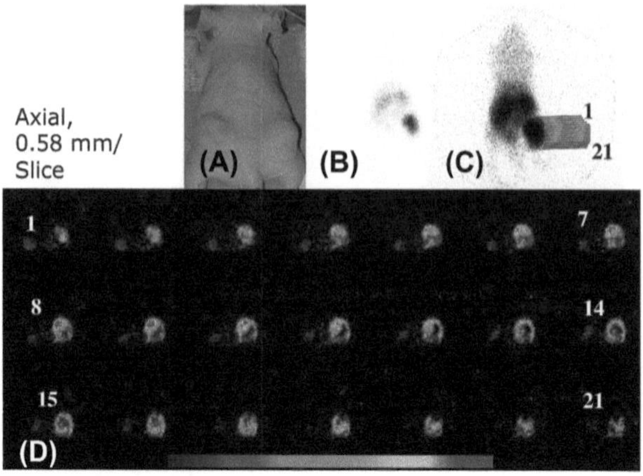


**Figure 4** Imaging of SSTR expression in (A) mice bearing A427 tumors, following intratumor injection of Ad5 encoding SSTR (right tumor) or Ad5 encoding mutant SSTR (left tumor, control), by (B,C) gamma camera imaging, and (D) SPECT. The images in (B) and (C) are scaled different.

## *SSTR Reporter Gene*

The human somatostatin receptor subtype II (SSTR) is an extracellular membrane receptor with variable levels of expression in several normal tissues (veins, gastrointestinal tract, pancreas, spleen, brain) and in a wide variety of cancers.[106] The latter observation has led to the development of many somatostatin analogs labeled with radioisotopes for identifying and treating SSTR-positive tumors.[83,92,93] The SSTR has been utilized as a reporter gene in several preclinical applications.[70,71,73,79,84,85,88,107–115]

Figure 4 presents images of a preclinical study with human nonsmall-cell lung A427 tumors after intratumoral injection of the Ad5 encoding SSTR (right tumor), or Ad5 encoding a mutant SSTR (left tumor, control). At 2 days after Ad5 injection the SSTR-avid $^{99m}$Tc-P2045 peptide was injected intravenously and 4 h later the animals were imaged with a planar gamma camera (Figure 4(B) and (C)) or by SPECT (Figure 4(D)). Figure 4(B) and (C) use different scaling, and show positive accumulation in the right tumor, and negative accumulation in the left control tumor. Figure 4(D) presents 0.58 mm axial slices showing the nonuniform distribution of the SSTR, as expected from the intratumor injection. Higher expression was found in the periphery of the tumor where growth was most active. Figure 5(A) and (B) present images of another animal for a SPECT/CT study with contrast (iohexol, omnipaque), and show excellent correspondence in the peripheral tumor area for perfusion (high level of CT contrast) and SSTR expression (Figure 5(B), arrows).

A single study that utilized adenoviral-mediated *SSTr* gene delivery in humans has been reported. A phase I clinical trial enrolling 12 patients with gynecologic (ovarian/endometrial) cancers utilized a replication-defective adenovirus encoding

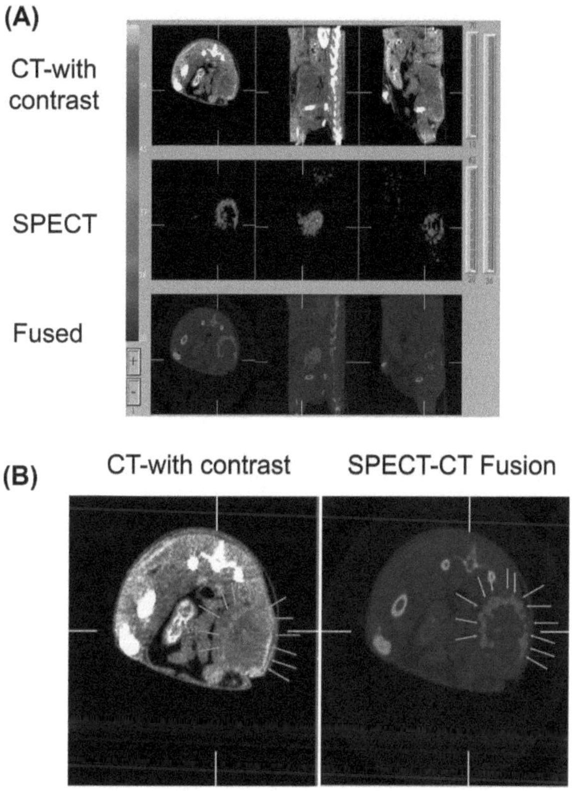


**Figure 5** SPECT/CT imaging of SSTR expression in A427 tumors, with (A) CT, SPECT, and SPECT/CT fused images of a tumor slice, and (B) zoomed area of CT and SPECT/CT fusion image. The mouse was injected with the CT contrast agent omnipaque.

genes for both *SSTr* and HSV1-*tk* in combination with ganciclovir prodrug therapy.[116] The virus capsid was modified to express the RGD sequence for receptor-mediated uptake via binding to cell surface integrin receptors as these receptors are commonly overexpressed in cancerous tissues. One day following the final of three daily intraperitoneal viral administrations, noninvasive SPECT imaging of SSTR expression was performed with $^{111}$In-pentetreotide. Although the results showed significant uptake of the SPECT probe relative to background levels in the groups receiving the highest viral dose ($10^{12}$), patients receiving lower doses did not show significantly increased uptake of the probe following viral administration relative to pretreatment levels. Ex vivo polymerase chain reaction (PCR) analysis of ascites samples from patients who received the mid-level viral doses ($5 \times 10^{10}$/dose), however, confirmed expression of HSV1-*tk* and *SSTr* gene transcripts, indicating that the SPECT probe employed was unable to detect low levels of gene expression. No direct correlation between therapeutic outcome and viral dose or gene expression was apparent from this study.

## Additional Reporter Genes

In addition to the reporter gene/reporter probe systems that have been demonstrated in human trials, several other targets have been evaluated for monitoring adenoviral-mediated transgene expression in preclinical studies. Viral transfection of the dopamine two receptor (D2R), which is not normally expressed in high levels outside of the brain and pituitary gland, has been assessed with PET probes (3-(2′-$^{18}$F-fluoroethyl)-spiperone, $^{11}$C-raclopride) that are capable of crossing the blood–brain barrier.[71,73,117–122] The norepinephrine transporter, which is restricted in expression to the nervous system, is another attractive reporter target for adenoviral applications and has been imaged with radioactive probes ($^{11}$C-*m*-hydroxyephedrine, $^{124/131}$I-*m*-iodobenzylguanidine) in cancer models.[70,123,124] The estrogen receptor ligand binding domain serves as a unique intracellular reporter target as it is limited in expression except within female reproductive organs. Successful adenoviral-mediated gene expression has been successfully visualized in vivo using 16α-$^{18}$F-fluoro-17β-estradiol.[125,126] Since the reporter probes utilized in the above preclinical studies have been previously validated for imaging endogenous receptor expression in humans, it is likely that the corresponding reporter genes will be evaluated in clinical adenoviral applications in the future.

Despite the relatively high sensitivity and quantitative characteristics of nuclear imaging with PET/SPECT tracers compared to other imaging modalities, nuclear probes *indirectly* allow determination of viral reporter gene expression. The probes must first reach the target cells before they interact with the gene products to allow visualization of vector expression. Insufficient delivery of the radioactive probe to the desired tissues would not allow detection of gene expression (regardless of expression level) and would result in false-negative findings. This problem is most common in brain imaging applications (e.g., glioma) with systemically administered agents that do not cross the blood–brain barrier (e.g., FIAU).[95,102] Nonspecific probe accumulation in nontarget tissues or slow pharmacokinetics must also be considered with nuclear reporter probes. High probe retention within normal tissues may lead to false-positive results or may not allow definitive verification of reporter gene expression due to high background activity.[96] To overcome these obstacles, reporter probe design with consideration of pharmacokinetics/half-life (radiochemical and biological) is crucial in utilizing established probes for new conditions and when developing novel probes for reporter genes.[97] Apart from probe design considerations, administration methods (e.g., systemic vs local) may also be important for adequate assessment of reporter gene expression.

The limited sensitivity of nuclear probes for detecting viral-mediated gene expression is a frequent observation noted in the studies above. While ex vivo techniques (PCR and immunohistochemical staining of biopsy samples) frequently indicated the presence of viral genome transcription, noninvasive imaging with the reporter probes failed to show the presence of the transgene.[103,116] Therefore, nuclear techniques may not adequately detect low levels of gene expression, even though the level of expression may be therapeutically relevant. This challenge regarding sensitivity limitations should be considered for future studies with probes for adenoviral reporter genes.

## 5.2 Optical Imaging Techniques

Optical detection of in vivo transgene expression can be achieved using fluorescence and BLI. Constitutive expression of optical reporters, such as fluorescent proteins or luciferases, allows for vector localization and monitoring over time. Optical reporters can also be linked to a specific promoter such that transcription of the reporter coding sequence is indicative of gene expression. There are four types of optical reporter genes that can be used to indirectly measure the expression of an endogenous gene or promoter. In general, these reporter genes encode (i) a fluorescence protein, (ii) a luciferase enzyme, (iii) an enzyme that can convert an exogenously added quiescent substrate into a fluorescent complex or induce a conformational change that modifies the fluorescence of the added substrate, or (iv) a fusion protein that uses a peptide linker to couple expressed transgenes with a fluorescent or bioluminescent reporter.[127]

Preclinically, optical imaging of adenoviral-mediated gene expression has been used to study a wide range of biological topics broadly related to tumor/cancer biology, cardiac applications, gene therapy, and cell trafficking. Several disadvantages of the optical reporter-based imaging, however, have thus far prevented the optical reporter gene approach from being clinically implemented. These disadvantages mainly involve the introduction of a foreign gene/probe into patients and the inability to perform deep tissue imaging due to autofluorescence, light scattering, and absorbance of light by hemoglobin at wavelengths below 600 nm. Given the significant and widespread success of in vivo optical imaging in the preclinical setting, however, many researchers are focused on developing optical detection systems and sophisticated probes that will enable optical imaging of gene therapy in humans.

### Fluorescent Proteins

The most widely used fluorescent reporter proteins are GFP, enhanced GFP, and red fluorescent protein. Numerous adenoviruses have been equipped with these fluorescent proteins to enable vector localization and quantification.[128–130] In addition, new fluorescent reporters that operate at wavelengths in the red and near-infrared window are being developed to maximize transmission, minimize absorption, and decrease background signals from tissue autofluorescence. These reporters have improved capabilities for in vivo fluorescence imaging and have the potential to be used as reporters of transgene expression.

A multitude of preclinical studies that describe the use of fluorescent reporters to monitor adenoviral gene transfer exist. For example, the cancer-specific promoter inhibition of differentiation-1 (Id1) has been used to derive the expression of an adenoviral-delivered mCherry fluorescent reporter for the detection and localization of breast and prostate cancer.[131,132] mCherry is a mutated variant of the widely used mRFP1 that has better tissue penetration and less autofluorescence due to its longer wavelength. In both human prostate cancer[131] and breast cancer[132] models, mCherry reporter expression was sensitively detected in situ using spectral imaging, and expression levels were shown to correlate with tumor levels of Id1.

OBP-301 (Telomelysin®) is a tumor-specific adenovirus that has demonstrated selective replication and antitumor effects in human cancer cells in vitro and in vivo.[133–136] Several clinical trials have also reported on the safety and potential therapeutic efficacy of Telomelysin® for oncolytic treatment of advanced solid tumors.[137,138] An animal study by Umeoka et al. evaluated a strategy for visualizing pleural dissemination of human nonsmall-cell lung cancer cells using OBP-301 in combination with a replication-deficient adenovirus expressing GFP (Ad-GFP). Locoregional coinjection of OBP-301 and Ad-GFP enabled real-time visualization of macroscopically invisible tumor tissues using optical CCD imaging, whereas no GFP signal was detected after coinjection into normal tissues or following local injection of only Ad-GFP.[139] To circumvent the need for intratumoral or local coinjections of the adenoviruses, this group subsequently modified the vector backbone of OBP-301 to include a cytomegalovirus (*CMV*)-controlled *GFP* gene. In theory, this modified adenovirus, named OBP-401, was designed to enable systemic administration leading to whole-body distribution of the vector for detection of distant metastatic lesions. The optical reporting capabilities of OBP-401 were evaluated in an orthotopic mouse model of human rectal cancer. Intratumoral administration of OBP-401 resulted in viral spread into the regional lymphatic area and hTERT-dependent replication and GFP expression in neoplastic lesions. Optical CCD imaging of exposed paraaortic lymph nodes allowed for direct visualization of micrometastases via GFP expression, which was shown to correlate histologically with the presence of adenocarcinoma cells.[140] OBP-401 has also been used clinically to identify circulating tumor cells in patients with various cancers.[128,141–143]

These examples highlight a few of the fluorescent imaging approaches that can be used for monitoring adenoviral reporter gene expression. Despite the clinically relevant information that can be extracted from these studies, the limitations associated with imaging fluorescent reporters are also evident. Most notably, the signal attenuation that occurs with increasing tissue depth restricts in vivo fluorescence imaging to surface-level detection whether it is directly beneath the skin or in dissected tissue (i.e., via surgical incision) or requires the use of endoscopic/laparoscopic techniques. The recent developments of endoscopic fluorescence systems and surgical imaging systems are helping in realizing the potential of high-resolution, real-time fluorescence imaging to aid in clinical decision-making. For example, fluorescent imaging of indocyanine green for evaluating tissue perfusion is now widely implemented in cardiac and surgical cases, and new NIR fluorescent dyes are being developed to aid in surgical resection of tumors. Advances in these areas of fluorescence imaging will likely pave the way for the translation of fluorescent reporter probes that enable real-time monitoring of gene therapies.

### *Luciferases*

Bioluminescence is a form of chemiluminescence in which visible light emission occurs within an organism as a result of a chemical reaction between a luciferase enzyme and its substrate (i.e., luciferin, coelenterazine). Bioluminescent light is produced when the electron from this reaction returns to its ground state; this light provides the imaging contrast necessary for BLI. Due to the lack of endogenous luciferases in mammalian tissues, the signal-to-noise ratio associated with BLI is extremely high, which enables very sensitive measurements.[4]

For molecular imaging, a reporter construct that contains the genetically encoded material necessary for the production of a luciferase enzyme is used. Luciferase reporter constructs are routinely controlled by a promoter, which can be constitutively expressed (i.e., always "on") in all tissues or expressed at varying levels dependent on the physiological status of the tissue. An example of the latter is the use of cyclooxygenase-2L (cox2L) as a promoter for controlling luciferase via an adenoviral vector (Ad-cox2L-luc). Since cox2L is "off" in the liver under normal conditions, luciferase was not expressed following intrahepatic injection of Ad-cox2L-luc. Injection of lipopolysaccharide (a known activator of cox2L) into transfected mice resulted in transient hepatic expression of luciferase.[4]

Numerous luciferases have been isolated from various species (mainly marine) and used as bioluminescent reporters of in vitro and in vivo biological phenomena. Among these, the most commonly used luciferases were isolated from beetles (firefly and click beetle—Coleoptera), jellyfish and sea pansies (Cnidaria), and bacteria (*Vibrio* and *Photorhabdus luminescens*). Firefly luciferase (FLuc) is the most common optical reporter for BLI. FLuc catalyzes the reaction between luciferin and oxygen in the presence of Mg-ATP; this process emits a yellow-green light (emission peak ~560 nm). Other luciferases also isolated from beetles emit light at red-shifted wavelengths (580–625 nm), which are attractive for in vivo imaging due to the advantage of better tissue penetration. Renilla luciferase (RLuc) emits blue-green light at ~480 nm and is attractive because it does not require ATP for enzymatic reaction with its substrate coelenterazine. Because luciferin and coelenterazine do not cross-react, utilization of both FLuc and RLuc reporters allows for simultaneous imaging of two different molecular events.[144,145]

The CMV promoter is commonly used in routine applications of bioluminescence since it is constitutively expressed and highly active in most tissues. In this way, luciferase reporter expression has been implemented in a variety of applications, most notably: (i) cancer biology for monitoring tumor growth, total tumor burden, metastasis, and angiogenesis, (ii) cell therapy for monitoring the migration of adoptively transferred cells, (iii) gene therapy for targeting vectors, (iv) infection for studying the clearance of pathogens, bacteria, etc., (v) protein–protein interactions and bioluminescence resonance energy transfer (BRET) for studying biological pathways, (vi) apoptosis detection for evaluating cancer therapies, (vii) viral replication for evaluating oncolytic therapies, (viii) signal transduction for real-time monitoring of biological pathways, and (ix) transgenic mice for understanding gene regulation and studying biological processes.[4]

In contrast to the CMV promoter, the use of specialized promoters to control luciferase reporter expression enables BLI to be used to noninvasively evaluate specific molecular pathways involved in disease pathogenesis. In addition, these selected molecular pathways can be monitored to determine the response to therapeutic intervention. Herein, a few studies that exemplify the use of adenoviral vectors to encode a luciferase reporter gene whose expression is driven by a disease-specific promoter are discussed.

Tumor-specific promoters enable targeted gene therapy approaches by preferentially localizing transgene expression to tumor cells and limiting therapeutic effects in normal cells. Many studies suggest that the human *survivin* gene, a member of the inhibitor of apoptosis family, is a favorable tumor-specific promoter since it is overexpressed in many cancers but absent in normal tissue. A study by Ray et al.[146] was performed to evaluate the efficiency and specificity of an adenoviral vector containing the survivin promoter, a therapeutic gene (TNF-alpha-related apoptosis inducing ligand; TRAIL), and a firefly luciferase reporter gene. To overcome the poor transcriptional efficiency of the survivin promoter, a bidirectional two-step transcriptional amplification process was used. This approach led to a 10-fold increase in therapeutic and reporter transgene expression following intratumoral injection as determined noninvasively and quantitatively with BLI.[146]

The ability of luciferin to pass through the blood–brain barrier has resulted in many studies that use luciferase reporters for studying neurodegenerative diseases. A number of studies, for example, have utilized luciferase expression in transgenic mice to study processes and pathways in Alzheimer disease, Parkinson disease, and amyotrophic lateral sclerosis.[147,148] Although most of these reports involve the use of animals that have been genetically manipulated to link luciferase expression to a promoter gene of interest, there is at least one study that utilized a luciferase-equipped adenovirus to correlate the location, magnitude, and duration of a therapeutic transgene.[149] In this study, an adenoviral vector containing both a yCD as a therapeutic transgene and an optical luciferase reporter gene was used to evaluate the antitumor efficacy of the prodrug 5-FC in an orthotopic 9L glioma model. The cytotoxicity of 5-FC is dependent on its enzymatic conversion by yCD into the antimetabolite 5-FU. Therefore, adequate expression of the yCD transgene is essential for the antitumor response of 5-FC. The location and magnitude of luciferase expression were monitored over time using BLI and correlated to yCD with histological observations. Anatomical and diffusion-weighted MRI were used to evaluate the efficacy of 5-FC treatment.[149] This study describes a therapeutic approach for optimizing the timing and dosage of 5-FC that could be envisioned for human use on replacement of the luciferase reporter with a more clinically relevant probe.

## 5.3 Magnetic Resonance-Based Techniques

The high spatial resolution and soft tissue contrast attainable with MR techniques are favorable for imaging adenoviral vector transduction efficiency and gene expression. Many applications have taken advantage of the enzymatic activity of reporter proteins to produce a unique spectroscopic signal detectable by MRS following administration of substrates containing naturally abundant $^{19}F$ or $^{31}P$; several $^{13}C$-labeled probes have been evaluated as well. As opposed to relying exclusively on sufficient probe accumulation within transduced tissues for detection, these MRS probes become "activated" by the reporter protein to generate the signal of interest. Creatine kinase (CK) and CD are two examples of reporter enzymes that have been utilized with probes for MR detection of transgene expression in preclinical adenoviral applications.

β-Galactosidase and various other kinases have also been proposed as reporter genes,[66,71] although they have not yet been assessed with adenoviral vectors in vivo.

CK, which catalyzes the phosphorylation of creatine, has been used as a reporter transgene in the liver, where it is not normally expressed. Zhou and coworkers[150] utilized $^{31}P$ MRS of phosphocreatine (PCr) and magnetization transfer to successfully evaluate CK activity in surgically exposed livers of mice following intravenous injection of adenovirus particles containing genes for CK and low-density lipoprotein receptor. In a related study, proliferation of transgenic hepatocytes expressing CK was evaluated by 3D $^{31}P$ MRS imaging of PCr in mice following adenoviral delivery of human growth factor to stimulate hepatocyte expansion.[151] This experiment did not require surgical exposure of tissues for successful MR analysis, indicating the viability of this technique for translational applications in the future.

As noted above, CD is a nonmammalian enzyme that catalyzes the conversion of 5-FC to the chemotherapeutic agent 5-FU. $^{19}F$ MRS was used to monitor the expression and activity of an adenovirally delivered gene containing CD fused to a translocation peptide sequence (vp22 from HSV-1).[152] 5-FU accumulation was apparent in subcutaneous 9L glioma tumors 1 day following intratumoral viral administration. When a nonfusion CD gene was employed in the study, no tumoral accumulation of 5-FU was apparent even though therapeutic efficacy (delayed tumor growth due to prodrug conversion) was observed. This suggests that the MRS threshold for 5-FU detection is above the level required for in vivo cytotoxicity using this tumor model.

A second strategy for MR detection of reporter gene expression relies exclusively on transgene-mediated accumulation of an endogenous or administered reporter probe rather than activation of the probe by the transgene. Paramagnetic metal ions or clusters are typically utilized as probes in these strategies. Ferritin has been one of the more frequently employed MR reporter genes in adenoviral applications. This metalloprotein sequesters endogenous iron ions in high densities, thus allowing favorable detection by MRI through T2 and T2* contrast without requiring administration of a reporter probe. Various constructs of ferritin (light subunit, heavy subunit, subunit fusions, mitochondrial) have been utilized for in vivo imaging in mice (e.g., brain striatum,[153] neuroblast migration,[154] olfactory sensory neurons[155]) following adenoviral delivery. Transferrin[156] and tyrosinase[157] have also been used as reporter ion transporters in preliminary applications with adenoviral vectors. While most studies have shown that metalloprotein-mediated iron accumulation does not induce toxicity in vitro or in vivo, the long-term retention of static iron deposits following initial viral administration may be detrimental for repetitive imaging of gene expression.

MRS has also been used to monitor changes in metabolic activity of natural substrates following adenoviral delivery of genes of interest. PCr/ATP ratios and fatty acid metabolism in rat hearts were assessed ex vivo by $^{31}P$ and $^{13}C$ NMR, respectively, after in vivo cardiac transfection with skeletal sarco(endo)plasmic reticulum $Ca^{2+}$-ATPase.[158] PCr/inorganic phosphate ratios in vivo were monitored by $^{31}P$ NMR in a rabbit ischemic hind-limb model after adenoviral delivery of growth factor genes.[159,160] Choi et al. used in vivo 3D $^{13}C$ NMR to monitor hepatic glycogen synthesis pathways in mice following transfection with a rat liver 6-phosphofructo-2-kinase/fructose-2,6-

bisphosphatase fusion construct.[161] These examples highlight the potential of MR technologies for evaluating functional enzymatic pathways within living tissues.

Despite the high spatial resolution and widespread utilization of MR techniques, these modalities suffer from low sensitivity and require high reporter concentrations (~μmol/liter) to allow adequate detection. This is a serious disadvantage for adenoviral vector imaging applications as the high reporter concentrations utilized may impair normal physiological function. These factors may prevent widespread clinical adoption of MR probes for directly imaging adenoviral transgene expression or localization. However, MR will remain an invaluable aspect for evaluating therapeutic efficacy and levels of natural, endogenous probes in adenoviral vector applications.

### 5.4 *Multimodal Imaging of Adenoviral Vectors*

The advances in instrumentation and computing power from 1995 to 2015 have brought multimodal imaging to the forefront of modern research and medical practices. Multimodal imaging instruments allow assessment of the subject with two (or more) different modalities in a single imaging session. Because a single instrument is used for image acquisition, there is essentially no delay or need to reposition the subject between different instruments while collecting the desired information for each modality. Typically, multimodal scanners pair complementary modalities together to maximize the information gained and offset the drawbacks associated with the individual techniques. For instance, CT (high tissue contrast, structural assessment) is frequently employed with nuclear imaging (high sensitivity, functional assessment). Fusion SPECT/CT, PET/CT, and PET/MRI scanners are now commonplace in the clinic and have greatly enhanced the potential for rapid diagnoses and patient throughput.

Many adenoviral vectors are now being designed with multimodal imaging applications in mind. Such vectors encode different reporter genes where each can be monitored individually by reporter probes from different modalities. Several preclinical applications have paired optical reporter genes with SPECT, PET, or MR reporter genes[71,73,79,81,87,88,112,156,162]; additional combinations (PET/MR, SPECT/MR) are also possible for applications with adenoviral vectors. As an alternative strategy, a single reporter gene can be visualized with probes for different modalities, such as assessment of TK expression by MRS and PET (using probes containing $^{31}P/^{19}F$ and $^{18}F$, respectively).[73] While an increasing number of multimodal adenoviral vectors are being explored in preclinical settings, these constructs have not yet been widely evaluated in human patients. Given the rapid growth of the field, however, it is likely that clinical assessments utilizing multimodal reporter genes will be attempted in the near future.

## 6. Conclusions

Imaging has served an important role in advancing human gene therapy since 2003 by allowing noninvasive assessment of novel therapeutic interventions in living subjects. This will continue in the future as imaging provides not only anatomical information but also real-time and repeated evaluations of function. Molecular imaging is expected to further enable adenoviral-based gene therapy in future years using recent advances in genetic reporter systems. These reporter systems have demonstrated great potential in preclinical studies.

Therapeutic strategies based on immune or stem cells represent an attractive growth area for both gene therapy and molecular imaging. Perhaps the adenoviral vectors could combine with cell-based therapies by transfer of the genetic reporter systems. Alternately, the adenoviral gene therapy vectors may target disease locations using a cell-based vehicle. Constitutively expressed genetic reporters offer opportunities for monitoring trafficking, homing–targeting, and persistence of administered cells, whereas inducible reporters could assess cell functionality and activation status. Clearly, the combination of molecular imaging and gene therapy offers great potential for improving treatments for a broad range of diseases.

## References


1. Zinn KR, Douglas JT, Smyth CA, Liu HG, Wu Q, Krasnykh VN, et al. Imaging and tissue biodistribution of 99mTc-labeled adenovirus knob (serotype 5). *Gene Ther* 1998;**5**(6):798–808.
2. Slomka PJ, Pan T, Berman DS, Germano G. Advances in SPECT and PET Hardware. *Prog Cardiovasc Dis* 2015;**57**(6):566–78.
3. Zinn KR, Szalai AJ, Stargel A, Krasnykh V, Chaudhuri TR. Bioluminescence imaging reveals a significant role for complement in liver transduction following intravenous delivery of adenovirus. *Gene Ther* 2004;**11**(19):1482–6.
4. Zinn KR, Chaudhuri TR, Szafran AA, O'Quinn D, Weaver C, Dugger K, et al. Noninvasive bioluminescence imaging in small animals. *ILAR J* 2008;**49**(1):103–15.
5. Shaner NC, Steinbach PA, Tsien RY. A guide to choosing fluorescent proteins. *Nat Methods* 2005;**2**(12):905–9.
6. Filonov GS, Piatkevich KD, Ting LM, Zhang J, Kim K, Verkhusha VV. Bright and stable near-infrared fluorescent protein for in vivo imaging. *Nat Biotechnol* 2011;**29**(8):757–61.
7. Borovjagin AV, McNally LR, Wang M, Curiel DT, MacDougall MJ, Zinn KR. Noninvasive monitoring of mRFP1- and mCherry-labeled oncolytic adenoviruses in an orthotopic breast cancer model by spectral imaging. *Mol Imaging* 2010;**9**(2):59–75.
8. Currie S, Hoggard N, Craven IJ, Hadjivassiliou M, Wilkinson ID. Understanding MRI: basic MR physics for physicians. *Postgrad Med J* 2013;**89**(1050):209–23.
9. Choquet P, Breton E, Goetz C, Marin C, Constantinesco A. Dedicated low-field MRI in mice. *Phys Med Biol* 2009;**54**(17):5287–99.
10. Zhai G, Kim H, Sarver D, Samuel S, Whitworth L, Umphrey H, et al. Early therapy assessment of combined anti-DR5 antibody and carboplatin in triple-negative breast cancer xenografts in mice using diffusion-weighted imaging and (1)H MR spectroscopy. *J Magn Reson Imaging* 2014;**39**(6):1588–94.
11. Soares DP, Law M. Magnetic resonance spectroscopy of the brain: review of metabolites and clinical applications. *Clin Radiol* 2009;**64**(1):12–21.
12. Mankoff DA. A definition of molecular imaging. *J Nucl Med: Official Publication, Society of Nuclear Medicine* 2007;**48**(6):18N, 21N.

13. Therasse P, Arbuck SG, Eisenhauer EA, Wanders J, Kaplan RS, Rubinstein L, et al. New guidelines to evaluate the response to treatment in solid tumors. European Organization for Research and Treatment of Cancer, National Cancer Institute of the United States, National Cancer Institute of Canada. *J Natl Cancer Inst* 2000;**92**(3):205–16.
14. Eisenhauer EA, Therasse P, Bogaerts J, Schwartz LH, Sargent D, Ford R, et al. New response evaluation criteria in solid tumours: revised RECIST guideline (version 1.1). *Eur J Cancer* 2009;**45**(2):228–47.
15. Khuri FR, Nemunaitis J, Ganly I, Arseneau J, Tannock IF, Romel L, et al. A controlled trial of intratumoral ONYX-015, a selectively-replicating adenovirus, in combination with cisplatin and 5-fluorouracil in patients with recurrent head and neck cancer. *Nat Med* 2000;**6**(8):879–85.
16. Reid T, Galanis E, Abbruzzese J, Sze D, Wein LM, Andrews J, et al. Hepatic arterial infusion of a replication-selective oncolytic adenovirus (dl1520): phase II viral, immunologic, and clinical endpoints. *Cancer Res* 2002;**62**(21):6070–9.
17. Reid TR, Freeman S, Post L, McCormick F, Sze DY. Effects of Onyx-015 among metastatic colorectal cancer patients that have failed prior treatment with 5-FU/leucovorin. *Cancer Gene Ther* 2005;**12**(8):673–81.
18. Sze DY, Freeman SM, Slonim SM, Samuels SL, Andrews JC, Hicks M, et al. Dr. Gary J. Becker Young Investigator Award: intraarterial adenovirus for metastatic gastrointestinal cancer: activity, radiographic response, and survival. *J Vasc Interv Radiol* 2003;**14**(3):279–90.
19. Sze DY, Iagaru AH, Gambhir SS, De Haan HA, Reid TR. Response to intra-arterial oncolytic virotherapy with the herpes virus NV1020 evaluated by [18F]fluorodeoxyglucose positron emission tomography and computed tomography. *Hum Gene Ther* 2012;**23**(1):91–7.
20. Nokisalmi P, Pesonen S, Escutenaire S, Sarkioja M, Raki M, Cerullo V, et al. Oncolytic adenovirus ICOVIR-7 in patients with advanced and refractory solid tumors. *Clin Cancer Res* 2010;**16**(11):3035–43.
21. Pesonen S, Diaconu I, Kangasniemi L, Ranki T, Kanerva A, Pesonen SK, et al. Oncolytic immunotherapy of advanced solid tumors with a CD40L-expressing replicating adenovirus: assessment of safety and immunologic responses in patients. *Cancer Res* 2012;**72**(7):1621–31.
22. Fujiwara T, Tanaka N, Kanazawa S, Ohtani S, Saijo Y, Nukiwa T, et al. Multicenter phase I study of repeated intratumoral delivery of adenoviral p53 in patients with advanced non-small-cell lung cancer. *J Clin Oncol* 2006;**24**(11):1689–99.
23. Hoos A, Eggermont AM, Janetzki S, Hodi FS, Ibrahim R, Anderson A, et al. Improved endpoints for cancer immunotherapy trials. *J Natl Cancer Inst* 2010;**102**(18):1388–97.
24. Wolchok JD, Hoos A, O'Day S, Weber JS, Hamid O, Lebbe C, et al. Guidelines for the evaluation of immune therapy activity in solid tumors: immune-related response criteria. *Clin Cancer Res* 2009;**15**(23):7412–20.
25. Tuma RS. New response criteria proposed for immunotherapies. *J Natl Cancer Inst* 2008;**100**(18):1280–1.
26. Sahani DV, Kalva SP, Hamberg LM, Hahn PF, Willett CG, Saini S, et al. Assessing tumor perfusion and treatment response in rectal cancer with multisection CT: initial observations. *Radiology* 2005;**234**(3):785–92.
27. Ehling J, Lammers T, Kiessling F. Non-invasive imaging for studying anti-angiogenic therapy effects. *Thromb Haemost* 2013;**109**(3):375–90.
28. Wahl RL, Jacene H, Kasamon Y, Lodge MA. From RECIST to PERCIST: evolving considerations for PET response criteria in solid tumors. *J Nucl Med* 2009;**50**(Suppl 1):122S–50S.

29. Yanagawa M, Tatsumi M, Miyata H, Morii E, Tomiyama N, Watabe T, et al. Evaluation of response to neoadjuvant chemotherapy for esophageal cancer: PET response criteria in solid tumors versus response evaluation criteria in solid tumors. *J Nucl Med* 2012;**53**(6):872–80.
30. Koski A, Ahtinen H, Liljenback H, Roivainen A, Koskela A, Oksanen M, et al. [(18)F]-fluorodeoxyglucose positron emission tomography and computed tomography in response evaluation of oncolytic adenovirus treatments of patients with advanced cancer. *Hum Gene Ther* 2013;**24**(12):1029–41.
31. Diaconu I, Cerullo V, Hirvinen ML, Escutenaire S, Ugolini M, Pesonen SK, et al. Immune response is an important aspect of the antitumor effect produced by a CD40L-encoding oncolytic adenovirus. *Cancer Res* 2012;**72**(9):2327–38.
32. Vardouli L, Lindqvist C, Vlahou K, Loskog AS, Eliopoulos AG. Adenovirus delivery of human CD40 ligand gene confers direct therapeutic effects on carcinomas. *Cancer Gene Ther* 2009;**16**(11):848–60.
33. Gomes EM, Rodrigues MS, Phadke AP, Butcher LD, Starling C, Chen S, et al. Antitumor activity of an oncolytic adenoviral-CD40 ligand (CD154) transgene construct in human breast cancer cells. *Clin Cancer Res* 2009;**15**(4):1317–25.
34. Grines CL, Watkins MW, Mahmarian JJ, Iskandrian AE, Rade JJ, Marrott P, et al. A randomized, double-blind, placebo-controlled trial of Ad5FGF-4 gene therapy and its effect on myocardial perfusion in patients with stable angina. *J Am Coll Cardiol* 2003;**42**(8):1339–47.
35. Hedman M, Hartikainen J, Syvanne M, Stjernvall J, Hedman A, Kivela A, et al. Safety and feasibility of catheter-based local intracoronary vascular endothelial growth factor gene transfer in the prevention of postangioplasty and in-stent restenosis and in the treatment of chronic myocardial ischemia: phase II results of the Kuopio Angiogenesis Trial (KAT). *Circulation* 2003;**107**(21):2677–83.
36. Gillies RJ, Bhujwalla ZM, Evelhoch J, Garwood M, Neeman M, Robinson SP, et al. Applications of magnetic resonance in model systems: tumor biology and physiology. *Neoplasia* 2000;**2**(1–2):139–51.
37. Evelhoch JL, Gillies RJ, Karczmar GS, Koutcher JA, Maxwell RJ, Nalcioglu O, et al. Applications of magnetic resonance in model systems: cancer therapeutics. *Neoplasia* 2000;**2**(1–2):152–65.
38. Hemminki O, Immonen R, Narvainen J, Kipar A, Paasonen J, Jokivarsi KT, et al. In vivo magnetic resonance imaging and spectroscopy identifies oncolytic adenovirus responders. *Int J Cancer* 2014;**134**(12):2878–90.
39. Begley JK, Redpath TW, Bolan PJ, Gilbert FJ. In vivo proton magnetic resonance spectroscopy of breast cancer: a review of the literature. *Breast Cancer Res* 2012;**14**(2):207.
40. Srivastava S, Roy R, Singh S, Kumar P, Dalela D, Sankhwar SN, et al. Taurine – a possible fingerprint biomarker in non-muscle invasive bladder cancer: a pilot study by 1H NMR spectroscopy. *Cancer Biomark* 2010;**6**(1):11–20.
41. Beckonert O, Monnerjahn J, Bonk U, Leibfritz D. Visualizing metabolic changes in breast-cancer tissue using 1H-NMR spectroscopy and self-organizing maps. *NMR Biomed* 2003;**16**(1):1–11.
42. De Micheli E, Alfieri A, Pinna G, Bianchi L, Colivicchi MA, Melani A, et al. Extracellular levels of taurine in tumoral, peritumoral and normal brain tissue in patients with malignant glioma: an intraoperative microdialysis study. *Adv Exp Med Biol* 2000;**483**:621–5.
43. Swanson MG, Vigneron DB, Tabatabai ZL, Males RG, Schmitt L, Carroll PR, et al. Proton HR-MAS spectroscopy and quantitative pathologic analysis of MRI/3D-MRSI-targeted postsurgical prostate tissues. *Magn Reson Med* 2003;**50**(5):944–54.

44. Liimatainen TJ, Erkkila AT, Valonen P, Vidgren H, Lakso M, Wong G, et al. 1H MR spectroscopic imaging of phospholipase-mediated membrane lipid release in apoptotic rat glioma in vivo. *Magn Reson Med* 2008;**59**(6):1232–8.
45. Delikatny EJ, Cooper WA, Brammah S, Sathasivam N, Rideout DC. Nuclear magnetic resonance-visible lipids induced by cationic lipophilic chemotherapeutic agents are accompanied by increased lipid droplet formation and damaged mitochondria. *Cancer Res* 2002;**62**(5):1394–400.
46. Bains LJ, Zweifel M, Thoeny HC. Therapy response with diffusion MRI: an update. *Cancer Imaging* 2012;**12**:395–402.
47. Kremser C, Judmaier W, Hein P, Griebel J, Lukas P, de Vries A. Preliminary results on the influence of chemoradiation on apparent diffusion coefficients of primary rectal carcinoma measured by magnetic resonance imaging. *Strahlenther Onkol* 2003;**179**(9):641–9.
48. Dzik-Jurasz A, Domenig C, George M, Wolber J, Padhani A, Brown G, et al. Diffusion MRI for prediction of response of rectal cancer to chemoradiation. *Lancet* 2002;**360**(9329):307–8.
49. Messiou C, Collins DJ, Giles S, de Bono JS, Bianchini D, de Souza NM. Assessing response in bone metastases in prostate cancer with diffusion weighted MRI. *Eur Radiol* 2011;**21**(10):2169–77.
50. Jang KM, Kim SH, Choi D, Lee SJ, Park MJ, Min K. Pathological correlation with diffusion restriction on diffusion-weighted imaging in patients with pathological complete response after neoadjuvant chemoradiation therapy for locally advanced rectal cancer: preliminary results. *Br J Radiol* 2012;**85**(1017):e566–72.
51. Tuppurainen L, Sallinen H, Hakkarainen H, Liimatainen TJ, Hassan M, Yla-Herttuala E, et al. Functional MRI measurements to predict early adenoviral gene therapy response in ovarian cancer mouse model. *J Genet Syndr Gene Ther* 2013;**4**(8):171–90.
52. Kiessling F, Jugold M, Woenne EC, Brix G. Non-invasive assessment of vessel morphology and function in tumors by magnetic resonance imaging. *Eur Radiol* 2007;**17**(8):2136–48.
53. Haney CR, Parasca AD, Fan X, Bell RM, Zamora MA, Karczmar GS, et al. Characterization of response to radiation mediated gene therapy by means of multimodality imaging. *Magn Reson Med* 2009;**62**(2):348–56.
54. Menendez MI, Clark DJ, Carlton M, Flanigan DC, Jia G, Sammet S, et al. Direct delayed human adenoviral BMP-2 or BMP-6 gene therapy for bone and cartilage regeneration in a pony osteochondral model. *Osteoarthr Cartil* 2011;**19**(8):1066–75.
55. Kurozumi K, Nakamura K, Tamiya T, Kawano Y, Kobune M, Hirai S, et al. BDNF gene-modified mesenchymal stem cells promote functional recovery and reduce infarct size in the rat middle cerebral artery occlusion model. *Mol Ther* 2004;**9**(2):189–97.
56. Horita Y, Honmou O, Harada K, Houkin K, Hamada H, Kocsis JD. Intravenous administration of glial cell line-derived neurotrophic factor gene-modified human mesenchymal stem cells protects against injury in a cerebral ischemia model in the adult rat. *J Neurosci Res* 2006;**84**(7):1495–504.
57. Liu H, Honmou O, Harada K, Nakamura K, Houkin K, Hamada H, et al. Neuroprotection by PlGF gene-modified human mesenchymal stem cells after cerebral ischaemia. *Brain* 2006;**129**(Pt 10):2734–45.
58. Kameda M, Shingo T, Takahashi K, Muraoka K, Kurozumi K, Yasuhara T, et al. Adult neural stem and progenitor cells modified to secrete GDNF can protect, migrate and integrate after intracerebral transplantation in rats with transient forebrain ischemia. *Eur J Neurosci* 2007;**26**(6):1462–78.
59. Chen C, Wang Y, Yang GY. Stem cell-mediated gene delivering for the treatment of cerebral ischemia: progress and prospectives. *Curr Drug Targets* 2013;**14**(1):81–9.

60. Horvath KA, Doukas J, Lu C-YJ, Belkind N, Greene R, Pierce GF, et al. Myocardial functional recovery after fibroblast growth factor 2 gene therapy as assessed by echocardiography and magnetic resonance imaging. *Ann Thorac Surg* 2002;**74**(2):481–6. discussion 87.
61. Horvath KA, Lu CY, Robert E, Pierce GF, Greene R, Sosnowski BA, et al. Improvement of myocardial contractility in a porcine model of chronic ischemia using a combined transmyocardial revascularization and gene therapy approach. *J Thorac Cardiovasc Surg* 2005;**129**(5):1071–7.
62. Guerrero M, Athota K, Moy J, Mehta LS, Laguens R, Crottogini A, et al. Vascular endothelial growth factor-165 gene therapy promotes cardiomyogenesis in reperfused myocardial infarction. *J Interv Cardiol* 2008;**21**(3):242–51.
63. Jin YN, Inubushi M, Masamoto K, Odaka K, Aoki I, Tsuji AB, et al. Long-term effects of hepatocyte growth factor gene therapy in rat myocardial infarct model. *Gene Ther* 2012;**19**(8):836–43.
64. Rissanen TT, Markkanen JE, Arve K, Rutanen J, Kettunen MI, Vajanto I, et al. Fibroblast growth factor 4 induces vascular permeability, angiogenesis and arteriogenesis in a rabbit hindlimb ischemia model. *FASEB J* 2003;**17**(1):100–2.
65. Tang C, Russell PJ, Martiniello-Wilks R, Rasko JE, Khatri A. Concise review: nanoparticles and cellular carriers-allies in cancer imaging and cellular gene therapy? *Stem Cells* 2010;**28**(9):1686–702.
66. Vande Velde G, Himmelreich U, Neeman M. Reporter gene approaches for mapping cell fate decisions by MRI: promises and pitfalls. *Contrast Media Mol Imaging* 2013;**8**(6):424–31.
67. Leong-Poi H. Molecular imaging using contrast-enhanced ultrasound: evaluation of angiogenesis and cell therapy. *Cardiovasc Res* 2009;**84**(2):190–200.
68. Sirsi SR, Flexman ML, Vlachos F, Huang J, Hernandez SL, Kim HK, et al. Contrast ultrasound imaging for identification of early responder tumor models to anti-angiogenic therapy. *Ultrasound Med Biol* 2012;**38**(6):1019–29.
69. Rix A, Lederle W, Siepmann M, Fokong S, Behrendt FF, Bzyl J, et al. Evaluation of high frequency ultrasound methods and contrast agents for characterising tumor response to anti-angiogenic treatment. *Eur J Radiol* 2012;**81**(10):2710–6.
70. Serganova I, Ponomarev V, Blasberg R. Human reporter genes: potential use in clinical studies. *Nucl Med Biol* 2007;**34**(7):791–807.
71. Waerzeggers Y, Monfared P, Viel T, Winkeler A, Voges J, Jacobs AH. Methods to monitor gene therapy with molecular imaging. *Methods* 2009;**48**(2):146–60.
72. Brader P, Wong RJ, Horowitz G, Gil Z. Combination of pet imaging with viral vectors for identification of cancer metastases. *Adv Drug Deliv Rev* 2012;**64**(8):749–55.
73. Collins SA, Hiraoka K, Inagaki A, Kasahara N, Tangney M. PET imaging for gene & cell therapy. *Curr Gene Ther* 2012;**12**(1):20–32.
74. Gil JS, Machado HB, Campbell DO, McCracken M, Radu C, Witte ON, et al. Application of a rapid, simple, and accurate adenovirus-based method to compare PET reporter gene/PET reporter probe systems. *Mol Imaging Biol* 2013;**15**(3):273–81.
75. Huh YM, Lee ES, Lee JH, Jun YW, Kim PH, Yun CO, et al. Hybrid nanoparticles for magnetic resonance imaging of target-specific viral gene delivery. *Adv Mater* 2007;**19**(20):3109.
76. Yun J, Sonabend AM, Ulasov IV, Kim DH, Rozhkova EA, Novosad V, et al. A novel adenoviral vector labeled with superparamagnetic iron oxide nanoparticles for real-time tracking of viral delivery. *J Clin Neurosci* 2012;**19**(6):875–80.
77. Rad AM, Iskander AS, Janic B, Knight RA, Arbab AS, Soltanian-Zadeh H. AC133+ progenitor cells as gene delivery vehicle and cellular probe in subcutaneous tumor models: a preliminary study. *BMC Biotechnol* 2009;**9**:28.

78. Varma NR, Barton KN, Janic B, Shankar A, Iskander A, Ali MM, et al. Monitoring adenoviral based gene delivery in rat glioma by molecular imaging. *World J Clin Oncol* 2013;**4**(4):91–101.
79. Dmitriev IP, Kashentseva EA, Kim KH, Matthews QL, Krieger SS, Parry JJ, et al. Monitoring of biodistribution and persistence of conditionally replicative adenovirus in a murine model of ovarian cancer using capsid-incorporated mCherry and expression of human somatostatin receptor subtype 2 gene. *Mol Imaging* 2014;**13**(8).
80. Mathis JM, Bhatia S, Khandelwal A, Kovesdi I, Lokitz SJ, Odaka Y, et al. Genetic incorporation of human metallothionein into the adenovirus protein IX for non-invasive SPECT imaging. *PLoS One* 2011;**6**(2):e16792.
81. Liu L, Rogers BE, Aladyshkina N, Cheng B, Lokitz SJ, Curiel DT, et al. Construction and radiolabeling of adenovirus variants that incorporate human metallothionein into protein IX for analysis of biodistribution. *Mol Imaging* 2014;**13**(7).
82. Blasberg R. PET imaging of gene expression. *Eur J Cancer* 2002;**38**(16):2137–46.
83. Xu C, Zhang H. Somatostatin receptor based imaging and radionuclide therapy. *Biomed Res Int* 2015;**2015**:917968.
84. Buchsbaum DJ, Chaudhuri TR, Yamamoto M, Zinn KR. Gene expression imaging with radiolabeled peptides. *Ann Nucl Med* 2004;**18**(4):275–83.
85. Buchsbaum DJ. Imaging and therapy of tumors induced to express somatostatin receptor by gene transfer using radiolabeled peptides and single chain antibody constructs. *Semin Nucl Med* 2004;**34**(1):32–46.
86. Min JJ, Gambhir SS. Gene therapy progress and prospects: noninvasive imaging of gene therapy in living subjects. *Gene Ther* 2004;**11**(2):115–25.
87. Serganova I, Blasberg R. Reporter gene imaging: potential impact on therapy. *Nucl Med Biol* 2005;**32**(7):763–80.
88. Jiang ZK, Sato M, Wu L. Chapter five – the development of transcription-regulated adenoviral vectors with high cancer-selective imaging capabilities. In: David TC, Paul BF, editors. *Advances in cancer research*, vol. 115. Academic Press; 2012. p. 115–46.
89. Yaghoubi SS, Campbell DO, Radu CG, Czernin J. Positron emission tomography reporter genes and reporter probes: gene and cell therapy applications. *Theranostics* 2012;**2**(4):374–91.
90. Brader P, Serganova I, Blasberg RG. Noninvasive molecular imaging using reporter genes. *J Nucl Med* 2013;**54**(2):167–72.
91. Gehrig S, Sami H, Ogris M. Gene therapy and imaging in preclinical and clinical oncology: recent developments in therapy and theranostics. *Ther Deliv* 2014;**5**(12):1275–96.
92. Johnbeck CB, Knigge U, Kjaer A. PET tracers for somatostatin receptor imaging of neuroendocrine tumors: current status and review of the literature. *Future Oncol* 2014;**10**(14):2259–77.
93. van Essen M, Sundin A, Krenning EP, Kwekkeboom DJ. Neuroendocrine tumours: the role of imaging for diagnosis and therapy. *Nat Rev Endocrinol* 2014;**10**(2):102–14.
94. Iyer M, Sato M, Johnson M, Gambhir SS, Wu L. Applications of molecular imaging in cancer gene therapy. *Curr Gene Ther* 2005;**5**(6):607–18.
95. Jacobs A, Braunlich I, Graf R, Lercher M, Sakaki T, Voges J, et al. Quantitative kinetics of [124I]FIAU in cat and man. *J Nucl Med* 2001;**42**(3):467–75.
96. Yaghoubi S, Barrio JR, Dahlbom M, Iyer M, Namavari M, Satyamurthy N, et al. Human pharmacokinetic and dosimetry studies of [$^{18}$F]FHBG: a reporter probe for imaging herpes simplex virus Type-1 thymidine kinase reporter gene expression. *J Nucl Med* 2001;**42**(8):1225–34.
97. Campbell DO, Yaghoubi SS, Su Y, Lee JT, Auerbach MS, Herschman H, et al. Structure-guided engineering of human thymidine kinase 2 as a positron emission tomography reporter gene for enhanced phosphorylation of non-natural thymidine analog reporter probe. *J Biol Chem* 2012;**287**(1):446–54.

98. Jacobs A, Voges J, Reszka R, Lercher M, Gossmann A, Kracht L, et al. Positron-emission tomography of vector-mediated gene expression in gene therapy for gliomas. *Lancet* 2001;**358**(9283):727–9.
99. Voges J, Reszka R, Gossmann A, Dittmar C, Richter R, Garlip G, et al. Imaging-guided convection-enhanced delivery and gene therapy of glioblastoma. *Ann Neurol* 2003;**54**(4):479–87.
100. Penuelas I, Mazzolini G, Boan JF, Sangro B, Marti-Climent J, Ruiz M, et al. Positron emission tomography imaging of adenoviral-mediated transgene expression in liver cancer patients. *Gastroenterology* 2005;**128**(7):1787–95.
101. Sangro B, Mazzolini G, Ruiz M, Ruiz J, Quiroga J, Herrero I, et al. A phase I clinical trial of thymidine kinase-based gene therapy in advanced hepatocellular carcinoma. *Cancer Gene Ther* 2010;**17**(12):837–43.
102. Dempsey MF, Wyper D, Owens J, Pimlott S, Papanastassiou V, Patterson J, et al. Assessment of $^{123}$I-FIAU imaging of herpes simplex viral gene expression in the treatment of glioma. *Nucl Med Commun* 2006;**27**(8):611–7.
103. Barton KN, Stricker H, Brown SL, Elshaikh M, Aref I, Lu M, et al. Phase I study of noninvasive imaging of adenovirus-mediated gene expression in the human prostate. *Mol Ther* 2008;**16**(10):1761–9.
104. Barton KN, Stricker H, Elshaikh MA, Pegg J, Cheng J, Zhang Y, et al. Feasibility of adenovirus-mediated hNIS gene transfer and $^{131}$I radioiodine therapy as a definitive treatment for localized prostate Cancer. *Mol Ther* 2011;**19**(7):1353–9.
105. Rajecki M, Kangasmaki A, Laasonen L, Escutenaire S, Hakkarainen T, Haukka J, et al. Sodium iodide symporter SPECT imaging of a patient treated with oncolytic adenovirus Ad5/3-Delta24-hNIS. *Mol Ther* 2011;**19**(4):629–31.
106. Reubi JC, Waser B, Schaer JC, Laissue JA. Somatostatin receptor sst1-sst5 expression in normal and neoplastic human tissues using receptor autoradiography with subtype-selective ligands. *Eur J Nucl Med* 2001;**28**(7):836–46.
107. Zinn KR, Chaudhuri TR, Krasnykh VN, Buchsbaum DJ, Belousova N, Grizzle WE, et al. Gamma camera dual imaging with a somatostatin receptor and thymidine kinase after gene transfer with a bicistronic adenovirus in mice. *Radiology* 2002;**223**(2):417–25.
108. Hemminki A, Zinn KR, Liu B, Chaudhuri TR, Desmond RA, Rogers BE, et al. In vivo molecular chemotherapy and noninvasive imaging with an infectivity-enhanced adenovirus. *J Natl Cancer Inst* 2002;**94**(10):741–9.
109. Rogers BE, Chaudhuri TR, Reynolds PN, Della Manna D, Zinn KR. Non-invasive gamma camera imaging of gene transfer using an adenoviral vector encoding an epitope-tagged receptor as a reporter. *Gene Ther* 2003;**10**(2):105–14.
110. Verwijnen SM, Sillevis Smith PA, Hoeben RC, Rabelink MJ, Wiebe L, Curiel DT, et al. Molecular imaging and treatment of malignant gliomas following adenoviral transfer of the herpes simplex virus-thymidine kinase gene and the somatostatin receptor subtype 2 gene. *Cancer Biother Radiopharm* 2004;**19**(1):111–20.
111. ter Horst M, Verwijnen SM, Brouwer E, Hoeben RC, de Jong M, de Leeuw BH, et al. Locoregional delivery of adenoviral vectors. *J Nucl Med* 2006;**47**(9):1483–9.
112. Chen R, Parry JJ, Akers WJ, Berezin MY, El Naqa IM, Achilefu S, et al. Multimodality imaging of gene transfer with a receptor-based reporter gene. *J Nucl Med* 2010;**51**(9):1456–63.
113. Singh SP, Han L, Murali R, Solis L, Roth J, Ji L, et al. SSTR2-based reporters for assessing gene transfer into non-small cell lung cancer: evaluation using an intrathoracic mouse model. *Hum Gene Ther* 2011;**22**(1):55–64.
114. Cotugno G, Aurilio M, Annunziata P, Capalbo A, Faella A, Rinaldi V, et al. Noninvasive repetitive imaging of somatostatin receptor 2 gene transfer with positron emission tomography. *Hum Gene Ther* 2011;**22**(2):189–96.

115. Ravoori MK, Han L, Singh SP, Dixon K, Duggal J, Liu P, et al. Noninvasive assessment of gene transfer and expression by in vivo functional and morphologic imaging in a rabbit tumor model. *PLoS One* 2013;**8**(6):e62371.
116. Kim KH, Dmitriev I, O'Malley JP, Wang M, Saddekni S, You Z, et al. A phase I clinical trial of Ad5.SSTR/TK.RGD, a novel infectivity-enhanced bicistronic adenovirus, in patients with recurrent gynecologic cancer. *Clin Cancer Res* 2012;**18**(12):3440–51.
117. Ogawa O, Umegaki H, Ishiwata K, Asai Y, Ikari H, Oda K, et al. In vivo imaging of adenovirus-mediated over-expression of dopamine D2 receptors in rat striatum by positron emission tomography. *Neuroreport* 2000;**11**(4):743–8.
118. Iyer M, Barrio JR, Namavari M, Bauer E, Satyamurthy N, Nguyen K, et al. 8-[$^{18}$F]fluoropenciclovir: an improved reporter probe for imaging HSV1-tk reporter gene expression in vivo using PET. *J Nucl Med* 2001;**42**(1):96–105.
119. Yaghoubi SS, Wu L, Liang Q, Toyokuni T, Barrio JR, Namavari M, et al. Direct correlation between positron emission tomographic images of two reporter genes delivered by two distinct adenoviral vectors. *Gene Ther* 2001;**8**(14):1072–80.
120. Liang Q, Satyamurthy N, Barrio JR, Toyokuni T, Phelps MP, Gambhir SS, et al. Noninvasive, quantitative imaging in living animals of a mutant dopamine D2 receptor reporter gene in which ligand binding is uncoupled from signal transduction. *Gene Ther* 2001;**8**(19):1490–8.
121. Liang Q, Gotts J, Satyamurthy N, Barrio J, Phelps ME, Gambhir SS, et al. Noninvasive, repetitive, quantitative measurement of gene expression from a bicistronic message by positron emission tomography, following gene transfer with adenovirus. *Mol Ther* 2002;**6**(1):73–82.
122. Chen IY, Wu JC, Min JJ, Sundaresan G, Lewis X, Liang QW, et al. Micro-positron emission tomography imaging of cardiac gene expression in rats using bicistronic adenoviral vector-mediated gene delivery. *Circulation* 2004;**109**(11):1415–20.
123. Buursma AR, Beerens AMJ, de Vries EFJ, van Waarde A, Rots MG, Hospers GAP, et al. The human norepinephrine transporter in combination with $^{11}$C-m-hydroxyephedrine as a reporter gene/reporter probe for PET of gene therapy. *J Nucl Med* 2005;**46**(12):2068–75.
124. Jia ZY, Deng HF, Huang R, Yang YY, Yang XC, Qi ZZ, et al. In vitro and in vivo studies of adenovirus-mediated human norepinephrine transporter gene transduction to hepatocellular carcinoma. *Cancer Gene Ther* 2011;**18**(3):196–205.
125. Lohith TG, Furukawa T, Mori T, Kobayashi M, Fujibayashi Y. Basic evaluation of FES-hERL PET tracer-reporter gene system for in vivo monitoring of adenoviral-mediated gene therapy. *Mol Imaging Biol* 2008;**10**(5):245–52.
126. Qin C, Lan X, He J, Xia X, Tian Y, Pei Z, et al. An in vitro and in vivo evaluation of a reporter gene/probe system hERL/(18)F-FES. *PLoS One* 2013;**8**(4):e61911.
127. Ronald J, Biswal S, Gambhir SS. In: Templeton NS, editor. *Monitoring gene and cell therapies in living subjects with molecular imaging technologies*. 4th ed. Boca Raton, FL: Taylor & Francis Group; 2015.
128. Yabusaki M, Sato J, Kohyama A, Kojima T, Nobuoka D, Yoshikawa T, et al. Detection and preliminary evaluation of circulating tumor cells in the peripheral blood of patients with eight types of cancer using a telomerase-specific adenovirus. *Oncol Rep* 2014;**32**(5):1772–8.
129. Ono HA, Le LP, Davydova JG, Gavrikova T, Yamamoto M. Noninvasive visualization of adenovirus replication with a fluorescent reporter in the E3 region. *Cancer Res* 2005;**65**(22):10154–8.
130. Quach CH, Jung KH, Paik JY, Park JW, Lee EJ, Lee KH. Quantification of early adipose-derived stem cell survival: comparison between sodium iodide symporter and enhanced green fluorescence protein imaging. *Nucl Med Biol* 2012;**39**(8):1251–60.

131. Richter JR, Mahoney M, Warram JM, Samuel S, Zinn KR. A dual-reporter, diagnostic vector for prostate cancer detection and tumor imaging. *Gene Ther* 2014;**21**(10):897–902.
132. Warram JM, Borovjagin AV, Zinn KR. A genetic strategy for combined screening and localized imaging of breast cancer. *Mol Imaging Biol* 2011;**13**(3):452–61.
133. Watanabe Y, Hashimoto Y, Kagawa S, Kawamura H, Nagai K, Tanaka N, et al. Enhanced antitumor efficacy of telomerase-selective oncolytic adenoviral agent OBP-301 with valproic acid (VPA) in human lung cancer cells. *J Gene Med* 2008;**10**(4):471–2.
134. Kawashima T, Kagawa S, Kobayashi N, Shirakiya Y, Umeoka T, Teraishi F, et al. Telomerase-specific replication-selective virotherapy for human cancer. *Clin Cancer Res* 2004;**10**(1 Pt 1):285–92.
135. Nakajima O, Matsunaga A, Ichimaru D, Urata Y, Fujiwara T, Kawakami K. Telomerase-specific virotherapy in an animal model of human head and neck cancer. *Mol Cancer Ther* 2009;**8**(1):171–7.
136. Huang P, Watanabe M, Kaku H, Kashiwakura Y, Chen J, Saika T, et al. Direct and distant antitumor effects of a telomerase-selective oncolytic adenoviral agent, OBP-301, in a mouse prostate cancer model. *Cancer Gene Ther* 2008;**15**(5):315–22.
137. Fujiwara T, Tanaka N, Nemunaitis JJ, Senzer NN, Tong A, Ichimaru D, et al. Phase I trial of intratumoral administration of OBP-301, a novel telomerase-specific oncolytic virus, in patients with advanced solid cancer: evaluation of biodistribution and immune response. *J Clin Oncol* 2008;**26**(15).
138. Nemunaitis J, Tong AW, Nemunaitis M, Senzer N, Phadke AP, Bedell C, et al. A phase I study of telomerase-specific replication competent oncolytic adenovirus (telomelysin) for various solid tumors. *Mol Ther* 2010;**18**(2):429–34.
139. Umeoka T, Kawashima T, Kagawa S, Teraishi F, Taki M, Nishizaki M, et al. Visualization of intrathoracically disseminated solid tumors in mice with optical imaging by telomerase-specific amplification of a transferred green fluorescent protein gene. *Cancer Res* 2004;**64**(17):6259–65.
140. Kishimoto H, Kojima T, Watanabe Y, Kagawa S, Fujiwara T, Uno F, et al. In vivo imaging of lymph node metastasis with telomerase-specific replication-selective adenovirus. *Nat Med* 2006;**12**(10):1213–9.
141. Igawa S, Gohda K, Fukui T, Ryuge S, Otani S, Masago A, et al. Circulating tumor cells as a prognostic factor in patients with small cell lung cancer. *Oncol Lett* 2014;**7**(5):1469–73.
142. Ito H, Inoue H, Sando N, Kimura S, Gohda K, Sato J, et al. Prognostic impact of detecting viable circulating tumour cells in gastric cancer patients using a telomerase-specific viral agent: a prospective study. *BMC Cancer* 2012;**12**:346.
143. Takakura M, Kyo S, Nakamura M, Maida Y, Mizumoto Y, Bono Y, et al. Circulating tumour cells detected by a novel adenovirus-mediated system may be a potent therapeutic marker in gynaecological cancers. *Br J Cancer* 2012;**107**(3):448–54.
144. Bhaumik S, Gambhir SS. Optical imaging of Renilla luciferase reporter gene expression in living mice. *Proc Natl Acad Sci USA* 2002;**99**(1):377–82.
145. Ray P. In: Rosenthal EL, Zinn KR, editors. *Illustrating molecular events with light: a perspective on optical reporter genes*. New York, NY: Springer-Verlag; 2009.
146. Ray S, Paulmurugan R, Patel MR, Ahn BC, Wu L, Carey M, et al. Noninvasive imaging of therapeutic gene expression using a bidirectional transcriptional amplification strategy. *Mol Ther* 2008;**16**(11):1848–56.
147. Hochgrafe K, Mandelkow EM. Making the brain glow: in vivo bioluminescence imaging to study neurodegeneration. *Mol Neurobiol* 2013;**47**(3):868–82.

148. Patterson AP, Booth SA, Saba R. The emerging use of in vivo optical imaging in the study of neurodegenerative diseases. *Biomed Res Int* 2014;**2014**:401306.
149. Rehemtulla A, Hall DE, Stegman LD, Prasad U, Chen G, Bhojani MS, et al. Molecular imaging of gene expression and efficacy following adenoviral-mediated brain tumor gene therapy. *Mol Imaging* 2002;**1**(1):43–55.
150. Li Z, Qiao H, Lebherz C, Choi SR, Zhou X, Gao G, et al. Creatine kinase, a magnetic resonance-detectable marker gene for quantification of liver-directed gene transfer. *Hum Gene Ther* 2005;**16**(12):1429–38.
151. Landis CS, Yamanouchi K, Zhou H, Mohan S, Roy-Chowdhury N, Shafritz DA, et al. Noninvasive evaluation of liver repopulation by transplanted hepatocytes using $^{31}$P MRS imaging in mice. *Hepatology* 2006;**44**(5):1250–8.
152. Lee KC, Hamstra DA, Bullarayasamudram S, Bhojani MS, Moffat BA, Dornfeld KJ, et al. Fusion of the HSV-1 tegument protein vp22 to cytosine deaminase confers enhanced bystander effect and increased therapeutic benefit. *Gene Ther* 2006;**13**(2):127–37.
153. Genove G, DeMarco U, Xu H, Goins WF, Ahrens ET. A new transgene reporter for in vivo magnetic resonance imaging. *Nat Med* 2005;**11**(4):450–4.
154. Iordanova B, Ahrens ET. In vivo magnetic resonance imaging of ferritin-based reporter visualizes native neuroblast migration. *Neuroimage* 2012;**59**(2):1004–12.
155. Iordanova B, Hitchens TK, Robison CS, Ahrens ET. Engineered mitochondrial ferritin as a magnetic resonance imaging reporter in mouse olfactory epithelium. *PLoS ONE* 2013;**8**(8):e72720.
156. Wang YF, Liu T, Guo YL, Gao FB. Construction and identification of the adenoviral vector with dual reporter gene for multimodality molecular imaging. *J Huazhong Univ Sci Technol Med Sci* 2013;**33**(4):600–5.
157. Yuan JP, Liang BL, Deng HR, Liu ZS, Bai SM, Zhong JL. Adenovirus mediated gene transfer of tyrosinase gene on HepG2 cell by magnetic resonance imaging. *Zhongguo Yi Xue Ke Xue Yuan Xue Bao* 2009;**31**(2):146–50.
158. O'Donnell JM, Fields A, Xu X, Chowdhury SA, Geenen DL, Bi J. Limited functional and metabolic improvements in hypertrophic and healthy rat heart overexpressing the skeletal muscle isoform of SERCA1 by adenoviral gene transfer in vivo. *Am J Physiol Heart Circ Physiol* 2008;**295**(6):H2483–94.
159. Korpisalo P, Rissanen TT, Bengtsson T, Liimatainen T, Laidinen S, Karvinen H, et al. Therapeutic angiogenesis with placental growth factor improves exercise tolerance of ischaemic rabbit hindlimbs. *Cardiovasc Res* 2008;**80**(2):263–70.
160. Niemi H, Honkonen K, Korpisalo P, Huusko J, Kansanen E, Merentie M, et al. HIF-1alpha and HIF-2alpha induce angiogenesis and improve muscle energy recovery. *Eur J Clin Invest* 2014;**44**(10):989–99.
161. Choi IY, Wu C, Okar DA, Lange AJ, Gruetter R. Elucidation of the role of fructose 2,6-bisphosphate in the regulation of glucose fluxes in mice using in vivo $^{13}$C NMR measurements of hepatic carbohydrate metabolism. *Eur J Biochem* 2002;**269**(18):4418–26.
162. Liu S-W, Cho K-H, Chen M-R, Yu H-C, Kao Y-Y, Tsou T-C, et al. Ferritin-red fluorescent protein fusion reporter for magnetic resonance and optical imaging. *Biomed Eng-App Bas C* 2012;**24**(04):333–41.

# Regulation of Adenoviral Vector-Based Therapies: An FDA Perspective

32

*Michael Havert, Rachael Anatol, Ilan Irony, Ying Huang*
Office of Cellular, Tissue, and Gene Therapies, Center for Biologics Evaluation and Research, Food and Drug Administration, Silver Spring, MD, USA

## 1. Introduction

The Center for Biologics Evaluation and Research (CBER) is one of seven centers at the United States Food and Drug Administration (FDA). CBER regulates a variety of biological products including allergenics, blood and blood components, medical devices and tests, gene therapies, human tissues, cellular therapies, vaccines, and xenotransplantation products. CBER is composed of six offices, three of which, the Office of Blood Research and Review, the Office of Vaccines Research and Review, and the Office of Cellular, Tissue, and Gene Therapies (OCTGT), are responsible for the regulation of medical products. OCTGT is charged with the regulation of gene therapies, cell-based products, therapeutic vaccines and cancer immunotherapies, human tissues for transplantation, xenotransplantation products, and certain combination products and medical devices.

In the United States, adenoviral vector-based therapies are regulated by the FDA as gene therapy products or vectored vaccines. They may be used only if licensed or under an Investigational New Drug application (IND). The first human gene therapy IND used a retroviral construct in patients with adenosine deaminase deficiency in 1990.[1] The first adenovirus-based gene therapy for cystic fibrosis followed soon thereafter.[2–4] Since that time the field of adenovirus gene therapy has grown and changed substantially.[5] Although the US FDA has not licensed any adenoviral-based gene therapy product, it is a product class with a large amount of investigational clinical experience and continues to be a robust area of research and development.[6] Approximately 200 adenovirus-based INDs have been submitted to OCTGT; the majority of these are for the treatment of cancer, and approximately 50 are currently active.

The requirements for an IND submission can be found in Title 21 Part 312 of the United States Code of Federal Regulations (21 CFR 312). These statutes apply to all INDs. This chapter describes the regulatory requirements and expectations for adenoviral vector-based IND submissions to OCTGT.

Historically, many developers of adenovirus-based gene therapy products have been clinical investigators who design, test, and produce a clinical product in their laboratory or their institution's core facility. Usually, this process relies heavily on funding from the National Institutes of Health (NIH) through competitive, investigator-initiated research grants or with assistance from program grants. Investigators who receive NIH funding must be in compliance with the NIH Guidelines for research involving recombinant nucleic acids and register their clinical protocol with the NIH/Office of Biotechnology Activities.[7,8] A review of oversight responsibilities for gene therapy products at the NIH and institutional level has been recently published and is beyond the scope of this chapter.[9]

This chapter provides OCTGT's perspective on the regulation of adenovirus-based vectors. The first section provides an overview of how these vectors are designed and implications for regulatory oversight. Design considerations have a significant impact on how the product will be manufactured, tested, and used in a clinical study. The remaining three sections describe more broadly how the scientific and medical disciplines in OCTGT review submissions for adenoviral vector-based products. This chapter describes general manufacturing requirements for products under IND, preclinical supporting information needed for initiating clinical studies, and considerations for clinical trial design and for investigators involved in treating patients with adenovirus-based investigational products. All of these requirements and recommendations have been published previously as FDA guidance or regulation, but here we present this information specifically in the context of developing adenovirus-based products.

## 2. Regulatory Considerations in the Design of Adenoviral Vector-Based Therapies

The first step in the development of an adenoviral vector-based therapy is the molecular design of a product for further study. In general, an understanding of the molecular biology of the virus, the transgene cassette, and its associated genetic elements influences the design and construction of adenovirus-vectored products. Researchers have found that the virus can accommodate a number of design configurations from which clinical grade production should be readily achievable. Ultimately, the design of these products is determined by the specific requirements of the user. The FDA recommends that construction and testing of these products should be done in consultation with the Agency before IND submission in what is called a preIND interaction.[10]

For many investigators who are developing adenoviral vector-based products, the selection of a transgene(s) is the first and most significant design consideration. Adenoviruses are very efficient at gene transfer and are capable of high-level expression of transgenes; however, the viral constructs are known to be immunogenic and usually have limited duration of gene expression in vivo. Therefore these vectors are, for the most part, best suited for applications in which the transgenes require only transient gene expression. Such applications would include expression of vaccine antigens, genes related to immune modulation, or cancer cell-specific cytotoxicity. All of these applications are being explored (some in combination) for cancer treatments and, after demonstrating a level of tumor response, may accelerate quickly through clinical development.[11]

The size of the transgene cassette must be taken into consideration because of size restrictions for the virus. Insertion capacity of the virus is limited by the amount of genomic DNA that can be efficiently packaged into the capsid.[12,13] Adenovirus-based vectors usually have at least one deletion to make space for the transgene cassette. The most common deletion is in the early region 1 (E1 region) of the virus. In addition to providing space for the transgene, an E1 deletion removes the principal activating proteins of the virus, thus rendering the adenovirus replication deficient. For so-called first generation constructs, the deletion of E1 significantly attenuates the virus and eliminates its ability to replicate. This feature restricts viral replication to cells capable of *trans*-complementing the deletion during manufacture (e.g., HEK293 or PER.c6 cells). It also simplifies safety studies needed to establish a basis for conducting a clinical trial. In most circumstances, the toxicity of a replication defective adenovirus can be evaluated in mouse model systems, even though the replication of human adenoviruses is greatly restricted in mice. Because of their transient expression, E1-deleted vector designs also limit the duration of shedding studies and long-term follow-up data collection in the clinical trial. Commercially available reagents may be used to generate these products, which also simplifies and standardizes the construction, and in turn enables faster initiation of clinical studies.

Incorporating additional deletions in second generation and high capacity (or gutless) vectors has allowed greater transgene sizes. These deletions eliminate expression of viral gene products that may be immunogenic or cytotoxic, which in turn may increase the duration of transgene expression. For example, high capacity vectors in which all the viral open reading frames have been removed are much less immunogenic and provide longer transgene expression times than first generation constructs (reviewed by Ventiri and Ng).[14] The elimination of additional viral genes required for replication, such as E2 or E4, will require a more specialized production cell line. In this situation, manufacturers must develop cells that are able to *trans*-complement the additional deletions, but many cell lines have already been established for this purpose (reviewed by Kovesdi and Hedley).[15] These cells that complement multiple deletions also have the advantage of significantly reducing the likelihood of replication-competent adenovirus (RCA) contamination in the product by the fact that additional homologous recombination events are required to generate an RCA.

Another significant design consideration is the selection of transcriptional and targeting control elements for the transgene. For most investigators, the selection of transcriptional control elements (i.e., promoter) is based on commercially available reagents. Strong promoters that are active in a variety of cell types are usually assumed to allow for the greatest amount of protein production in vivo. This is a reasonable starting point, although expression levels may vary from cell line to cell line and may not necessarily correlate with activity in vivo. Promoter selection is a particularly important manufacturing consideration when expression of the transgene can interfere with virus replication. Transgene interference may decrease yields and impose a negative selection pressure against expression, leading to genetic instability.[16,17] In these situations, inducible promoters for transgene expression or promoters that are active in restricted cell types may be required for production. Cell-restricted promoters may also improve the safety profile of a product. For example, the safety profile of a virus expressing a cytotoxic gene for cancer gene therapy may be improved with the addition of a tumor-specific promoter. In this case, establishing the safety profile for clinical use would include a demonstration of promoter selectivity.

The designers of a new adenovirus-based product may also choose to incorporate a variety of capsid serotypes or capsid modifications that can direct cellular targeting of their product. Many designers choose a serotype 5 adenovirus capsid based on the convenience of commercially available reagents. A large body of knowledge has been generated regarding the pharmacokinetics of serotype 5 adenovirus after administration in humans via a number of routes and at a number of dose levels. However, some investigators are exploring pseudotyping, retargeting, or use of novel serotypes as a way to improve target cell uptake. These approaches are being rationally designed as we understand more about the biology of adenoviruses and their interaction with the human host. Specific manufacturing concerns regarding these targeted adenoviruses include the efficiency of assembly of an altered capsid and potential selection against the intended modification during production and the stability of the modified virus capsid during manufacturing and storage, all of which will be taken into consideration during the development process for these products.

For investigators who are developing adenovirus vector-based products that are replication selective, the mutation or mechanism by which the virus is made to be replication selective is a significant design consideration that affects manufacturing, preclinical safety testing, and clinical trial design. Most of these vectors are being developed as oncolytic products to allow limited or selective replication in tumor cells, but there has also been interest in developing replication-competent vectored vaccines.[18–20] For replication-competent oncolytic products, tumor cell-specific replication is desirable. Complete infection of all tumor cells may not be achievable using the standard serotype 5 adenovirus vector. Low levels of coxsackievirus and adenovirus receptor expression on tumor cells and preexisting serotype 5 immunity limit transduction Low viral transduction limits the ability of the virus to replicate and spread in tumor cells.[21–23] Advances are being made in the development of vectors that target and replicate in tumors. Complete infection of all tumor cells may not be required because host immune responses to virus-infected cells also generate anticancer effects.[24–26] The oncolytic methods currently being explored for adenovirus-based therapies include transcriptional targeting of E1 gene expression and removal of tumor suppressor binding domains in the E1 genes (reviewed in Kruyt and Curiel).[27] In addition, replication-competent viruses from different serotypes with low seroprevalence are being investigated because they are unimpeded by preexisting immunity, as well as constructs selected for the ability to lyse cancer cell lines in vitro.[28,29] The degree to which these viruses are replication selective is an important consideration to be established prior to use in human clinical studies.[30] In addition, steps to assess shedding and the possible transmission from treated to untreated individuals may be required. Recommendations regarding shedding studies have been published in FDA guidance (FDA, *Draft Guidance for Industry: Design and Analysis of Shedding Studies for Virus or Bacteria-Based Gene Therapy and Oncolytic Products*) and discussed publicly in a recent Cellular, Tissue, and Gene Therapies Advisory Committee (CTGTAC) meeting.[31]

Finally, another significant design consideration is the compatibility between the adenovirus vector and the delivery devices used to administer the product. For some adenoviral products, delivery to a specific organ or tissue may require a specialized delivery device. Medical devices are cleared by the FDA for their intended use, which may include placement in a specific tissue, accessing a particular body compartment, or injecting/infusing an approved drug. Adenovirus vectors have known stability problems when they come into contact with certain materials, including some materials that are commonly used in catheters.[32–34] Contact surfaces or leachables from these surfaces may bind or inactivate the virus in as little as 10 min and may result in suboptimal delivery of the product. Therefore delivery devices should be carefully considered and assessed for their biocompatibility with adenoviruses before use in clinical studies.

## 3. Chemistry, Manufacturing, and Control Requirements

A summary of regulatory requirements and considerations for specific manufacturing steps for adenoviral vector-based products is presented in the following section. This section describes measures taken to qualify cell substrates, viral banks, purification strategies, and reagents and raw materials used in production. More specific information can be found in relevant guidance documents published by the Agency and referenced in this section.

### *3.1 Establishing a Cell Bank*

Cells used for production are a critical reagent in the manufacturing of clinical grade adenoviral vector-based products. The properties of the production cells and the potential adventitious contaminants harbored by them influence the quality of the viral product. As with any biological product, cells used for production should be characterized with respect to history, in vitro growth characteristics, and presence of detectable microbial agents (21 CFR 610.18 (c)). There are limitations when selecting a production cell line based on the design requirements of the adenoviral vector. For replication-defective viruses in which coding sequences needed for viral replication have been removed, viral proteins must be provided *in trans* by the production cell line. Cells suitable for the generation of E1-deleted replication-defective adenovirus vectors contain and express the E1 region of the viral genome. The two most common are cell lines derived from embryonic human kidney fibroblasts (HEK 293) or retinoblasts (PER.C6). In HEK 293 cells, the integrated E1 sequences contain enough flanking adenovirus sequence such that homologous recombination with the vector will occur, resulting in contamination of the product with E1-containing RCA.[35] PER.C6 cells were engineered to contain minimal regions of homology to prevent recombination and reduce the occurrence of RCA.[36] PER.C6 and other similar cells may be advantageous; however, there have also been reports of a rare RCA-like material known as helper-dependent E1-positive virus particle during production with PER.C6 cells.[37,38] For more complicated vector designs with additional essential viral genes missing, manufacturers may have to develop custom cells that express additional viral genes[15] and/or qualify additional reagents (i.e., a helper virus bank) to provide viral genes necessary for replication. For production of adenovirus gene therapy products in which no complementation is required (e.g., conditionally replicating vectors having modified E1 regions), a cell line should be chosen that does not contain adenovirus sequences in order to prevent homologous recombination.

Cell lines used for adenovirus vector production can be easily frozen and a cell banking system can be established to ensure consistency and control of production cells. With an appropriate cell bank system, production with the same stock of cells can occur for the lifetime of the product. The Agency has published guidance documents with recommendations for the qualification of cell banks.[39,40] The recommendations include a description of the history of the cell line and the banking system used for storage. This history should include a reference to the cell source and where the cells were first obtained. For the derivation of new cell lines from primary human tissue the history should include whether the requirements for screening and testing of human donors were met (21 CFR 1271). Characterization tests are performed when the bank is developed, and include tests for phenotype, genotype, and cellular isoenzyme expression to establish identity. In addition to consistency and control, a cell bank is tested to ensure safety of the production cell lines. The Agency recommends that testing be performed to demonstrate that a master cell bank (MCB) is free from *Mycoplasma*, endotoxin, bacteria, fungi, and adventitious agents (through in vitro and in vivo assays). Depending on its species of origin, the MCB should be tested for species-specific pathogens. For human adenoviruses made in human cells, this will include CMV, HIV1/2, HTLV1/2, EBV, B19, HBV, and HCV. The Agency recommends that sponsors summarize the analytical testing in the regulatory file and establish acceptance criteria for these tests of the MCB. Many cell lines used in the production of adenoviral gene therapies are known to be tumorigenic in animal models and therefore additional testing to verify a tumorigenic phenotype is not usually required for qualification of a cell bank.[40] In these cases, the manufacturers will demonstrate in final product testing that the product is free of the transforming agent (e.g., host cell DNA). More on host cell DNA is presented later in this chapter.

The MCB is usually the first tier of a two-tier cell bank system; the second is the working cell bank (WCB). Usually, the purpose of the WCB is to extend the supply of the qualified MCB and it is created by expansion of one or more vials of the MCB. The amount of information for characterizing a WCB is generally less extensive because it is generated from the fully qualified MCB, usually at the same facility, in a similar manner, and using the same, or similarly qualified, raw materials and ancillary agents. Therefore, WCB qualification is typically limited to tests for sterility, *Mycoplasma*, adventitious agents (by the in vitro assay) and identity.

## 3.2 Establishing a Virus Bank

The generation of a master viral bank (MVB, sometimes referred to as the master virus seed) is the next stage in development after the adenovirus product has been designed and a cell bank has been established. A banking system is needed to ensure the control and consistency of a product and, like the MCB, the MVB is expected to last for the lifetime of a product. The MVB requires proper storage, identification, and appropriate record keeping (21 CFR 610.18). OCTGT has published guidance documents that review recommendations for the qualification of an MVB under IND. These recommendations include an assessment of the history of the virus stock and a description of the banking system used for storage.[39] The MVB is an important control point that allows thorough characterization and qualification of virus stocks used for production. Characterization and safety tests for qualification of the MVB involve some of the same testing recommended for the MCB and final product. For most manufacturers, the history of the virus can be fully documented and viral stocks have only been in contact with qualified cells. However, in situations where the initial virus construct was made using nonqualified cells or under poorly defined laboratory conditions, it would be appropriate to clone the virus by plaque purification or limiting dilution to remove any potential contaminants before being expanded into an MVB. Because of the redundancy in testing for the MVB and final clinical product, investigators for most first-in-human trials sometimes combine testing to qualify a new master virus bank and the release of a final product for clinical use at the same time. In these situations, after the first production run is completed, a fraction of the final vialed product is used for clinical release and the remaining vials are stored as the MVB. Combining tests for the MVB and clinical lot has a cost advantage.

As an important control point in manufacturing, the MVB is used for full sequence analysis and RCA testing. All adenovirus vectors for gene therapy should be sequenced prior to being used in clinical trials, based on recommendations in the November 2000 Advisory Committee and published in guidance.[39,41] Most vectors have been derived from serotype 5 and can be compared to the complete sequence in GenBank such as the adenovirus reference material (ARM) sequence (Accession Number AY339865). Some minor changes may be detected based on passage history and plasmids used during construction. The virus has an error-checking polymerase and a relatively low mutation rate. As such, evaluation of sequence stability of the transgene in the MVB and during production is not usually recommended for early-phase trials. However, some vector constructs have been reported to be unstable; examples include those that have repetitive sequences, which might allow for recombination, or those with transgene deleterious to the virus.[42] Testing the MVB for the presence of sequence changes in these situations may be recommended at early stages of development.

Another factor that may have a significant impact on the qualification of the virus bank is the presence of RCA. RCA arises by homologous recombination during replication of E1-deleted constructs in a complementing cell line. Recombination occurs between the left end of the vector and the adenovirus sequence in the cell, resulting in the acquisition of E1 by the vector. The effect of RCA on production and control of the product can be significant. If RCA-contaminated stocks are expanded, RCA levels can increase at the expense of the intended product after multiple passages.[35] The presence of RCA can interfere with in vitro adventitious agents testing performed for lot release. RCA-contaminated products in animals have been demonstrated to have reduced transgene expression and increased inflammatory responses.[43] High levels of RCA in the final product are also a potential safety consideration that may necessitate additional toxicology studies or additional precautions taken in the design of the clinical trial. In general RCA levels can be reduced by limiting the amount of homologous sequence in the vector and producer cell line.[36,44–47] The Agency has recommended that products contain no more than one RCA in $3 \times 10^{10}$ vp.[39] This number was chosen based on an estimation of the level of purity that typical manufacturers were able to achieve. If a manufacturer is not able to meet this recommendation, a new production system with a lower chance of producing RCA should be considered.

## 3.3 Production and Purification Strategies

Most gene therapy applications require relatively high doses of a pure product. High titers are usually needed to target sufficient cells for gene transfer, and impurities may limit transgene expression by inducing immune response or inflammation. Adenovirus products are attractive candidate vector systems because of the ability to manufacture a high titer and high purity product. High titers are possible because the virus is efficient at creating progeny. One infected cell can result in 10,000 new virus particles and yields of about $1 \times 10^{11}$ vp per milliliter of cell culture are commonly achieved.[48] The typical purification process involves harvest of infected cells and a lysis step to release cell-associated virus. This is followed by clarification of cellular debris, digestion of nonencapsidated DNA, separation step(s), and buffer exchange for final formulation. The stable, nonenveloped nature of the capsid allows for robust enrichment methods, and the typical approaches include density gradient ultracentrifugation or anion exchange chromatography. Both allow purification of virus particles from cellular components and adventitious agents. Density gradient ultracentrifugation has the advantage of removing empty capsids based on the lower density of empty capsids, but this technique cannot be performed as a closed system and is difficult to scale-up. In the past, many investigators started with purification by ultracentrifugation in early-phase studies, and then switched to chromatographic separation later in development. The Agency advises IND sponsors to discuss major manufacturing changes, including scale-up, before implementation. Prior to initiation of Phase 3 studies the Agency will consider the potential impact manufacturing changes have on the safety profile of the product. After Phase 3 studies have been initiated, manufacturing changes are reviewed for potential impacts on the efficacy and safety of the product.

Product-related impurities include noninfectious particles, aggregates, free viral capsid proteins, and vector nucleic acid. Process-related impurities include host cell protein, host cell DNA, residual culture media components, residual solvents, additives, antifoaming agents, or enzymes used in production. The Agency recommends that manufacturers develop quantitative lot release assays for purity prior to initiating clinical studies (e.g., relative amount of infectious particles expressed as a ratio compared to viral particles).[39] Other tests should be developed during early-phase studies and be in place by the start of Phase 3. Probably the most significant product-related impurity at the production and purification stage is DNA. The greatest by-product of replication is viral nucleic acid, and both viral and cellular DNA may be present as impurities in the final product. Steps to remove DNA are needed for many reasons. During manufacturing the presence of DNA can interfere with chromatography, and thus digestion of DNA can reduce fouling of columns. DNA may also interfere with quantitation methods for calculating virus particle levels, and the presence of low levels of DNA has also been associated with vector aggregation.[49] For all these reasons, removal of DNA is an important process step during purification. The presence of DNA from tumorigenic cell lines in the final product is also a safety concern. For gene therapy vectors, it has been recommended that products contain not greater than 10 ng host cell DNA per dose.[50] This is consistent with general guidelines for vaccines and other biologicals prepared in continuous cell lines.[51] A size limit for host cell DNA is also recommended since it is likely that reduction of DNA fragment size reduces the risk from DNA, as the smaller the DNA fragments are, the lower the probability that intact oncogenes and other functional sequences would be present. DNA fragments smaller than 200 bp will give substantial safety margins for products that meet the 10 ng per dose limit.[52] Therefore host cell DNA is recommended to be between 100 and 200 bp.

Process-related impurities also include chemicals such as plasticizers and plastic additives that may leach from contact surfaces into the adenovirus product during production or storage. Leachables may pose safety concerns or interfere with product performance, although they are usually present only at very low levels. Another significant concern is that leachables can also interfere with analytical test methods. Some contaminants have been identified that absorb strongly at 260 nm and can interfere with viral quantification methods, leading to inaccurate measurements of viral dose.[53] Careful study of leachables is usually performed during late stages of development. Prior to this stage, manufacturers should carefully qualify and control the plastic material used in production that comes into contact with the viral product.

### 3.4 Qualification of Raw Materials and Reagents Used in Production

Various raw materials are used in the production of adenoviruses. Use of these materials may affect the quality, safety, and efficacy of the final product. The quality of the culture media and raw materials should be documented in the regulatory file with respect to identity, purity, sterility, biologic activity, and absence of adventitious agents. Typically trypsin and serum are the only animal-derived components commonly used in adenovirus gene therapy production. For all animal-derived components, the manufacturer should provide information on the source organism, supplier/vendor, country of origin, infectious agent testing, and stage of manufacture in which the component is used. This information is collected to mitigate potential concerns regarding transmission of bovine spongiform encephalopathy and transmissible spongiform encephalopathy and other infectious agents. Trypsin and serum should be tested by vendors to ensure the absence of bovine viruses and porcine parvoviruses. If these are not documented in the certificates of analysis, these tests will be required to qualify the MCB and/or MVB. Animal-derived materials should be compliant with the requirements for the ingredients of animal origin used for the production of biologics (9 CFR 113.53). Serum and trypsin may also harbor infectious agents not detected by qualification tests. Because these reagents may potentially introduce contaminants into the production process that may not be removed during purification and may not be detected during final product testing, nonanimal-derived, alternative synthetic, or gamma-irradiated materials may be advantageous. If proprietary serum-free medium is used, the sponsor must provide a letter of cross-reference for the material.

# 4. Manufacturing Control

## *4.1 Product Testing*

Testing on the adenoviral vector-based product should include the routine safety testing outlined in the general biological standards (21 CFR 610) and applicable guidance documents.[39] These include tests for sterility, *Mycoplasma*, endotoxin, and adventitious agents (including RCA). In addition to these assays, additional customized assays will be required over the course of clinical development. Common virological assays are used throughout development for quantification of the virus and these assays play an important role in the analytical characterization and release testing of adenoviral vector-based products. Testing the final product relies on measures of both physical and biological properties of the viral particles. Viral particles are assessed through detection of particle components such as DNA or capsid proteins. The most common method for calculating virus particles is an absorbance measurement of lysed particles at OD 260.[54,55] The data are then converted to a particle number based on the extinction coefficient for a 36 kb adenovirus, or based on an extinction coefficient specifically determined for the genome size of the adenovirus product. The OD 260 method is commonly used to determine dose for clinical-grade preparations of adenovirus. To facilitate the standardization of this method (as well as other methods noted below), an ARM was established in 2002. This material serves as a quantitative and qualitative reference of serotype 5 adenovirus, and is commercially available from ATCC to be used by manufacturers in their qualification program.[56]

Another common assay for calculating adenovirus virus particles is anion exchange high-performance liquid chromatography (AEX-HPLC). AEX-HPLC is useful for analyzing both crude and pure samples.[57,58] Virus particle quantification by AEX-HPLC relies on an absorbance measurement after elution from a positively charged matrix. Pure adenovirus can be eluted at a nearly symmetrical peak and can be distinguished from residual DNA, hexon protein, and cellular debris. The viral capsids remain intact during chromatography, and empty and full capsids are not separated. Virus particle concentration is derived from a calibration curve that reports the known virus particle concentrations against the corresponding absorbance area of the viral peak using a reference standard. This approach is more robust than the OD260 method because it can be used to measure virus particle concentration in samples of nonpurified cell lysate.[57] This method can also measure the overall purity of a preparation by integrating all of the absorbance peaks and determining the percentage of the viral peak.

Physical particle counts alone are not sufficient to measure the activity of a virus preparation because they are unable to distinguish functional from nonfunctional particles. In order to assess the infectivity of a preparation (or number of infectious particles), cell culture infectivity assays are performed. These types of assays report the infectivity of the virus by assessing the ability of the virus particles to be taken up by cells and execute one or more steps in the virus life cycle. A large number of assays can be used to measure infectivity. The most common infectivity assays are based on cell killing or CPE of the infected cell and these include plaque assays and limiting dilution assays. The theory behind the limiting dilution assay is that, at limiting dilution, each CPE-positive well will represent a single infectious particle, distributed according to the Poisson distribution. This method is more standardizable than the plaque assay, but as with any biological assay the results are highly variable. Under typical conditions, only a small percentage of virions actually infect the cell monolayer, and therefore the results are often corrected for the theoretical diffusion rate of the virus. This calculation is often referred to as the NAS (normalized and adjusted standard) method.[59] NAS titers can correct for differences in incubation time and incubation volume that may otherwise affect the results of this assay. A number of other infectious titer assays have been developed, including focus-forming assays and polymerase chain reaction (PCR)- or flow cytometry-based assays (reviewed in Dormond et al.).[60] All of these are acceptable approaches for measuring infectious virus. A well-characterized reference standard is an important control to include in these assays to understand assay variation. The ARM has been useful in qualifying a manufacturer's in-house reference standard and standardizing infectious titer results between manufacturers.[56]

The Agency has developed recommendations to ensure that preparations of adenovirus have a minimum activity in terms of infectivity per physical particle.[39] This ratio was designed to be a measure of the specific activity of a preparation and is expressed as a ratio of particles to infectious units. The current recommendation is that the particle to infectious unit ratio be no greater than 30:1. The previous recommendation of 100:1 plaque-forming units was developed shortly after the first adenovirus gene therapy trials were initiated, and remained in effect until 2000.[50] After review of data received in response to the March 6, 2000, Gene Therapy letter, it was apparent that almost all vector lots have a ratio of less than 30:1 particle to infectious units.[61] However this ratio can be highly variable, depending on the assay used to measure infectivity. Using the NAS titer can help overcome some of the variability of different infectivity assays. However, it may not be possible to manufacture some viral constructs, including those from different serotypes, at this level of specific activity. This recommendation was developed based on experience with the serotype 5 viruses and is applied generally as a target level for early development. Ultimately, manufacturers will have to set a specification for the particle–infectious unit ratio based on manufacturing experience.

Investigators will also have to develop specific custom assays for a product over the course of clinical development. These include the development of methods for assessment of potency, identity, and purity. The development of specifications for each of these parameters is an important part of product development and characterization. Potency assays are probably the most challenging custom assays to develop. Tests for potency should consist of either in vitro or in vivo tests (or both) that measure a biological activity(ies) linked to how the product functions.[62] Acceptability will be determined on a case by case basis, but these tests must (1) measure the biological activity of active components, (2) be available for lot release, (3) provide a quantitative readout, (4) meet predefined acceptance/rejection criteria, (5) include appropriate reference controls, (6) be stability indicating, and (7) be validated before licensure.

Identity assays are also specific for each product. Before initiation of Phase 1 studies, an identity assay is recommended that will adequately identify the product and distinguish it from any other product being processed in the same facility.[39] For most adenoviruses, this would include restriction enzyme digestion patterns and PCR. An appropriately designed PCR should be specific enough to satisfy this recommendation, but may need to be updated if other similar products are made in the same facility over time.

Tests for purity may vary based on the purification methods used in the manufacturing process. For example, process-related impurities such as CsCl, detergents, or Benzonase® are necessary only when these substances are used during the purification. Not all tests necessarily need to be developed into lot release tests with specifications. Rather some may be used mainly in situations when more extensive testing is required, such as process validation and comparability testing. Custom tests may be designed with a specific understanding of the product and manufacturing process. For example, an improved manufacturing process to remove human SET and Nucleolin was able to provide higher purity vector preparations.[63] Tests to measure specific protein impurities can be very sensitive, and these tests may be helpful for supplementing more general methods to assess viral purity.

Normal adenovirus replication results in the production of a small proportion of empty capsids.[64] Empty capsids may copurify with viral particles using anion exchange chromatography based on a similar capsid structure and charge.[65] Empty capsids are immature viral particles that contain no DNA when measured by the OD 260 assay, and therefore the extinction coefficient for empty capsids is very different than that of intact capsids. It can be difficult to accurately measure the percentage of empty capsids through direct measurements. Measurements of empty capsids are not required for initiation of Phase 1 studies, but may be included as a part of product characterization later in development. Direct measurements of empty capsids usually involve analytical ultracentrifugation or transmission electron microscopy. However, Vellekamp et al. have identified a unique approach for measuring a marker of empty capsids, a capsid precursor protein (pVIII).[65] Thus quantification of pVIII by SDS-PAGE or RP-HPLC may help simplify empty capsid measurements for some vectors.

Like other highly purified, highly concentrated protein preparations, adenovirus vectors may aggregate under certain circumstances. Aggregation might be visible to the naked eye as cloudiness or a white precipitate, but aggregation can occur that is not visible. A simple spectrophotometer-based static light scattering assay has been reported and is suitable for routine lot release; however, it is somewhat qualitative and not very sensitive. More sophisticated assays use disc centrifuge or dynamic light scattering, which can measure particle diameter and dispersity.[66,67] Finally, analytical ultracentrifugation can provide detailed information about particle sizes.[68] Development tests for aggregation may be more sensitive in detecting loss of infectious particles than infectivity assays, and therefore perhaps may serve as a sensitive indicator of batch-to-batch activity.

### 4.2 Quality and Current Good Manufacturing Practices

A goal of the Agency's current good manufacturing practices (cGMPs) is to ensure quality through control and regulation at each step of a manufacturing process. The cGMPs are directed at the commercial manufacturer for large, repetitive commercial batch production. Therefore, full application of cGMPs may not be achievable for the manufacturing of most investigational drugs used for Phase 1 clinical trials. In 2008, the FDA amended the cGMP regulations under 21 CFR part 211 to exempt most products made for use in Phase 1 clinical trials and published the Guidance for Industry "CGMP for Phase I Investigational Drugs". The Agency provides recommendations in this guidance document that manufacturers of Phase 1 investigational drugs can use to comply with cGMP requirements of section 501(a)(2)(B) of the Food, Drug and Cosmetics Act. The adherence to cGMP during the manufacture of Phase 1 investigational drugs occurs through having well-defined, written procedures, and having adequately controlled equipment and an adequately controlled manufacturing environment. Consistent with the cGMPs, the Phase 1 manufacturing program should include accurate and systematically recorded data from manufacturing (including process and final product testing). In addition, a quality control (QC) program, which is separate from manufacturing, should be in place at the earliest phase of product manufacture. This QC unit should be responsible for ensuring the quality of the product and for product release for clinical use. The role of the QC unit is described in Part 211 of Section 21 of the CFR.

## 5. Preclinical Evaluation of Adenoviral Vector-Based Therapies

Advances in science and technology have resulted in the development of many types of adenovirus vector-based therapies, intended to treat a diversity of medical diseases and conditions. However, prior to administration of such products in humans, "adequate information about pharmacological and toxicological studies…on the basis of which the sponsor has concluded that it is reasonably safe to conduct the proposed clinical investigations" is needed (21 CFR 312.23 (a)(8)). The preclinical program for each adenovirus-based therapy should thus be designed to evaluate the benefit of product administration in relation to the risk of administration in the identified patient population. This section will provide an overview of preclinical evaluation considerations for adenoviral vector-based products to support administration in clinical trials. For a more comprehensive discussion, refer to the FDA/CBER guidance titled, *Guidance for Industry: Preclinical Assessment of Investigational Cellular and Gene Therapy Products* (herein referred to as the "preclinical guidance").[69]

### 5.1 Proof-of-Concept Studies

The proof-of-concept (POC) studies, the initial translational step from the discovery stage, help establish the feasibility of use of a specific investigational product in a particular disease/condition. The data generated for the adenoviral-based gene therapy can (1) help define a pharmacologically active dose range, (2) determine a potentially optimal route and anatomic location for product administration, and (3) inform a possible dosing regimen for the clinical trial. In addition, a preliminary biodistribution (BD) profile of the administered vector can help delineate potential target and nontarget tissues.

The POC program can consist of both in vitro and in vivo studies. In vitro studies provide additional insight into the biological activity and mechanism of action of the gene therapy product and help support the biological relevancy of animal species/model used, or may be used for POC if there is not an applicable animal model. The use of animal models of disease/injury allows for potential further insight and understanding of the pharmacological profile of the administered product. For example, antitumor activity may be evaluated in syngeneic murine tumor models, murine:human hybrid tumor xenografts, transgenic models, and animals with spontaneously occurring cancers.

Considerations when selecting an appropriate animal species can include (1) assessment of the animal's pharmacological response to the clinical transgene delivered by the adenoviral vector or to the ex vivo adenoviral transduced cells, (2) permissiveness/susceptibility of the species to vector transduction or to virus replication, and (3) the comparative physiology and pathophysiology of the animal species to the proposed human population.[70] These considerations also affect the use of a particular model of disease/injury to study the adenoviral vector-based therapy.

## 5.2 Safety Testing

The overall goal of the toxicology studies is to further understand the benefit:risk ratio for use of a specific investigational product in a defined clinical trial. Characterizing the vector-based product proposed for clinical use (e.g., vector backbone, replication status, expressed transgene) and knowing the administration route/anatomic location for product delivery are considerations in the overall paradigm for safety assessment. Additional factors are the BD, viral replication, and potential persistence of the vector in nontarget tissues and the potential for inappropriate immune activation in the host. Another important element is the expressed transgene, which can link to risks such as undesired immunogenicity and toxicity from overexpression in target as well as nontarget tissues.

The design of studies to evaluate the safety of gene therapies, including adenoviral vector-based products, should consider the biological relevancy and disease status of the animal species, the route of administration, and other factors, as discussed in the preclinical guidance.

### 5.2.1 Selection of Animal Species

Appropriate scientific justification should accompany the selection of the relevant animal species/models used in the safety assessment of the adenoviral vector-based therapy. Consideration of anatomy, physiology, age, clinical delivery system and procedure, and the anatomic site of product delivery is important. The inclusion of multiple animal species (e.g., rodent and nonrodent) for safety testing of gene therapy products is not a default position.

In addition, selection of a particular animal species/strain may affect the interpretation of resulting study data. For example, intravenous injection of adenovirus vectors in rats resulted in induction of shock mediated by platelet-activating factor.[71] Adenovirus interaction with a variety of host proteins, in particular coagulation Factor X, has also been identified as part of the inherent effects on liver tropism, which, combined with macrophage scavenging, may impact viral activity and induce potential adverse effects.[72,73] In many instances, toxicities related to the vector itself (e.g., inflammatory reaction to adenovirus capsid proteins) can occur.

In some cases, the expressed human transgene may be biologically active only in humans and in nonhuman primates. In such cases, consideration is given to other animal species, such as (1) rodents responsive to adenovirus that are subsequently "humanized" to express the human target receptor(s) and (2) to vector expressing an analogous animal transgene.

The use of large, nonrodent animal species may be appropriate in certain instances, such as for the evaluation of a new delivery system and novel anatomic site proposed for the clinical trial; however, providing appropriate rationale for the species is important.

If feasible, safety endpoints should be added to POC studies to facilitate assessment of the disease-related pathology to any toxicity observed. However, when analyzing the resulting data, the underlying pathology associated with the disease state may be a possible confounder.

### 5.2.2 Route of Administration and Dosing Regimen

It is important that the route of administration, delivery device, and the delivery procedure for the adenoviral vector-based therapy reflect the clinical plan to the extent possible. Although the dosing regimen should mimic the intended clinical trial regimen, this may be difficult to achieve in rodents. A modified regimen may be substituted if it reflects a worst case scenario in terms of the frequency of exposure and total exposure levels that are equivalent in humans. The adequacy of the modified regimen should be based on the data derived from BD studies.

### 5.2.3 Dose Level Selection

Knowing the pharmacologically effective dose level range obtained from POC studies is an important element when designing the toxicology studies. It is important that dose levels bracketing this range be administered in the safety studies in order to determine a possible threshold at which toxicity is observed. The highest dose level administered may be restricted due to animal size, route of administration, tissue volume/size, and/or product manufacturing capacity. The dose level at which no biologically or statistically significant increase in the severity or frequency of adverse findings in safety parameters such as clinical observations, clinical pathology, histopathology, or other observation parameters selected with the specific product in evaluation, as compared to appropriate concurrent controls, is observed is termed a "no-observed-adverse effect dose level" (NOAEL).

The extrapolation of the NOAEL in animals to the starting clinical dose level for most gene therapy products can be determined based on body weight, if the product is delivered systemically or results in systemic exposure, organ mass or volume if the product is delivered locally, or other factors. It is important to justify the extrapolation method used. In all preclinical studies appropriate concurrent controls, such as animals that are untreated, given vehicle alone, or administered null vector, should also be included.

## 5.3 BD Assessment

Studies evaluating the BD of the in vivo administered vector in "expected" target tissues, "unexpected" nontarget tissues, and biological fluids (e.g., blood, semen) help inform the design of the toxicology studies (e.g., dose levels, dosing regimen, route of administration). Adenovirus vectors can remain in the host tissues indefinitely following administration. Thus, the data generated from analysis of blood samples obtained at various time points are not necessarily indicators of the virus level in tissues; therefore, multiple sacrifice intervals are usually needed in order to obtain a comprehensive BD profile.

The BD profile for the adenoviral vector-based therapy is characterized prior to initiation of the initial clinical trial. For subsequent clinical trials, additional BD studies may be needed. Examples include a change in product (e.g., formulation), route of administration, or dosing schedule.

Although the risk of inadvertent gene transfer to germ cells or vertical transmission of the foreign gene is not significant for adenoviral vectors, if the BD data indicate high levels of vector DNA in the reproductive tissues and germ cells, then reproductive/developmental toxicity concerns may need to be assessed prior to administration of the product to humans. Further discussion on this aspect can be found in the preclinical guidance.

In addition, the guidance document titled, *Guidance for Industry: Gene Therapy Clinical Trials—Observing Subjects for Delayed Adverse Events* contains recommendations regarding tissues to be collected and analyzed via a quantitative PCR assay. Depending on the route of administration used, additional tissues may need to be collected and analyzed. Vector presence in tissues or biological fluids may trigger further analysis to determine transgene levels.

## 5.4 Good Laboratory Practice

Preclinical toxicology studies should be conducted in compliance with Good Laboratory Practice (GLP) as set forth in 21 CFR Part 58, in order to ensure the quality and integrity of the safety data that are generated. However, some toxicology studies do not fully comply with GLP. In such cases, an explanation for this noncompliance is expected in the final study report and aspects of the study that deviate from the protocol and the potential impact of these deviations on study integrity should be also included in the report. It is important that all preclinical studies be conducted according to a prospectively designed protocol, be performed in as nonbiased a manner as possible, and have appropriate record keeping and documentation of all data. In addition, oversight of the conduct of all non-GLP toxicology studies and each resulting final study report by a Quality Assurance unit/person that is independent of the personnel responsible for the conduct of the study (21 CFR Part 58.35) is recommended. Such oversight is important to ensure study conduct according to sound procedures and to ensure the quality and integrity of the resulting data.[69]

### 5.5 *Considerations for Late-Phase Clinical Trials*

As the product development program progresses into late-phase clinical trials, the need for additional in vitro and/or in vivo preclinical studies depends on factors such as (1) significant changes in product manufacturing or final product formulation, (2) changes in the administration route, process, or anatomical site, (3) dosing schedule modifications, (4) modified patient population, (5) safety concerns that arise in the clinical trial, or (6) reason to conduct reproductive/developmental toxicity evaluations. Discussions with CBER to address such concerns for an adenoviral vector-based therapy are recommended.

## 6. Introduction to Clinical Testing

The goal of clinical testing is to provide information about a product's safety and effectiveness and, ultimately, allow new products to come to the marketplace. The principles described below are not unique to gene therapy vectors in general nor to adenoviral vectors in particular.

### 6.1 *Phases of Clinical Development*

Premarket clinical testing proceeds in a stepwise manner, often referred to as Phases 1, 2, and 3 of clinical development, although the phases are not always discrete. Phase 4 studies are those performed after a marketing authorization is obtained. Each phase of product clinical testing has its series of goals or objectives.

The primary goals of Phase 1 testing are to learn about the product's safety and pharmacokinetic profile, and to identify a safe dose or doses for further study. Phase 1 studies involve small numbers of study participants who are closely monitored for the drug effects. A common Phase 1 design is a single dose, ascending dose, cohort study. Escalation to the next dose cohort occurs after sufficient safety assessment of the preceding cohort. The starting dose and dose-escalation scheme employed depend on the data gleaned in preclinical testing, and the choices may also be influenced by other clinical data, if available (e.g., from closely related products or from the same product studied in a different population). Dose escalation usually proceeds until a defined endpoint, such as a near maximal tolerated dose or an optimal biologic dose, is reached.

Phase 1 studies for some kinds of drugs might be conducted in healthy volunteers when anticipated adverse effects of the product are expected to be minimal and transient and the target population (those with the disease or condition of interest) has high background rates of adverse events, making it difficult to tease out the safety profile of the product. However, for many classes of drugs and biologicals, including adenovirus gene therapy products, the potential short- and long-term adverse effects generally make their risks unacceptable for testing in healthy volunteers.

The next phases of clinical testing, Phases 2 and 3, build on the information generated from the prior studies. The goal of Phase 2 testing is to gain preliminary or further evidence of the product's activity in the disease or condition of interest and to begin to characterize that activity. Phase 2 is the ideal time to optimize the dose and/or dosing regimen, determine the specific subset of the target population expected to respond to the product, the response parameters that are most likely to reflect clinical benefit, and to build on the safety database. Phase 2 trials often are randomized, controlled, and conducted in multiple centers.

Phase 3 of clinical testing includes clinical studies to establish the product's effectiveness, as well as to continue to ascertain the product safety in the setting of a larger study population. The number of efficacy trials, trial design(s), and size of the safety database necessary to determine net clinical benefit depend on a number of factors, including but not limited to the class of product under development, the condition or disease being studied, and the availability of other therapies.

Phase 4 of clinical testing comprises studies conducted after market approval, usually referred to as postmarketing studies. Common reasons for conducting these studies are to acquire more complete safety data, to address questions that arose during the premarketing investigations, or to evaluate the product in other related settings, such as children, the elderly, or people with more advanced stages of the disease. The design of a postmarketing study (such as a randomized controlled clinical trial or a registry) depends on the questions to be addressed.

## 6.2 Good Clinical Practices

Good clinical practice (GCP) is a set of principles and procedures intended to preserve and protect the rights and confidentiality of human research subjects and to ensure, to the extent possible, that the clinical research generates valid scientific data. The origins of a code of conduct to protect human subjects in clinical research date back to the Nuremberg war trials and the Declaration of Helsinki. In 1996, the FDA, under the auspices of the International Conference on Harmonization (ICH), published the guidance document titled: *Guidance for Industry—E6 Good Clinical Practice: Consolidated Guideline*. Basic principles of GCP will be discussed below; the reader is referred to the CBER Website http://www.fda.gov/Drugs/GuidanceComplianceRegulatoryInformation/Guidances/ucm065004.htm for the full document.[74]

### 6.2.1 Responsibilities of Sponsor and Investigators

The sponsor has responsibility over all clinical studies conducted under the IND application and communicates with the FDA. As set forth in regulations at 21 CFR 312

Subpart D and in the ICH GCP guidance, the oversight function includes selecting study investigators, reporting safety information to the FDA, and providing accurate and timely information to all investigators. In some cases, a sponsor may transfer all or some of its obligations to a contract research organization, although the sponsor retains ultimate responsibility for the IND.

Clinical investigators also have specific obligations, delineated in 21 CFR 312 Subpart D and in the ICH GCP guidance. Investigators are responsible for selecting study participants based on eligibility requirements of the protocol, and for obtaining the protocol-specified evaluations. The investigator is responsible for the welfare of the study subject at his/her clinical site. This includes collecting safety data and reporting safety information to the IND sponsor. The investigator also must account for all investigational medical product, maintain accurate records, provide annual updates to the Institutional Review Board (IRB), and obtain informed consent from all study participants.

Individual physicians who assume the roles of both sponsor and investigator (sponsor–investigator) should be familiar with guidances and federal regulations that set out the respective duties of the sponsor and the investigator.

Where the sponsor and investigator are distinct, their separate roles, with the former overseeing the latter, incorporate the checks and balances that minimize bias, and maximize patient safety and trial validity. Under these conditions, a sponsor must provide his or her investigator(s) with an Investigator's Brochure, containing important information about the product (including any nonclinical and clinical studies conducted using the product) to allow the investigator to recognize adverse events and protect subjects in the study.

### 6.2.2 Adverse Event Reporting

Adverse event collection and reporting is a fundamental aspect of drug development and of human subjects protection. The clinical investigator is the individual who identifies, evaluates, and documents adverse events experienced by study participants at his or her site and who is responsible for updating the IND sponsor and the IRB as appropriate, as set forth in federal regulations (at 21 CFR 312.64).

The sponsor is responsible for submitting safety information to the FDA. The timing and reporting format will depend on the nature of the adverse event. The sponsor must report to the FDA in writing every serious and unexpected suspected adverse reaction (SUSAR). An adverse event is considered serious if it results in death, a life-threatening event, inpatient hospitalization or prolongation of hospitalization, persistent or significant incapacity or substantial disruption of the ability to conduct normal life functions, or a congenital anomaly/birth defect. An adverse event is a suspected adverse reaction if there is evidence to suggest a causal relationship between the drug and the adverse event. It is unexpected if it is not listed in an investigator's brochure or in the risk information described in the general investigational plan, or, if listed, is not listed at the specificity or severity observed. The sponsor must report a SUSAR within 15 calendar days of determining that the event qualifies for reporting. Any unexpected life-threatening or fatal SUSAR must be reported by telephone (or facsimile) within 7 calendar days of receipt of the information (as per 21 CFR 312.32). The telephone and written reports constitute expedited reports. Although causality assessment is integral to expedited reporting, a determination that a given investigational product caused or was associated with an adverse event in the course of a clinical study is not always easy. In 2010, the FDA modified the safety reporting regulations to, among other objectives, reduce the number of safety reports that were uninformative. The causality standard was clarified to explain that a *reasonable possibility* that the experience was caused by the drug meant that there was evidence to suggest a causal relationship. In evaluating and reporting an individual adverse event, the sponsor also needs to identify previous similar SUSARs and analyze the event in light of similar reports or relevant information. Additional information about IND safety reporting requirements can be found in the FDA publication titled *Guidance for Industry and Investigators—Safety Reporting Requirements for INDs and BA/BE Studies* (2012), which is available at http://www.fda.gov/BiologicsBloodVaccines/GuidanceComplianceRegulatoryInformation/Guidances/General/ucm239423.htm.

The sponsor is also required to submit to the IND an annual report that includes a summary of the most frequent and the most serious adverse events (21 CFR 312.33). The ICH E3 guidance titled *Guideline for Industry—Structure and Content of Clinical Study Reports* (1996) and a companion guidance titled *Guidance for Industry—E3 Structure and Content of Clinical Study Reports—Questions and Answers (R1)* (2013) describe the manner in which safety data for individual studies should be organized and presented to regulatory authorities in marketing applications.[75,76] A marketing application includes an integrated summary of the entire safety experience for the product. The FDA, as part of the ICH process, is developing a guideline titled *The Common Technical Document for the Registration of Pharmaceuticals for Human Use* that addresses, among other items, formatting of integrated safety data.[77] Once a product is marketed, a passive surveillance system allows for the continued collection and reporting of safety information.[78] For some products, such as ones that pose unique long-term risks, a more active type of postmarketing follow-up might be required as a condition of approval.

### 6.2.3 Consent and Vulnerable Populations

In general, prospective participants cannot be enrolled into a trial without their consent. Elements of the consent form and the informed consent process are set forth in 21 CFR 50.25. Before consenting, study participants must be informed of known and potential toxicities that may occur from participation in a trial of an investigational product, even if the likelihood of toxicity is remote. The IRB for each institution participating in a study must review and approve the consent form and the clinical research protocol before the study can be initiated at that institution. The composition and duties of the IRB are described in the ICH GCP guideline and in 21 CFR 56.

For some of the disorders that are targets of gene therapy, such as inborn errors of metabolism, the affected population will be pediatric subjects. Mechanisms exist to strengthen the human subject protections for study participants who may be particularly vulnerable, such as children, who cannot give valid consent. When a child is to be enrolled in a research study, the parent or legal guardian consents (gives permission) for the child. Additional required safeguards for children in clinical investigations are described in 21 CFR 50, Subpart D. Similar protections apply also to individuals who cannot provide informed consent due to neurologic or psychiatric disabilities.

### 6.2.4 Monitoring and Auditing

Monitoring and auditing are fundamental aspects of GCP. Although their purposes are similar (to ensure appropriate trial conduct and data validity), the approaches differ. As stated in the ICH GCP document, monitoring is "the act of overseeing the progress of the clinical trial and ensuring that it is conducted, recorded, and reported in accordance with the protocol, standard operating procedures, GCP, and applicable regulatory requirements." Medical monitors, usually employees of the sponsor, perform on-site (and, if applicable, off-site) evaluations of trial-related activities. The extent and frequency of monitoring should be appropriate for the length, complexity, and other particulars of the trial. Among the functions of the monitor is identification of deviations in protocol conduct so that the sponsor may take appropriate corrective steps, for example, retraining investigators, closing out certain sites, etc.

An additional measure of human subject protection is use of a Data Monitoring Committee (DMC) to evaluate accumulating data from a clinical trial.[79] Generally, the sponsor establishes the DMC, including selecting the members and devising their charter. The DMC members should be independent of the sponsor and clinical investigators. The role of the DMC varies according to the charter and the nature of the study. The DMC is usually empowered to recommend study modifications to enhance safety of participants; in some cases, a DMC may recommend that a study be stopped if accruing data indicate a major safety concern. Of note, DMCs review data submitted to them but do not visit sites to directly ensure that the data are accurate, the protocol is followed, consent is documented, etc. Thus, a DMC does not perform the functions of or obviate the need for study monitors. Further discussion of DMCs and recommendations regarding their composition, responsibilities, and operation can be found in the FDA guidance document titled *Guidance for Clinical Trial Sponsors—Establishment and Operation of Clinical Trial Data Monitoring Committees* (2006).[80]

Auditing is defined in the ICH GCP document as "the systematic and independent examination of trial-related activities and documents." The audit is usually conducted at the conclusion of the trial. The sponsor may hire auditors who document findings in a written report to the sponsor. FDA field inspectors also conduct independent study audits. Traditionally, the purpose of the FDA audits has been to verify the data submitted to the FDA in support of a marketing application. However, the FDA and the sponsor may conduct "for cause" or directed audits at any stage of clinical investigation if there is reason to suspect a problem with trial conduct or data integrity.

## 6.3 Clinical Safety of Adenoviral Vector Products

Most of the completed and ongoing adenoviral vector clinical trials are in early or middle phases of development, with a few products being investigated in Phase 3 trials, and a preliminary picture of the general safety of adenoviral vector products has begun to emerge. An overwhelming systemic inflammatory response, to which has been attributed, at least in part, the death of a volunteer in a trial of ornithine transcarbamylase deficiency who received intrahepatic artery injection of a high dose of adenovirus-containing product, has not been observed in other clinical trials, including those that employ systemic administration of similar doses of adenovirus vector. Although individuals have experienced systemic reactions such as fever, chills, hypotension, and laboratory findings of mild, transient elevated liver aminotransferases,[81] the safety profile of adenoviral vector products has generally been well tolerated, so that issues related to the intrinsic safety of adenovirus as a vector or as oncolytic therapy have not been seen as a significant obstacle to the continued investigation and development of these products.

Preexisting antibody to adenovirus, or the development of an antibody response following administration of an adenovirus-containing product, may play a role in product safety, although a clear relationship has not been established.[82] The limited data available have not suggested a correlation between high baseline levels of neutralizing antibody and adenovirus toxicity (or activity). Products intended to be injected directly into a tumor would be expected to be even less susceptible to baseline presence of neutralizing antibody. Moreover, in a study that involved repeat administration of an adenovirus-containing product, participants developed large spikes in levels of neutralizing antibody after the initial dose. However, the toxicity profiles of the first and subsequent doses were similar, again suggesting a lack of correlation. It may be of value for clinical investigators to characterize the immune status of study participants at baseline and following adenovirus vector administration, and attempt to correlate adverse events with levels of, or changes in, antibody titer. Ultimately, such information might have relevance in patient selection criteria or in clinical monitoring to enhance safety and effectiveness.

Concerns about late adverse sequelae, such as new malignancies, occurring years or decades following administration of replication-competent, integrating viruses resulted in FDA guidance regarding the recommended duration to follow subjects who have received gene therapies [FDA, *Guidance for Industry: Gene Therapy Clinical Trials—Observing Subjects for Delayed Adverse Events* (2006)]. Because the usual modifications used to create adenovirus vectors do not result in the virus being integrating or replication competent, the FDA is not currently recommending that patients exposed to such products be followed for the 15 year term that is being asked for retroviruses. Protocols for adenovirus product typically follow subjects for 5 years.

Shedding refers to the release of oncolytic or gene therapy vectors from the patient. Shedding raised the possibility of transmission of product-based vectors from treated to untreated individuals. Shedding studies are typically done by incorporating additional specimen collection and testing into studies that are part of the usual clinical development program for the product. Recommendations regarding shedding studies have been published in FDA guidance (FDA, *Guidance for Industry: Design and Analysis of Shedding Studies for Virus or Bacteria-Based Gene Therapy and Oncolytic Products*), which has been discussed publically in a recent CTGTAC meeting.

For the typical adenovirus product that is not replication competent, the recommendation is that shedding data should be collected during Phase 2 trials, after a dose and dosing regimen have been determined.

## 6.4 Bioactivity of Adenoviral Vector Products

A goal of Phase 1 and Phase 2 testing is to determine if the adenovirus containing the transgene of interest is bioactive, and, if so, to determine whether the observed activity findings, together with the safety profile, warrant further clinical testing. Bioactivity measures may be laboratory findings, clinical outcomes (i.e., outcomes affecting how the patient feels, functions, or survives), or a combination of the two. One measure of bioactivity for gene therapy products is detection of gene transfer and gene expression. Documentation of clinical or surrogate outcomes or alternative assessments (e.g., pharmacodynamic measurements), and correlations, if any, to levels of gene expression, can be very helpful in early product development. The extent to which the generation of such data will be feasible depends on, among other factors, the nature of the product, the clinical population in the study, and the state of the science regarding assays to detect the transgene.

The majority of the clinical investigations with adenoviral vectors to date target patients with cancer. In the oncology setting, studies that are in Phase 2 of development are usually designed to capture data on tumor responses (complete and partial response rates). The demonstration that the adenovirus gene therapy product results in a certain level of tumor response, and the characterization of those responses (rates of complete and partial responses, duration of response, etc.), along with an acceptable safety profile, will usually be sufficient evidence of activity to warrant efficacy trials.

## 6.5 Clinical Efficacy of Adenoviral Vector Products

FDA grants market approval for products that are shown to be safe and effective. The efficacy standard, applicable to all drug and biologic products, as stated in section 505(d) of the Food, Drug, and Cosmetic Act, is *substantial evidence*, defined as "evidence consisting of adequate and well-controlled investigations, including clinical investigations, by experts qualified by scientific training and experience to evaluate the effectiveness of the drug involved, on the basis of which it could be fairly and responsibly concluded by such experts that the drug will have the effect it purports or is represented to have under the conditions of use prescribed, recommended, or suggested in the labeling or proposed labeling thereof." The following paragraphs address the issues of the quality and quantity of clinical investigations that can provide "substantial evidence."

### 6.5.1 Choice of Control

An "adequate and well-controlled" investigation is one whose design and execution produces valid scientific data. Clinical investigations intended to show efficacy must be controlled so that the effect(s) of the intervention can be distinguished from other influences, such as spontaneous change, placebo effect, or biased observation. In Phase 2 testing, controlled trials are helpful in teasing out adverse events and in assessing the magnitude and variability of the effect relative to the control group. Such information may be useful for sample size calculations for the efficacy trial(s).

The choice of control (e.g., historical, active, placebo, etc.) depends on the clinical setting. The Agency has approved products for marketing based on studies with various types of control groups. Each type of control has its advantages and limitations. The reader is referred to the ICH guidance titled *Guidance for Industry—E 10 Choice of Control Group and Related Issues in Clinical Trials* (2001) for an extensive discussion on this topic.[83]

A few adenovirus products are currently in Phase 3 testing in patients with malignancies. Most are designed as "add-on" trials, that is, chemotherapy + gene product versus chemotherapy + placebo (or no additional treatment if a placebo is not feasible). If a trial is not blinded, such as would be the case if the control arm could not receive a placebo, it will be important to utilize objective outcome measures and to control use of concomitant therapies. If measures are not resistant to bias, then blinded third party assessors may be useful.

### 6.5.2 Endpoint Selection

Trials intended to provide substantial evidence of efficacy must be "adequate" in addition to "well-controlled." They must be conducted according to GCP (as discussed in Section 6.2) to maximize human subject protection and data validity. They must also be designed with appropriate, relevant endpoints that either reflect clinical outcomes or are acceptable surrogate endpoints.

Surrogate endpoints are laboratory or other measurements not directly indicating clinical benefit but that are reasonably expected to predict clinical benefit. Surrogate endpoints are usually easier to measure than clinical endpoints and occur earlier in the course of the disease, allowing for shorter, smaller, and thus, less expensive studies. Their major disadvantage is the uncertainty surrounding whether and to what extent the effect on the surrogate reflects the true clinical benefit. Thus, if the FDA bases important regulatory decisions regarding product licensure on a surrogate, and the medical community bases practice decisions on data generated from trials using surrogates, it is critical that the surrogate be valid for the particular treatment and disease. Once a surrogate is validated for one treatment and disease using a particular product, the extent to which that validation applies to other products in the same class and across product classes could become important, particularly as one might define a product class in the context of adenoviral-containing products. In earlier phases of clinical testing, use of surrogate endpoints may serve useful and potentially less problematic roles. For instance, during product development, a surrogate may be used to assess dose–response and thus provide the rationale for dose selection for later trials, or may be used as initial POC to base decisions about further clinical development. Several excellent papers provide more in-depth discussions about surrogates and validation of surrogates.[84,85]

Where the disease is serious or life threatening and without acceptable alternative treatments, it may be possible to establish efficacy and receive FDA approval based on trials employing a surrogate endpoint that is not yet validated but reasonably likely to predict clinical benefit. If a product is marketed based on an effect on such a surrogate endpoint, postmarketing studies are required to verify the clinical benefit. These provisions are set forth in 21 CFR 601.40 Subpart E. Oncology and AIDS are two areas where this provision has been used with some frequency.

The number of adequate and well-controlled trials that will be necessary to make a determination of substantial evidence of effectiveness has been discussed in FDA guidance.[86] Sponsors should meet with the Agency at the end of Phase 2 to discuss critical product development issues such as the number and types of clinical trials and the size of a safety database necessary to file a marketing application.

## 7. Sponsor Outreach and Education

CBER has routinely been involved in educational and training activities aimed at sponsors and investigators who are involved in gene transfer research. The Agency recognizes the need to inform potential sponsors of not only the issues specific to the conduct of gene therapy studies but also on the issues involved in the design of a clinical program and the elements of GCP. Education sessions have taken place at various venues, including annual meetings of various professional societies involved in drug development in general and in gene therapy in particular. The FDA has published numerous guidance documents relevant to various aspects of biologics development that are available at: http://www.fda.gov/BiologicsBloodVaccines/GuidanceComplianceRegulatoryInformation/Guidances/General/default.htm.

CBER has additional guidance documents addressing issues relating specifically to gene therapy, which are available at: http://www.fda.gov/BiologicsBloodVaccines/GuidanceComplianceRegulatoryInformation/Guidances/CellularandGeneTherapy/default.htm.

The Office of Cellular, Tissue, and Gene Therapies has made available a series of prerecorded Web-based presentations that cover several topics of relevance to developers of gene therapy products. The presentations are available at: http://www.fda.gov/BiologicsBloodVaccines/NewsEvents/ucm232821.htm.

## Acknowledgments

The authors would like to acknowledge Andrew Byrnes for very helpful advice and discussion in the preparation of this manuscript. The following individuals also provided comments on this document: Denise Gavin, Robert Aksamit, Kimberly Benton, John Hyde, Mercedes Serabian, Wilson Bryan, and Stephanie Simek.

## References


1. Anderson WF, Blaese RM, Culver K. The ADA human gene therapy clinical protocol: Points to Consider response with clinical protocol. *Hum Gene Ther* July 6, 1990;**1**(3):331–62.
2. Crystal RG, McElvaney NG, Rosenfeld MA, Chu CS, Mastrangeli A, Hay JG, et al. Administration of an adenovirus containing the human CFTR cDNA to the respiratory tract of individuals with cystic fibrosis. *Nat Genet* 1994;**8**(1):42–51.
3. Zabner J, Couture LA, Gregory RJ, Graham SM, Smith AE, Welsh MJ. Adenovirus-mediated gene transfer transiently corrects the chloride transport defect in nasal epithelia of patients with cystic fibrosis. *Cell* 1993;**75**(2):207–16.
4. Zuckerman JB, Robinson CB, McCoy KS, Shell R, Sferra TJ, Chirmule N, et al. A phase I study of adenovirus-mediated transfer of the human cystic fibrosis transmembrane conductance regulator gene to a lung segment of individuals with cystic fibrosis. *Hum Gene Ther* 1999;**10**(18):2973–85.
5. Crystal RG. Adenovirus: the first effective in vivo gene delivery vector. *Hum Gene Ther* 2014;**25**(1):3–11.
6. Smith JG, Wiethoff CM, Stewart PL, Nemerow GR. Adenovirus. *Curr Top Microbiol Immunol* 2010;**343**:195–224.
7. *NIH Guidelines for research.* http://osp.od.nih.gov/office-biotechnology-activities/biosafety/nih-guidelines [accessed 20.04.15].
8. *NIH Oversight of human gene transfer research.* http://osp.od.nih.gov/office-biotechnology-activities/biomedical-technology-assessment/hgt [accessed 20.04.15].
9. *Oversight and review of clinical gene transfer protocols: assessing the role of the recombinant DNA advisory committee.* Washington, DC: National Academies Press; 2014.
10. FDA. *Guidance for industry: formal meetings between the FDA and sponros or applicants.* 2009.
11. FDA. *Guidance for industry: expedited programs for serious conditions - drugs and biologics.* 2014.
12. Bett AJ, Prevec L, Graham FL. Packaging capacity and stability of human adenovirus type 5 vectors. *J Virol* 1993;**67**(10):5911–21.
13. Parks RJ, Graham FL. A helper-dependent system for adenovirus vector production helps define a lower limit for efficient DNA packaging. *J Virol* 1997;**71**(4):3293–8.
14. Vetrini F, Ng P. Gene therapy with helper-dependent adenoviral vectors: current advances and future perspectives. *Viruses* 2010;**2**(9):1886–917.
15. Kovesdi I, Hedley SJ. Adenoviral producer cells. *Viruses* 2010;**2**(8):1681–703.
16. Cottingham MG, Carroll F, Morris SJ, Turner AV, Vaughan AM, Kapulu MC, et al. Preventing spontaneous genetic rearrangements in the transgene cassettes of adenovirus vectors. *Biotechnol Bioeng* 2012;**109**(3):719–28.
17. Gall JG, Lizonova A, EttyReddy D, McVey D, Zuber M, Kovesdi I, et al. Rescue and production of vaccine and therapeutic adenovirus vectors expressing inhibitory transgenes. *Mol Biotechnol* 2007;**35**(3):263–73.
18. Gurwith M, Lock M, Taylor EM, Ishioka G, Alexander J, Mayall T, et al. Safety and immunogenicity of an oral, replicating adenovirus serotype 4 vector vaccine for H5N1 influenza: a randomised, double-blind, placebo-controlled, phase 1 study. *Lancet Infect Dis* 2013;**13**(3):238–50.
19. Lubeck MD, Davis AR, Chengalvala M, Natuk RJ, Morin JE, Molnar-Kimber K, et al. Immunogenicity and efficacy testing in chimpanzees of an oral hepatitis B vaccine based on live recombinant adenovirus. *Proc Natl Acad Sci USA* 1989;**86**(17):6763–7.

20. Lubeck MD, Natuk R, Myagkikh M, Kalyan N, Aldrich K, Sinangil F, et al. Long-term protection of chimpanzees against high-dose HIV-1 challenge induced by immunization. *Nat Med* 1997;**3**(6):651–8.
21. Kim M, Zinn KR, Barnett BG, Sumerel LA, Krasnykh V, Curiel DT, et al. The therapeutic efficacy of adenoviral vectors for cancer gene therapy is limited by a low level of primary adenovirus receptors on tumour cells. *Eur J Cancer* 2002;**38**(14):1917–26.
22. Molnar-Kimber KL, Sterman DH, Chang M, Kang EH, ElBash M, Lanuti M, et al. Impact of preexisting and induced humoral and cellular immune responses in an adenovirus-based gene therapy phase I clinical trial for localized mesothelioma. *Hum Gene Ther* 1998;**9**(14):2121–33.
23. Xiang Z, Gao G, Reyes-Sandoval A, Cohen CJ, Li Y, Bergelson JM, et al. Novel, chimpanzee serotype 68-based adenoviral vaccine carrier for induction of antibodies to a transgene product. *J Virol* 2002;**76**(6):2667–75.
24. Tuve S, Liu Y, Tragoolpua K, Jacobs JD, Yumul RC, Li ZY, et al. In situ adenovirus vaccination engages T effector cells against cancer. *Vaccine* 2009;**27**(31):4225–39.
25. Hemminki O, Parviainen S, Juhila J, Turkki R, Linder N, Lundin J, et al. Immunological data from cancer patients treated with Ad5/3 E2F Delta24 GMCSF suggests utility for tumor immunotherapy. *Oncotarget* 2015;**6**(6):4467–81.
26. Cerullo V, Pesonen S, Diaconu I, Escutenaire S, Arstila PT, Ugolini M, et al. Oncolytic adenovirus coding for granulocyte macrophage colony-stimulating factor induces antitumoral immunity in cancer patients. *Cancer Res* 2010;**70**(11):4297–309.
27. Kruyt FA, Curiel DT. Toward a new generation of conditionally replicating adenoviruses: pairing tumor selectivity with maximal oncolysis. *Hum Gene Ther* 2002;**13**(4):485–95.
28. Kuhn I, Harden P, Bauzon M, Chartier C, Nye J, Thorne S, et al. Directed evolution generates a novel oncolytic virus for the treatment of colon cancer. *PLoS One* 2008;**3**(6):e2409.
29. Silver J, Mei YF. Transduction and oncolytic profile of a potent replication-competent adenovirus 11p vector (RCAd11pGFP) in colon carcinoma cells. *PLoS One* 2011;**6**(3):e17532.
30. International Conference on Harmonization. *Considerations. Oncolytic viruses.* http://www.ema.europa.eu/docs/en_GB/document_library/Scientific_guideline/2009/09/WC500003915.pdf [accessed 20.04.15].
31. *Cellular, tissue, and gene therapies advisory committee meeting: design and analysis of shedding studies for viruses or bacteria-based gene therapy and oncolytic products.* http://www.fda.gov/AdvisoryCommittees/CommitteesMeetingMaterials/BloodVaccinesandOtherBiologics/CellularTissueandGeneTherapiesAdvisoryCommittee/ucm380047.htm [accessed 20.4.15].
32. Naimark WA, Lepore JJ, Klugherz BD, Wang Z, Guy TS, Osman H, et al. Adenovirus-catheter compatibility increases gene expression after delivery to porcine myocardium. *Hum Gene Ther* 2003;**14**(2):161–6.
33. Marshall DJ, Palasis M, Lepore JJ, Leiden JM. Biocompatibility of cardiovascular gene delivery catheters with adenovirus vectors: an important determinant of the efficiency of cardiovascular gene transfer. *Mol Ther* 2000;**1**(5 Pt 1):423–9.
34. Bilbao R, Reay DP, Koppanati BM, Clemens PR. Biocompatibility of adenoviral vectors in poly(vinyl chloride) tubing catheters with presence or absence of plasticizer di-2-ethylhexyl phthalate. *J Biomed Mater Res A* 2004;**69**(1):91–6.
35. Lochmuller H, Jani A, Huard J, Prescott S, Simoneau M, Massie B, et al. Emergence of early region 1-containing replication-competent adenovirus in stocks of replication-defective adenovirus recombinants (delta E1 + delta E3) during multiple passages in 293 cells. *Hum Gene Ther* 1994;**5**(12):1485–91.

36. Fallaux FJ, Bout A, van der Velde I, van den Wollenberg DJ, Hehir KM, Keegan J, et al. New helper cells and matched early region 1-deleted adenovirus vectors prevent generation of replication-competent adenoviruses. *Hum Gene Ther* 1998;**9**(13):1909–17.
37. Murakami P, Pungor E, Files J, Do L, van RR, Vogels R, et al. A single short stretch of homology between adenoviral vector and packaging cell line can give rise to cytopathic effect-inducing, helper-dependent E1-positive particles. *Hum Gene Ther* 2002;**13**(8):909–20.
38. Murakami P, Havenga M, Fawaz F, Vogels R, Marzio G, Pungor E, et al. Common structure of rare replication-deficient E1-positive particles in adenoviral vector batches. *J Virol* 2004;**78**(12):6200–8.
39. FDA. *Guidance for FDA review staff and sponors: content and review of Chemistry, manufacturing, and control (CMC) information for human gene therapy investigational new drug applications (INDs).* http://www.fda.gov/BiologicsBloodVaccines/GuidanceCompliance RegulatoryInformation/Guidances/CellularandGeneTherapy/ucm072587.htm≥; 2008 [accessed 20.04.15].
40. FDA. *Guidance for industry: characterization and qualification of cell substrates and other biological materials used in the production of viral vaccines for infectious disease indications.* http://www.fda.gov/downloads/biologicsbloodvaccines/guidancecompliance regulatoryinformation/guidances/vaccines/ucm202439.pdf≥; 2010 [accessed 20.04.15].
41. BRMAC Meeting #28. *Briefing document: structural characterization of gene transfer vectors. FDA Dockets Home Page.* http://www.fda.gov/ohrms/dockets/ac/00/backgrd/3664b1.htm; 2000 [accessed 20.04.15].
42. Small JC, Kurupati RK, Zhou X, Bian A, Chi E, Li Y, et al. Construction and characterization of E1- and E3-deleted adenovirus vectors expressing two antigens from two separate expression cassettes. *Hum Gene Ther* 2014;**25**(4):328–38.
43. Hermens WT, Verhaagen J. Adenoviral vector-mediated gene expression in the nervous system of immunocompetent Wistar and T cell-deficient nude rats: preferential survival of transduced astroglial cells in nude rats. *Hum Gene Ther* 1997;**8**(9):1049–63.
44. Schiedner G, Hertel S, Kochanek S. Efficient transformation of primary human amniocytes by E1 functions of Ad5: generation of new cell lines for adenoviral vector production. *Hum Gene Ther* 2000;**11**(15):2105–16.
45. Kim JS, Lee SH, Cho YS, Park K, Kim YH, Lee JH. Development of a packaging cell line for propagation of replication-deficient adenovirus vector. *Exp Mol Med* 2001;**33**(3):145–9.
46. Hehir KM, Armentano D, Cardoza LM, Choquette TL, Berthelette PB, White GA, et al. Molecular characterization of replication-competent variants of adenovirus vectors and genome modifications to prevent their occurrence. *J Virol* 1996;**70**(12):8459–67.
47. Gao GP, Engdahl RK, Wilson JM. A cell line for high-yield production of E1-deleted adenovirus vectors without the emergence of replication-competent virus. *Hum Gene Ther* 2000;**11**(1):213–9.
48. Nadeau I, Kamen A. Production of adenovirus vector for gene therapy. *Biotechnol Adv* 2003;**20**(7–8):475–89.
49. Konz JO, Pitts LR, Sagar SL. Scaleable purification of adenovirus vectors. *Methods Mol Biol* 2008;**434**:13–23.
50. Bauer S, Pilaro AM, Weiss KD. Testing of adenoviral gene therapy products: FDA expectations. 2002;(21):615–54.
51. Sheng-Fowler L, Lewis Jr AM, Peden K. Issues associated with residual cell-substrate DNA in viral vaccines. *Biologicals* 2009;**37**(3):190–5.
52. Peden K, Sheng L, Pal A, Lewis A. Biological activity of residual cell-substrate DNA. *Dev Biol (Basel)* 2006;**123**:45–53.

53. Vellekamp G, Ravindran S, Lin M, Sluzky V, Lehmberg E. A contaminant in the adenovirus reference material. *Bioprocess J* 2002 (Fall):57–61.
54. Sweeney JA, Hennessey Jr JP. Evaluation of accuracy and precision of adenovirus absorptivity at 260 nm under conditions of complete DNA disruption. *Virology* 2002;**295**(2):284–8.
55. Maizel Jr JV, White DO, Scharff MD. The polypeptides of adenovirus. II. Soluble proteins, cores, top components and the structure of the virion. *Virology* 1968;**36**(1):126–36.
56. Simek S, Byrnes AP, Bauer SR. FDA Perspectives on the use of the adenovirus reference material. *BioProcess J* 2002 (Fall):40–2.
57. Shabram PW, Giroux DD, Goudreau AM, Gregory RJ, Horn MT, Huyghe BG, et al. Analytical anion-exchange HPLC of recombinant type-5 adenoviral particles. *Hum Gene Ther* 1997;**8**(4):453–65.
58. Shabram P, Scandella C, Vellekamp G. Purification of adenovirus. 2002;(7):167–203.
59. Nyberg-Hoffman C, Shabram P, Li W, Giroux D, Aguilar-Cordova E. Sensitivity and reproducibility in adenoviral infectious titer determination. *Nat Med* 1997;**3**(7):808–11.
60. Dormond E, Perrier M, Kamen A. From the first to the third generation adenoviral vector: what parameters are governing the production yield? *Biotechnol Adv* 2009;**27**(2):133–44.
61. BRMAC Meeting #30. *Breifing document: adenovirus titer measurements and RCA levels. FDA Dockets home page.* http://www.fda.gov/ohrms/dockets/ac/01/briefing/3768b1_01.pdf≥; 2001 [accessed 20.04.15].
62. FDA. *Guidance for industry: potency tests for cellular and gene therapy products.* 2011.
63. Riske F, Berard N, Albee K, Pan P, Henderson M, Adams K, et al. Development of a platform process for adenovirus purification that removes human SET and nucleolin and provides high purity vector for gene delivery. *Biotechnol Bioeng* 2013;**110**(3):848–56.
64. Burlingham BT, Brown DT, Doerfler W. Incomplete particles of adenovirus. I. Characteristics of the DNA associated with incomplete adenovirions of types 2 and 12. *Virology* 1974;**60**(2):419–30.
65. Vellekamp G, Porter FW, Sutjipto S, Cutler C, Bondoc L, Liu YH, et al. Empty capsids in column-purified recombinant adenovirus preparations. *Hum Gene Ther* 2001;**12**(15):1923–36.
66. Shih S-J, Yagami M, Tseng W-J, Lin A, Validation of a quantitative method for detection of adenovirus aggregation. *Bioprocess J* 2011;**9**(2):25–33.
67. Rexroad J, Evans RK, Middaugh CR. Effect of pH and ionic strength on the physical stability of adenovirus type 5. *J Pharm Sci* 2006;**95**(2):237–47.
68. Berkowitz SA, Philo JS. Monitoring the homogeneity of adenovirus preparations (a gene therapy delivery system) using analytical ultracentrifugation. *Anal Biochem* 2007;**362**(1):16–37.
69. FDA. *Guidance for industry: preclinical assessment of investigational cellular and gene therapy products.* http://www.fda.gov/BiologicsBloodVaccines/GuidanceComplianceRegulatoryInformation/Guidances/CellularandGeneTherapy/ucm376136.htm; 2013 [accessed 20.04.15].
70. Serabian M, Huang Y. Preclinical safety evaluation of gene therapy products. 2008;(32):739.
71. Xu Z, Smith JS, Tian J, Byrnes AP. Induction of shock after intravenous injection of adenovirus vectors: a critical role for platelet-activating factor. *Mol Ther* 2010;**18**(3):609–16.
72. Duffy MR, Parker AL, Bradshaw AC, Baker AH. Manipulation of adenovirus interactions with host factors for gene therapy applications. *Nanomedicine (Lond)* 2012;**7**(2):271–88.
73. Xu Z, Qiu Q, Tian J, Smith JS, Conenello GM, Morita T, et al. Coagulation factor X shields adenovirus type 5 from attack by natural antibodies and complement. *Nat Med* 2013;**19**(4):452–7.

74. International Conference on Harmonization. *Guidance for industry. E6: good clinical practice: consolidated guidance*. http://www.fda.gov/Drugs/GuidanceComplianceRegulatoryInformation/Guidances/ucm065004.htm≥; 1996 [accessed 20.04.15].
75. International Conference on Harmonization. *Guidance for industry. E3: structure and content of the final study report*. http://www.fda.gov/Drugs/GuidanceComplianceRegulatoryInformation/Guidances/ucm065004.htm [accessed 20.04.15].
76. FDA. *Guideline for industry – structure and content of clinical study reports*. http://www.fda.gov/Drugs/GuidanceComplianceRegulatoryInformation/Guidances/ucm065004.htm≥; 1996 [accessed 20.04.15].
77. International Conference on Harmonization. The Common Technical Document.
78. Kessler DA, Kennedy DL. MedWatch: FDA's new medical products reporting program. *J Clin Eng* 1993;**18**(6):489–92.
79. Wittes J. Behind closed doors: the data monitoring board in randomized clinical trials. *Stat Med* 1993;**12**(5–6):419–24.
80. FDA. *Guidance for clinical trial sponsors – establishment and operation of clinical trial data monitoring committees*. www.fda.gov/downloads/drugs/guidancecompliance-regulatoryinformation/guidances; 2006 [accessed 20.04.15].
81. Harvey BG, Maroni J, O'Donoghue KA, Chu KW, Muscat JC, Pippo AL, et al. Safety of local delivery of low- and intermediate-dose adenovirus gene transfer vectors to individuals with a spectrum of morbid conditions. *Hum Gene Ther* 2002;**13**(1):15–63.
82. Kafri T, Morgan D, Krahl T, Sarvetnick N, Sherman L, Verma I. Cellular immune response to adenoviral vector infected cells does not require de novo viral gene expression: implications for gene therapy. *Proc Natl Acad Sci USA* 1998;**95**(19):11377–82.
83. FDA. *Guidance for industry – e 10 choice of control group and related issues in clinical trials*. http://www.fda.gov/Drugs/GuidanceComplianceRegulatoryInformation/Guidances/ucm065004.htm≥; 2001 [accessed 20.04.15].
84. Fleming TR. Surrogate markers in AIDS and cancer trials. *Stat Med* 1994;**13**(13–14):1423–35.
85. Fleming TR, DeMets DL. Surrogate end points in clinical trials: are we being misled? *Ann Intern Med* 1996;**125**(7):605–13.
86. FDA. *Guidance for industry: providing clinical evidence of effectiveness*. www.fda.gov/downloads/Drugs/.../Guidances/ucm078749.pdf; 1998 [accessed 20.04.15].

# Index

*Note*: Page numbers followed by "f" indicate figures and "t" indicate tables.

## C

## D



## E

## W



## X



## Y



## Z

FSC
www.fsc.org
MIX
Papier | Fördert gute Waldnutzung
FSC® C083411

Kolibri 360 GmbH
Ferdinand-Jühlke-Straße 7
99095 Erfurt, Deutschland
produktsicherheit@kolibri360.de